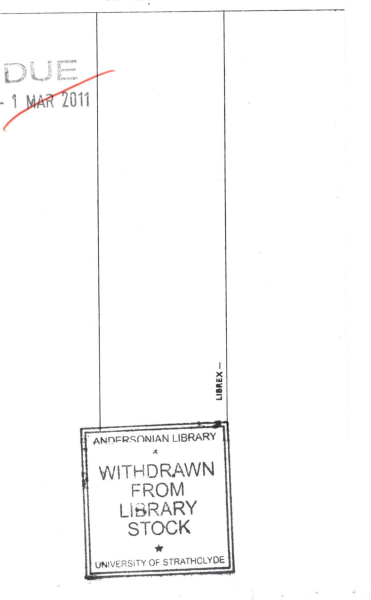

HANDBOOK OF

FOOD ENGINEERING

SECOND EDITION

FOOD SCIENCE AND TECHNOLOGY

HANDBOOK OF
FOOD
ENGINEERING

SECOND EDITION

EDITED BY

DENNIS R. HELDMAN
DARYL B. LUND

CRC Press
Taylor & Francis Group
Boca Raton   London   New York

CRC Press is an imprint of the
Taylor & Francis Group, an informa business

CRC Press
Taylor & Francis Group
6000 Broken Sound Parkway NW, Suite 300
Boca Raton, FL 33487-2742

© 2007 by Taylor and Francis Group, LLC
CRC Press is an imprint of Taylor & Francis Group, an Informa business

No claim to original U.S. Government works
Printed in the United States of America on acid-free paper
10 9 8 7 6 5 4 3 2 1

International Standard Book Number-10: 0-8247-5331-3 (Hardcover)
International Standard Book Number-13: 978-0-8247-5331-3 (Hardcover)
Library of Congress Card Number 2006010365

Library of Congress Cataloging-in-Publication Data

Handbook of food engineering / [edited by] Dennis R. Heldman and Daryl B. Lund. -- 2nd ed.
    p. cm. -- (Food science and technology ; 161)
  Includes bibliographical references and index.
  ISBN-13: 978-0-8247-5331-3 (acid-free paper)
  ISBN-10: 0-8247-5331-3 (acid-free paper)
  1. Food industry and trade--Handbooks, manuals, etc. I. Heldman, Dennis R. II. Lund, Daryl B. III. Series: Food science and technology (Taylor & Francis) ; 161.

TP370.4.H36 2007
664--dc22
                                      2006010365

Visit the Taylor & Francis Web site at
http://www.taylorandfrancis.com

and the CRC Press Web site at
http://www.crcpress.com

Preface

The primary mission of the second edition of the *Handbook of Food Engineering* is the same as the first. The most recent information needed for efficient design and development of processes used in the manufacturing of food products has been assembled, along with the traditional background on these processes. The audience for this handbook includes three groups: (1) practicing engineers in the food and related industries, (2) the student preparing for a career as a food engineer, and (3) other scientists and technologists seeking information about processes and the information needed in design and development of these processes. For the practicing engineer, the handbook assembles information needed for the design and development of a given process. For the student, the handbook becomes the primary reference needed to supplement textbooks used in the teaching of process design and development concepts. Other scientists and technologists should use the handbook to locate important information and physical data related to foods and food ingredients.

As in the first edition, the handbook assembles the most recent information on thermophysical properties of foods, rate constants about changes in food components during a process, and illustrations of the use of these properties and constants in process design. Researchers will be able to use the information as a guide in establishing the direction of future research on thermophysical properties and rate constants. In this edition, an appendix has been created to assemble tables and figures containing property data needed for the design of processes described in various chapters of the handbook.

Although the first three chapters focus primarily on properties of food and food ingredients, the chapters that follow are organized according to traditional unit operations associated with the manufacturing of foods. Two key chapters cover the basic concepts of transport and storage of liquids and solids, and the heating and cooling of foods and food ingredients. An additional background chapter focuses on basic concepts of mass transfer in foods. More specific unit operations on freezing, concentration, dehydration, thermal processing, and extrusion are discussed and analyzed in separate chapters. The chapter on membrane processes deals with liquid food concentration but provides the basis for other applications of membranes in food processing. The final chapters of the handbook cover the important topics of packaging and cleaning and sanitation.

The editors of this handbook hope that the information presented will continue to contribute to the evolution of food engineering as an interface between engineering and other food sciences. As demands for safe, high quality, nutritious and convenient foods continue to increase, the needs for the concepts presented will become more critical. In the near future, the applications of new science from molecular biology, nanotechnology, and nutritional biochemistry in food manufacturing will increase, and the role of engineering in process design and scale-up will be even more visible. At the same time, new process technologies will continue to emerge and require input from engineers for application, design, and development in food manufacturing. Ultimately, the use of engineering concepts should lead to the highest quality food products at the lowest possible cost.

The editors wish to acknowledge the authors and their significant contributions to the second edition of this handbook. These authors are among the leading scientists and engineers in the field of food engineering. We are pleased to be associated with their contributions to this field and to the handbook.

Dennis R. Heldman
Daryl B. Lund

Editors

Dennis R. Heldman is a principal of Heldman Associates in Weston, Florida. He has been professor of food process engineering at Rutgers, The State University of New Jersey, the University of Missouri and Michigan State University. In addition, he has industry experience at the Campbell Soup Company, the National Food Processors Association and the Weinberg Consulting Group. Dr. Heldman is the author or co-author of over 140 journal articles, and the author, co-author or editor of over 10 textbooks, handbooks and encyclopedias. He is a fellow of the Institute of Food Technologists and the American Society of Agricultural Engineers. He served as president of the IFT, the Society for Food Science and Technology, an organization with over 20,000 members, from 2006–2007, and was elected fellow in the International Academy of Food Science & Technology in 2006. Dr. Heldman was awarded a BS (1960) and an MS (1962) from The Ohio State University, and a PhD (1965) from Michigan State University.

Darryl B. Lund earned a BS (1963) in mathematics and a PhD (1968) in food science with a minor in chemical engineering at the University of Wisconsin-Madison. During 21 years at the University of Wisconsin, he was a professor of food engineering in the food science department serving as chair of the department from 1984–1987. He has contributed over 150 scientific papers, edited 5 books, and co-authored one major textbook in the area of simultaneous heat and mass transfer in foods, kinetics of reactions in foods, and food processing.

In 1988 he continued his administrative responsibilities by chairing the Department of Food Science at Rutgers University, and from December 1989 through July 1995 served as the executive dean of Agriculture and Natural Resources with responsibilities for teaching, research and extension at Rutgers University. In that position, among other achievements, he initiated a rigorous strategic planning process for Cook College and the New Jersey Agricultural Experiment Station, streamlined administrative services, fostered a review of the undergraduate curriculum and encouraged the faculty to develop a social contact for undergraduate instruction.

In August 1995, he joined the Cornell University faculty as the Ronald P. Lynch Dean of Agriculture and Life Sciences. During his tenure as dean of CALS, he initiated a strategic positioning process for the college that guided the college through 20% downsizing, promoted the Agriculture Initiative to gain increased state support for the Agricultural Experiment Station and Cooperative Extension, supported an initiative in genomics and overhaul of the biological sciences, fostered a review of undergraduate programs that led to major changes, and supported the adoption of electronic technologies for undergraduate teaching and distance education. In July 2000, Dr. Lund returned to the Department of Food Science as professor of food engineering.

In January 2001, Dr. Lund became the executive director of the North Central Regional Association of State Agricultural Experiment Station Directors. In this position he facilitates interstate collaboration on research and a greater integration between research and extension in the twelve-state region.

Among many awards in recognition of personal achievement, he is a recipient of the ASAE/DFISA Food Engineering Award, the IFT International Award and Carl R. Fellers Award, and the Irving Award from the American Distance Education Consortium. He is an elected fellow of the Institute of Food Technologists, elected fellow of the Institute of Food Science and Technology (UK), and charter inductee in the International Academy of Food Science and Technology.

Contributors

Osvaldo Campanella
Purdue University
West Lafayette, Indiana

Munir Cheryan
University of Illinois at Urbana-Champaign
Urbana, Illinois

Hulya Dogan
The State University of New Jersey, Rutgers
New Brunswick, New Jersey

Vassilis Gekas
Technical University of Crete
Chania, Greece

Albrecht Graßhoff
Bundesanstalt für Milchforschung
Kiel, Germany

Bengt Hallström
University of Lund
Lund, Sweden

Richard W. Hartel
University of Wisconsin
Madison, Wisconsin

James G. Hawkes
Nutriscience Technologies, Inc.
Naperville, Illinois

Dennis R. Heldman
Heldman Associates
Weston, Florida

Jozef L. Kokini
The State University of New Jersey, Rutgers
New Brunswick, New Jersey

John M. Krochta
University of California
Davis, California

Leon Levine
Leon Levine and Associates, Inc.
Albuquerque, New Mexico

Robert C. Miller
Consulting Engineer
Auburn, New York

Ken R. Morison
University of Canterbury
Christchurch, New Zealand

Ganesan Narsimhan
Purdue University
West Lafayette, Indiana

Martin R. Okos
Purdue University
West Lafayette, Indiana

Erwin A. Plett
Sello Verde Ingenieria Ambiental S.A.
Santiago, Chile

M.A. Rao
Cornell University
Ithaca, New York

Anne Marie Romulus
Université Paul Sabatier
Toulouse, France

Yrjö H. Roos
University College
Cork, Ireland

R. Paul Singh
University of California
Davis, California

Rakesh K. Singh
Purdue University
West Lafayette, Indiana

Ingegerd Sjöholm
University of Lund
Lund, Sweden

Arthur Teixeira
University of Florida
Gainesville, Florida

Ricardo Villota
Kraft Foods
Glenview, Illinois

A.C. Weitnauer
Purdue University
West Lafayette, Indiana

Table of Contents

1 Rheological Properties of Foods

Hulya Dogan and Jozef L. Kokini

CONTENTS

1.1 INTRODUCTION

Rheological properties are important to the design of flow processes, quality control, storage and processing stability measurements, predicting texture, and learning about molecular and conformational changes in food materials (Davis, 1973). The rheological characterization of foods provides important information for food scientists, ingredient selection strategies to design, improve, and optimize their products, to select and optimize their manufacturing processes, and design packaging and storage strategies. Rheological studies become particularly useful when predictive relationships for rheological properties of foods can be developed which start from the molecular architecture of the constituent species.

Reliable and accurate steady rheological data are necessary to design continuous-flow processes, select and size pumps and other fluid-moving machinery and to evaluate heating rates during engineering operations which include flow processes such as aseptic processing and concentration (Holdsworth, 1971; Sheath, 1976), and to estimate velocity, shear, and residence-time distribution in food processing operations including extrusion and continuous mixing.

Viscoelastic properties are also useful in processing and storage stability predictions. For example, during extrusion, viscoelastic properties of cereal flour doughs affect die swell and extrudate expansion. In batch mixing, elasticity is responsible for the *rod climbing phenomenon*, also known as the *Weissenberg effect* (Bird et al., 1987). To allow for elastic recovery of dough during cookie making, the dough is cut in the form of an ellipse which relaxes into a perfect circle.

Creep and small-amplitude oscillatory measurements are useful in understanding the role of constituent ingredients on the stability of oil-in-water emulsions. Steady shear and creep measurements help identify the effect of ingredients that have stabilizing ability, such as gums, proteins, or other surface-active agents (Fischbach and Kokini, 1984).

Dilute solution viscoelastic properties of biopolymeric materials such as carbohydrates and protein can be used to characterize their three-dimensional configuration in solution. Their configuration affects their functionality in many food products. It is possible to predict better and improve the flow behavior of food polymers through an understanding of how the molecular structure of polymers affects their rheological properties (Liguori, 1985). Examples can be found in the improvement of

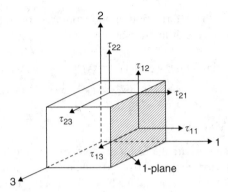

FIGURE 1.1   Stress components on a cubical material element.

the consistency and stability of emulsions by using polymers with enhanced surface activity and greater viscosity and elasticity.

This chapter will review recent advances in basic rheological concepts, methods of measurement, molecular theories, linear and nonlinear constitutive models, and numerical simulation of viscoelastic flows.

1.2 BASIC CONCEPTS

1.2.1 STRESS AND STRAIN

Rheology is the science of the deformation and flow of matter. Rheological properties define the relationship between stress and strain/strain rate in different types of shear and extensional flows. The stress is defined as the force F acting on a unit area A. Since both force and area have directional as well as magnitude characteristics, stress is a second order tensor and typically has nine components. Strain is a measure of deformation or relative displacement and is determined by the displacement gradient. Since displacement and its relative change both have directional properties, strain is also a second order tensor with nine components.

A rheological measurement is conducted on a given material by imposing a well-defined stress and measuring the resulting strain or strain rate or by imposing a well-defined strain or strain rate and by measuring the stress developed. The relationship between these physical events leads to different kinds of rheological properties.

When a force F is applied to a piece of material (Figure 1.1), the total stress acting on any infinitesimal element is composed of two fundamental classes of stress components (Darby, 1976):

Normal stress components, applied perpendicularly to the plane (τ_{11}, τ_{22}, τ_{33})
Shear stress components, applied tangentially to the plane (τ_{12}, τ_{13}, τ_{21}, τ_{23}, τ_{31}, τ_{32})

There are a total of nine stress components acting on an infinitesimal element (i.e., two shear components and one normal stress component acting on each of the three planes). Individual stress components are referred to as τ_{ij}, where i refers to the plane the stress acts on, and j indicates the direction of stress component (Bird et al., 1987). The stress tensor can be written as a matrix of nine components as follows:

$$\tau = \begin{bmatrix} \tau_{11} & \tau_{12} & \tau_{13} \\ \tau_{21} & \tau_{22} & \tau_{23} \\ \tau_{31} & \tau_{32} & \tau_{33} \end{bmatrix}$$

In general, the stress tensor in the deformation of an incompressible material is described by three shear stresses and two normal stress differences:

Shear stresses: $\tau_{12}(=\tau_{21})$ $\tau_{13}(=\tau_{31})$ $\tau_{23}(=\tau_{32})$

Normal stress differences: $N_1 = \tau_{11} - \tau_{22}$ $N_2 = \tau_{22} - \tau_{33}$

1.2.2 CLASSIFICATION OF MATERIALS

Rheological properties of materials are the result of their stress-strain behavior. Ideal solid (elastic) and ideal fluid (viscous) behaviors represent two extreme responses of a material (Darby, 1976).

An ideal solid material deforms instantaneously when a load is applied. It returns to its original configuration instantaneously (complete recovery) upon removal of the load. Ideal elastic materials obey Hooke's law, where the stress (τ) is directly proportional to the strain (γ). The proportionality constant (G) is called the modulus.

$$\tau = G\gamma$$

An ideal fluid deforms at a constant rate under an applied stress, and the material does not regain its original configuration when the load is removed. The flow of a simple viscous material is described by Newton's law, where the shear stress (τ) is directly proportional the shear rate ($\dot{\gamma}$). The proportionality constant (η) is called the Newtonian viscosity.

$$\tau = \eta\dot{\gamma}$$

Most food materials exhibit characteristics of both elastic and viscous behavior and are called viscoelastic. If viscoelastic properties are strain and strain rate independent, then these materials are referred to as linear viscoelastic materials. On the other hand if they are strain and strain rate dependent, than they are referred to as nonlinear viscoelastic materials (Ferry, 1980; Bird et al., 1987; Macosko, 1994).

A simple and classical approach to describe the response of a viscoelastic material is using mechanical analogs. Purely elastic behavior is simulated by springs and purely viscous behavior is simulated using dashpots. The Maxwell and Voigt models are the two simplest mechanical analogs of viscoelastic materials. They simulate a liquid (Maxwell) and a solid (Voigt) by combining a spring and a dashpot in series or in parallel, respectively. These mechanical analogs are the building blocks of constitutive models as discussed in Section 1.4 in detail.

1.2.3 TYPES OF DEFORMATION

1.2.3.1 Shear Flow

One of the most useful types of deformation for rheological measurements is simple shear. In simple shear, a material element is placed between two parallel plates (Figure 1.2) where the bottom plate is stationary and the upper plate is displaced in x-direction by Δx by applying a force F tangentionally to the surface A. The velocity profile in simple shear is given by the following velocity components:

$$v_x = \dot{\gamma}y, \quad v_y = 0, \quad \text{and} \quad v_z = 0$$

The corresponding shear stress is given as:

$$\tau = \frac{F}{A}$$

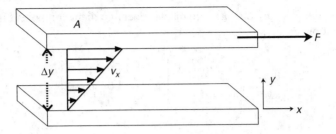

FIGURE 1.2 Shear flow.

If the relative displacement at any given point Δy is Δx, then the shear strain is given by

$$\gamma = \frac{\Delta x}{\Delta y}$$

If the material is a fluid, force applied tangentially to the surface will result in a constant velocity v_x in x-direction. The deformation is described by the strain rate ($\dot{\gamma}$), which is the time rate of change of the shear strain:

$$\dot{\gamma} = \frac{d\gamma}{dt} = \frac{d}{dt}\left(\frac{\Delta x}{\Delta y}\right) = \frac{dv_x}{dy}$$

Shear strain defines the displacement gradient in simple shear. The displacement gradient is the relative displacement of two points divided by the initial distance between them. For any continuous medium the displacement gradient tensor is given as:

$$\frac{\partial u_i}{\partial x_j} = \begin{bmatrix} \dfrac{\partial u_1}{\partial x_1} & \dfrac{\partial u_1}{\partial x_2} & \dfrac{\partial u_1}{\partial x_3} \\[2mm] \dfrac{\partial u_2}{\partial x_1} & \dfrac{\partial u_2}{\partial x_2} & \dfrac{\partial u_2}{\partial x_3} \\[2mm] \dfrac{\partial u_3}{\partial x_1} & \dfrac{\partial u_3}{\partial x_2} & \dfrac{\partial u_3}{\partial x_3} \end{bmatrix}$$

A nonzero displacement gradient may represent pure rotation, pure deformation, or both (Darby, 1976). Thus, each displacement component has two parts:

$$\frac{\partial u_i}{\partial x_j} = \underbrace{\frac{1}{2}\left(\frac{\partial u_i}{\partial x_j} + \frac{\partial u_j}{\partial x_i}\right)}_{\text{Pure deformation}} + \underbrace{\frac{1}{2}\left(\frac{\partial u_i}{\partial x_j} - \frac{\partial u_j}{\partial x_i}\right)}_{\text{Pure rotation}}$$

Then the strain tensor (e_{ij}) can be defined as:

$$e_{ij} = \left(\frac{\partial u_i}{\partial x_j} + \frac{\partial u_j}{\partial x_i}\right)$$

Similarly, the rotation tensor (r_{ij}) can be defined as:

$$r_{ij} = \left(\frac{\partial u_j}{\partial x_i} - \frac{\partial u_i}{\partial x_j}\right)$$

In simple shear, there is only one nonzero displacement gradient component that contributes to both strain and rotation tensors.

$$\frac{\partial u_i}{\partial x_j} = \begin{bmatrix} 0 & \frac{\partial u_x}{\partial y} & 0 \\ 0 & 0 & 0 \\ 0 & 0 & 0 \end{bmatrix} = \frac{du_x}{dy} \begin{bmatrix} 0 & 1 & 0 \\ 0 & 0 & 0 \\ 0 & 0 & 0 \end{bmatrix}$$

The time derivative of the strain tensor gives the rate of strain tensor (Δ_{ij}):

$$\Delta_{ij} = \frac{\partial}{\partial t}(e_{ij}) = \frac{\partial}{\partial t}\left(\frac{\partial u_i}{\partial x_j} + \frac{\partial u_j}{\partial x_i}\right) = \frac{\partial v_i}{\partial x_j} + \frac{\partial v_j}{\partial x_i}$$

Similarly, time derivative of the rotation tensor gives the vorticity tensor (Ω_{ij}):

$$\Omega_{ij} = \frac{\partial}{\partial t}(r_{ij}) = \frac{\partial v_j}{\partial x_i} - \frac{\partial v_i}{\partial x_j}$$

Simple shear flow, or viscometric flow, serves as the basis for many rheological measurement techniques (Bird et al., 1987). The stress tensor in simple shear flow is given as:

$$\tau = \begin{bmatrix} 0 & \tau_{12} & 0 \\ \tau_{21} & 0 & 0 \\ 0 & 0 & 0 \end{bmatrix}$$

There are three shear rate dependent material functions used to describe material properties in simple shear flow:

Viscosity: $$\mu(\dot{\gamma}) = \frac{\tau_{12}}{\dot{\gamma}}$$

First normal stress coefficient: $$\psi_1(\dot{\gamma}) = \frac{\tau_{11} - \tau_{22}}{\dot{\gamma}^2} = \frac{N_1}{\dot{\gamma}^2}$$

Second normal stress coefficient: $$\psi_2(\dot{\gamma}) = \frac{\tau_{22} - \tau_{33}}{\dot{\gamma}^2} = \frac{N_2}{\dot{\gamma}^2}$$

Among the viscometric functions, viscosity is the most important parameter for a food material. In the case of a Newtonian fluid, both the first and second normal stress coefficients are zero and the material is fully described by a constant viscosity over all shear rates studied. First normal stress data for a wide variety of food materials are available (Dickie and Kokini, 1982; Chang et al., 1990; Wang and Kokini 1995a). Well-known practical examples demonstrating the presence of normal stresses are the Weissenberg or road climbing effect and the die swell effect. Although the exact molecular origin of normal stresses is not well understood, they are considered to be the result of the elastic properties of viscoelastic fluids (Darby, 1976) and are a measure of the elasticity of the fluids. Figure 1.3 shows the normal stress development for butter at 25°C. Primary normal stress coefficients vs. shear rate plots for various semisolid food materials on log-log coordinates are shown in Figure 1.4 in the shear rate range 0.1 to 100 sec^{-1}.

1.2.3.2 Extensional (Elongational) Flow

Pure extensional flow does not involve shearing and is referred to as shear-free flow (Bird et al., 1987; Macosko, 1994). Extensional flows are generically defined by the following velocity

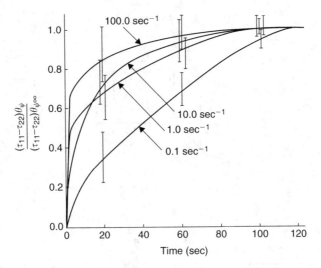

FIGURE 1.3 Normal stress development for butter at 25°C. (Reproduced from Kokini, J.L. and Dickie, A., 1981, *Journal of Texture Studies*, 12: 539–557. With permission.)

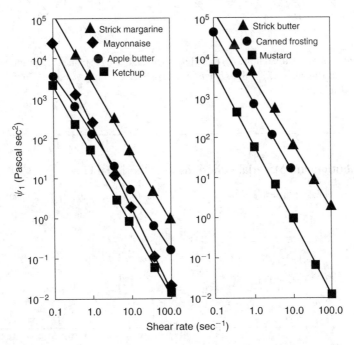

FIGURE 1.4 Steady primary normal stress coefficient ψ_1 vs. shear rate for semisolid foods at 25°C. (Reproduced from Kokini, J.L. and Dickie, A., 1981, *Journal of Texture Studies*, 12: 539–557. With permission.)

field:

$$v_x = -\tfrac{1}{2}\dot{\varepsilon}(1 + b)x$$
$$v_y = -\tfrac{1}{2}\dot{\varepsilon}(1 - b)y$$
$$v_z = +\dot{\varepsilon}\, z$$

where $0 \leq b \leq 1$ and $\dot{\varepsilon}$ is the elongation rate (Bird et al., 1987).

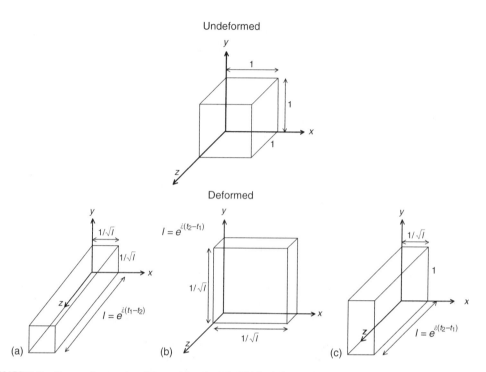

FIGURE 1.5 Types of extensional flows (a) uniaxial, (b) biaxial, and (c) planar. (Reproduced from Bird, R.B., Armstrong, R.C., and Hassager, O., 1987, *Dynamics of Polymeric Liquids*, 2nd ed., John Wiley & Sons Inc., New York. With permission.)

TABLE 1.1
Velocity Distribution and Material Functions in Extensional Flow

	Uniaxial ($b = 0$, $\dot{\varepsilon} > 0$)	Biaxial ($b = 0$, $\dot{\varepsilon} < 0$)	Planar ($b = 1$, $\dot{\varepsilon} > 0$)
Velocity distribution	$v_x = -\frac{1}{2}\dot{\varepsilon}x$	$v_x = +\dot{\varepsilon}x$	$v_x = -\dot{\varepsilon}x$
	$v_y = -\frac{1}{2}\dot{\varepsilon}y$	$v_y = -2\dot{\varepsilon}x$	$v_y = 0$
	$v_z = +\dot{\varepsilon}z$	$v_z = +\dot{\varepsilon}z$	$v_z = +\dot{\varepsilon}z$
Normal stress differences	$\sigma_{11} - \sigma_{22}$ and $\sigma_{11} - \sigma_{33}$	$\sigma_{11} - \sigma_{22}$ and $\sigma_{33} - \sigma_{22}$	$\sigma_{11} - \sigma_{22}$
Viscosity	$\eta_E = \dfrac{\sigma_{11} - \sigma_{22}}{\dot{\varepsilon}} = \dfrac{\sigma_{11} - \sigma_{33}}{\dot{\varepsilon}}$	$\eta_B = \dfrac{\sigma_{11} - \sigma_{22}}{\dot{\varepsilon}} = \dfrac{\sigma_{33} - \sigma_{22}}{\dot{\varepsilon}}$	$\eta_P = \dfrac{\sigma_{11} - \sigma_{22}}{\dot{\varepsilon}}$

There are three basic types of extensional flow: uniaxial, planar, and biaxial as shown in Figure 1.5. When a cubical material is stretched in one or two direction(s), it gets thinner in the other direction(s) as the volume of the material remains constant. During uniaxial extension the material is stretched in one direction which results in a corresponding size reduction in the other two directions. In biaxial stretching, a flat sheet of material is stretched in two directions with a corresponding decrease in the third direction. In planar extension, the material is stretched in one direction with a corresponding decrease in thickness while the height remains unchanged.

The velocity distribution in Cartesian coordinates and the resulting normal stress differences and viscosities for these three extensional flows are given in Table 1.1 (Bird et al., 1987).

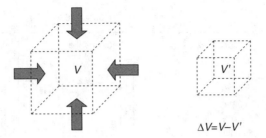

FIGURE 1.6 Volumetric strain.

The concept of extensional flow measurements goes back to 1906 with measurements conducted by Trouton. Trouton established a mathematical relationship between extensional viscosity and shear viscosity. The dimensionless ratio known as the Trouton number (N_T) is used to compare relative magnitude of extensional (η_E, η_B, or η_P) and shear (η) viscosities:

$$N_T = \frac{\text{extensional viscosity}}{\text{shear viscosity}}$$

The Trouton ratio for a Newtonian fluid is 3, 6, and 4 in uniaxial, biaxial, and planar extensions, respectively (Dealy, 1984).

$$\eta = \frac{\eta_E}{3} = \frac{\eta_B}{6} = \frac{\eta_P}{4}$$

1.2.3.3 Volumetric Flows

When an isotropic material is subjected to identical normal forces (e.g., hydrostatic pressure) in all directions, it deforms uniformly in all axes resulting in a uniform change (decrease or increase) in dimensions of a cubical element (Figure 1.6). In response to the applied isotropic stress, the specimen changes its volume without any change in its shape. This uniform deformation is called volumetric strain. An isotropic decrease in volume is called a compression, and an isotropic increase in volume is referred to as dilation (Darby, 1976). In this case all shear stress components will be zero and the normal stresses will be constant and equal:

$$\sigma_{ij} = \sigma \begin{bmatrix} 1 & 0 & 0 \\ 0 & 1 & 0 \\ 0 & 0 & 1 \end{bmatrix}$$

The bulk elastic properties of a material determine how much it will compress under a given amount of isotropic stress (pressure). The modulus relating hydrostatic pressure and volumetric strain is called the bulk modulus (K), which is a measure of the resistance of the material to the change in volume (Ferry, 1980). It is defined as the ratio of normal stress to the relative volume change:

$$K = \frac{\sigma}{\Delta V / V}$$

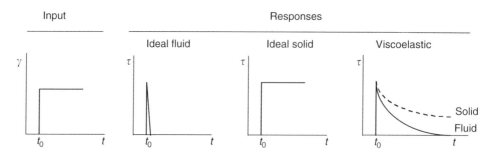

FIGURE 1.7 Response of ideal fluid, ideal solid, and viscoelastic materials to imposed step strain. (From Darby, R., 1976, *Viscoelastic Fluids: An Introduction to Their Properties and Behavior*, Dekker Inc., New York.)

1.2.4 RESPONSE OF VISCOUS AND VISCOELASTIC MATERIALS IN SHEAR AND EXTENSION

Viscoelastic properties can be measured by experiments which examine the relationship between stress and strain and strain rate in time dependent experiments. These experiments consist of (i) stress relaxation, (ii) creep, and (iii) small amplitude oscillatory measurements. Stress relaxation (or creep) consists of instantaneously applying a constant strain (or stress) to the test sample and measuring change in stress (or strain) as a function of time. Dynamic testing consists of applying an oscillatory stress (or strain) to the test sample and determining its strain (or stress) response as a function of frequency. All linear viscoelastic rheological measurements are related, and it is possible to calculate one from the other (Ferry, 1980; Macosko, 1994).

1.2.4.1 Stress Relaxation

In a stress relaxation test, a constant strain (γ_0) is applied to the material at time t_0, and the change in the stress over time, $\tau(t)$, is measured (Darby, 1976; Macosko, 1994). Ideal viscous, ideal elastic, and typical viscoelastic materials show different responses to the applied step strain as shown in Figure 1.7. When a constant stress is applied at t_0, an ideal (Newtonian) fluid responds with an instantaneous infinite stress. An ideal (Hooke) solid responds with instantaneous constant stress at t_0 and stress remains constant for $t > t_0$. Viscoelastic materials respond with an initial stress growth which is followed by decay in time. Upon removal of strain, viscoelastic fluids equilibrate to zero stress (complete relaxation) while viscoelastic solids store some of the stress and equilibrate to a finite stress value (partial recovery) (Darby, 1976).

The relaxation modulus, $G(t)$, is an important rheological property measured during stress relaxation. It is the ratio of the measured stress to the applied initial strain at constant deformation. The relaxation modulus has units of stress (Pascals in SI):

$$G(t) = \frac{\tau}{\gamma_0}$$

A logarithmic plot of $G(t)$ vs. time is useful in observing the relaxation behavior of different classes of materials as shown in Figure 1.8. In glassy polymers, there is a little stress relaxation over many decades of logarithmic time scale. cross-linked rubber shows a short time relaxation followed by a constant modulus, caused by the network structure. Concentrated solutions show a similar qualitative response but only at very small strain levels caused by entanglements. High molecular weight concentrated polymeric liquids show a nearly constant equilibrium modulus followed by a sharp fall at long times caused by disentanglement. Molecular weight has a significant impact on relaxation time, the smaller the molecular weight the shorter the relaxation time. Moreover,

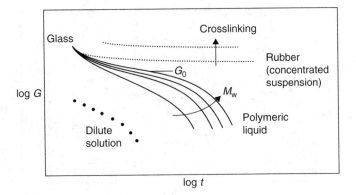

FIGURE 1.8 Typical relaxation modulus data for various materials. (Reproduced from Macosko, C.W., 1994, *Rheology: Principles, Measurements and Applications*, VCH Publishers, Inc., New York. With permission.)

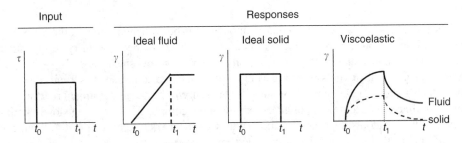

FIGURE 1.9 Response of ideal fluid, ideal solid, and viscoelastic materials to imposed instantaneous step stress. (From Darby, R., 1976, *Viscoelastic Fluids: An Introduction to Their Properties and Behavior*, Dekker Inc., New York.)

a narrower molecular weight distribution results in a much sharper drop in relaxation modulus. Uncross-linked polymers, dilute solutions, and suspensions show complete relaxation in short times. In these materials, $G(t)$ falls rapidly and eventually vanishes (Ferry, 1980; Macosko, 1994).

1.2.4.2 Creep

In a creep test, a constant stress (τ_0) is applied at time t_0 and removed at time t_1, and the corresponding strain $\gamma(t)$ is measured as a function of time. As in the case with stress relaxation, various materials respond in different ways as shown by typical creep data given in Figure 1.9. A Newtonian fluid responds with a constant rate of strain from t_0 to t_1; the strain attained at t_1 remains constant for times $t > t_1$ (no strain recovery). An ideal (Hooke) solid responds with a constant strain from t_0 to t_1 which is recovered completely at t_1. A viscoelastic material responds with a nonlinear strain. Strain level approaches a constant rate for a viscoelastic fluid and a constant magnitude for a viscoelastic solid. When the imposed stress is removed at t_1, the solid recovers completely at a finite rate, but the recovery is incomplete for the fluid (Darby, 1976).

The rheological property of interest is the ratio of strain to stress as a function of time and is referred to as the creep compliance, $J(t)$.

$$J(t) = \frac{\gamma(t)}{\tau_0}$$

The compliance has units of Pa^{-1} and describes how compliant a material is. The greater the compliance, the easier it is to deform the material. By monitoring how the strain changes as a function

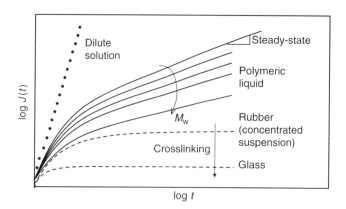

FIGURE 1.10 Typical creep modulus data for various materials. (From Ferry, J., 1980, *Viscoelastic Properties of Polymers*, 3rd ed., John Wiley & Sons, New York.)

of time, the magnitude of elastic and viscous components can be evaluated using available viscoelastic models. Creep testing also provides means to determine the zero shear viscosity of fluids such as polymer melts and concentrated polymer solutions at extremely low shear rates.

Creep data are usually expressed as logarithmic plots of creep compliance vs. time (Figure 1.10). Glassy materials show a low compliance due to the absence of any configurational rearrangements. Highly crystalline or concentrated polymers exhibit creep compliance increasing slowly with time. More liquid-like materials such as low molecular weight or dilute polymers show higher creep compliance and faster increase in $J(t)$ with time (Ferry, 1980).

1.2.4.3 Small Amplitude Oscillatory Measurements

In small amplitude oscillatory flow experiments, a sinusoidal oscillating stress or strain with a frequency (ω) is applied to the material, and the oscillating strain or stress response is measured along with the phase difference between the oscillating stress and strain. The input strain (γ) varies with time according to the relationship

$$\gamma = \gamma_0 \sin \omega t$$

and the rate of strain is given by

$$\dot{\gamma} = \gamma_0 \omega \cos \omega t$$

where γ_0 is the amplitude of strain.

The corresponding stress (τ) can be represented as

$$\tau = \tau_0 \sin(\omega t + \delta)$$

where τ_0 is the amplitude of stress and δ is shift angle (Figure 1.11).

$$\delta = 0 \qquad \text{for a Hookean solid}$$

$$\delta = 90^\circ \qquad \text{for a Newtonian fluid}$$

$$0 < \delta < 90^\circ \quad \text{for a viscoelastic material}$$

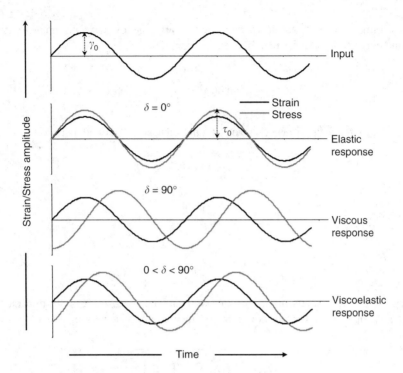

FIGURE 1.11 Input and response functions differing in phase by the angle δ. (From Darby, R., 1976, *Viscoelastic Fluids: An Introduction to Their Properties and Behavior*, Dekker Inc., New York.)

A perfectly elastic solid produces a shear stress in phase with the strain. For a perfectly viscous liquid, stress is 90° out of phase with the applied strain. Viscoelastic materials, which have both viscous and elastic properties, exhibit an intermediate phase angle between 0 and 90°. A solid like viscoelastic material exhibits a phase angle smaller than 45°, while a liquid like viscoelastic material exhibits a phase angle greater than 45°.

Two rheological properties can be defined as follows:

$$G'(\omega) = \frac{\tau_0}{\gamma_0} \cos \delta$$

$$G''(\omega) = \frac{\tau_0}{\gamma_0} \sin \delta$$

The storage modulus, G', is related to the elastic character of the fluid or the storage energy during deformation. The loss modulus, G'', is related to the viscous character of the material or the energy dissipation that occurs during the experiment. Therefore, for a perfectly elastic solid, all the energy is stored, that is, G'' is zero and the stress and the strain will be in phase. However, for a perfect viscous material all the energy will be dissipated that is, G' is zero and the strain will be out of phase by 90°.

By employing complex notation, the complex modulus, $G^*(\omega)$, is defined as

$$G^*(\omega) = G'(\omega) + iG''(\omega)$$

or

$$G^*(\omega) = \sqrt{(G'(\omega))^2 + (G''(\omega))^2}$$

Another commonly used dynamic viscoelastic property, the loss tangent, $\tan \delta(\omega)$, denotes ratio of viscous and elastic components in a viscoelastic behavior:

$$\tan \delta(\omega) = \frac{G''}{G'}$$

For fluid-like systems, appropriate viscosity functions can be defined as follows:

$$\eta' = \frac{G''}{\omega}$$

and

$$\eta'' = \frac{G'}{\omega}$$

where η' represents the viscous or in-phase component between stress and strain rate, while η'' represents the elastic or out-of-phase component. The complex viscosity η^* is equal to

$$\eta^* = \sqrt{\left(\frac{G'}{\omega}\right)^2 + \left(\frac{G''}{\omega}\right)^2}$$

The quantities of G', G'', and η', η'' collectively enable the rheological characterization of a viscoelastic material during small amplitude oscillatory measurements. The objective of oscillatory shear experiment is to determine these material specific moduli (G' and G'') over a wide range of frequency, temperature, pressure, or other material affecting parameters. Because of experimental constraints (e.g., weak torque values at low frequencies or large slip and inertial effects at high frequencies) it is usually impossible to measure $G'(\omega)$ and $G''(\omega)$ over 3 to 4 decades of frequency. However, the frequency range can be extended to the limits which are not normally experimentally attainable by the time–temperature superposition technique (Ferry, 1980).

Some rheologically simple materials obey the time–temperature superposition principle where time and temperature changes are equivalent (Ferry, 1980). Frequency data at different temperatures are superimposed by simultaneous horizontal and vertical shifting at a reference temperature. The resulting curve is called a master curve which is used to reduce data obtained at various temperatures to one general curve as shown in Figure 1.12. The time–temperature superposition technique allows an estimation of rheological properties over many decades of time.

The shift factor (a_T) for each curve has different values, which is a function of temperature. There are different methods to describe the temperature dependence of the horizontal shift factors. The William–Landel–Ferry (WLF) equation is the most widely accepted one (Ferry, 1980). The WLF equation enables to calculate the time (frequency) change at constant temperature, which is equivalent to temperature variations at constant time (frequency).

$$\log \left| \frac{\eta(T)}{\eta(T_{\text{ref}})} \right| = \log a_T = \frac{-C_1(T - T_{\text{ref}})}{C_2 + T - T_{\text{ref}}}$$

where $\eta(T)$ and $\eta(T_{\text{ref}})$ are viscosities at temperature T and T_{ref}, respectively. C_1 and C_2 are WLF constants for a given relaxation process.

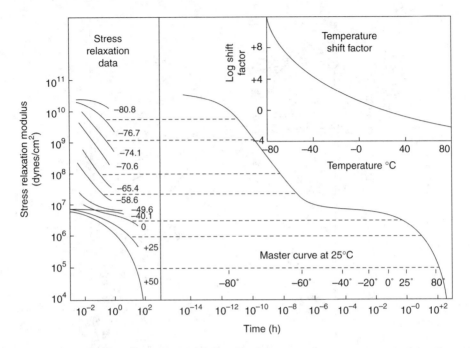

FIGURE 1.12 Construction of master curve using time–temperature superposition principle. (Reproduced from Sperling, L.H., 2001, *Introduction to Physical Polymer Science*, John Wiley and Sons, Inc., New York. With permission.)

1.2.4.4 Interrelations between Steady Shear and Dynamic Properties

Steady shear rheological properties and small amplitude oscillatory properties of fluid materials can be related. The steady viscosity function, η, can be related to the complex viscosity, η^*, and the dynamic viscosity function, η', while the primary normal stress coefficient, ψ_1, can be related to η''/ω. The Cox–Mertz rule (1954) suggests a way of obtaining a relation between the linear viscoelastic properties and the viscosity. It predicts that the magnitude of complex viscosity is equal to the viscosity at corresponding values of frequency and shear rate (Bird et al., 1987):

$$\eta^*(\omega) = \eta(\dot{\gamma})|_{\dot{\gamma}=\omega}$$

Figure 1.13 shows data to compare small amplitude oscillatory properties (η^*, η', and η''/ω) and steady rheological properties (η and ψ_1) for 0.50% and 0.75% guar (Mills and Kokini, 1984). Guar suspensions tend to a limiting Newtonian viscosity at low shear rates as is typical of many polymeric materials. At small shear rates η^* and η' are approximately equal and are very close in magnitude to the steady viscosity η. At higher shear rates η' and η^* diverge while η^* and η converge.

When the out-of-phase component of the complex viscosity is divided by frequency (η''/ω) it has the same dimensions as the primary normal stress coefficient, ψ_1. Both η''/ω and ψ_1, in the region where data could be obtained, are also plotted vs. shear rate/frequency as in Figure 1.13; η''/ω and ψ_1 curves show curvature at low shear rates. This is also consistent with observations in other macromolecular systems (Ferry, 1980; Bird et al., 1987). Moreover, the rate of change in the magnitude of ψ_1 closely follows that of η''/ω.

A second example is shown for 3% gum karaya, which is a more complex material (Figure 1.14). Both steady and dynamic properties of gum karaya deviate radically from the rheological behavior observed with guar gum. First, within the shear rate range studied, η^* was higher than η. This is in

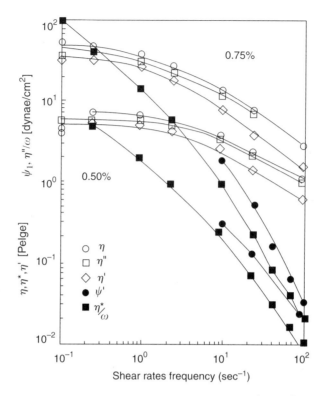

FIGURE 1.13 Comparison of small amplitude oscillatory properties (η^*, η', and η''/ω) and steady rheological properties (η and ψ_1) for 0.5% and 0.75% guar. (Reproduced from Mills, P.L. and Kokini, J.L., 1984, *Journal of Food Science*, 49: 1–4, 9. With permission.)

contrast to the behavior observed with guar, where η was either equal to nor higher than η^*. Second, none of the three viscosities approached a zero shear viscosity in the frequency/shear rate range studied. Third, the steady viscosity function η was closer in magnitude to η' than η^* and seemed to be nonlinearly related to both η' and η^*. Finally, values of ψ_1 were smaller than values of η''/ω, in contrast to observations with guar where ψ_1 was larger than η''/ω (Mills and Kokini, 1984).

There are several theories (Spriggs et al., 1966; Carreau et al., 1968; Chen and Bogue, 1972) which essentially predict two major kinds of results for the interrelationship between steady and dynamic macromolecular systems. These results can be summarized as follows:

$$\eta'(\omega) = \eta(\dot{\gamma})|_{\dot{\gamma}=\omega}$$

$$\frac{2\eta^*(\omega)}{\omega} = \psi_1(\dot{\gamma})|_{\dot{\gamma}=\omega}$$

$$\eta'(c\omega) = \eta(\dot{\gamma})|_{\dot{\gamma}=c\omega}$$

$$\frac{\eta''(c\omega)}{\omega} = \psi_1(\dot{\gamma})|_{\dot{\gamma}=c\omega}$$

These equations are strictly applicable at small shear rates. At large shear rates the Cox–Mertz rule applies. For guar in the range of shear rates between 0.1 and 10 sec^{-1}, η^* is equal to η. Similarly, in the zero shear region $2\eta''/\omega$ is approximately equal to ψ_1. In the case of gum karaya, on the other

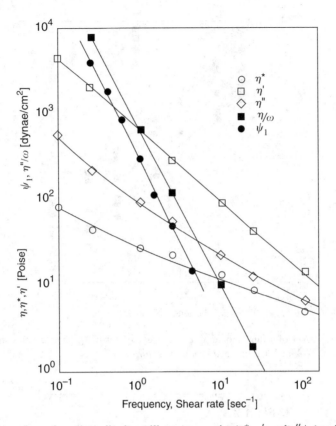

FIGURE 1.14 Comparison of small amplitude oscillatory properties (η^*, η', and η''/ω) and steady rheological properties (η and ψ_1) for 3% karaya. (Reproduced from Mills, P.L. and Kokini, J.L., 1984, *Journal of Food Science*, 49: 1–4, 9. With permission.)

hand, nonlinear relationships are needed as follows (Mills and Kokini, 1984):

$$\eta^* = c[\eta(\dot{\gamma})]^{\alpha}|_{\dot{\gamma}=\omega}$$

$$\eta' = c'[\eta(\dot{\gamma})]^{\alpha'}|_{\dot{\gamma}=\omega}$$

$$\frac{\eta''(\omega)}{\omega} = c[\psi_1(\dot{\gamma})]^{\alpha}|_{\dot{\gamma}=\omega}$$

Similar results are obtained in the case of semisolid food materials, as shown in Figure 1.15 (Bistany and Kokini, 1983b). Values for the constants c and α, c', and α' for a variety of food materials are shown in Tables 1.2 and 1.3. It can be seen from these figures and tables that semisolid foods follow the above relationships.

A dimensional comparison of the primary normal stress coefficient ψ_1 and G'/ω^2 shows that these quantities are dimensionally consistent, both possessing units of Pa sec^2. The primary normal stress coefficient ψ_1 and G'/ω^2 vs. frequency followed power law behavior as seen in Figure 1.16. As with viscosity, a nonlinear power law relationship can be formed between G'/ω^2 and ψ_1,

$$\frac{G'}{\omega^2} = c^*[\psi_1(\dot{\gamma})]^{\alpha*}|_{\dot{\gamma}=\omega}$$

The values for the constants c^* and α^* for a variety of foods are given in Table 1.4.

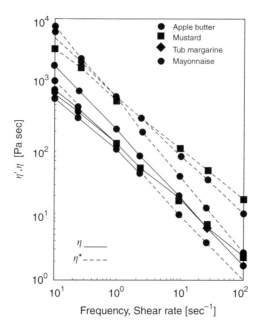

FIGURE 1.15 Comparison of η^* and η for apple butter, mustard, tub margarine, and mayonnaise. (Reproduced from Bistany, K.L. and Kokini, J.L., 1983a, *Journal of Texture Studies*, 14: 113–124. With permission.)

TABLE 1.2

Empirical Constants for $\eta^* = c[\eta(\dot{\gamma})]^\alpha|_{\dot{\gamma}=\omega}$

Food	α	C	R^2
Whipped cream cheese	0.750	93.21	0.99
Cool whip	1.400	50.13	0.99
Stick butter	0.986	49.64	0.99
Whipped butter	0.948	43.26	0.99
Stick margarine	0.934	35.48	0.99
Ketchup	0.940	13.97	0.99
Peanut butter	1.266	13.18	0.99
Squeeze margarine	1.084	11.12	0.99
Canned frosting	1.208	4.40	0.99
Marshmallow fluff	0.988	3.53	0.99

Source: Bistany, K.L. and Kokini, J.L., 1983b, *Journal of Rheology*, 27: 605–620.

1.3 METHODS OF MEASUREMENT

There are many test methods used to measure rheological properties of food materials. These methods are commonly characterized according to (i) the nature of the method such as fundamental and empirical; (ii) the type of deformation such as compression, extension, simple shear, and torsion; (iii) the magnitude of the imposed deformation such as small or large deformation (Bird et al., 1987; Macosko, 1994; Steffe, 1996; Dobraszczyk and Morgenstern, 2003).

TABLE 1.3
Empirical Constants for $\eta' = c'[\eta(\dot{\gamma})]^{\alpha'}|_{\dot{\gamma}=\omega}$

Food	α'	c'	R^2
Whipped cream cheese	0.847	9.52	0.98
Cool whip	1.732	6.16	0.99
Whipped butter	1.082	5.84	0.99
Ketchup	0.897	5.14	0.99
Peanut butter	1.272	4.78	0.99
Squeeze margarine	1.042	3.57	0.99
Marshmallow fluff	1.078	1.22	0.99
Stick margarine	1.202	1.02	0.99
Stick butter	1.339	0.94	0.99
Canned frosting	1.520	0.16	0.95

Source: Bistany, K.L. and Kokini, J.L., 1983b, *Journal of Rheology*, 27: 605–620.

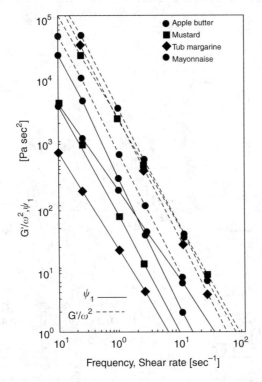

FIGURE 1.16 Comparison of G'/ω and ψ_1 for apple butter, mustard, tub margarine, and mayonnaise. (Reproduced from Bistany, K.L. and Kokini, J.L., 1983a, *Journal of Texture Studies*, 14: 113–124. With permission.)

1.3.1 SHEAR MEASUREMENTS

Steady shear rheological properties of semisolid foods have been studied by many laboratories (Kokini et al., 1977; Kokini and Dickie, 1981; Rao et al., 1981; Barbosa-Canovas and Peleg, 1983; Dickie and Kokini, 1983; Kokini et al., 1984a; Rahalkar et al., 1985; Dervisoglu and Kokini,

TABLE 1.4
Empirical Constants for $G'/\omega^2 = c^*[\psi_1(\dot{\gamma})]^{\alpha^*}|_{\dot{\gamma}=\omega}$

Food	α^*	c^*	R^2
Squeeze margarine	1.022	52.48	0.99
Whipped butter	1.255	33.42	0.99
Ketchup	1.069	14.15	0.99
Whipped cream cheese	1.146	13.87	0.99
Cool whip	1.098	6.16	0.99
Canned frosting	1.098	4.89	0.99
Peanut butter	1.124	1.66	0.99
Stick margarine	1.140	1.28	0.99
Marshmallow fluff	0.810	1.26	0.99
Stick butter	1.204	0.79	0.99

Source: Bistany, K.L. and Kokini, J.L., 1983b, *Journal of Rheology*, 27: 605–620.

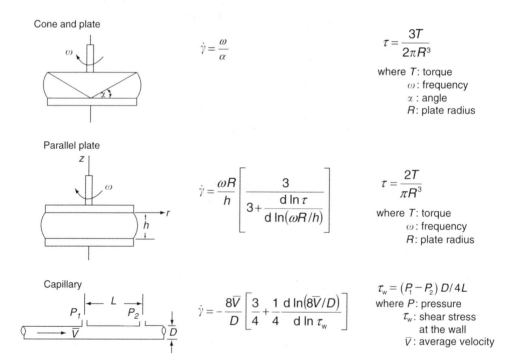

Cone and plate

$$\dot{\gamma} = \frac{\omega}{\alpha}$$

$$\tau = \frac{3T}{2\pi R^3}$$

where T: torque
ω: frequency
α: angle
R: plate radius

Parallel plate

$$\dot{\gamma} = \frac{\omega R}{h}\left[\frac{3}{3+\dfrac{d\ln\tau}{d\ln(\omega R/h)}}\right]$$

$$\tau = \frac{2T}{\pi R^3}$$

where T: torque
ω: frequency
R: plate radius

Capillary

$$\dot{\gamma} = -\frac{8\bar{V}}{D}\left[\frac{3}{4}+\frac{1}{4}\frac{d\ln(8\bar{V}/D)}{d\ln\tau_w}\right]$$

$\tau_w = (P_1 - P_2)D/4L$
where P: pressure
τ_w: shear stress
at the wall
$\bar{V}$: average velocity

FIGURE 1.17 Commonly used geometries for shear stress and shear rate measurements.

1986a and 1986b; Kokini and Surmay, 1994; Steffe, 1996; Gunasekharan and Ak, 2000). The most commonly used experimental geometries for achieving steady shear flow are the capillary, cone and plate, parallel-plate, and couette geometries referred to as narrow gap rheometers and are shown in Figure 1.17 with appropriate equations to estimate shear stresses and shear rates.

The use of narrow gap rheometers is limited to relatively small shear rates. At high shear rates, end effects arising from the inertia of the sample make measurements invalid (Walters, 1975). The edge

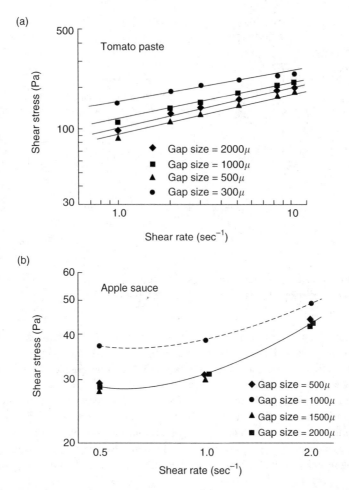

FIGURE 1.18 Effect of gap size on measurement of shear stress as a function of shear rate for (a) tomato paste and (b) applesauce using the parallel plate geometry. (Reproduced from Dervisoglu, M. and Kokini, J.L., 1986b, *Journal of Food Science*, 51: 541–546, 625. With permission.)

and end effects result mainly from the fracturing of the sample at high shear rates. At high rotational speeds, secondary flows are generated, making rheological measurements invalid. Another limitation of narrow-gap rheometers results from the fact that some suspensions contain particles comparable in size to the gap between the plates (Mitchell and Peart, 1968; Bongenaar et al., 1973; Dervisoglu and Kokini, 1986b). This limitation is most pronounced in cone and plate geometry, where the tip of the cone is almost in contact with the plate. In cases where the particle size is comparable to the gap between the plates, large inaccuracies are introduced due to particle–plate contact. In parallel plate geometry this limitation may be improved to a certain extent by increasing gap size. However, the gap size selected should still be much smaller than the radius of the plate.

An example for the case of tomato paste is shown in Figure 1.18a. With tomato paste, the effect of particle-to-plate contact was observed for gap sizes smaller than 500 μm. At gap sizes larger than 500 μm, measured shear stresses increased with increasing gap size. This is thought to be due to the dependence of shear stress values on structure breakdown during loading. As the gap is increased, structure breakdown due to loading decreases since the sample is not squeezed as much. A second example for applesauce is shown in Figure 1.18b. At the smallest gap size of 500 μm, shear stress values are largest, suggesting that particle-to-plate contact controls the resistance to flow. For gap sizes larger than 1000 μm, shear stress measurements no longer depend on gap size.

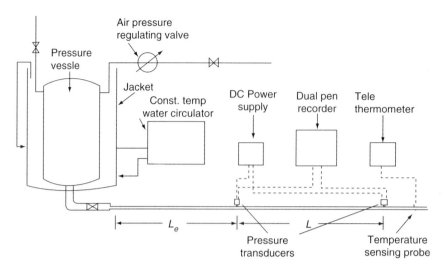

FIGURE 1.19   Schematic diagram of the capillary set-up. (Reproduced from Dervisoglu, M. and Kokini, J.L., 1986b, *Journal of Food Science*, 51: 541–546, 625. With permission.)

In capillary flow, shear stresses and shear rates are calculated from the measured volumetric flow rates and pressure drops as well as the dimensions of the capillary, as shown in Figure 1.17 (Toledo, 1980). There are, however, two important effects that need to be considered with non-Newtonian materials: the entrance effect and the wall effects.

The entrance effect in capillary flow is due to abrupt changes in the velocity profile when the material is forced from a large diameter reservoir into a capillary tube. This effect can be effectively eliminated by using a long entrance region and by determining the pressure drop as the difference of two pressure values measured in the fully developed laminar flow region (i.e., away from the entrance region). Dervisoglu and Kokini (1986b) developed the rheometer shown in Figure 1.19 based on these ideas.

When the entrance effects cannot be eliminated, Bagley's procedure (1957) allows for correction of the data. In this procedure the entrance effects are assumed to increase the length of the capillary because streamlines are stretched so that the true shear stress is considered equal to:

$$\tau = \frac{\Delta P R}{2(L + eR)}$$

where ΔP is the total pressure drop, L and R are the length and the radius of the capillary, and e is Bagley end correction factor. Rearranging this equation the more useful form is obtained

$$\Delta P = 2\tau \frac{L}{R} + 2\tau e$$

Plotting ΔP vs. L/R allows estimation of the true shear stress through the slope of the line, and e is estimated through the value of L/R where $\Delta P = 0$. This estimation procedure is shown in Figure 1.20.

The wall effect in capillary flow results from interactions between the wall of the capillary and the liquid in the vicinity of the wall. In many polymer solutions and suspensions the velocity gradient near the wall may induce some preferred orientation of polymeric molecules or drive suspended particles away from the wall generating effectively a slip like phenomenon (Skelland, 1967). The suspended particles tend to move away from the wall region, leaving a low viscosity thin layer adjacent to the wall (Serge and Silberberg, 1962; Karnis et al., 1966). This in turn causes higher flow rates at a given

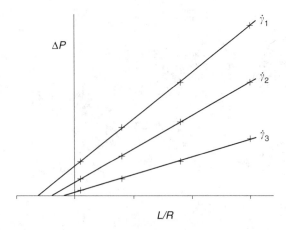

FIGURE 1.20 Bagley plot for entry pressure drop at different shear rates (P = pressure, R = radius, L = length, $\dot{\gamma}$ = shear rate).

pressure drop as if there were an effective slip at the wall surface. The wall effect associates with the capillary flow of polymer solutions, and suspensions can therefore be characterized by a slip velocity at the wall (Oldroyd, 1949; Jastrzebsky, 1967; Kraynik and Schowalter, 1981).

If the slip coefficient is defined as $\beta_c = v_s R / \tau_w$ (Jastrzebsky, 1967; Kokini and Dervisoglu, 1990) then it can be shown that:

$$\frac{Q}{\pi R^3 \tau_w} = \frac{\beta_c}{R^2} + \frac{1}{\tau_w^4} \int_0^{\tau_w} \tau^2 f(\tau) d\tau$$

where Q is the flow rate. Plotting $Q / \pi R^3 \tau_w$ vs. $1/R^2$ at constant τ_w gives β_c as the slope of the line. Corrected flow rates can now be calculated using (Goto and Kuno, 1982):

$$Q_c = Q - \pi R \tau_w \beta_c$$

and the true shear rate at the wall is given by

$$\dot{\gamma} = \left(\frac{3n+1}{n} \right) \frac{Q}{\pi R^3}$$

An example of such data is shown for apple sauce in Figure 1.21 as a function of tube diameter. The data clearly indicates a strong dependency of flow behavior on tube diameter. Smaller shear stress values are observed for smaller tube diameters. The wall effect is also greater at the smaller shear rates.

When $Q / \pi R^3 \tau_w$ calculated at constant wall shear stresses are plotted against $1/R^2$ as in Figure 1.22, the corrected slip coefficients, β_c, can be calculated from the slopes of the resulting lines at specific values of τ_w. The corresponding true shear rates can then be calculated. The different flow curves obtained with different tube diameters can then be used to generate a true flow curve after being corrected for apparent slip as shown in Figure 1.21 for applesauce.

Narrow gap geometries give the rheologist a lot of flexibility in terms of measuring rheological properties at different shear rate ranges and to be used for different purposes. For example, when data are necessary at small shear rates, the cone and plate or parallel plate geometry can be used. This would be particularly useful in understanding structure–rheology relationships. A capillary rheometer can be used if flow data at high shear rates of most processing operations are needed.

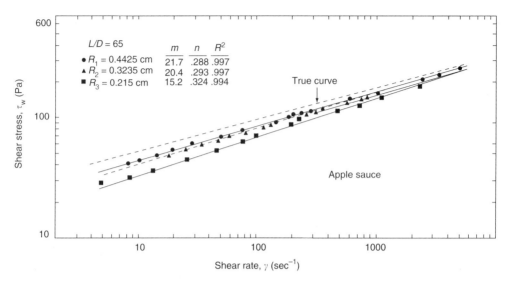

FIGURE 1.21 Effect of tube diameter on measurement of wall shear stress as a function of wall shear rate for applesauce using capillary rheometer. (Reproduced from Kokini, J.L. and Dervisoglu, M., 1990, *Journal of Food Engineering*, 11: 29–42. With permission.)

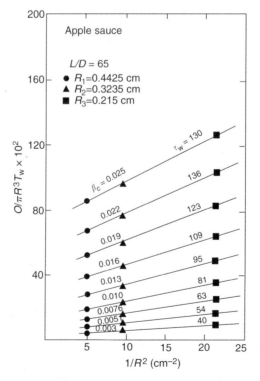

FIGURE 1.22 Determination of slip coefficients β_c at constant wall shear stress through plots of $Q/\pi R^3 \tau_w$ vs. $1/R^2$. The slope of the line is equal to β_c. (Reproduced from Kokini, J.L. and Dervisoglu, M., 1990, *Journal of Food Engineering*, 11: 29–42. With permission.)

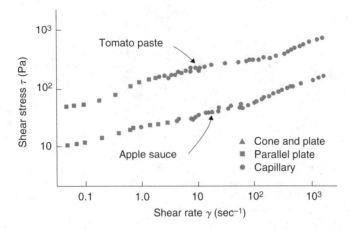

FIGURE 1.23 Superposition of cone and plate, parallel plate and capillary, and shear stress-shear rate data for tomato paste and applesauce. (Reproduced from Dervisoglu, M. and Kokini, J.L., 1986b, *Journal of Food Science*, 51: 541–546, 625. With permission.)

When rheological measurements are conducted with knowledge of their limitations, and appropriate corrections are made, superposition of cone and plate, parallel plate, and capillary flow measurements can be obtained. Examples of such superpositions are given for ketchup and mustard in Figure 1.23.

1.3.2 SMALL AMPLITUDE OSCILLATORY MEASUREMENTS

Small amplitude oscillatory measurements have become very popular for a lot of foods that are shear sensitive and are not well suited for steady shear measurements. These include hydrocolloid solutions, doughs, batter, starch solutions, and fruit and vegetable purees among many others. One of the major advantages of this method is that it provides simultaneous information on the elastic (G') and viscous (G'') nature of the test material. Due to its nondestructive nature, it is possible to conduct multiple tests on the same sample under different test conditions including temperature, strain, and frequency (Gunasekaran and Ak, 2000; Dobraszczyk and Morgenstern, 2003).

During dynamic testing samples held in various geometries are subjected to oscillatory motion. A sinusoidal strain is applied on the sample, and the resulting sinusoidal stress is measured or vice versa. The cone and plate or parallel plate geometries are usually used. The magnitude of strain used in the test is very small, usually in the order of 0.1–2%, where the material is in the linear viscoelastic range.

Typical experimentally observed behavior of η^*, G', and G'' for a dilute hydrocolloid solution, a hydrocolloid gel, and a concentrated hydrocolloid solution are shown in Figure 1.24 (Ross-Murphy, 1988). In dilute hyrocolloid solutions (Figure 1.24a), storage of energy is largely by reversible elastic stretching of the chains under applied shear, which results in conformations of higher free energy, while energy is lost in the frictional movement of the chains through the solvent. At low frequencies the principal mode of accommodation to applied stress is by translational motion of the molecules, and G'' predominates, as the molecules are not significantly distorted. With increasing frequency, intramolecular stretching and distorting motions become more important and G' approaches G''.

By contrast, hydrocolloid gels are interwoven networks of macromolecules and would be subject primarily to intramolecular stretching and distorting. The network bonding forces prevent actual transnational movement; therefore, these materials show properties approaching those of an elastic solid (Figure 1.24b). G' predominates over G'' at all frequencies and neither shows any appreciable frequency dependence. For concentrated solutions at high frequencies (Figure 1.24c), where inter-chain entanglements do not have sufficient time to come apart within the period of one oscillation,

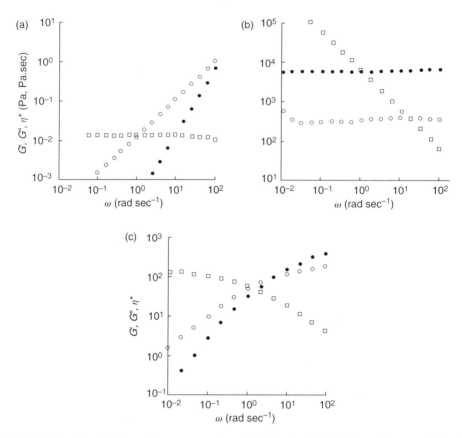

FIGURE 1.24 Small amplitude oscillatory properties, G', G'', and η^* for (a) dilute dextran solution, (b) an agar gel of 1 g/dL concentration, and (c) an λ-carrageenan solution of 5 g/dL concentration. (Reproduced from Ross-Murphy, S.B., 1988, Small deformation measurements. In: *Food Structure — Its Creation and Evaluation*, J.M.V. Blanshard and J.R. Mitchell, Eds, Butterworth Publishing Co., London. With permission.)

the concentrated solution begins to approximate the behavior of a network and higher G' values are obtained. When the frequency is so high that translational movements are no longer possible, they start behaving similarly as to true gels, with G' greater than G'' and showing little change with frequency (Ross-Murphy, 1988).

Small amplitude oscillatory measurements have been used to study the rheological properties of many foods and in particular wheat flour doughs. Smith et al. (1970) showed that as protein content increased in a protein (gluten)–starch–water system the magnitude of both the storage and loss moduli increased. Dus and Kokini (1990) used the Bird–Carreau model used to predict the steady viscosity (η), the primary normal stress coefficient (ψ_1), and the small amplitude oscillatory properties (η' and η''/ω) for a hard wheat flour dough containing 40% total moisture in the region of frequencies of 0.01 to 100 rad/sec for the dynamic viscoelastic properties and a region of 10^{-5} through 10^3 sec^{-1} for steady shear properties.

Small amplitude oscillatory measurements have the limitation of not being appropriate in practical processing situations due to the rates at which the test is applied. Typical examples include dough mixing and expansion and oven rise during baking. The extension rates of expansion during fermentation and oven rise are in the range of 5×10^{-3} and 5×10^{-4} sec^{-1} and are several orders of magnitude smaller than the rates applied during small amplitude oscillatory measurements (Bloksma, 1990). Small strains are also not comparable to the actual strain levels encountered during dough expansion. Strain in gas expansion during proofing is reported to be in the region of several hundred percent (Huang and Kokini, 1993). In such cases, tests resulting in large deformation levels such as extensional methods are applied.

1.3.3 EXTENSIONAL MEASUREMENTS

Extensional flow is commonly encountered in many food processing such as dough sheeting, sheet stretching, drawing and spinning, CO_2 induced bubble growth during dough fermentation, die swell during extrudate expansion due to vaporization of water, squeezing to spread a product (Padmanabhan, 1995; Brent et al., 1997; Huang and Kokini, 1999; Gras et al., 2000; Charalambides et al., 2002a; Nasseri et al., 2004; Sliwinski et al., 2004a, 2004b). Extensional flow is also associated with mixing, particularly dough mixing (Bloksma, 1990; Dobraszczyk et al., 2003). It is an important factor in the human perception of texture with regard to the mouthfeel and swallowing of fluid foods (Kokini, 1977; Dickie and Kokini, 1983; Elejalde and Kokini, 1992a; Kampf and Peleg, 2002).

Shear and extensional flow have a different influence on material behavior since the molecules orient themselves in different ways in these flow fields. Presence of velocity gradients in shear flow causes molecules to rotate (Darby, 1976). Rotation action reduces the degree of stretching. However, in extensional flow the molecules are strongly oriented in the direction of the flow field since there are no forces to cause rotation. Long chain high molecular weight polymer melts are known to behave differently in shear and extensional fields (Dobraszczyk and Morgenstern, 2003). The nature of the molecule influences its flow behavior in extension significantly. Linear molecules align themselves in the direction of extensional flow more easily than branched molecules. Similarly, stiffer molecules are more quickly oriented in an extensional flow field. The molecular orientations caused by extensional flow leads to the development of final products with unique textures (Padmanabhan, 1995).

While shear rheological properties of food materials have been studied extensively, there are a limited number of studies on extensional properties of food materials due to the difficulty in generating controlled extensional flows with foods. Several studies have been done to investigate the extensional properties of wheat flour dough in relation to bread quality, which usually involve empirical testing devices such as alveograph, extensigraph, mixograph, and farinograph. Brabender Farinograph is the first special instrument designed for the physical testing of doughs in about the 1930s (Janssen et al., 1996b). Then the National Mixograph, the Brabender Extensigraph, and the Chopin Alveograph were developed. The farinograph and mixograph record the torque generated during dough mixing. In the extensigraph, doughs are subjected to a combination of shear and uniaxial extension, while in the alveograph, doughs are subjected to biaxial extension.

Empirical test are widely used in routine analysis, usually for quality control purposes, since they are easy to perform, provide useful practical data for evaluating the performance of dough during processing. In these empirical tests the sample geometry is variable and not well-defined; the stress and strain are not controllable and uniform throughout the test. Since the data obtained cannot be translated into a well-defined physical quantity the fundamental interpretation of the experimental results is extremely difficult.

There are several fundamental rheological tests that have been developed for measuring the extensional properties of polymeric liquids over the last 30 years. Some of these techniques are used to measure the extensional behavior of food materials. Macosko (1994) classified the extensional flow measurement methods in several geometries as shown in Table 1.5. Mathematical equations to convert measured forces and displacements into stresses and strains, which are in turn used to calculate extensional material functions, are given in detail in Macosko (1994). The strengths and weaknesses of each method are also discussed in detail in this excellent text of rheology. Readers should also refer to an extensive review on the fiber wind-up, the entrance pressure drop technique for high viscosity liquids, and the opposed jets device for low viscosity liquids (Padmanabhan and Bhattacharya, 1993b; Padmanabhan, 1995).

TABLE 1.5
Extensional Flow Measurement
Methods

Simple extension
　End clamps
　Rotating clamps
　Buoyancy baths
　Spinning drop
Lubricated compression
Sheet stretching, multiaxial extension
　Rotating clamps
　Inflation methods
Fiber spinning (tubeless siphon)
Bubble collapse
Stagnation flows
　lubricated and unlubricated dies
　opposed nozzles
Entrance flows

It has long been recognized that baking performance and bread quality are strongly dependent on the rheological properties of the dough used (Huang and Kokini, 1993; Janssen, 1996b; Huang, 1998; Dobraszczyk et al., 2003). The extensional behavior of wheat dough is of special interest since it relates directly to deformations during mixing and bubble growth during fermentation and baking (Dobraszczyk et al., 2003). Extensional viscosity data of wheat doughs are useful in predicting the functional properties of bread, such as loaf volume. Gas cell expansion leading to loaf volume development during baking is a largely biaxial stretching flow (Bloksma and Nieman, 1975; de Bruijne et al., 1990). Bloksma (1990) estimated that extensional rates during bread dough fermentation range from 10^{-4} to 10^{-3} sec^{-1} and that during oven rise are approximately 10^{-3} sec^{-1}. Experimental measurements have to be performed in these ranges of extension rate in order to predict the performance of these processes accurately.

Extensional deformation of dough has been widely studied using the mechanical testing apparatus (Gras et al., 2000; Newberry et al., 2002; Sliwinski et al., 2004a, 2004b), bubble inflation technique (Hlynka and Barth, 1955; Joye et al., 1972; Launay et al., 1977; Huang and Kokini, 1993; Charalambides et al., 2002a, 2002b; Dobraszczyk et al., 2003), lubricated squeezing flow technique (Huang and Kokini, 1993; Janssen et al., 1996a and 1996b; Nasseri et al., 2004).

The bubble inflation method is the most popular in the dough industry as it simulates the expansion of gas cells during proof and oven rise (Bloksma, 1990; Huang and Kokini, 1993; Charalambides et al., 2002a, 2002b). In this technique, a thin circular material sheet is clamped around its perimeter and inflated using pressurized air (Figure 1.25). The thickness of bubble wall during bubble inflation varies, with a maximum deformation near the pole and a minimum at the rim (Figure 1.26). Bloksma (1957) derived an analysis that takes into account this nonuniformity in thickness where the wall thickness distribution of the inflating bubble is given as:

$$t = t_0 \left\{ \frac{a^4 + s^2 h^2(t)}{a^2[a^2 + h^2(t)]} \right\}$$

where t_0 is the original sample thickness, a is the original sample radius, and h is the bubble height. With the knowledge of the thickness around the bubble, Launay et al. (1977) calculated the strain

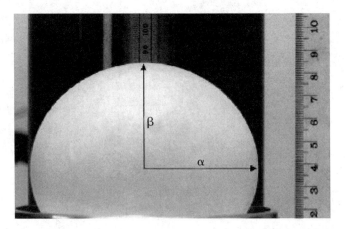

FIGURE 1.25 Inflated sample. (Reproduced from Charalambides, M.N., Wanigasooriya, L., Williams, J.G., and Chakrabarti, S., 2002a, *Rheologica Acta*, 41: 532–540. With permission.)

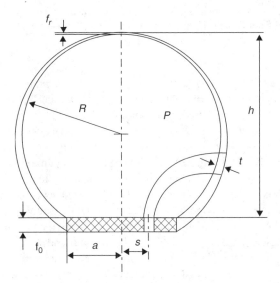

FIGURE 1.26 Geometry of bubble inflation. (Reproduced from Charalambides, M.N., Wanigasooriya, L., Williams, J.G., and Chakrabarti, S., 2002a, *Rheologica Acta*, 41: 532–540. With permission.)

in the axisymmetric direction as:

$$\varepsilon_B = -\frac{1}{2} \ln \left[\frac{t}{t_0} \right]$$

In extensional flow experiments, maintaining steady extensional flow in the tested sample for a sufficient time to determine the steady extensional viscosity is the basic problem (Jones et al., 1987). Huang and Kokini (1993) used the biaxial extensional creep first developed by Chatraei et al. (1981). They measured the biaxial extensional viscosity of wheat flour doughs using a lubricated squeeze film apparatus with an extension rate of 0.011 sec^{-1}. Obtaining steady extensional flow necessitated 10 to 200 sec depending on the magnitude of normal stresses which ranged from 5.018 to 0.361 kPa. Doughs with different protein contents (13.2%, 16.0%, and 18.8%) showed different biaxial extensional viscosities. The extensional viscosity of doughs increased with increasing protein

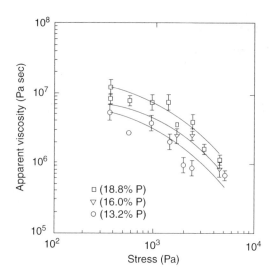

FIGURE 1.27   The effect of protein content of wheat doughs on the biaxial extensional viscosities at different stress levels. (Reproduced from Huang, H. and Kokini, J.L., 1993, *Journal of Rheology*, 37: 879–891. With permission.)

content (Figure 1.27). Results showed that different wheat doughs can be prepared by manipulating the protein contents to maintain desired extensional properties which are required for processing of specific baked foods (such as pasta, bread, and cookie).

During the last two decades several extensive experimental and theoretical studies on the rheology of polymer melts during extensional deformation were performed. Techniques have been developed to study the behavior of polymers in equal biaxial (equibiaxial) and uniaxial extensional flows. In the early 1970s, Meissner developed practical methods for extensional flow measurements based on the fixed rotating clamp (Meissner et al., 1981). Chatraei et al. (1981) developed and used the lubricated squeezing flow technique, which provides a simple way to perform extensional measurements, where the sample is squeezed between a moving and a fixed plate. Since then the technique has been used by various research groups to measure the rheology of soft solids and structured fluids (Kompani and Venerus, 2000). Although squeeze flow is a simple and convenient method, it is used less frequently than conventional methods due to its transient nature and complications described by Meeten (2002). Recently, Campanella and Peleg (2002) reviewed extensively the theory, applications, and artifacts of squeezing flow viscometry for semiliquid foods.

The lubricated squeezing flow technique has been used for the characterization of a growing number of semiliquid and soft solid foods extensively due to its simplicity and versatility. Suwonsichon and Peleg (1999a), and Kamf and Peleg (2002) used imperfect squeezing flow method as a tool to assess the consistency of mustards with seeds and chickpea pastes (humus), which gives complementary information to steady shear measurements. Suwonsichon and Peleg (1999b, 1999c, 1999d) also worked on rheological characterization of commercial refried beans, stirred yogurt, and ricotta cheese by squeezing flow viscometry. Corradini et al. (2000) used Teflon coated parallel plates to generate lubricated squeezing flow which allowed to calculate the elongational viscosity of commercial tomato paste, low fat mayonnaise, and mustard samples as a function of biaxial strain rate.

1.3.4   STRESS RELAXATION

From the theory of linear viscoelasticity, the linear response to any type of deformation can be predicted using the relaxation modulus, $G(t)$, in the linear viscoelastic region. Constitutive equations, such as generalized Maxwell model, can be used to simulate the linear relaxation modulus.

The relaxation modulus with N Maxwell elements is given as:

$$G(t) = \sum_{i=1}^{N} G_i \, e^{t/\lambda_i}$$

Nonlinear regression of experimental data provides N sets of relaxation times (λ_i) and moduli (G_i). Among the discrete spectra, the G_i–λ_i pair with longest value of λ_i dominates the $G(t)$ behavior in the terminal zone (Huang, 1998).

If the spectrum of relaxation times is continuous instead of discrete, the relaxation modulus can be defined in integral form. The continuous relaxation spectrum, $H(\lambda)$, can be obtained from relaxation modulus $G(t)$:

$$G(t) = \int_{-\infty}^{+\infty} H(\lambda) \, e^{-t/\lambda} \, d(\ln \lambda)$$

The relaxation time spectrum contains the complete information on the distribution of relaxation times which is very useful in describing a material's response to a given deformation history. The linear viscoelastic material functions can simply be calculated from the relaxation time spectrum. Sets of G_i–λ_i pairs need to be obtained from the simulation of the experimental data in order to convert the measured dynamic material functions into the relaxation time spectrum of the test sample (Ferry, 1980; Orbey and Dealy, 1991; Mead, 1994). However, calculation of relaxation time spectrum from material functions has many numerical difficulties. Inversion of the integrals may result in extremely unstable problems, which are called ill-posed problems (Honerkamp and Weese, 1989). There can be infinite numbers of solutions that can fit the criterion of the nonlinear regression method used (Honerkamp and Weese, 1990). Furthermore, the relaxation moduli G_i depend strongly on the initial choice of relaxation time λ_i (Dealy and Wissbrun, 1990).

A mathematical solution is considered an ill-posed problem if a function $f(t)$ cannot be measured experimentally but can be related to experimental data $g(t)$ using:

$$g(t) = K[f(t)] + \varepsilon(t)$$

where K is the operator that relates $f(t)$ to $g(t)$, and $\varepsilon(t)$ is the error function. To obtain $f(t)$, it is necessary to use the inverse relation:

$$f = K^{-1}[g]$$

This inverse problem is an ill-posed problem as small errors in $g(t)$ will result in large errors in $f(t)$. The problem is tackled by supplying a regularization parameter which adds additional constraints to the solution of $f(t)$ (Roths et al., 2001). Many methods have been developed for solving such ill-posed problems most of which are regularization methods. Tikhonov regularization is one of the oldest and most common techniques (Honerkamp and Weese, 1989; Weese, 1992).

In a particular study where the relaxation processes of wheat flour dough of various protein contents were studied, Huang (1998) reported that infinite solutions of discrete generalized Maxwell model elements obtained by conventional nonlinear regression method using experimental linear relaxation moduli had large regression standard errors. In this study, experimental $G(t)$ and the corresponding data errors were used to calculate the small space relaxation spectra $H(\lambda)$ where the relationship between $G(t)$ and $H(\lambda)$ can be written as (Weese, 1991):

$$G(t) = \int_{-\infty}^{\infty} e^{-t/\lambda} H(\lambda) d(\ln \lambda) + \sum_{j=1}^{m} a_j b_j(t)$$

where the set of m coefficients $a_1, a_2, a_3, \ldots, a_m$ is related to corresponding experimental errors.

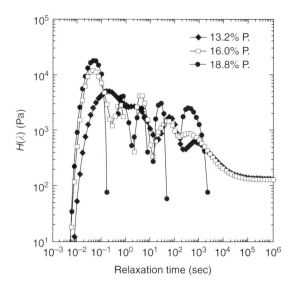

FIGURE 1.28 Relaxation spectra $H(\lambda)$ for wheat flour doughs with three different protein levels obtained using Tikhonov regularization method and the generalized Maxwell model. (Reproduced from Huang, H., 1998, Shear and extensional rheological measurements of hard wheat flour doughs and their simulation using Wagner constitutive model, Ph.D. Thesis, Rutgers University. With permission.)

The simulation accuracy of the continuous relaxation spectrum, $H(\lambda)$, is strongly determined by the second term $\sum_{j=1}^{m} a_j b_j(t)$, which makes the problem ill-posed. With Tikhonov regularization an estimate for the function $H(\lambda)$ and coefficients $a_1, a_2, a_3, \ldots, a_m$ is obtained from experimental $g_1^{\sigma}, g_2^{\sigma}, g_3^{\sigma}, \ldots, g_n^{\sigma}$ for $G(t_1), G(t_2), G(t_3), \ldots, G(t_n)$ with errors $\sigma_1, \sigma_2, \sigma_3, \ldots, \sigma_n$ by minimizing

$$L(\zeta) = \sum_{i=1}^{n} \frac{1}{\sigma_i^2} \left[g_i^{\sigma} \left(\int_{-\infty}^{\infty} e^{-t/\lambda} \cdot H(\lambda) d(\ln \lambda) + \sum_{j=1}^{m} a_j b_j(t) \right) \right]^2 + \zeta \| O \cdot H(\lambda) \|^2$$

where O is an operator, and ζ is the so-called regularization parameter. With an appropriate value for the regularization parameter, the first term on the right hand side of the above equation forces the result to be compatible with the experimental $G(t)$ (Weese, 1991).

Huang (1998) used a FORTRAN program (FTIKREG) with the classical Tikhonov regularization technique developed by Honerkamp and Weese (1989) to calculate continuous relaxation spectra $H(\lambda)$ from experimental linear relaxation moduli. Unlike the regression method, the accuracy of the simulated relaxation spectrum, $H(\lambda)$, was not affected by the total number of relaxation time. Figure 1.28 shows the relaxation spectra of doughs with different protein content within the relaxation times of 10^{-3} to 10^5 sec. The simulated linear relaxation moduli were calculated from the simulated relaxation spectrum, $H(\lambda)$. The wheat flour doughs showed jagged relaxation spectra. The physical–chemical interactions of starch/starch, starch/protein and protein/protein might have affected the relaxation process of the protein molecules and given the jagged relaxation spectra that are not often seen in polymer melts.

In this study, vital wheat gluten was added to regular hard wheat flour (13.2% protein content) to reinforce the gluten strength in 16.0 and 18.8% protein flour doughs. This extra added gluten protein benefited the dough relaxation by both increasing the relaxation process contributed by both gliadin and glutenin molecules, and at the same time decreasing the volume percentage of starch particles in the network. The simulated linear relaxation moduli superimposed very well for all three doughs. Figure 1.29 shows the experimental and simulated $G(t)$ of the tested wheat flour doughs. There was

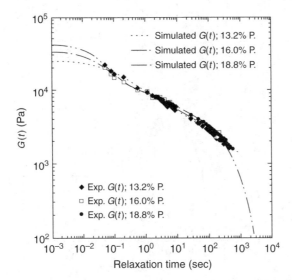

FIGURE 1.29 Comparison of experimental linear shear relaxation moduli $G(t)$ and predicted moduli using the generalized Maxwell model for wheat flour doughs with three different protein levels at 27°C. (Reproduced from Huang, H., 1998, Shear and extensional rheological measurements of hard wheat flour doughs and their simulation using Wagner constitutive model, Ph.D. Thesis, Rutgers University. With permission.)

no terminal relaxation time observed up to a relaxation time of 1000 sec. This dough acted like a polymer with a broad molecular weight distribution.

In order to understand the effect of gluten protein on wheat dough relaxation process, Huang (1998) used the linear relaxation moduli of gluten dough with 55% moisture and high purity gliadin dough with 35% moisture to calculate their relaxation spectra $H(\lambda)$ by the Tikhonov regularization method. Simulated relaxation moduli G_i of gluten dough vanished at a time of 5×10^4 sec as shown in Figure 1.30. Similarly, simulated relaxation moduli G_i of gliadin dough completely relaxed at an even shorter time at around 4×10^2 sec after suddenly imposing step strain. Gluten dough with highly extensible gliadin molecular or with interchangeable disulfide bond in the glutenin might have helped higher protein flour dough to accelerate the relaxation process at long relaxation time. The simulated relaxation moduli G_i of the gluten and gliadin doughs are shown in Figure 1.31.

Stress relaxation experiments have been widely used with many foods but have found a lot of intense applications with wheat flour dough since slower relaxation times are associated with good baking quality (Bloksma, 1990; Wang and Sun, 2002). Measurements of large-deformation creep and stress relaxation properties were found to be useful to distinguish between different wheat varieties (Safari-Ardi and Phan-Thien, 1998; Edwards et al., 2001; Wang and Sun, 2002; Keentook et al., 2002). Safari-Ardi and Phan-Thien (1998) studied the relaxation properties of weak, medium, strong, and extra strong wheat doughs at strain amplitudes between 0.1 and 29%. Oscillatory testing did not distinguish between the types of dough. However, the relaxation modulus of dough behaved quite distinctly at high strains. The magnitude of the modulus was found to be in the order of extra strong > strong > medium > weak dough, indicating higher levels of elasticity in stronger doughs.

Bekedam et al. (2003) studied the dynamic and relaxation properties of strong and weak wheat flour dough and their gluten components. They observed that sample preparation method, testing fixture, and sample age had a significant influence on the results. Sample age affected the shape of the relaxation modulus curve, which developed a terminal plateau upon aging. The relaxation spectrum for hard and soft wheat dough and their gluten fractions were obtained from the dynamic data using a Tikhonov regularization algorithm. The relaxation spectra obtained were consistent with the molecular character of the protein and previous studies suggesting dominant low and high molecular components.

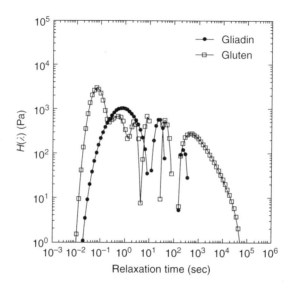

FIGURE 1.30   Relaxation spectrum $H(\lambda)$ for 35% moisture gliadin and 55% moisture gluten doughs obtained using the Tikhonov regularization method and the generalized Maxwell model. (Reproduced from Huang, H., 1998, Shear and extensional rheological measurements of hard wheat flour doughs and their simulation using Wagner constitutive model, Ph.D. Thesis, Rutgers University. With permission.)

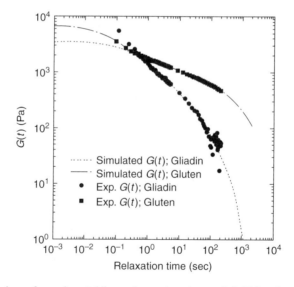

FIGURE 1.31   Comparison of experimental linear shear relaxation moduli $G(t)$ and predicted moduli using the generalized Maxwell model for 35% moisture gliadin and 55% moisture gluten doughs at 25°C. (Reproduced from Huang, H., 1998, Shear and extensional rheological measurements of hard wheat flour doughs and their simulation using Wagner constitutive model, Ph.D. Thesis, Rutgers University. With permission.)

1.3.5 CREEP RECOVERY

There are several studies on rheological characterization of food materials using creep-recovery technique since the 1930s. Creep-recovery tests are sometimes preferred over stress relaxation tests due to the ease of sample loading and the creeping flows which do not significantly change the food structure.

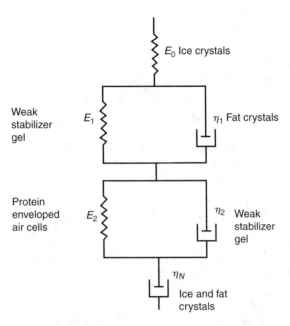

FIGURE 1.32   Six-element model for frozen ice cream showing rheological associations with structural components. (Reproduced from Shama, F. and Sherman, P., 1966a, *Journal of Food Science*, 31: 699–706. With permission.)

Shama and Sherman (1966a and 1966b) developed a mechanical model for ice cream based on the rheological properties. They presented the creep behavior of frozen ice cream by a six-element model which is composed of a spring in series with a dashpot (Maxwell body) and two units each comprising a spring in parallel with a dashpot (Voigt body) as shown in Figure 1.32. The parameters involved are the instantaneous elasticity (E_0), two elastic moduli (E_1 and E_2), and two viscosity components (η_1 and η_2) associated with retarded elasticity and a Newtonian viscosity (η_N). Shama and Sherman (1966a) assigned various model parameters to the structural parameters of the ice cream by examining the relative effect of fat, overrun, and temperature on rheological properties. They studied the creep behavior of several ice cream recipes at various temperatures. Typical creep curve for 10% fat ice cream is shown in Figure 1.33. From the effect of fat, overrun, and temperature on the magnitude of rheological parameters, it is suggested that E_0 is affected primarily by ice crystals, E_1 and η_2, by the weak stabilizer-gel network, E_2, by protein-enveloped air cells, η_1, by the fat crystals, and η_N by both fat and ice crystals.

Carillo and Kokini (1988) studied the effect of egg yolk powder and egg yolk powder and salt, on the stability of xanthan gum and propylene glycol alginate gum-stabilized o/w model salad dressing using creep test, steady shear test, and particle size analysis. Results showed that the magnitude of creep compliance, $J(t)$, increased as aging time increased. The added ingredients decreased compliance values indicating more viscous and stable emulsion (Figure 1.34). Data clearly showed that increased egg yolk or salt concentration resulted in an increase in increasing levels of structure formation in emulsions (Figure 1.35). At all salt concentrations, creep compliance increased significantly with increasing storage time indicating more liquid like structure development over the storage time (Figure 1.36). Increasing amount of additives becomes more effective on emulsion stability as storage time increases.

Edwards et al. (1999) applied creep test on durum wheat cultivars of varying gluten strength (Wascana, Kyle, AC Melita, and Durex), a parameter affecting extrusion properties and pasta cooking quality. Differences in creep parameters were significant at different absorption levels and among the cultivars at a given absorption level (Figure 1.37). Wascana was consistently the most extensible

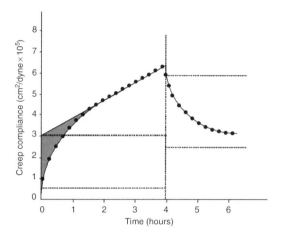

FIGURE 1.33 Typical creep curve for 10% fat ice cream. (Reproduced from Shama, F. and Sherman, P., 1966a, *Journal of Food Science*, 31: 699–706. With permission.)

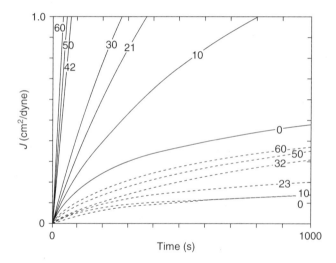

FIGURE 1.34 Averaged creep curves for 0% egg yolk emulsions (———) and 1% egg + 1% salt emulsions (-----) at different aging times. (Reproduced from Carrillo, A.R. and Kokini, J.L., 1988, *Journal of Food Science*, 53: 1352–1366. With permission.)

cultivar while AC Melita and Durex were the least extensible at all absorption levels. Increasing water absorption increased maximum strain attained, expectedly, since water addition facilitates flow.

1.3.6 TRANSIENT SHEAR STRESS DEVELOPMENT

Shear stress overshoot at the inception of steady shear flow is frequently observed with many semisolid food materials. These overshoots can range anywhere from 30 to 300% of their steady-state value, depending on the particular shear rate and material used. These stresses are of particular importance when the relaxation time of the material is larger or comparable to the time scale of the experiment. They become significant in assessments of the textural attributes, spreadability (Kokini and Dickie, 1982), and thickness (Dickie and Kokini, 1983) and also in the startup of flow equipment.

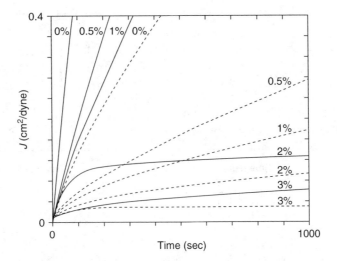

FIGURE 1.35 Averaged creep curves for different concentrations of egg yolk emulsions (———) and different concentration of egg yolk + 2% salt emulsions (- - - - -) after 30 days of aging. (Reproduced from Carrillo, A.R. and Kokini, J.L., 1988, *Journal of Food Science*, 53: 1352–1366. With permission.)

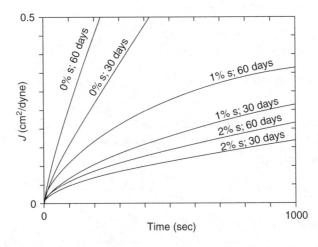

FIGURE 1.36 Effect of aging time on averaged creep data curves for 1% egg yolk and 0%, 1% and 2% salt emulsions. (Reproduced from Carrillo, A.R. and Kokini, J.L., 1988, *Journal of Food Science*, 53: 1352–1366. With permission.)

Several first attempts have been made to develop an equation capable of predicting transient shear stress growth in food materials. Elliot and Green (1972) have modeled transient shear stress growth in several foods assuming that these foods could be simulated by a Maxwell element coupled with a yield element. This analysis, although fundamentally very enlightening, did not account for nonlinear viscoelastic behavior frequently observed with most foods. It is, nevertheless, a first, very worthwhile attempt at explaining shear stress overshoots in materials that portray yield stresses, such as foods.

Dickie and Kokini (1982) have simulated shear stress growth in 15 foods using an empirical equation developed by Leider and Bird (1974). This equation has the following form:

$$\tau_{\theta\phi} = m(\dot{\gamma})^n[1 + (b\dot{\gamma}t - 1)e^{-t/an\lambda}]$$

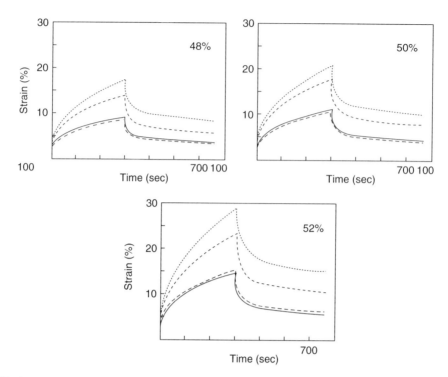

FIGURE 1.37 Creep-recovery of four durum wheat cultivars at different water absorption levels (······ Wascana, ------- Kyle, ——— AC Melita, – – – Durex). (Reproduced from Edwards, N.M., Dexter, J.E., Scanlon, M.G., and Cenkowski, S., 1999, *Cereal Chemistry*, 76: 638–645. With permission.)

where $\tau_{\theta\phi}$ is shear stress, m and n are limiting viscous power law parameters, $\dot{\gamma}$ is the shear rate, t is time, a and b are adjustable parameters, and λ is the time constant.

$$\lambda = \left(\frac{m'}{2m}\right)^{1/(n'-n)}$$

with m' and n' first normal stress power law parameters. In this model it is assumed that both shear stress and first normal stress differences are simulated using the power law behavior:

$$\tau_{12} = m(\dot{\gamma})^n$$

$$\tau_{11} - \tau_{22} = m'(\dot{\gamma})^{n'}$$

where τ_{12} is shear stress, $\tau_{11} - \tau_{22}$ is the first normal stress difference, m, n, m', and n' are the power law parameters, and $\dot{\gamma}$ is the shear rate. These parameters for 15 typical food materials are shown in Table 1.6.

A distinct convenience of this equation is that at long times it converges to the power law behavior observed with a large number of food materials (Rao, 1977; Rha, 1978). An example of the ability of this equation to fit transient shear stress growth data is shown in Figure 1.38 for peanut butter. The equation was found to predict peak shear stresses and peak times fairly well but failed to predict transient decay accurately. Although the model is able to account for nonlinear behavior, one of its more serious shortcomings is a single exponential term to simulate the relaxation part of the data. Time constants for 15 typical food systems are shown in Table 1.6. To account for this limitation, a family of empirical models was developed (Mason et al., 1983). These models are an

TABLE 1.6
Power Law Parameters of Various Foods

Products	m (Pa sec^n)	n	R^2	m' (Pa sec$^{n'}$)	n'	R^2	λ (sec)
Apple butter	222.90	0.145	0.99	156.03	0.566	0.99	8.21×10^{-2}
Canned frosting	355.84	0.117	0.99	816.11	0.244	0.99	2.90×10^{0}
Honey	15.39	0.989	—	—	—	—	—
Ketchup	29.10	0.136	0.99	39.47	0.258	0.99	4.70×10^{-2}
Marshmallow cream	563.10	0.379	0.99	185.45	0.127	0.99	1.27×10^{3}
Mayonnaise	100.13	0.131	0.99	256.40	−0.048	0.99	2.51×10^{-1}
Mustard	35.05	0.196	0.99	65.69	0.136	0.99	2.90×10^{0}
Peanut butter	501.13	0.065	0.99	3785.00	0.175	0.99	1.86×10^{5}
Stick butter	199.28	0.085	0.99	3403.00	0.398	0.99	1.06×10^{3}
Stick margarine	297.58	0.074	0.99	3010.13	0.299	0.99	1.34×10^{3}
Squeeze margarine	8.68	0.124	0.99	15.70	0.168	0.99	9.93×10^{-2}
Tube margarine	106.68	0.077	0.99	177.20	0.353	0.99	5.16×10^{1}
Whipped butter	312.30	0.057	0.99	110.76	0.476	0.99	1.61×10^{-2}
Whipped cream cheese	422.30	0.058	0.99	363.70	0.418	0.99	8.60×10^{-2}
Whipped dessert topping	35.98	0.120	0.99	138.00	0.309	0.99	3.09×10^{1}

Source: Dickie, A.M., 1982, Predicting the spreadability and thickness of foods from time dependent viscoelastic rheology. M.S. Thesis, Rutgers University, New Brunswick, NJ.

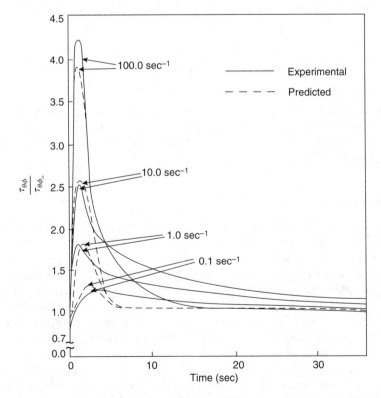

FIGURE 1.38 Shear stress development of peanut butter at 25°C and comparison of the Bird–Leider equation with experimental data. (Reproduced from Dickie, A.M. and Kokini, J.L., 1982, *Journal of Food Process Engineering*, 5: 157–174. With permission.)

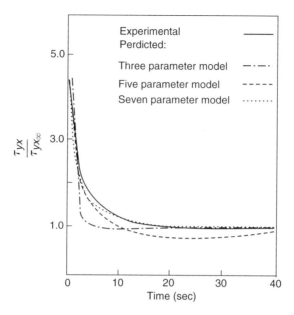

FIGURE 1.39 Comparison of the predictions of the three-, five-, and seven-parameters models with experimental data for stick butter at a shear rate of $100\,\text{sec}^{-1}$ (Reproduced from Mason, P.L., Puoti, M.P., Bistany, K.L., and Kokini, J.L., 1983, *Journal of Food Process Engineering*, 6: 219–233. With permission.)

extension of the earlier model developed by Leider and Bird (1974) and contain several relaxation terms:

$$\tau_{yx} = m(\dot{\gamma})^n \left[1 + (b_0 \dot{\gamma} t - 1) \frac{\sum b_i e^{-t/\lambda_i}}{\sum b_i} \right]$$

where m and n are power law parameters, $\dot{\gamma}$ is the shear rate, t time, λ_i are tine constants, and b_0 and b_i are constants. In Figure 1.39 for stick butter at a shear rate of $10\,\text{sec}^{-1}$, it can be seen that the seven-parameter model predicts shear stress growth better than does the three-parameter Bird–Leider equation (Mason et al., 1983).

1.3.7 YIELD STRESSES

Many semisolid food materials portray yield stresses. Yield stresses can be measured with a variety of techniques. These include measuring the shear stress at vanishing shear rates, extrapolation of data using rheological models that include yield stresses, and stress relaxation experiments, among others (Barbosa-Canovas and Peleg, 1983). One particularly useful technique is plotting viscosity vs. shear stress (Dzuy and Boger, 1983). In this form the viscosity tends to infinity when the yield stress value is reached. This technique gives one of the most accurate values for yield stress. Figure 1.40 and Figure 1.41 show such graphs for guar gum and gum karaya, respectively (Mills and Kokini, 1984). Guar gum did not show yield stresses as viscosity tends to a constant value. However, in case of gum karaya, viscosity tends to large values as a limiting value of shear stress is reached, signifying the presence of a yield stress. Guar gum is a linear polysaccharide which readily disperses in aqueous solutions. Dispersions of gum karaya, on the other hand, are formed by deformable particles that swell to many times their original size and are responsible for the observed yield stresses. Similar data are obtained for mustard, whereas viscosity tended to large values as the yield stress was approached (Figure 1.42).

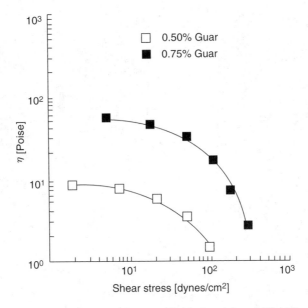

FIGURE 1.40 Viscosity vs. shear stress curve for guar. (Reproduced from Mills, P.L. and Kokini, J.L., 1984, *Journal of Food Science*, 49: 1–4, 9. With permission.)

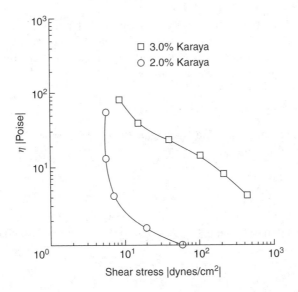

FIGURE 1.41 Viscosity vs. shear stress curve for gum karaya. (Reproduced from Mills, P.L. and Kokini, J.L., 1984, *Journal of Food Science*, 49: 1–4, 9. With permission.)

1.4 CONSTITUTIVE MODELS

A growing field of importance in food rheology is the development of constitutive models that describe the behavior of food materials in all components of stress, strain, and strain rates. Constitutive models predict rheological properties through mathematical formalism which makes fundamental assumptions about the structure and molecular properties of materials (Kokini, 1993 and 1994). Relating rheological measurements to molecular structures and conformations of food polymers and food systems in general is a goal of considerable importance. Constitutive models are

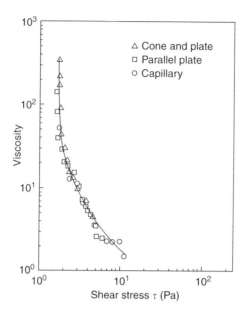

FIGURE 1.42   Viscosity vs. shear stress data for mustard. (Reproduced from Kokini, J.L., 1992, *Handbook of Food Engineering*, D.R. Heldman and D.B. Lund, Eds, Marcel Dekker, Inc., New York. With permission.)

gaining importance in food science research because of their applications in predictive rheological modeling and also because of their use in numerical simulation of unit operations such as dough sheeting, extrusion, which can provide insight into design and scale-up (Kokini, 1993 and 1994; Kokini et al., 1995b; Dhanasekharan and Kokini, 2003).

1.4.1 SIMULATION OF STEADY RHEOLOGICAL DATA

There are several basic models available to simulate the flow behavior of semisolid food materials. These include the power law model,

$$\tau = m(\dot{\gamma})^n$$

where τ is shear stress, $\dot{\gamma}$ shear rate, and m and n are power law parameters. A special case where $n = 1$ reduces this equation to Newton's law. Other models include the Bingham model:

$$\tau = \tau_0 + \mu\dot{\gamma}$$

where τ_0 is the yield stress described before the Casson model:

$$\tau^{1/2} = \tau_0^{1/2} + \mu(\dot{\gamma})^{1/2}$$

A general model to describe the flow behavior of inelastic time-independent fluids is that proposed by Herschel and Bulkley:

$$\tau = \tau_0 + m(\dot{\gamma})^n$$

The power law, Newtonian, and Bingham plastic models are all special cases of the Herschel–Bulkley model. The literature is abundant with other models, but the Herschel–Bulkley

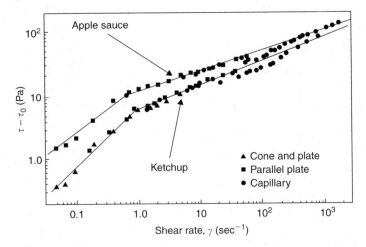

FIGURE 1.43 Log $(\tau - \tau_0)$ vs. log (shear rate) for applesauce and ketchup. (Reproduced from Dervisoglu, M. and Kokini, J.L., 1986b, *Journal of Food Science*, 51: 541–546, 625. With permission.)

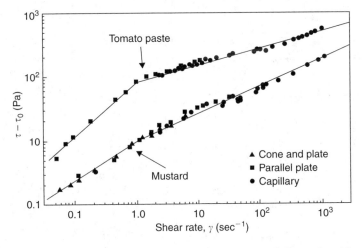

FIGURE 1.44 Log $(\tau - \tau_0)$ vs. log (shear rate) for tomato paste and mustard. (Reproduced from Dervisoglu, M. and Kokini, J.L., 1986b, *Journal of Food Science*, 51: 541–546, 625. With permission.)

model is the one most commonly used. A convenient way of linearizing the Herschel–Bulkley model is by subtracting τ_0 from shear stresses τ and to plot $\tau - \tau_0$ vs. shear rate on logarithmic coordinates. Examples of such plots are shown in Figure 1.43 for applesauce and ketchup and in Figure 1.44 for tomato paste and mustard. All of the flow curves portray a gradual transition from a less shear thinning behavior to a more shear thinning behavior with increasing shear rate. At lower shear rates the time of shear is comparable to the time necessary to reform aggregates, and the forces exerted are small compared to the overall force necessary to achieve extensive breakdown. Consequently, the effective rate of breakdown is smaller than that observed in the larger shear rates. As a result of this gradual transition, two clearly different regions become evident. For all of the materials, the less shear thinning region is observed for shear rates approximately less than $1.0 \ sec^{-1}$. This is consistent with observations on tomato juice obtained by DeKee et al. (1983).

Steffe et al. (1986) has compiled a large amount of food data using the Herschel–Bulkley model as a basis. Additional information, such as shear rate range, total solids, and temperature are also given. Steffe's compilation for fruit and vegetable products is given in Table 1.7. Additional information,

TABLE 1.7
Properties of Fruit and Vegetables

Product	Total solids (%)	Temp. (°C)	n	m (Pa.sec^n)	π_y (Pa)	Shear rate ranges (sec^{-1})
APPLE						
Pulp	—	25.0	0.084	65.03	—	—
Sauce	—	23.8	0.645	0.50	—	—
Sauce	—	23.8	0.408	0.66	—	—
Sauce	—	—	0.470	5.63	58.6	—
Sauce	—	20.0	0.302	16.68	—	3.3–530
Sauce + 12.5% water	—	25.0	0.438	2.39	—	0.1–1.1
Sauce	11.6	27.0	0.28	12.7	—	160–340
	11.0	30.0	0.30	11.6	—	5–50
	11.0	82.2	0.30	9.0	—	5–50
Sauce	10.5	26.0	0.45	7.32	—	0.78–1260
	9.6	26.0	0.45	5.63	—	0.78–1260
	8.5	26.0	0.44	4.18	—	0.78–1260
APRICOT						
Puree	17.7	26.6	0.29	5.4	—	—
	23.0	26.6	0.35	11.2	—	—
	41.4	26.6	0.35	54.0	—	—
	44.3	26.6	0.37	56.0	—	0.5–80
	51.4	26.6	0.36	108.0	—	0.5–80
	55.2	26.6	0.34	152.0	—	0.5–80
	59.3	26.6	0.32	300.0	—	0.5–80
Reliable, conc., green	27.0	4.4	0.25	170.0	—	3.3–137
	27.0	25.0	0.22	141.0	—	3.3–137
Reliable, conc., ripe	24.1	4.4	0.25	67.0	—	3.3–137
	24.1	25.0	0.22	54.0	—	3.3–137
Reliable, conc., ripened	25.6	4.4	0.24	85.5	—	3.3–137
	25.6	25.0	0.26	71.0	—	3.3–137
Reliable, conc., overripe	26.0	4.4	0.27	90.0	—	3.3–137
	26.0	25.0	0.30	67.0	—	3.3–137
BANANA						
Puree A	—	23.8	0.458	6.5	—	—
Puree B	—	23.8	0.333	10.7	—	—
Puree (17.7 brix)	—	22.0	0.283	107.3	—	28–200
BLUEBERRY						
Pie Filling	—	20.0	0.426	6.08	—	3.3–530
CARROT						
Puree	—	25.0	0.228	24.16	—	—
GREEN BEAN						
Puree	—	25.0	0.246	16.91	—	—
GUAVA						
Puree (10.3 brix)	—	23.4	0.494	38.98	—	15–400
MANGO						
Puree (9.3 brix)	—	24.2	0.334	20.58	—	15–1000

(Continued)

TABLE 1.7
Continued

Product	Total solids (%)	Temp. (°C)	n	m (Pa·sec^n)	π_y (Pa)	Shear rate ranges (sec^{-1})
ORANGE JUICE CONCENTRATE						
Hamlin, early (42.5 brix)	—	25.0	0.585	4.121	—	0–500
	—	15.0	0.602	5.973	—	0–500
	—	0.0	0.676	9.157	—	0–500
	—	−10.0	0.705	14.255	—	0–500
Hamlin, late (41.1 brix)	—	25.0	0.725	1.930	—	0–500
	—	15.0	0.560	8.118	—	0–500
	—	0.0	0.620	1.754	—	0–500
	—	−10.0	0.708	13.875	—	0–500
Pineapple, early (40.3 brix)	—	25.0	0.643	2.613	—	0–500
	—	15.0	0.587	5.887	—	0–500
	—	0.0	0.681	8.938	—	0–500
	—	−10.0	0.713	12.184	—	0–500
Pineapple, late (41.8 brix)	—	25.0	0.532	8.564	—	0–500
	—	15.0	0.538	13.432	—	0–500
	—	0.0	0.636	18.584	—	0–500
	—	−10.0	0.629	36.414	—	0–500
Valencia, early (43.0 brix)	—	25.0	0.538	5.059	—	0–500
	—	15.0	0.609	6.714	—	0–500
	—	0.0	0.622	14.036	—	0–500
	—	−10.0	0.619	27.16	—	0–500
Valencia, late (41.9 brix)	—	25.0	0.538	8.417	—	0–500
	—	15.0	0.568	11.802	—	0–500
	—	0.0	0.644	18.751	—	0–500
	—	−10.0	0.628	41.412	—	0–500
Naval (65.1 brix)	—	−18.5	0.71	39.2	—	—
	—	−14.1	0.76	14.6	—	—
	—	−9.3	0.74	10.8	—	—
	—	−5.0	0.72	7.9	—	—
	—	−0.7	0.71	5.9	—	—
	—	10.1	0.73	2.7	—	—
	—	19.9	0.72	1.6	—	—
	—	29.5	0.74	0.9	—	—
PAPAYA						
Puree (7.3 brix)	—	26.0	0.528	9.09	—	20–450
PEACH						
Pie Filling	—	20.0	0.46	20.22	—	0.1–140
Puree	10.9	26.6	0.44	0.94	—	—
	17.0	26.6	0.55	1.38	—	—
	21.9	26.6	0.55	2.11	—	—
	26.0	26.6	0.40	13.4	—	80–1000
	29.6	26.6	0.40	18.0	—	80–1000
	37.5	26.6	0.38	44.0	—	—
	40.1	26.6	0.35	58.5	—	2–300
	49.8	26.6	0.34	85.5	—	2–300
	58.4	26.6	0.34	440.0	—	—
Puree	11.7	30.0	0.28	7.2	—	5–50
	11.7	82.2	0.27	5.8	—	5–50
	10.0	27.0	0.34	4.5	—	160–3200

(Continued)

TABLE 1.7
Continued

Product	Total solids (%)	Temp. (°C)	n	m (Pa.sec^n)	π_y (Pa)	Shear rate ranges (sec^{-1})
PEAR						
Puree	15.2	26.6	0.35	4.3	—	—
	24.3	26.6	0.39	5.8	—	—
	33.4	26.6	0.38	38.5	—	80–1000
	37.6	26.6	0.38	49.7	—	—
	39.5	26.6	0.38	64.8	—	2–300
	47.6	26.6	0.33	120.0	—	0.5–10
	49.3	26.6	0.34	170.0	—	—
	51.3	26.6	0.34	205.0	—	—
	45.8	32.2	0.479	35.5	—	—
	45.8	48.8	0.477	26.0	—	—
	45.8	65.5	0.484	20.0	—	—
	45.8	82.2	0.481	16.0	—	—
	14.0	30.0	0.35	5.6	—	5–50
	14.0	82.2	0.35	4.6	—	5–50
PLUM						
Puree	14.0	30.0	0.34	2.2	—	5–50
	14.0	8.2	0.34	2.0	—	5–50
	—	25.0	0.222	5.7	—	—
SQUASH						
Puree A	—	25.0	0.149	20.65	—	—
Puree B	—	25.0	0.281	11.42	—	—
TOMATO						
Juice Concentrate	5.8	32.2	0.590	0.223	—	500–800
	5.0	48.8	0.540	0.27	—	500–800
	5.8	65.5	0.470	0.37	—	500–800
	12.8	32.2	0.430	2.00	—	500–800
	12.8	48.8	0.430	1.88	—	500–800
	12.8	65.5	0.340	2.28	—	500–800
	12.8	82.2	0.350	2.12	—	500–800
	16.0	32.2	0.450	3.16	—	500–800
	16.0	48.8	0.450	2.77	—	500–800
	16.0	65.5	0.400	3.18	—	500–800
	16.0	82.2	0.380	3.27	—	500–800
	25.0	32.2	0.410	12.9	—	500–800
	25.0	48.8	0.420	10.5	—	500–800
	25.5	65.5	0.430	8.0	—	500–800
	25.0	82.2	0.430	6.1	—	500–800
	30.0	32.2	0.400	18.7	—	500–800
	30.0	48.8	0.420	15.1	—	500–800
	30.0	65.5	0.430	11.7	—	500–800
	30.0	82.2	0.450	7.9	—	500–800
Ketchup	—	25.0	0.27	18.7	32	10–560
	—	45.0	0.29	16.0	24	10–560
	—	65.0	0.29	11.3	14	10–560
	—	95.0	0.253	7.45	10.5	10–560
Puree	—	25.0	0.236	7.78	—	—
	—	47.7	0.550	1.08	2.04	—

Source: Steffe, J.F., Mohamed, I.O., and Ford, E.W., 1986, Rheological properties of fluid foods: data compilation. In: *Physical and Chemical Properties of Food*, M.E. Okos, Ed., ASAE Publications.

TABLE 1.8

Properties of Apple and Grape Juice Concentrates

	°Brix	η_0 (Pa sec)	E_a kcal/gmole	Temp range (°C)
Apple juice concentrate	45.1	3.394×10^{-7}	6.0	−5 to 40
(from McIntosh apples)	50.4	1.182×10^{-7}	6.9	−10 to 40
	55.2	2.703×10^{-9}	9.4	−15 to 40
	60.1	3.935×10^{-10}	10.9	−15 to 40
	64.9	7.917×10^{-12}	13.6	−15 to 40
	68.3	1.156×10^{-12}	15.3	−15 to 40
Grape juice concentrate	43.1	8.147×10^{-8}	7.0	−5 to 40
(from concord grapes)	49.2	1.074×10^{-8}	8.5	−10 to 40
	54.0	9.169×10^{-8}	10.3	−15 to 40
	59.2	1.243×10^{-10}	11.8	−15 to 40
	64.5	1.340×10^{-10}	12.3	−15 to 40
	68.3	6.086×10^{-12}	14.5	−15 to 40

Source: Rao, M.A., Cooley, H.J., and Vitali, A.A., 1984. Flow properties of concentrated juices at low temperatures, *Food Technology*, 38: 113–119.

such as shear rate range, total solids, and temperature are also given (Table 1.7). Additional data for apple and juice concentrates reported by Rao et al. (1984) are given in Table 1.8.

The temperature dependence in most cases is considered to be an Arrhenius one given by

$$\eta = \eta_0 \exp\left(\frac{E_a}{RT}\right)$$

where η is the viscosity in Pa · sec, η_0 the viscosity at a reference temperature, E_a, the activation energy, T, the absolute temperature, and R, the gas constant. Data for meat, fish, and dairy products are given in Table 1.9, and data for oils and other products are given in Table 1.10.

1.4.2 LINEAR VISCOELASTIC MODELS

Linear viscoelasticity is observed when the deformations encountered by food polymers are small enough that the polymeric material is negligibly disturbed from its equilibrium state (Bird et al., 1987). The level of deformation where linear viscoelasticity is observed depends on the molecular architecture of the food polymer molecules and structure of the food. For example, for high viscosity concentrated dispersions, linear viscoelastic behavior is observed when the deformation occurs very slowly, as in creep tests or small amplitude oscillatory tests at very low frequencies. When the flow is slow, Brownian motion can return the deformed molecule to its original state before the next molecule tends to deform it again, and the viscoelastic material is the linear range.

Linear viscoelastic properties are very useful in terms of elucidating structural characteristics of polymeric materials. In the linear viscoelastic region, moreover, the measured rheological properties are independent of the magnitude of the applied strain or stress. However, linear viscoelastic properties are of little value in terms of predicting the deformation behavior of the materials during many food processing operations which occur in the large strains (Table 1.11).

Constitutive equations enable the simulation of a wide range of rheological data obtained by a variety of experiments. These models necessitate rheological constants, which are determined either from molecular properties or from an independent set of experiments. The simplest constitutive

TABLE 1.9
Properties of Meat, Fish and Dairy Products

Product	Total solids (%)	Temp. (°C)	n	m (Pa.sec^n)	π_y (Pa)	Shear rate ranges (sec^{-1})
CREAM						
10% Fat	—	40	1.0	0.00148	—	—
	—	60	1.0	0.00107	—	—
	—	80	1.0	0.00083	—	—
20% Fat	—	40	1.0	0.00238	—	—
	—	60	1.0	0.00171	—	—
	—	80	1.0	0.00129	—	—
30% Fat	—	40	1.0	0.00395	—	—
	—	60	1.0	0.00289	—	—
	—	80	1.0	0.00220	—	—
40% Fat	—	40	1.0	0.00690	—	—
	—	60	1.0	0.00510	—	—
	—	80	1.0	0.00395	—	—
FISH						
Minced paste	—	3–6	0.91	8.55	1600	0.7–238
MEAT						
Raw comminated batters						
% Fat % Prot. % MC						
15.0 13.0 66.8	—	15	0.156	639.3	1.53	300–500
18.7 12.9 65.9	—	15	0.104	858.0	0.28	300–500
22.5 12.1 63.2	—	15	0.209	429.5	0.00	300–500
30.0 10.4 57.5	—	15	0.341	160.2	27.80	300–500
33.8 9.5 54.5	—	15	0.390	103.3	17.90	300–500
45.0 6.9 45.9	—	15	0.723	14.0	2.30	300–500
45.0 6.9 45.9	—	15	0.685	17.9	27.60	300–500
67.3 28.9 1.8	—	15	0.205	306.8	0.00	300–500
MILK						
Homogenized	—	20	1.0	0.00200	—	—
	—	30	1.0	0.00150	—	—
	—	40	1.0	0.00110	—	—
	—	50	1.0	0.00095	—	—
	—	60	1.0	0.00078	—	—
	—	70	1.0	0.00070	—	—
	—	80	1.0	0.00060	—	—
Raw	—	0	1.0	0.00344	—	—
	—	5	1.0	0.00305	—	—
	—	10	1.0	0.00264	—	—
	—	15	1.0	0.00231	—	—
	—	20	1.0	0.00199	—	—
	—	25	1.0	0.00170	—	—
	—	30	1.0	0.00149	—	—
	—	35	1.0	0.00134	—	—
	—	40	1.0	0.00123	—	—

(Continued)

TABLE 1.9
Continued

Product	Total solids (%)	Temp. (°C)	n	m (Pa.sec^n)	π_y (Pa)	Shear rate ranges (sec^{-1})
WHOLE SOYBEAN						
7% Soy Cotyledon Solids	—	10	0.85	0.0640	—	0–1300
7% Soy Cotyledon Solids	—	20	0.84	0.0400	—	0–1300
7% Soy Cotyledon Solids	—	30	0.80	0.0400	—	0–1300
7% Soy Cotyledon Solids	—	40	0.81	0.0330	—	0–1300
7% Soy Cotyledon Solids	—	50	0.82	0.0270	—	0–1300
7% Soy Cotyledon Solids	—	60	0.83	0.0240	—	0–1300
4.9% Soy Cotyledon Solids	—	25	0.90	0.0187	—	0–1300
6.2% Soy Cotyledon Solids	—	25	0.85	0.0415	—	0–1300
7.2% Soy Cotyledon Solids	—	25	0.84	0.0665	—	0–1300
8.1% Soy Cotyledon Solids	—	25	0.78	0.1171	—	0–1300
9.0% Soy Cotyledon Solids	—	25	0.76	0.2133	—	0–1300
10.2% Soy Cotyledon Solids	—	25	0.71	0.4880	—	0–1300

Source: Steffe, J.F., Mohamed, I.O., and Ford, E.W., 1986, Rheological properties of fluid foods: data compilation. In: *Physical and Chemical Properties of Food*, M.E. Okos, Ed., ASAE Publications.

theories are Newton's law for purely viscous fluids,

$$\tau = \mu\dot{\gamma}$$

and Hooke's law for purely elastic materials

$$\tau = G\gamma$$

A classical approach to describe the response of materials which exhibit combined viscous and elastic properties is based upon an analogy with the response of springs and dashpots arranged in series or in parallel representing purely elastic and purely viscous properties (Figure 1.45).

1.4.2.1 Maxwell Model

The Maxwell element consists of a Hookean spring and a Newtonian dashpot combined in series, representing the simplest model for the flow behavior of viscoelastic fluids. In this model, both spring and dashpot are subjected to the same stress. The total strain in the Maxwell element is equal to the sum of the strains in the spring and dashpot.

$$\gamma = \gamma_{\text{spring}} + \gamma_{\text{dashpot}}$$

The governing differential equation for Maxwell fluid model is (Darby, 1976):

$$\tau + \lambda\dot{\tau} = \mu\dot{\gamma}$$

TABLE 1.10
Properties of Oils and Miscellaneous Products

Product	Total solids (%)	Temp. (°C)	n	m (Pa.sec^n)	π_y (Pa)	Shear rate ranges (sec^{-1})
CHOCOLATE						
Melted	—	46.1	0.574	0.57	1.16	—
HONEY						
Buckwheat	18.6	24.8	1.0	3.86	—	—
Golden rod	19.4	24.3	1.0	2.93		—
Sage	18.6	25.9	1.0	8.88		—
Sweet clover	17.0	24.7	1.0	7.2		—
White clover	18.2	25.0	1.0	4.8		—
MAYONNAISE	—	25.0	0.55	6.4	—	30–1300
	—	25.0	0.54	6.6	—	30–1300
	—	25.0	0.60	4.2	—	40–1100
	—	25.0	0.59	4.7	—	40–1100
MUSTARD	—	25.0	0.39	18.5	—	30–1300
	—	25.0	0.39	19.1	—	30–1300
	—	25.0	0.34	27	—	40–1100
	—	25.0	0.28	33	—	−40–1100
OILS						
Castor	—	10.0	1.0	2.42	—	—
	—	30.0	1.0	0.451	—	—
	—	40.0	1.0	0.231	—	—
	—	100.0	1.0	0.0169	—	—
Corn	—	38.0	1.0	0.0317	—	—
	—	25.0	1.0	0.0565	—	—
Cottonseed	—	20.0	1.0	0.0704	—	—
	—	38.0	1.0	0.0386	—	—
Linseed	—	50.0	1.0	0.0176	—	—
	—	90.0	1.0	0.0071	—	—
Olive	—	10.0	1.0	0.1380	—	—
	—	40.0	1.0	0.0363	—	—
	—	70.0	1.0	0.0124	—	—
Peanut	—	25.0	1.0	0.0656	—	—
	—	38.0	1.0	0.0251	—	—
	—	21.1	1.0	0.0647	—	0.32–64
	—	37.8	1.0	0.0387	—	0.32–64
	—	54.4	1.0	0.0268	—	0.32–64
Rapeseed	—	0.0	1.0	2.530	—	—
	—	20.0	1.0	0.163	—	—
	—	30.0	1.0	0.096	—	—
Safflower	—	38.0	1.0	0.0286	—	—
	—	25.0	1.0	0.0922	—	—
Sesame	—	38.0	1.0	0.0324	—	—
Soybean	—	30.0	1.0	0.0406	—	—
	—	50.0	1.0	0.0206	—	—
	—	90.0	1.0	0.0078	—	—
Sunflower	—	38.0	1.0	0.0311	—	—

Source: Steffe, J.F., Mohamed, I.O., and Ford, E.W., 1986, Rheological properties In: *Physical and Chemical Properties of Food*, M.E. Okos, Ed., ASAE Publications.

TABLE 1.11

Typical Shear Rates Involved in Some Processes

Operation or equipment	Shear rate (sec^{-1})
Particle sedimentation	10^{-6}–10^{-3}
Flow under gravity	10^{-1}–10^{1}
Chewing and swallowing	10^{1}–10^{2}
Mixing	10^{1}–10^{3}
Pipe flow	10^{0}–10^{3}
Plate heat exchanger	10^{2}–10^{3}
Scrape surface heat exchanger	10^{1}–5×10^{2}
Extruder	10^{2}–5×10^{4}

Source: Lagarrigue, S. and Alvarez, G., 2001, *Journal of Food Engineering*, 50: 189–202.

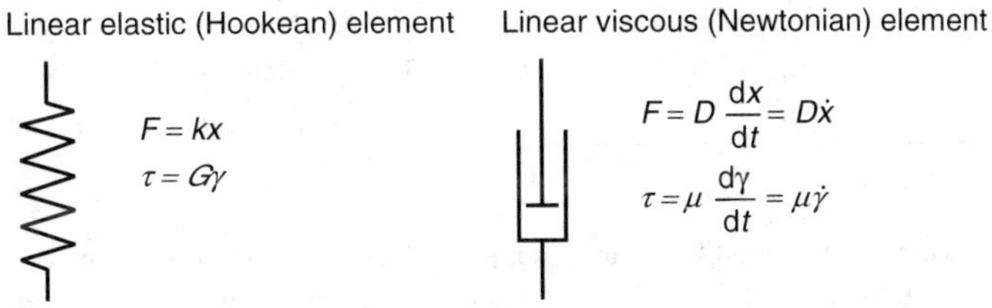

Linear elastic (Hookean) element Linear viscous (Newtonian) element

$$F = kx$$
$$\tau = G\gamma$$

$$F = D\frac{dx}{dt} = D\dot{x}$$
$$\tau = \mu\frac{d\gamma}{dt} = \mu\dot{\gamma}$$

FIGURE 1.45 Linear elastic and viscous mechanical elements.

where the relaxation time (λ) is given by

$$\lambda = \frac{\mu}{G}$$

During stress relaxation test, where a constant shear strain (γ_0) is instantly applied at $t = 0$ and maintained constant for times $t > 0$, the resulting stress for a Maxwell fluid as a function of time is given by

$$\tau(t) = G\gamma_0 e^{-t/\lambda} = \tau_0 e^{-t/\lambda}$$

The initial response is purely elastic, that is, $\tau \to G\gamma_0$ as $t \to 0^+$, due to the initial extension of the spring element, then it decays exponentially with time reaching 37% of its initial value at $t = \lambda$ (Figure 1.46b) (Darby, 1976).

Another test that distinguishes relative viscous and elastic behavior is the creep test. When a constant shear stress (τ_0) is instantly applied at $t = 0$ and maintained constant for times $t < t_1$, the resulting deformation observed as a function of time is given as (Darby, 1976):

$$\gamma(t) = \frac{\tau_0}{\mu}\{t + \lambda - [(t - t_1) + \lambda]U(t - t_1)\}$$

where $U(t - t_1)$ is the unit step function.

As shown in Figure 1.47b, the initial response is elastic, followed by a purely viscous flow response with a slope τ_0/μ. When the stress is removed, the material again shows an elastic response, indicating a recoverable strain of τ_0/G (Darby, 1976). This is also known as recoil or memory effect.

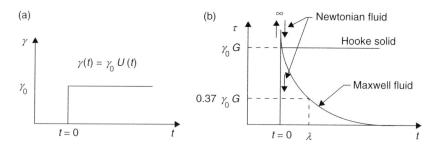

FIGURE 1.46 Behavior of a Maxwell fluid during stress relaxation (a) input function, (b) material response. (Reproduced from Darby, R., 1976, *Viscoelastic Fluids: An Introduction to Their Properties and Behavior*, Dekker Inc., New York. With permission.)

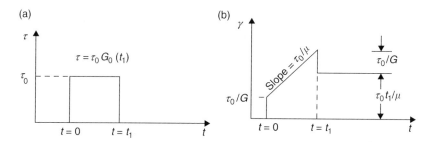

FIGURE 1.47 Behavior of a Maxwell fluid during creep test (a) input function, (b) material response. (Reproduced from Darby, R., 1976, *Viscoelastic Fluids: An Introduction to Their Properties and Behavior*, Dekker Inc., New York. With permission.)

1.4.2.2 Voigt Model

The Voigt or Kelvin element consists of a Hookean spring and a Newtonian dashpot combined in parallel. It is the simplest model for a viscoelastic solid. Due to parallel arrangements, both spring and dashpot in the Voigt element are constrained to deform the same amount, and the total stress is equal to the sum of the stress in the spring and dashpot.

$$\tau = \tau_{\text{spring}} + \tau_{\text{dashpot}}$$

The governing differential equation relating stress and strain is

$$\tau = G\gamma + \mu\dot{\gamma}$$

which can also be written

$$\frac{\tau}{G} = \lambda'\dot{\gamma} + \gamma$$

where the retardation time (λ') is given as

$$\lambda' = \frac{\mu}{G}$$

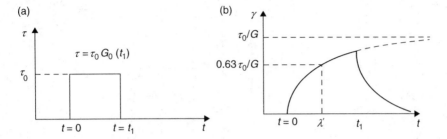

FIGURE 1.48 Behavior of a Voigt solid during creep test (a) Input function, (b) material response. (Reproduced from Darby, R., 1976, *Viscoelastic Fluids: An Introduction to Their Properties and Behavior*, Dekker Inc., New York. With permission.)

The strain response of a Voigt solid to creep test is calculated as (Darby, 1976):

$$\gamma(t) = \frac{\tau_0}{G}[(1 - e^{-t/\lambda'}) - (1 - e^{-(t-t_1)/\lambda'})U(t - t_1)]$$

As shown in Figure 1.48b, the strain initially increases exponentially and reaches an equilibrium strain (τ_0/G) asymptotically. The Hookean solid component of Voigt element retards the rate at which the equilibrium strain is approached, and 63% of the final equilibrium value is attained at $t = \lambda'$. The quantity λ' represents a characteristic time of the material and is called as the retardation time of the viscoelastic solid.

The response of a Voigt solid to stress relaxation test is:

$$\tau(t) = \gamma_0[G + \mu\delta(t)]$$

where $\delta(t)$ represents the Dirac delta or impulse function, which has an infinite magnitude at $t = 0$ but is zero at $t \neq 0$ (Darby, 1976)

$$\delta(t) = \begin{cases} \infty & \text{at } t = 0 \\ 0 & \text{for } t \neq 0 \end{cases}$$

Response function to stress relaxation shows that viscous component relaxes infinitely fast in Voigt solid, whereas the elastic component does not relax at all (Darby, 1976). Voigt solid shows incomplete instantaneous relaxation, which is in contrast with the stress relaxation properties of the Maxwell fluid shown in Figure 1.46.

1.4.2.3 Multiple Element Models

Although the Voigt and Maxwell elements are the building blocks for linear viscoelasticity, they are inadequate for modeling real material behavior except for very simple fluid and solid materials. More complex models are formulated by combining springs, dashpots, Voigt, and Maxwell elements in a variety of mechanical analogs, in order to simulate the flow behavior of a specific viscoelastic material.

An improvement over the simple viscoelastic fluids is obtained by using generalized models. The generalized Maxwell model involves n number of Maxwell elements in parallel. Figure 1.49 shows the mechanical analog of the generalized Maxwell model.

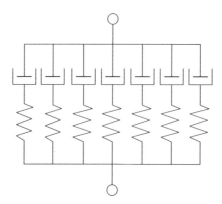

FIGURE 1.49 Mechanical analog of the generalized Maxwell model. (Reproduced from Kokini, J.L., Wang, C.F., Huang, H., and Shrimanker, S., 1995b, *Journal of Texture Studies*, 26: 421–455. With permission.)

The total stress of this model is the sum of the individual stresses in each element (Darby, 1976):

$$\tau = \sum_{p=1}^{n} \tau_p$$

For each Maxwell element in the generalized model, τ_p is associated with a viscosity μ_p and a relaxation time λ_p. Then the constitutive equation for each element in the generalized Maxwell model can be formulated as follows:

$$\tau_p + \lambda \dot{\tau}_p = \mu_p \dot{\gamma}$$

where p ranges from 1 to n for n elements.

Similarly, the generalized Kelvin model consists of Voigt elements arranged in series. For any possible combination of Maxwell and Voigt elements in series and/or in parallel, the constitutive behavior of the elements can be modeled in the form of an ordinary differential equation of the nth order:

$$\tau + p_1 \dot{\tau} + p_2 \ddot{\tau} + p_3 \dddot{\tau} + \cdots + p_m \tau^{(m)} = q_0 \gamma + q_1 \dot{\gamma} + q_2 \ddot{\gamma} + q_3 \dddot{\gamma} + \cdots + q_n \gamma^{(n)}$$

One to one correspondence exists between the parameters associated with the springs and dashpots of the mechanical analog and the coefficients (p and q) of the associated governing equations.

A linear viscoelastic constitutive model is then an equation that describes all components of stress and strain in all types of linear behavior. To develop such an equation, the "Boltzmann superposition" principle is used. The superposition principle assumes that stresses resulting from strains at different times can simply add on stresses resulting from strains at different times.

$$\sigma(t) = \sum_{i=1}^{n} G(t - t_i)\delta\gamma(t_i)$$

where $\delta\gamma(t_i)$ is the incremental strain applied at time t_i and $G(t - t_i)$ is the influence function which links stress strain behavior. The integral form of this equation when $\delta\gamma(t_i) \rightarrow 0$ is:

$$\sigma(t) = \int_0^t G(t - t')\mathrm{d}\gamma(t')$$

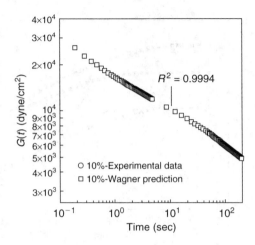

FIGURE 1.50 Relaxation modulus, $G(t)$, of 55% moisture gluten dough using 12 element generalized Maxwell model. (Reproduced from Kokini, J.L., Wang, C.F., Huang, H., and Shrimanker, S., 1995b, *Journal of Texture Studies*, 26: 421–455. With permission.)

It is necessary to determine the relaxation modulus $G(t)$ in order to relate all components of stress to all components of strain and strain rate. The relaxation modulus for Maxwell model element given by:

$$G(t) = G_0 \exp(-t/\lambda)$$

and the linear integral constitutive model is given by:

$$\tau_{ij}(t) = \int_{-\infty}^{t} G_0 \{\exp[-(t - t')/\lambda]\} \dot{\gamma}_{ij}(t') dt'$$

The generalized Maxwell with n elements leads to the following integral model:

$$\tau_{ij}(t) = \int_{-\infty}^{t} \sum_{k=1}^{n} G_k \{\exp[-(t - t')/\lambda_k]\} \dot{\gamma}_{ij}(t') dt'$$

where G_k and λ_k are the appropriate moduli and relaxation times of the Maxwell element.

The behavior of the relaxation modulus at sufficiently long times is dominated by the relaxation time with the largest value and is called the "longest relaxation time" or "terminal relaxation time." Simulation of the relaxation modulus using the generalized Maxwell model for wheat flour dough is shown in Figure 1.50 (Kokini et al., 1995b).

The Boltzmann superposition principle can also be used in dynamic measurements to obtain equations for the storage and loss moduli when a generalized Maxwell model is used to represent the relaxation modulus:

$$G'(\omega) = \sum_{i=1}^{n} \frac{G_i(\omega\lambda_i)^2}{[1 + (\omega\lambda_i)^2]}$$

$$G''(\omega) = \sum_{i=1}^{n} \frac{G_i\omega\lambda_i}{[1 + (\omega\lambda_i)^2]}$$

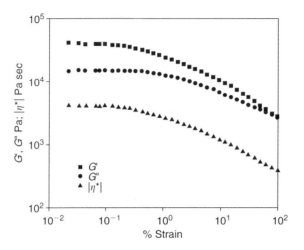

FIGURE 1.51 Dynamic measurements vs. strain for 40% moisture hard wheat flour dough sample at testing frequency of 10 rad/sec. (Reproduced from Dus, S.J. and Kokini, J.L., 1990, *Journal of Rheology*, 34: 1069–1084. With permission.)

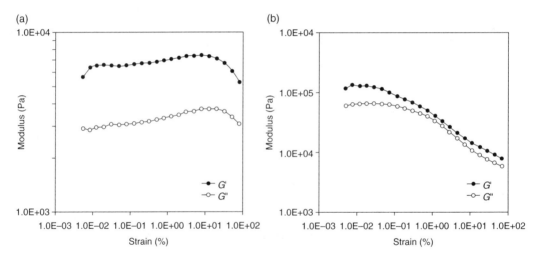

FIGURE 1.52 Dynamic moduli as function of shear strain for 55% moisture gluten at testing frequency of (a) 1.6 Hz and (b) 10 Hz. (Reproduced from Dhanasekharan, M., 2001, Dough rheology and extrusion: Design and scaling by numerical simulation, Ph.D. Thesis, Rutgers University. With permission.)

Identifying a linear viscoelastic range is a challenge with many food materials. In particular, dough has been the subject of many studies (Dus and Kokini, 1990; Wang and Kokini, 1995a, 1995b; Phan-Thien et al., 1997). It has been generally agreed that the wheat flour doughs exhibit linear behavior until a strain of O(0.001) (Dus and Kokini, 1990; Phan-Thien et al., 1997) as shown in Figure 1.51. Wang (1995) reported a linear viscoelastic strain limit of O(0.1) for gluten doughs. Dhanasekharan (2001) found that the linear viscoelastic strain limit for gluten doughs is dependent on the testing frequency (Figure 1.52). At low testing frequencies 10 rad/sec as used by Wang and Kokini (1995b) a linear viscoelastic strain limit of O(0.1) is observed. At testing frequencies of 10 Hz as used by Phan-Thien et al. (1997) for wheat flour dough, a viscoelastic strain limit of O(0.001) is observed.

1.4.2.4 Mathematical Evolution of Nonlinear Constitutive Models

Linear viscoelastic models have limited applicability due to the nonlinear nature of a majority of viscoelastic materials at realistic levels of applied strain. However, they are critical to the evolution of nonlinear models, since any material, even the most nonlinear, exhibit essentially linear behavior when subjected to a sufficiently small deformation. For food materials, more complicated nonlinear viscoelastic models are needed.

Constitutive equations of linear viscoelasticity can be evolved into nonlinear models by replacing the tensors as shown below (Bird et al., 1987):

	Tensors in linear viscoelasticity	Tensors in nonlinear viscoelasticity
Time derivatives of the rate of strain tensor	$\dfrac{\partial^n \gamma}{\partial t^n}$	$\gamma_{(n)}$
Time derivative of the stress tensor	$\dfrac{\partial \tau}{\partial t}$	$\tau_{(1)}$
Strain tensor at t' referred to state at t	$\gamma(t, t')$	$\gamma^{[0]}(t, t'), \gamma_{[0]}(t, t')$

Linear viscoelastic models are modified to nonlinear differential constitutive equations by replacing the time derivatives of rate-of-strain tensor and stress tensor by convected derivatives.

The convected time derivatives of rate-of-strain tensor are given as follows:

$$\gamma_{(1)} = \dot{\gamma}$$

$$\gamma_{(n+1)} = \frac{D}{Dt}\gamma_{(n)} - \{(\nabla v)^\dagger \cdot \gamma_{(n)} + \gamma_{(n)} \cdot (\nabla v)\}$$

where $\gamma_{(n+1)}$ is called as the nth convected derivative of the rate-of-strain tensor $\gamma_{(1)}$.

The convected time derivative of the stress tensor is similarly given as follows:

$$\tau_{(1)} = \frac{D}{Dt}\tau - \{(\nabla v)^\dagger \cdot \tau + \tau \cdot (\nabla v)\}$$

Integral constitutive equations are the integral form of differential linear viscoelastic models. They involve the use of memory functions. Modification of general linear viscoelastic models to nonlinear models is done by replacing the infinitesimal strain tensor $\gamma(t, t')$ with relative strain tensors $\gamma_{[0]}(t, t')$.

Table 1.12 shows the classical evolution of nonlinear models (Bird et al., 1987). Many of the nonlinear models have resulted from the rewriting of the Maxwell model in convected coordinates. Nonlinear viscoelastic fluids exhibit dependence of stress not only on the instantaneous rate of strain but also on the strain history.

Description of the flows with large displacement gradients necessitates evolution of linear constitutive models to growing complexities to accurately represent real material behavior. Below is an example of evolution of linear models to quasilinear model and then to nonlinear models:

Maxwell equation is given as

$$\tau + \lambda \dot{\tau} = \mu \dot{\gamma} \quad \text{or} \quad \tau + \lambda \frac{\partial \tau}{\partial t} = \mu \dot{\gamma}$$

TABLE 1.12
Mathematical Evolution of Nonlinear Models

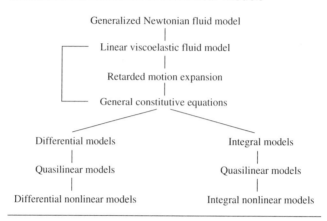

By introducing the time derivative of $\dot\gamma$ into the above equation and replacing μ by η_0 (the zero-shear-rate viscosity) we get Jeffrey's model of the form:

$$\tau + \lambda_1 \frac{\partial \tau}{\partial t} = -\eta_0 \left(\dot\gamma + \lambda_2 \frac{\partial \dot\gamma}{\partial t} \right)$$

where λ_1 is a relaxation and λ_2 is a retardation time.

By replacing the partial time derivatives with the convected time derivatives we generate a quasilinear model known as Oldroyd's fluid B model:

$$\tau + \lambda_1 \tau_{(1)} = -\eta_0(\gamma_{(1)} + \lambda_2\gamma_{(2)})$$

where $\tau_{(1)}$ is the convected time derivative of stress tensor, $\gamma_{(1)}$ is the convected time derivative of the rate of strain tensor, and $\gamma_{(2)}$ is the second convected derivative of the rate of strain tensor. When retardation time, $\lambda_2 = 0$, Oldroyd's B equation reduces to "convected Maxwell" model:

$$\tau + \lambda\tau_{(1)} = -\eta_0\gamma_{(1)}$$

where λ is the relaxation time. This is one of the simplest models, which can be used to characterize nonlinear viscoelastic effects. However, this model is primarily applicable to small strains because it is a quasi-linear viscoelastic model.

1.4.3 NONLINEAR CONSTITUTIVE MODELS

1.4.3.1 Differential Constitutive Models

Nonlinear differential models are of particular interest in numerical simulations for process design, optimization and scale-up. In a differential viscoelastic constitutive equation, the extra-stress tensor (τ_p) is related to the rate of deformation tensors ($\dot\gamma$) by means of a differential equation. The total stress tensor, τ, is given as the sum of the viscoelastic component, τ_p, and the purely Newtonian component, τ_s, as:

$$\tau = \tau_p + \tau_s$$

where $\tau_s = 2\eta\dot{\gamma}$. Differential viscoelastic constitutive models that are frequently used for characterizing the rheological properties of food materials are presented below.

1.4.3.1.1 The Giesekus Model

The Giesekus model considers polymer molecules as unbranched or branched chains of structural elements, which can be viewed as beads, joined either by elastic springs or rigid rods and subjected to Brownian motion forces (Dhanasekaran and Kokini, 2001). Entanglement loss and regeneration process cause the relative emotion of the beads with respect to the same or neighboring molecules. The relationship between this relative motion and the generating force is described by a configuration-dependent nonisotropic mobility tensor (Giesekus, 1982). The constitutive equation has the following form:

$$\left[I + \alpha\frac{\lambda}{\eta_1}\tau\right]\tau + \lambda\tau_{(1)} = 2\eta_2\dot{\gamma}$$

with a purely Newtonian component $\tau_p = 2\eta_2\dot{\gamma}$, where $\dot{\gamma}$ is the strain rate tensor, and $\tau_{(1)}$ is the upper-convected derivative of the stress tensor of the viscoelastic component, I is the unit tensor, and λ and η are the relaxation times and the viscosity factors. Parameter α controls the shear thinning properties and extensional viscosity as well as the ratio of second normal stress difference to the first one, when $\alpha > 0$ shear thinning behavior is always obtained. The term involving α is the "mobility factor" that can be associated with anisotropic Brownian motion and anisotropic hydrodynamic drag on the constituent polymer molecules.

Material functions for the Giesekus model in steady shear flow are (Bird et al., 1987):

$$\frac{\eta}{\eta_0} = \frac{\lambda_2}{\lambda_1} + \left(1 - \frac{\lambda_2}{\lambda_1}\right)\frac{(1-f)^2}{1+(1-2\alpha)f}$$

$$\frac{\psi_1}{2\eta_0(\lambda_1 - \lambda_2)} = \frac{f(1-\alpha f)}{(\lambda_1\dot{\gamma})^2\alpha(1-f)}$$

$$\frac{\psi_2}{\eta_0(\lambda_1 - \lambda_2)} = \frac{-f}{(\lambda_1\dot{\gamma})^2}$$

where

$$f = \frac{1-\chi}{1+(1-2\alpha)\chi}$$

$$\chi^2 = \frac{(1+16\alpha(1-\alpha)(\lambda_1\dot{\gamma})^2)^{1/2}-1}{8\alpha(1-\alpha)(\lambda_1\dot{\gamma})^2}$$

and ψ_1 and ψ_2 are the first and second normal stress coefficients, respectively.

Material functions in small amplitude oscillatory flow are:

$$\frac{\eta'}{\eta_0} = \frac{1+\lambda_1\lambda_2\omega^2}{1+\lambda_1^2\omega^2}$$

$$\frac{\eta''}{\eta_0\omega} = \frac{(\lambda_1 - \lambda_2)}{1+\lambda_1^2\omega^2}$$

1.4.3.1.2 The White–Metzner Model

The White–Metzner (1963) model is derived from the network theory of polymers developed by Lodge (1956) and Yamomoto (1956). The theory assumes a flowing polymer system consists of long chain molecules connected in a continuously changing network structure with temporary junctions. The viscoelastic differential constitutive model is given by:

$$\tau + \lambda \tau_{(1)} = 2\eta\dot{\gamma}$$

η is obtained from the experimental shear viscosity curve, and the function λ is obtained from the experimental first normal stress difference experimental curve. Both parameters, η and λ, can be obtained using Constant, Power law, or Bird–Carreau type dependences.

Using the Bird–Carreau type of dependence, for instance, we get the shear viscosity of the following form:

$$\eta = \eta_\infty + (\eta_0 - \eta_\infty)(1 + \lambda_v^2\dot{\gamma}^2)^{(n_v-1)/2}$$

The dependence of relaxation time on shear rate is found by fitting the experimental first normal stress difference using a Bird–Carreau type model as:

$$\lambda = \lambda_0(1 + \lambda_r^2)\dot{\gamma}^{(n_r-1)/2}$$

where η_0 is zero shear rate viscosity, η_∞ is infinite shear rate viscosity, λ_v and λ_r are the natural time (i.e., inverse of the shear rate at which fluid changes from Newtonian to power-law behavior), and n_v and n_r are the power-law index.

So the first normal stress coefficient is given by:

$$\psi_1 = 2\eta\lambda$$

The transient properties are given by:

$$\eta^+ = \eta(1 - e^{-t/\lambda})$$

$$\psi_1^+ = \psi_1 \left(1 - e^{-t/\lambda} - \frac{t}{\lambda}e^{-t/\lambda}\right)$$

where η^+ the transient viscosity and ψ^+ is the first normal stress coefficient.

1.4.3.1.3 Phan-Thien–Tanner Model

Weilgel (1969) proposed an alternative approach to Lodge and Yamamoto's network theory similar to that of Boltzmann's kinetic theory of gases. The stress tensor was shown to assume a Boltzmann integral form. Phan-Thien and Tanner (1977) used this approach to show that the stress tensor can be explicitly written in terms of an effective Finger tensor. They assumed specific forms for the creation and destruction rates of the network junctions and derived a constitutive equation containing two adjustable parameters ε and ξ. The final form of the constitutive equation is:

$$\exp\left[\varepsilon\frac{\lambda}{\eta}tr(\tau)\right]\tau + \lambda\left[\left(1 - \frac{\xi}{2}\right)\overset{\nabla}{\tau} + \frac{\xi}{2}\overset{\Delta}{\tau}\right] = 2\eta\dot{\gamma}$$

where the parameters η and λ are the partial viscosity and relaxation time, respectively, measured from the equilibrium relaxation spectrum of the fluid. They are not considered as adjustable parameters of the model. The parameter ξ can be obtained using the dynamic viscosity (η')-shear viscosity (η) shift according to:

$$\eta'(x) = \eta\left(\frac{x}{\sqrt{\xi(2-\xi)}}\right)$$

The shear viscosity, η, is given by:

$$\eta = \sum_{i=1}^{n} \frac{G_i \lambda_i}{1 + \xi(2 - \xi)\lambda_i^2 \dot{\gamma}^2}$$

where the summation of n refers to the number of nodes. The first normal stress difference (ψ_1) is given by:

$$\psi_1 = 2 \sum_{i=1}^{n} \frac{G_i \lambda_i^2}{1 + \xi(2 - \xi)\lambda_i^2 \dot{\gamma}^2}$$

The transient shear properties are obtained numerically due to the nonlinear nature of the model. The models can be used in multiple modes. This means that relaxation spectra can be chosen instead of a single relaxation time and relaxation modulus. This enables good prediction of the oscillatory shear properties. Figure 1.53a shows the predictions of the Giesekus, White–Metzner, and

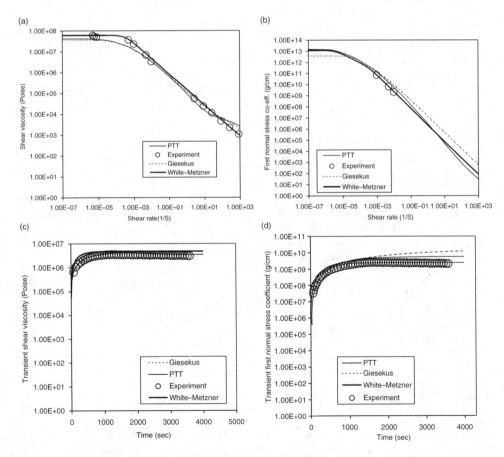

FIGURE 1.53 Prediction of (a) the steady shear viscosity, (b) the first normal stress coefficient, (c) the transient shear viscosity, and (d) the transient first normal stress coefficient of gluten dough using Giesekus, Phan-Thien–Tanner and White–Metzner model. (Reproduced from Kokini, J.L., Dhanasekharan, M., Wang, C.-F., and Huang, H., 2000, *Trends in Food Engineering*, J.E. Lozano, C. Anon, E. Parada-Arias, and G.V. Barbosa-Canovas, Eds, Technomics Publishing Co. Inc., Lancaster, PA. With permission.)

Phan-Thien–Tanner models for the shear viscosity of gluten dough (Dhanasekharan et al., 2001). The White–Metzner model resulted in the most accurate estimated values for shear viscosity using Bird–Carreau type model which has a power law parameter to predict shear viscosity in the shear thinning regime and the zero shear viscosity in the constant viscosity regime at low shear rates. Figure 1.53b shows the predictions of first normal stress coefficient for gluten dough using three different models. The White–Metzner model again provided the best fit for the first normal stress co-efficient. Figure 1.53c and Figure 1.53d show the predictions of the transient shear properties of gluten dough. The White–Metzner model under-predicted the observed transient properties while the Phan-Thien–Tanner model provided the best fit for the transient shear viscosity and the transient first normal stress coefficient.

Dhanasekharan et al. (1999) used the same three models to predict the steady shear and transient shear properties of 50% hard wheat flour/water dough. The White–Metzner model gave the best overall prediction of the observed results, as shown in Figure 1.54. However, this model exhibited asymptotic behavior at biaxial extension rates greater than 0.01 sec^{-1} and therefore is not well suited for predicting extensional flows. The Giesekus and Phan-Thien–Tanner models over-predicted the steady shear viscosity in the shear-thinning region (Figure 1.54a), the first normal stress coefficient, the transient properties (Figure 1.54b) and the biaxial viscosity (Figure 1.54d) but accurately predicted

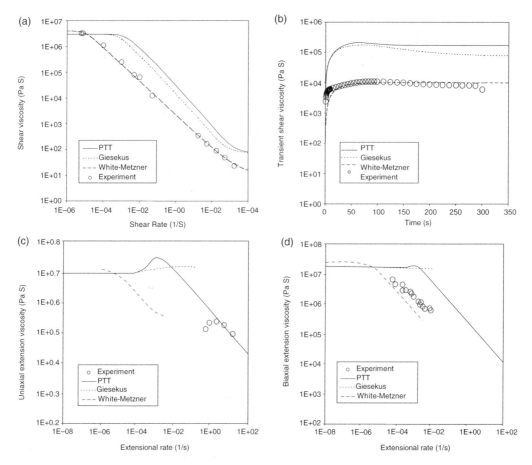

FIGURE 1.54 Comparison of (a) shear viscosity, (b) transient shear viscosity, (c) uniaxial extension, and (d) biaxial extension data for hard wheat flour dough with the predictions of nonlinear differential viscoelastic models. (Reproduced with permission from Dhanasekharan, M., Huang, H., and Kokini, J.L., 1999, *Journal of Texture Studies*, 30: 603–623. With permission.)

the dynamic properties. Only the Phan-Thien–Tanner model was able to give a good prediction of the uniaxial extensional viscosity (Figure 1.54c), with the Giesekus model requiring a higher than has been reported mobility factor (α) in order to give good results.

1.4.3.2 Integral Constitutive Models

Nonlinear integral constitutive models evolve from the general linear viscoelastic models as well. Since linear viscoelastic models are based on the infinitesimal strain tensor which specifically applies to flows with small displacement gradients, it has to be generalized to describe flows with large deformation levels. The infinitesimal strain tensor is replaced by the finite deformation tensor, a mathematical operator that transforms material displacement vectors from their past to their present state.

Finite deformation tensor (F) is used to describe the present (deformed) state in terms of the past (undeformed) state:

$$\mathrm{d}x = F \cdot \mathrm{d}x'$$

where x and x' indicate the present and past states, respectively.

Finite deformation tensor F_{ij} describes the state of deformation and rotation at any point and it depends on both the current and past state of deformation (Macosko, 1994):

$$F_{ij} = \begin{bmatrix} \dfrac{\partial x_1}{\partial x_1'} & \dfrac{\partial x_1}{\partial x_2'} & \dfrac{\partial x_1}{\partial x_3'} \\ \dfrac{\partial x_2}{\partial x_1'} & \dfrac{\partial x_2}{\partial x_2'} & \dfrac{\partial x_2}{\partial x_3'} \\ \dfrac{\partial x_3}{\partial x_1'} & \dfrac{\partial x_3}{\partial x_2'} & \dfrac{\partial x_3}{\partial x_3'} \end{bmatrix}$$

There are two types of finite deformation tensors: Cauchy (C_{ij}) and Finger (B_{ij}) tensors, which are the measures of finite strain.

$$\text{Cauchy tensor:} \quad C_{ij} = F^T \cdot F$$

$$\text{Finger tensor:} \quad B_{ij} = F \cdot F^T$$

where F^T is the transpose of finite deformation tensor. Physically Finger tensor describes the local change in area within the sample, whereas the Cauchy tensor expresses deformation in terms of length change.

The Finger tensor has three scalar invariants for a given deformation, a specific property of a second order tensor. These invariants are as follows:

$$I_1(B_{ij}) = B_{11} + B_{22} + B_{33}$$

$$I_2(B_{ij}) = C_{11} + C_{22} + C_{33}$$

$$I_3(B_{ij}) = 1$$

The Boltzmann superposition principle is generalized using the Finger tensor to formulate a theory of nonlinear viscoelasticity as follows:

$$\tau_{ij}(t) = \int_{-\infty}^{t} m(t - t') B_{ij}(t, t') \mathrm{d}t'$$

where $m(t - t')$ is the memory function. This is the equation for a rubber-like liquid developed by Lodge (1964). The constitutive equation that results from Lodge's network theory is:

$$\tau_{ij}(t) = \int_{-\infty}^{t} \frac{G_i}{\lambda_i} \exp\left[-\frac{(t-t')}{\lambda_i}\right] B_{ij}(t,t') dt'$$

The rubber-like liquid theory is of limited applicability since it predicts that the viscosity and first normal stress coefficient are independent of shear rate which is not the case with most food materials. Based on the concepts originally used in the development of the theory of rubber viscoelasticity Bernstein, Kearsley and Zapas (Bernstein et al., 1964) proposed an equation known as the BKZ equation to predict nonlinear viscoelastic behavior of materials:

$$\tau_{ij} = \int_{-\infty}^{t} \left[2\frac{\partial \mu}{\partial I_1} C_{ij}(t,t') - 2\frac{\partial \mu}{\partial I_2} B_{ij}(t,t')\right] dt'$$

where μ is a time-dependent elastic energy potential function given by:

$$\mu = \mu(I_1, I_2, t - t')$$

and I_1 and I_2 are the first and second invariants of the Finger tensor. A more practical form of the BKZ equation involves a product of a time-dependent and a strain-dependent term:

$$\mu = \mu(I_1, I_2, t - t') = m(t - t')U(I_1, I_2)$$

Wagner (1976) further simplified the equation and proposed the following factorable model of the form:

$$M[(t - t'), I_1, I_2] = m(t - t')h(I_1, I_2)$$

where $h(I_1, I_2)$ is called the damping function. This is a form of the memory function, which is separable and factorable and leads to the Wagner constitutive equation:

$$\tau(t) = \int_{-\infty}^{t} m(t - t')h(I_1, I_2)B_{ij}(t,t') dt'$$

The Wagner equation is not a complete constitutive equation since it contains the unknown $h(I_1, I_2)$ which has to be determined experimentally. There are several approximations proposed for damping functions which have all been shown to be valid in shear flows:

Wagner (1976)	$h(\gamma) = \exp(-n\gamma)$
Osaki (1976)	$h(\gamma) = a\exp(-n_1\gamma) + (1 - a)\exp(-n_2\gamma)$
Zapas (1966)	$h(\gamma) = \dfrac{1}{1 + a\gamma^2}$
Soskey and Winter (1984)	$h(\gamma) = \dfrac{1}{1 + a\gamma^b}$

where γ is the shear strain, n, n_1, n_2, and a and b are fitting parameters.

Similarly, damping functions are proposed for extensional flows as well (Meissner, 1971):

$$h(\varepsilon) = \{a[\exp(2\varepsilon)] + (1 - a)\exp(k\varepsilon)\}^{-1}$$

Damping functions have been obtained for food materials and in particular for gluten and wheat flour doughs (Wang, 1995; Kokini et al., 1995b; Huang, 1998; Kokini et al., 2000). The form proposed by Osaka was found to be the most successful in simulating the experimental data (Figure 1.55 and Figure 1.56).

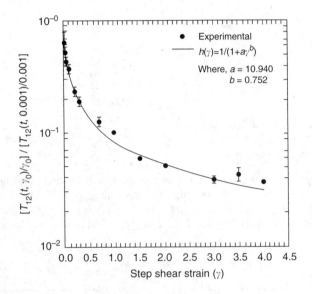

FIGURE 1.55 Simulation of shear damping function for 18.8% protein flour dough. (Reproduced from Kokini, J.L., Dhanasekharan, M., Wang, C.-F., and Huang, H., 2000, *Trends in Food Engineering*, J.E. Lozano, C. Anon, E. Parada-Arias, and G.V. Barbosa-Canovas, Eds, Technomics Publishing Co. Inc., Lancaster, PA. With permission.)

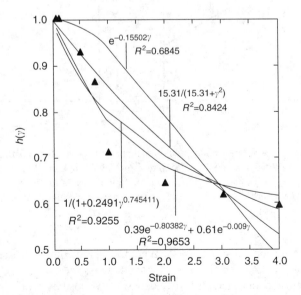

FIGURE 1.56 Simulation of the damping function $h(\gamma)$ using four types of mathematical models for 55% moisture gluten dough at 25°C. (Reproduced from Kokini, J.L., Wang, C.F., Huang, H., and Shrimanker, S., 1995b, *Journal of Texture Studies*, 26: 421–455. With permission.)

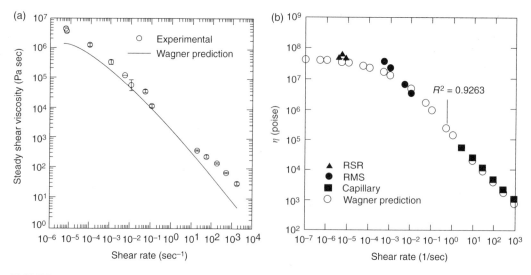

FIGURE 1.57 Comparison of Wagner model prediction of the steady shear viscosities with experimental data for (a) 18.8% protein flour dough and (b) 55% moisture gluten. (Reproduced from Kokini, J.L., Dhanasekharan, M., Wang, C.-F., and Huang, H., 2000, In: *Trends in Food Engineering*, J.E. Lozano, C. Anon, E. Parada-Arias, and G.V. Barbosa-Canovas, Eds, Technomics Publishing Co. Inc., Lancaster, PA. With permission.)

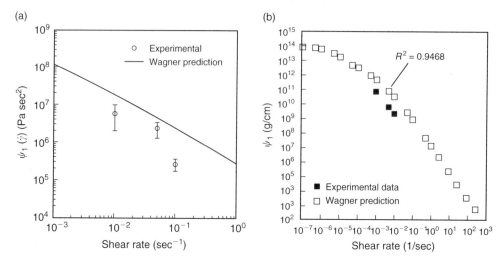

FIGURE 1.58 Comparison of Wagner model prediction of the first normal stress coefficients with experimental data for (a) 18.8% protein flour dough and (b) 55% moisture gluten dough. (Reproduced from Kokini, J.L., Dhanasekharan, M., Wang, C.-F., and Huang, H., 2000, In: *Trends in Food Engineering*, J.E. Lozano, C. Anon, E. Parada-Arias, and G.V. Barbosa-Canovas, Eds, Technomics Publishing Co. Inc., Lancaster, PA. With permission.)

Figure 1.57 shows the comparison of the Wagner model prediction of the steady shear viscosities with the experimental data for wheat flour and gluten doughs. Shear viscosity predictions using the Wagner model showed an under-prediction of steady shear viscosities in the experimental shear rate range of 1×10^{-6} to 1×10^{-1} sec^{-1} (Kokini et al., 2000). Higher differences between experimental and simulated steady shear viscosities were observed in the shear rate region where viscosities were measured using a capillary rheometer. Figure 1.58 shows the predictions of first normal stress coefficient of wheat flour and gluten doughs using the Wagner model. The model over-predicted the first normal stress coefficient values. The high volume percentage of starch fillers in the dough violates the core assumptions included in the development of the Wagner model, which may account for the discrepancy between simulated and experimental results (Kokini et al., 2000).

1.5 MOLECULAR INFORMATION FROM RHEOLOGICAL MEASUREMENTS

1.5.1 DILUTE SOLUTION MOLECULAR THEORIES

Molecular models of rheology aim at quantitatively linking rheological properties to molecular structure and use rheological data as a diagnostic tool to understand the molecular conformation of food polymers and the structural organization of complex materials. In order to achieve this goal, idealizations of molecular architecture or conformation are necessary. Such idealizations lead to molecular theories of rheology.

The simplest polymer systems are for a dilute solution of linear flexible polymers. The molecular evolution of molecular theories started by considering dilute solutions of high molecular weight polymeric materials. These theories (Rouse, 1953; Zimm, 1956; Marvin and McKinney, 1965) are useful in characterizing the effect of long-range conformation on the flexibility of some carbohydrates and proteins. Dilute solution molecular theories have further evolved to predict rheological properties of concentrated polymeric systems. They are based on key assumptions pertaining to network formation and dissolution which occur during deformation processes. There are many other constitutive models, which count for the effect of entanglements or cross-links. The models that have an accurate molecular and conformational basis enable us to predict rheological properties from the detailed understanding of the molecular structure.

The Rouse (1953) and Zimm (1956) theories provide a basis for quantitative prediction of linear viscoelastic properties for linear high molecular weight polymers in dilute solutions. The Rouse model is based on the assumption that large polymer molecules can be simulated using straight segments that act as simple linear elastic springs. The springs are connected by beads which give rise to viscous resistance. The combination of elastic and viscous effects develops viscoelastic behavior (Labropoulos et al., 2002a). The equations to predict the reduced storage and loss moduli of flexible random coil molecules of the Rouse and Zimm type are given below:

$$[G']_R = \sum_{p=1}^{n} \frac{\omega^2 \tau_p^2}{(1 + \omega^2 \tau_p^2)}$$

$$[G'']_R = \sum_{p=1}^{n} \frac{\omega \tau_p}{(1 + \omega^2 \tau_p^2)}$$

where $[G']_R$ is the reduced intrinsic storage modulus and $[G'']_R$ is the reduced intrinsic loss modulus, τ_p is the spectrum of relaxation time and ω is the frequency of the applied oscillatory deformation, and p is an index number.

Estimation of intrinsic moduli $[G']$ and $[G'']$ necessitates measurement of the storage modulus G' and the loss modulus G'' at several concentrations in dilute solution region. When (G') and $(G'' - \omega \eta_s)$ are plotted against concentration, the intercept at zero concentration gives:

$$[G'] = \lim_{c \to 0} \frac{G'}{c} \quad \text{and} \quad [G''] = \lim_{c \to 0} \frac{G'' - \omega \eta_s}{c}$$

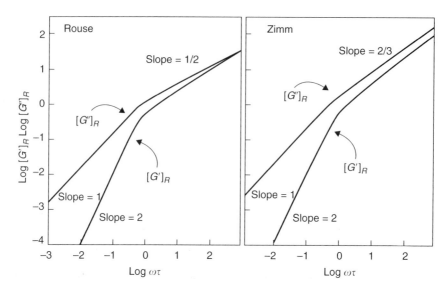

FIGURE 1.59 Prediction of reduced moduli for flexible random coils as proposed by Rouse (1953) and Zimm (1956). (Reproduced from Ferry, J., 1980, *Viscoelastic Properties of Polymers*, 3rd ed., John Wiley & Sons, New York. With permission.)

Then the reduced moduli are calculated as

$$[G']_R = \frac{[G']M}{RT} \quad \text{and} \quad [G'']_R = \frac{[G'']M}{RT}$$

where c is the polymer concentration, M is the polymer molecular weight, T is the temperature, and R is the gas constant.

The difference in the reduced moduli between Rouse and Zimm type of molecules is in the calculation of relaxation time. Calculated theoretical values of $[G']_R$ and $[G'']_R$ for each model are given by Ferry (1980). The predicted reduced moduli from the theories of Rouse and Zimm for random coils as a function of $\omega\tau$ are plotted in Figure 1.59. At high frequencies the reduced moduli of the Rouse theory become equal and increase together with a slope of $1/2$, while those in the Zimm theory remain unequal and increase in a parallel manner with a slope of $2/3$.

A number of theories have been developed for dilute solutions of elongated rigid rod-like macromolecules. The main feature of rod-like models is the prediction of an end-to-end rotation relaxation time (Labropoulos et al., 2002a) which can be related to the relaxation behavior of clusters in solution. The reduced storage and loss moduli and the spectrum of relaxation time can be generalized as follows:

$$[G']_R = \frac{m_1 \omega^2 \tau^2}{(1 + \omega^2 \tau^2)}$$

$$[G'']_R = \omega\tau \left[\frac{m_1}{(1 + \omega^2 \tau^2)} + m_2 \right]$$

$$\tau = \frac{m[\eta]\eta_s M}{RT}$$

where

$$m = (m_1 + m_2)^{-1}$$

TABLE 1.13

Geometrical Constants m_1 and m_2 for the Elongated Rigid Rod Model

Model	m_1	m_2	m
Cylinder	0.60	0.29	1.15
Cylinder	0.46	0.16	1.61
Rigid dumbell	0.60	0.40	1.00
Prolate ellipsoid	0.60	0.24	1.19
Shishkebob	0.60	0.20	1.25

Source: From Ferry, J., 1980, *Viscoelastic Properties of Polymers*, 3rd ed., John Wiley & Sons, New York. With permission.

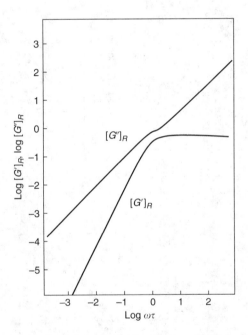

FIGURE 1.60 Prediction of reduced moduli for the rigid rod theory of Marvin and McKinney (1965). (Reproduced from Kokini, J.L., 1993, In: *Plant Polymeric Carbohydrates*, F. Meuser, D.J. Manners, and W. Siebel, Eds, Royal Society of Chemistry, Cambridge. With permission.)

where ω is the frequency, τ is relaxation time, $[\eta]$ is the intrinsic viscosity of the solution, and η_s is the viscosity of the solvent. m_1 and m_2 are constants for different geometrical variations such as a cylinder of a dumbell for the elongated rigid rod model (Ferry, 1980). Table 1.13 shows the values of the geometrical constants calculated using different rigid-rod models.

The predicted reduced moduli from the theory of Marvin and McKinney (1965) for rigid rods as a function of $\omega\tau$ are given in Figure 1.60 (Kokini, 1993).

Dilute solution theories found some applications in food polymer rheology. Chou and Kokini (1987) and Kokini and Chou (1993) studied the rheological properties of dilute solutions of hot break and cold break tomato, commercial citrus and apple pectins. Tomato processing was found to have a significant effect on the chain length and rheological properties of tomato pectins. Tomato pectin from cold break tomato paste had intrinsic viscosity value three times lower than that of tomato pectin from hot break paste, suggesting that cold break processing affected the chain length of tomato pectins through the action of pectic enzymes. Consistent with the viscosity data, the weight-average molecular weight of cold break tomato pectin was found to be 38 times lower than that of hot break tomato pectin. Kokini and Chou (1993) studied the conformation of tomato, apple, and citrus pectins as a function of the degree of esterification using constitutive models. The fit of the experimental $[G']_R$ and $[G'']_R$ with the theoretical rigid model of Marvin and McKinney for apple pectin of degree of methylation of 73.5% is shown in Figure 1.61a. The graph clearly shows that this apple pectin does not follow rod-like behavior. Experimental reduced moduli were also compared with the predictions of Rouse and Zimm models. The Rouse model gave a slightly better

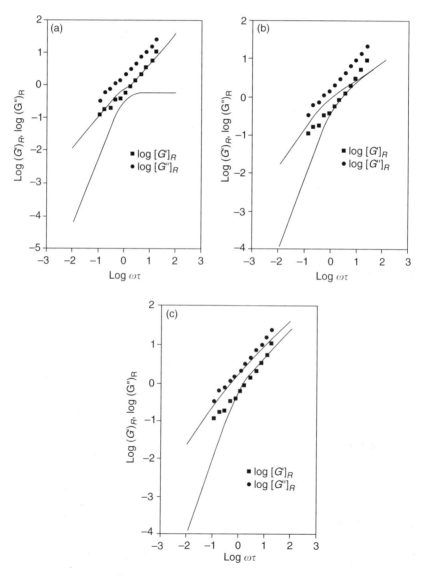

FIGURE 1.61 Comparison of experimental reduced moduli of apple pectin with (a) rod model, (b) Rouse model, and (c) Zimm model. (Reproduced from Kokini, J.L., 1994, *Carbohydrate Polymers*, 25: 319–329. With permission.)

agreement with the experimental data compared to rod-like model (Figure 1.61b) but still not well approximated by the flexible random coil theory. Among the dilute solution theories, the random coil theory of Zimm best explained the experimental data (Figure 1.61c) and suggested a certain level of intermolecular interaction present in the dilute pectin. This interaction is expected since opposite charges on the molecule will tend to attract providing an environment for considerable intermolecular interactions.

1.5.2 CONCENTRATED SOLUTION THEORIES

1.5.2.1 The Bird–Carreau Model

The rheological properties of concentrated dispersions cannot be predicted accurately using dilute solution theories due to the fundamental conformational differences between dilute polymer solutions and undiluted polymers. In a concentrated solution, the polymer chain cannot freely move sideways and its principal motion is in the direction of the chain backbone. James (1947) was the first to develop a mathematical model for the statistical properties of a molecular network, which consist of physically cross-linked polymer chains, forming a macromolecular structure. This theory was expanded by Kaye, Lodge, and Yamamoto to better explain viscoelastic behavior by assuming that deformation creates and destroys temporary cross-links (Leppard, 1975).

The Carreau constitutive model is an integral model that incorporates the entire deformation history of a material. The model can describe non-Newtonian viscosity, shear-rate dependent normal stresses, frequency-dependent complex viscosity, stress relaxation after large deformation shear flow, recoil, and hysteresis loops (Bird and Carreau, 1968). The Bird–Carreau model employs the use of zero-shear-rate limiting viscosity, η_0, and the time constants, λ_1 and λ_2, and α_1 and α_2.

The prediction for η is (Bird et al., 1987):

$$\eta = \sum_{p=1}^{\infty} \frac{\eta_p}{1 + (\lambda_{1p}\dot{\gamma})^2}$$

and large shear rates above equation is approximated by

$$\eta = \frac{\pi \eta_0}{Z(\alpha_1) - 1} \cdot \frac{(2^{\alpha_1}\lambda_1\dot{\gamma})^{(1-\alpha_1)/\alpha_1}}{2\alpha_1 \sin[((1 - \alpha_1)/(2\alpha_1)) \cdot \pi]}$$

where

$$\lambda_{1p} = \lambda_1 \left[\frac{2}{p+1} \right] \alpha_1$$

$$\eta_p = \eta_0 \frac{\lambda_{1p}}{\sum_{p=1}^{\infty} \lambda_{1p}}$$

$$z(\alpha_1) = \sum_{k=1}^{\infty} K^{-\alpha_1}$$

The Bird–Carreau prediction for η' is:

$$\eta' = \sum_{p=1}^{\infty} \frac{\eta_p}{1 + (\lambda_{2p}\omega)^2}$$

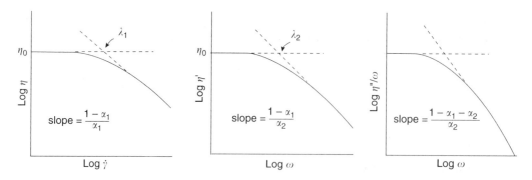

FIGURE 1.62 Determination of the Bird–Carreau constants λ_1, λ_2, α_1, and α_2. (Reproduced from Bird, R.B., Armstrong, R.C., and Hassager, O., 1987, *Dynamics of Polymeric Liquids*, 2nd ed., John Wiley & Sons Inc., New York. With permission.)

and at high frequencies it is approximated by

$$\eta' = \frac{\pi\eta_0}{z(\alpha_1) - 1} \frac{(2\alpha_2\lambda_2\omega)^{(1-\alpha_1)/\alpha_2}}{2\alpha_2 \sin[((1+2\alpha_2-\alpha_1)/(2\alpha_2))\pi]}$$

Finally, the prediction for η''/ω is:

$$\eta''/\omega = \sum_{p=1}^{\infty} \frac{\eta_p\lambda_{2p}}{1 + (\lambda_{2p}\omega)^2}$$

and at high frequencies it converges to

$$\eta''/\omega = \frac{2^{\alpha_2}\lambda_2\pi\eta_0}{z(\alpha_1) - 1} \frac{(2^{\alpha_2}\lambda_2\omega)^{(1-\alpha_1-\alpha_2)/\alpha_2}}{2\alpha_2 \sin[((1+\alpha_2-\alpha_1)/(2\alpha_2))\pi]}$$

where

$$\lambda_{2p} = \lambda_2 \left[\frac{2}{p+1}\right]^{\alpha_2}$$

The empirical model constants are obtained from steady shear and oscillatory shear experiments: η_0, λ_1, and α_1 are determined from a logarithmic plot of η vs. γ, while λ_2 and α_2 are obtained from a logarithmic plot of η' vs. ω (Figure 1.62).

η_0 is readily obtained by extrapolating the steady shear viscosity to low shear rates. The time constant λ_1 represents the characteristic time for the onset of non-Newtonian behavior under steady shear conditions. λ_1 values are taken as the inverse of the shear rate at the intersection of the line extending from η_0 to the line tangent to the high-shear-rate non-Newtonian region of the log η vs. log $\dot{\gamma}$ curve. The time constant λ_2 represents the characteristic time for the onset of non-Newtonian behavior under oscillatory shear conditions and is determined by the same procedures as for λ_1, where ω replaces $\dot{\gamma}$.

The constant α_1 is obtained from the slope of the non-Newtonian region of the log η vs. log $\dot{\gamma}$ curve as follows:

$$\text{slope of } \eta = \frac{1 - \alpha_1}{\alpha_1}$$

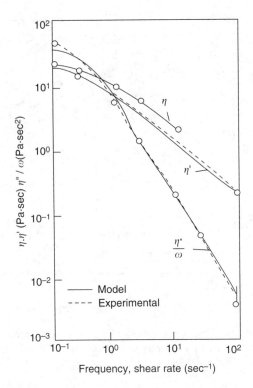

FIGURE 1.63 Comparison of predictions of the Bird–Carreau constitutive model and experimental data for 1% guar. (Reproduced from Kokini, J.L., 1994, *Carbohydrate Polymers*, 25: 319–329. With permission.)

where slope η is the slope of the steady shear non-Newtonian region; α_2 is then determined from either the slope of the non-Newtonian region of the log η' vs. log ω curve or the slope of the high-frequency region of the log η''/ω vs. log ω curve as follows:

$$\text{slope of } \eta' = \frac{1 - \alpha_1}{\alpha_2}$$

and

$$\text{slope of } \frac{\eta''}{\omega} = \frac{1 - \alpha_1 - \alpha_2}{\alpha_2}$$

where the slopes η' and η''/ω are the slopes of the log η' vs. log ω curve and log η''/ω vs. log ω, respectively.

The semiempirical nature of the Bird–Carreau model facilitates the estimation of parameters and makes the models easily applicable to a variety of materials such as concentrated dispersions of polysaccharides including guar gum and CMC (Plutchok and Kokini, 1986; Kokini and Plutchok, 1987b), protein networks as well as doughs (Dus and Kokini, 1990; Cocero and Kokini, 1991). As an example, Figure 1.63 shows that the Bird–Carreau model was able to predict η, η' and η''/ω in the high and low frequency regions for 1% guar solution.

Such constitutive models can also be used to predict the rheological properties of concentrated gum blend systems. Plutchok and Kokini (1986) developed empirical equations capable of predicting η_0, λ_1, and λ_2, as well as the slope of non-Newtonian region of η and η', using concentration and molecular weight data. A generalized correlation to predict rheological constants from concentration

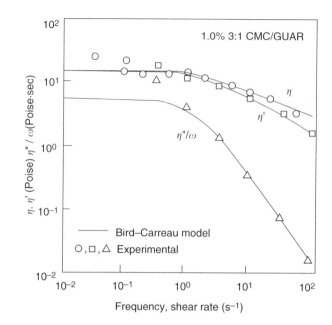

FIGURE 1.64 Comparison of predictions of the Bird–Carreau constitutive model and experimental data for a 3:1 CMC/guar blend at a combined 1% concentration. (Reproduced from Kokini, J.L., 1994, *Carbohydrate Polymers*, 25: 319–329. With permission.)

and molecular weight of the following form was used:

$$f(c_{\text{blend}}, \bar{M}_{w,\text{blend}}) = p_0 (c_{\text{blend}})^{p_1} (\bar{M}_{w,\text{blend}})^{p_2}$$

where

p_0, p_1, and p_2	parameters to be determined
c	concentration (g/100 ml)
$\bar{M}_w$	weight-average molecular weight
$f(c_{\text{blend}}, M_{w,\text{blend}})$	η_0, λ_1, λ_2, slope of η and η'

The rheological properties of guar gum-CMC blends at several proportions were predicted using these empirical equations (Plutchok and Kokini, 1986). In the case of 3:1 CMC-guar gum blend, the Bird–Carreau model explained steady-shear and dynamic properties very well in the higher shear rate or frequency region of 1 to 100 $\sec^{-1}$. However, η''/ω does not tend to a zero shear constant value (Figure 1.64).

In the case of cereal biopolymers, the rheological properties at moderate to low moisture contents are highly significant. Proteins exist in any amorphous metastable glassy state which is very sensitive to changes in moisture, temperature, and processing history. Cocero and Kokini (1991) showed that both gluten and its high molecular weight component glutenin are plastizable polymers. Dus and Kokini (1990) used the Bird–Carreau model to predict the rheology of gluten and glutenin. The model successfully predicted the apparent steady shear viscosity for 40% moisture glutenin at 25°C (Figure 1.65). The Bird–Carreau parameters suggested that 40% moisture glutenin is indeed in the free-flow region. Since glutenin is the principal protein component of wheat flour dough, the presence of disulfide bonds and noncovalent interactions determine the density of entanglements. 40% moisture glutenin at 25°C experienced rubbery flow, where the entanglements slip so that

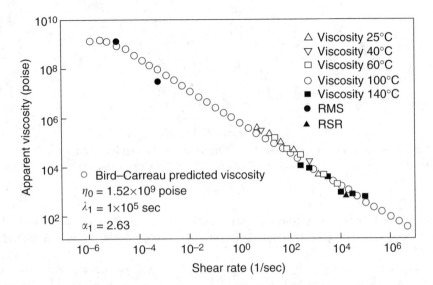

FIGURE 1.65 Bird–Carreau prediction of the steady shear viscosity for 40% moisture glutenin at 25°C. (Reproduced from Kokini, J.L., 1994, *Carbohydrate Polymers*, 25: 319–329. With permission.)

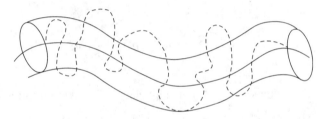

FIGURE 1.66 A schematic representation of a worm-like polymer chain (dashed line) surrounded by an outer tube-like cage. (Reproduced from Doi, M. and Edwards, S.F., 1978a, *Journal of Chemical Society, Faraday Transactions II*, 74: 1789–1801.)

configurational rearrangements of segments separated by entanglements can take place (Kokini, 1993 and 1994).

1.5.2.2 The Doi–Edwards Model

Doi and Edwards viscoelasticity is explained by considering entanglements within the polymer network (Doi and Edwards, 1978a,b). Accordingly, a model chain (or primitive path) is constructed which describes molecular motions in a densely populated system assuming that each polymer chain moves independently in the mean field imposed by the other chains. The mean field is represented by a three-dimensional cage. In this cage each polymer is confined in a tube-like region surrounding it as shown in Figure 1.66. The primitive chain can move randomly forward or backward only along itself.

A sliplink network concept is introduced to define dynamic properties under flow. The junctions of sliplinks are assumed not to be permanent cross-links but small rings through which the chain can pass freely as shown in Figure 1.67. In highly entangled polymer systems, the molecular motion of a single chain can be divided into two types: (i) the small-scale wiggling motion which does not alter the topology of the entanglement and (ii) the large-scale diffusive motion which changes the topology. The time scale of the first motion is essentially the Rouse relaxation time (Shrimanker, 1989).

FIGURE 1.67 A schematic representation of the sliding motion of the wiggling chain through sliplinks (small circles). (Reproduced from Doi, M. and Edwards, S.F., 1978b, *J. Chem. Soc., Faraday Trans. II.*, 74: 1802–1817. With permission.)

The Doi–Edwards theory is only concerned with motion of the second type. The time scale of the second motion is a renewal proportion of the topology of a single chain is proportional to M^3 (Doi and Edwards, 1978a,b).

The theory has been modified by Rahalkar et al. (1985) for a polydisperse system. The following results are relevant to the storage and loss moduli (G' and G'') of a monodispersed polymer.

$$G'(\omega) = \frac{8}{\pi^2} G_N^0 \sum_{p=1,\text{odd}}^{\infty} [(\omega T_1)^2/p^6]/[1 + (\omega T_1)^2/p^4]$$

$$G''(\omega) = \frac{8}{\pi^2} G_N^0 \sum_{p=1,\text{odd}}^{\infty} [(\omega T_1)/p^6]/[1 + (\omega T_1)^2/p^4]$$

where G_N^0 is the plateau modulus obtained at high frequency, T_1 is the extra stress tensor and p is an integer.

For a polydisperse polymer with a molecular weight distribution of $f(\mu)$, the weight fraction of chains with molecular weight between M and $M + dM$ is given by $W(M)dM$, where

$$W(M) = 1/M_n f(\mu)$$

and where μ is the dimensionless molecular weight ($=M/M_n$).

For this case, the storage and loss modulus are by:

$$G'(\omega) = G_N^0 \int_0^\infty \frac{8}{\pi^2} \sum_{p=1,\text{odd}}^{\infty} [(\omega T_1)^2 \mu^6 f(\mu)/p^6]/[1 + (\omega T_1)^2 \mu^6/p^4]d\mu$$

and

$$G''(\omega) = G_N^0 \int_0^\infty \frac{8}{\pi^2} \sum_{p=1,\text{odd}}^{\infty} [(\omega T_1)\mu^3 f(\mu)/p^6]/[1 + (\omega T_1)^2 \mu^6/p^4]d\mu$$

where G_N^0, and the plateau modulus, is given by

$$G_N^0 = G_{0\text{ave}}/5$$

Shrimanker (1989) used the Doi–Edwards theory to predict G' and G'' values for a 5% apple pectin dispersion, assuming both monodisperse and polydisperse polymer. Figure 1.68 shows the plot

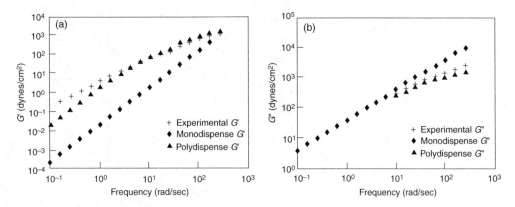

FIGURE 1.68 G' and G'' values for 5% pectin solution predicted by the Doi–Edwards model. (Reproduced from Kokini, J.L., 1994, *Carbohydrate Polymers*, 25: 319–329. With permission.)

of predicted values along with the experimental values for the simulation of $G'(\omega)$ and $G''(\omega)$. The polydisperse model explained the experimental data better than the monodisperse model, expectedly, since apple pectin is highly polydisperse with a reported polydispersity ratio (Mw/Mn) of 15 to 45.

Although the constitutive models discussed above provide major clues in designing food molecules with desired rheological properties, they usually do not permit prediction of the rheological properties of complex mixtures. For structurally complex materials it is difficult to describe the viscoelastic behavior with just one polymer model. Agar gel is a typical example for such a case. At high temperatures, the rheological behavior of agar sols is similar to dilute solutions of linear polymers. On the other hand, at low temperatures below the gelation point, their behavior is similar to that of cross-linked polymers. In the temperature range where the sol-gel transition occurs, the situation is further complicated. Moreover, the rheological properties of agar gels depend on their thermal history (Labropoulos et al., 2002a, 2002b).

Labropoulos et al. (2002a) developed a theoretical rheological model for agar gels, based on the bead-spring model for linear flexible random coils and the model for cross-linked polymers. A temperature dependence was introduced into the proposed model to determine the fraction of molecules that undergo gelation and thus to predict the gelation behavior of agar gels as a function of time and temperature. At high temperatures, agar molecules take on a random coil conformation. During cooling, agar molecules associate with each other forming double helices and higher order assemblies. At temperatures below gelation temperature the rheological behavior of agar gel is dominated by contributions from an agar network.

The proposed model was successfully fitted to experimental gelation curves obtained over a wide range of cooling rates (0.5–20°C/min) and agar concentrations in the range of (1–3%w), demonstrating a good flexibility of the model to fit a wide range of thermal histories. Figure 1.69 shows dynamic moduli as a function of time for a 2% (w/w) agar cooled from 90 to 25°C at 0.5°C/min. Solid lines represent the theoretical predictions of the model. Similar results were obtained for other agar concentrations and cooling rates. The theoretical predictions for G' and G'' are very close to the experimental data, and the theoretical $G' - G''$ crossover matches the experimental one closely.

1.5.3 UNDERSTANDING POLYMERIC PROPERTIES FROM RHEOLOGICAL PROPERTIES

1.5.3.1 Gel Point Determination

Crosslinking polymers undergo phase transitions from liquid to solid at a critical extent of reaction, which is called gelation. Gel point is defined as the moment at which a polymer/biopolymer system

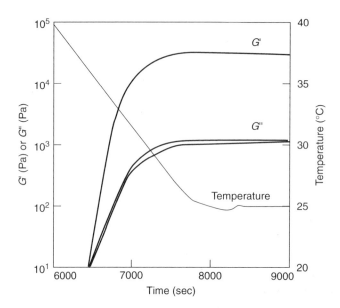

FIGURE 1.69 Dynamic storage (G') and loss (G'') moduli as a function of time for a 2% agar cooled from 90 to 25°C at 0.5°C/min. (Reproduced from Labropoulos, K.C., Niesz, D.E., Danforth, S.C., and Kevrekidis, P.G., 2002b, *Carbohydrate Polymers*, 50: 407–415. With permission.)

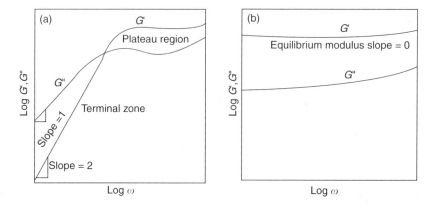

FIGURE 1.70 Dynamic mechanical spectra of storage and loss modulus for (a) an entanglement network system and (b) a covalently cross-linked network. (Reproduced from Ross-Murphy, S.B., 1995b, *Journal of Rheology*, 39: 1451–1463. With permission.)

changes from a viscous liquid (sol) to an elastic solid (gel) (Ross-Murphy, 1995a, 1995b). It can be determined from rheological properties such as steady shear viscosity for the liquid state and equilibrium shear modulus for the solid state (Gunasekaran and Ak, 2000). The polymer is considered to be at the gel point where its steady shear viscosity is infinite and its equilibrium modulus is zero (Winter and Chambon, 1986).

Small amplitude oscillatory measurements have been widely used for determining gel point and properties of the final gel network. Dynamic measurements provide continuous rheological data for the entire gelation process in contrast to steady rheological measurements. This is extremely import-ant due to lack of singularity in the gelation process. Two commonly used rheological measures to detect gel point are the cross-over point between G' and G'' (Figure 1.70) and the point when loss

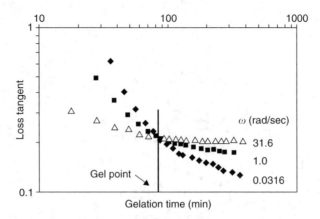

FIGURE 1.71 Gel point determination using Winter–Chambon criterion. (Reproduced from Gunasekaran, S. and Ak, M.M., 2000, *Trends in Food Science & Technology*, 11: 115–127. With permission.)

tangent (tan δ) becomes frequency independent (Figure 1.71), also known as the Winter–Chambon method (Gunasekaran and Ak, 2000). The cross-over method is a special case of the Winter–Chambon method. The gelation time determined by these two methods does not necessarily match in a single frequency experiment (Winter and Chambon, 1986). The cross-over method depends on the frequency of the oscillation depending on the gel strength. Entangled polymer network systems (weak gels) show a strong frequency dependence, that is, G' increases with increasing test frequency as shown in Figure 1.70a, while the cross-linked network gels (strong gels or chemical gels) show very little frequency dependence (Figure 1.70b).

Both cross-over and Winter–Chambon methods have been extensively used for gel point determination of biopolymers. Svegmark and Hermansson (1991) reported that cross-over criterion becomes difficult to use in complex mixed systems such as potato, wheat, and maize starch dispersions. Lopes da Silva and Goncalves (1994) studied rheological properties of curing high methoxyl pectin/ sucrose gels at different temperatures using small amplitude oscillatory experiments. They observed that the time of $G' - G''$ crossover point is dependent on the oscillation frequency (Figure 1.72). Thus, the $G' - G''$ crossover method could not be used as a criterion to identify the gel point; they instead applied the Winter–Chambon criterion.

Jauregui et al. (1995) studied the viscoelastic behavior of two commercial hydroxyl ethers of potato starch with different degrees of substitution. They reported three different viscoelastic behaviors of hydroxyethylated starch aqueous systems at different concentrations as shown in Figure 1.73:

I. Fluid-like behavior at low concentrations: G'' is greater than G', $G' \propto \omega^2$, and $G'' \propto \omega$ as predicted by the general linear viscoelastic model.
II. Fluid-gel transition zone at intermediate concentrations: G'' is still greater than G' but both moduli are proportional to frequency as $\omega^{0.5}$.
III. Gel-like behavior at concentrations of $>30\%$: G' is greater than G'' and is independent of frequency at low frequencies.

The loss tangent (tan δ) vs. frequency plots at different starch concentrations (Figure 1.74) confirmed the existence of the three viscoelastic behaviors defined as fluid-like, fluid-gel transition, and gel-like zones. When the system is not a gel, tan δ decreases as the frequency increases, as is typical for a viscoelastic liquid. However, when gelation takes place, the loss factor increases with frequency indicating that the system has changed into the viscoelastic solid state. An intermediate behavior is observed for 25% w/w, which gives rise to an almost frequency-independent tan δ, as corresponds to the transition region.

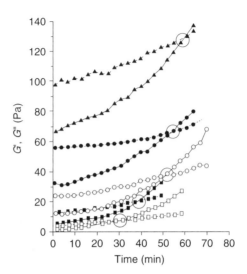

FIGURE 1.72 Storage modulus (———) and loss modulus (------) recorded at different oscillatory frequencies for 1% high methoxyl pectin (60% sucrose, pH 3) (□) 0.50 rad/sec; (■) 1.58 rad/sec; (○) 5.0 rad/sec; (●) 15.8 rad/sec; (△) 50.0 rad/sec. (Reproduced from Lopes da Silva, J.A. and Goncalves, M.P., 1994, *Carbohydrate Polymers*, 24: 235–245. With permission.)

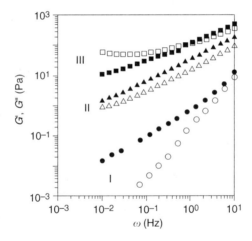

FIGURE 1.73 Storage moduli (G'; open symbols) and loss moduli (G''; solid symbols) of aqueous starch systems as a function of frequency. I, II, and III correspond to 15, 25, and 30% polymer concentrations, respectively. (Reproduced from Jauregui, B., Muñoz, M.E., and Santamaria, A., 1995, *International Journal of Biological Macromolecules*, 17: 49–54. With permission.)

Labropoulos and Hsu (1996) studied the gel forming ability of whey protein isolate (WPI) dispersions subjected to different processing variables (e.g., temperature, pH, and concentration) using small amplitude oscillatory measurements. They observed a wide range of gelation times from 12 to 164 min depending on the experimental conditions when the Winter–Chambon method was applied. A frequency-independent $\tan\delta$ was determined from a multifrequency scan of $\tan\delta$ vs. gelling time at the gel point (Figure 1.75). The rheological data demonstrated a power-law frequency dependence of the viscoelastic functions G' and G'' (i.e., $G'(\omega) = A\omega^n$ and $G''(\omega) = B\omega^n$). A unique power low exponent n at the gel point was obtained from linear regression fits of $\log G'$ and $\log G''$ vs. $\log\omega$. The experimental results showed that high correlations between the applied processing

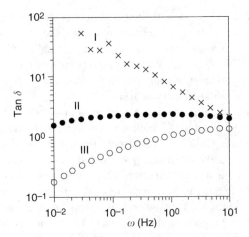

FIGURE 1.74 Loss factors plotted against frequency. I, II, and III correspond to 15, 25, and 30% polymer concentrations, respectively. (Reproduced from Jauregui, B., Mufioz, M.E., and Santamaria, A., 1995, *International Journal of Biological Macromolecules*, 17: 49–54. With permission.)

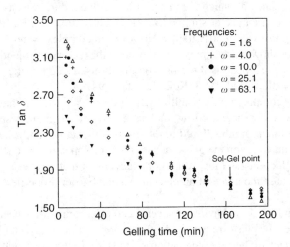

FIGURE 1.75 Time evolution of tan δ during gelation for different frequencies, showing critical gel points of WPI. (Reproduced from Labropoulos, A.E. and Hsu, S.-H., 1996, *Journal of Food Science*, 61: 65–68. With permission.)

conditions and resulting gelling times would serve as valuable tools for controlling the variables during gelation of WPI dispersions.

1.5.3.2 Glass Transition Temperature and the Phase Behavior

Synthetic amorphous polymers exhibit five regions of time-dependent viscoelastic behavior: glassy zone, glass transition zone, rubbery zone, rubbery flow region, and free flow region. Amorphous materials undergo transition from a solid glassy state to viscous liquid state at a material specific temperature called the glass transition temperature. In complex systems, such as food formulations, this transition occurs over a wide range of temperature although it is usually referred to as a single temperature value (Cocero and Kokini, 1991; Madeka and Kokini, 1996; Ross et al., 1996; Morales and Kokini, 1997; Toufeili et al., 2002). Molecular mobility and physicochemical properties change dramatically over the temperature range of glass transition. Understanding the thermal behavior of food biopolymers and mapping the changes in their rheological properties resulting from plasticization and other processing parameters are very important to control the final quality of the food products. Glass transition has a great effect on processing, properties, quality, safety, and stability of foods (Ross et al., 1996). It affects the physical and textural properties of foods (e.g., stickiness, viscosity, brittleness, crispness, or crunchiness), the rates of deteriorative changes, such as enzymatic reactions, nonenzymatic browning, oxidation, and crystallization.

State transitions and chemical reactions in food systems can be identified and characterized using differential scanning calorimetry, rheometry, dilatometry, thermal expansion measurements, or dielectric constants measurements (Kokini et al., 1994). During transition from glassy to rubbery state the properties such as heat capacity, thermal expansion and dielectric constant show a discontinuity, which is used as the basis for most of the experimental techniques for T_g measurements. Differential scanning calorimetry (DSC) and rheometry, in particular small amplitude oscillatory measurements, are the most common techniques used to study the glass transition of biopolymers.

In the glassy state, the storage modulus, G', is in the range 10^9–10^{11} Pa. At the glass-to-rubber transition, a characteristic drop of 10^3–10^5 Pa in G' is observed, reflecting the change in the rheological properties. The experimental T_g can be determined from the change in storage moduli as function of temperature either as the onset of drop in storage modulus (G') or as the peak of loss modulus (G'') as shown Figure 1.76. When the material is at rubbery plateau region, G' shows little dependence on the frequency at which the material is oscillated during measurements, whereas the loss modulus, G'', shows a characteristic maximum which is considered as the T_g. The tan δ peak (tan $\delta = G''/G'$) is also used to identify the T_g. However, in complex systems the tan δ peak may be very broad and does not show a single maximum. Among the techniques mentioned, the temperature corresponding to G' or tan δ peak is the most commonly used marker of T_g (Cocero and Kokini, 1991; Kalichevsky and Blanshard, 1993; Kokini et al., 1994).

Molecular weight, composition, crystallinity, and chemical structure alter the glass transition temperature of materials significantly. Low molecular weight compounds such as water act as an effective plasticizer by lowering the T_g of biopolymers. Kalichevsky and Blanshard (1993) studied the effect of fructose and water on the glass transition of amylopectin and observed that the fructose has more significant effect on T_g at low water contents. Gontard et al. (1993) reported on the strong plasticizing effect of water and glycerol on mechanical and barrier properties of edible wheat gluten films.

Glass transition and phase behavior of several cereal proteins have been studied extensively. Kokini et al. investigated the phase transitions of gliadin, zein, glutenin, 7S and 11S soy globulins, and gluten to map the changes in their rheological properties as a function of moisture and temperature. The state diagrams of glutenin (Cocero and Kokini, 1991), gliadin (Madeka and Kokini, 1994),

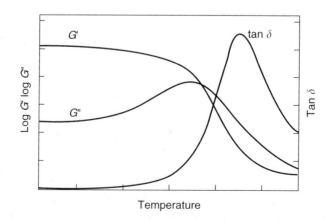

FIGURE 1.76 Determination of glass transition temperature from storage modulus, loss modulus, and tan δ.

zein (Madeka and Kokini, 1996), 7S and 11S soy globulins (Morales and Kokini, 1997, 1999), and gluten (Toufeili et al., 2002) are given Figure 1.77 through Figure 1.80.

Cocero and Kokini (1991) demonstrated the plasticizing effect of water, as measured by the storage modulus (G'), on the glutenin component of wheat proteins. Small amplitude measurements showed that hydrated glutenin between 4 and 14% moisture content showed a wide range of glass transition temperatures between 132 and 22°C. The temperature and frequency dependency of storage (G') and loss (G'') moduli were obtained to characterize the physical states of gliadin (Madeka and Kokini, 1994). Morales and Kokini (1997) studied the glass transition of soy 7S and 11S globulin fractions as a function of moisture content.

Moraru et al. (2002) studied the effect of plasticizers on the mechanical properties and glass transitions of meat–starch extruded systems. Water, in general, decreased the mechanical properties and glass transition temperatures. However, at low moisture content, the addition of water caused an increase in mechanical properties, interpreted as antiplasticization effect. Peleg (1996) reported

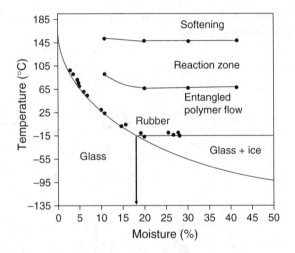

FIGURE 1.77 State diagram for Glutenin. (Reproduced from Kokini, J.L., Cocero, A.M., Madeka, H., and de Graaf, E., 1994, *Trends in Food Science and Technology*, 5: 281–288. With permission.)

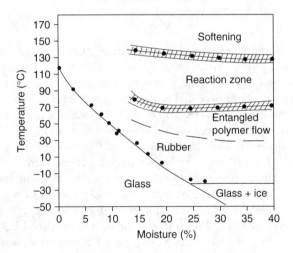

FIGURE 1.78 State diagram for Gliadin. (Reproduced from Madeka, H. and Kokini, J.L., 1994, *Journal of Food Engineering*, 22: 241–252. With permission.)

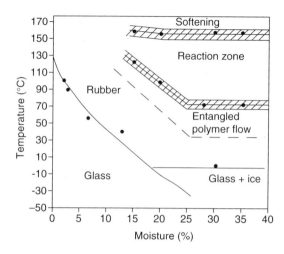

FIGURE 1.79 State diagram for Zein. (Reproduced from Kokini, J.L., Cocero, A.M., and Madeka, H., 1995a, *Food Technology*, 49: 74–81. With permission.)

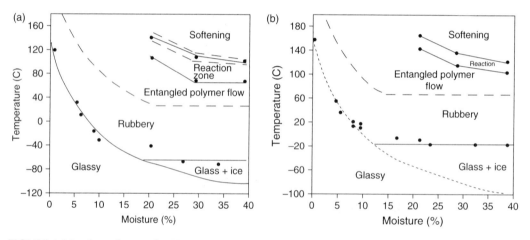

FIGURE 1.80 State diagram for 7S and 11S soy globulins. (Reproduced from Morales, A. and Kokini, J.L., 1999, *Journal of Rheology*, 43: 315–325. With permission.)

that the addition of low molecular weight diluents, such as fructose and glycerol, to glassy polymers lowers T_g but at the same time exerts an antiplasticizing effect on the mechanical properties.

Predicting the changes in rheological properties that occur as a result of plasticization with water or of processing conditions is central to the ability to predict the physical properties and the resulting quality and stability of a food (Kokini et al., 1994). Knowledge of the rheological behavior of food products is essential for process design and evaluation, and quality control and consumer acceptability (Dervisoglu and Kokini, 1986b; Kokini and Plutchok, 1987a; Slade and Levine, 1987; Roos and Karel, 1991). For instance, rheological property changes encountered by wheat dough during baking affect the final texture of breads, cookies, and snacks. The state diagrams allow the prediction of the material phases that can be expected during processes such as baking and extrusion. The state diagrams also describe the moisture content and temperature region at which the material will undergo appropriate reactions. For example, during extrusion and baking the protein phase is expected to undergo crosslinking reactions to generate appropriate texture in the extrudate or on the crumb of the baked product. The physical states of a material during wetting, heating and cooling/drying stages of extrusion cooking is shown in a hypothetical diagram in Figure 1.81.

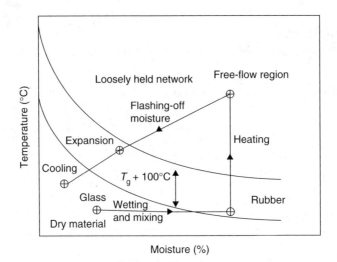

FIGURE 1.81 Hypothetical phase diagram showing the physical states of material during wetting, heating and cooling/drying stages of extrusion cooking. (Reproduced with permission from Kokini, J.L., Cocero, A.M., Madeka, H., and de Graaf, E., 1994, *Trends in Food Science and Technology*, 5: 281–288. With permission.)

1.5.3.3 Networking Properties

Molecular weight between entanglements (Me) or cross-links (Mc), and the slope and magnitude of the storage modulus (G') are the most commonly used rheological measures for quantifying the network formation in biopolymers. When a polymer undergoes cross-linking, this molecular orientation results in a major increase in its solid-like properties. As the network density increases the molecular weight between entanglements/cross-links decreases, and G' increases and remains approximately constant with frequency.

The theory of elasticity can be used to estimate the molecular weight cross-links (Mc) and the number of cross-links (Nc). The rubber elasticity theory explains the relationships between stress and deformation in terms of numbers of active network chains and temperature (Sperling, 2001). The dependence of the stress necessary to deform the amorphous cross-linked polymers above T_g is on the cross-links density and temperature. The statistical theory of rubber elasticity is based on the concept of an entropy driven restraining force. The shear modulus, G, is affected by the work of deformation and the total change in free energy of the deformed network due to the deformation (Treloar, 1975). The resulting equation is given as:

$$G = NRT$$

where R is the ideal gas constant and T is the absolute temperature. The number of chains per unit volume (N) is given as:

$$N = \rho/Mc$$

where ρ is density in g/cm^3, and Mc is the average molecular weight of each chain segment in the network in g/mol. Then the molecular weight between cross-links can be calculated using the equation:

$$Mc = \frac{\rho RT}{G'}$$

TABLE 1.14
Calculated Values for Cross-Linked Waxy Maize Starches

Starch	Degree of cross-link	M_c (g/mol)	Swell factor (ml/g)	N_c ($Mw = 5 \times 10^7$)
Cleargel S	Low	2.7×10^6	18.0	9
W-13	Low	2.6×10^6	16.2	10
400S	Moderate	2.5×10^6	14.5	10
WNA	High	1.2×10^6	12.5	20
W-11	High	1.2×10^6	11.6	21

Source: From Gluck-Hirsh and Kokini, J.L., 1997, *Journal of Rheology*, 41(1): 129–139. With permission.

Gluck–Hirsh and Kokini (1997) used the theory of elasticity for solvent swollen rubbers to calculate the average molecular weight of chain between cross-links of five waxy maize starches with different degrees of cross-linking. This work is the first study in which the cross-link densities of swollen deformable starch granules were quantified. It was hypothesized that a starch with a high degree of cross linking swells less than its lightly cross-linked counterpart. Highly cross-linked starches therefore require a higher concentration to reach maximum packing. In the regime above the threshold concentration of maximum packing, when the granules become tightly packed, rubberlike behavior occurs. A rubbery plateau is achieved, whereby storage modulus (G') remains approximately constant with frequency. Above this critical starch concentration, the interior of the granules controls the rheological behavior of the starch suspensions. Calculated M_c values based on the maximum packing (plateau modulus G') are shown in Table 1.14. As expected, a lower degree of cross-linking resulted in a higher molecular weight between the covalent bonds.

Morales (1997) studied thermally induced phase transitions of 7S and 11S soy globulins, main soy storage proteins, as a function of moisture by monitoring their rheological and calorimetric properties. Pressure rheometry and DSC were used to characterize the denaturation and completing reactions as a function of moisture and temperature. The frequency dependence of G', G'', and tan δ was monitored to identify phase behavior of soy proteins: the rubbery zone, the entangled polymer flow region, and the reaction zone. Figure 1.82 shows $G'(\omega)$ and $G''(\omega)$ for the 7S globulin fraction with 30% moisture (Morales and Kokini, 1998). At 65°C, G' and G'' were both frequency dependent, and their relative values were very close to each other. The slope of the logarithmic plots of G' vs. ω was 0.30, high enough to suggest that the material had a nonnetwork structure capable of experiencing flow, that is, entangled polymer flow. At 70°C, G' and G'' were farther apart from each other than they were at 65°C. Both moduli became less frequency dependent as well. The slope of G' vs. ω was 0.19. At the highest ω values G' and G'' became frequency independent, reaching a plateau and suggesting shorter range networking compared to 65°C. At 115°C, G' and G'' were almost frequency independent in the whole frequency range that was evident by a slope of 0.07. The value of G' was about eight times larger than that at 70°C, and the relative difference between G' and G'' was larger than at 70°C as well indicating all characteristics of a cross-linked polymer (Ferry, 1980). At 138°C, G' and G'' decreased, suggesting depolymerization of the cross-linked network. G' and G'' continued to show little frequency dependence.

The molecular weight between cross-links of the network under different time and temperature conditions were calculated to study the kinetics of complexing reactions. The cross-linked process was shown to be time dependent in the temperature reaction zone. The evolution of the molecular weight between cross-links (M_c) of the 7S and 11S fraction subjected to different time–temperature conditions is shown in Figure 1.83. M_c continued to decrease from the initial M_c (i.e., $M_{c,0}$) during

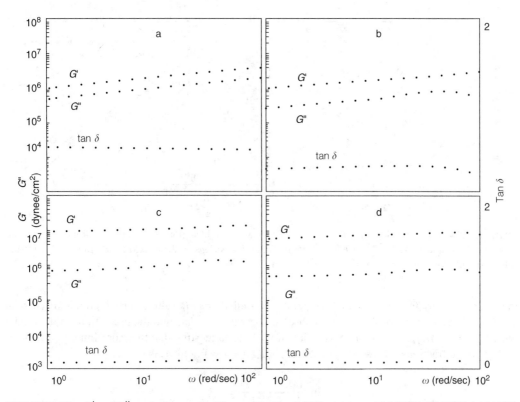

FIGURE 1.82 G' and G'' vs. frequency for 7S soy globulins at 30% moisture at (a) 65°C, (b) 70°C, (c) 115°C, (d) 138°C (Reproduced from Morales, A.M. and Kokini, J.L., 1998, In: *Phase/State Transitions in Foods*, M.A. Roa and R.W. Hartel, Eds, Marcel Dekker, Inc., New York. With permission.)

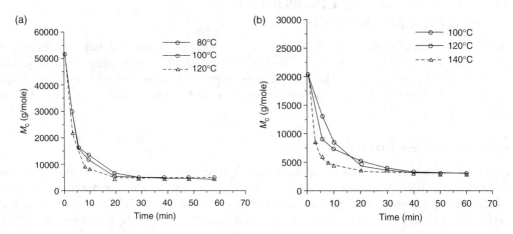

FIGURE 1.83 Molecular weight between cross-links (M_c) vs. time for (a) 7S and (b) 11S globulin fractions. (Reproduced from Morales, A.M. and Kokini, J.L., 1998, In: *Phase/State Transitions in Foods*, M.A. Roa and R.W. Hartel, Eds, Marcel Dekker, Inc., New York. With permission.)

the complexing reactions of the globulins, which is an indication of the existence of an increasingly cross-linked network.

Significant differences were observed in the decreasing rate of M_c at different temperatures. The higher the treatment temperature, the higher the rate of M_c reduction, which is related to

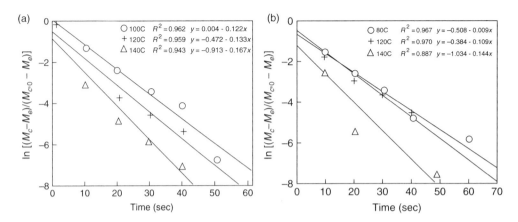

FIGURE 1.84 Cross-linking kinetics for the (a) 7S and (b) 11S globulin fractions. (Reproduced from Morales, A.M. and Kokini, J.L., 1998, In: *Phase/State Transitions in Foods*, M.A. Roa and R.W. Hartel, Eds, Marcel Dekker, Inc., New York. With permission.)

the rate of network formation. At all temperatures studied, the M_c values of each protein reached an equilibrium. Good correspondence was observed between the data and the predictions of the model when $\ln[(M_c - M_e)/(M_{c,0} - M_e)]$ was plotted vs. time suggesting that the cross-linking process of both globulin fractions follows first-order reaction kinetics (Figure 1.84).

$$\ln \frac{(M_c - M_e)}{(M_{c,0} - M_e)} = -kt$$

1.6   USE OF RHEOLOGICAL PROPERTIES IN PRACTICAL APPLICATIONS

The rheology of food products influences their sensory properties and plays a major role in texture and texture–taste interactions. Changes in rheology is a strong indicator of changes in food quality during its shelf life and finally most food engineering operations need be designed to with knowledge of rheological properties of foods. We will give examples of the role of rheology in some of these applications.

1.6.1   SENSORY EVALUATIONS

Texture is a key quality factor for acceptability of food materials. Quality attributes such as thickness, spreadability, and creaminess are extremely important to the acceptance of semisolid food products by consumers. Rheological behavior is associated directly with texture, taste and mouth feel (Kokini et al., 1977, 1984b; Kokini and Cussler, 1983; Elejalde and Kokini, 1992a, 1992b).

Subjective viscosity is the most studied sensory attribute in fluid foods, since it is generally recognized that the rheological properties liquid food materials have a profound impact on the perceived texture by the consumers (Shama and Sherman, 1973; Shama et al., 1973; Kokini et al., 1977). Early studies by Shama et al. (1973) initiated the first semiquantitative design rules in reference to liquid and semisolid food materials. These were then followed by mathematical models that are able to predict liquid perception in the mouth, developed by Kokini et al. (1977).

Psychophysical models have been used to evaluate the effect of external stimulus on the impression of subjective intensity. According to the psychophysical power law model, the sensation magnitude, ψ, grows as a power function of the stimulus magnitude, ϕ (Stevens, 1975).

$$\psi = a\phi^b$$

The constant a depends on the units of measurement. The value of exponent b serves as a signature that may differ from one sensory continuum to another.

The exponent of the power function determines its curvature:

$b \sim 1.0$ sensation varies directly with the intensity of stimulus
$b > 1.0$ concave upward, sensation grows more and more rapidly as the stimulus increases
$b < 1.0$ downward curvature, sensation grows less and less rapidly with increasing stimulus

The linear form of the power law model gives the simple relation between stimulus and sensory response:

$$\log \psi = \log a + b \log \phi$$

According to this linear relationship, equal stimulus ratios produce equal subjective ratios, which means a constant percentage change in stimulus produces a constant percentage change in the sensed effect. Once the appropriate sensory perception mechanisms are identified they can be linked to the operating conditions of each sensory test through psychophysical models.

The sensory thickness is one of the most important textural attributes of semisolid foods. To develop predictive correlation between thickness and rheological properties of foods, it is necessary to understand the deformation process in the mouth. Kokini (1977) estimated sensory viscosity of liquid foods in the mouth from the fundamental physical properties of these fluids using the lubrication theory. Kokini et al. (1977) showed that sensory thickness was perceived as the shear stress between the tongue and the roof of the mouth, smoothness as the inverse of the boundary force, and slipperiness as the average of the reciprocal boundary friction and hydrodynamic forces.

Elejalde and Kokini (1992b) approximated the roof of the mouth and the tongue to squeeze flow solution assuming parallel plate geometry to estimate the sensory viscosity in the mouth (Figure 1.85). The proposed psychophysical model is:

$$\text{Subjective viscosity} = a \, (\text{Shear stress in the mouth})^b$$

Elejalde and Kokini (1992b) estimated the sensory viscosity of low calorie viscoelastic syrups in the mouth, while pouring out of a bottle, and spreading over a flat surface from the fundamental physical properties of these fluids. In order to estimate the sensory viscosity during pouring, the flow conditions were approximated by an inclined trough, with circular channel profile identical to that on the neck of the bottle (Figure 1.86) with incompressible, steady and fully developed flow. The following psychophysical model was proposed:

$$\text{Subjective viscosity} = a(A_c)^b$$

where A_c is the degree of fill of the flow channel, or the cross-sectional area of the neck of the bottle that fills up when a given amount of syrup is being poured.

In a third study, Elejalde and Kokini (1992b) approximated the flow during spreading by a squeeze flow solution, where the height of liquid under gravitational forces provides the squeezing force at any instant. The squeezing force is equal to the hydrostatic force exerted by the height of the syrup in the puddle (Figure 1.87). Thus a transient force exists. The proposed psychophysical model is

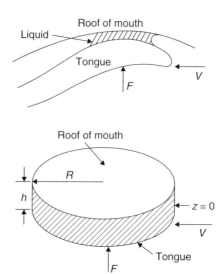

FIGURE 1.85 A model geometry of the mouth. (Reproduced from Kokini, J.L., Kadane, J.B., and Cussler, E.L., 1977, *Journal of Texture Studies*, 8: 195–218. With permission.)

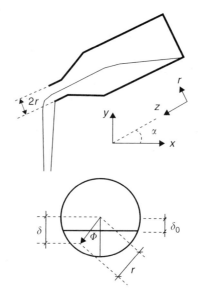

FIGURE 1.86 A model geometry for flow out of a bottle. (Reproduced from Elejalde, C.C. and Kokini, J.L., 1992b, *Journal of Texture Studies*, 23: 315–336. With permission.)

therefore:

$$\text{Subjective viscosity} = a \, (1/\text{Radial Growth of Syrup Puddle})^b$$

All of the sensory cues were found to be appropriate in estimating the sensory response of subjective viscosity in the mouth, pouring out of the bottle and spreading over a flat surface. Oral sensory viscosity correlated with the shear stress in the mouth ($R^2 = 0.96$); pouring sensory viscosity correlated well with the cross-sectional area filled by the fluid at the neck of the bottle ($R^2 = 0.86$); and spreading sensory viscosity correlated inversely with the radial growth of the spreading fluid

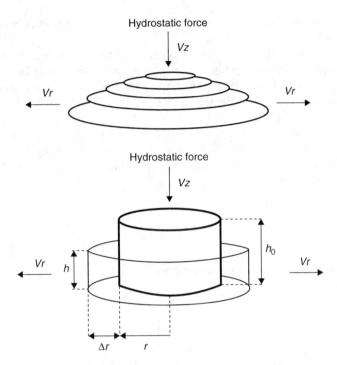

FIGURE 1.87 A model geometry for flow during spreading over a flat surface. (Reproduced from Elejalde, C.C. and Kokini, J.L., 1992b, *Journal of Texture Studies*, 23: 315–336. With permission.)

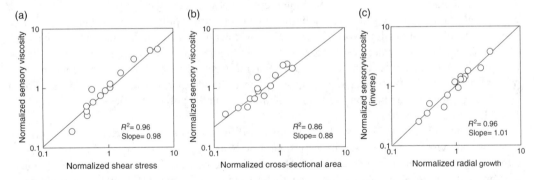

FIGURE 1.88 Normalized sensory viscosity vs. (a) normalized shear stress in the mouth, (b) normalized cross-sectional area of bottle neck filled with syrup during pouring out of a bottle, and (c) normalized radial growth of syrup puddle during spreading over a flat surface. (Reproduced from Elejalde, C.C. and Kokini, J.L., 1992b, *Journal of Texture Studies*, 23: 315–336. With permission.)

puddle ($R^2 = 0.96$). Figure 1.88 shows the correlations between sensory and experimental measures (Elejalde and Kokini, 1992b).

1.6.2 MOLECULAR CONFORMATIONS

Similar to the synthetic polymers, functional and rheological properties of food biopolymers (proteins and polysaccharides) are directly related to their structure and conformation. Consistency is a major quality factor in many semisolid foods such as purees and pastes. Polymer concentration and intermolecular interactions that are most likely to occur at high concentrations are two important factors involved in viscosity development.

Food producers are continuously seeking economical food ingredients which will impart the same level of quality to the final product as would the expensive ones. Two ingredients having similar chemical structures may behave differently. Therefore, when there is a need to replace one ingredient with another one it is extremely critical to monitor the behavior of all ingredients and compare their performance.

Two commonly used generalization techniques η_{sp} vs. $c[\eta]$, and η_{sp} vs. cMw allow comparison of the rheological properties of polymers, where η_{sp} is the specific viscosity, $[\eta]$ is intrinsic viscosity, c is the concentration and Mw is the molecular weight (Chou and Kokini, 1987).

Specific viscosity is defined as:

$$\eta_{sp} = \frac{\eta - \eta_s}{\eta_s}$$

where η and η_s are the viscosity of the solution and the solvent, respectively. Intrinsic viscosity is then calculated using:

$$[\eta] = \lim_{c \to 0} \frac{\eta_{sp}}{c}$$

Intrinsic viscosity is a measure of the hydrodynamic volume occupied by a molecule. The nondimensional parameter $c[\eta]$ can be taken as a measure of the extent of overlapping between polymer molecules (Morris and Ross-Murphy, 1981). When polymer coils start to overlap, molecules will start free draining behavior, and frictional interactions between neighboring polymers generate the major contribution to the viscosity. In addition, entanglement coupling may occur and the solution behaves like a cross-linked network (Ferry, 1980). At the onset of the molecular contact, the slope of the η_{sp} vs. $c[\eta]$ curve increases sharply as shown in Figure 1.89. The concentration at which this transition from dilute to concentrated solution behavior occurs is called as the critical concentration ($c*$). This behavior is also typical to random coil polysaccharides. Critical concentration varies from system to system, depending on the hydrodynamic volume of polymer molecules (Morris et al., 1981; Lazaridou et al., 2003).

Chou and Kokini (1987) showed that pectins of different plant origins behave similarly when their intrinsic viscosities are taken into consideration When η_{sp} data were plotted against $c[\eta]$ for citrus pectin, apple pectin, hot break, and cold break tomato pectins all data points fell on one curve as shown in Figure 1.89. It was concluded that tomato, citrus, and apple pectins all have a random coil conformation because their common transition from dilute to concentrated solution region occurs at a common $c[\eta]$ value. Lazaridou et al. (2003) studied the molecular weight effects on solution rheology of pullulan and observed a systematic increase in c^* with increasing molecular weight (Mw) of the polysaccharide.

The second way of superposition of viscosity data to compare the rheological properties of polymers is by plotting η_{sp} vs. cM_w. In this case it is assumed that only molecules with the same approximate shape and conformation will superimpose. Such a curve is proven to be useful in terms of identifying polymers of similar solution properties (Chou and Kokini, 1987; Kokini and Chou, 1993). The slope of the concentrated solution region is also a useful indication of the conformation of biopolymers in solution. For flexible random coil the slope of $\log \eta_{sp}$ vs. $\log cM_w$ curve gives exponents around 3.5 while for stiffer chains molecules it is around 8.

The steady shear viscosity data of biopolymers can also be superposed if η/η_0 is plotted vs. $\tau\dot{\gamma}$, where τ is the characteristic relaxation time, η_0 is the zero-shear viscosity, and $\dot{\gamma}$ is the shear rate (Chou and Kokini, 1987). For flexible monodisperse random coil molecules the Rouse relaxation

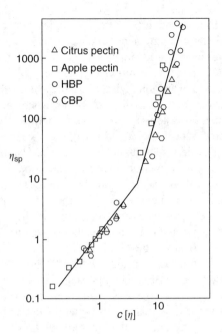

FIGURE 1.89 Specific viscosity η_{sp} vs. $c[\eta]$ for tomato, citrus, and apple pectins. (Reproduced from Chou, T.C. and Kokini, J.L., 1987, *Journal of Food Science*, 52: 1658–1664. With permission.)

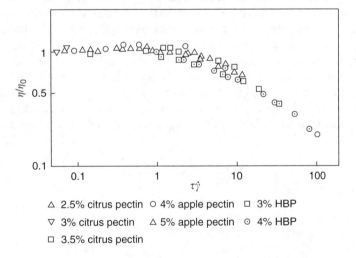

FIGURE 1.90 η/η_0 vs. $\tau\dot{\gamma}$ for tomato, citrus, and apple pectins. (Reproduced from Chou, T.C. and Kokini, J.L., 1987, *Journal of Food Science*, 52: 1658–1664. With permission.)

time provides a good approximation in the semidilute solution region:

$$\tau_R = (6/\pi^2)\frac{[\eta]\eta_s M}{RT}$$

where $[\eta]$ is intrinsic viscosity, η_s is solvent viscosity, M is the molecular weight, R is gas constant and T is absolute temperature. Chou and Kokini (1987) superimposed the steady shear viscosities of tomato, citrus and apple pectins measured at several concentrations (Figure 1.90). The slope of limiting non-Newtonian region was found to be −0.6, which is typical for random coil molecules.

Branching is an important factor that affects the rheological properties of synthetic polymers. Side branches lead intermolecular entanglements in concentrated systems, which result in unique rheological properties. The presence of side branches are known to influence the intrinsic viscosity, zero shear viscosity, shear rate dependence of viscosity, temperature dependence of viscosity, zero shear recoverable compliance, and extensional viscosity (Cogswell, 1981).

Gelling is an important attribute of carbohydrate polymers, where the elastic properties determine the overall quality of the gels in food systems such as jams and jellies. The clear understanding on the contribution of side branches to the elasticity is of critical importance in designing the gelling systems constituted by carbohydrate polymers such as polysaccharides. Hwang and Kokini (1991, 1992) investigated the contribution of side branches to rheological properties of carbohydrate polymers using apple pectins which naturally posses significant branching size. It was observed that the side branches of apple pectins greatly influence steady shear rheological properties such as zero-shear viscosity and shear rate dependence of viscosity. Based on the rheological theories developed in synthetic polymers, the results suggested that side branches of pectins exist as significant entangled states in concentrated solutions (Hwang and Kokini, 1991).

The rheological data were superimposed using a variety of generalization curves such as η_{sp} vs. $c[\eta]$, and η_{sp} vs. cM_w. Increased degree of branching resulted in higher η_0 and increased shear rate dependence of viscosity. The gradients of η_{sp} vs. $c[\eta]$ in the concentrated region ($c > c^*$) were dependent upon the degree of branching, that is, the higher the branching, the higher the gradients, whereas there was no significant difference in the dilute region ($c < c^*$) irrespective of the degree of branching. Circular dichroism (CD) studies of pectins showed that the conformation of pectin molecules was not affected by the degree of branching. It is concluded that side branches of pectins can result in significant entanglements in concentrated solutions (Hwang and Kokini, 1992).

Branching is also known to affect the elastic properties of synthetic polymers. Hwang and Kokini (1995) studied the branching effects on dynamic viscoelastic properties of carbohydrate polymers using apple pectins with varying branching degrees. The storage and loss moduli of apple pectin solutions were measured in the range of 2–6% pectin concentration. A typical dynamic moduli vs. frequency profile for low (sample-I) and high (sample-II) branched pectin samples at 4% concentration is shown in Figure 1.91. Both G' and G'' increased with increasing pectin concentration suggesting that increasing intermolecular entanglements enhance the elasticity as well as the viscosity. G'' was observed to be higher than G' for both samples indicating the predominant liquid-like behavior of

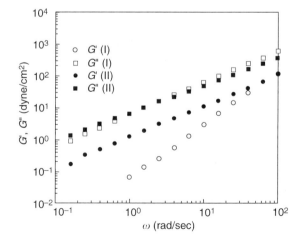

FIGURE 1.91 Dynamic storage modulus (G') and loss modulus (G'') vs. frequency of 4% apple pectins. Sample II posses twice as much side branches as sample I. (Reproduced from Hwang, J. and Kokini, J.L., 1995, *The Korean Journal of Rheology*, 7: 120–127. With permission.)

pectin solutions. Moreover, G' of more branched pectin sample was observed to be higher than that of less branched sample reflecting the positive contribution of side branches to the elasticity of pectin solutions.

The frequency dependence of G' and G'' was expressed using a power-law type relation (i.e., $G' \propto \omega^\alpha$ and $G'' \propto \omega^\beta$). Pectins of high degree of branching were found to give lower α and β values than less branched samples indicating lower frequency dependence of loss and storage moduli. Zero-shear recoverable compliance, a useful parameter of fluid elasticity, was also calculated to confirm the findings of branching effects on elastic properties (Figure 1.92). Experimental data showed the same trend with observed storage modulus indicating the positive contribution of side branches to the elastic properties of pectins.

1.6.3 PRODUCT AND PROCESS CHARACTERIZATION

Food products are complex mixtures of several ingredients where individual ingredients are mixed together to produce a particular finished product. Each ingredient and its interactions have a strong influence on the finished product characteristics. The behavior of each component has to be monitored under the test conditions that mimic the processing, storage, and handling conditions that the product will be subjected to.

A small change in the amount of certain ingredients such as stabilizers and emulsifiers can have a dramatic effect on the final product's characteristics. Food researchers continuously seek alternative food ingredients due to both cost and health/nutrition considerations. It is extremely important to fully compare the rheological behavior of alternative ingredients with the conventional ones both during processing and storage before switching formulations. Two ingredients having similar chemical structures and conformation may behave differently in processing. It is also critical to adjust the processing conditions accordingly to achieve desired performance from the end product.

Rheological measurements are useful in storage stability predictions of emulsion-based products such as ice cream, margarine, butter, beverages, sauces, salad dressings, and mayonnaise. These measurements also allow for a better understanding of how various emulsifiers/stabilizers interact to stabilize emulsions. The relationship between rheology and processing and formulation of emulsions has been studied extensively. Goff et al. (1995) studied the effects of temperature, polysaccharide stabilizing agents, and overrun on the rheological properties of ice cream mix and ice cream using dynamic rheological techniques. Storage and loss moduli and $\tan \delta$ decreased significantly with

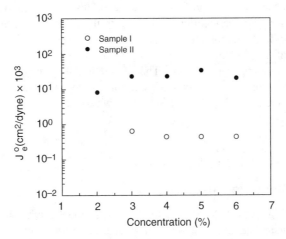

FIGURE 1.92 Zero-shear recoverable compliance (J_e^0) vs. concentration of apple pectin samples I and II (Reproduced from Hwang, J. and Kokini, J.L., 1995, *The Korean Journal of Rheology*, 7: 120–127. With permission.)

increasing temperature. Unstabilized samples demonstrated significantly higher G' and G'' and $\tan \delta$ than stabilized samples. Experimental results indicated the importance of considering both ice and unfrozen phases in determining the impact of stabilizers on ice cream rheology. Dickinson and Yamamoto (1996) investigated the effect of lecithin addition on the rheological properties of heat-set β-lactoglobulin emulsion gels. The storage and loss modulus data showed that the lecithin-containing emulsion gels behave like a strong gel as indicated by less frequency-dependence as compared to the gels without lecithin.

Among many food materials, wheat dough is one of the most complex and also the most widely studied one. The gas cell expansion during proofing and baking has been shown to be closely related to biaxial stretching flow (Bloksma, 1990; Huang and Kokini, 1993; Dobraszczyk and Morgenstern, 2003). Wang and Sun (2002) studied the creep recovery of different wheat flour doughs and its relation to bread-making performance. The maximum recovery strain of doughs has been observed to be highly correlated to bread loaf volume. Stress relaxation has also been widely used to study the viscoelastic behavior of wheat flour doughs. Bagley and Christianson (1986) reported that the stress relaxation behavior is closely related to bread volume. Slow relaxation is usually associated with better baking quality since the strength of gas cells are important in maintaining stability against premature failure during baking (Dobraszczyk and Morgenstern, 2003).

Understanding the flow and deformation behavior of biopolymers is important both in designing the process equipments and setting appropriate parameters during several food processing operations. Rheological techniques can be used to predict the performance of a material during mixing, extrusion, sheeting, baking, etc. For instance, mixing is a critical step in bread making as it develops the viscoelastic properties of gluten which dictates the bread quality. Dough development and protein networking depend on the right balance of mechanical work and temperature rise within the mixer as well as the entrainment of air (Prakash and Kokini, 2000). Knowledge of the velocity profiles and velocity gradients across the mixer makes it possible to design equipment for optimum mixing, correct mixing deficiencies and to set scale up criteria for rheologically complex fluids like wheat flour dough. The shear rate in Brabender Farinograph was found to be a function of blade geometry, blade position, and location (Prakash and Kokini, 2000). The equations developed for shear rate were suggested to be used as powerful design tools to predict the state of gluten development in real time nonintrusively, helping in better process control and enhanced finished product quality.

Madeka and Kokini (1994) used small amplitude oscillatory measurements to monitor chemical reactions zones, which served as a basis for the construction of phase diagrams. During the processing of wheat doughs gluten proteins (gliadin and glutenin) undergo physical and chemical changes due to applied heat and shear. Storage and loss moduli were observed to increase significantly at the process conditions where the chemical reactions lead to the formation of higher molecular weight products (Figure 1.93). In the temperature range of 50–75°C the storage modulus (G') was roughly equal to the loss modulus (G''). As the material is heated above 75°C, G' increased almost 100 fold during heating throughout the reaction zone. The storage modulus reached a peak at 115°C where loss modulus G'' made a minimum indicating maximum structure build up. When the reaction was complete, the expected temperature-induced softening was observed.

1.7 NUMERICAL SIMULATION OF FLOWS

1.7.1 Numerical Simulation Techniques

Mathematical simulations provide a very effective way to probe the dynamics of a process and learn about what goes on inside the material being processed nonintrusively (Puri and Anantheswaran, 1993). Numerical simulations have a wide range of applications in equipment design, process optimization, trouble shooting, and scale up and scale down in many food processing operations. The geometrical complexities of process equipments and the nonlinear viscoelastic properties of food

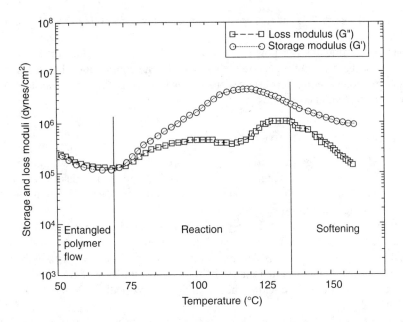

FIGURE 1.93 Temperature sweep of 25% moisture gliadin showing different reaction zones. (Reproduced from Madeka, H. and Kokini, J.L., 1994, *Journal of Food Engineering*, 22: 241–252. With permission.)

materials makes it a necessity to invest in numerical simulation if appropriate progress is to be made in improving food operations (Connelly, 2004).

Computational fluid dynamics (CFD) offers a powerful design and an investigative tool to process engineers. The advent of powerful computers and work stations has provided the opportunity to simulate various real-world processes. CFD has only recently been applied to food processing applications. It assists in a better understanding of the complex physical mechanisms that govern the operations of food processes, such as mechanical and thermal effects during processing.

When simulating the processing of food products, it is necessary to take the rheological nature of a food into account as this will dictate its flow behavior. There are many CFD approaches to discretizing the equations of conservation of momentum, mass, and energy, together with the constitutive equation that defines the rheology of the fluid being modeled and the boundary and initial conditions that govern the flow behavior, in particular geometries such as extruders and mixers (Connelly and Kokini, 2003, 2004; Dhanasekharan and Kokini, 2003). The most important of these are finite difference (FDM), finite volume (FVM), and finite element (FEM) methods. Other CFD techniques can be listed as spectral schemes, boundary element methods, and cellular automata, but their use is limited to special classes of problems.

The use of the finite element method (FEM) as a numerical procedure for solving differential equations in physics and engineering has increased considerably. The finite-element method has various advantages contributing to this popularity: Spatial variations of material properties can be handled with relative ease; irregular regions can be modeled with greater accuracy; element size can be easily varied; it is better suited to nonlinear problems; and mixed-boundary value problems are easier to handle (de Baerdemaeker et al., 1977). The major disadvantage of the method is that it is numerically intensive and can therefore take high CPU time and memory storage space.

In FEM there are three primary steps: The domain under consideration is divided into small elements of various shapes called finite elements. All elements are connected at nodal points located throughout the domain and along the boundaries, and the collection of elements is called the mesh. Over each element, the solution is approximated as a linear combination of nodal values and approximation functions, and then algebraic relations are derived between physical quantities and the nodal values. Finally the elements are assembled in order to obtain the solution to the whole (Reddy, 1993).

FEM numerical simulation of flow processes is conducted by simultaneously solving the FEM representations of the continuum equations that describe the conservation laws of momentum and energy, with a rheological equation of state (constitutive models) of the food material to be processed, along with boundary/initial conditions.

1.7.2 SELECTION OF CONSTITUTIVE MODELS

Constitutive models play a significant role in the accuracy of the predictions by numerical simulations. A proper choice of a constitutive model that describes the behavior of the material under investigation is important. Three classes of flow models are being used in numerical simulations: Newtonian, generalized Newtonian, and viscoelastic.

Differential viscoelastic models have generally been more popular than integral models in numerical developments (Crochet, 1989). Nonlinear differential models are of particular interest in numerical simulations for process design, optimization, and scale-up. This is because integral viscoelastic models are not well suited for use in numerical simulation of complex flows due to high computational costs involved in tracking the strain history, particularly in three-dimensional flows (Dhanasekharan, 2001).

Dhanasekharan et al. (1999, 2001, 2003) focused on the proper choice of constitutive models for wheat flow doughs for the design and scaling of extrusion by numerical simulation. The flow in an extruder is shear dominant, and therefore two groups of models which give a good prediction of shear properties of dough were tested: Generalized Newtonian models (Newtonian fluid, power-law fluid, Hershel–Bulkley fluid, and Morgan fluid) and differential viscoelastic models (Phan-Thien Thanner, White–Metzner, and Giesekus–Leonov model).

1.7.3 FINITE ELEMENT SIMULATIONS

For an incompressible fluid, the stress tensor (σ) is given as the sum of an isotropic pressure (p) component and an extra stress tensor (T). The extra stress tensor is obtained using the constitutive models as shown in Section 1.4.

$$\sigma = -pI + T$$

The conservation of linear momentum is then given by:

$$\nabla \cdot \sigma + \rho f = \rho \left(\frac{\partial v}{\partial t} + v \cdot \nabla v \right)$$

where ρ is the fluid density and f is the external body force per unit mass. For incompressible fluids, conservation of mass yields the continuity equation:

$$\nabla \cdot v = 0$$

and the conservation of energy equation is given as:

$$\rho C(T) \cdot \left(\frac{\partial T}{\partial t} + v \cdot \nabla T \right) = T : \nabla v + r - \nabla \cdot q$$

where $C(T)$ is the heat capacity as a function of temperature, r is the given volumetric heat source, q is the heat flux, and $T : \nabla v$ is the viscous heating term. These equations together with constitutive models form a complete set of governing equations. The solutions of these equations give velocity and temperature profiles for a particular problem. In most cases, the solution of these equations requires numerical methods, such as the finite element method. An abundance of software tools are available in the market using finite element methods to solve flow problems.

1.7.3.1 FEM Techniques for Viscoelastic Fluid Flows

A variety of numerical methods based on finite element methodology are available for use with viscoelastic fluids. One of the formulations is the so-called weak formulation. In this method, the momentum equation and the continuity equation are weighted with fields V and P and integrated over the domain Ω. The finite element formulations are given by:

$$\int_\Omega (-\nabla p + \nabla \cdot T + f) \cdot u \, d\Omega = 0, \quad \forall u \in V$$

$$\int_\Omega (\nabla \cdot v) \cdot q \, d\Omega = 0, \quad \forall q \in P$$

where T is the extra stress tensor, V and P denoting the velocity and pressure fields, respectively. The domain Ω, is discretized using finite elements covering a domain, Ω^h on which the velocity field and pressure fields are approximated using v^h and p^h. The superscript h refers to the discretized domain. The approximations are obtained using:

$$v^h = \sum V^i \psi_i, \quad p^h = \sum p^i \pi_i$$

where V^i and p^i are nodal variables and ψ_i and π_i are shape functions. The unknowns V^i and p^i are calculated by solving the weak forms of equations of motion and the continuity equation, along with the formulations for the constitutive models, using two basic approaches.

The first approach, also known as the coupled method, is the mixed or stress-velocity-pressure formulation. The primary unknown, the stress tensor, is formulated using an approximation T^h with:

$$T^h = \sum T^i \phi_i$$

where T^i are nodal stresses while ϕ_i are shape functions. This procedure is normally used with differential models. The main disadvantage of this method is the large number of unknowns and hence high computational costs for typical flow problems.

The second approach, called the decoupled scheme, uses an iterative method. The computation of the viscoelastic extra-stress is performed separately from that of flow kinematics. The stress field is calculated from flow kinetics. In this approach, the number of variables is much lower than in the mixed method, but the number of iterations is much larger.

A straightforward implementation of these two approaches gives an instability and divergence of the numerical algorithms for viscoelastic problems. FEM solvers use a variety of numerical methods to circumvent convergence problems for viscoelastic flows as explained below.

Viscoelastic fluids exhibit normal stress differences in simple shear flow. Early attempts to simulate viscoelastic flows numerically were restricted to very moderate Weissenberg numbers (i.e., a nondimensional measure of fluid elasticity) as the solutions invariably became unstable at unrealistically low *Wi* values. This problem is called the "high Weissenberg number problem," and it is mostly due to the hyperbolic part of the differential constitutive equations. Numerical methods were unable to handle flows at *Wi* values sufficiently high to make comparisons with the experimental results. Progress has been made by the use of central numerical methods, such as central finite differences or Galerkin finite elements, by which small Weissenberg numbers are attainable. More insight into the type of the system of differential equations led to the development of upwind schemes, such as the Streamline Upwind (SU) by Marchal and Crochet (1987) and streamline integration method by Luo and Tanner (1986a, 1986b). Furthermore, the Streamline Upwind/Petrov-Galerkin (SUPG) method was developed by rewriting the set of partial differential equations in the explicit elliptic momentum equation form. The SUPG method is considered more accurate compared to the SU method but it is only applicable to smooth geometries.

In order to ease the problems caused by the high stress gradients, viscoelastic extra-stress field interpolation techniques, which include biquadratic and bilinear subelements, are used. Marchal and Crochet (1987) introduced the use of 4×4 subelements for the stresses. These bilinear subelements smoothed the mixed method solution of the Newtonian stick-slip problem, as well as aided in the convergence of the viscoelastic problem. Perera and Walters (1977) introduced a method known as Elastic Viscous Stress Splitting (EVSS) by splitting the stress tensor into an elastic part and a viscous part, which stabilizes the behavior of the constitutive equations.

1.7.3.2 FEM Simulations of Flow in an Extruder

Dhanasekharan and Kokini (1999) characterized the 3D flow of whole flour wheat dough using three nonlinear differential viscoelastic models, Phan-Thien–Tanner, the White–Metzner and the Giesekus models. The Phan-Thien–Tanner (PTT) model gave good predictions for transient shear and extensional properties of wheat flour doughs as shown in Figure 1.53. Based on the rheological studies using differential viscoelastic models, it was concluded that the PTT model was most suitable for numerical simulations (Dhanasekharan et al., 1999).

Dhanasekharan and Kokini (2000) modeled the 3D flow of a single mode PTT fluid in the metering zone of completely filled single-screw extruder. The modeling was done by means of a stationary screw and rotating barrel. The pressure build up for the PTT model was found to be smaller than the Newtonian case, which is explained by the shear-thinning nature incorporated into the differential viscoelastic model. The velocity profile generated using the viscoelastic model, however, was found to be very close to the Newtonian case.

A fundamental analysis was done using two important dimensionless numbers, Deborah number (De) and Weissenberg number (Wi). For the chosen flow conditions and the extruder geometry Deborah and Weissenberg numbers were reported to be 0.001 and 5.22, respectively. $De = 0.001$ explained the velocity profile predictions close to Newtonian case, as $De \to 0$ indicates a viscous liquid behavior. When the relaxation processes are of the same order of magnitude of the residence time of flow (i.e., $De \sim 1$), the impact of viscoelasticity on the flow becomes significant. $Wi = 5.22$ indicated "high Weissenberg number problem." In spite of the difficulties in convergence due to high Wi, these results provided a starting point for further simulations of viscoelastic flow using more realistic parameters.

Dhanasekharan and Kokini (2003) proposed a computational method to obtain simultaneous scale-up of mixing and heat transfer in single screw extruders by several parametric 3D nonisothermal numerical simulations. The finite element meshes used for numerical simulations are shown in Figure 1.94. The numerical experiments of flow and heat transfer modeling studies were conducted using the Mackey and Ofoli (1990) viscosity model for low to intermediate moisture wheat doughs in the metering section of a single screw extruder. In order to develop the trend charts, numerical simulations of nonisothermal flow were conducted by varying screw geometric variables such as helix angle (θ), channel depth (H), screw diameter to channel depth ratio (D/H), screw length to screw diameter ratio (L/D), and the clearance between the screw flights and barrel (ε). The nonisothermal flow model included viscous dissipation and the complete three-dimensional flow geometry including leakage flows without any simplifications such as unwinding the screw.

The down channel velocity profile, temperature profile in the flow region between the screw root and the barrel, pressure along the axial distance, and local shear rate along the axial distance were predicted under nonisothermal conditions. Residence time distribution (RTD) and specific

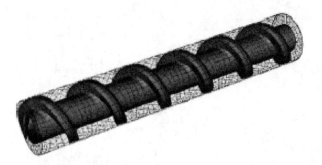

FIGURE 1.94 A typical screw geometry and FEM mesh used for the nonisothermal simulation. (Reproduced from Dhanasekharan, K.M. and Kokini, J.L., 2003, *Journal of Food Engineering*, 60: 421–430. With permission.)

mechanical energy (SME) were chosen as the design parameters for the scale-up of mixing and heat transfer, respectively. Effect of helix angle, clearance, channel depth, and L/D on RTD was studied with various screw geometries (Figure 1.95). Increasing D/H at constant helix angle shifted the RTD curve to the right and increased the peak. Increasing helix angle while keeping D/H constant shifted the RTD curve to the left and decreased the peak, while decreasing channel depth at constant D/H and helix angle reduced RTD peak. The clearance between the flights and the barrel did not have any significant impact on the RTD curve. Decreasing L/D at constant helix angle and D/H decreased RTD because of smaller channel volume.

Numerical simulations showed that similar residence time distributions can be maintained by decreasing D/H and helix angle and decreasing the channel depth. Two differently sized extruders that would have the same SME input were chosen following these scaling rules to illustrate the effect of screw geometries on RTD (Figure 1.96). Two different geometries (geometry I and II) had a scale-up of about 10 times based on throughput rates as calculated from the design charts. The throughput rates are 1.6 and 17.4 kg/h for the small and big extruder, respectively. The results obtained for these two extruders gave SME values of 164.4 kJ/kg and 152.6 kJ/kg and average residence times for the two extruders were 31.2 and 31.8 sec, respectively. Figure 1.97 shows the RTD distribution for the two geometries.

The computational method used was capable of taking viscoelastic effects and the three-dimensional nature of the flow in the extruder into consideration. SME and RTD curves vs. screw parameters developed from the numerical simulations provided powerful tools for accurate extrusion design and scaling.

1.7.3.3 FEM Simulations of Flow in Model Mixers

Research on mixing flows can be classified according to the complexity of the geometries that have been studied. These include:

1. Studies using classical geometries such as eccentric cylinder, flow past cylinder, or sphere and lid-driven cavity mixers (Anderson et al., 2000a, 2000b; Fan et al., 2000)
2. Studies involving simple model mixer geometries including stirred tank reactors and couette geometries (Alvarez et al., 2002; Binding et al., 2003)
3. Mixing research on complex geometries such as twin-screw continuous mixers, batch Farinograph and helical mixers (Bertrand et al., 1999; Connelly and Kokini, 2003, 2004, 2006b, 2006c)

The classical geometries such as contraction flows, flow past a cylinder in a channel, flow past a sphere in a tube, and flow between eccentrically rotating cylinders have been traditionally used as benchmark

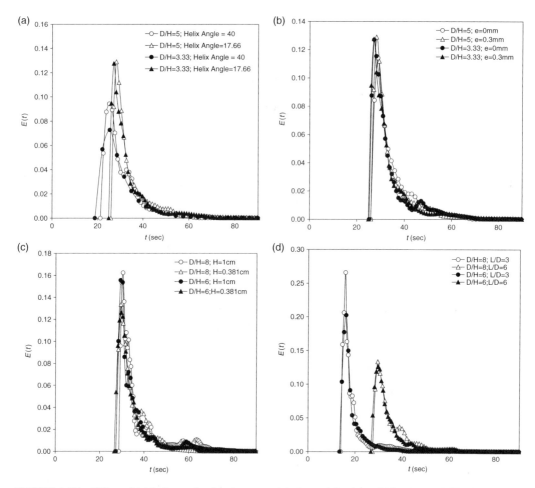

FIGURE 1.95 Effect of (a) helix angle, (b) clearance, (c) channel depth, and (d) L/D on RTD. Other screw parameters are $\varepsilon = 0.3$ mm, $H = 0.381$ cm, $L/D = 6$, and $\theta = 17.66°$. (Reproduced from Dhanasekharan, K.M. and Kokini, J.L., 2003, *Journal of Food Engineering*, 60: 421–430. With permission.)

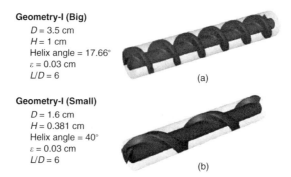

Geometry-I (Big)
$D = 3.5$ cm
$H = 1$ cm
Helix angle = 17.66°
$\varepsilon = 0.03$ cm
$L/D = 6$

(a)

Geometry-I (Small)
$D = 1.6$ cm
$H = 0.381$ cm
Helix angle = 40°
$\varepsilon = 0.03$ cm
$L/D = 6$

(b)

FIGURE 1.96 Screw geometry and FEM meshes (a) big and (b) small extruder. (Reproduced from Dhanasekharan, K.M. and Kokini, J.L., 2003, *Journal of Food Engineering*, 60: 421–430. With permission.)

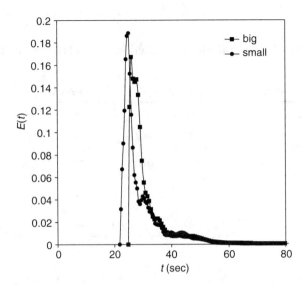

FIGURE 1.97 RTD curve comparison between scaling geometries I and II. (Reproduced from Dhanasekharan, K.M. and Kokini, J.L., 2003, *Journal of Food Engineering*, 60: 421–430. With permission.)

problems for testing new techniques and understanding fundamental effects involved in mixing. Studies involving simple model mixer geometries have been done to understand mixing phenomena in mixers with geometries closer to industrial mixers. Only recently mixing in complex geometries such as the twin-screw continuous mixers and batch Farinograph mixers has been addressed utilizing new advances in numerical simulation techniques and computational capabilities (Connelly and Kokini, 2004).

Good progress has been made in understanding the effects of rheology and geometry on the flow and mixing in batch and continuous mixers, as well as in identifying conditions necessary for efficient mixing (Connelly and Kokini, 2003, 2004). The finite element method (FEM) was used for numerical simulations of the flow of dough-like fluids in model batch and continuous dough mixing geometries. Several FEM techniques, such as elastic viscous stress splitting (EVSS), Petrov-Galerkin (PG), 4×4 subelements, streamline upwind (SU) and Streamline Upwind/Petrov-Galerkin (SUPG) were used for differential viscoelastic models. The mixing of particles was analyzed statistically using the segregation scale and cluster distribution index. Efficiency of mixing was evaluated using lamellar model and dispersive mixing. Series of strategies to systematically increase the complexity were used to encounter the flows in commercial dough mixers properly as discussed below.

Connelly and Kokini (2004) explored the viscoelastic effects on mixing flows obtained with kneading paddles in a single screw continuous mixer. A simple 2D representation of a single paddle in a fully filled, rotating cylindrical barrel with a rotating reference frame was used as a starting point to evaluate the FEM techniques. The single screw mixer was modeled by taking the kneading paddle as the point of reference, fixing the mesh in time. Here, either the paddle turns clockwise with a stationary wall in a reference frame or the wall moves counterclockwise in the rotating reference frame originating from the center of the paddle.

The single-mode, nonlinear Phan-Thien–Tanner differential viscoelastic model was used to simulate the mixing behavior of dough-like materials. Different numerical simulation techniques including EVSS SUPG, 4×4 SUPG, EVSS SU, and 4×4 SU were compared for their ability to simulate viscoelastic flows and mixing. Mesh refinement and comparison between methods were also done based on the relaxation times at 1 rpm and the Deborah number (*De*) to find the appropriate mesh size and the best technique to reach the desired relaxation time of 1000 sec. The limits of the *De* that were reachable in this geometry with the PTT model are listed in Table 1.15. The coarser meshes

TABLE 1.15
Limits of *De* Reached by Several Methods Used in Viscoelastic Simulations during Mesh Refinement at 1 rpm

Mesh size	EVSS SUPG		4 × 4 SUPG		EVSS SU		4 × 4 SU	
	λ (1 rpm)	*De*	λ (1 rpm)	*De*	λ (1 rpm)	*De*	λ (1 rpm)	*De*
360 elements	0.327	0.034	0.23	0.024	651.04	68.20	1000	104.7
600 elements	0.178	0.019	1.04	0.109	14.12	1.47	23.40	2.45
1480 elements	0.089	0.009	0.089	0.009	0.73	0.076	131.78	13.8
2080 elements	—	—	0.066	0.007	0.79	0.082	543.58	56.9
3360 elements	—	—	—	—	0.58	0.061	110.32	11.6

Source: From Connelly, R.K., and Kokini, J.L., 2003, *Advances in Polymer Technology*, 22(1): 22–41. With permission.

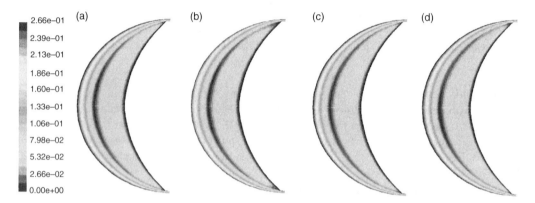

FIGURE 1.98 Velocity magnitude distribution at 1 rpm of (a) Newtonian (λ = 0 sec), (b) Oldroyd-B (λ = 0.5 sec), (c) Bird–Carreau Viscous (λ = 60 sec), and (d) PTT (λ = 100 sec) where the units of velocity are cm/sec. (Reproduced from Connelly, R.K., 2004, Numerical simulation and validation of the mixing of dough-like materials in model batch and continuous dough mixers, Ph.D. Thesis, Rutgers University. With permission.)

allowed convergence at higher *De* since the high gradients at the discontinuity are smoothed in the boundary layers.

The SUPG technique and less computationally intensive EVSS technique were found to be in adequate for this geometry. Only the 4 × 4 SU technique was able to attain *De* values representative of the level of viscoelasticity closer to dough viscoelasticity. Even with this technique it was not possible to reach the desired relaxation time of 1000 sec at low rpm values. High rpm values are more representative of the actual conditions found in this type of mixer. At high rpm levels the instabilities in the calculations were found to disappear.

The effect of shear thinning and viscoelastic flow behavior on mixing was systematically explored using the Newtonian, Bird–Carreau viscous, Oldroyd B, and Phan-Thien–Tanner models using single screw simulations with the rotating reference frame approach. For the application of these techniques the rheological data and nonlinear viscoelastic models for wheat flour doughs previously studied by Dhanasekharan et al. (1999), Wang and Kokini (1995a, 1995b) were utilized. Comparison of the predictions by these viscoelastic models with experimental data showed that viscoelastic flow predictions differ significantly in shear and normal stress predictions resulting in a loss of symmetry in velocity (Figure 1.98) and pressure profiles (Figure 1.99) in the flow region. Introduction of shear

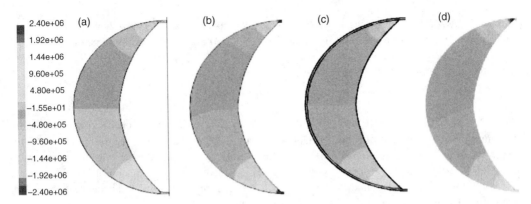

FIGURE 1.99 Pressure distributions at 1 rpm of (a) Newtonian ($\lambda = 0$ sec), (b) Oldroyd-B ($\lambda = 0.5$ sec), (c) Bird–Carreau Viscous ($\lambda = 60$ sec), and (d) PTT ($\lambda = 100$ sec) where the units of pressure are dyne/cm^2 (Reproduced from Connelly, R.K., 2004, Numerical simulation and validation of the mixing of dough-like materials in model batch and continuous dough mixers, Ph.D. Thesis, Rutgers University. With permission.)

thinning behavior resulted in a decrease in the magnitude of the pressure and stress and an increase the size of low velocity or plug flow regions.

Remeshing or a moving mesh technique is required to model the flow in mixers that contain more than one mixing element or a nonsymmetrical geometry that does not contain a reference point from which the mesh can be fixed. The mesh superposition technique is very useful since it allows the use of a periodically changing moving element without remeshing (Avalosse, 1996; Avalosse and Rubin, 2000).

Connelly (2004) compared the mesh superposition and rotating reference frame techniques using a generalized Newtonian dough model. The first step in mesh superposition techniques is to mesh the flow domains and moving elements separately. Then the meshes are superimposed as they would be positioned at a given time interval. The mesh superposition technique (Polyflow, 2001b) uses a penalty force term, $H(v - v_p)$, that modifies the equation of motion as follows:

$$H(v - v_p) + (1 - H)\left[-\nabla p + \nabla \cdot T + \rho f - \rho \frac{Dv}{Dt}\right] = 0$$

where v_p is the velocity of the moving part. H is zero outside the moving part and 1 within the moving part (Connelly and Kokini, 2006a). When $H = 0$, the normal Navier–Stokes equations are left, but when $H = 1$ the equation degenerates into $v = v_p$.

The results of the comparison between the rotating reference frame and mesh superposition technique show relatively good agreement between the velocities from the rotating reference frame and the mesh superposition technique, except near the wall where there is uncertainty in the exact shape that is dependant on the mesh discretization. This uncertainty also leads to a significant number of material points bleeding into the paddles during particle tracking.

1.7.3.4 FEM Simulations of Mixing Efficiency

There are various parameters such as segregation scale, cluster distribution index, length of stretch, and efficiency of mixing which are used to characterize the nature and efficiency of mixing (Connelly and Kokini, 2006a).

The Manas-Zloczower mixing index (Cheng and Manas-Zloczower, 1990), which is also known as the flow number, is used for analysis of the dispersive mixing ability and the type of the flow:

$$\lambda_{MZ} = \frac{|D|}{|D| + |\Omega|}$$

where D is rate of strain tensor and Ω is vorticity tensor. The flow number characterizes the extent of elongation and rotational flow components with values from 0 to 1.0 (0 for pure rotation, 0.5 for simple shear, and 1 for pure elongation). High flow number values combined with high shear rates have been shown to indicate areas of highly effective dispersive mixing (Yang and Manas-Zloczower, 1992), although the results from different reference frames cannot be compared because the measure is reference frame dependant (Manas-Zloczower, 1995).

The cluster distribution index (ε) is defined as follows (Yang and Manas-Zloczower, 1994):

$$\varepsilon = \frac{\int_0^\infty [c(r) - c(r)_{\text{ideal}}]^2 \mathrm{d}r}{\int_0^\infty [c(r)_{\text{ideal}}]^2 \mathrm{d}r}$$

where $c(r)$ is the coefficient of probability density function. This index is used to measure the difference of the current distribution of particles that were initially in a noncohesive cluster from an ideal random distribution.

Li and Manas-Zloczower (1995) proposed a similar approach based on the correlation coefficient of the length of stretch experienced by particles in a noncohesive cluster:

$$G(\lambda, t) = \frac{2M(\lambda, t)}{\sum_{j=1}^I N_j(N_j - 1)} = g(\lambda, t)\Delta\lambda$$

where $G(\lambda, t)$ and $g(\lambda, t)$ are the length of stretch correlation functions and λ is length of stretch.

In a random mixing process of two components, the maximum attainable uniformity is given by the binomial distribution. A quantitative measure of the binomial distribution is the scale of segregation, L_s, which is defined as:

$$L_s = \int_0^\xi R(|r|)\mathrm{d}|r|$$

where $R(|r|)$ is the Eulerian coefficient of correlation between concentration of pairs of points and it is given as:

$$R(|r|) = \frac{\sum_{j=1}^M (c_j' - \bar{c}) \cdot (c_j'' - \bar{c})}{MS^2}$$

where c_j' and c_j'' are concentration of the pairs in the jth pair while $\bar{c}$ is the average concentration, M is number of pairs, and S is sample variance.

Another model developed by Ottino et al. (1979, 1981) gives a kinematic approach to modeling distributive mixing by tracking the amount of deformation experienced by fluid elements. The length of stretch of an infinitely small material line is defined as:

$$\lambda = \frac{|\mathrm{d}x|}{|\mathrm{d}X|}$$

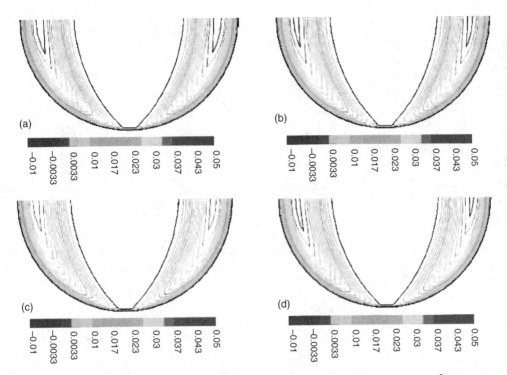

FIGURE 1.100 Velocity profiles generated for different fluid models as stream lines (cm^2/sec) at 1 rpm. (a) Newtonian, (b) Oldroyd-B, (c) Bird–Carreau viscous, and (d) Phan-Thien–Tanner fluid. (Reproduced from Connelly, R.K. and Kokini, J.L., 2004, *Journal of Non-Newtonian Fluid Mechanics*, 123: 1–17. With permission.)

The local efficiency of mixing is defined as:

$$e_\lambda = \frac{\dot{\lambda}/\lambda}{(D:D)^{1/2}} = \frac{-D:\hat{m}\hat{m}}{(D:D)^{1/2}} = \frac{D\ln\lambda/Dt}{(D:D)^{1/2}}$$

where D is the rate of strain tensor, and $\hat{m}$ the current orientation unit vector.

Connelly and Kokini (2004) used the simulated flow profiles (Figure 1.100) in a model 2D mixer they developed for purely viscous, shear thinning inelastic and viscoelastic fluids to calculate the trajectories of initially randomly placed neutral material points. Then the effect of viscoelasticity on mechanism and efficiency of mixing was explored. Mixing parameters such as segregation scale, cluster distribution index, length of stretch, and efficiency of mixing were used to characterize the nature and the effectiveness of mixing.

The mechanism of dispersive mixing within the mixer in a rotating reference frame environment for different fluid models was mapped using the flow number as shown in Figure 1.101. The flow number results indicate that the mixing is primarily due to shearing mechanism. The effect of the variation in rheology is also evident in the simulations as depicted by the increase in the size of regions dominated by elongational flow. Larger areas of high dispersive mixing flow number values were observed with the presence of viscoelasticity. Moreover, low values of dispersive mixing flow number values were obtained with an increase in the intensity of shear thinning as depicted by the increase in the size of the poorly mixed plug flow regions.

Experimental results indicated that shear thinning is detrimental to dispersion, since it decreases the magnitude of the shear stress and increases the sizes of dead zones. However, the effect of viscoelasticity on the overall dispersive ability of the mixer was observed to depend on whether or

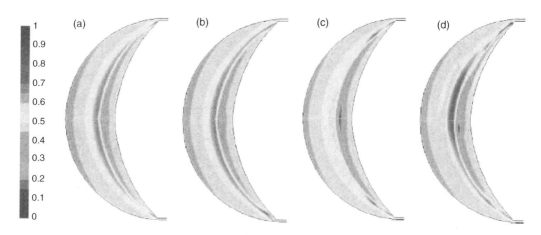

FIGURE 1.101   Flow number distribution at 1 rpm of (a) Newtonian, (b) Oldroyd-B, (c) Bird–Carreau viscous, and (d) Phan-Thien–Tanner fluid models (Reproduced from Connelly, R.K. and Kokini, J.L., 2004, *Journal of Non-Newtonian Fluid Mechanics*, 123: 1–17. With permission.)

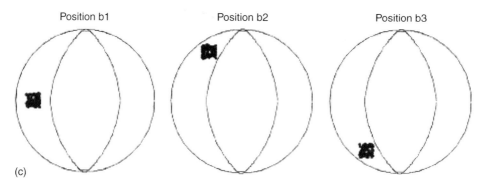

FIGURE 1.102   Initial positions of fixed 0.5×0.5 cm boxes containing 100 randomly placed points. (Reproduced from Connelly, R.K. and Kokini, J.L., 2004, *Journal of Non-Newtonian Fluid Mechanics*, 123: 1–17. With permission.)

not it increases the coincidence of elongational flow. This elongational flow includes high enough shear stresses that would overcome the cohesive or surface forces in clumps and immiscible droplets in a given situation.

The ability of the mixer to distribute clusters of material is analyzed statistically by comparing the distance between pairs of points at each recorded time step of clusters of material points placed initially in one of three boxes in the flow domain as shown in Figure 1.102 (Connelly and Kokini, 2004).

The ability of this mixer to distribute noncohesive clusters of 100 material points was studied by positioning the cluster at the center and upper and lower corners of flow region as shown in Figure 1.102. The effect of fluid rheology on mixing efficiency was evidenced by superimposing the cluster positions in Figure 1.103 over the streamlines shown in Figure 1.100. The streamlines indicates that the center of rotation is not centered in the cluster but is near the left edge. This causes most of the particles to move up towards the back of the blade. A small fraction of the material points located on the left edge of the cluster move slowly down toward the front of the blade. Also, some particles move faster than others due to the velocity gradients, causing the points to spread out.

After 1 revolution.

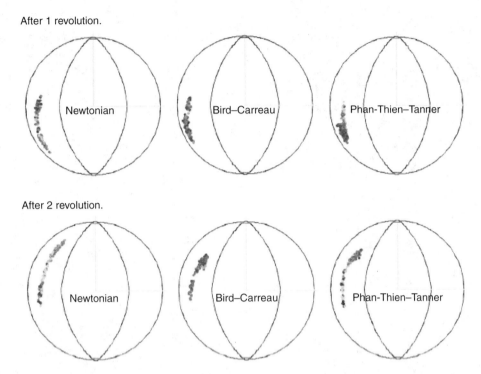

After 2 revolution.

FIGURE 1.103 Distribution of 100 particles in cluster b1 after one and two revolutions while mixing at 1 rpm (Reproduced from Connelly, R.K. and Kokini, J.L., 2004, *Journal of Non-Newtonian Fluid Mechanics*, 123: 1–17. With permission.)

The points in the Newtonian fluid are farther along in the circulation pattern than those in the inelastic Bird–Carreau fluid, with the PTT fluid points falling in the middle. Shear thinning causes irregularity in the shape of the cluster when it is moving towards the back of the blade tip. Circulation of the points caught in the plug flow region will be retarded, allowing all the points to become more spread out over time. It is also apparent that there is no mechanism for moving particles out of the circular streamlines that are present in this region. In order for the distributive mixing to be improved, a mechanism to fold the fluid is required.

Both the scale of segregation and the cluster distribution index showed dependence of rheology on the period of circulation. The length of stretch and efficiency of mixing showed some stretching near the walls. The overall efficiency decreased with increasing mixing times since there is no mechanism to reorient the material lines in this geometry. The secondary flow pattern caused the material to circulate around a central point, and it was shifted in the presence of viscoelasticity. This circulation dominated the mixing, with a period of circulation of approximately two revolutions that was dependant on the fluid rheology. Material is trapped within the circular streamlines, except very near the gap and did not distribute effectively in this geometry (Figure 1.104).

The positions of color-coded particles after 1, 5, and 10 revolutions during mixing at 1 rpm for a Newtonian, inelastic Bird–Carreau and a PTT fluid are shown in Figure 1.105. Clusters of 1000 material points initially were placed randomly in the flow domain. The concentration of neutral material points randomly distributed throughout the flow domain are arbitrarily set to a value of 1, while the concentration of the rest of the neutral material points are set to 0. Then the positions of the particles at any given time are used to calculate the value of the scale of segregation at that point in time. The calculation is done at each recorded time step in order to track the evolution of this parameter over time. After 1 revolution, the particles are still segregated between the upper and lower halves, except near the wall and paddle surfaces. After five revolutions, there are still

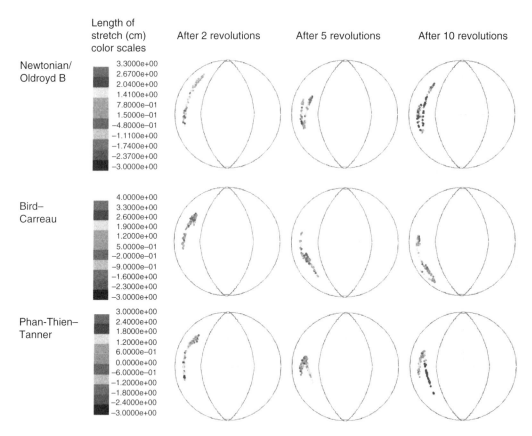

FIGURE 1.104 Distribution of 100 particles initially in cluster b1 after 2, 5, and 10 revolutions while mixing at 1 rpm with length of starch scales. (Reproduced from Connelly, R.K. and Kokini, J.L., 2004, *Journal of Non-Newtonian Fluid Mechanics*, 123: 1–17. With permission.)

considerable amounts of segregated regions with all three fluid models. However, their interfaces and positions are not located in similar manner indicating that the circulation time is rheology dependent. After 10 revolutions, the size of the segregated regions has been reduced significantly, with some randomness in the distribution of particles apparent near the wall. The material is observed to flow through the gap and along the blade surfaces near the walls. However, the material in the center of the flow region that was originally segregated between the upper and lower halves is unable to be redistributed randomly with the flow pattern. It is also evident that the sizes of the central segregated regions are larger with the PTT viscoelastic fluid, likely due to the asymmetry of the velocity distribution (Connelly and Kokini, 2004).

Finite element simulations were also performed in 2D co-rotating twin paddles in a figure eight shaped barrel using the viscous Bird–Carreau dough model of Dhanasekharan et al. (1999) to compare the effectiveness of single and twin screw mixers. Flow profiles were generated from the FEM simulations and particle tracking was conducted to analyze for measures of mixing efficiency. The mixing ability of the single screw and twin screw mixers were then compared. Although the 2D single screw mixer had limited mixing capability, particularly in distributing clumps of material in the upper and lower halves, the 2D twin screw mixer had greater mixing ability with the length of stretch increasing exponentially leading to positive mixing efficiencies over time (Figure 1.106). The results from the 2D twin screw simulation also showed the presence of dead zones in the twin-screw mixer.

The studies mentioned above demonstrate the effectiveness of numerical simulation in studying the flow of materials with different rheological properties in different mixer geometries

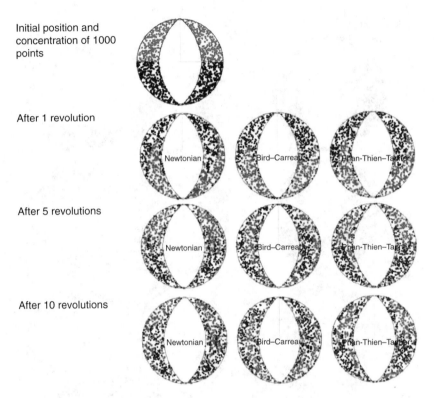

Initial position and concentration of 1000 points

After 1 revolution

Newtonian Bird–Carreau Phan-Thien–Tanner

After 5 revolutions

Newtonian Bird–Carreau Phan-Thien–Tanner

After 10 revolutions

Newtonian Bird–Carreau Phan-Thien–Tanner

FIGURE 1.105 Distribution of 1000 massless particles with concentration of 1 (blue) and 0 (red) initially and after 1, 5, and 10 revolutions. (Reproduced from Connelly, R.K. and Kokini, J.L., 2004, *Journal of Non-Newtonian Fluid Mechanics*, 123: 1–17. With permission.)

nonintrusively. Numerical simulations were clearly shown to serve as valuable tools for process and design engineers to examine the flow behavior of materials of different rheological characteristics. It is also a very effective way to test new ideas to see if they will actually improve a specific food process application without having to build the process equipment in question.

1.7.4 VERIFICATION AND VALIDATION OF MATHEMATICAL SIMULATIONS

The first step in the verification and validation of a numerical simulation is to determine the potential sources of error or uncertainty in the simulation. There are two basic types of uncertainties in the simulations: numerical or physical (Karniadakis, 2002). Numerical uncertainty includes discretization error, round-off error, programming bugs, solution instability, and incomplete convergence. Physical uncertainty includes insufficient knowledge of the geometry, bad assumptions in the development of the physics, simplifications, approximate constitutive laws, unknown boundary conditions, imprecise parameter values, etc. that are the inputs for the simulations.

Several measurement and visualization techniques have been utilized to experimentally validate numerical simulation results and gain a deeper understanding of the processes involved in flow and mixing such as measurements of velocities, at either specific points or through an entire plane, pressure, and residence time. Flow visualization can be achieved using acid-base reactions or the diffusion of a dye in a flow and then using imaging techniques to capture the flow patterns. Velocity measurement has traditionally been carried out at point locations using laser doppler velocimetry (LDA). Velocity measurements through entire planar cross sections are done using particle image or tracking velocimetry (PIV) and planar laser-induced fluorescence (PLIF).

Single screw   **Twin screw**

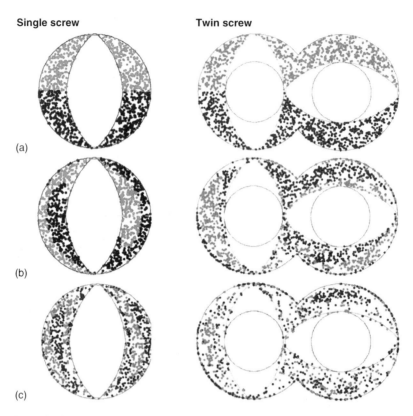

(a)

(b)

(c)

FIGURE 1.106 Distributive mixing between upper and lower halves of single and twin screw mixers at 100 rpm (a) initial position of blades and particles, (b) particle positions after 1 revolution, and (c) particle positions after 10 revolutions. (Reproduced from Connelly, R.K. and Kokini, J.L., 2006a, *Journal of Food Engineering* (accepted). With permission.)

LDA has been used extensively by various authors to estimate velocities at point locations in 2D or 3D flows. Prakash et al. (Prakash, 1996; Prakash and Kokini, 1999, 2000; Prakash et al., 1999) used the LDA to measure velocity distribution in a twin sigma blade mixer (Brabender Farinograph) and estimated the shear rate and various mixing parameters such as instantaneous area stretch efficiency, time averaged efficiency of mixing, strain rate, vorticity rate, dispersive mixing index, and lineal stretch ratio using the velocity vectors. Connelly and Kokini (2006b) compared these LDA results to validate numerical simulations of the flow and mixing in a Brabender Farinograph mixer using exact representations of the blade geometry utilizing the mesh superposition technique. Two positions (180°/270° and 270°/405°) were undertaken for three experimental fluids particle tracking. As an illustration, the comparison of the experimental shear rates and mixing index with numerical simulation results from the Farinograph are shown in Figure 1.107 and Figure 1.108, respectively for three different fluid rheologies.

Moreover, Connelly and Kokini (2006c) simulated the flow of a viscous Newtonian fluid during a complete cycle of the blades positions in a sigma blade mixer. The distributive mixing and overall efficiency of the mixer over time were analyzed using particle tracking. The differential in the blade speeds was observed to allow an exchange of material between the blades with a circulation pattern of material moving up toward the top. The fast blade pushes material towards the slow blade near the bottom of the mixer. The zone in the center of the mixer between the two blades is shown to have excellent distributive and dispersive mixing ability with high shear rates and mixing index values. In contrast, the area away from the region swept by the blades that is generally not filled during normal use of this mixer demonstrates very slow mixing that is made worse by the presence of shear thinning.

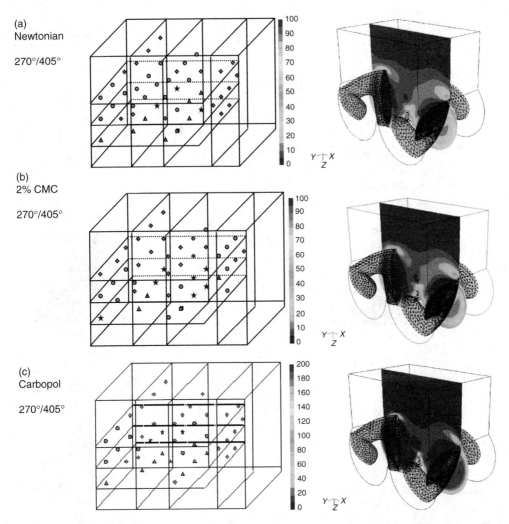

FIGURE 1.107 Simulated shear rates on plane across center of bowl and compared with the experimental LDA results mapped in 3D across the flow domain at the 270°/405° position. Shear rate (sec^{-1}): ♦: 0–50; ●: 50–100; ★: 100–150; ▲: 150–200; ■: above 200 (figures at the left reproduced from Prakash, S. and Kokini, J.L., 2000, *Journal of Food Engineering*, 44: 135–148. With permission; figures at the right reproduced from Connelly, R.K. and Kokini, J.L., 2005b, *Advanced Polymer Technology* (in review). With permission.).

The length of stretch calculated for material points in the Newtonian fluid increased exponentially, indicating effective mixing of the majority of material points. In the area swept by the blades, the highest values of the length of stretch are generally located near the blade edges or in the area swept by the blade edges. High points, however, are also found outside these zones in a more random position. The instantaneous efficiency indicates which blade positions are the most and least effective at applying energy to stretch rather than displace material points. The efficiency was found to be the lowest when the flattened central sections of both blades are horizontal, while the most effective mixing occurred when the flattened section of the fast blade is vertical. The mean time averaged efficiency was found to remain above zero while its standard deviation reduces over time, indicating that the majority of the points are experiencing equivalent levels of stretching over time.

Mixing analysis results reported by Connelly and Kokini (2006b and 2006c) demonstrate how CFD numerical simulations can be used to examine flow and mixing and mixing efficiency in model

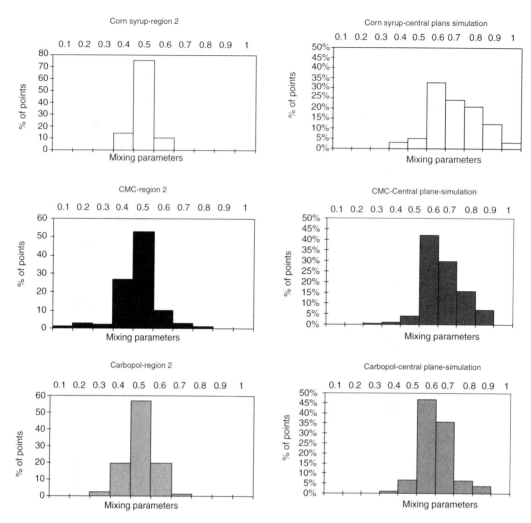

FIGURE 1.108 Distribution of mixing index values at the 270°/405° position for (a) the portion of the ~44 points between the blades. (Reproduced from Prakash, S. and Kokini, J.L., 1999, *Advances in Polymer Technology*, 18: 208–224. With permission.) and (b) the 456 nodes from the simulation on the vertical center plane between the blades. (Reproduced from Connelly, R.K. and Kokini, J.L., 2006b, *Advanced Polymer Technology* (in review). With permission.)

food mixers. The continuous improvements both in hardware and software capabilities will allow process engineers to better predict the behavior of complex materials in complex mixing geometries. Advances in numerical simulations will ultimately lead to better understanding, and control in mixing process thus will allow the design of systems which facilitate effective mixing.

1.8 CONCLUDING REMARKS

Food rheology has grown as a useful tool for many applications in food processing, food handling and storage during the last 20 years. As the chapter points out, better and well understood measurement techniques are currently available with a strong body of work to interpret experimental data. Only 20 years ago the first serious viscoelasticity papers had begun to be published. Today 100s of excellent papers with excellent interpretations from laboratories around the world are available. The quality, novelty, and creativity of the work is of such high level that a significant part of worldwide advances are published in prestigious journals such as the *Journal of Rheology*, *Rheology Acta*, the *Journal of Non Newtonian Mechanics*, and the premier food rheology journal, the *Journal of Texture Studies*.

Rheology has become a routine research and quality control tool for companies around the world. Many companies have been able to develop on-line or at-line measurement programs making rheology an integral part of their process control and monitoring tools. Rheology has also found many applications in the sensory evaluation of texture. With the availability of reliable psychophysical models rheology is being successfully and reliably used for texture design. The ability of rheology to provide information on polymeric properties of food materials, in particular, about enabling the industry to conduct reliable measurements on the glass to rubber transition has given a powerful tool for shelf life characterization of many foods. Food rheology has also provided the necessary constitutive models applicable to various food materials that can be used in numerical simulation of various complex processes such as extrusion, mixing, dough sheeting, and others allowing better understanding of these processes and the improvement of their design based on sound understanding. Clearly rheology has delivered on its promise.

The advances in modern food rheology should further motivate the scientific community to expand the development of more powerful rheological tools. For example, rheology is showing promise as a very accurate and quantitative analytical tool to determine the weight average molecular distribution of many polymers. It has the potential to extend these capabilities to the measurement of particle size distributions in suspensions and droplet size distribution in emulsions. With advances in imaging including Atomic Force Microscopy it is possible to study the nano-scale rheological properties of food molecules and obtain thorough evidence on their conformation and aggregation properties. Molecular models including constitutive models with a molecular basis have not yet found their way in food rheology. Only very simple attempts have been made so far, and this is an area where remarkable predictability can be gained about molecular structure and structural changes during processing. With the advent of a variety of nonthermal processing technologies, electrorheology and the effect of very high pressures on rheological properties are needed, and this area should be a fertile ground for research in the next decade.

In this chapter, we have made an attempt to give an overview of the most recent advances in food rheology with a goal to enable the practitioner to find key ideas and then to expand further into the field with the available references. We hope that this chapter will serve as a useful reference to those who want to capture some of the key advances in the field.

REFERENCES

Alvarez, M.M., Zalc, J.M., Shinbrot, T., Arratia, P.E., and Muzzio, F.J., 2002, Mechanisms of mixing and creation of structure in laminar stirred tanks, *AIChE Journal*, 48: 2135–2148.

Anderson, P.D., Galaktionov, O.S., Peters, G.W.M., van de Vosse, F.N., and Meijer, H.E.H., 2000a, Chaotic fluid mixing in non-quasi-static time-periodic cavity flows, *International Journal of Heat & Fluid Flow*, 21: 176–185.

Anderson, P.D., Galaktionov, O.S., Peters, G.W.M., van de Vosse, F.N., and Meijer, H.E.H., 2000b, Mixing of non-Newtonian fluids in time-periodic cavity flows, *Journal of Non-Newtonian Fluid Mechanics*, 93: 265–286.

Avalosse, Th., 1996, Numerical simulation of distributive mixing in 3D flows, *Macromolecules Symposium*, 112: 91–98.

Avalosse, Th. and Rubin, Y., 2000, Analysis of mixing in corotating twin screw extruders through numerical simulation, *International Polymer Process*, XV: 117–123.

Baaijens, F.P.T., 1998, Mixed finite element methods of viscoelastic flow analysis: a review, *Journal of Non-Newtonian Fluid Mechanics*, 79: 361–385.

Bagley, E.B., 1957, End corrections in the capillary flow of polyethylene, *Journal of Applied Physics*, 28: 624–627.

Bagley, E.B., 1992, Constitutive models for dough. In: *Food Extrusion Science and Technology*, pp. 203–212, J.L. Kokini, C.T. Ho, and M.V. Karwe, Eds, Marcel Dekker, Inc, New York.

Bagley, E.B. and Christianson, D.D., 1986, Response of chemically leavened doughs to uniaxial compression. In: *Fundamentals of Dough Rheology*, H. Faridi and J.M. Faubion, Eds, AACC Publications, Minnesota.

Barbosa-Canovas, G.V. and Peleg, M., 1983, Flow parameters of selected commercial semi-liquid food products, *Journal of Texture Studies*, 14: 213–234.

Bekedam, K, Chambon, L., Ashokan,B., Dogan, H., Moraru, C.I., and Kokini, J.L., 2003, Spectra of relaxation times of wheat flour doughs and their proteins: Molecular origin and measurements, ICEF-9 International Conference of Food Engineering and Food, France.

Bernstein, B., Kearsley, E.A., and Zapas, L.J., 1964, Thermodynamics of perfect elastic fluids, *Journal of Research of the National Bureau of Standards*, 68B: 103–113.

Bertrand, F., Tanguy, P.A., de la Fuente, B., and Carreau, P., 1999, Numerical modeling of the mixing flow of second-order fluids with helical ribbon impellers, *Computational Methods in Applied Mechanics and Engineering*, 180: 267–280.

Binding, D.M., Couch, M.A., Sujatha, K.S., and Webster, M.F., 2003, Experimental and numerical simulation of dough kneading in filled geometries, *Journal of Food Engineering*, 58: 111–123.

Bird, R.B. and Carreau, P.J., 1968, A nonlinear viscoelastic model for polymer solutions and melts-I, *Chemical Engineering Science*, 23: 427–434.

Bird, R.B., Armstrong, R.C., and Hassager, O., 1987, *Dynamics of Polymeric Liquids*, 2nd ed., John Wiley & Sons Inc., New York.

Bistany, K.L., 1984, Viscoelastic properties of guar gum and constitutive considerations, Master Thesis, Rutgers University.

Bistany, K.L. and Kokini, J.L., 1983a, Comparison of steady shear rheological properties and small amplitude dynamic viscoelastic properties of fluid food materials, *Journal of Texture Studies*, 14: 113–124.

Bistany, K.L. and Kokini, J.L., 1983b, Dynamic viscoelastic properties of foods in texture control, *Journal of Rheology*, 27: 605–620.

Bloksma, A.H., 1957, A calculation of the shape of the alveograms of some rheological model substances, *Cereal Chemistry*, 34: 126–136.

Bloksma, A.H., 1990, Rheology of the breadmaking process, *Cereal Foods World*, 35: 228–236.

Bloksma, A.H. and Nieman, W., 1975, The effect of temperature on some rheological properties of wheat flour dough, *Journal of Texture Studies*, 6: 343–361.

Bongenaar, J.J.T.M., Kossen, N.W.F., and Meijboom, F.W., 1973, A method for characterizing the rheological properties of viscous fermentation broths, *Bioengineering*, 15: 201–206.

Brent, J.L., Mulvaney, S.J., Cohen C., and Bartsch, J.A., 1997, Viscoelastic Properties of Extruded Cereal Melts, *Journal of Cereal Science*, 26: 313–328.

Campanella, O.H. and Peleg, M., 2002, Squeezing flow viscometry for nonelastic semiliquid foods — Theory and applications, *Critical Reviews in Food Science and Nutrition*, 42: 241–264.

Carreau, P.J., MacDonald, I.F., and Bird, R.B., 1968, A nonlinear viscoelastic model for polymer solutions and melts, *Chemical Engineering Science*, 23: 901–911.

Carrillo, A.R. and Kokini, J.L., 1988, Effect of egg yolk and egg yolk + salt on rheological properties and particle size distribution of model oil-in-water salad dressing emulsions, *Journal of Food Science*, 53: 1352–1366.

Chang, C.N., Dus, S., and Kokini, J.L., 1990, Measurement and interpretation of batter rheological properties. In: *Batters and Breadings in Food Processing*, K. Kulp and R. Loewe, Eds, American Association of Cereal Chemists Inc., Minnesota.

Charalambides, M.N., Wanigasooriya, L., Williams, J.G., and Chakrabarti, S., 2002a, Biaxial deformation of dough using the bubble inflation technique. I. Experimental, *Rheologica Acta*, 41: 532–540.

Charalambides, M.N., Wanigasooriya, L., Williams, J.G., and Chakrabarti, S., 2002b, Biaxial deformation of dough using the bubble inflation technique. II. Numerical modeling, *Rheologica Acta*, 41: 541–548.

Charm, S.E., 1960, Viscometry of non-Newtonian food materials, *Food Research*, 25: 351.

Chatraei, S.H., Macosko, C.W., and Winter, H.H., 1981, Lubricated squeezing flow: a new biaxial extensional rheometer, *Journal of Rheology*, 25: 433–443.

Chen, I. and Bogue, D.C., 1972, Time dependent stress in polymer melts and review of viscoelastic theory, *Transactions of the Society of Rheology*, 16: 59–78.

Cheng, J.J. and Manas-Zloczower, I., 1990, Flow field characterization in a banbury mixer. *International Polymer Processing*, V: 178–183.

Chou, T.C. and Kokini, J.L., 1987, Rheological properties and conformation of tomato paste pectins, citrus and apple pectins, *Journal of Food Science*, 52: 1658–1664.

Cocero, A.M. and Kokini, J.L., 1991, The study of the glass transition of glutenin using small amplitude oscillatory rheological measurements and differential scanning calorimetry, *Journal of Rheology*, 35: 257–270.

Cogswell, F.N., 1981, *Polymer Melt Rheology: A Guide for Industrial Practice*, John Wiley & Sons, Inc., New York.

Connelly, R.K., 2004, Numerical simulation and validation of the mixing of dough-like materials in model batch and continuous dough mixers, Ph.D. Thesis, Rutgers University.

Connelly, R.K. and Kokini, J.L., 2001, Analysis of mixing in a model dough mixer using numerical simulation with particle tracking. In: *Proceedings of the Seventh Conference of Food Engineering, A Topical Conference of the AIChE Annual Meeting*, Reno, NV.

Conelly, R.K. and Kokini, J.L., 2003, 2D Numerical Simulation of Differential Viscoelastic Fluids in a Single-Screw Continuous Mixer: Application of Viscoelastic Finite Element Methods, *Advances in Polymer Technology*, 22: 22–41.

Conelly, R.K. and Kokini, J.L., 2004, The effect of shear thinning and differential viscoelasticity on mixing in a model 2D mixer as determined using FEM with particle tracking, *Journal of Non-Newtonian Fluid Mechanics*, 123: 1–17.

Connelly, R.K. and Kokini, J.L., 2006a, Examination of the mixing ability of single and double screw mixers using 2D Finite Element Method simulation with particle tracking, *Journal of Food Engineering* accepted.

Connelly, R.K. and Kokini, J.L., 2006b, 3D numerical simulation of the flow of viscous Newtonian and shear thinning fluids in a twin sigma blade mixer, *Advanced Polymer Technology* (in review).

Connelly, R.K. and Kokini, J.L., 2006c, Simulation and analysis of mixing of a viscous Newtonian liquid in a twin sigma blade mixer, *AIChE Journal* (in review).

Corradini, M.G., Stern, V., Suwonsichon, T., and Peleg, M., 2000, Squeezing flow of semi liquid foods between parallel Teflon coated plates, *Rheologica Acta*, 39: 452–460.

Cox, W.P. and Mertz, E.H., 1954, Correlation of dynamic and steady flow viscosities, *Journal of Polymer Science*, 28: 619–622.

Crochet, M.J., 1989, Numerical simulation of viscoelastic flow: a review, *Rubber Chemistry and Technology*, 62: 426–455.

Darby, R., 1976, *Viscoelastic Fluids: An Introduction to Their Properties and Behavior*, Dekker Inc., New York.

Davis, S.S., 1973, Rheological properties of semi-solid foodstuffs: viscoelasticity and its role in quality control, *Journal of Texture Studies*, 4: 15–40.

de Baerdemaeker, J., Singh, R.P., and Segerlind, L.J., 1977, Modeling heat transfer in foods using the finite element method, *Journal of Food process Engineering*, 1: 37–50.

de Bruijne, D.W., de Loof, J., and van Eulem, A., 1990, The rheological properties of breads dough and their relation to baking. In: *Rheology of Food, Pharmaceutical and Biological Materials with General Rheology*, R.E. Carter, Ed., Elsevier Applied Science, London, UK.

Dealy, J.M., 1984, Official nomenclature for material functions describing the response of a viscoelastic fluid to various shearing and extensional deformations, *Journal of Rheology*, 28: 181–185.

Dealy, J.M. and Wissbrun, K.F., 1990, *Melt Rheology and Its Role in Plastics Processing: Theory and Application*, VNR, New York.

DeKee, D., Code, R.K., and Turcotte, G., 1983, Flow properties of time dependent foodstuffs, *Journal of Rheology*, 27: 581–604.

Dervisoglu, M. and Kokini, J.L., 1986a, Effect of different tube materials on the steady shear tube flow semi-solid foods, *Journal of Food Process Engineering*, 8: 137–146.

Dervisoglu, M. and Kokini, J.L., 1986b, The steady shear rheology and fluid mechanics of four semi-solid foods, *Journal of Food Science*, 51: 541–546, 625.

Dhanasekharan, K.M., 2001, Dough rheology and extrusion: Design and scaling by numerical simulation, Ph.D. Thesis, Rutgers University.

Dhanasekharan, K.M. and Kokini, J.L., 1999, A study of viscoelastic flows in the extrusion of wheat flour doughs. In: *Proceedings of the Sixth Conference of Food Engineering, A Topical conference of the AIChE Annual Meeting*, Dallas, TX.

Dhanasekharan, K.M. and Kokini, J.L., 2003, Design and scaling of wheat dough extrusion by numerical simulation of flow and heat transfer, *Journal of Food Engineering*, 60: 421–430.

Dhanasekharan, K.M., Huang, H., and Kokini, J.L., 1999, Comparison of observed rheological properties of hard wheat flour dough with predictions of the Giesekus-Leonov, White-Metzner and Phan-Thien–Tanner models, *Journal of Texture Studies*, 30: 603–623.

Dhanasekharan, K.M., Wang, C.F., and Kokini, J.L., 2001, Use of nonlinear differential viscoelastic models to predict the rheological properties of gluten dough, *Journal of Food Process Engineering*, 24: 193–216.

Dhanasekharan, K.M. and Kokini, J.L., 2000, Viscoelastic flow modeling in extrusion of a dough-like fluid *Journal of Food Process Engineering*, 23: 237–247.

Dickie, A.M. and Kokini, J.L., 1982, Use of Bird–Leider equation in food rheology, *Journal of Food Process Engineering*, 5: 157–174.

Dickie, A.M. and Kokini, J.L., 1983, An improved model for food thickness from non-Newtonian fluid mechanics in the mouth, *Journal of Food Science*, 48: 57–61.

Dickinson, E. and Yamamoto, Y., 1996, Effect of lecithin on the viscoelastic properties of β-lactoglobulin-stabilized emulsion gels, *Food Hydrocolloids*, 10: 301–307.

Dobraszczyk, B.J. and Morgenstern, M.P., 2003, Rheology and the breadmaking process, *Journal of Cereal Science*, 38: 229–245.

Dobraszczyk, B.J., Smewing, J., Albertini, M., Maesmans, G., and Schofield, J.D., 2003, Extensional rheology and stability of gas cell walls in bread doughs at elevated temperatures in rations to breadmaking performance, *Cereal Chemistry*, 80: 218–224.

Doi, M. and Edwards, S.F., 1978a, Dynamics of concentrated polymer systems Part I, *Journal of Chemecial Society, Faraday Transactions II*, 74: 1789–1801.

Doi, M. and Edwards, S.F., 1978b, Dynamics of concentrated polymer systems Part II, *J. Chem. Soc., Faraday Trans. II.*,74: 1802–1817.

Dus, S.J. and Kokini, J.L., 1990, Prediction of the nonlinear viscoelastic properties of a hard wheat flour dough using the Bird–Carreau constitutive model, *Journal of Rheology*, 34: 1069–1084.

Dzuy, N.Q. and Boger, D.V., 1983, Yield stress measurements of concentrated suspension, *Journal of Rheology*, 27: 321–349.

Edwards, N.M., Dexter, J.E., Scanlon, M.G., and Cenkowski, S., 1999, Relationship of creep-recovery and dynamic oscillatory measurements to durum wheat physical dough properties, *Cereal Chemistry*, 76: 638–645.

Edwards, N.M., Peressini, D., Dexter, J.E., and Mulvaney, S.J., 2001, Viscoelastic properties of drum wheat and common wheat dough of different strengths, *Rheologica Acta*, 40: 142–153.

Elejalde, C.C. and Kokini, J.L., 1992a, Identification of key textural attributes of viscoelastic syrups by regression analysis, *Journal of Food Science*, 57: 167–171.

Elejalde, C.C. and Kokini, J.L., 1992b, The psychophysics of pouring, spreading and in-mouth viscosity, *Journal of Texture Studies*, 23: 315–336.

Elliott, J.H. and Green, C.E. 1972, Modification of food characteristics with cellulose hydrocolloids. II. The modified Bingham body — A useful rheological model. *J. Texture Studies*, 3: 194–205.

Fan, Y.R., Tanner, R.I., and Phan-Thien, N., 2000, A numerical study of viscoelastic effects in chaotic mixing between eccentric cylinders. *Journal of Fluid Mechanics*, 412: 197–225.

Ferry, J., 1980, *Viscoelastic Properties of Polymers*, 3rd ed., John Wiley & Sons, New York.

Fischbach, E.R. and Kokini, J.L., 1984, Comparison of yield stresses of semi-solid foods obtained using constant stress and constant strain rate techniques, *Paper presented at the 44th Annual Meeting of the Institute of Food Technologists, Anaheim, CA.*

Giesekus, H., 1982, A simple constitutive equation for polymeric fluids based on the concept of deformation dependent tensorial mobility, *Journal of Non-Newtonian Fluid Mechanics*, 11: 69–109.

Gluck-Hirsh and Kokini, J.L., 1997, Determination of the molecular weight between crosslinks of waxy maize starches using the theory of rubber elasticity, *Journal of Rheology*, 41: 129–139.

Goff, H.D., Freslon, B., Sahagian, M.E., Hauber, T.D., Stone, A.P., and Stanley, D.W., 1995, Structural development in ice cream-Dynamic rheological measurements, *Journal of Texture Studies*, 26: 517–536.

Gontard, N., Guilbertand, S., and Cuq, J.L., 1993, Water and glycerol as plasticizers affect mechanical and water barrier properties of an edible wheat gluten film, *Journal of Food Science*, 58: 206–211.

Goto, H. and Kuno, H., 1982, Flow of suspensions containing particles of different sizes through a capillary tube, *Rheology*, 26: 387–398.

Gras, P.W., Carpenter, H.C., and Anderssen, R.S., 2000, Modelling the developmental rheology of wheat-flour dough using extension tests, *Journal of Cereal Science*, 31: 1–13.

Gunasekaran, S. and Ak, M.M., 2000, Dynamic oscillatory shear testing of foods — selected applications, *Trends in Food Science & Technology*, 11: 115–127.

Herum, F.L., Isaacs, G.W., and Peart, R.M., 1966, Flow properties of highly viscous organic pastes and slurries, *Transactions American Society of Agriculture Engineers*, 9: 45–51.

Hlynka, I. and Barth, F.W., 1955, Chopin alveograph studies I. Dough resistance at constant sample deformation, *Cereal Chemistry*, 32: 463–471.

Holdsworth, S.D., 1971, Applicability of rheological models to the interpretation of flow and process behavior of fluid food products, *Journal of Texture Studies*, 2: 393–396.

Honerkamp, J. and Weese, J., 1989, Determination of the relaxation spectrum by a regularization method. *Macromolecules*, 22: 4372–4377.

Honerkamp, J. and Weese, J., 1990, Tikhonov regularization method for ill-posed problems. *Continuum Mechanics and Thermodynamics*, 2: 17–30.

Huang, H., 1998, Shear and extensional rheological measurements of hard wheat flour doughs and their simulation using Wagner constitutive model, Ph.D. Thesis, Rutgers University.

Huang, H. and Kokini, J.L., 1993, Measurement of biaxial extensional viscosity of wheat flour doughs, *Journal of Rheology*, 37: 879–891.

Huang, H. and Kokini, J.L. 1994. Steady shear and extensional rheological measurements of hard wheat flour doughs and their simulation using Wagner constitutive model. In: *Progress and Trends in Rheology IV, Proceedings of the Fourth European Rheology Conference*, Sevilla, Spain.

Huang, H. and Kokini, J.L., 1999, Prediction of dough volume development which considers the biaxial extensional growth of cells. In: *Bubbles in Food*, G.M. Campbell, C. Webb, S.S. Pandiella, and K. Niranjan, Eds, American Association of Cereal Chemists Inc., Minnesota.

Hwang, J. and Kokini, J.L., 1991, Structure and rheological function of side branches of carbohydrate polymers, *Journal of Texture Studies*, 22: 123–137.

Hwang, J. and Kokini, J.L., 1992, Contribution of side branches to theological properties of pectins, *Carbohydrate Polymers*, 19: 41–50.

Hwang, J. and Kokini, J.L., 1995, The branching effects of pectic polysaccharides on viscoelastic properties, *The Korean Journal of Rheology*, 7: 120–127.

James, H.M., 1947, Properties of networks on flexible chains, *Journal of Chemical Physics*, 15: 651–668.

Janssen, A.M., van Vliet, T., and Vereijken, J.M., 1996a, Rheological behaviour of wheat glutens at small and large deformations. Effect of gluten composition, *Journal of Cereal Science*, 23: 33–42.

Janssen, A.M., van Vliet, T., and Vereijken, J.M., 1996b, Fundamental and empirical rheological behavior of wheat flour doughs and comparison with bread making performance, *Journal of Cereal Science*, 23: 43–54.

Jastrzebsky, Z.D., 1967, Entrance effects and wall effects in an extrusion rheometer during the flow of concentrated suspensions, *Industrial and Engineering Chemistry Fundamentals*, 6: 445–454.

Jauregui, B., Mufioz, M.E., and Santamaria, A., 1995, Rheology of hydroxyethylated starch aqueous systems: Analysis of gel formation, *International Journal of Biological Macromolecules*, 17: 49–54.

Joye, D.D., Poehlein, G.W., and Denson, C.D., 1972, A bubble inflation technique for the measurement of viscoelastic properties in equal biaxial extensional flow, *Transactions of the Society of Rheology*, 16: 421–455.

Kalichevsky, M.T. and Blanshard, J.V.M., 1993, The effect of fructose and water on the glass transition of amylopectin, *Carbohydrate Polymers*, 20: 107–113.

Kampf, N. and Peleg, M., 2002, Characterization of chick pea (*Cicer arietum L*) pastes using squeezing flow viscometry, *Rheologica Acta*, 41: 549–556.

Karniadakis, G.E., 2002, Quantifying uncertainty in CFD, *Journal of Fluids Engineering*, 24: 2–3.

Karnis, A., Goldsmith, H.L., and Mason, S.G., 1966, The flow of suspensions through tubes. V. Inertial effects, *Canadian Journal of Chemical Engineering*, 44: 181–193.

Keentook, M., Newberry, M.P., Gras, P., and Tanner, R.I., 2002, The rheology of bread dough made from four commercial flours, *Rheologica Acta*, 41: 173–179.

Kokini, J.L., 1977, Predicting liquid food texture from fluid dynamics and lubrication theory, Ph.D. Thesis, Carnegie-Mellon University, Pittsburg, PA.

Kokini, J.L., 1985, Fluid and semi-solid food texture and texture-taste interactions, *Food Technology*, 39: 86–92, 94.

Kokini, J.L., 1992, Theological properties of Foods. In: *Handbook of Food Engineering*, D.R. Heldman and D.B. Lund, Eds, Marcel Dekker, Inc., New York.

Kokini, J.L., 1993, Constitutive models for dilute and concentrated food biopolymers systems. In: *Plant Polymeric Carbohydrates*, F. Meuser, D.J. Manners, and W. Siebel, Eds, Royal Society of Chemistry, Cambridge.

Kokini, J.L., 1994, Predicting the rheology of biopolymers using constitutive models, *Carbohydrate Polymers*, 25: 319–329.

Kokini, J.L. and Chou, T.C., 1993, Comparison of the conformation of tomato pectins with apple and citrus pectins, *Journal of Texture Studies*, 24: 117–137.

Kokini, J.L. and Cussler, E.L., 1983, Predicting the texture of liquid and melting semi-solid foods, *Journal of Food Science*, 48: 1221–1225.

Kokini, J.L. and Dervisoglu, M., 1990, Wall effects in the laminar pipe flow of four semi-solid foods, *Journal of Food Engineering*, 11: 29–42.

Kokini, J.L. and Dickie, A., 1981, An attempt to identify and model transient viscoelastic flow in foods, *Journal of Texture Studies*, 12: 539–557.

Kokini, J.L. and Dickie, A., 1982, A model food spreadability from fluid mechanics, *Journal of Texture Studies*, 13: 211–227.

Kokini, J.L. and Plutchok, G., 1987a, Viscoelastic properties of semisolid foods and their biopolymeric components, *Food Technology*, 41: 89–95.

Kokini, J.L. and Plutchok, G., 1987b, Predicting steady and oscillatory shear rheological properties of CMC/guar blends using the Bird–Carreau constitutive model, *Journal of Texture Studies*, 18: 31–42.

Kokini, J.L. and Surmay, K., 1994, Steady shear viscosity first normal stress difference and recoverable strain in carboxymethyl cellulose, sodium alginate and guar gum, *Carbohydrate Polymers*, 22: 241–252.

Kokini, J.L., Bistany, K.L., and Mills, P.L., 1984a, Predicting steady shear and dynamic viscoelastic properties of guar and carrageenan using Bird–Carreau constitutive model, *Journal of Food Science*, 49: 1569–1576.

Kokini, J.L., Cocero, A.M., and Madeka, H., 1995a, State Diagrams help predict rheology of cereal proteins, *Food Technology*, 49: 74–81.

Kokini, J.L., Cocero, A.M., Madeka, H., and de Graaf, E., 1994, The development of state diagrams for cereal proteins, *Trends in Food Science and Technology*, 5: 281–288.

Kokini, J.L., Dhanasekharan, M., Wang, C.-F., and Huang, H., 2000, Integral and differential linear and non-linear constitutive models for rheology of wheat flour doughs. In: *Trends in Food Engineering*, J.E. Lozano, C. Anon, E. Parada-Arias, and G.V. Barbosa-Canovas, Eds, Technomics Publishing Co. Inc., Lancaster, PA.

Kokini, J.L., Kadane, J.B., and Cussler, E.L., 1977, Liquid texture perceived in the mouth, *Journal of Texture Studies*, 8: 195–218.

Kokini, J.L., Poole, M., Mason, P., Miller, S. and Stier, E., 1984b, Identification of key textural attributes of fluid and semi-solid foods using regression analysis, *Journal of Food Science*, 49: 47–51.

Kokini, J.L., Wang, C.F., Huang, H., and Shrimanker, S., 1995b, Constitutive models of foods, *Journal of Texture Studies*, 26: 421–455.

Kompani, M. and Venerus, D.C., 2000, Equibiaxial extensional flow of polymer melts via lubricated squeezing flow I. Experimental analysis, *Rheologica Acta*, 39: 444–451.

Kraynik, A.M. and Schowalter W.R., 1981, Slip at wall and extrudate roughness with aqueous solutions of polyvinylalcohol and sodium berate, *Journal of Rheology*, 25: 95–114.

Labropoulos, A.E. and Hsu, S.-H., 1996, Viscoelastic behavior of whey protein isolates at the sol-gel transition point, *Journal of Food Science*, 61: 65–68.

Labropoulos, K.C., Niesz, D.E., Danforth, S.C., and Kevrekidis, P.G., 2002a, Dynamic rheology of agar gels: theory and experiments. Part I. Development of a rheological model, *Carbohydrate Polymers*, 50: 393–406.

Labropoulos, K.C., Niesz, D.E., Danforth, S.C., and Kevrekidis, P.G., 2002b, Dynamic rheology of agar gels: theory and experiments. Part II. Gelation behavior of agar sols and fitting a theoretical rheological model, *Carbohydrate Polymers*, 50: 407–415.

Lagarrigue, S. and Alvarez, G., 2001, The rheology of starch dispersions at high temperatures and high shear rates: a review, *Journal of Food Engineering*, 50: 189–202.

Launay, B., Bure, J., and Praden, J., 1977, Use of Chopin alveograph as a rheological tool I. Dough deformation measurements, *Cereal Chemistry*, 54: 1042–1048.

Lazaridou, A., Biliaderis, C.G., and Kontogiorgos, V., 2003, Molecular weight effects on solution rheology of pullulan and mechanical properties of its films, *Carbohydrate Polymers*, 52: 151–166.

Leider, P.I. and Bird, R.B., 1974, Squeezing flow between parallel disks I. Theoretical analysis. *Industrial and Engineering Chemistry Fundamentals*, 13: 336–341.

Leppard, W.R., 1975, Viscoelasticity: Stress measurements and constitutive theory, Ph.D. Thesis, University of Utah.

Li, T. and Manas-Zloczower, I., 1995, Evaluation of distributive mixing efficiency in mixing equipment, *Chemical Engineering Communications*, 139: 223–231.

Liguori, C.A., 1985, The relationship between the viscoelastic properties and the structure of sodium alginate and propylene glycol alginate, M.S. Thesis, Rutgers University.

Lodge, A.S., 1956, A network theory of flow birefringence and stress in concentrated polymer solutions, *Transactions of the Faraday Society*, 52: 120–130.

Lodge, A.S., 1964, *Elastic Liquids*, Academic Press, New York.

Lopes da Silva, J.A. and Goncalves, M.P., 1994, Rheological study into the aging process of high methoxyl pectin/sucrose aqueous gels, *Carbohydrate Polymers*, 24: 235–245.

Luo X.-L. and Tanner, R.I., 1986a. A streamline element scheme for solving viscoelastic flow problems Part I: Differential constitutive models, *Journal of Non-Newtonian Fluid Mechanics*, 21: 179–199.

Luo X.-L. and Tanner, R.I., 1986b. A streamline element scheme for solving viscoelastic flow problems Part II: Integral constitutive models, *Journal of Non-Newtonian Fluid Mechanics*, 22: 61–89.

Ma, L. and Barbosa-Canovas, G.V., 1995, Rheological characterization of mayonnaise, Part II. Flow and viscoelastic properties at different oil and xanthan gum concentrations, *Journal of Food Engineering*, 25: 409–425.

Mackey, K.L. and Ofoli, R.Y., 1990, Rheology of low to intermediate moisture whole wheat flour dough, *Cereal Chemistry*, 67: 221–226.

Macosko, C.W., 1994, *Rheology: Principles, Measurements and Applications*, VCH Publishers, Inc., New York.

Madeka, H. and Kokini, J.L., 1994, Changes in rheological properties of gliadin as a function of temperature and moisture: development of a state diagram, *Journal of Food Engineering*, 22: 241–252.

Madeka, H. and Kokini, J.L., 1996, Effect of glass transition and cross-linking on rheological properties of zein: development of a preliminary state diagram, *Cereal Chemistry*, 73: 433–438.

Marchal J.M. and Crochet, M.J., 1987, A new mixed finite element for calculating viscoelastic flow, *Journal of Non-Newtonian Fluid Mechanics*, 26: 77–114.

Marvin, R.S. and McKinney, J.E., 1965. In: *Physical Acoustics*, W.P. Mason, Ed., Vol. B. Academic Press, New York.

Mason, P.L., Puoti, M.P., Bistany, K.L., and Kokini, J.L., 1983, A new empirical model to simulate transient shear stress growth in semi-solid foods, *Journal of Food Process Engineering*, 6: 219–233.

Matsumura, Y., Kang, I.J., Sakamoto, H., Motoki, M. and Mori, T., 1993, Filler effects of emulsion gels, *Food Hydrocolloids*, 7: 227–240.

Mead, D.W., 1994, Determination of molecular weight distributions of linear flexible polymers from linear viscoelastic material functions, *Journal of Rheology*, 38: 1797–1827.

Meeten, G.H., 2002, Constant-force squeeze flow of soft solids, *Rheologica Acta*, 41: 557–566.

Meissner, J., 1971, Dehnungsverhalten von Polyathylen-Schmelzen, *Rheologica Acta*, 10: 230–240.

Meissner, J., Raible, R., and Stephenson, S.E., 1981, Rotary clamp in uniaxial and biaxial extension rheometry of polymer melts, *Journal of Rheology*, 25: 1–28.

Mills, P.L. and Kokini, J.L., 1984, Comparison of steady shear and dynamic viscoelastic properties of guar and karaya gums, *Journal of Food Science*, 49: 1–4, 9.

Mitchell, B.W. and Peart, R.M., 1968, Measuring apparent viscosity of organic slurries, *Transaction ASAE*, 11: 523–.

Morales, A. and Kokini, J.L., 1997, Glass transition of soy globulins using differential scanning calorimetry and mechanical spectroscopy, *Biotechnology. Progress*, 13: 624–629.

Morales, A. and Kokini, J.L., 1999, State diagrams of soy globulins, *Journal of Rheology*, 43: 315–325.

Morales, A.M., 1997, Rheological properties and phase transitions of soy globulins, Ph.D. Thesis, Rutgers University.

Morales, A.M. and Kokini, J.L., 1998, Understanding phase transitions and chemical complexing reactions in 7S and 11S soy protein fractions. In: *Phase/State Transitions in Foods*, M.A. Roa and R.W. Hartel, Eds, Marcel Dekker, Inc., New York.

Moraru, C.I., Lee, T.-C., Karwe, M.V., and Kokini, J.L., 2002, Plasticizing and antiplasticizing effects of water and polyols on a meat-starch extruded matrix, *Journal of Food Engineering*, 67: 3026–3032.

Morris, E.R. and Ross-Murphy, S.B., 1981, Chain flexibility of polysaccharides and glycoproteins from viscosity measurements, *Techniques in Carbohydrate Metabolism*, 201–246.

Morris, E.R., Coulter, A.N., Ross-Murphy, S.B., Rees, D.A., and Price, J., 1981, Concentration and shear rate dependence of viscosity in random coil polysaccharide solutions, *Carbohydrate Polymers*, 1: 5–21.

Munoz, J. and Sherman, P., 1990, Dynamic viscoelastic properties of some commercial salad dressings, *Journal of Texture Studies*, 21: 411–426.

Nasseri, S., Bilston, L., Fasheun, B., and Tanner, R., 2004, Modelling the biaxial elongational deformation of soft solids, *Rheologica Acta*, 43: 68–79.

Newberry, M.P., Phan-Thien, N., Larroque, O.R., Tanner, R.I., and Larsen, N.G., 2002, Dynamic and elongation rheology of yeasted bread doughs, *Cereal Chemistry*, 79: 874–879.

Oldroyd, J.G., 1949, The interpretation of observed pressure gradients in laminar flow of non-Newtonian liquids through tubes, *Journal of Colloid Science*, 4: 333–342.

Orbey, N. and Dealy, J.M., 1991, Determination of the relaxation spectrum from oscillatory shear data, *Journal of Rheology*, 35: 1035–1049.

Osaki, K., 1976, Proc. VIIth Intern. Congr. Rheol., p. 104. Gothenburg.

Ottino, J.M., Ranz, W.E., and Macosko, C.W., 1979, A lamellar model for analysis of liquid-liquid mixing, *Chemical Engineering Science*, 34: 877–890.

Ottino, J.M., Ranz, W.E., and Macosko C.W., 1981, A framework for description of mechanical mixing of fluids, *AIChE Journal*, 27: 565–577.

Padmanabhan, M., 1995, Measurement of extensional viscosity of viscoelastic liquid foods, *Journal of Food Engineering*, 25: 311–327.

Padmanabhan, M. and Bhattacharya, M., 1991, Rheological measurement of fluid elasticity during extrusion cooking, *Trends in Food Science & Technology*, 2: 149–151.

Padmanabhan, M. and Bhattacharya, M., 1993a, Effect of extrusion processing history on the theology of corn meal, *Journal of Food Engineering*, 18: 335–349.

Padmanabhan, M. and Bhattacharya, M., 1993b, Planar extensional viscosity of corn meal dough, *Journal of Food Engineering*, 18: 389–411.

Papanastasiou, A.C., Scriven, L.E., and Macosko, C.W., 1983, An integral constitutive equation for mixed flows: viscoelastic characterization, *Journal of Rheology*, 27: 387–410.

Peleg, M., 1995, A note on the tan δ peak as a glass transition indicator in biosolids, *Rheologica Acta*, 34: 215–220.

Peleg, M., 1996, On modeling changes in food and biosolids at and around their glass transition temperature range, *Critical Reviews in Food Science and Nutrition*, 36: 49–67.

Perera, M.G.N. and Walters, K., 1977, Long-range memory effects in flows involving abrupt changes in geometry. Part I: Flows associated with L-shaped and T-shaped geometries, *Journal of Non-Newtonian Fluid Mechanics*, 2: 49–81.

Phan-Thien, N. and Tanner, R.I., 1977, A new constitutive equation derived from network theory, *Journal of Non-Newtonian Fluid Mechanics*, 2: 353–365.

Phan-Thien, N., Safari-Ardi, M., and Marales-Patino, A., 1997, Oscillatory and simple shear flows of a flour-water dough: a constitutive model, *Rheologica Acta*, 36: 38–48.

Plutchok, G. and Kokini, J.L., 1986, Predicting steady and oscillatory shear rheological properties of CMC/guar blends using Bird–Carreau constitutive model, *Journal of Food Science*, 51: 1284–1288.

Polyflow. 1997, *Notes on Mixing*. Place del'Universite 16, B-1348 Louvain-la-Neuve, Belgium.

Polyflow. 2001a, *Theoretical Background, Version* 3.8.0. Place del'Universite 16, B-1348 Louvain-la-Neuve, Belgium.

Polyflow. 2001b, *User's Manual, Version* 3.8.0. Place del'Universite 16, B-1348 Louvain-la-Neuve, Belgium.

Prakash, S., 1996, Characterization of shear rate distribution in a model mixer using Laser Doppler Anemometry, Ph.D. Thesis, Rutgers University.

Prakash, S. and Kokini, J.L., 1999, Determination of mixing efficiency in a model food mixer, *Advances in Polymer Technology*, 18: 208–224.

Prakash, S. and Kokini, J.L., 2000, Estimation and prediction of shear rate distribution as a model mixer, *Journal of Food Engineering*, 44: 135–148.

Prakash, S., Karwe, M.V., and Kokini, J.L., 1999, Measurement of velocity distribution in the Brabender farinograph as a model mixer using Laser Doppler Anemometry, *Journal of Food Process Engineering*, 22: 435–454.

Puri, V.M. and Anantheswaran, R.C., 1993, The finite element method in food processing: a review, *Journal of Food Engineering*, 19: 247–274.

Rahalkar, R.R., Lavanaud, C., Richmond, P., Melville and Pethrick, R., 1985, Oscillatory shear measurements on concentrated Dextran solutions Comparison with Doi and Edwards theory of reptation, *Journal of Rheology*, 129: 955–970.

Rao, M.A., 1977, Rheology of liquid foods: a review, *Journal of Texture Studies*, 5: 135–168.

Rao, M.A., Bourne, M.C., and Cooley, H.J., 1981, Flow properties of tomato concentrates, *Journal of Texture Studies*, 12: 521–538.

Rao, M.A., Cooley, H.J., and Vitali, A.A., 1984, Flow properties of concentrated juices at low temperatures, *Food Technology*, 38: 113–119.

Rauwendaal, C., 1986, *Polymer Extrusion*, Hanser Publishers, New York.

Reddy, 1993, *An Introduction to Finite Element Method*, 2nd ed., McGraw-Hill, Inc, New York.

Rha, C.K., 1978, Rheology of fluid foods, *Food Technology*, 7: 32–35.

Roos, Y.H. and Karel, M., 1991, Applying state diagrams to food processing and development, Food Technology, 45: 66, 68–71, 107.

Ross, Y.H., Karel, M., and Kokini, J.L., 1996, Glass transition in low moisture and frozen foods: effects on shelf life and quality, *Food Technology*, 50: 95–108.

Ross-Murphy, S.B., 1995a, Rheological characterization of gels, *Journal of Texture Studies*, 26: 391–400.

Ross-Murphy, S.B., 1988, Small deformation measurements. In: *Food Structure-Its Creation and Evaluation*, J.M.V. Blanshard and J.R. Mitchell Eds, Butterworth Publishing Co., London.

Ross-Murphy, S.B., 1995b, Structure-property relationships in food biopolymer gels and solutions, *Journal of Rheology*, 39: 1451–1463.

Roths, T., Marth, M., Weese, J., and Honerkamp, J., 2001, A generalized regularization method for nonlinear ill-posed problems enhanced for nonlinear regularization terms, *Computer Physics Communications*, 139: 279–296.

Rouse, P.E., 1953, *Journal of Chemical Physics*, 21: 1272–1280.

Safari-Ardi, M. and Phan-Thien, N., 1998, Stress relaxation and oscillatory tests to distinguish between doughs prepared from wheat flour of different varietal origin, *Cereal Chemistry*, 75: 80–84.

Seow, C.C., Cheah, P.B., and Chang, Y.P., 1999, Antiplasticization by water in reduced moisture food systems, *Journal of Food Science*, 64: 576–581.

Segre, G. and Silberberg, A., 1962, Behavior of macroscopic rigid spheres in Poseuille flow, *Journal of Fluid Mechanics*, 14: 115–135.

Shama, F. and Sherman, P., 1966a, Texture of ice cream 2. Rheological properties of frozen ice cream, *Journal of Food Science*, 31: 699–706.

Shama, F. and Sherman, P., 1966b, Texture of ice cream 3. Rheological properties of mix and melted ice cream, *Journal of Food Science*, 31: 707–716.

Shama, F. and Sherman, P., 1973, Identification of stimuli controlling the sensory evaluation of viscosity, II. Oral methods, *Journal of Texture Studies*, 4: 111–118.

Shama, F., Parkinson, C., and Sherman, P., 1973, Identification of stimuli controlling the sensory evaluation of viscosity, I. Non-oral methods, *Journal of Texture Studies*, 4: 102–110.

Sheth, B.B., 1976, Viscosity measurements and interpretation of viscosity data, *Journal of Texture Studies*, 7: 157–

Shrimanker, S.H., 1989, Evaluation of the Doi-Edwards theory for predicting viscoelastic properties of food biopolymer. M.S. Thesis, Rutgers University.

Skelland, A.H.P., 1967, *Non-Newtonian Flow and Heat Transfer*. Wiley, New York.

Slade, L. and Levine, H., 1987, Recent advances in starch retrogradation. In: *Industrial Polyscchar-ides*, S.S. Stivade, V. Crescenzi, and I.C.M. Deo, Eds, pp. 387–430. Gordon and Buech Science Publishers, NY.

Sliwinski, E.L., Kolster, P., and van Vliet, T., 2004a, Large-deformation properties of wheat dough in uni- and biaxial extension. Part I. Flour dough, *Rheologica Acta*, 43: 306–320.

Sliwinski, E.L., van der Hoef, F., Kolster, P., van Vliet, T., 2004b, Large-deformation properties of wheat dough in uni- and biaxial extension. Part II. Gluten dough, *Rheologica Acta*, 43: 321–332.

Smith, J.R., Smith, T.L., and Tschoegl, N.W., 1970, Rheological properties of wheat flour doughs III. Dynamic shear modulus and its dependence on amplitude, frequency and dough composition, *Rheologica Acta*, 9: 239–252.

Soskey, P.R. and Winter, H.H., 1984, Large step shear strain experiments with parallel-disk rotational rheometers., *Journal of Rheology*, 28: 625–645.

Sperling, L.H., 2001, *Introduction to Physical Polymer Science*, John Wiley and Sons, Inc, New York.

Spriggs, T.W., Huppler, J.P., and Bird, R.B., 1966, An experimental appraisal of viscoelastic models, *Transactions of the Society of Rheology*, 10: 191–213.

Steffe, J.F., 1996, *Rheological Methods in Food Process Engineering*, 2nd ed., Freeman Press, East Lansing.

Steffe, J.F., Mohamed, I.O., and Ford, E.W., 1986, Rheological properties of fluid foods: data compilation. In: *Physical and Chemical Properties of Food*, M.E. Okos, Ed., ASAE Publications.

Stevens, S.S., 1975, *Psychophysics, Introduction to Its Perceptual and Social Perspectives*, John Wiley & Sons, Inc., New York.

Suwonsichon, T. and Peleg, M., 1999a, Imperfect squeezing flow viscometry of mustards with suspended particulates, *Journal of Food Engineering*, 39: 217–226.

Suwonsichon, T. and Peleg, M., 1999b, Imperfect squeezing flow viscometry for commercial refried beans, *Food Science and Technology International*, 5: 159–166.

Suwonsichon, T. and Peleg, M., 1999c, Rheological characterization of almost intact and stirred yogurt by imperfect flow viscometry, *Journal of the Science of Food and Agriculture*, 79: 911–921.

Suwonsichon, T. and Peleg, M., 1999d, Rheological characterization of ricotta cheeses by imperfect flow viscometry, *Journal of Texture Studies*, 30: 89–103.

Svegmark, K. and Hermansson, A.M., 1991, Changes induced by shear and gel formation in the viscoelastic behavior of potato, wheat and maize starch dispersions, *Carbohydrate Polymer*, 15: 151–169.

Toledo, R., 1980, *Fundamentals of Food Process Engineering*, AVI Publishing Co., New York.

Toufeili, I., Lambert, I.A., and Kokini, J.L., 2002, Effect of glass transition and cross-linking on rheological properties of gluten: Development of a preliminary state diagram, *Cereal Chemistry*, 79: 138–142.

Treloar, L.R.G., 1975, *The Physics of Rubber Elasticity*, Claredon Press, Oxford.

Wagner, M.H., 1976, Analysis of time-dependent nonlinear stress-growth data for shear and elongational flow of a low-density branched polyethylene melt, *Rheologica Acta*, 15: 136.

Walters, K., 1975, *Rheometry*, John Wiley & Sons, Inc., New York.

Wang, C.F., 1995, Simulation of the linear and non-linear rheological properties of gluten and gliadin using the Bird–Carreau and the Wagner constitutive models, Ph.D. Thesis, Rutgers University.

Wang, C.F. and Kokini, J.L., 1995a, Prediction of the nonlinear viscoelastic properties of gluten doughs, *Journal of Food Engineering*, 25: 297–309.

Wang, C.F. and Kokini, J.L., 1995b, Simulation of the nonlinear rheological properties of gluten dough using the Wagner constitutive model, *Journal of Rheology*, 39: 1465–1482.

Wang, F.C. and Sun, X.S., 2002, Creep-recovery of wheat flour doughs and relationship to other physical dough tests and breadmaking performance, *Cereal Chemistry*, 79: 567–571.

Weese, J., 1991, *FTIKREG: A Program for the Solution of Fredholm Integral Equations of the First Kind, User Manual*, Freiburger Materialforschungszentrum, FRG.

Weese, J., 1992, A reliable and fast method for the solution of Fredholm integral equations of the first kind based on Tikhonov regularization, *Computer Physics Communications*, 69: 99–111.

White, J.L. and Metzner, A.B., 1963, Development of constitutive equations for polymeric melts and solutions, *Journal of Applied Polymer Science*, 7: 1867–1889.

Winter, H.H. and Chambon, F., 1986, Analysis of linear viscoelasticity of a crosslinking polymer at the gel point, *Journal of Rheology*, 30: 367–382.

Yamamoto, M., 1956, The viscoelastic properties of network structure I. General formalism, *Journal of the Physical Society*, 11: 413–421.

Yang, H.-H. and Manas-Zloczower, I., 1994, Analysis of mixing performance in a VIC mixer, *Internernational Polymer Processing*, IX: 291–302.

Zapas, L.J., 1966, Viscoelastic behavior under large deformations, *J. Res. Nat. Bur. Stds, 70A*, 525.

Zimm, B.H., 1956, Dynamics of polymer molecules in dilute solution viscoelasticity, flow birefringence and dielectric loss, *Journal of Chemical Physics*, 24: 269–278.

2 Reaction Kinetics in Food Systems

Ricardo Villota and James G. Hawkes

CONTENTS

2.1 INTRODUCTION

Advances in more efficient and versatile methods of food processing and preservation have been occurring exponentially over the past few decades in order to meet the continually increasing population and consumer demands for quality foods with particular focus on their nutritional aspects. The quality of processed foods depends not only upon the initial integrity of the raw materials but also on the changes occurring during processing and subsequent storage that may result in potential losses and decreased bioavailability. Emphasis is growing in the area of nutraceuticals and fortified high-energy foods (Giese, 1995; Molyneau and Lee, 1998; Sloan, 1999, 2005). Trends also indicate that fortification of food products has increased tremendously in the past years in multiple categories, including beverages, meals, biscuits, etc. Not only is the nutritional quality important to the food processor, but also the general appearance of the food, its flavor, color, and texture, factors which are highly dependent upon the target consumer. It is, therefore, of critical importance to the food industry to minimize losses of quality in food products during processing and subsequent storage. It is through the development of mathematical models to predict behavior of food components and optimization of processes for maximum product quality that continued advancement can be achieved. To obtain these goals, extensive information is needed on the rates of destruction of quality parameters and their dependence on variables such as temperature, pH, light, oxygen, and moisture content. A food engineer can then develop new processing techniques to achieve optimum product quality based on an understanding of reaction rates and mechanisms of destruction of individual quality factors combined with heat and mass transfer information. The need for this type of information has become critically important to the food industry with required nutritional labeling for food products.

Chemical kinetics encompasses the study of the rates at which chemical reactions proceed. The area of kinetics in food systems has received a great deal of attention in past years, primarily due to efforts to optimize or at least maximize the quality of food products during processing and storage. Moreover, a good understanding of reaction kinetics can provide a better idea of how to formulate or fortify food products in order to preserve the existing nutrients or components in a food system or, on the other hand, minimize the appearance of undesirable breakdown products. Unfortunately, limited kinetic information is available at present for food systems or ingredients that would facilitate the development of food products with improved stability or the optimization of processing conditions. A major consideration, however, is that indirectly some of the information available may be used to predict kinetic trends and, thus, establish major guidelines in formulation, storage, and process conditions. Thus, it is within the scope of this chapter to (a) present a general discussion on kinetics, outlining some of the fundamental principles and (b) provide information on a variety of food systems, indicating their reactivity and reported kinetic behavior. It is considered that a better understanding of kinetics in food systems will facilitate the development of a more complete and sound database.

It should be emphasized that the level of accuracy of kinetic data is dependent upon its final application. For instance, if mathematical models are developed to optimize the retention of a particular attribute in a given food system, it is evident that more sophisticated techniques are required than for those where the measurement is used for routine quality control. This is an area where careful judgement needs to be exerted. In fact, many of the major drawbacks existing in current kinetic data have originated in the analytical technique selected to compile kinetic information. Not only sensitivity but also selectivity in the assay procedure needs to be taken into account when monitoring individual compounds in highly complex food systems.

There are three main areas of concern when dealing with reaction kinetics: (1) the stoichiometry, (2) the order and rate of reaction, and (3) the mechanism. For simple reactions, the stoichiometry is probably the first consideration. Once this is clarified or elucidated, the mechanisms involved in the reaction are determined. It should be mentioned that based on kinetic data, our idea of the stoichiometry may change. In highly complex reactions, as in the case of many reactions occurring in food systems, a great deal of overlap exists among the three aforementioned areas. Thus, it is of critical importance to take a close and analytical look at the overall system to be able to characterize reaction pathways.

2.2 BASIC PRINCIPLES OF KINETICS

2.2.1 ORDER OF REACTION

Of particular significance to the area of kinetics, is the determination of the order of the reaction. Understanding of the mechanisms involved in the reaction is important to properly obtain and report meaningful kinetic information, select reaction conditions leading to a desired end product, and/or minimize the appearance of undesirable compounds. Unfortunately, very seldom has effort been made to clearly understand the mechanisms involved in the reaction in complex systems, as in the cases with food and biological materials. Most information available has been oversimplified. In fact, most investigators have often tried to adapt fairly simple zero- or first-order reaction kinetics to complex situations without trying to understand the actual pathways involved. Although from a practical point of view it is clear that simplifications may be taken, applicability of the information may be restricted only to the conditions encompassed by the experimental design, and thus, one may incorrectly predict trends by directly extrapolating reported information.

The reaction pathway, also called reaction mechanism, may be determined through proper experimentation. A chemical reaction may take place in a single step, as in the case of elementary reactions, or in a sequence of steps, as would be the case of most reactions occurring in food systems. Conditions such as temperature, oxygen availability, pressure, initial concentration, and the overall composition of the system may affect the mechanism of the reaction. For instance, the degradation of folic acid and ascorbic acid can be affected by the presence of oxygen, resulting in modification of the reaction pathway and, thus, the type of breakdown products. Moreover, the rate at which these parent compounds disappear may be highly influenced by the presence and concentration of the breakdown products generated. It is true that the level of complexity involved in these reactions may be of such magnitude that a complete understanding of the mechanism of deterioration cannot always be easily determined or identified or, worse, may hinder the development of simple techniques to rapidly evaluate the stability of a given system. Nevertheless, it should be stressed that more reliable information is obtained when understanding of the reaction pathways is achieved.

A basic approach for the determination of the reaction order for a simple reaction, taking into consideration its initial rate is as follows:

$$-\frac{dC}{dt} = kC^n \tag{2.1}$$

which after taking the natural logarithm on both sides of the equation results in:

$$\ln\left(-\frac{dC}{dt}\right) = \ln k + n \ln C \tag{2.2}$$

where C is the concentration, k is the reaction rate constant, n is the order of the reaction, and t is time. According to this approach, a plot of the $\ln(-dC/dt)$ vs. $\ln C$ will give a straight line, whose slope corresponds to the reaction order (n), as shown in Figure 2.1. Although the intercept should correspond to the reaction rate constant (k), it is normally considered that, for the sake of accuracy, this would not be the preferred approach for its estimation. Rather, once the reaction order has been determined, the rate constant can be calculated by applying the corresponding equation for that reaction order.

The method of least squares can also be used to determine the order of the reaction with respect to the reactants and products involved in the reaction. For instance, for a given reaction $A + B \rightarrow P$, where A and B are the reactants and P is the product, the reaction rate (r) can be defined by an equation such that:

$$-r = k[A]^a[B]^b \tag{2.3}$$

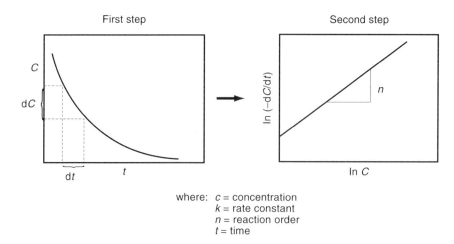

where: c = concentration
        k = rate constant
        n = reaction order
        t = time

FIGURE 2.1   Graphical representation for determining reaction order (n) of a reaction.

where [A] and [B] are the respective concentrations, and a and b are the respective reaction orders of the reactants A and B. By taking the natural log on each side, the equation becomes:

$$\ln(-r) = \ln k + a \ln[A] + b \ln[B] \tag{2.4}$$

which is of the form:

$$y = a_0 + a x_1 + b x_2 \tag{2.5}$$

where a_0, a, and b are constants. Thus by the least squares approach, the coefficients a and b, corresponding to the order of the reaction, can be determined.

For the particular case of complex reactions, such as in the case of lipid oxidation, many investigators have tried to apply simple reaction kinetics to describe their behavior. In this particular situation it is obvious that a clear understanding of the overall chain reaction including initiation, propagation and termination stages may reach levels of complexity, totally undesirable from a practical point of view and extremely difficult to express in simple mathematical terms. Complex reaction kinetics will be discussed in a subsequent section.

2.2.2 REACTION RATE

When dealing with food systems, a common approach to report reaction rates is as the change in concentration of a reactant as a function of time. The reaction rate thus provides a measurement of the reactivity and stability of a given system. A number of variables have been observed to influence the reaction rate. Major factors include: (a) concentration of reactants, products, and catalysts; (b) environmental factors such as temperature, pressure, and oxygen availability; (c) wavelength and intensity of light; and (d) physicochemical properties such as viscosity, ionic strength, and conductivity. Depending upon the type of reaction and the components, other factors will also be influential in controlling reaction kinetics.

Although traditionally one can apply reaction kinetics to monitor chemical changes occurring in a system, other physicochemical changes may also be described using a kinetic approach. For instance, textural and color changes occurring in food systems can be described using reaction rates. It is obvious that the numbers obtained represent the final effect caused by other complex reaction mechanisms leading to an overall result. For instance, color changes in a product containing carotenoids may be an indication of the stability of the system, and in particular, stability of the carotenoids as related to environmental conditions. Another example is the textural changes in starch-based systems as a function of time, which may be the result of starch retrogradation mechanisms as well as lipid-amylose interactions as influenced by environmental conditions.

How to properly measure the changes occurring in a system as influenced by different factors becomes a key issue. Changes may be measured by monitoring, for instance, the disappearance of a compound, the appearance of a breakdown product, or changes in the physicochemical properties of the system such as in thermal conductivity. Colored species may be monitored through their appearance and disappearance by using spectrophotometric techniques. Depending very much on the final application of the kinetic data, one may need to actually monitor the reaction rate in such a way that the reaction does not proceed to any significant extent during the analytical test. Quenching of the reaction may be accomplished in many different ways, such as by lowering the temperature of the system or by addition of the reaction mixture to a system that provides stability. For most cases in food systems, the rates of chemical reactions that proceed slowly can be easily studied through convenient methods. Since nutrient retention is of primary concern in food systems subject to deleterious conditions, a great deal of attention has been given to the study of vitamin degradation. A number of different techniques have been developed for their analyses, a summary of which is presented in Table 2.1. Many of these techniques, however, suffer from the lack of differentiation of intermediate compounds that may form, resulting in erroneous conclusions. For instance, the transformation of

---

TABLE 2.1
Methods Used for Vitamin Assays

Vitamin C:	Indophenol Method (Using Tillman's Reagent — titration or photometric); monitors color changes due to 2,6-dichloroindophenol reduction
	2,4-dinitrophenylhydrazine (DNPH), monitors total ascorbic acid (AA) based on oxidation of AA to dehydroascorbic acid (DHAA) followed by coupling with DNPH to form red-colored osazones
	Microfluorometric (based on reaction of DHAA with o-phenylenediamine)
	HPLC (direct measurement)
	Polarographic
	Oxidation-Reduction Methods (with iodine, bromine, iron, copper, mercury or selenious acid)
	Capillary Zone Electrophoresis (separation of L- and D-isoascorbic acids)
Vitamin B_1:	Fluorometric Thiochrome (alkaline oxidation of thiamine to fluorescent thiochrome)
	HPLC (direct measurement)
	Microbiological (*Lactobacillus viridescens* [intact thiamine-specific])
	Animal (rat, pigeon, chick)
Vitamin B_2:	Fluorometric (native fluorescence of riboflavin)
	Polarographic
	Microbiological (*Lactobacillus casei* [non-specific]; *Tetrahymena pyriformis* [B_2-specific])
	HPLC (direct measurement)
	Animal (rat, chick)
Vitamin B_6:	Microbiological (*Saccharomyces uvarum*)
	HPLC (fluorescence detection)
	Animal (chick and rat growth)
Vitamin B_{12}:	Microbiological (*Lactobacillus leichmennii*)
	Radioassays (competitive inhibition assay based on isotope-dilution principle)
	Protozoan assays (*Euglena gracilis*, *Ochromonas malhamensis*)
Folates:	Microbiological (*Lactobacillus casei*, *Streptococcus faecalis*, *Pediococcus cerivisiae*)
	Radiometric Assay (competitive binding radioassay using liquid or dry skim milk as binder)
	Electrophoretic (polyglutamyl chain-length determination)
	HPLC (direct measurement)

**TABLE 2.1
Continued**

	Stable isotope LC-MS method
	Polarographic
	Animal (chick and rat)
Niacin:	Microbiological (*Lactobacillus plantarum*)
	Radiometric Assay
	HPLC (direct measurement)
	Spectrophotometric:
	based on König reaction (niacin + cyanogen bromide → pyridinium compound →
	+ aromatic amine → glutaconic dialdehyde derivative [colored])
	niacinimide in potassium dihydrogen phosphate + cyanogen bromide + barbituric acid →
	purple color
	Metabolite Measurement:
	N^1-methyl nicotinamide (NMN) and 6-pyridone of N^1-methyl nicotinamide
	NMN + ketones in alkali solution → green fluorescent compound
	Animal (dog, chick, weanling rat)
Pantothenic acid:	Fluorometric: alkaline hydrolysis + *o*-phthalaldehyde + 2-mercaptoethanol in boric acid
	solution → fluorogenic compound
	Microbiological (*Lactobacillus plantarum, Saccharomyces carlsbergensis, S. cerivisiae*)
	Radioimmunoassay
	Enzyme-linked immunosorbent assay (ELISA)
	Animal (chick)
Vitamin A:	Spectrophotometric
(Carotenoids):	Fluorometric
	HPLC (direct measurement)
	Supercritical Fluid Chromatography (direct measurement)
Vitamin D:	Spectrophotometric (UV)
	Colorimetric (Vit. D + $SbCl_3$ in ethylene dichloride → pink complex)
	HPLC
	GLC
	Animal (rat [line test]; chick [bone ash])
Vitamin E:	Spectrophotometric (UV)
	Colorimetric (based on Emmerie & Engel's Reaction:
	tocopherols + bathophenanthroline + $FeCl_3$ → + orthophosphoric acid → pink-colored
	complex)
	Spectrofluorometric
	HPLC (direct measurement)
	GLC (native or trimethylsilyl derivatives)
	Animal (rat, chick, duckling)
Vitamin K:	GLC
	Spectrophotometric
	Reduction-Oxidation Method
	Animal (chick prothrombin time determinations)
	Ethylcyanoacetate Method
	2,4-Dinitrophenylhydrazine Method
Biotin:	Microbiological (*L. plantarum*)
	Animal (chick, rat)
	Radiometric (not for general use)
	Fluorometric (for non-biological materials)
	HPLC/avidin binding assay

trans-carotenoids into their *cis*-form is difficult to detect unless a very specific technique is utilized. Since iomerization causes losses of the provitamin activity, it is critical to be able to identify the different isomers. Most techniques utilized thus far to monitor kinetics of carotenoid degradation have not taken this factor into account. A great deal of effort has recently been directed towards more conclusive methods such as high-performance liquid chromatography (HPLC) and supercritical fluid chromatography, which afford separation of the individual compounds, thus enabling the collection of proper kinetic information to elucidate mechanisms of deterioration and characterize kinetic parameters. Other methods include the use of liquid chromatography in combination with mass spectrometry (LC-MS) for the quantitative determination of 5-methyltetrahydrofolic and folic acids (Pawlosky and Flanagan, 2001; Thomas et al., 2003), and capillary zone electrophoresis for the separation of *L*-ascorbic and *D*-isoascorbic acids (Liao et al., 2000).

Since reactions in food systems are normally complex and are a combination of several elementary steps, additional basic information may be necessary to postulate reaction rate expressions. Identification of intermediates and previous knowledge of rate equations to fit data for other systems may provide assistance in properly characterizing a given reaction. Additional factors that may affect reaction rates are the type of energy and/or conditions surrounding the process to which the food is subjected such as those associated with nontraditional techniques including high pressure, irradiation, ohmic, and pulsed electric field processing.

In the following paragraphs, a short summary of the mathematical description of concentration vs. time for single irreversible and complex reactions is presented. The most commonly found reactions in food and biological systems are the zero-, first-, and second-order reactions.

2.2.3 TYPES OF REACTIONS

2.2.3.1 Zero-Order Reactions

In zero-order reactions, the rate is independent of the concentration. This may occur in two different situations: (a) when intrinsically the reaction rate is independent of the concentration of reactants and (b) when the concentration of the reacting compound is so large that the overall reaction rate appears to be independent of its concentration. Many catalyzed reactions fall in the category of zero-order reactions with respect to the reactants. On the other hand, the reaction rate may depend upon the catalyst concentration or other factors unrelated to the concentration of the compound under investigation.

Thus for a zero-order reaction at constant density, the overall expression would be as follows:

$$-\frac{dC}{dt} = k_0 \tag{2.6}$$

where C is the concentration, t is time, and k_0 is the zero-order reaction rate constant, which by integration would result in:

$$C_0 - C = k_0 t \tag{2.7}$$

where C is the concentration at time (t), and C_0 is the initial concentration. According to this mathematical expression, a distinguishing feature of this type of reaction is a linear decrease in concentration as a function of time, as illustrated in Figure 2.2.

Typical reactions that have been represented by zero-order reactions include some of the autooxidation and nonenzymatic browning reactions. It is clear that zero-order reactions do not appear to occur as frequently in food systems as other reaction orders. In most cases, it is evident that the most common situation for this type of reaction is when the concentration of the reactants is so large that the system appears to be independent of concentration.

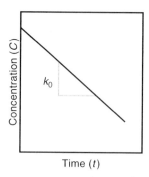

FIGURE 2.2   Graphical representation for the determination of a zero-order rate constant (k_0).

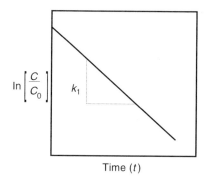

FIGURE 2.3   Graphical representation for the determination of a first-order rate constant (k_1).

2.2.3.2 First-Order Reactions

A large number of reactions occurring in food systems appear to follow a first-order reaction. A mathematical expression for this behavior would be as follows:

$$-\frac{dC}{dt} = k_1 C \tag{2.8}$$

where k_1 is the first-order reaction rate constant. By integration, this equation becomes:

$$-\ln\left(\frac{C}{C_0}\right) = k_1 t \tag{2.9}$$

Thus, according to this mathematical expression $\ln C$ vs. time will be a linear function where the slope corresponds to $-k_1$ as shown in Figure 2.3.

The half-life ($t_{1/2}$) is given by:

$$k_1 t_{1/2} = -\ln(1/2) \tag{2.10}$$

$$t_{1/2} = \ln 2 / k_1 \tag{2.11}$$

The mathematical expressions above clearly indicate that the half-life and the reaction rate for a true first-order reaction are independent of the initial concentration. However, although in a number of systems this may be the case, formulated products will not follow true first-order reaction kinetics, but rather a pseudo-first-order reaction. In fact, in formulated systems the presence of breakdown products may strongly influence the order of the reaction. However, for only a given value of initial concentration the reaction may follow apparent first-order kinetics. To determine if a given reaction does indeed follow a pseudo-first-order kinetics, conditions for the kinetic study can be chosen to follow the technique of flooding. Through this approach, all but one of the concentrations are set sufficiently high such that compared to the one reagent present at lower concentration the others are effectively constant during the time of the experiment. Since only one of the concentrations changes appreciably during the run, the effective kinetic order is reduced to the reaction order with respect to that one substance. If the order of the reaction is determined to be one, the reaction is said to follow a pseudo-first-order reaction. The degradation of ascorbic acid, for instance, has been primarily found to follow first-order kinetics in food systems. On the contrary, degradation of ascorbic acid in model systems has frequently been found to follow pseudo-first-order kinetics. It appears that the presence of breakdown products modifies the kinetics of deterioration of ascorbic acid and, thus, its initial concentration will influence its rate of degradation.

2.2.3.3 Second-Order Reactions

Two types of second-order reaction kinetics are of importance:

$$\text{Type I:} \quad A + A \rightarrow P$$

$$-\frac{dC_A}{dt} = k_2 C_A{}^2 \tag{2.12}$$

and

$$\text{Type II:} \quad A + B \rightarrow P$$

$$-\frac{dC_A}{dt} = k_2 C_A C_B \tag{2.13}$$

where C_A is the concentration of reactant species (A) at time (t), C_B is the concentration of reactant species (B) at time (t), and k_2 is the second-order reaction rate constant. For Type I, the integrated kinetic expression yields:

$$\frac{1}{C_A} - \frac{1}{C_{A_0}} = k_2 t \tag{2.14}$$

which in terms of the half-life becomes:

$$t_{1/2} = \frac{1}{k_2 C_{A_0}} \tag{2.15}$$

For Type II, the integrated form yields:

$$k_2 t = \frac{1}{C_{A_0} - C_{B_0}} \ln \left(\frac{C_{B_0} C_A}{C_{A_0} C_B} \right) \tag{2.16}$$

where C_{A_0} and C_{B_0} are the respective initial concentrations and C_A and C_B are the respective concentrations at time (t). It should be stressed, however, that Type II reactions do not necessarily have to follow a second-order reaction. For instance, for the particular case where component A is present in large amounts as compared with component B, the reaction may follow first-order kinetics with respect to B. A typical plot of second-order kinetics is presented in Figure 2.4.

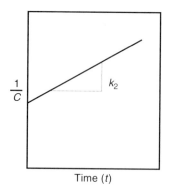

FIGURE 2.4 Graphical representation for the determination of a second-order rate constant (k_2) for a Type I reaction.

2.2.3.4 Nonelementary Reactions

To characterize the kinetics of nonelementary reactions, one can assume a series of individual elementary reactions taking place. In these reactions, intermediates may not be observed or quantitated, either because they are present in very small amounts or because they are unstable. Such reactions would fall under three main categories: (1) consecutive or series reactions, (2) reversible or opposing reactions, which attain a finite equilibrium, and (3) parallel or competitive reactions. The types of intermediates postulated may fall in any one of the following categories, namely, (a) free radicals, (b) ions and polar substances, (c) molecules, and (d) transition complexes (chain reactions and nonchain reactions). The following are examples of the various mechanisms proposed.

2.2.3.4.1 Intermediates
2.2.3.4.1.1 Free radicals
Free atoms or fragments of stable molecules containing one or more unpaired electrons are called free radicals. For these compounds, a standard convention is to designate the unpaired electron by a dot ($\cdot$). Some of these radicals are stable as in the case of ascorbic acid, where its hydroxyl on the C-3 readily ionizes ($pK_1 = 4.04$ at 25°C) and may undergo degradation to dehydroascorbic acid in the presence of oxygen. Other free radicals are unstable as in the case of lipid oxidation. For instance, when a hydroperoxide decomposes to form $RO\cdot$ radicals, such compounds can participate in other reactions, thus being capable of continuing the chain propagation process and forming several products. Hydroxy acids, keto acids, and aldehydes have been isolated from oxidizing lipid systems. The formation of protein free radicals in systems undergoing lipid oxidation is another example of reactions involving unstable radicals. Free radicals may form on the α-carbons of the proteins, while cysteil radicals may form in proteins containing cysteine or cystine (Karel, 1973).

2.2.3.4.1.2 Ions and polar substances
Electrically charged atoms, molecules or fragments of molecules such as Na^+, NH_4^+, I^- and NO_2^- are called ions, which may serve as reactive intermediates in a variety of reactions.

2.2.3.4.1.3 Molecules
In reactions such as: $A \rightarrow B \rightarrow C$, where compound B is highly reactive or its concentration in a given reaction mixture is very small, such compound B acts as an intermediate.

2.2.3.4.1.4 Transition Complexes

Chain-reactions
Catalyzed reactions may fall in the category of chain reactions. In these reactions an intermediate is formed in the initiation step. Such an intermediate then interacts with the reactant to obtain a product

and more intermediate to participate in the reaction. A key consideration is that the intermediate may catalyze a series of reactions before being destroyed.

A classical example for the case of chain reactions is the degradation of β-carotene, an autooxidation reaction involving three main periods: (1) induction (formation of free radicals), (2) propagation (free radical-chain reactions), and (3) termination (formation of nonradical products). According to this pathway, the reaction may be expressed as follows:

$$\left.\begin{array}{ll} 2RH & \xrightarrow{k_i} \quad 2R^{\cdot} \\[2mm] R^{\cdot} + O_2 & \underset{}{\overset{k}{\rightleftharpoons}} \quad RO_2^{\cdot} \end{array}\right\} \quad (1)\ \text{induction}$$

$$\left.\begin{array}{ll} RO_2^{\cdot} + RH & \xrightarrow{k_p} \quad R^{\cdot} + ROOH \\[2mm] R^{\cdot} + R'H & \xrightarrow{k_p'} \quad R^{\cdot} + RH \end{array}\right\} \quad (2)\ \text{propagation}$$

$$\left.\begin{array}{ll} R^{\cdot} + RO_2^{\cdot} & \xrightarrow{k_t} \quad products \\[2mm] R^{\cdot} + R^{\cdot} & \xrightarrow{k_t'} \quad products \end{array}\right\} \quad (3)\ \text{termination}$$

Although different approaches have been taken to mathematically describe the kinetic behavior of β-carotene, a simplified free-radical recombination has been suggested by a number of authors (Alekseev et al., 1968; Gagarina et al., 1970: Finkel'shtein et al., 1973, 1974).

According to this approach, the rate of consumption of a hydrocarbon, that is, carotenes, in a chain process with a second-order chain termination can be described by:

$$-\frac{dC}{dt} = aC\sqrt{w_i} \tag{2.17}$$

By replacing the value corresponding to w_i, the rate of formation of free radicals, the rate of consumption of the hydrocarbon can be expressed as being:

$$-\frac{dC}{dt} = aC\sqrt{b_0C + b(C_0 - C)} \tag{2.18}$$

where C is the carotenoid concentration at time (t), C_0 is the initial carotenoid concentration, a is the constant derived from Equation 2.19, b is the initiation rate constant of the products, b_0 is the initiation rate constant of unoxidized carotenoids, and t is time.

The initiation rate is considered to be the sum of the initiation rates of radicals formed by the unreacted carotene and by the intermediate products. The value of constant "a" can be represented by:

$$a = \frac{k_p}{\sqrt{k_t}}\sqrt{kK_SP_{O_2}} \tag{2.19}$$

where k_p, k_t, and k are rate constants as previously shown; K_S is the solubility coefficient of oxygen in carotenoids; and P_{O_2} is the partial pressure of oxygen. Integrating Equation 2.19 using the dimensionless variables suggested by Gagarina et al. (1970) yields:

$$\ln\left(\frac{1 + \sqrt{1 - C/C_0}}{1 - \sqrt{1 - C/C_0}}\right) = a\sqrt{(b - b_0)}\sqrt{C_0}t \tag{2.20}$$

If all the constants on the right side of the previous equation are lumped together to denote the effective rate constant, namely (σ),

$$\sigma = a\sqrt{(b - b_0)}\sqrt{C_0} \qquad (2.21)$$

Equation 2.20 can be simplified to yield:

$$\ln\left(\frac{1 + \sqrt{1 - C/C_0}}{1 - \sqrt{1 - C/C_0}}\right) = \sigma t \qquad (2.22)$$

which corresponds to the equation for a straight line. It is evident that the slope of a graph of $\ln[(1 + \sqrt{1 - C/C_0})/(1 - \sqrt{1 - C/C_0})]$ vs. t will give the value corresponding to the effective rate constant.

The aforementioned simplified models have been successfully used by various investigators to describe the reaction kinetics of carotenoids assuming a scheme of an unbranched chain as previously described. It is obvious that in systems where oxygen accessibility is limited due to the density of the material, higher rates of oxidation will take place close to the surface. Hence, different rates of degradation of the carotenoids will take place simultaneously, highly complicating the analysis of the kinetics for the overall system.

Nonchain reactions

Nonchain catalyzed reactions may involve the interaction of the substrate with the catalyst to form a complex, followed by its decomposition to form the product. Upon decomposition, the catalyst is then regenerated and is capable of taking part in the reaction once again. A typical example of this behavior is the acid-catalyzed hydrolysis of pyranosides according to the pathways presented in Figure 2.5. According to this diagram, the mechanism involves rapid reversible protonation of the glycosidic oxygen atom to produce a protonated oligosaccharide (2), which undergoes a slow unimolecular decomposition to a stable monosaccharide and an acyclic carbonium ion (3). It is

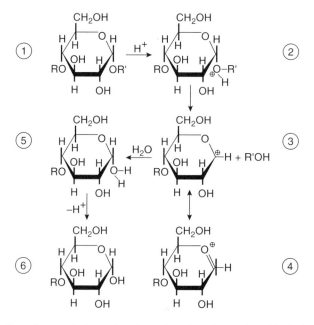

FIGURE 2.5   Illustration of a non-chain catalyzed reaction: the acid-catalyzed hydrolysis of pyranosides.

considered that the carbonium ion is stabilized by resonance with the oxonium ion (4). Nucleophilic addition of water would yield a protonated reducing sugar (5), which through the loss of a proton would result in the hydrolytic products (6) and reappearance of the catalyst.

Enzyme kinetics

Another example of transition complex reactions is that of enzyme-catalyzed reactions. It should be mentioned that most of the reactions occurring in biological systems are catalytic in nature. The basic principles for enzyme-catalyzed reactions have been presented by Michaelis–Menten, who proposed the theory of complex formation according to the following equation:

$$E + S \quad \underset{k_{-1}}{\overset{k_1}{\rightleftarrows}} \quad ES \quad \underset{k_{-2}}{\overset{k_2}{\rightleftarrows}} \quad E + P$$

where E is the enzyme, S is the substrate, ES is the enzyme-substrate complex, and P is the product. Both reactions are considered to be reversible. In this equation, k_1, k_{-1}, k_2, and k_{-2} are the specific constants for the designated reactions.

Although the general principle of chemical kinetics may apply to enzymatic reactions, the phenomenon of saturation with substrate is unique to enzymatic reactions. In fact, at low substrate concentrations the reaction velocity is proportional to the substrate concentration, and thus the reaction is first-order with respect to the substrate. As the substrate concentration increases, the reaction progressively decreases, being no longer proportional to the concentration of the substrate and deviating from any first-order kinetics. The reaction follows zero-order reaction kinetics, due to saturation with the substrate (Figure 2.6).

For the particular case of enzyme kinetics, the cases of competitive and noncompetitive inhibition need to be considered. In the first case, the competitive inhibitor (I) is able to interact with the enzyme to generate a complex (EI), according to the reaction:

$$E + I \quad \underset{k_{-1}}{\overset{k_1}{\rightleftarrows}} \quad EI$$

In this type of reaction, the complex EI does not break down to create products. However, the reaction can be reversed by increasing the substrate concentration.

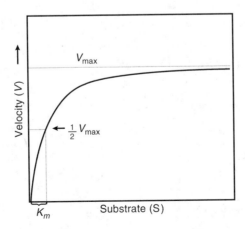

FIGURE 2.6 Illustration of reaction rate dependence on substrate concentration for the initial phase of an enzyme-catalyzed reaction, indicating maximum velocity (V_{max}), $\frac{1}{2} V_{max}$ and Michaelis constant (K_m).

On the other hand, in the case of noncompetitive inhibition, the inhibitor may bind to the enzyme on a locus different from the active site of the enzyme, and thus, it may bind to the free enzyme or to the complex according to the equations presented below:

$$E + I \; \leftrightarrows \; EI$$

$$ES + I \; \leftrightarrows \; ESI$$

where the forms EI and ESI are inactive.

The most common type of competitive inhibition is that created by compounds that bind reversibly with the sulfhydryl groups of cysteine residues that are essential for the catalytic activity of some enzymes. Such groups may be located at or near the active site. In this second case, the catalytic activity may be related to stearic hindrance or the ability to maintain the three-dimensional conformation of the enzyme.

It is of critical importance to be able to identify competitive and irreversible inhibition. For the particular case of irreversible inhibition, the inhibitor binds to the enzyme irreversibly, and some may modify its molecular structure. It is obvious that for this particular case the use of the Michaelis–Menten equation is not possible since this approach assumes that the interaction between the enzyme and the inhibitor is reversible. A typical case of irreversible reactions would be the case of the trypsin inhibitors found in soybeans, namely, the Kunitz and the Bowman–Birk inhibitors. Chymotrypsin has been found to be strongly inhibited by the Bowman–Birk inhibitor, while only weakly inhibited by the Kunitz inhibitor. Both inhibitors have also been shown to be active against bovine trypsin. The activity of human trypsin has been observed to be inhibited to a significant extent by the Kunitz inhibitor.

2.2.3.5 Consecutive Reactions

Consecutive reactions form another category of reactions of importance in food products. Intermediates are formed in such reactions that may decompose or react to create other compounds. In many cases, the intermediate may have a short life, and thus, simplifications may be taken to describe their kinetics. Since decomposition of the intermediates may proceed under different reaction kinetics, complex situations may arise.

For the particular case of reactions in sequence, following first-order or pseudo first-order kinetics:

$$A \; \xrightarrow{k_1} \; B \; \xrightarrow{k_2} \; C$$

and for the particular case where the aforementioned reactions are not reversible, the rate of disappearance of A can be expressed as follows:

$$-\frac{d[A]}{dt} = k_1[A] \tag{2.23}$$

which becomes:

$$[A] = [A_0] \exp(-k_1 t) \tag{2.24}$$

The concentration of the intermediate can be determined by:

$$\frac{d[B]}{dt} = k_1[A] - k_2[B] \tag{2.25}$$

By substituting Equation 2.24 into Equation 2.25 and multiplying each term by exp (k_2t), the following expression is derived:

$$\exp(k_2t)\left(\frac{d[B]}{dt}\right) + [\exp(k_2t)][B]k_2 = k_1[A_0][\exp(k_2 - k_1)t] \tag{2.26}$$

Integration of this equation with $[B] = 0$ at $t = 0$ yields:

$$[B] = \frac{k_1[A_0]}{k_2 - k_1}[\exp(-k_1t) - \exp(-k_2t)] \tag{2.27}$$

It can be easily illustrated that the final product (C) does not form immediately as A is decomposed due to the formation of B. This period is normally termed an *induction period*.

It is evident that the aforementioned case is a simple case for these particular types of reactions. More complicated situations correspond to the cases for consecutive reactions with a reversible step or when the individual steps do not follow the same order reaction kinetics. If the rate of disappearance of B is fairly rapid, it is evident that if one monitors the appearance of C only based on the concentration of A, inaccuracies may be introduced in the kinetic analysis.

2.2.3.6 Reversible First-Order Reactions

For the most part, we have considered reactions whose rate constant consists of a single value with an integral reaction order. However, now we will consider the reaction:

$$A \underset{k_{-1}}{\overset{k_1}{\rightleftarrows}} B$$

in which the rate of disappearance of A is given by:

$$-\frac{d[A]}{dt} = k_1[A] - k_{-1}[B] \tag{2.28}$$

To solve this equation, two considerations are to be taken into account: (1) from the stoichiometry:

$$[A_0] + [B_0] = [A_\infty] + [B_\infty] = [A] + [B] \tag{2.29}$$

and (2) from the condition $-d[A]/dt = 0$ at equilibrium:

$$k_1[A_\infty] = k_{-1}[B_\infty] \tag{2.30}$$

Through substitution and rearrangement:

$$-\frac{d[A]}{dt} = (k_1 + k_{-1})([A] - [A_\infty]) \tag{2.31}$$

Integration between the corresponding limits will give:

$$\ln\left(\frac{[A] - [A_\infty]}{[A_0] - [A_\infty]}\right) = -(k_1 + k_{-1})t \tag{2.32}$$

According to this equation, a plot of $\ln([A] - [A_\infty])$ vs. time will be a straight line, whose slope corresponds to $-(k_1 + k_{-1})$. Figure 2.7 describes the concentration of reactant A as a function of time for two different situations. In the first case, the reaction will reach a certain equilibrium with

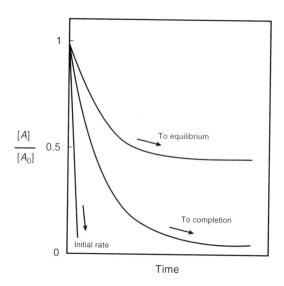

FIGURE 2.7   Illustration of time dependence of a hypothetical reversible first-order reaction.

retention values leveling off. On the other hand, if a reactant is added to rapidly consume B, this will prevent its return to A. In this situation only the forward reaction controls $-d[A]/dt$ with a continuous depletion of reactant A. For both cases, the initial rate of the reaction is the same. Although this type of behavior may be possible in food systems, as in the interconversion of pyridoxamine and pyridoxal, where these compounds may undergo further degradation, most information reported in the literature will consider the rate for the overall reaction with no simultaneous information for the forward and the reverse reactions. Huang and von Elbe (1985), however, developed a kinetic reaction model accounting for the forward and reverse reactions for the degradation and regeneration of betanine in solution and its degradation products, betalamic acid and cyclodopa-5-O-glycoside. This model could predict the amount of betanine remaining before and after regeneration of the pigment under different experimental conditions. Since in most practical situations dealing with food products the majority of investigators have not explored the mechanisms of degradation of the compound under question to minimize the amount of work involved, it becomes simple to visualize the limitations of available kinetic information if one tries to extrapolate to other systems.

2.2.3.7   Simultaneous Competitive Reactions

In reference to complex reactions, another case of significance corresponds to the degradation of a single compound to different products, following different pathways. The degradation of ascorbic acid is a typical example of a complex reaction where several pathways may operate simultaneously. Seldom have investigators taken the time to identify the contribution of each pathway to the degradation of this vitamin but rather have reported an overall value. In the case of chlorophylls, Heaton et al. (1996a, 1996b) developed a general mechanistic model for rates of chlorophyll degradation to pheophorbides via either pheophytins or chlorophyllides. Depending upon the contribution of each pathway when dealing with competitive reactions, variable levels of inaccuracy may result since each reaction will proceed at a different rate. Thus if the reaction is expressed by the following mechanism:

$$A \begin{cases} \xrightarrow{k_1} & P_1 \\ \xrightarrow{k_2} & P_2 \\ \xrightarrow{k_n} & P_n \end{cases}$$

where $P_1, P_2, \ldots, P_n$ = products, $k_1, k_2, \ldots, k_n$ = rate constants. The rate of disappearance of A is given by:

$$-\frac{d[A]}{dt} = (k_1 + k_2 + \cdots + k_n)[A] \tag{2.33}$$

which by integration results in:

$$[A] = [A_0]\exp\left(-\sum k_n t\right) \tag{2.34}$$

where $\sum k_n = k_1 + k_2 + \cdots + k_n$.

Since the products are generated with different yields, it is obvious that information on product formation is needed to evaluate the rates of the reaction. Moreover, it is evident that solely monitoring the rate of disappearance of A and assigning an overall reaction rate may be highly inaccurate, since each pathway may have a different reaction order and a different associated rate constant. This approach has been commonly taken by many investigators in determining reaction rates for the degradation of vitamins, pigments, etc. This discussion should emphasize the need for a better understanding of the reaction mechanisms to properly report kinetics. Since this is a more time-consuming approach, it is understandable why many investigators circumvented the complexity of the reaction kinetics.

2.2.4 EFFECT OF TEMPERATURE

When considering reaction rates, it is clear that these values may be influenced by a large number of parameters, including temperature and pressure. In fact, equilibrium yields, chemical reaction rates, and product distribution may be drastically influenced by temperature. Since chemical reactions are accompanied by heat effects, if these are large enough to cause a significant change in temperature of the reaction mixture, these effects also need to be taken into consideration. This would be particularly important in reactor design. The effect of temperature for an elementary process may follow, in most cases, the Arrhenius equation:

$$k = k_o e^{-E_a/RT} \tag{2.35}$$

where k_o is the frequency or collision factor; E_a is the activation energy; R is the gas constant (1.987 cal/mol $\cdot$ $^\circ$K); and T is the absolute temperature ($^\circ$K). It is obvious that if the frequency factor and the activation energy could be evaluated from the molecular properties of the reactants, it would be possible to estimate the values corresponding to the reaction rate. Unfortunately, our knowledge of kinetics is limited, particularly for complex systems, as would be the case of food systems or products.

It is, however, important to mention the collision theory as an approach to deal with kinetics. In Figure 2.8, the energy levels involved in a reaction are illustrated. According to the collision theory, upon the collision of reactive molecules, enough energy is generated to provide the necessary activation energy. Such a theory was used as the foundation for the determination of rate expressions based on the frequency of molecular collision required to generate a minimum energy.

Another theory, the activated-complex or transition-state theory, has also been suggested. According to this approach which still relies on reactions occurring due to collision between reactive molecules, an activated complex is formed from the reactants, which eventually decomposes to generate products. The activated complex is in thermodynamic equilibrium with the reactants. Complex decomposition is then the limiting step. Regardless of the theory considered, these approaches

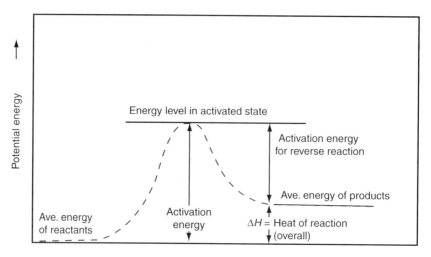

FIGURE 2.8 Representation of potential energy levels during the process of a given endothermic reaction.

do not provide the means to rapidly and easily calculate activation energies from simple thermo-dynamic information. Thus in practical terms, one has to obtain basic kinetic information to be able to determine the effect of temperature as affecting reaction kinetics. Based on the Arrhenius equation it is clear that if one plots the $\ln k$ vs. $1/T$, the slope would correspond to the activation energy divided by the gas constant. Moreover, this value will not provide by itself any idea of the reactivity of a given system, only information on temperature dependence of the reaction.

Although the Arrhenius equation is commonly used to describe temperature dependence of the reaction rate in most food systems, deviations may occur as reported by several authors including Labuza and Riboh (1982) and Taoukis et al. (1997). In fact, a large number of factors may contribute to deviations. Changes in reaction mechanisms may occur for a large temperature range. For instance, it is highly possible that mechanisms of deterioration may change at conditions below freezing point due to a concentration effect. On the other hand, at high temperatures, changes in the physical state of some compounds, including fats and sugars, may occur. Lipids may change from a solid to a liquid state, while sugars may change from an amorphous to a crystalline or to a liquid state. Because of the high complexity of food systems, it is also possible that when various mechanisms of deterioration operate simultaneously, the effect of temperature may alter the rates of one, thus causing inhibition or catalysis in the other mechanisms. Finally, irreversible changes such as starch hydrolysis or protein denaturation may occur due to temperature, thus modifying the reactivity of the system. In fact, although enzyme-catalyzed reactions will have an increasing reaction rate upon an increase in temperature, a decrease will be observed beyond a certain temperature due to enzyme inactivation. Typical values for activation energies for a number of reactions are summarized in Table 2.2.

2.2.5 EFFECT OF PRESSURE

Traditionally, processing of foods has involved thermal treatments with the specific goal of making foods microbiologically safe. With greater concerns over the nutritional benefits of foods as well as qualitative aspects such as texture and color, investigation into advanced food processing techniques has resulted in the emergence of various new technologies within the food industry. One area of recent interest is the use of ultra-high pressure (UHP) processing of foods, also referred to as high hydrostatic pressure (HHP) and high pressure processing (HPP). According to the Le Chatelier principle, any reaction, conformational change, or phase change that is accompanied by a decrease

TABLE 2.2

Activation Energies for Selected Reactions in Food Related Systems

Reaction	Activation energy (Kcal/mole)
Enzyme reactions	0–8
Chlorophyll degradation	5–27
Ascorbic Acid	5–40
Anthocyanins	7–30
Alpha-tocopherol	9–13
Trans-retinol	9–29
Betalains	9–29
Non-enzymatic browning	9–40
Hydrolysis of disaccharides	10–15
Carotenoid oxidation	10–22
Lipid oxidation	10–25
Spore destruction	60–80
Vegetative cell destruction	50–150
Protein denaturation	80–120

in volume will be favored at high pressure, while reactions involving an increase in volume will be inhibited (Williams, 1994).

The kinetics of reactions as influenced by pressure may be best approached from the Eyring (activated complex or absolute theory) relation where reaction rates are based on the formation of an unstable intermediate complex, which is in quasi-equilibrium with the reactants. For instance in a bimolecular reaction, reactants A and B form an intermediary complex $(AB)^*$, with an equilibrium rate constant (k_1), which may further decompose at a rate constant (k_2) to form product(s).

$$A + B \quad \overset{k_1}{\underset{}{\rightleftharpoons}} \quad (AB)^* \quad \overset{k_2}{\Longrightarrow} \quad P$$

The overall reaction rate is therefore controlled by the rate of formation of the activated complex which is a function of the change in Gibbs free energy (ΔG) going from the normal to the activated state, similar to the previous discussion (Section 2.2.3.4.1.). In the case of effect of changing temperature, the relationship was given by the Arrhenius equation, assuming pressure was held constant. The influence of pressure on reaction rate, however, may be described by the basic thermodynamic relationship, as shown in Equation 2.36 as applied to the Eyring Equation 2.37:

$$\left(\frac{\mathrm{d}\Delta G^\circ}{\mathrm{d}P} \right)_T = \Delta V^\circ \tag{2.36}$$

where ΔG° is the standard free energy associated with the formation of one mole of substance at 25°C and 1 atm (molal free energy), P is pressure and ΔV° is the associated volume at constant temperature (T) and, the Eyring equation, where $\Delta G^* = \Delta H^* - \Delta S^* T$:

$$k = \frac{k_B T}{h} \exp \left(\frac{\Delta S^*}{R} \right) \exp \left(-\frac{\Delta H^*}{RT} \right) = \frac{k_B T}{h} \exp \left(-\frac{\Delta G^*}{RT} \right) \tag{2.37}$$

where ΔG^* is the change in free energy, ΔS^* is the entropy, ΔH^* is the enthalpy, k_B is the Boltzman constant, h is Plank's constant, and R is the gas constant. Combining these equations results in the following expression at constant temperature:

$$\left(\frac{d \ln k}{dP}\right)_T = -\frac{\Delta V^*}{RT} \tag{2.38}$$

By integration this gives the expression for the rate constant, k:

$$\ln k = \ln k_o - \frac{\Delta V^*}{RT}P \tag{2.39}$$

where k_o is a constant dependant on the system, ΔV^* is the volume of activation, and T is temperature ($^\circ$K). The activation volume relates the change in volume between that of the reactants and that of the activated complex. Basically this expression indicates that the rate constant increases with increasing pressure if ΔV^* is negative. In other words, the molar volume of the activated complex is smaller than that of the reactants together. From a practical point of view, the activation volume can be determined from the slope $(-\Delta V^*/RT)$ of the plot of $\ln k$ vs. P at constant temperature. This is similar to the method of determination of the activation energy (E_a) from the slope $(-E_a/R)$ from a plot of $\ln k$ vs. $1/T$ at constant pressure. It is also important for the rate constant (k) to be measured above the "critical" or "threshold" pressure in order for activation volume constants to be meaningful.

It should be pointed out that in the process of pressure treating foods, there is an increase in temperature due to the work of compression. It is therefore, critical to maintain a constant temperature during the pressure treatment in order to obtain meaningful kinetic data. Farkas and Hoover (2000) pointed out several critical process factors to take into account when conducting pressure related studies. These factors include maintaining constant composition, pH, water activity, come-up-times and pressure release times, change in temperature due to compression, and in the case of microorganism testing, the type, age, culturing, and growth conditions should all be kept the same for comparison.

Limited work on the effect of pressure on kinetics of vitamin and pigment degradation has been reported in the literature. Thus far, much of the emphasis has been placed on the microbiological aspects, as would be expected. A further discussion on effects of pressure will be addressed in a later section of this chapter.

In the following section, a brief discussion of the mechanisms of deterioration of food components including vitamins and pigments, and some of the most relevant kinetic information are presented.

2.3 KINETICS OF FOOD COMPONENTS

Several reviews on the stability of various nutrients and pigments have been published (Harris and von Loesecke, 1960; Harris and Karmas, 1975; DeRitter, 1976; Archer and Tannenbaum, 1979; Thompson, 1982; Villota and Hawkes, 1986; Clydesdale et al., 1991; Delgado-Vargas et al., 2000). It is evident by analyzing these reviews that, although a large number of kinetic studies have been reported in the literature on the stability of nutrients, pigments, textural properties, etc., systematic studies geared to elucidate mechanisms of reaction and to provide information to develop comprehensive kinetic models are still scarce. Moreover, it is also clear that although work carried out in model systems greatly contribute to our understanding of reactions, the actual kinetic information obtained may not be readily applicable to highly complex systems such as foods. Thus, information compiled for this review considers primarily real food products, with limited emphasis given to model systems.

2.3.1 VITAMINS

2.3.1.1 Water-Soluble Vitamins

2.3.1.1.1 Vitamin C (Ascorbic Acid)

Ascorbic acid, chemically known as *L*-3-keto-*threo*-hexuronic acid lactone, is naturally found as the *L*-isomer in various citrus fruits, hip berries and fresh tea leaves. It also exists in a stereoisomeric form referred to as *D*-isoascorbic acid (also called *D*-araboascorbic or erythorbic acid) and has only a twentieth of the bioavailability of *L*-ascorbic acid. *L*-ascorbic acid can reversibly convert to dehydroascorbic acid in the presence of mild oxidants, and may subsequently and irreversibly convert to 2,3-diketogulonic acid, which has no bioavailability. This makes it important for proper differentiation during its analysis for nutritional purposes. Considerable amounts of ascorbic acid may be lost during processing and storage of food products. In fact, ascorbic acid is readily destroyed by heating and oxidation, and its protection is particularly difficult to achieve. Other factors influencing the degradation of this vitamin include water activity or moisture content, pH, and metal traces, especially copper and iron. In general, it has been observed for a wide number of products containing ascorbic acid that the reaction appears to follow first-order kinetics. It should be mentioned, however, that different pathways exist for the degradation of ascorbic acid. Such pathways give origin to different breakdown products, and therefore, affect the overall rates of vitamin degradation. In fact the reaction may proceed under aerobic or anaerobic conditions, or through catalyzed or uncatalyzed aerobic pathways (Figure 2.9). Understanding of the mechanisms involved facilitates the handling of kinetic data. However, because of the fact that many parameters influence the kinetics of ascorbic acid decomposition, it is difficult to establish a precursor-product relationship, except for the earlier part of the reactions. For instance in stored canned products, the reaction may occur at the beginning through catalyzed or uncatalyzed aerobic mechanisms. Upon storage after the disappearance of the free oxygen, subsequent losses may be due to anaerobic decomposition of the compound. It is also possible that various mechanisms of deterioration can operate simultaneously, thus highly complicating the treatment of the kinetic data. For instance, in the presence of oxygen, it is possible that the anaerobic pathway will take place, although its contribution appears to be less significant than even the pathway for uncatalyzed degradation. Eison–Perchonok and Downes (1982) used second-order kinetics to describe the degradation of ascorbic acid under limiting oxygen concentrations. Finholt et al. (1963) indicated a maximum rate for anaerobic degradation of ascorbic acid at pH 4 and attributed this behavior to a 1:1 complex of ascorbic acid molecules and hydrogen ascorbate ions which would be present at the highest concentration at a pH near 4.0. Since at normal temperatures the anaerobic degradation proceeds at low rates, it is considered that normally its contribution can be considered insignificant in the presence of excess oxygen.

It is a well-known fact that the autooxidation of ascorbic acid to dehydroascorbic acid is a reversible process, which takes place in two steps with the formation of a free radical as an intermediate. Early studies of ascorbic acid radicals, reported by Yamazaki et al. (1959, 1960) and Yamazaki and Piette (1961), indicated its decay according to second-order reaction kinetics. Bielski et al. (1971) indicated that complex reactions occur for the decay of ascorbic acid radicals, despite the fact that the decay follows strictly second-order reaction kinetics. Huelin (1953) investigated the stability of ascorbic acid in a variety of food products. The author observed that decomposition of the compound proceeded faster in the pH range of 3–4. Under anaerobic conditions, it has been observed that a maximum degradation occurs at approximately pH 4, followed by a rate decrease upon a decrease in pH to 2.0 (Huelin et al., 1971). More recently, the affinity of ascorbic acid for free radicals was shown by Giroux et al. (2001) to improve the color stability of beef during storage, following gamma

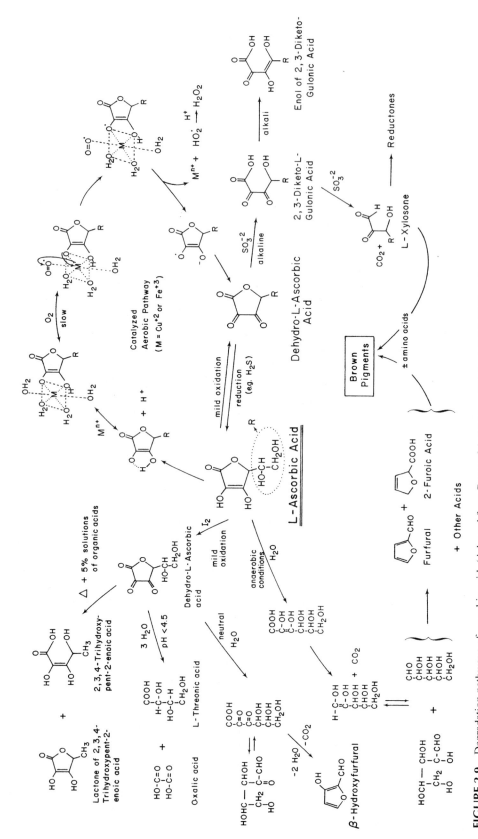

FIGURE 2.9 Degradation pathways of ascorbic acid. (Adapted from Bauernfeind J.C. and Pinkert, D.M. 1970. *Advances in Food Research*, 18: 219–315. Tannenbaum, S.R., Archer, M.C., and Young, V.R. 1985. Vitamins and minerals. Ch. 7, pp. 477–544. In: *Food Chemistry*, 2nd ed., Fennema, O.R. (ed.), Marcel Dekker, Inc., NY.)

irradiation treatment. The vitamin C-scavenging of free radicals (such as $\cdot OH$ and $\cdot SH$) showed a reduction of pigment oxidation in meats. Similarly, addition of ascorbic acid was found to improve the stability of the color of dry chilli powder (carotenoids) during storage at different water activities (Bera et al., 2001).

With regard to metal-catalyzed reactions of ascorbic acid, it has been found that the reaction can be catalyzed by transition metals such as copper, iron, and vanadium, both free and bound in complex compounds (Khan and Martell, 1967a, 1967b, 1968). The reaction may also be catalyzed by the copper containing metal enzyme, ascorbate oxidase. Two different mechanisms have been proposed for the nonenzyme-catalyzed degradation of ascorbic acid by free and complex ions. Differences in their reaction order, rate limiting step, and products of oxygen reduction were observed. Schwertnerová et al. (1976) reported that the autooxidation of ascorbic acid, catalyzed by copper ions, followed the Michaelis–Menten law in the presence of an inhibitor. Pekkarinen (1974) observed that the autooxidation of ascorbic acid catalyzed by iron salts in citric acid solutions had a clear induction period, particularly at lower concentrations. Although the addition of copper salt decreased the induction period, it was also observed that the rate of the reaction slowed down at a later stage. It is possible, according to the author, that the copper salt may destroy radicals produced by the reaction catalyzed by the iron salts. Spanyár and Kevei (1963) had previously reported that copper was very destructive to ascorbic acid in air but had an insignificant effect in nitrogen. The authors also indicated that although iron had a prooxidant activity, iron and copper combined accelerated the degradation of ascorbic acid at a lower rate than either of the two acting alone.

In most situations it has been observed that an increase in temperature also increases the destruction rates of ascorbic acid. On the other hand, contradicting results have been suggested for conditions at sub-freezing temperatures. Grant and Alburn (1965a, 1965b) studied the degradation of ascorbic acid in frozen and unfrozen solutions containing 10^{-4} M ascorbic acid in a 0.02M acetate buffer at pH values of 5.0 and 5.5. At both pH values, the rates of ascorbic acid oxidation were observed to be significantly higher for the $-11°C$ as compared to the system at $1°C$ and more pronounced for the system at pH 5.5. A number of factors have been suggested as being responsible for the observed trends. An increase in concentration of the reactants occurs upon freezing of the system (Pincock and Kiovsky, 1966). Other factors, such as a catalytic effect exerted by the ice crystals, favorable orientation of the reactants in the partially frozen system, and a decrease in dielectric constant or an increase in proton mobility, have also been given as possible reasons for the enhanced rates of ascorbic acid degradation at sub-freezing temperatures. It is also possible that oxygen availability is another factor to be taken into consideration. Results reported by Thompson and Fennema (1971) indicated that, in fact, an increase in reactant concentration and pH and oxygen availability could explain higher rates of degradation or a smaller than expected decrease in rate in partially frozen systems. Changes of pH as a result of freezing should also be taken into consideration. In fact, Van den Berg and Rose (1959) reported significant changes in pH during the freezing of well-buffered phosphate solutions upon freezing from 0 to $-10°C$.

In general, taking into consideration the possible reaction pathways, as summarized by Bauernfeind and Pinkert (1970), the specific kinetic parameters may be dependent on more factors than normally accounted for. In fact, it is a common assumption to measure the rate of degradation of ascorbic acid in food products, natural or formulated, for only a given initial concentration. Work done by various investigators indicates that for the case of this particular vitamin the degradation may follow pseudo first-order reaction kinetics. In fact, the presence of breakdown products will alter the kinetic rates and, possibly, the mechanisms of degradation (Villota, 1979). In fact, depending on the composition of the system and considering that some of the steps involved in the degradation of ascorbic acid are reversible, it is clear that although the reaction may still follow first-order reaction kinetics the rates of degradation will vary depending on the equilibrium created in the system and the levels of breakdown products. Lavelli and Giovanelli (2003) further exemplified the effect of initial concentration in tomato products. They found higher rates of degradation of ascorbic acid at lower initial concentration levels (Table 2.3).

TABLE 2.3   Kinetic Parameters for Vitamin Degradation during Thermal Processing or Storage

Commodity	Process / Conditions	r^2	k_T value (min^{-1})	E_a (kcal/mol)	Reaction Order	r^2	Temp. Range (°C)	$t_{1/2}$ (min)	Reference
Vitamin C (ascorbic acid)			**Water Soluble Vitamins:**						
Apricots	Canned	0.94	$k_{26.7} = 53.236 \times 10^{-8}$		1			1.302×10^6	Cameron, et al., 1955
		0.82	$k_{18.3} = 9.079 \times 10^{-8}$	23.7		0.91	10 - 26.7	7.635×10^6	
		0.84	$k_{10} = 5.043 \times 10^{-8}$					13.745×10^6	
Asparagus	Canned	0.75	$k_{26.7} = 12.967 \times 10^{-8}$		1			5.345×10^6	Cameron, et al., 1955
		0.72	$k_{18.3} = 7.769 \times 10^{-8}$	8.1		0.97	10 - 26.7	8.922×10^6	
		0.85	$k_{10} = 5.808 \times 10^{-8}$					11.951×10^6	
	Raw, whole in crushed ice; room temperature	0.91	$k_0 = 0.0001065$	-	1	-	~0	6,508	Cameron, et al., 1955
		0.92	$k_{20} = 0.0002394$	-	1	-	~20	2,895	
Beef									
Model system (40% soy flour, 25% beef, 20% sucrose, 10% propylene glycol)	$a_w = 0.69$	-	-	3.3	0	-	61 - 100		Laing et al. 1978
	$a_w = 0.80$	-	-	4.1	0	-	61 - 100	-	
	$a_w = 0.90$	-	-	3.8	0	-	61 - 100	-	
Breakfast cereal	Packaging: 208 x 006 TDT cans								Kirk et al., 1977
Model system (soy protein/ fat/ carbohydrate/ salt/sugar)	$a_w = 0.10$	-	$k_{37} = 0.681 \times 10^{-5}$		1			10.18×10^4	
		-	$k_{30} = 0.632 \times 10^{-5}$	8.1		0.956	10 - 37	10.97×10^4	
		-	$k_{20} = 0.313 \times 10^{-5}$					22.15×10^4	
		-	$k_{10} = 0.215 \times 10^{-5}$					32.24×10^4	
Total ascorbic acid	$a_w = 0.24$	-	$k_{37} = 3.479 \times 10^{-5}$		1			1.99×10^4	
		-	$k_{30} = 1.236 \times 10^{-5}$	15.9		0.971	10 - 37	5.61×10^4	
		-	$k_{20} = 0.660 \times 10^{-5}$					10.50×10^4	
		-	$k_{10} = 0.257 \times 10^{-5}$					26.97×10^4	

a_w	Rate constant					$\times 10^4$	Reference
$a_w = 0.40$	$k_{37} = 4.882 \times 10^{-5}$		1	0.997	10 - 37	1.42×10^4	Kirk et al., 1977
	$k_{30} = 2.174 \times 10^{-5}$	17.6				3.19×10^4	
	$k_{20} = 0.889 \times 10^{-5}$					7.80×10^4	
	$k_{10} = 0.292 \times 10^{-5}$					23.74×10^4	
$a_w = 0.50$	$k_{37} = 6.417 \times 10^{-5}$		1	0.986	10 - 37	1.08×10^4	
	$k_{30} = 2.771 \times 10^{-5}$	19.2				2.50×10^4	
	$k_{20} = 0.778 \times 10^{-5}$					8.91×10^4	
	$k_{10} = 0.340 \times 10^{-5}$					20.39×10^4	
$a_w = 0.65$	$k_{37} = 10.931 \times 10^{-5}$		1	0.984	10 - 37	0.634×10^4	
	$k_{30} = 3.313 \times 10^{-5}$	19.2				2.09×10^4	
	$k_{20} = 1.000 \times 10^{-5}$					6.93×10^4	
	$k_{10} = 0.347 \times 10^{-5}$					19.98×10^4	
Reduced ascorbic acid $a_w = 0.10$	$k_{37} = 0.854 \times 10^{-5}$		1	0.979	10 - 37	8.12×10^4	
	$k_{30} = 0.771 \times 10^{-5}$	7.1				8.99×10^4	
	$k_{20} = 0.451 \times 10^{-5}$					15.37×10^4	
	$k_{10} = 0.299 \times 10^{-5}$					23.18×10^4	
$a_w = 0.24$	$k_{37} = 3.083 \times 10^{-5}$		1	0.985	10 - 37	2.25×10^4	
	$k_{30} = 1.597 \times 10^{-5}$	14.2				4.34×10^4	
	$k_{20} = 0.931 \times 10^{-5}$					7.45×10^4	
	$k_{10} = 0.312 \times 10^{-5}$					22.22×10^4	
$a_w = 0.40$	$k_{37} = 5.465 \times 10^{-5}$		1	0.994	10 - 37	1.27×10^4	
	$k_{30} = 2.667 \times 10^{-5}$	17.8				2.60×10^4	
	$k_{20} = 1.174 \times 10^{-5}$					5.90×10^4	
	$k_{10} = 0.326 \times 10^{-5}$					21.26×10^4	
$a_w = 0.50$	$k_{37} = 6.181 \times 10^{-5}$		1	0.989	10 - 37	1.12×10^4	
	$k_{30} = 3.251 \times 10^{-5}$	18.1				2.16×10^4	
	$k_{20} = 0.903 \times 10^{-5}$					7.68×10^4	
	$k_{10} = 0.403 \times 10^{-5}$					17.20×10^4	
$a_w = 0.65$	$k_{37} = 11.701 \times 10^{-5}$		1	0.989	10 - 37	0.592×10^4	
	$k_{30} = 3.674 \times 10^{-5}$	21.1				1.89×10^4	
	$k_{20} = 1.340 \times 10^{-5}$					5.17×10^4	
	$k_{10} = 0.382 \times 10^{-5}$					18.154×10^4	

(Continued)

TABLE 2.3 Continued

Commodity	Process / Conditions	r^2	k_T value (min^{-1})	E_a (kcal/mol)	Reaction Order	r^2	Temp. Range (°C)	$t_{1/2}$ (min)	Reference
Breakfast cereal (cont.)	Packaging: cardboard boxes /w/ wax paper liners								Kirk et al., 1977
Total ascorbic acid	$a_w = 0.10$	-	$k_{30} = 1.701 \times 10^{-5}$	-	1	-	30	4.07×10^4	
	$a_w = 0.40$	-	$k_{30} = 2.521 \times 10^{-5}$	-	1	-	30	2.75×10^4	
	$a_w = 0.85$	-	$k_{30} = 4.868 \times 10^{-5}$	-	1	-	30	1.45×10^4	
Reduced ascorbic acid	$a_w = 0.10$	-	$k_{30} = 1.847 \times 10^{-5}$	-	1	-	30	3.75×10^4	
	$a_w = 0.40$	-	$k_{30} = 2.618 \times 10^{-5}$	-	1	-	30	2.65×10^4	
	$a_w = 0.85$	-	$k_{30} = 4.903 \times 10^{-5}$	-	1	-	30	1.41×10^4	
	Packaging: 303 cans /w/ XS headspace								
Total ascorbic acid	$a_w = 0.24$	-	$k_{30} = 2.507 \times 10^{-5}$	-	1	-	30	2.76×10^4	
	$a_w = 0.40$	-	$k_{30} = 5.944 \times 10^{-5}$	-	1	-	30	1.17×10^4	
Reduced ascorbic acid	$a_w = 0.24$	-	$k_{30} = 2.625 \times 10^{-5}$	-	1	-	30	2.64×10^4	
	$a_w = 0.40$	-	$k_{30} = 5.313 \times 10^{-5}$	-	1	-	30	1.30×10^4	
Breakfast cereal model system [similar to Kirk et al. (1977)]	Packaging: 303 cans								Dennison and Kirk, 1978
Total ascorbic acid	$a_w = 0.10$	-	$k_{37} = 1.229 \times 10^{-5}$					5.64×10^4	
		-	$k_{30} = 0.944 \times 10^{-5}$	10.7	1	0.963	10 - 37	7.34×10^4	
		-	$k_{20} = 0.597 \times 10^{-5}$					11.61×10^4	
		-	$k_{10} = 0.229 \times 10^{-5}$					30.27×10^4	
	$a_w = 0.24$	-	$k_{20} = 0.764 \times 10^{-5}$	-	1	-	20	9.07×10^4	
	$a_w = 0.40$	-	$k_{37} = 5.035 \times 10^{-5}$					1.38×10^4	
		-	$k_{30} = 2.542 \times 10^{-5}$	16.0	1	0.998	10 - 37	2.73×10^4	
		-	$k_{20} = 1.014 \times 10^{-5}$					6.84×10^4	
		-	$k_{10} = 0.417 \times 10^{-5}$					16.62×10^4	

Sample	Treatment	a_w	Rate constant						Reference
	$a_w = 0.65$	-	$k_{37} = 8.382 \times 10^{-5}$	18.3	1	0.999	10 - 37	0.827×10^4	Dennison and Kirk. 1978
		-	$k_{30} = 3.854 \times 10^{-5}$					1.80×10^4	
		-	$k_{20} = 1.410 \times 10^{-5}$					4.92×10^4	
		-	$k_{10} = 0.479 \times 10^{-5}$					14.47×10^4	
Reduced ascorbic acid	Packaging: 303 cans /w/ XS headspace								
	$a_w = 0.10$	-	$k_{37} = 1.129 \times 10^{-5}$	10.7	1	0.977	10 - 37	5.36×10^4	
		-	$k_{30} = 0.896 \times 10^{-5}$					7.74×10^4	
		-	$k_{20} = 0.583 \times 10^{-5}$					11.89×10^4	
		-	$k_{10} = 0.236 \times 10^{-5}$					29.377×10^4	
	$a_w = 0.24$	-	$k_{20} = 0.715 \times 10^{-5}$	-	1	-	20	9.69×10^4	
	$a_w = 0.40$	-	$k_{37} = 5.271 \times 10^{-5}$	15.5	1	0.984	10 - 37	1.32×10^4	
		-	$k_{30} = 2.167 \times 10^{-5}$					3.20×10^4	
		-	$k_{20} = 1.021 \times 10^{-5}$					6.79×10^4	
		-	$k_{10} = 0.438 \times 10^{-5}$					15.83×10^4	
	$a_w = 0.65$	-	$k_{37} = 8.104 \times 10^{-5}$	16.9	1	0.994	10 - 37	0.885×10^4	
		-	$k_{30} = 3.549 \times 10^{-5}$					1.95×10^4	
		-	$k_{20} = 1.417 \times 10^{-5}$					4.89×10^4	
		-	$k_{10} = 0.563 \times 10^{-5}$					12.31×10^4	
Corn (yellow)	Canned	0.92	$k_{26.7} = 18.768 \times 10^{-8}$	9.7	1	0.99	10 - 26.7	3.693×10^6	Cameron et al., 1955
		0.74	$k_{18.3} = 10.867 \times 10^{-8}$					6.378×10^6	
		0.76	$k_{10} = 7.183 \times 10^{-8}$					9.650×10^6	
Grapefruit segments	Canned	0.99	$k_{26.7} = 76.141 \times 10^{-8}$	20.7	1	0.99	10 - 26.7	0.910×10^6	Cameron et al., 1955
		0.90	$k_{18.3} = 22.558 \times 10^{-8}$					3.073×10^6	
		0.73	$k_{10} = 9.810 \times 10^{-8}$					7.066×10^6	
Grapefruit juice Solids	Thermal concentration 11.2 °Bx	0.937	$k_{96} = 0.002642$	4.98	1 (apparent)	0.998	61 - 96	262	Saguy et al., 1978b
		0.996	$k_{95} = 0.002503$					277	
		0.966	$k_{80} = 0.001899$					365	
		0.937	$k_{61} = 0.001276$					543	

(Continued)

TABLE 2.3 Continued

Commodity	Process / Conditions	r^2	k_T value (min^{-1})	E_a (kcal/mol)	Reaction Order	r^2	Temp. Range (°C)	$t_{1/2}$ (min)	Reference
Grapefruit juice (cont.)	31.2 °Bx	0.978	k_{91} = 0.002701	5.33	1	0.997	60 - 91	257	Saguy et al., 1978b
		0.956	k_{82} = 0.002165					320	
		0.974	k_{75} = 0.001874					370	
		0.925	k_{60} = 0.001349					514	
	47.1 °Bx	0.935	k_{96} = 0.003777	6.69	1	0.998	61 - 96	184	
		0.964	k_{90} = 0.003121					222	
		0.976	k_{80} = 0.002460					282	
		0.962	k_{61} = 0.001430					485	
	55.0 °Bx	0.988	k_{91} = 0.004712	8.60	1	0.999	61 - 91	147	
		0.972	k_{81} = 0.003348					207	
		0.970	k_{75} = 0.002715					255	
		0.951	k_{61} = 0.001618					428	
	62.5 °Bx	0.924	k_{96} = 0.01068	11.30	1	0.998	68 - 96	69	
		0.945	k_{81} = 0.005365					129	
		0.929	k_{76} = 0.004561					163	
		0.947	k_{68} = 0.003022					229	
Green beans	Canned	0.89	$k_{26.7}$ = 25.832 x 10^{-8}	7.5	1	0.99	10 - 26.7	2.683 x 10^6	Cameron et al., 1955
		0.86	$k_{18.3}$ = 17.819 x 10^{-8}					3.890 x 10^6	
		0.81	k_{10} = 12.317 x 10^{-8}					5.628 x 10^6	
	Raw, whole, sieved by mesh size								
	No. 5	0.96	k_{20} = 0.0002454	-	1	-	20	2,825	
	No. 4	0.95	k_{20} = 0.0002337	-	1	-	20	2,966	
	Nos. 2 and 3	0.85	k_{20} = 0.0003789	-	1	-	20	1,829	
	Stored at room temp. (18.9-21.1°C)								
Green beans	Frozen	-	k_{-5} = 22.920 x 10^{-6}	24.3	1	.967	-5 to -20	3.024 x 10^4	Giannakourou et al., 2003
	Blanch: 90°C/2min	-	k_{-10} = 9.627 x 10^{-6}					7.200 x 10^4	
	IQF: -22°C/2min	-	k_{-15} = 3.946 x 10^{-6}					17.568 x 10^4	
			k_{-20} = 1.548 x 10^{-6}					44.784 x 10^4	

Food	Treatment	k	value	R^2	(value)	n	R^2	Temp. range	values	Reference
Infant formula [15% protein (milk-based); 24% lipids; 57% carbohydrates; 4% vitamins, minerals, water]	With ferrous sulfate	k_{45} =	4.24×10^{-6}	0.920	9.0	1	0.998	20 - 45	1.63×10^5	Galdi et al., 1989
		k_{37} =	2.80×10^{-6}	0.916					2.48×10^5	
		k_{20} =	1.25×10^{-6}	0.954					5.55×10^5	
	With ferric glycinate	k_{45} =	3.24×10^{-6}	0.959	8.7	1	0.992	20 - 45	2.14×10^5	
		k_{37} =	2.04×10^{-6}	0.932					3.40×10^5	
		k_{20} =	0.972×10^{-6}	0.926					7.13×10^5	
Okra	Frozen	k_{-5} =	12.034×10^{-6}	-	25.2	1	0.868	-5 to -20	5.760×10^4	Giannakourou et al., 2003
	Blanch: 90°C/2m	k_{-10} =	4.912×10^{-6}	-					14.112×10^4	
	IQF: -22°C/2m	k_{-15} =	1.933×10^{-6}	-					35.856×10^4	
		k_{-20} =	0.729×10^{-6}						95.040×10^4	
Oranges (fresh squeezed / filtered) Thermal treatment: pH 3.5	In capillary tubes (1.15 x 150 mm) - vac - 0-150 min	k_{150} =	0.0967	0.99	28.1	1	0.998	120 - 150	7.2	Van den Broeck, et al., 1998
		k_{140} =	0.0480	0.98					14.4	
		k_{130} =	0.0205	0.98					33.8	
		k_{120} =	0.0076	0.99					91.2	
Thermal/pressure treatment:	8500 bar	k_{80} =	0.010289	0.98	20.1	1	0.999	65 - 80	67.4	
	0.3 ml / N₂ flush	k_{70} =	0.004338	0.98					159.8	
		k_{65} =	0.002897	0.99					239.3	
Orange juice	Canned	$k_{26.7}$ =	68.467×10^{-8}	0.98	24.3	1	0.99	10 - 26.7	1.012×10^6	Cameron et al., 1955
		$k_{18.3}$ =	22.738×10^{-8}	0.98					3.048×10^6	
		k_{10} =	6.171×10^{-8}	0.81					11.232×10^6	
Peaches	Canned	$k_{26.7}$ =	56.939×10^{-8}	0.97	-	1	-	26.7	1.217×10^6	Cameron et al., 1955
Peas (sweet)	Canned	$k_{26.7}$ =	17.954×10^{-8}	0.92	9.0	1	0.99	10 - 26.7	3.860×10^6	Cameron et al., 1955
		$k_{18.3}$ =	10.794×10^{-8}	0.98					6.422×10^6	
		k_{10} =	7.386×10^{-8}	0.82					9.385×10^6	
Peas (green)	Frozen	k_{-5} =	20.060×10^{-6}	-	23.4	1	0.958	-5 to -20	3.456×10^4	Giannakourou et al., 2003
	Blanch: 90°C/2m	k_{-10} =	8.596×10^{-6}	-					8.064×10^4	
	IQF: -22°C/2m	k_{-15} =	3.647×10^{-6}	-					19.008×10^4	
		k_{-20} =	1.481×10^{-6}						46.800×10^4	

(Continued)

TABLE 2.3 Continued

Commodity	Process / Conditions	r^2	k_T value (min^{-1})		E_a (kcal/mol)	Reaction Order	r^2	Temp. Range (°C)	$t_{1/2}$ (min)	Reference
Peas (green) (cont.) Predicted	$T_{eff} = -3.8°C$	-	$k_{eff} =$	24.24×10^{-6}	-	1	-	-5 to -20	2.86×10^4	Giannakourou et al., 2003
	Temp. Cycle: -3°C/72 h -5°C/24 h -8°C/12 h	-	$k_{exp} =$	21.88×10^{-6}	-	1	0.981	-3 to -8°C	3.17×10^4	
Peas (sweet)	Dehydro-frozen (50% H_2O) Storage losses (in air)									Neumann et al., 1965 (from Labuza, 1972)
	$a_w = 0.90$	-	$k_{-7} =$	3.19×10^{-5}	46.0	1	-	-15 to -7	2.17×10^4	
	$a_w = 0.90$	-	$k_{-15} =$	0.214×10^{-5}					32.39×10^4	
Peas (sweet)	Canned	-	$k_{132.2} =$	0.00900					77	Lathrop and Leung, 1980
		-	$k_{126.7} =$	0.00430					161	
		-	$k_{121.1} =$	0.00250	39.3	1	0.984	110 - 132	277	
		-	$k_{115.6} =$	0.00140					495	
		-	$k_{110.0} =$	0.00046					1,507	
Pineapple slices	Canned	0.98	$k_{26.7} =$	67.577×10^{-8}					1.026×10^6	Cameron et al., 1955
		0.87	$k_{18.3} =$	26.975×10^{-8}	11.5	1	0.88	10 - 26.7	2.570×10^6	
		0.70	$k_{10} =$	21.456×10^{-8}					3.231×10^6	
Spinach	Canned	0.81	$k_{26.7} =$	17.475×10^{-8}					3.967×10^6	Cameron et al., 1955
		0.80	$k_{18.3} =$	11.574×10^{-8}	6.3	1	0.96	10 - 26.7	5.989×10^6	
		0.95	$k_{10} =$	9.347×10^{-8}					7.4165×10^6	
Spinach	Frozen Blanch: 90°C/2m IQF: -22°C/2m	-	$k_{-5} =$	60.169×10^{-6}					1.152×10^4	Giannakourou et al., 2003
		-	$k_{-10} =$	24.068×10^{-6}	26.8	1	0.992	-5 to -20	2.880×10^4	
		-	$k_{-15} =$	8.752×10^{-6}					7.920×10^4	
			$k_{-20} =$	3.146×10^{-6}					22.032×10^4	

System	Condition		k				Temp (°C)		Reference
Predicted	$T_{eff} = -1.7°C$ Temp. Cycle: -1°C/72 h, -4°C/24 h, -7°C/12 h	-	$k_{eff} = 119.9 \times 10^{-6}$	-	1	-	-5 to -20	0.578×10^4	Haralampu and Karel, 1983
		-	$k_{exp} = 113.9 \times 10^{-6}$	-	1	0.917	-1 to -7°C	0.609×10^4	
Sweet potato (flour)	Freeze-dried/ ground/ rehumidified								
	$a_w = 0.020$	0.629	$k_{40} = 1.58 \times 10^{-5}$	-	1	-	40	43.87×10^3	
	$a_w = 0.058$	0.843	$k_{40} = 2.53 \times 10^{-5}$	-	1	-	40	27.40×10^3	
	$a_w = 0.112$	0.973	$k_{40} = 2.67 \times 10^{-5}$	-	1	-	40	25.96×10^3	
	$a_w = 0.316$	0.999	$k_{40} = 6.57 \times 10^{-5}$	-	1	-	40	10.55×10^3	
	$a_w = 0.484$	0.986	$k_{40} = 10.77 \times 10^{-5}$	-	1	-	40	6.44×10^3	
	$a_w = 0.555$	0.960	$k_{40} = 16.48 \times 10^{-5}$	-	1	-	40	4.21×10^3	
	$a_w = 0.661$	0.985	$k_{40} = 27.67 \times 10^{-5}$	-	1	-	40	2.51×10^3	
	$a_w = 0.747$	1.000	$k_{40} = 31.83 \times 10^{-5}$	-	1	-	40	2.18×10^3	
Tomatoes	Canned	0.94	$k_{26.7} = 32.380 \times 10^{-8}$	13.3	1	0.82	10 - 26.7	2.141×10^6	Cameron et al., 1955
		0.70	$k_{18.3} = 9.865 \times 10^{-8}$					7.026×10^6	
		0.70	$k_{10} = 8.578 \times 10^{-8}$					8.081×10^6	
Tomatoes (fresh sqeezed / filtered) Thermal treatment: Variety A	pH 4.5 In capillary tubes (1.15 x 150 mm) - vac - 0-150 min	0.99	$k_{150} = 0.0487$	25.2	1	0.999	120 - 150	14.2	Van den Broeck, et al., 1998
		0.98	$k_{140} = 0.0245$					60.3	
		0.98	$k_{130} = 0.0115$					28.3	
		0.99	$k_{120} = 0.0049$					141.5	
Thermal/pressure treatment:	8500 bar 0.3 ml / N₂ flush	0.99	$k_{80} = 0.005744$	17.8	1	0.965	65 - 80	120.7	
		0.98	$k_{70} = 0.003235$					214.3	
		0.98	$k_{65} = 0.001789$					387.4	
Thermal treatment: Variety B		0.99	$k_{150} = 0.0864$	27.5	1	0.998	120 - 150	8	
		0.99	$k_{140} = 0.0411$					16.9	
		0.99	$k_{130} = 0.0183$					37.9	
		0.98	$k_{120} = 0.0071$					97.6	

(Continued)

TABLE 2.3 Continued

Commodity	Process / Conditions	r^2	k_T value (min⁻¹)	E_a (kcal/mol)	Reaction Order	r^2	Temp. Range (°C)	$t_{1/2}$ (min)	Reference
Tomatoes (fresh sqeezed / filtered) cont. - Thermal treatment:									Van den Broeck, et al., 1998
Phosphate buffer	pH 4.0	-	$k_{140} = 0.01302$	-	1	-	140	53.2	
	pH 7.0	-	$k_{140} = 0.00606$	-	1	-	140	114.4	
	pH 8.0 (0 - 360 min)	-	$k_{140} = 0.00227$	-	1	-	140	305.4	
Tomato juice	Heat processed, stored in 8 oz cans + citrate buffer								Lee et al., 1977
	pH 3.53	-	$k_{37.8} = 0.128 \times 10^{-5}$	4.5	1	-	10 - 37.8	54.15×10^4	
	pH 3.78	-	$k_{37.8} = 0.158 \times 10^{-5}$	4.0	1	-	10 - 37.8	43.87×10^4	
	pH 4.06	-	$k_{37.8} = 0.172 \times 10^{-5}$	3.3	1	0.999	10 - 37.8	40.30×10^4	
		-	$k_{29.4} = 0.147 \times 10^{-5}$					47.15×10^4	
		-	$k_{18.3} = 0.121 \times 10^{-5}$					57.28×10^4	
		-	$k_{10} = 0.101 \times 10^{-5}$					68.63×10^4	
	pH 4.36	-	$k_{37.8} = 0.158 \times 10^{-5}$	3.8	1	-	37.8	43.87×10^4	
Tomato juice	Freeze-dried / rehydrated								Riemer and Karel, 1977
	$a_w = 0$	0.99	$k_{51} = 2.083 \times 10^{-5}$	18.8	1	0.971	20 - 51	3.32×10^4	
		0.99	$k_{37} = 0.340 \times 10^{-5}$	$(24.6)^a$				20.37×10^4	
		0.74	$k_{20} = 0.090 \times 10^{-5}$					77.02×10^4	
	$a_w = 0.11$	0.99	$k_{51} = 6.944 \times 10^{-5}$	17.0	1	0.974	20 - 51	0.981×10^4	
		0.98	$k_{37} = 1.389 \times 10^{-5}$	$(22.3)^a$				4.99×10^4	
		0.95	$k_{20} = 0.410 \times 10^{-5}$					16.91×10^4	
	$a_w = 0.32$	0.99	$k_{51} = 19.444 \times 10^{-5}$	20.3	1	0.999	20 - 51	0.357×10^4	
		0.99	$k_{37} = 4.861 \times 10^{-5}$	$(20.2)^a$				1.43×10^4	
		0.99	$k_{20} = 0.694 \times 10^{-5}$					9.99×10^4	

Tomato juice - cont.	$a_w = 0.57$	$k_{51} = 48.611 \times 10^{-5}$	16.0	1			0.143×10^4	Riemer and Karel, 1977
		$k_{37} = 13.889 \times 10^{-5}$	$(18.1)^{[a]}$				0.499×10^4	
		$k_{20} = 3.472 \times 10^{-5}$			0.997	20 - 51	2.00×10^4	
	$a_w = 0.75$	$k_{51} = 81.944 \times 10^{-5}$	14.6	1			0.0846×10^4	
		$k_{37} = 38.194 \times 10^{-5}$	$(16.2)^{[a]}$				0.181×10^4	
		$k_{20} = 7.639 \times 10^{-5}$			0.986	20 - 51	0.907×10^4	
Tomato paste Total Ascorbic acid	Storage: 90 d 130 g Al-tubes	$k_{50} = 9.51 \times 10^{-6}$	5.84	1			7.29×10^4	Lavelli and Giovanelli, 2003
	$a_w = 0.94$	$k_{40} = 6.67 \times 10^{-6}$	$(5.64)^{[a]}$		0.985	30 - 50	10.40×10^4	
	$a_w = 0.69$	$k_{30} = 5.21 \times 10^{-6}$					13.31×10^4	
Tomato pulp	Storage: 90 d 450 g cans	$k_{50} = 14.93 \times 10^{-6}$	25.90	1			4.64×10^4	
	$a_w = 0.84$	$k_{40} = 3.47 \times 10^{-6}$	$(25.1)^{[a]}$		0.995	30 - 50	19.96×10^4	
	$a_w = 0.77$	$k_{30} = 1.04 \times 10^{-6}$					66.54×10^4	
Tomato pulp /w/ XS heat	$a_w = 0.94$	$k_{40} = 11.11 \times 10^{-6}$	-	1	-	40	6.24×10^4	
Tomato puree	$a_w = 0.94$	$k_{40} = 7.43 \times 10^{-6}$	-	1	-	40	9.33×10^4	

B-Vitamins

Vitamin B₁ (thiamine)

Breakfast cereal model system (Vit. B₁)	Packaged in: TDT cans							Dennison, et al., 1977
	$a_w = 0.10$	$k_{45} = 0.00066$	-	1	-	45	1,050	
	$a_w = 0.25$	$k_{45} = 0.00091$	-	1	-	45	762	
	$a_w = 0.40$	$k_{45} = 0.000675$	-	1	-	45	103	
	$a_w = 0.50$	$k_{45} = 0.01101$	-	1	-	45	63	
	$a_w = 0.65$	$k_{45} = 0.00867$	-	1	-	45	80	
Breakfast cereal model system (Vit. B₁ + Vit. C + Vit. A)	$a_w = 0.10$	$k_{45} = 0.00014$	-	1	-	45	4951	
	$a_w = 0.25$	$k_{45} = 0.00065$	-	1	-	45	1066	
	$a_w = 0.40$	$k_{45} = 0.00648$	-	1	-	45	107	
	$a_w = 0.50$	$k_{45} = 0.00927$	-	1	-	45	75	
	$a_w = 0.65$	$k_{45} = 0.00948$	-	1	-	45	73	

(Continued)

TABLE 2.3 Continued

Commodity	Process / Conditions	r²	k_T value (min⁻¹)	E_a (kcal/mol)	Reaction Order	r²	Temp. Range (°C)	$t_{1/2}$ (min)	Reference
Carrots (puree)		-	[a]k_{149} = 0.1669					4.2	Feliciotti and Esselen, 1957
		-	k_{139} = 0.0711					9.7	
		-	k_{129} = 0.0285	28.3	1	0.999	109 - 149	24	
		-	k_{119} = 0.0120					58	
		-	k_{109} = 0.0049					141	
Green beans (puree)		-	k_{149} = 0.1744					4.0	Feliciotti and Esselen, 1957
		-	k_{139} = 0.0717					9.7	
		-	k_{129} = 0.0311	28.6	1	0.999	109 - 149	22	
		-	k_{119} = 0.0122					57	
		-	k_{109} = 0.0049					141	
Meats Beef heart (puree)		-	k_{149} = 0.2171					3.2	Feliciotti and Esselen, 1957
		-	k_{139} = 0.0161	22.1	1	0.687	109 - 149	43	
		-	k_{129} = 0.0392	27.9	1	0.999	109 - 149 (-k_{139})	18	
		-	k_{119} = 0.0157					44	
		-	k_{109} = 0.0068					102	
Beef liver (puree)		-	k_{149} = 0.2326					3.0	Feliciotti and Esselen, 1957
		-	k_{139} = 0.0892					7.8	
		-	k_{129} = 0.0364	28.5	1	0.996	109 - 149	19	
		-	k_{119} = 0.0147					47	
		-	k_{109} = 0.0067					103	
Beef (puree)		-	$k_{137.8}$ = 0.03673					19	Mulley et al., 1975a
		-	$k_{132.2}$ = 0.02509					28	
		-	$k_{126.7}$ = 0.01436	27.5	1	0.996	121 - 138	48	
		-	$k_{121.1}$ = 0.00906					77	

Sample							
Lamb (puree)		k_{149} = 0.1935				3.6	Feliciotti and Esselen, 1957
		k_{139} = 0.0814				8.5	
		k_{129} = 0.0377	27.7	1	0.998	109 - 149	18
		k_{119} = 0.0138				50	
		k_{109} = 0.0062				112	
Pork (puree)		k_{149} = 0.1693				4.1	Feliciotti and Esselen, 1957
		k_{139} = 0.0717				9.7	
		k_{129} = 0.0288	27.4	1	0.998	109 - 149	24
		k_{119} = 0.0129				54	
		k_{109} = 0.0055				126	
Pork (puree)		[a]$k_{137.8}$ = 0.0840				8.3	Lenz and Lund, 1977b
		$k_{126.7}$ = 0.0385	25.6	1	0.997	115 - 138	18
		$k_{115.6}$ = 0.0140				50	
Pork (puree)		-	18.4[a]	1	-	99 - 126.5	-	Greenwood et al., 1944
Meat loaf (ground meat, potato starch and bread crumbs)	baked in convection oven	[a]k_{98} = 0.002511				276	Skjöldebrand, et al., 1983
		$k_{85.5}$ = 0.001244	27.1	1	0.955	70.5 - 98	557
		$k_{70.5}$ = 0.000139				4,987	
Milk infant formula [15% protein, (milk-based); 24% lipids; 57% carbohydrates; 4% vitamins, minerals, water]	Flexible plastic containers, no headspace						Galdi et al., 1989
	With ferrous sulfate	0.992 k_{45} = 2.18 x 10^{-6}				3.18×10^5	
		0.915 k_{37} = 1.27 x 10^{-6}	7.6	1	0.950	20 - 45	5.46×10^5
		0.984 k_{20} = 0.741 x 10^{-6}				9.35×10^5	
	With ferric glycinate	0.920 k_{45} = 1.74 x 10^{-6}				3.98×10^5	
		0.967 k_{37} = 0.794 x 10^{-6}	10.6	1	0.933	20 - 45	9.07×10^5
		0.920 k_{20} = 0.370 x 10^{-6}				18.73×10^5	

(Continued)

TABLE 2.3 Continued

Commodity	Process / Conditions	r^2	k_T value (min^{-1})	E_a (kcal/mol)	Reaction Order	r^2	Temp. Range (°C)	$t_{1/2}$ (min)	Reference
Milk (3.5% fat)	Pasteurized/UHT stored in 0.5-L glass bottles / dark	-	k_{85} = 13.68 x 10^{-4}	30.5	2	0.970	35 - 85	507	Fink and Kessler, 1985
		-	k_{72} = 2.69 x 10^{-4}					2,580	
		-	k_{50} = 0.248 x 10^{-4}					27,950	
		-	k_{35} = 0.0396 x 10^{-4}					175,000	
Pasta (enriched)	Steady-state conditions								
	a_w = 0.44	0.992	k_{55} = 1.087 x 10^{-5}	27.7	1	0.976	25 - 55	0.638 x 10^5	Kamman et al., 1981
		0.982	k_{45} = 0.154 x 10^{-5}					4.50 x 10^5	
		0.931	k_{35} = 0.0451 x 10^{-5}					15.37 x 10^5	
		0.875	k_{25} = 0.0138 x 10^{-5}					50.23 x 10^5	
	a_w = 0.54	0.995	k_{55} = 1.406 x 10^{-5}	29.0	1	0.996	25 - 55	0.493 x 10^5	
		0.994	k_{45} = 0.267 x 10^{-5}					2.60 x 10^5	
		0.979	k_{35} = 0.066 x 10^{-5}					10.50 x 10^5	
		0.828	k_{25} = 0.0153 x 10^{-5}					45.30 x 10^5	
	a_w = 0.65	0.995	k_{55} = 1.809 x 10^{-5}	24.2	1	0.992	25 - 55	0.383 x 10^5	
		0.997	k_{45} = 0.435 x 10^{-5}					1.59 x 10^5	
		0.983	k_{35} = 0.127 x 10^{-5}					5.46 x 10^5	
		0.941	k_{25} = 0.0424 x 10^{-5}					16.35 x 10^5	
	Square-wave fluctuations								
	a_w = 0.44								
	25 - 55 °C	0.930	k = 0.463 x 10^{-5}	-	1	-		1.50 x 10^5	
	25 - 45 °C	0.973	k = 0.0807 x 10^{-5}	-	1	-		8.59 x 10^5	
	a_w = 0.54								
	25 - 55 °C	0.954	k = 0.589 x 10^{-5}	-	1	-		1.18 x 10^5	
	25 - 45 °C	0.960	k = 0.132 x 10^{-5}	-	1	-		5.25 x 10^5	
	a_w = 0.65								
	25 - 55 °C	0.989	k = 0.194 x 10^{-5}	-	1	-		3.57 x 10^5	
	25 - 45 °C	0.989	k = 0.882 x 10^{-5}	-	1	-		0.786 x 10^5	

Product		k						Reference
Peas								
Brine packed (whole)	-	^{a}k$_{132.2}$ = 0.0351					20	Bendix et al., 1951
	-	k$_{126.7}$ = 0.0286					24	
	-	k$_{118.3}$ = 0.0122	20.5	1	0.976	104 - 132	57	
	-	k$_{104.4}$ = 0.0058					120	
Vacuum packed (whole)	-	k$_{132.2}$ = 0.0351					20	Bendix et al., 1951
	-	k$_{126.7}$ = 0.0226	21.9	1	0.996	104 - 132	31	
	-	k$_{118.3}$ = 0.0142					49	
	-	k$_{104.4}$ = 0.0046					151	
Peas (puree)	-	k$_{149}$ = 0.1659					4.2	Feliciotti and Esselen, 1957
	-	k$_{139}$ = 0.0708					9.8	
	-	k$_{129}$ = 0.0276	28.1	1	0.998	109 - 149	25	
	-	k$_{119}$ = 0.0114					61	
	-	k$_{109}$ = 0.0051					136	
Peas (puree)	-	^{a}k$_{137.8}$ = 0.04350					16	Lenz and Lund, 1977b
	-	k$_{126.7}$ = 0.02100	23.2	1	0.998	115 - 138	33	
	-	k$_{115.6}$ = 0.00860					81	
Peas (puree)	-	k$_{137.8}$ = 0.03757					18	Mulley et al., 1975a
	-	k$_{132.2}$ = 0.02206	27.8	1	0.964	121 - 138	31	
	-	k$_{126.7}$ = 0.01164					60	
	-	k$_{121.1}$ = 0.009328					74	
Peas (puree, in brine)	-	k$_{137.8}$ = 0.03897					18	
	-	k$_{132.2}$ = 0.03026	27.1	1	0.977	121 - 138	23	
	-	k$_{126.7}$ = 0.01584					44	
	-	k$_{121.1}$ = 0.010160					68	
Spinach (puree)	-	k$_{149}$ = 0.2280					3.0	Feliciotti and Esselen, 1957
	-	k$_{139}$ = 0.0825					8.4	
	-	k$_{129}$ = 0.0336	28.2	1	0.993	109 - 149	21	
	-	k$_{119}$ = 0.0143					48	
	-	k$_{109}$ = 0.0067					103	

(Continued)

TABLE 2.3 Continued

Commodity	Process / Conditions	r^2	k_T value (min^{-1})	E_a (kcal/mol)	Reaction Order	r^2	Temp. Range (°C)	$t_{1/2}$ (min)	Reference
Vitamin B$_2$ (Riboflavin)									
Breakfast cereal model system (B$_2$)	Packaged in TDT cans								Dennison, et al., 1977
	$a_w = 0.10$	-	$^{a,b}k_{37} = 0.00023$	-	1	-	37	3,010	
	$a_w = 0.25$	-	$k_{37} = 0.00188$	-	1	-	37	369	
	$a_w = 0.40$	-	$k_{37} = 0.00263$	-	1	-	37	264	
	$a_w = 0.50$	-	$k_{37} = 0.00411$	-	1	-	37	169	
	$a_w = 0.65$	-	$k_{37} = 0.00503$	-	1	-	37	138	
	Packaged in paperboard boxes								
	$a_w = 0.10$	-	$k_{30} = 0.0044$	-	1	-	30	158	
	$a_w = 0.40$	-	$k_{30} = 0.0043$	-	1	-	30	161	
	$a_w = 0.85$	-	$k_{30} = 0.0043$	-	1	-	30	161	
Macaroni Light exposure (lumens/m^2)	First phase			$0.6 - 2.0^a$	1	-	25 - 55		Woodcock et al., 1982
27.87	$a_w = 0.32$	-	$^{a,b}k_{55} = 1.38$					0.502	
		-	$k_{25} = 1.26$					0.550	
27.87	$a_w = 0.44$	-	$k_{55} = 1.32$					0.525	
		-	$k_{35} = 1.52$					0.546	
		-	$k_{25} = 1.56$					0.444	
18.58	$a_w = 0.44$	-	$k_{55} = 1.72$					0.403	
		-	$k_{35} = 1.27$					0.546	
		-	$k_{25} = 1.30$					0.533	
9.29	$a_w = 0.44$	-	$k_{55} = 1.32$					0.525	
		-	$k_{35} = 1.53$					0.453	
		-	$k_{25} = 1.37$					0.506	

Sample	Condition	a_w / corr.	k	Value	E_a	n	r^2	Temp (°C)	Reference
Light exposure (lumens/m²)	Second phase								Woodcock et al., 1982
27.87		$a_w = 0.32$	$k_{55} = 0.55$	1.260					
			$k_{25} = 0.78$	0.889					
27.87		$a_w = 0.44$	$k_{55} = 0.87$	0.797					
			$k_{35} = 0.55$	1.260					
			$k_{25} = 0.57$	1.220	1.9 - 4.3[a]	1	-	25 - 55	
18.58		$a_w = 0.44$	$k_{55} = 1.10$	0.630					
			$k_{35} = 0.85$	0.815					
			$k_{25} = 0.84$	0.825					
9.29		$a_w = 0.44$	$k_{55} = 1.33$	0.521					
			$k_{35} = 0.80$	0.866					
			$k_{25} = 0.70$	0.990					
Macaroni	Light exposure = 150 ft-c at 4°C								Furuya et al., 1984
Whole	First phase	0.83	$k_4 = 0.326 \times 10^{-3}$	2.13×10^3	-	1	-	4	
	Second phase	0.80	$k_4 = 0.764 \times 10^{-5}$	90.73×10^3	-	1	-	4	
Particulate	First phase	0.88	$k_4 = 0.386 \times 10^{-3}$	1.80×10^3	-	1	-	4	
	Second phase	0.75	$k_4 = 1.181 \times 10^{-5}$	58.69×10^3	-	1	-	4	
Milk Infant formula	Flexible plastic containers (no head space)								Galdi et al., 1989
[15% protein (milk-based); 24% lipids; 57% cabohydrates; 4% vitamins, minerals, water]	With ferrous sulfate	0.952	$k_{45} = 3.06 \times 10^{-6}$	2.27×10^5					
		0.939	$k_{37} = 2.50 \times 10^{-6}$	2.77×10^5	8.0	1	0.986	20 - 45	
		0.930	$k_{20} = 1.07 \times 10^{-6}$	6.48×10^5					
	With ferric glycinate	0.968	$k_{45} = 2.186 \times 10^{-6}$	3.18×10^5					
		0.796	$k_{37} = 0.949 \times 10^{-6}$	7.30×10^5	27.0	1	0.994	20 - 45	
		0.907	$k_{20} = 0.058 \times 10^{-6}$	120×10^5					

(Continued)

TABLE 2.3 Continued

Commodity	Process / Conditions	r^2	k_T value (min^{-1})	E_a (kcal/mol)	Reaction Order	r^2	Temp. Range (°C)	$t_{1/2}$ (min)	Reference
Milk (whole) (1-gal samples)	Container type: Glass								Singh et al., 1975
Light intensity:	150 ft-c	-	$k_{10} = 1.83 \times 10^{-5}$	6.6	1	0.098	1.7 - 10	37.88×10^3	
		-	$k_{4.4} = 1.46 \times 10^{-5}$					47.48×10^3	
		-	$k_{1.7} = 1.28 \times 10^{-5}$					54.15×10^3	
	300 ft-c	-	$k_{10} = 5.37 \times 10^{-5}$	8.1	1	0.999	1.7 - 10	12.91×10^3	
		-	$k_{4.4} = 4.04 \times 10^{-5}$					17.16×10^3	
		-	$k_{1.7} = 3.48 \times 10^{-5}$					19.92×10^3	
	450 ft-c	-	$k_{10} = 5.98 \times 10^{-5}$	20.8	1	-	4.4 - 10	11.52×10^3	
		-	$k_{4.4} = 5.55 \times 10^{-5}$					18.02×10^3	
	Blow-molded polyethylene (BMP)								
Light intensity:	150 ft-c	-	$k_{10} = 1.76 \times 10^{-5}$	16.2	1	0.999	1.7 - 10	3.94×10^4	
		-	$k_{4.4} = 0.998 \times 10^{-5}$					6.95×10^4	
		-	$k_{1.7} = 0.735 \times 10^{-5}$					9.43×10^4	
	300 ft-c	-	$k_{10} = 5.19 \times 10^{-5}$	11.4	1	0.962	1.7 - 10	1.34×10^4	
		-	$k_{4.4} = 3.79 \times 10^{-5}$					1.83×10^4	
		-	$k_{1.7} = 2.75 \times 10^{-5}$					2.52×10^4	
	450 ft-c	-	$k_{10} = 8.75 \times 10^{-5}$	11.9	1	-	4.4 - 10	0.792×10^4	
		-	$k_{4.4} = 5.70 \times 10^{-5}$					1.22×10^4	
	Gold-pigmented BMP								
Light intensity:	150 ft-c	-	$k_{10} = 0.655 \times 10^{-5}$	11.6	1	0.749	1.7 - 10	10.58×10^4	
		-	$k_{4.4} = 0.570 \times 10^{-5}$					12.16×10^4	
		-	$k_{1.7} = 0.327 \times 10^{-5}$					21.20×10^4	
	300 ft-c	-	$k_{10} = 1.53 \times 10^{-5}$	20.5	1	0.773	1.7 - 10	4.53×10^4	
		-	$k_{4.4} = 1.16 \times 10^{-5}$					5.98×10^4	
		-	$k_{1.7} = 0.453 \times 10^{-5}$					15.30×10^4	
	450 ft-c	-	$k_{10} = 0.920 \times 10^{-5}$	15.5	1	-	4.4 - 10	7.53×10^4	
		-	$k_{4.4} = 0.527 \times 10^{-5}$					13.15×10^4	

System / Condition		Rate constant	E_a	n		T (°C)	D	Reference
Light intensity:								Singh et al., 1975
Paperboard 150 ft-c		$k_{10} = 0.453 \times 10^{-5}$	14.2	1	0.313	1.7 - 10	15.30×10^{4}	
		$k_{4,4} = 0.648 \times 10^{-5}$					10.70×10^{4}	
		$k_{1,7} = 0.170 \times 10^{-5}$					40.77×10^{4}	
300 ft-c		$k_{10} = 1.75 \times 10^{-5}$	49.8	1	0.916	1.7 - 10	3.96×10^{4}	
		$k_{4,4} = 1.547 \times 10^{-5}$					12.67×10^{4}	
		$k_{1,7} = 0.103 \times 10^{-5}$					67.30×10^{4}	
450 ft-c		$k_{10} = 0.302 \times 10^{-5}$	23.0	1	-	4.4 - 10	22.95×10^{4}	
		$k_{4,4} = 0.690 \times 10^{-5}$					10.05×10^{4}	
Milk (skim) Liquid — Light exposure = 150 ft-c at 4°C								Furuya et al., 1984
First phase	0.97	$k_{4} = 0.834 \times 10^{-3}$	-	1	-	4	831	
Milk (nonfat dry milk powder)								
First phase	0.90	$k_{4} = 0.190 \times 10^{-3}$	-	1	-	4	3,648	
Second phase	0.68	$k_{4} = 2.50 \times 10^{-3}$	-	1	-	4	27,730	
Vitamin B$_6$								Evans et al., 1981
Breakfast cereal model system — Toasted		[a]$k_{200} = 0.4895$	30.0	1	0.999	155 - 200	1.4	
		$k_{185} = 0.1688$					4.1	
Pyridoxine		$k_{170} = 0.0522$					13	
		$k_{155} = 0.0174$					40	
Casein-based liquid model system								Gregory and Hiner, 1983
Pyridoxine	0.98	$k_{133} = 0.0083$	28.6	1	0.995	105 - 133	84	
	0.98	$k_{118} = 0.0025$					277	
	0.98	$k_{105} = 0.0006$					1,155	
Pyridoximine	0.99	$k_{133} = 0.0187$	23.8	1	0.999	105 - 133	37	
	0.98	$k_{118} = 0.0064$					108	
	0.98	$k_{105} = 0.0021$					330	
Pyridoxal	0.94	$k_{133} = 0.0266$	20.7	1	0.998	105 - 133	26	
	0.98	$k_{118} = 0.0092$					75	
	0.96	$k_{105} = 0.0040$					173	

(Continued)

TABLE 2.3 Continued

Commodity	Process / Conditions	r^2	k_T value (min^{-1})	E_a (kcal/mol)	Reaction Order	r^2	Temp. Range (°C)	$t_{1/2}$ (min)	Reference
Cauliflower (puree)									
Overall B$_6$	Heated in 125-ml flasks	-	[a]$k_{137.7}$ = 0.01145	13.8	1	0.990	105.9 - 137.7	61	Navankasattusas and Lund, 1982
		-	$k_{125.6}$ = 0.00652	(27±2)[a]	pseudo-1[a]			106	
		-	$k_{114.6}$ = 0.00453					153	
		-	$k_{105.9}$ = 0.00265					262	
Vitamin B$_{12}$									
Milk (cow)									Watanabe et al., 1998
Conventional boil 30 min / CUT = ~7.5 min		0.990	k_{100} = 0.01828	-	1	-	100	37.9	
Microwave 6 min / CUT = ~ 2 min		0.999	k_{100} = 0.09155	-	1	-	100	7.6	
Folates									
Folic acid									
Apple juice	pH 3.4	-	[a]k_{140} = 0.01370	19.9	1	0.998	100 - 140	51	Mnkeni and Beveridge, 1982
Method: *L. casei*		-	k_{130} = 0.00792					88	
		-	k_{121} = 0.00468					148	
		-	k_{110} = 0.00202					343	
		-	k_{100} = 0.00105					660	
Tomato juice	pH 4.3	-	k_{140} = 0.012050					58	
		-	k_{130} = 0.006567					106	
		-	k_{121} = 0.003417	19.9	1	0.996	100 - 140	203	
		-	k_{110} = 0.001733					400	
		-	k_{100} = 0.000933					743	
Citrate buffer	pH 3.0	-	k_{140} = 0.039330					18	
		-	k_{130} = 0.027670					25	
		-	k_{121} = 0.012400	22.4	1	0.989	100 - 140	56	
		-	k_{110} = 0.004883					142	
		-	k_{100} = 0.002417					287	
	pH 4.0	-	k_{140} = 0.014000					50	
		-	k_{130} = 0.009667					72	
		-	k_{121} = 0.003900	19.4	1	0.982	100 - 140	178	
		-	k_{110} = 0.002133					325	
		-	k_{100} = 0.001233					562	

System / Conditions		Rate constants							Reference
Citrate buffer cont. -	pH 5.0	k_{140} = 0.003450	-	201					Mnkeni and Beveridge, 1982
		k_{130} = 0.001867	-	371					
		k_{121} = 0.0008833	-	785	17.8	1	0.975	100 - 140	
		k_{110} = 0.0005167	-	1,341					
		k_{100} = 0.0003500	-	1,980					
Model system (solid) Avicel / glycerol (60:40) Initial folate conc. = 0.2 mg/g solids HPLC	Heat in 6 x 50 mm glass tubes Moisture content (g H₂O/100g solids)								Hawkes and Villota, 1989a
	6.8	0.997 k_{80} = 0.0005109		1,357					
		0.994 k_{70} = 0.0003386		2,047	9.47	1	0.998	50 - 80	
		0.961 k_{60} = 0.0002298		3,016					
		0.984 k_{50} = 0.0001436		4,827					
	11.0	0.986 k_{80} = 0.0005210		1,330					
		0.977 k_{70} = 0.0003386		2,047	9.39	1	0.998	50 - 80	
		0.998 k_{60} = 0.0002365		2,931					
		0.952 k_{50} = 0.0001475		4,699					
	18.0	0.995 k_{80} = 0.0006502		1,066					
		0.990 k_{70} = 0.0004199		1,651	8.65	1	0.995	50 - 80	
		0.990 k_{60} = 0.0003033		2,285					
		0.997 k_{50} = 0.0002025		3,423					
	27.0	0.993 k_{80} = 0.0007266		954					
		0.993 k_{70} = 0.0005189		1,336	8.82	1	0.998	50 - 80	
		0.984 k_{60} = 0.0003357		2,065					
		0.999 k_{50} = 0.0002294		3,022					
	40.0	0.998 k_{80} = 0.0008392		826					
		0.998 k_{70} = 0.0005628		1,232	9.66	1	0.996	50 - 80	
		0.992 k_{60} = 0.0003463		2,002					
		0.998 k_{50} = 0.0002379		2,914					
Phosphate buffer pH 7.0 HPLC	Capillary tubes: 1.5 x 150 mm	k_{160} = 0.0047	-	147					Nguyen et al., 2003
		k_{140} = 0.0013	-	550	[a]12.35	1	0.8	120 - 160	
		k_{120} = 0.0010	-	666					

(Continued)

TABLE 2.3 Continued

Commodity	Process / Conditions	r^2	k_T value (min^{-1})	E_a (kcal/mol)	Reaction Order	r^2	Temp. Range (°C)	$t_{1/2}$ (min)	Reference
Strawberries (fresh / whole - *Zephyr*)									Strälsjö et al., 2003
Total folate	Stored: 0 - 9 days	0.998	k_4 = 0.000026	-	1	-	4	26,925	
	Open air	0.996	k_{20} = 0.000104	-	1	-	20	6,677	
Swiss chard	Storage conditions:								Gami and Chen, 1985
	Plastic bag	0.998	k_{21} = 0.000135	-	1	-	21	5,134	
	Moist conditions	0.996	k_{21} = 0.000149	-	1	-	21	4,652	
	Open air	-	k_{40} = 0.001645					421	
		-	k_{35} = 0.001042	22.9	1	0.996	4 - 40	665	
		-	k_{21} = 0.000205					3,381	
		-	k_4 = 0.00001433					48,370	
5-Methyltetrahydrofolate									
Apple juice Method: *L. casei*	pH 3.4 (unlimited O_2)	0.97	k_{70} = 0.249					2.8	Mnkeni and Beveridge, 1983
		0.95	k_{60} = 0.193	7.9	1	0.99	50 - 70	3.6	
		0.96	k_{50} = 0.123					5.6	
	pH 3.4 (limited O_2, at 5.3 ppm)	0.89	k_{70} = 0.200					3.5	
		0.92	k_{60} = 0.126	9.5	1	0.98	50 - 70	5.5	
		0.99	k_{50} = 0.089					7.8	
Tomato juice	pH 4.3 (unlimited O_2)	0.93	k_{130} = 1.065					0.65	
		0.92	k_{121} = 0.792	10.6	1	0.99	100 - 130	0.88	
		0.95	k_{110} = 0.508					1.4	
		0.99	k_{100} = 0.374					1.9	
	pH 4.3 (limited O_2, at 5.3 ppm)	0.98	k_{130} = 0.488					1.4	
		0.95	k_{121} = 0.353	10.8	1	0.99	100 - 130	2.0	
		0.97	k_{110} = 0.259					2.7	
		0.98	k_{100} = 0.160					4.3	
Phosphate buffer	Capillary tubes: 1.5 x 150 mm pH 7.0	-	k_{90} = 0.0683					10.2	Nguyen et al., 2003
		-	k_{80} = 0.0281	19.1	1	0.993	65 - 90	24.6	
		-	k_{70} = 0.0131					53.1	
		-	k_{65} = 0.0097					71.2	

System / Conditions	k						Reference
Phosphate buffer							Viberg et al., 1997
Micro-scale UHT simulation pH 7.0							
anaerobic O$_2$ at 0.3 ppm	k_{150} = 1.630					0.425	
	k_{140} = 1.120					0.619	
	k_{120} = 0.488					1.420	
	k_{110} = 0.242	14.9	1	0.989	110 - 150	2.864	
aerobic O$_2$ at 6.8 ppm	k_{150} = 3.180					0.218	
	k_{140} = 1.650					0.132	
	k_{120} = 0.638	18.3	1	0.994	110 - 150	0.207	
	k_{110} = 0.300	(25.5)	(2)	0.998		0.690	
Model system (solid)							Hawkes and Villota, 1989a
Avicel / glycerol (60:40)							
Initial folate conc. = 0.02 mg/g solids							
HPLC							
Moisture content g H$_2$O/100g solids							
5.0	k_{80} = 0.007075			0.974		98.0	
6.8	k_{80} = 0.007098			0.986		97.7	
11.0	k_{80} = 0.007249	-	1	0.994	80	95.6	
18.0	k_{80} = 0.009377			0.970		73.9	
27.0	k_{80} = 0.011590			0.980		59.8	
40.0	k_{80} = 0.014780			0.987		46.9	
Pantothenic acid (PA)							Hamm and Lund, 1978
Meat puree (beef)							
Free PA pH 5.4	-	[a]20	1	-	118 - 143	-	
Total PA pH 5.4	-	25	1	-	118 - 143	-	
Pea puree							
Free PA pH 7.0	-	38	1	-	118 - 143	-	
Total PA pH 7.0	-	36	1	-	118 - 143	-	
Buffered PA pH 4.0	-	20	1	-	118 - 143	-	
pH 5.0	-	22	1	-	118 - 143	-	
pH 6.0	-	27	1	-	118 - 143	-	

(Continued)

TABLE 2.3 Continued

Commodity	Process / Conditions	r^2	k_T value (min⁻¹)	E_a (kcal/mol)	Reaction Order	r^2	Temp. Range (°C)	$t_{1/2}$ (min)	Reference
Vitamin A									
Beef liver puree	*Trans*-retinol	-	$k_{126.7}$ = 0.09738					7.1	Wilkinson
		-	$k_{122.1}$ = 0.05766					12	et al., 1981
		-	$k_{118.3}$ = 0.04080	26.9	1	0.995	102.9 - 126.7	17	
		-	$k_{111.0}$ = 0.02316					30	
		-	$k_{102.9}$ = 0.01074					65	
			Fat Soluble Vitamins:						
Beef liver model systems	(% fat/ protein/ moisture/ ash + carbohydrate)								
Sample 1:	10.9/ 21.6/ 63.2/ 4.3	[a]0.992	k_{122} = 0.006162					112	Wilkinson
		0.993	k_{112} = 0.002694	26.4	1	0.999	102 - 122	257	et al., 1982
		0.986	k_{102} = 0.001026					676	
Sample 2:	27.9/ 11.7/ 56.0/ 4.4	0.993	k_{122} = 0.003480					199	
		0.984	k_{112} = 0.001392	24.2	1	0.993	102 - 122	498	
		0.990	k_{102} = 0.000672					1,031	
Sample 3:	10.0/ 13.0/ 72.2/ 4.8	0.996	k_{122} = 0.007704					90	
		0.997	k_{112} = 0.003696	23.9	1	0.998	102 - 122	188	
		0.999	k_{102} = 0.001524					455	
Sample 4:	26.1/ 20.2/ 51.7/ 2.0	0.979	k_{122} = 0.002718					255	
		0.925	k_{112} = 0.000780	27.0	1	0.953	102 - 122	889	
		0.998	k_{102} = 0.000432					1,605	
(15% protein/ 17-30% fat)	ppm Cu^{+2} / pH / % moisture								
Sample 1:	6/ 5.6/ 55	[a]0.999	k_{122} = 0.002730	[a]21.8	1	-	102 - 122	254	
		0.974	k_{102} = 0.000666					1,041	
Sample 2:	30/ 5.6/ 55	0.994	k_{122} = 0.001213	8.6	1	-	102 - 122	571	
		0.994	k_{102} = 0.000774					896	

	Condition	r	k	Ea		n	T (°C)		Reference
Sample 3:	6/ 7.0/ 55	0.975	k_{122} = 0.002781	28.0	-	1	102 - 122	255	Wilkinson et al., 1982
		0.980	k_{102} = 0.000498					1,392	
Sample 4:	30/ 7.0/ 55	0.971	k_{122} = 0.001368	8.6	-	1	102 - 122	507	
		0.998	k_{102} = 0.000666					1,041	
Sample 5:	6/ 5.6/ 68	0.998	k_{122} = 0.011020	28.2	-	1	102 - 122	63	
		0.996	k_{102} = 0.001650					420	
Sample 6:	30/ 5.6/ 68	0.999	k_{122} = 0.009210	23.2	-	1	102 - 122	75	
		0.988	k_{102} = 0.002270					305	
Sample 7:	6/ 7.0/ 68	0.998	k_{122} = 0.010690	28.9	-	1	102 - 122	65	
		0.996	k_{102} = 0.001542					450	
Sample 8:	30/ 7.0/ 68	0.993	k_{122} = 0.005028	17.2	-	1	102 - 122	138	
		0.999	k_{102} = 0.001458					475	
Carotene (crystalline)	Heated dry in 2-ml glass vials (50-150°C / 10-30 min)								Chen et al., 1994
All-*trans*-α-Carotene	HPLC analysis dissolved in hexane for analysis								
⇌ 13-*cis*-α-carotene		0.994							
Forward reaction (k_f)			[a]k_{150} = 0.0323 →	-		1	150	-	
Reverse reaction (k_r)			k_{150} = 0.0998 ↓	-		1	150	-	
⇌ 9-*cis*-α-carotene		0.980							
Forward reaction (k_f)			k_{150} = 0.0038 →	-		1	150	-	
Reverse reaction (k_r)			k_{150} = 0.0449 ↓	-		1	150	-	
⇌ 15-*cis*-α-carotene		0.990							
Forward reaction (k_f)			k_{150} = 0.0067 →	-		1	150	-	
Reverse reaction (k_r)			k_{150} = 0.0407 ↓	-		1	150	-	

(Continued)

(Continued)

TABLE 2.3 Continued

Commodity	Process / Conditions	r^2	k_T value (min⁻¹)	E_a (kcal/mol)	Reaction Order	r^2	Temp. Range (°C)	$t_{1/2}$ (min)	Reference
All-*trans*-β-Carotene	dissolved in MeOH: CHCl₃ (45:55) for analysis								Chen et al., 1994
⇌ 13-*cis*-β-carotene									
Forward reaction (k_f)		0.996	k_{150} = 0.0125 →	-	1	-	150	-	
Reverse reaction (k_1)			k_{150} = 0.0339 ←	-	1	-	150	-	
⇌ 9-*cis*-β-carotene									
Forward reaction (k_f)		0.990	k_{150} = 0.0043 →	-	1	-	150	-	
Reverse reaction (k_1)			k_{150} = 0.0184 ←	-	1	-	150	-	
Corn flakes (fortified)	Storage:								Kim et al., 2000
Vitamin A palmitate (15% RDI)	23 - 45°C, up to 16 wks	0.960	ᵃk_{45} = 1.2639 x 10⁵	-	apparent-2	-	23 - 45	5.484 x 10⁴	
		0.954	k_{23} = 1.0208 x 10⁵	-	apparent-2	-	23 - 45	6.790 x 10⁴	
Vitamin A palmitate	Packaging:								
(plus B₁, B₆, B₁₂, C, D₂ & D₃ - 15% RDI each)	cardboard box with plastic liner	0.960	k_{45} = 17.5694 x 10⁵	-	apparent-1	-	23 - 45	0.395 x 10⁴	
		0.793	k_{23} = 4.5139 x 10⁵	-	apparent-1	-	23 - 45	1.536 x 10⁴	
Enteral feeding formula (*5.5% protein, 3.6% lipid, 11.4% CHO*) Reconstitution; presterilization (UHT: 136°C / 3-4 s); homogenize; add vitamins; pack in glass; sterilize (118°C / min); store 4-30°C / 0-9 mo.									Frias and Vidal-Valverde, 2001
All-*trans*-retinol		0.780	k_{30} = 5.628 x 10⁻⁶					1.23 x 10⁵	
		0.840	k_{20} = 5.859 x 10⁻⁶	1.06	1	0.708	4 - 30	1.18 x 10⁵	
		0.829	k_4 = 4.849 x 10⁻⁶					1.43 x 10⁵	
13-*cis*-retinol		0.945	k_{30} = 3.853 x 10⁻⁶					1.80 x 10⁵	
		0.881	k_{20} = 2.766 x 10⁻⁶	2.47	1	0.784	4 - 30	2.51 x 10⁵	
		0.809	k_4 = 2.538 x 10⁻⁶					2.73 x 10⁵	
Vitamin A activity		0.831	k_{30} = 11.21 x 10⁻⁶					0.619 x 10⁵	
		0.844	k_{20} = 5.581 x 10⁻⁶	5.19	1	0.783	4 - 30	1.24 x 10⁵	
		0.830	k_4 = 4.664 x 10⁻⁶					1.49 x 10⁵	

Description	r²	k	E_a	n	r²	Temp (°C)	value	Reference
Milk / infant (stored 12 months)								
Liquid: 115°C/15 min sealed in brown glass jars	-	k_{37} = 7.87 x 10^{-7}	-	1	-	37	8.81 x 10^5	Albalá-Hurtado et al., 2000
Powdered: 40% conc. / 60°C; 70-72°C/15 sec; spray dry 80°C; agglomerated	-	k_{37} = 9.44 x 10^{-7}	-	1	-	37	7.34 x 10^5	
Milk								
Flexible plastic containers, no headspace								Galdi et al., 1989
infant formula [15% protein, (milk-based); 24% lipids; 57% carbohydrates; 4% vitamins, minerals, water]								
With ferrous sulfate	0.978	k_{45} = 2.92 x 10^{-6}	7.6	1	0.991	20 - 45	2.37 x 10^5	
	0.993	k_{37} = 2.36 x 10^{-6}					2.94 x 10^5	
	0.935	k_{20} = 1.07 x 10^{-6}					6.48 x 10^5	
With ferric glycinate	0.991	k_{45} = 2.39 x 10^{-6}	8.5	1	0.909	20 - 45	2.90 x 10^5	
	0.970	k_{37} = 1.18 x 10^{-6}					5.87 x 10^5	
	0.929	k_{20} = 0.694 x 10^{-6}					9.99 x 10^5	
Butternut squash Carotene	-	[a]k_{80} = 0.00110	11.8 (13.5)[a]	1	0.998	60 - 80	630	Stefanovich and Karel, 1982
	-	k_{70} = 0.00070					990	
	-	k_{60} = 0.00040					1,733	
Yellow corn Carotene	-	k_{80} = 0.00135	20.5 (4.8)[a]	1	0.735	60 - 80	513	
	-	k_{70} = 0.00023					3,014	
	-	k_{60} = 0.00023					3,014	
Model system Avicel-PH102 / β-carotene	-	k_{80} = 0.02390	21.8 (21.7)[a]	pseudo-1	0.984	60 - 80	29	
	-	k_{70} = 0.01190					58	
	-	k_{60} = 0.00370					187	
Sweet potato	-	k_{80} = 0.00116	10.5 (10.6)[a]	1	0.936	60 - 80	598	
	-	k_{70} = 0.00061					1,136	
	-	k_{60} = 0.00047					1,475	

(Continued)

TABLE 2.3   Continued

Commodity	Process / Conditions	r^2	k_T value (min^{-1})	E_a (kcal/mol)	Reaction Order	r^2	Temp. Range (°C)	$t_{1/2}$ (min)	Reference
Sweet potato									Haralampu and Karel. 1983
β-carotene: test 1	Freeze-dried / ground / rehumidified								
	g H$_2$O / g solids								
a$_w$ = 0.020	0.008	0.978	k$_{40}$ = 1.01 x 10^{-4}	-	pseudo-1	-	40	6.88 x 10^3	
a$_w$ = 0.058	0.026	0.999	k$_{40}$ = 0.638 x 10^{-4}	-	pseudo-1	-	40	10.86 x 10^3	
a$_w$ = 0.112	0.034	0.993	k$_{40}$ = 0.570 x 10^{-4}	-	pseudo-1	-	40	12.16 x 10^3	
a$_w$ = 0.316	0.068	0.993	k$_{40}$ = 0.423 x 10^{-4}	-	pseudo-1	-	40	16.39 x 10^3	
a$_w$ = 0.484	0.085	0.992	k$_{40}$ = 0.382 x 10^{-4}	-	pseudo-1	-	40	18.15 x 10^3	
a$_w$ = 0.555	0.101	0.996	k$_{40}$ = 0.313 x 10^{-4}	-	pseudo-1	-	40	22.15 x 10^3	
a$_w$ = 0.661	0.116	0.981	k$_{40}$ = 0.347 x 10^{-4}	-	pseudo-1	-	40	19.98 x 10^3	
a$_w$ = 0.747	0.204	0.995	k$_{40}$ = 0.320 x 10^{-4}	-	pseudo-1	-	40	21.66 x 10^3	
β-carotene: test 2	g H$_2$O / g solids								
a$_w$ = 0.020	0.030	0.994	k$_{40}$ = 2.15 x 10^{-4}	-	pseudo-1	-	40	3.22 x 10^3	
a$_w$ = 0.058	0.058	0.996	k$_{40}$ = 1.16 x 10^{-4}	-	pseudo-1	-	40	5.98 x 10^3	
a$_w$ = 0.112	0.064	0.998	k$_{40}$ = 0.789 x 10^{-4}	-	pseudo-1	-	40	8.79 x 10^3	
a$_w$ = 0.316	0.094	0.997	k$_{40}$ = 0.653 x 10^{-4}	-	pseudo-1	-	40	10.61 x 10^3	
a$_w$ = 0.484	0.110	0.994	k$_{40}$ = 0.577 x 10^{-4}	-	pseudo-1	-	40	12.01 x 10^3	
a$_w$ = 0.555	0.133	0.972	k$_{40}$ = 0.417 x 10^{-4}	-	pseudo-1	-	40	16.62 x 10^3	
a$_w$ = 0.661	0.147	0.999	k$_{40}$ = 0.368 x 10^{-4}	-	pseudo-1	-	40	18.84 x 10^3	
a$_w$ = 0.747	0.190	0.995	k$_{40}$ = 0.423 x 10^{-4}	-	pseudo-1	-	40	16.39 x 10^3	
Tomato paste	Storage 0-90 days	0.98	k$_{50}$ = 7.43 x 10^{-6}	4.97	1	0.917	30 - 50	1.35 x 10^5	Lavelli and Giovanelli, 2003
(reported as reduction in total antioxidant activity - β-carotene and lycopene)		0.91	k$_{40}$ = 5.07 x 10^{-6}	(4.83)[a]				1.37 x 10^5	
		0.89	k$_{30}$ = 4.44 x 10^{-6}					1.56 x 10^5	

Description		r^2	k					Temp (°C)	rate	Reference
Tomato pulp		0.96	k_{50}	$= 5.28 \times 10^{-6}$	5.49	1	0.795	30 - 50	1.31×10^5	Lavelli and Giovanelli, 2003
		0.89	k_{40}	$= 3.13 \times 10^{-6}$	$(5.31)^{a}$				2.22×10^5	
		0.94	k_{30}	$= 2.99 \times 10^{-6}$					2.32×10^5	
Tomato pulp /w/ XS heat (additional 98°C / 50 min)		0.97	k_{40}	$= 3.33 \times 10^{-6}$	-	1	-	40	2.08×10^5	
Tomato puree		0.980	k_{40}	$= 2.71 \times 10^{-6}$	-	1	-	40	2.56×10^5	

Vitamin D

Vitamin D_2 (ergocalciferol)

Description		r^2	k					Temp (°C)	rate	Reference
Model system (12% water / 88% acetone)	Air-tight serum bottles /w/ light									Li and Min, 1998
Phase 1:										
Initial concentration: 1000 ppm D_2	0 ppm B_2	0.99	k_{25}	$= 1.12 \times 10^{-4}$	-	1	-	25 - 60	6.21×10^3	
		0.91	k_{60}	$= 3.24 \times 10^{-4}$	-	1	-	25 - 60	2.14×10^3	
	15 ppm B_2	0.90	k_{25}	$= 5.48 \times 10^{-4}$	-	1	-	25 - 60	1.27×10^3	
		0.94	k_{60}	$= 6.96 \times 10^{-4}$	-	1	-	25 - 60	0.996×10^3	
Phase 2:	0 ppm B_2	0.86	k_{25}	$= 0.217 \times 10^{-4}$	-	1	-	25 - 60	31.9×10^3	
		0.96	k_{60}	$= 0.200 \times 10^{-4}$	-	1	-	25 - 60	34.7×10^3	
	15 ppm B_2	0.87	k_{25}	$= 0.583 \times 10^{-4}$	-	1	-	25 - 60	11.9×10^3	
		0.83	k_{60}	$= 0.250 \times 10^{-4}$	-	1	-	25 - 60	27.7×10^3	

Vitamin E

Enteral feeding formula (5.5% protein, 3.6% lipid, 11.4% CHO)
Reconstitution; presterilization (UHT: 136°C / 3-4 s); homogenize; add vitamins; pack in glass; sterilize (118°C / 9 min); store 4-30°C / 0-9 mo.

Description	r^2	k					Temp (°C)	rate	Reference
α-Tocopherol	0.874	k_{30}	$= 1.835 \times 10^{-6}$	2.46	1	0.999	4 - 30	3.78×10^5	Frias and Vidal-Valverde, 2001
	0.608	k_{20}	$= 1.598 \times 10^{-6}$					4.34×10^5	
	0.611	k_{4}	$= 1.250 \times 10^{-6}$					5.55×10^5	
γ-Tocopherol	0.874	k_{30}	$= 2.300 \times 10^{-6}$	4.11	1	0.999	4 - 30	3.01×10^5	
	0.608	k_{20}	$= 1.799 \times 10^{-6}$					3.85×10^5	
	0.611	k_{4}	$= 1.210 \times 10^{-6}$					5.73×10^5	

(Continued)

TABLE 2.3 Continued

Commodity	Process / Conditions	r^2	k_T value (min^{-1})	E_a (kcal/mol)	Reaction Order	r^2	Temp. Range (°C)	$t_{1/2}$ (min)	Reference
Enteral feeding formula (5.5% protein, 3.6% lipid, 11.4% CHO) (cont.)									Frias and Vidal-Valverde, 2001
δ-Tocopherol		0.874	k_{30} = 1.943 x 10^{-6}					3.57 x 10^5	
		0.608	k_{20} = 1.572 x 10^{-6}	2.82	1	0.985	4 - 30	4.41 x 10^5	
		0.611	k_4 = 1.242 x 10^{-6}					5.58 x 10^5	
Vitamin E activity		0.939	k_{30} = 1.844 x 10^{-6}					3.76 x 10^5	
		0.936	k_{20} = 1.603 x 10^{-6}	2.50	1	0.999	4 - 30	4.32 x 10^5	
		0.915	k_4 = 1.249 x 10^{-6}					5.55 x 10^5	
Enteral feeding formula (3.7% protein, 3.9% lipid, 12.5% CHO)									
α-Tocopherol		0.952	k_{30} = 2.126 x 10^{-6}					3.26 x 10^5	
		0.946	k_{20} = 1.774 x 10^{-6}	2.51	1	0.989	4 - 30	3.91 x 10^5	
		0.938	k_4 = 1.427 x 10^{-6}					4.86 x 10^5	
γ-Tocopherol		0.873	k_{30} = 2.471 x 10^{-6}					2.81 x 10^5	
		0.775	k_{20} = 1.879 x 10^{-6}	3.21	1	0.964	4 - 30	3.69 x 10^5	
		0.772	k_4 = 1.474 x 10^{-6}					4.70 x 10^5	
δ-Tocopherol		0.920	k_{30} = 2.473 x 10^{-6}					2.80 x 10^5	
		0.885	k_{20} = 2.057 x 10^{-6}	2.68	1	0.993	4 - 30	3.37 x 10^5	
		0.891	k_4 = 1.620 x 10^{-6}					4.28 x 10^5	
Vitamin E activity		0.952	k_{30} = 2.134 x 10^{-6}					3.25 x 10^5	
		0.944	k_{20} = 1.777 x 10^{-6}	2.54	1	0.989	4 - 30	3.90 x 10^5	
		0.936	k_4 = 1.428 x 10^{-6}					4.85 x 10^5	
Milk									
Infant formula [15% protein (milk-based); 24% lipids; 57% carbohydrates; 4% vitamin, minerals, water]	Flexible plastic containers, no headspace. With ferrous sulfate	0.988	k_{45} = 6.25 x 10^{-6}					1.11 x 10^5	Galdi et al., 1989
		0.900	k_{37} = 2.59 x 10^{-6}	9.4	1	0.998	20 - 45	2.68 x 10^5	
		0.861	k_{20} = 1.78 x 10^{-6}					3.89 x 10^5	

System	Conditions	a_w / rate constants		Q_{10}		Correlation	Temp (°C)	k values	Reference
Infant formula cont. -	With ferric glycinate	0.958	$k_{45} = 3.82 \times 10^{-6}$	9.6	1	0.999	20 - 45	1.81×10^5	Galdi et al., 1989
		0.932	$k_{37} = 2.50 \times 10^{-6}$					2.77×10^5	
		0.926	$k_{20} = 1.04 \times 10^{-6}$					6.66×10^5	
Model system: α-Tocopherol (48% starch, 34% corn syrup solids, 10.2% soy isolate, 5.1% sucrose, 2.0% NaCl)	Freeze-dried / rehumidified Packaging: 303 x 406 cans (XS headspace)	$a_w = 0.10$	$k_{37} = 0.557 \times 10^{-5}$ $k_{30} = 0.345 \times 10^{-5}$ $k_{20} = 0.225 \times 10^{-5}$	9.5	1	0.977	20 - 37	1.24×10^5 2.01×10^5 3.08×10^5	Widicus et al., 1980
		$a_w = 0.24$	$k_{37} = 0.892 \times 10^{-5}$ $k_{30} = 0.432 \times 10^{-5}$ $k_{20} = 0.310 \times 10^{-5}$	10.8	1	0.894	20 - 37	0.777×10^5 1.60×10^5 2.24×10^5	
		$a_w = 0.40$	$k_{37} = 1.031 \times 10^{-5}$ $k_{30} = 0.436 \times 10^{-5}$ $k_{20} = 0.363 \times 10^{-5}$	10.4	1	0.788	20 - 37	0.672×10^5 1.59×10^5 1.91×10^5	
		$a_w = 0.65$	$k_{37} = 1.107 \times 10^{-5}$ $k_{30} = 0.494 \times 10^{-5}$ $k_{20} = 0.360 \times 10^{-5}$	11.4	1	0.871	20 - 37	0.626×10^5 1.40×10^5 1.93×10^5	
	Packaging: 208 x 006 TDT (no headspace)	$a_w = 0.10$	$k_{37} = 0.599 \times 10^{-5}$ $k_{30} = 0.326 \times 10^{-5}$ $k_{20} = 0.224 \times 10^{-5}$	10.1	1	0.936	20 - 37	1.16×10^5 2.12×10^5 3.09×10^5	
		$a_w = 0.24$	$k_{37} = 0.790 \times 10^{-5}$ $k_{30} = 0.410 \times 10^{-5}$ $k_{20} = 0.226 \times 10^{-5}$	13.1	1	0.978	20 - 37	0.877×10^5 1.69×10^5 3.07×10^5	

(Continued)

TABLE 2.3 Continued

Commodity	Process / Conditions	r^2	k_T value (min⁻¹)	E_a (kcal/mol)	Reaction Order	r^2	Temp. Range (°C)	$t_{1/2}$ (min)	Reference
Model system: α -Tocopherol cont. -									Widicus et al., 1980
	$a_w = 0.40$	-	$k_{37} = 0.913 \times 10^{-5}$					0.795×10^5	
		-	$k_{30} = 0.419 \times 10^{-5}$	10.8	1	0.861	$20 - 37$	1.65×10^5	
		-	$k_{20} = 0.315 \times 10^{-5}$					2.20×10^5	
	$a_w = 0.65$	-	$k_{37} = 0.931 \times 10^{-5}$					0.745×10^5	
		-	$k_{30} = 0.451 \times 10^{-5}$	8.8	1	0.791	$20 - 37$	1.54×10^5	
		-	$k_{20} = 0.385 \times 10^{-5}$					1.80×10^5	
Seaweed (*Ascophyllum nodosum*) α -Tocopherol Moisture content:	Air-dried / ground								Jenson, 1969
	10%	0.954	$k_{25} = 0.441 \times 10^{-5}$					1.57×10^5	
		0.883	$k_{15} = 0.296 \times 10^{-5}$	10.1	1	0.880	$4 - 25$	2.34×10^5	
		0.881	$k_{10} = 0.256 \times 10^{-5}$					2.71×10^5	
		0.987	$k_4 = 0.111 \times 10^{-5}$					6.24×10^5	
	15%	0.923	$k_{25} = 0.538 \times 10^{-5}$					1.29×10^5	
		0.980	$k_{15} = 0.354 \times 10^{-5}$	12.5	1	0.893	$4 - 25$	1.96×10^5	
		0.926	$k_{10} = 0.261 \times 10^{-5}$					2.66×10^5	
		0.965	$k_4 = 0.0999 \times 10^{-5}$					6.94×10^5	
	25%	0.983	$k_{25} = 0.833 \times 10^{-5}$					0.832×10^5	
		0.902	$k_{15} = 0.561 \times 10^{-5}$	6.3	1	0.998	$4 - 25$	1.24×10^5	
		0.885	$k_{10} = 0.464 \times 10^{-5}$					1.49×10^5	
		0.902	$k_4 = 0.373 \times 10^{-5}$					1.86×10^5	

[a] Values reported by authors.

When dealing with vitamin C, it is important to understand the bioavailability and kinetics of retention in processing and storage of related compounds. Ascorbic acid and dehydroascorbic acid are the chemical forms with primary vitamin activity. Dehydroascorbic acid, however, is not stable to heat and is rapidly hydrolyzed to 2,3-diketogulonic acid and further breakdown products. Other compounds such as ascorbate-2-phosphate, a fully active compound (Liao and Seib, 1988); ascorbigen, a form of ascorbic acid bound to phenols, with 10 to 20% bioavailability in guinea pigs (Matano and Kato, 1967); and isoascorbic acid (erythorbic acid) with about 5% vitamin activity and with limiting effect on the absorption of ascorbic acid (Hornig et al., 1974) may need some consideration as well.

Taking into account the stability of ascorbic acid in food systems during processing, vitamin C as well as thiamine and folic acid are normally considered to be good indicators of the severity of a food process. If these vitamins are well retained, we may safely assume that all other nutrients are well retained during processing. Based on the recent FDA dietary recommendations and admonitions to increase fruit and vegetable, as well as whole grain consumption, the significance of understanding vitamin retention during processing and its bioavailability have become increasingly important.

Since fruits and vegetables are major contributors to micronutrients, in particular vitamin C, a vast number of studies have been published reporting kinetic information on the stability of this compound in blanching, canning, dehydration, high pressure sterilization and freezing operations (Selman, 1994; Martins and Silva, 2003; Van den Broeck, et al., 1998; Giannakourou and Taouki, 2003; etc). Killeit (1994), for example, reviewed vitamin retention during extrusion and pointed out mostly destructive effects, with vitamin C being the most sensitive. He also reported that through modification of the vitamin molecule, there was potential for improved stability. An example cited was the commercially available L-ascorbyl-2-polyphosphate (AsPP), a modified form of ascorbic acid, used for feed applications. An extruded feed for catfish containing AsPP showed improved stability through the process as compared to a traditionally used ethylcellulose-coated ascorbic acid (83% retention vs. 39%) and no losses during storage as compared with 78% loss of the traditionally coated ascorbic acid (Robinson et al., 1989). Reports indicate complete bioavailability of this compound for the animals (Grant et al., 1989).

As summarized by Clydesdale et al. (1991), other components present in food systems may have major impact on the kinetics of ascorbic acid degradation by changing the reactivity of the system, thus eventually impacting its bioavailability. For instance, minerals such as iron and copper, vitamin E, flavonoids, amino acids, and sugars can significantly change the retention of ascorbic acid during processing and storage, as well as its bioavailability. Van den Broeck et al. (1998) reported that ascorbic acid found in real food systems such as orange juice and tomatoes had less stability than buffered solutions of ascorbic acid (pH 4–8) subjected to heat (120–150°C) or combined pressure/thermal treatment (8.5 kbar [850 MPa]/65–80°C). Lavelli and Giovanelli (2003) reported pseudo first-order kinetics of ascorbic acid degradation in tomato products as related to stability of various carotenoids and phenolics during storage (30–50°C/90 days) and found degradation of ascorbic acid even at 30°C.

Light-induced degradation of ascorbic acid in the presence of riboflavin has been a subject of investigation. Several studies suggest that riboflavin is first excited by visible light, and then, through an excitation transfer, ascorbic acid is oxidized. The reaction mechanism proposes the formation of H_2O_2 (Şahbaz and Somer, 1993; Şansal and Somer, 1997).

2.3.1.1.2 Vitamin B_1 (Thiamine)

Thiamine, also known as Vitamin B_1, thiamine chloride, aneurin, and antiberiberi vitamin, occurs either in the free thiamine form, as a protein complex, as mono-, di-, or triphosphate esters, or as a phosphorous protein complex. Foods considered to be rich in vitamin B_1 include nuts, pork, yeast, and cereal germ. Thiamine is also one of the least stable of the vitamins. Thiamine has been found unstable in neutral and alkaline solutions, while it can withstand up to 120°C for one hour in acidic solutions, although it is susceptible to cleavage by sulfites even in an acidic environment. In the dry form it is stable to oxidation, but in solution it is very unstable to oxidation and reduction reactions. Since thiamine can exist in many forms, it is obvious that its stability and its kinetics of degradation are greatly affected by the relative concentrations of the different forms (Farrer, 1955). Feliciotti and Esselen (1957), studying the thermal destruction of thiamine in pureed meats and vegetables, suggested that its destruction in foods was dependent on the interrelationship of pH and the relative proportions of the free and the combined forms of the vitamin. It has been observed that the enzyme-bound forms, cocarboxylases, appear to be less stable than the free forms. Mulley et al. (1975b) also reported that under identical conditions, cocarboxylase is destroyed faster than thiamine hydrochloride. It appears that the faster destruction of the cocarboxylase may be due to the pyrophosphoric acid group which is the basic difference between the two molecules, and that appears to cause additional reactivity or stress on the cocarboxylase molecule. The authors also reported that the presence of the cocarboxylase form does not affect the destruction of the free thiamine up to concentration levels of 35%. Since this appears to be the situation in most food products, it is expected that the cocarboxylase form will not interfere with the kinetics of degradation.

A number of factors will be highly influential in the stability of thiamine, including water activity, pH, temperature, ionic strength, and the presence of other compounds. Some of the suggested mechanisms for thiamine degradation are presented in Figure 2.10. The particular instability of thiamine to heat under neutral and alkaline conditions has resulted in various studies on the chemistry of thiamine degradation. However, due to the high complexity of food materials, a number of studies have been carried out in model systems in order to clarify the mechanisms involved in thiamine degradation. Several authors such as Farrer (1955), Beadle et al. (1943) and Greenwood et al. (1943) established that the thermal destruction of thiamine in aqueous and buffered solutions followed a first-order reaction. Farrer and Morrison (1949) studied the thermal degradation of thiamine in buffered solutions and determined that the Arrhenius equation could be used to describe the effect of temperature. Two possible reactions have been considered leading to the degradation of thiamine, namely, (a) the breaking of the "CH-bridge" leaving the pyrimidine and thiazole moieties and (b) the breakdown of the thiazole ring with the production of hydrogen sulfite. Limited efforts have been addressed to determine the governing mechanism of thiamine degradation in food systems and rather an overall response is commonly monitored. In fact, the lack of comprehensive kinetic data has limited our understanding of the significance of the different mechanisms involved. Dwivedi and Arnold (1973) summarized the most important aspects affecting the degradation of thiamine in food products and model systems.

Of great significance in the area of stability of thiamine has been the observations of several investigators that thiamine in natural foods is more heat-resistant than in aqueous and buffered systems. Thus, it appears that certain factors will influence the stability of the vitamin. For instance, Frost and McIntire (1944) indicated that α- and β-amino acids and some of their derivatives had a significant stabilizing effect upon thiamine at pH 6.0. In general, this effect became noticeable at pH values above the range 4.5 to 5.0. Other compounds such as proteins and starch have been found to improve the thermal stability of thiamine. The exact mechanisms involved are not well elucidated.

A controversy still exists on the role that oxygen plays in the thermal stability of thiamine. In fact, results presented by Williams and Spies (1938), Farrer (1955), and Mulley et al. (1975a) have demonstrated that the thermal degradation of thiamine can be described by a first-order kinetics and that the reaction was not oxidative in nature. On the other hand, other authors have suggested that in the case of products containing oxygen the reaction became a true first-order reaction upon the disappearance of oxygen (Farrer and Morrison, 1949). Fink and Kessler (1985) studied the retention of thiamine in milk in the temperature range from 4 to 150°C. The authors reported that for the range 35–50°C and 72–85°C, the reaction followed second-order kinetics. With regard to the effect

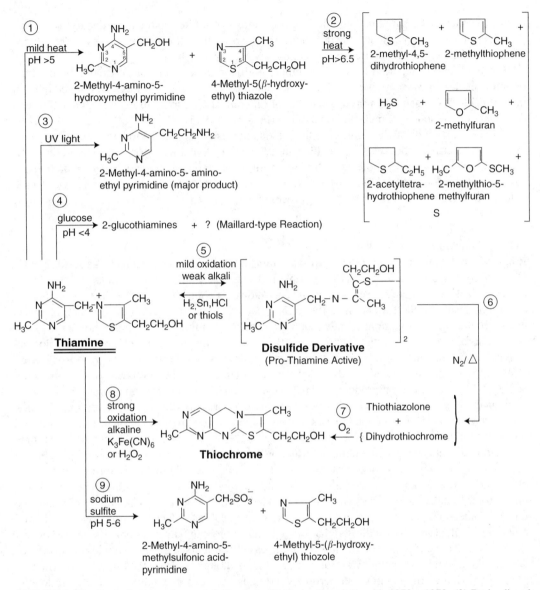

FIGURE 2.10 Degradation pathways of thiamine. (1) Dwivedi and Arnold, 1972a; 1973; (2) Dwivedi and Arnold, 1973; (3) Kawasaki and Daira, 1963; (4) Lhoest, 1958; (5) Zima and Williams, 1940; (6) Sykes and Todd, 1951; (7) Sykes and Todd, 1951; (8) Barger et al., 1935; (9) Metzler, 1960; Dwivedi and Arnold, 1972b.

of oxygen, the authors did not observe any effect on the rate of thiamine losses. Dennison et al. (1977), working with a dehydrated food system, also indicated that the presence of oxygen did not significantly affect the degradation of thiamine.

In model systems, looking at the effect of different solutes, Fernández et al. (1986) reported that the degradation of thiamine followed a first-order reaction and that the degradation was affected by the type of solute used in the formulation, increasing in the order sodium chloride, potassium chloride, glycerol, and sodium sulfate. Thus, the rate of degradation was affected not only by water activity but also by the specific solute.

It has been determined that in food products, compounds such as sulfites, phenols, amino acids, and proteins, as well as lipids, may have a significant effect on thiamine degradation and its associated kinetic parameters. Of particular significance is the effect of sulfites on thiamine due to the nucleophilicity of the sulfite ion. Hence, the destruction of thiamine by sulfite becomes a key issue in foods claiming to be a significant source of this vitamin (Vanderveen, 1988).

Fox et al. (1997) reported losses of thiamine in ground pork as a result of irradiation but little or no losses during conventional cooking, heat denaturation, or storage. It was also pointed out that the exclusion of oxygen may improve stability of thiamine during irradiation and that stability was dependant on the source of meat and type of cut (Fox et al., 1995).

As previously indicated, retention of thiamine is normally considered to be an indicator of the intensity of thermal processes such as blanching, canning, freezing, extrusion, dehydration, etc. (Selman, 1994; Ilo and Berghofer, 1998). Further information on kinetic destruction of thiamine is required for less conventional food processes such as gamma irradiation, high pressure, pulse electric fields, and microwave and radio frequency processing.

2.3.1.1.3 Vitamin B₂ (Riboflavin)

Riboflavin, or vitamin B_2, (also referred to as vitamin G, lactoflavin, and chemically as 7,8-dimethyl-10-($1'$-ribityl) isoalloxazine) is relatively stable in foods under ordinary conditions. Its stability is pH dependent, being more stable under acidic conditions. Its photochemical cleavage under alkaline conditions results in the formation of the highly reactive compound, lumiflavin, which mediates the destruction of other vitamins. Under neutral and acidic conditions, this vitamin loses the ribityl side chain forming lumichrome (Figure 2.11). Both lumichrome and lumiflavin have no biological activity. Moreover, the photolysis reaction is irreversible. Work carried out by Woodcock et al. (1982) in pasta products indicated that lumichrome, a photolysate of riboflavin, was not the only one or the final degradation product. In fact, for these types of products, 60% of the losses were accounted for by the presence of lumichrome and varied according to the process conditions. Palanuk and Warthesen (1988) studied the kinetics of degradation of riboflavin and lumichrome in milk and observed that the rate of degradation of riboflavin was 2.8 times greater than the rate of lumichrome formation and that the rate of lumichrome formation was 6.3 times greater than the rate of lumichrome degradation. The combined effect was such that after an increase in lumichrome formation, leveling off of the reaction took place. According to the authors' model, 23.4% of the riboflavin degraded to lumichrome, indicating that either riboflavin degraded to other products or that lumichrome became bound to other components in the system, becoming unavailable for determination. Results reported by Furuya et al. (1984) also indicated that lumichrome content in buffer systems leveled off during storage, while the concentration of riboflavin decreased continuously. Thus, simple monitoring of the formation of this compound will not be a reliable method to measure the losses of the vitamin. The degradation of riboflavin under aerobic conditions has been generally reported as following first-order reaction kinetics.

A list of some of the most important studies with the corresponding rates of degradation is presented in Table 2.3. Although the degradation of the vitamin has often been categorized as being a first-order reaction, the exact approach for the monitoring of the degradation will play a significant role. Woodcock et al. (1982) indicated that lumichrome production in pasta products followed two basic steps. The first stage followed first-order reaction kinetics that proceeded at a fast rate and a second stage that involved the disappearance of lumichrome and could not be easily described in kinetic terms. The photodegradation of riboflavin in milk has been determined to follow first-order kinetics (Singh et al., 1975; Allen and Parks, 1979). Kinetic parameters were found to be influenced by temperature and the presence of light. From the point of view of kinetics, limited work has been carried out to clearly determine the mechanisms involved in the degradation of riboflavin and their contribution to the overall kinetic values. Based on some of the work reported by several authors, it is evident that the concentration of lumiflavin and lumichrome will influence the kinetic values reported, if these products of the reaction are the ones to be monitored as a measurement of stability.

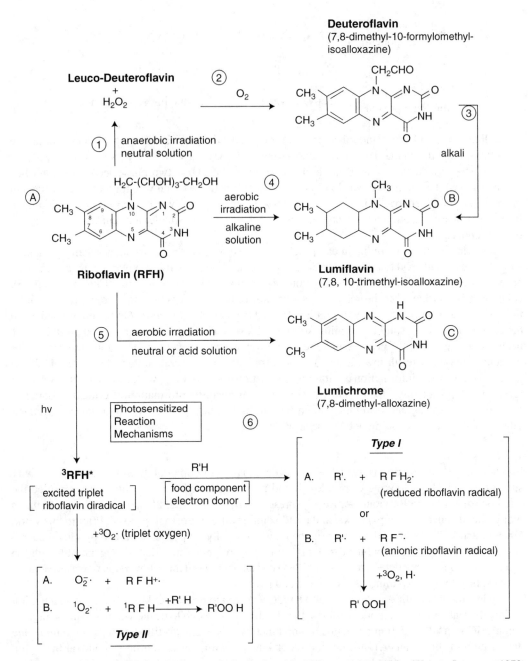

FIGURE 2.11 Degradation pathways of riboflavin. (1) Kuhn et al., 1933; Wagner-Jauregg, 1972; (2) Holmström and Oster, 1961; Kuhn et al., 1933; (3) Holmström and Oster, 1961; (4) Holmström and Oster, 1961; (5) Karrer et al., 1934; (6) Choe et al., 2005.

Studies presented by Toyosaki et al. (1988) indicate that the photolysis mechanism can be described according to the following reactions:

$$\text{Riboflavin} + h\nu \underset{k_{-1}}{\overset{k_1}{\rightleftharpoons}} \text{riboflavin-H} \overset{k_2}{\longrightarrow} \text{riboflavin decomposition}$$

$$\text{Riboflavin} + hv \atop \text{(multicomponents)} \quad \underset{k_{-1}}{\overset{k_1}{\rightleftharpoons}} \quad \text{riboflavin-H} + \text{active oxygen}$$

$$\overset{k_2}{\longrightarrow} \quad \text{riboflavin decomposition}$$

$$\text{Riboflavin} \atop \text{(multicomponents)} \; + \; \text{active oxygen} \quad \overset{k_3}{\longrightarrow} \quad \text{riboflavin decomposition}$$

According to the authors, standard riboflavin proceeded by one-phase decomposition under all the conditions studied. Standard riboflavin underwent photolysis in which no active oxygen was produced. On the other hand, milk serum riboflavin was photolyzed by a two-phase decomposition when the intensity of the irradiation was low. The presence of active oxygen was found to be involved in the reaction. Increased irradiation was reported to change the photolysis of the milk serum riboflavin from a two-phase to a one-phase decomposition mechanism, with smaller amounts of active oxygen being produced.

Photosensitization of riboflavin can produce reactive oxygen species such as superoxide anion, singlet oxygen, hydroxyl radical, and hydrogen peroxide. These reactive oxygen species and radicals have been found to affect the decomposition of proteins, lipids, and vitamins (Choe et al., 2005). Riboflavin is considered to be relatively stable during food processing and storage, except under light, where absorption of light will produce excited triplet state riboflavin. On the other hand, as in the case of most water-soluble vitamins, substantial losses of riboflavin in many food products may be due to leaching into the process and/or cooking water. Thus, blanching operations and cooking of vegetables normally result in extensive losses of this vitamin. Several authors (Petrou et al., 2002) have reported kinetic information of vitamin degradation upon cooking, which in the light of the most recent developments in terms of nutritional claims has become information of critical importance to product processing and development. In fact, to make nutritional claims in a finished product, vitamin losses during cooking need to be accounted for.

2.3.1.1.4 Vitamin B$_6$

Limited kinetic information is available for the degradation of compounds with vitamin B$_6$ activity such as pyridoxal, pyridoxine or pyridoxol, and pyridoxamine. Several authors have reported the interconversion of the B$_6$ vitamers. Results presented by Gregory and Hiner (1983) indicate that a bidirectional conversion of pyridoxal and pyridoxamine during processing exists. Prior to this work, Gregory and Kirk (1978) had reported on the rapid conversion of pyridoxamine to pyridoxal during storage at low moisture content in systems containing protein and reducing sugars, although the reverse reaction was not detected. Yonker (1984) had also observed interconversion of pyridoxamine-5′-phosphate to pyridoxamine to pyridoxine and pyridoxal.

Under storage conditions, Gregory and Kirk (1978) found first-order kinetics for the degradation of the different vitamers in model systems. Pyridoxamine appeared to be the vitamer with the highest complexity in its degradation mechanisms and kinetics. In systems fortified with pyridoxamine, the degradation of this vitamer followed first-order kinetics with conversion into pyridoxal through a transamination process. The limiting factor in this case was the degradation of pyridoxal since the reverse reaction, pyridoxal conversion into pyridoxamine, was not significant.

Navankasattusas and Lund (1982) reported on the stability of vitamin B$_6$ vitamers in phosphate buffer solutions and cauliflower puree at high temperature, in the range 110 to 140°C. The authors reported that the thermal degradation of pyrydoxamine followed pseudo first-order reaction kinetics, whereby the degradation of the vitamer appeared to be slightly dependent on the initial concentration, while the degradation of pyridoxine and pyridoxal followed 1.5 and second-order kinetics, respectively. In the case of cauliflower puree, the kinetics of degradation were observed to deviate from first-order reaction kinetics throughout the entire heating time. Working with a model food system simulating a ready-to-eat breakfast cereal, Evans et al. (1981) determined that the degradation of pyridoxine followed first-order kinetics for the range 155 to 200°C.

Limited kinetic information is available with regard to the influence of pH on the stability of the different vitamers, although alkali pH has been shown to enhance the decomposition of all the B_6 vitamers. Saidi and Warthesen (1983) evaluated the effect of pH as well as water activity, light, and temperature on B_6 vitamer model systems. Pyridoxine was found to be very stable in the pH range 4.0 to 7.0 when held at 40 and $60°C$ for up to 140 days. In the case of pyridoxamine under the same conditions, the authors indicated that the reaction followed first-order kinetics under the conditions causing more vitamin losses. Pyridoxamine degradation appeared to follow the trend of higher degradation upon an increase in pH. The authors, however, indicated that the effect of pH was not totally clear. Experiments carried out in model systems indicated that pyridoxal was much more light sensitive than pyridoxine or pyridoxamine. In general, the stability of vitamin B_6 has been observed to be influenced by pH, light, and temperature; however, additional information is needed to clearly characterize the effect of these parameters on the kinetics of degradation and on the mechanisms involved. Kinetic information on vitamin B_6 stability in food products is reported in Table 2.3.

Information reported by Paul and Southgate (1978) indicated a substantial loss ($\sim 40\%$) of vitamin B_6 during the cooking of vegetables. Losses were significant in both root and leafy vegetables. Their information points out the significant losses of thiamine, folate, vitamin C, and vitamin B_6 expected during industrial blanching operations.

2.3.1.1.5 Vitamin B_{12} (Cyanocobalamin)

Vitamin B_{12} is chemically the largest and most complex of all vitamins, composed of a central cobalt atom planarly coordinated via nitrogen atoms to a porphyrin-like group referred to as corrin with axial coordination sites occupied by a 5,6-dimethylbenzimidazole base and a cyano group. Vitamin B_{12} is also referred to as cyanocobalamin and part of a larger group collectively known as cobalamins. One of the predominant forms of this vitamin is referred to as coenzyme B_{12}, where the cyano group at the sixth coordination position is substituted by 5-deoxyadenosine, attached to the cobalt atom via a methylene group. Another common form found in foods is the hydroxo-cobalamin (also called vitamin B_{12a} or hydroxo-B_{12}), where the cyano group is replaced with a hydroxy group (Farquharson and Adams, 1976; Schneider, 1987). Other cobalamins include methyl- (CH_3), nitrito- (NO_2), and sufito- (HSO_3) groups substituting for the cyano group. B_{12} analogues refer to a group where the 5,6-dimethylbenzimidazole base is replaced with other substituent groups (Figure 2.12).

In general, limited kinetic information is available on the stability of vitamin B_{12}. It has been reported, however, that this vitamin is slightly unstable in mild acid or alkaline solutions. In the pH range 4.0 to 7.0, this vitamin appears to have good stability. The presence of compounds such as ascorbic acid have been reported to influence the destruction of this vitamin by authors such as Herbert and Jacob (1974), while others such as Newmark et al. (1976) did not appear to detect any significant difference in food systems containing ascorbic acid. Iron, either ionic or in complex forms, has been found to provide vitamin stability in the presence of ascorbic acid in liver extracts and pharmaceutical formulations (Shenoy and Ramasarma, 1955). It is considered that the stability of vitamin B_{12} in food systems is very different from that found in pharmaceutical formulations or model systems, since vitamin B_{12} in foods is tightly bound. For instance in liver, cobalamin is present in the form of a coenzyme bound to a liver protein. Stability has been attributed to reduced accessibility of the vitamin to chemical attack.

Information on vitamin B_{12} retention when food products are heated or processed using microwaves is limited. Watanabe et al. (1998) reported that appreciable losses (~ 30–40%) of vitamin B_{12} occurred when microwave heating raw beef, pork, and milk. The authors also reported that when microwave heating hydroxo vitamin B_{12}, which predominates in foods, two biologically inactive degradation compounds were identified. Very limited information is available on other food-occurring vitamin B_{12} analogues with different β-ligands, such as methyl vitamin B_{12} and $5'$-deoxyadenosyl vitamin B_{12}.

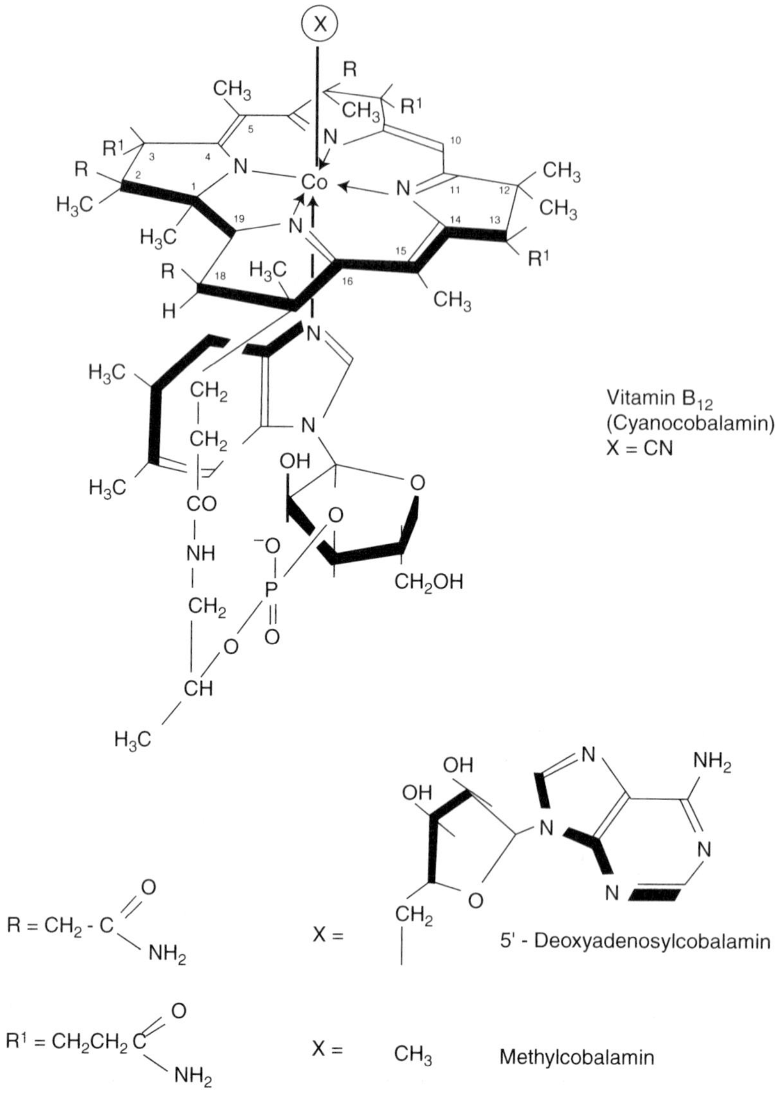

FIGURE 2.12 Structures of vitamin B_{12} and cobalamin cofactors. (Adapted from Matthews, R.G. 1984. Methionine biosynthesis. Ch. 13, pp. 497–553, In: *Folates and Pterins, Vol. 1. Chemistry and Biochemistry of Folates*, Blakley, R.L. and Benkovic, S.J. (eds.), John Wiley & Sons, NY.)

2.3.1.1.6 Folates (Pteroylpolyglutamates)

Folates or folacin refer to a large group of heterocyclic derivatives with similar biological function and a common basic structure, N-[4-[{(2-amino-1, 4-dihydro-4-oxo-6-pteridinyl)-methyl} amino] benzoyl] glutamic acid, with or without additional L-glutamic acid residues conjugated via peptide linkages through the γ-carboxyl groups of succeeding glutamate molecules. Since its discovery as an important dietary factor in the early 1930s, it has undergone a series of name changes including vitamin M, vitamin U, vitamin B_c, and *L. casei* factor. Folate deficiencies have increasingly become a worldwide concern at all socioeconomic levels. It is a common cause of megaloblastic anemia and is either directly or indirectly responsible for the defective synthesis of nucleic acids, frequently occurring in newborn infants. This is usually found due to a folate deficiency during gestation often subsequently resulting in mental retardation. This group of compounds, of great nutritional significance, has not received adequate attention from the point of view of kinetics. In fact, a large number of parameters can affect the stability of folates, including pH, water activity, temperature, oxygen availability, light, metal traces, etc. Moreover, the stability is dependent upon the particular vitamer under consideration. Some of the most important derivatives of this group include 5-methyltetrahydrofolic, tetrahydrofolic, dihydrofolic, 5-formyltetrahydrofolic and folic acids. Due to complications in the separation of this extremely large number of individual derivatives in this group of vitamins, a thorough analysis of the stability of the different vitamers has seldom been carried out. However, because of their individual biological activity, they each need to be taken into consideration. In fact, the literature is very scarce when it comes to the stability of folates in foods other than the parent compound, folic acid. Most information corresponds to model systems or buffer solutions, except for cases such as apple and tomato juices, for which kinetic information on 5-methyltetrahydrofolate (Mnkeni and Beveridge, 1983) as well as on folic acid (Mnkeni and Beveridge, 1982) has been reported. Several authors, including Hawkes and Villota (1986), working with model systems, indicated that folic acid had greater stability as compared to tetrahydrofolic acid and 5-methyltetrahydrofolic acid, also biologically available compounds. It was also indicated that the degradation of these three folates followed first-order kinetics within narrow temperature ranges.

Oxidation of tetrahydrofolate (THF) or dihydrofolate (DHF) generally results in the loss of the side chain, especially at neutral and low pH (Maruyama et al., 1978). Tetrahydrofolate has been shown to follow a number of degradation pathways in the presence of air, where both the rate and the mechanisms involved are highly dependant on the pH of the system (Reed and Archer, 1980). It should be mentioned that under neutral and acidic conditions, tetrahydrofolate is degraded to p-aminobenzoyl glutamates and pterin products with no vitamin activity. At higher pH, dihydrofolate is a product of the reaction with vitamin activity but undergoes further oxidation to compounds without any activity. On the other hand, as summarized by Hawkes and Villota (1989a), folic acid is stable under anaerobic conditions in alkaline environment, although as reported by Temple et al. (1981), opening of the pyrimidine ring forming 2-pyrazine carboxylic acid will occur over long periods of storage. Under aerobic conditions, however, degradation will result in cleavage of the side chain to p-aminobenzoyl glutamic acid plus pterin-6-carboxylic acid. Acid hydrolysis, on the other hand, in the presence of oxygen yields a 6-methyl pterin. Hawkes and Villota (1986) indicated that the stability of folic acid, tetrahydrofolic acid and 5-methyltetrahydrofolic acid decreased with a decrease in pH for the range 7.0 to 2.0. Folic acid solutions have also been shown to be sensitive to light and may undergo photodecomposition to p-aminobenzoylglutamic acid plus pterin-6-carboxylic acid (Lowry et al., 1949).

Most of the literature information appears to indicate that the degradation of folates follows a true first-order reaction. However, the effect of initial concentration has almost never been monitored. Studies have indicated that initial concentration is an important consideration for the kinetics of folate degradation, thus, following pseudo-first-order reaction kinetics. It has also been pointed out that the temperature range studied will affect the mechanism of degradation of the folates, resulting in different energies of activation (Hawkes, 1988).

Based on the work presented by different authors and summarized by Hawkes and Villota (1989a), it is clear that the presence of oxygen affects the specific pathways of degradation of folic acid. Considering the oxidative pathways for folic acid, it is obvious that aerobic and anaerobic degradation of the vitamin may occur simultaneously (Figure 2.13). This is of particular importance when fortifying food products subjected to various deleterious processing techniques such as in the case with spray dried fortified formulations (Hawkes and Villota, 1989b). Because of the high complexity existing in food systems, seldom has a clear characterization of the mechanisms involved been reported along with kinetic information.

The degradation of folic acid and its derivatives due to oxygen is well documented, although our understanding of the kinetic mechanisms involved is limited. Ruddick et al. (1980) reported that the degradation of 5-methyltetrahydrofolate in phosphate buffer pH 7.3 followed initially pseudo-first-order reaction kinetics, while following a second-order reaction with limited oxygen supply. Similar findings were reported on the methyl derivative in pH 7.0 phosphate buffer under simulated UHT

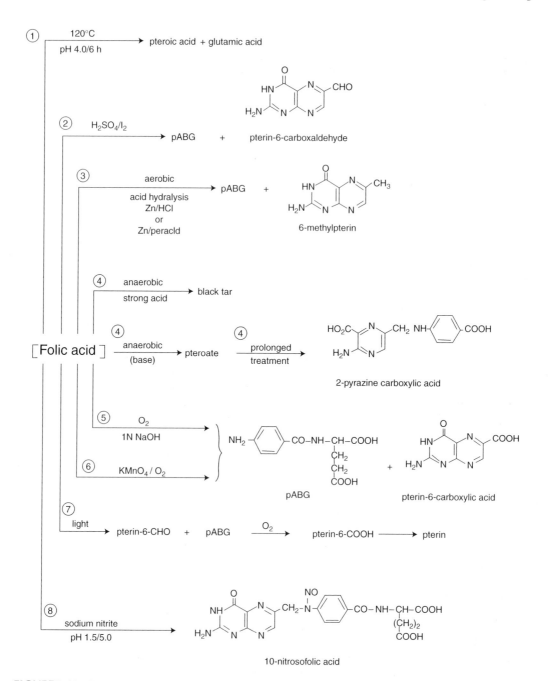

FIGURE 2.13 Degradation pathways of folic acid. (1) Hutchings et al., 1948; (2) Waller et al., 1950; (3) Baugh et al., 1979; (4) Temple et al., 1981; (5) Stokstad et al., 1948; (6) Brown et al., 1974; Maruyama et al., 1978; (7) Lowry et al., 1949; (8) Reed and Archer, 1979. (From Hawkes, J.G. and Villota, R. 1989a. *CRC Critical Reviews in Food Science and Nutrition*, 28: 439–538.)

heating conditions in the presence of controlled oxygen concentrations (Viberg et al., 1997). Day and Gregory (1983) have also reported that for folic acid and 5-methyl tetrahydrofolate the reaction was a second-order for the case of phosphate buffer at pH 7.0. With the advent of high pressure processing, Nguyen et al. (2003) reported kinetics of folic acid and 5-methyltetrahydrofolic acid degradation as affected by the combined treatment of temperature and pressure. This is discussed further in a separate section on the effect of high pressure on the quality of foods (Section 2.4.1).

For the particular case of folate degradation, an added problem corresponds to the methodology by which folates are determined. At the time of the first edition of this chapter, it was pointed out that since microbiological tests were more widely accepted because of their ability to monitor the bioavailability of the different folates, less valid kinetic information was available for the stability of these individual vitamers. Unfortunately, over the past decade, not much has changed in that respect. Rader et al. (2000) point out in a discussion on compliance for mandatory folic acid fortification of enriched cereal grain products in the U.S. that a more accurate methodology for free folic acid is needed over traditional microbiological assays. This concern is targeted particularly at distinguishing between specific forms of folate and total versus free folates. The need for better methodology impacts not only health concerns and proper nutritional labeling, but also costs and problems associated with over fortification of food products. It should be stressed that from the point of view of kinetics, the specificity and accuracy of the assay procedure is of critical importance to mathematically describe and understand the mechanisms of deterioration involved in the reaction. Moreover, the fact that free and bound folates have different stability, accurate techniques to monitor both are required. Unfortunately, a great deal of technical work is still needed to measure folates in food systems using selective and accurate techniques such as HPLC-MS. Release of the bound folates also needs to be properly controlled. Current methods for the measurement of bound folates have not yet reached the level of development needed to use the information in kinetic studies. Some more recent studies have shown some progress in the methodology of folate analysis through a stable isotope liquid chromatography-mass spectrometric (LC-MS) method for the quantitation of folic acid and 5-methyltetrahydrofolic acid of food systems (Pawlosky and Flanagan, 2001; Doherty and Beecher, 2003; Thomas et al., 2003).

In summary, although some information is available for the stability and kinetics of folates as reported in Table 2.3, a better understanding of the mechanisms taking place in food systems is needed.

2.3.1.1.7 Pantothenic Acid

Pantothenic acid, D $(+)$-N-(2,4-dihydroxy-3,3-dimethylbutyryl)-β-alanine, is a member of the B-complex vitamins, also referred to as vitamin B_5, "chick antidermatitis factor", or "yeast growth factor". It is present in almost all plant and animal tissue and is essential for the biosynthesis of coenzyme A. Prevalent sources of pantothenic acid are liver, jelly of the queen bee, rice bran, and molasses. Only the natural dextrorotary form has vitamin activity. Pantothenic acid is most stable in the pH range 4–7. It undergoes alkaline hydrolysis to yield pantoic acid and β-alanine (Frost, 1943), or γ–lactone and pantoic acid under acid hydrolysis. This vitamin has also been reported to be susceptible to thermal decomposition. Frost and McIntire (1944) determined that hydrolysis of pantothenic acid in the temperature range 10–100°C and in the pH range 3.7–4.0 followed first-order kinetics. Hamm and Lund (1978) working with buffer systems and meat and pea purees indicated that pantothenic acid was more stable in food products than in model systems, clearly indicating that the stability of this vitamin improved due to the presence of other compounds in food products. The degradation appeared to follow first-order kinetics (Table 2.3). The authors also indicated that for the systems studied, the vitamin was quite heat stable, contrary to other results reported in the literature indicating vitamin instability during processing (Schroeder, 1971). Cheng and Eitenmiller (1988) also reported that steam blanching, water blanching, canning, and frozen storage caused losses of pantothenic acid in spinach and broccoli to a different degree. Water blanching, in particular, resulted in large losses of pantothenic acid in both spinach and broccoli. In general, however, it appears that due to problems associated with the assay procedures, kinetic data are limited and inconclusive. A calcium salt of pantothenic acid is commercially available and commonly used in extruded feeds for animals.

2.3.1.1.8 Biotin

Biotin, also known as vitamin H, "egg white injury factor," coenzyme R, and chemically as hexahydro-2-oxo-1-*H*-thiene[3, 4-d]imidazole-4-pentanoic acid, is a highly biologically active growth factor found in all living cells (Figure 2.14). It plays an important role in carboxylation, transcarboxylation and decarboxylation reactions and functions in vitally important metabolic processes of glucose and fat synthesis, a deficiency of which has been found to result in dermatitis and perosis in chicks and poults and basically reduces the activity of biotin-dependent enzymes (Dobson, 1970). Some of the richest sources are liver, kidney, pancreas, yeast, and milk. It occurs naturally as the *d*-isomer and is present in a conjugated or bound form to proteins and polypeptides. In fact, raw egg whites produce a biotin deficiency due to the glycoprotein, avidin, which forms a complex with biotin, rendering it unavailable. However, since egg white is heat labile, prolonged heating of egg white denatures avidin and destroys its biotin-binding capacity. The *d*-isomer has about twice the biological activity of the *d*, *l*-isomer, where the *l*-isomer is biologically inactive. Biotin is reported as stable to heat, oxygen, in moderately acid (pH 4.0), and neutral solutions up to pH 9.0 but less stable under alkaline conditions (Merck, 2001). Hoppner and Lampi (1993) reported on relative losses of biotin and pantothenic acid in various legumes and found biotin was significantly more stable. After 24 h soaking, followed by conventional cooking for 20 min, they found 90% retention of biotin compared to only 44% pantothenic acid, but no mention was made from a kinetic point of view. It is expected that the cooking and processing of foods can convert biotin to the oxidized forms. It has been reported that the different biotin derivatives such as desthiobiotin, oxybiotin, biotinol, norbiotin, biotin sulfoxide, and biotin sulphone have different biological activities. However, limited information is available on the stability of biotin and its derivatives. It was not until 1966 that biotin became more important commercially, particularly in the fortification of feeds for livestock such as poultry and swine as well as pets. Watson and Marsh (2001) developed a patent for a biotin supplement for animals to withstand extrusion processing. Gyorgy and Langer (1968) have reviewed the chemistry of this vitamin. It is clear that more information is needed in this area.

2.3.1.2 Fat-Soluble Vitamins

2.3.1.2.1 Vitamin A

Vitamin A is generally classified into two main groups possessing biological activity: (a) C20 unsaturated hydrocarbons including retinol and its derivatives from animal origin and (b) C40 unsaturated hydrocarbons including carotene and a number of other provitamin A carotenoids of plant origin. Vitamin A is a generic descriptor for all β-ionone derivatives with the biological activity of all-*trans*-retinol (also referred to as vitamin A alcohol or vitamin A_1). Provitamin A carotenoid is a generic descriptor for all carotenoids with the qualitative activity of β-carotene. Natural forms of vitamin A predominantly occur in the more stable form of all-*trans*-retinyl esters, along with small levels of 13-*cis*-retinol as found in fish livers. Other natural retinyl derivatives present include esters of 3-dehydroretinol (vitamin A_2, with ~40% retinol activity) and retinal (vitamin A aldehyde with ~90% retinol activity). Commercially available forms of synthetic vitamin A may be found as either retinol acetate or palmitate and can be supplied in crystalline form, or as concentrates in oil, emulsions, or in encapsulated forms. Similarly, different forms of provitamin A carotenoids are available. Some of the carotenoids with significant provitamin A activity include β-carotene (100%); 3,4-dehydro-β-carotene (75%); β-apo-8'-carotenal (72%); β-apo-12'-carotenal (120%); 3-hydroxy-β-carotene (50–60%); α-carotene (50–54%); and γ-carotene (42–50%) (Bauernfeind, 1972). Because of the wide variety of forms of vitamin A and provitamin A carotenoids, labeling requirements report total vitamin A activity in terms of "retinol equivalents" (RE), where one RE is equivalent to 1 μg retinol, 6 μg β-carotene, and 12 μg other provitamin carotenoids. In terms of international units (IU), one RE = 3.33 IU retinol or 10 IU β-carotene (NRC, 1980).

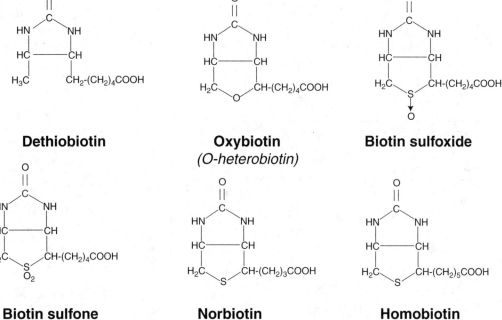

Biotin

(Hexahydro-2-oxo-1-H-thieno [3,4-d-] imidazole-4-pentanoic acid)

Dethiobiotin

Oxybiotin
(O-heterobiotin)

Biotin sulfoxide

Biotin sulfone

Norbiotin

Homobiotin

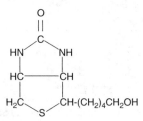

Biotinol

FIGURE 2.14 Structures of biotin and derivatives. (From Scheiner, J. 1985. Biotin. ch. 21, pp. 535–553, In: *Methods of Vitamin Assays, 4th Edition*, Augustin, J., Klein, B.P., Becker, D., and Venugopal, P.B. (eds.), John Wiley & Sons, NY.)

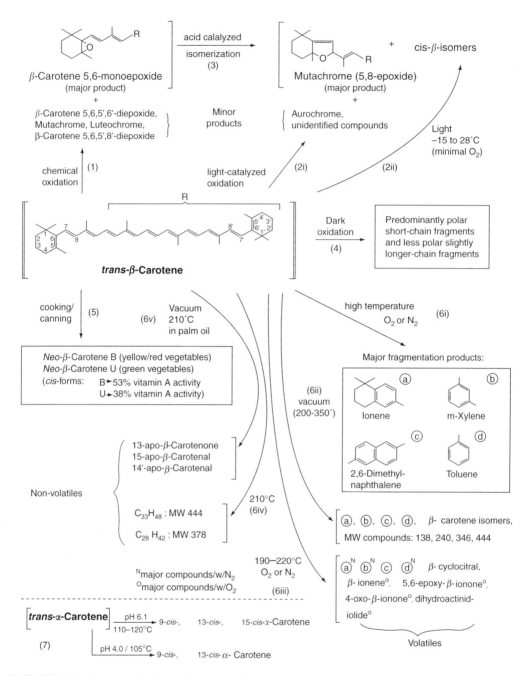

FIGURE 2.15 Some mechanisms of β-carotene degradation. (1) Seely and Myer, 1971; (2i) Seely and Myer, 1971; (2ii) Pesek and Warthesen, 1988; (3) Seely and Myer, 1971; (4) Walter et al., 1970; (5) Sweeney and Marsh, 1971, 1970; (6i) Ishiwatari, 1980; Mader, 1964; Day and Erdman, 1963; (6ii) Ishiwatari, 1980; (6iii) Schreir et al., 1979; (6iv) Onyewu et al., 1982; (6v) Ouyang et al., 1980; (7) Chen et al., 1995.

A variety of pathways have been proposed to describe the destruction or autooxidation of carotenoids, depending on the process conditions, the presence of light, the presence of oxygen, and the composition of the system including peroxidizing lipids or enzymatic activity. A summary of the most important mechanisms is presented in Figure 2.15. It appears that at high temperatures, the destruction of carotenoids results in fragmentation including the formation of aromatic compounds.

In the absence of oxygen, *trans-cis* isomerization seems to be one of the most important mechanisms of deterioration. Light catalyzed oxidation appears to result primarily in the formation of mutachrome. In general, the degradation of carotenoids has been considered to be an autooxidation reaction involving the formation of free radicals, thus giving origin to a propagation reaction and finally a termination stage. A variety of approaches have been taken to describe the kinetics of carotenoid degradation. For instance, Ramakrisnan and Francis (1979a), using a microcrystalline cellulose/starch model system, stated that since this is an autooxidative process, the two principal reactants are oxygen and carotenoids, and since oxygen is in excess, the reaction is expected to follow first-order reaction kinetics. Other authors such as Chou and Breene (1972), Baloch et al. (1977c), Stefanovich and Karel (1982), Goldman et al. (1983), and Pesek and Warthesen (1987, 1988) have also reported that a first-order reaction would describe, within limits, the degradation of carotenoids in model systems. On the other hand, authors such as Quackenbush (1963) working with corn, Baloch et al. (1977a, 1977b) working with carrots, and Stefanovich and Karel (1982) working with butternut squash, sweet potato, and yellow corn also concluded that first-order kinetics could describe the degradation of carotenoids, although the model did not take into account an induction period. Haralampu and Karel (1983) working with dehydrated sweet potato indicated that the degradation of β-carotene was described with the use of pseudo first-order reaction kinetics. Taking into consideration that carotenoid degradation is a chain reaction, other authors such as Alekseev et al. (1968), Finkel'shtein et al. (1974), Gagarina et al. (1970), Goldman et al. (1983), Stefanovich and Karel (1982), and Smith–Molina (1983) have applied a simplified free radical recombination model to describe the autooxidation of carotenoids. Finkel'shtein et al. (1973) indicated that the autooxidation of β-carotene followed the same basic trends regardless of the conditions, namely, (a) an induction period, (b) an acceleration period, (c) a stationary induction period, and (d) a retardation period.

Limited information is available with regard to the effect of initial concentration on the degradation of carotenoids. Budowski and Bondi (1960), working with model systems, indicated that the higher the initial concentration of carotenoids, the shorter the induction period, with an associated faster reaction rate. Gagarina et al. (1970) also indicated that when working with β-carotene in chloroform, the time required for complete consumption of the carotene was shorter upon an increase in the initial concentration. Similarly, Stefanovich and Karel (1982), working with β-carotene in a model system, observed that the kinetics of degradation were dependent on the initial concentration. The authors indicated that the dependence was related to the thickness of the carotene layer since the diffusion of oxygen becomes a limiting factor in the oxidation reaction. Smith–Molina (1983) indicated that in liquid systems the rate of degradation of carotenoids increased with an increase in the initial concentration. The high mobility of the reactants including free radicals was assumed to be the reason for the observed trends. The author also reported that when taking into consideration the history of the sample, systems where degradation of carotenoids had proceeded to a larger extent were also more reactive. Similarly, Goldman et al. (1983) observed that carotene degradation was strongly affected by the presence of free-radical initiators.

Although a certain amount of information is available for the kinetics of degradation of carotenoids, limited work has been done in trying to follow the most important mechanisms of deterioration on the degradation of carotenoids in food systems (Table 2.3). Moreover, analytical techniques have been limited in their ability to monitor isomerization of the carotenoids, which is expected to be one of the most critical changes occurring as a result of processing.

Working at high temperatures, Wilkinson et al. (1981) indicated that the destruction of vitamin A in beef liver puree, measured as *trans*-retinol, followed a first-order reaction. Wilkinson et al. (1982) indicated that increased concentrations of copper increased the losses of vitamin A, whereas increased pH (5.6–7.0) resulted in a decrease. The authors appeared to believe that changes in the copper concentration modified the mechanism by which the vitamin was lost.

Chen et al. (1995) reported effects of different processing techniques on the stability of various carotenoids in carrot juice including α-carotene, β-carotene, and lutein. Depletion and/or conversion of the *trans*-isomeric forms to various *cis*-forms of the vitamers were monitored by HPLC. It was reported that acidification of fresh carrot juice to pH 4.0 followed by heating to 105°C/25 sec showed little change. HTST heating at 110°C/30 sec and 120°C/30 sec showed progressively higher levels of loss of the *trans*-carotenoids with the predominant *cis*-isomers being 13-*cis*-β-carotene, followed by 13-*cis*-lutein and 15-*cis*-α-carotene. Retort processing (121°C/30 min) showed the highest level of carotenoid destruction with formation of 13, 15-di-*cis*-β-carotene. Color changes as monitored by *Lab*-values showed decreases that paralleled losses of the *trans*-forms and increase in *cis*-forms. Lin and Chen (2005) indicated that in tomato juice during storage both temperature and light would influence the proportion of isomers that predominate over time (4–35°C/12 wks). For instance, they found that all-*trans*-β-carotene degraded to di-*cis*-, 9-*cis*-, and 13-*cis*-β-carotene isomers after storage under light, depending on temperature. In the absence of light, degradation products included 5-*cis*-, 9-*cis*-, and 13-*cis*-β-carotene, again, depending on temperature.

In food systems the effect of water on carotene oxidation appears to be dependent on composition (Kanner et al., 1978). As reported by Arya et al. (1979) and Maloney et al. (1966), an increase in water content could mobilize the pro-oxidant factors in the matrix or expose new sites in the matrix resulting in accelerated oxidation. On the other hand, Haralampu and Karel (1983) indicated that in the case of sweet potato flour, the degradation of β-carotene was inversely proportional to the water activity. With respect to the photosensitized oxidation of β-carotene, mutachrome has been identified to be the most important oxidation product, although other compounds such as aurochrome and a number of compounds absorbing in the violet and near ultraviolet region have also been detected. The 5,6-monoepoxide was not detected in significant amounts, although this compound is unstable and can be converted into mutachrome by acid traces and catalysts (Seely and Meyer, 1971). The authors determined that 5,6-monoepoxide was not the first product of photochemical oxidation. Pesek and Warthesen (1988) working with model dispersions indicated that the photodegradation of β-carotene followed a first-order reaction that was affected by temperature; the physical state of the sample, frozen vs. liquid; and the microenvironment. The authors also indicated that the presence of *cis*-isomers and the rate of their formation was larger than that of their degradation. Moreover, it had been previously shown by Zechmeister (1944) that the *cis*-isomers may convert back to the all-*trans* form, which is more stable, or undergo further degradation.

Heat stability studies of α-carotene and β-carotene indicate that β-carotene is about 1.9 times more susceptible to heat damage than α-carotene during normal blanching and cooking operations (Baloch et al., 1977a). With regard to cooking losses, Sweeney and March (1971) indicated that heating promotes the *cis*-*trans* isomerization of carotenoids in vegetables, with an increase of the *cis* isomers. In general, literature reports clearly indicate the instability of carotenoids at high temperatures such as those encountered in canning and drying operations, particularly in high temperature-long time type of processes, while freezing and low temperature processing normally results in much lower losses.

Of critical importance in processing, particularly fruits and vegetables, would be the isomerization of all-*trans*-β-carotene sensitized by chlorophylls. As reported by O'Neil and Schwartz (1995), the photoisomerization of β-carotene sensitized by chlorophyll will result in 9-, 13-, and 15-*cis*-β-carotene primarily, with a higher ratio of the 9-*cis*-isomer.

2.3.1.2.2 Vitamin D

Originally in the early 1920s, the term vitamin D was given to the active component present in cod liver oil, which could cure or prevent rickets, a disease resulting in weakness and deformation of the bones. The most important forms of vitamin D are D_2 (ergocalciferol) and D_3 (cholecalciferol). Vitamin D_2 is formed by ultraviolet irradiation of the provitamin ergosterol. Similarly, D_3 is formed from the provitamin 7-dehydrocholesterol. The following are also considered to be provitamin D compounds: 22,23-dehydroergosterol, 7-dehydrositosterol, 7-dehydrostigmasterol and 7-dehydro-campesterol. The D provitamins do not have any vitamin activity unless the B ring is opened between carbons 9 and 10, and a double bond is formed between carbons 10 and 19 forming the 3(beta)-hydroxy-9,

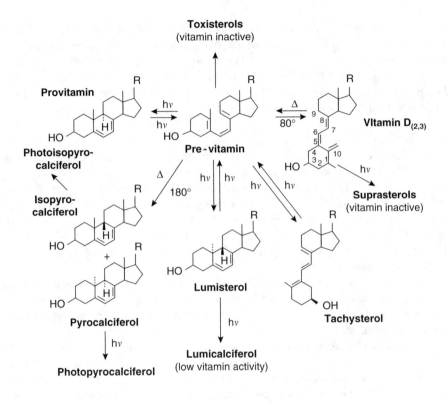

FIGURE 2.16 Some reaction pathways as affected by light or heat for the D vitamers. (Adapted from Miller, B.E. and Norman, A.W. 1984. Vitamin D. ch. 2, pp. 45–97, In: *Handbook of Vitamins. Nutritional, Biochemical, and Clinical Aspects*, Machlin, L.J. (ed.), Marcel Dekker, Inc., NY. and Li, T.-L., King, J.M., and Min, D.B. 2000. *Journal of Food Biochemistry*, 24: 477–492.)

10-seco-5, 7,10(19)-triene derivative (Figure 2.16). Limited information is available on the stability of the provitamin D compounds and the active derivatives, although this vitamin has been reported to be susceptible to oxygen and light. Photochemical transformations of provitamin D give origin to the *anti*-9:10 isomers, while thermal isomerization yields the *syn*-9:10 isomers, procalciferol and isopyrocalciferol (Sebrell and Harris, 1971). Yamada et al. (1983) indicated that vitamin D undergoes 1,4-cycloaddition and ene-type reactions with singlet oxygen, generating: (a) two carbon (6) epimers of 6,19-epidioxyvitamin D (55–65% yields) and (b) two carbon (6) epimers of the $\Delta^{4,7,10(19)}$ 6-hydroperoxide (15–25% yields). Figure 2.16 presents some of the reaction pathways for Vitamin D as affected by light and heat. With regard to kinetic information, our understanding of the stability of vitamin and provitamin D in food systems is almost non-existent. Li and Min (1998), however, reported first-order rate constants (2 phases) for the degradation of vitamin D_2 in model systems as a function of riboflavin concentration in the presence of light. The authors indicated riboflavin acted as a photosensitizer and accelerated the oxidation of vitamin D_2 by singlet oxygen under light but had no affect on stability in the absence of light. Further studies showed that the presence of carotenoids could quench singlet oxygen activity and provide stability to vitamin D_2 in model systems. Li et al. (2000) reported quenching rate constants for retinol, retinyl acetate, fucoxanthin and β-carotene (1.22×10^8, 5.98×10^8, 1.78×10^9, 5.00×10^9 $M^{-1}sec^{-1}$, respectively) and indicated that with increasing number of carotenoid double bonds (5, 6, 10, 11, respectively), the quenching rate constant of carotenoid increased.

2.3.1.2.3 Vitamin E

Tocopherols or compounds with vitamin E activity are methyl-substituted hydroxychromans with an isoprenoid side chain. Tocopherols are composed of two homologous series: (a) tocopherols with a saturated side chain and (b) tocotrienols with a side chain unsaturated between carbons $3'$ and $4'$, $7'$ and 8, and $11'$ and $12'$ (Parrish, 1980). It is considered that α-tocopherol is the compound with the most significant biological activity of all E vitamers, although it is also the least resistant to oxidation. Since tocopherols are monoethers of a hydroquinone, they can be easily oxidized. A comprehensive summary of the most influential factors on the stability of tocopherols has been presented by Bauernfeind (1977, 1980).

Storage studies carried out by Widicus et al. (1980) indicated that the degradation of α-tocopherol followed a first-order reaction in model systems not containing fat. Although the presence of an autooxidation mechanism was not observed, the involvement of oxygen in the degradation of this compound was determined.

A number of studies have indicated that α-tocopherol is highly susceptible to degradation depending on moisture content, temperature, light, alkali, and the presence of metal ions such as iron and copper. Moreover, tocopherols have been found to be more unstable in peroxidizing systems. The decomposition products of oxidized tocopherols include dimers, trimers, dihydroxy compounds, and quinones (Csallany and Draper, 1963; Skinner and Parkhurst, 1964; Csallany et al., 1970). Some pathways for the degradation of α-tocopherol are illustrated in Figure 2.17.

Due to the high instability of this vitamin, processing in particular has been reported to be detrimental to the stability of tocopherols. For instance, Thomas and Calloway (1961) indicated severe α-tocopherol losses (41–65%) in various meat products as a result of canning. Livingston et al. (1968) reported losses of α-tocopherol ranging from 5 to 33% during the drying of alfalfa.

Widicus et al. (1980) indicated that for a nonlipid containing model system the degradation of α-tocopherol was directly related to the water activity, suggesting that the system was highly dependant on the rate of diffusion of the reactants. The reaction followed first-order kinetics. Similarly, Jensen (1969) reported higher stabilities of α-tocopherol during the storage of seaweed meal at lower moisture contents for the range 10–25%. Frias and Vidal–Valverde (2001) reported on the stability of α-, β-, and δ-tocopherols, vitamin A and thiamine during storage of enteral feeding formulas. Analysis of their data is presented in Table 2.3, along with some of the most relevant information on vitamin E stability in food systems.

Considerable losses of tocopherols have been reported during the storage of oils, as influenced by temperature, time, and the presence of other antioxidants. Since tocopherols are highly susceptible to free radical oxidation, it is obvious that their stability is influenced by the levels of lipid oxidation taking place in a food system. It has also been found that the relative stabilities of natural tocopherols may vary according to the biological source. For instance, Chow and Draper (1974) found that both vitamin E oxidation and peroxide formation occurred more rapidly in corn oil than in soybean oil.

With regard to the influence of metal traces, Cort et al. (1978) reported that both $\alpha-$ and γ-tocopherols were degraded by Fe^{3+} and Cu^{2+}. Chelating compounds such as ascorbic acid and EDTA appeared to inhibit the Cu^{2+} oxidation, while ascorbic acid prevented the Fe^{3+} oxidation of tocopherols in alcohol solutions.

Tocopherols have been reported to combine with various proteins and amino acids, thus modifying their stability. The conjugate appears to be a protein-tocopherol linkage without the involvement of lipids (Voth and Miller, 1958). Binding affinities of proteins and free amino acids have shown a

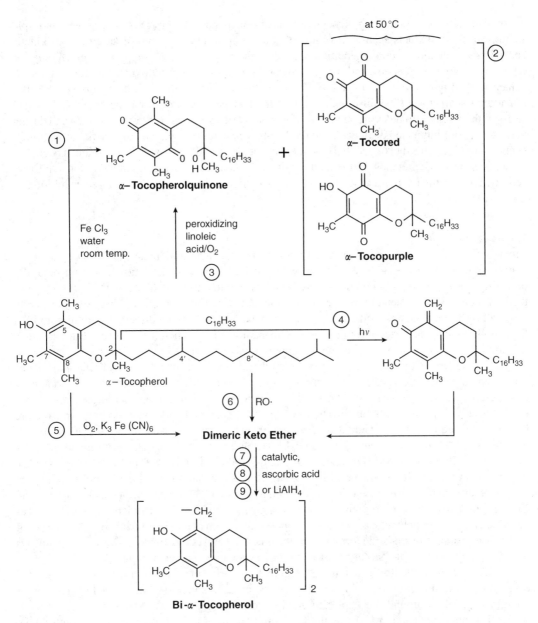

FIGURE 2.17   Some mechanisms of α-tocopherol degradation. (1) John et al., 1939; John and Emte, 1941; (2) Frampton et al., 1960, 1954; (3) Knapp and Tappel, 1961; (4) Knapp and Tappel, 1961; (5) Skinner and Alaupovic, 1963; Nelan and Robeson, 1962; (6) Dürckheimer and Cohen, 1962; Schnudel et al., 1972; (7) Schnudel et al., 1972; (8) Nelan and Robeson, 1962; (9) Csallany and Draper, 1963.

similar behavior, whereby an increase in the negative charge resulted in an increase in the binding of tocopherols. Thus, relatively positive amino acids such as lysine, arginine, and histidine do not participate in the binding of the tocopherols or modify the affinity of proteins for binding (Voth and Miller, 1958).

Fortification of foods with Vitamin E has become increasingly more important. Vitamin E is important in human nutrition since it has potent antioxidant activity, thereby preventing the damage of cells through the inactivation of free radicals and oxygen species (Diplock, 1994). Due to its antioxidant activity, vitamin supplementation has been found to be effective on pigment and lipid stability in food products such as frozen beef. Lanari et al. (1994) demonstrated through kinetic analysis that vitamin E supplementation stabilized the oxymyoglobin complex by enhancing the deoxymyoglobin oxygenation and by decreasing the oxymyoglobin autoxidation rate. Vitamin E enhanced the pigment and lipid stability of frozen beef, stored in the dark or under constant illumination. Lanari et al. (1993) also indicated the significance of dietary supplementation of Holstein steers with vitamin E in delaying surface discoloration after repeated freeze-thaw cycles and during dark storage or illuminated display. Similarly, Houben et al. (2000) studied the benefits of vitamin E supplementation to the diet of beef bulls on the color stability and lipid oxidation of minced beef. The authors corroborated previous studies on the proposed mechanisms of the vitamin E color stabilizing activity, which calls for indirectly delaying the oxidation of oxymyoglobin via direct inhibition of lipid oxidation.

2.3.1.2.4 Vitamin K

Vitamin K is a generic term referring to a group of bicyclic naphthoquinones participating as activators of prothrombin precursor activity in warm-blooded animals. The most common naturally occurring are known as phylloquinone (K_1) and farnoquinone (K_2) along with numerous synthetic variations such as menadione, acetomenaphthone, menadione sodium diphosphate, menadione sodium bisulfite, menadione sodium bisulfite complex, and menadione dimethylpyrimidinol bisulfite. Phylloquinone, also referred to as 3-phytylmenadione, has a chemical structure established as 2-methyl-3-phytyl-1, 4-naphthoquinone. Although relatively heat stable, vitamin K_1 is sensitive to light and oxidation. Common sources of K_1 include alfalfa, spinach, cabbage, kale, lettuce, cauliflower, and other leafy vegetables. Because it is difficult and costly to synthesize, it is not generally considered a practical source for fortification. Vitamin K_2 is found in various microorganisms, particularly bacteria; small amounts can also be found in milk, eggs, and pork liver. Its chemical structure has been defined as either 2-methyl-3-*all-trans*-farnesyl-farnesyl-1, 4-naphthoquinone [$K_{2(30)}$] or 2-methyl-3-*all-trans*-farnesyl-geranylgeranyl-1, 4-naphthoquinone [$K_{2(35)}$]. Very limited information is available for the kinetics of degradation and stability of vitamin K in foods. Indyk (1988), working with vitamin K_1 dissolved in hexadecane, observed stability of this vitamin to mild heat treatment, even in the presence of oxygen. Vitamin K was found to be susceptible to degradation and isomerization in the presence of light even at low intensity. The loss of the isomers, *cis* and *trans*, was described by zero-order kinetics, indicating the possibility of an autooxidation mechanism. Losses of either isomer did not result in the formation of the other. Several competing decomposition pathways have been proposed for the photolysis of vitamin K depending upon conditions.

Because of the difficulty in obtaining vitamin K, a series of synthetic forms were developed over time, each progressively improved over its predecessor. The basis for the activity of vitamin K revolves around the naphthoquinone nucleus (Figure 2.18). Menadione (2-methyl-1, 4-naphthoquinone or vitamin K_3) was the first commercially available product with actually three times the biological activity of K_1. Unfortunately, it had some serious side effects and is no longer in use. Menadione formed the basis, however, for the later versions including menadione sodium bisulfite (MSB) and the more recent version, menadione sodium bisulfite complex (MSBC). It is reportedly water-soluble, stable to light and air but not heat. Minimal information is available on kinetic stability.

2.3.2 PIGMENTS

Color is one of the first notable characteristics of food that often predetermines a consumer's judgement of food quality. Pigments are compounds that absorb light in the wavelength of the visible region. Obviously, color stability is an extremely important factor in foods. Griffiths (2005) reviewed currently acceptable synthetic and natural colors used in the US food industry and indicated a trend

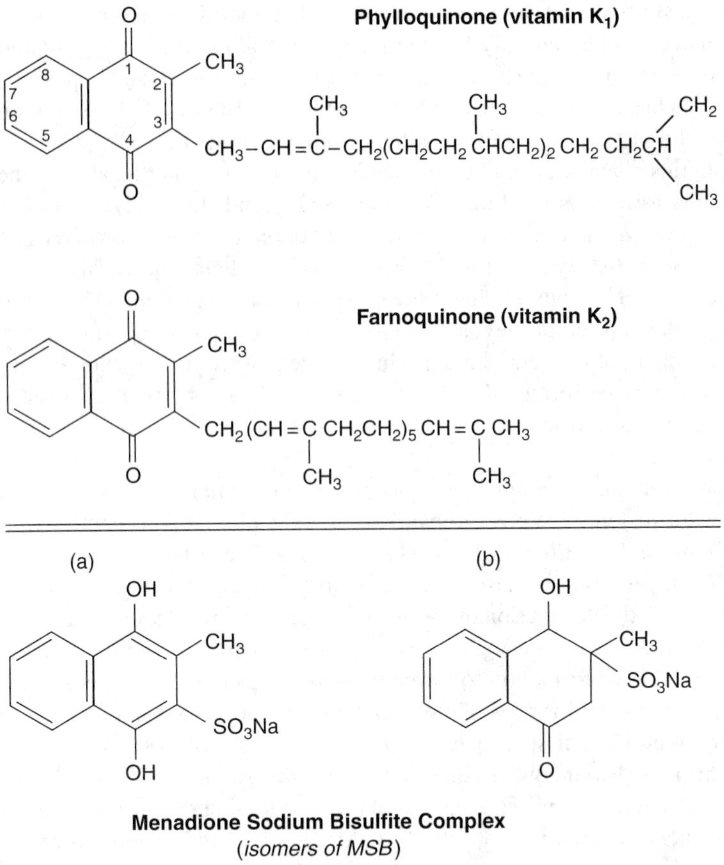

Phylloquinone (vitamin K₁)

Farnoquinone (vitamin K₂)

(a) (b)

Menadione Sodium Bisulfite Complex
(isomers of MSB)

FIGURE 2.18 Structures of vitamin K and derivatives. (Adapted from Berruti, R. 1985. Vitamin K. ch. 11, pp. 285–302, In: *Methods of Vitamin Assays, 4th Edition*, Augustin, J., Klein, B.P., Becker, D., and Venugopal, P.B. (eds.), John Wiley & Sons, NY.)

for an increased use of colors particularly in novelty snacks, desserts, and beverages, further emphasizing the importance of color stability in foods. Colorants such as carotenoids including β-carotene, annatto, paprika, and particularly lycopene are known to exhibit antioxidant activity. Flavonoids, including the anthocyanin group, have also been attributed to having health benefits such as antioxidant properties, anti-inflammatory effects, lowered blood pressure, and anti-tumor properties. Another group of colorants, known as the curcuminoids found in turmeric, are also found to have similar health-related properties as well as antithrombic effects and antimicrobial activity (Taylor, 1996). Overall changes in color may be due to a number of reactions such as pigment degradation or polymerization, interaction with other components in the food product, nonenzymatic browning, oxidation of tannins, and other reactions. The following section discusses some of the major pigments present in foods and their relative stability to processing and/or storage conditions.

2.3.2.1 Chlorophylls

Chlorophylls refer collectively to a group of pigments providing color to green plant tissues. They range in color from a bright green to a dull olive-brown and are often used as indicators of product quality of processed green vegetables, as measured by the intensity of their green color. The predominant green colored pigments include chlorophylls *a* and *b* at a reported ratio of about 3:1 as naturally occuring in plants. Both are derivatives of a tetrapyrrole phorbin (porphyrin ring with C9-C10 iso-cyclic ring) chelated with a centrally located magnesium atom and a C7 20-carbon phytol chain (Figure 2.19). Their main differences are their substituent groups at the C3 position and perceived color; chlorophyll a has a methyl group and is blue-green, and chlorophyll b has a formyl group with a yellow-green color (Belitz and Grosch, 1987). Isomeric forms may also exist as chlorophylls a' and b' or pheophytins a' and b', due to epimerization at the C10 center located on the isocyclic ring. Other less common forms that exist include chlorophyll c and chlorophyll d, isolated from marine algae. Chlorophyllides a and b are the respective acid derivatives of chlorophylls a and b resulting from enzymatic (e.g., naturally occurring chlorophyllase) or chemical hydrolysis of the C7 propion-ate ester and cleavage of the phytol chain; they too posses a green color. The main transformation or degradation products of chlorophylls a, b are, respectively, pheophytin a, b, which are formed through the replacement of the central magnesium of the porphyrin ring with hydrogen atoms. Also pheophorbides a, b may be formed through the removal of the phytol chain from the pheophytins or through magnesium loss from the chlorophyllides. These degradation products all exhibit a dull olive-brown color.

It has long been reported by many investigators that chlorophylls are susceptible to thermal treat-ment, being transformed into the predominantly dull green pheophytins a and b (Schwartz and von Elbe, 1983; LaBorde and von Elbe, 1994; Steet and Tong, 1996a, 1996b; Heaton et al., 1996a, 1996b; Gunawan and Barringer, 2000). These compounds may also further degrade to pyropheophytin or other products through the destruction of the porphyrin ring. Schwartz and von Elbe (1983) working with spinach indicated that pyropheophytin was a predominant product of the thermal breakdown of chlorophylls and that its formation followed first-order kinetics. Heaton et al. (1996a) developed a general mechanistic model for rates of chlorophyll degradation to pheophytin, chlorophyllide and pheophorbide in green plant tissue, including models such as coleslaw, pickles, and olives. Their claim was that this model could discriminate between pathways of degradation and enable quantitat-ive definition on which pathways were operational or predominate under different conditions. This would also allow better comparison of rates of chlorophyll degradation between various commod-ities. For instance, Heaton et al. (1996b) found no significant change over time with chlorophyllide in coleslaw, but with pickles and olives, the formation of chlorophyllide with further degradation to pheophorbide was a predominant reaction pathway. Some of this variation may be due to the relative activity levels of chlorophyllase present, as well as pH and other environmental factors. This type of approach is important in understanding the mechanisms for discoloration and should aid the food processor in determining optimum shelf life.

Other factors influencing the stability of chlorophylls include light, oxygen, water activity, irra-diation, pH, presence of metal traces, and enzymatic activity. Lajolo and Lanfer Marquez (1982) indicated higher rates of degradation with water activity in a spinach model system at 38.6°C. Sim-ilarly, the authors observed an increase in the rates of chlorophyll degradation upon a decrease in pH for the range 5.9–6.8. These results confirm the well-known and most common mechanism for chlorophyll degradation through its acid-catalyzed transformation into pheophytin (Figure 2.19). This reaction has been reported by several authors to follow first-order kinetics. The mechanisms by which chlorophylls degrade, of course, depend upon the process under consideration. For instance, Minguez–Mosquera et al. (1989) found that chlorophillides were intermediary products in the fer-mentation of olives and that the ratio of the various degradation products, including chlorophyllides a, b; pheophytins a, b; and pheophorbides a, b, were very dependent upon the pH of the system (Figure 2.19).

Gunawan and Barringer (2000) studied the effect of acid (pH 3–8) and microbial growth on the stability of the green color of blanched broccoli under low temperature storage (7°C). Through HPLC determination, they found only conversion of chlorophylls to pheophytins. This conversion was greater at lower pH and fit a first order kinetic model. Some isomers were also isolated, including chlorophylls a' and b', present in blanched broccoli, and pheophytins a' and b' after acidification. The authors also found that chlorophyll degradation was dependent on the type of acid used. Acids

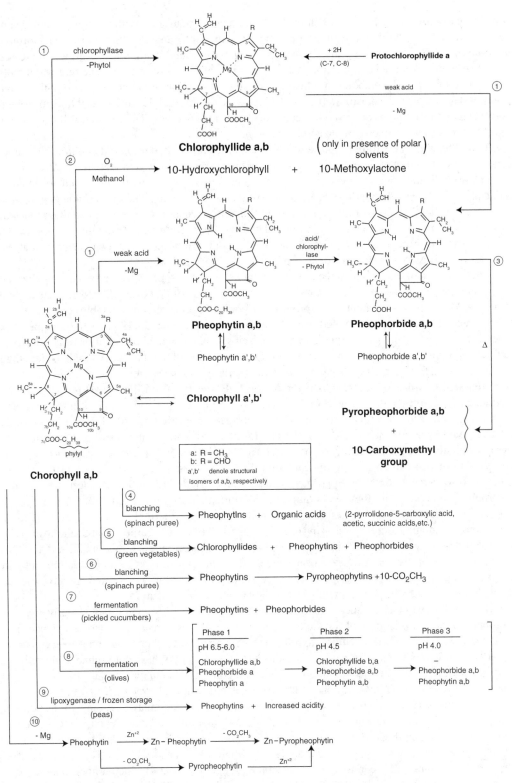

FIGURE 2.19 Selected mechanisms of chlorophyll degradation. (1) Aronoff, 1966; (2) Schaber et al., 1984; (3) Seely, 1966; (4) Clydesdale et al., 1972; (5) Jones et al., 1963; (6) Schwartz and von Elbe, 1983; (7) Jones et al., 1962; (8) Minguez-Mosquera et al., 1989; (9) Wagenknecht et al., 1952; (10) Canjura et al., 1999.

containing a benzene ring resulted in more rapid color change than acids with a simple carbon chain; perhaps due to the hydrophobicity of the aromatic acids, they were able to diffuse more easily through the lipid membrane surrounding the chloroplasts. They also found that microbial growth increased loss of color and proposed two possible mechanisms by which this may occur. The first was simply that production of acid metabolite products would lower the pH and, thereby decrease chlorophyll stability. The second mechanism was due to the breakdown of the cellular structure of the broccoli, as evidenced by surface holes observed by scanning electron microscopy. This could result in exposing the chloroplasts more directly to the acidic medium.

Ryan–Stoneham and Tong (2000) developed a mathematical model to predict chlorophyll concentration as a function of time, temperature, and pH using pea puree as a model. Since pH naturally lowers during heating due to acid formation, the authors used a specially designed reactor to automatically adjust pH of the medium to keep it constant during heating. They found that degradation of both chlorophylls a and b followed first order kinetics. Reaction rate constants and energies of activation are presented in Table 2.4, as calculated by the conventional Arrhenius equation. The authors reported through the use of their modified model, factoring in pH as a variable, that the energies of activation were independent of pH.

The formation of more stable green metallocomplexes of chlorophyll derivatives during thermal processing has been of interest for years in the canning industry (Jones et al., 1977; Tonucci and von Elbe, 1992). In fact a patent for improving the color of canned green vegetables (known as the "Veri-Green" process) was developed over 20 years ago by Segner et al. (1984). Although the FDA amended the standard of identity for canned green beans at that time to allow addition of $ZnCl_2$ to the optional ingredient list, the provision was that the concentration of the metal salt should be less than 75 ppm (Federal Register, 1986). In many cases (particularly with peas), much higher concentrations were required in order to yield a satisfactory color. LaBorde and von Elbe (1994) investigated the degradation of chlorophyll and zinc complexation with chlorophyll derivatives in processed pea puree containing added Zn^{+2} as a function of pH (4–10) during heating at 121°C/up to 150 min. They found relatively rapid degradation of chlorophyll a and formation of zinc complexes of pheophytin a and pyropheophytin a at lower pH levels. At pH above 8.0, chlorophyll a was retained with less Zn-complexing of pheophytin and pyropheophytin. The authors suggested that any improvement in green color occurring at elevated pH may only be temporary due to the lower stability of natural chlorophyll compared with the metal complexes. A similar study was carried out by Canjura et al. (1999) using whole fresh and frozen peas with $ZnCl_2$ subjected to thermal treatment using a particle cell reactor, simulating aseptic processing (121–145°C/0–20 min). The authors found zinc absorption into the pea tissue was dependent upon available Zn^{+2} ion concentration, temperature, and duration of reaction. The reaction pathway proposed suggested that at the lower temperatures, Zn-pheophytin was formed and at the higher temperature range Zn-pyropheophytin was formed. In any case, as compared with controls, improvement was found with addition of metal ions.

Much of the work mentioned above includes complete analysis of individual derivatives of chlorophylls, which is important for the elucidation of mechanistic pathways. However, there is a fair amount of kinetic work on chlorophyll losses, as reported by total color changes over time of a given process without regard to any specific mechanistic pathway. These color changes have been monitored with L-a-b-values as measured with either a Hunter, Minolta, or similar type colorimeter, where L-value represents lightness or brightness, a-value measures degree of red ($+$) to green ($-$), and b-value measures degree of yellow ($+$) to blue ($-$). Various mathematical combinations of these values have also been used. For instance, Lau et al. (2000) monitored kinetic changes of both color and texture in asparagus. Color was measured by the reflectance of the surface of the whole product. Changes were reported in terms of hue angle (h), where $h = \tan^{-1}(b/a)$; this value takes into account both yellow and green tonal changes in the product (Table 2.4). The authors found energy of activation for color change at 13.1 kcal/mole and 24.0 kcal/mole for changes in texture over a temperature range of 70–98°C, and time 5–120 min. Steet and Tong (1996a) found that using a-value (greenness) from a tristimulus colorimeter to monitor changes in color of pea puree during thermal treatment followed first order kinetics. They based their kinetics on a fractional conversion format. These authors measured simultaneous destruction of chlorophylls a and b using HPLC. Activation energies were reported as 18.2 for change in "greenness" (a-value) and 19.5 and 17.1 kcal/mole for chlorophyll a and b, respectively (70–90°C, up to 600 min). These authors also studied color changes resulting from pheophytinization and nonenzymatic browning reactions in pea puree at higher temperatures, 100–120°C, up to about 200 min (Steet and Tong, 1996b). Again they found first order kinetics to describe loss of green value, but nonenzymatic browning was found to follow zero-order kinetics after an initial lag period, with E_a's equal to 20.4 and 22.3 kcal/mole, respectively. Rate constants are presented in Table 2.4.

Similarly, Sánchez et al. (1991) measured surface color and textural changes of pickled green olives during pasteurization 70–100°C up to 60 min. They measured color by both a tristimulus colorimeter (L-a-b-values) and a reflectance spectrophotometer at specific wavelengths (Table 2.4). Shin and Bhowmik (1995) studied changes in color of pea puree and reported D- and z-values for several different mathematical combinations of L-a-b-values at several temperatures from 110–125°C for up to 20 minutes. They found that the mathematical expressions $-La/b$ and $-a/b$ provided the best fit as applied to a first order reaction with an E_a of 16.2 kcal/mole.

Other workers have used various mathematical combinations of L-a-b-values to describe losses of color during thermal treatment. Ahmed et al. (2000, 2002b) investigated color and rheological changes in green chilli. Color changes were reported as loss of green (a-value) and total color change ($L \times a \times b$-values), both following first-order kinetics (50–90°C). Similarly, Ahmed et al. (2002a) found first order kinetics with green color loss in purees of spinach, mustard leaves, and a blend of mustard, spinach and fenugreek during heating at 75–115°C, 0–20 min. It was reported that activation energies for green color loss were consistently higher than those for total color change and proposed that green color could be used for on-line quality monitoring of these types of green leafy vegetables during thermal processing.

Schwartz and Lorenzo presented a general review of the reactivity and analysis of chlorophylls in foods in 1990. Since then, a great deal of information has been reported and now needs to be critically reviewed. Van Boekel (2000), for instance, reviewed a case study on kinetic modeling of chlorophyll degradation in olives. Selected kinetic information on chlorophyll stability is reported in Table 2.4.

2.3.2.2 Anthocyanins

Anthocyanins are derivatives of the basic C15 flavylium cation structure with a chromane ring bearing a second aromatic ring B in position 2 (C6-C3-C6) with one or more sugar molecules bonded at different hydroxylated positions (Figure 2.20). This group of water-soluble colored compounds is found in a wide variety of fruits, flowers, and vegetables. Over 240 anthocyanins have been reported, varying in the number of hydroxyl groups, the degree of methylation, the nature and number of sugars esterified, their position of attachment, and the nature and number of aliphatic or aromatic acids attached to the sugar molecules. The term, "anthocyanidin" refers to the basic C15 structure with various specific R-substitutions, of which there are at least 17 known combinations, but six of them are the most important. These six anthocyanidins include pelargonidin, cyanidin, delphinidin, peonidin, petunidin, and malvidin (Harborne and Grayer, 1988). These anthocyanidins are then esterified to one or more sugar molecules (e.g., glucose, rhamnose, xylose, galactose, arabinose, or fructose) forming the individual anthocyanin pigments, hence the large variety and variation in color dependent upon the combination of these substitutions. In addition, some of these anthocyanins have ester bonds between sugars and organic acids, including coumaric, caffeic, ferulic, p-hydroxy benzoic, synapic, malonic, acetic, succinic, oxalic, and malic acids (Francis, 1989, 1985). The degree of substitution of hydroxyl or methoxy groups influences the color of the anthocyanin. An increase in hydroxyl groups tends to deepen the color to a bluish tone and the more the methoxy groups, the greater the increase in redness.

TABLE 2.4 Kinetic Parameters for Pigment Degradation / Formation during Thermal Processing or Storage

Commodity	Process / Conditions	r^2	k_T value (min^{-1})	E_a (kcal/mol)	Reaction Order	r^2	Temp. Range (°C)	$t_{1/2}$ (min)	Reference
Chlorophyll									
Asparagus									
Fresh / whole bud segment	Heated in distilled water (5 - 120 min)	0.95	k_{98} = 0.0130	12.9	1	-	70 - 98	53	Lau et al., 2000
		0.91	k_{90} = 0.0087					80	
	Color measure: Lab-values / hue angle (h)	0.98	k_{80} = 0.0050					139	
		0.96	k_{70} = 0.0029					239	
Fresh / whole butt segment		-	k_{98} = 0.0167	13.2	1	-	70 - 98	42	
		-	k_{90} = 0.0069					100	
		-	k_{80} = 0.0054					128	
		-	k_{70} = 0.0032					217	
Broccoli juice	Fresh broccoli liquified								Weemaes et al., 1999b
Chlorophyll a	Heat in 800 µl sealed vials	-	k_{120} = 0.1224					5.7	
		-	k_{110} = 0.0611					11.3	
	(0 - 180 min)	-	k_{100} = 0.0284	17.0	1	-	80 - 120	24.4	
		-	k_{90} = 0.0187					37.1	
	HPLC	-	k_{80} = 0.0101					68.6	
Chlorophyll b		-	k_{120} = 0.0564					12.3	
		-	k_{110} = 0.0270					25.7	
		-	k_{100} = 0.0128	16.0	1	-	80 - 120	54.2	
		-	k_{90} = 0.0083					83.5	
		-	k_{80} = 0.0055					126.0	
Total chlorophyll		-	k_{120} = 0.0943					7.4	
		-	k_{110} = 0.0489					14.2	
		-	k_{100} = 0.0229	16.5	1	-	80 - 120	30.3	
		-	k_{90} = 0.0149					46.5	
		-	k_{80} = 0.0085					81.5	

Product	Method	r^2	k		n	r^2	Temp. range		Reference
Brussel sprouts Whole or halves Total chlorophyll									Dietrich and Neumann, 1965
	Water blanched, wire mesh immersion	0.999	$k_{100} = 0.56390$					12.29	
		0.996	$k_{93.3} = 0.04603$	12.9	1	0.949	87.8 - 100	15.06	
		0.980	$k_{87.8} = 0.03113$					22.27	
	Steam blanched	0.998	$k_{115.6} = 0.12000$					5.78	
		0.991	$k_{100} = 0.06890$	12.3	1	0.987	87.8 - 115.6	10.06	
		0.993	$k_{93.3} = 0.05027$					13.79	
		0.996	$k_{87.8} = 0.03476$					19.94	
Green beans Whole (cross-cut) Total chlorophyll									Dietrich et al., 1959
	Water blanched, wire mesh immersion	0.998	$k_{100} = 0.04872$					14.23	
		0.999	$k_{93.3} = 0.04293$	5.2	1	0.999	87.8 - 100	16.15	
		0.994	$k_{87.8} = 0.03847$					18.02	
	Steam blanched	0.996	$k_{100} = 0.07599$					9.12	
	Spectrophotometric A_{534} / A_{556}	0.996	$k_{93.3} = 0.05322$	9.2	1	-	87.8 - 100	13.02	
		0.956	$k_{87.8} = 0.005027$					138.00	
Green beans (cross-cut)	Water blanched	-	$k_{100} = 0.0931$	-	1	-	100	-	Walker, 1964
Olives (pickled) Heated water bath (0 - 60 min) Surface color:									Sánchez et al., 1991
S-value (560, 590, 635μ)	Color via Spec-20	0.974	$k_{90} = 0.004488$					157.00	
		0.988	$k_{80} = 0.003707$	7.2	1	0.970	70 - 90	187.00	
		0.942	$k_{70} = 0.002528$					274.00	
L-value	Color via Hunter	0.914	$k_{90} = 0.003200$					217.00	
		0.972	$k_{80} = 0.001757$	10.5	1	0.937	70 - 90	394.00	
		0.946	$k_{70} = 0.001367$					506.00	
b-value	Color via Hunter	0.963	$k_{90} = 0.003465$					200.00	
		0.964	$k_{80} = 0.003329$	5.1	1	0.836	70 - 90	208.00	
		0.921	$k_{70} = 0.002309$					300.00	
Peas, whole Blanched Total chlorophyll	Packed in 2% salt in cans	-	$k_{137.8} = 0.382$	16.1	1	0.999	115.6 - 137.8	1.81	Gold and Weckel, 1959
		-	$k_{126.7} = 0.219$	$(16.1)^{a}$				3.17	
		-	$k_{115.6} = 0.124$					5.59	

(Continued)

TABLE 2.4 Continued

Commodity	Process / Conditions	r²	k_T value (min⁻¹)	E_a (kcal/mol)	Reaction Order	r²	Temp. Range (°C)	$t_{1/2}$ (min)	Reference
Peas, whole (cont.) Unblanched	Spectrophotometric analysis, Hunter colorimeter	-	$k_{137.8}$ = 0.312	14.3	1	0.992	115.6 - 137.8	2.22	Gold and Weckel, 1960
		-	$k_{126.7}$ = 0.208	(12.6)[a]				3.33	
		-	$k_{115.6}$ = 0.115					6.03	
Peas, puree	pH 6.5	-	-	22[a]	1	-	79.4 - 137.8	-	Lenz and Lund, 1977b
Peas, puree Chlorophyll *a*	Freeze-dried/ rehydrated pH 5.5 (w/control)	-	k_{100} = 0.160	16.3	1	0.994	80 - 100	4.3	Ryan-Stoneham and Tong, 2000
		-	k_{90} = 0.080					8.7	
		-	k_{80} = 0.046					15.1	
	pH 6.2 (w/control)	-	k_{100} = 0.082	17.2	1	0.997	80 - 100	8.5	
		-	k_{90} = 0.046					15.1	
		-	k_{80} = 0.022					31.5	
	pH 6.8 (w/control)	-	k_{100} = 0.034	18.1	1	0.996	80 - 100	20.4	
		-	k_{90} = 0.016					43.3	
		-	k_{80} = 0.009					81.6	
	pH 7.5 (w/control)	-	k_{100} = 0.017	18.9	1	0.998	80 - 100	40.8	
		-	k_{90} = 0.008					86.6	
		-	k_{80} = 0.004					173.0	
Chlorophyll *b*	pH 5.5 (w/control)	-	k_{100} = 0.077	16.4	1	0.996	80 - 100	9.0	
		-	k_{90} = 0.039					17.8	
		-	k_{80} = 0.022					31.5	
	pH 6.2 (w/control)	-	k_{100} = 0.031	14.8	1	0.969	80 - 100	22.4	
		-	k_{90} = 0.015					46.2	
		-	k_{80} = 0.010					69.3	
	pH 6.8 (w/control)	-	k_{100} = 0.013	17.1	1	0.986	80 - 100	53.3	
		-	k_{90} = 0.006					115.5	
		-	k_{80} = 0.004					198.0	
	pH 7.5 (w/control)	-	k_{100} = 0.008	18.1	1	0.950	80 - 100	86.6	
		-	k_{90} = 0.003					223.6	
		-	k_{80} = 0.002					346.6	

Peas, puree							Steet and Tong, 1996a
Chlorophyll a	HPLC	$^{a}k_{90}$ = 0.0344				20	
		k_{90} = 0.0376				18	
		k_{80} = 0.0170	19.5	1	70 - 90	41	
		k_{80} = 0.0175				40	
		k_{70} = 0.0075				92	
		k_{70} = 0.0074				94	
Chlorophyll b	HPLC	k_{90} = 0.0152				46	
		k_{90} = 0.0160				43	
		k_{80} = 0.0080	17.1	1	70 - 90	87	
		k_{80} = 0.0086				81	
		k_{70} = 0.0039				178	
		k_{70} = 0.0039				94	
Total green color	Lab color (a-value)	k_{90} = 0.0184				38	
		k_{90} = 0.0187				37	
		k_{80} = 0.0092	18.2	1	70 - 90	75	
		k_{80} = 0.0087				80	
		k_{70} = 0.0043				161	
		k_{70} = 0.0042				165	
Peas, puree							Steet and Tong, 1996b
Chlorophyll a	HPLC	$^{a}k_{120}$ = 0.2672				2.59	
		k_{120} = 0.2536				2.73	
		k_{110} = 0.1370	20.5	0.999	100 - 120	5.06	
		k_{110} = 0.1324				5.24	
		k_{100} = 0.0652				10.63	
		k_{100} = 0.0630				11.00	
Chlorophyll b	HPLC	k_{120} = 0.1070				6.48	
		k_{120} = 0.1007				6.88	
		k_{110} = 0.0537	18.0	0.993	100 - 120	12.91	
		k_{110} = 0.0557				12.44	
		k_{100} = 0.0311				22.29	
		k_{100} = 0.0284				24.41	

(Continued)

TABLE 2.4 Continued

Commodity	Process / Conditions	r^2	k_1 value (min^{-1})	E_a (kcal/mol)	Reaction Order	r^2	Temp. Range (°C)	$t_{1/2}$ (min)	Reference
Peas (cont.)									
Total green color	Lab color (a-value)	-	$k_{120} = 0.1540$					4.50	Steet and
		-	$k_{120} = 0.1547$					4.48	Tong, 1996b
		-	$k_{110} = 0.0774$	20.3	0.999		100 - 120	8.96	
		-	$k_{110} = 0.0766$					9.05	
		-	$k_{100} = 0.0383$					18.10	
		-	$k_{100} = 0.0381$					18.19	
Peas									
Whole	Packed in distilled water in No. 303 cans (Hunter colorimeter)	-	-	17.5[a]	1	-	98.9 - 126.7	-	Rao et al., 1981
Spinach									
Clorophyll a	Pureed in pyrex tubes pH 6.5	-	$k_{148.9} = 0.6580$					1.05	Gupte et al., 1964
		-	$k_{143.3} = 0.5099$					1.36	
		-	$k_{137.8} = 0.3947$	15.4 (143)[a]	1	0.999	126.7 - 148.9	1.76	
		-	$k_{132.6} = 0.3056$					2.27	
		-	$k_{126.7} = 0.2365$					2.93	
Clorophyll b	pH 5.5	-	$k_{148.9} = 0.3024$					2.29	
		-	$k_{143.3} = 0.2667$					2.60	
		-	$k_{137.8} = 0.2350$	7.6 (35.2)[a]	1	0.999	126.7 - 148.9	2.95	
		-	$k_{132.6} = 0.2072$					3.35	
		-	$k_{126.7} = 0.1828$					3.79	
Spinach									
Clorophyll a	Pureed in cans (natural pH)	0.992	[a]$k_{126} = 0.2666$					2.60	Schwartz and von Elbe, 1983
		0.994	$k_{121} = 0.1777$	27.3 (25.2)[a]	1	0.998	116 - 126	3.90	
	HPLC analysis	0.984	$k_{116} = 0.1100$					6.30	
Clorophyll b		0.982	$k_{126} = 0.1195$					5.80	
		0.998	$k_{121} = 0.0845$	24.7 (22.5)[a]	1	0.995	116 - 126	8.20	
		0.996	$k_{116} = 0.0537$					12.91	

Pheophytin a

$k_{126} = 0.07877$	24.5	1	-	8.80
$k_{121} = 0.05545$	$(20.7)^a$	0.996	116 - 126	12.50
$k_{116} = 0.03555$		1		19.50

Pheophytin b

$k_{126} = 0.10350$	16.9	1	-	6.70
$k_{121} = 0.07877$	$(15.7)^a$	0.999	116 - 126	8.80
$k_{116} = 0.05975$		1		11.60

Spinach, puree Clorophyll a

pH 5.9 / no glycerol

a_w	Blanched / freeze-dried / rehumidified $g\,H_2O\,/\,100\,g$						
0.11	-	$k_{56.7} = 2.00 \times 10^{-5}$	62.0	1	-	46 - 56.7	34.7×10^3
		$k_{46} = 0.083 \times 10^{-5}$					325×10^3
0.32	5.9	$k_{56.7} = 13.95 \times 10^{-5}$	34.0	1	0.999	38.6 - 56.7	5.0×10^3
	6.0	$k_{46} = 2.30 \times 10^{-5}$					30.1×10^3
	6.4	$k_{38.6} = 0.717 \times 10^{-5}$					96.7×10^3
0.52	8.3	$k_{56.7} = 17.67 \times 10^{-5}$	21.0	1	0.997	38.6 - 56.7	3.9×10^3
	10.9	$k_{46} = 5.58 \times 10^{-5}$					12.4×10^3
	12.2	$k_{38.6} = 2.82 \times 10^{-5}$					24.6×10^3
0.75	17.7	$k_{32} = 6.25 \times 10^{-5}$	9.7	1	-	32 - 38.6	11.1×10^3
	29.2	$k_{38.6} = 13.40 \times 10^{-5}$					5.2×10^3
	38.0	$k_{32} = 18.83 \times 10^{-5}$					3.7×10^3

pH 5.9 /w/ glycerol

	a_w	$g\,H_2O\,/\,100\,g$						
60%	0.32	8.5	$k_{38.6} = 7.20 \times 10^{-5}$	-	1	-	38.6	9.6×10^3
35%	0.32	6.1	$k_{38.6} = 3.85 \times 10^{-5}$	-	1	-	38.6	18.0×10^3
	0.52	14.1	$k_{38.6} = 10.28 \times 10^{-5}$	-	1	-	38.6	6.7×10^3
15%	0.32	-	$k_{38.6} = 1.83 \times 10^{-5}$	-	1	-	38.6	37.9×10^3
	0.52	12.6	$k_{38.6} = 6.58 \times 10^{-5}$	-	1	-	38.6	10.5×10^3
10%	0.32	5.4	$k_{38.6} = 1.03 \times 10^{-5}$	-	1	-	38.6	67.3×10^3
	0.52	10.2	$k_{38.6} = 6.63 \times 10^{-5}$	-	1	-	38.6	10.5×10^3
	0.75	27.7	$k_{38.6} = 51.87 \times 10^{-5}$	-	1	-	38.6	1.3×10^3

Lajolo and Lanfer Marquez, 1982

(Continued)

TABLE 2.4 Continued

Commodity	Process / Conditions		r^2	k_T value (min^{-1})	E_a (kcal/mol)	Reaction Order	r^2	Temp. Range (°C)	$t_{1/2}$ (min)	Reference
Spinach, puree (cont.) Clorophyll a	a_w	$g\,H_2O/100\,g$								Lajolo and Lanfer Marquez, 1982
0% glycerol	0.32	5.0	–	$k_{38.6} = 1.00 \times 10^{-5}$	–	1	–	38.6	69.3×10^3	
	0.52	10.2	–	$k_{38.6} = 5.18 \times 10^{-5}$	–	1	–	38.6	13.4×10^3	
	0.75	27.7	–	$k_{38.6} = 21.78 \times 10^{-5}$	–	1	–	38.6	3.2×10^3	
Anthocyanins										
Blackberry juice (cyanidin-3-glucoside)										Debicki-Pospišil et al., 1983
Control			–	$k_{70} = 0.001178$	15.0	1			588	
			–	$k_{50} = 0.0003500$	$(14.8)^a$		0.998	24 - 70	1,980	
			–	$k_{24} = 0.0000395$					17,548	
With furfural			–	$k_{70} = 0.001395$	13.2	1			497	
			–	$k_{50} = 0.0005067$	$(13.0)^a$		0.996	24 - 70	1,368	
			–	$k_{24} = 0.0000717$					9,667	
HMF			–	$k_{70} = 0.001478$	13.1	1			469	
			–	$k_{50} = 0.0005783$	$(12.7)^a$		0.992	24 - 70	1,199	
			–	$k_{24} = 0.0000780$					8,887	
benzaldehyde			–	$k_{70} = 0.001927$	11.3^a	1	–	24 - 70	360	
formaldehyde			–	$k_{70} = 0.003367$	6.8^a	1	–	24 - 70	206	
Cyanidin-3-glucoside										Debicki-Pospišil et al., 1983
Control (in citrate buffer)			–	$k_{70} = 0.000960$	20.5	1			722	
			–	$k_{50} = 0.0001823$	$(20.1)^a$		0.998	24 - 70	3,802	
			–	$k_{24} = 0.00000933$					74,290	
With furfural			–	$k_{70} = 0.001217$	17.9	1			570	
			–	$k_{50} = 0.0003067$	$(17.4)^a$		0.996	24 - 70	2,260	
			–	$k_{24} = 0.0000217$					31,940	
HMF			–	$k_{70} = 0.001525$	18.8	1			455	
			–	$k_{50} = 0.0003683$	$(14.9)^a$		0.995	24 - 70	1,882	
			–	$k_{24} = 0.00002183$					31,750	
Boysenberry juice (A_{520}/A_{420})			0.907	$k_{100} = 0.001546$	$(20)^a$	1	–	20 - 120	488	Ponting et al., 1960

Cherries Fresh fruit in sucrose soln./ packed in glass jars (0.073m dia x 0.012m h)	Total anthocyanins Pasteurized: 90°C / 20 min Store: 10 mo	R^2	k	value		n	R^2	T	value	Reference
Sour (*Prunus cerasus*)	With light	0.986[a]	k_{20}	$= 3.194 \times 10^{-5}$	-	1	-	20	2.170×10^4	Ochoa et al., 2001
	No light	0.995	k_{20}	$= 2.410 \times 10^{-5}$	-	1	-	20	2.876×10^4	
Sweet (*Prunus avium*)	With light	0.992	k_{20}	$= 2.500 \times 10^{-5}$	-	1	-	20	2.773×10^4	
	No light	0.982	k_{20}	$= 2.083 \times 10^{-5}$	-	1	-	20	3.328×10^4	
	Store: 5 mo									
Sour (*Prunus cerasus*)	With light	0.975	k_{40}	$= 3.729 \times 10^{-6}$		1			1.859×10^5	
		0.998	k_{20}	$= 1.729 \times 10^{-6}$	8.49		0.998	4 - 40	4.009×10^5	
		0.991	k_4	$= 0.764 \times 10^{-6}$					9.073×10^5	
Sweet (*Prunus avium*)	With light	0.963	k_{40}	$= 4.826 \times 10^{-6}$		1			1.436×10^5	
		0.991	k_{20}	$= 2.694 \times 10^{-6}$	8.31		0.999	4 - 40	2.573×10^5	
		0.957	k_4	$= 0.889 \times 10^{-6}$					7.797×10^5	
Cherry juice (sour)										Cemeroglu et al., 1994
15° Brix	Heated: 20 ml Pyrex tubes /w/ minimal headspace Max. 48 h	0.982	k_{80}	$= 5.661 \times 10^{-4}$		1			1.22×10^3	
		0.937	k_{70}	$= 2.048 \times 10^{-4}$	16.37		0.936	50 - 80	3.38×10^3	
		0.960	k_{60}	$= 0.875 \times 10^{-4}$					7.92×10^3	
		0.949	k_{50}	$= 0.665 \times 10^{-4}$					10.42×10^3	
45° Brix		0.976	k_{80}	$= 9.532 \times 10^{-4}$		1			0.727×10^3	
		0.982	k_{70}	$= 4.052 \times 10^{-4}$	18.13		0.997	50 - 80	1.71×10^3	
		0.972	k_{60}	$= 1.832 \times 10^{-4}$					3.78×10^3	
		0.916	k_{50}	$= 0.858 \times 10^{-4}$					8.08×10^3	
71° Brix		0.996	k_{80}	$= 16.192 \times 10^{-4}$		1			0.428×10^3	
		0.927	k_{70}	$= 6.745 \times 10^{-4}$	19.14		0.999	50 - 80	1.03×10^3	
		0.931	k_{60}	$= 3.125 \times 10^{-4}$					2.22×10^3	
		0.951	k_{50}	$= 1.250 \times 10^{-4}$					5.55×10^3	
45° Brix	Pasteurized / stored up to 160 days	0.958	k_{37}	$= 1.281 \times 10^{-5}$		1			0.541×10^5	
		0.910	k_{20}	$= 0.367 \times 10^{-5}$	15.58		0.992	5 - 37	1.89×10^5	
		0.906	k_5	$= 0.694 \times 10^{-6}$					9.99×10^5	
71° Brix		0.970	k_{37}	$= 1.659 \times 10^{-5}$		1			0.418×10^5	
		0.904	k_{20}	$= 0.455 \times 10^{-5}$	18.02		0.981	5 - 37	1.524×10^5	
		0.949	k_5	$= 0.569 \times 10^{-6}$					12.17×10^5	

(Continued)

TABLE 2.4 Continued

Commodity	Process / Conditions	r^2	k_T value (min^{-1})		E_a (kcal/mol)	Reaction Order	r^2	Temp. Range (°C)	$t_{1/2}$ (min)	Reference
Cherry juice model system										
Pelargonidin-3, 5-diglucoside	pH 2.5	-	k_{108}	= 0.04260	27.4	1	0.999	78 - 108	16	Ioncheva and Tanchev, 1974
		-	k_{98}	= 0.01608					43	
		-	k_{88}	= 0.00600					116	
		-	k_{78}	= 0.001896					366	
	pH 3.5	-	k_{108}	= 0.03612	27.8	1	0.998	78 - 108	19	
		-	k_{98}	= 0.01500					46	
		-	k_{88}	= 0.004596					151	
		-	k_{78}	= 0.001650					420	
	pH 4.5	-	k_{108}	= 0.02928	26.7	1	0.999	78 - 108	24	
		-	k_{98}	= 0.01206					57	
		-	k_{88}	= 0.00396					175	
		-	k_{78}	= 0.001482					468	
Cyanidin-3, 5-diglucoside	pH 2.5	-	k_{108}	= 0.03570	23.5	1	0.999	78 - 108	19	
		-	k_{98}	= 0.01626					43	
		-	k_{88}	= 0.00657					106	
		-	k_{78}	= 0.002532					274	
	pH 3.5	-	k_{108}	= 0.02880	25.4	1	0.996	78 - 108	24	
		-	k_{98}	= 0.01296					53	
		-	k_{88}	= 0.00417					166	
		-	k_{78}	= 0.001746					397	
	pH 4.5	-	k_{108}	= 0.03228	24.2	1	0.999	78 - 108	21	
		-	k_{98}	= 0.01392					50	
		-	k_{88}	= 0.005466					127	
		-	k_{78}	= 0.002112					328	
Peonidin-3, 5-diglucoside	pH 2.5	-	k_{108}	= 0.03822	27.2	1	0.998	78 - 108	18	
		-	k_{98}	= 0.01566					44	
		-	k_{88}	= 0.004950					140	
		-	k_{78}	= 0.001842					376	

Compound	pH	Rate constant				Temp (°C)		Reference
Peonidin-3, 5-diglucoside (cont.)	pH 3.5	$k_{108} = 0.03156$					22	Ioncheva and Tanchev, 1974
		$k_{98} = 0.01338$	24.3	1	0.934	78 - 108	52	
		$k_{88} = 0.004548$					152	
		$k_{78} = 0.002136$					325	
	pH 4.5	$k_{108} = 0.02952$					23	
		$k_{98} = 0.01161$	24.3	1	0.999	78 - 108	60	
		$k_{88} = 0.004986$					139	
		$k_{78} = 0.001848$					375	
Petunidin-3, 5-diglucoside	pH 2.5	$k_{108} = 0.05070$					14	
		$k_{98} = 0.02532$	19.4	1	0.999	78 - 108	27	
		$k_{88} = 0.01200$					58	
		$k_{78} = 0.00573$					121	
	pH 3.5	$k_{108} = 0.03846$					18	
		$k_{98} = 0.01836$	19.6	1	0.999	78 - 108	38	
		$k_{88} = 0.009180$					76	
		$k_{78} = 0.004152$					167	
	pH 4.5	$k_{108} = 0.02910$					24	
		$k_{98} = 0.01296$	20.4	1	0.998	78 - 108	53	
		$k_{88} = 0.00600$					116	
		$k_{78} = 0.002904$					239	
Malvidin-3, 5-diglucoside	pH 2.5	$k_{108} = 0.05130$					14	
		$k_{98} = 0.01986$	26.5	1	0.999	78 - 108	35	
		$k_{88} = 0.007620$					91	
		$k_{78} = 0.002544$					272	
	pH 3.5	$k_{108} = 0.02628$					26	
		$k_{98} = 0.01140$	26.0	1	0.998	78 - 108	61	
		$k_{88} = 0.003816$					182	
		$k_{78} = 0.001458$					475	
	pH 4.5	$k_{108} = 0.02244$					31	
		$k_{98} = 0.00942$	26.2	1	0.998	78 - 108	74	
		$k_{88} = 0.003090$					224	
		$k_{78} = 0.001212$					572	

(Continued)

TABLE 2.4 Continued

Commodity	Process / Conditions	r^2	k_T value (min^{-1})		E_a (kcal/mol)	Reaction Order	r^2	Temp. Range (°C)	$t_{1/2}$ (min)	Reference
Cranberries	Pigment extracted and concentrated solution in pH 2.5 phosphate buffer									Attoe and von Elbe, 1981
Cyanidin-3-arabinoside	No light	-	k_{55}	= 0.000283	26.8	1	-	40 - 55	2,450	
		-	k_{40}	= 0.000395					17,500	
	With light (400 ft-c)	-	k_{55}	= 0.000435	8.7	1	-	40 - 55	1,590	
		-	k_{40}	= 0.000230					3,010	
Cyanidin-3-galactoside	No light	-	k_{55}	= 0.000267	26.7	1	-	40 - 55	2,600	
		-	k_{40}	= 0.0000373					18,600	
	With light (400 ft-c)	-	k_{55}	= 0.000363	7.7	1	-	40 - 55	1,900	
		-	k_{40}	= 0.000207					3,350	
Peonidin-3-arabinoside	No light	-	k_{55}	= 0.000287	24.9	1	-	40 - 55	2,420	
		-	k_{40}	= 0.0000458					15,100	
	With light (400 ft-c)	-	k_{55}	= 0.000422	7.6	1	-	40 - 55	1,640	
		-	k_{40}	= 0.000242					2,860	
Peonidin-3-galactoside	No light	-	k_{55}	= 0.000265	26.4	1	-	40 - 55	2,620	
		-	k_{40}	= 0.0000380					18,200	
	With light (400 ft-c)	-	k_{55}	= 0.000337	5.4	1	-	40 - 55	2,060	
		-	k_{40}	= 0.0000227					30,500	
18 Anthocyanins (averaged)										Tanchev, 1983
Fruit juice		-	k_{108}	= 0.01925	22.2	1	0.999	78 - 108	36	
		-	k_{98}	= 0.008351					83	
		-	k_{88}	= 0.003667					189	
		-	k_{78}	= 0.001561					444	
Citrate buffer		-	k_{108}	= 0.02666	25.1	1	0.999	78 - 108	26	
		-	k_{98}	= 0.01005					69	
		-	k_{88}	= 0.004101					169	
		-	k_{78}	= 0.001540					450	

Grape juice	(Total Anthocyanins)									Ponting
Alicante bouschet (A)	Evelyn colorimeter (A_{520}/A_{420})	0.991	k_{100} = 0.002822	-	-	1	-	100	246	et al., 1960
Carignane (B)		0.943	k_{100} = 0.003438	-	-	1	-	100	202	
Zinfandel (C)		0.913	k_{100} = 0.003443	-	-	1	-	100	201	
Grape blend (45A:45B:10C)		0.996	k_{100} = 0.002592	$(28)^{a}$	-	1	-	20 - 120	267	
Concord grape pigments										Sastry and
Buffer solution (McIlvaine)	pH 3.4	0.674	k_{121} = 0.018790	13.1	1	0.999	76.7 - 121	37		Tischer, 1952
		0.880	$k_{98.9}$ = 0.007430						93	
		0.838	$k_{76.7}$ = 0.002263						306	
Concord grape pigments (*V. labrusca*)										Calvi and
CON - (Control: 0.1M citrate-phosphate buffer)	pH 3.2	0.958	k_{95} = 0.005467	18.9	1	0.991	85 - 95	127		Francis, 1978
		0.990	k_{90} = 0.003600						193	
		0.992	k_{85} = 0.002650						262	
GLU - (buffer + 15% glucose)	pH 3.2	0.990	k_{95} = 0.005383	19.7	1	0.999	85 - 95	129		
		0.992	k_{90} = 0.003700						187	
		0.984	k_{85} = 0.002533						274	
SUC - (buffer + 15% sucrose)	pH 3.2	0.994	k_{95} = 0.008233	23.8	1	0.995	80 - 95	84		
		0.998	k_{90} = 0.004783						145	
		0.978	k_{85} = 0.003300						210	
		0.990	k_{80} = 0.002000						347	
FJD - (buffer + 15% sucrose + 10% white grape juice)	pH 3.2	0.994	k_{95} = 0.008783	17.9	1	0.993	85 - 95	79		
		0.996	k_{90} = 0.006600						105	
		0.990	k_{85} = 0.004433						156	
CON	pH 2.8	-	k_{90} = 0.004400	-	1	-	90	158		
GLU	pH 2.8	-	k_{90} = 0.003983	-	1	-	90	174		
SUC	pH 2.8	-	k_{90} = 0.004667	-	1	-	90	149		
CON	pH 3.6	-	k_{90} = 0.003867	-	1	-	90	179		
GLU	pH 3.6	-	k_{90} = 0.003517	-	1	-	90	197		
SUC	pH 3.6	-	k_{90} = 0.005133	-	1	-	90	135		

(Continued)

TABLE 2.4　Continued

Commodity	Process / Conditions	r^2	k_T value (min⁻¹)	E_a (kcal/mol)	Reaction Order	r^2	Temp. Range (°C)	$t_{1/2}$ (min)	Reference
Plum juice									
Cyanidin-3-rutinoside	pH 2.5	-	k_{108} = 0.02028	21.8	1	0.991	78 - 108	34	Tanchev and Joncheva, 1973
		-	k_{98} = 0.00768					90	
		-	k_{88} = 0.00348					199	
		-	k_{78} = 0.00170					408	
	pH 3.5	-	k_{108} = 0.02052	20.6	1	0.984	78 - 108	34	
		-	k_{98} = 0.00762					91	
		-	k_{88} = 0.00370					187	
		-	k_{78} = 0.00196					354	
	pH 4.5	-	k_{108} = 0.02664	22.6	1	0.981	78 - 108	26	
		-	k_{98} = 0.00876					79	
		-	k_{88} = 0.00395					175	
		-	k_{78} = 0.00201					345	
Peonidin-3-rutinoside	pH 2.5	-	k_{108} = 0.02904	29.9	1	0.999	78 - 108	24	
		-	k_{98} = 0.01014					68	
		-	k_{88} = 0.00327					212	
		-	k_{78} = 0.000996					696	
	pH 3.5	-	k_{108} = 0.03000	23.4	1	0.997	78 - 108	23	
		-	k_{98} = 0.01254					55	
		-	k_{88} = 0.00477					145	
		-	k_{78} = 0.00219					317	
	pH 4.5	-	k_{108} = 0.02988	22.9	1	0.995	78 - 108	23	
		-	k_{98} = 0.01428					49	
		-	k_{88} = 0.00509					136	
		-	k_{78} = 0.00239					290	

Sample	k values						Half-life	Reference
Pomegranate juice (mostly delphinidin-3, 5-diglucoside) Total anthocyanins	k_{92} = 0.00180	-					385	Mishkin and Saguy, 1982
	k_{90} = 0.00088	-	25.0	1	0.945	70 - 92	788	
	k_{80} = 0.00054	-					1,284	
	k_{70} = 0.00015	-					4,621	
Raspberries Total anthocyanins								Ochoa et al., 2001
Fresh fruit in sucrose soln./ packed in glass jars (0.073m dia x 0.012m h) Pasteurized: 90°C / 20 min Store: 10 mo With light	k_{20} = 3.819 x 10^{-5}	[a]0.975	-	1	-	20	1.815 x 10^{4}	
No light	k_{20} = 1.528 x 10^{-5}	0.995	-	1	-	20	4.537 x 10^{4}	
Store: 5 mo With light	k_{40} = 4.931 x 10^{-6}	0.975					1.406 x 10^{5}	
	k_{20} = 2.347 x 10^{-6}	0.998	6.24	1	0.998	4 - 40	2.953 x 10^{5}	
	k_{4} = 1.389 x 10^{-6}	0.991					4.990 x 10^{5}	
Raspberry juice Total anthocyanins								Tanchev, 1972
Maling Promis pH 3.2	k_{108} = 0.01986	-					35	
	k_{98} = 0.008520	-	22.0	1	0.999	78 - 108	81	
	k_{88} = 0.003960	-					175	
	k_{78} = 0.001620	-					428	
New Burg pH 3.4	k_{108} = 0.01638	-					42	
	k_{98} = 0.006480	-	24.1	1	0.999	78 - 108	107	
	k_{88} = 0.002874	-					241	
	k_{78} = 0.001050	-					660	
Bulgarian Ruby pH 3.3	k_{108} = 0.02040	-					34	
	k_{98} = 0.008940	-	21.2	1	0.998	78 - 108	78	
	k_{88} = 0.003510	-					197	
	k_{78} = 0.001950	-					355	

(Continued)

Handbook of Food Engineering

TABLE 2.4 Continued

Commodity	Process / Conditions	r^2	k_T value (min^{-1})	E_a (kcal/mol)	Reaction Order	r^2	Temp. Range (°C)	$t_{1/2}$ (min)	Reference
Raspberry juice (with added sugar)									
Maling Promis	pH 3.2	-	k_{108} = 0.01710					41	Tanchev, 1972
		-	k_{98} = 0.006150	23.0	1	0.988	78 - 108	113	
		-	k_{88} = 0.003450					201	
		-	k_{78} = 0.001164					595	
New Burg	pH 3.4	-	k_{108} = 0.01416					49	
		-	k_{98} = 0.006180	23.6	1	0.999	78 - 108	112	
		-	k_{88} = 0.002652					261	
		-	k_{78} = 0.000972					713	
Bulgarian Ruby	pH 3.3	-	k_{108} = 0.02052					34	
		-	k_{98} = 0.008670	23.6	1	0.999	78 - 108	80	
		-	k_{88} = 0.003480					199	
		-	k_{78} = 0.001434					483	
Strawberry juice									
In oxygen	pH 3.05	-	k_{45} = 5.95 x 10^{-4}	-	1	-	45	1,165	Lukton et al., 1956
	pH 3.55	-	k_{45} = 10.83 x 10^{-4}	-	1	-	45	640	
	pH 4.30	-	k_{45} = 15.33 x 10^{-4}	-	1	-	45	452	
In nitrogen	pH 3.05	-	k_{45} = 0.85 x 10^{-4}	-	1	-	45	8,155	
	pH 3.55	-	k_{45} = 1.00 x 10^{-4}	-	1	-	45	6,966	
	pH 4.30	-	k_{45} = 1.24 x 10^{-4}	-	1	-	45	5,590	
Strawberry juice	Spectrophotometer (A_{490} / A_{420})	-	-	(19)[a]	1	-	20 - 120	-	Ponting et al., 1960

System	Conditions								Reference
Wine model systems									Baranowski and Nagle, 1983
Malvidin-3-glucoside (M-3-G)	Glass tubes, O$_2$-free atmos., HPLC	-	k_{52} = 6.60 x 10^{-5}	16.9	1	0.778	22 - 52	1.05 x 10^4	
		-	k_{42} = 9.60 x 10^{-4}	(28)[a]				0.0722 x 10^4	
		-	k_{32} = 4.08 x 10^{-6}					16.99 x 10^4	
		-	k_{22} = 4.38 x 10^{-6}					15.83 x 10^4	
M-3-G + d-catechin		-	k_{52} = 3.24 x 10^{-5}	12.0	1	0.971	22 - 52	2.14 x 10^4	
		-	k_{42} = 1.98 x 10^{-5}	(11)[a]				3.50 x 10^4	
		-	k_{32} = 1.26 x 10^{-5}					5.50 x 10^4	
		-	k_{22} = 4.68 x 10^{-6}					14.81 x 10^4	
M-3-G + d-catechin + equimolar acetaldehyde		-	k_{52} = 2.94 x 10^{-5}	13.0	1	0.832	22 - 52	2.36 x 10^4	
		-	k_{42} = 2.04 x 10^{-5}	(10)[a]				3.40 x 10^4	
		-	k_{32} = 1.68 x 10^{-5}					4.13 x 10^4	
		-	k_{22} = 3.30 x 10^{-6}					21.00 x 10^4	
M-3-G + d-catechin + XS acetaldehyde		-	k_{52} = 9.60 x 10^{-5}	12.7	1	0.994	22 - 52	0.722 x 10^4	
		-	k_{42} = 5.94 x 10^{-5}	(13)[a]				1.17 x 10^4	
		-	k_{32} = 2.88 x 10^{-5}					2.41 x 10^4	
		-	k_{22} = 1.32 x 10^{-5}					5.25 x 10^4	
Wine model systems (grape skin pigments in potassium hydrogen tartrate, pH 3.5)	Storage: 50 ml vials / air / dark / up to 140 days HPLC *PA / TA Ratio*								Romero and Bakker, 2000
Malvidin-3-glucoside	0	0.984	k_{32} = 59.03 x 10^{-6}	0.32	1	-	10 - 32	1.17 x 10^{-4}	
		0.984	k_{20} = 29.17 x 10^{-6}					2.38 x 10^{-4}	
		0.963	k_{15} = 5.56 x 10^{-6}					12.48 x 10^{-4}	
		0.975	k_{10} = 4.86 x 10^{-6}					14.26 x 10^{-4}	
PA = pyruvic acid	300	0.975	k_{32} = 39.58 x 10^{-6}	0.24	1	-	10 - 32	1.75 x 10^{-4}	
TA = total anthocyanins		0.981	k_{20} = 20.14 x 10^{-6}					3.44 x 10^{-4}	
		0.987	k_{15} = 8.33 x 10^{-6}					8.32 x 10^{-4}	
		0.986	k_{10} = 4.86 x 10^{-6}					14.26 x 10^{-4}	

(Continued)

TABLE 2.4 Continued

Commodity	Process / Conditions	r^2	k_T value (min^{-1})		E_a (kcal/mol)	Reaction Order	r^2	Temp. Range (°C)	$t_{1/2}$ (min)	Reference
Wine model systems (cont.)										
Malvidin-3-acetylglucoside	0	0.933	k_{32}	= 54.86 x 10^{-6}	0.31	1	-	10 - 32	1.17 x 10^{-4}	Romero and
		0.994	k_{20}	= 34.03 x 10^{-6}					2.38 x 10^{-4}	Bakker, 2000
		0.919	k_{15}	= 5.56 x 10^{-6}					12.48 x 10^{-4}	
		0.973	k_{10}	= 4.86 x 10^{-6}					14.26 x 10^{-4}	
	300	0.927	k_{32}	= 41.67 x 10^{-6}	0.23	1	-	10 - 32	1.66 x 10^{-4}	
		0.921	k_{20}	= 18.06 x 10^{-6}					3.84 x 10^{-4}	
		0.933	k_{15}	= 11.81 x 10^{-6}					5.87 x 10^{-4}	
		0.875	k_{10}	= 5.56 x 10^{-6}					12.48 x 10^{-4}	
Malvidin-3p-coumaryl-glucoside	0	0.984	k_{32}	= 104.9 x 10^{-6}	0.37	1	-	10 - 32	0.661 x 10^{-4}	
		0.997	k_{20}	= 53.47 x 10^{-6}					1.30 x 10^{-4}	
		0.933	k_{15}	= 8.33 x 10^{-6}					8.32 x 10^{-4}	
		0.906	k_{10}	= 5.56 x 10^{-6}					12.48 x 10^{-4}	
	300	0.963	k_{32}	= 67.36 x 10^{-6}	0.30	1	-	10 - 32	1.03 x 10^{-4}	
		0.927	k_{20}	= 26.39 x 10^{-6}					2.63 x 10^{-4}	
		0.938	k_{15}	= 10.42 x 10^{-6}					6.65 x 10^{-4}	
		0.925	k_{10}	= 5.56 x 10^{-6}					12.48 x 10^{-4}	
Betalains										
Beets										
Beet puree (Betaine)	Natural pH (electrophoretic separation)	0.995	[a]k_{116}	= 0.01419	[a]8.7 (10 ± 2)[a]	1	0.996	102 - 116	49	Von Elbe et al., 1974
		0.996	[a]k_{110}	= 0.01219					57	
		0.950	[a]k_{102}	= 0.009333					74	
Beet juice (Betanine)	pH 3.0	-	k_{100}	= 0.079	-	1	-	100	8.8	
	pH 5.0	-	k_{100}	= 0.024	-	1	-	100	29.0	
	pH 7.0	-	k_{100}	= 0.135	-	1	-	100	5.1	
Betanine solution (citric-phosphate buffer)	pH 3.0	-	k_{100}	= 0.094	-	1	-	100	7.4	
	pH 4.0	-	k_{100}	= 0.051	-	1	-	100	13.6	

Sample	k	z	n	R^2	Temp		Reference
Betanine solution (cont.)							Von Elbe et al., 1974
pH 5.0	– $k_{100} = 0.048$		1	–	–	14.4	
pH 5.0	– $k_{75} = 0.007800$	[a]12.6	1	0.976	25 - 100	89.0	
pH 5.0	– $k_{50} = 0.002200$	10.5	1	0.998	25 - 75	315	
pH 5.0	– $k_{25} = 0.000610$	14.7	1	0.979	50 - 100	1136	
pH 6.0	– $k_{100} = 0.079$		1	–	100	8.8	
pH 7.0	– $k_{100} = 0.118$		1	–		5.9	
pH 7.0	– $k_{75} = 0.035000$	[a]8.5	1	0.974	25 - 100	20.0	
pH 7.0	– $k_{50} = 0.013800$	7.1	1	0.992	25 - 75	50.0	
pH 7.0	– $k_{25} = 0.006200$	10.1	1	0.986	50 - 100	112	
Beet powder Betanine Dry powders sealed in glass vials							Kopelman and Saguy, 1977
Drum-dried (4% MC)	– $k_{45} = 3.04 \times 10^6$					2.28×10^5	
	– $k_{40} = 2.65 \times 10^6$					2.62×10^5	
	– $k_{35} = 2.37 \times 10^6$	5.9	1	0.992	25 - 45	2.92×10^5	
	– $k_{31} = 1.99 \times 10^6$					3.48×10^5	
	– $k_{25} = 1.63 \times 10^6$					4.25×10^5	
Air-dried (4% MC)	– $k_{45} = 3.49 \times 10^6$					1.99×10^5	
	– $k_{40} = 3.28 \times 10^6$					2.11×10^5	
	– $k_{35} = 2.79 \times 10^6$	6.6	1	0.949	25 - 45	2.48×10^5	
	– $k_{31} = 2.10 \times 10^6$					3.30×10^5	
	– $k_{25} = 1.84 \times 10^6$					3.77×10^5	
Vulgaxanthin Drum-dried (4% MC)	– $k_{45} = 3.01 \times 10^6$					2.30×10^5	
	– $k_{40} = 2.68 \times 10^6$					2.59×10^5	
	– $k_{35} = 2.19 \times 10^6$	5.6	1	0.935	25 - 45	3.17×10^5	
	– $k_{31} = 1.80 \times 10^6$					3.85×10^5	
	– $k_{25} = 1.75 \times 10^6$					3.96×10^5	
Air-dried (4% MC)	– $k_{45} = 3.23 \times 10^6$					2.15×10^5	
	– $k_{40} = 2.90 \times 10^6$					2.39×10^5	
	– $k_{35} = 2.38 \times 10^6$	6.5	1	0.986	25 - 45	2.91×10^5	
	– $k_{31} = 1.96 \times 10^6$					3.54×10^5	
	– $k_{25} = 1.67 \times 10^6$					4.15×10^5	

(Continued)

TABLE 2.4 Continued

Commodity	Process / Conditions	r^2	k_T value (min^{-1})	E_a (kcal/mol)	Reaction Order	r^2	Temp. Range (°C)	$t_{1/2}$ (min)	Reference
Beet powder	Freeze-dried/ ground/ rehumidified to 0.75 aw								Cohen and Saguy, 1983
Betanine	$g\ H_2O\ /\ 100\ g\ solids$								
No additives:	25.8	0.982	k_{35} = 5.80 x 10^{-5}	-	1	-	35	1.20 x 10^4	
Beet:CMC (3:1)	20.1	0.984	k_{35} = 5.35 x 10^{-5}	-	1	-	35	1.30 x 10^4	
Beet:CMC (1:1)	14.9	0.984	k_{35} = 2.92 x 10^{-5}	-	1	-	35	2.37 x 10^4	
Beet:pectin (1:1)	13.6	0.992	k_{35} = 3.01 x 10^{-5}	-	1	-	35	2.30 x 10^4	
Vulgaxanthin-I	$g\ H_2O\ /\ 100\ g\ solids$								Cohen and Saguy, 1983
No additives:	25.8	0.986	k_{35} = 4.47 x 10^{-5}	-	1	-	35	1.55 x 10^4	
Beet:CMC (3:1)	20.1	0.972	k_{35} = 4.04 x 10^{-5}	-	1	-	35	1.72 x 10^4	
Beet:CMC (1:1)	14.9	0.992	k_{35} = 2.45 x 10^{-5}	-	1	-	35	2.83 x 10^4	
Beet:pectin (1:1)	13.6	0.460	k_{35} = 0.150 x 10^{-5}	-	1	-	35	13.86 x 10^4	
Beet juice									Saguy, 1979
Betanine									
	pH 4.8	-	k_{100} = 0.1130					6.2	
		-	$k_{85.5}$ = 0.0405	18.2	1	0.991	61.5 - 100	17.0	
		-	$k_{75.5}$ = 0.0243					29.0	
		-	$k_{61.5}$ = 0.0063					110	
	pH 5.2	-	k_{100} = 0.0980					7.1	
		-	$k_{85.5}$ = 0.0374	18.6	1	0.999	61.5 - 100	19.0	
		-	$k_{75.5}$ = 0.0165					42.0	
		-	$k_{61.5}$ = 0.0056					124	
	pH 5.8	-	k_{100} = 0.0946					7.3	
		-	$k_{85.5}$ = 0.0320	19.6	1	0.999	61.5 - 100	22.0	
		-	$k_{75.5}$ = 0.0146					47.0	
		-	$k_{61.5}$ = 0.0045					154	
	pH 6.2	-	k_{100} = 0.1177					5.9	
		-	$k_{85.5}$ = 0.0405	19.9	1	0.999	61.5 - 100	17.0	
		-	$k_{75.5}$ = 0.0168					41.0	
		-	$k_{61.5}$ = 0.0055					126	

Vulgaxanthin-1	pH 4.8	-	k_{100} = 0.1337	15.4	1	0.997	61.5 - 100	5.2	Saguy, 1979
		-	$k_{85.5}$ = 0.0560					12.0	
		-	$k_{75.5}$ = 0.0341					20.0	
		-	$k_{61.5}$ = 0.0119					58.0	
	pH 5.2	-	k_{100} = 0.1204	16.4	1	0.999	61.5 - 100	5.8	
		-	$k_{85.5}$ = 0.0497					14.0	
		-	$k_{75.5}$ = 0.0251					28.0	
		-	$k_{61.5}$ = 0.0095					73.0	
	pH 5.8	-	k_{100} = 0.1146	16.5	1	0.999	61.5 - 100	6.0	
		-	$k_{85.5}$ = 0.0456					15.0	
		-	$k_{75.5}$ = 0.0234					30.0	
		-	$k_{61.5}$ = 0.0088					79.0	
	pH 6.2	-	k_{100} = 0.1239	16.9	1	0.999	61.5 - 100	5.6	
		-	$k_{85.5}$ = 0.0493					14.0	
		-	$k_{75.5}$ = 0.0234					30.0	
		-	$k_{61.5}$ = 0.0091					76.0	
Beet slices Betanine	Partially freeze-dried/ rehumidified $g\ H_2O\,/\,100\ g\ solids$								Saguy et al., 1980
	0.03	0.99	k_{90} = 0.00297	9.1	1	0.99	70 - 90	233	
		0.98	k_{80} = 0.00214	$(13.1)^a$				324	
		0.98	k_{70} = 0.00142					488	
	0.41	0.99	k_{90} = 0.00397	10.4	1	0.99	70 - 90	175	
		0.98	k_{80} = 0.00283	$(14.9)^a$				245	
		0.99	k_{70} = 0.00172					403	
	0.56	0.99	k_{90} = 0.00733	10.8	1	0.97	70 - 90	95	
		0.99	k_{80} = 0.00546	$(15.5)^a$				127	
		0.98	k_{70} = 0.00306					227	
	4.26	0.99	k_{90} = 0.00872	11.3	1	0.96	70 - 90	79	
		0.98	k_{80} = 0.00659	$(16.2)^a$				105	
		0.99	k_{70} = 0.00350					198	
	6.67	0.97	k_{90} = 0.00984	12.0	1	0.98	70 - 90	70	
		0.98	k_{80} = 0.00690	$(17.4)^a$				100	
		0.98	k_{70} = 0.00370					187	

(Continued)

TABLE 2.4 Continued

Commodity	Process / Conditions	r^2	k_T value (min^{-1})		E_a (kcal/mol)	Reaction Order	r^2	Temp. Range (°C)	$t_{1/2}$ (min)	Reference
Beet slices (cont.) Vulgaxanthin-I	*g H$_2$O / 100 g solids*									Saguy et al., 1980
	0.03	0.99	k_{90}	= 0.00081	5.3	1	0.98	70 - 90	856	
		0.99	k_{80}	= 0.00069	$(7.3)^a$				1005	
		0.98	k_{70}	= 0.00053					1308	
	0.41	0.99	k_{90}	= 0.00194	10.4	1	0.86	70 - 90	357	
		0.98	k_{80}	= 0.00096	$(15.0)^a$				722	
		0.99	k_{70}	= 0.00083					835	
	0.56	0.99	k_{90}	= 0.00456	13.0	1	0.99	70 - 90	152	
		0.98	k_{80}	= 0.00292	$(16.7)^a$				237	
		0.99	k_{70}	= 0.00159					436	
	4.26	0.95	k_{90}	= 0.00598	13.4	1	0.99	70 - 90	116	
		0.97	k_{80}	= 0.00336	$(18.8)^a$				206	
		0.99	k_{70}	= 0.00207					335	
	6.67	0.99	k_{90}	= 0.00718	13.4	1	0.99	70 - 90	97	
		0.99	k_{80}	= 0.00403	$(19.2)^a$				172	
		0.98	k_{70}	= 0.00243					285	
Betanine → Betalamic acid	pH 5.5 / spectrophotometric	-	k_{86}	= 0.04610	20.4	1	0.999	60 - 86	15	Saguy et al., 1978c
		-	k_{81}	= 0.02960	$(20.4)^a$				23	
	A$_{max}$ = 535 nm	-	k_{75}	= 0.01750					40	
		-	k_{60}	= 0.00490					141	
Betalamic acid → Betalamic acid brown compounds		-	k_{86}	= 0.00341	20.3	1	0.909	60 - 86	203	
		-	k_{81}	= 0.00261	$(20.7)^a$				266	
	pH 5.5	-	k_{75}	= 0.00081					856	
	A$_{max}$ = 430 nm	-	k_{60}	= 0.00390					178	

				Order	r^2	Temp. (°C)		Reference
Betanine	0.1M citrate-phosphate buffer: pH 5.0							Huang and von Elbe, 1985
Forward reaction:	2-ml glass vials, in N_2 atmos.	$^a k_{90}$ = 6.3 ± 0.3					0.11	
		k_{85} = 4.5 ± 0.2					0.15	
		k_{75} = 2.1 ± 0.1	17.9 $(17.3)^a$	1	0.999	65 - 90	0.33	
		k_{65} = 1.01 ± 0.05					0.69	
Reverse reaction:		k_{90} = 86.7					0.0080	
		k_{85} = 85.6	0.66 $(0.64)^a$	1	0.999	65 - 90	0.0081	
		k_{75} = 83.3					0.0083	
		k_{65} = 81.0					0.0086	
Betalamic acid		k_{90} = 1.6 ± 0.2					0.43	
		k_{85} = 1.1 ± 0.1	17.7 $(18.2)^a$	1	0.999	65 - 90	0.63	
		k_{75} = 0.53 ± 0.06					1.30	
		k_{65} = 0.26 ± 0.02					2.70	
Cyclodopa-5-0-glycoside		k_{90} = 0.22 ± 0.03					3.20	
		k_{85} = 0.15 ± 0.02	25.4 $(29)^a$	1	0.998	65 - 90	4.60	
		k_{75} = 0.048 ± 0.005					14.00	
		k_{65} = 0.017 ± 0.002					41.00	
Cranberries	Pigment extracted and concentrated solution in pH 5.0 phosphate buffer							Attoe and von Elbe, 1981
Betanine	No light	k_{55} = 0.00368					188	
		k_{40} = 0.000642	25.1	1	0.999	25 - 55	1,080	
		k_{25} = 0.0000765					9,060	
	With light (400 ft-c)	k_{55} = 0.00468					148	
		k_{40} = 0.001170	19.3	1	0.999	25 - 55	592	
		k_{25} = 0.000238					2,910	
Prickly pear fruit Betacyanine	10-ml aqueous solution of extracted pigment							Merin et al., 1987
	Dilute	k_{90} = 0.05711					12.1	
		k_{70} = 0.01076	15.7 $(7.7)^a$	pseudo-1	0.973	50 - 90	64.4	
		k_{50} = 0.00380					182.4	
	Concentrated (x 10)	k_{90} = 0.02990					23.2	
		k_{70} = 0.00690	21.9 $(10.7)^a$	pseudo-1	0.992	50 - 90	100.4	
		k_{50} = 0.00070					990.0	

(Continued)

TABLE 2.4 Continued

Commodity	Process / Conditions	r^2	k_T value (min^{-1})	E_a (kcal/mol)	Reaction Order	r^2	Temp. Range (°C)	$t_{1/2}$ (min)	Reference
Carotenoids (as pigment)									
Blue crab									
Astaxanthin	XS-water cook	-	k_{100} = 3.4700					0.20	Himelbloom et al., 1983
		-	$k_{93.9}$ = 2.8530					0.24	
	color measured by a-value (colorimeter)	-	$k_{87.8}$ = 1.0510	22.5	1	0.941	76.6 - 100	0.66	
		-	$k_{82.2}$ = 0.7674					0.90	
		-	$k_{76.6}$ = 0.5377					1.29	
Paprika (Total carotenoids)		0.998	k_{125} = 0.003786	-	1	-	125 - 150	183	Ramakrishnan and Francis, 1973
Salmon	Freeze-dried/ rehydrated								Martinez and Labuza, 1968
Astacene	a_w = 0	-	k_{37} = 1.83×10^{-5}					3.79×10^4	
	a_w = 0.11	-	k_{37} = 1.67×10^{-5}	-	1	-	37	4.15×10^4	
	a_w = 0.32	-	k_{37} = 0.83×10^{-5}					8.35×10^4	
	a_w = 0.40	-	k_{37} = 0.17×10^{-5}					40.77×10^4	
Tomato juice Lycopene loss	Heat: 0 - 7 min	0.896	k_{130} = 0.024105					28.8	Miki & Akatsu, 1970 (from Shi & LeMaguer, 2000a)
		0.885	k_{127} = 0.020146					34.4	
		0.884	k_{124} = 0.017068					40.6	
		0.882	k_{121} = 0.014357					48.3	
		0.683	k_{118} = 0.010680	21.2	1	0.977	90 - 130	64.9	
		0.917	k_{115} = 0.009462					73.3	
		0.822	k_{110} = 0.005581					124.2	
		0.756	k_{100} = 0.002107					329.0	
		0.855	k_{90} = 0.001647					420.9	

Tomato paste
Overall color change (Lycopene)
9 ml glass vials / 70, 80, 90, 100°C / 5-90 min

Color measure: Gardner XL-23

Effect of color parameter on determination of reaction order and E_a

Parameter	r^2	k	order	R^2	Temp (°C)	E_a	Reference
ΔE	0.992	$k_{100} = 0.458000$	0	0.977	70 - 100	1.51	Barreiro et al., 1997
	0.991	$k_{70} = 0.122000$				5.68	
L-value: phase 1	0.986	$k_{100} = 0.015000$	1	0.996	70 - 100	46.20	
	0.988	$k_{70} = 0.003600$				192.54	
L-value: phase 2	0.960	$k_{100} = 0.001460$	1	0.979	70 - 100	474.76	
	0.968	$k_{70} = 0.000762$				909.64	
a-value	0.996	$k_{100} = 0.017200$	apparent 1	0.983	70 - 100	40.30	
	0.979	$k_{70} = 0.005000$				138.63	
b-value	0.958	$k_{100} = 0.003210$	apparent 1	0.952	70 - 100	215.93	
	0.902	$k_{70} = 0.000240$				2888.11	
a/b	0.964	$k_{100} = 0.011000$	apparent 1	0.986	70 - 100	69.31	
	0.962	$k_{70} = 0.004600$				150.68	
hue angle	0.960	$k_{100} = 0.009240$	apparent 1	0.993	70 - 100	75.02	
	0.966	$k_{70} = 0.003840$				180.51	
SI	0.998	$k_{100} = 0.014400$	apparent 1	0.989	70 - 100	48.14	
	0.986	$k_{70} = 0.004060$				170.73	

Tomato pulp
Lycopene loss
Heat: 0 - 3 hr

Parameter	r^2	k	order	R^2	Temp (°C)	E_a	Reference
Dark / CO_2	0.633	$k_{100} = 0.000264$	1	-	100	2631.0	Cole & Kapur, 1957b (from Shi & LeMaguer, 2000)
Dark / O_2	0.986	$k_{100} = 0.001990$	1	-	100	348.3	
Light / CO_2	0.978	$k_{100} = 0.000657$	1	-	100	1055.0	
Light / O_2	0.990	$k_{100} = 0.002237$	1	-	100	309.9	

Lycopene Model
(Hexane / light petroleum solution)

Parameter	r^2	k	order	R^2	Temp (°C)	E_a	Reference
Lycopene	0.956	$k_{100} = 0.002341$	1	-	100	296.1	Cole & Kapur, 1957a (from Shi & LeMaguer, 2000)
	0.950	$k_{65} = 0.001613$	1	-	65	429.6	
Lycopene + Cu-stearate	0.985	$k_{100} = 0.012972$	1	-	100	53.4	
	0.999	$k_{65} = 0.005098$	1	-	65	136.0	

(Continued)

TABLE 2.4 Continued

Commodity	Process / Conditions	r^2	k_T value (min^{-1})	E_a (kcal/mol)	Reaction Order	r^2	Temp. Range (°C)	$t_{1/2}$ (min)	Reference
Tomato puree (fresh) Lycopene	Heat: 90-150°C, up to 6 hr								Shi et al., 2003
	all-*trans*-lycopene → mono & poly-cis isomers	-	k_{150} = 0.014638					47.4	
		-	k_{120} = 0.007148	7.98	1	0.959	90 - 150	97.0	
		-	k_{110} = 0.004250	^{a}(0.982)				163.1	
		-	k_{90} = 0.003223					215.1	
	cis isomers → oxidized by-products	-	k_{150} = 0.014455					48.0	
		-	k_{120} = 0.005965	11.80	1	0.967	90 - 150	116.2	
		-	k_{110} = 0.002633	^{a}(0.891)				263.2	
		-	k_{90} = 0.001522					455.5	
	all-*trans*-lycopene → oxidized by-products	-	k_{150} = 0.025783					26.9	
		-	k_{120} = 0.014617	7.00	1	0.996	90 - 150	47.4	
		-	k_{110} = 0.011383	^{a}(1.507)				60.9	
		-	k_{90} = 0.006483					106.9	
Tomato puree (fresh)	Storage: 25°C / 1 -6 days								
	Light: 400 µmol/m^{-2}s^{-1}								
	all-*trans*-lycopene → mono & poly-cis isomers	-	k_{25} = 0.000058	-	1	-	25	12025.7	
	cis isomers → oxidized by-products	-	k_{25} = 0.002708	-	1	-	25	256.0	
	all-*trans*-lycopene → oxidized by-products	-	k_{25} = 0.002826	-	1	-	25	245.2	
	Light: 500 µmol/m^{-2}s^{-1}								
	all-*trans*-lycopene → mono & poly-cis isomers	-	k_{25} = 0.000083	-	1	-	25	8317.8	
	cis isomers → oxidized by-products	-	k_{25} = 0.004384	-	1	-	25	158.1	
	all-*trans*-lycopene → oxidized by-products	-	k_{25} = 0.004938	-	1	-	25	140.4	
	Light: 600 µmol/m^{-2}s^{-1}								
	all-*trans*-lycopene → mono & poly-cis isomers	-	k_{25} = 0.000126	-	1	-	25	5484.2	
	cis isomers → oxidized by-products	-	k_{25} = 0.004883	-	1	-	25	142.0	
	all-*trans*-lycopene → oxidized by-products	-	k_{25} = 0.006119	-	1	-	25	113.3	

Myoglobin

Beef (ground)

Lean cut — Change in red color ("a")

	Condition		k								Reference
Control	PVC foil / O_2 perm	0.898	k_7	=	1.242×10^{-4}	-	1	7		5.58×10^3	Houben et al., 2000
	Modified atmos.	0.998	k_7	=	1.236×10^{-4}	-	1	7		5.61×10^3	
Vit. E supplement	PVC foil / O_2 perm	0.928	k_7	=	1.022×10^{-4}	-	1	7		6.78×10^3	
	Modified atmos.	0.942	k_7	=	0.618×10^{-4}	-	1	7		11.21×10^3	

Fat cut

	Condition		k								
Control	PVC foil / O_2 perm	0.636	k_7	=	1.137×10^{-4}	-	1	7		6.10×10^3	
	Modified atmos.	0.988	k_7	=	1.430×10^{-4}	-	1	7		4.85×10^3	
Vit. E supplement	PVC foil / O_2 perm	0.637	k_7	=	1.022×10^{-4}	-	1	7		6.78×10^3	
	Modified atmos.	0.892	k_7	=	1.127×10^{-4}	-	1	7		6.15×10^3	

Browning

Annato — Aqueous solutions heated up to 450 min

COMCOR 1500 Plus

Color measure:

		k							Reference
L-value	0.721	k_{140}	= 4.49×10^{-4}					15.43×10^2	Ferreira et al., 1999
	0.925	k_{120}	= 3.15×10^{-4}	11.3	0.969	90 - 140	1	22.00×10^2	
	0.755	k_{100}	= 1.13×10^{-4}					61.34×10^2	
	0.624	k_{90}	= 0.732×10^{-4}					94.69×10^2	
a-value	0.721	k_{140}	= 1.59×10^{-2}					0.44×10^2	
	0.925	k_{120}	= 0.204×10^{-2}	24.4	0.979	90 - 140	1	3.40×10^2	
	0.755	k_{100}	= 4.60×10^{-4}					15.07×10^2	
	0.624	k_{90}	= 2.60×10^{-4}					26.66×10^2	
b-value	0.721	k_{140}	= 15.30×10^{-4}					4.53×10^2	
	0.925	k_{120}	= 2.95×10^{-4}	11.0	0.966	90 - 140	1	23.50×10^2	
	0.755	k_{100}	= 1.07×10^{-4}					64.78×10^2	
	0.624	k_{90}	= 0.694×10^{-4}					99.88×10^2	
ΔE	0.721	k_{140}	= 2.74×10^{-2}					0.25×10^2	
	0.925	k_{120}	= 2.75×10^{-2}	12.0	0.883	90 - 140	1	0.25×10^2	
	0.755	k_{100}	= 0.722×10^{-2}					0.96×10^2	
	0.624	k_{90}	= 0.422×10^{-2}					1.64×10^2	

(Continued)

TABLE 2.4 Continued

Commodity	Process / Conditions	r^2	k_T value (min⁻¹)	E_a (kcal/mol)	Reaction Order	r^2	Temp. Range (°C)	$t_{1/2}$ (min)	Reference
Browning (cont.)									
Norbixin salt	Spectrophotometer: 453 nm	0.721	k_{140} = 5.82 x 10⁻²					0.12 x 10²	Ferreira et al., 1999
		0.925	k_{120} = 1.10 x 10⁻²	23.9	1	0.971	90 – 140	0.63 x 10²	
		0.755	k_{100} = 0.151 x 10⁻²					4.59 x 10²	
		0.624	k_{90} = 0.128 x 10⁻²					5.42 x 10²	
Apple juice									
Maillard reaction	Natural pH	–	k_{130} = 0.0170					41	Herrmann, 1970
		–	k_{110} = 0.0037					187	
		–	k_{100} = 0.0014					495	
		–	k_{90} = 0.00060					1,160	
		–	k_{80} = 0.00024	24.0	1	0.99	40 – 130	2,890	
		–	k_{70} = 0.00009					7,700	
		–	k_{60} = 0.000036					19,300	
		–	k_{50} = 0.000010					69,300	
		–	k_{40} = 0.0000028					248,000	
Nonenzymatic		–	k_{130} = 0.0100					69	
		–	k_{120} = 0.0060					116	
		–	k_{110} = 0.0040					173	
		–	k_{100} = 0.0018					385	
		–	k_{90} = 0.0008	19.8	1	0.99	40 – 130	866	
		–	k_{80} = 0.0004					1,730	
		–	k_{70} = 0.00019					3,650	
		–	k_{60} = 0.00007					9,900	
		–	k_{50} = 0.00003					23,100	
		–	k_{40} = 0.000008					86,600	

	solids:								Toribo and Lozano, 1984
Apple juice Granny Smith	65° Bx	0.991	k_{37} = 7.36 x 10^{-6}	19.5	1	0.997	5 - 37	9.42 x 10^4	
		0.972	k_{20} = 1.38 x 10^{-6}					5.02 x 10^5	
		0.983	k_5 = 1.92 x 10^{-7}					3.61 x 10^6	
	70° Bx	0.975	k_{37} = 8.26 x 10^{-6}	17.8	1	0.999	5 - 37	8.39 x 10^4	
		0.976	k_{20} = 1.57 x 10^{-6}					4.41 x 10^5	
		0.991	k_5 = 2.99 x 10^{-7}					2.32 x 10^6	
	75° Bx	0.964	k_{37} = 8.82 x 10^{-6}	16.7	1	0.999	5 - 37	7.86 x 10^4	
		0.971	k_{20} = 1.82 x 10^{-6}					3.81 x 10^5	
		0.982	k_5 = 3.90 x 10^{-7}					1.78 x 10^6	
Red Delicious	65° Bx	0.985	k_{37} = 3.46 x 10^{-6}	19.4	1	0.997	5 - 37	2.00 x 10^5	
		0.979	k_{20} = 4.72 x 10^{-7}					1.47 x 10^6	
		0.983	k_5 = 9.30 x 10^{-8}					7.45 x 10^6	
	70° Bx	0.973	k_{37} = 3.73 x 10^{-6}	17.8	1	0.987	5 - 37	1.86 x 10^5	
		0.968	k_{20} = 4.97 x 10^{-7}					1.39 x 10^6	
		0.988	k_5 = 1.33 x 10^{-7}					5.21 x 10^6	
	75° Bx	0.981	k_{37} = 3.79 x 10^{-6}	16.7	1	0.977	5 - 37	1.83 x 10^5	
		0.961	k_{20} = 5.22 x 10^{-7}					1.33 x 10^6	
		0.990	k_5 = 1.68 x 10^{-7}					4.13 x 10^6	
Aspartame / Glucose	a_w = 0.8	-	k_{100} = 0.2100	22.6	1	0.998	70 - 100	3.30	Stamp and Labuza, 1983
		-	k_{90} = 0.0870	(22)[a]				8.00	
		-	k_{80} = 0.0340					20.00	
		-	k_{70} = 0.01480					47.00	
Glycine / Glucose	a_w = 0.8	-	k_{100} = 0.9950	16.2	1	0.999	80 - 100	0.70	
		-	k_{90} = 0.5483	(15.5)[a]				1.26	
		-	k_{80} = 0.2883					2.40	

(Continued)

TABLE 2.4 Continued

Commodity	Process / Conditions	r^2	k_T value (min^{-1})	E_a (kcal/mol)	Reaction Order	r^2	Temp. Range (°C)	$t_{1/2}$ (min)	Reference
Glycine / Glucose Models (0.2 M Gly + Glu / phos. Buffer / pH 6.8) 16 x 160 mm vials									Martins and VanBoekel, 2005
Non-linear regression / determinant criterion									
	Pathways:								
	1. D-Glu + Gly→ E₁	-	-	23.1	-	-	80 - 120	-	
	2. D-Glu→ Fru	-	-	29.3	-	-	80 - 120	-	
	3. D-Fru→ D-Glu	-	-	22.3	-	-	80 - 120	-	
AA = acetic acid	4. E₁ → Gly + 3-DG	-	-	23.2	-	-	80 - 120	-	
Cn = CHO fragments	5. 3-DG→ FA	-	-	7.1	-	-	80 - 120	-	
DFG = N-(1-deoxy-D-fructos-1-yl)-glycine	6. Cn → MG + FA + AA	-	-	29.8	-	-	80 - 120	-	
1-DG = 1-deoxyglucosone	7. E₂ → Gly + 1-DG	-	-	25.6	-	-	80 - 120	-	
3-DG = 3-deoxyglucosone	8. 1-DG→ AA	-	-	18.1	-	-	80 - 120	-	
E1, E2 = intermediates	9. Cn + Gly→ Mel	-	-	22.8	-	-	80 - 120	-	
FA = formic acid	10. DFG→ E₂	-	-	56.6	-	-	80 - 120	-	
Fru = fructose									
Glu = glucose	MG = methylglyoxal								
Gly = glycine	OA = organic acids								
Mel = melanoidins									
Glucose:Glycine Solutions	Heat up to 40 hr								Barbanti et al., 1990
	Molar ratio:								
Solids concentration:	1:1								
20%		-	k₉₀ = 0.2070		-	-		3.35	
		-	k₈₀ = 0.0833		0	-	70 - 90	8.32	
		-	k₇₀ = 0.0383			-		18.10	
	2:1	-	k₉₀ = 0.1700		-	-		4.08	
		-	k₈₀ = 0.0383		0	-	70 - 90	18.10	
		-	k₇₀ = 0.0250			-		27.73	

	5:1	k_{90} = 0.0250	-	-		27.73
		k_{80} = 0.0067	0	-	70 - 90	103.45
		k_{70} = 0.0017	-	-		407.73
30%	1:1	k_{90} = 0.2517	-	-		2.75
		k_{80} = 0.1050	0	-	70 - 90	6.60
		k_{70} = 0.0600	-	-		11.55
	2:1	k_{90} = 0.1750	-	-		3.96
		k_{80} = 0.0717	0	-	70 - 90	9.67
		k_{70} = 0.0300	-	-		23.10
	5:1	k_{90} = 0.1167	-	-		5.94
		k_{80} = 0.0200	0	-	70 - 90	34.66
		k_{70} = 0.0117	-	-		59.41
30%	1:1	k_{90} = 0.4783	-	-		1.45
		k_{80} = 0.2167	0	-	70 - 90	3.20
		k_{70} = 0.0717	-	-		9.67
	2:1	k_{90} = 0.3533	-	-		1.96
		k_{80} = 0.1133	0	-	70 - 90	6.12
		k_{70} = 0.0550	-	-		12.60
	5:1	k_{90} = 0.2983	-	-		2.32
		k_{80} = 0.0633	0	-	70 - 90	10.94
		k_{70} = 0.0317	-	-		21.89

Cabbage	Blanched/ freeze-dried/ ground/ rehumidified $g H_2O$ / 100 g solids					Mizrahi et al., 1970
Moisture content:	18.0	28	0	-	30 - 52	-
	11.7	30	0	-	30 - 52	-
	8.9	34	0	-	30 - 52	-
	5.6	35	0	-	30 - 52	-
	3.2	38	0	-	30 - 52	-
	1.5	40	0	-	30 - 52	-

(Continued)

TABLE 2.4 Continued

Commodity	Process / Conditions	r^2	k_T value (min^{-1})	E_a (kcal/mol)	Reaction Order	r^2	Temp. Range (°C)	$t_{1/2}$ (min)	Reference
Flour dough Maillard Browning	Baked in 1 x 4 cm steel cylinders *% Moisture:*								Herrmann and Nour, 1977
	6	-	k_{190} = 0.02647					26	
		-	k_{170} = 0.009597	21.3		0.999	150 - 190	72	
		-	k_{150} = 0.002962					234	
	14	-	k_{190} = 0.04516					15	
		-	k_{170} = 0.01181	23.4		0.992	150 - 190	59	
		-	k_{150} = 0.00405					171	
	22	-	k_{190} = 0.02414					29	
		-	k_{170} = 0.007637	22.3		0.999	150 - 190	91	
		-	k_{150} = 0.002445					283	
	30	-	k_{190} = 0.01476					47	
		-	k_{170} = 0.005222	22.4		0.999	150 - 190	133	
		-	k_{150} = 0.001472					471	
Goat milk	pH 6.5 - 6.6								Burton, 1963
Homogenized		-	-	27.0[a]	1	-	93.3 - 121	-	(from Lund, 1975)
Unhomogenized		-	-	27.0	1	-	93.3 - 121	-	
Grapefruit juice Lag period	Thermal concentration								Saguy et al., 1978a
Solids:	11.2° Bx	-	k_{95} = 0.004451					156	
		-	k_{80} = 0.003230	8.19	1	0.978	61 - 95	215	
		-	k_{61} = 0.001446					479	
	31.2° Bx	-	k_{91} = 0.006320					110	
		-	k_{82} = 0.004430	9.86	1	0.999	60 - 91	156	
		-	k_{75} = 0.003412					203	
		-	k_{60} = 0.001770					392	

47.1° Bx	k_{96} = 0.013110						53
	k_{90} = 0.010530						66
	k_{80} = 0.00498	} 15.3	1	0.989	60 - 96		154
	k_{60} = 0.001461						474
55.0° Bx	k_{91} = 0.023340						30
	k_{81} = 0.010010	} 21.6	1	0.999	61 - 91		69
	k_{75} = 0.005910						117
	k_{61} = 0.001590						436
62.5° Bx	k_{96} = 0.164300						4.2
	k_{81} = 0.013260	} 30.4	1	0.923	68 - 96		52
	k_{76} = 0.009083						76
	k_{68} = 0.006055						114
Postlag period Solids:						Saguy et al., 1978a	
11.2° Bx	k_{95} = 0.02450						28
	k_{80} = 0.01100	} 15.10	1	0.998	61 - 95		63
	k_{61} = 0.00300						231
31.2° Bx	k_{91} = 0.02050						34
	k_{82} = 0.01730	} 15.70	1	0.964	60 - 91		40
	k_{75} = 0.00960						72
	k_{60} = 0.00300						231
47.1° Bx	k_{96} = 0.06330						11
	k_{90} = 0.02840	} 17.2	1	0.973	61 - 96		24
	k_{80} = 0.01650						42
	k_{61} = 0.00470						147
55.0° Bx	k_{91} = 0.05100						14
	k_{81} = 0.02360	} 19.7	1	0.999	60 - 91		29
	k_{75} = 0.01450						48
	k_{60} = 0.00440						158
62.5° Bx	k_{96} = 0.14810						4.7
	k_{81} = 0.04250	} 23.8	1	0.994	68 - 96		16
	k_{76} = 0.02130						33
	k_{68} = 0.01070						65

(Continued)

TABLE 2.4 Continued

Commodity	Process / Conditions	r^2	k_T value (min^{-1})	E_a (kcal/mol)	Reaction Order	r^2	Temp. Range (°C)	$t_{1/2}$ (min)	Reference
Sugar solutions	"Caramelization": (phosphate buffer, NaCl, 0.90 a$_w$)								Buera et al., 1987a
Fructose (0.27M)	pH 6.0	-	k_{65} = 0.000292					2,370	
		-	k_{55} = 0.000093	24.6	0	0.98	45 - 65	7,450	
		-	k_{50} = 0.000042					16,500	
		-	k_{45} = 0.000032					21,700	
Lactose (0.27M)		-	k_{65} = 0.000250					2,770	
		-	k_{55} = 0.000045	35.2	0	0.99	45 - 65	15,400	
		-	k_{45} = 0.0000092					75,300	
Lactose (0.27M)		-	-	48.4	1	0.99	45 - 65	-	
Maltose (0.27M)		-	k_{65} = 0.000208					3,330	
		-	k_{55} = 0.000023	39.7	0	0.98	45 - 65	30,100	
		-	k_{45} = 0.0000050	$(40.2)^a$				139,000	
Sucrose (0.27M)		-	-	57.6	0	0.99	45 - 65	-	
Sugar solutions (0.27M sugar / 0.67M glycine)	Maillard reaction: (phosphate buffer, NaCl, 0.90 a$_w$)								Buera et al., 1987b
Glucose	pH 6.0	-	k_{65} = 14.28					4.85 x 10^{-2}	
		-	k_{60} = 11.37					6.10 x 10^{-2}	
		-	k_{55} = 4.92	27.20	0	0.96	45 - 65	14.10 x 10^{-2}	
		-	k_{50} = 1.93					35.91 x 10^{-2}	
		-	k_{45} = 1.43					48.47 x 10^{-2}	
	pH 6.0	-	k_{65} = 94.48					0.73 x 10^{-2}	
		-	k_{60} = 58.95					1.18 x 10^{-2}	
		-	k_{55} = 22.95	25.30	1	0.68	45 - 65	3.02 x 10^{-2}	
		-	k_{50} = 52.15					1.33 x 10^{-2}	
		-	k_{45} = 5.28					13.13 x 10^{-2}	
	pH 5.0	-	k_{55} = 1.37	-	0	-	55	0.51	
	pH 5.0	-	k_{55} = 9.05	-	1	-	55	0.08	
	pH 4.0	-	k_{55} = 0.45	-	0	-	55	1.54	
	pH 4.0	-	k_{55} = 2.98	-	1	-	55	0.23	

Sample	Condition	Rate constants			Temp (°C)	R^2		Reference
Fructose	pH 6.0	$k_{65} = 4.90$					0.14	Buera et al., 1987b
		$k_{55} = 1.65$					0.42	
		$k_{50} = 0.91$	29.40	0	45 - 65	0.97	0.76	
		$k_{45} = 0.28$					2.48	
	pH 5.0	$k_{55} = 1.33$	-	0	55	-	0.52	
	pH 4.0	$k_{55} = 0.9$	-	0	55	-	0.77	
Sucrose	pH 6.0	$k_{65} = 2.08$					0.33	
		$k_{55} = 0.23$	36.6	0	45 - 65	0.97	3.01	
		$k_{45} = 0.067$					10.3	
	pH 6.0	$k_{65} = 17.2$					0.04	
		$k_{55} = 4.43$	32.8	1	45 - 65	0.99	0.16	
		$k_{45} = 0.8$					0.87	
	pH 5.0	$k_{55} = 0.28$	-	0	55	-	2.48	
	pH 5.0	$k_{55} = 4.45$	-	1	55	-	0.16	
	pH 4.0	$k_{55} = 0.87$	-	0	55	-	0.80	
	pH 4.0	$k_{55} = 7.37$	-	1	55	-	0.09	
Peaches (fresh puree) 0 ml vials 5 temps. / up to 150 min Color: Minolta CR-300								Ávila and Silva, 1999
	L-value	$^{a}k_{122.5} = 2.9 \times 10^{-3}$	$^{a}25.7$	1	110 - 135	0.999	0.24×10^{3}	
	b-value	$k_{122.5} = 4.0 \times 10^{-3}$	26.2	1	110 - 135	0.999	0.17×10^{3}	
	a-value	$k_{122.5} = 0.03 \times 10^{-3}$	25.3	1	110 - 135	0.990	23.1×10^{3}	
	La/b	$k_{122.5} = 0.026 \times 10^{-3}$	25.3	1	110 - 135	0.993	26.7×10^{3}	
(Total color difference)	TCD $(=\Delta E)$	$k_{122.5} = 0.0085 \times 10^{-3}$	28.4	1	110 - 135	0.986	81.5×10^{3}	
Pear puree (11° Brix) Heat up to 500 min Color measure:	UV – 420 nm							Ibarz et al., 1999
	0.901	$^{a}k_{98} = 5.1 \times 10^{-4}$	15.0	0	80 - 98	0.650	1.4×10^{3}	
	0.841	$k_{95} = 7.0 \times 10^{-4}$					0.99×10^{3}	
	0.902	$k_{90} = 7.0 \times 10^{-4}$					0.99×10^{3}	
	0.863	$k_{85} = 3.8 \times 10^{-4}$					1.8×10^{3}	
	0.625	$k_{80} = 2.0 \times 10^{-4}$					3.5×10^{3}	

(Continued)

TABLE 2.4 Continued

Commodity	Process / Conditions	r^2	k_T value (min^{-1})	E_a (kcal/mol)	Reaction Order	r^2	Temp. Range (°C)	$t_{1/2}$ (min)	Reference
Pear puree (cont.)	Macbeth CE3000								Ibarz et al., 1999
	a-value								
	[a]k_{98}	-	$= 3.01 \times 10^{-2}$					23	
	k_{95}	-	$= 2.08 \times 10^{-2}$					33	
	k_{90}	-	$= 1.38 \times 10^{-2}$	24.4	0	0.987	80 - 98	50	
	k_{85}	-	$= 0.76 \times 10^{-2}$					91	
	k_{80}	-	$= 0.57 \times 10^{-2}$					122	
	ΔE								
	[a]k_{98}	-	$= 5.22 \times 10^{-2}$					13	
	k_{95}	-	$= 3.99 \times 10^{-2}$					17	
	k_{90}	-	$= 3.49 \times 10^{-2}$	14.9	0	0.970	80 - 98	20	
	k_{85}	-	$= 2.65 \times 10^{-2}$					26	
	k_{80}	-	$= 1.71 \times 10^{-2}$					41	
	L-value								
	[a]k_{98}	-	$= 6.5 \times 10^{-4}$					1.07×10^3	
	k_{95}	-	$= 5.3 \times 10^{-4}$					1.31×10^3	
	k_{90}	-	$= 3.8 \times 10^{-4}$	18.3	1	0.964	80 - 98	1.82×10^3	
	k_{85}	-	$= 2.2 \times 10^{-4}$					3.15×10^3	
	k_{80}	-	$= 2.0 \times 10^{-4}$					3.47×10^3	
Analysis:									
HPLC / RI	Sucrose degradation	-	-	17.7	1	0.720	80 - 98	-	
HPLC / 254 nm	HMF formation	-	-	27.1	1	0.922	80 - 98	-	
Potato strips	Deep-fat fried								Moyano et al., 2002
(Pretreatment: phosphate dip / blanched)									
Control	Color: Minolta CR200b (ΔE)								
	Moisture content: g H_2O / 100 g solids								
Color measured as ΔE	140	-	-	36.29	1	-	160 - 180	-	
	120	-	-	24.33	1	-	160 - 180	-	
	100	-	-	19.61	1	-	160 - 180	-	
	80	-	-	17.81	1	-	160 - 180	-	
	60	-	-	15.65	1	-	160 - 180	-	
	40	-	-	14.76	1	-	160 - 180	-	
	30	-	-	11.73	1	-	160 - 180	-	

NaCl dip (3%)

Moisture content: g H₂O / 100 g solids

Moisture content							Reference
140	-	-	41.73	1	160 - 180	-	Moyano et al., 2002
120	-	-	28.78	1	160 - 180	-	
100	-	-	21.49	1	160 - 180	-	
80	-	-	17.95	1	160 - 180	-	
60	-	-	15.82	1	160 - 180	-	
40	-	-	14.92	1	160 - 180	-	
30	-	-	13.35	1	160 - 180	-	

Corn syrup DE-42 / NaCl dip (50% / 3%)

Moisture content: g H₂O / 100 g solids

Moisture content							Reference
140	-	-	24.54	1	160 - 180	-	
120	-	-	19.65	1	160 - 180	-	
100	-	-	16.36	1	160 - 180	-	
80	-	-	13.31	1	160 - 180	-	
60	-	-	11.50	1	160 - 180	-	
40	-	-	10.74	1	160 - 180	-	
30	-	-	10.62	1	160 - 180	-	

White potato dice   Air dehydration

Moisture content: g H₂O / 100 g solids

Moisture content							Reference
370	-	-	26.0	1	65 - 99.5	-	Hendel et al., 1955
110	-	-	25.0	1	65 - 99.5	-	
33	-	-	25.0	1	65 - 99.5	-	
15	-	-	28.0	1	65 - 99.5	-	
9.4	-	-	32.0	1	65 - 99.5	-	
4.9	-	-	37.0	1	65 - 99.5	-	

Wine model system

(K-tartrate / KOH / aqueous   UV / HPLC
ethanol / catechin / catechin / ascorbic   (440 nm)
acid)   45°C / up to 24 days

							Reference
Catechin alone	0.967	k_{45} = 0.401 x 10^{-6}	-	0	45	17.29 x 10^5	Bradshaw et al., 2001
Catechin + Vit. C	0.995	k_{45} = 2.981 x 10^{-6}	-	0	45	2.33 x 10^5	

[a] Values as reported by authors

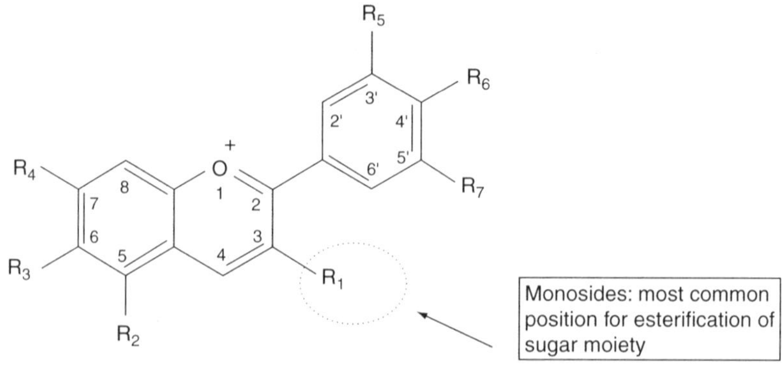

Basic structure of anthocyanidin pigments

Anthocyanidin	Group Positions						
	R1	R2	R3	R4	R5	R6	R7
Cyanidin	OH	OH	H	OH	OH	OH	H
Delphinidin	OH	OH	H	OH	OH	OH	OH
Malvidin	OH	OH	H	OH	OCH_3	OH	OCH_3
Pelargonidin	OH	OH	H	OH	H	OH	H
Peonidin	OH	OH	H	OH	OCH_3	OH	H
Petunidin	OH	OH	H	OH	OCH_3	OH	OH

FIGURE 2.20   Basic structure of some of the most significant anthocyanidins.

The high content of anthocyanins and proanthocyanins (colorless) in fruits such as blueberries and raspberries has recently received a great deal of attention since these compounds appear to have high potential in improving both motor and cognitive functioning in humans. It has also been suggested that hydroxycinnamates may work together with anthocyanins to accomplish all these benefits. Proanthocyanidins are found in the juice, as well as in fresh, frozen and dried cranberries and blueberries. They can be readily converted to their corresponding anthocyanidin when heated in the presence of acid. Zheng and Wang (2003), for instance, have found high levels of antioxidant activity in anthocyanidins and various phenolics in blueberries, cranberries, and lingonberries. Others have investigated the radical scavenging capabilities of anthocyanidins and anthocyanins found in wine, such as pelargonidin, cyanidin, peonidin, delphinidin, and malvidin as well as some of their substituted glycosides. They found antioxidant activity to be dependent upon pH (Borkowski et al., 2005). Kähkönen and Heinonen (2003) found higher levels of antioxidant activity in anthocyanidins than their glycosides in an aqueous environment, but the reverse was found in an oil system. Other workers have monitored the metabolism of various anthocyanins in rats fed an anthocyanin-fortified diet and found elevated concentrations of the additive in the brain, liver, and kidney (Talavéra et al., 2005). Thus, although the retention of anthocyanins is key to the general quality and appearance of food products, their medical significance seems to have become as important. A great deal of emphasis has recently been placed on the health benefits of these natural pigments (Boyd, 2000).

A large number of factors have been reported as influencing the stability of anthocyanins in food products, including the presence of oxygen (Daravingas and Cain, 1968; Starr and Francis, 1968;

Clydesdale et al., 1978); light (Van Buren et al., 1969; Palmidis and Markakis, 1975; Sweeny et al., 1981); ascorbic acid (Sondheimer and Kertesz, 1953; Sistrunk and Cash, 1970); enzymatic action (Peng and Markakis, 1963); pH (Lukton et al., 1956; Brouillard and Delaporte, 1977; Timberlake, 1980; Brouillard, 1982); metal traces (Sistrunk and Cash, 1970; Starr and Francis, 1973; Francis, 1977); additives such as sulfite or sulfur dioxide (Timberlake and Bridle, 1967a, 1967b); and heat (Markakis, 1974; Clydesdale et al., 1978; Main et al., 1978; Timberlake, 1980) as well as condensation reactions (Jurd, 1969). The final breakdown of anthocyanins results in either the formation of brown-colored substances or bleaching of the system. Since anthocyanins are the most important natural colorant in many fruits, as well as reported as having various health benefits, its stability is of prime concern during processing and storage. General discussions on the stability of anthocyanins from a chemistry point of view or as affected in food systems have been presented by Timberlake and Bridle (1975), Markakis (1982), and Delgado–Vargas et al. (2000). Some of the suggested mechanisms of decomposition are presented in Figure 2.21.

Debicki–Pospišil et al. (1983), working with blackberry juice and a citrate buffer model solution, indicated that the rate of degradation of cyanidin-3-glucoside followed first-order kinetics for the range 24 to 70°C. Russu and Valuiko (1980) reported that the thermal decomposition of anthocyanins during the heating of mash at temperatures ranging from 20 to 100°C followed first-order kinetics. Cemeroglu et al. (1994) also reported first-order kinetics for degradation of anthocyanins in sour cherry juice concentrates, both during storage (-18 to 37°C/up to 180 days) and processing (50–80°C/up to 48 h), Table 2.4.

Torskangerpoll and Anderson (2005) studied the effect of pH on anthocyanins. They presented an extensive set of tabulated data on change in color of three selected anthocyanins at two initial concentrations and 11 different pH levels (1.1–10.5). The anthocyanins chosen were cyanidin 3-glucoside [from black rice]; cyanidin 3-(2″-glucosylglucoside)-5-glucoside [from red cabbage]; and cyanidin 3-(2″-(2‴-sinapoylglucosyl)-6″-sinapoylglucoside)-5-glucoside [from red cabbage]. Color values were tabulated from L-a-b colorimeter values, calculated as chroma (C), hue (h_{ab}), and L-value. Their focus was on the impact of the various substituents and aromatic acylations on color and stability of anthocyanins at different pH. Sarni–Manchado et al. (1996) reported on the stability of anthocyanin-derived pigments from wine. They reported increased stability of vinylphenol anthocyanin-3-glucoside adducts compared with anthocyanin-3-glucosides both under adverse pH conditions and in the presence of SO_2, which is known to bleach anthocyanin pigments in wine.

Ascorbic acid has been reported to induce the destruction of anthocyanins both aerobically and anaerobically (Sondheimer and Kertesz, 1953). Meschter (1953) also found that dehydroascorbic acid resulted in decolorization of anthocyanins, although at a slower rate. The formation of an intermediate peroxide even at low pH has been considered to be responsible, at least in part with the presence of oxygen, for anthocyanin degradation. The rates of destruction appeared to be enhanced by the presence of cupric ions. In the absence of oxygen, however, other mechanisms involving ascorbic acid and anthocyanins must be involved. Jurd (1972) suggested a condensation reaction between ascorbic acid and anthocyanins, which results in an unstable complex that degrades to a colorless compound. Later work by López–Serrano and Ros Barceló (1999) showed that peroxide stimulates oxidation of pelargonidin-3-glucoside, the main anthocyanin in ripe strawberries. They suggested a coupled mechanism for glucosidase and peroxidase to be responsible for oxidation of anthocyanins. Others have investigated stability of co-pigmented anthocyanins and ascorbic acid in red grape model systems using water-soluble polyphenolic cofactors isolated from *Rosmarinus officinalis* (Brenes et al., 2005). They reported first-order degradation kinetics for both anthocyanins and ascorbic acid during storage (dark/25–35°C/30 days), following pasteurization (85°C/30 min/pH 3.5). The authors found no effect of co-pigmentation on stability of anthocyanins without the presence of ascorbic acid. However, in the presence of ascorbic acid, they found higher anthocyanin content. This was attributed to a delayed conversion of L-ascorbic acid to dehydroascorbic acid, which leads to the destruction of anthocyanins. Malien–Aubert et al. (2001) also proposed the protective

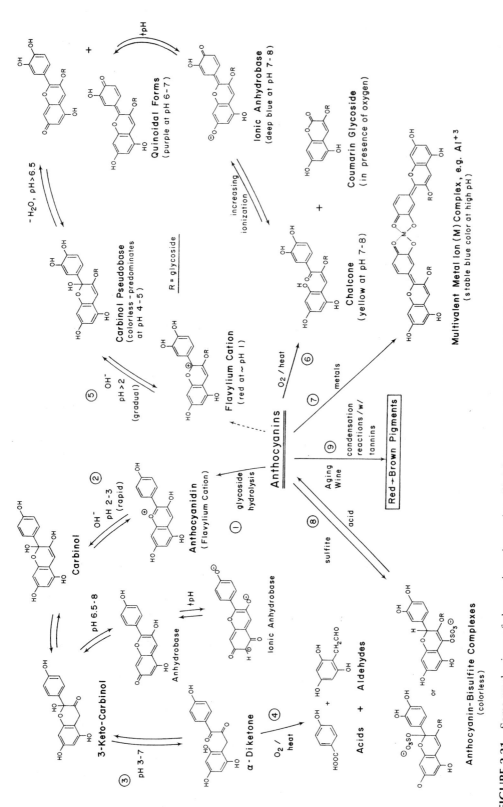

FIGURE 2.21   Some mechanisms of changes in anthocyanins. (1) Adams, 1973; (2) Harper, 1968; Timberlake and Bridle, 1966; (3) Hrazdina, 1981; Harper, 1968; 1967; Jurd, 1972; (4) Jurd, 1972; (5) Timberlake and Bridle, 1967a; (6) Jurd, 1972; Hrazdina, 1971; (7) Asen et al., 1969; Jurd and Asen, 1966; (8) Jurd, 1972; 1964; (9) Ribeéreau-Gayon, 1982; Somers, 1971; Jurd, 1969.

effects of co-pigmentation of anthocyanins. They found that colorants rich in flavonols and high co-pigment/pigment ratio show improved stability with acylated anthocyanins.

With regard to the effect of light, Palamidis and Markakis (1975) indicated a marked increase in the loss of grape anthocyanins during storage due to their exposure to fluorescence and day light. In model systems, Attoe and von Elbe (1981) determined that at high temperature (55°C), light had minimal effect on the degradation of the major cranberry anthocyanins in model systems. At lower temperatures (40°C), on the other hand, light was responsible for most of the losses. The authors also reported that fluorescent light had limited effect on the stability of anthocyanins, unless minimum concentrations of oxygen were available. Degradation of anthocyanins in the dark and in the presence of light was reported to follow first-order kinetics.

Merin et al. (1987) observed that the degradation rates of prickly-pear-fruit were influenced by the initial concentration of the pigment, being slower for higher concentrations. The presence of oxygen appeared to have only a marginal effect. On the other hand, Kallio et al. (1986) indicated that the stability of the 12 anthocyanins in crowberry juice was improved upon oxygen removal, increasing by a factor of 3 to 4. Lin et al. (1989) observed that carbon dioxide levels in package-modified atmosphere greatly destabilized cyanidin-3-galactoside, cyanidin-3-arabinoside, and other unidentified cyanidin arabinosides in Starkrimson apples. Peonidin and malvidin, anthocyanins containing the ring B substituted with only one hydroxyl group, were reported to be the most stable when crowberry juice was supplemented with Fe^{+3} ions (Kallio et al., 1986). In general, however, all anthocyanins were found to have improved stability with the addition of Fe^{+3} and Al^{+3}. The formation of stable complexes of anthocyanins with tin, copper, and iron have been suggested as a means for increased stability of these compounds in the presence of metals (Sarma, et al., 1997). Flavonoids such as flavones, isoflavones and aurone sulfonates have been found to increase the photostability of anthocyanins (Francis, 1989). Sucrose addition has been found to protect anthocyanins in quick frozen strawberries due to inhibition of degradative enzymes and steric interference with condensation reactions as reported by Wrolstad et al. (1990). Moreover, enzymes present in plant tissue such as glycosidases that convert anthocyanins into anthocyanidins and sugars, polyphenoloxidases that catalyze the oxidation of o-dihydrophenols to o-quinones, and peroxidases can promote significant loss of these pigments (Francis, 1989). It is clear from the aforementioned studies that formulation or composition of food products will have a major impact on the stability of anthocyanins during processing and storage. Anthocyanins can be easily destroyed during the processing of fruits and vegetables due to high temperatures, pH, or the presence of ascorbic acid. Kinetic data on anthocyanin stability are presented in Table 2.4.

2.3.2.3 Betalains

Another group of water-soluble pigments of great significance to the food industry corresponds to betalains. Two major categories, which constitute this group of pigments, are the betacyanins (red colored resonating compounds) and betaxanthins (yellow colored non-resonating compounds). Betalains are basically immonium derivatives of betalamic acid, with their chromophore described as a protonated 1,2,4,7,7-penta-substituted 1,7-diazaheptamethin system (Piatelli, 1976, 1981; Strack et al., 1993). All betalain molecules have a betalamic acid moiety present with attached R-groups extending from the N-1; the nature of those substituent groups will determine whether or not the molecule belongs to a betacyanin or a betaxanthin group (Figure 2.22). Betanidin (found in red beets) is an aglycone of betacyanin and is actually the most basic structural unit of most betacyanins, followed by the C-15 epimer, isobetanidin (Piatelli, 1981). Differentiation of betacyanins is based on the glycosidation of one of the hydroxyl groups at the 5- and 6-position. Betaxanthins, on the other hand, comprise different proteinogenic and nonproteinogenic amino acids or biogenic amino-conjugated moieties of betalamic acids, typical examples of which are indicaxanthin, found in prickly pear cactus fruits and vulgaxanthins-1 and -2 from beets.

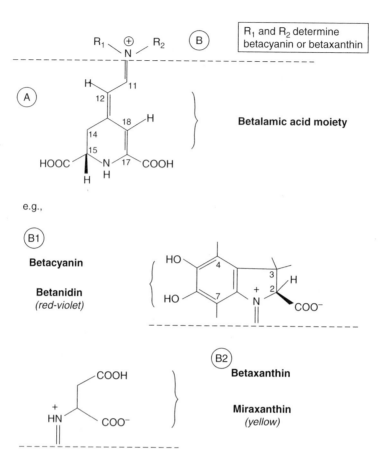

FIGURE 2.22 Basic structural differences between betacyanins and betaxanthins. (Adapted from Delgado-Vargas, F., Jiménez, A.R., and Paredes-López, O. 2000. *Critical Reviews in Food Science and Nutrition*, 40: 173–289; Böhm, H. and Rink, E. 1988. Betalains. pp. 449–463. In: *Cell Culture and Somatic Cell Genetics of Plants*, Academic Press, NY; Piatelli, M. 1981. The betalains: structure, biosynthesis and chemical taxonomy, pp. 557–575, In: *The Biochemistry of Plants: A Comprehensive Treatise, Vol. 17*, Conn, E.E. (ed.), Academic Press, NY; and Strack, D., Steglich, W., and Wray, V. 1993. Betalains. pp. 421–50, In: *Methods in Plant Biochemistry*, Vol.8, Academic Press, Orlando, FL.)

This group of approximately 70 compounds has been found to be susceptible to environmental conditions including temperature, oxygen, light, and pH (Huang and Von Elbe, 1987). Lashley and Wiley (1979), and Lee and Wiley (1981) indicated the presence of decolorizing enzymes in beet tissues, thus making this an important consideration for storage conditions. Von Elbe et al. (1974) found the rate of betanin degradation increased after exposure to daylight at 15°C. They reported that degradation of betalains exposed to fluorescent light followed first-order kinetics, with higher rate constants at pH 3.0 than pH 5.0.

Experiments carried out by Saguy et al. (1980) indicated that the degradation of betanin and vulgaxanthin I in beet slices could be described by first-order kinetics for the temperature range 70 to 90°C. With respect to moisture content, increased rates of degradation were observed for both pigments with increased water levels. Saguy et al. (1984) working with beet powders observed that the stability of betanin and vulgaxanthin was markedly enhanced at water activity levels below 0.5 in a nitrogen environment. At water activity levels above 0.5, a transition in the mechanism of beet pigment deterioration was shown to be related to oxygen concentration. In the presence of Al_2O_3 and at water activities allowing reactant mobility, the formation of alumina-oxygen-pigment complexes and free radicals or activated oxygen species were suggested as possible mechanisms accelerating betalain degradation. Šimon et al. (1993) found similar dependency of betanine stability in water–alcohol model systems with increasing rates of degradation at higher water activities, perhaps due to greater mobility of reactants or increased oxygen solubility. More recently, Wybraniec (2005) investigated the mechanisms of degradation of extracts of *B. vulgaris* L. roots (red beet). Samples were heated in aqueous or ethanolic solutions at 75 to 80°C/60 to 180 min. and analyzed by HPLC with tandem mass spectrometry and diode-array detection. Degradation products were a mixture of mono-, bi-, and tri-decarboxylated betacyanins along with their corresponding neobetacyanins. Some specific identification of breakdown products included 17-decarboxy-betacyanins and 2-decarboxy-betacyanins, 2,17-bidecarboxybetanin, its isoform and 14,15-dehydrogenated (neobetacyanin) derivatives of all the decarboxylated betacyanins.

Attoe and von Elbe (1981) reported that at high temperatures the degradation of betanin in model systems was primarily induced by heat. However, at low temperatures (25°C), the influence of light became predominant. In both situations, degradation of betanin was determined to follow first-order kinetics. Their results also indicated that molecular oxygen was necessary for the photocatalyzed destruction of this pigment. The exact mechanism of oxygen involvement was not elucidated. However, Attoe and von Elbe (1985) indicated that the inefficiency of anti-oxidants, capable of interacting with free radicals, to improve the stability of betalains suggested that the mechanism of betanin oxidation did not involve free radical chain reactions. Kinetic information on the stability of some betalain-containing systems is presented in Table 2.4.

Their sensitivity to environmental factors such as temperature, light, oxygen, and humidity has limited their application as food colorants. A better understanding of their kinetic stability will facilitate their wider application in the food industry and may potentially displace the application of synthetic dies. On the other hand, betalains are compounds with antioxidant activity, and thus, have gained great importance in human nutrition. Betanin and betanidin in very small concentrations have been found to inhibit lipid peroxidation and heme decomposition (Kanner et al., 2001). Red beet products in the diet have been suggested as having protection against oxidative stress-related disorders in humans, since they are good electron donors. Stintzing et al. (2005) substantiated the antioxidant properties of betalains, as found in cactus pear clones. The authors pointed out that this plant was a good economical source of betalains but content was dependent upon individual plant species. Other sources of betalains as food colorants have been investigated. Cai and Cork (1999), for instance, carried out a comparative stability of *Amaranthus* betacyanin pigments against a radish anthocyanin and a synthetic FDA Red No. 3 as added to different food types such as ice cream, jelly, and model beverage, with pH adjusted to 5 to 6.0. They found similar stability of the amaranth to the anthocyanins but less stability compared to the synthetic dye. In general, there has been a great deal of interest in betalains, not only as a source of natural colorant but also as a nutraceutical. Pharmaceutical interest in red pigments from beets has not only been reported as antioxidants, but also as having anticarcinogenic properties (Kapadia et al., 2003). Currently, extracts from beets are the only betalain colorants permitted in the United States. Commercial beet powders generally contain about 0.4 to 1.0% pigment, 80% sugar, 8% ash, and 10% protein with citric acid and/or ascorbic acid as a preservative (Francis, 2000a).

2.3.2.4 Carotenoids

Although carotenoids (as vitamin A) have been previously discussed from a nutritional point of view, it should be briefly mentioned that there is also significant interest in carotenoids as coloring agents. Carotenoids can be divided into two major classes. The first class includes a large group of yellow and red-pigmented unsaturated hydrocarbons such as lycopene, α-, β-, γ-, and ξ-carotenes. The second group is the oxygenated derivatives referred to as xanthophylls (e.g., β-cryptoxanthin, lutein, and zeaxanthin). Commercial interest in these pigments lies in their use as colorants in oil-based food systems such as margarine, butter, cheese, ice cream, meats, soups, beverages, and confectionery

(Francis, 2000b). These pigments may be obtained from annatto (bixin), red peppers (capasanthin), tomatoes (lycopene), saffron, and paprika.

Carotenoids may be complexed with sugars, such as di-gentiobiose in α-crocin, the main pigment in saffron. Proteins may also be complexed with carotenoids, such as in the case of astaxanthin, commonly found in crustacea such as crab and lobster. The natural bluish-green color of these crustaceans is due to the astaxanthin-protein complex, which upon heating results in denaturation of the protein portion of the complex giving a reddish-orange color.

Recent clinical evidence supports the role of carotenoids as important micronutrients. For example, lycopene has been reported to provide protection against multiple types of cancer (Levy et al., 1995). Lycopene functions as an antioxidant and exhibits a high quenching rate constant for singlet oxygen, reportedly twice that of β-carotene and 10 times more than that of α-tocopherol (DiMascio et al., 1989; and Shi et al., 2003). Lycopene exists in nature as the all-*trans*-form and can isomerize to the higher energy and more reactive mono- or poly-*cis*-forms under the influence of heat, light, or certain chemical reactions. However, it has no provitamin A activity due to the lack of a β-ionone ring structure.

Several approaches have been taken to describe the kinetics of discoloration of carotenoids as previously discussed in the section corresponding to vitamin A. In fact, several investigators have observed apparent first-order kinetics. This is definitely difficult to apply to a number of systems, where an induction period is present. Martinez and Labuza (1968) proposed that the deterioration of astacene in freeze-dried salmon followed a first-order reaction. Other investigators have successfully used the free-radical recombination approach to describe the kinetics of carotenoid decoloration. Since the reaction is a chain reaction in which initiators and inhibitors are involved, it is obvious that a more complex approach may be required to accurately characterize the reaction kinetics of carotenoid decoloration. Working with model systems simulating dehydrated foods, Goldman et al. (1983) indicated that the decoloration of β-carotene followed three periods, namely, an induction period, a fast main period, and a retardation period, typical of an autocatalytic radical reaction. Saguy et al. (1985) successfully predicted losses of β-carotene under dynamic conditions using kinetic information obtained under static conditions through the use of the free radical recombination approach.

Mortensen and Skibsted (2000) described degradation of carotenoids under mild acid conditions to follow pseudo zero-order kinetics, and their reaction rates were very much dependent upon the individual carotenoid under consideration. For instance, carotenoids with carbonyl groups (e.g., astaxanthin and canthaxanthin) showed slower rates of degradation than β-carotene and zeaxanthin. The mechanism suggested carotenoids containing carbonyl groups are preferentially protonated at this site and not on a carbon atom of the conjugated system.

Kanner et al. (1978) monitored the bleaching of carotenoids in powdered paprika and reported that the kinetics were complex and did not follow a simple first-order or pseudo first-order reaction. The authors suggested that variations in the prooxidant-antioxidant balance in different media would explain differences in the results for various products. Chen and Gutmanis (1968) had previously indicated that an autocatalytic mechanism would describe the bleaching of carotenoids in pepper and in other products. A second-order reaction was found to describe the autooxidation process in dry chili peppers during storage. Chou and Breene (1972) also indicated that the decoloration of β-carotene was an autooxidative reaction. The authors reported that the reaction could be described by a first-order or pseudo first-order reaction when oxygen was not a limiting factor. Ramakrishnan and Francis (1973) monitored color changes in paprika, subjected to heat from 125 to 150°C. Simultaneous carotenoid color reduction and browning development were measured. Texeira Neto et al. (1981) studied the decoloration of β-carotene in a dehydrated food model system stored at 37°C. The authors observed that first-order kinetics described the reaction. However, values corresponding to oxygen uptake measurements indicated that six to seven molecules of oxygen were consumed per mole of oxidized carotene. These authors' results seemed to be in agreement with previous results presented by Walter and Purcell (1974) who suggested that due to the high level of unsaturation of the carotenes more than one site is available for oxidative attack. Walter and Purcell (1974) had also suggested that some of the oxidative products may undergo further oxidation. Texeira Neto et al. (1981) were unable to confirm this claim using their model system. The authors developed a computer program to predict β-carotene decoloration by measuring oxygen uptake or to determine oxygen uptake from decoloration experimental data.

Barreiro et al. (1997) observed different reaction order kinetics, depending on the color parameter used to quantify the change in color during heating. These authors monitored color changes in double concentrated tomato paste heated 70–100°C/up to 90 min. Color measurements were made with a tristimulus colorimeter, and the usual L-, a-, and b-values were reported along with calculated values including color difference (ΔE), where $\Delta E = \sqrt{(L_0 - L)^2 + (a_0 - a)^2 + (b_0 - b)^2}$; saturation index (SI), where $\text{SI} = \sqrt{a^2 + b^2}$; and a/b ratio, where hue angle $= \tan^{-1} a/b$. The authors reported that all the color parameters followed pseudo first-order kinetics except for ΔE, which showed a zero-order behavior ($E_a = 10.2$ kcal/mole). L-value was defined as two consecutive first-order reactions with E_a values of 11.5 and 5.73 kcal/mole for phase one and phase two, respectively. Energies of activation for a-, b-, a/b, and SI-values were reported as 9.79, 20.5, 6.86, and 10.1 kcal/mole.

Many reports have been published on the stability of carotenoids during food processing and storage. Although the overall amount of carotenoids may remain similar after processing, isomerization has been identified as one of the main pathways for carotenoid losses (Waché et al., 2003 and Chen et al., 1994). This is of importance since isomers have different biological potencies. In general, heat, oxygen, and light will have a significant impact on carotenoid stability. Baloch et al. (1987) reported on the beneficial effect of sulfiting on the stability of carotenoids in blanched and unblanched carrots, while Nutting et al. (1970) reported on the beneficial effects in parsley. Canning of fruits and vegetables has been identified as causing *cis-trans* isomerization (Weckel et al., 1962 and Ogunlesi and Lee, 1979).

With the recent reports on health benefits of lycopene, in particular, a great deal of research has been directed toward this carotenoid. Although lycopene exists in nature as the all-*trans* form, it will readily isomerize to various *cis*-forms under the influence of heat, light, acid, or certain chemical reactions (Shi and Le Maguer, 2000). Shi et al. (2003) presented a study on lycopene in tomato puree as subjected to various processing conditions (90–150°C/up to 6 hrs) and reported rate constants and energies of activation for the individual consecutive reactive pathways. They included (a) *trans*-lycopene isomerizing to *cis*-isomer (k_1), (b) *cis*-isomer oxidation (k_3), and (c) the predominant reaction of all-*trans*-lycopene oxidation (k_4), Table 2.4. The reaction rate constant (k_2) for the reversible reaction of *cis*-isomers to *trans*-isomers was shown to be slow under the thermal and light irradiation treatments of the experiments. Lin and Chen (2005) found significant losses of both all-*trans*- and *cis*-forms of lycopene, lutein and β-carotene during storage of tomato juice (4–35°C/up to 12 weeks). The presence of light during storage enhanced degradation of all carotenoids. All-*trans*-lycopene showed the highest degradation loss, followed by β-carotene and lutein. However, more *cis*-isomers were generated during storage than either lutein or β-carotene.

Shi and LeMaguer (2000) have reviewed the physical properties and reported findings on the health aspects of lycopene as affected by processing. It has been demonstrated that the biopotency of lycopene is dependent on the level of isomerization and oxidation. For instance, the bound chemical form of lycopene in tomatoes when converted to the *cis* form by the processing temperatures makes it more easily absorbable by the body. In general, however, further characterization of lycopene as well as other carotenoid isomers is highly needed in order to provide a better understanding of their biopotency and health related benefits. Kinetic data for the destruction of carotenoids from the point of view of color are presented in Table 2.4.

2.3.2.5 Myoglobin

A major pigment that has been often overlooked in terms of kinetic stability either during thermal processing or during storage is myoglobin. Myoglobin is responsible for the visual appeal whether as freshly wrapped meat on the shelf in the market or after cooking. Since most of the iron from hemoglobin is removed from the animal after slaughter, myoglobin retains about 95% or more of the remaining iron. This myoglobin pigment is actually a complex muscle protein comprising a protein moiety (globin) and a nonpeptide portion referred to as the heme consisting of a central iron atom within a porphyrin ring. The color of myoglobin is actually purple but may be oxygenated to a bright red oxymyoglobin with a reduced ferrous state or may be oxidized to metmyoglobin with the oxidized ferric state.

Most investigations involving myoglobin have been involved with storage as affected by oxygen, various types of packaging materials, and pretreatments to maintain a desirable red color while in the market (Huffman, 1980; Kropf, 1980; Griffin et al., 1982; Koohmaraie et al., 1983). However, due to myoglobin's sensitivity not only to oxygen but also to heat, enzymes, metal ions, light, alcohols, acids, etc., it is of particular interest to study these variables in relation to kinetic parameters of myoglobin denaturation and coloration. Another important area of study with respect to color changes in myoglobin is the area of cured meats. In the presence of nitric oxide and heat, the cured meats as affected by different processing techniques at low temperature has been reported by Kamarei et al. (1979, 1981) and Fox et al. (1967) as well as several reviews on meat pigment chemistry (Fox, 1966; Fox and Ackerman, 1968). Some mechanisms of pigment reactions in both fresh and cured meats are presented in Figure 2.23.

Kamarei et al. (1981) studied the color of nitrate-cured samples exposed to gamma radiation. The authors reported that contrary to prior reports nitrate did not affect color development or post-irradiation fading of pork semimembranosus muscles. Radiation-reduced pigments were oxidized by air to brown globin myohemochromogen as previously reported by Kamarei et al. (1979). The subsequent pigment fading of cured samples as due to photooxidation or autooxidation are commonly accepted mechanisms.

Fox et al. (1967) reported on color development during frankfurter processing and indicated that the presence of oxygen, the addition of ascorbate or cysteine, and temperature were critical to the rate of color development, the levels of cured meat pigment formed, and the levels of color maintained during storage.

Limited information is available concerning the kinetics of color changes of myoglobin and related compounds, either due to processing or storage. Since the formation of pigments responsible for the bright red color of uncooked cured meats, namely, nitric oxide myoglobin, and nitric oxide hemoglobin has been reported to be associated with chemical, nonenzymatic as well as enzymatic reaction pathways, it is clear that kinetic characterization of the color changes can be very complex, depending on the system under investigation. Moreover as pointed out by Lougovois and Houston (1989), mechanisms for the interaction of ferrous and ferric myoglobin and hemoglobin with nitrite to produce pigments have not been well elucidated.

On the other hand, variation in the color of precooked meat products to the same internal temperature has been reported to be a problem as evidenced by variations in redness of highly pigmented muscles such as beef muscle (Anonymous, 1983) and variations in pinkness in the less pigmented poultry muscles (Cornforth et al., 1986). Two possible explanations have been reported, namely, (a) conversion of the myoglobin to a pink hemochrome during heating and (b) incomplete denaturation of the myoglobin. The formation of nitrosylhemochrome as in the case of cured meats may be the result of nitrate, nitrite, or nitrous oxide present during cooking. It has also been suggested that under certain conditions denatured myoglobin may react with certain amino acids, denatured proteins, and other nitrogen containing compounds to produce pink hemochromes (Drabkin and Austin, 1935; Barron, 1937; Dymicky et al., 1975). Cornforth et al. (1986) suggested that the pink hemochrome was formed under reducing conditions between the heme from myoglobin and the nicotinamide normally present in the muscle. Trout (1989) reported that in fully cooked high pH meat products the pink color was due to (a) incomplete denaturation of myoglobin at low temperatures ($\leq 76°C$) and (b) formation of a hemochrome at higher temperatures ($\geq 76°C$). The presence of sodium chloride and sodium tripolyphosphate, additives that were

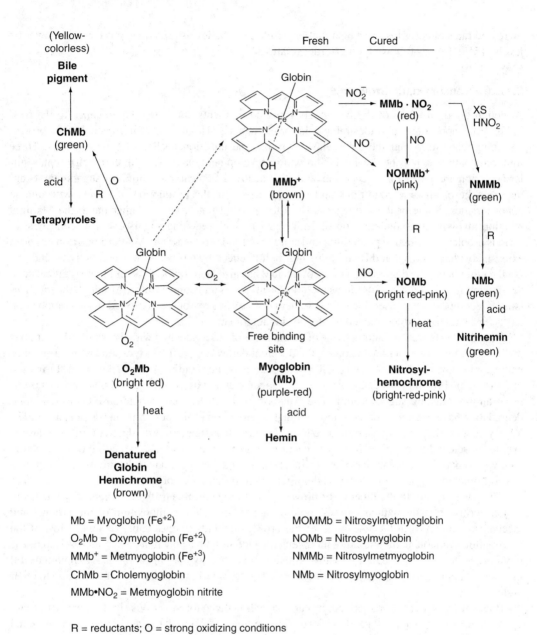

Mb = Myoglobin (Fe^{+2})

O$_2$Mb = Oxymyoglobin (Fe^{+2})

MMb$^+$ = Metmyoglobin (Fe^{+3})

ChMb = Cholemyoglobin

MMb•NO$_2$ = Metmyoglobin nitrite

MOMMb = Nitrosylmetmyoglobin

NOMb = Nitrosylmyglobin

NMMb = Nitrosylmetmyoglobin

NMb = Nitrosylmyoglobin

R = reductants; O = strong oxidizing conditions

FIGURE 2.23 Some mechanisms of myoglobin reactions in fresh and cured meats. (Adapted from Fox, J.B. Jr. 1966. *Journal of Agricultural and Food Chemistry*, 14: 207–210 and Giroux, M., Yefsah, R., Smoragiewicz, W., Saucier, L., and Lacroix, M. 2001. *Journal of Agricultural and Food Chemistry*, 49: 919–925.)

found to alter the denaturation of myoglobin, also resulted in increased pinkness of the cooked products.

As previously indicated, a number of other factors play an important role in the stability of color in meat and meat products. For instance Houben et al. (2000) reported on the effect of vitamin E stabilizing color of minced beef by indirectly delaying oxidation of oxymyoglobin via direct inhibition of lipid oxidation. Vitamin E was added to the animal's diet. The reported effect was more significant in lean meat. Similar trends were reported by Bhattacharya and Hanna (1989) who found increased rates of color losses in beef patties with higher fat levels (30%) vs. product with lower fat levels (15%). Losses followed a first-order relationship.

2.3.2.6 Nonenzymatic Browning

Although the destruction of the previously reported pigments is of great significance to the food industry, the appearance of undesirable color or pigments is also of great importance. In fact, browning compounds may alter the color, flavor, aroma, and nutritional value of food products. Three important pathways may be involved in browning development, namely, sugar caramelization, Maillard reaction and oxidation of ascorbic acid. The heat-induced caramelization of sugars may occur under acidic or alkaline conditions and is associated with the production of flavors with unique characteristics. Some of these flavors have bitter or burnt notes. On the other hand, the Maillard reaction involving the condensation of amino groups with reducing sugars can also contribute to serious problems during the processing and storage of food products. The Maillard reaction has been extensively characterized and found to be highly influenced by temperature, water activity, and pH. On the other hand, the degradation of ascorbic acid has also been found to be of great significance in the darkening of a number of products including fruit juices and concentrates. Although the mechanism of decomposition of ascorbic acid is rather complex as previously discussed, its decomposition has been found to be accompanied by the production of carbon dioxide.

As indicated by various authors, one of the main problems associated with the Maillard reaction is the lack of knowledge on how to control the different pathways, particularly when trying to enhance pathways leading to desirable aroma or flavor formation, while avoiding those leading to the formation of undesirable brown pigments, as well as carcinogenic and toxic substances. Many reactions occur in an amino-reducing sugar system. The Amadori pathway as well as acid/base catalyzed sugar degradation have been considered important in the generation of flavor compounds (Lu et al., 1997). Yaylayan et al. (2003) suggested the use of phosphorylated sugars in providing control over aroma profiles developed through the Maillard reaction. In general, a better understanding of the reaction pathways to avoid the appearance of undesirable compounds and/or enhance the presence of desirable flavor/aroma during food processing is definitely an important area of research.

Thus, with regard to the kinetics of browning development, it is difficult to generalize its behavior in terms of mathematical models, unless knowledge of the composition of the product and mechanisms involved in the formation of the brown pigments is reasonably defined. Most of the information available in the literature has been collected in such a way that an overall reaction rate is monitored when characterizing browning kinetics. It is obvious that this approach, although useful in practical applications, has serious limitations when trying to extrapolate this information to other systems.

Buera et al. (1987a) characterized the caramelization behavior of various single sugars as a function of temperature (45–65°C) and pH (4–6). The authors indicated that a zero-order reaction model for fructose, xylose, and maltose, and a mixed-order reaction model for glucose, lactose, and sucrose, described the kinetics of sugar caramelization. A lag period was observed for glucose, lactose, and sucrose. Fructose and xylose solutions exhibited faster rates of browning as compared with maltose, glucose, lactose, or sucrose. The authors attributed the reported trends to the relative structural stability of the sugars, including their tendency for mutarotation, opening of the hemiacetal ring, and enolization. Lowering of the pH was observed to strongly influence the kinetics of caramelization, resulting in decreased caramelization upon a decrease in pH. It is obvious that depending on the pH involved, additional mechanisms, such as sugar hydrolysis, may play an important role. Activation energies in the range of 25 to 30 kcal/mole for fructose and xylose and in the range of 35 to 48 kcal/mole for lactose and maltose were determined.

It is clear that when systems containing reducing sugars and amino groups are heated, caramelization and the Maillard reaction may take place simultaneously. Hence, it becomes important to characterize the contribution of each pathway to properly establish kinetic models.

Buera et al. (1987b) reported on the nonenzymatic browning of sugar-glycine liquid model systems at high water activity heated at 45 to 65°C. The authors indicated that the kinetic behavior of fructose-glycine solutions was described by a zero-order reaction, while those of xylose-, glucose-, lactose-, maltose-, and sucrose-glycine solutions followed a fractional order kinetic model (~0.5). It was also reported that the contributions of caramelization and the Maillard reaction were dependant upon composition, pH, water activity, and temperature. Based on their results, the contribution to caramelization increased with temperature. Hence, it becomes more evident by looking at the information presented for model systems, that various mechanisms contribute to the formation of browning, and thus its kinetics cannot always be described by simple mathematical expressions. Mundt and Wedzichz (2003) clearly pointed out, working with model systems, that the Maillard reaction in a fructose–glucose–glycine system can be described as a combination of two parallel reactions of glucose+glycine and fructose+glycine. These reactions share a common intermediate, and thus a synergistic behavior of the two sugars is observed. Similarly, it is expected that in complex food systems containing mixtures of sugars, some independent pathways may exist, although most likely there will be interaction between the various reaction pathways, thus highly complicating any type of kinetic analysis based on reaction mechanisms.

Martins and Van Boekel ((2005) have strongly suggested the use of multiresponse modeling to take simultaneously into account all measured reactant-, intermediate-, and end-product concentration changes, as opposed to only one response in simple kinetics when characterizing the Maillard reaction. The authors investigated a kinetic model for the glucose/glycine Maillard reaction, indicating that the reaction constants followed an Arrhenius type temperature dependence and that the model performed well for the temperature range studied (80–120°C). The multiresponse modeling was found to be very useful in unraveling complicated reaction pathways as normally found in Maillard reactions, where multiple parallel reactions proceed at the same time (Table 2.4).

Although a large number of kinetic studies have been conducted on the kinetics of browning, the mathematical description of the reaction has not always been simple. A more traditional approach, as presented by Haugaard et al. (1951), studying the kinetics of browning formation of D-glucose and glycine at reflux temperatures, concluded that the rate of browning formation (dB/dt) was proportional to the square of the amino acid concentration (A) and the concentration of the reducing sugar (R), according to an overall equation:

$$-\frac{dB}{dt} = kA^2Rt \qquad (2.40)$$

where k is the rate constant and t is time.

Labuza (1970) reported that the Maillard reaction could be considered to follow zero-order kinetics when reactant concentrations are not limited. In fact, several investigators have found zero-order reaction models to be the easiest to describe the kinetics of browning with reasonable accuracy. Labuza and Saltmarch (1981), in their review on the non-enzymatic browning reaction as affected by water in foods, indicated that for most dry foods a zero-order reaction had been found to describe the kinetics of browning. Mizrahi et al. (1970) working with dehydrated cabbage observed that the kinetics of browning formation could be described by a zero-order reaction for the moisture content range investigated (1.32–17.9 g water/100 g solids). Warmbier et al. (1976a), working on the effect of glycerol on nonenzymatic browning in a solid intermediate moisture model system containing casein and glucose, reported zero-order kinetics for Maillard browning formation after a period of induction, although the initial loss of available lysine and glucose followed first-order kinetics. Warmbier et al. (1976b) also observed that the browning rates increased as the glucose concentration increased possibly due to the closer proximity with the amino groups, thus overriding diffusional problems of the reactants due to the viscosity of the system. Waletzko and Labuza (1976), working with intermediate moisture foods, and Labuza and Saltmarch (1981), working with whey powders, have reported zero-order reactions to describe the kinetics of nonenzymatic browning development. Petriella et al. (1985), working with model systems at high water activity containing lysine and glucose, also reported zero-order kinetics after a short initial induction period in the range 45 to 55°C but with deviations at lower temperatures. In fact, at 35°C the order of the reaction was between zero and one. The authors reported a strong influence of pH and temperature on the rate of brown color development and a lack of influence of water activity for that particular narrow range (0.90–0.95). Working with model systems, Stamp and Labuza (1983) reported that browning of aspartame followed zero-order kinetics. It is considered that aspartame, a dipeptide, can undergo nonenzymatic browning in the presence of reducing sugars.

Saguy et al. (1978a) reported on the kinetics of browning in grapefruit juice during thermal and concentration processes and observed that two stages were present. First-order reaction kinetics were used to describe the rates of browning formation in both stages. The first stage or lag period had an activation energy in the range 8 to 30 kcal/mole, while the second stage proceeded more rapidly and had an activation energy between 15 and 24 kcal/mole.

Toribio and Lozano (1984) investigated the rate of browning in Red Delicious and Granny Smith apple juice concentrates during storage. The authors reported a first-order reaction for the solid content under investigation (65–75° Brix) in the range 5 to 37°C.

Kinetics of color changes during extrusion of yellow maize grits were reported as a function of processing parameters, including barrel temperature (140–180°C), moisture (13–17%), feed rate (38–52 kg/h) and screw speed (60–80rpm). Color changes were monitored using an L-a-b colorimeter, specifically lightness (L-value) and redness (a-value). Major parameters affecting color were product temperature and feed moisture content. Activation energies were calculated for L-value and a-value as 16 and 18 kcal/mole, respectively. L-value was found to be the best indicator for modeling browning kinetics of the extruded product (Ilo and Berghofer, 1999). Further studies by the authors (Ilo and Berghoffer, 2003) using lysine-fortified maize grits indicated that lysine degradation followed a first-order reaction kinetics. Cystine and arginine losses also followed first-order kinetics. Activation energies were found to be 30.4, 16.3, and 18.2 kcal/mole for lysine, arginine, and cystine, respectively. Shear stress appeared to significantly affect the rate constants of amino acid degradation.

Kinetics of color changes for the thermal processing of onion and garlic paste were reported by Ahmed and Shivare (2001a, 2001b). Color changes followed first-order reaction kinetics for both products as described by the tristimulus color value combination La/b, with activation energies of 3.9 and 3.3 kcal/mole, respectively. Ávila and Silva (1999) determined that the retention of peach puree total color difference (TCD) and the La/b may be used as quality indicators for sterilization processing conditions. The peach puree color thermal degradation followed a first-order kinetics with an Arrhenius model well describing the temperature dependence. The activation energies for a, La/b, and TCD were found to be 25.3, 25.3, and 28.4 kcal/mole, respectively. In general, different approaches have been taken to monitor color changes as due to nonenzymatic browning, primarily looking at L-a-b values and their corresponding kinetic changes. The basic problem with this approach is that most of these models are not based on a true understanding of the mechanisms of browning formation or pigment losses but rather a composite final effect as measured by a colorimetric response. Thus, although this information has some value as it may help optimize process or storage conditions, limited information can be extrapolated to other formulations or systems due to lack of understanding on how to avoid undesirable or favor desirable reactions associated with nonenzymatic browning.

In the following paragraphs, we will address kinetic information available in relatively novel processing techniques such as high pressure, irradiation, and ohmic heating/pulsed electric field.

2.4 KINETICS ASSOCIATED WITH ALTERNATIVE PROCESSING TECHNOLOGIES

2.4.1 HIGH PRESSURE PROCESSING

In response to the ever-increasing demands by the consumer for better quality and nutritious foods, the food industry continually searches for new techniques to bring these demands to realization. One of the new emerging technologies that has received a great deal of attention in the past 10–15 years is that of high-pressure processing (HPP) technology. Pressure is applied via a pressure-transferring medium such as water or other fluid, and can be effective at ambient temperature, thereby decreasing the amount of thermal energy needed during conventional processing (Hashizume et al., 1995; Palou et al., 1997). Reports have shown that HPP treatments are independent of product size and geometry, and their effect is uniform and instantaneous, unlike conventional heat transfer systems (Knorr, 1993; Zimmerman and Bergman, 1993; Alemán et al., 1996). During high-pressure processing, foods are generally subjected to pressures in a range of 100–600 MPa at around room temperature.

The following are considered the major effects of high pressure on food systems. First and foremost is (1) proper inactivation of microorganisms/yeast (Hoover et al., 1989; Zook et al., 1999) followed by (2) enzyme inactivation or activation (Morild, 1981; Indrawati et al., 2001; and Nienaber and Shellhammer, 2001a, 2001b). Other areas that are highly significant are (3) modification of biopolymers such as protein denaturation (Heremans, 1982; Heremans et al., 1999), or gel formation (Cheftel, 1991); (4) product functionality (e.g., density changes, freezing, and melting temperatures) or texture attributes (Farr, 1990; Deuchi and Hayashi, 1991; Eshtiaghi and Knorr, 1993); and last but certainly not the least is (5) quality retention such as flavor and color (Hayashi, 1989; Cheftel, 1991; Weemaes et al, 1999a; Suthanthangjai et al., 2005) as well as nutritional value (Sancho et al., 1999). Most work up until now has been generally focussed on inactivation of microorganisms (Farkas and Hoover, 2000). Of course, the key to the success of this HPP technology will be the inactivation of microorganisms with the simultaneous retention of nutritive and organoleptic characteristics of a given product. The attainment of reliable and consistent data on the destruction of microorganisms has yet to be accomplished, although great strides have been made. There are many reports on the effect of HPP on reduction of microorganisms but few authors have actually reported kinetics, which is essential to the prediction of destruction of pathogenic bacteria and spores. Some investigators have reported first-order kinetics on destruction of bacteria and yeast as a function of pressure (Butz and Ludwig, 1986; Smelt and Rijke, 1992; Carlez et al., 1993; Hashizume et al., 1995; Palou et al., 1997); however, other researchers have reported a possible two-phase inactivation with the first population inactivated rapidly and the second found to be more resistant (Cheftel, 1995). It is important to determine inactivation kinetics as a function of pressure, temperature, and medium composition (Ludwig et al., 1992). Many of the authors who have determined first-order kinetics have attempted to apply the traditional concepts of D- and z-values; however, other authors have shown evidence that death is not first-order, indicating the invalidity of D- and z-values. Similarly to studies on thermal processing, a major drawback is the incomplete reporting of necessary conditions when collecting kinetic data. In the case of HPP processing this is even more crucial with added variables such as rate of pressure increase and decrease, the come-up time for the pressure, the actual isostatic pressure, along with the initial microbial population concentration. Their pressure resistance depends on the type of microorganism and suspension media composition, temperature of applied pressure, gas solubility, ionic strength, and pH. Recovery after release of pressure treatment is also an important consideration for proper assessment of the treatment.

Another area on the effect of high-pressure treatment that has received a fair amount of attention is inactivation of enzymes. Indrawati et al. (2001) reported that the inactivation of lipoxygenase in green pea juice and whole peas as a function of thermal (60–70°C) or combined thermal-pressure effects (–15–70°C/0–600 MPa) followed first-order kinetics. They reported energies of activation of 129 and 140 kcal/mole for green pea juice and whole peas, respectively. Nienaber and Shellhammer (2001b) also reported first-order kinetics for the inactivation of pectin methylesterase in orange juice when subjected to combined high pressure and thermal treatments (400–600 MPa/25–50°C/0–30 min). The same authors also looked at the effect of HPP/thermal treatment (500–800 MPa/25–50°C/1 min) on the shelf life of orange juice in terms of not only enzyme activity but also ascorbic acid and color loss (Nienaber and Shellhammer, 2001a). Processing at 800 MPa/25°C/1 min resulted in less than 20% loss of ascorbic acid over a period of 3 months at 4°C or 2 months at 15°C with stable color (as measured by Lab-values) at all conditions.

Due to the complexity of food systems, however, it is somewhat difficult to exactly predict the effects of increasing pressure on food products. For instance, there have been reports that lipid oxidation may increase with application of high pressure in pork fat and meat (Cheah and Ledward, 1995, 1996, 1997) but browning may decrease at elevated pressures (Tamaoka et al., 1991). Other investigators have found that the effect of pressure on nonenzymatic browning in glucose–lysine model systems was dependent on pH; increasing pressure suppressed browning at pH $\leq$ 8.0 but accelerated browning at pH 10.2 (Moreno et al., 2003). In theory, due to the lower temperatures required for HPP treatment of foods, nutritional quality should be protected via this type of processing. HPP has been reported to keep covalent bonds intact and affects only noncovalent ones, thus resulting in improvement in overall quality characteristics (Hayashi, 1989). For instance, Horie et al. (1991) reported 95% retention of vitamin C in strawberry jam after HPP processing, and Sancho et al. (1999) reported 89% retention of ascorbic acid in a strawberry puree model system after treatment at 400 MPa/30 min at ambient temperature.

Work is relatively scarce in the area of effect of HPP treatment on the nutritional quality of foods. There is some significant work reported, however. Nguyen et al. (2003), for instance, studied the effect of pressure and temperature on the degradation of 5-methyltetrahydrofolic and folic acid model systems. They reported first-order kinetics for both folate derivatives at ambient pressure in a temperature range of 65 to 165°C with energies of activation of 19.1 and 12.3 kcal/mole, respectively. High-pressure treatment combined with heat (0–800 MPa/10–65°C) showed little effect on folic acid but increased 5-methyltetrahydrofolic acid degradation with increasing pressure, particularly at temperatures > 40°C. The authors developed a predictive model describing the combined effects of pressure and temperature on the 5-methyltetrahydrofolic acid rate constants (Table 2.5a).

Van den Broeck et al. (1998) reported first-order kinetics for the destruction of L-ascorbic acid in orange and tomato juices subjected to a single pressure level of 850 MPa at 65–80°C. Both energies of activation and z-values were reported; parameters are previously reported in Section 2.3.1.1 on vitamin C. Sanchez–Moreno et al. (2003) reported on the effect of specific combinations of HPP and thermal/time treatments on vitamin C, provitamin A, and other carotenoids in orange juice after initial processing and during storage at 4°C up to 10 days. The authors found an 8–10% total loss of vitamin C during processing depending upon the designated treatment with up to 60% loss during subsequent storage. Vitamin A, on the other hand, showed increases in β-carotene immediately following process treatment (from 6 to 28%) with increasing concentrations as pressure increased from 100 to 400 MPa. Losses occurred, however, during storage up to 21%. Some of the increase in β-carotene concentration immediately following pressure treatment was attributed to a structural change allowing greater extractability of the carotenoids. Nevertheless, it appeared that the effect of high pressure treatment did not adversely affect vitamin A.

Other reports have shown effects of high pressure treatment on color stability. For instance, Weemaes et al. (1999a) investigated the effect of HPP on degradation of green color in broccoli juice (0.1–850 MPa/70–90°C/up to 180 min). Loss of color was measured as change in a-value. The authors found that two consecutive first-order degradation steps using the Arrhenius equation with energies of activation increasing with increasing pressure could model loss of color. However, the Eyring equation did not effectively describe their dependency on pressure (rate constants are presented in Table 2.5a for reference). Suthanthangjai et al. (2005) also investigated the effect of HPP on color of red raspberry puree, as monitored by degradation of two predominant anthocyanins, cyanidin-3-glucoside and cyanidin-3-sophoroside as measured by HPLC. The ranges of pressure and temperature treatments were 200, 400, 600, and 800 MPa at 18–22°C for 15 min, followed by storage from 4–30°C, up to 9 days. The highest stability for red color was reported to occur at 200 and 800 MPa with storage at 4°C. The authors surmised that there may be higher remaining enzyme activity of β–glucosidase, peroxidase, and polyphenolase at those intermediary pressures. This trend was observed at all storage temperatures. Similar enzyme activity as a function of pressure has been reported by Chéret et al. (2005) in the case of certain proteolytic enzymes in sea bass. This phenomenon is of critical importance when modeling effect of pressure on overall quality parameters.

TABLE 2.5a Kinetic Parameters for Losses during Nonconventional Processing

Commodity	Process / Conditions		r^2	k_T value (min^{-1})		E_a (kcal/mol)	Reaction Order	r^2	Temp. Range (°C)	$t_{1/2}$ (min)	Reference
High Pressure Processing:											
Vitamins											
Ascorbic acid											
Orange juice (fresh squeezed / filtered / C-fortified)	Process: 850 MPa / 0 - 400 min 0.3 ml flexible microtubes / N₂ flush		0.935	k_{80} = 0.010289						67.4	Van den Broeck, et al., 1998
			0.976	k_{70} = 0.004338		20.10	1	0.999	65 - 80	159.8	
			0.893	k_{65} = 0.002897						239.3	
Orange juice (fresh squeezed w/ pulp)	Process: 800 MPa / 25°C / 1 min Storage: 4-37°C / up to 14 days		0.986	k_{37} = 9.9834x10⁻⁶						6.95x10⁴	Nienaber and Shellhammer, 2001a
			0.862	k_{26} = 3.7958x10⁻⁶		8.11	1	0.907	4 - 37	18.26x10⁴	
			0.987	k_{15} = 2.7434x10⁻⁶						25.27x10⁴	
			0.933	k_{4} = 1.9145x10⁻⁶						36.21x10⁴	
Tomato juice (fresh squeezed / filtered / C-fortified)	Process: 850 MPa / 0 - 400 min 0.3 ml flexible microtubes / N₂ flush		-	k_{80} = 0.005744						120.7	Van den Broeck, et al., 1998
			-	k_{70} = 0.003235		17.82	1	0.965	65 - 80	214.2	
			-	k_{65} = 0.001789						387.4	
Folates - 5-Methyltetrahydrofolate	Pilot scale high-pressure unit /w/ flexible 500 μl tubes										Nguyen et al., 2003
	Pressure rate ↑ : 125-150 MPa/ 5 min equilibration	P = 0.1 MPa	0.97	k_{90} = 0.06831						10.15	
	Process up to 130 min (10 μl / ml in phos buffer / pH 7)		0.99	k_{80} = 0.02814		19.12	1	0.99	65 - 90	24.63	
			0.96	k_{70} = 0.01306						53.07	
			0.96	k_{65} = 0.00973						71.24	
	HPLC analysis	P = 100 MPa	0.93	k_{65} = 0.02508						27.64	
			0.91	k_{60} = 0.02421		18.88	1	0.91	65 - 50	28.63	
			0.98	k_{50} = 0.00738						93.92	

5-Methyltetrahydrofolate (cont.)

P = 200 MPa	0.99	k_{65} = 0.03812						18.18
	0.99	k_{60} = 0.03519		17.18	1	0.98	65 - 40	19.70
	0.99	k_{50} = 0.01174						59.04
	0.88	k_{40} = 0.00566						122.46
P = 400 MPa	0.98	k_{65} = 0.07899						8.78
	0.95	k_{60} = 0.07012						9.89
	0.80	k_{50} = 0.02298		19.38	1	0.99	65 - 30	30.16
	0.99	k_{40} = 0.01103						62.84
	0.94	k_{30} = 0.00292						237.38
P = 600 MPa	0.93	k_{60} = 0.10634						6.52
	0.98	k_{50} = 0.02596						26.70
	0.98	k_{40} = 0.01825		23.95	1	0.96	60 - 20	37.98
	0.97	k_{30} = 0.00451						153.69
	0.90	k_{20} = 0.00055						1260.27
P = 800 MPa	0.98	k_{60} = 0.15037						4.61
	0.98	k_{50} = 0.05441						12.74
	0.95	k_{40} = 0.01832		21.54	1	0.99	60 - 20	37.84
	0.93	k_{30} = 0.00733						94.56
	0.99	k_{20} = 0.00160						433.22

Reference: Nguyen et al., 2003

See Table 2.3, ambient pressure

5-Methyltetrahydrofolate Temperature	V_a (cm^3/mol)	r^2	Pressure Range (MPa)	Reference
65	-13.95	0.94	0.1 - 400	Nguyen et al., 2003
60	-7.23	0.94	100 - 800	
50	-7.05	0.96	100 - 800	
40	-5.24	0.90	200 - 800	
30	-5.79	0.99	400 - 800	

(Continued)

TABLE 2.5a Continued

Commodity	Process / Conditions	r^2	k_T value (min^{-1})		E_a (kcal/mol)	Reaction Order	r^2	Temp. Range (°C)	$t_{1/2}$ (min)	Reference
Chlorophyll - Broccoli Juice 2-Step Consecutive Pathway: *1. Chlorophyll → Peophytin* (measured as change in *a*-value)				**Pigments**						Weemaes et al, 1999a
	P = 0.1 MPa	-	[a]k_{90} = 0.02630						26.36	
		-	k_{80} = 0.01575		[a]13.9	1	[a]0.969	70 - 90	44.01	
		-	k_{75} = 0.01002						69.18	
		-	k_{70} = 0.00908						76.34	
	P = 50 MPa	-	k_{80} = 0.02065						33.57	
		-	k_{75} = 0.01493		13.77	1	0.993	70 - 80	46.43	
		-	k_{70} = 0.01165						59.5	
	P = 100 MPa	-	k_{80} = 0.02892						23.97	
		-	k_{75} = 0.02341		18.23	1	0.943	70 - 80	29.61	
		-	k_{70} = 0.01358						51.04	
	P = 150 MPa	-	k_{80} = 0.02552						27.16	
		-	k_{75} = 0.01588		15.48	1	0.926	70 - 80	43.65	
		-	k_{70} = 0.01340						51.73	
	P = 300 MPa	-	k_{80} = 0.05602						12.37	
		-	k_{75} = 0.01925		32.27	1	0.891	70 - 80	36.01	
		-	k_{70} = 0.01462						47.41	
	P = 500 MPa	-	k_{80} = 0.10574						6.56	
		-	k_{75} = 0.04825		44.02	1	0.995	70 - 80	14.37	
		-	k_{70} = 0.01702						40.73	
	P = 700 MPa	-	k_{80} = 0.09671						7.17	
		-	k_{75} = 0.04032		38.71	1	0.997	70 - 80	17.19	
		-	k_{70} = 0.01937						35.78	
	P = 850 MPa	-	k_{80} = 0.10239		-	1	-	70 - 80	6.77	
		-	k_{70} = 0.01697						40.85	

2. Peophytin → pyropheophytin

P							Value	Reference
P = 0.1 MPa	$^{a}k_{90}$ = 0.00269						26.36	Weemaes et al., 1999a
	k_{80} = 0.00085						44.01	
	k_{75} = 0.00057	^{a}22.14	1	^{a}0.954	70 - 90		69.18	
	k_{70} = 0.00047						76.34	
P = 50 MPa	k_{80} = 0.00082						33.57	
	k_{75} = 0.00079	22.46	1	0.791	70 - 80		46.43	
	k_{70} = 0.00032						59.5	
P = 100 MPa	k_{80} = 0.00109						23.97	
	k_{75} = 0.00058	25.19	1	0.984	70 - 80		29.61	
	k_{70} = 0.00038						51.04	
P = 150 MPa	k_{80} = 0.00084						27.16	
	k_{75} = 0.00053	36.85	1	0.955	70 - 80		43.65	
	k_{70} = 0.00018						51.73	
P = 300 MPa	k_{80} = 0.00214						12.37	
	k_{75} = 0.00067	-	1	-	75 - 80		36.01	
P = 500 MPa	k_{80} = 0.00221						6.56	
	k_{75} = 0.00137	-	1	-	75 - 80		14.37	
P = 700 MPa	k_{80} = 0.00199						7.17	
	k_{75} = 0.00096	50.55	1	0.973	70 - 80		17.19	
	k_{70} = 0.00024						35.78	
P = 850 MPa	k_{80} = 0.00218						6.77	
	k_{70} = 0.00038	-	1	-	70 - 80		40.85	

Anthocyanins - Raspberry puree
Cyanidin-3-glucoside

Process: 200-800 MPa / 15 min / 18 - 22°C

Storage: 0 - 9 days / 4 - 30°C

P								Value	Reference
P = 0 MPa	0.935	k_{30} = 11.139 x 10^{-5}						6223	Suthanthangjai et al., 2005
	0.976	k_{20} = 4.066 x 10^{-5}	11.45	1	0.957	4 - 30		17048	
	0.893	k_{4} = 1.760 x 10^{-5}						39373	
P = 200 MPa	0.974	k_{30} = 8.251 x 10^{-5}						8401	
	0.896	k_{20} = 3.909 x 10^{-5}	10.24	1	0.988	4 - 30		17730	
	0.933	k_{4} = 1.626 x 10^{-5}						42621	

(Continued)

TABLE 2.5a Continued

Commodity	Process / Conditions	r^2	k_T value (min^{-1})	E_a (kcal/mol)	Reaction Order	r^2	Temp. Range (°C)	$t_{1/2}$ (min)	Reference
Cyanidin-3-glucoside (cont.)	P = 400 MPa	0.962	k_{30} = 8.170 x 10^{-5}	4.47	1	0.979	4 - 30	8484	Suthanthangjai et al., 2005
		0.995	k_{20} = 6.977 x 10^{-5}					9935	
		0.938	k_4 = 4.133 x 10^{-5}					16770	
	P = 600 MPa	0.958	k_{30} = 8.917 x 10^{-5}	6.29	1	0.994	4 - 30	7773	
		0.940	k_{20} = 6.699 x 10^{-5}					10347	
		0.877	k_4 = 3.385 x 10^{-5}					20477	
	P = 800 MPa	0.949	k_{30} = 8.689 x 10^{-5}	10.46	1	0.996	4 - 30	7977	
		0.880	k_{20} = 5.2809 x 10^{-5}					13129	
		0.688	k_4 = 1.728 x 10^{-5}					40120	
Cyanidin-3-sophoroside	P = 0 MPa	0.989	k_{30} = 11.249 x 10^{-5}	12.24	1	0.878	4 - 30	6162	
		0.987	k_{20} = 11.002 x 10^{-5}					6300	
		0.936	k_4 = 1.867 x 10^{-5}					37120	
	P = 200 MPa	0.982	k_{30} = 14.051 x 10^{-5}	9.22	1	0.999	4 - 30	4933	
		0.966	k_{20} = 8.247 x 10^{-5}					8405	
		0.757	k_4 = 3.334 x 10^{-5}					20793	
	P = 400 MPa	0.986	k_{30} = 17.696 x 10^{-5}	4.41	1	0.919	4 - 30	3917	
		0.965	k_{20} = 11.367 x 10^{-5}					6098	
		0.956	k_4 = 8.615 x 10^{-5}					8046	
	P = 600 MPa	0.966	k_{30} = 13.832 x 10^{-5}	10.71	1	0.974	4 - 30	5011	
		0.993	k_{20} = 9.757 x 10^{-5}					7104	
		0.847	k_4 = 2.720 x 10^{-5}					25482	
	P = 800 MPa	0.991	k_{30} = 9.7409 x 10^{-5}	10.59	1	0.965	4 - 30	7116	
		0.968	k_{20} = 7.196 x 10^{-5}					9632	
		0.879	k_4 = 1.963 x 10^{-5}					35310	

TABLE 2.5b Kinetic Parameters for Losses during Nonconventional Processing

Commodity	Irradiation Dose (kGy)	% Retention	r^2	k_T value (kGy^{-1})	Reaction Order	Temp. Range (°C)	Reference
Vitamins							
Irradiation							
Thiamin							
Chicken	0.5	100.0					Hannis et al., 1989
	1.0	81.2					(from Kilcast, 1994)
	2.5	70.9	0.932	k_{10} = 0.08181	1	10	
	5.0	56.4					
	10.0	42.7					
	0.5	100.0					
	1.0	88.8					
	2.5	77.6	0.949	k_{-15} = 0.05665	1	-15	
	5.0	72.0					
	10.0	55.2					
Thiamin							
Beef	30.0	90.0					Wierbicki et al., 1970
	60.0	75.0	-	k_{-80} = 0.00608	1	-80	(from Karel, 1975)
	30.0	10.0				5	
Ham	40.0	69.0					
	50.0	88.0	-	k_{-80} = 0.02432	1	-80	
	40.0	4.0				20	
Thiamin							
Pork (ground)	Cs γ-ray, N$_2$ atmosphere						
Raw / Irradiated	0 - 10.0		-	$^b k_5$ = 0.100	1	5	Fox et al., 1997
Denatured / Irradiated	0 - 10.0		-	k_5 = 0.170	1	5	
Irradiated / Denatured	0 - 10.0		-	k_5 = 0.155	1	5	
Riboflavin							
Pork (ground)							
Raw / Irradiated	0 - 10.0		-	$^b k_5$ = 0.0067	1	5	Fox et al., 1997
Denatured / Irradiated	0 - 10.0		-	k_5 = 0.0235	1	5	
Irradiated / Denatured	0 - 10.0		-	k_5 = 0.0095	1	5	

[b] average of 2 replicates

In summary, in order to favor the expansion of this technology, it is important to continue studying the effects of high pressure on various quality factors of foods to assure the best quality products.

2.4.2 IRRADIATION

Similar to conventional thermal processing, the concept of irradiation is to temporarily raise the energy level of a system sufficiently to result in the death of microorganisms, thus, extending the shelf life of a product. In the case of irradiation, however, instead of raising the temperature of the system for a given period of time, the system is exposed to a source of radiation, resulting in ionization of individual atoms or molecules to produce an electron and a positively charged atom. The energy of radiation is often measured in electron volts (eV), where one electron volt is the energy acquired by an electron falling through a potential of 1 V (or $1/eV = 1.602 \times 10^{-12}$ erg). The total effect of the radiation upon a given material is dependent on the energy of each photon, its source, as well as the total number of photons impinging on the material. The relative ionization of electrons varies with the depth of absorption in a given material (absorber). Using irradiation as a technique for preservation allows the foods to be kept cold or frozen during treatment, thus, potentially allowing greater stability of quality factors of the food product.

From a microbiological safety viewpoint, some issues that are raised on the effects of irradiation of foods are whether or not they may result in the mutation of microorganisms which may lead to more virulent pathogens, and if there is reduction of the spoilage microorganisms, whether or not there will result as a consequence the ability of pathogens to grow undetected without competition. FDA, however, has indicated that radiation-induced mutation is not a concern with respect to increased virulence or increased heat resistance. In fact, Farkas (1989) has indicated that radiation is more likely to reduce virulence of surviving pathogens. Other issues of concern involve losses of nutrients. There has been general evidence, however, that indicates conventional cooking alters the nutrient quality much more than irradiation. Macronutrients, including proteins, lipids, and carbohydrates are not significantly affected up to doses of 10 kGy with only minor changes at sterilization doses of 50 kGy (Diehl, 1995). There is evidence that vitamins may degrade with irradiation, which would be expected of any process that elevates the energy level of the individual food constituents. The degree of degradation will depend on a number of factors such as the dose of radiation, the food type, the temperature at which irradiation occurs, and the presence of oxygen. Generally low temperature radiation in the absence of oxygen reduces significantly the losses of vitamins as well as maintains storage at low temperature and in sealed containers (WHO, 1994). Similarly to thermal processing, it has been found that thiamine is the most labile of the water soluble vitamins in the presence of irradiation processing, whereas, vitamin E has been shown to be the most susceptible of the oil soluble vitamins to degradation from irradiation (Kilcast, 1994; Fox et al., 1995, 1997). FDA requires that the vitamins most affected by irradiation are not a significant source from that particular processed food product in the overall diet.

Overall quality of irradiated foods may be affected by: (1) radiation dose; (2) dose rate; (3) temperature and atmospheric conditions during irradiation (e.g., presence of oxygen); (4) temperature and environmental conditions following irradiation during storage; and (5) development of radiolytic products (Thayer, 1990). The presence of radiolytic products can result in oxidation of myoglobin and fat, which may result in discoloration and rancidity or other off-odor and/or off-flavor development. For instance, ozone, which is a strong oxidizer produced during irradiation in the presence of oxygen, may oxidize myoglobin resulting in a bleached appearance. Color changes may be influenced by the packaging environment. It has been reported that irradiated vacuum packaged meats develop a fairly stable bright pink or red color in turkey breasts, pork, or beef (Niemand et al., 1983; Lebepe et al., 1990; Lynch et al., 1991). This stresses the importance of elimination of oxygen before irradiation. Irradiation in the frozen state minimizes movement of free radicals to react throughout the food system and, thus, can minimize sensory quality issues.

According to Kropf et al. (1995) and Luchsinger et al. (1996), any irradiation-induced off-odors may be removed during conventional cooking; however, studies are ongoing for investigation. Crone et al. (1992) detected the formation of 2-alkyl-cyclobutanone formed from fatty acids in irradiated but not cooked foods.

From a microbiological point of view, the predominant spoilage organisms are Gram-negative psychrotrophic microorganisms, which are very susceptible to irradiation (Monk et al., 1995). It has been shown that doses of about 1 kGy virtually eliminate Gram-negative microorganisms. However, it is not as effective on Gram-positive lactic acid-producing microorganisms. Nevertheless, refrigerated storage of meats has increased dramatically as a result of irradiation. Lambert et al. (1992) found that pork loin slices packaged under nitrogen and irradiated to 1 kGy had an extension of 21 days beyond the control at 5°C. As with thermal death time (TDT) in conventional thermal processing, the death of a microorganism resulting from exposure to radiation can also be evaluated by plotting the logarithm of the surviving fraction, in this case, against dose. With thermal sterilization, the effect of kill is not solely dependent on the quantity of heat absorbed by the cell but also on the intensity factor (temperature) and on time. Radiation sterilization, on the other hand, is actually less complicated since the intensity factor is called "dose rate" or the amount of radiation absorbed by the cell per unit time. Although dose rate has some lethal effects, it is possible to relate radiation effects to dose alone according to the following equation: $N = N_0 e^{-D/D_0}$, where N is the number of live organisms after irradiation, N_0 is the initial number of microorganisms, D is the dose of radiation received, and D_0 is the constant dependent upon organism type and environmental factors (Figure 2.24). As commercial use of irradiation becomes more viable, it will become increasingly

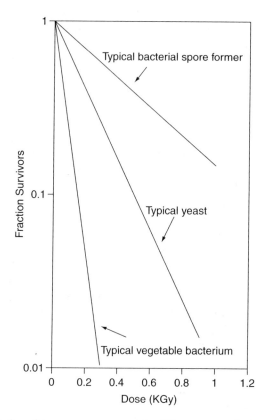

FIGURE 2.24 Representation of dose-response curves for different types of microorganisms. (Adapted from Karel, M. 1975. Radiation preservation of foods, ch. 4, pp. 93–130, In: *Principles of Food Science. Part II. Physical Principles of Food Preservation*, Fennema, O.W. (ed.), Marcel Dekker, NY.)

important to provide additional kinetic data on the stability of various quality factors as affected by treatment (Table 2.5b).

2.4.3 PROCESSES ASSOCIATED WITH ELECTROMAGNETIC FIELDS

Other areas of alternative technologies for processing of foods include the direct transfer of energy from an electromagnetic source without the use of traditional heat transfer surfaces, providing the advantage of high-energy utilization. Considering the electromagnetic spectrum, there are three main frequency areas for direct heating of foods: (1) 50/60 Hz or typical household power as used in resistance or ohmic heating, with direct immersion of electrodes; (2) 10–60 MHz (high frequency), where the food material acts as a conductor between two electrodes; and (3) 1–3 GHz (microwave region), where energy is transferred to the food through air by guided waves controlled by an electromagnetic device (Ohlsson, 1999). Two areas that have recently received a great deal of attention include ohmic heating and pulsed electric field (PEF). Ohmic (or Joule) heating was one of the earliest forms of electricity applied to food pasteurization and is reemerging as a viable process (Allen et al., 1996). Its mode of operation is by direct passage of electric current through the food product, which generates heat as a result of electrical resistance. Ohmic heating reduces the time of heating as compared to conventional heat transfer by convection or conduction. Its advantages are (1) rapid and uniform heating, (2) less thermal damage to products, (3) decreased operational costs, and (4) absence of hot surfaces and therefore reduced fouling, as occurs in UHT processing (Scott, 1995; Reznick, 1996). Ohmic heating is of particular interest to aseptic processing, keeping in mind factors such as particle size, particle density, carrier viscosity and composition, and electrical conductivity (Zoltai and Swearingen, 1996; Kim et al., 1996). Leizerson and Shimoni (2005) compared ultra-high temperature ohmic processing of fresh orange juice (90–150°C/1.13–0.68 sec) to conventional thermal processing (90°C/50 sec) and found complete inactivation of bacteria, yeast, and mold with either treatment and superior organoleptic properties and less loss of vitamin C with the high temperature ohmic processed juice.

In most cases up until the present, inactivation of microorganisms with electrical energy has been considered to be due mainly to thermal effects, utilizing traditional thermal death time (TDT) equations. However, there are considerations that may need to be taken into account for nonthermal effects, particularly when these technologies are applied at low temperatures. One such technology includes pulsed electric fields (PEF) processing. This process involves the application of pulses of high voltage (20–80 kV/cm) to foods placed between two electrodes. PEF may be applied in the form of exponentially decaying, square wave, bipolar, or oscillatory pulses at ambient, sub-ambient, or slightly above ambient temperatures for less than 1 sec (Barbosa-Cánovas et al., 2000). This would obviously reduce the potential of heat losses present in traditional thermal processing. By increasing the electric field intensity, the frequency of the number of pulses, and their duration, greater inactivation of microorganisms can be achieved (Benz and Zimmermann, 1980; Tsong, 1990; Knorr et al., 1994; Larkin and Spinak, 1996). This, again, is dependent on other factors such as treatment temperature, pH, ionic strength, and conductivity of the fluid medium (i.e., high lipid containing fluids are less conductive than aqueous media).

Similarly to high pressure processing, the effectiveness of applying electrical energy for the inactivation of microorganisms will depend on processing above the threshold (or critical) electric field intensity for a target microorganism, below which inactivation does not occur. A somewhat similar model was proposed for the influence of electric field intensity on reducing a microorganism population as that with the influence of pressure (Hülsheger and Niemann, 1980):

$$\ln(\mathbf{S}) = -b_E(E - E_c) = \ln(N/N_0) \tag{2.41}$$

where, $\mathbf{S}$ is the survival ratio, N_0 and N are the initial and final microbial populations, respectively; b_E is the regression coefficient (microbial media constant); E is the applied electric field; and E_c is the critical electric field (based on an extrapolated value of E for 100% survival) and is dependent on cell size and pulse width. This relationship was later revised by Hülsheger et al. (1981) relating survival of microorganisms with PEF treatment time (t):

$$\ln(\mathbf{S}) = -b_t \ln(t/t_c) \quad \text{or} \quad \mathbf{S} = (t/t_c)^{-([E-E_c]/K)} \tag{2.42}$$

where, b_t is the regression coefficient; t_c is the extrapolated value of t for 100% survival; and K is a kinetic constant.

Vega-Mercado et al. (1999) and Barbosa-Cánovas et al. (2000) have reviewed some of the most recent advances in the area of high-intensity pulsed electric fields. It is clear that this is still a developing technology but warrants mention as a new technology of the future in food processing. Nevertheless, a great deal of work needs to be accomplished in the area of determining the kinetics of destruction of microorganisms as a function of PEF, determination of uniform delivery of treatment, the impact of temperature, pH, moisture and lipid content on the process, and the influence of food additives. There is also little known on the influence of PEF on vitamin retention. PEF has been predominantly used for the preservation of bread, milk, orange juice, liquid eggs, and apple juice. Zang et al. (1997), for instance, evaluated shelf life of reconstituted orange juice treated with a PEF pilot plant system, maintaining ambient temperature and using different ware-shaped pulses. They confirmed that square wave was the most effective pulse shape, with a reduction of aerobic counts by 3–4 log cycles under 32 kV/cm, similar reduction to that by conventional thermal processes. Less vitamin C losses were observed with PEF compared with conventionally heat-processed juice when stored at 4°C for up to 90 days.

2.5 SUMMARY

After reviewing the published literature, it is evident that a major effort is still needed to attain reliable kinetic information for nutrient or quality retention in food materials or products applicable to the optimization of processing and storage conditions. A better understanding of the mechanisms involved in the reactions leading to the destruction of a compound or a quality parameter should facilitate the development of kinetic information with wider applicability. In addition, this approach will help in providing guidelines to determine formulation or fortification protocols of products leading to a higher nutritional value or quality. The accuracy, sensitivity and specificity of the methodology applied to quantify the specific changes are also of crucial relevance to the quality of the kinetic information.

It should be kept in mind that some of the new emerging technologies in food processing such as high pressure, irradiation, microwave and radio frequency processing, pulsed electric fields, and ultrasound will require more basic studies in order to predict retention of nutrients such as vitamins, and organoleptic properties, as well as establishing experimental protocols to obtain reliable kinetic parameters to describe survival curves for microbial populations. Efforts such as those set forth by a joint effort between the Food and Drug Administration of the U.S. Department of Health and Human Sciences in conjunction with the Institute of Food Technologists in 2000 are critical in assessing alternate processing technologies and identifying research needs. This joint organized effort was put together in order to provide a scientific review and analyze the important issues in food safety, food processing, and human health.

With the growing concerns over health-related issues and consumer desire for better quality foods, combined with the increase in potential new technologies for alternative food processing, it is evident that there is an even greater challenge to establish appropriate kinetic models for process optimization.

NOMENCLATURE

SIMPLE REACTIONS/REVERSIBLE FIRST-ORDER REACTIONS

A_0 Initial concentration of reactant species A

A_∞ Concentration of reactant species (A) at time infinity

B_0 Initial concentration of reactant species (B)

B_∞ Concentration of reactant species (B) at time infinity

C Concentration at time (t)

C_0 Initial concentration

C_A Concentration of reactant species (A) at time (t)

C_B Concentration of reactant species (B) at time (t)

C_{A_0} Initial concentration of reactant species (A)

C_{B_0} Initial concentration of reactant species (B)

k reaction rate constant

k_0 Zero-order reaction rate constant

k_1 First-order reaction rate constant

k_{-1} Reverse first-order reaction rate constant

k_2 Second-order reaction rate constant

n Order of the reaction

r Reaction rate

t Time

$t_{1/2}$ Half-life

CHAIN REACTIONS

a Constant

b Initiation rate constant of the products

b_0 Initiation rate constant of unoxidized carotenoids

C Carotenoid concentration at time (t)

C_0 Initial carotenoid concentration

K_S Solubility coefficient of oxygen in carotenoids

$k, k_i, k_p, k_p', k_t, k_t'$ Rate constants

P_{O_2} Partial pressure of oxygen

t Time

w_i Rate of formation of free radicals

σ Effective rate constant

ENZYME KINETICS/SIMULTANEOUS COMPETITIVE REACTIONS

E   Enzyme

EI   Enzyme-inhibitor complex

ES   Enzyme-substrate complex

ESI   Enzyme-substrate-inhibitor complex

I   Competitive inhibitor

$k_1, k_{-1}, k_2, k_{-2}, k_n$ Rate constants

P, P_1, P_2, P_n Products

S   Substrate

EFFECT OF TEMPERATURE/ARRHENIUS

E_a Activation energy (cal/mole)

k_0 Frequency or collision factor

R Gas constant ($1.987\,\mathrm{cal/mol} \cdot {}^\circ\mathrm{K}$)
T Absolute temperature (${}^\circ\mathrm{K}$)

EFFECT OF PRESSURE/EYRING

ΔG Gibb's free energy
ΔH^* Activation enthalpy
ΔS^* Activation entropy
h Planck's constant ($1.584 \times 10^{-23}\,\mathrm{cal} \cdot \mathrm{s}$)
k Rate constant
P Pressure
k_B Boltzmann's constant ($3.301 \times 10^{-22}\,\mathrm{cal/}{}^\circ\mathrm{K}$)
ΔV^* Activation volume (V_a)

NONENZYMATIC BROWNING

dB/dt Rate of browning formation
A   Amino acid concentration
R   Reducing sugar concentration

IRRADIATION

D Dose of irradiation received [Units: 1 gray (Gy) = 100 rad]
D_0 Constant dependent upon microorganism type and environmental factors
N Number of live microorganisms after irradiation
N_0 Initial number of microorganisms

ELECTROMAGNETIC FIELDS

b_E, b_t Regression coefficients (microbial media constants)
E Applied electric field
E_c Critical electric field
K Kinetic constant
S   Survival ratio (N/N_0)
t Treatment time
t_c Critical time (extrapolated for 100% survival)

REFERENCES

Adams, J.B. 1973. Thermal degradation of anthocyanins with particular reference to the 3-glycosides of cyanidin. I. In acidified aqueous solution at 100°C. *Journal of the Science of Food and Agriculture*, 24(7):747–762.

Ahmed, J., Kaur, A. and Shivhare, U. 2002a. Color degradation kinetics of spinach, mustard leaves, and mixed puree. *Journal of Food Science*, 67(3):1088–1091.

Ahmed, J. and Shivhare, U.S. 2001a. Thermal kinetics of color change, rheology, and storage characteristics of garlic puree/paste. *Journal of Food Science*, 66(5):754–757.

Ahmed, J. and Shivhare, U.S. 2001b. Thermal kinetics of color degradation and storage characteristics of onion paste. *Lebensmittel-Wissenschaft und-Technologie*, 34(6):380–383.

Ahmed, J., Shivhare, U.S. and Debnath, S. 2002b. Colour degradation and rheology of green chilli pee during thermal processing. *International Journal of Food Science and Technology*, 37(1):57–63.

Ahmed, J., Shivhare, U.S. and Raghavan, G.S.V. 2000. Rheological characteristics and kinetics of colour degradation of green chilli puree. *Journal of Food Engineering*, 44(4):239–244.

Albalá-Hurtado, S., Veciana-Nogués, M.T., Vidal-Carou, M.C. and Mariné-Font, A. 2000. Stability of vitamins A, E, and B complex in infant milks claimed to have equal final composition in liquid and powdered form. *Journal of Food Science*, 65(6):1052–1055.

Alekseev, É. V., Gagarina, A.B., Etveeva, N.M., Vakulova, L.A., Samokhvalov, G.I. and Émanuél, N.M. 1968. Kinetic principles of the oxidation of polyenic hydrocarbons. Communication 1. Decomposition of β-carotene in the presence of free radical initiators. *Bulletin of the Academy of Sciences of the USSR. Division of Chemical Sciences*, 11:2342–2347. (translated from *Izvestiya Akademii Nauk SSSR, Seriya Khimicheskaya*).

Alemán, G.D., Ting, E.Y., Mordre, S.C., Hawes, A.C.O., Walker, M., Farkas D.F. and Torres, J.A. 1996. Pulsed ultra high pressure treatments for pasteurization of pineapple juice. *Journal of Food Science*, 61(2):388–390.

Allen, C. and Parks, O.W. 1979. Photodegradation of riboflavin in milks exposed to fluorescent light. *Journal of Dairy Science*, 62(9):1377–1379.

Allen, K., Eidman, V. and Kinsey, J. 1996. An economic engineering study of ohmic food processing. *Food Technology*, 50(5):269–273.

Anonymous. 1983. Red meat mystery. *Meat Industry*, 29(10):76.

Archer, M.C. and Tannenbaum, S.R. 1979. Vitamins. ch. 3, p. 47–95, In: *Nutritional and Safety Aspects of Food Processing*. Tannenbaum, S.R. (ed.), Marcel Dekker, Inc., NY.

Aronoff, S. 1966. The chlorophylls — An introductory survey. ch. 1, p. 3–20, In: *The Chlorophylls*. Vernon, L.P. and Seely, G.R. (eds.), Academic Press, NY.

Arya, S.S., Natesan, V., Parihar, D.B. and Vijayaraghavan, P.K. 1979. Stability of carotenoids in dehydrated carrots. *Journal of Food Technology*, 14(6):579–586.

Asen, S., Norris, K.H. and Stewart, R.N. 1969. Absorption spectra and color of aluminum-cyanidin 3-glucoside complexes as influenced by pH. *Phytochemistry*, 8(3):653–659.

Attoe, E.L. and von Elbe, J.H. 1985. Oxygen involvement in betanine degradation: Effect of antioxidants. *Journal of Food Science*, 50(1):106–110.

Attoe, E.L. and von Elbe, J.H. 1981. Photochemical degradation of betanine and selected anthocyanins. *Journal of Food Science*, 46(6):1934–1937.

Ávila, I.M.L.B. and Silva, C.L.M. 1999. Modelling kinetics of thermal degradation of colour in peach puree. *Journal of Food Engineering*, 39(2):161–166.

Baloch, A.K., Buckle, K.A. and Edwards, R.A. 1977a. Effect of processing variables on the quality of dehydrated carrot. I. Leaching losses and carotenoid content. *Journal of Food Technology*, 12(3):285–293.

Baloch, A.K., Buckle, K.A. and Edwards, R.A. 1977b. Effect of processing variables on the quality of dehydrated carrot. II. Leaching losses and stability of carrot during dehydration and storage. *Journal of Food Technology*, 12(3):295–307.

Baloch, A.K., Buckle, K.A. and Edwards, R.A. 1977c. Stability of β-carotene in model systems containing sulphite. *Journal of Food Technology*, 12(3):309–316.

Baloch, A.K., Buckle, K.A. and Edwards, R.A. 1987. Effect of sulphur dioxide and blanching on the stability of carotenoids of dehydrated carrots. *Journal of the Science of Food and Agriculture*, 40:179–187.

Baranowski, E.S. and Nagel, C.W. 1983. Kinetics of malvidin-3-glucoside condensation in wine model systems. *Journal of Food Science*, 48(2):419–421, 429.

Barbanti, D., Mastrocola, D. and Lerici, C.R. 1990. Early indicators of chemical changes in foods due to enzymic or nonenzymic browning reactions. Part II: Colour changes in heat treated model systems. *Lebensmittel-Wissenschaft und-Technologie*, 23(6):494–498.

Barbosa-Cánovas, G.V., Pierson, M.D., Zhang, Q.H. and Schaffner, D.W. 2000. Pulsed electric fields. *Journal of Food Science–Special Supplement–Kinetics of Microbial Inactivation for Alternative Food Processing Technologies*, 65(4):65–77.

Barger, G., Bergel, F. and Todd, A.R. 1935. Über das Thiochrom aus Vitamin B_1 (Antineurin). *Berichte der Deutschen Chemischen Gesellschaft*, 68(12):2257–2262.

Barreiro, J.A., Milano, M. and Sandoval, A.J. 1997. Kinetics of colour change of double concentrated tomato paste during thermal treatment. *Journal of Food Engineering*, 33(3–4):359–371.

Barron, E.S.G. 1937. Studies in biological oxidations. IX. The oxidation-reduction potentials of blood hemin and its hemochromogens. *Journal of Biological Chemistry*, 121(1):285–312.

Bauernfeind, J. 1980. Tocopherols in food. ch.4, p.99–167, In: *Vitamin E — A Comprehensive Treatise*, Machlin, L.J. (ed.), Marcel Dekker, Inc., NY.

Bauernfeind, J.C. 1972. Carotenoid vitamin A precursors and analogs in foods and feeds. *Journal of Agricultural and Food Chemistry*, 20(3):456–473.

Bauernfeind, J.C. 1977. The tocopherol content of food and influencing factors. *CRC Critical Reviews in Food Science and Nutrition*, 8(4):337–382.

Bauernfeind, J.C. and Pinkert, D.M. 1970. Food processing with added ascorbic acid. *Advances in Food Research*, 18:219–315.

Baugh, C.M., May, L., Braverman, E. and Nair, M.G. 1979. Determination of the gammaglutamyl chain lengths in the folates by a combined zinc/acid peracid procedure, p. 219–224. In: *Chemistry and Biology of Pteridines*, Kisliuk, R.L. and Brown, G.M., (eds.), Elsevier North-Holland, Inc., Amsterdam.

Beadle, B.W., Greenwood, D.A. and Kraybill, H.R. 1943. Stability of thiamine to heat. I. Effect of pH and buffer salts in aqueous solutions. *Journal of Biological Chemistry*, 149(2):339–347.

Belitz, H.D. and Grosch, W. (eds.) 1987. Vegetables and their products. p. 549–576. In: *Food Chemistry*, D. Hadziyev (ed.), Springer-Verlag, New York.

Bendix, G.H., Herberlein, D.G., Ptak, L.R. and Clifcorn, L.E. 1951. Factors influencing the stability of thiamine during heat sterilization. *Food Research*, 16(6):494–503.

Benz, R. and Zimmermann, U. 1980. Pulse-length dependence of the electrical breakdown in lipid bilayer membranes. *Biochimica et Biophysica Acta*, 597(3):637–642.

Bera, M.B., Singh, C.J., Shrivastava, D.C., Kumar, K.S. and Sharma, Y.K. 2001. Storage stability of colour substance in thermally processed dry chilli powder. *Journal of Food Science*, 38(1):8–11.

Berruti, R. 1985. Vitamin K. ch. 11, p. 285–302, In: *Methods of Vitamin Assays, 4rth Edition*, Augustin, J., Klein, B.P., Becker, D. and Venugopal, P.B. (eds.), John Wiley & Sons, NY.

Bhattacharya, M. and Hanna, M.A. 1989. Kinetics of drip loss, cooking loss and color degradation in frozen ground beef during storage. *Journal of Food Engineering*, 9(2):83–96.

Bielski, B.H.J., Comstock, D.A. and Bowen, R.A. 1971. Ascorbic acid free radicals. I. Pulse radiolysis study of optical absorption and kinetic properties. *Journal of the American Chemical Society*, 93(22):5624–5629.

Böhm, H. and Rink, E. 1988. Betalains. p. 449–463. In: *Cell Culture and Somatic Cell Genetics of Plants*, Academic Press, NY.

Borkowski, T., Szymusiak, H., Gliszczyñska-Świgto, A., Rietjens, I.M.C.M. and Tyrakowska, B. 2005. Radical scavenging capacity of wine anthocyanins is strongly pH-dependent. *Journal of Agricultural and Food Chemistry*, 53(14):5526–5534.

Boyd, W. 2000. Natural colors as functional ingredients in healthy foods. *Cereal Foods World*, 45(5):221–222.

Bradshaw, M.P., Prenzler, P.D. and Scollary, G.R. 2001. Ascorbic acid-induced browning of (+)-catechin in a model wine system. *Journal of Agricultural and Food Chemistry*, 49(2):934–939.

Brenes, C.H., Del Pozo-Insfran, D. and Talcott, S.T. 2005. Stability of copigmented anthocyanins and ascorbic acid in a grape juice model system. *Journal of Agricultural and Food Chemistry*, 53(1):49–56.

Brouillard, R. 1982. Chemical structure of anthocyanins. ch. 1, p.1–40. In: *Anthocyanins as Food Colors*, Markakis, P. (ed.), Academic Press, NY.

Brouillard, R. and Delaporte, B. 1977. Chemistry of anthocyanin pigments. 2. Kinetic and thermodynamic study of proton transfer, hydration, and tautomeric reactions of malvidin 3-glucoside. *Journal of the American Chemical Society*, 99(26):8461–8468.

Brown, J.P., Davidson, G.E. and Scott, J.M. 1974. The identification of the forms of folate found in the liver, kidney and intestine of the monkey and their biosynthesis from exogenous pteroylglutamate (folic acid). *Biochimica et Biophysica Acta*, 343(1):78–88.

Budowski, P. and Bondi, A. 1960. Autoxidation of carotene and Vitamin A. Influence of fat and antioxidants. *Archives of Biochemistry and Biophysics*, 89(1):66–73.

Buera, M.P., Chirife, J., Resnik, S.L. and Lozano, R.D. 1987a. Nonenzymatic browning in liquid model systems of high water activity: Kinetics of color changes due to carmelization of various single sugars. *Journal of Food Science*, 52(4):1059–1062, 1073.

Buera, M.P., Chirife, J., Resnik, S.L. and Wetzler, G. 1987b. Nonenzymatic browning in liquid model systems of high water acitvity: Kinetics of color changes due to Maillard's reaction between different single sugars and glycine and comparison with carmelization browning. *Journal of Food Science*, 52(4):1063–1067.

Burton, H. 1963. A note on the effect of heat on the colour of goat's milk. *Journal of Dairy Research*, 30(2):217–222.

Butz, P. and Ludwig, H. 1986. Pressure inactivation of microorganisms at moderate temperatures. *Physica B&C*, 139–140B:875–877.

Cai, Y. and Corke, H. 1999. *Amaranthus* betacyanin pigments applied in model systems. *Journal of Food Science*, 64(5):869–873.

Calvi, J.P. and Francis, F.J. 1978. Stability of Concord grape (*V. labrusca*) anthocyanins in model systems. *Journal of Food Science*, 43(5):1448–1456.

Cameron, E.J., Clifcorn, L.E., Esty, J.R., Feaster, J.F., Lamb, F.C., Monroe, K.H. and Royce, R. (eds.) 1955. *Retention of Nutrients During Canning*. Bulletin. Research Laboratories. National Canners Association.

Canjura, F.L., Watkins, R.H. and Schwartz, S.J. 1999. Color improvement and metallo-chlorophyll complexes in continuous flow aseptically processed peas. *Journal of Food Science*, 64(6):987–990.

Carlez, A., Rosec, J-P., Richard, N. and Cheftel, J-C. 1993. High pressure inactivation of *Citrobacter freundii*, *Pseudomonas fluorescens* and *Listeria innocua* in inoculated minced beef muscle. *Lebensmittel-Wissenschaft und-Technologie*, 26(4):357–363.

Cemeroglu, B., Velioglu, S. and Isik, S. 1994. Degradation kinetics of anthocyanins in sour cherry juice and concentrate. *Journal of Food Science*, 59(6):1216–1218.

Cheah, P.B. and Ledward, D.A. 1995. High pressure effects on lipid oxidation. *Journal of the American Oil Chemist's Society*, 72(9):1059–1063.

Cheah, P.B. and Ledward, D.A. 1996. High pressure effects on lipid oxidation in minced pork. *Meat Science*, 43(2):123–134.

Cheah, P.B. and Ledward, D.A. 1997. Catalytic mechanism of lipid oxidation following high pressure treatment in pork fat and meat. *Journal of Food Science*, 62(6):1135–1138, 1141.

Cheftel, J-C. 1991. Applications des hautes pressions en technologie alimentaire. *Actualité des Industries Alimentaires et Agro-Alimentaires*, 108(3):141–153.

Cheftel, J-C. 1995. Review: High pressure, microbial inactivation and food preservation. *Food Science and Technology International*, 1(2–3):75–90.

Chen, B.H., Peng, H.Y. and Chen, H.E. 1995. Changes of carotenoids, color, and vitamin A contents during processing of carrot juice. *Journal of Agriculture and Food Chemistry*, 43(7):1912–1918.

Chen, B.H., Chen, T.M. and Chien, J.T. 1994. Kinetic model for studying the isomerization of α– and β–carotene during heating and illumination. *Journal of Agricultural and Food Chemistry*, 42(11):2391–2397.

Chen, S.L. and Gutmanis, F. 1968. Auto-oxidation of extractable color pigments in chili pepper with special reference to ethoxyquin treatment. *Journal of Food Science*, 33(3):274–280.

Cheng, T.S. and Eitenmiller, R.R. 1988. Effects of processing and storage on the pantothenic acid content of spinach and broccoli. *Journal of Food Processing and Preservation*, 12(2):115–123.

Chéret, R., Delbarre-Ladrat, C., DeLamballerie-Anton, M. and Verrez-Bagnis, V. 2005. High-pressure effects on the proteolytic enzymes of sea bass (*Dicentrarchus labrax* L.) fillets. *Journal of Agricultural and Food Chemistry*, 53(10):3969–3973.

Choe, E., Huang, R. and Min, D.B. 2005. Chemical reactions and stability of riboflavin in foods. *Journal of Food Science*, 70(1):R28-R36.

Chou, H. and Breene, W. 1972. Oxidative decoloration of β-carotene in low-moisture model systems. *Journal of Food Science*, 37(1):66–68.

Chow, C.K. and Draper, H.H. 1974. Oxidative stability and antioxidant activity of the tocopherols in corn and soybean oils. *International Journal of Vitamin and Nutrition Research*, 44(3):396–403.

Clydesdale, F.M., Ho, C-T., Lee, C.Y., Mondy, N.I. and Shewfelt, R.L. 1991. The effects of postharvest treatment and chemical interactions on the bioavailability of ascorbic acid, thiamin, vitamin A, carotenoids, and minerals. *Critical Reviews in Food Science and Nutrition*, 30(6):599–638.

Clydesdale, F.M., Lin, Y.D. and Francis, F.J. 1972. Formation of 2-pyrrolidone-5-caboxylic acid from glutamine during processing and storage of spinach puree. *Journal of Food Science*, 37(1):45–47.

Clydesdale, F.M., Main, J.H., Francis, F.J. and Damon, R.A., Jr. 1978. Concord grape pigments as colorants for beverages and gelatin desserts. *Journal of Food Science*, 43(6):1687–1692, 1697.

Cohen, E. and Saguy, I. 1983. Effect of water activity and moisture content on the stability of beet powder pigments. *Journal of Food Science*, 48(3):703–707.

Cole, E.R. and Kapur, N.S. 1957a. The stability of lycopene. I. Degradation by oxygen. *Journal of the Science of Food and Agriculture*, 8(6):360–365.

Cole, E.R. and Kapur, N.S. 1957b. The stability of lycopene. II. Oxidation during heating of tomato pulps. *Journal of the Science of Food and Agriculture*, 8(6):366–368.

Cornforth, D.P., Vahabzadeh, F., Carpenter, C.E. and Bartholomew, D.T. 1986. Role of reduced hemochromes in pink color defect of turkey rolls. *Journal of Food Science*, 51(5):1132–1135.

Cort, W.M., Mergens, W. and Greene, A. 1978. Stability of alpha- and gamma-tocopherol: Fe^{3+} and Cu^{2+} interactions. *Journal of Food Science*, 43(3):797–798.

Crone, A.V.J., Hamilton, J.T.G. and Stevenson, M.H. 1992. Effects of storage and cooking on the dose response of 2-dodecylcyclobutanone, a potential marker for irradiated chicken. *Journal of the Science of Food and Agriculture,* 58(2):249–252.

Csallany, A.S., Chiu, M. and Draper, H.H. 1970. Oxidation products of α-tocopherol formed in autoxidizing methyl linoleate. *Lipids*, 5(1):63–70.

Csallany, A.S. and Draper, H.H. 1963. The structure of a dimeric metabolite of d-α-tocopherol isolated from mammalian liver. *Journal of Biological Chemistry*, 238(9):2912–2918.

Daravingas, G. and Cain, R.F. 1968. Thermal degradation of black raspberry anthocyanin pigments in solution. *Journal of Food Science*, 33(1):138–142.

Day, B.P.F. and Gregory, J.F., III. 1983. Thermal stability of folic acid and 5-methyltetrahydrofolic acid in liquid model food systems. *Journal of Food Science*, 48(2):581–587,599.

Day, W.C. and Erdman, J.G. 1963. Ionene: A thermal degradation product of β-carotene. *Science*, 141(3583):808.

Debicki-Pospišil, J., Lovrić, T., Trinajstić, N. and Sabljić, A. 1983. Anthocyanin degradation in the presence of furfural and 5-hydroxymethyl furfural. *Journal of Food Science*, 48(2):411–416.

Delgado-Vargas, F., Jiménez, A.R. and Paredes-López, O. 2000. Natural pigments: carotenoids, anthocyanins, and betalains–characteristics, biosynthesis, processing, and stability. *Critical Reviews in Food Science and Nutrition*, 40(3):173–289.

Dennison, D.B. and Kirk, J.R. 1978. Oxygen effect on the degradation of ascorbic acid in a dehydrated food system. *Journal of Food Science*, 43(2):609–612, 618.

Dennison, D., Kirk, J., Bach, J., Kokoczka, P. and Heldman, D. 1977. Storage stability of thiamin and riboflavin in a dehydrated food system. *Journal of Food Processing and Preservation*, 1(1):43–54.

De Ritter, E. 1976. Stability characteristics of vitamins in processed foods. *Food Technology*, 30(1): 48–51, 54.

Deuchi, T. and Hayashi, R. 1991. Pressure-application to thawing of frozen foods and to food preservation under sub-zero temperature. p. 101–110, In: *High Pressure Science for Food*, Hayashi, R. (ed.), San-ei Pub. Co., Kyoto, Japan.

Diehl, J.F. 1995. Nutritional adequacy of irradiated foods. In: *Safety of Irradiated Foods*, 2nd ed., Marcel Dekker, Inc., NY.

Dietrich, W.C. and Neumann, H.J. 1965. Blanching Brussels sprouts. *Food Technology*, 19(7):1174–1177.

Dietrich, W.C., Olson, R.L., Nutting, M.-D., Neumann, H.J. and Boggs, M.M. 1959. Time-temperature tolerance of frozen foods. XVIII. Effect of blanching conditions on color stability of frozen beans. *Food Technology*, 13(5):258–261.

DiMascio, P., Kaiser, S. and Sies, H. 1989. Lycopene as the most efficient biological carotenoid singlet oxygen quencher. *Archives of Biochemistry and Biophysics*, 274(2):532–538.

Diplock, A.T. 1994. Antioxidants and disease prevention. *Molecular Aspects of Medicine*, 15:293–376.

Dobson, D.C. 1970. Biotin requirement of turkey poults. *Poultry Science*, 49(2):546–553.

Doherty, R.F. and Beecher, G.R. 2003. A method for the analysis of natural and synthetic folate in foods. *Journal of Agricultural and Food Chemistry*, 51(2):354–361.

Drabkin, D.L. and Austin, J.H. 1935. Spectrophotometric studies. IV. Hemochromogens. *Journal of Biological Chemistry*, 112(1):89–104.

Dürckheimer, W. and Cohen, L.A. 1962. Mechanisms of α-tocopherol oxidation: Synthesis of highly labile 9-hydroxy-α-tocopherone. *Biochemistry and Biophyics Research Communications*, 9(3):262–265.

Dymicky, M., Fox, J.B. and Wasserman, A.E. 1975. Color formation in cooked model and meat systems with organic and inorganic compounds. *Journal of Food Science*, 40(2):306–309.

Dwivedi, B.K. and Arnold, R.G. 1973. Chemistry of thiamine degradation in food products and model systems: A review. *Journal of Agricultural and Food Chemistry*, 21(1):54–60.

Dwivedi, B.K. and Arnold, R.G. 1972a. Chemistry of thiamine degradation. Mechanisms of thiamine degradation in a model system. *Journal of Food Science*, 37(6):886–888.

Dwivedi, B.K. and Arnold, R.G. 1972b. Gas chromatographic estimation of thiamine. *Journal of Food Science*, 37(6):889–891.

Eison-Perchonok, M.H. and Downes, T.W. 1982. Kinetics of ascorbic acid and autoxidation as a function of dissolved oxygen concentration and temperature. *Journal of Food Science*, 47(3):765–767, 773.

Eshtiaghi, M.N. and Knorr, D. 1993. Potato cubes response to water blanching and high hydrostatic pressure. *Journal of Food Science*, 58(6):1371–1374.

Evans, S.R., Gregory, J.F. III and Kirk, J.R., 1981. Thermal degradation kinetics of pyridoxine hydrochloride in dehydrated model food systems. *Journal of Food Science*, 46(2):555–558, 563.

Farkas, D.F. and Hoover, D.G. 2000. High pressure processing. *Journal of Food Science–Special Supplement–Kinetics of Microbial Inactivation for Alternative Food Processing Technologies*, 65(4):47–64.

Farkas, J. 1989. Microbiological safety of irradiated foods. *Review. International Journal of Food Microbiology*, 9(1):1–15.

Farquharson, J. and Adams, J.F. 1976. The forms of vitamin B_{12} in foods. *British Journal of Nutrition*, 36(1):127–136.

Farr, D. 1990. High pressure technology in the food industry. *Trends in Food Science and Technology*, 1(1):14–16.

Farrer, K.T.H. and Morrison, P.G. 1949. The thermal destruction of vitamin B_1. 6. The effect of temperature and oxygen on the rate of destruction of aneurin. *The Australian Journal of Experimental Biology and Medical Science*, 27(5):517–522.

Farrer, K.T. 1955. The thermal destruction of vitamin B_1 in foods. *Advances in Food Research*, 6:257–312.

Federal Register. 1986. Canned green beans deviating from identity standard; extension and amendment of temporary permit for market testing. *Fed. Regist.*, 51 (March 13), 49.

Feliciotti, E. and Esselen. 1957. Thermal destruction rates of thiamine in pureed meats and vegetables. *Food Technology*, 11(2):77–84.

Fernández, B., Mauri, L.M., Resnik, S.L. and Tomio, J.M. 1986. Effect of adjusting the water activity to 0.95 with different solutes on the kinetics of thiamin loss in a model system. *Journal of Food Science*, 51(4):1100–1101.

Ferriera, V.L.P., Teixeira Neto, R.O. and Moura, S.C. and Silva, M.S. 1999. Kinetics of color degradation of water-soluble commercial annatto solutions under thermal treatments. (Cinética da degradação da cor de solução hidrossolúvel comercial de urucum, submetida a tratamentos térmicos.) *Ciência e Tecnologia de Alimentos*, 19(1):37–42.

Finholt, P., Paulssen, R.B. and Higuchi, T. 1963. Rate of anaerobic degradation of ascorbic acid in aqueous solution. *Journal of Pharmaceutical Sciences*, 52(10):948–954.

Fink, R. and Kessler, H.G. 1985. Reaction kinetics study of thiamine losses in stored UHT milk. *Milchwissenschaft*, 40(12):709–712.

Finkel'shtein, E.I., Alekseev, É. V. and Kozlov, É. I. 1973. The kinetics of β-carotene solid film autooxidation. *Doklady Akademii Nauk SSSR.*, 208(6):1408–1411.

Finkel'shtein, E.I., Alekseev, É. V. and Kozlov, É. I. 1974. Kinetic relationships of the solid state autooxidation of β-carotene. *Zhurnal Organicheskoi Khimii.*, 10(5):1027–1034.

Fox, J.B. Jr. 1966. The chemistry of meat pigments. *Journal of Agricultural and Food Chemistry*, 14(3):207–210.

Fox, J.B. Jr. and Ackerman, S.A. 1968. Formation of nitric oxide myoglobin: Mechanisms of the reaction with various reductants. *Journal of Food Science*, 33(4):364–370.

Fox, Jr., J.B., Lakritz, L., Hampson, J., Richardson, R., Ward, K. and Thayer, D.W. 1995. Gamma irradiation effects on thiamin and riboflavin in beef, lamb, pork, and turkey. *Journal of Food Science*, 60(3):596–603.

Fox, Jr., J.B., Lakritz, L. and Thayer, D.W. 1997. Thiamin, riboflavin and α–tocopherol retention in processed and stored irradiated pork. *Journal of Food Science*, 62(5):1022–1025.

Fox, J.B., Townsend, W.E., Ackerman, S.A. and Swift, C.E. 1967. Cured color development during frankfurter processing. *Food Technology*, 21(3A):386–392.

Frampton, V.L., Skinner, Jr., W.A., Bailey, P.S. 1954. The product of tocored upon the oxidation of *dl-α*-tocopherol with ferric chloride. *Journal of the American Chemical Society*, 76(1):282–284.

Frampton, V.L., Skinner, W.A., Cambour, P. and Bailey, P.S. 1960. α-Tocopurple, an oxidation product of α-tocopherol. *Journal of the American Chemical Society*, 82(16):4632–4634.

Francis, F.J. 2000a. Anthocyanins and betalains: composition and applications. *Cereal Foods World*, 45(5):208–213.

Francis, F.J. 2000b. Carotenoids as food colorants. *Cereal Foods World,* 45(5):198–203.

Francis, F.J. 1989. Food colorants: anthocyanins. *Critical Reviews in Food Science and Nutrition*, 28(4):273–317.

Francis, F.J. 1985. Pigments and other colorants. ch. 8, p.545–584. In: *Food Chemistry*, 2nd ed., Fennema, O.R. (ed.), Marcel Dekker, Inc., NY.

Francis, F.J. 1977. Anthocyanins. p. 19–27. In: *Current Aspects of Food Colorants*, Furia, T.E. (ed.), CRC Press, Boca Raton, FL.

Frias, J. and Vidal-Valverde, C. 2001. Stability of thiamine and vitamins E and A during storage of enteral feeding formula. *Journal of Agricultural and Food Chemistry*, 49(5):2313–2317.

Frost, D.V. 1943. Pantothenic acid. Optical rotation as a measure of stability. *Industrial and Engineering Chemistry. Analytical Edition*. Washington, D.C., 15(5):306–310.

Frost, D.V. and McIntire, F.C. 1944. The hydrolysis of pantothenate: A first-order reaction. Relation to thiamin stability. *Journal of the American Chemical Society*, 66(3):425–427.

Furuya, E.M., Warthesen, J.J. and Labuza, T.P. 1984. Effects of water activity, light intensity, and physical structure of food on the kinetics of riboflavin photodegradation. *Journal of Food Science*, 49(2):525–528.

Gagarina, A.B., Kasaikina, O.T. and Émanuél, N.M. 1970. Kinetics of autooxidation of polyene hydrocarbons in aromatic solvents. *Doklady Akademii Nauk SSSR.*, 195(2):387–390.

Galdi, M., Carbone, N. and Valencia, M.E. 1989. Comparison of ferric glycinate to ferrous sulfate in model infant formulas: Kinetics of vitamin losses. *Journal of Food Science*, 54(6):1530–1533, 1539.

Gami, D.B. and Chen, T.S. 1985. Kinetics of folacin destruction in Swiss chard during storage. *Journal of Food Science*, 50(2):447–449, 453.

Giannakourou, M.C. and Taoukis, P.S. 2003. Kinetic modelling of vitamin C loss in frozen green vegetables under variable storage conditions. *Food Chemistry*, 83(1):33–41.

Giese, J. 1995. Vitamin and mineral fortification of foods. *Food Technology*, 49(5):109–122.

Giroux, M., Yefsah, R., Smoragiewicz, W., Saucier, L. and Lacroix, M. 2001. *Journal of Agricultural and Food Chemistry*, 49(2):919–925.

Garrett, E.R. 1956. Prediction of stability in pharmaceutical preparations. II. Vitamin stability in liquid multivitamin preparations. *Journal of Pharmaceutical Sciences*, 45(3):171–178.

Gold, H.J. and Weckel, K.G. 1959. Degradation of chlorophyll to pheophytin during sterilization of canned green peas by heat. *Food Technology*, 13(5):281–286.

Goldman, M., Horev, B. and Saguy, I. 1983. Decolorization of β-carotene in model systems simulating dehydrated foods. Mechanism and kinetic principles. *Journal of Food Science*, 48(3):751–754.

Goodhue, C.T. and Risley, H.A. 1965. Reactions of vitamin E with peroxides. II. Reaction of benzoyl peroxide with d-α-tocopherol in alcohols. *Biochemistry*, 4(5):854–858.

Grant, B.F., Seib, P.A., Liao, M-L. and Corpron, K.E. 1989. Polyphosphorylated ascorbic acid: a stable form of vitamin C for aquaculture feeds. *Journal of the World Aquaculture Society*, 20(3):143–157.

Grant, N.H. and Alburn, H.E. 1965a. Fast reactions of ascorbic acid and hydrogen peroxide in ice, a presumptive early environment. *Science*, 150(3703):1589–1590.

Grant, N.H. and Alburn, H.E. 1965b. Transfer reactions in ice. Inhibition of nonenzymatic hydroxylaminolysis of amino acid esters by structural analogs. *Biochemistry*, 4(10):1913–1916.

Greenwood, D.A., Beadle, B.W. and Kraybill, H.R. 1943. Stability of thiamine to heat. II. Effect of meat-curing ingredients in aqueous solutions and in meat. *Journal of Biological Chemistry*, 149(2):349–354.

Greenwood, D.A., Kraybill, H.R., Feaster, J.F. and Jackson, J.M. 1944. Vitamin retention in processed meat. *Industrial and Engineering Chemistry. Industrial Edition*, 36(10):922–927.

Gregory, J.F. III and Hiner, M.E. 1983. Thermal stability of vitamin B_6 compounds in liquid model food systems. *Journal of Food Science*, 48(4):1323–1327, 1339.

Gregory, J.F. and Kirk, J.R. 1978. Assessment of storage effects on vitamin B_6 stability and bioavailability in dehydrated food systems. *Journal of Food Science*, 43(6):1801–1808, 1815.

Griffin, D.B., Savell, J.W., Smith, G.C., Vanderzant, C., Terrell, R.N., Lind, K.D. and Galloway, D.E. 1982. Centralized packaging of beef loin steaks with different oxygen barrier films: Physical and sensory characteristics. *Journal of Food Science*, 47(4):1059–1069.

Griffiths, J.C. 2005. Coloring foods and beverages. *Food Technology*, 59(5):38–44.

Gunawan, M.I. and Barringer, S.A. 2000. Green color degradation of blanched broccoli *(Brassica oleracea)* due to acid and microbial growth. *Journal of Food Processing and Preservation*, 24(3):253–263.

Gupte, S.M., El-Bisi, H.M. and Francis, F.J. 1964. Kinetics of thermal degradation of chlorophyll in spinach puree. *Journal of Food Science*, 29(4):379–382.

Gyorgy, P. and Langer, B. 1968. Biotin–Chemistry. p. 263. In: *The Vitamins, Vol. 2*, Sebrell, W.H. and Harris, R.S. (eds.), Academic Press, New York.

Hamm, D.J. and Lund, D.B. 1978. Kinetic parameters for thermal inactivation of pantothenic acid. *Journal of Food Science*, 43(2):631–633.

Hanis, T., Jelen, P., Klir, P., Mnuková, J., Perez, B. and Pesek, M. 1989. Poultry meat irradiation–effect of temperature on chemical changes and inactivation of microorganisms. *Journal of Food Protection*, 52(1):26–29.

Haralampu, S.G. and Karel, M. 1983. Kinetic models for moisture dependence of ascorbic acid and β-carotene degradation in dehydrated sweet potato. *Journal of Food Science*, 48(6): 1872–1873.

Harborne, J.B. and Grayer, R.J. 1988. The anthocyanins. p. 1–20, In: *The Flavonoids*, Harborne, J.B. (ed.), Chapman and Hall, Ltd., London.

Harper, K.A. 1967. Structural changes of flavylium salts. III. Polarographic and spectrometric examination of 3,7,4'-trihydroxyflavylium perchlorate. *Australian Journal of Chemistry*, 20(12): 2691–2700.

Harper, K.A. 1968. Structural changes of flavylium salts. IV. Polarographic and spectrometric examination of pelargonidin chloride. *Australian Journal of Chemistry*, 21(1):221–227.

Harris, R.S. and von Loesecke, H. (eds.) 1960. *Nutritional Evaluation of Food Processing*. John Wiley and Sons, Inc., NY.

Harris, R.S. and Karmas, E. (eds.) 1975. *Nutritional Evaluation of Food Processing*. 2nd. ed. AVI Publishing Co., Inc., Westport, CT.

Hashizume, C., Kimura, K. and Hayashi, R. 1995. Kinetic analysis of yeast inactivation by high pressure treatment at low temperatures. *Bioscience, Biotechnology and Biochemistry*, 59(8): 1455–1458.

Haugaard, G., Tumerman, L. and Silvestri, H. 1951. A study on the reaction of aldoses and amino acids. *Journal of the American Chemical Society*, 73(10):4594–4600.

Hawkes, J.G. and Villota, R. 1986. Kinetics of folate degradation during food processing. ch. 30, p. 323–333. In: *Food Engineering and Process Applications, Volume I–Transport Phenomena*. LeMaguer, M. and Jelen, P. (eds.), Elsevier Applied Science, NY.

Hawkes, J.G. 1988. *Kinetics of Folate Degradation*. M.S. Thesis. University of Illinois, Urbana, IL.

Hawkes, J.G. and Villota, R. 1989a. Folates in foods: Reactivity, stability during processing, and nutritional implications. *CRC Critical Reviews in Food Science and Nutrition*, 28(6):439–538.

Hawkes, J.G. and Villota, R. 1989b. Prediction of folic acid retention during spray dehydration. *Journal of Food Engineering*, 10(4):287–317.

Hayashi, R. 1989. Application of high pressure to food processing and preservation: philosophy and development, p.815–826 In: *Engineering and Food, Vol.2*, Spiess, W.E.L. and Schubert, H. (eds.), Elsevier Applied Science, London.

Heaton, J.W., Lencki, R.W. and Marangoni, A.G. 1996a. Kinetic model for chlorophyll degradation in green tissue. *Journal of Agricultural and Food Chemistry*, 44(2):399–402.

Heaton, J.W., Yada, R.Y. and Marangoni, A.G. 1996b. Discoloration of coleslaw is caused by chlorophyll degradation. *Journal of Agricultural and Food Chemistry*, 44(2):395–398.

Hendel, C.E., Silveira, V.G. and Harrington, W.O. 1955. Rates of non-enzymatic browning of white potato during dehydration. *Food Technology*, 9(9):433–438.

Herbert, V. and Jacob, E. 1974. Destruction of vitamin B$_{12}$ by ascorbic acid. *Journal of the American Medical Association*, 230(2):241–242.

Heremans, K. 1982. High pressure effects on proteins and other biomolecules. *Annual Review of Biophysics and Bioengineering*, 11(1):1–21.

Heremans, K., Meersman, F., Rubens, P., Smeller, L., Snauwaert, J. and Vermeulen, G. 1999. A comparison between pressure and temperature effects on food constituents. ch. 15, p. 269–280, In: *Processing of Foods–Quality Optimization and Process Assessment*, Oliveira, F.A.R. and Oliveira, J.C. (eds.), CRC Press, Boca Raton, FL.

Herrmann, J. 1970. Calculation of the chemical and sensory alterations in food during heating and storage processes. *Ernaehrungsforschung*, 15(4):279–299.

Herrmann, J. and Nour, S. 1977. Modelluntersuchungen zur Bildung von Carbonylverbindungen und Bräunung durch die Maillard-Reaktion beim Backprozeß. *Die Nahrung*, 21(4):319–330.

Himelbloom, B.H., Rugledge, J.E. and Biede, S.L. 1983. Color changes in blue crabs (*Callinectes sapidus*) during cooking. *Journal of Food Science*, 48(2):652–653.

Holmström, B. and Oster, G. 1961. Riboflavin as an electron donor in photochemical reactions. *Journal of the American Chemical Society*, 83(8):1867–1871.

Hoover, D.G., Metrick, C., Papineau, A.M., Farkas, D.F. and Knorr, D. 1989. Biological effects of high hydrostatic pressure on food microorganisms. *Food Technology*, 43(3):99–107.

Hoppner, K. and Lampi, B. 1993. Pantothenic acid and biotin retention in cooked legumes. *Journal of Food Science*, 58(5):1084–1085, 1089.

Horie, Y., Kimura, K., Ida, M., Yoshida, Y. and Ohki, K. 1991. Jam preparation by pressurization. *Nippon Nogeikagaku Kaishi*, 65(6):975–980.

Hornig, D., Weber, F. and Wiss, O. 1974. Influence of erythorbic acid on vitamin C status in guinea pigs. *Experientia*, 30(2):173–174.

Houben, J.H., VanDijk, A., Eikelenboom, G. and Hoving-Bolink, A.H. 2000. Effect of dietary vitamin E supplementation, fat level and packaging on colour stability and lipid oxidation in minced beef. *Meat Science*, 55(3):331–336.

Hrazdina, G. 1971. Reactions of the anthocyanidin-3,5-diglucosides: Formation of 3,5-di-(O-β-D-glucosyl)-7-hydroxy coumarin. *Phytochemistry*, 10(5):1125–1130.

Hrazdina, G. 1981. Anthocyanins and their role in food products. *Lebensmittel-Wissenschaft und -Technologie*, 14(6):283–286.

Huang, A.S. and von Elbe, J.H. 1985. Kinetics of the degradation and regeneration of betanine, *Journal of Food Science*, 50(4):1115–1120, 1129.

Huang, A.S. and von Elbe, J.H. 1987. Effect of pH on the degradation and regeneration of betanine, *Journal of Food Science*, 52(6):1689–1693.

Huelin, F.E. 1953. Studies on the anaerobic decomposition of ascorbic acid. *Food Research*, 18(6):633–639.

Huelin, F.E., Coggiola, I.M., Sidhu, G.S. and Kennett, B.H. 1971. The anaerobic decomposition of ascorbic acid in the pH range of foods and in more acid solutions. *Journal of the Science of Food and Agriculture*, 22(10):540–542.

Huffman, D.L. 1980. Processing effects on fresh and frozen meat color. *Proceedings. 33rd Annual Reciprocal Meat Conference of the American Meat Science Association*, Perdue University, W. Lafayette, IN, Jun 22–25, 1980, 33:4–14.

Hülsheger, H. and Nieman, E.G. 1980. Lethal effects of high-voltage pulses on *E. coli* K12. *Radiation and Environmental Biophysics*, 18(4):281–288.

Hülsheger, H., Pottel, J. and Nieman, E.G. 1981. Killing of bacteria with electric pulses of high field strength. *Radiation and Environmental Biophysics*, 20(1):53–65.

Hutchings, B.L., Stokstad, E.L.R., Mowat, J.H., Boothe, J.H., Waller, C.W., Angier, R.B., Semb, J. and SubbaRow, Y. 1948. Degradation of the fermentation *L. casei* factor. II., *Journal of the American Chemical Society*, 70(1):10–13.

Ibarz, A., Pagán, J. and Garza, S. 1999. Kinetic models for colour changes in pear puree during heating at relatively high temperatures. *Journal of Food Engineering*, 39(4):415–422.

Ilo, S. and Berghofer, E. 1998. Kinetics of thermomechanical destruction of thiamin during extrusion cooking. *Journal of Food Science*, 63(2):312–316.

Ilo, S. and Berghofer, E. 1999. Kinetics of colour changes during extrusion of maize grits. *Journal of Food Engineering*, 39(1):73–80.

Ilo, S. and Berghofer, E. 2003. Kinetics of lysine and other amino acids loss during extrusion cooking of maize grits. *Journal of Food Science*, 68(2):496–502.

Indrawati, A.M., VanLoey, L.R., Ludikhuyze, L.R. and Hendrickx, M.E. 2001. Pressure-temperature inactivation of lipoxygenase in green peas (*Pisum sativum*): a kinetic study. *Journal of Food Science*, 66(5):686–693.

Indyk, H. 1988. Liquid chromatographic study of vitamin K1 degradation: possible nutritional implications in milk. *Milchwissenschaft*, 43(8):503–506.

Inglett, G.E. and Mattill, H.A. 1955. Oxidation of hindered 6-hydroxychromans. *Journal of the American Chemical Society*, 77(24):6552–6554.

Ioncheva, N. and Tanchev, S. 1974. Kinetics of thermal degradation of some anthocyanidin-3,5-diglucosides. *Zeitschrift für Lebensmittel-Untersuchung und-Forschung*, 155(5):257–262.

Ishiwatari, M. 1980. Thermal reaction of β-carotene. Part I. *Journal of Analytical and Applied Pyrolysis*, 2(2):153–167.

Jensen, A. 1969. Tocopherol content of seaweed and seaweed meal. III. Influence of processing and storage on the content of tocopherols, carotenoids and ascorbic acid in seaweed meal. *Journal of the Science of Food and Agriculture*, 20(10):622–626.

John, W., Dietzel, E. and Emte, W. 1939. Über einige Oxydationsprodukte der Tokopherole und analoger einfacher Modellkörper. 6. Mitteilung über Antisterilitätsfaktoren (Vitamin E). *Hoppe-Seyler's Zeitschrift für Physiologische Chemie*, 257(5/6):173–189.

John, W. and Emte, W. 1941. Über einige neue Oxydationsprodukte der Tokopherole. 8. Mitteilung über Antisterilitätsfaktoren (Vitamin E). *Hoppe-Seyler's Zeitschrift für Physiologische Chemie*, 268(3/4): 85–103.

Jones, I.D., White, R.C. and Gibbs, E. 1962. Some pigment changes in cucumbers during brining and brine storage. *Food Technology*, 16(3):96–102.

Jones, I.D., White, R.C. and Gibbs, E. 1963. Influence of blanching or brining treatments on the formation of chlorophyllides, pheophytins and pheophorbides in green plant tissue. *Journal of Food Science*, 28(4):437–439.

Jones, I.D., White, R.C., Gibbs, E., Butler, L.S. and Nelson, L.A. 1977. Experimental formation of zinc and copper complexes of chlorophyll derivatives in vegetable tissue by thermal processing. *Journal of Agricultural and Food Chemistry*, 25(1):149–153.

Jurd, L. 1964. Reactions involved in sulfite bleaching of anthocyanins. *Journal of Food Science*, 29(1):16–19.

Jurd, L. and Asen, S. 1966. The formation of metal and "co-pigment" complexes of cyanidin 3-glucoside. *Phytochemistry*, 5(6):1263–1271.

Jurd, L. 1969. Review of polyphenol condensation reactions and their possible occurrence in the aging of wine. *American Journal of Enology and Viticulture*, 20(3):191–195.

Jurd, L. 1972. Some advances in the chemistry of anthocyanin-type pigments. p. 123–142, In: *The Chemistry of Plant Pigments*, Chichester, C.O. (ed.), Academic Press, NY.

Kähkönen, M.P. and Heinonen, M. 2003. Antioxidant activity of anthocyanins and their aglycons. *Journal of Agricultural and Food Chemistry*, 51(3):628–633.

Kallio, H., Pallasaho, S., Kärppä, J. and Linko, R.R. 1986. Comparison of the half-lives of the anthocyanins in the juice of crowberry, *Empetrum nigrum. Journal of Food Science*, 51(2):408–410, 430.

Kamarei, A.R., Karel, M. and Wierbicki, E. 1979. Spectral studies on the role of ionizing radiation in color changes of radappertized beef. *Journal of Food Science*, 44(1):25–32.

Kamarei, A.R., Karel, M. and Wierbicki, E. 1981. Color stability of radappertized cured meat. *Journal of Food Science*, 46(1):37–40.

Kamman, J.F., Labuza, T.P. and Warthesen, J.J. 1981. Kinetics of thiamin and riboflavin loss in pasta as a function of constant and variable storage conditions. *Journal of Food Science*, 46(5):1457–1461.

Kanner, J., Harel, S. and Granit, R. 2001. Betalains–A new class of dietary cationized antioxidants. *Journal of Agricultural and Food Science*, 49(11):5178–5185.

Kanner, J., Mendel, H. and Budowski, P. 1978. Carotene oxidizing factors in red pepper fruits (*Capsicum annuum L.*): Oleoresin-cellulose solid model. *Journal of Food Science*, 43(3):709–712.

Kapadia, G.J., Azuine, M.A., Sridhar, R., Okuda, Y., Tsuruta, A., Ichiishi, E., Mukainake, T., Takasaki, M., Konoshima, T., Nishino, H. and Tokuda, H. 2003. Chemoprevention of DMBA-induced UV-B promoted, NOR-1-induced TPA promoted skin carcinogenesis, and DEN-induced phenobarbitol promoted liver tumors in mice by extract of beetroot. *Pharmacological Research*, 47(2):141–148.

Karel, M. 1973. Symposium: Protein interactions in biosystems. Protein-lipid interactions. *Journal of Food Science*, 38(5):756–763.

Karel, M. 1975. Radiation preservation of foods. ch. 4, p. 93–130, In: *Principles of Food Science. Part II. Physical Principles of Food Preservation*, Fennema, O.W. (ed.), Marcel Dekker, NY.

Karrer, P., Salomon, H., Schöpp, K., Schlittler, E. and Fritzsche, H. 1934. Ein neues Bestrahlungsprodukt des Lactoflavins: Lumichrom. *Helvetica Chimica Acta*, 17:1010–1013.

Kawasaki, C. and Daira, I. 1963. Decomposition of thiamine derivatives by ultraviolet degradation. *Journal of Nutritional Science and Vitaminology*, 9:264–268.

Kearsley, M.W. and Katasaboxakis, K.Z. 1980. Stability and use of natural colours in foods (Red beet powder, copper chlorophyl powder, and cochineal). *Journal of Food Technology*, 15(5):501–514.

Khan, M.M.T. and Martell, A.E. 1967a. Metal ion and metal chelate catalyzed oxidation of ascorbic acid by molecular oxygen. I. Cupric and ferric ion catalyzed oxidation. *Journal of the American Chemical Society*, 89(16):4176–4185.

Khan, M.M.T. and Martell, A.E. 1967b. Metal ion and metal chelate catalyzed oxidation of ascorbic acid by molecular oxygen. II. Cupric and ferric chelate catalyzed oxidation. *Journal of the American Chemical Society*, 89(26):7104–7111.

Khan, M.M.T. and Martell, A.E. 1968. Kinetics of metal ion and metal chelate catalyzed oxidation of ascorbic acid. III. Vanadyl ion catalyzed oxidation. *Journal of the American Chemical Society*, 90(22):6011–6017.

Kilcast, D. 1994. Effect of irradiation on vitamins. *Food Chemistry*, 49(2):157–164.

Killeit, U. 1994. Vitamin retention in extrusion. *Food Chemistry*, 49(2):149–155.

Kim, H-J., Choi, Y-M., Yang, T.C.S., Taub, I.A., Tempest, P., Skudder, P., Tucker, G. and Parrott, D.L. 1996. Validation of ohmic heating for quality enhancement of food products. *Food Technology*, 50(5):253–261.

Kim, Y-S., Strand, E., Dickmann, R. and Warthesen, J. 2000. Degradation of vitamin A palmitate in corn flakes during storage. *Journal of Food Science*, 65(7):1216–1219.

Kirk, J., Dennison, D., Kokoczka, P. and Heldman, D. 1977. Degradation of ascorbic acid in a dehydrated food system. *Journal of Food Science*, 42(5):1274–1279.

Knapp, F.W. and Tappel, A.L. 1961. Some effects of γ-radiation or linoleate peroxidation on α-tocopherol. *Journal of the American Oil Chemists' Society*, 38:151–156.

Knorr, D. 1993. Effects of high-hydrostatic pressure process on food safety and quality. *Food Technology*, 47(6):156–161.

Knorr, D., Geulen, M., Grahl, T. and Sitzmann, W. 1994. Food application of high electric field pulses. *Trends in Food Science and Technology*, 5(3):71–75.

Koohmaraie, M., Kennick, W.H., Elgasim, E.A., Dickson, R.L. and Sandine, W.E. 1983. Effect of previgor pressurization on the retail characteristics of beef. *Journal of Food Science*, 48(3):998–999.

Kopelman, I.J. and Saguy, I. 1977. Color stability of beet powders. *Journal of Food Processing and Preservation*, 1(3):217–224.

Kropf, D.H. 1980. Effects of retail display conditions on meat color. *Proceedings. 33rd Annual Reciprocal Meat Conference of the American Meat Science Association*, Perdue University, W. Lafayette, IN, Jun 22–25, 1980, 33:15–32.

Kropf, D.H., Hunt, M.C., Castner, C.L. and Luchsinger, S.E. 1995. Palatability, color, and shelf-life of low-dose irradiated beef. *Proceedings of 1995 International Congress of Meat Science and Technology*, San Antonio, TX.

Kuhn, R., Rudy, H. and Wagner-Jauregg, T. 1933. Über lacto-flavin (Vitamin B$_2$). *Berichte der Deutschen Chemischen Gesellschaft*, 66(12):1950–1956.

LaBorde, L.F. and Von Elbe, J.H. 1994. Chlorophyll degradation and zinc complex formation with chlorophyll derivatives in heated green vegetables. *Journal of Agricultural and Food Chemistry*, 42(5):1100–1103.

Labuza, T.P. 1970. Properties of water as related to the keeping quality of foods. *Proceedings SOS/70 Third International Congress Food Science and Technology*, p. 618–635.

Labuza, T.P. 1972. Nutrient losses during drying and storage of dehydrated foods. *CRC Critical Review in Food Technology*, 3(2):217–240.

Labuza, T.P. and Saltmarch, M. 1981. Kinetics of browning and protein quality loss in whey powders during steady state and nonsteady state storage conditions. *Journal of Food Science*, 47(1):92–96, 113.

Labuza, T.P. and Riboh, D. 1982. Theory and application of Arrhenius kinetics to the prediction of nutrient losses in foods. *Food Technology*, 36(10):66–74.

Laing, B.M., Schlueter, D.L. and Labuza, T.P. 1978. Degradation kinetics of ascorbic acid at high temperature and water activity. *Journal of Food Science*, 43(5):1440–1443.

Lajolo, F.M. and Lanfer Marquez, U.M. 1982. Chlorophyll degradation in a spinach system at low and intermediate water activities. *Journal of Food Science*, 47(6):1995–1998, 2003.

Lambert, A.D., Smith, J.P., and Dodds, K.L. 1992. Physical, chemical and sensory changes in irradiated fresh pork packaged in modified atmosphere. *Journal of Food Science*, 57(6):1294–1299.

Lanari, M.C., Cassens, R.G., Schaefer, D.M. and Scheller, K.K. 1993. Dietary vitamin E enhances color and display life of frozen beef from Holstein steers. *Journal of Food Science*, 58(4):701–704.

Lanari, M.C., Cassens, R.G., Schaefer, D.M. and Scheller, K.K. 1994. Effect of dietary vitamin E on pigment and lipid stability of frozen beef: A kinetic analysis. *Meat Science*, 38(1):3–15.

Larkin, J.W. and Spinak, S.H. 1996. Regulatory aspects of new / novel technologies. p.86, In: *New Processing Technologies Yearbook*, Chandarana, D.I. (ed.), National Food Processors Assoc. (NFPA), Washington, D.C.

Lashley, D. and Wiley, R.C. 1979. A betacyanine decolorizing enzyme found in red beet tissue. *Journal of Food Science*, 44(5):1568–1569.

Lathrop, P.J. and Leung, H.K. 1980. Rates of ascorbic acid degradation during thermal processing of canned peas. *Journal of Food Science*, 45(1):152–153.

Lau, M.H., Tang, J. and Swanson, B.G. 2000. Kinetics of textural and color changes in green asparagus during thermal treatments. *Journal of Food Engineering*, 45(4):231–236.

Lavelli, V. and Giovanelli, G. 2003. Evaluation of heat and oxidative damage during storage of processed tomato products. II. Study of oxidative damage indices. *Journal of the Science of Food and Agriculture*, 83(9):966–971.

Lebepe, N., Molins, R.A., Charoen, S.P., Iv, H.F. and Showronski, R.P. 1990. Changes in microflora and other characteristics of vacuum-packaged pork loins irradiated at 3.0 kGy. *Journal of Food Science*, 55(4):918–924.

Lee, Y.C., Kirk, J.R., Bedford, C.L. and Heldman, D.R. 1977. Kinetics and computer simulation of ascorbic acid stability of tomato juice as functions of temperature, pH and metal catalyst. *Journal of Food Science*, 42(3):640–644, 648.

Lee, Y.N. and Wiley, R.C. 1981. Betalaine yield from a continuous solid-liquid extraction system as influenced by raw product, post-harvest and processing variables. *Journal of Food Science*, 46(2):421–424.

Leizerson, S. and Shimoni, E. 2005. Effect of ultrahigh-temperature continuous ohmic heating treatment on fresh orange juice. *Journal of Agricultural and Food Chemistry*, 53(9):3519–3524.

Lenz, M.K. and Lund, D.B. 1977a. The Lethality-Fourier number method: Experimental verification of a model for calculating temperature profiles and lethality in conduction-heating, canned foods. *Journal of Food Science*, 42(4):989–996, 1001.

Lenz, M.K. and Lund, D.B. 1977b. The Lethality-Fourier number method: Experimental verification of a model for calculating average quality factor retention in conduction-heating canned foods. *Journal of Food Science*, 42(4):997–1001.

Levy, J., Bosin, E., Feldman, B., Giat, Y., Miinster, A., Danilenko, M. and Sharoni, Y. 1995. Lycopene is a more potent inhibitor of human cancer cell proliferation than either α-carotene or β-carotene. *Nutrition and Cancer*, 24(3):257–266.

Lhoest, W. 1958. Application of chromatography to the study of incompatibilities of thiamine. The Maillard reaction. *J. Pharm. Belg*, 13:519–533. *Chemical Abstracts*, 53:15144g.

Li, T-L. and Min, D.B. 1998. Stability and photochemistry of vitamin D_2 in model system. *Journal of Food Science*, 63(3):413–417.

Li, T-L., King, J.M. and Min, D.B. 2000. Quenching mechanisms and kinetics of carotenoids in riboflavin photosensitized singlet oxygen oxidation of vitamin D_2. *Journal of Food Biochemistry*, 24(6):477–492.

Liao, M.-L. and Seib, P.A. 1988. Chemistry of *L*-ascorbic acid related to foods. *Food Chemistry*, 30(4): 289–312.

Liao, T., Wu, S-B.J., Wu, M-C. and Chang, H-M. 2000. Epimeric separation of *L*-ascorbic acid and *D*-isoascorbic acid by capillary zone electrophoresis. *Journal of Agricultural and Food Chemistry*, 48(1):37–41.

Lin, C.H. and Chen, B.H. 2005. Stability of carotenoids in tomato juice during storage. *Food Chemistry*, 90(4):837–846.

Lin, Y.D., Clydesdale, F.M. and Francis, F.J. 1971. Organic acid profiles of thermally processed, stored spinach puree. *Journal of Food Science*, 36(2):240–242.

Lin, T.Y., Koehler, P.E. and Shewfelt, R.L. 1989. Stability of anthocyanins in the skin of Starkrimson apples stored unpackaged, under heat shrinkable wrap and in-package modified atmosphere. *Journal of Food Science*, 54(2):405–407.

Livingston, A.L., Nelson, J.W. and Kohler, G.O. 1968. Stability of α-tocopherol during alfalfa dehydration and storage. *Journal of Agricultural and Food Chemistry*, 16(3):492–495.

López-Serrano, M. and Ros Barceló, A. 1999. H_2O_2-mediated pigment decay in strawberry as a model system for studying color alterations in processed plant foods. *Journal of Agricultural and Food Chemistry*, 47(3):824–827.

Lougovois, V. and Houston, T.W. 1989. Kinetic study of the anaerobic formation of nitric oxide haemoglobin. *Food Chemistry*, 32(1):47–57.

Lowry, O.H., Bessey, O.A. and Crawford, E.J. 1949. Photolytic and enzymatic transformations of pteroyl-glutamic acid. *Journal of Biological Chemistry*, 180(1):389–398.

Lu, G., Yu, T-H. and Ho, C-T. 1997. Generation of flavor compounds by the reaction of 2-deoxyglucose with selected amino acids. *Journal of Agricultural and Food Chemistry*, 45(1):233–236.

Ludwig, H., Bieller, C., Hallbauer, K. and Scigalla, W. 1992. Inactivation of microorganisms by hydrostatic pressure. p. 25, In: *High Pressure and Biotechnology, Vol. 224*, Balny, C., Hayashi, R., Heremans, K. and Masson, P. (eds.), Colloque INSERM, John Libbey Eurotext, Montrouge, France.

Lukton, A., Chichester, C.O. and Mackinney, G. 1956. The breakdown of strawberry anthocyanin pigment. *Food Technology*, 10(9):427–432.

Lund, D.B. 1975. Heat transfer in foods. ch. 2, p. 11–30, In: *Principles of Food Science. Part II. Physical Principles of Food Preservation*, Fennema, O.W. (ed.), Marcel Dekker, NY.

Luchsinger, S.E., Kropf, D.H., Garciá Zepeda, C.M., Chambers, E. IV, Hollingsworth, M.E., Hunt, M.C., Marsden, J.L., Kastner, C.L. and Kuecker, W.G. 1996. Sensory analysis and consumer acceptance of irradiated boneless pork chop. *Journal of Food Science*, 61(6):1261–1266.

Lynch, J.A., MacFie, H.J.H. and Mead, G.C. 1991. Effect of irradiation and packaging type on sensory quality of chilled-stored turkey breast fillets. *International Journal of Food Science and Technology*, 26(6): 653–668.

Mader, I. 1964. Beta-carotene: thermal degradation. *Science*, 144(3618):533–534.

Main, J.H., Clydesdale, F.M. and Francis, F.J. 1978. Spray drying anthocyanin concentrates for use as food colorants. *Journal of Food Science*, 43(6):1693–1694, 1697.

Malien-Aubert, C., Dangles, O. and Amiot, M.J. 2001. Color stability of commercial anthocyanin-based extracts in relation to the phenolic composition. Protective effects by intra- and intermolecular copigmentation. *Journal of Agricultural and Food Chemistry*, 49(1):170–176.

Maloney, J.F., Labuza, T.P., Wallace, D.H. and Karel, M. 1966. Autoxidation of methyl linoleate in freeze-dried model systems. I. Effect of water on the autocatalyzed oxidation. *Journal of Food Science*, 31(6):878–884.

Markakis, P. 1974. Anthocyanins and their stability in foods. *CRC Critical Reviews in Food Technology*, 4(4):437–456.

Markakis, P. 1982. Stability of anthocyanins in foods. ch.6, p.163–180. In: *Anthocyanins as Food Colors*, Markakis, P. (ed.), Academic Press, NY.

Martinez, F. and Labuza, T.P. 1968. Rate of deterioration of freeze-dried salmon as a function of relative humidity. *Journal of Food Science*, 33(3):241–247.

Martins, R.C. and Silva, C.L.M. 2003. Kinetics of frozen stored green bean (*Phaseolus vulgaris* L.) quality changes: texture, vitamin C, reducing sugars, and starch. *Journal of Food Science*, 68(7):2232–2237.

Martins, S.I.F.S. and VanBoekel, M.A.J.S. 2005. A kinetic model for the glucose/glycine Maillard reaction pathways. *Food Chemistry*, 90(1–2):257–269.

Maruyama, T., Shiota, T. and Krumdieck, C.L. 1978. The oxidative cleavage of folates–A critical study. Analytical Biochemistry, 84(1):277–295.

Matano, K. and Kato, N. 1967. Studies on synthetic ascorbigen as a source of vitamin C for guinea pigs. *Acta Chemica Scandinavica*, 21(10):2886-2887.

Matthews, R.G. 1984. Methionine biosynthesis. ch. 13, p. 497–553, In: *Folates and Pterins, Vol. 1. Chemistry and Biochemistry of Folates*, Blakley, R.L. and Benkovic, S.J. (eds.), John Wiley & Sons, NY.

Merck Index, 13th Edition, 2001. Merck and Co., Inc., Whitehouse Station, NJ.

Merin, U., Gagel, S., Popel, G., Bernstein, S. and Rosenthal, I. 1987. Thermal degradation kinetics of prickly-pear-fruit red pigment. *Journal of Food Science*, 52(2):485–486.

Meschter, E.E. 1953. Fruit color loss. Effect of carbohydrates and other fractions on strawberry products. *Journal of Agricultural and Food Chemistry*, 1(8):574–579.

Metzler, D.E. 1960. Thiamine coenzymes. ch.9, p.295–337, In: *The Enzymes Vol. 2*, Boyer, P.D., Lardy, H. and Myrbäck, K. (eds.), Academic Press, New York.

Miki, N. and Akatsu, K. 1970. Effect of heating sterilization on color of tomato juice. Nippon Shokuhn Kogyo Gakkai-Shi [Japan], 17(5):175–181.

Miller, B.E. and Norman, A.W. 1984. Vitamin D. ch. 2, p. 45–97, In: *Handbook of Vitamins. Nutritional, Biochemical, and Clinical Aspects*, Machlin, L.J. (ed.), Marcel Dekker, Inc., NY.

Minguez-Mosquera, M.I., Garrido-Fernández, J. and Gandul-Rojas, B. 1989. Pigment changes in olives during fermentation and brine storage. *Journal of Agricultural and Food Chemistry*, 37(1):8–11.

Mishkin, M. and Saguy, I. 1982. Thermal stability of pomegranate juice. *Zeitschrift für Lebensmittel-Untersuchung und -Forschung*, 175(6):410–412.

Mizrahi, S., Labuza, T.P. and Karel, M. 1970. Feasibility of accelerated tests for browning in dehydrated cabbage. *Journal of Food Science*, 35(6):804–807.

Mnkeni, A.P. and Beveridge, T. 1982. Thermal destruction of pteroyl glutamic acid in buffer and model food systems. *Journal of Food Science*, 47(6):2038–2041, 2063.

Mnkeni, A.P. and Beveridge, T. 1983. Thermal destruction of 5-methyltetrahydrofolic acid in buffer and model systems. *Journal of Food Science*, 48(2):595–599.

Molyneaux, M. and Lee, C.M. 1998. The U.S. market for marine nutraceutical products. *Food Technology*, 52(6):56–57.

Monk, J.D., Beuchat, L.R. and Doyle, M.P. 1995. Irradiation inactivation of food-borne microorganisms. *Journal of Food Protection*, 58:197–208.

Moreno, F.J., Molina, E., Olano, A. and López-Fandiño, R. 2003. High-pressure effects on Maillard reaction between glucose and lysine. *Journal of Agricultural and Food Chemistry*, 51(2): 394–400.

Morild, E. 1981. The theory of pressure effects on enzymes. p. 93, In: *Advances in Protein Chemistry, Vol. 34*, Anfinsen, C.B., Edsall, J.T. and Richards, F.M. (eds.), Academic Press, Inc., London, GB.

Mortensen, A. and Skibsted, L.H. 2000. Kinetics and mechanism of the primary steps of degradation of carotenoids by acid in homogeneous solution. *Journal of Agricultural and Food Chemistry*, 48(2): 279–286.

Moyano, P.C., Rioseco, V.K. and Gonzalez, P.A. 2002. Kinetics of crust color changes during deep-fat frying of impregnated french fries. *Journal of Food Engineering*, 54(3):249–255.

Mulley, E.A., Stumbo, C.R. and Hunting, W.M. 1975a. Kinetics of thiamine degradation by heat. A new method for studying reaction rates in model systems and food products at high temperatures. *Journal of Food Science*, 40(5):985–988.

Mulley, E.A., Stumbo, C.R. and Hunting, W.M. 1975b. Kinetics of thiamine degradation by heat. Effect of pH and form of the vitamin on its rate of destruction. *Journal of Food Science*, 40(5):989–992.

Mundt, S. and Wedzicha, B.L. 2003. A kinetic model for glucose-fructose-glycine browning reaction. *Journal of Agricultural and Food Chemistry*, 51(12):3651–3655.

National Research Council, Food and Nutrition Board. 1980. *Recommended Daily Allowances*, 9th ed. National Academy of Sciences, Washington, D.C.

Navankasattusas, S. and Lund, D.B. 1982. Thermal destruction of vitamin B_6 vitamers in buffer solution and cauliflower puree. *Journal of Food Science*, 47(5):1512–1518.

Nelan, D.R. and Robeson, C.D. 1962. The oxidation product from α-tocopherol and potassium ferricyanide and its reaction with ascorbic and hydrochloric acids. *Journal of the American Chemical Society*, 84(15):2963–2965.

Neumann, H.J., Shepherd, A.D., Dietrich, W.C., Guadagni, D.G., Harris, J.G. and Durkee, E.L. 1965. Effect of drying temperatures on initial quality and storage stability of dehydrofrozen peas. *Food Technology*, 19(11):125–128.

Newmark, H.L., Scheiner, J., Marcus, M. and Prabhudesai, M. 1976. Stability of vitamin B_{12} in the presence of ascorbic acid. *The American Journal of Clinical Nutrition*, 29(6):645–649.

Nguyen, M.T., Indrawati and Hendrickx, M. 2003. Model studies on the stability of folic acid and 5-methyltetrahydrofolic acid degradation during thermal treatment in combination with high hydrostatic pressure. *Journal of Agriculture and Food Chemistry*, 51(11):3352–3357.

Niemand, J.G., Van der Linde, H.J. and Holzapfel, W.H. 1983. Shelf-life extension of minced beef through combined treatments involving radurization. *Journal of Food Protection*, 46(9):791–796.

Nienaber, U. and Shellhammer, T.H. 2001a. High-pressure processing of orange juice: Combination treatments and a shelf life study. *Journal of Food Science*, 66(2):332–336.

Nienaber, U. and Shellhammer, T.H. 2001b. High-pressure processing of orange juice: kinetics of pectinmethylesterase inactivation. *Journal of Food Science*, 66(2):328–331.

Nutting, M.-D., Neumann, H.J. and Wagner, J.R. 1970. Effects of processing variables on the stability of β-carotenes and xanthophylls of dehydrated parsley. *Journal of the Science of Food and Agriculture*, 21(4):197–202.

Ochoa, M.R., Kesseler, A.G., DeMichelis, A., Mugrid, A. and Chaves, A.R. 2001. Kinetics of colour change of raspberry, sweet (*Prunus avium*) and sour (*Prunus cerasus*) cherries preserves packed in glass containers: light and room temperature effects. *Journal of Food Engineering*, 49(1):55–62.

Ogunlesi, A.T. and Lee, C.Y. 1979. Effect of thermal processing on the stereoisomerization of major carotenoids and vitamin A value of carrots. *Food Chemistry*, 4(4):311–318.

Ohlsson, T. 1999. Minimal processing of foods with electric heating methods. ch. 6, p. 97–105, In: *Processing Foods–Quality Optimization and Process Assessment*, Oliveira, F.A.R. and Oliveira, J.C. (eds.), CRC Press, Boca Raton, FL.

O'Neil, C.A. and Schwartz, S.J. 1995. Photoisomerization of β-carotene by photosensitization with chlorophyll derivatives as sensitizers. *Journal of Agricultural and Food Chemistry*, 43(3):631–635.

Onyewu, P.N., Daun, H. and Ho, C.-T. 1982. Formation of two thermal degradation products of β-carotene. *Journal of Agricultural and Food Chemistry*, 30(6):1147–1151.

Ouyang, J.M., Daun, H., Chang, S.S. and Ho, C.-T. 1980. Formation of carbonyl compounds from β-carotene during palm oil deodorization. *Journal of Food Science*, 45(5):1214–1217, 1222.

Palamidis, N. and Markakis, P. 1975. Stability of grape anthocyanin in a carbonated beverage. *Journal of Food Science*, 40(5):1047–1049.

Palanuk, S.L. and Warthesen, J.J. 1988. The kinetics of lumichrome in skim milk using nonlinear regression analysis. *Food Chemistry*, 27(2):115–121.

Palou, E., López-Malo, A., Barbosa-Cánovas, G.V., Welti-Chanes, J. and Swanson, B.G. 1997. Kinetic analysis of *Zygosaccharomyces bailii* by high hydrostatic pressure. *Lebensmittel-Wissenschaft und-Technologie*, 30(7):703–708.

Parkhurst, R.M., Skinner, W.A. and Sturm, P.A. 1968. The effect of various concentrations of tocopherols and tocopherol mixtures on the oxidative stability of sample lard. *Journal of the American Oil Chemists' Society*, 45(10):641–642.

Parrish, D.B. 1980. Determination of vitamin E in foods–A review. *CRC Critical Reviews in Food Science and Nutrition*, 13(2):161–187.

Paul, A.A. and Southgate, D.A.T. 1978. *McCance and Widdowson's The Composition of Foods, 4rth ed*. Ministry of Agriculture, Fisheries and Food, HMSO, London, UK.

Pawlosky, R.J. and Flanagan, V.P. 2001. A quantitative stable-isotope LC-MS method for the determination of folic acid in fortified foods. *Journal of Agriculture and Food Chemistry*, 49(3):1282–1286.

Pekkarinen, L. 1974. The mechanism of the autoxidation of ascorbic acid catalyzed by iron salts in citric acid solution. *Finnish Chemical Letters*, 7:233–236.

Peng, C.Y. and Markakis, P. 1963. Effect of phenolase on anthocyanins. *Nature*, 199(4893):597–598.

Pesek, C.A. and Warthesen, J.J. 1987. Photodegradation of carotenoids in a vegetable juice system. *Journal of Food Science*, 52(3):744–746.

Pesek, C.A. and Warthesen, J.J. 1988. Characterization of the photodegradation of β-carotene in aqueous model systems. *Journal of Food Science*, 53(5):1517–1520.

Petriella, C., Resnik, S.L., Lozano, R.D. and Chirife, J. 1985. Kinetics of deteriorative reactions in model food systems of high water activity: Color changes due to nonenzymatic browning. *Journal of Food Science*, 50(3):622–626.

Petrou, A.L., Roulia, M. and Tampouris, K. 2002. The use of the Arrhenius equation in the study of deterioration and of cooking of foods–some scientific and pedagogic aspects. *Chemistry Education: Research and Practice in Europe*, 3(1):87–97.

Piatelli, M. 1976. Betalains. p. 560–596, In: *Chemistry and Biochemistry of Plant Pigments, Vol. 1.*, Goodwin, T.W. (ed.), Academic Press, New York.

Piatelli, M. 1981. The betalains: structure, biosynthesis and chemical taxonomy. p. 557–575, In: *The Biochemistry of Plants: A Comprehensive Treatise, Vol. 17*, Conn, E.E. (ed.), Academic Press, New York.

Pincock, R.E. and Kiovsky, T.E. 1966. Kinetics of reactions in frozen solutions. *Journal of Chemical Education*, 43(7):358–360.

Ponting, J.D., Sanshuck, D.W. and Brekke, J.E. 1960. Color measurement and deterioration in grape and berry juices and concentrates. *Food Research*, 25(4):471–478.

Quackenbush, F.W. 1963. Corn carotenoids: effects of temperature and moisture on losses during storage. *Cereal Chemistry*, 40(3):266–269.

Rader, J.I., Weaver, C.M. and Angyal, G. 2000. Total folate in enriched cereal-grain products in the United States following fortification. *Food Chemistry*, 70(3):275–289.

Ramakrishnan, T.V. and Francis, F.J. 1973. Color and carotenoid changes in heated paprika. *Journal of Food Science*, 38(1):25–28.

Ramakrishnan, T.V. and Francis, F.J. 1979a. Stability of carotenoids in model aqueous systems. *Journal of Food Quality*, 2(3):177–189.

Ramakrishnan, T.V. and Francis, F.J. 1979b. Coupled oxidation of carotenoids in fatty acid esters of varying unsaturation. *Journal of Food Quality*, 2(4):277–287.

Rao, M.A., Lee, C.Y., Katz, J. and Cooley, H.J. 1981. A kinetic study of the loss of vitamin C, color, and firmness during thermal processing of canned peas. *Journal of Food Science*, 46(2): 636–637.

Reed, L.S. and Archer, M.C. 1979. Action of sodium nitrite on folic acid and tetrahydrofolic acid. *Journal of Agricultural and Food Chemistry*, 27(5):995–999.

Reed, L.S. and Archer, M.C. 1980. Oxidation of tetrahydrofolic acid by air. *Journal of Agricultural and Food Chemistry*, 28(4):801–805.

Reznick, D. 1996. Ohmic heating of fluid foods. *Food Technology*, 50(5):250–251.

Ribéreau-Gayon, P. 1982. The anthocyanins of grapes and wines. ch. 8, p.209–244, In: *Anthocyanins as Food Colors*, Markakis, P. (ed.), Academic Press, New York.

Riemer, J. and Karel, M. 1977. Shelf-life studies of vitamin C during food storage: prediction of L-ascorbic acid retention in dehydrated tomato juice. *Journal of Food Processing and Preservation*, 1(4):293–312.

Robinson, E.H., Brent, J.R. and Grabtree, J.T. 1989. AsPP, an ascorbic acid, resists oxidation in fish feed. *Feedstuffs*, 61:64–66.

Romero, C. and Bakker, J. 2000. Effect of storage temperature and pyruvate on kinetics of anthocyanin degradation, vitisin A derivative formation, and color characteristics of model solutions. *Journal of Agricultural and Food Chemistry*, 48(6):2135–2141.

Ruddick, J.E., Vanderstoep, J. and Richards, J.F. 1980. Kinetics of thermal degradation of methyltetrahydrofolate. *Journal of Food Science*, 45(4):1019–1022.

Russu, S.I. and Valuiko, G.G. 1980. [Decomposition of anthocyanins during heating of mash]. *Vinodelie i Vinogradarstvo SSSR*, 8:25–28.

Ryan-Stoneham, T. and Tong, C-H. 2000. Degradation kinetics of chlorophyll in peas as a function of pH. *Journal of Food Science*, 65(8):1296–1302.

Saguy, I., Kopelman, I.J. and Mizrahi, S. 1978a. Extent of nonenzymatic browning in grapefruit juice during thermal and concentration processes: kinetics and prediction. *Journal of Food Processing and Preservation*, 2(3):175–184.

Saguy, I., Kopelman, I.J. and Mizrahi, S. 1978b. Simulation of ascorbic acid stability during heat processing and concentration of grapefruit juice. *J. Food Process Engineering*, 2(3):213–225.

Saguy, I., Kopelman, I.J. and Mizrahi, S. 1978c. Thermal kinetic degradation of betanine and betalamic acid. *Journal of Agricultural and Food Chemistry*, 26(2):360–362.

Saguy, I. 1979. Thermostability of red beet pigments (betanine and vulgaxanthin-I): influence of pH and temperature. *Journal of Food Science*, 44(5):1554–1555.

Saguy, I., Kopelman, I.J. and Mizrahi, S. 1980. Computer-aided prediction of beet pigment (betanine and vulgaxanthin-1) retention during air-drying. *Journal of Food Science*, 45(2):230–235.

Saguy, I., Goldman, M., Bord, A. and Cohen, E. 1984. Effect of oxygen retained on beet powder on the stability of betanine and vulgaxanthine I. *Journal of Food Science*, 49(1):99–101, 113.

Saguy, I., Goldman, M. and Karel, M. 1985. Prediction of beta-carotene decolorization in model system under static and dynamic conditions of reduced oxygen environment. *Journal of Food Science*, 50(2):526–530.

Şahbaz, F. and Somer, G. 1993. Photosensitized decomposition of ascorbic acid in the presence of riboflavin. *Food Chemistry*, 46(2):177–182.

Saidi, B. and Warthesen, J.J. 1983. Influence of pH and light on the kinetics of vitamin B_6 degradation. *Journal of Agricultural and Food Chemistry*, 31(4):876–880.

Sánchez, A.H., Rejano, L. and Montaño, A. 1991. Kinetics of the destruction by heat of colour and texture of pickled green olives. *Journal of the Science Food Agriculture*, 54(3):379–385.

Sánchez-Moreno, C., Plaza, L., DeAncos, B. and Pilar Cano, M. 2003. Vitamin C, provitamin A carotenoids, and other carotenoids in high-pressurized orange juice during refrigerated storage. *Journal of Agricultural and Food Chemistry*, 51(3):647–653.

Sancho, F., Lambert, Y., Demazeau, G., Largeteau, A., Bouvier, J-M. and Narbonne, J-F. 1999. Effect of ultra-high hydrostatic pressure on hydrosoluble vitamins. *Journal of Food Engineering*, 39(3):247–253.

Şansal, Ü. and Somer, G. 1997. The kinetics of photosensitized decomposition of ascorbic acid and the determination of hydrogen peroxide as a reaction product. *Food Chemistry*, 59(1):81–86.

Sarma, A.D., Sreelakshmi, Y. and Sharma, R. 1997. Antioxidant ability of anthocyanins against ascorbic acid oxidation. *Phytochemistry*, 45(4):671–674.

Sarni-Manchado, P., Fulcrand, H., Souquet, J-M., Cheynier, V. and Moutounet, M. 1996. Stability and color of unreported wine anthocyanin-derived pigments. *Journal of Food Science*, 61(5):938–941.

Sastry, L.V.L. and Tischer, R.G. 1952. Behavior of anthocyanin pigments in Concord grapes during heat processing and storage. *Food Technology*, 6(3):82–86.

Schaber, P.M., Hunt, J.E., Fries, R. and Katz, J.J. 1984. High-performance liquid chromatographic study of the chlorophyll allomerization reaction. *Journal of Chromatography*, 316:25–41.

Scheiner, J. 1985. Biotin. ch. 21, p. 535–553, In: *Methods of Vitamin Assays, 4rth Edition*, Augustin, J., Klein, B.P., Becker, D. and Venugopal, P.B. (eds.), John Wiley & Sons, NY.

Schneider, Z. 1987. Cobalamin in food and feeding stuff. p.194–198. In: *Comprehensive B₁₂*. Schneider, Z. and Stroiński, A. (eds.), de Gruyter: Berlin.

Schnudel, P., Mayer, H. and Isler, O. 1972. Tocopherols. II. Chemistry. p.168–218. In: *The Vitamins. Chemistry, Physiology, Pathology, Methods, Vol. V*, Sebrell, W.H., Jr. and Harris, R.S. (eds.), Academic Press, NY.

Schreir, P., Drawert, F. and Bhiwapurkar, S. 1979. Volatile compounds formed by thermal degradation of β-carotene. *Chemie Mikrobiologie Technologie der Lebensmittel*, 6(3):90–91.

Schroeder, H.A. 1971. Losses of vitamins and trace minerals resulting from processing and preservation of foods. *American Journal of Clinical Nutrition*, 24(5):562–573.

Schwartz, S.J. and von Elbe, J.H. 1983. Kinetics of chlorophyll degradation to pyropheophytin in vegetables. *Journal of Food Science*, 48(4):1303–1306.

Schwartz, S.J. and Lorenzo, T.V. 1990. Chlorophylls in foods. *CRC Critical Reviews in Food Science and Nutrition*, 29(1):1–17.

Schwertnerová, E., Wagnerová, D.M. and Vepřek-Šiška, J. 1976. Catalytic effect of copper ions and chelates on the oxidation of ascorbic acid. *Collection of Czechoslovak Chemical Communications*, 41(9):2463–2472.

Scott, E. 1995. Ohmic heating hits commercial scale. *Food Technology*, NZ30(7):8.

Sebrell, W.H., Jr. and Harris, R.S. (eds.) 1971. Vitamin D group. ch. 7, p. 156–301, In: *The Vitamins. Chemistry, Physiology, Pathology, Methods Vol. 3*, Academic Press, NY.

Seely, G.R. 1966. The structure and chemistry of functional groups. ch. 3, p. 67–109, In: *The Chlorophylls*, Vernon, L.P. and Seely, G.R. (eds.), Academic Press, NY.

Seely, G.R. and Meyer, T.H. 1971. The photosensitized oxidation of β-carotene. *Photochemistry and Photobiology*, 13(1):27–32.

Segner, W.P., Ragusa, T.J., Nank, W.K. and Hoyle, W.C. 1984. Process for the preservation of green color in canned vegetables. U.S. Patent 4,473,591. Continental Can Co., Inc., Stamford, CT.

Selman, J.D. 1994. Vitamin retention during blanching of vegetables. *Food Chemistry*, 49(2):137–147.

Shenoy, K.G. and Ramasarma, G.B. 1955. Iron as a stabilizer of vitamin B_{12} activity in liver extracts and the nature of so-called alkali stable factor. *Archives of Biochemistry and Biophysics*, 55(1): 293–295.

Shi, J. and LeMaguer, M. 2000. Lycopene in tomatoes: chemical and physical properties affected by food processing. *Critical Reviews in Food Science and Nutrition*, 40(1):1–42.

Shi, J., LeMaguer, M., Bryan, M. and Kakuda, Y. 2003. Kinetics of lycopene degradation in tomato puree by heat and light irradiation. *Journal of Food Process Engineering*, 25(6):485–498.

Shin, S. and Bhowmik, S.R. 1995. Thermal kinetics of color changes in pea puree. *Journal of Food Engineering*, 24(1):77–86.

Šimon, P., Drdák, M. and Altamirano, R.C. 1993. Influence of water activity on the stability of betanin in various water/alcohol model systems. *Food Chemistry*, 46(2):155–158.

Singh, R.P., Heldman, D.R. and Kirk, J.R. 1975. Kinetic analysis of light-induced riboflavin loss in whole milk. *Journal of Food Science*, 40(1):164–167.

Singh, R.P., Heldman, D.R. and Kirk, J.R. 1976. Kinetics of quality degradation: ascorbic acid oxidation in infant formula during storage. *Journal of Food Science*, 41(2):304–308.

Sistrunk, W.A. and Cash, J.N. 1970. The effect of certain chemicals on the color and polysaccharides of strawberry puree. *Food Technology*, 24(4):473–477.

Skinner, W.A. and Alaupovic, P. 1963. Oxidation products of vitamin E and its model, 6-hydroxy-2,2,5,7,8-pentamethylchroman. V. Studies of the products of alkaline ferricyanide oxidation. *Journal of Organic Chemistry*, 28(10):2854–2858.

Skinner, W.A. and Parkhurst, R.M. 1964. Oxidation products of vitamin E and its model, 6-hydroxy-2,2,5,7,8-pentamethylchroman. VII. Trimer formed by alkaline ferricyanide oxidation. *Journal of Organic Chemistry*, 29(12):3601–3603.

Skjöldebrand, C., Anäs, A., Öste, R. and Sjödin, P. 1983. Prediction of thiamine content in convective heated meat products. *Journal of Food Technology*, 18(1):61–73.

Sloane, A.E. 1999. The new market: foods for the not-so-healthy. *Food Technology*, 53(2):54–60.

Sloane, A.E. 2005. Healthy vending and other emerging trends. *Food Technology*, 59(2):26–35.

Smelt, J. and Rijke, G. 1992. High pressure treatment as a tool for pasteurization of foods. p. 361, In: *High Pressure and Biotechnology, Vol. 224*, Balny, C., Hayashi, R., Heremans, K. and Masson, P. (eds.), Colloque INSERM, John Libbey Eurotext, Montrouge, France.

Smith-Molina, Antonio. 1983. *Kinetics of Beta-Carotene Degradation.* M.S. Thesis, University of Illinois, Urbana, IL.

Somers, T.C. 1971. The polymeric nature of wine pigments. Phytochemistry, 10(9):2175–2186.

Sondheimer, E. and Kertesz, Z.I. 1953. Participation of ascorbic acid in the destruction of anthocyanin in strawberry juice and model systems. *Food Research*, 18(5):475–479.

Spanyár, P. and Kevei, E. 1963. Über die Stabilisierung von Vitamin C in Lebensmitteln. I. Mitteilung. *Zeitschrift für Lebensmittel-Untersuchung und -Forschung*, 120(1):1–17.

Stamp, J.A. and Labuza, T.P. 1983. Kinetics of the Maillard reaction between aspartame and glucose in solution at high temperatures. *Journal of Food Science*, 48(2):543–544, 547.

Starr, M.S. and Francis, F.J. 1973. Effect of metallic ions on color and pigment content of cranberry juice cocktail. *Journal of Food Science*, 38(6):1043–1046.

Starr, M.S. and Francis, F.J. 1968. Oxygen and ascorbic acid effect on the relative stability of four anthocyanin pigments in cranberry juice. *Food Technology*, 22(10):1293–1295.

Steet, J.A. and Tong, C-H. 1996a. Degradation kinetics of green color and chlorophylls in peas by colorimetry and HPLC. *Journal of Food Science*, 61(5):924–927, 931.

Steet, J.A. and Tong, C-H. 1996b. Quantification of color change resulting from pheophytinization and non-enzymatic browning reactions in thermally processed green peas. *Journal of Agricultural and Food Chemistry*, 44(6):1531–1537.

Stefanovich, A.F. and Karel, M. 1982. Kinetics of beta-carotene degradation at temperatures typical of air drying of foods. *Journal of Food Processing and Preservation*, 6(4):227–242.

Stintzing, F.C., Herbach, K.M., Mosshammer, M.R., Carle, R., Yi, W., Sellappan, S., Akoh, C.C., Bunch, R. and Felker, P. 2005. Color, betalain pattern, and antioxidant properties of cactus pear (*Opuntia* spp.) clones. *Journal of Agricultural and Food Science*, 53(2):442–451.

Stokstad, E.L.R., Hutchings, B.L., Mowat, J.H., Boothe, J.H., Waller, C.W., Angier, R.B., Semb, J. and SubbaRow, Y. 1948. The degradation of the fermentation *L. casei* factor. I. *Journal of the American Chemical Society*, 70(1):5–9.

Strack, D., Steglich, W. and Wray, V. 1993. Betalains. p. 421–450, In: *Methods in Plant Biochemistry, Vol. 8*, Academic Press, Orlando, FL.

Strålsjö, L.M., Witthöft, C.M., Sjöholm, I.M. and Jägerstad, M.I. 2003. Folate content in strawberries (*Fragaria ananassa*): effects of cultivar, ripeness, year of harvest, storage, and commercial processing. *Journal of Agricultural and Food Chemistry*, 51(1):128–133.

Suthanthangjai, W., Kajda, P. and Zabetakis, I. 2005. The effect of high hydrostatic pressure on the anthocyanins of raspberry (*Rubus idaeus*). *Food Chemistry*, 90(1–2):193–197.

Sweeney, J.P. and Marsh, A.C. 1970. Vitamins and other nutrients. Separation of carotene stereoisomers in vegetables. *Journal of the Association of Official Analytical Chemists*, 53(5):937–940.

Sweeney, J.P. and Marsh, A.C. 1971. Effect of processing on provitamin A in vegetables. *Journal of the American Dietetic Association*, 59(3):238–243.

Sweeny, J.G., Wilkinson, M.M. and Iacobucci, G.A. 1981. Effect of flavonoid sulfonates on the photobleaching of anthocyanins in acid solution. *Journal of Agricultural and Food Chemistry*, 29(3): 563–567.

Sykes, P. and Todd, A.R. 1951. Aneurin. Part X. The mechanism of thiochrome formation from aneurin and aneurin disulphide. *Journal of the Chemical Society*, Part I: 534–544.

Talavéra, S., Felgines, C., Texier, O., Besson, C., Gil-Izquierdo, A., Lamaison, J-L. and Rémésy, C. 2005. Anthocyanin metabolism in rats and their distribution to digestive area, kidney, and brain. *Journal of Agricultural and Food Chemistry*, 53(10):3902–3908.

Tamaoka, T., Itoh, N. and Hayashi, R. 1991. High pressure effect on Maillard reaction. *Agricultural and Biological Chemistry*, 55(8):2071–2074.

Tanchev, S. 1983. Kinetics of thermal degradation of anthocyanins. Research in Food Science and Nutrition. Vol. 2. Basic Studies in Food Science. *Proceedings of the 6th International Congress of Food Science and Technology*, 2:96.

Tanchev, S.S. 1972. Kinetics of the thermal degradation of anthocyanins of the raspberry. *Zeitschrift für Lebensmittel-Untersuchung und -Forschung*, 150(1):28–30.

Tanchev, S.S. and Joncheva, N. 1973. Kinetics of the thermal degradation of cyanidin-3-rutinoside and peonidin-3-rutinoside. *Zeitschrift für Lebensmittel-Untersuchung und -Forschung*, 153(1):37–41.

Tannenbaum, S.R., Archer, M.C. and Young, V.R. 1985. Vitamins and minerals. ch. 7, p.477–544. In: *Food Chemistry*, 2nd ed., Fennema, O.R. (ed.), Marcel Dekker, Inc., NY.

Taoukis, P.S., Labuza, T.P. and Saguy, I.S. 1997. Kinetics of food deterioration and shelf-life prediction. ch. 9, p. 361–403, In: *Handbook of Food Engineering Practice*, Valentas, K.J., Rotstein, E. and Singh, R.P. (eds.), CRC Press, Boca Raton, FL.

Taylor, B. 1996. Natural food colorants as nutraceuticals. Paper presented at the INF/COL II Symposium, January 23–26, Hamden, CT, sponsored by The Hereld Organization, p. 83.

Teixeira Neto, R.O., Karel, M., Saguy, I. and Mizrahi, S. 1981. Oxygen uptake and β-carotene decoloration in a dehydrated food model. *Journal of Food Science*, 46(3):665–669, 676.

Temple, C., Jr., Rose, J.D. and Montgomery, J.A. 1981. Chemical conversion of folic acid to pteroic acid. *Journal of Organic Chemistry*, 46(18):3666–3667.

Thayer, D.W. 1990. Food irradiation: benefits and concerns. *Journal of Food Quality*, 13(3):147–169.

Thomas, M.H. and Calloway, D.H. 1961. Nutritional value of dehydrated foods. *Journal of the American Dietetic Association*, 39(2):105–116.

Thomas, P.T., Flanagan, V.P. and Pawlosky, R.J. 2003. Determination of 5-methyltetrahydrofolic acid and folic acid in citrus juices using stable isotope dilution–mass spectrometry. *Journal of Agricultural and Food Chemistry*, 51(5):1293–1296.

Thompson, D.R. 1982. The challenge in predicting nutrient changes during food processing. *Food Technology*, 36(2):97–108, 115.

Thompson, L.U. and Fennema, O. 1971. Effect of freezing on oxidation of L-ascorbic acid. *Journal of Agricultural and Food Chemistry*, 19(1):121–124.

Timberlake, C.F. 1980. Anthocyanins — Occurrence, extraction and chemistry. *Food Chemistry*, 5(1): 69–80.

Timberlake, C.F. and Bridle, P. 1975. The anthocyanins. ch.5, p.214–266. In: *The Flavonoids*, Harborne, J.B., Mabry, T.J. and Mabry, H. (eds.), Chapman and Hall, London.

Timberlake, C.F. and Bridle, P. 1966. Spectral studies of anthocyanin and anthocyanidin equilibria in aqueous solutions. *Nature*, 212(5058):158–159.

Timberlake, C.F. and Bridle, P. 1967a. Flavylium salts, anthocyanidins and anthocyanins. I. Structural transformations in acid solutions. *Journal of the Science of Food and Agriculture*, 18(10):473–478.

Timberlake, C.F. and Bridle, P. 1967b. Flavylium salts, anthocyanidins and anthocyanins. II. Reactions with sulphur dioxide. *Journal of the Science of Food and Agriculture*, 18(10):479–485.

Tonucci, L.H. and von Elbe, J.H. 1992. Kinetics of the formation of zinc complexes of chlorophyll derivatives. *Journal of Agricultural and Food Chemistry*, 40(12):2341–2344.

Toribio, J.L. and Lozano, J.E. 1984. Nonenzymatic browning in apple juice concentrate during storage. *Journal of Food Science*, 49(3):889–892.

Torskangerpoll, K. and Anderson, Ø.M. 2005. Colour stability of anthocyanins in aqueous solutions at various pH values. *Food Chemistry*, 89(3):427–440.

Toyosaki, T., Yamamoto, A. and Mineshita, T. 1988. Kinetics of photolysis of milk riboflavin. *Milchwissenschaft*, 43(3):143–146.

Trout, G.R. 1989. Variation in myoglobin denaturation and color of cooked beef, pork, and turkey meat as influenced by pH, sodium chloride, sodium tripolyphosphate, and cooking temperature. *Journal of Food Science*, 54(3):536–540, 544.

Tsong, T.Y. 1990. Review On electroporation of cell membranes and some related phenomena. *Bioelectrochemistry and Bioenergetics*, 24(3):271–293.

Van Boekel, M.A.J.S. 2000. Kinetic modelling in food science: a case study on chlorophyll degradation in olives. *Journal of the Science of Food and Agriculture*, 80(1):3–9.

Van Buren, J.P., Bertino, J.J. and Robinson, W.B. 1969. The stability of wine anthocyanins on exposure to heat and light. *American Journal of Enology and Viticulture*, 19(3):147–154.

Van den Berg, L. and Rose, D. 1959. Effect of freezing on the pH and composition of sodium and potassium phosphate solutions: the reciprocal system KH_2PO_4-Na_2HPO_4-H_2O. *Archives of Biochemistry and Biophysics*, 81(2):319–329.

Van den Broeck, I., Weemaes, L.C., Van Loey, A. and Hendrickx, M. 1998. *Journal of Agricultural and Food Chemistry*, 46(5):2001–2006.

Vanderveen, J.E. 1988. Interactions of food additives and nutrients. ch. 14, In: *Nutrient Interactions*, Bodwell, C.E. and Erdman, J.W. (eds.), Marcel Dekker, New York.

Vega-Mercado, H., Góngora-Nieto, M.M., Barbosa-Cánovas, G.V. and Swanson, B.G. 1999. Nonthermal preservation of liquid foods using pulsed electric fields. ch. 17, p. 487–520, In: *Handbook of Preservation*, Rahman, M.S. (ed.), Marcel Dekker, Inc., NY.

Viberg, U., Jägerstad, M., Öste, R. and Sjöholm, I. 1997. Thermal processing of 5-methyltetrahydrofolic acid in the UHT region in the presence of oxygen. *Food Chemistry*, 59(3):381–386.

Villota, R. 1979. *Ascorbic Acid Degradation upon Air-Drying in Model Systems*. Ph.D. Thesis. Massachusetts Institute of Technology, Cambridge, MA.

Villota, R. and Hawkes, J.G. 1986. Kinetics of nutrients and organoleptic changes in foods during processing. p. 266–366, In: *Physical and Chemical Properties of Food*, Okos, M.R. (ed.), American Society of Agricultural Engineers, St. Joseph, MI.

Von Elbe, J.H., Maing, I.-Y., and Amundson, C.H. 1974. Color stability of betanine. *Journal of Food Science*, 39(2):334–337.

Voth, O.L. and Miller, R.C. 1958. Interactions of tocopherol with proteins and amino acids. Archives of Biochemistry and Biophysics, 77(1):199–205.

Waché, Y., Bosser-DeRatuld, A., Lhuguenot, J.-C., and Belin, J.-M. 2003. Effect of *cis / trans* isomerism of β–carotene on the ratios of volatile compounds produced during oxidative degradation. *Journal of Agricultural and Food Chemistry*, 51(7):1984–1987.

Wagenknecht, A.C., Lee, F.A. and Boyle, F.P. 1952. The loss of chlorophyll in green peas during frozen storage and analysis. *Food Research*, 17(4):343–350.

Wagner-Jauregg, T. 1972. Riboflavin. II. Chemistry. p.3–43. In: *The Vitamins. Chemistry, Physiology, Pathology, Methods. Vol. V*, Sebrell, W.H., Jr. and Harris, R.S. (eds.), Academic Press, NY.

Waletzko, P. and Labuza, T.P. 1976. Accelerated shelf-life testing of an intermediate moisture food in air and in an oxygen-free atmosphere. *Journal of Food Science*, 41(6):1338–1344.

Walker, G.C. 1964. Color deterioration in frozen french beans (*Phaseolus vulgaris*). 2. The effect of blanching. *Journal of Food Science*, 29(4):389–392.

Waller, C.W., Goldman, A.A., Angier, R.B., Boothe, J.H., Hutchings, B.L., Mowat, J.H. and Semb, J. 1950. 2-Amino-4-hydroxy-6-pteridinecarboxaldehyde. *Journal of the American Chemical Society*, 72(10):4630–4633.

Walter, W.M. Jr., Purcell, A.E. and Cobb, W.Y. 1970. Fragmentation of β-carotene in autoxidizing dehydrated sweet potato flakes. *Journal of Agricultural and Food Chemistry*, 18(5):881–885.

Walter, W.M. Jr. and Purcell, A.E. 1974. Lipid autoxidation in precooked dehydrated sweet potato flakes stored in air. *Journal of Agricultural and Food Chemistry*, 22(2):298–302.

Warmbier, H.C., Schnickels, R.A. and Labuza, T.P. 1976a. Effect of glycerol on nonenzymatic browning in a solid intermediate moisture model food system. *Journal of Food Science*, 41(3):528–531.

Warmbier, H.C., Schnickels, R.A. and Labuza, T.P. 1976b. Nonenzymatic browning kinetics in an intermediate moisture model system: Effect of glucose to lysine ratio. *Journal of Food Science*, 41(5):981–983.

Watanabe, F., Abe, K., Fujita, T., Goto, M., Hiemori, M. and Nakano, Y. 1998. Effects of microwave heating on the loss of vitamin B_{12} in foods. *Journal of Agricultural and Food Chemistry*, 46(1):206–210.

Watson, T.D.G. and Marsh, K.A. 2001. Biotin and B vitamins containing pet food. US Patent 6,177,107 B1, Mars UK Limited, Jan 23.

Weckel, K.G., Santos, B., Hernan, E., Laferriere, L. and Gabelman, W.H. 1962. Carotene components of frozen and processed carrots. *Food Technology*, 16(8):91–94.

Weemaes, C.A., Ooms, V., Indrawati, Ludikhuyze, I., Van den Broeck, A., VanLoey, A.M. and Hendrickx, M.E. 1999a. Pressure-temperature degradation of green color in broccoli juice. *Journal of Food Science*, 64(3):504–508.

Weemaes, C.A., Ooms, V., VanLoey, A.M. and Hendrickx, M.E. 1999b. Kinetics of chlorophyll degradation and color loss in heated broccoli juice. *Journal of Agricultural and Food Chemistry*, 47(6):2404–2409.

WHO, 1994. *Safety and Nutritional Adequacy of Irradiated Food*, World Health Organization, Geneva.

Widicus, W.A., Kirk, J.R. and Gregory, J.F. 1980. Storage stability of α-tocopherol in a dehydrated model food system containing no fat. *Journal of Food Science*, 45(4):1015–1018.

Wierbicki, E., Anellis, A., Killoran, J.J., Johnson, E.L., Thomas, M.H. and Josephson, E.S. 1970. High dose radiation processing of meat, poultry and seafood products. US Army Natick Labs, Natick, MA, paper presented at Third International Congress of Food Science and Technology, Washington, D.C.

Wilkinson, S.A., Earle, M.D. and Cleland, A.C. 1981. Kinetics of vitamin A degradation in beef liver puree on heat processing. *Journal of Food Science*, 46(1):32–33, 40.

Wilkinson, S.A., Earle, M.D. and Cleland, A.C. 1982. Effects of food composition, pH, and copper on the degradation of vitamin A in beef liver puree during heat processing. *Journal of Food Science*, 47(3):844–848.

Williams, A. 1994. New technologies in food preservation and processing: Part II. *Nutrition and Food Science*, 94(1):20–23.

Williams, R.R. and Spies, T.D. 1938. *Vitamin B₁(Thiamin) and Its Use in Medicine*. Minot, G.B. (ed.) The MacMillan Company, New York.

Woodcock, E.A., Warthesen, J.J. and Labuza, T.P. 1982. Riboflavin photochemical degradation in pasta measured by high performance liquid chromatography. *Journal of Food Science*, 47(2):545–555.

Wrolstad, R.E., Skrede, G., Lea, P. and Enersen, G. 1990. Influence of sugar on anthocyanin pigment stability in frozen strawberries. *Journal of Food Science*, 55(4):1064–1065, 1072.

Wybraniec, S. 2005. Formation of decarboxylated betacyanins in heated purified betacyanin fractions from red beet root (*Beta vulgaris* L.) monitored by LC-MS/MS. *Journal of Agricultural and Food Chemistry*, 53(9):3483–3487.

Yamada, S., Nakayama, K. and Takayama, H. 1983. Studies of vitamin D oxidation. 3. Dye-sensitized photooxidation of vitamin D and chemical behavior of vitamin D 6,19-epoxides. *Journal of Organic Chemistry*, 48(20):3477–3483.

Yamazaki, I., Mason, H.S. and Piette, L. 1959. Identification of intermediate substrate free-radicals formed during peroxidatic oxidations, by electron paramagnetic resonance spectroscopy. *Biochemical and Biophysical Research Communications*, 1(6):336–337.

Yamazaki, I., Mason, H.S. and Piette, L. 1960. Identification, by electron paramagnetic resonance spectroscopy, of free radicals generated from substrates by peroxidase. *Journal of Biological Chemistry*, 235(8):2444–2449.

Yamazaki, I. and Piette, L.H. 1961. Mechanism of free radical formation and disappearance during the ascorbic acid oxidase and peroxidase reactions. *Biochimica et Biophysica Acta*, 50(1):62–69.

Yaylayan, V.A., Machiels, D. and Istasse, L. 2003. Thermal decomposition of specifically phosphorylated *D*-glucoses and their role in the control of the Maillard reaction. *Journal of Agricultural and Food Chemistry*, 51(11):3358–3366.

Yonker, C.B. 1984. *Determination of B-6 Vitamer Degradation Kinetics by High Performance Liquid Chromatography*. Ph.D. Thesis, Rutgers University, New Brunswick, NJ.

Zechmeister, L. 1944. *Cis-trans* isomerization and stereochemistry of carotenoids and diphenylpolyenes. *Chemical Reviews*, 34(2):267–344.

Zhang, Q.H., Qiu, X. and Sharma, S.K. 1997. Recent development in pulsed electric field processing. Washington, D.C. National Food Processors Association. *New Technologies Yearbook*, 31–42.

Zheng, W. and Wang, S.Y. 2003. Oxygen radical absorbing capacity of phenolics in blueberries, cranberries, chokeberries, and lingonberries. *Journal of Agricultural and Food Chemistry*, 51(2):502–509.

Zima, O. and Williams, R.R. 1940. Über ein antineuritisch wirksames Oxydationsprodukt des Aneurins. *Berichte der Deutschen Chemischen Gesellschaft*, 73(9):941–949.

Zimmerman, F. and Bergman, C. 1993. Isostatic high-pressure equipment for food preservation. *Food Technology*, 47(6):162–163.

Zoltai, P. and Swearingen, P. 1996. Product development considerations for ohmic processing. *Food Technology*, 50(5):263–266.

Zook, C.D., Parish, M.E., Braddock, R.J. and Balaban, M.O. 1999. High pressure inactivation kinetics of *Saccharomyces cerevisiae* ascospores in orange and apple juices. *Journal of Food Science*, 64(3):533–535.

3 Phase Transitions and Transformations in Food Systems

Yrjö H. Roos

CONTENTS

3.1 INTRODUCTION

Phase transitions govern changes in the physical state of all materials, including food components. Phase transitions of pure materials occur at temperatures, which are pressure-dependent and specific to each material. In foods, internal and external pressure and temperature conditions contribute to their physical state during processing, storage, and consumption. Most phase transitions in foods are phase changes that occur in their main components: carbohydrates, lipids, proteins, and water. The physical state and engineering properties of most foods are defined by the physical state of their major components, especially that of lipids and water.

The basic theories of equilibrium thermodynamics can be applied in studies of most first-order transitions of pure compounds. Food materials and their component compounds exhibit both equilibrium and nonequilibrium states within a complex, often multiphase structure, which makes their phase behavior complicated and close to that of many noncrystalline synthetic polymers (Slade and Levine, 1990). The physical state of a chemically pure material can be described in terms of temperature, T, volume, V, and pressure, p. The relationships between these variables and the physical state of a material is usually described by a three dimensional phase diagram showing equilibrium lines for temperature, volume, and pressure.

Equilibrium thermodynamics of pure single-component systems are used to describe phase transitions in terms of chemical potential, μ, Gibbs energy, G, enthalpy, H, entropy, S, and temperature, T. Unfortunately, foods are not chemically pure systems, and their phase transitions and transition temperatures often depend on their composition. Furthermore, foods often exist in nonequilibrium, metastable, amorphous states which exhibit time-dependent properties. The thermodynamic characteristics of the phase behavior of amorphous food components are related to their free volume (Slade et al., 1989; Roos and Karel, 1990, 1991a). The free volume theory of synthetic polymers has been well documented (Eyring, 1936; Fox and Flory, 1950; Ferry, 1980; Tant and Wilkes, 1981). It seems that the free volume theory can be applied and used to describe the general behavior of many amorphous food systems (Slade and Levine, 1991).

Water is probably the most common and most important component of all foods which affects their physical, chemical, microbial, organoleptical, and other important properties. Most fresh foods, excluding some cereals, contain 60 to 95% (w/w) liquid water. The only other major components, which may exist in their liquid state in foods are oils and fats. However, these lipids and water do not mix well and they often exist in separate phases. Consequently, carbohydrates and proteins as well as minerals are the main food solids that have significant interactions with water, that is, phase transitions of both nonlipid solids and water are affected by their concentration in the nonlipid phase. Water in foods may exist in all its physical states, that is, ice, liquid, and vapor, at typical temperatures of food processing and storage. Hence, water in foods shows transitions between solid ice, liquid water, and gaseous vapor. Water is important to all physical properties of food materials, and its most important role in affecting phase transitions of other food components is its ability to act as a solvent or a plasticizer. Water affects significantly, for example, protein denaturation (Hägerdal and Martens, 1976; Wright, 1982) and starch gelatinization (Lund, 1984), and it may strongly influence state transitions of amorphous food components (Slade and Levine, 1988a; Roos and Karel, 1990, 1991a).

In their chemically pure state, most carbohydrates and proteins are crystalline or amorphous solids at room temperature. Therefore, in the absence of water, they may exist in completely crystalline, semicrystalline, partially crystalline, and amorphous states. Thermodynamical properties and phase and state transitions of food solids are extremely important to food dehydration and dried food stability which are closely related to their water content and sorption behavior (White and Cakebread, 1966; Karel, 1973; Levine and Slade, 1986; Roos, 1987; Roos and Karel, 1990). In the nonaqueous state, decomposition of a number of food components may occur before transition temperatures are reached which makes experimental determination of high-temperature phase and state transitions impossible.

Classification of phase transitions into first-order, second-order, and higher-order transitions on a thermodynamic basis was published by Ehrenfest (1933). Transitions between solid crystals and liquid, and liquid and gaseous states are thermodynamically first-order phase transitions. Other important first-order transitions are transitions between solid and liquid states of fats. Starch gelatinization and protein denaturation also show first-order thermodynamic characteristics, but these transformations may be associated with other changes which make them more complicated and different from pure first-order phase transitions. Melting of lipids has been studied extensively and thermodynamical data for transitions between various polymorphic states of mostly pure triacylglycerols have been reported in a number of studies (Hagemann, 1988). Most of the studies reported thermodynamical data for phase transitions of lipids (Hagemann, 1988), protein denaturation (Wright, 1982), and starch gelatinization (Donovan, 1979; Biliaderis et al., 1980; Lund, 1984) obtained by using differential scanning calorimetry (DSC). Use of DSC in food analysis has been reviewed by numeros authors, for example, by Biliaderis (1983) and Lund (1983), and in the analysis of polymers by Wunderlich (1981).

Phase and state transitions of amorphous food components often dominate in dehydrated, low moisture and frozen foods. In these foods, the properties of food solids are typical of other amorphous materials (Kauzmann, 1948; White and Cakebread, 1966; Levine and Slade, 1986; Roos 1987; Roos and Karel, 1990; Roos, 1995). The most typical and important state transition in amorphous systems is the transition where a glass is transformed to a rubber during heating over a glass transition temperature, T_g, range. This transition has some second-order characteristics, although it occurs in amorphous systems between different nonequilibrium states. Although the glass transition is not a true second-order transition, the amorphous state of synthetic polymers has been well characterized (Tant and Wilkes, 1981). The amorphous state of food materials and its impact on food processing and storage stability has also been recognized (Biliaderis et al., 1986; Roos, 1987; Zeleznak and Hoseney, 1987; Simatos and Karel, 1988; Roos and Karel, 1990, 1991a; Roos 1995). In foods containing amorphous components both transport phenomena and relaxation phenomena may define the state of the system at a given time and temperature. The amorphous state of food components increases the complexity of food characterization and the need for understanding the effects of the metastable, nonequilibrium properties of food solids on various physical and chemical changes occurring in food systems (Slade and Levine, 1990; Roos, 1995; Slade and Levine, 1995; Le Meste et al., 2002).

The gaseous state is not typical of carbohydrates, lipids, and proteins. However, foods contain many volatile compounds which are essential to their flavor and palatability. Transition of water from liquid to gas is the basis of food concentration and drying which makes the gaseous state important to all food preparation and processing. The gaseous state of water is also important in defining the amount of water sorbed by foods at various relative humidities.

Transition temperatures of food components and the compositional effects on phase transitions can be used in the formulation of food products, in the design of food processes, and in the evaluation of storage conditions necessary for maximum food stability. All compounds exhibit phase transitions which depend on pressure and temperature and other compounds present in food. Most transitions occur at atmospheric pressures, but high pressure technology has advanced possibilities to manipulate transitions by using both pressure and temperature (LeBail et al., 2003). Because of the complexity of both phase transitions in various media and the number of compounds only typical transitions of carbohydrates, lipids, proteins, and water mainly at normal pressures will be discussed in the present chapter. The main purpose is to describe phase and state transitions which are of importance to food product development, processing, and storage.

3.2 THERMODYNAMIC ASPECTS OF PHASE TRANSITIONS

3.2.1 Basic Thermodynamics and Equilibrium States

Thermodynamics is the study of transformations of energy. Energy transformation in relation to phase transitions may occur within a system or between a system and its surroundings. A closed system has no transfer of matter between the system and its surroundings. An open system exhibits transfer of matter through a boundary between the system and its surroundings. Both closed and open systems can transfer energy between the systems and their surroundings. If there is transfer of neither energy nor matter between a system and its surroundings, the system is an isolated system. The state of a single system can be characterized according to its internal energy, U, temperature, T, volume, V, pressure, p, number of moles, N, and mass, M. Internal energy, volume, number of moles, and mass are extensive functions of state, which means that they are proportional to the amount of matter. Temperature and pressure are independent of the amount of matter and they are defined as intensive functions of state. The basic thermodynamics of phase transitions can be found in books of Physical Chemistry, such as Atkins and de Paula (2006), and an excellent summary is available in Singh and Heldman (2001).

3.2.1.1 The First Law of Thermodynamics

The internal energy, U, is the sum of all forms of energy within a system, that is, all kinetic and potential energy of all molecules within the system. A change in internal energy, ΔU, may occur as a result of energy transfer and the amount of internal energy changes from the initial state of the system, U_i, to a final state, U_f. The internal energy is a state function. This means that the internal energy is dependent on the state of the system, but independent of how that state may have been achieved. State functions are properties which are dependent on state variables, for example, pressure. The internal energy of a system may change as a result of transfer of heat, q, or work, w, with the surroundings of the system. This is quantified by Equation 3.1, which is also known as the *First Law of Thermodynamics*.

$$\Delta U = q + w \tag{3.1}$$

If a system has a constant volume, a change in internal energy may involve only transfer of heat between the system and its surroundings, that is, $dU = dq + 0$ and $dw = 0$, because there is no pressure–volume work done on the system. If the volume of the system changes, the change in internal energy is the sum of the amount of heat transferred between the system and its surroundings and pressure–volume work, because $dU \neq dq$ and $dw \neq 0$. When a system has a constant pressure, the amount of energy that may be transferred between the material and its surroundings is equal to the change in internal energy of the material and the work corresponding to the change in volume, according to Equation 3.2.

$$H = U + pV \tag{3.2}$$

Equation 3.2 is a state function, which defines that the enthalpy, H, of a system as the sum of its internal energy, U, and pressure–volume work, pV, which the system has done on its surroundings. At atmospheric conditions, changes occur at a constant pressure and a change in enthalpy equals the amount of heat exchanged with the surroundings, that is, $dH = dq$. A change in enthalpy is defined by Equation 3.3 which applies also when there is a change in pressure and, hence, the system exchanges additional work with its surroundings.

$$dH = dU + pdV + Vdp \tag{3.3}$$

The internal energy of a substance increases with temperature, that is, heat is either removed from the material or transferred into it as temperature changes. A plot of the internal energy against temperature describes the change. The slope of the curve at any temperature gives the heat capacity of the system at that temperature. However, heat capacity can be defined for a substance at a constant pressure or a constant volume. At a constant volume, there is no expansion and the pressure of the system changes with temperature. Heat capacity at a constant volume is given by Equation 3.4.

$$C_V = \left(\frac{\partial U}{\partial T}\right)_V \tag{3.4}$$

Heat capacity is an extensive property of a material. It depends on the amount of the material, and it can be expressed as a molar heat capacity, that is, heat capacity per mole or as the specific heat capacity or specific heat, which refers to heat capacity per weight in grams.

The heat capacity at a constant volume gives a quantitative relationship between the change in internal energy and temperature 3.5.

$$dU = C_V dT \tag{3.5}$$

Heat capacity at a constant pressure takes into account both the change in the internal energy of the material and the pressure–volume work exchanged with surroundings. Hence, heat capacity at a constant pressure, C_p, is the slope of a plot of enthalpy against temperature at a constant pressure 3.6.

$$C_p = \left(\frac{\partial H}{\partial T}\right)_p \tag{3.6}$$

The heat capacity at constant pressure relates a change in enthalpy with a change in temperature according to Equation 3.7.

$$dH = C_p dT \tag{3.7}$$

The heat capacity of a system at a constant pressure is often larger than heat capacity at a constant volume. The difference between the heat capacity of a substance at a constant pressure and constant volume can be expressed by Equation 3.8.

$$C_p - C_V = \left(\frac{\partial H}{\partial T}\right)_p - \left(\frac{\partial U}{\partial T}\right)_V \tag{3.8}$$

A change in temperature at a constant pressure results in a change in volume. The effect of temperature on the volume is defined by the thermal expansion coefficient, α, given by Equation 3.9.

$$\alpha = \frac{1}{V}\left(\frac{\partial V}{\partial T}\right)_p \tag{3.9}$$

The thermal expansion coefficient is derived from Equation 3.10 which gives the temperature dependence of internal energy.

$$dU = \left(\frac{\partial U}{\partial V}\right)_T dV + C_V dT \tag{3.10}$$

The differential $(\partial U/\partial V)_T$ is defined as the *internal pressure*, π_T, of the system. Equation 3.10 may then be written to apply at constant pressure 3.11 from which α can be defined.

$$\left(\frac{\partial U}{\partial T}\right)_p = \left(\frac{\partial U}{\partial V}\right)_T \left(\frac{\partial V}{\partial T}\right)_p + C_V \tag{3.11}$$

$$\left(\frac{\partial U}{\partial T}\right)_p = \alpha \pi_T + C_V \tag{3.12}$$

Equation 3.12 gives the relationship between internal energy and temperature at a constant pressure. The enthalpy change of a system at a constant pressure is related to the heat capacity of the system. At a constant volume, the enthalpy change of a system is defined by Equation 3.13.

$$dH = \left(\frac{\partial H}{\partial p}\right)_T dp + C_p dT \tag{3.13}$$

Equation 3.13 may be used to obtain Equation 3.14 which uses the isothermal compressibility, κ_T, and the Joule–Thomson coefficient, μ, to relate enthalpy changes with temperature. The coefficients are related to the state functions according to Equation 3.15 and Equation 3.16, respectively.

$$\left(\frac{\partial H}{\partial T}\right)_{\mathrm{V}} = \left(1 - \frac{\alpha\mu}{\kappa_{\mathrm{T}}}\right) C_{\mathrm{p}} \tag{3.14}$$

$$\kappa_{\mathrm{T}} = -\frac{1}{V}\left(\frac{\partial V}{\partial p}\right)_{\mathrm{T}} \tag{3.15}$$

$$\mu = \left(\frac{\partial T}{\partial p}\right)_{\mathrm{H}} \tag{3.16}$$

The use of the thermal expansion coefficient and isothermal compressibility leads to the relationship given by Equation 3.17.

$$C_{\mathrm{p}} - C_{\mathrm{V}} = \frac{\alpha^2 TV}{\kappa_{\mathrm{T}}} \tag{3.17}$$

The thermal expansion of many liquids and solids is relatively small and the heat capacities are about the same, that is, $C_{\mathrm{p}} \approx C_{\mathrm{V}}$.

3.2.1.2 The Second Law of Thermodynamics

Spontaneity of chemical and physical changes can be analyzed using entropy and Gibbs energy of the system. The second law of thermodynamics explains when processes occur spontaneously or nonspontaneously. The first law of thermodynamics defines that the amount of energy in an isolated system is constant while the second law of thermodynamics takes into account the distribution of energy within the system.

The second law of thermodynamics uses a state function, *entropy*, S, to describe the direction of spontaneous changes. Entropy is a measure of the dispersion of energy in a system which means that in an isolated system, entropy increases as a result of a spontaneous change, that is, $\Delta S_{\mathrm{tot}} > 0$. The thermodynamic definition of entropy uses the quantity of heat exchanged in a chemical or physical change to quantify the change in distribution of energy in the process according to Equation 3.18.

$$\mathrm{d}S = \frac{\mathrm{d}q_{\mathrm{rev}}}{T} \tag{3.18}$$

When the system and its surroundings are at the same temperature, it follows that $\mathrm{d}S + \mathrm{d}S_{\mathrm{sur}} \geq 0$. This takes into account that the entropy of the surroundings, $\mathrm{d}S_{\mathrm{sur}}$ must change with an equal or smaller, but opposite amount of entropy.

The first law of thermodynamics states that the internal energy of a system is the sum of heat and work, that is, $\mathrm{d}U = \mathrm{d}q + \mathrm{d}w$. In a reversible change of a system with a constant composition and with no nonexpansion work $\mathrm{d}w_{\mathrm{rev}} = -p\mathrm{d}V$ and $\mathrm{d}q_{\mathrm{rev}} = T\mathrm{d}S$, giving Equation 3.19.

$$\mathrm{d}U = T\mathrm{d}S - p\mathrm{d}V \tag{3.19}$$

Equation 3.19 applies to any change of a closed system with no nonexpansion work, and it is known as the fundamental equation in combining the first and second laws of thermodynamics.

3.2.1.3 The Gibbs and Helmholtz Energies

Heat exchange between a system and its surroundings at a constant pressure requires that $\mathrm{d}q_{\mathrm{p}} = \mathrm{d}H$, which means that $T\mathrm{d}S \geq \mathrm{d}H$ at constant p, with no nonexpansion work applies. This relationship is

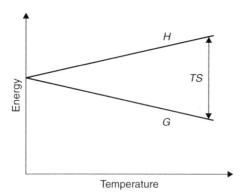

FIGURE 3.1 Change of enthalpy, H, and Gibbs energy, G, as function of temperature. The energy difference between enthalpy and Gibbs energy is obtained as $H = G + TS$ (T temperature, S entropy).

used in the definition of Gibbs energy, G, and a change of G in a process can be defined according to Equation 3.20.

$$dG = dH - T dS \qquad (3.20)$$

Since the absolute temperature, T, is always positive, dS of a spontaneous change is positive, and $T dS \geq dH$, the change in Gibbs energy of a spontaneous change is negative, that is, $dG_{T \cdot p} \leq 0$ (Figure 3.1).

The importance of Gibbs energy in normal processes is that it defines the spontaneous nature of a process in terms of pressure and temperature, which are the two variables that can be controlled. The variation of the Gibbs energy with pressure and temperature, in agreement with Equation 3.18, can be expressed by two exact differentials.

$$\left(\frac{\partial G}{\partial T}\right)_p = -S \quad \left(\frac{\partial G}{\partial p}\right)_T = V \qquad (3.21)$$

These differentials are of great importance in understanding phase transitions. They show that Gibbs energy is a function of temperature at a constant volume defined by entropy or a function of pressure at a constant temperature defined by volume. The entropy increases with temperature resulting in a decrease in Gibbs energy. The change in entropy in the gaseous, liquid, and solid states is considerably different, and the gaseous state with the highest temperature dependence of entropy is most sensitive to changes in temperature. At a constant volume, an increase in pressure increases the Gibbs energy, and the increase is highest for the state with the largest volume.

Equation 3.17 can be rearranged to Equation 3.22. If a change occurs at a constant volume, there is no expansion work and the amount of heat exchanged with the surroundings is equal to the change in internal energy, that is, $dq_V = dU$, and Equation 3.23 applies.

$$dS - \frac{dq}{T} \geq 0 \qquad (3.22)$$

$$dS - \frac{dU}{T} \geq 0 \qquad (3.23)$$

The relationship of Equation 3.23 can be rewritten to Equation 3.24, which defines that at a constant internal energy and volume, entropy increases in a spontaneous change.

$$T\mathrm{d}S \geq \mathrm{d}U \quad (constant\, V, \text{no nonexpansion work})$$ (3.24)

A change in Helmholz free energy, A, is defined by Equation 3.25, which gives the amount of internal energy in a change that is available for the system to do work.

$$\mathrm{d}A = \mathrm{d}U - T\mathrm{d}S$$ (3.25)

In a spontaneous change at a constant temperature and volume the change in Helmholtz energy is negative, that is, $\mathrm{d}A_{T,V} \leq 0$.

3.2.1.4 Enthalpy and Entropy Changes in a Physical Change

A change in enthalpy often occurs at standard conditions, for example, at atmospheric pressure and a given temperature at which transition occurs. Commonly, the standard state is that of a substance at the transition temperature in its pure form at a pressure of 1 bar. The standard enthalpy change of a change in physical state of a material is the standard enthalpy of transition, $\Delta H*_{trs}$. The standard enthalpy of transition applies to changes between the physical states of materials, for example, crystallization, fusion, and vaporization. Changes between the various states may occur following different paths, but the final, total change in enthalpy is always independent of the path. For example, transformation of ice to vapor may occur directly by sublimation or first to liquid and then to water vapor. The total enthalpy change, however, is the same. Furthermore, a reverse change has an equal but opposite value for change in enthalpy.

A change in entropy is obvious in a change in the phase of a system, that is, the molecular disorder in the gaseous, liquid, and solid states differ. At the transition temperature, at least two of the phases may coexist at equilibrium, and heat exchange between the system and its surroundings is reversible. At a constant pressure, the heat exchange is equal to the change in enthalpy of the transition, that is, $q = \Delta_{trs}H$. The transition occurs at a constant temperature, T_{trs}, and the change in entropy, $\Delta_{trs}S$, is given by Equation 3.26.

$$\Delta_{trs}S = \frac{\Delta_{trs}H}{T_{trs}}$$ (3.26)

An empirical rule, known as the Trouton's rule, states that liquids have approximately the same standard entropy of vaporization. There are differences in standard entropies of vaporization between liquids, but many liquids have about a constant standard entropy of vaporization of about $85\ JK^{-1}\ mol^{-1}$.

3.2.2 PHYSICAL STATE AND PHASE DIAGRAMS

3.2.2.1 Physical State

Phase transitions of pure substances result in a change of the physical state of a system without any change in its chemical composition. A *phase* of a material is uniform and homogeneous in chemical composition and physical state, that is, a solid, liquid, or gaseous phase. A phase transition may be defined as a spontaneous transformation of one phase into another phase. A phase transition occurs at a constant, well-defined pressure and temperature, for example, boiling of water at 1 bar at 100°C. At a transition temperature, T_{trs}, the chemical potentials of the two phases of the material must be the same at equilibrium, allowing these phases to coexist.

The temperature dependence of the Gibbs energy of a substance at constant pressure is defined by $(\partial G/\partial T)_p = -S$. The chemical potential, μ, of a pure substance is the same as its molar Gibbs energy, and Equation 3.21 can be rewritten to Equation 3.27 defining the relationship between chemical potential and molar entropy, S_m.

$$\left(\frac{\partial \mu}{\partial T}\right)_p = -S_m \tag{3.27}$$

Equation 3.27 shows that a plot of chemical potential against temperature has a negative slope, because $S_m < 0$ applies in all cases. The molar entropy of the three phases at a constant pressure differs in the order $S_m(g) > S_m(l) > S_m(s)$ giving a different slope for each phase. The stable phase is always the phase with the lowest chemical potential, and a phase transition occurs at the temperature at which the chemical potentials of the two phases are the same.

Most substances have a higher melting temperature with increasing pressure. One exception, however, is water, which melts at a lower temperature at a higher pressure. A higher melting temperature favors the lower density liquid phase which does not apply to water, because the density of liquid water is higher than that of ice. The variation of chemical potential with pressure is defined by Equation 3.28.

$$\left(\frac{\partial \mu}{\partial p}\right) = V_m \tag{3.28}$$

Equation 3.28 states that a plot of chemical potential against pressure has a slope, which is equal to the molar volume of the substance. The chemical potential increases with increasing pressure $(V_m > 0)$ and for most substances $V_m(l) > V_m(s)$ with some exceptions, for example, water, as discussed above and also by LeBail et al. (2003).

The temperature at which the solid and liquid states of a material at a constant pressure coexist at equilibrium is the melting temperature. The thermodynamical requirements for melting and freezing temperatures are the same. However, formation of nuclei preceding crystallization in cooling requires some supercooling to a temperature below the equilibrium crystallization temperature at which the liquid and solid phases may coexist. Thereafter the actual change in phase occurs at the transition temperature. A phase diagram of a pure substance shows in addition to phase boundaries a point at which the phase boundaries meet, that is, the solid, liquid, and gaseous states have the same chemical potential and the three phases may coexist at equilibrium. That pressure–temperature point is known as the triple point of the material. The triple point pressure is also the lowest pressure at which a liquid phase of a material may exist.

An increase in the temperature of a liquid at a given external pressure in an open system, for example, normal atmospheric pressure, increases the vapor pressure of the liquid. When the vapor pressure becomes equal to the external pressure, vaporization occurs throughout the liquid and the liquid starts to boil. Hence, the temperature at which the vapor pressure of a liquid is equal to the external pressure is the boiling temperature of the liquid at that pressure. In a closed container, the vapor pressure of a liquid increases with temperature and an increasing amount of the liquid is transformed into the vapor phase. This increases the vapor density and the increasing temperature decreases the density of the remaining liquid phase. Liquids also have a critical temperature, T_c, at which the densities of the liquid and vapor phases become the same and a surface between these two phases disappears. Hence, the two phases form a uniform phase which is called a supercritical fluid.

3.2.2.2 Phase Diagrams

Phase diagrams are important tools, or maps, which describe the equilibrium state of a material at any combination of pressure, temperature, and volume. A two dimensional phase diagram may show regions of pressure and temperature at which various phases are thermodynamically stable. Phase boundaries in a phase diagram are lines, which describe the pressure–temperature combinations at which two phases may coexist at equilibrium.

When two or more phases of a substance coexist, their chemical potentials must be the same, for example, the liquid (l) and solid (s) phases at the same pressure and temperature have the same chemical potential, $\mu_s(p, T) = \mu_l(p, T)$. The equilibrium can be maintained by moving along a phase boundary, that is, changing the pressure and temperature at the same time to maintain an equal chemical potential for the two phases. The change in chemical potential is defined by Equation 3.29, which at equilibrium is equal for each phase.

$$d\mu = -S_m dT + V_m dp \tag{3.29}$$

Considering that the changes in entropy and volume in the phase change between two phases, α and β, are given by the relationships $\Delta_{trs}S = S_{l,m} - S_{s,m}$ and $\Delta_{trs}V = V_{l,m} - V_{s,m}$, respectively, gives the Clapeyron Equation 3.30.

$$\frac{dp}{dT} = \frac{\Delta_{trs}S}{\Delta_{trs}V} \tag{3.30}$$

The Clapeyron equation is the exact definition of the phase boundary in a phase diagram. A phase transition involves a change in molar enthalpy, $\Delta_{trs}H$, and the transition occurs at a constant temperature, T. Hence, the change in molar entropy is given by $\Delta_{trs}S = \Delta_{trs}H/T$, and the Clapeyron equation can be written into the form of Equation 3.31.

$$\frac{dp}{dT} = \frac{\Delta_{trs}H}{T\Delta_{trs}V} \tag{3.31}$$

Equation 3.31 can be used for a solid–liquid transformation (fusion) by assuming that the changes in enthalpy and volume are very small and, therefore, the quantities can be considered as constants. For a solid–liquid transformation, the boundary can be described by the approximate Equation 3.32, where the initial pressure and temperature are referred as p^0 and T^0, respectively.

$$p = p^0 + \frac{\Delta_{fus}H}{\Delta_{fus}V} \ln\left(\frac{T}{T^0}\right) \tag{3.32}$$

In case T and T^0 do not differ greatly, an assumption that $\ln(T/T^0) \approx T - T^0/T^0$ can be made, and the relationship of Equation 3.33 is obtained.

$$p \approx p^0 + \frac{(T - T^0)\Delta_{fus}H}{T^0\Delta_{fus}V} \tag{3.33}$$

Equation 3.33 describes the solid–liquid boundary in a phase diagram when pressure is plotted against temperature.

The Clapeyron equation for vaporization is given in Equation 3.34,

$$\frac{dp}{dT} = \frac{\Delta_{vap}H}{T\Delta_{vap}V} \tag{3.34}$$

The volume change in transformation from liquid into a gas is large and the transformation temperature is highly pressure-dependent. It may be assumed that $\Delta_{vap}V \approx V_m(g)$ and $V_m(g) = RT/p$.

Equation 3.34 may then be written to the form of Equation 3.35.

$$\frac{\mathrm{d}p}{\mathrm{d}T} = \frac{\Delta_{\mathrm{vap}}H}{T(RT/p)} \tag{3.35}$$

Furthermore, Equation 3.35 can be written to the form of the Clausius–Clapeyron Equation 3.36, which describes the temperature dependence of vapor pressure in a liquid–gas transformation.

$$\frac{\mathrm{d}\ln p}{\mathrm{d}T} = \frac{\Delta_{\mathrm{vap}}H}{RT^2} \tag{3.36}$$

The integrated form of Equation 3.36 with the assumption that $\Delta_{\mathrm{vap}}H$ is independent of temperature gives the relationship between pressure and temperature in a liquid–gas transformation 3.37.

$$p = p^0 e^{-\chi} \quad \chi = \frac{\Delta_{\mathrm{vap}}H}{R}\left(\frac{1}{T} - \frac{1}{T^0}\right) \tag{3.37}$$

Equation 3.37 gives the phase boundary of a liquid–gas transformation in a phase diagram showing pressure against temperature. The same equation can also be used in sublimation, that is, transformation of a solid directly into the gaseous state.

3.2.3 Classification of Phase Transitions

Ehrenfest (1933) classified phase transitions to first-order, second-order, and higher-order transitions (Figure 3.2). Ehrenfest (1933) used the chemical potential of substances in his classification of changes in phase into first-order, second-order, and higher-order transitions. The Ehrenfest classification of phase transitions is based on the use of Equation 3.27 and Equation 3.28, which define that either a plot of chemical potential against temperature (constant pressure) or chemical potential against pressure (constant volume) shows a change in slope at a phase transition temperature or pressure, respectively. The first derivatives of the chemical potential, that is, enthalpy, entropy, and

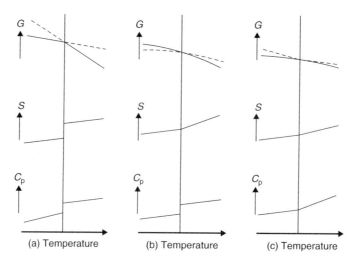

FIGURE 3.2 Classification of phase transitions according to Ehrenfest (1933), and their effect on specific heat, c_p, entropy, S, and Gibbs energy, G. (a) First order transition. (b) Second order transition. (c) Third order transition.

volume show a discontinuity for a first-order transition. The second-order and higher-order transitions are defined as those that show discontinuity for the second or higher derivatives of chemical potential at the transition temperature or pressure. For second- and higher-order transitions, the enthalpy, entropy, and volume do not change at the transition. In a second-order transition, there is a discontinuity in the heat capacity and thermal expansion coefficient of the substance.

3.2.3.1 First-Order Transitions

First-order phase transitions govern the changes of the physical state between solid, liquid, and gaseous states. At a first-order transition temperature, for example, melting, crystallization, condensation, and evaporation temperature, the change of the physical state at atmospheric pressure occurs isothermally, and a given amount of heat is either released or required as the latent heat for the transition. In first order phase transitions, Gibbs energy is the same in both phases $\Delta G = 0$, but the entropy and volume are different in the two phases. Gibbs energy is a continuous function of temperature and pressure, but it suffers a break at the transition temperature (Figure 3.2). Therefore, at least one of the first derivatives of Gibbs energy shows a discontinuous change at the transition temperature or pressure, and the transition can be noticed from a discontinuity in enthalpy, entropy, volume, and other thermodynamic functions.

Most of the latent heats of first-order transitions of compounds other than water in foods have been obtained using differential scanning calorimetry by the integration of the first-order transition peak which gives the enthalpy change of the transition. In melting, energy is required for the transition (ΔH_m, enthalpy of melting or latent heat of melting) and in crystallization the same amount of heat (ΔH_{cr}, latent heat of crystallization) is released ($\Delta H_m = -\Delta H_{cr}$) (Figure 3.3). At a constant pressure a change in enthalpy applies to all changes in phase.

3.2.3.2 Second-Order Transitions

Thermodynamically well-defined second-order transitions are not typical of food solids. However, all noncrystalline, amorphous solids exhibit a glass transition when they are transformed from a glassy solid to a supercooled liquid state. The glass transition includes no latent heat for the transition, but a change in heat capacity and thermal expansion coefficient giving to it thermodynamic characteristics of a second-order transition. The glass transition is a change within a nonequilibrium system which makes it complicated and related to a number of time-dependent phenomena (Sperling, 1992; Roos, 1995).

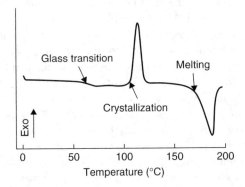

FIGURE 3.3 Typical phase transitions of amorphous, crystallizable compounds shown for amorphous sucrose as an example. Amorphous sucrose in its anhydrous state has glass transition, T_g, > 62°C, instant crystallization, T_{cr}, at 103°C, and melting at 185°C. Crystallization may occur at any temperature between T_g and T_{cr} depending on holding time. The heats of crystallization (exothermal, ΔH_{cr}) and melting (endothermal, ΔH_m) have the same but opposite values.

In the glassy state, the internal state variables can be considered to be frozen in below the glass transition temperature to a nonequilibrium, solid state with a higher energy and volume relative to the corresponding, crystalline equilibrium state (Tant and Wilkes, 1981). All amorphous materials have a random structure, and they resemble liquids above their glass transition temperature. Below the glass transition, unlike liquids, glasses are unable to change their molecular macroconformations, and the molecular motions are often limited to molecular rotations and vibration. Translational mobility of molecules in amorphous systems appear above the glass transition (Sperling, 1992).

In second-order phase transitions, both Gibbs energy and its first derivatives are continuous functions of temperature or pressure. At least one of the second derivatives of G, given in Equation 3.38, has a discontinuity at the second-order transition temperature (Figure 3.2).

$$\left(\frac{\partial^2 G}{\partial T^2}\right) = -\frac{C_p}{T} \quad \left(\frac{\partial^2 G}{\partial p \partial T}\right) = V\alpha \quad \left(\frac{\partial^2 G}{\partial p^2}\right) = V\beta \quad (3.38)$$

where α is thermal expansion coefficient and β is isothermal compressibility.

3.3 GLASS TRANSITION

3.3.1 PROPERTIES OF GLASS TRANSITION

Glass transition is the change in the physical state of amorphous materials which involves transformation of a nonequilibrium solid to a viscous liquid (rubbery, leathery, syrup, etc.) state. Glass transition takes place also in cooling when highly supercooled liquids vitrify to the nonequilibrium, solid glassy state. The glass transition is often observed at about 100°C below the equilibrium melting temperature, T_m, of the crystalline substance. However, there is a significant variation between the observed temperature range differentiating T_g and T_m. The ratio, T_m/T_g, is often a useful parameter in the characterization of noncrystalline materials (Slade and Levine, 1991, 1995; Roos, 1993, 1995).

The glass transition has a significant effect on relaxation times of various changes in material properties. The change of heat capacity, ΔC_p, or specific heat, Δc_p, as well as most other relaxations associated with glass transition occur over a temperature range. The broadness of the transition may differ largely for various food components (Roos, 1995). For example, a temperature range of 10 to 20°C applies to many low molecular weight amorphous sugars, whereas the broadness of the transition for carbohydrate polymers and proteins, such as starch (Zeleznak and Hoseney, 1987) and gluten (Hoseney et al., 1986), may extend to tens of degrees (Roos, 1995).

According to Wunderlich (1981), the most precise determination of the glass transition temperature of polymeric materials is obtained by the cooling of a melt at a specified rate and determining the transition temperature using heat capacity, expansion coefficient or compressibility measurement (Figure 3.4). However, this is not always possible for food materials which are substantially plasticized by water and in some cases decomposed below the melting temperature (Roos, 1995; Slade and Levine, 1995). Moreover, different cooling and heating rates give different T_g values for all amorphous systems (Sperling, 1992), including foods, as shown in Figure 3.5, and the transition temperature within a temperature range may be taken using various criteria. There is also a variation of T_g caused by thermal history which is often neglected, but the hysteresis in glass transition can be used to gain information of the thermal history of the material. Glasses can also be annealed to obtain various types of glassy solids, which may give different endothermal and exothermal enthalpy relaxations around the glass transition as well as relaxations in other thermodynamic properties (Wunderlich, 1981; Roos, 1995; Roos, 2002). For food materials there is, however, very little information available about the effects of thermal and moisture history on the state and relaxations of amorphous components. A single glass transition temperature gives a reference temperature for the glass transition temperature range, but relaxation times at that temperature may vary depending on the method and criteria of its determination.

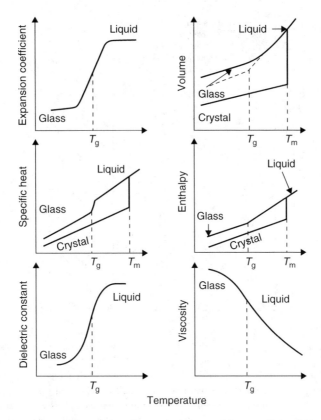

FIGURE 3.4 Schematic presentation of the effects of glass transition on dielectric constant, specific heat, thermal expansion coefficient, viscosity, enthalpy, and volume when an amorphous glass is heated over its glass transition temperature region.

The most important effect of glass transition on physical properties of food materials is the increase in mobility above the glass transition in the rubbery state which may affect rates of various physical and chemical deteriorative changes (White and Cakebread, 1966; Flink, 1983; Simatos and Karel, 1988; Slade and Levine, 1990, 1995; Roos, 1995). Hence, many amorphous foods must be processed and stored in their glassy state to maintain quality and avoid rapid deterioration (Roos, 1995, 2002; Slade and Levine, 1995).

3.3.2 Theories of Glass Transition

3.3.2.1 Free-Volume Theory

The basic principle of the free-volume theory is that molecular mobility requires vacancies or holes in the bulk state allowing molecules to move from one position to another between the holes. The free-volume theory, which was first used to describe transport properties related to viscosity and diffusivity, has been widely used to describe second-order transitions in polymers (Shen and Eisenberg, 1967; Tant and Wilkes, 1981; Sperling, 1992). Its importance in the analysis of amorphous food materials has been emphasized and discussed by Slade and Levine (1991, 1995).

The free-volume theory recognizes that glass transition temperature can be taken as the temperature at which the thermal expansion coefficient of a material is altered. It also assumes that the free volume of amorphous materials is constant at the glass transition (Sperling, 1992). The theory utilizes a single parameter, free volume, V_f, in addition to temperature and pressure to describe

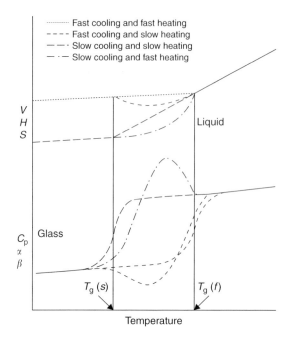

FIGURE 3.5 Effect of thermal history on thermal compressibility coefficient, β, thermal expansion coefficient, α, specific heat, c_p, entropy, S, enthalpy, H, and volume, V, on their values at the glass transition temperature region, and the possible thermal hysteresis effects observed as a glass is heated over its glass transition temperature. The same cooling and heating rate leads to no observed hysteresis but the T_g values determined using slow or fast rates may become slightly different. Fast cooling produces a glass with extra volume and leads to release of energy during slow heating over T_g region. Slow cooling leads to formation of a low energy state glass and requires extra energy as heated over the T_g region. The size of an endotherm observed in DSC curves also increases with increasing annealing or aging time below the glass transition temperature. (Adapted from Weitz A and Wunderlich B. *J Polym Sci Polym Phys* 12, 2473–2491, 1974. With permission.)

the nonequilibrium amorphous state which can be defined according to Equation 3.39. The volume occupied by molecules, V_0, includes the volume within the Van der Waals radii and volume associated with molecular vibrations.

$$V_f = V - V_0 \qquad (3.39)$$

where V is the macroscopic volume of the material and V_0 is volume occupied by molecules.

Fractional free volume, f, is defined by Equation 3.40, and it can be related to the glass transition temperature, as discussed in more detailed in the polymer literature (e.g., Ferry, 1980; Sperling, 1992), according to Equation 3.41. The free volume is also related to the thermal expansion of the material.

$$f = \frac{V}{V_f} \qquad (3.40)$$

$$f = f_g + \alpha(T - T_g) \qquad (3.41)$$

where f_g is factional free volume at T_g and α_f is coefficient of expansion of free volume.

Free volume is proportional to inverse molecular weight, and low molecular weight plasticizers, such as water in amorphous food materials, increase free volume (Slade and Levine, 1990, 1991; Sperling 1992). The free-volume theory has been successfully applied to predict enthalpy changes or enthalpy relaxations at T_g resulting from differences in thermal history. According to Tant and

Wilkes (1981), the free volume theory has been useful qualitatively, but it has shown only limited success as a quantitative tool to predict nonequilibrium phenomena.

3.3.2.2 Free Volume and Molecular Mobility

Williams et al. (1955) found that a number of glass forming materials exhibited almost universal changes in relaxation times in their glass transition. This finding gave the well-known but empirical Williams–Landel–Ferry (WLF) Equation 3.42, which relates relaxation times of mechanical properties to a reference temperature above the glass transition temperature. In later studies, the WLF equation has been derived from thermodynamics (Shen and Eisenberg, 1967; Ferry, 1980; Tant and Wilkes, 1981), and it can also be derived from the free-volume theory (Bauwens, 1986; Sperling, 1992).

Williams et al. (1955) related the ratio of relaxation times, A_t, at an observation temperature, θ, to relaxation time at a reference temperature, θ_0, according to Equation 3.42.

$$A_t = \frac{\theta}{\theta_0} \tag{3.42}$$

A_t has been shown to relate to a number of time-dependent quantities at the glass transition and at another temperature (Sperling, 1992). The most common quantity related to free volume changes above the glass transition is viscosity, η. Williams et al. (1955) suggested that Equation 3.43 can be used to model changes in relaxation times, for example, viscosity, above glass transition.

$$A_t = \frac{-C_1(T - T_0)}{C_2 + (T - T_0)} \tag{3.43}$$

where C_1 and C_2 are constants, T is observation temperature, and T_0 is a reference temperature.

An analysis of time-dependent changes of a number of inorganic and organic glass forming materials showed that when the glass transition temperature, T_g, was taken as the reference temperature, C_1 and C_2 had their universal values of -17.44 and 51.6, respectively (Williams et al., 1955). The universal values with the glass transition temperature are often used to model relaxation times above the glass transition. The use of the universal values, however, was not recommended by Williams et al. (1955), and they may not always apply to food systems (Peleg, 1992).

The free-volume theory has related the constants of the WLF equation to fractional free volume and thermal expansion. Hence, the theoretical form of the WLF equation can be written in the form of Equation 3.43.

$$\ln A_T = \frac{-(B/f_0)(T - T_0)}{f_0/\alpha_f + (T - T_0)} \tag{3.44}$$

where B is a constant, f_0 is fractional free volume at T_0 and α_f is expansion coefficient of the free volume.

According to Equation 3.44, when the glass transition temperature is used as the reference temperature with the universal WLF constants, the free volume at the glass transition of any polymer is 2.5% (Sperling, 1992).

The WLF equation can be written to give the temperature dependence of viscosity above T_g (Soesanto and Williams, 1981; Angell et al., 1982). The viscosity decreases above T_g as shown in Figure 3.6. The universal WLF constants were also used to model time to crystallization of amorphous sugars above their T_g by Roos and Karel (1990, 1991a). The WLF equation applies approximately over the temperature range from T_g to $T_g + 100°C$. Below T_g and above $T_g + 100°C$ Arrhenius type temperature dependence often applies. [The most drastic changes of the mechanical properties occur at temperatures up to $T_g + 50°C$, as shown in Figure 3.6.]

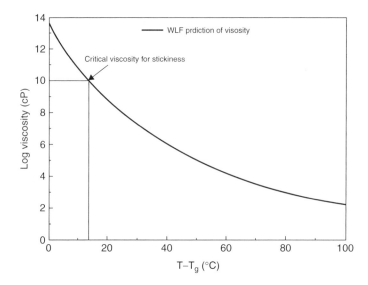

FIGURE 3.6 Temperature dependence of viscosity of amorphous materials above their glass transition temperature as predicted using the Williams–Landel–Ferry equation (Equation 3.43) (Williams et al., 1955). The equation applies usually at a temperature range of T_g to $T_g + 100°C$. Downton et al. (1982) reported viscosities below 10^{10} cP to lead to stickiness of dehydrated materials. The viscosity of an amorphous material heated over its glass transition temperature decreases to 10^{10} cP in the vicinity of the endpoint of the transition.

3.3.2.3 Kinetic and Thermodynamic Properties of Glass Transition

The free-volume theory has been successful in relating time and temperature of mechanical changes above the glass transition, but it cannot explain changes in observed glass transition temperature at different heating rates and the second-order phase transition characteristics of the transition (Sperling, 1992). Hence, a number of other theories have been developed to explain the kinetic and thermodynamic characteristics of the glass transition.

The kinetic theory of glass transition considers changes in the number of holes around the glass transition. The change in the number of holes corresponds to the change in heat capacity. The glass transition temperature is defined as the temperature at which the relaxation time for segmental motions of a polymer chain has the same time scale as the experiment (Sperling, 1992). The kinetic theory explains the change in heat capacity and changes of T_g with the time scale of the experiment.

The thermodynamic theory of glass transition aims at confirming the second-order thermodynamic properties of the glass transition. It assumes that the true second-order characteristics and an equilibrium state are approached at an infinite observation time. The theory explains changes in T_g with molecular weight and plasticizer content, but the true second order transition temperature is not well defined (Sperling, 1992). However, it seems that the nonequilibrium, noncrystalline materials cannot exhibit true second-order characteristics at realistic time scales, although a true equilibrium, noncrystalline state could probably be achieved at an infinite time.

3.3.3 Material Properties in Relation to Glass Transition

3.3.3.1 Relaxations and Time Dependence

Freezing of molecules of amorphous solids at temperatures below glass transition retains the nonequilibrium state, although changes in the solid state occur slowly with time. The term *transition* refers to a change in state resulting from a change of pressure or temperature (Sperling, 1992), but it does not consider the time dependence of the nonequilibrium state. The term *relaxation* is used to refer to the time required by a nonequilibrium system to respond to a change in pressure or temperature. Relaxation times in amorphous solids are extremely long. The main relaxations observed below glass transition are molecular vibrations and sidechain rotations of polymer molecules (Sperling, 1992). The glass transition can also be defined as a series of time-dependent relaxations. These can be recorded at decreasing temperatures with increasing time of observation.

Relaxations in amorphous materials can be observed from changes in dielectric, mechanical, and thermal properties below and around the glass transition. The dielectric and mechanical spectroscopic methods are extremely sensitive in observing relaxations as a function of frequency and temperature. These relaxations involve both sub-glass transition relaxations, which are referred as β and γ relaxations, and the glass transition which is the main, α denoted relaxation. Although there is some disagreement in relating rates of chemical reactions to the relaxations of amorphous food systems, it seems obvious that mechanical properties of amorphous foods are significantly affected by the glass transition.

Relaxation times of amorphous materials may be considered to change over five regions of their viscoelastic behavior, as described by changes in modulus against frequency or temperature. The five regions are (i) the glassy region, (ii) the glass transition region, (iii) the rubbery plateau region, (iv) the rubbery flow region, and (v) the liquid flow region (Sperling, 1992). A comprehensive discussion of the properties of amorphous solids and liquids is available in Ferry (1980) and Sperling (1992).

In the glassy region, amorphous polymers are solid and brittle, and their modulus is fairly constant at 3×10^9 Pa (Sperling, 1992). Within the glass transition region, the modulus decreases by a factor of 10^3 over a temperature range of 20–30°C. The stiffness of amorphous materials in the glass transition region is very sensitive to changes in temperature, but the extent of the change in modulus is dependent on a number of factors, such as molecular weight, crystallinity, extent of cross-linking, etc. The glass transition region represents the onset of long-range, coordinated molecular motions. The rubbery plateau region has not received much attention in the characterization of food materials. It follows the glass transition of polymers as their modulus levels off at an almost constant value of 2×10^6 Pa. For linear polymers, often the higher the molecular weight the broader is the rubbery plateau, although modulus increases with increasing crystallinity. Many amorphous, low molecular weight sugars and foods are transformed very rapidly to viscous liquids, syrups, or molasses and they exhibit no rubbery plateau (Talja and Roos, 2001). Cross-linking improves rubber elasticity of polymers and extends their rubbery plateau. When polymers reach their rubbery flow region, they exhibit rubber elasticity at very short experimental times but flow in a long experiment. This region is followed by the liquid flow region in which polymers behave like molasses (Sperling, 1992). Obviously, low molecular weight food components are transformed extremely rapidly from the solid glassy state to the liquid flow region.

3.3.3.2 Differential Scanning Calorimetry

Differential scanning calorimetry (DSC) is the most common thermal analytical technique for observing phase and state transitions, including glass transition in foods. Most DSC instruments can be operated in a dynamic or isothermal mode. This allows observation of phase transitions from exothermal and endothermal changes when the heat flow to a sample is compared with a reference. First-order transitions exhibit an exotherm, for example, crystallization, when heat is released or an endotherm, for example, melting, when heat is required for the change in phase. Because of its second-order phase transition characteristics, a glass transition in a heating scan produces a step change in heat flow to a sample resulting from the change in heat capacity over the transition temperature range (Figure 3.7). The glass transition temperature can be taken from the onset or midpoint temperature of the thermal event, although the onset temperatures seem to be more important, because a number of changes in food properties occur rapidly above the onset of the transition (Roos, 1995). Both endothermal and exothermal enthalpy relaxations are often associated with the glass transition

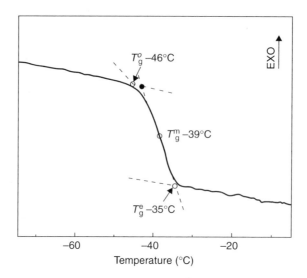

FIGURE 3.7 Determination of glass transition temperatures from differential scanning calorimetric data shown for 80% (w/w) sucrose solution. The temperature values shown are onset temperature, T_g^o, midpoint temperature, T_g^m, and endpoint temperature, T_g^e, of the glass transition temperature range. Both T_g^o and T_g^m values for glass transitions are used as glass transition temperature values in polymer and food literature. The temperature range of the glass transition region ($\Delta T = T_g^e - T_g^o$) is often 10–20°C.

of food systems (Roos, 2002). These may be eliminated or reduced in an immediate rescan of the sample.

3.3.3.3 Mechanical Thermal Analysis

Mechanical thermal analysis systems include mechanical spectroscopy and various dynamic mechanical analysis methods. The basic principle of the measurements is to apply an oscillating mechanical stress at a given frequency to cause a nondestructive strain in the sample and detect its recovery. A portion of the applied energy is stored in the material and used in recovery while some of the applied energy is converted to heat and lost. The corresponding Young's or shear moduli are referred to as storage modulus (E' or G') and loss modulus (E'' or G''). The angle, δ, between the in-phase and out-of-phase components of the cyclic motions is related to the ratio of the loss and storage modulus according to Equation 3.45.

$$\tan \delta = \frac{E''}{E'} \tag{3.45}$$

The α-relaxation or glass transition is observed from changes in E', E'', and tan δ. The storage modulus decreases significantly over the glass transition, while both the loss modulus and loss tangent exhibit a peak around the glass transition. Mechanical spectrometers and dynamic mechanical analysis equipment may be use to collect both isothermal and dynamic data on mechanical properties in tension, compression, shear, bending, and other tests over a broad frequency range.

3.3.3.4 Dielectric Analysis

Dielectric analysis uses samples placed between parallel plate capacitors; an alternating electric field is applied at given frequencies, and the dielectric permittivity, ε', and dielectric loss factor, ε'', of the sample are measured. The results are analogous to dynamic mechanical measurements with similar sub-T_g and α-relaxation behavior. The phase loss angle, tan δ, is defined as the ratio of the dielectric permittivity and loss factor. Modern dielectric analyzers may also be operated isothermally or in a dynamic mode to collect data at various frequencies.

3.3.3.5 Annealing

Annealing is used to observe time-dependent changes in a state which may occur in crystallization and glass formation. In studies involving food materials, annealing is used to analyze crystallization properties of lipids, starch gelatinization, protein denaturation, ice formation in concentrated solutions, and relaxations associated with the glass transition. According to Struik (1978), the basic property that changes upon annealing of polymers below T_g is the segmental mobility of molecules. Since molecules are in a nonequilibrium state, there exists a thermodynamic potential as a driving force further packing or conformational rearrangement. The driving force for nonequilibrium processes can be derived from the total change of Gibbs energy. Sub-T_g annealing causes a decrease of enthalpy and free volume (Tant and Wilkes, 1981; Matsuoka et al., 1985). Annealing of starch-water suspensions below gelatinization temperature leads to increased gelatinization temperatures (Lund, 1984).

Since the excess enthalpy decreases during annealing of glasses, the enthalpy recovered during heating over the glass transition region increases with increasing annealing time (Figure 3.5). This is observed in a DSC scan at the glass transition as an endothermic peak, which becomes correspondingly large. Annealing below T_g leads also to increased density, tensile and flexural yield stress and elastic modulus, decreased impact strength, fracture energy, ultimate elongation and creep rate, and transition from ductile to brittle behavior (Tant and Wilkes, 1981).

3.3.3.6 Aging

Aging of amorphous materials below their T_g was studied extensively by Struik (1978). He showed that in all glassy materials, including low molecular weight carbohydrates, and other natural amorphous materials, aging proceeds similarly. Aging as a phenomenon is a thermoreversible process that affects the properties of glasses primarily by changes in the relaxation times, all of them in the same way. Relaxation times were found to increase in proportion with aging time (Struik, 1978). Aging affects such material properties as density, heat capacity, enthalpy, modulus and stress (Figure 3.4). Struik (1978) stated that aging below T_g is a continuous process in which decreasing free volume leads to decreased mobility. This indicated aging to proceed nonlinearly and to be caused by decreasing segmental mobility of molecules. Molecular mobility depends mainly on the configurational free volume, which decreases with physical aging, and affects the structural state. The structural state is affected by time, temperature, and thermal and mechanical histories. As the molecular mobility and free volume decrease with increasing aging time, glassy materials become more brittle as a result of aging (Tant and Wilkes, 1981; Bauwens, 1986).

Aging occurs in amorphous materials to an extent which decreases with increasing crystallinity in partially crystalline polymers (Tant and Wilkes, 1981). Aging is important to the characteristics of polymeric materials and their changes as a function of time. No studies or attempts have been made to study the effect of aging of amorphous glassy food materials on their properties. However, aging may be important to the quality of hard glassy sugar candies or products which have an amorphous glassy sugar coating. It may also affect properties of food powders and affect crispness of snack foods. It should also be noticed that the aging of amorphous foods is affected by water plasticization. Hence, fluctuations in storage relative humidity and temperature may accelerate structural changes and aging of food systems and have an affect on their quality attributes, such as crispness and flavor retention. Furthermore, aging may be an important factor affecting properties of edible and biodegradable films.

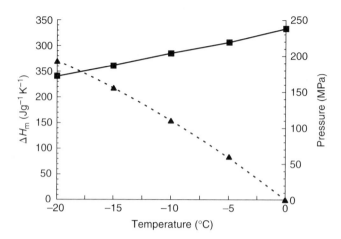

FIGURE 3.8   Latent heat of melting, ΔH_m, of ice at various pressure-dependent melting temperatures.

3.4   PHASE TRANSITIONS OF WATER IN FOODS

In food freezing, a substantial amount of heat, released from water as a result of ice formation and to decrease the temperature of ice and an unfrozen solids phase, has to be removed. In thawing, melting of ice crystals requires an equal amount of heat which must be supplied as frozen foods are melted. Freezing of water in foods results in crystallization of pure water separated as ice crystals within the food microstructure, and the latent heat for ice melting is the same as that of pure ice (334 kJ/kg). Although, the latent heat of melting is often considered to be constant and independent of temperature (Simatos et al., 1975; Roos, 1995), it varies with melting temperature according to the thermodynamic requirements for a total enthalpy change required for ice melting and increasing the temperature of ice and water to a given level. The latent heat of ice melting at various pressure-dependent melting temperatures is shown in Figure 3.8.

In evaporation and drying, water has to be transformed from liquid into vapor. Sublimation of ice or evaporation of water occur at all practical temperatures, and they may lead to quality changes during food storage. The enthalpy of evaporation is a function of temperature (Table 3.1 and Table 3.2), and its temperature dependence can be predicted using the Clausius–Clapeyron equation (Equation 3.46). The latent heat of ice sublimation is 2826 J/g at 0°C. The sublimation of ice is the basis of freeze-drying (King, 1970; Mellor, 1978). Vapor pressure values of ice and water are given in Table 3.1 and Table 3.2.

$$\ln \frac{p_2}{p_1} = -\frac{\Delta H_\mathrm{V}}{R} \left(\frac{1}{T_2} - \frac{1}{T_1} \right) \tag{3.46}$$

where p_1 and p_2 is pressure at temperature T_1 and T_2, respectively, ΔH_v is latent heat of vaporization, and R is the gas constant.

3.4.1   PHASE DIAGRAM OF WATER

The phase diagram of water shows the phase boundaries at which two phases may coexist and the triple point at which all three phases coexist. The two dimensional pressure–temperature phase diagram of water is shown in Figure 3.9. The phase diagram of water is of great importance to food engineering as water often dominates the thermodynamic properties of food due to their high water content. The mixing behavior of liquid systems is another fundamental factor affecting properties and phase equilibria in food systems. Often the stability of foods is achieved by manipulating and understanding the behavior of mixed systems. The chemical potential of a liquid, for example water,

TABLE 3.1
Vapor Pressure of Ice, p_i, and Water, p_u, Vapor Volume, V_v, and Heat of Sublimation, ΔH_s at Temperatures Below 0°C

T (°C)	p_i (mbar)	p_u (mbar)	V_v (m³/Kg)	ΔH_s (J/g)
−98	0.00002			
−90	0.00009			
−80	0.00053			
−76	0.00103		908600	
−70	0.00259			
−65	0.00500		191600	
−60	0.01077		98110	
−55	0.02106		50260	
−50	0.03939		25760	
−45	0.07265		15030	
−40	0.12876			
−36	0.20088		5472	
−30	0.38110		2800	
−25	0.63451			
−20	1.03441	1.25323		
−15	1.65425	1.91419		
−10	2.59935	2.86462		
−5	4.01633	4.21628		
0	6.10381	6.10381		2834

changes as a result of introducing and mixing another component into the liquid phase. If two phases of a substance coexist at equilibrium, the chemical potentials of the two phases need to be equal. For example, the chemical potential of liquid water and water vapor are the same at equilibrium, which is used in the definition of the water activity concept.

3.4.2 BOILING TEMPERATURE ELEVATION AND FREEZING TEMPERATURE DEPRESSION

Several properties of dilute solution differ from those of pure water, which is well known for food systems. These properties are referred to as colligative properties, because they change as a result of collection of other substances in the system. The changes in colligative properties resulting from changes in the concentration of dissolved substances in food systems are extremely common.

3.4.2.1 Raoult's Law

The vapor pressure of pure water, A, at a constant temperature has a definite value, p_A^0, and its chemical potential at atmospheric (1 bar) pressure is given by Equation 3.47. At equilibrium, the chemical potential of water and its vapor phase is the same, that is, $\mu_A^0 = \mu_A^0(g) = \mu_A^0(l)$.

$$\mu_A^0 = \mu_A^* + RT \ln p_A^0 \tag{3.47}$$

where μ_A^* is the chemical potential at a standard state.

In the presence of another component in water, the vapor pressure of the liquid is changed to p_A and the chemical potential of water, μ_A, is given by Equation 3.48. The relationships of

TABLE 3.2
Vapor Pressure,p_A^0, Temperature, T, and Corresponding Latent Heat of Vaporization, $\Delta_{vap}H$, and Volume, V_{vap}, of Vapor

p_A^0 (mbar)	T (°C)	V_{vap} (m³/kg)	$\Delta_{vap}H$ (kJ/kg)
10	6.98	129.21	2485
20	17.51	67.01	2460
30	24.09	45.69	2445
40	28.97	34.82	2433
50	32.89	28.20	2424
60	36.18	23.74	2416
70	39.02	20.53	2409
80	41.54	18.10	2403
90	43.79	16.20	2398
100	45.84	14.67	2393
150	54.00	10.02	2373
200	60.09	7.650	2359
250	64.99	6.206	2346
300	69.12	5.231	2336
400	75.88	3.995	2319
500	81.34	3.241	2305
600	85.95	2.732	2294
700	89.96	2.365	2283
800	93.51	2.087	2274
900	96.71	1.869	2266
1000	99.63	1.694	2258
1500	111.4	1.159	2226
2000	120.2	0.8853	2202
2500	127.4	0.7184	2181
3000	133.5	0.6056	2163
4000	143.6	0.4623	2133
5000	151.8	0.3747	2108
6000	158.8	0.3155	2085
7000	164.9	0.2727	2065
8000	170.4	0.2403	2047
9000	175.4	0.2148	2030
10000	179.9	0.1943	2014
15000	198.3	0.1316	1945
20000	212.4	0.09952	1889
25000	223.9	0.07990	1839
30000	233.8	0.06663	1794
40000	250.3	0.04975	1713
50000	263.9	0.03943	1640

Equation 3.47 and Equation 3.48 give Equation 3.49.

$$\mu_A = \mu_A^* + RT \ln p_A \tag{3.48}$$

$$\mu_A = \mu_A^0 + RT \ln \left(\frac{p_A}{p_A^0} \right) \tag{3.49}$$

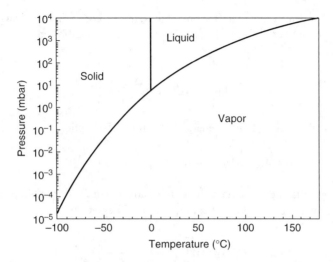

FIGURE 3.9 Phase diagram of water. The triple point of water is at $0.0099°C$ and 6.104 mbar. In food materials the vapor pressure curve of ice is shifted to lower pressure values because of other compounds present.

The chemical potential of water in a mixture is a function of its concentration, which in Equation 3.49 is related to p_A/p_A^0. This relationship leads to Equation 3.50, which is generally known as Raoult's law.

$$p_A = x_A p_A^0 \tag{3.50}$$

The vapor pressure of water in a solution is always lower than the vapor pressure of pure water, that is, $p_A < p_A^0$. Raoult's law applies to dilute solutions, for example, dilute salt and sugar solutions, and it is an important relationship in estimating freezing and boiling temperatures of solutions.

When a solute is mixed with water, there is a reduction in the chemical potential of water, because the chemical potential of pure water, μ_A^0, decreases with the quantity of $RT \ln x_A$ (x_A is mole fraction of water). However, the solute molecules appear only in the liquid, because the vapor and solid states contain only water molecules. The change in the chemical potential of water means that the transformation to ice at a constant pressure occurs at a lower temperature and its transformation to vapor occurs at a higher temperature.

3.4.2.2 Freezing Temperature Depression

The freezing temperature depression by solutes in water affects the freezing behavior of foods. The freezing and melting of water in a food system is much more complicated and the freezing temperature depression only refers to the temperature at which the last ice crystals have the same vapor pressure as unfrozen water. Hence, the solid phase contains only the solvent molecules and its chemical potential at equilibrium with the unfrozen water is given by Equation 3.51.

$$\mu_A^0(s) = \mu_A^0(l) + RT \ln x_A \tag{3.51}$$

The mole fraction of water can be written as $x_A = 1 - x_B$, which gives Equation 3.52.

$$\ln(1 - x_B) = \frac{\mu_A^0(s) - \mu_A^0(l)}{RT} \tag{3.52}$$

The change of chemical potential in freezing, that is, the difference in chemical potential resulting from freezing at a pressure of 1 bar, is the same as the change in Gibbs energy of freezing, $\Delta_{fng}G$, which gives Equation 3.53.

$$\ln(1 - x_B) = -\frac{\Delta_{fng}G}{RT} \tag{3.53}$$

According to the definition of Gibbs energy, Equation 3.54 applies.

$$\Delta_{fng}G = \Delta_{fng}H - T\Delta_{fng}S \tag{3.54}$$

In dilute solutions, the change of the values are relatively small and the temperature dependence of $\Delta_{fng}H$ and $\Delta_{fng}S$ may be ignored. Hence, Equation 3.55 can be written.

$$\ln(1 - x_B) = -\left(\frac{\Delta_{fng}H}{RT} - \frac{\Delta_{fng}S}{R}\right) \tag{3.55}$$

At the freezing temperature of pure water, T^*, the solid phase contains only water molecules giving $x_A = 1$ and $x_B = 0$, and the following relationship applies:

$$\ln 1 = -\left(\frac{\Delta_{fng}H}{RT*} - \frac{\Delta_{fng}S}{R}\right) \tag{3.56}$$

Equation 3.55 and Equation 3.56 may be combined according to Equation 3.57 to obtain Equation 3.58 which estimates the freezing temperature depression of a dilute solution.

$$\ln(1 - x_B) = -\left(\frac{\Delta_{fng}H}{RT} - \frac{\Delta_{fng}S}{R}\right) + \left(\frac{\Delta_{fng}H}{RT^*} - \frac{\Delta_{fng}S}{R}\right) \tag{3.57}$$

$$\ln(1 - x_B) = \frac{\Delta_{fng}H}{R}\left(\frac{1}{T^*} - \frac{1}{T}\right) \tag{3.58}$$

The freezing temperature depression may also be derived from the temperature at which the vapor pressure of ice and unfrozen water in the food are equal (Roos, 1995). The freezing temperature depression is discussed in more detail by Heldman (2006) in this book.

Equation 3.58 states that the freezing temperature depression is a linear function of the solute mole fraction, because all other parameters and, therefore, $RT^{*2}/\Delta_{fng}H$ have a constant value. This constant value can be calculated for any solvent and it is referred to as cryoscopic constant. The freezing temperature depression at a pressure of 1 bar is not dependent on the solute and, if a food can be considered as a dilute system, it may be assumed to have an average solute mole fraction that results in a freezing temperature depression of $1.86°C$ for each mole of solute in 1 L of water. The cryoscopic method in measuring the freezing temperature of milk, and therefore, the solid content or dilution resulting from added water is used as a standard rapid method in the analysis of the quality of milk supplied by farmers to dairies.

3.4.2.3 Boiling Temperature Elevation

The elevation of boiling temperature by solutes is a property of all solute–solvent systems. In foods, the elevation of the boiling temperature of water is common even in a simple solution of salt and water. As water in a solution exhibits a lower vapor pressure than pure water, the vapor pressure may overcome the atmospheric pressure only at a higher temperature. The requirement for boiling is also that the vapor pressure of water in the solution and in the vapor phase above the liquid phase be the same, that is, $\mu_A^0(g) = \mu(l)$. As the vapor phase consists only of water, A, Equation 3.59 applies.

$$\mu_A^0(g) = \mu_A^0(l) + RT \ln x_A \tag{3.59}$$

The mole fraction of water can be written as $x_A = 1 - x_B$, and Equation 3.59 can be written into the form of Equation 3.60

$$\ln(1 - x_B) = \frac{\mu_A^0(g) - \mu_A^0(l)}{RT} \tag{3.60}$$

The change of chemical potential in vaporization, that is, the difference in chemical potential resulting from vaporization at a pressure of 1 bar is the same as the change in Gibbs energy of vaporization, $\Delta_{vap}G$, which gives Equation 3.61. $\Delta_{vap}G$ is defined by Equation 3.62.

$$\ln(1 - x_B) = \frac{\Delta_{vap}G}{RT} \tag{3.61}$$

$$\Delta_{vap}G = \Delta_{vap}H - T\Delta_{vap}S \tag{3.62}$$

Over a large temperature range which may apply to some foods and many concentrates, there may be a large change in the heat and entropy of vaporization. However, for a dilute solution, the changes of these values are relatively small and the temperature dependence of $\Delta_{vap}H$ and $\Delta_{vap}S$ may be ignored. Then Equation 3.63 is obtained.

$$\ln(1 - x_B) = \frac{\Delta_{vap}H}{RT} - \frac{\Delta_{vap}S}{R} \tag{3.63}$$

At the boiling temperature of pure water, T^*, the vapor phase contains only the solvent molecules giving $x_A = 1$ and $x_B = 0$, and the relationship of Equation 3.64 applies.

$$\ln 1 = \frac{\Delta_{vap}H}{RT^*} - \frac{\Delta_{vap}S}{R} \tag{3.64}$$

Equation 3.63 and Equation 3.64 may be combined to obtain Equation 3.65 that reduces to Equation 3.66.

$$\ln(1 - x_B) = \frac{\Delta_{vap}H}{RT} - \frac{\Delta_{vap}S}{R} - \left(\frac{\Delta_{vap}H}{RT^*} - \frac{\Delta_{vap}S}{R} \right) \tag{3.65}$$

$$\ln(1 - x_B) = \frac{\Delta_{vap}H}{R} \left(\frac{1}{T} - \frac{1}{T^*} \right) \tag{3.66}$$

Equation 3.66 can be used to evaluate the boiling temperature elevation of food systems.

3.4.3 FREEZING, FREEZE-CONCENTRATION AND MELTING

The latent heat of ice melting is relatively high, being 333.5 J/g (6003 J/mol). Below 0°C pure water (triple point 0.0099°C and 6.104 mbar) can exist only as ice or vapor (Figure 3.9). The freezing temperature of water in foods is always lower than that of pure water, as defined by the freezing temperature depression. Most materials expand 1 to 20% in volume during the phase change from a crystal to a melt. Water, however, is an unusual exception showing an increase in the volume of ice, which is 8 to 10% larger than the volume of liquid water.

Water in high moisture foods begins to crystallize as pure ice at temperatures slightly below $0°C$. The amount of ice crystallized at a particular temperature depends on food composition, because various dissolved substances depress the freezing temperature of water. As ice forms, freeze-concentration of solids occurs simultaneously with ice formation and decreases the freezing temperature of the remaining unfrozen water. Between the temperature at which the first ice crystals melt in a maximally freeze-concentrated system, defined as antemelting (T_{am}) (Luyet and Rasmussen, 1967, 1968; Rasmussen and Luyet, 1969; Simatos et al., 1975; LeMeste and Simatos, 1980) or as onset of ice melting (T'_m) (Roos and Karel, 1991b), and the temperature at which all ice is finally converted to water (equilibrium melting temperature, T_m), the equilibrium amount of unfrozen water and ice are controlled by their chemical potential or vapor pressures. In an equilibrium state, the vapor pressure of the unfrozen water (p_u) in food is the same as that of ice (p_i) at a constant temperature. Values of $p_u < p_i$ lead to melting and $p_u > p_i$ to ice formation. It should be noted that freezing is often not completed at normal temperatures used for industrial food freezing and frozen storage, and some of the water remains unfrozen even at very low temperatures.

The water activity of frozen food can be related to the vapor pressure of the unfrozen water, p_u, which in frozen food should be equal to the vapor pressure of ice, that is, $p_u = p_i$. Hence, the water activity of ice at a given temperature needs to be the same as the water activity of food at the same temperature. The water activity of frozen food decreases with temperature and is given by Equation 3.67.

$$A_w = \frac{p_u}{p_A^0} \tag{3.67}$$

Freeze-concentration of nonaqueous food constituents increases with decreasing temperature, which also affects pH, titratable acidity, ionic strength, viscosity, freezing temperature, surface and interfacial tension, and oxidation-reduction potential of the remaining unfrozen phase (Fennema, 1985). At low temperatures, the unfrozen matrix is solidified as an amorphous glass (vitrification) (Rey and Bastien, 1962; Rasmussen and Luyet, 1969; Bellows and King, 1973; Simatos et al., 1975; Levine and Slade, 1988; Slade and Levine, 1990; Roos and Karel, 1991b; Roos, 1995). Foods, however, do not have exact melting points, and they show a wide melting range.

Although various forms of ice crystals can be formed at different pressures, the melting temperature of water is not significantly affected by pressure at traditional food processing and storage conditions. New possibilities in manipulating the freezing and thawing temperatures of water in foods have been provided by the developments in high pressure food processing technology (Kalichevsky et al., 1995; Li and Sun, 2002; LeBail et al., 2003). The freezing temperature of ordinary ice decreases when pressure is increased to $-22°C$ at 207.5 MPa (Kalichevsky et al., 1995). High pressure technology allows cooling without ice formation well below the freezing temperature at normal pressures (Figure 3.10). When pressure is released, rapid nucleation results in uniform formation of small ice crystals. In high pressure thawing, frozen food is exposed to high pressure and a sufficient amount of heat is supplied to allow the change in phase (Figure 3.10). The main advantage of high pressure thawing is the greater driving force (higher temperature above melting temperature resulting from decreasing melting temperature under high pressure) for thawing and a more rapid process (Kalichevsky et al., 1995; Li and Sun, 2002; LeBail et al., 2003).

3.4.3.1 Freezing and Melting of Eutectic Solutions

Eutectic solutes in water, for example, dissolved salts, may crystallize at their eutectic temperatures (Figure 3.11). During freezing of eutectic solutions, a freeze-concentrated saturated solution is formed, as the solvent water is removed from the solution to the solid ice phase. At the saturation concentration, the solute crystallizes resulting in full crystallization of both the solvent and solute. The two step melting of frozen eutectic solutions can be determined using thermal analytical

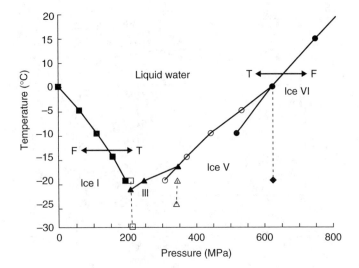

FIGURE 3.10 Phase diagram of water showing the pressure-dependence of ice melting temperature for various pressure-dependent polymorphic forms of ice.

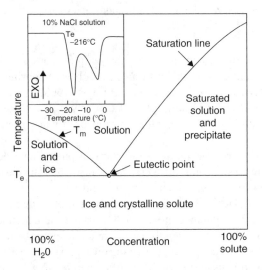

FIGURE 3.11 Phase diagram of materials showing eutectic behavior with water. The inset figure shows two step melting of a 5% NaCl solution observed during heating at 5°C/min in a DSC scan. The first peak is due to eutectic melting and the second peak to ice melting.

methods, for example, DSC (Rey, 1960; Roos, 1995). The two step melting endotherm determined by differential scanning calorimetry for a NaCl solution is shown in Figure 3.11. The phase behavior of foods containing salts having low eutectic points, for example, $CaCl_2$ is complicated, because many food components probably solidify with the unfrozen water as a glass before the eutectic point is achieved. Hence, many eutectic solutes may not crystallize in foods, because of their complex composition and high viscosity at low temperatures.

3.4.3.2 Freeze-Concentrated Systems and Ice Melting in Foods

Freeze-concentration of foods results from ice formation, which increases concentration of potential reactants which in some conditions favors increasing reaction rates in frozen foods. At low

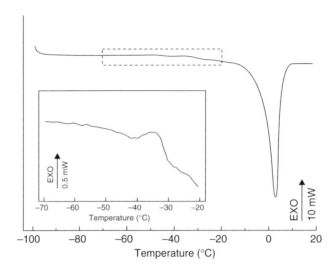

FIGURE 3.12 Differential scanning calorimetry thermogram of 20% sucrose solution determined by heating the sample from -100 to $20°C$ at $5°C/min$. The ice melting endotherm dominates over the whole temperature range but an expanded thermogram from $-70°$ to $-20°C$ (inset thermogram) shows a glass transition, T_g, crystallization exotherm for ice formation (or devitrification), and subsequent onset of melting, T_m.

temperatures the viscosity of the concentrated unfrozen phase often becomes a limiting factor for additional ice formation and leads to various transitions observed during rewarming of frozen solutions (Luyet and Rasmussen, 1967; Rasmussen and Luyet, 1969; Simatos et al., 1975; Levine and Slade, 1986; Roos, 1987; Roos and Karel, 1991b; Roos, 1995). The physical state of the amorphous, unfrozen phase containing concentrated solids and unfrozen water often defines rates of physical changes.

Freeze-concentrated matrices exhibit various thermal phenomena depending on composition, initial concentration, temperature and time (Franks et al., 1977). The basic thermal events observed during heating of frozen binary solutions of sugars and water are glass transition of the freeze-concentrated matrix, T_g, and ice melting, T_m. Freeze-concentration also increases the viscosity of the unfrozen matrix with decreasing temperature which leads to time-dependent and probably diffusion controlled ice formation. Therefore, rapidly frozen materials may show ice crystallization during rewarming (devitrification) at a temperature allowing diffusion (Figure 3.12 and Figure 3.13). For the same reason, thermal treatments or annealing are often applied to achieve maximum ice formation. Growth of ice crystals during freezing may also be substantially delayed by high molecular weight food components, which increase the viscosity of the unfrozen solution (Muhr and Blanshard, 1986).

Ice formation during freezing of dilute solutions occurs fairly freely. In a DSC heating scan the transitions observed are the following, (i) glass transition, (ii) onset of ice melting, and (iii) the main melting endotherm. Melting curves for 20% and 65% sucrose solutions are shown in Figure 3.12 through Figure 3.14. As the concentration of the initial solution is increased, the viscosity of the initial solution as well as at any stage of freeze-concentration decreases the rate of ice formation. Limited ice formation may also occur in rapid cooling and rapidly cooled solutions, often exhibit ice formation (devitrification) on rewarming. This produces an exotherm above the T_g of the partially freeze-concentrated matrix at a given temperature corresponding to a viscosity allowing ice formation within a time period compatible with the rewarming rate (Figure 3.12 and Figure 3.13). The T_g observed occurs at a lower temperature than that of a maximally freeze-concentrated 20% sucrose solution, because more unfrozen water is plasticizing the unfrozen matrix, thus, lowering the T_g. Isothermal holding at a certain temperature below the onset temperature of ice melting, T_m' of the maximally freeze-concentrated system, but above T_g', allows time-dependent ice formation. T_g' is the glass

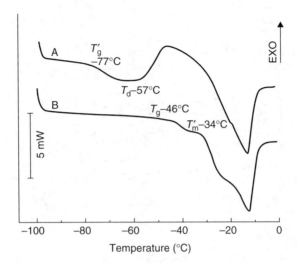

FIGURE 3.13 Thermal behavior of 65% sucrose solutions observed in nonannealed and annealed samples. (A) Thermogram of solution cooled to −100°C and heated to 0°C at 5°C/min. The transitions observed are glass transition of the supercooled solution, T_g, devitrification at the onset of ice formation exotherm, T_d, which is followed by ice melting endotherm. (B) Thermogram of solution cooled to −100°C, heated to −35°C at 10°C/min, held isothermally 30 min at −35°C to allow time for ice crystallization at a temperature above T_g' but below T_m', cooled to −100°C at 10°C/min and scanned from −100°C to 0°C at 5°C/min. Annealing leads to maximal freeze-concentration, and the transitions observed in the annealed sample are glass transition of the maximally freeze-concentrated solids, T_g', and onset of ice melting, T_m', followed by the ice melting endotherm. The effect of freeze-concentration is observed as an increased glass transition temperature, disappearance of the ice formation exotherm, and increased size of the ice melting endotherm.

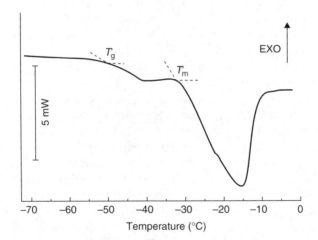

FIGURE 3.14 Phase transitions of 65% sucrose solution (w/w) observed during heating from −100° to 0°C at 5°C/min using differential scanning calorimetry. The transitions observed are the glass transition of the maximally freeze-concentrated solution, T_g', and onset of ice melting, T_m', followed by the melting endotherm of ice. The peak area is proportional to the latent heat of melting of ice, ΔH_m, in the sample.

transition of the maximally freeze-concentrated unfrozen matrix. No further unfrozen water is able to crystallize below T_g', and, therefore, depending on viscosity in the vicinity of T_g', the ice melts at the same temperature as it was formed. Hence, the theoretical values for the T_g' and T_m' are the same. However, partial softening of the glass is required before time-dependent ice formation

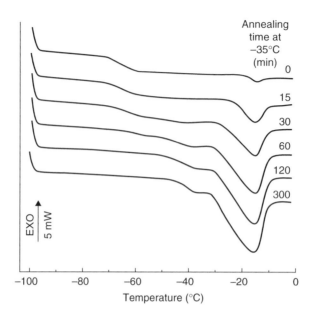

FIGURE 3.15 Effect of annealing on ice formation in 68% sucrose solution. The annealing was done as given in Figure 3.14 for varying times at $-35°C$. The ice formation is governed by viscosity and only a small exotherm followed by immediate melting is observed in a nonannealed sample. The glass transition temperature increases, the ice formation exotherm decreases and the size of the ice melting endotherm increases with increasing annealing time. The sample annealed for 300 min has no exothermal change above glass transition temperature due to ice formation, and it shows the T_g' and T_m' values of sucrose.

can occur. This is shown by the maximally freeze-concentrated solutions of most sugars, which often show onset of melting above the T_g' and in some cases above the endpoint temperature of the glass transition range. The melting behavior of a maximally freeze-concentrated annealed solution is shown in Figure 3.14 and Figure 3.15.

At a given extent of freeze-concentration, the melting temperature of ice and the T_g theoretically reach the same temperature (T_g' for T_g or T_m' for T_m). The unfrozen water content in this system (W_g') is the amount of water that remains unfrozen, because of the high viscosity of the maximally freeze-concentrated unfrozen phase. At this temperature, the extremely high viscosity becomes a kinetic barrier for further ice formation (Roos and Karel, 1991b). At a temperature below the T_g' of the maximally freeze-concentrated matrix, the amorphous unfrozen phase vitrifies and exists as a glassy solid within the ice crystals. Levine and Slade (1986) defined an endothermal transition, which is often observed between antemelting, T_{am}, and incipient melting, T_{im}, as the glass transition temperature of the maximally freeze-concentrated amorphous matrix, T_g'. Other authors including the present author, however, have suggested that this transition considered as a glass transition by Levine and Slade (1986) results from melting of ice at T_m' (LeMeste and Simatos, 1980; Izzard et al., 1991; Roos and Karel, 1991b; Roos, 1995; Talja and Roos, 2001).

Freezing and melting of water is important to such food processes as freeze-concentration, freeze-drying, and freezing. The T_m' defines a structural collapse (a material flows and cannot support its own weight) temperature for various materials during their freeze-drying (Roos, 1997), and above T_m' increasing water content leads to rapid decrease of viscosity, because of dilution of the food solids (increasing plasticization by unfrozen water) and concomitant decrease of the viscosity controlling T_g. Below T_m' the viscosity of frozen materials decreases until a glass is formed at T_g'. At temperatures below T_m', if the maximum amount of ice has formed, several frozen foods should exhibit improved stability (Slade and Levine, 1990, 1995).

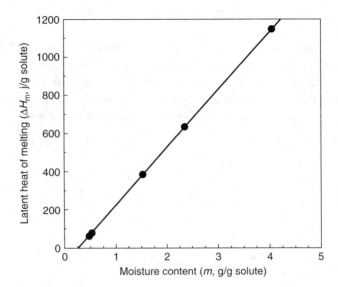

FIGURE 3.16 Latent heat of melting, ΔH_m, of annealed sucrose solutions obtained by integration of the melting endotherm above $-35°C$ against their moisture content, m. Extrapolation of the line gives a maximally freeze-concentrated sucrose concentration ($\Delta H_m = 0$ J/g solute) of 78%.

The concentration of the maximally freeze-concentrated solution can be obtained from a state diagram based on glass transition data at various water contents. Hence, the water content or solid content corresponding to the glass transition of the maximally freeze-concentrated solids give the most precise estimate for the composition of the unfrozen, maximally freeze-concentrated phase. Another possibility is to plot the latent heat of melting of ice in solutions with various initial water contents against the water content. This produces a straight line, which can be extrapolated to obtain $\Delta H_m = 0$ J/g at W'_g (Figure 3.16).

Glucose, fructose, sucrose, and lactose belong to the most important natural carbohydrates in foods. They are also important cryoprotectants in the freezing of biologically active materials. The various phenomena related to freezing and melting of these carbohydrate solutions have been studied intensively (Rasmussen and Luyet, 1969), but some of the basic phenomena have been understood only recently (Slade and Levine, 1988b; Izzard et al., 1990; Roos and Karel, 1991b; Talja and Roos, 2001; Roos, 2002), although there is still some disagreement on the true nature of the transitions (Goff et al., 2003). The complex transitions in frozen systems are time-dependent and include both equilibrium and nonequilibrium phenomena, including the viscosity controlled nature of ice formation at low temperatures. Therefore, only annealing of solutions at an appropriate temperature may favor maximum amount of ice formation, which is extremely important to freeze-drying. It has been observed that, for example, flavor retention during freeze-drying is significantly improved if the material is slowly frozen (Thijssen, 1971; Karel and Flink, 1973), and, therefore, allowed to form a greater amount of ice. Crystallization and recrystallization of ice and other compounds in partially freeze-concentrated solutions is an important factor affecting quality of frozen foods. These processes are also related to the glass transition and dilution of freeze-concentrated systems (Roos, 1995; Hartel, 2001). T'_g and T'_m values for selected food materials are given in Table 3.3.

3.4.4 EVAPORATION OF WATER IN FOODS

The temperature of food materials with high water content is often controlled by evaporation of water which requires 2255.3 J/g at 100°C. The latent heat of evaporation decreases with increasing boiling temperature (Table 3.2). The critical temperature of water vapor is 374.15°C, and pressure

TABLE 3.3
Melting Temperatures, T_m, Glass Transition Temperatures, T_g, Change in Heat Capacity Over the Glass Transition, ΔC_p, Glass Transition Temperatures of Maximally Freeze-Concentrated Solutes, T'_g, and Onset Temperatures of Ice Melting of Freeze-Concentrated Matrices, T_m', for Selected Carbohydrates, k is Constant for Gordon-Taylor Equation 3.69

Compound	T_m (°C)	T_g (°C)	ΔC_p	T_m/T_g	k	T'_g (°C)	T'_m (°C)
Altrose	107	10.5					−44
Arabinose	150	−2	0.66	1.56	3.55	−66	−53
Fructose	124	5	0.75	1.37	3.76	−57	−46
Fucose		26		1.36	4.37	−62	−48
Galactose	170	30	0.50	1.44	4.49	−56	−45
Glycerol	18	−93					−65
Glucose	158	31	0.63	1.37	4.52	−57	−46
Isomalt		64					
Lactose	214	107			6.56	−41	−30
Lactulose		79	0.45		5.92	−42	−32
Lyxose	115	8					−47.5
Maltitol		39	0.56	1.32	4.75	−47	−37
Maltose	129	87	0.61		6.15	−42	−32
Maltotriose	133.5	76					−23.5
Mannitol	170	15		1.55			
Mannobiose	205	90					−30.5
Mannose	139.5	25	0.72	1.32	4.34	−58	−45
Melibiose		85	0.58		6.10	−42	−32
Raffinose		70	0.45		5.66	−36	−28
Rhamnose		−7	0.69		3.40	−60	−47
Ribose	87	−20	0.67	1.36	3.02	−67	−53
Sorbitol	111	−9	0.96	1.36	3.35	−63	−49
Sorbose		19	0.69	1.46	4.17	−57	−44
Sucrose	192	>62	0.60	1.33	5.42	−46	−34
Talose	140	11.5					−44
Trehalose	203	107	0.55		6.54	−40	−30
Turanose	177	52					−31
Xylitol	94	−29	1.02	1.48	2.76	−72	−57
Xylose	153	6	0.66	1.49	3.78	−65	−53

Source: From Levine H and Slade L. *Cryo-Lett* 9, 21–63, 1988; Roos Y and Karel M. *Biotehcnol Prog* 6, 159–163, 1990; Orford PD, Parker R, and Ring SG. *Carbohydr Res* 196, 11–18, 1990; and Slade L and Levine H. *CRC Crit Rev Food Sci Nutr* 30, 115–360, 1990.

221.4 bar. Vapor pressure over ice, supercooled water, and water at some temperatures are given in Table 3.1 and Table 3.2. Evaporation of water occurs usually in all food processing. The boiling point and also the latent heat of evaporation are strong functions of pressure (Figure 3.9). Evaporation of water is fastest at boiling temperature, which is the highest possible temperature of liquid water at a given pressure. Many food materials are heat sensitive (for example, heating results in protein denaturation, change of flavor, change of color, etc.), and reduced evaporation pressures are used to depress the boiling point of water to 40–60°C.

Evaporation of water occurs also in drying. However, in drying processes the amount of energy needed is substantially higher than the energy needed for evaporation, because of poor energy

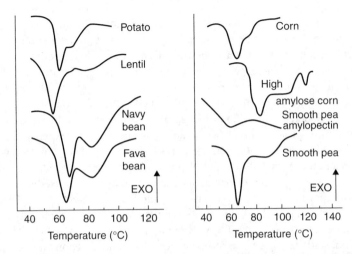

FIGURE 3.17 Gelatinization thermograms of starches of various origin. (Adapted from Biliaderis CG, Maurice TJ, and Vose JR. *J Food Sci* 45, 1669–1680, 1980. With permission.)

efficiency in drying equipment. Therefore, foods are usually concentrated before drying to improve efficiency. The importance of phase and state transitions in dehydration of biomaterials is discussed by Achanta and Okos (1996) and Roos (2004).

3.5 PHASE TRANSITIONS OF CARBOHYDRATES, LIPIDS, AND PROTEINS

3.5.1 PHASE TRANSITIONS OF STARCH

Native starches are glucose polymers, which are present in starch granules of plants and include cereal, legume, and tuber starches. The structure of starch granules has been discussed by Blanshard (1987), Slade and Levine (1988a), Lund (1989) and Autio and Salmenkallio-Marttila (2003). Most starches contain about 25% amylose and 75% amylopectin. Amylose is a mostly linear polysaccharide composed of about 4000 (1-4 linked) glucose units, and amylopectin is a highly branched (1-4-6 linked branch every 20–25 residues) polysaccharide with about 100,000 glucose units (Ring, 1985; Morris, 1990).

Donovan (1979) studied various thermal transitions occurring during heating of starch water mixtures. At high water levels, a single endotherm was observed, and it was referred to as the gelatinization endotherm. At intermediate moisture contents two separate endotherms were observed. At low moisture contents this endotherm was shifted to higher temperatures with decreasing water content. Also the enthalpy of the transition decreased with decreasing water content. Similar behavior was reported to apply to starches of various origins (Biliaderis et al., 1980). The typical phase transitions of starch are glass transition of the amorphous components, melting of crystallites and melting of amylose–lipid complexes (Biliaderis et al., 1980; Eliasson, 2003). The most important phase transition of starches in foods is gelatinization (Lund, 1989). Typical DSC thermograms for starches are shown in Figure 3.17.

3.5.1.1 Gelatinization of Starch

Atwell et al. (1988) defined gelatinization as collapse (disruption) of molecular orders within the starch granule manifested by irreversible changes, such as granular swelling, native crystallite melting, loss of birefringence, and starch solubilization. The temperature of initial gelatinization, and the range over which it occurs, was stated to be governed by starch concentration, method of observation, granule type, and heterogeneities within the granule population under observation. The basic methods to detect starch gelatinization are determinations of turbidity, swelling, absorption of dyes, x-ray diffraction, loss of birefringence, enzymatic digestibility, light scattering, NMR spectra, and enthalpy change using differential scanning calorimetry. Gelatinization usually occurs over a temperature range of 5–10°C. Starch gelatinization may be treated as a melting transition of a semicrystalline synthetic polymer (Biliaderis et al., 1986). Levine and Slade (1990) described starch gelatinization as a nonequilibrium melting process, which occurs during the heating of starch in the presence of water. Gelatinization of starch starts in amorphous regions of starch granules. Starch shows annealing and recrystallization phenomena typical of water plasticizable polymers (Biliaderis et al., 1986; Levine and Slade, 1990). Lipids often interact with starches and they can be used to influence swelling, solubilization, skinning, and stickiness properties of starches and cereal grains (Eliasson, 2003).

The thermodynamic relationship between the melting point of a crystalline polymer and its diluent concentration can be expressed by the Flory–Huggins equation (Flory, 1953). The Flory–Huggins equation (Equation 3.20) has been used to analyze the melting phenomenon of starch as a first-order phase transition between crystalline and amorphous states (Lelievre, 1976; Donovan, 1979; Biliaderis et al., 1980). The equation relates melting point of a polymer, its heat of fusion, and molar volumes of polymer repeating unit and diluent, and it is useful in describing the water content dependence of melting (Biliaderis et al., 1986).

$$\frac{1}{T_m} = \frac{R}{\Delta H_m} \frac{V_u}{V_1}(1 - \chi_1)v_1 + \frac{1}{T_m^0} \tag{3.68}$$

where T_m is the observed melting temperature (undiluted polymer), T_m^0 is the melting temperature of the polymer, ΔH_m is the latent heat of melting of a polymer repeating unit, V_u and V_1 are the molar volumes of the repeating unit and water, respectively, v_1 is volume fraction of water, and χ_1 is Flory–Huggins polymer–diluent interaction parameter.

Although the equation fits well to melting of crystals in starch (Donovan, 1979; Roos, 1995), Levine and Slade (1990) stated that the equation should not be used to gelatinization and melting processes of starch, because the transitions cannot be assumed as equilibrium transitions.

3.5.1.2 Gelatinization Temperature

Starch granules are insoluble in cold water, but they swell upon heating and crystallites become disordered. Starch gelatinization occurs during heating of starch-water mixtures. The gelatinization of starch also requires water, and the gelatinization temperature depends on moisture content. Water in the amorphous parts of starch acts as a plasticizer. This decreases the glass transition temperature of the noncrystalline regions of native starch and leads to melting of the crystalline parts as temperature is increased.

High-amylose corn starch may contain up to 85% amylose. Amylose has a high crystallinity and high temperatures are needed for its gelatinization. Waxy corn, barley, and rice starches contain mainly amylopectin which gives them ability to form clear stable pastes. Usually gelatinization requires a minimum of about 6% starch in water. Gelatinization temperatures and latent heats of gelatinization for selected starches are given in Table 3.4 and Table 3.5.

In partially crystalline polymers, only amorphous regions are plasticized by water. Biliaderis et al. (1980) suggested that when excess water is present, hydration and swelling of the amorphous regions of starch facilitate melting of the crystallites during heating, and a single endotherm is obtained. In concentrated starch water mixtures, this destabilizing effect of the amorphous regions decreases, and only partial melting of crystallites occurs. Subsequent redistribution of the water around remaining crystallites assists their melting during further heating. Gelatinization temperatures

TABLE 3.4
Gelatinization Temperature Range, Amylase Content, Crystallinity, and Latent Heat of Gelatinization, ΔH_{gel}, for Various Starches Obtained by DSC and Microscopy

Starch	Amylose content (%)	Crystallinity (%)	Method	Temperature range (°C)	ΔH_{gel} (J/g)
Barley	22		Hot stage	51–60	
Corn			Hot stage	65–76	
			DSC	65–77	13.8
	23–28	40	DSC	62–76	13.8–20.5
	52	15–22	DSC	67–86	28.0
	1	40	DSC	63–80	16.7–20.1
Oats	23–24	33	DSC	52–64	9.2
Pea	29		DSC	62–	12.5
Potato			Hot stage	59–68	
			DSC	57–95	21–23
	19–23	28	DSC	58–71	17.6–18.8
Rice			Hot stage	72–79	
			DSC	68–82	13.0
	17-21	38	DSC	68–82	13.0–16.3
Rye	27	34	DSC	49–70	10.0
Sorghum	25	37	Hot stage	68–78	
Tapioca	17–18	38	DSC	63–80	15.1–16.7
Triticale	23–24		Hot stage	55–62	
Wheat			Hot stage	55–66	
	23–26	36	DSC	52–66	9.7–12.0
			Amylograph	54–67	
			Turbidity	55–100	

Source: From Lund D. *CRC Crit Rev Food Sci Nutr* 20, 249–273, 1984; and Roos YH. *Phase Transitions in Foods*. San Diego CA: Academic Press, 1995, p. 361.

and latent heats of gelatinization at minimum water content and maximum gelatinization water content are given in Table 3.6.

Annealing of starch-water mixtures below gelatinization temperature has been found to cause an increase of gelatinization temperature but narrowing of gelatinization temperature range (Wirakartakusumah, 1981; Lund, 1984). Annealing leads also to lower gelatinization enthalpies (Table 3.7). Krueger et al. (1987) found that commercially prepared starches were closer to annealed starches than to native starches, probably because the wet-milling process may effectively anneal starch. They pointed out that the gelatinization temperature and heat of gelatinization values of native, laboratory isolated starches were quite different from values obtained for starch that was annealed by heating in water at subgelatinization temperatures. Knutson (1990) reported that annealing close to gelatinization temperature resulted in partial gelatinization with increasing time.

3.5.1.3 Effect of Added Compounds on Starch Gelatinization

Sugars decrease the rate of starch gelatinization and increase gelatinization temperatures (Table 3.8). Chungcharoen and Lund (1987) reported addition of sucrose and sodium chloride to increase gelatinization temperatures and to decrease enthalpy values of gelatinization. Levine and Slade (1990)

TABLE 3.5
Gelatinization Temperatures (T_1 Onset Temperature of Gelatinization, T_2 Temperature at Peak of Gelatinization, T_3 Endset Temperature of Gelatinization), and Latent Heat of Gelatinization, ΔH_{gel}, of Starches from Various Origins

Starch origin	Concentration (%)	Melting (°C) T_1	T_2	T_3	ΔH_{gel} (J/g)
Potato					
amylose	12	143.2	153.1	164.2	39.3
Amylopectin					
Faba bean	25	40.1	59.5	71.1	20
Maize native	25	45.3	58.7	64.7	28.5
pregel.	25	47.7	60.2	67.5	26.8
Potato	25	42.3	59.3	73.2	20.5
Rice	25	40.7	56.7	68.7	25.1
Tapioca	25	41.7	57.5	65.8	13
Wheat	25	44.2	54.2	64.0	30.1

Source: From Ring SG, Colonna P, I'Anson KJ, Kalichevsky MT, Miles MJ, Morris VJ, and Orford PD. *Carbohydr Res* 162, 277–293, 1987.

TABLE 3.6
The Minimum and Maximum Values of Heat of Gelatinization, ΔH_{gel}, and Amount of Water Needed for Gelatinization of Starches, and Water Content Needed for Maximum Gelatinization

Starch	Minimum Amount	Maximum water(%)	ΔH_{gel} (J/g) min.	max.	Reference
Bean			11.7	21.8	Biliaderis et al., 1980
Corn	35	60			Reid and Charoenrein, 1985
Lentil			5.86	14.2	Biliaderis et al., 1980
Pea			11.3	14.7	Biliaderis et al., 1980
Potato	31	58			Collison and Chilton, 1974
Wheat	33		0	13.8	Eliasson, 1980
			0.84	19.7	Wootton and Bamunuarachchi, 1979

postulated that sugar solutions decrease the T_g of starch less than water alone which leads to increased gelatinization temperatures. This assumption was based on the effect of increase in the molecular weight of the plasticizer (sucrose + water > water).

Wootton and Banmunuarachchi (1980) reported that sodium chloride had the maximum effect on gelatinization at concentrations of 6 to 9% (Table 3.9). Other salts including sodium sulfate and sodium hydrogen phosphate also increased gelatinization temperature (Evans and Haisman, 1982). The gelatinization temperature was found to first decrease at a low concentration and then increase with increasing concentration when calcium chloride was added. The effects of surfactants on starch

TABLE 3.7

Effect of Water Content and Annealing on the Heat of Gelatinization, ΔH_{gel}, of Waxy Maize Starch

Water (%)	Annealing temperature (°C)	Time (min)	ΔH_{gel} (J/g)
65		0	16
	62	15	11
	70	15	1.7
	80	15	0
55		0	16
	65	30	13
	70	10	8
	80	15	0.7
	90	15	0
45		0	16
	65	15	14
	75	15	12
	80	15	10
	95	15	7
	110	15	0

Source: From Maurice TJ, Slade L, Sirett RR, and Page CM. In: Simatos D and Multon JL, Eds, *Properties of Water in foods in Relation to Quality and Stability*. Dordrecht. Netherlands: Martinus Nijhoff Publishers, 1985, pp. 211–227.

TABLE 3.8

Effect of Sucrose Concentration on Gelatinization Temperature (T_1 Onset Temperature of Gelatinization, T_2 Temperature at Gelatinization Peak, T_3 Endset Temperature of Gelatinization) and Latent Heat of Gelatinization, ΔH_{gel}, of Wheat Starch

Sucrose (% in H_2O)	ΔH_{gel} (J/g)	Gelatinized (% of starch)	Gelatinization temperature T_1	T_2	T_3
0	19.7	100	50	68	86
15	13.4	68	50	70	86
30	11.7	60	50	73	86
45	9.6	49	50	75	86

Source: From Wootton M and Bamunuarachchi A. *Staerke* 32, 126–129, 1980.

TABLE 3.9
Effect of NaCl Concentration on Gelatinization Temperature (T_1 Onset Temperature of Gelatinization, T_2 Temperature at Gelatinization Peak, T_3 Endset Temperature of Gelatinization) and Latent Heat of Gelatinization, ΔH_{gel}, of Wheat Starch

NaCl	ΔH_{gel}	Gelatinized	Gelatinization temperature (°C)		
(% in H$_2$O)	(J/g)	(% of starch)	T_1	T_2	T_3
0	19.7	100	50	68	86
3	11.3	57	58	71	88
6	10.5	53	64	75	88
9	10.9	55	68	78	88
12	11.3	57	65	77	88
15	11.3	57	65	77	88
21	11.7	60	61	80	90
30	13.8	70	59	79	91

Source: From Wootton M and Bamunuarachchi A. *Staerke* 32, 126–129, 1980.

TABLE 3.10
Gelatinization Temperatures and Latent Heats of Gelatinization, ΔH_{gel}, for Wheat and Potato Starch in Presence of Various Compounds

	Wheat starch			Potato starch		
Additive	T_1 (°C)	T_2 (°C)	ΔH_{gel} (J/g)	T_1 (°C)	T_2 (°C)	ΔH_{gel} (J/g)
No	57.0	61.3	12.7	58.2	63.7	17.0
Sodium dodecyl sulfate	54.7	60.1	9.5	55.4	60.8	13.97
Cetyltrimethyl Ammoniumbromide	57.6	61.7	8.5	58.6	63.9	14.1
saturated monoglycerides	56.7	60.5	12.8	58.8	63.3	17.8
Lysolecithin	55.7	60.6	7.5	57.8	63.4	15.8
Lecithin	56.7	60.8	12.2	59.1	63.7	18.1

Source: From Eliasson A-C. *Carbohydr Polym* 6, 464–476, 1986.

gelatinization are given in Table 3.10. Allen et al. (1982) reported that chlorine treatments of wheat flour did not alter the gelatinization temperature and heat of gelatinization of starch.

3.5.1.4 Melting of Amylose-Lipid Complexes

Lipids form complexes with amylose in which the lipid molecules form the core of an amylose helix. A differential scanning calorimetric study by Kugimiya et al. (1980) showed that an endothermal transition typical of cereal starches near 100°C was due to melting of amylose-lipid complexes.

Formation of the complexes can be observed as an exothermal transition at a temperature range of 60 to 80°C when starch free of lipid was gelatinized in the presence of lipids. However, the observed melting temperatures were dependent on the water content (Biliaderis et al., 1985). The size of the melting endotherm of the amylose-lipid complex can be used to estimate the amylose content of starch.

Amylose has been shown to form inclusion complexes also with other long chain aliphatic compounds (Kowblansky, 1985). The melting temperature and latent heat of melting of such complexes formed with straight chain compounds varied with the number of carbons in the aliphatic chain, and the melting properties vary with water content (Eliasson, 2003).

3.5.1.5 Starch Retrogradation

Retrogradation of starches is a complex phenomenon involving rapid crystallization of amylose and less rapid crystallization of amylopectin from a solution or gel. This leads to textural changes known as staling. Amylopectin crystallization has been described as a nucleation limited growth process that occurs above the glass transition temperature of the amorphous starch in the gel network plasticized by water (Slade and Levine, 1988a). Indeed, a strong relationship between the extent and rate of crystallization with glass transition in corn starch was found by Jouppila and Roos (1997, 1998). The extent of recrystallization increases with increasing water content in the range of 27 to 50% and then decreases with further increase in water content. However, this behavior seems to depend on water content and the temperature difference above the glass transition, $T - T_g$ (Jouppila et al., 1998). Sugars have been shown to act as antistaling agents, probably because of the increase of glass transition temperature caused by increased average molecular weight of the solution (Slade and Levine, 1990). Mestres et al. (1988) reported that pasting, drum drying, and extrusion lead to different types of crystallization phenomena of amylose and amylopectin.

3.5.2 Protein Denaturation

Kauzmann (1959) defined denaturation of proteins as a process in which the spatial arrangement of the polypeptide chains within the molecule were changed from that typical of the native protein to a more disordered arrangement. Protein denaturation in other words is a transition from a native, folded structure to the denaturated, unfolded structure. Protein denaturation occurs over a temperature range and denaturation temperature is often defined as a temperature at which 50% of the protein is denaturated (Foegeding, 1988). Protein denaturation involves a heat of denaturation, but it is often an irreversible process that occurs in native proteins during heating at a temperature typical of each protein. The amount of energy needed for protein denaturation is very small, usually about 20 J/g of protein. However, the denaturation temperature and heat of denaturation depend on pH, ionic strength, and heating rate (Wright, 1982; Kinsella and Whitehead, 1989) as well as on previous processing of the protein (Murray et al., 1985). A typical thermogram of food protein denaturation is shown in Figure 3.18.

Rüegg et al. (1975) studied denaturation of β-lactoglobulin. The denaturation temperature was at 80.3°C, and it was dependent on water content below 0.7 gH_2O/g but not at water contents higher than 0.7 gH_2O/g. Heat of denaturation, ΔH_d, was found to be constant at high water contents being 14.4 J/g. At water contents below 0.7 g/g ΔH_d decreased to a value as low as 7.5 J/g. Denaturation temperatures and latent heat values for denaturation of some food proteins are given in Table 3.11. Water affects the denaturation temperature and latent heat of denaturation values as shown in Table 3.12. The denaturation kinetics has been reported to follow pseudo first-order or second-order processes (Park and Lund, 1984; Bernal and Jelen, 1985; Harwalkar, 1986; O'Neill and Kinsella, 1988).

Donovan et al. (1975) and Donovan and Mapes (1976) studied denaturation of egg white proteins using differential scanning calorimetry. Egg white had typically three denaturation endotherms

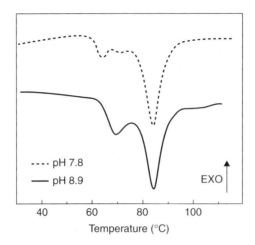

FIGURE 3.18   Denaturation thermogram of egg white proteins at two different pH values. (Adapted from Wright DJ. In: Hudson BJF, Ed. *Developments in Food Proteins-1*. London: Applied Science Publishers, 1982, pp. 61–89. With permission.)

TABLE 3.11

Denaturation Temperatures, T_d, and Heats of Denaturation, ΔH_d, of Food Proteins

Protein	T_d (°C)	ΔH_d (J/g)	Reference
Actin	83.5	14.5	Wright et al. (1977)
Avidin	85	298[a]	Donovan and Ross (1973)
Avidin-biotin complex	131	1065[a]	Donovan and Ross (1973)
Bovine serum albumin	64	12.2	deWit and Klarenbeek (1984)
Conalbumin	61	15.2	Donovan et al. (1975)
Egg globulins	92.5	11.8	Donovan et al. (1975)
Fababean protein	88	18.4	Arntfield and Murray (1981)
Field pea protein	86	15.6	Arntfield and Murray (1981)
Immunoglobulins	72	13.9	deWit and Klarenbeek (1984)
α-lactalbumin	62	17.8	deWit and Klarenbeek (1984)
β-lactoglobulin	78	16.9	deWit and Klarenbeek (1984)
Lysozyme	75	28.2	Donovan et al. (1975)
Myofibrils	59.5 and 74.5	22.6	Wright et al. (1977)
Myosin	55	13.9	Wright et al. (1977)
Oats	112	18.8	Arntfield and Murray (1981)
Ovomucoid	79	21.9	Donovan and Beardslee (1975)
Ovalbumin	84	15.2	Donovan et al. (1975)
Sarcoplasmic protein	63, 67, and 75	16.5	Wright et al. (1977)
Soybean protein	93	14.6	Arntfield and Murray (1981)
Soybean trypsin inhibitor (STI)	76	110[a]	Donovan and Beardslee (1975)
β-trypsin	72	194[a]	Donovan and Beardslee (1975)
Wheat gluten	88.4 and 101.4	0.25 and 0.13	Eliasson and Hegg (1980)
Whey proteins	62–78	11.5	deWit and Klarenbeek (1984)

[a] J/mol.

TABLE 3.12
Effect of Moisture Content on Myoglobin Denaturation Temperature and Heat of Denaturation, ΔH_d

Moisture content (%)	Denaturation temperature (°C)	ΔH_d (J/g)
2.3	122	1.7
9.5	89	6.7
15.6	82	11.3
20.6	79	15.1
35.2	75	19.7

Source: From Hägerdal B and Martens H. *J Food Sci* 41, 933–937, 1976.

equivalent to those of ovotransferring at 64°C, lysozyme at 72°C and ovalbumin at 84°C. During the storage of eggs, ovalbumin was found to be converted to S-ovalbumin, which resulted in the development of a denaturation endotherm at 92.5°C. Sucrose was found to stabilize egg-white proteins against denaturation. Sucrose has also been reported to increase the denaturation temperature of horseradish peroxidase from 89 to 101.5°C at 60% sucrose concentration (Chang et al., 1988).

3.5.3 Melting of Oils and Fats

The successful manufacture of industrial lipid-based food products is based on the manipulation of the composition or triglyceride properties of a fat blend to achieve desired physical and chemical properties and to prevent undesirable changes during processing and storage (Birker and Padley, 1987). Most phase transitions of lipids are first-order transitions, that is, melting, crystallization, and recrystallization transitions, between their different polymorphic crystal forms (sub-α, α, β', β) and the liquid state. The transition between sub-α and α modifications is a second-order phase transition. Unfortunately, most of the studies reporting melting temperatures and latent heats of melting of the various crystalline forms have been done using pure triacylglycerides or simple mixtures, and published data on phase transitions of natural oils and fats are limited. Crystallization and the crystal forms of cocoa butter has received extensive attention because of its importance to chocolate quality and stability (Hartel, 2001).

Mono- and diglycerides have two and one free hydroxyl groups, respectively. They do not occur naturally in appreciable quantities except in fats and oils that have undergone partial hydrolysis. Individual fats and oils vary over relatively large ranges in the proportions of the component fatty acids. Most fatty acids that occur in natural fats and oils are straight chain acids which contain an even number of carbon atoms.

Edible oils and fats usually consist of more than 95% of a complex mixture of triacylglycerols. The fatty acid composition of some natural fats and oils are given in Table 3.13. Typically an edible oil or fat contains more than 500 different triacylglycerols. These mixtures do not have exact melting points but have a melting range. In industrial products a fat with an optimal melting range for a particular application is obtained by the blending of natural and modified oils and fats.

TABLE 3.13
Positional Distribution of Fatty Acids in Triacylglycerols of Some Natural Fats

Source	Position	4:0	6:0	8:0	10:0	12:0	14:0	16:0	18:0	18:1	18:2	18:3	20:0	20:1	22:0	24:0
Cow's milk	1	5	3	1	3	3	11	36	15	21	1					
	2	3	5	2	6	6	20	33	6	14	3					
	3	43	11	2	4	3	7	10	4	15	0.5					
Coconut	1		1	4	4	39	29	16	3	4	2					
	2		0.3	2	5	78	8	1	0.5	3	2					
	3		3	32	13	38	8	1	0.5	3	1					
Cocoa butter	1							34	50	12	1					
	2							2	2	87	9					
	3							37	53	9						
Corn	1							18	3	28	50					
	2							2		27	70					
	3							14	31	52	1					
Soybean	1							14	6	23	48	9				
	2							1		22	70	7				
	3							13	6	28	45	8				
Olive	1							13	3	72	10	0.6				
	2							1		83	14	0.8				
	3							17	4	74	5	1				
Peanut	1							14	5	59	19		1	1		1
	2							2		59	39					
	3							11	5	57	10		4	3	6	0.5
Beef	1						4	41	17	20	4	1				3
	2						9	17	9	41	5	1				
	3						1	22	24	37	5	1				
Pig	1						1	10	30	51	6					
	2						4	72	2	13	3					
	3								7	73	18					

Column group header: Fatty acid (mol %)

Source: From Nawar WW. Lipids. In: Fennema OR, Ed., *Food Chemistry* 2nd ed., New York: Marcel Dekker, 1985, pp. 139–244.

3.5.3.1 Polymorphism

The existence of a number of alternative crystal structures is a characteristic property of many lipids (alkanes, fatty acids, soaps, methyl esters of fatty acids, and triacylglycerols). The study of fat crystallization involves determination of crystal polymorphism and solid content in a fat blend. The methods commonly used are dilatometry, temperature controlled X-ray diffraction, NMR and differential scanning calorimetry (Birker and Padley, 1987).

The α modification is the least stable polymorphic form of oils and fats, which occurs in edible fats during their preparation. It is usually converted to β' in some minutes. In commercial mixtures of fats and oils β' transformation to β may be delayed for months or years, and it may have an important effect on the quality of fat spreads (Wesdorp, 1990). β crystals occur in edible fats if they have isomorphous triacylglycerides like hardened rape-seed oils and cocoa butter. Transformation of α crystals to β crystals occurs always through the β'-form. During crystallization of lipids α form is obtained first, and this is recrystallized into β' and finally into β form. For example, palm oil crystallizes to α form at 10°C, to β' form at 25°C and to β form at 32°C. The coexistence of α and β'-forms is also possible. The thermodynamical stability of the various crystalline states is determined by Gibbs energy, and crystal forms with higher Gibbs energies possess lower melting points (Figure 3.19).

Lipids may also show time-dependence in transitions between the various crystalline structures, and, therefore, exhibit thermal hysteresis mostly because of supercooling (Hagemann, 1988). Stable crystals of fats and oils are usually not formed close to the melting points of the polymorphic forms, and lipids often exist in metastable, supercooled states before crystallization. In DSC curves, recrystallization, and melting endotherms of the polymorphic forms often overlap which makes it difficult to analyze transition temperatures and latent heats of the various crystal forms. Heat of fusion data on β-forms predominate in the literature, because α- and β'-forms are less stable and difficult to prepare. Rapid cooling generally produces only α-forms, and ΔH_m (latent heat of melting) values can be obtained from cooling exotherms.

Hagemann (1988) reviewed the thermal behavior and polymorphism of acylglycerides. In DSC analysis of solidified triglycerides, a reheating thermogram shows an endotherm for the α-form followed immediately by an exotherm resulting from the transformation of α to β-form. Annealing

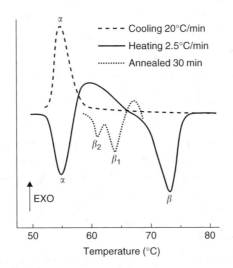

FIGURE 3.19 Heating and melting thermograms of tristearin showing crystallization to α-form, melting of α and β-forms, and effect of annealing for the formation of β'-forms. (Adapted from Hagemann JW. In: Garti N and Sato K, Eds, *Crystallization and Polymorphism of Fats and Fatty Acids*. New York: Marcel Dekker, 1988, pp. 9–95. With permission.)

TABLE 3.14
Melting Points of Fatty Acids, Mono-, Di-, and Triglycerides

Fatty acid	Carbons	Melting point (°C)			
		Acid	1-mono-glyceride	1,3-di-glyceride	tri-glyceride
Butyric	4	−7.9			
Valeric	5	−33.5			
Caproic	6	−3.4	19.4		−25
Heptanoic	7	−7.1			
Caprylic	8	16.7			8.3
Pelargonic	9	12.5			
Capric	10	31.6	53	44.5	31.5
Undecanoic	11	28.7	56.5	49	30.5
Lauric	12	44.2	63	57.8	46.4
Tridecanoic	13	41.4	65	59.5	44.0
Myristic	14	54.4	70.5	66.8	57.0
Pentadecanoic	15	52.1	72	68.5	54.0
Palmitic	16	62.9	77	76.3	63.5
Heptadecanoic	17	61.3	77	74.5	63.5
Stearic	18	69.6	81.5	79.4	73.1
Nonadecanoic	19	68.7			
Arachidic	20	75.4	84		
Heneicosanoic	21	74.3			
Behenic	22	80.0			
Tricosanoic	23	79.1			
Lignoceric	24	84.2			
Oleic	18:1 cis	16.3	35.2	21.5	5.5
Elaidic	18:1 trans	43.7	58.5	55	42
Linoleic	18:2 cis	−6.5	12.3	−2.6	−13.1
Linolenic	18:3 cis	−12.8	15.7	−12.3	−24.2
Ricinoleic OH	18:1 cis	5.5			
α-Eleostearic	conj. 18:3	49			
β-Eleostearic	conj. 18:3	72			
Erucic	22:1	33.4	50	46.6	30

Source: From Formo MW. In: Swern D, Ed., *Bailey's Industrial Oil and Fat Products*, Vol I, 4th ed. New York: Wiley, 1979, pp. 177–232.

at the α-endotherm leads to formation of β'-form. The required annealing time depends on the chain length of the fatty acids and it increases with increasing chain length. Further heating leads to the appearance of the melting endotherm of the β crystals. One of the most studied unsaturated monoacid triglycerides is triolein. Trierucin, triolein, and trilinolein each exhibits three different β'-forms.

3.5.3.2 Melting of Fats and Oils

The most important phase transition characteristic of fats and oils is their melting or softening point. Generally melting temperatures of fatty acids increase with increasing chain length and decrease with increasing unsaturation (Formo, 1979). Melting points of some fatty acids and their mono-, di-, and triglycerides are given in Table 3.14. Melting temperatures are affected not only by the type of fat or oil but also by the origin of the lipid, processing, various treatments, and seasonal variation.

TABLE 3.15
Molecular Weight, M, Melting Points, mp, and Heats of Fusion, ΔH_m, of Saturated Monoacid Triglycerides

Fatty acid chain length	M	α mp (°C)	α ΔH_m (kJ/mol)	β' mp (°C)	β' ΔH_m (kJ/mol)	β mp (°C)	β ΔH_m (kJ/mol)
8	470.69	−54	17.2	−19		11	69.5
9	512.77	−30	35.6	7	49.4	10	54.8
10	554.85	−10	56.5	13		33	92.1
11	596.93	4	55.3	28	71.6	31	84.6
12	639.01	14	72.0	34	81.2	46	116.4
13	681.09	23	74.5	40	83.7	42	99.6
14	723.17	31	85.0	45	100.5	56	136.9
15	765.25	41	97.1	54	108.8	58	144.4
16	807.33	46	103.0	57	131.5	66	165.8
17	849.41	50	109.7	60	128.5	65	169.1
18	891.49	55	112.2	64	142.8	73	191.8
19	933.57	60	120.6	65	136.1	72	186.7
20	975.65	64	122.2	69	160.4	78	220.6
21	1017.73	66	127.7	70	149.9	76	180.9
22	1059.82	69	143.2	74	152.4	83	224.0
23	1101.90	72	140.7	75	159.5	80	191.3
24	1143.98	74	160.8	79	155.7	86	226.5
26	1228.14	78	162.4	82	159.5	89	234.5
28	1312.30	80	113.9	91	173.8		
30	1396.46	79	96.3	93	150.3		

Source: From Lutton ES and Fehl AJ. *Lipids* 5, 90–99, 1970; Ollivon M and Perron R. *Thermochimica Acta* 53, 183–194, 1982; and Hagemann JW and Rothfus JA. *J Am Oil Chem Soc* 60, 1308–1314, 1983.

TABLE 3.16
Latent Heating of Melting, ΔH_m, of Monoacid Triglycerides Estimated from Experimental Data

Equation	Reference
$\Delta H_m(\alpha) = 0.616\,n - 7.20$ kcal/mol	Ollivon and Perron (1982)
$\Delta H_m(\beta') = 0.93\,n - 13.3$ kcal/mol	Ollivon and Perron (1982)
$\Delta H_m(\beta) = 1.023\,n - 7.79$ kcal/mol	Timms (1978)

Melting temperatures and latent heats of melting of the different polymorphic forms for some saturated monoacid triacylglycerides are given in Table 3.15. The melting point and latent heats of melting are increased with the increasing chain length of the fatty acid (Table 3.16). Mixed saturated and unsaturated triglycerides have been investigated by many workers, because palmitic, stearic, and oleic combinations occur commonly in edible fats (Table 3.17). In these structures, few degrees separate the melting temperatures of β'- and β-forms. The main interest in the phase behavior of the mixed-acid triglycerides systems lies in the contrasting effects of saturated and unsaturated acid chains and the influence of the positional isomerism on the behavior (Hagemann, 1988).

TABLE 3.17
Stable Crystalline Form, Typical Melting Temperature, T_m, and Composition of Natural Fats and Oils

Fat or oil	Stable form	T_m (°C)	Saturated acids (%, w/w)		Unsaturated acids (%, w/w)	
			Palmitic	Stearic	Oleic	Linoleic
Butterfat	β'	32.2	22.5	8.4	21.1	3.5
Lard oil	β	30.5	22.1	10.6	32.2	5.7
Cocoa butter	β	34.1	19.6	26.1	27.6	2.1
Coconut oil	β	25.1	9.5	2.2	7.0	
Corn oil	β	−20.0	9.3	2.9	33.2	25.5
Cottonseed oil	β'	−1.0	19.0	1.1	18.6	32.3
Olive oil	β	−6.0	6.5	2.2	45.8	4.4
Palm oil	β'	35.0	28.6	5.2	29.9	9.3
Palmkernel oil	β	24.1	8.1	1.3	15.6	0.7
Peanut oil	β	3.0	7.7	3.0	35.9	20.6
Rapeseed oil	β'	−10.0	1.0		24.2	13.0
Sesame oil	β	−6.0	8.3	4.1	31.2	28.8
Soybean oil	β	−16.0	8.9	2.3	22.4	33.6
Sunflower oil	β	−17.0	5.3	2.2	20.1	39.8

Thermodynamical data for some mixed acid triglycerides were reported by Hagemann (1988).

Almost all fats and fatty acids have two or more solid phases under given thermodynamical conditions. Most triglycerides may exist in one of the three different crystalline forms depending on temperature, time, and composition. Soybean, peanut, corn, olive, coconut and safflower oils, cocoa butter, and lard tend to crystallize in β form, and cottonseed, palm, rapeseed oils, milk fat tallow, modified lard, and most natural fats and fat blends have the β' modification as their stable solid form (Nawar, 1985; Birker and Padley, 1987). The sub-α and α modifications are always unstable.

The solid content is an important characteristic of any fat and lipid product. The solid fat content of fats is measured using dilatometry to monitor volume contraction during crystallization and by wide-line NMR, which can be used to determine differences in molecular mobility in liquid and solid triacylglycerols. Dilatation values at a certain temperature, ΔT, are related to the difference in volume between a completely liquid fat and stabilized crystallized fat at the same temperature, and it is expressed in mm^3 25 g of fat. Solid fat index (SFI) is calculated as $\Delta T/25$. Pulse-NMR has mostly replaced the dilatometric methods (Birker and Padley, 1987). The dilatometric and NMR data can be used to establish phase diagrams, which show the solid content as a function of temperature or composition. Iso-solids and iso-dilatation diagrams are used to show lines indicating constant dilatation values or solid contents as a function of concentration (Figure 3.20), and they indicate interaction of two triacylglycerols or fats or fat blends. Properties of various fats and their manufacture has been discussed by various authors in Hui (1996).

3.5.3.3 Phase Behavior of Cocoa Butter

Cocoa butter is probably the most studied natural fat. The various polymorphic forms, transition temperatures, and latent heats are well documented (Schlichter-Aronhime and Garti, 1988; Minifie, 1989; Hartel, 2001). Cocoa butter has four polymorphic forms: α, β, β', and γ. The γ form is obtained by very fast cooling, and its melting point is 17°C. The γ form is the most unstable polymorphic form of cocoa butter. The α form is produced by fast cooling, and it melts at 21–24°C. The α-form is

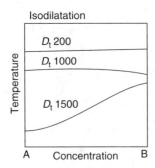

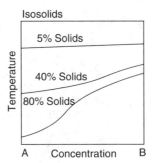

FIGURE 3.20 Schematic isodilation and isosolids diagrams of fat blends. The lines indicate temperatures at which varying concentrations of fats in binary blends have the same dilatation number or solid content. (Adapted from Birker PJMWL and Padley FB. In: Hamilton RJ and Bhati A, Eds, *Recent Advances in Chemistry and Technology of Fats and Oils*. London: Elsevier, 1987, pp. 1–11. With permission.)

TABLE 3.18
Temperatures and Enthalpies of Crystallization, ΔH_{cr}, and Melting, ΔH_m, of Cocoa Butter Crystallized at Different Cooling Rates

Rate	ΔH_{cr}	T_{cr}	ΔH_m	T_m
0.3	89	13.2	95	22.8
0.2	94	14.2	99	23.7
0.1	101	15.5	104	24.6
0.05	92	16.8	107	25.3
0.02	83	19.2	106	26.3

Source: From Schlichter-Aronhime J and Garti N. In: Garti N and Sato K, Eds, *Crystallization and Polymorphism of Fats and Fatty Acids*. New York: Marcel Dekker, 1988, pp. 363–393.

transformed to β' form at normal storage temperatures. The β' form melts at 27–29°C. The β form is stable, and it melts at 34–35°C. Effect of annealing on cocoa butter melting is given in Table 3.18.

3.6 STATE TRANSITIONS AND WATER IN FOOD SYSTEMS

3.6.1 WATER PLASTICIZATION OF FOOD COMPONENTS

Water sorption by food solids is an important property affecting glass transition, melting, crystallization, and other phase and state transitions of food materials. Amorphous foods are highly water plasticizable, similar to water plasticizable polymers (White and Cakebread, 1966; Levine and Slade, 1986; Roos and Karel 1990, 1991a, 1991b; Roos, 1995). Ellis (1988) reported that 1% water in water plasticizable polymers may induce a 15–20°C reduction of T_g compared to a typical value of 4–5°C for common polymers plasticized by organic diluents. The plasticization of amorphous food materials is of similar magnitude, and even traces of water may significantly alter the T_g value of the nonaqueous

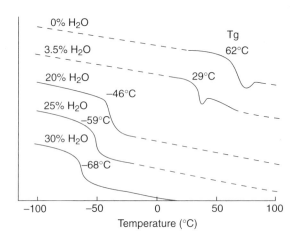

FIGURE 3.21 Water plasticization of amorphous sucrose as detected in DSC thermograms (heating rate 5°C/min) as decreasing values of the glass transition temperature.

material (Roos, 1987; Roos and Karel, 1991a; Roos, 1995). The effect of water plasticization on amorphous food materials is observed as decreasing T_g values with increasing water content (Rasmussen and Luyet, 1969; Simatos et al., 1975; Roos, 1987; Zeleznak and Hoseney, 1987; Orford et al., 1989; Levine and Slade, 1990; Roos and Karel, 1991a). The water plasticization effect on T_g is shown in Figure 3.21. Water plasticization decreases also crystallization temperatures of amorphous crystallizable materials and the melting temperatures of the crystals (Roos, 1995; Hartel, 2001).

3.6.1.1 Prediction of T_g at Varying Relative Humidities

At low and intermediate moisture contents the T_g values of amorphous food materials decrease linearly with increasing water activity. This linear relationship is of practical importance since it allows prediction of glass transitions temperatures for materials exposed to various relative humidities (Roos, 1987; Roos and Karel, 1991a; Roos, 1995). The dependence of the T_g values of maltodextrins of various molecular weights on water activity is shown in Figure 3.22. Roos and Karel (1991c) showed that the glass transition temperature of all maltodextrins studied decreased with increasing water content and water activity. The decrease of T_g with increasing water activity was similar for all maltodextrins showing a constant slope (−150) in the linear relationship. The linearity between T_g and A_w allows prediction of T_g (Figure 3.22) over a wide range of relative humidities (RH) if the slope, temperature, and RH are known for at least one point.

3.6.1.2 Prediction of T_g Using Mixing Equation

It has been shown (Kelley et al., 1987; Ellis, 1988) that the effect of water plasticization on the glass transition temperature of water plasticizable polymers can be predicted using the Gordon and Taylor equation (Equation 3.69) which has been used to predict glass transition temperatures of polymer mixtures (Gordon and Taylor, 1952; Couchman, 1978). The equation has been later modified by Coachman (1978) to include the specific heat change at the T_g to give constant $k = \Delta c_{p2}/\Delta c_{p1}$ (Equation 3.70).

$$T_g = \frac{w_1 T_{g1} + kw_2 T_{g2}}{w_1 + kw_2} \tag{3.69}$$

$$T_g = \frac{w_1 \Delta c_{p1} T_{g1} + w_2 \Delta c_{p2} T_{g2}}{w_1 \Delta c_{p1} + w_2 \Delta c_{p2}} \tag{3.70}$$

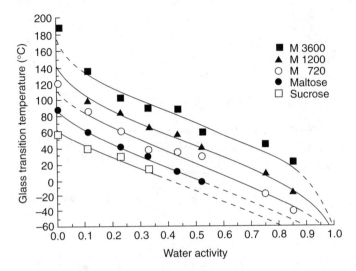

FIGURE 3.22 Glass transition temperatures, T_g, of maltodextrins of varying molecular weights (Maltrin M040, M150, and M250), maltose and sucrose as a function of water activity, A_w. The relationship between T_g and A_w is sigmoid but shows linearity at a practical a_w range of 0.1 to 0.9. The slope of the lines is -150. The effect of molecular weight on T_g at the linear range is obtained by the Fox and Flory (1950) equation: $T_g = T_g(\infty) - 1/M_w$. $T_g(\infty)$ is the T_g value of highest molecular weight, M_w. (Adapted from Roos Y and Karel M. *Biotechnol Prog* 7, 49–53, 1991c. With permission.)

where Δc_{p1} and Δc_{p2} are specific heat change of component 1 and 2 over the glass transition, T_g is glass transition temperature of mixture, T_{g1} and T_{g2} are T_g of component 1 and 2, w_1 and w_2 are weight fractions of component 1 and 2, and k is a constant.

Various glass transition temperatures have been reported for amorphous water around $-135°C$ (Johari et al., 1987). The Δc_p for amorphous water at the glass transition region was reported by Sugisaki et al. (1968) to be 1.94 J/gK, although other measured values vary significantly (Roos, 1995). Attempts to derive k values from the change of the specific heat of components have failed (Orford et al., 1990), but empirical k values may lead to good correlation with experimental data (Roos and Karel, 1991c). The glass transition temperature of water is theoretically reached as the weight fraction of water becomes 1. However, water in food and other biological materials is a crystallizable plasticizer (Levine and Slade, 1988), and the more concentrated the system studied the lower is the melting or freezing temperature of water. In foods, crystallization of water leads to freeze-concentration of the food solids, and thus to a gradually decreasing melting temperature and eventually to a constant T_g of the unfrozen matrix, which remains at a constant solute concentration.

3.6.2 STATE DIAGRAMS

A state diagram proposed by Franks et al. (1977) represents various states in which a system can exist as a function of temperature, concentration, time, and pressure (Levine and Slade, 1990). A simplified state diagram shows the state of a material as a function of concentration and moisture content (Roos, 1995; Slade and Levine, 1995). The state diagram is of significance in the characterization of the state of food materials having intermediate and low moisture contents, and frozen foods (Blond, 1989; Slade and Levine, 1990; Roos, 1995).

A typical state diagram of a water plasticizable material is that of amorphous sucrose shown in Figure 3.23. The T_g line of sucrose is based on the Gordon and Taylor equation (Equation 3.18). This line indicates the transformation of a glass to a rubber at any given concentration. The effect of freeze-concentration is shown by the fact that at water concentrations above 20% ice formation

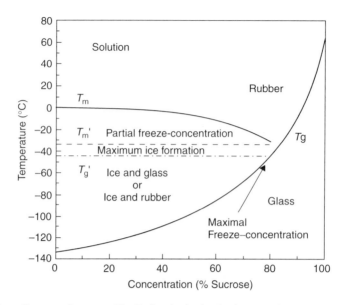

FIGURE 3.23   State diagram of sucrose. The T_g line is obtained using experimental T_g values, and T_g values predicted by the Gordon and Taylor equation (Equation 3.69) with an empirical value of $k = 4.8$. The T_g' line indicates the T_g of maximally freeze-concentrated solutions, and it intercepts with the T_g at the maximally freeze-concentrated concentration. The sucrose concentration of the maximally freeze-concentrated matrix is 80%, which agrees with the T_g' at $-46°C$.

leads to a constant T_g of the maximally freeze-concentrated solution. The ice melting has an onset temperature of T_m' which is in the vicinity of the end point of the glass transition region. The state diagram can be used to analyze both processing and storage conditions of various food materials. The use of state diagrams has been discussed for example by Slade and Levine (1990), Roos (1995), Slade and Levine (1995), and Roos et al. (1996).

3.6.3   PHASE AND STATE TRANSITIONS OF AMORPHOUS FOOD COMPONENTS

Carbohydrates show both first-order and second-order phase transitions. Typical first-order transitions of carbohydrates are melting (Table 3.19), crystallization, and gelatinization of starch. Most carbohydrates are also found to exist in amorphous forms. The amorphous state of low molecular weight sugars are of importance to all dry and frozen food behavior. Starch exists in both amorphous and crystalline forms. Usually amylopectin exists in an amorphous state and amylose in a partially crystalline state. Cellulose exists mostly as crystals but wood lignin and hemicellulose may exist in amorphous forms and become plasticized by water (Kelley et al., 1987).

3.6.3.1   Phase and State Transitions of Amorphous Sugars

Manufacturing of noncrystalline solid sugar materials is the basic technology needed for hard candy production, and the existence of amorphous lactose has been reported since the development of dry milk products (Troy and Sharp, 1930; Sharp and Doob, 1941; Bushill et al., 1965; Roos, 2002). White and Cakebread (1966) listed processes, which normally favor formation of amorphous structures in sugar containing products. The methods reported were concentration of solutions at high temperatures and rapid cooling, roller, spray- and freeze-drying of solutions, rapid freezing of solutions, and fusion of crystals and rapid cooling. They also pointed out that the main criterion for

TABLE 3.19
Melting Temperatures (Peak Temperature of DSC Melting Endotherm) and Latent Heats of Melting, ΔH_m, of Sugars and Sugar Alcohols

Compound	Melting temperature (°C)	ΔH_m (J/g)
Monosaccharides		
L-Arabinose	155	260
L-Fucose	130	190
D-Fructose	115	180
D-Galactose	165	280
D-Galacturonic acid	105	70
D-Glucoheptose	180	270
α-D-Glucose	150	180
α-D-Glucose $\times$ H_2O	75	60
β-D-Glucose	150	150
D-Glucuronic acid	140	280
D-Glucurono-δ-lactone	165	120
L-Rhamnose $\times$ H_2O	100	210
D-Ribose	90	150
L-Sorbose	160	250
D-Xylose	150	280
Disaccharides		
Cellobiose	220	160
α-Lactose	195	250
β-Lactose	220	250
Lactulose	155	110
Sucrose	185	120
Turanose	165	150
Oligosaccharides		
Melezitose $\times$ $2H_2O$	165	140
Raffinose $\times$ $5H_2O$	85	150
Polyols		
L-Arabinitol	105	230
Galactitol	190	330
Isomaltitol	175	180
D-Mannitol	170	290
Maltitol	150	150
Meso-erythritol	125	330
Ribitol	110	250
Sorbitol	95	150
Xylitol	100	250
Other		
Inulin	150	40
Myo-inositol	230	260

Source: From Raemy A and Schweizer TF. *J Thermal Anal* 28, 95–108, 1983.

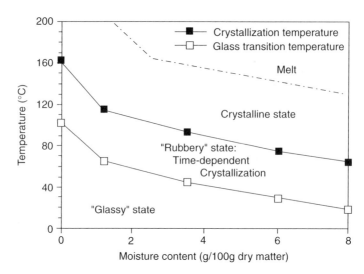

FIGURE 3.24   Glass transition, crystallization, and melting temperatures of amorphous lactose as a function of moisture content. Amorphous crystallizable sugars show time-dependent crystallization above their glass transition temperature.

stability of food materials containing amorphous components is their storage at temperatures below T_g of the amorphous matrix.

A typical thermogram of an amorphous sugar is shown in Figure 3.3. Water as a plasticizer of the amorphous food components decreases their glass transition temperatures, T_g, crystallization temperatures of crystallizable amorphous compounds, T_{cr}, and melting temperatures, T_m. T_g, T_{cr} and T_m values of lactose at different water contents as shown in Figure 3.24.

3.6.3.2 Crystallization of Amorphous Sugars

Makower and Dye (1956) found that crystallization of amorphous sucrose and glucose occurred as function of time, and time to crystallization was dependent on moisture content. At low relative humidities no crystallization was observed during three years of observation. Karel (1973) showed that the crystallization data reported by Makower and Dye (1956) could be used to establish a sorption isotherm showing time-dependent crystallization during water sorption. Similarly, Roos and Karel (1990) showed that crystallization of amorphous lactose was time-dependent at different relative humidities (Figure 3.25). They showed that time to crystallization at various moisture contents depended on the temperature difference to the T_g, and the time to crystallization was only slightly affected by humidity at the same $T - T_g$ value. Therefore, time to crystallization was a function of $T - T_g$, showing WLF type temperature dependence. Because amorphous sugars are plasticized by water, their T_g values decrease with increasing water content, and thus time to crystallization decreases with increasing moisture content at isothermal conditions. Crystallization is not possible below T_g because of the high viscosity of the glassy state, and amorphous sugars are not able to crystallize below their T_g. Hence, food products containing amorphous sugars, for example, dried milk products, have to be stored below fairly low relative humidity and temperature values.

Crystallization of amorphous sugars is also related to stability of products containing encapsulated flavors or oils in the amorphous matrix (Gejl-Hansen and Flink, 1977; To and Flink, 1978). Shimada et al. (1991) showed that crystallization of amorphous lactose containing encapsulated methyl linoleate lead to the release of the encapsulated oil and subsequent rapid oxidation.

Crystallization of amorphous sugars can be delayed by the addition of high molecular weight substances (Iglesias and Chirife, 1978; Roos and Karel, 1990, 1991a) or other sugars (Herrington

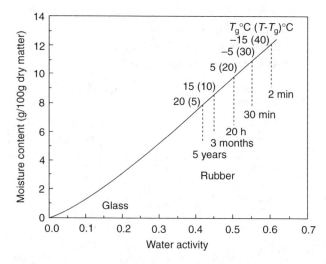

FIGURE 3.25 Time-dependence of crystallization of amorphous lactose at 25°C shown as a function of water activity and moisture content. Time to crystallization decreases with increasing moisture content because of decreasing glass transition temperature. Crystallization is not probable if the glass transition temperature is higher than 25°C. (Adapted from Roos Y and Karel M. *Biotehcnol Prog* 6, 159–163, 1990. With permission.)

and Branfield, 1984). Small amounts of added polymers affect the T_g values of amorphous sugars only slightly but the crystallization is delayed significantly (Roos and Karel, 1991a, 1991c; Roos, 1995; Hartel, 2001).

3.6.3.3 Glass Transition of Proteins

Proteins in common with other biopolymers are often found to exist as amorphous polymers. The most studied amorphous proteins include gelatin, elastin (Kakivaya and Hoeve, 1975), lysozyme (Morozov and Gevorkian, 1985), gluten (Hoseney et al., 1986), glutenin (Cocero and Kokini, 1991), zein (Madeka and Kokini, 1996), and soy proteins (Morales-Diaz and Kokini, 1998). The glass transition temperatures of dry proteins are relatively high which makes their analysis difficult (Table 3.20). However, similar to most amorphous food components they are significantly plasticized by water (Slade and Levine, 1990).

The glass transition temperature of amorphous food proteins is important to the structure formation of many cereal products. Some diffusion controlled reactions may be related to the temperature difference to the glass transition temperature. Fujio and Lim (1989) reported a critical temperature of color change in gluten to be related to its glass transition temperature. Madeka and Kokini (1996) and Morales-Diaz and Kokini (1998) developed state diagrams, which show various transition and reaction zones for amorphous protein systems.

3.6.3.4 Effect of Composition on Glass Transition Temperature

Food materials are often mixtures of various compounds of which several may exist in the amorphous state. As has been clearly shown for polymers, the composition of mixtures of amorphous miscible compounds affects the glass transition temperature of the mixture. The glass transition temperature of the mixture can be predicted using the Gordon and Taylor equation.

Many amorphous low molecular weight sugars are extremely hygroscopic, and their processing and storage is difficult because of their low glass transition temperatures. Therefore, various high molecular weight compounds have been used to improve their processability and storage stability. In spray drying industry, materials with high sugar content especially fruit juices have been difficult

TABLE 3.20
Glass Transition Temperatures, T_g, of Anhydrous Food Polymers

Material	T_g (°C)
Proteins	
α-Casein	152
β-Casein	148
κ-Casein	156
Collagen	197
Gelatin	207
Gliadin	179
Glutenin	189
Elastin	208
α-Lactalbumin	159
β-Lactoglobulin	146
Legumin	171
Lyzozyme	165
Myoglobin	149
Ovalbumin	157
Pea 7S-globulin	149
Pea 11S-globulin	172
Soybean 2S-globin	173
Soybean 11S-globulin	187
Vicilin	146
Zein	165
Polysaccharides	
Amylose	302
Amylopectin	294
Dextran	259
Pea amylose	332
Potato starch	316
Pullulan	263 (215)
Waxy maize starch	285

Source: From Bizot H, le Bail P, Leroux B, Davy J, Roger P, and Buleon A. *Carbohydr Polym* 32, 33–50, 1997; Matveev YuI, Slade L, and Levine H. *Food Hydrocolloids* 13, 381–388, 1999; and Matveev YuI, Grinberg V Ya, and Tolstoguzov VB. *Food Hydrocolloids* 14, 425–437, 2000.

to dry because of their stickiness, and maltodextrins or other high molecular weight materials have been used as drying aids to increase the sticky point of the materials. The effect of composition on the glass transition temperature of sucrose maltodextrin mixtures is shown in Figure 3.26.

3.6.4 PHYSICAL, STRUCTURAL AND CHEMICAL CHANGES

3.6.4.1 Flow and Structural Changes

The existence of second-order transitions in food materials has been known a long time, and many foods have been reported to be in an amorphous phase after dehydration (Troy and Sharp, 1930;

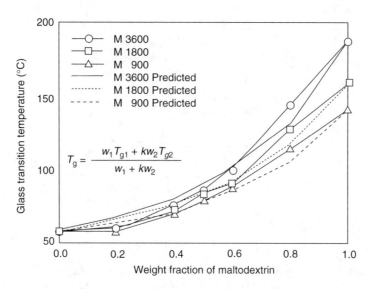

$$T_g = \frac{w_1 T_{g1} + kw_2 T_{g2}}{w_1 + kw_2}$$

FIGURE 3.26 Glass transition temperature of maltodextrin (Maltrin M040, M100, and M200) sucrose mixtures. The glass transition increases with increasing maltodextrin concentration and molecular weight. Fairly high amounts of maltodextrins are needed to increase the glass transition temperature. (Adapted from Roos Y and Karel M. *Biotechnol Prog* 7, 49–53, 1991c. With permission.)

Bushill et al., 1965; Karel, 1973; Alexander and King, 1985; Roos, 1987). White and Cakebread (1966) discussed various effects of second-order transitions on food processing and storage stability. The amorphous metastable system has a constant driving force to a more stable state, as can be expected from the higher energy state as compared to the crystalline structure (Figure 3.4).

A liquid whose viscosity is greater than 10^{12} Pa s is capable of supporting its own weight which is typical of the glassy state. Above T_g, amorphous materials are not able to support their own weight, and thus in foods glass transition governs various collapse phenomena and stickiness of dehydrated food powders. Cooling of an amorphous material through the glass transition into the glassy state causes a rapid decrease in molecular mobility and a sudden increase in viscosity (Williams et al., 1955), which in foods can be related to most mobility related phenomena (Simatos and Karel, 1988; Roos and Karel, 1990, 1991a, 1991b). However, the importance of the physical state to processability and storage behavior of both dehydrated and frozen food materials has not been studied until recently (Roos, 1987; Levine and Slade, 1988; Simatos and Karel, 1988; Roos and Karel, 1991a; Roos, 1995; Le Meste et al., 2002).

The glass transition temperatures of many food components in their dry state are relatively high, but almost all food materials are significantly plasticized by water which decreases both first- order and second-order transition temperatures (Levine and Slade, 1986, 1988, 1990; Roos and Karel, 1990, 1991a). Due to the decrease of viscosity above T_g (Williams et al., 1955; Soesanto and Williams, 1981) it controls also the crystallization of amorphous components, especially that of low molecular weight carbohydrates (Roos and Karel, 1990) and crystallization of amorphous starch (Slade and Levine, 1990; Jouppila and Roos, 1997; Jouppila et al., 1998). Second-order transitions of food materials are important to processes which cause a rapid decrease of the amount of water, and thus not allowing crystallization during the process, for example, dehydration, extrusion and freezing, and to the stability of these products. The physical changes occurring at the glass transition temperature have been reviewed, for example, by Slade and Levine (1990, 1995), Roos (1995), and Le Meste et al. (2002).

Roos and Karel (1991a) showed that the traditional sticky point determined by the method of Lazar (Lazard et al., 1956) was dependent on the glass transition temperature of the material,

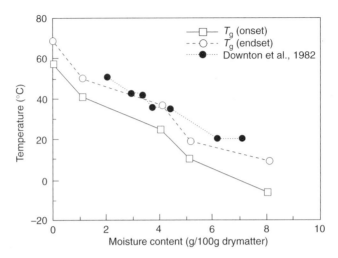

FIGURE 3.27   Relationship between glass transition temperature, T_g, and sticky point, T_s, of sucrose-fructose (7:1) mixture at varying water contents. The sticky point is related to the decrease of viscosity (Downton et al., 1982) which decreases according to the WLF temperature dependence (Soesanto and Williams, 1980). The T_s values correlate with the end point temperature of the glass transition range. (Adapted from Roos Y and Karel M. *J Food Sci* 56, 38–43, 1991a. With permission.)

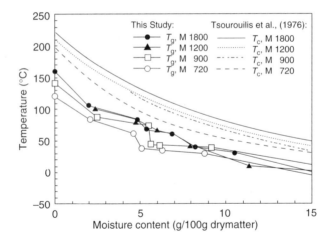

FIGURE 3.28   Glass transition temperatures, T_g, of maltodextrins of varying molecular weights (Maltrin M100, M150, M200, M250), and the corresponding collapse temperatures, T_c, as a function of moisture content. Collapse occurs time dependently above the T_g (Tsourouflis et al., 1976; To and Flink, 1978), and the instant collapse temperature is about 50 to 70°C above the corresponding T_g value. (Adapted from Roos Y and Karel M. *Biotechnol Prog* 7, 49–53, 1991c. With permission.)

and that stickiness occurred immediately after the completion of the transition (Figure 3.6 and Figure 3.27). Later it was shown that the collapse temperature of dried amorphous powders determined by Tsourouflis et al. (1976) and To and Flink (1978) was also dependent on the glass transition temperature of the materials, and occurred above the T_g (Figure 3.28) (Roos and Karel, 1991c). In this case, the flow where the material is no more able to carry its own weight is measured. However, as the flow rate depends on viscosity and temperature, collapse is a time-dependent phenomenon (Flink, 1983). The critical viscosity of stickiness that occurs is 10^7 Pa s (Downton et al., 1982) which agrees well with the correlation with the end point temperature of glass transition noticed by Roos and Karel (1991a).

Above glass transition temperatures of various materials, the diffusion coefficient is often increased (Vrentas and Duda, 1978; Vrentas et al., 1982). Amorphous carbohydrates have very low diffusion coefficients in their glassy state. This is observed during drying as decreased diffusion of water and selective retention of volatile compounds which show a dramatic decrease in diffusion (King, 1983). The hindered diffusion in the glassy state may also protect encapsulated oils from oxidation (Gejl-Hansen and Flink, 1977; To and Flink, 1978; Shimada et al., 1990).

NOMENCLATURE

A	Helmholz free energy
a_w	Water activity
a_T	Relaxation time
C_g'	Solute concentration in maximally freeze concentrated matrices
C_1	Constant
C_2	Constant
c	Constant
C_p	Heat capacity (constant pressure)
C_V	Heat capacity (constant volume)
c_p	Specific heat (constant pressure)
Δc_p	Change of specific heat
D_t	Dilatation value
F	Hemholtz free energy
f	Fractional free volume
f_g	Fractional free volume at T_g
G	Gibbs energy
H	Enthalpy
ΔH_{cr}	Latent heat of crystallization
ΔH_d	Heat of denaturation
ΔH_{gel}	Heat of gelatinization
ΔH_f	Heat of fusion
ΔH_m	Latent heat of melting
ΔH_s	Latent heat of sublimation
ΔH_v	Heat of vaporization
k	Constant
M	Mass
M	Molecular weight
M_w	Molecular weight
m	Moisture content
mp	Melting point
N	Number of moles
p	Pressure
p_i	Vapor pressure of ice
p_u	Vapor pressure of unfrozen water
p_w	Vapor pressure of water
q	Heat content
R	Gas constant
RH	Relative humidity
S	Entropy
ΔS_m	Entropy change at melting point
T	Temperature

T_{am}	Ante-melting temperature
T_c	Collapse temperature
T_{cr}	Crystallization temperature
T_d	Devitrification temperature
T_d	Denaturation temperature
T_g	Glass transition temperature
$T_g(\infty)$	Highest glass transition temperature of homopolymers
T_g^o	Onset temperature of glass transition
T_g^m	Midpoint temperature of glass transition
T_g^e	Endpoint temperature of glass transition
T_{im}	Incipient melting temperature of ice
T_m	Melting point
T_m'	Onset temperature of ice melting
T_m^0	Equilibrium melting point of undiluted polymer
T_s	Sticky point
T_1	Onset temperature of gelatinization
T_2	Peak temperature of gelatinization endotherm
T_3	Endpoint temperature of gelatinization
U	Internal energy
V	Volume
V_0	Volume occupied by molecules
V_f	Free volume
V_u/V_1	Ratio of molar volume of repeating unit to that of water
V_v	Vapor volume
v_1	Volume fraction of water
w	Work
W_g'	Unfrozen water of maximally freeze-concentrated matrices
w	Weight fraction
x_1	Flory–Huggins polymer–diluent interaction parameter

GREEK LETTERS

α	Thermal expansion coefficient
α_f	Coefficient of expansion of free volume
β	Isothermal compressibility
κ_T	Isothermal compressibility
μ	Chemical potential
η	Viscosity
η_g	Viscosity at glass transition temperature
θ_{cr}	Time to crystallization

sub-α, α, β', β, and γ refer to crystal forms of lipids

REFERENCES

Achanta S and Okos MR. Predicting the quality of dehydrated foods and biopolymers — research needs and opportunities. *Drying Technol* 14, 1329–1368, 1996.

Alexander K and King CJ. Factors governing surface morphology of spray-dried amorphous substances. *Drying Technol* 3, 321–348, 1985.

Allen JE, Sherbon JW, Lewis BA, and Hood LF. Effect of chlorine treatment on wheat flour and starch, Measurement of thermal properties by differential scanning calorimetry. *J Food Sci* 47, 1508–1511, 1982.

Angell CA, Stell RC, and Sichina W. Viscosity-temperature function for sorbitol from combined viscosity and differential scanning calorimetry studies. *J Phys Chem* 86, 1540–1542, 1982.

Arntfield SD and Murray ED. The influence of processing parameters on food protein functionality I. Differential scanning calorimetry as an indicator of protein denaturation. *Can Inst Food Sci Technol J* 14, 289–294, 1981.

Atkins PW and de Paula J. *Physical Chemistry*. 8th Ed. Oxford: Oxford University press, London, p. 1094, 2006.

Atwell WA, Hood CF, Lineback DR, Varriano-Marston E, and Zobel HF. The terminology and methodology associated with basic starch phenomenon. *Cereal Foods World* 33, 306–311, 1988.

Autio K and Salmenkallio-Mattila M. Understanding microstructural changes in biopolymers using light and electron microscopy. In: Kaletunç G and Breslauer KJ, Eds, *Characterization of Cereals and Flours*. New York: Marcel Dekker, 2003, pp. 387–408.

Bauwens JC. Physical aging, Relation between volume and plastic deformation. In: Brostow W and Corneliussen RD, Eds, *Failure of Plastics*. Munich: Hanser Publishers, 1986, pp. 235–258.

Bellows RJ and King CJ. Product collapse during freeze-drying of liquid foods. *AIChE Symp Ser* 69, 33–41, 1973.

Bernal V and Jelen P. Thermal stability of whey proteins — a calorimetric study. *J Dairy Sci* 68, 2847, 1985

Biliaderis CG. Differential scanning calorimetry in food research — a review. *Food Chem* 10, 239–265, 1983.

Biliaderis CG, Maurice TJ, and Vose JR. Starch gelatinization phenomena studied by differential scanning calorimetry. *J Food Sci* 45, 1669–1680, 1980.

Biliaderis CG, Page CM, Slade L, and Sirett RR. Thermal behavior of amylose-lipid complexes. *Carbohydr Polym* 5, 367–389, 1985.

Biliaderis CG, Page CM, Maurice TJ, and Juliano B. Thermal characterization of rice starches, A polymeric approach to phase transitions of granular starch. *J Agric Food Chem* 34, 6–14, 1986.

Birker PJMWL and Padley FB. Physical properties of fats and oils. In: Hamilton RJ and Bhati A, Eds, *Recent Advances in Chemistry and Technology of Fats and Oils*. London: Elsevier, 1987, pp. 1–11.

Bizot H, le Bail P, Leroux B, Davy J, Roger P, and Buleon A. Calorimetric evaluation of the glass transition in hydrated, linear and branched polyanhydroglucose compounds. *Carbohydr Polym* 32, 33–50, 1997.

Blanshard JMV. Starch granule structure and function, A physicochemical approach. In: Gilliard T, Ed., Critical Reports on Applied Chemistry Vol. 13. *Starch, Properties and Potential*. New York: Wiley, 1987, pp. 17–54.

Blond G. Water-galactose system, Supplemented state diagram and unfrozen water. *Cryo-Letters* 10, 299–308, 1989.

Bushill JH, Wright WB, Fuller CHF, and Bell AV. The crystallization of lactose with particular reference to its occurrence in milk powder. *J Sci Food Agric* 16, 622–628, 1965.

Chang BS, Park KH, and Lund DB. Thermal inactivation kinetics of horseradish peroxidase. *J. Food Sci* 53, 920–923, 1988.

Chungcharoen A and Lund DB. Influence of solutes and water on rice starch gelatinization. *Cereal Chem* 64, 240–243, 1987.

Cocero AM and Kokini JL. The study of the glass transition of Osborne glutenin using small amplitude oscillatory rheological measurements and differential scanning calorimetry. *J Rheol* 35, 257–270, 1991.

Collison R and Chilton WG. Starch gelation as a function of water content. *J Food Technol* 9, 309, 1974.

Couchman PR. Compositional variation of glass transition temperatures. 2. Application of the thermodynamic theory to compatible polymer blends. *Macromolecules* 11, 1156–1161, 1978.

De Wit JN and Klarenbeek G. Effects of various heat treatments on structure and solubility of whey proteins. *J Dairy Sci* 67, 2701–2710, 1984.

Donovan JW. Phase transitions of the starch-water system. *Biopolymers* 18, 263–275, 1979.

Donovan JW and Beardslee RA. Heat stabilization produced by protein-protein association. *J Biol Chem* 250, 1966–1971, 1975.

Donovan JW and Mapes CJ. A differential scanning calorimetric study of conversion of ovalbumin to S-ovalbumin in eggs. *J Sci Food Agric* 27, 197–204, 1976.

Donovan JW and Ross KD. Increase in the stability of avidin produced by binding biotin. A DSC study of denaturation by heat. *Biochemistry* 12, 512–517, 1973.

Donovan JW, Mapes CJ, Davis JG, and Garibaldi JA. A differential scanning calorimetric study of the stability of egg white to heat denaturation. *J Sci Food Agric* 26, 73–83, 1975.

Downton GE, Flores-Luna JL, and King CJ. Mechanism of stickiness in hygroscopic, amorphous powders. *Ind Eng Chem Fundam* 21, 447–451, 1982.

Ehrenfest P. 1933. Proc Acad Sci, Amsterdam 36, 153.

Eliasson A-C. On the effects of surface active agents on the gelatinization of starch — a calorimetric investigation. *Carbohydr Polym* 6, 464–476, 1986.

Eliasson A-C. Utilization of thermal properties for understanding baking and staling processes. In: Kaletunç G and Breslauer KJ, Eds, *Characterization of Cereals and Flours*. New York: Dekker, 2003, pp. 65–115.

Eliasson A-C and Hegg P-O. Thermal stability of wheat gluten. *Cereal Chem* 57, 436–437, 1980.

Ellis TS. Moisture-induced plasticization of amorphous polyamides and their blends. *J Appl Polym Sci* 36, 451–466, 1988.

Evans ID and Haisman DR. The effect of solutes on the gelatinization temperature range of potato starch. *Staerke* 34, 224–231, 1982.

Eyring H. Viscosity, plasticity, and diffusion as examples of absolute reaction rates. *J Chem Phys* 4, 283–291, 1936.

Fennema OR. Water and ice. In: Fennema OR, Ed., *Food Chemistry*. 2nd ed. New York: Marcel Dekker, 1985, pp. 23–67.

Ferry JD. *Viscoelastic Properties of Polymers*. 3rd ed. New York: Wiley, 1980, p. 641.

Flink JM. Structure and structure transitions in dried carbohydrate material. In: Peleg M and Bagley EB, Eds, *Physical Properties of Foods*. Westport: AVI, 1983, pp. 473–521.

Flory PJ. *Principles of Polymer Chemistry*. Ithaca NY: Cornell University Press, 1953, p. 672.

Foegeding EA. Thermally induced changes in muscle proteins. *Food Technol* 42, 58–64, 1988.

Formo MW. Physical properties of fats and fatty acids. In: Swern D, *Bailey's Industrial Oil and Fat Products*, Vol I, 4th ed. New York: Wiley, 1979, pp. 177–232.

Fox TG and Flory PJ, Second-order transition temperatures and related properties of polystyrene. I. Influence of molecular weight. *J Appl Phys* 21, 581–591, 1950.

Franks F, Asquith MH, Hammond CC, Skaer HB, and Echlin P. Polymeric cryoprotectants in the preservation of biological ultrastructure. I. Low temperature states of aqueous solutions of hydrophilic polymers. *J Microsc* 110, 223–238, 1977.

Fujio Y and Lim J-K. Correlation between the glass transition point and color change of heat treated gluten. *Cereal Chem* 66, 268–270, 1989.

Gejl-Hansen F and Flink JM. Freeze-dried carbohydrate containing oil-in-water emulsions, Microstructure and fat distribution. *J Food Sci* 42, 1049–1055, 1977.

Goff HD, Verespej E, and Jermann D. Glass transitions in frozen sucrose solutions are influenced by solute inclusions within ice crystals. *Thermochimica Acta* 399, 43–55, 2003.

Gordon M and Taylor JS. Ideal copolymers and the second-order transitions of synthetic rubbers. I. Non-crystalline copolymers. *J Appl Chem* 2, 493–500, 1952.

Hagemann JW. Thermal behavior and polymorphism of acylglycerides. In: Garti N and Sato K, Eds, *Crystallization and Polymorphism of Fats and Fatty Acids*. New York: Marcel Dekker, 1988, pp. 9–95.

Hagemann JW and Rothfus JA. Computer modeling of theoretical structures of monoacid triglyceride α-forms in various subcell arrangements. *J Am Oil Chem Soc* 60, 1308–1314, 1983.

Hägerdal B and Martens H. Influence of water content on the stability of myoglobin to heat treatment. *J Food Sci* 41, 933–937, 1976.

Hartel RW. *Crystallization in Foods*. Gaithersburg ML, Aspen, 2001, p. 325.

Harwalkar VR. Kinetic study of thermal denaturation of proteins in whey. *Milchwissenschaft* 41, 206, 1986.

Herrington TM and Branfield AC. Physico-chemical studies on sugar glasses. I. Rates of crystallization. *J Food Technol* 19, 409–425, 1984.

Hoseney RC, Zeleznak K, and Lai CS. Wheat gluten, a glassy polymer. *Cereal Chem* 63, 285–286, 1986.

Hui YH. *Bailey's Industrial Oil and Fat Products*, 5th ed, Vol. 1–5. New York: Wiley, 1996.

Iglesias HA and Chirife J. Delayed crystallization of amorphous sucrose in humidified freeze dried model systems. *J Food Technol* 13, 137–144, 1978.

Izzard MJ, Ablett S, and Lillford PJ. A calorimetric study of the glass transition occurring in sucrose solutions. In: Dickinson E, Ed., *Food Polymers, Gels and Colloids*. Cambridge: The Royal Society of Chemistry, 1991, pp. 289–300.

Johari GP, Hallbrucker A, and Mayer E. The glass-liquid transition of hyperquenched water. *Nature* 330, 552–553, 1987.

Jouppila K and Roos YH. The physical state of amorphous corn starch and its impact on crystallization. *Carbohydr Polym* 32, 95–104, 1997.

Jouppila K, Kansikas J, and Roos YH. Factors affecting crystallization and crystallization kinetics in amorphous corn starch. *Carbohydr Polym* 36, 143–149, 1998.

Kakivaya SR and Hoeve CAJ. The glass point of elastin. *Proc Nat Acad Sci* 72, 3505–3507, 1975.

Kalichevsky MT, Knorr D, and Lillford PJ. Potential food applications of high-pressure effects on ice-water transitions. *Trends Food Sci Technol* 6, 253–259, 1995.

Karel M. Recent research and development in the field of low-moisture and intermediate-moisture foods. *CRC Crit Rev Food Technol* 3, 329–373, 1973.

Karel M and Flink JM. Influence of frozen state reactions on freeze-dried foods. *J Agric Food Chem* 21, 16–21, 1973.

Katz JR. Gelatinization and retrogradation of starch in relation to the problem of bread staling. In: Walton RP, Ed., *Comprehensive Survey of Starch Chemistry* Vol I. New York: Chemical Catalog Co, 1928, p. 68.

Kauzmann W. Some factors in the interpretation of protein denaturation. *Adv Prot Chem* 14, 1–63, 1959.

Kauzmann W. The nature of the glassy state and the behavior of liquids at low temperatures. *Chem Rev* 43, 219–256, 1948.

Kelley SS, Rials TG, and Glasser, WG. Relaxation behavior of the amorphous components of wood. *J Mat Sci* 22, 617–624, 1987.

King CJ. Physical and chemical properties governing volatilization of flavor and aroma components. In: Peleg M and Bagley EB, Eds, *Physical Properties of Foods*. Westport: AVI, 1983, pp. 399–421.

King CJ. Freeze-drying of foodstuffs. *CRC Crit Rev Food Technol* 1, 379–451, 1970.

Kinsella JE and Whitehead DM. proteins in whey, Chemical, physical, and functional properties. *Adv Food Nutr Res* 33, 343–438, 1989.

Knutson CA. Annealing of maize starches at elevated temperatures. *Cereal Chem* 67, 376–384, 1990.

Kowblansky M. Calorimetric investigation of inclusion complexes of amylose with long-chain aliphatic compounds containing different functional groups. *Macromolecules* 18, 1776–1779, 1985.

Krueger BR, Knutson CA, Inglett GE, and Walker CE. A differential scanning calorimetry study on the effect of annealing on gelatinization behavior of corn starch. *J Food Sci* 52, 715–718, 1987.

Kugimiya M, Donovan JW, and Wong RY. Phase transitions of amylose-lipid complexes in starches, A calorimetric study. *Staerke* 36, 265, 1980.

Lazar ME, Brown AH, Smith GS, Wong FF, and Lindquist FE. Experimental production of tomato powder by spray drying. *Food Technol* 10, 129–134, 1956.

LeBail A, Boillereaux L, Davenel A, Hayert M, Lucas T, and Monteau JY. Phase transition in foods, effect of pressure and methods to assess or control phase transition. *Innov Food Sci Emerging Technol* 4, 15–24, 2003.

Lelievre J. Theory of gelatinization in a starch-water-solute system. *Polymer* 17, 854–858, 1976.

Le Meste M and Simatos D. Use of electron spin resonance for the study of the "ante-melting" phenomenon, observed in sugar solutions by differential scanning calorimetry. *Cryo-Lett* 1, 402–407, 1980.

Le Meste M, Champion D, Roudaut G, Blond G, and Simatos D. Glass transition and food technology, a critical appraisal. *J Food Sci* 67, 2444–2458, 2002.

Levine H and Slade L. A polymer physico-chemical approach to the study of commercial starch hydrolysis products (SHPs). *Carboh Polym* 6, 213–244, 1986.

Levine H and Slade L. Principles of "cryostabilization" technology from structure/property relationships of carbohydrate/water systems — a review. *Cryo-Lett* 9, 21–63, 1988.

Levine H and Slade L. Influences of the glassy and rubbery states on the thermal, mechanical and structural properties of doughs and baked products. In: Faridi H and Faubion JM, Eds, *Dough Rheology and Baked Product Texture*. New York: AVI, 1990, pp. 157–330.

Li B and Sun D-W. Novel methods for rapid freezing and thawing of foods — a review. *J Food Eng* 54, 175–182, 2002.

Lund DB. Starch gelatinization. In: Singh RP and Medina AG, Eds, *Food Properties and Computer-Aided Engineering of Food Processing Systems*. Kluwer Academic Publishers, 1989, pp. 299–311.

Lund D. Influence of time, temperature, moisture, ingredients and processing conditions on starch gelatinization. *CRC Crit Rev Food Sci Nutr* 20, 249–273, 1984.

Lund DB. Applications of differential scanning calorimetry in foods. In: Peleg M and Bagley EB, Eds, *Physical Properties of Foods*. Westport: AVI, 1983, pp. 125–143.

Luyet B and Rasmussen D. Study by differential thermal analysis of the temperatures of instability of rapidly cooled solutions of glycerol, ethylene glycol, sucrose and glucose. *Biodynamica* 10, 167–191, 1968.

Luyet B and Rasmussen D. Study by differential thermal analysis of the temperatures of instability in rapidly cooled solutions of polyvinyl pyrrolidone. *Biodynamica* 10, 137–147, 1967.

Lutton ES and Fehl AJ. The polymorphism of odd and even saturated single acid triglycerides, C8–C22. *Lipids* 5, 90–99, 1970.

Madeka H and Kokini JL. Effect of glass transition and cross-linking on rheological properties of zein, Development of a preliminary state diagram. *Cereal Chem* 73, 433–438, 1996.

Makower B and Dye WB. Equilibrium moisture content and crystallization of amorphous sucrose and glucose. *J Agric Food Chem* 4, 72–77, 1956.

Matsuoka S, Williams G, Johnson GE, Anderson EW, and Furukawa T. Phenomenological relationship between dielectric relaxation and thermodynamic recovery processes near the glass transition. *Macromolecules* 18, 2652–2663, 1985.

Matveev YuI, Slade L, and Levine H. Determination of the main technological parameters of food substances by means of the additive contribution method. *Food Hydrocolloids* 13, 381–388, 1999.

Matveev YuI, Grinberg VYa, and Tolstoguzov VB. The plasticizing effect of water on proteins, polysaccharides and their mixtures. Glassy state of biopolymers, food and seeds. *Food Hydrocolloids* 14, 425–437, 2000.

Maurice TJ, Slade L, Sirett RR, and Page CM. Polysaccharide-water interactions — thermal behavior of rice starch. In: Simatos D and Multon JL, Eds, *Properties of Water in foods in Relation to Quality and Stability*. Dordrecht. Netherlands: Martinus Nijhoff Publishers, 1985, pp. 211–227.

Mellor JD. *Fundamentals of Freeze-Drying*. London: Academic Press, 1978, p. 386.

Mestres C, Colonna P, and Buleon A. Gelation and crystallization of maize starch after pasting, drum-drying or extrusion cooking. *J Cereal Sci* 7, 123–134, 1988.

Minifie BW. *Chocolate, Cocoa and Confectionary, Science and Technology*. 3rd ed. Westport: AVI, 1989, p. 904.

Morales-Diaz A and Kokini JL. Understanding phase transitions and chemical complexing reactions in the 7S and 11S soy protein fractions. In: Rao MA and Hartel RW, Eds, *Phase/State Transitions in Foods*. New York: Marcel Dekker, 1998, pp. 273–311.

Morozov VN and Gevorkian SG. Low-temperature glass transition in proteins. *Biopolymers* 24, 1785–1799, 1985.

Morris VJ. Starch gelation and retrogradation. *Trends Food Sci Technol* 1, 2–6, 1990.

Muhr AH and Blanshard JMV. Effect of polysaccharide stabilizers on the rate of growth of ice. *J Food Technol* 21, 683–710, 1986.

Murray ED, Arntfield SD, and Ismond MAH. The influence of processing parameters on food protein functionality II. Factors affecting thermal properties as analyzed by differential scanning calorimetry. *Can Inst Food Sci Technol J* 18, 158–162, 1985.

Nawar WW. Lipids. In: Fennema OR, Ed., *Food Chemistry* 2nd ed. New York: Marcel Dekker, 1985, pp. 139–244.

Ollivon M and Perron R. Measurements of enthalpies and entropies of unstable crystalline forms of saturated even monoacid triglycerides. *Thermochimica Acta* 53, 183–194, 1982.

O'Neill T and Kinsella JE. Effect of heat treatment and modification on conformation and flavor binding by β-lactoglobulin. *J Food Sci* 53, 906–909, 1988.

Orford PD, Parker R, Ring SG, and Smith AC. Effect of water as a diluent on the glass transition behaviour of malto-oligosaccharides, amylose and amylopectin. *Int J Biol Macrom* 11, 91–96, 1989.

Orford PD, Parker R, and Ring SG. Aspects of the glass transition behaviour of mixtures of carbohydrates of low molecular weight. *Carbohydr Res* 196, 11–18, 1990.

Park KH and Lund DB. Calorimetric study of thermal denaturation of β-lactoglobulin. *J Dairy Sci* 67, 1699–1706, 1984.

Parks GS, Huffmann HM, and Caffoir FR. Glass. II. The transition between glassy and liquid states in the case of glucose. *J Phys Chem* 32, 1366–1379, 1928.

Peleg M. On the use of the WLF model in polymers and foods. *Crit Rev Food Sci Nutr* 32, 59–66, 1992.

Raemy A and Schweizer TF. Thermal behaviour of carbohydrates studied by heat flow calorimetry. *J Thermal Anal* 28, 95–108, 1983.

Rasmussen D and Luyet B. Complementary study of some non-equilibrium phase transitions in frozen solutions of glycerol, ethylene glycol, glucose and sucrose. *Biodynamica* 10, 319–331, 1969.

Reid DS and Charoenrein S. DSC studies of the starch water interaction in the gelatinization process. Proceedings of 14th Natas Conference, San Francisco, 1985, pp. 335–340.

Rey LR. Thermal analysis of eutectics in freezing solutions. *Ann NY Acad Sci* 85, 510–534, 1960.

Rey LR and Bastien MC. Biophysical aspects of freeze-drying. In: Fisher FR, Ed., *Freeze-Drying of Foods*. Washington DC: National Academy of Sciences — National Research Council, 1962, pp. 25–42.

Ring SG. Observations on the crystallization of amylopectin from aqueous solution. *Int J Biol Macromol* 7, 253–254, 1985.

Ring SG, Colonna P, I'Anson KJ, Kalichevsky MT, Miles MJ, Morris VJ, and Orford PD. The gelation and crystallization of amylopectin. *Carbohydr Res* 162, 277–293, 1987.

Roos Y. Effect of moisture on the thermal behavior of strawberries studied using differential scanning calorimetry. *J Food Sci* 52, 146–149, 1987.

Roos Y. Melting and glass transitions of low molecular weight carbohydrates. *Carbohydr. Res.* 238: 39–48, 1993.

Roos YH. *Phase Transitions in Foods*. San Diego ČA: Academic Press, 1995, p. 361.

Roos YH. Frozen state transitions in relation to freeze drying. *J Thermal Anal* 48, 535–544, 1997.

Roos YH. Thermal analysis, state transitions and food quality. *J Thermal Anal Calorim* 71, 197–203, 2002.

Roos YH. Phase and state transitions in dehydration of biomaterials and foods. In: Mujumdar AS, Ed. *Dehydration of Products of Biological Origin*. Enfield NH: Science Publishers, 2003, In press.

Roos Y and Karel M. Differential scanning calorimetry study of phase transitions affecting quality of dehydrated materials. *Biotehcnol Prog* 6, 159–163, 1990.

Roos Y and Karel M. Plasticizing effect of water on thermal behavior and crystallization of amorphous food models. *J Food Sci* 56, 38–43, 1991a.

Roos Y and Karel M. Phase transitions of amorphous sucrose and frozen sucrose solutions. *J Food Sci* 56, 266–267, 1991b.

Roos Y and Karel M. Phase transitions of mixtures of amorphous polysaccharides and sugars. *Biotechnol Prog* 7, 49–53, 1991c.

Roos YH, Karel M, and Kokini JL. Glass transitions in low moisture and frozen foods, Effects on shelf life and quality. *Food Technol* 50, 95–108, 1996.

Rüegg M, Moor U, and Blanc B. Hydration and thermal denaturation of β-lactoglobulin. A calorimetric study. *Biochim Biophys Acta* 400, 334–342, 1975.

Schlichter-Aronhime J and Garti N. Solidification and polymorphism in cocoa butter and blooming problems. In: Garti N and Sato K, Eds, *Crystallization and Polymorphism of Fats and Fatty Acids*. New York: Marcel Dekker, 1988, pp. 363–393.

Sharp PF and Doob H. Effect of humidity on moisture content and forms of lactose in dried whey. *J Dairy Sci* 24, 679–690, 1941.

Shen MC and Eisenberg A. Glass transitions in polymers. In: Reiss H, Ed. *Progress in Solid State Chemistry*, Vol. 3. London: Pergamon Press, 1967, pp. 407–481.

Shimada Y, Roos Y, and Karel M. Oxidation of methyl linoleate encapsulated in amorphous lactose-based food model. *J Agric Food Chem* 39, 637–641, 1991.

Simatos D, Faure M, Bonjour E, and Couach M. The physical state of water at low temperatures in plasma with different water contents as studied by differential thermal analysis and differential scanning calorimetry. *Cryobiology* 12, 202–208, 1975.

Simatos D and Karel M. Characterization of the condition of water in foods — Physico-chemical aspects. In: Seow CC, Ed. *Food Preservation by Water Activity Control*. Amsterdam: Elsevier 1988, pp. 1–41.

Singh RP and Heldman DR. Introduction to Food Engineering, Third Edition. Academic Press, New York. p. 620, 2001.

Slade L and Levine H. Non-equilibrium melting of native granular starch, Part I. Temperature location of the glass transition associated with gelatinization of A-type cereal starches. *Carbohydr Polym* 8, 183–208, 1988a.

Slade L and Levine H. Non-equilibrium behavior of small carbohydrate-water systems. *Pure Appl Chem* 60, 1841–1864, 1988b.

Slade L and Levine H. Beyond water activity, Recent advances based on an alternative approach to the assessment of food quality and safety. *CRC Crit Rev Food Sci Nutr* 30, 115–360, 1990.

Slade L and Levine H. Beyond Water Activity: Recent Advances Based on an Alternative Approach to the Assessment of Food Quality and Safety. *CRC Crit Revs Food Sci Nutr* 30: 115–360, 1991.

Slade L and Levine H. Glass transitions and water-food structure interactions. *Adv Food Nutr Res* 38, 103–269, 1995.

Slade L, Levine H and Finley JW. Protein-Water Interactions: Water as a Plasticizer of Gluten and Other Protein Polymers. In Protein Quality and the Effects of Processing, Phillips RD and Finley JW, Eds Marcel Dekker, New York, 9–124, 1989.

Smith JM and Van Ness HC. *Introduction to Chemical Engineering Thermodynamics*. 4th ed. New York: McGraw-Hill, 1987, p. 698.

Soesanto T and Williams MC. Volumetric interpretation of viscosity for concentrated and dilute sugar solutions. *J Phys Chem* 85, 3338–3341, 1981.

Sperling LH. *Introduction to Physical Polymer Science*, 2nd ed. New York: Wiley, 1992, p. 594.

Struik LCE. *Physical Aging in Amorphous Polymers and Other Materials*. Amsterdam: Elsevier, 1978, p. 229.

Sugisaki M, Suga H, and Seki S. Calorimetric study of the glassy state. IV. Heat capacities of glassy water and cubic ice. *Bull Chem Soc Japan* 41, 2591–2599, 1968.

Talja RA and Roos YH. Phase and state transition effects on dielectric, mechanical, and thermal properties of polyols. *Thermochimica Acta* 380, 109–121, 2001.

Tant MR and Wilkes GL. An overview of the nonequilibrium behavior of polymer glasses. *Polym Eng Sci* 21, 874–895, 1981.

Thijssen HAC. Flavor retention in drying preconcentrated food liquids. *J Appl Chem Biotechnol* 21, 372–377, 1971.

Timms RE. Heats of fusion of glycerides. *Chem Phys Lipids* 21, 113–129, 1978.

To EC and Flink JM. "Collapse," a structural transition in freeze-dried carbohydrates. III. Prerequisite of recrystallization. *J Food Technol* 13, 583–594, 1978.

Troy HC and Sharp PF. α and β lactose in some milk products. *J Dairy Sci* 13, 140–157, 1930.

Tsourouflis S, Flink JM, and Karel M. Loss of structure in freeze-dried carbohydrates solutions, Effect of temperaure, moisture content and composition. *J Sci Food Agric* 27, 509–519, 1976.

Vrentas JS and Duda JL. A free volume interpretation of the influence of the glass transition on diffusion in amorphous polymers. *J Appl Polym Sci* 22, 2325–2339, 1978.

Vrentas JS, Duda JL, and Lau MK. Solvent diffusion in molten polyethylene. *J Appl Polym Sci* 27, 3987–3997, 1982.

Weitz A and Wunderlich B. Thermal analysis and dilatometry of glasses formed under elevated pressure. *J Polym Sci Polym Phys* 12, 2473–2491, 1974.

Wesdorp LH. Liquid–multiple solid phase equilibria in fats. Theory and experiments. Technische Universiteit Delft, Doctoral Thesis, 1990, p. 253.

White GW and Cakebread SH. The glassy state in certain sugar-containing food products. *J Food Technol* 1, 73–82, 1966.

Williams ML, Landel RF, and Ferry JD. The temperature dependence of relaxation mechanisms in amorphous polymers and other glass-forming liquids. *J Am Chem Soc* 77, 3701–3707, 1955.

Wirakartakusumah MA. Kinetics of starch gelatinization and water absorption in rice. Department of Food Science. University of Wisconsin-Madison, Ph.D. Thesis, 1981.

Wootton M and Bamunuarachchi A. Application of differential scanning calorimetry to starch gelatinization. II. Effect of heating rate and moisture level. *Staerke* 31, 262–264, 1979.

Wootton M and Bamunuarachchi A. Application of differential scanning calorimetry to starch gelatinization. III. Effect of sucrose and sodium chloride. *Staerke* 32, 126–129, 1980.

Wright DJ. Application of scanning calorimetry to the study of protein behaviour in foods. In: Hudson BJF, Ed., *Developments in Food Proteins-1*. London: Applied Science Publishers, 1982, pp. 61–89.

Wright DJ, Leach IB, and Wilding P. Differential scanning calorimetric studies of muscle and its constituent proteins. *J Sci Food Agric* 28, 557–564, 1977.

Wunderlich B. The basis of thermal analysis. In: Turi'EA, Ed. *Thermal Characterization of Polymeric Materials*. New York: Academic Press, 1981, pp. 91–234.

Zeleznak KJ and Hoseney RC. The glass transition in starch. *Cereal Chem* 64, 121–124, 1987.

4 Transport and Storage of Food Products

M.A. Rao

CONTENTS

4.1 INTRODUCTION

The transportation and storage of food products are two very important unit operations in the food processing industry. Because of the biological and fragile nature of foods, and the ever-present threat of attack by insects and microorganisms, the design of transportation and storage systems poses special challenges and problems. Therefore, sanitary and microbiological considerations play important roles and the design considerations include the use of stainless steel and other approved materials of construction, the use of sanitary fittings, the design of pipelines to eliminate stagnation zones and to facilitate drainage of the foods, the control of temperature and humidity of storage, and minimal contact with oxygen to minimize degradation reactions. Foods are stored either for relatively short periods of times in food processing plants when used as ingredients for processed foods, or they are stored for extended periods in warehouses, where they are held as part of the distribution sector. In this chapter the emphasis is on in-plant storage, although some of the principles are also applicable to storage of foods on a large scale in warehouses.

4.2 LIQUID-INGREDIENT STORAGE

Design of liquid-ingredient storage tanks depends to some extent on the product to be stored. For foods such as dairy products that pose high microbiological risk, 3-A standards (discussed later) need to be consulted with regard to considerations not only for the design of storage tanks but also with respect to the design of the components of the associated transfer systems. For the most part, vertical tanks are used for storage of liquid ingredients. The design considerations with respect to radii of corners and welds discussed for solid bins are also applicable to liquid storage tanks. In addition, the bottoms of vertical tanks must be sloped at the rate of 3/4 in./ft for small tanks and 1 in./ft for large tanks. Horizontal tanks must have a pitch of 1/8 in./ft toward (he outlet (Imholte, 1984).Tanks for liquid sugar and syrups can be constructed out of mild steel or they could be fiberglass tanks made out of food-grade resins. Oil tanks can also be constructed out of mild steel.

4.2.1 PROPERTIES FOR STORAGE AND TRANSPORTATION OF LIQUID FOODS

For storage of liquid foods, density of the foods is a property that must be known in order to estimate the volume of storage tanks. The density and viscosity (or apparent viscosity) are important properties in the drainage of liquid food storage tanks and in the transportation of liquid foods. Because many foods are non-Newtonian fluids, their rheological behavior and properties play an important role in the transportation of liquid foods. The rheological properties of fluid foods were discussed in Chapter 1.

4.2.2 DENSITIES OF LIQUID FOODS

The density of many liquid foods can be calculated from empirical equations compiled from the literature by Choi and Okos (1986). For fruit juices, the density vs. the index of refraction (n) data of sugar solutions developed by Riedel (1949) may be used:

$$\frac{n^2 - 1}{n^2 + 2} \times \frac{62.4}{0.206} \times 16.0185 \tag{4.1}$$

The density of whole and skim milk as a function of temperature (°C) can be calculated from equations developed by Short (1955):

$$\text{Whole milk}: \quad \rho = 1035.0 - 0.358\,T + 0.0049\,T^2 - 0.00010\,T^3 \tag{4.2}$$

$$\text{Skim milk}: \quad \rho = 1036.6 - 0.146\,T + 0.0023\,T^2 - 0.00016\,T^3 \tag{4.3}$$

Phipps (1969) reported an equation for the estimation of the density of cream as a function of temperature (T) and fat content (X_f) with an accuracy of $\pm 0.45\%$:

$$\rho = 1038.2 - 0.17\,T - 0.003\,T^2 - \left(133.7 - \frac{475.5}{T}\right) X_f \tag{4.4}$$

Roy et al. (1971) reported equations for estimating the density of fat from buffalo and cow's milk:

$$\text{Buffalo milk}: \quad \rho = 923.84 - 0.44\,T \tag{4.5}$$

$$\text{Cows's milk}: \quad \rho = 923.51 - 0.43\,T \tag{4.6}$$

For the density of tomato juice, Choi and Okos (1983) developed the predictive equation based on the water (X_W) and solids fraction (X_S):

$$\rho = \rho_W X_W + \rho_S X_S \tag{4.7}$$

where

$$\rho_W = 9.9989 \times 10^2 - 6.0334 \times 10^{-2}\,T - 3.6710 \times 10^{-3}\,T^2 \tag{4.8}$$

and

$$\rho_S = 1.4693 \times 10^3 + 5.4667 \times 10^{-1}\,T - 6.9646 \times 10^{-3}\,T^2 \tag{4.9}$$

Choi and Okos (1986) presented a model applicable to many liquid foods as a function of composition and temperature:

$$\rho = \Sigma \rho_i x_i \tag{4.10}$$

where x_i is the weight fraction of each component of the liquid food (water, protein, fat, carbohydrate, and ash) that can be obtained from tabulations in handbooks such as that of Watt and Merrill (1975), and ρ_i is the density of the component at the temperature selected. The standard errors between the predicted and experimental values of densities were found to be between 2.14 and 3.15%.

4.3 SANITARY STANDARDS

Because of microbiological considerations, ordinary pipe and fittings cannot be used for transporting liquid foods in food processing facilities; instead, stainless steel fittings should be used. Foods such as liquid sugars, syrups, and honey are generally not affected by bacteria, but they can support mold and yeast growth; therefore, the piping used should be of stainless steel. Transportation systems for liquid foods, such as dairy and egg. products, sensitive to microbial spoilage Table 4.1 and Table 4.2 must be designed according the 3-A and E-3-A standards, listed in Table 4.1 and Table 4.2, respectively. Some of the considerations include (Imholte, 1984): (1). self-draining installation of pipe systems without sags where the product could accumulate; (2) for clean-in-place systems, sanitary weld fittings with standard take-apart fittings must be provided at pumps, valves, tanks, and

---

TABLE 4.1
Published 3-A Sanitary Standards[a]

Number	Title	Effective date
01-08	Storage Tanks for Milk and Milk Products	20-Nov-2001
02-09	Centrifugal and Positive Rotary Pumps for Milk and Milk Products	01-Nov-1996
04-04	Homogenizers and Reciprocating Pumps	01-Nov-1996
05-15, replaces 05-14	Stainless Steel Automotive Milk and Milk Product Transportation Tanks for Bulk Delivery and/or Farm Pick-Up Service	01 -Nov-2002
10-04	Sanitary Standards for Filters Using Single Service Filter Media	12-Nov-2000
11-06	Sanitary Standards for Plate-Type Heat Exchangers	20-Nov-2001
12-06	Sanitary Standards for Tubular Heat Exchangers	31-May-2002
13-09	Farm Milk Cooling and Holding Tanks	01-Nov-1993
16-05	Milk and Milk Products Evaporators and Vacuum Pans	01-Aug-1985
17-10, replaces 17-09	Formers, Fillers, and Sealers of Single-Service Containers for Fluid Milk and Fluid Milk Products	01-Nov-2002
18-03	Multiple-Use Rubber and Rubber-Like Materials Used as Product Contact Surfaces in Dairy Equipment	01-Aug-1999
19-05	Batch and Continuous Freezers for Ice Cream, Ices, and Similarly Frozen Dairy Foods	01-Nov-1999
21-00	Sanitary Standards for Centrifugal Separators and Clarifiers	24-Nov-2002
20-22, replaces 20-21	Sanitary Standards for Multiple-Use Plastic Materials Used as Product Contact Surfaces for Dairy Equipment	01-Jun-2003
22-07	Silo-type Storage Tanks for Milk and Milk Products	01-Nov-1996
23-04 replaces 23-03 12	Sanitary Standards for Equipment for Packaging Viscous Products	24-Nov-2002
24-02	Non-Coil Type Batch Pasteurizers for Milk and Milk Products	01-Nov-1989
25-03, replaces 25-02	Non-Coil Type Batch Processors for Milk and Milk Products	01-Nov-2002
26-04	Sanitary Standards for Sifters for Dry Products	20-Nov-2001
27-05 replaces 27-04 01	Equipment for Packaging Dry Milk and Dry Milk Products	01-Nov-2002
28-03	Flow Meters for Milk and Milk Products	19-Jun-2002
29-02	Air Eliminators for Milk and Milk Products	12-Nov-2000
30-01	Farm Milk Storage Tanks	01-Sep-1984
31-03	Sanitary Standards for Scraped Surface Heat Exchangers	12-Nov-2000
32-02	Uninsulated Tanks for Milk and Milk Products	01-Aug-1994
33-01	Polished Metal for Milk and Milk Products	01-Nov-1994
34-02	Portable Bins for Dry Milk and Dry Milk Products	01-Sep-1992
35-00	Continuous Blenders	01-Aug-1997
36-00	Colloid Mills	01-Aug-1997
38-00	Cottage Cheese Vats	01-Aug-1980

(Continued)

TABLE 4.1
Continued

Number	Title	Effective date
39-00	Pneumatic Conveyors for Dry Milk and Dry Milk Products	01-Sep-1982
40-02, replaces 40-01	Bag Collectors for Dry Milk and Dry Milk Products	24-Nov-2002
41-01	Mechanical Conveyors for Dry Milk and Dry Milk Products	01-Nov-1996
42-01	In-Line Strainers for Milk and Milk Products	01-Nov-1997
43-00	Wet Collectors for Dry Milk and Dry Milk Products	01-Sep-1986
44-03	Sanitary Standards for Diaphragm Pumps	20-Nov-2001
45-01	Crossflow Membrane Modules	01-Nov-1999
46-03, replaces 46-02	Refractometers and Energy-Absorbing Optical Sensors for Milk and Milk Products	24-Nov-2002
47-00	Centrifugal and Positive Rotary Pumps for Pumping Cleaning and Sanitizing Solutions	01-Nov-1996
49-01	Sanitary Standards for Air-Driven Sonic Horns for Dry Products	20-Nov-2001
50-01	Sanitary Standards for Level Sensing Devices for Dry Products	20-Nov-2001
51-01	Plug-Type Valves for Milk and Milk Products	01-Nov-1998
52-02	Plastic Plug-Type Valves for Milk and Milk Products	01-Nov-1998
53-02	Compression-Type Valves for Milk and Milk Products	01-Apr-2002
54-02	Diaphragm-Type Valves for Milk and Milk Products	01-Nov-1997
55-01	Boot Seal-type Valves for Milk and Milk Products	01-Nov-1996
56-00	Inlet and Outlet Leak-Protector Plug-Type Valves for Milk and Milk Products	01-May-1993
57-01	Tank Outlet Valves for Milk and Milk Products	01-Nov-1996
58-00	Vacuum Breakers and Check Valves for Milk and Milk Products	01-Jun-1992
59-00	Automatic Positive Displacement Samplers for Fluid Milk and Fluid Milk Products	01-Nov-1993
60-00	Rupture Discs for Milk and Milk Products	01-Sep-1983
61-00	Steam Injection Heaters for Milk and Milk Products	01-Nov-1994
62-01	Hose Assemblies for Milk and Milk Products	01-Nov-1996
63-03, replaces 63-02	Sanitary Fittings for Milk and Milk Products	24-Nov-2002
64-00	Pressure Reducing and Back Pressure Regulating Valves for Milk and Milk Products	01-Nov-1993
65-00	Sight and/or Light Windows and Sight Indicators in Contact with Milk and Milk Products	01-Nov-1994
66-00	Caged-Ball Valves for Milk and Milk Products	01-Nov-1995
68-00	Ball-Type Valves for Milk and Milk Products	01-Nov-1996
70-01, replaces 70-00	Italian-Type Pasta Filata Style Cheese Cookers	24-Nov-2002
71-01, replaces 71-00	Italian-Type Pasta Filata Style Cheese Molders	24-Nov-2002
72-01	Italian-Type Pasta Filata Style Molded Cheese Chillers	24-Nov-2002
73-00	Shear Mixers, Mixers, and Agitators	01-Nov-1996
74-02	Sanitary Standards for Sensors and Sensor Fittings and Connections Used on Milk and Milk Products Equipment	24-Nov-2002
75-00	Belt-Type Feeders	01-Nov-1998
78-00	Spray Devices to Remain in Place	01-Nov-1998
81-00	Auger-Type Feeders	01-Nov-1998
82-00	Sanitary Standards for Pulsation Dampening Devices	24-Nov-2002
	E-3-A Sanitary Standards (Eggs)	
E-1500	Shell Egg Washer	01-Jan-1977
E-600	Egg Breaking and Separating Machines	01-Jan-1977

[a] A current list of 3-A standards can be obtained at the above website.

Source: From www.3-A.org.

TABLE 4.2
Published 3-A Accepted Practices

Number	Title	Effective date
603-06	Sanitary Construction, Installation, Testing, and Operation of High-Temperature Short-Time and Higher-Heat Shorter-time Pasteurizer Systems	01-Dec-1992
604-04	Supplying Air Under Pressure in Contact with Milk, Milk Products, and Product Contact Surfaces	01-Nov-1994
05-04	Permanently Installed Product and Solution Pipelines and Cleaning Systems Used in Milk and Milk Product Processing Plants	01-Aug-1994
606-04	Replaced by 3A 606–05 Design, Fabrication, and Installation of Milking and Milk Handling Equipment	01-Nov-1996
607-04	Milk and Milk Products Spray Drying Systems	01- Nov-1998
608-02	Accepted Practices for Instantizing Systems	20-Nov-2001
609-02	Method of Producing Steam of Culinary Quality	01-Nov-1996
610-00	Sanitary Construction, Installation, and Cleaning of Crossflow Membrane Processing Systems for Milk and Milk Products	01-Sep-1990
611-00	Farm Milk Cooling and Storage Systems	01-Nov-1994

[a] A current list of 3-A standards can be obtained at the above website.

Source: From www.3-A.org

other points of connection; and (3) pipelines must be rigidly supported. For liquid sugars and syrups, piping systems must be generously oversized; Imholte (1984) suggested a 3-in. diameter pipe as the minimum size. For oil-handling systems, one may use mild steel pipe. Finally, pipeline strainers and magnets must be used on the suction side of pumps.

4.3.1 3-A STANDARDS

Whereas the sizes of storage tanks are calculated from simple arithmetic expressions, the sanitary aspects of storage of milk and egg products are very important and must be based on 3-A standards for storage tanks for milk and milk products (No. 01-06) and for egg products (No. E-0100). Because the movement of fluid foods in processing plants requires pumps, piping, and other auxiliary equipment, one must be aware of the 3-A standards for various equipments also.

The International Association of Dairy and Milk Inspectors in the late 1920s established a committee on Dairy and Milk Plant Equipment and recommended efforts in the area of standards for dairy and milk plant equipment. The 3-A sanitary standards refer to the development by the original three parties: International Association of Milk Dealers (now the Milk Industry Foundation), the Dairy and Ice Cream Machinery Supply Association [now the Dairy and Food Industries Supply Association (DFISA)], and the people associated with city, state, and federal enforcement. While the title has been retained, it now refers to three different groups: (1), the International Association of Milk, Food, and Environmental Sanitarians; (2) USPHS/FDA; and (3) the Dairy Industry Committee (DIC), which represents a group of eight trade associations.

The 3-A sanitary standards set forth the criteria for (1) the material used in the construction of a piece of dairy equipment, (2) the fabrication and design of such material, and (3) its construction, including such things as the finish of the material, which are considered essential from a sanitary standpoint in the use, performance, and maintenance of such equipment. The E-3-A sanitary standards set forth the criteria for egg processing equipment. Each standard was developed through the joint collaboration of (1) manufacturers of the equipment, (2) users of the equipment, (3) the International Association of Milk, Food, and Environmental Sanitarians' (IAMFES) Committee on Sanitary Procedure, (4) Poultry and Egg Institute of America, and (5) representatives of the U.S.

Public Service/FDA, U.S. Department of Agriculture. The first rough equivalent of a 3-A sanitary standard was developed in 1929 and applied to sanitary fittings used in milk plants. Table 4.1 is a list of the published 3-A standards and Table 4.2 is a list of E-3-A standards. New standards are needed from time to time, so ensure that one has an as to up-to-date list of standards by contacting either IAMFES or the Journal of Food Protection, 502 E. Lincoln Way, Ames, IA 50010.

Labrie (1987) suggested that in addition to the 3-A standards, the requirements according to Pasteurized Milk Ordinance and the FDA's Current Good Manufacturing Practices also be considered. Labrie (1987) also suggested pointers to provide a clean operating production line (1) one must think small when it comes to wet process areas; (2) to form a project team consisting of the project engineer, state health inspector, and personnel from plant production, maintenance, and sanitation, and representatives of the equipment manufacturer and the installation contractor, (3) for the processor to participate in equipment design; and (4) that for essential communication, flow sheets, and sanitary details of installation drawings be used.

4.4 TRANSPORTATION OF FLUID FOODS

4.4.1 MECHANICAL ENERGY BALANCE EQUATION

The energy required to pump a liquid food through a pipeline can be calculated from the mechanical energy balance equation (MEBE). The MEBE can be used to analyze pipe flow systems. For the steady-state flow of an incompressible fluid, the mechanical energy balance can be written as follows (Brodkey, 1967; Heldman and Singh, 1981):

$$gZ_1 + \frac{P_1}{\rho} + \frac{u_1^2}{\alpha} - W = gZ_2 + \frac{P_2}{\rho} + \frac{u_2^2}{\alpha} + E_f \tag{4.11}$$

where Z is the height above a reference point, P is the pressure, u is the fluid velocity, W is the work output per unit mass, E_f is the energy loss per unit mass, α is the kinetic energy correction factor, ρ is the density, and the subscripts 1 and 2 refer to two, points in the pipe system. The velocities at the entrance and exit of the system can be calculated from the respective diameters of the tanks or pipes and the volumetric flow rate of the food. The energy loss term E_f consists of losses due to friction in pipe and that due to friction in valves and fittings:

$$E_f = \frac{2fu^2L}{D} + \sum_1^b \frac{k_f u^2}{2} \tag{4.12}$$

where f is the friction factor, u the velocity, L the length of straight pipe of diameter D, k_f the friction coefficient for a fitting, and b the number of valves or fittings. It is emphasized that k_f is unique to a particular fitting and that different values of u, k_f, and f may be required when the system contains pipes of different diameters. Further, losses due to special equipment, such as heat exchangers, must be added to E_f (Steffe and Morgan, 1986).

4.4.2 FRICTION LOSSES IN PIPES

Because many fluid foods are non-Newtonian in nature, estimation of friction losses for these fluids in straight pipes and in fittings is of interest. One can estimate the friction losses in straight pipes and tubes from the magnitude of the Fanning friction factor, f:

$$\frac{\Delta P_f}{\rho} = \frac{2fLu^2}{D} \tag{4.13}$$

where f is the friction factor. For laminar flow conditions, Garcia and Steffe (1987) suggested that based on the work of Hanks (1978), the friction factor can be calculated from a single general relationship for fluids that can be described by Newtonian power-law (Equation 4.15), Bingham plastic (Equation 4.16), and Herschel–Bulkley fluid (Equation 4.14) models.

$$\sigma = \sigma_0 + k_H \dot{\gamma}^{n_H} \tag{4.14}$$

$$\sigma = K \dot{\gamma}^n \tag{4.15}$$

$$\sigma = \sigma_0 + \eta \dot{\gamma} \tag{4.16}$$

For the laminar flow of a Herschel–Bulkley fluid (Equation 4.14), the friction factor can be written as

$$f = \frac{16}{\Psi \cdot \mathrm{GRe}} \tag{4.17}$$

where GRe is the generalized Reynolds number, defined as

$$\mathrm{GRe} = \frac{D^n u^{2-n} \rho}{8^{n-1} K} \left(\frac{4n}{3n+1} \right) \tag{4.18}$$

and

$$\Psi = (3n+1)^n (1-\xi_0)^{1+n} \left[\frac{(1-\xi_0)^2}{3n+1} + \frac{2\xi_0(1-\xi_0)}{2n+1} + \frac{\xi_0^2}{n+1} \right]^n \tag{4.19}$$

where ξ_0 is the dimensionless unsheared plug radius:

$$\xi_0 = \frac{\sigma_0}{\sigma_w} = \frac{\sigma_0}{D\Delta P_f\, 4L} = \frac{\sigma_0}{f\rho u^2/2} \tag{4.20}$$

ξ_0 can be calculated as an implicit function of GRe and the generalized Hedstrom number GHe:

$$\mathrm{Re} = 2\mathrm{GHe} \left(\frac{n}{3n+1} \right)^2 \left(\frac{\Psi}{\xi_0} \right)^{(2/n)-1} \tag{4.21}$$

where

$$\mathrm{GHe} = \frac{D^2 \rho}{K} \left(\frac{\sigma_0}{K} \right)^{(2/n)-1} \tag{4.22}$$

For power-law and Newtonian fluids, the friction factor can be estimated directly from Equation 4.17 because $\xi_0 = 0$ and $\Psi = 1$ when $\sigma_0 = 0$. For Bingham plastic and H-B fluids, ξ_0 is calculated through iteration of Equation 4.21 using Equation 4.17 to Equation 4.19.

Because of the highly viscous nature of non-Newtonian foods, laminar flow conditions are likely to be encountered more often than turbulent conditions. Nevertheless, it is important to be aware of developments with respect to the prediction of friction factors in turbulent flow of non-Newtonian foods. For turbulent flow, except for Newtonian fluids, the predicted magnitudes for non-Newtonian fluids may differ greatly depending on the relationship employed (Garcia and Steffe, 1987). The relationships of Dodge and Metzner (1959), Clapp (1961), and Hanks and Ricks (1975) for power-law fluids were found to predict similar magnitudes of the friction factors. For the Herschel–Bulkley model used when non-Newtonian foods exhibit yield stress, the analysis of Hanks (1978) was found

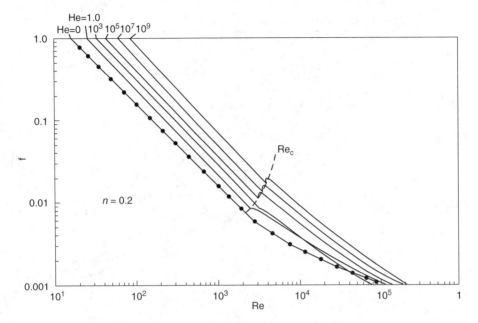

FIGURE 4.1 Friction factor for a fluid food that follows the Herschel–Bulkley model, for flow behavior index (n) = 0.2 as a function of the generalized Reynolds number (GRe) and the generalized Hedstrom number (GHe). Re_c is the critical Reynolds number. (From Garcia, E.J. and Steffe, J.F. 1987. *J. Food Process Eng.*, 9: 93–120. With permission.)

to be the most comprehensive, but to use the derived relationship, it is necessary to perform a numerical integration and several iterations. Figure 4.1 for $n = 0.2$ and Figure 4.2 for $n = 0.5$ can be used to calculate the friction factor for H-B fluids over a wide range of generalized Reynolds and Hedstrom numbers.

4.4.3 KINETIC ENERGY LOSSES

Kinetic energy losses can be calculated easily provided that the kinetic energy correction factor α can be determined. In turbulent flow, $\alpha = 2$. When the flow is laminar, α may be determined from Figure 4.3 or it may also be calculated from the following analytical expression (Osorio and Steffe, 1984):

$$\alpha = \{2(1 + 3n + 2n^2 + 2n^2\zeta_0 + 2n^2\xi_0 + 2n^2\xi_0^2)^3(3n + 2)(5n + 3)(4n + 3)\}$$
$$\times \{[(2n + 1)^2(3n + 1)^2][18 + n(105 + 66\zeta_0) + n^2(243 + 306\xi_0 + 85\xi_0^2)$$
$$+ n^3(279 + 522\xi_0 + 350\xi_0^2) + n^4(159 + 390\xi_0 + 477\xi_0^2) + n^5(36 + 108\xi_0 + 216\xi_0^2)]\}^{-1}$$

(4.23)

In Figure 4.3, α is presented as a function of the flow behavior index and the stress ratio defined as the ratio of the yield stress to the shear stress at the wall:

$$\xi_0 = \frac{\sigma_0}{\sigma_w} = \frac{\sigma_0 4L}{D\Delta P}$$

(4.24)

4.4.4 FRICTION LOSS COEFFICIENTS FOR FITTINGS

Steffe et al. (1984) determined magnitudes of the coefficient of for a fully open plug valve, a tee with flow from line to branch and a 90° short elbow as a function of GRe using applesauce as the

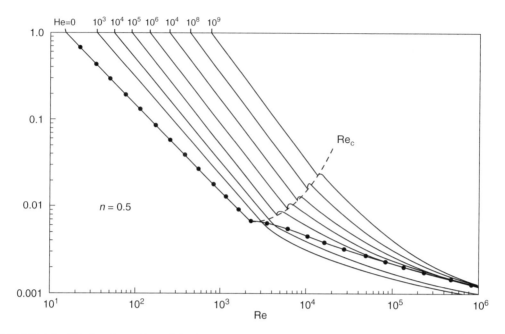

FIGURE 4.2 Friction factor for a fluid food that follows the Herschel–Bulkley model, for flow Behavior index (*n*) = 0.5 as a function of the generalized Reynolds number (GRe) and the generalized Hedstrom number (GHe). Re$_c$ is the critical Reynolds number. (From Garcia, E.J. and Steffe, J.F. 1987. *J. Food Process Eng.*, 9: 93–120. With permission.)

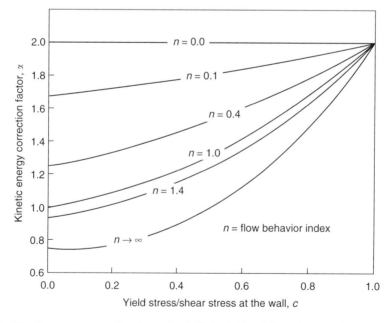

FIGURE 4.3 Kinetic energy correction factor for fluid foods that follow the Herschel–Bulkley model as a function of the ratio (yield stress/shear stress at the wall) and the flow behavior index (*n*). (From Osorio, F.A. and Steffe, J.F. 1984. *J. Food Sci.* 49: 1295–1296, 1315. With permission.)

test fluid. They found that as for Newtonian fluids, k_f increases with decreasing values of GRe. The regression equations for the three fittings were:

$$\text{Three-way plug valve:} \quad k_f = 30.3 \, \text{GRe}^{-0.492} \tag{4.25}$$

$$\text{Tee:} \quad k_f = 29.4 \, \text{GRe}^{-0.504} \tag{4.26}$$

$$\text{Elbow:} \quad k_f = 191.0 \, \text{GRe}^{-0.896} \tag{4.27}$$

In many instances, the practice is to employ values determined for Newtonian fluids, such as those in *Perry's Chemical Engineer's Handbook* (Perry and Green, 1984).

4.4.5 PUMP SELECTION AND PIPE SIZING

Steffe and Morgan (1986) discussed in detail the selection of pumps and the sizing of pipes for non-Newtonian fluids. Preliminary selection of a pump is based on the volumetric pumping capacity only from data provided by the manufacturers of pumps. Effective viscosity η_e was defined by Skelland (1967) as the viscosity that is obtained assuming that the Hagen–Poiseuille equation for laminar flow of Newtonian fluids is applicable:

$$\eta_e = \frac{D\Delta P/4L}{\pi D^3/32Q} \tag{4.28}$$

An alternative form of Equation 4.28 in terms of the mass flow rate m and the friction factor f is

$$\eta_e = \frac{fm}{4\pi D} \tag{4.29}$$

In calculating η_e from Equation 4.28, either the port size of a pump or the dimensions of the assumed pipe size can be used. Based on the magnitude of η_e, the suitability of the pump volumetric size must be verified from plots of effective viscosity vs. volumetric flow rate. It is emphasized that a pump size is assumed based on the volumetric pumping requirements and the assumption is verified by performing detailed calculations.

A comprehensive example for sizing a pump and piping for a non-Newtonian fluid whose rheological behavior can be described by the Herschel–Bulkley model (Equation 4.14) was developed by Steffe and Morgan (1986) for the system shown in Figure 4.4 and it is summarized in the

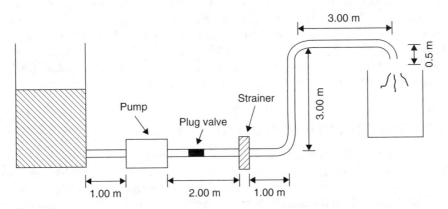

FIGURE 4.4 Flow system for the illustration of the mechanical energy balance equation. (From Steffe, J.F. and Morgan, R.G. 1986. *Food Technol.* 40(12): 78–85. With permission.)

following. The Herschel–Bulkley parameters, density, and other data for the example are: yield stress $= 157$ Pa, flow behavior index $= 0.45$, consistency index $= 5.20$ Pav·sec^n, density $1,250$ kg/m^3, flow rate $= 1.57 \times 10^{-3}$ m^3/sec, mass average velocity $= 1.66$ m/sec, and internal pipe diameter $= 0.0348$ m. The magnitudes of the generalized Hedstrom (GHe) number, the generalized Reynolds (GRe) number, and the ratio $\xi_0 = (\sigma_0/\sigma_w)$ were 36, 430, 323, and 0.585, respectively. Based on GRe and GHe, from Figure 4.2, the magnitude of the friction factor f was 0.156. The friction loss coefficients for the elbow and the gate valve, based on data for Newtonian fluids, were taken from *Perry's Chemical Engineer's Handbook* (Perry and Green, 1984) to be 0.45 and 9.0, respectively.

4.4.6 PUMP DISCHARGE PRESSURE

The discharge pressure of the pump can be calculated by applying the MEBE equation between the pump discharge and the exit point of the system so that the upper seal pressure limits are not exceeded. The MEBE for this purpose can be written as:

$$P_1 = \left[g(Z_2 - Z_1) + \frac{P_2}{\rho} + E_f \right] \rho \qquad (4.30)$$

The energy loss due to friction in the pipe, valve, and fittings was estimated to be 329 0 J/kg, and the discharge pressure of the pump, P_1, was estimated to be 4.42×10^5 Pa.

4.4.7 POWER REQUIREMENTS

The total power requirements for pumping are calculated by adding the hydraulic and viscous power requirements. The former can be estimated from the MEBE written for the work input. $-W$. We note that $P_1 = P_2$ and that E_f includes not only the friction losses on the discharge section but also the inlet section: As stated earlier, the former was estimated to be 329.0 J/kg and the latter was estimated to be 24.7 J/kg by applying the MEBE between the exit of the tank and the pump inlet. The work input $(-W)$ was estimated to be 380.0 J/kg, and because the mass flow rate was 1 97 kg/s the hydraulic power input was estimated to be 749.0 J/sec or 0.749 kW. For estimating the viscous power requirements due to energy losses in the pump due to friction, the operating speed and the effective viscosity of the fluid food must be calculated. The former can be calculated from the displacement volume per revolution of the pump and the required volumetric flow, while the latter can be calculated from Equation 4.28. For a size 30 Waukesha pump, the volumetric displacement per revolution was 2.27×10^{-4} m^3/s, and hence the pump speed was 417 rpm, while the equivalent viscosity for the fluid food under consideration was 0.703 Pa·sec (Steffe and Morgan, 1986). The energy losses in the pump were estimated from the manufacturer's data to be 0.835 kW. Therefore, the sum of the hydraulic and the viscous losses were 1.58 kW or 2.12 hp. These data allow selection of a suitable motor and drive system. In this example, pipe size was based on the pump port's diameter. It may also be based on plans for future expansion and ease of cleaning.

4.5 MIXING OF NON-NEWTONIAN FLUID FOODS

Mixing, also called agitation, of fluid foods is an important operation in food processing plants. The goals of a mixing operation include: homogenization, dispersion, suspension, blending, and heat exchange. Several agitators are used in the food industry and many, undoubtedly, are proprietary designs. Kalkschmidt (1977) listed several types of agitators used in the dairy industry that were classified under, propellers: screw, edge, and ring; under stirrers: disc, cross bar, paddle, anchor, blade, gate-paddle, spiral, and finger-paddle, and moving cutters. In Figure 4.5, three agitators: an

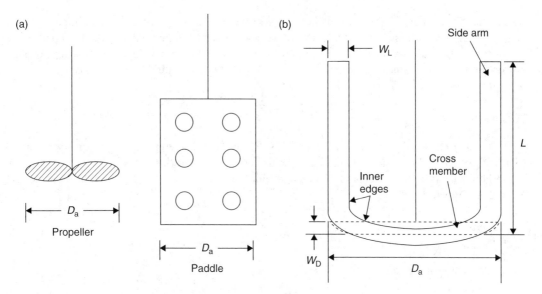

FIGURE 4.5   Schematic of agitators: propeller, paddle, and anchor.

anchor, a propeller, and a paddle are shown. Sometimes, vertical baffles placed along the circumference of a mixing tank are used to avoid vortex formation at high rotational speeds in low-viscosity foods.

Agitators may also be classified according to: top-entry, side-entry, and clamp-on mixers (Anonymous, 2002). Top-entry mixers are used for mixing viscous foods. Side-entry units are popular in the wine and beverage industry. Clamp-on agitators are ideal for mixing contents of small tanks and open drums.

It is important to match the agitator and agitation conditions to the characteristics of the product. For example, agitators for intact fruit must not shear or damage the product. Even in a low-viscosity fluid, like milk, the type of agitator, and its dimensions and rotational speed are important. Miller (1981) studied mixing efficiency of two top-entering agitators: straight paddle and pitched blade impeller at various rotational speeds in milk storage tanks. A simple two-bladed paddle was recommended for use in cylindrical vessels. The dimensions of the mixing systems, expressed in terms of vessel diameter (D_T), being: paddle diameter, 0.3 D_T, blade width, 0.06 D_T; clearance between paddle and vessel base, 0.10 D_T; offset from vessel axis, 0.08 D_T. For continuous operation a rotational speed of 35 rpm was recommended. Since damage to milk fat globules can occur at high rotational speeds (e.g., 150 rpm), the lowest speed capable of providing the required mixing effect should be selected.

4.5.1 POWER CONSUMPTION IN AGITATION

The effectiveness of and energy consumption in agitation depend on the basic principles of fluid mechanics; however, the flow patterns are much too complex for their rigorous application. Therefore, empirical relationships based on dimensionless groups are used. Here, because most fluid foods are non-Newtonian in nature, the discussion emphasizes in-tank agitation of such fluids using top-entering agitators.

In agitation of a non-Newtonian food, one has to relate shear rate vs. shear stress data to the flow field in an agitated vessel. The main complication is that the shear rate ($\dot{\gamma}$) is not uniform in an agitated vessel. For example, it has highest value at the point of highest fluid velocity; such a point occurs at the tip of the rotating agitator and decreases with increasing distance. It should be noted that while some impellers, such as propellers, act at or near the axis of a cylindrical tank and mix the bulk of the fluid, others, such as the anchor, act near the tank wall. However, helical impellers can be designed to act either near the axis or the vessel wall or can be combined to act at both places.

Based on dimensional analysis, Henry Rushton and coworkers developed the concept of the power number, Po, for studying mixing of fluids that for Newtonian fluids is defined as:

$$Po = \frac{p}{D_a^5 N_a^3 \rho} \tag{4.31}$$

where, p is power ($P = 2\pi N_a \times T$), (Jsec^{-1}), D_a is agitator diameter (m), T is torque on agitator (N m), N_a is agitator speed (sec^{-1}), and ρ is the density of the food (kg m^{-3}).

In laminar mixing conditions of Newtonian fluids, Po is linearly related to agitator rotational Reynolds number, Re_a:

$$Po = \frac{A}{\text{Re}_a} \tag{4.32}$$

where, A is a constant and,

$$\text{Re}_a = \frac{D_a^2 N_a \rho}{\eta} \tag{4.33}$$

The value of the constant A depends on the type of agitator. Laminar mixing conditions are encountered as long as Re_a is less than about 10.

For non-Newtonian foods, such as FCOJ, the viscosity is not constant, but depends on the shear due to agitation. Therefore, for FCOJ and other non-Newtonian fluids we would like to define an agitator Reynolds number that can be used in place of Re_a to estimate Po from data presented in Figure 4.6. A widely accepted procedure (Metzner and Otto, 1957) assumes that the average shear rate during mixing is directly proportional to the agitator rotational speed, N_a, that is,

$$\dot{\gamma} = k_s N_a \tag{4.34}$$

Further, the apparent viscosity is given by:

$$\eta_a = K(k_s N_a)^{n-1} \tag{4.35}$$

Substituting the expression for apparent viscosity in the rotational Reynolds number for Newtonian fluids, $\text{Re}_a = \frac{D_a^2 N_a \rho}{\eta}$, the power law agitation Reynolds number is:

$$\text{Re}_{pl} = \frac{D_a^2 N_a \rho}{K(k_s N_a)^{n-1}} \tag{4.36}$$

The curves of Re_{pl} vs. Po for several agitators, adapted from Skelland (1967), are shown in Figure 4.6. It is emphasized that each line in Figure 4.6 is valid for a specific agitator, its orientation and dimensions, and the mixing tank dimensions, as well as the configuration of the tank's baffles. Because of proprietary agitator designs and mixing tanks, it would be advisable to develop Re_{pl} vs. Po data for agitation systems being used for a specific food.

Once the magnitude of the power law agitation Reynolds number (Re_{pl}) is known, assuming that it is equal to Re_a, the corresponding value of the Power number (Po) can be determined from the applicable curve for the specific agitator, such as Figure 4.6 or similar data. From the known values of Po, the diameter of the agitator (D_a), agitator rotational speed (N_a), and the density (ρ) of the food, the power required (p) for agitation can be calculated.

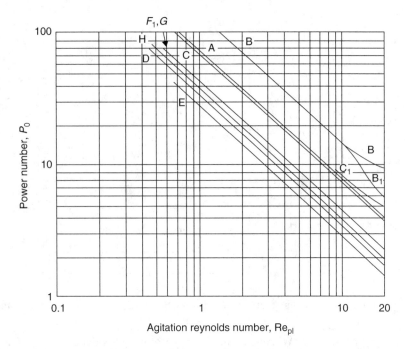

FIGURE 4.6   Curves of power law agitation Reynolds number (Re_{pl}) vs. power number (Po) for several agitators: curve A — single turbine with 6 flat blades, B — two turbines with 6 flat blades, C — a fan turbine with 6 blades at 45°, D and E — square-pitch marine propeller with 3 blades with shaft vertical and shaft 10° from vertical, respectively, F and G — square-pitch marine propeller with shaft vertical and with 3 blades and 4 blades, respectively, and H — anchor agitator. (Adapted from Skelland, A.H.P. 1967. *Non-Newtonian Flow and Heat Transfer.* John-Wiley, New York.)

4.5.2   ESTIMATION OF k_s OF AN IMPELLER

Procedures for determining k_s in Equation 4.34 of a specific agitator and mixing tank can be found in (Rao and Cooley, 1984). In one procedure (Rieger and Novak, 1973; Rao and Cooley, 1984), the constant k_s was determined from a plot of $\log [p/KN_a^{n+1}D_a^3]$ vs. $(1-n)$; the slope of the line is equal to $-\log k_s$. For a given agitator, tests must be conducted such that the following data are obtained: p, the power ($p = 2\pi N_a \times T)(Jsec^{-1})$, D_a, agitator diameter (m), T is torque on agitator ($N \cdot m$), N_a is agitator speed (sec^{-1}), and the power law rheological parameters of test fluids so that a wide range of $(1-n)$ values are obtained.

Typical values of the proportionality constant k_s for chemical industry impellers range from about 10 to 13 (Skelland, 1967). However, higher values have been reported for an anchor, $k_s = 24.5$ for $D_a/D_T = 0.98$, and a helical-ribbon, $k_s = 29.4$, $D_a/D_T = 0.96$; $Pi/D_a = 1$ (Pi is the pitch) impellers (Wilkens et al., 2003). Cantu-Lozano et al. (2000) reported a value of 17.8 for a helical impeller for $D_a/D_T = 0.77$; $Pi/D_a = 0.89$.

4.5.3   SCALE-UP CONSIDERATIONS FOR MIXING VESSELS

In scale-up of mixing vessels, the objective is to predict the rotational speed, N_{a2}, in Scale 2 that will duplicate the performance in Scale 1 due to agitation at a speed of N_{a1} (Figure 4.7). An important assumption in scale-up is geometric similarity that is achieved when all corresponding linear dimensions in Scale 1 and Scale 2 have a constant ratio. One popular scale-up criterion is based on equal power per volume, p/V, because it is understandable and practical. Other scale-up criteria, include (Wilkens et al., 2003): equal agitation Reynolds number, equal impeller tip speed, equal bulk fluid

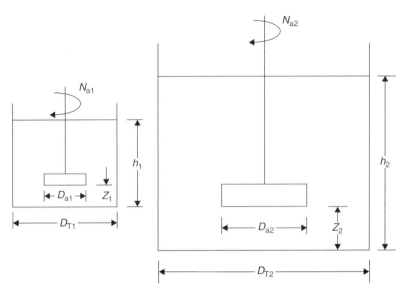

FIGURE 4.7 Illustration of scale-up of mixing vessels from Scale 1 to Scale 2 and geometric similarity.

velocity, and equal blend time. Wilkens et al. (2003) recommended that the rotational speed N_{a2} be estimated using each criterion and select that which predicts a high value that is economical to implement in view of high capital and operating costs of high-speed mixing systems.

4.6 STORAGE OF SOLIDS

Both transportation and the storage of solid foods are important operations in a food processing plant. A number of topics have been very well covered in other handbooks. For example, conveying of bulk solids, storage and weighing of solids in bulk, packaging and handling of solid and liquid products, and transportation of solids were well covered in the sixth edition of *Perry's Chemical Engineer's Handbook* (Perry and Green, 1984). The storage and conveying of grains are well covered in ASAE Standards (Hahn and. Rosentreter, 1987). Therefore, these topics will not be covered here. In this section those features applicable to food processing plants are discussed.

4.6.1 SOLID STORAGE BINS

Bins for storage can take many shapes and sizes, but many storage tanks are circular in cross section with storage capacities ranging from several thousand pounds to over 1 million bushels. For small and intermediate-sized bins, Imholte (1984) suggested that stainless steel is the material of choice when frequent wet cleaning is necessary and when the bins can be used for a variety of purposes. This is true for use bins and/or holding bins that are used to hold a product for a short period of time.

For bulk storage of solids, the bins can be built from mild steel, and they should not be located in wet manufacturing areas. These bins are of welded construction with continuously welded butt seams and corner welds ground to have a 3/16 to 1/4 in. radius. The interior areas of the bins should be free from any horizontal ledges where the product may accumulate. The discharge hoppers must be either properly sloped, as discussed later, or suitable mechanical unloading devices must be provided to provide for complete product discharge.

Use/holding bins must also be constructed in a manner similar to bulk bins, except that the material of construction is stainless steel. Imholte (1984) suggested a 2B mill finish for the interior and a number 4 finish for the exterior. Square or rectangular bins must have corner radii of about 1/2 in. for small bins and 1 to 1/2 in. for larger bins.

For both use and storage bins, access doors measuring a minimum of 18 in. must be provided. These doors must be fabricated with quick-release fasteners that do not require tools. Mild steel surfaces must be painted with food-grade epoxy enamel.

4.6.2 DESIGN OF STORAGE TANKS

The volume of storage tanks can be calculated from either the volume or the mass of the solid or liquid food that needs to be stored. When the mass is known, it can be converted to volume by dividing it by the density of the food. The equation relating the volume V on one hand and the diameter D and height h on the other for cylindrical tanks is $V = h(\pi/4)D^2$, and it can be employed to calculate either the height or the diameter from the known value of the other.

4.6.3 BULK DENSITIES OF SOLID FOODS

The bulk densities of several solid foods are given in Table 4.3. Caution must be exercised in using these figures because considerable variation in the magnitudes of bulk densities can result due to vibration, different particle sizes, and other factors. For example, bulk density of roasted and ground coffee is lower in smaller diameter vessels (e.g., less than 30 cm) than in larger diameter vessels due to greater wall support in the smaller diameter vessel and also finer grinds form denser beds (Sivetz and Desrosier, 1979).

4.7 BASIC CONCEPTS OF SOLID FRICTION

The ratio of friction force, F, and the force normal to the surface of contact, W, is given by the relationship

$$f = \frac{F}{W} \qquad (4.37)$$

TABLE 4.3
Densities and Bulk Densities of Some Food Powders

Food	Solid density, ρ_S (g/cm^3)	Bulk density, ρ_B (g/cm^3)
Wheat flour	1.45–1.49	0.55–0.65
Rye flour	1.45	0.45–0.70
Corn flour	1.54	0.50–0.70
Corn starch	1.62	0.55
Potato starch	1.65	0.65
Rice, polished	1.37–1.39	0.7–0.8
Cocoa powder 10% fat	1.45	0.35–0.40
Cocoa powder 22% fat	1.42	0.40–0.55
Sucrose	1.60	0.85–1.05
Instant dried whole milk	1.30–1.45	0.45–0.55
Instant dried skim milk	1.20–1.40	0.25–0.55
Roast ground coffee	—	0.31–0.40
Instant coffee powder	—	0.20–0.43

Source: Data from Schubert, H. 1987a. *J. Food Eng.*, 6: 1–32, Schubert, H. 1987b. *J. Food Eng.*, 6: 83–102, and Sivetz, M. and Desrosier, N. 1979. *Coffee Technology.* AVI, Westport, CT.

where f is the coefficient of friction. The commonly accepted concepts of friction are (Mohsenin, 1986):

1. The friction force can be defined as the force acting in a plane containing the contact points that resists relative motion of the contact surfaces.
2. The friction force can be divided into two main components: (a) a force required to deform and sometimes shear the asperities of the contacting surface and (b) a force required to overcome adhesion or cohesion of surfaces.
3. The friction force is directly proportional to the actual contact area.
4. The friction force depends on the sliding velocity of the contacting surfaces.
5. The friction force depends on the nature of materials in contact.
6. The friction force is not dependent on the surface roughness except in the case of very fine and very rough surfaces.

Friction phenomenon can be considered to be the sum of shearing force S and plowing force P:

$$F = S + P \tag{4.38}$$

which can be expanded to

$$F = \frac{W_s}{p_m} + AP_d \tag{4.39}$$

where s is the shearing stress of the softer material. From Equation 4.37 and Equation 4.39 it can be concluded that the coefficient of static friction f_s, is virtually independent of the area of contact. When the plowing term in Equation 4.39 is negligible, the coefficient of friction may be expressed in terms of the mechanical properties of the softer material:

$$f = \frac{s}{p_m} = \frac{\text{shear strength}}{\text{yield pressure of softer material}} \tag{4.40}$$

A number of factors affect friction; these include sliding velocity, water film, and surface roughness. In general, at low velocities, the coefficient of friction increases with velocity, and at high velocities, friction either remains constant or decreases. Under certain conditions, increase in moisture may cause an increase in friction due to an increase in adhesion. For many materials, however, addition of moisture results in lubrication (i.e., reduction in friction) (Mohsenin, 1986). There are considerable data on the coefficient of friction of agricultural materials such as chopped grass, corn silage, chopped alfalfa, shelled corn, and other grains, and they can be found in the text of Mohsenin (1986) and in ASAE Standards (Hahn and Rosentreter, 1987).

4.8   ROLLING RESISTANCE OF MATERIALS

In applications such as the gravity conveying of fruits and vegetables, the rolling resistance or the maximum angle of stability in rolling of foods with rounded shapes may be useful information. For the case of a cylindrical or a spherical object of radius r and weight W rolling over a horizontal surface with a force F, the coefficient of rolling resistance c can be defined as (Mohsenin, 1986):

$$c = \frac{Fr}{W} \tag{4.41}$$

TABLE 4.4
Coefficient of Rolling Resistance and Maximum Stability Angle of Apples and Tomatoes

Surface	Coefficient of friction		Rolling resistance (deg)	
	Static (f_s)	Kinetic (f_k)	Static	Kinetic
Apples (six different varieties)				
Plywood	0.32–0.44	0.24–0.33	12–18	2.5–4.5
Galvanized steel	0.38–0.46	0.28–0.36	13–18	2.5–4.0
Rigid foam	0.34–0.44	0.28–0.38	13–18	2.5–4.0
Soft foam	0.72–0.93	0.55–0.75	11–16	4.0–5.0
Canvas	0.36–0.44	0.25–0.36	12–16	4.0–5.0
Tomatoes (four different varieties)				
Sheet aluminum	0.33–0.52	0.28–0.40	7–11	3.6–4.8
Plywood	0.41–0.60	0.41–0.56	9–14	3.6–4.8
Rigid foam	0.44–0.56	0.48–0.56	11–13	4.2–4.8
Soft foam	0.77–0.83	0.68–0.79	11–13	4.8–4.8
Canvas	0.48–0.75	0.49–0.67	13–14	4.8–7.0

Source: From Mohsenin, N.N. 1986. *Physical Properties of Plant and Animal Materials.* Gordon and Breach, New York.

From Equation 4.41 the rolling resistance F is directly proportional to the weight of the object and the coefficient of rolling resistance and inversely proportional to the radius r of the rolling object. Coefficients of friction for apples and tomatoes are given in Table 4.4.

4.9 ANGLE OF INTERNAL FRICTION AND ANGLE OF REPOSE OF GRANULAR MATERIALS

The coefficient of friction between granular materials is equal to the tangent of the angle of internal friction of the material. The angle of repose is the angle made by a material with respect to the horizontal when piled. While it is generally assumed that the angle of friction and the angle of repose are approximately equal, for some materials, such as sorghum, the magnitudes of the two angles can be different (Mohsenin, 1986). There are two angles of repose, a static angle of repose taken up by a granular solid that is about to slide upon itself and a dynamic angle of repose that arises in all cases where the bulk of the material is in motion, such as during discharging of solids from bins and hoppers.

Magnitudes of angle of repose increase with moisture content. Mohsenin credited Fowler and Wyatt with developing an equation for calculating the angle of repose of wheat, sand, canary seed, and other solids:

$$\tan \Phi_r = an^2 + b\frac{M}{D_{av}} + cs_g + D \qquad (4.42)$$

where Φ_r is the angle of repose, n the shape factor based on the specific surface, M the percent moisture content, D_{av} the average screen particle diameter, s_g the specific gravity, and a, b, c, and d are constants. The magnitudes of the coefficients were determined to be $a = 0.4621$, $b = 0.0342$, $c = 0.0898$, and $d = 0.0978$, with a magnitude of $R = 0.97$. Data presented by Bhattacharya et al. (1972) on rice over the range of moisture content of 12 to 26% also showed that the angle of repose, measured indirectly as the height at the back of a box, increased with increase in moisture content.

4.10 ANGLE OF INTERNAL FRICTION

The angle of internal friction is needed for calculating the lateral pressure on a wall of a storage bin or in the design of gravity flow bins and hoppers (Mohsenin, 1986). For example, the Rankine equation is used to calculate at a point the lateral pressure against the wall, σ_3:

$$\sigma_3 = wy \tan^2(45 - \Phi_i/2) \tag{4.43}$$

where y is the distance below the top of the wall, w the density of the material, and Φ_i the angle of internal friction. In the design of deep bins, one needs the ratio, k, of the lateral pressure σ_3 to the vertical pressure σ_1. This quantity can also be calculated using the angle of internal friction:

$$k = \frac{1 - \sin \Phi_i}{1 + \sin \Phi_i} \tag{4.44}$$

The horizontal pressure against the wall can be estimated from the known value of the magnitude of k and the vertical pressure.

For any given magnitude of vertical pressure, the horizontal pressure against the wall can be calculated from the known value of the magnitude of k. The vertical pressure causes a column action on the wall, while the lateral pressure causes a bending action on the wall. In grain bins when the height of the material exceeds about twice the bin diameter, no increase in bottom pressure can be detected with increasing depth of grain. The pressure ratio k can be obtained either directly from pressure measurements in full- or model-sized bins or by the use of triaxial compression chamber and the Mohr's circle.

Several factors affect the lateral pressure in bins, as indicated by Janssen's equation (Mohsenin, 1986):

$$\sigma_3 + \frac{wR}{f_s}\left[1 - \exp\left(\frac{-kf_s h}{R}\right)\right] \tag{4.45}$$

where R is the hydraulic radius (ratio of area of cross section to circumference), w the density of the material, f_s the static friction of the material against the wall, and h the depth of material.

4.11 FLOW OF POWDERS AND GRANULAR SOLIDS

Although there are some similarities in the storage of food grains and food powders and many properties are common to both, it appears that their flow properties have been studied along separate paths. Further, it is clear that the major impetus to the study of the flow properties of all granular materials was the pioneering study of Jenike at the University of Utah during the decade of the 1960s. The mechanisms of liquid and solid flow are substantially different (Peleg, 1977). The two major differences are: (1) in liquid flow, the flow rate is proportional to the square root of the liquid head above the outlet, and in granular solids, the flow rate is independent, or nearly independent, of the head when the solid bed height is at least 2.5 times the outlet diameter; and (2) particulate solid materials can support considerable shear stresses or form stable structures that will prevent flow despite the existence of a head.

4.11.1 FACTORS AFFECTING FLOW OF POWDERS

The flow of a granular solid is affected by several forces: gravitational forces, friction, cohesion (interparticle attraction), and adhesion (particle–wall attraction). The formation of a stable solid arch above the aperture is also possible. Gravity is the natural driving force of unaided flow that can also

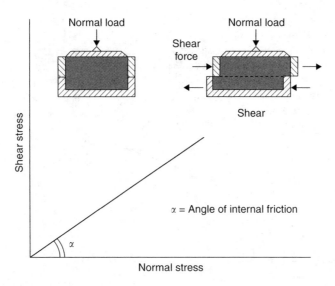

FIGURE 4.8 Plot of normal stress vs. shear stress for a noncohesive powder. (From Peleg, M. 1977. *J. Food Process Eng.* 1: 303–328. With permission.)

cause considerable compaction of the bed. Consequently, the bed, due to enhanced cohesive forces, will have measurable mechanical properties, such as tensile strength and compressive breaking strength. Therefore, flow takes place due to solid failure.

Noncohesive or free-flowing powders are those in which interparticle forces are negligible and the major obstruction to flow is internal friction. Referring to Figure 4.8 the condition for flow to occur is:

$$\tau > \mu\sigma \tag{4.46}$$

where τ is the shear stress, μ the coefficient of friction, and σ the normal stress. We note here that σ is the symbol for normal stress in the literature of food powders, and this notation will be used in this section; it must not be confused with the shear stress of liquid foods.

Interparticle forces can develop under special conditions, such as due to moisture absorption, elevated temperature, or static pressure (Peleg, 1977), and they can reduce the flowability, stop it altogether, or form stable bridges (agglomeration). The latter phenomenon is usually referred to as a caking problem, characterized by the formation of soft lumps to total solidification. Unlike noncohesive powders, the shear yield stress vs. normal stress data for cohesive powders are characterized by (1) a family of yield loci curves (i.e., at each consolidation level there is a different curve), and (2) the curves do not pass through the origin. Figure 4.9 is a schematic diagram at two consolidation levels. At zero normal stress, the compacted powder has nonzero shear strength called cohesion whose magnitude depends on the properties of the powder and the consolidation conditions (C_1, and C_2 in Figure 4.9). The yield loci intercepts with the normal stress axis indicate the tensile strength (T_1 and T_2 in Figure 4.9). The tensile strength provides a direct indication of the interparticle forces and its magnitude depends on the consolidation stress. For many powders, the yield loci can be described by the Warren Spring equation (Peleg, 1977):

$$\left(\frac{\tau}{c}\right)^n = \frac{\sigma + T}{T} \tag{4.47}$$

From an energy balance of a compacted powder being sheared, one can show that cohesion is proportional to the tensile strength.

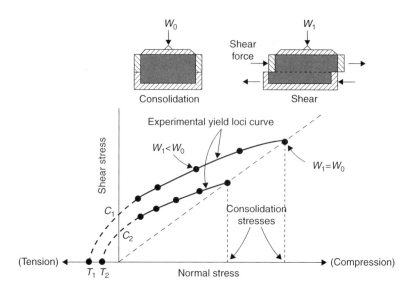

FIGURE 4.9 Plot of normal stress vs. shear stress for a cohesive powder showing a family of yield loci curves, cohesion (C_1 and C_2) and tensile strength (T_1 and T_2). (From Peleg, M. 1977. *J. Food Process Eng.* 1: 303–328. With permission.)

The angle of repose is a simple technical test in which the angle the powder forms with the horizontal is determined. Irrespective of the method of measurement of the angle of repose, it can be assumed that a small angle indicates a free-flowing granular solid. One thumb rule is that powders with an angle of repose of less than 40° are free flowing, while those exhibiting angles of 50° or more are likely to cause flow problems (Peleg, 1977). Magnitudes of angles of repose determined by different techniques will differ from each other and cannot be compared with each other. Because irregular cone angles are formed in the case of cohesive powders, the measurement of the angle becomes difficult.

4.11.2 FLOWABILITY OF POWDERS

Jenike (1970) established fundamental methods for determining flowability characteristics of powders. The experimental yield loci are obtained by plotting the normal stress vs. the shear stress at a particular powder porosity. The normal and shear stresses are obtained in a Jenike shear cell. The stress condition of each point on the yield locus may be described by Mohr stress circles. The yield loci curves of many food materials can be approximated by a straight line (Peleg, 1977), although it is expected to be slightly concave for all granular materials. Two Mohr stress semicircles that characterize two important properties are shown in Figure 4.10. The larger semicircle characterizes the stress conditions during steady-state flow since it is passing through the point of consolidation conditions at which point steady-state flow is reached and no changes in stress or volume take place (Schubert, 1987a, 1987b). σ_1 is the major consolidation stress. The smaller semicircle is drawn tangential to the yield locus and passing through the origin; this gives f_c, the unconfined yield stress.

The unconfined yield stress (f_c and the major consolidation stress (σ_1) are important parameters in design problems. The former is a measure of the solid's strength at a free surface, while the latter is related to the pressure applied to the solid if it were compressed in a cylinder with frictionless rigid wall. The angle of internal friction of the solid is the slope of the yield locus and it varies from point to point due to the curvature of the yield locus (Mohsenin, 1986). The flow function ff_c is defined as:

$$ff_c = \frac{\sigma_1}{f_c} \qquad (4.48)$$

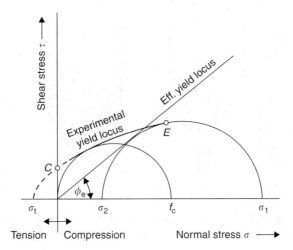

FIGURE 4.10 Illustration of two Mohr semicircles that define the major consolidation stress (σ_1) and the unconfined yield stress (f_c). (From Schubert, H. 1987a. *J. Food Eng.* 6: 1–32; 1987b. *J. Food Eng.* 6: 83–102. With permission.)

TABLE 4.5
Flowability of Powders According to Jenike's Flow Function

$ff_c < 2$	Very cohesive, nonflowing	Cohesive powders
$2 < ff_c < 4$	Cohesive	
$4 < ff_c < 10$	Easy flowing	Noncohesive powders
$10 < ff_c$	Free flowing	

The flow function, ff_c characterizes the flowability of powders, an important property in designing bins and hoppers. Table 4.5 contains a classification of powders according to their flowability based on the magnitude of the flow function, ff_c. While the flow function, ff_c, is a useful guide to the flowability of powders, a complete description of the flowability of particulate solids can be obtained only by measuring the yield loci at different magnitudes of porosities (Schubert, 1987a, 1987b).

Cohesion C of powders described earlier is not as good an index of flowability as the flow function, ff_c The tensile strength, s_1, also defined earlier, may. be used to interpolate the yield locus in the region in which the Mohr circle is plotted to determine f_c accurately. This is of particular importance in the dispensing of slightly cohesive instant food powders in vending machines. The tangent to the major stress circle passing through the origin is called the effective yield locus. The ratio of the two principal stresses σ_1, and σ_2 and the effective angle of friction Φ_e are related by the general equation describing the steady-state flow of powders:

$$\frac{\sigma_1}{\sigma_2} = \frac{1 + \sin \Phi_e}{1 - \sin \Phi_e} \tag{4.49}$$

4.11.3 DESIGN OF BINS AND HOPPERS

Jenike's theory and the associated measurements are applied to the design of bins and hoppers. Mass flow bins in which the entire bulk powder is in motion (Figure 4.11a) during discharge are preferable to plug flow bins in which a portion of the powder at the sides of the bin is stationary (Figure 4.11b). For foods with limited shelf life, funnel flow, also known as plug flow bins, must be

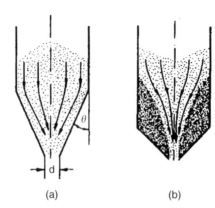

(a) (b)

FIGURE 4.11 Illustration of mass flow (a) and plug flow (b) in a bin; θ is the angle of the cone and d is the diameter of bin outlet. (From Schubert, H. 1987a. *J. Food Eng.* 6: 1–32; 1987b. *J. Food Eng.* 6: 83–102. With permission.)

avoided because in the dead regions the material can remain for long periods of time. By using a procedure suggested by Jenike, the critical cone angle θ_c, which is a function of the effective friction angle Φ_e and the friction angle Φ_W between bulk material and the wall may be taken from diagrams in Jenike (1970) that contain values of θ_c as a function of θ, Φ_e, and Φ_W.

Mass flow occurs when the cone angle $q \leq \theta_c$. For the paniculate solids to flow without the formation of a stable arch, the outlet must have a minimum diameter, d_c. A stable arch is possible only for $\sigma_1' < f_c$; σ_1' is the major principal stress acting at the abutment of an arch and its magnitude may be calculated from:

$$\sigma_1' = \frac{\sigma_1}{\text{ff}} = \frac{\sigma_1}{f(\theta, \Phi_c, \Phi_w)} \tag{4.50}$$

The flow factor, ff, as a function of θ, Φ_e, and Φ_W may be taken from Jenike (1970). The critical diameter d_c can be calculated by first assuming the condition $\sigma_1' = f_c$, which allows calculation of the critical stress σ_{1c}' which acts at the abutment of an arch:

$$d_c = \frac{\sigma_{1c}' H(\theta)}{\rho_B g} \tag{4.51}$$

where ρ_B is bulk density, g the acceleration due to gravity, and $H(\theta)$ a function of the geometry of the hopper given in diagrams by Jenike (1970). When $d > d_c$, the formation of a stable arch that in turn would prevent the discharge of a powder can be avoided.

4.11.4 APPLICATION OF JENIKE'S THEORY

Schubert (1987a, 1987b) illustrated the application of Jenike's theory for two batches of instant cocoa mix. The yield loci for one of the batches are shown in Figure 4.12 The f_c curves for the two batches of instant cocoa mix are shown in Figure 4.13, from which it can be recognized that data on shear forces must be obtained at extremely small normal stresses. The figure also illustrates the difference in flowability in food powders arising due to normal variations of process conditions. Because the f_c curve is passing through the origin, batch I powder has ideal flow properties. The flow function (ff$_c$ = σ_1/f_c) is independent of σ_1' and it has a magnitude equal to 14. From Table 4.5, batch I is noncohesive and hence free flowing. For batch II it was determined that ff$_c$ = 2; therefore, from Table 4.5, this powder was deemed to be cohesive.

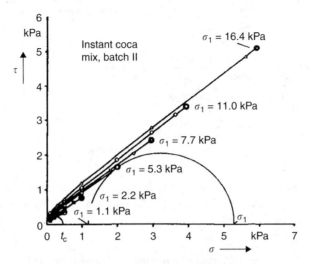

FIGURE 4.12 Yield loci and a Mohr stress semicircle for a batch of instant cocoa mix. (From Schubert, H. 1987a. *J. Food Eng.* 6: 1–32; 1987b. *J. Food Eng.* 6: 83–102. With permission.)

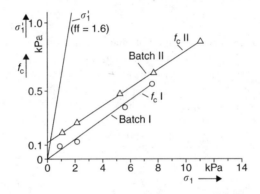

FIGURE 4.13 Unconfined yield stresses of two batches of instant cocoa mix. Also shown is the major principal stress (σ_1') acting on the abutment of an arch. Batch I has ideal flow properties. (From Schubert, H. 1987a. *J. Food Eng.* 6: 1–32; 1987b. *J. Food Eng.* 6: 83–102. With permission.)

4.12 FLUIDIZATION AND HYDRAULIC TRANSPORT OF FOOD PIECES

The bulk fluidization characteristics of agricultural and food materials are important in the design of conveying systems for the materials.

4.12.1 RELATIONSHIP BETWEEN PARTICLE PROPERTIES

When a fluid is passed vertically upward through a bed of particles, the pressure drop ΔP will initially rise as the velocity U increases and eventually reaches a constant value (Figure 4.14). The minimum velocity at which fluidization takes place is designated as U_o. The Carmen–Kozeny equation for fixed beds will be applicable in this linear region. A plot of the logarithm of the fluid velocity vs. the logarithm of the bed voidage results in a straight line whose slope for spherical particles depends on the particle/column ratio, d/D, and the particle Reynolds number.

The terminal velocity U_t is the velocity at which particles would be transported from the system and hence is the upper limit to avoid particle entrainment; its magnitude can be approximated by the

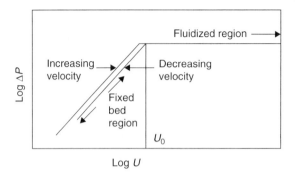

FIGURE 4.14 Plot of the logarithm of velocity (U) vs. the logarithm of pressure drop (ΔP) for fixed and fluidized-bed regions. (From McKay, G., Murphy, W.R., and Jodierie-Dabbaghzadeh, S. 1987. *J. Food Eng.* 6: 377–399. With permission.)

terminal or the free-fall settling velocity of the particles (McKay et al., 1987):

$$U_t = \left[\frac{4 g d_p (\rho_p - \rho) \times 10^{-3}}{3 \rho C_d} \right]^{0.5} \tag{4.52}$$

where d_p is the nominal diameter, g the gravitational constant, ρ the fluid density, ρ_p the particle density, and C_d the particle drag coefficient. The magnitude of U_t can also be predicted by extrapolating the log ε vs. log U plots to a log ε value of unity. The ratios of the terminal velocity to the incipient fluidization velocity, U_t/U_0, for laminar (Re < 0.2) and turbulent (Re > 500) are given by Equation 4.53 and Equation 4.54, respectively:

$$\frac{U_t}{U_0} = \frac{150(1 - \varepsilon_0)}{18 \varepsilon_0^3} \tag{4.53}$$

$$\frac{U_t}{U_0} = \left(\frac{5.3}{\varepsilon_0^3} \right)^{0.5} \tag{4.54}$$

Fluidization of small solids ($d_P < 1.0$ mm) is a well-established technology and the necessary theory is well documented in several texts (Davidson and Harrison, 1967; Kunii and Levenspiel, 1969). Fluidization of carrot pieces by water in a glass test section was investigated by McKay et al. (1987). The effects of length, diameter, L/D ratio, density, and the solids settling characteristics were studied. Specifically, cylinders of carrots (specific gravity = 1033 to 1035), nylon 6,6 (specific gravity = 1145), and PVC (specific gravity = 1145) having L/D ratios from 0.63 to 2.55 were employed. Water was the fluidizing medium. Figure 4.15 is a plot of logarithm of bed voidage against superficial velocity. The transport velocities can be predicted by extrapolating the plots to a log (voidage) value of zero.

4.12.2 Friction Factor

The friction factor, f_v, representing the ratio of pressure drop to the viscous energy term was proposed by Ergun; it is defined as:

$$f_v = \frac{\Delta p_0 d_e^2 \varepsilon_0^3}{Z_0 \eta U_0 (1 - \varepsilon_0)^2} \tag{4.55}$$

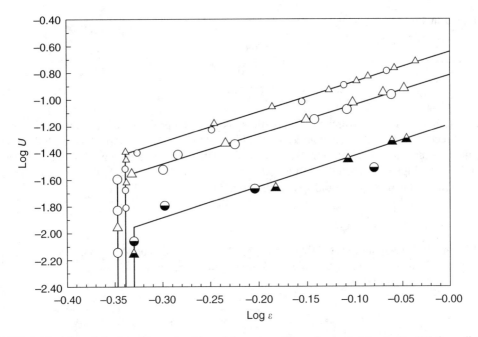

FIGURE 4.15 Plot of the logarithm of voidage (ε) vs. logarithm of superficial velocity (U) for cylinders of carrots, nylon 66, and PVC. Temperature 18°C; length 12.5 mm; aspect ratio 1:1. Small open circles and triangles for PVC increasing and decreasing order of velocities, respectively; large open circles and triangles for nylon 66 increasing and decreasing order of velocities, respectively; and shaded triangles and circles for carrots increasing and decreasing order of velocities, respectively. The transport velocities are obtained by extrapolation of (he log voidage value to zero). (From McKay, G., Murphy, W.R., and Jodierie-Dabbaghzadeh, S. 1987. *J. Food Eng.* 6: 377–399. With permission.)

where ΔP_0 is the pressure drop across the bed of particles at minimum fluidization conditions (N/m^2), d_e the diameter of the sphere having the same volume/surface ratio as the actual particle, ε_0 the bed voidage at minimum fluidization conditions, Z_0 the height of the fluidized bed at minimum fluidization conditions, U_0 the superficial flow velocity at minimum fluidization conditions, and η the fluid viscosity (Pa·sec). The friction factor can be correlated against the term $\mathrm{Re}_{de}/(1 - \varepsilon_0)$. The following correlation was developed by regression analysis of the experimental data:

$$\log f_v = 1.439 + 0.53 \log \frac{\mathrm{Re}_{de}}{1 - \varepsilon_0} \tag{4.56}$$

The authors also cited a correlating equation from an earlier study for cuboids having a sphericity factor, $\Phi_s = 0.66$; the sphericity factor Φ_s defined as the ratio of the surface area of a sphere of the same volume as the particle to the surface area of the particle.

4.12.3 Resistance Coefficient

McKay et al. (1987) also determined the correlation between the resistance coefficient, defined by Equation 4.58, and the sphericity factor, Φ_s. The resistance coefficient was used, instead of the drag coefficient, because it provided higher values of correlation coefficient. The correlation for the cylinders was given by Equation 4.59, while that for plastic cuboids was given by Equation 4.60.

$$C_r = \frac{2v(\rho_p - \rho)g}{AU_t^2 \rho} \tag{4.57}$$

$$\log C_r = 0.77 + 4.75 \log \Phi_s. \tag{4.58}$$

In Equation 4.57, v is the volume of the particle and A is the projected area of the particle normal to the direction of flow. For magnitudes of L/D less than 1:1, the cylindrical solids fell with their circular projected area normal to the vertical, while with an L/D ratio above 1:1, the particles fell with their rectangular projected area normal to the vertical. At $L/D = 1$, there was no preferred orientation and the particles would spin as they fell.

4.12.4 FLUIDIZATION OF GRAINS

Mohsenin (1986) presented an equation derived by Roberts for calculating the force (pounds) required to move a column of grain L inches high in a tube against wall friction:

$$f = \frac{w\pi R^3}{2fk[\exp(2fkL/P) - 1]} \tag{4.59}$$

where w is bulk density in pounds per cubic inch, R the radius of the tube in inches, and P the pressure acting on a perforated piston in inches; in addition, the coefficient of friction f of the grain against the tube wall and the pressure ratio k were constant. When compared to pneumatic conveying, dense-phase fluidization requires more power. For example, in tests performed by the U.S. Department of Agriculture on seed fluidization (Mohsenin, 1986), the horsepower requirements for seed transport in the dense phase was about 3.5 compared with a horsepower of 2.5 in the lean-phase operation. The major advantage of dense-phase transport was the low-velocity seed flow in flexible tubes. Mohsenin (1986) also presented a table of physical properties (bulk density, angle of repose, and coefficient of friction) of several agricultural materials that can be used for the design of fluidized transport systems based on the work of Roberts. In addition, the table contains the expected rating of the fluidization characteristics of the materials.

4.13 STORAGE OF COMMODITIES

A very important aspect of food storage is that foods, particularly commodities, require different temperature and humidity conditions for maximum shelf lives. The commodities require precooling or prechilling prior to storage. The methods of precooling and prechilling and the refrigeration load calculations are covered in detail in the *Applications Handbook* of ASHRAE. The storage behavior of different cultivars of fruits and vegetables, and species of fish are also covered in the *Handbook*. Pertinent information on storage also can be found in texts devoted to commodities: fruits and vegetables (Ryall and Lipton, 1979, 1984), cereal grains and their products (Christensen, 1953), potatoes (Rastovski and van Es, 1981), and in a manual prepared by the Tropical Stored Products Centre (Anonymous, 1970). However, because storage of foods is an important subject, the temperature and humidity conditions of food storage and the expected storage life are covered in this subsection. In addition, where available, space requirements and other practical information will be provided.

A substantial amount of information on storage on the temperature and humidity conditions, and the expected storage life is given in Table 4.6 (ASHRAE, 1982). It should be noted that Table 4.6 also contains data on the properties of many perishable foods. The space, weight, and density data for commodities stored in refrigerated warehouses are given in Table 4.7. These data are in English units because are used for the dimensions of packages and their weights.

TABLE 4.6
Storage Requirements and Properties of Perishable Products

Commodity	Storage Temp. (°C)	Relative humidity (%)	Approximate storage life[a]	Water content (%)	Highest freezing (°C)	Specific heat above freezing[b] (J/kg ·°C)	Specific heat below freezing[b] (J/kg ·°C)	Latent heat[c] (J/kg)
			Vegetables[d]					
Artichokes								
Globe	0	95	2 weeks	84	−1.2	3.651	1.892	280.18
Jerusalem	0	90 to 95	5 months	80	−2.5[b]	3.517	1.842	266.84
Asparagus	0 to 2	95	2 to 3 weeks	93	−0.6	3.952	2.005	310.20
Beans								
Snap or green	4 to 7	90 to 95	7 to 10 days	89	−0.7	3.818	1.955	296.86
Lima	0 to 4	90 to 95	3 to 5 days	67	−0.6	3.081	1.679	223.48
Dried	10	70	6 to 8 months	11		1.206	0.975	
Beets								
Roots	0	95 to 100	4 to 6 months	88	−0.9	3.785	1.943	293.52
Bunch	0	95	10 to 14 days		−0.4			
Broccoli	0	95	10 to 14 days	90	−0.6	3.852	1.968	300.20
Brussels sprouts	0	95	3 to 5 weeks	85	−0.8	3.684	1.905	283.52
Cabbage late	0	95 to 100	5 to 6 months	92	−0.9	3.919	1.993	306.87
Carrots								
Topped, immature	0	98 to 100	4 to 6 weeks	88	−1.4	3.785	1.943	293.52
Topped, mature	0	98 to 100	5 to 9 months	88	−1.4	3.785	1.943	293.52
Cauliflower	0	95	2 to 4 weeks	92	−0.8	3.919	1.993	306.87
Celeriac	0	95 to 100	3 to 4 months	88	−0.9	3.785	1.943	293.52
Celery	0	95	1 to 2 months	94	−0.5	3.986	2.018	313.54
Collards	0	95	10 to 14 days	87	−0.8	3.751	1.930	290.19
Corn, sweet	0	95	4 to 8 days	74	−0.6	3.316	1.767	246.83
Cucumbers	10 to 13	90 to 95	10 to 14 days	96	−0.5	4.053	2.043	320.21
Eggplant	7 to 10	90 to 95	7 to 10 days	93	−0.8	3.952	2.005	310.20
Endive (escarole)	0	95	2 to 3 weeks	93	−0.1	3.952	2.005	310.20
Frozen vegetables	−23 to −18		6 to 12 months					
Garlic, dry	0	65 to 70	6 to 7 months	61	−0.8	2.880	1.604	203.46
Greens, leafy	0	95	10 to 14 days	93	−0.3	3.952	2.005	310.20
Horseradish	−1 to 0	95 to 100	10 to 12 months	75	−1.8	3.349	1.779	250.16
Kale	0	95	3 to 4 weeks	87	−0.5	3.751	1.930	290.19
Kohlrabi	0	95	2 to 4 weeks	90	−1.0	3.852	1.968	300.20
Leeks, green	0	95	1 to 3 months	85	−0.7	3.684	1.905	283.52
Lettuce, head	0 to 1	95 to 100	2 to 3 weeks	95	−0.2	4.019	2.031	316.87
Mushrooms	0	90	3 to 4 days	91	−0.9	3.885	1.980	303.53
Okra	7 to 10	90 to 95	7 to 10 days	90	−1.8	3.852	1.968	300.20
Onions								
Greens	0	95	3 to 4 weeks	89	−0.9	3.818	1.955	296.86
Dry and onion sets	0	65 to 75	1 to 8 months	88	−0.8	3.785	1.943	293.52

(Continued)

TABLE 4.6
Continued

Commodity	Storage Temp. (°C)	Relative humidity (%)	Approximate storage life[a]	Water content (%)	Highest freezing (°C)	Specific heat above freezing[b] (J/kg ·°C)	Specific heat below freezing[b] (J/kg ·°C)	Latent heat[c] (J/kg)
Parsley	0	95	1 to 2 months	85	−1.1	3.684	1.905	283.52
Parsnips	0	98 to 100	4 to 6 months	79	−0.9	3.483	1.830	263.50
Peas								
Green	0	95	1 to 3 weeks	74	−0.6	3.316	1.767	246.83
Dried	10	70	6 to 8 months	12		1.239	0.988	
Peppers								
Dried	0 to 10	60 to 70	6 months	12		1.239	0.988	
Sweet	7 to 10	90 to 95	2 to 3 weeks	92	−0.7	3.919	1.993	306.87
Potatoes								
Early	10 to 13	90		81	−0.6	3.550	1.855	270.18
Main crop	3 to 10	90 to 95	5 to 8 months	78	−0.7	3.450	1.817	260.17
Sweet	13 to 16	85 to 90	4 to 7 months	69	−1.3	3.148	1.704	230.15
Pumpkins	10 to 13	70 to 75	2 to 3 months	91	−0.8	3.885	1.980	303.53
Radishes								
Spring	0	95	3 to 4 weeks	95	−0.7	4.019	2.031	316.87
Winter	0	95 to 100	2 to 4 months	95	−0.7	4.019	2.031	316.87
Rhubarb	0	95	2 to 4 weeks	95	−0.9	4.019	2.031	316.87
Rutabagas	0	98 to 100	4 to 6 months	89	−1.1	3.818	1.955	296.86
Salsify	0	98 to 100	2 to 4 months	79	−1.1	3.483	1.830	263.50
Seed, vegetable	0 to 10	50 to 65	10 to 12 months	7 to 15		1.206	0.976	036.69
Spinach	0	95	10 to 14 days	93	−0.3	3.952	2.005	310.20
Squash								
Acorn	7 to 10	70 to 75	5 to 8 weeks		−0.8			
Summer	0 to 10	85 to 95	5 to 14 days	94	−0.5	3.986	2.018	313.54
Winter	10 to 13	70 to 75	4 to 6 months	85	−0.8	3.684	1.905	283.52
Tomatoes								
Mature green	13 to 21	85 to 90	1 to 3 weeks	93	−0.6	3.952	2.005	310.20
Firm, ripe	7 to 10	85 to 90	4 to 7 days	94	−0.5	3.986	2.018	313.54
Turnips								
Roots	0	95	4 to 5 months	92	−1.1	3.919	1.993	306.87
Greens	0	95	10 to 14 days	90	−0.2	3.852	1.968	300.20
Watercress	0	95	3 to 4 days	93	−0.3	3.952	2.005	310.20
Yams	16	85 to 90	3 to 6 months	74		3.316	1.767	246.83
Fruits and Melons[d]								
Apples	−1 to 4	90	3 to 8 months	84	−1.1	3.651	1.892	280.18
Dried	0 to 5	55 to 60	5 to 8 months	24		1.641	1.139	
Apricots	0	90	1 to 2 weeks	85	−1.1	3.684	1.905	283.52
Avocados	4 to 13	85 to 90	2 to 4 weeks	65	−0.3	3.014	1.654	216.81
Bananas	[d]	85 to 95	[d]	75	−0.8	3.349	1.779	250.16

(Continued)

TABLE 4.6
Continued

Commodity	Storage Temp. (°C)	Relative humidity (%)	Approximate storage life[a]	Water content (%)	Highest freezing (°C)	Specific heat above freezing[b] (J/kg · °C)	Specific heat below freezing[b] (J/kg · °C)	Latent heat[c] (J/kg)
Blackberries	−0.5 to 0	95	3 days	85	−0.8	3.684	1.905	283.52
Blueberries	−1 to 0	90 to 95	2 weeks	82	−1.6	3.584	1.867	273.51
Cantaloupes	2 to 4	90 to 95	5 to 15 days	92	−1.2	3.919	1.993	306.87
Cherries								
Sour	−1 to 0	90 to 95	3 to 7 days	84	−1.7	3.651	1.892	280.18
Sweet	−1	90 to 95	2 to 3 weeks	80	−1.8	3.517	1.842	266.84
Casaba melons	7 to 10	85 to 95	4 to 6 weeks	93	−1.1	3.952	2.005	310.20
Cranberries	2 to 4	90 to 95	2 to 4 months	87	−0.9	3.751	1.930	290.19
Currants	−0.5 to 0	90 to 95	10 to 14 days	85	−1.0	3.684	1.905	283.52
Dates, cured	−18 or 0	75 or less	6 to 12 months	20	15.7	1.507	1.089	66.71
Dewberries	−1 to 0	90 to 95	3 days	85	−1.3	3.684	1.905	283.52
Figs								
Dried	0 to 4	50 to 60	9 to 12 months	23		1.608	1.126	76.72
Fresh	−1 to 0	85 to 90	7 to 10 days	78	−2.4	3.450	1.817	260.17
Frozen fruits	−23 to −18	90 to 95	6 to 12 months					
Gooseberries	−1 to 0	90 to 95	2 to 4 weeks	89	−1.1	3.818	1.955	296.86
Grapefruit	10 to 16	85 to 90	4 to 6 weeks	89	−1.1	3.818	1.955	296.86
Grapes								
American	−1 to 0	85 to 90	2 to 8 weeks	82	−1.6	3.584	1.867	273.51
Vinifera	−1	90 to 95	3 to 6 months	82	−2.1	3.584	1.867	273.51
Guavas	7 to 10	90	2 to 3 weeks	83		3.617	1.880	276.85
Honeydew melons	7 to 10	90 to 95	3 to 4 weeks	93	−0.9	3.952	2.005	310.20
Lemons	0 or 10 to 14[e]	85 to 90	1 to. 6 months	89	−1.4	3.818	1.955	296.86
Limes	9 to 10	85 to 90	6 to 18 weeks	86	−1.6	3.718	1.918	286.85
Mangoes	13	85 to 90	2 to 3 weeks	81	−0.9	3.550	1.855	270.18
Nectarines	−0.5 to 0	90	2 to 4 weeks	82	−0.9	3.584	1.867	273.51
Olives, fresh	7 to 10	85 to 90	4 to 6 weeks	75	−1.4	3.349	1.779	250.16
Oranges	0 to 9	85 to 90	3 to 12 weeks	87	−0.8	3.751	1.930	290.19
Papayas	7	85 to 90	1 to 3 weeks	91	−0.8	3.885	1.980	303.53
Peaches	−0.5 to 0	90	2 to 4 weeks	89	−0.9	3.818	1.955	296.86
Dried	0 to 5	55 to 60	5 to 8 months	25		1.675	1.151	
Pears	−1.6 to −0.5	90 to 95	2 to 7 months	83	−1.6	3.617	1.880	276.85
Persian melons	7 to 10	90 to 95	2 weeks	93	−0.8	3.952	2.005	310.20
Persimmons	−1	90	3 to 4 months	78	−2.2	3.450	1.817	260.17
Pineapples, ripe	7	85 to 90	2 to 4 weeks	85	−1.0	3.684	1.905	283.52
Plums	−1 to 0	90 to 95	2 to 4 weeks	86	−0.8	3.718	1.918	286.85
Pomegranates	0	90	2 to 4 weeks	82	−3.0	3.584	1.867	273.51

(Continued)

TABLE 4.6
Continued

Commodity	Storage Temp. (°C)	Relative humidity (%)	Approximate storage life[a]	Water content (%)	Highest freezing (°C)	Specific heat above freezing[b] (J/kg ·°C)	Specific heat below freezing[b] (J/kg ·°C)	Latent heat[c] (J/kg)
Prunes								
Fresh	−1 to 0	90 to 95	2 to 4 weeks	86	−0.8	3.718	1.918	286.85
Dried	0 to 5	55 to 60	5 to 8 months	28		1.775	1.189	
Quinces	−1 to 0	90	2 to 3 months	85	−0.2	3.684	1.905	283.52
Raisins				18		1.440	1.063	
Raspberries								
Black	−0.5 to 0	90 to 95	2 to 3 days	81	−1.1	3.550	1.855	270.18
Red	−0.5 to 0	90 to 95	2 to 3 days	84	−0.6	3.651	1.892	280.18
Strawberries	−0.5 to 0	90 to 95	5 to 7 days	90	−0.8	3.852	1.968	300.20
Tangerines	0 to 3	85 to 90	2 to 4 weeks	87	−1.1	3.751	1.930	290.19
Watermelons	4 to 10	80 to 90	2 to 3 weeks	93	−0.4	3.952	2.005	310.20
Seafood (Fish)[d]								
Haddock cod, perch	−1 to 1	95 to 100	12 days	81	−2.2	3.550	1.855	270.17
Hake, whiting	0 to 1	95 to 100	10 days	81	−2.2	3.550	1.855	270.17
Halibut	−1 to 1	95 to 100	18 days	75	−2.2	3.349	1.779	250.16
Herring								
Kippered	0 to 2	80 to 90	10 days	61	−2.2	2.880	1.604	203.46
Smoked	0 to 2	80 to 90	10 days	64	−2.2	2.981	1.641	213.47
Mackerel	0 to 1	95 to 100	6 to 8 days	65	−2.2	3.014	1.654	216.81
Menhaden	1 to 5	95 to 100	4 to 5 days	62	−2.2	2.914	1.615	206.80
Salmon	−1 to 1	95 to 100	18 days	64	−2.2	2.981	1.641	213.47
Tuna	0 to 2	95 to 100	14 days	70	−2.2	3.182	1.717	233.49
Frozen fish	−29 to −18	90 to 95	6 to 12 months					
Seafood (Shellfish)[d]								
Scallop meat	0 to 1	95 to 100	12 days	80	−2.2	3.517	1.842	266.84
Shrimp	−1 to 1	95 to 100	12 to 14 days	76	−2.2	3.383	1.792	253.50
Lobster, American	5 to 10	In sea water	Indefinitely in sea water	79	−2.2	3.483	1.830	263.50
Oysters, clams (meat and liquor)	0 to 2	100	5 to 8 days	87	−2.2	3.751	1.930	290.19
Oyster in shell	5 to 10	95 to 100	5 days	80	−2.8	3.517	1.842	266.84
Frozen shellfish	−29 to −18	90 to 95	3 to 8 months					
Meat (Beef)[d]								
Beef, fresh, average	0 to 1	88 to 92	1 to 6 weeks	62 to 77	−2.2 to −1.7[f]	2.914 to 3.426	1.616 to 1.804	206.80 to 256.83

(Continued)

TABLE 4.6
Continued

Commodity	Storage Temp. (°C)	Relative humidity (%)	Approximate storage life[a]	Water content (%)	Highest freezing (°C)	Specific heat above freezing[b] (J/kg ·°C)	Specific heat below freezing[b] (J/kg ·°C)	Latent heat[c] (J/kg)
Carcass								
Choice, 60% lean	0 to 4	85 10 90	1 to 3 weeks	49	−1.7	2.478	1.453	163.44
Prime, 54% lean	0 to 1	85	1 to 3 weeks	45	−2.2	2.345	1.403	150.10
Sirloin cut (choice)	0 to 1	85	1 to 3 weeks	56		2.713	1.541	186.79
Round cut (choice)	0 to 1	85	1 to 3 weeks	67		3.081	1.679	223.48
Dried, chipped	10 to 15	15	6 to 8 weeks	48		2.445	1.440	160.10
Liver	0 to I	90	1 to 5 days	70	−1.7	3.182	1.717	233.48
Veal, 81% lean	0 to 1	90	1 to 7 days	66		3.048	1.666	220.14
Beef, frozen	−23 to −18	90 to 95	9 to 12 months					
Meat (Pork)[d]								
Pork, fresh, average	0 to 1	85 to 90	3 to 7 days	32 to 44	−2.2 to −2.7[f]	1.909 to 2.311	1.239 to 1.390	106.74 to 146.76
Carcass, 47% lean	0 to 1	85 to 90	3 to 5 days	37		2.077	1.302	123.41
Bellies, 33% lean	0 to 1	85	3 to 5 days	30		1.842	1.214	100.06
Backfat, 100% fat	0 to 1	85	3 to 7 days	8		1.105	0.938	
Shoulder, 67% lean	0 to 1	85	3 to 5 days	49	−2.2[f]	2.478	1.453	163.44
Pork, frozen	−23 to −18	90 to 95	4 to 6 months					
Ham								
74% lean	0 to 1	80 to 95	3 to 5 days	56	−1.7[f]	2.713	1.541	186.79
Light cure	3 to 5	80 to 85	1 to 2 weeks	57		2.746	1.553	190.12
Country cure	10 to 15	65 to 70	3 to 5 months	42		2.244	1.365	140.09
Frozen	−23 to-18	90 to 95	6 to 8 months					
Bacon								
Medium fat class	3 to 5	80 to 85	2 to 3 weeks	19		1.474	1.076	63.37
Cured, farm style	16 to 18	85	4 to 6 months	13–20		1.273 to 1.507	1.001 to 1.088	43.46 to 66.71
Cured, packer style	1 to 4	85	2 to 6 weeks					
Frozen	−23 to −18	90 to 95	4 to 6 months					
Sausage								
Links or bulk	0 to 1	85	1 to 7 days	38		2.110	1.315	126.75

(Continued)

TABLE 4.6
Continued

Commodity	Storage Temp. (°C)	Relative humidity (%)	Approximate storage life[a]	Water content (%)	Highest freezing (°C)	Specific heat above freezing[b] (J/kg ·°C)	Specific heat below freezing[b] (J/kg ·°C)	Latent heat[c] (J/kg)
Country, smoked	0	85	1 to 3 weeks	50	−3.9	2.512	1.465	166.78
Frankfurters average	0	85	1 to 3 weeks	56	−1.7	2.713	1.541	186.79
Polish style	0	85	1 to 3 weeks	54		2.646	1.516	180.12
			Meat (Lamb)[d]					
Fresh, average	0 to 1	85 to 90	5 to 12 days	60 to 70	−2.2 to −1.7[f]	2.847 to 3.182	1.591 to 1.717	200.01 to 233.48
Choice, 67% lean	0	85	5 to 12 days	61	−1.9	2.880	1.604	203.47
Leg, choice, 83% lean	0	85	5 to 12 days	65		3.014	1.654	216.81
Frozen	−23 to −18	90 to 95	8 to 10 months					
			Meat (Poultry)[d]					
Poultry, fresh, average	0	85 to 90	1 week	74	−2.8	3.316	1.767	246.83
Chicken, all classes	0	85	1 week	74	−2.8	3.316	1.767	246.83
Turkey, all classes	0	85	1 week	64	−2.8	2.981	1.641	213.47
Duck	0	85	1 week	69	−2.8	3.148	1.704	230.15
Poultry, frozen	−23 to −18	90 to 95	8 to 12 months					
			Meat (Miscellaneous)[d]					
Rabbits, fresh	0 to 1	90 to 95	1 to 5 days	68		3.115	1.691	226.81
			Dairy Products[d]					
Butter	4	75 to 85	1 month	16	−20 to −0.6	1.373	1.038	53.37
Butter, frozen	−23	70 to 85	12 months					
Cheese								
Cheddar, long storage	−1 to 1	65 to 70	18 months	37	−13.3	2.077	1.302	123.41
Cheddar, short storage	4.4	65 to 70	6 months	37	−13.3	2.077	1.302	123.41
Cheddar, processed	4.4	65 to 70	12 months	39	−7.2	2.143	1.327	130.08
Cheddar, grated	4.4	60 to 70	12 months	31		1.876	1.227	103.40
Ice cream, 10% fat	−29 to −26		3 to 23 months	63	−5.6	2.948	1.629	210.14
Milk								
Whole, pasteurized grade A	0 to 1.1		2 to 4 months	87	−0.56	3.751	1.930	290.19

(Continued)

TABLE 4.6
Continued

Commodity	Storage Temp. (°C)	Relative humidity (%)	Approximate storage life[a]	Water content (%)	Highest freezing (°C)	Specific heat above freezing[b] (J/kg ·°C)	Specific heat below freezing[b] (J/kg ·°C)	Latent heat[c] (J/kg)
Dried, whole	21	Low	6 to 9 months	2		0.904	0.862	66.71
Dried, nonfat	7 to 21	Low	16 months	3		0.938	0.895	10.01
Evaporated,	4		24 months	74	−1.4	3.316	1.767	246.83
Evaporated, unsweetened	21		12 months	74	−1.4	3.316	1.767	246.83
Condensed, sweetened	4		15 months	27	−15	1.742	1.176	90.06
Whey, dried	21	Low	12 months	5		1.005	0.900	16.68
Poultry Products[d]								
Eggs								
Shell	−2 to 0[g]	80 to 85	5 to 6 months	66	−2.2[f]	3.048	1.666	220.14
Shell, farm cooler	10 to 13	70 to 75	2 to 3 weeks	66	−2.2[f]	3.048	1.666	220.14
Frozen, whole	−18 or below		1 year plus	74		3.316	1.767	246.83
Frozen, yolk	−18 or below		1 year plus	55		2.680	1.528	183.45
Frozen, white	−18 or below		1 year plus	88		3.785	1.943	293.52
Whole egg solids	2 to 4	Low	6 to 12 months	2 to 4		0.938	0.875	10.01
Yolk solids	2 to 4	Low	6 to 12 months	3 to 5		0.972	0.888	13.34
Flake albumen solids	Room	Low	1 year plus	12 to 16		1.306	1.013	46.70
Dry spray albumen solids	Room	Low	1 year plus	5 to 8		1.055	0.919	21.68
Candy[d]								
Milk chocolate	−18 to 1.1	40	6 to 12 months	1		0.871	0.850	03.34
Peanut brittle	−18 to 1.1	40	1.5 to 6 months	2		0.904	0.862	06.67
Fudge	−18 to 1.1	65	5 to 12 months	10		1.172	0.963	33.35
Marshmallows	−18 to 1.1	65	3 to 9 months	17		1.407	1.051	56.70
Miscellaneous[d]								
Alfalfa meal	−18 or below	70 to 75	1 year plus					
Beer								
Keg[d]	2 to 4		3 to 8 weeks	90	−2.2[f]	3.852	1.968	300.20
Bottles and cans	2 to 4	65 or below	3 to 6 months	90				
Bread[d]	−18		3 to 13 weeks	32 to 37		1.993	1.271	106.74 to 123.41
Canned goods	0 to 16	70 or lower	1 year					
Cocoa	0 to 4	50 to 70	1 year plus					
Coconuts	0 to 2	80 to 85	1 to 2 months	47	−0.9	2.412	1.428	156.77

(Continued)

TABLE 4.6
Continued

Commodity	Storage Temp. (°C)	Relative humidity (%)	Approximate storage life[a]	Water content (%)	Highest freezing (°C)	Specific heat above freezing[b] (J/kg ·°C)	Specific heat below freezing[b] (J/kg ·°C)	Latent heat[c] (J/kg)
Coffee, green	2 to 3	80 to 85	2 to 4 months	10 to 15		1.172 to 1.340	0.962 to 1.026	033.36 to 050.03
Fur and fabrics	1 to 4	45 to 55	Several years					
Honey	below 10		1 year plus	17		1.407	1.051	056.70
Hops	−2 to 0	50 to 60	Several months					
Lard								
(without	7	90 to 95	4 to 8 months	0				
antioxidant)	−18	90 to 95	12 to 14 months	0				
Maple syrup				33		1.943	1.252	110.07
Nuts	0 to 10	65 to 75	8 to 12 months	3 to 6		0.938 to 1.038	0.875 to 0.913	010.01 to 020.01
Oil, vegetable, salad	21		1 year plus	0				
Oleomargarine	2	60 to 70	1 year plus	16		1.372	1.038	053.37
Orange juice	−1 to 2		3 to 6 weeks	89		3.818	1.955	296.86
Popcorn, unpopped	0 to 4	85	4 to 6 weeks	10		1.172	0.963	033.36
Yeast baker's, compressed	−0.6 to 0			71		3.215	1.729	236.82
Tobacco								
Hogshead	10 to 18	50 to 55	1 year					
Bales	2 to 4	70 to 85	1 to 2 years					
Cigarettes	2 to 8	50 to 55	6 months					
Cigars	2 to 10	60 to 65	2 months					

[a]Storage life is not based on maintaining nutritional value.

[b]Calculated by Siebel's *formula and converted* to Sf units. For values below freezing, specific heat in kJ/kg·°C (Btu/lb·°F) = $0.0355a + 0.8374(0.008a + 0.20)$.

For values below freezing, specific heat in kJ/kg·°C (Btu/lb·°F) = $0.0126a + 0.8374(0.003a + 0.20)$. Seibel's formula is not very accurate in the frozen region, because foods are not simple mixtures of solids and liquids.

[c]Values for latent heat in kJ/kg were calculated by multiplying the percentages of water content by the latent heat of fusion of water.

[d]More specific information is available in the commodity chapters of the *ASHRAE Applications Handbook*.

[e]Lemons stored in production areas for conditioning are held at 12.8 to 14.4°C, but sometimes at 0°C.

[f]Average freezing point.

Source: From ASHRAE. 1982. *ASHRAE Handbook: Applications.* American Society of Heating, Refrigerating, and Air-Conditioning Engineers, Atlanta. GA.

TABLE 4.7
Space, Weight, and Density Data for Commodities Stored in Refrigerated Warehouses

Commodity	Type of package	Outside dimensions of package, (in.)	Avg gross wt of pkg (lb)	Avg net wt mdse (lb)	Avg gross wt density (lb/cf)	Avg net wt density (lb/cf)
Apples	Wood box					
	Northwestern	$19\frac{1}{2} \times 11 \times 12\frac{3}{16}$	50	42	33.1	27.8
	Fiber tray carton	$20\frac{1}{2} \times 12\frac{1}{2} \times 13\frac{1}{4}$	$46\frac{3}{4}$	43	23.8	21.9
	Fiber master carton	$22\frac{1}{2} \times 12\frac{1}{2} \times 13$	$44\frac{3}{4}$	41	21.2	19.4
	Fiber bulk carton	$19 \times 12\frac{1}{2} \times 13$	$44\frac{3}{4}$	41	25.0	22.9
	Pallet box	$47 \times 47 \times 30$	1030	900	6.9	23.5
Beef						
Boneless	Fiber carton	$28 \times 18 \times 6$	146	140	83.4	80.0
Fores	Loose					22.2
Hinds	Loose					22.2
Celery	Wirebound crates	$20\frac{1}{4} \times 16 \times 9\frac{3}{4}$	60	55	32.8	30.0
	Fiber carton	$16 \times 11 \times 10$	36	32	35.4	31.4
Cheese	Hoops	$16 \times 16 \times 13$	84	78	43.6	40.5
	Wood, export	$17 \times 17 \times 14$	87	76	37.1	32.5
Cheese, Swiss	Wheels	$32\frac{1}{2} \times 32\frac{1}{2} \times 7$		171		40.0
Chili peppers	Bags	$45 \times 21 \times 26$	234	229	16.5	16.1
Citrus fruits						
Oranges	Box	$12\frac{1}{8} \times 13\frac{1}{4} \times 26\frac{1}{4}$	77	69	31.5	28.3
	Bruce box	$13 \times 11 \times 26\frac{1}{4}$	88	83	40.5	38.2
	Pallet, 40 cartons	$40 \times 48 \times 58\frac{1}{2}$	1690	1480	26.0	22.8
California oranges	Fiber carton	$16\frac{3}{8} \times 10\frac{1}{16} \times 10\frac{1}{2}$	40	37	38.0	35.2
Florida oranges	Fiber carton	$19\frac{1}{4} \times 12\frac{1}{4} \times 8$	45	37	41.3	33.9
Lemons	Fiber carton	$16\frac{3}{8} \times 10\frac{1}{16} \times 10\frac{1}{2}$	40	37	40.0	37.0
Grapefruit	Fiber carton	$19\frac{1}{4} \times 12\frac{1}{4} \times 8$	40	38	36.7	34.9
Coconut, shredded	Bags	$38 \times 18\frac{1}{2} \times 8$	101	100	31.0	30.7
Cranberries	Fiber carton	$15\frac{3}{4} \times 11\frac{1}{4} \times 10\frac{1}{2}$	26	24	24.1	22.2
Cream	Tins	$12 \times 12 \times 14$	$52\frac{3}{4}$	50	45.2	42.9
Dried fruit	Wood box	$15\frac{1}{2} \times 10 \times 6\frac{1}{2}$	$26\frac{1}{2}$	25	45.4	42.9
Dates	Fiber carton	$14 \times 14 \times 11$	32	30	25.7	24.0
Raisins, prunes, figs, peaches	Fiber carton	$15 \times 11 \times 7$	32	30	47.9	44.9
Eggs, shell	Wood cases	$26 \times 12 \times 13$	55	45	23.4	19.1
Eggs, frozen	Cans	$10 \times 10 \times 12\frac{1}{2}$	32	30	44.2	41.5
Frozen fishery products						
Blocks	$4/13\frac{1}{2}$-lb carton	$20\frac{3}{4} \times 12\frac{1}{8} \times 6\frac{3}{4}$	56	54	57.0	55.0
	$4/16\frac{1}{2}$-lb carton	$19\frac{3}{4} \times 10\frac{3}{4} \times 11\frac{1}{4}$	68	66	49.2	47.8
Filets	12/16-oz carton	$12\frac{3}{4} \times 8\frac{5}{8} \times 3\frac{13}{16}$	13.5	12	55.8	49.6
	10/5-lb carton	$14\frac{1}{2} \times 10 \times 14$	52.25	50	44.6	42.7

(Continued)

TABLE 4.7
Continued

Commodity	Type of package	Outside dimensions of package, (in.)	Avg gross wt of pkg (lb)	Avg net wt mdse (lb)	Avg gross wt density (lb/cf)	Avg net wt density (lb/cf)
	5/10-lb carton	$14\frac{1}{2} \times 10 \times 14$	52.2	50	44.5	42.7
Fish sticks	12/8-oz carton	$11 \times 8\frac{3}{8} \times 3\frac{7}{8}$	6.9	6	33.6	29.3
	24/8-oz carton	$16\frac{7}{16} \times 8\frac{5}{16} \times 4\frac{5}{8}$	13.8	12	37.8	32.9
Panned fish	None, glazed	Wooden boxes				35.0
portions	2-, 3-, 5-, and 6-lb carton	Custom packing				29–33
Round ground fish	None, glazed	Stacked loose				33–35
Round Halibut	None, glazed	Wooden box, loose				30–35
		Stacked loose				38.0
Round salmon	None, glazed	Stacked loose				33–35
Shrimp	$2\frac{1}{2}$- and 5-lb cartons	Custom packing				35.0
Steaks	1-, 5-. or 10-lb packages	Custom packing				50–60
Frozen fruits, juices, vegetables						
Asparagus	24/12-oz carton	$13\frac{1}{2} \times 11\frac{3}{4} \times 8\frac{1}{4}$	21	18	27.7	23.8
Beans, green	36/10-oz carton	$12\frac{1}{2} \times 11 \times 8$	$25\frac{1}{2}$	$22\frac{1}{2}$	40.1	35.3
Blueberries	24/12-oz carton	$12 \times 11\frac{1}{2} \times 8$	20	18	31.3	28.2
Broccoli	24/10-oz carton	$12\frac{1}{2} \times 11\frac{1}{2} \times 8\frac{1}{2}$	$18\frac{1}{2}$	15	26.2	21.2
Citrus concentrates	Fiber carton 48/6 oz	$13 \times 8\frac{3}{4} \times 7\frac{1}{2}$	27	26	54.7	52.7
Peaches	24/1-lb carton	$13\frac{1}{2} \times 11\frac{1}{4} \times 7\frac{1}{2}$	27	24	41.0	36.4
Peas	6/5-1b carton	$17 \times 11 \times 9\frac{1}{2}$	32	30	31.1	28.2
	48/12-oz carton	$21\frac{1}{2} \times 8\frac{1}{2} \times 12\frac{1}{2}$	38	36	28.7	27.2
Potatoes, french fries	12/16-oz carton					28.6
	24/9-oz carton					24.0
Spinach	24/14-oz carton	$12\frac{1}{2} \times 11 \times 8\frac{1}{2}$	24	21	35.5	31.0
Strawberries	30-lb can	$12\frac{1}{2} \times 10 \times 10$	32	30	44.2	41.5
	24/1–1b carton	$13 \times 11 \times 8$	28	24	42.3	36.2
	450-lb barrel	$35 \times 25 \times 25$		450		35.5
Grapes, California	Wood lug box	$6\frac{1}{2} \times 15 \times 18$	31	28	32.4	29.2
Lamb, boneless	Fiber box	$20 \times 15 \times 5$	57	53	65.7	61.0
Lard (2/28 1b)	Wood export box	$18 \times 13\frac{1}{4} \times 7\frac{3}{4}$	64	56	59.8	52.5
Lettuce, head	Fiber carton	$20\frac{1}{2} \times 13\frac{1}{2} \times 9\frac{1}{2}$	$37\frac{1}{2}$	35	24.7	
	Fiber carton	$21\frac{1}{2} \times 14\frac{1}{4} \times 10\frac{1}{2}$	45–55	42–52	26.9	25.2
	Pallet, 30 cartons	$42 \times 50 \times 66$	1350	1170	16.8	14.6
Milk, condensed	Barrels	$35 \times 25\frac{1}{2} \times 25\frac{1}{2}$	670	600	50.9	45.6
Nuts						
Almonds, in shell	Sacks	$24 \times 15 \times 33$	$91\frac{1}{2}$	90	13.3	13.1

(Continued)

TABLE 4.7
Continued

Commodity	Type of package	Outside dimensions of package, (in.)	Avg gross wt of pkg (lb)	Avg net wt mdse (lb)	Avg gross wt density (lb/cf)	Avg net wt density (lb/cf)
Almonds, shelled	Cases	$6\frac{3}{4} \times 23\frac{1}{2} \times 11$	32	28	31.7	27.7
English walnuts, in shell	Cases	$25 \times 11 \times 31$	103	100	20.9	20.3
English walnuts, shelled	Fiber carton	$14 \times 14 \times 10$	27	25	23.8	22.0
Peanuts, shelled	Burlap bag	$35 \times 10 \times 15$	127	125	39.2	38.6
Pecans, in shell	Burlap bag	$35 \times 22 \times 12$	$126\frac{1}{2}$	125	23.7	23.4
Pecans, shelled	Fiber carton	$13 \times 13 \times 11$	32	30	29.8	27.9
Peaches	3/4 bushel	$16\frac{7}{8}$ top dia	41	48	43.9	40.7
	1/2 bushel	$14\frac{1}{2}$ top dia	28	25	45.0	40.2
	Wirebound crate	$19 \times 11\frac{3}{4} \times 11\frac{1}{8}$	42	38	29.2	26.4
	Wood lug box	$18\frac{1}{8} \times 11\frac{1}{2} \times 5\frac{3}{4}$	26	23	38.0	33.1
Pears	Wood box	$8\frac{1}{2} \times 11\frac{1}{2} \times 18$	52	48	51.0	47.1
Pears, place pack	Fiber carton	$18\frac{1}{2} \times 12 \times 10$	52	46	40.5	35.6
Pork						
Bundle bellies	Bundles	$23\frac{1}{2} \times 10\frac{1}{2} \times 7$	57	57	57.0	57.0
Loins (regular)	Wood box	$28 \times 10 \times 10$	60	54	37.0	33.3
Loins (boneless)	Fiber box	$20 \times 15 \times 5$	57	52	65.7	59.9
Potatoes	Sack	$33 \times 17\frac{1}{2} \times 11$	101	100	27.5	27.2
Poultry, fresh (eviscerated)						
Fryers, whole, 24–30 to pkg.	Wirebound crate	$24 \times 10 \times 7$	65	60	27.5	25.4
Fryer parts Poultry, frozen (eviscerated)	Wirebound crate	$17\frac{3}{4} \times 10 \times 12\frac{1}{2}$	54	50	42.1	38.9
Ducks, 6 to pkg.	Fibei carton	$22 \times 16 \times 4$	$32\frac{1}{2}$	31	39.9	38.0
Fowl, 6 to pkg.	Fiber canon	$20\frac{3}{4} \times 18 \times 5\frac{1}{2}$	$33\frac{1}{2}$	31	28.2	26.1
Fryers, cut up, 12 to pkg.	Fiber. carton	$17\frac{1}{4} \times 15\frac{3}{4} \times 4\frac{1}{4}$	$30\frac{1}{2}$	28	45.4	41.7
Roasters, 8 to pkg.	Fiber carton	$20\frac{3}{4} \times 18 \times 5\frac{1}{2}$	$32\frac{1}{2}$	30	27.3	25.2
Turkeys,						
3–6 lb, 6 to pkg.	Fiber carton	$21 \times 17 \times 6\frac{1}{2}$	30	27	22.5	20.1
6–10 lb, 6 to pkg.	Fiber canon	$26 \times 21\frac{1}{2} \times 7$	$52\frac{1}{2}$	48	23.3	21.2
10–13 lb, 4 to pkg.	Fiber carton	$26\frac{1}{2} \times 16 \times 7\frac{1}{2}$	50	46	27.2	25.0
13–16 lb, 4 to pkg.	Fiber carton	$29 \times 18\frac{1}{2} \times 9$	$67\frac{1}{2}$	62	24.2	22.2
16–20 lb, 2 to pkg.	Fiber carton	$17 \times 16 \times 9$	39	36	27.7	25.4
20–24 lb, 2 to pkg.	Fiber carton	$19 \times 16\frac{1}{2} \times 9\frac{1}{2}$	$47\frac{1}{2}$	44	27.6	25.5
Tomatoes						
Florida	Fiber carton	$19 \times 10\frac{7}{8} \times 10\frac{3}{4}$	43	40	33.3	31.10
	Wirebound crate	$18\frac{3}{4} \times 11\frac{15}{16} \times 11\frac{15}{16}$	64	60	41.3	38.7
California	Wood lug box	$17\frac{1}{2} \times 14 \times 7\frac{3}{4}$	34	30	30.9	27.3
Texas	Wood lug box	$17\frac{1}{2} \times 14 \times 6\frac{5}{8}$	34	30	36.2	31.9
Veal (boneless)	Fiber carton	$20 \times 15 \times 5$	57	53	65.7	61.0

Source: From ASHRAE. 1982. *ASHRAE Handbook: Applications*. American Society of Heating, Refrigerating, and Air-Conditioning Engineers, Atlanta, GA.

NOMENCLATURE

A	Area; projected area of particle normal to direction of flow, Equation 4.57
b	Number of fittings
c	Coefficient of rolling resistance, Equation 4.41 cohesion, Equation 4.47
d_p	Particle diameter
c_d	Particle drag coefficient
c_r	Resistance coefficient
d_e	Diameter of sphere having same volume/surface ratio of actual particle
D	Diameter
D_a	Agitator diameter (m)
D_T	Tank diameter (m)
D_{av}	Average screen particle diameter, Equation 4.42
E_f	Energy loss term
f	Friction factor in pipe for fluid flow; force to move a column of grain, Equation 4.60
f_c	Unconfined yield stress, Equation 4.48
f_s	Static friction
f_v	Friction factor for hydraulic transport of solid foods
ff_c	Powder flow function, Equation 4.48
F	Friction force
g	Acceleration due to gravity
h	Height; depth of a material
$H(q)$	A function of hopper geometry given by Jenike
k	Ratio of lateral to vertical pressure
k_f	Friction coefficient for pipe fittings
K, K_H	Consistency index
k_s	Agitator constant (dimensionless)
L	Length of pipe; length of particle
m	Mass flow rate
M	Percent moisture
n	Index of refraction, Equation 4.1; flow behavior index, Equation 4.15; shape factor, Equation 4.42
N_a	Agitator speed ($\sec^{-1}$)
p	Power ($J\sec^{-1}$)
p_m	Yield pressure
p	Pressure, Equation 4.11; plowing force
Po	Power number (dimensionless)
Q	Volumetric flow rate
r	Radius
R	Hydraulic radius, Equation 4.45; radius of tube, Equation 4.60
Re_a	Agitator rotational Reynolds number (dimensionless)
Re_a	Power law agitation Reynolds number (dimensionless)
s	Shearing stress
s_g	Specific gravity
S	Shearing force
T	Temperature; Equation 4.2 and Equation 4.3; tensile stress, Equation 4.47
T	Torque on agitator (N m)
u	Average fluid velocity
U	Fluid velocity in packed bed
U_0	Minimum fluidization velocity

U_t	Terminal velocity to avoid particle entrainment
v	Volume of particle
V	Volume
w	Density of material
W	Power, Equation 4.11; force normal to contact, Equation 4.37 weight, Equation 4.41
X_s	Solids fraction
X_w	Water fraction
Z	Height above a reference point
Z_0	Height of fluidized bed at minimum fluidization conditions

DIMENSIONLESS NUMBERS

$\mathrm{GRe} = \dfrac{D^n u^{2-n} \rho}{8^{n-1} K} \dfrac{4n}{3n+1}$	Generalized Reynolds number
$\mathrm{Re}_{de} = \dfrac{d_e u \rho}{\eta}$	Reynolds number based on d_e
$\mathrm{Po} = \dfrac{p}{D_a^5 N_a^3 \rho}$	Power number in mixing
$\mathrm{Re}_a = \dfrac{D_a^2 N_a \rho}{\eta}$	Newtonian agitation Reynolds number
$\mathrm{Re}_{pl} = \dfrac{D_a^2 N_a \rho}{K(k_s N_a)^{n-1}}$	Power law agitation Reynolds number

GREEK LETTERS

α	Kinetic energy correction factor
ΔP	Pressure drop
ε	Bed voidage fraction
ε_0	Bed voidage at minimum fluidization condition
η	Viscosity
η_e	Effective viscosity
Φ_r	Angle of repose
Φ_s	Sphericity factor = ratio of surface area of a sphere of same volume as particle to surface area of the particle
Φ_w	Friction angle between bulk material and wall
Φ_e	Effective angle of friction
Φ_i	Angle of internal friction
$\dot{\gamma}$	Shear rate
μ	Coefficient of friction
ρ	Density of a fluid
ρ_p	Density of a particle
ρ_b	Bulk density
θ	Angle
θ_c	Critical cone angle
ξ	Unsheared plug radius
σ	shear stress in liquid, Equation 4.14 through Equation 4.16; normal stress in powders, Equation 4.46
σ_1	Consolidation stress in powders
σ_1'	Major principal stress at abutment of arch
σ_0	Yield stress
σ_w	Wall stress
s	Yield stress

τ	Shear stress in solids
Ψ	Function in Equation 4.17
SUBSCRIPTS	
1,2	Position in system
s	Solids
w	Water

REFERENCES

Anonymous. 1970. *Food Storage Manual*. World Food Programme, FAO, Rome.

Anonymous, 2002. Stir it up. *Food Technol. New-Zealand*, 37: 31.

ASHRAE. 1982. *ASHRAE Handbook: Applications*. American Society of Heating, Refrigerating, and Air-Conditioning Engineers, Atlanta, GA.

Bhattacharya, K.R., Sowbhagya, C.M., and Indudhara Swamy, Y.M. 1972. Some physical properties of paddy and rice. *J. Sci. Food Agric.*, 23: 171–186.

Brodkey, R.S. 1967. *The Phenomena of Fluid Motions*. Addition-Wesley, Reading, Mass.

Cantú-Lozano, D., Rao, M.A., and Gasparetto, C.A. 2000. Rheological properties of non-cohesive apple dispersion with helical and vane impellers: effect of concentration and particle size. *J. Food Process Eng.*, 23: 373–385.

Choi, Y. and Okos, M.R. 1983. The thermal properties of tomato juice. *Trans. ASAE*, 26: 305–311.

Choi, Y. and Okos, M.R. 1986. Thermal properties of liquid foods: review. In *Physical and Chemical Properties of Food*, M.R. Okos (ed.). American Society of Agricultural Engineers. St. Joseph, Mich., pp. 35–77.

Christensen, C.M. 1953. *Storage of Cereal Grains and Their Products*. American Association of Cereal Chemists, St. Paul, Minn.

Clapp, R.M. 1961. Turbulent heat transfer in pseudoplastic non-Newtonian fluids. *International Developments in Heat Transfer*. ASME, Part III, Sec. A., pp. 652–661 [cited in Garcia and Steffe (1987)].

Davidson, J.F. and Harrison, D. 1967. *Fluidisation*. Academic Press, New York.

Dodge, D.W. and Metzner, A.B. 1959. Turbulent flow of non-Newtonian systems. *AIChE J.*, 5: 189–204.

Garcia, E.J. and Steffe, J.F. 1987. Comparison of friction factor equations for non-Newtonian fluids in pipe flow. *J. Food Process Eng.*, 9: 93–120.

Hahn, R.H. and Rosentreter, E.E. 1987. *Standards 1987*. American Society of Agricultural Engineers, St. Joseph, Mich.

Hanks, R.W. 1978. Low Reynolds number turbulent pipeline flow of pseudohomogeneous slurries. In *Proc. 5th Int. Conf. Hydraulic Transport of Solids in Pipes (Hydrotransport)* May 8–11, Hanover, West Germany, Paper C2, pp. C2–23 to C2–34 [cited in Garcia and Steffe (1987)].

Hanks, R.W. and Ricks, B.L. 1974. Laminar-turbulent transition in flow of pseudoplastic fluids with yield stress. *J. Hydronaut.*, 8: 163–166 [cited in Garcia and Steffe (1987)].

Heldman, D.R. and Singh, R.P. 1981. *Food Process Engineering*, 2nd ed. AVI, Westport, Conn. IAMFES. 1988. International Association of Milk, Food, and Environmental Sanitarians, Ames, Iowa.

Imholte, T.J. 1984. *Engineering for Food Safety and Sanitation: A Guide to the Sanitary Design of Food Plants and Food Plant Equipment*. Technical Institute for Food Safely, Crystal, Minn.

Jenike, A.W. 1970. *Storage and Flow of Solids*, Bulletin 123 of the Utah Engineering Experiment Station, 4th printing (revised). University of Utah, Salt Lake City.

Kalkschmidt-J. 1977. Ruehr- und Mischeinrichtungen - unter besonderer Beruecksichtigung der Milchwirtschaft [Stirring and mixing equipment for the dairy industry.]. *Fette,-Seifen,-Anstrichmittel*, 77: 357–359. Food Science & Technology Abstract 76–04-P0664.

Kunii, D. and Levenspiel, O. 1969. *Fluidization Engineering*. John-Wiley, New York.

Labrie, J. 1987. Correct the equipment design—don't fix it on the job or how you can provide a clean operating production line. In *FPEI Workshop on Engineering for Sanitation*. American Society of Agricultural Engineers, St. Joseph, Mich.

Lutz, J.M. and Hardenburg, R.E. 1968. *The Commercial Storage of Fruits, Vegetables, and Florist and Nursery Stocks*, USDA Agriculture Handbook 66. U.S. Government Printing Office, Washington, D.C.

McKay, G., Murphy, W.R., and Jodieri-Dabbaghzadeh, S. 1987. Fluidization and hydraulic transport of carrot pieces. *J. Food Eng.*, 6: 377–399.

Metzner, A.B. and Otto, R.E. 1957. Agitation of non-Newtonian fluids. *AIChE J.*, 3: 3–10.

Miller, E.J. 1981. The design and operation of agitators for use in whole milk storage vessels. *New-Zealand-Journal-of-Dairy-Science-and-Technology*, 16: 221–229. Food Science & Technology Abstract 80–10-P1720.

Mohsenin, N.N. 1986. *Physical Properties of Plant and Animal Materials*. Gordon and Breach, New York.

Osorio, F.A. and Steffe, J.F. 1984. Kinetic energy calculations for non-Newtonian fluids in circular tubes. *J. Food Sci.*, 49: 1295–1296, 1315.

Peleg, M. 1977. Flowability of food powders and methods for its evaluation. *J. Food Process Eng.*, 1: 303–328.

Perry, R.H. and Green, D. 1984. *Perry's Chemical Engineer's Handbook*. McGraw-Hill, New York.

Phipps, L.W. 1969. The interrelationship of viscosity, fat content, and temperature of cream between 40°C and 80°C. *J. Dairy Res.* 36: 417–426 [cited in Choi and Okos (1986)].

Rao, M.A. and Cooley, H.J. 1984. Determination of effective shear rates of complex geometries. *J. Texture Studies*, 15: 327–335.

Rastovski, A. and van Es, A. 1981. *Storage of Potatoes*. Centre for Agricultural Publishing and Documentation, Wageningen, The Netherlands.

Riedel, L. 1949. Thermal conductivity measurement on sugar solutions, fruit juices and milk. *Chem. Ing. Tech.*, 21: 340–341 [cited in Choi and Okos (1986)].

Rieger, F. and Novak, V. 1973. Power consumption of agitators in highly viscous non-Newtonian liquids. *Trans. IChem. E.*, 51: 105–111.

Roy, N.K., Yadav, P.L., and Dixit, R.N. 1971. Density of buffalo milk fat. II. Centrifuged fat. *Milchwissenschaft*, 26: 735–738 [cited in Choi and Okos (1986)].

Ryall, A.L. and Lipton, W.J. 1979. *Handling, Transportation and Storage of Fruits and Vegetables*. Vol. 1, *Vegetables and Melons*, 2nd ed. AVI, Westport, Conn.

Ryall, A.L. and Lipton, W.J. 1984. *Handling, Transportation and Storage of Fruits and Vegetables*, Vol. 2, *Fruits and Nuts*, 2nd ed. AVI, Westport, Conn.

Schubert, H. 1987a. Food particle technology. I. Properties of particles and participate food systems. *J. Food Eng.*, 6: 1–32.

Schubert, H. 1987b. Food particle technology. II. Some specific cases. *J. Food Eng.*, 6: 83–102.

Short, A.L. 1955. The temperature coefficient of expansion of raw milk. *J. Dairy Res.*, 22: 69 [cited in Choi and Okos (1986)].

Sivetz, M. and Desrosier, N. 1979. *Coffee Technology*. AVI, Westport, Conn.

Skelland, A.H.P. 1967. *Non-Newtonian Flow and Heat Transfer*. John-Wiley, New York.

Steffe, J.F. and Morgan, R.G. 1986. Pipeline design and pump selection for non-Newtonian fluid foods. *Food Technol.*, 40: 78–85.

Steffe, J.F., Mohamed, I.O., and Ford, E.W. 1984. Pressure drop across valves and fittings for pseudoplastic fluids in laminar flow. *Trans. ASAE*, 27: 616–619.

Watt, B.K., and Merrill, A.L. 1975. *Composition of Foods*, USDA Handbook 8. U.S. Government Printing Office, Washington, D.C.

Wilkens, R.J., Henry, C., and Gates, L.E. 2003. How to scale-up mixing processes in non-Newtonian fluids, *Chem. Eng. Progress*, 99: 44–52.

5 Heating and Cooling Processes for Foods

R. Paul Singh

CONTENTS

5.1 INTRODUCTION

Heating and cooling processes are important part of food processing operations. Many desirable changes, as well as undesirable reactions, occur in foods when they are heated or cooled. The rate and extent of these reactions can be controlled by controlling the rate of heat transfer. Thus, the heating and cooling characteristics of foods must be well understood to bring about intended changes in foods during processing.

As a food material is heated or cooled, there is an initial period of unsteady state when the temperature at a given location in the material is changing with time. After a certain time has elapsed, the rate of heat transfer reaches a steady state where the temperature may vary from one location to another, but at any given location, there is no change in temperature with time.

In this chapter a mathematical description of both steady-state and unsteady-state heat transfer in foods is presented. A description of important thermal properties of foods is given with particular emphasis on mathematical models. Knowledge of these thermal properties is essential to the general study of heat transfer.

The topic of heat transfer covers extensive material. Several excellent textbooks give detailed coverage of this area. This chapter provides a summary of different modes of heat transfer relevant to food processing. Several mathematical expressions will be given without derivations. For additional background material, the reader is referred to books by Holman (1997), Chapman (1974), and Kreith and Black (1980).

5.2 THERMAL PROPERTIES OF FOODS

5.2.1 THERMAL CONDUCTIVITY

Thermal conductivity k is the rate of heat transfer q through a unit cross-sectional area A when a unit temperature difference $(T_1 - T_2)$ is maintained over a unit distance L:

$$k = \frac{qL}{A(T_1 - T_2)} \tag{5.1}$$

The definition above, which implies steady-state heat transfer conditions, has been used to design experiments for measuring thermal conductivity of foods. In addition, transient techniques are also used for more rapid determination of thermal conductivity. These experimental methods have been reviewed by Choi and Okos (1986) and Reidy and Rippen (1971). The steady-state methods include the guarded hot-plate method, the concentric cylinder method, and the concentric sphere method. The transient methods include the Fitch method, the line heat source or probe method, and the plate heat source method. Experimental data on thermal conductivities measured for various food groups have been expressed by mathematical relationships. These models are useful in estimating thermal conductivity of food materials. Some of the commonly used models are presented in the following.

Riedel (1949) presented the following model to predict thermal conductivity of fruit juices, sugar solutions, and milk:

$$k = (326.58 + 1.0412T - 0.00337T^2) \times (0.46 + 0.54X_w) \times 1.73 \times 10^{-3} \tag{5.2}$$

It was estimated that between 0 and 180°C there was an error of 1% when this model was used.

Sweat (1974) suggested the following equation, obtained with regression analysis of data on thermal conductivities of several fruits and vegetables:

$$k = 0.148 + 0.00493 \times (\%\text{water}) \tag{5.3}$$

Equation 5.3 should predict thermal conductivity within $\pm 15\%$ of experimental values for fruits and vegetables with moisture content greater than 60%. This model is unsuitable for low-density products, or foods with void spaces, such as apples. Experimental values of thermal conductivity of selected foods are given in Table 5.1.

Choi and Okos (1986) have suggested the following model for liquid foods based on the food composition:

$$k = \Sigma_i k_i X_i^v \tag{5.4}$$

where the estimated volume fraction, $X_i^v = (X_i^w/\rho_i)/\Sigma(X_i^w/\rho_i)$.

The values of thermal conductivities of pure components of liquid foods are given in Table 5.2. For porous foods, a review of thermal conductivity values and mathematical models was given by Wallapapan et al. (1986).

Thermal conductivities of anisotrophic materials vary with the direction of heat transfer. For example, for meats the thermal conductivity along the meat fibers is different from what it is across the fibers. These differences were considered by Kopelman (1966). His models for thermal conductivity are presented by Heldman and Singh (1981).

5.2.2 DENSITY

Density of a food material is the mass of the sample divided by its volume. Experimental determination of density can be done using a pycnometer, an air comparison pycnometer, and or a platform scale method (Choi and Okos, 1986). Mathematical models of density of foods have been developed for prediction purposes. For fruit juices, Riedel (1949) suggested measuring the index of refraction of the juice, s, and using the following relationship:

$$\rho = \frac{s^2 - 1}{s^2 + 2} \times \frac{62.4}{0.206} \times 16.0185 \tag{5.5}$$

Other researchers have reported specific models for milk, cream, and tomato juice (Short, 1955; Phipps, 1969; Choi and Okos, 1983).

Using the compositional information of liquid foods, Choi and Okos (1966) have suggested the following model:

$$\rho = \frac{1}{\sum (X_i^w/\rho_i)} \tag{5.6}$$

The density values of pure components are given in Table 5.1.

5.2.3 SPECIFIC HEAT

Specific heat of a food material is a measure of the amount of energy required by a unit mass to raise its temperature by a unit degree. Specific heat, or the mass heat capacity of food materials, has been determined experimentally by several methods, including the method of mixtures, method of guard plate, and using a differential scanning calorimeter (Choi and Okos, 1986).

TABLE 5.1

Thermal Conductivity and Specific Heat of Selected Food Products

Product	Moisture content (%)	Temperature (°C)	Thermal conductivity (W/m·K)
Apple	85.6	2 to 36	0.393
Applesauce	78.8	2 to 36	0.516
Beef, freeze dried			
1000 mmHg pressure	—	0	0.065
0.001 mmHg pressure	—	0	0.037
Beef, lean			
Perpendicular to fibers	78.9	7	0.476
	78.9	62	0.485
Parallel to fibers	78.7	8	0.431
	78.7	61	0.447
Beef fat	—	24 to 38	0.19
Butter	15	46	0.197
Cod	83	2.8	0.544
Corn, yellow dent	0.91	8 to 52	0.141
	30.2	8 to 52	0.172
Egg, Frozen whole	—	−10 to −6	0.97
Egg, white	—	36	0.577
Egg. yolk	—	33	0.338
Fish muscle	—	0 to 10	0.557
Grapefruit, whole	—	30	0.45
Honey	12.6	2	0.502
	80	2	0.344
	14.8	69	0.623
	80	69	0.415
Juice, apple	87.4	20	0.559
	87.4	80	0.632
	36.0	20	0.389
	36.0	80	0.436
Lamb			
Perpendicular to fiber	71.8	5	0.45
		61	0.478
Parallel to fiber	71.0	5	0.415
		61	0.422
Milk	—	37	0.530
Milk, condensed	90	24	0.571
	—	78	0.641
	50	26	0.329
	—	78	0.364
Milk, skimmed	—	1.5	0.538
	—	80	0.635
Milk, nonfat dry	4.2	39	0.419
Olive oil	—	15	0.189
	—	100	0.163
Oranges, combined	—	30	0.431
Peas, black-eyed	—	3 to 17	0.312
Pork			
Perpendicular to fibers	75.1	6	0.488
		60	0.54

(Continued)

TABLE 5.1
Continued

Product	Moisture content (%)	Temperature (°C)	Thermal conductivity (W/m·K)
Parallel lo fibers	75.9	4	0.443
		61	0.489
Pork fat	—	25	0.152
Potato, raw flesh	81.5	1 to 32	0.554
Potato, starch gel	—	1 to 67	0.04
Poultry, broiler muscle	69.1 to 74.9	4 to 27	0.412
Salmon			
Perpendicular to fibers	73	4	0.502
Salt	—	87	0.247
Sausage mixture	64.72	24	0.407
Soybean oil meal	13.2	7 to 10	0.069
Strawberries	—	−14 to 25	0.675
Sugars	—	29 to 62	0.087 to 0.22
Turkey, breast			
Perpendicular to fibers	74	3	0.502
Parallel to fibers	74	3	0.523
Veal			
Perpendicular to fibers	75	6	0.476
		62	0.489
Parallel to fibers	75	5	0.441
		60	0.452
Vegetable and animal oils	—	4 to 187	0.169
Wheat flour	8.8	43	0.45
		65.5	0.689
		1.7	0.542
Whey		80	0.641

Specific Heat

Product	Water (%)	Specific heat, experimental (kJ/kg·K)
Beef (hamburger)	68.3	3.52
Butter	15.5	2.051–2.135
Milk, whole pasteurized	87.0	3.852
Skim milk	90.5	3.977–4.019
Egg yolk	49.0	2.810
Fish, fresh	76.0	3.600
Beef, lean	71.7	3.433
Potato	79.8	3.517
Apple, raw	84.4	3.726–4.019
Bacon	49.9	2.01
Cucumber	96.1	4.103
Potato	75.0	3.517
Veal	68.0	3.223
Fish	80.0	3.60
Cheese, cottage	65.0	3.265
Shrimp	66.2	3.014
Sardines	57.4	3.014
Beef, roast	60.0	3.056
Carrot, fresh	88.2	3.81–3.935

Source: Reidy, G.A. 1968. Thermal properties of foods and methods of their determination. M.S. Thesis, Food Science Dept., Michigan State University.

TABLE 5.2
Thermal Property Models

Thermal property	Major component	Group model temperature function	Standard error	Standard % error
a. Major food components				
k(W/m·°C)	Protein	$k = 1.7881 \times 10^{-1} + 1.1958 \times 10^{-3}T - 2.7178 \times 10^{-6}T^2$	0.012	5.91
	Fat	$k = 1.8071 \times 10^{-1} - 2.7604 \times 10^{-3}T - 1.7749 \times 10^{-7}T^2$	0.0032	1.95
	Carbohydrate	$k = 2.0141 \times 10^{-1} + 1.3874 \times 10^{-3}T - 4.3312 \times 10^{-6}T^2$	0.0134	5.42
	Fiber	$k = 1.8331 \times 10^{-1} + 1.2497 \times 10^{-3}T - 3.1683 \times 10^{-6}T^2$	0.0127	5.55
	Ash	$k = 3.2962 \times 10^{-1} + 1.4011 \times 10^{-3}T - 2.9069 \times 10^{-6}T^2$	0.0083	2.15
α(m²/sec)	Protein	$\alpha = 6.8714 \times 10^{-2} + 4.7578 \times 10^{-4}T - 1.4646 \times 10^{-6}T^2$	0.0038	4.50
	Fat	$\alpha = 9.8777 \times 10^{-2} - 1.2569 \times 10^{-4}T - 3.8286 \times 10^{-8}T^2$	0.0020	2.15
	Carbohydrate	$\alpha = 8.0842 \times 10^{-2} + 5.3052 \times 10^{-4}T - 2.3218 \times 10^{-6}T^2$	0.0058	5.84
	Fiber	$\alpha = 7.3976 \times 10^{-2} + 5.1902 \times 10^{-4}T - 2.2202 \times 10^{-6}T^2$	0.0026	3.14
	Ash	$\alpha = 1.2461 \times 10^{-1} + 3.7321 \times 10^{-4}T - 1.2244 \times 10^{-6}T^2$	0.0022	1.61
ρ(kg/m³)	Protein	$\rho = 1.3299 \times 10^3 - 5.1840 \times 10^{-1}T$	39.9501	3.07
	Fat	$\rho = 9.2559 \times 10^2 - 4.1757 \times 10^{-1}T$	4.2554	0.47
	Carbohydrate	$\rho = 1.5991 \times 10^3 - 3.1046 \times 10^{-1}T$	93.1249	5.98
	Fiber	$\rho = 1.3115 \times 10^3 - 3.6589 \times 10^{-1}T$	8.2687	0.64
	Ash	$\rho = 2.4238 \times 10^3 - 2.8063 \times 10^{-1}T$	2.2315	0.09
C_p(kJ/kg·°C)	Protein	$C_p = 2.0082 + 1.2089 \times 10^{-3}T - 1.3129 \times 10^{-6}T^2$	0.1147	5.57
	Fat	$C_p = 1.9842 + 1.4733 \times 10^{-3}T - 4.8008 \times 10^{-6}T^2$	0.0236	1.16
	Carbohydrate	$C_p = 1.548.8 + 1.9625 \times 10^{-3}T - 5.9399 \times 10^{-6}T^2$	0.0986	5.96
	Fiber	$C_p = 1.8459 + 1.8306 \times 10^{-3}T - 4.6509 \times 10^{-6}T^2$	0.0293	1.66
	Ash	$C_p = 1.0926 + 1.8896 \times 10^{-3}T - 3.6817 \times 10^{-6}T^2$	0.0296	2.47
		Temperature functions[a]	Standard error	Standard % error
b. Water and ice as a function of temperature				
Water		$k_w = 5.7109 \times 10^{-1} + 1.7625 \times 10^{-3}T - 6.7036 \times 10^{-6}T^2$	0.0028	0.45
		$\alpha_w = 1.3168 \times 10^{-1} + 6.2477 \times 10^{-4}T - 2.4022 \times 10^{-6}T^2$	0.002×10^{-6}	1.44
		$\rho_w = 9.9718 \times 10_2 + 3.1439 \times 10^{-3}T - 3.7574 \times 10^{-3}T^2$	2.1044	0.22
		$C_{pw1}\cdot = 4.0817 - 5.3062 \times 10^{-3}T + 9.9516 \times 10^{-4}T^2$	0.0988	2.15
		$C_{pw2}\cdot = 4.1762 - 9.0864 \times 10^{-5}T + 5.4731 \times 10^{-6}T^2$	0.0159	0.38
Ice		$k_l = 2.2196 - 6.2489 \times 10^{-3}T + 1.0154 \times 10^{-4}T^2$	0.0079	0.79
		$\alpha_l = 1.1756 - 6.0833 \times 10^{-3}T + 9.5037 \times 10^{-5}T^2$	0.0044×10^{-6}	0.33
		$\rho_l = 9.1689 \times 10^2 - 1.3071 \times 10^{-1}T$	0.5382	0.06
		$C_{pl} = 2.0623 + 6.0769 \times 10^{-3}T$	0.0014	0.07

[a] C_{pw1} = for the temperature range −40 to 0°C.
C_{pw2} = for the temperature range 0 to 150°C.

Source: Choi, Y. and Oleos, M. R. 1986. Effects of temperature and Composition on the thermal properties of foods. In *Food Engineering and Process Applications*, Vol. 1, *Transport Phenomenon*, L. Maguer and P. Jelen (eds.). Elsevier, New York. pp. 93–101.

For high-moisture foods, above the freezing point, Siebel (1982) developed the following equation:

$$c_p = 0.837 + 3.349 X_w \tag{5.7}$$

A similar equation was suggested by Dickerson (1969) for high-moisture foods.

Charm (1971) suggested the following model:

$$c_p = 2.093X_F + 1.256X_s + 4.187X_w \tag{5.8}$$

Using a similar approach, Choi and Okos (1986) have suggested the following model for the specific heat for liquid foods:

$$c_p = \Sigma c_{pi} X_i^w \tag{5.9}$$

where the specific heat of pure components is given in Table 5.2.

A review of specific heat values for porous foods is given by Wallapapan and Sweat (1986).

5.2.4 THERMAL DIFFUSIVITY

Thermal diffusivity, α, can be expressed in terms of thermal conductivity, density, and specific heat as:

$$\alpha = \frac{k}{\rho c_p} \tag{5.10}$$

If the values of properties on the right-hand side of the equation are known, thermal diffusivity can be calculated. Most researchers use this procedure to determine thermal diffusivity. There are a few direct methods for experimental determination, such as the use of a cylinderical object and time-temperature data, the use of a spherical object and time-temperature data, and the use of a thermal conductivity probe (Choi and Okos, 1986).

Thermal diffusivity is strongly influenced by the water content as shown by the following models developed by Dickerson (1969) and Martens (1980), respectively:

$$\alpha = 0.088 \times 10^{-6} + (\alpha_w - 0.088 \times 10^{-6})X_w \tag{5.11}$$

$$\alpha = 0.057363X_w + 0.00028(T + 273) \times 10^{-6} \tag{5.12}$$

Based on the composition of liquid foods, Choi and Okos (1986) suggest the following model:

$$\alpha = \Sigma \alpha_i X_i^v \tag{5.13}$$

where the values of thermal diffusivity for pure components are given in Table 5.2.

Information on thermal properties of porous foods is presented in a review paper by Wallapapan et al. (1983). Thermal diffusivity values of selected foods are given in Table 5.3.

Example A new engineered food has the following composition: water 79.4%, protein 2%, fat 0.1 %, carbohydrate 17.6%, and ash 0.9%. Estimate thermal conductivity, density, specific heat, and thermal diffusivity at 20°C.

Solution The thermal properties will be determined using the models suggested by Choi and Okos (1986). The models shown in Table 5.2 may be programmed in a spreadsheet to yield the following results:

Thermal conductivity: $k = 0.54$ W/m·°C
Density: $\rho = 1082$ kg/m^3
Specific heat: $c_p = 3.65$ kJ/kg·°C
Thermal diffusivity: $\alpha = 0.137 \times 10^{-6}$ m^2/sec

TABLE 5.3
Thermal Diffusivity Some Foodstuffs

Product	Water content (wt %)	Temperature[a] (°C)	Thermal diffusivity ($\times 10^{-7}$ m^2/sec)
Fruits, vegetables, and by-products			
Apple, whole, Red Delicious	85	0–30	1.37
Applesauce	37	5	1.05
	37	65	1.12
	80	5	1.22
	80	65	1.40
	—	26–129	1.67
Avocado, flesh	—	24,0	1.24
Seed	—	24,0	1.29
Whole	—	41,0	1.54
Banana, flesh	76	5	1.18
	76	65	1.42
Beans, baked	—	4–122	1.68
Cherries, tart, flesh	—	30,0	1.32
Grapefruit, Marsh, flesh	88.8	—	1.27
Grapefruit, Marsh, albedo	72.2	—	1.09
Lemon, whole	—	40,0	1.07
Lima bean, pureed	—	26–122	1.80
Pea, pureed	—	26–128	1.82
Peach, whole	—	27,4	1.39
Potato, flesh	—	25	1.70
Potato, mashed, cooked	78	5	1.23
	78	65	1.45
Rutabaga	—	48,0	1.34
Squash, whole	—	47,0	1.71
Strawberry, flesh	92	5	1.27
Sugarbeet	—	14,60	1.26
Sweet potato, whole	—	35	1.06
	—	55	1.39
	—	70	1.91
Tomato, pulp	—	4.26	1.48
Fish and meat products			
Codfish	81	5	1.22
	81	65	1.42
Corned beef	65	5	1.32
	65	65	1.18
Beef, chuck[b]	66	40–65	1.23
Beef, round	71	40–65	1.33
Beef, tongue	68	40–65	1.32
Halibut	76	40–65	1.47
Ham, smoked	64	5	1.18
Ham, smoked	64	40–65	1.38
Water	—	30	1.48
	—	65	1.60
Ice	—	0	11.82

[a]Where two temperatures, separated by a comma, are given, the first is the initial temperature of the sample, and the second is that of the surroundings.

[b]Data are applicable only where juices that exuded during heating remain in the food samples.

Source: Singh, R.P. 1982. Thermal diffusivity in food processing. *Food Technology.*, 36: 87–91.

5.3 STEADY-STATE HEATING AND COOLING OF FOODS

5.3.1 CONDUCTION HEAT TRANSFER

The conduction mode of heat transfer involves energy transfer from regions of higher to lower temperatures. This energy transfer occurs mainly by the contact of matter at a given location with adjacent matter. There is no physical movement of the mass from one location to another. The rate of heat transfer due to conduction was described by Fourier using the following equation, also called Fourier's law:

$$\frac{q}{A} = -k\frac{dT}{dx} \tag{5.14}$$

From Equation 5.14 it is evident that the rate of heat transfer per unit area is proportional to the temperature gradient along the x-axis. The negative sign indicates that heat flow occurs from a hotter region to a colder region. The thermal conductivity is a unique property of the material. Thermal conductivity of food materials may exhibit a strong dependence on temperature and location. A further description of this property is given in Section 5.2.1.

Fourier's law may be solved for a rectangular, cylindrical, or spherical coordinate system, depending on the geometrical shape of the object being studied. Some commonly used solutions of Fourier's law are presented below.

1. Conduction heat transfer in a flat plate (Figure 5.1a):

$$q_x = -kA\frac{\Delta T}{\Delta x} = -kA\frac{T_2 - T_1}{x_2 - x_1} \tag{5.15}$$

2. Conductive heat transfer in a hollow pipe (Figure 5.1b):

$$q = \frac{2\pi Lk(T_2 - T_1)}{\ln(r_2/r_1)} \tag{5.16}$$

3. Conductive heat transfer in a three-layered wall where thermal conductivities of the three layers are k_A, k_B, and k_C; and thickness of each layer is $x_2 - x_1$, $x_3 - x_2$, and $x_4 - x_3$, respectively (Figure 5.1c):

$$q = \frac{A(T_4 - T_1)}{(x_2 - x_1)/k_A + (x_3 - x_2)/k_B + (x_4 - x_3)/k_C} \tag{5.17}$$

4. Conductive heat transfer in a three-layered composite cylindrical tube (Figure 5.1d):

$$q_r = \frac{2\pi L(T_4 - T_1)}{\ln(r_2/r_1)/k_A + \ln(r_3/r_2)/k_B + \ln(r_4/r_3)/k_C} \tag{5.18}$$

5. Conductive heat transfer in a hollow sphere:

$$q_r = -4\pi kr_1 r_2\frac{T_2 - T_1}{r_2 - r_1} \tag{5.19}$$

6. Steady-state heat transfer with heat generation:
   Certain foods, especially fruits and vegetables, and cereal grains respire during storage. The respiration process results in evolution of heat. The heat of respiration is a function

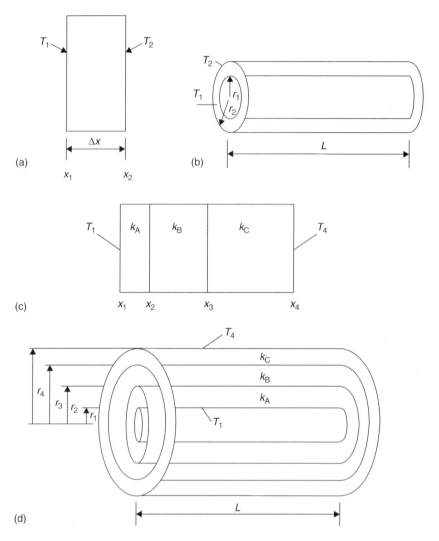

FIGURE 5.1   Schematic of (a) a plane wall, (b) a cylinder, (c) a composite wall, and (d) a composite cylinder.

of temperature, increasing with higher temperatures. Heat of respiration values for a wide variety of commodities from 0 to 20°C are shown in Table 5.4. Products such as strawberries, snap beans, mushrooms, and parsley evolve considerably more heat during storage than commodities such as apples, onions, and turnips.

For steady state conditions, the following equations may be used to calculate surface temperature, T_s, of objects with internal heat generation, q, if the surrounding temperature is T_∞, and h is the surface heat transfer coefficient incorporating convection and radiation. For a large plane wall of thickness $2L$

$$T_s = T_\infty + \frac{qL}{h} \tag{5.20}$$

For a long solid cylinder of radius r_1:

$$T_s = T_\infty + \frac{qr_1}{2h} \tag{5.21}$$

TABLE 5.4
Heat of Respiration of Selected Fruits and Vegetables

Commodity	Heat of respiration, m W/kg				
	0°C	5°C	10°C	15°C	20°C
Apples	6.8–12.1	15–21.3	—	40.3–91.7	50–103.8
Apricots	15.5–17	18.9–26.7	33–55.8	63–101.8	87.3–155.2
Asparagus	81–237.6	162–404.5	318.1–904	472.3–971.4	809.4–1484
Avocados	*	*	—	183.3–465.6	218.7–1029.1
Bananas - green	*	*	—	59.7–130.9	87.3–155.2
Bananas - ripening	*	*	—	37.3–164.9	97–242.5
Beans, Snap	*	101.4–103.8	162–172.6	252.2–276.4	350.6–386
Berries, Blueberries	6.8–31	27.2–36.4	—	101.4–183.3	153.7–259
Berries, Cranberries	*	12.1–13.6	—	—	32.5–53.8
Berries, Strawberries	36.4–52.4	48.5–98.4	145.5–281.3	210.5–273.5	303.1–581.0
Broccoli, Sprouting	55.3–63.5	102.3–474.8	—	515–1008.2	824.9–1011.1
Cabbage, White, Winter	14.5–24.2	21.8–41.2	36.4–53.3	58.2–80	106.7–121.2
Carrots, Roots, Imperator, Texas	45.6	58.2	93.1	117.4	209
Cauliflower, Texas	52.9	60.6	100.4	136.8	238.1
Cherries, Sweet	12.1–16	28.1–41.7	—	74.2–133.4	83.4–94.6
Grapes, Vinifera, Thompson Seedless	5.8	14.1	22.8	—	—
Grapefruit, Calif. Marsh	*	*	*	34.9	52.4
Kiwi Fruit	8.3	19.6	38.9	—	51.9–57.3
Lemons, Eureka, Calif	*	*	*	47	67.4
Lettuce, Head, Calif	27.2–50	39.8–59.2	81–118.8	114.4–121.2	178
Lettuce, Romaine	—	61.6	105.2	131.4	203.2
Mangoes	*	*	—	133.4	222.6–449.1
Melons, Cantaloupes	*	25.7–29.6	46.1	99.9–114.4	132.4–191.6
Melons, Honeydew	—	*	23.8	34.9–47	59.2–70.8
Melons, Watermelon	*	*	22.3	—	51.4–74.2
Mushrooms	83.4–129.2	210.5	—	—	782.2–938.9
Onions, Dry, White Bermuda	8.7	10.2	21.3	33	50.0
Oranges, W. Navel Calif	*	18.9	40.3	67.4	81
Papayas	*	*	33.5	44.6–64.5	—
Parsley	98–136.5	195.9–252.3	388.8–486.7	427.4–661.9	581.7–756.8
Peaches, Elberta	11.2	19.4	46.6	101.8	181.9
Pears, Bartlett	9.2–20.4	15–29.6	—	44.6–178	89.2–207.6
Peppers, Sweet	*	*	42.7	67.9	130
Pineapple, Ripening	*	*	22.3	53.8	118.3
Plums, Wickson	5.8–8.7	11.6–26.7	26.7–33.9	35.4–36.9	53.3–77.1
Potatoes, White	*	17.5–20.4	19.7–29.6	19.7–34.9	19.7–47

(Continued)

TABLE 5.4
Continued

Commodity	Heat of respiration, m W/kg				
	0°C	**5°C**	**10°C**	**15°C**	**20°C**
Rose Mature					
Radishes, with Tops	43.2–51.4	56.7–62.1	91.7–109.1	207.6–230.8	368.1–404.5
Radishes, topped	16–17.5	22.8–24.2	44.6–97	82.4–97	141.6–145.5
Spinach, Texas	—	136.3	328.3	530.5	682.3
Tomatoes, Texas, Mature Green	*	*	*	60.6	102.8
Tomatoes, Texas, Ripening	*	*	*	79.1	120.3
Tomatoes, Calif., Mature Green	*	*	*	—	71.3–103.8
Turnip, Roots	25.7	28.1–29.6	—	63.5–71.3	71.3–74.2

*denotes chilling injury temperature

Source: Adapted from American Society of Heating, Refrigerating and Air-Conditioning Engineers. With permission of the American Society of Heating, Refrigerating, and Air-conditioning Engineers, Atlanta, Georgia, (1985)

For a sphere of radius r_1:

$$T_s = T_\infty + \frac{qr_1}{3h} \tag{5.22}$$

5.3.2 Convection Heat Transfer

Convection heat transfer is the major mode of heat transfer between the surface of a solid material and the surrounding fluid. The rate of convective heat transfer depends on the properties of the fluid and the fluid flow characteristics. Originally suggested by Prandtl, the resistance to heat transfer may be considered to be localized in a boundary layer within the fluid present at the surface of the solid material. Although this concept is for ideal situations, it has been widely used in studying convective heat transfer.

Using the boundary- layer concept, the rate of convective heat transfer may be written as:

$$q = \frac{k}{\delta}A(T_s - T_\infty) \tag{5.23}$$

Since δ, the thickness of the boundary layer, cannot be measured, the quantity k/δ is expressed by h, the convective heat transfer coefficient. Then

$$q = hA(T_s - T_\infty) \tag{5.24}$$

The convective heat transfer coefficient, h, has been measured by numerous researchers for a variety of different conditions. This coefficient is dependent on fluid properties, such as k, ρ, c_p, and μ, velocity of flow, and the geometrical shape of the object undergoing heating or cooling. Using dimensional analysis, correlations have been developed to determine the convective heat transfer coefficient. Some of these relationships that are important to food processing are presented in the following section.

There are two modes of convective heat transfer depending on the fluid flow characteristics. The first mode of heat-transfer is called forced convection. The fluid flow is artificially induced, such as blowing air with a fan or a blower or pumping liquid on a heating (or cooling) surface. On the other hand, if the fluid flow is due primarily to changes in fluid density that are caused by differences in temperature, heat transfer occurs by free (or natural) convection.

5.3.2.1 Forced Convection in Newtonian Fluids

The convective heat transfer coefficient, under forced convection conditions, has been measured experimentally for a variety of different conditions. The experimental results are typically presented by correlations developed using dimensional analysis. Some of these correlations are presented in the following.

Laminar flow inside circular pipe (horizontal or vertical):

$$N_{Nu} = 1.86(N_{Gz})^{1/3} \left(\frac{\mu}{\mu_w} \right)^{0.14} \quad \text{for } N_{Re} < 2100 \tag{5.25}$$

Transitional flow inside circular pipe:

$$N_{Nu} = 0.116[(N_{Re})^{0.667} - 125](N_{Pr})^{1/3} \left[1 + \left(\frac{d}{L} \right)^{0.667} \right] \left(\frac{\mu}{\mu_w} \right)^{0.14}, \quad \text{for } 2100 < N_{Re} < 10,000 \tag{5.26}$$

Turbulent flow inside a circular pipe:

$$N_{Nu} = 0.023(N_{Re})^{0.8} (N_{Pr})^{0.667} \left(\frac{\mu_b}{\mu_w} \right)^{0.14} \tag{5.27}$$

Flow of liquid normal to a single cylinder:

$$N_{Nu} = [0.35 + 0.56(N_{Re})^{0.52}](N_{Pr})^{0.3} \quad \text{for } N_{Re} = 0.1\text{--}300 \tag{5.28}$$

Flow of gases past a sphere:

$$N_{Nu} = 2 + 0.6(N_{Re})^{0.5} (N_{Pr})^{0.33} \quad \text{for } N_{Re} < 325 \tag{5.29}$$

and

$$N_{Nu} = 0.4(N_{Re})^{0.6} (N_{Pr})^{0.33} \quad \text{for } 325 < N_{Re} < 70,000 \tag{5.30}$$

Flow of gases past a sphere:

$$N_{Nu} = [0.97 + 0.68N_{Re}^{0.52}](N_{Pr})^{0.3} \tag{5.31}$$

5.3.2.2 Free Convection in Newtonian Fluids

In the free-convection mode of heat transfer, the temperature of the fluid affects its density, which causes buoyant forces to develop. The following functional relationship is used to determine the convective heat transfer coefficient:

$$N_{Nu} = f(N_{Gr}, N_{Pr}) \tag{5.32}$$

TABLE 5.5
Constants a and b for Equation 5.33

Configuration	$N_{Gr}N_{Pr}$	a	b
Vertical plates and cylinders			
Length > 1 m			
Laminar	$< 10^4$	1.36	1/5
Laminar	$10^4 > N_{Gr}N_{Pr} < 10^9$	0.55	1/4
Turbulent	$> 10^9$	0.13	1/3
Spheres and horizontal cylinders			
Diameter < 0.2 m			
Laminar	$10^3 > N_{Gr}N_{Pr} < 10^9$	0.53	1/4
Turbulent	$> 10^9$	0.13	1/3
Horizontal plates			
Heated plate facing up (or cooled plate facing down)			
Laminar	$10^5 < N_{Gr}N_{Pr} < 2 \times 10^7$	0.54	1/4
Turbulent	$2 \times 10^7 < N_{Gr}N_{Pr} < 3 \times 10^{10}$	0.14	1/3
Heated plate facing down (or cooled plate facing up)			
Laminar	$3 \times 10^5 < N_{Gr}N_{Pr} < 3 \times 10^{10}$	0.27	1/4

Experimentally obtained results have been expressed using the following equation:

$$N_{Nu} = a(N_{Gr}N_{Pr})^b \qquad (5.33)$$

where a and b are evaluated from Table 5.5 for respective conditions.

5.3.2.3 Convective Heat Transfer in Non-Newtonian Fluids

Piston Flow: For a Graetz number larger than 500, the following expression presented by Metzner et al. (1959) is useful:

$$N_{Nu} = \frac{8}{\pi} + \frac{4}{\pi}(N_{GZ})^{0.5} \qquad (5.34)$$

Fully Developed Velocity Profile: For a power-law fluid, the following equation may be used:

$$N_{Nu} = 1.75 \left(\frac{3n+1}{4n} \right)^{1/3} (N_{Gz})^{1/3} \qquad (5.35)$$

Another expression that is useful for a fully developed velocity profile was proposed by Charm and Merrill (1959):

$$N_{Nu} = 2(N_{Gz})^{1/3} \left[\frac{m_b}{m_s} \frac{3n+1}{2(3n-1)} \right]^{0.14} \qquad (5.36)$$

5.3.3 Radiation Heat Transfer

The study of heat transfer by radiation includes three important properties of food materials: emissivity, ε; absorptivity, α; and transmittance, τ.

The energy emitted from a surface can be described using the Stefan–Boltzmann law:

$$q = \sigma A \varepsilon T_A^4 \tag{5.37}$$

where $\sigma = 5.67 \times 10^{-8}$ W/m$^2 \cdot$ K^4 and T_A = absolute temperature (K).

Equation 5.37 can be used determine radiative energy exchange between a surface A of a body and the surroundings at temperature T_2 that envelope the body. An example is baking bread inside an oven.

$$q_{1-2} = \sigma A_1 (\varepsilon_1 T_{A1}^4 - \phi_{1-2} T_{A2}^4) \tag{5.38}$$

According to Kirchhoff's law, the emissivity of a body is equal to its absorptivity for the same wavelength.

An example of radiative heat exchange that is commonly encountered in food processing is radiation between two parallel gray surfaces:

$$q_{12} = \frac{\sigma A_1 (T_1^4 - T_2^4)}{1/\,\epsilon_1 + 1/\,\epsilon_2 - 1} \tag{5.39}$$

where A_1 is the smaller area. In heat transfer calculations, it is desirable to determine whether radiative heat exchange is appreciable or small enough to be considered negligible. Similar to convective heat transfer efficient, a radiative heat transfer coefficient may be expressed as

$$q = h_r A (T_1 - T_2) \tag{5.40}$$

The radiative heat transfer coefficient can be estimated using Figure 5.2.

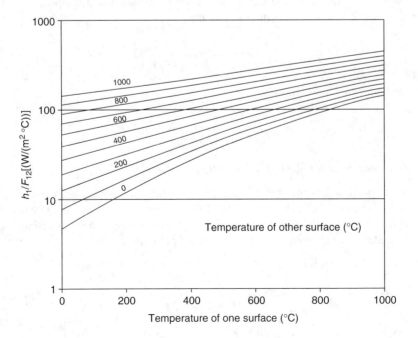

FIGURE 5.2 Radiative heat transfer coefficient as a function of temperature.

5.4   UNSTEADY-STATE HEATING AND COOLING OF FOODS

Often, it is vital to know the change in temperature with time during the unsteady-state period of heating or cooling of foods. For example, in processes involving food sterilization, the temperature history during heating and cooling periods must be known so that the lethal effect of the thermal process on microbial population can be accurately determined.

The governing equation describing unsteady-state heat transfer is:

$$\frac{\partial T}{\partial t} = \alpha \left(\frac{\partial^2 T}{\partial x^2} + \frac{\partial^2 T}{\partial y^2} + \frac{\partial^2 T}{\partial z^2} \right) \tag{5.41}$$

Before considering the solution of the governing equation, it is desirable to determine the relative importance of internal versus external resistance to heat transfer. For this purpose, a dimensionless number, called the Biot number, N_{Bi}, is useful:

$$N_{Bi} = \frac{hD}{k} = \frac{D/k}{1/h} = \frac{\text{internal resistance}}{\text{external resistance}} \tag{5.42}$$

A high Biot number (greater than 40) implies that external resistance to heat transfer is large (e.g., steam condensation on the surface of the object). A low Biot number (smaller than 0.2) means that the product has very small internal resistance to heat transfer, the thermal conductivity of the object is high.

Between a Biot number of 0.2 and 40 there is a finite resistance to heat transfer both internally and at the surface of the object undergoing heating or cooling.

Mathematical expressions useful in estimating temperature histories for the three cases above are presented in the following.

5.4.1   NEGLIGIBLE INTERNAL RESISTANCE TO HEAT TRANSFER

The dimensionless temperature ratio, also called the unaccomplished temperature ratio, can be expressed with the following exponential equation:

$$\frac{T - T_m}{T_0 - T_m} = \exp \left(\frac{-hA_s}{\rho c_p V} \right) t \tag{5.43}$$

In a completely dimensionless form,

$$\frac{T - T_m}{T_0 - T_m} = \exp(-N_{Bi} N_{Fo}) \tag{5.44}$$

5.4.2   NEGLIGIBLE SURFACE RESISTANCE TO HEAT TRANSFER

The governing Equation 5.41 may be solved analytically for some regular shaped objects such as an infinite plate, infinite cylinder and a sphere. The following set of equations allows calculation of the variable temperature, T, with time at any location within the object, when the initial temperature is uniform, T_0, and the surface temperature, T_s is constant (Crank, 1975).

For a plane wall where thickness is much smaller than length and height,

$$\frac{T - T_s}{T_0 - T_s} = 1 - \frac{4}{\pi} \sum_{n=0}^{\infty} \frac{(-1)''}{2n+1} \exp \left(-\frac{(2n+1)^2}{4} \pi^2 N_{Fo} \right) \cos \frac{(2n+1)\pi x}{2L} \tag{5.45}$$

where the wall's thickness is $2L$, and x is the variable distance from the center axis.

For an infinite cylinder

$$\frac{T - T_s}{T_0 - T_s} = 1 - \frac{2}{a} \sum_{n=1}^{\infty} \frac{\exp(-\alpha \mu_n^2 t) J_0(r \mu_n)}{\mu_n J_1(a \mu_n)} \tag{5.46}$$

where a is the radius of the cylinder, r is the variable distance from the axis, α is the thermal diffusivity, t is the time, and J_0 and J_1 are Bessel functions of the first kind of zero and first order, respectively. The discrete values of μ_n are the roots of the transcendental equation.

$$J_0(a \mu_n) = 0 \tag{5.47}$$

For a sphere

$$\frac{T - T_s}{T_0 - T_s} = 1 + \frac{2a}{\pi r} \sum_{n=1}^{\infty} \frac{(-1)^n}{2n + n} \sin \frac{n \pi r}{a} \exp(-n^2 \pi^2 N_{F0}) \tag{5.48}$$

where a is the radius of sphere, r is variable distance from the center. Temperature at the center axis is given by the limit at $r \to 0$

$$\frac{T - T_s}{T_0 - T_s} = 1 + 2 \sum_{n=1}^{\infty} (-1)^n \exp(-n^2 \pi^2 N_{F0}) \tag{5.49}$$

5.4.3 FINITE SURFACE AND INTERNAL RESISTANCE TO HEAT TRANSFER

The preceding equations describe temperature distribution in an object when the surface temperature is constant. However, many cases involve convection at the boundary between the solid object and the surrounding fluid. The following expressions are useful to calculate the temperature, T, anywhere in the object with a uniform initial temperature T_0, when immersed in a fluid of temperature, T_m, and a convection boundary condition.

Plane wall with convection heat transfer at the surface (where thickness is much smaller than length and height):

$$\frac{T - T_m}{T_0 - T_m} = \sum_{n=1}^{\infty} \frac{4 \sin \mu_n}{2 \mu_n + \sin(2 \mu_n)} \exp(-\mu_n^2 N_{F0}) \cos(\mu_n x^*) \tag{5.50}$$

where $N_{F0} = \frac{\alpha t}{L^2}$ thickness of plane wall= $2L, x^* = x/L$, at the centerline, $x^*=0$, and the eigenvalues are positive roots of the transcendental equation

$$\mu_n \tan \mu_n = N_{Bi} \tag{5.51}$$

Infinite cylinder with convection at the surface:

$$\frac{T - T_m}{T_0 - T_m} = \sum_{n=1}^{\infty} \frac{2 J_1(\mu_n)}{\mu_n (J_0^2(\mu_n) + J_1^2(\mu_n))} \exp(-\mu_n^2 N_{F0}) J_0(\mu_n r^*) \tag{5.52}$$

where $r^* = \frac{r}{R}$ and the discrete values of μ_n are the positive roots of the transcendental equation:

$$\mu_n \frac{J_1(\mu_n)}{J_0(\mu_n)} = N_{\text{Bi}} \tag{5.53}$$

where J_0 and J_1 are Bessel functions of first kind and order zero and one, respectively.

Sphere with convection at the surface:

$$\frac{T - T_m}{T_0 - T_m} = \sum_{n=1}^{\infty} \frac{4[\sin(\mu_n) - \mu_n \cos(\mu_n)]}{2\mu_n - \sin(2\mu_n)} \exp(-\mu_n^2 N_{F0}) \frac{1}{\mu_n r^*}(\mu_n r^*) \tag{5.54}$$

where $r^* = \frac{r}{R}$, and the discrete values of μ_n are the positive roots of the transcendental equation

$$1 - \mu_n \cot \mu_n = N_{\text{Bi}} \tag{5.55}$$

Roots of the transcendental equation are tabulated in mathematical handbooks.

5.4.4 Use of Charts to Estimate Temperature History During Unsteady-State Heating or Cooling

The use of analytical solutions to determine temperature history is cumbersome because of the need to evaluate numerous terms in the series. These solutions have been reduced to charts that are much easier to use. These charts are presented in Figure 5.3 through Figure 5.5 for infinite slab, infinite cylinder, and a sphere, respectively. The use of these charts in solving for unsteady-state heat transfer is illustrated in the following example.

Example   Apples are being cooled to 8°C from an initial temperature of 20°C using water at 5°C. The water flow over the surface of the apple creates a convective heat transfer coefficient of 10 W/m²·K. Assume that the apple can be described by a sphere with an 8-cm diameter and the geometric center is to be reduced to 8°C. The properties of apple include thermal conductivity of 0.4 W/m·K, specific heat of 3.8 kJ/kg·K, and density of 960 kg/m³. Determine how long the apples must be exposed to water.

Solution   Calculate the temperature ratio:

$$\frac{T_a - T_f}{T_a - T_i} = \frac{5 - 8}{5 - 20} = 0.2$$

Calculate the Biot number:

$$N_{\text{Bi}} = \frac{hr_0}{k} = \frac{(10)(4 \times 10^{-2})}{0.4} = 1.0$$

$$k/hr_0 = 1.0.$$

From Figure 5.5

$$N_{\text{Fo}} = \frac{\alpha t}{r^2} = 0.78$$

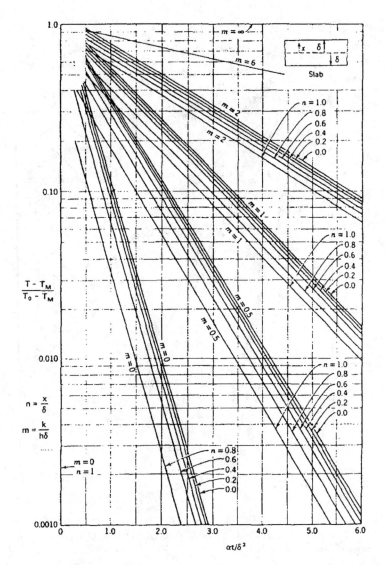

FIGURE 5.3 Unsteady-state temperature distributions in an infinite slab. (From Foust, A.S., Wenzel, L.A., Clump, C.W., Mans, L., and Anderson, L.B., 1960. *Principles of Unit Operations*. John Wiley and Sons, NY. With permission.)

Then

$$t = \frac{0.78 r^2}{\alpha} = \frac{(0.78)(0.04)^2}{(0.4)/(960)(3800)} = 11,382 \text{ sec or } 3.16 \text{ h}$$

5.5 HEAT EXCHANGERS

5.5.1 TUBULAR HEAT EXCHANGER

The most common type of heat exchanger used in the food industry is a tubular double-pipe heat exchanger. Typically, liquid food that needs to be heated or cooled is pumped into the inner pipe and

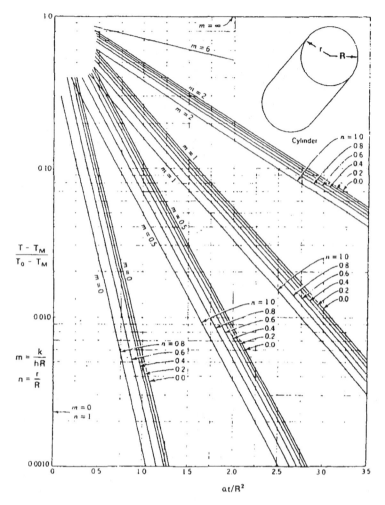

FIGURE 5.4 Unsteady-state temperature distributions in an infinite cylinder. (From Foust, A.S., Wenzel, L.A., Clump, C.W., Mans, L, and Anderson, L.B., 1960. *Principles of Unit Operations.* John Wiley and Sons, NY. With permission.)

the heating or cooling medium is pumped into the annular space formed by the concentric pipes. The fluid flow inside the heat exchanger can be either concurrent or countercurrent.

Typical temperature profiles inside a heat exchanger are shown in Figure 5.6. The rate of heat transfer between the two fluids is:

$$q = UA(\Delta T)_{lm} \qquad (5.56)$$

where

$$(\Delta T)_{lm} = \frac{(\Delta T)_2 - (\Delta T)_1}{\ln(\Delta T_2/\Delta T_1)} \qquad (5.57)$$

5.5.2 TRIPLE-TUBE HEAT EXCHANGERS

Triple-tube heat exchangers are used in the food industry for both heating and cooling applications. Typically, heating or cooling medium flows in the inner tube and outermost annular space,

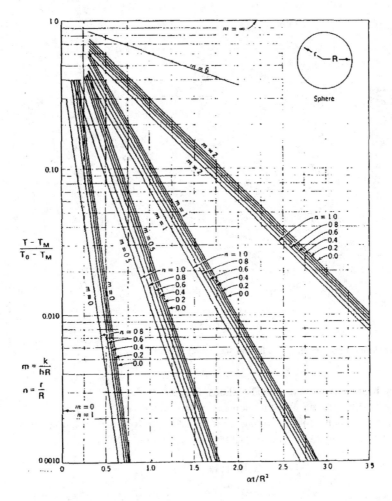

FIGURE 5.5 Unsteady-state temperature distributions in a sphere. (From Foust, A.S., Wenzel, L.A., Clump, C.W., Mans, L, and Anderson, L.B., 1960. *Principles of Unit Operations*. John Wiley and Sons, NY. With permission.)

whereas product flows in the middle annular space. Expressions for estimating convective heat transfer coefficient for both inner and outer heat transfer surfaces were suggested by Jacob (1949).

$$N_{Nu} = \left(\frac{D_2}{D_1}\right)^{0.8} (N_{GZ})^{0.45} N_{Gr}^{0.05} \tag{5.58}$$

5.5.3 PLATE HEAT EXCHANGERS

Plate heat exchangers are commonly used in heating and cooling applications in the dairy and food beverage industry. A schematic diagram of a plate heat exchanger is shown in Figure 5.7. The heat exchanger consists of closely spaced metal plates parallel to each other and held securely in a metal frame. The plates are often corrugated to induce turbulence in the flowing liquid. The ports and edges of the plate are sealed with gaskets to prevent intermixing of the liquid streams. Since the food product flows in a thin film over the heat transfer area, the retention time is small, thus reducing thermal damage to the product. The fluids can be pumped in concurrent or countercurrent flow with

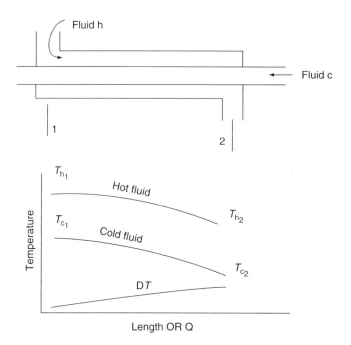

FIGURE 5.6   Temperature profile in a tubular heat exchanger.

respect to the heat transfer medium. The plates can easily be removed for cleaning or changing the surface area; thus the desired heat flux can easily be obtained.

For liquid foods with non-Newtonian characteristics the following relationship was proposed by Skelland (1967) to estimate the convective heat transfer coefficient:

$$N_{Nu} = \frac{6}{\zeta} \frac{(n+1)}{(2n+1)}$$

$$\zeta = \frac{5}{4} - \frac{2n}{2n+1} + \frac{3n}{4n+1} - \frac{n}{5n+1}$$

(5.59)

This relationship is valid for constant physical properties and when there exists a cubic polynomial temperature distribution in the fluid. For Newtonian liquids, the results of Nunge et al. (1967), may be used to calculate the rate of heat transfer.

5.5.4 SCRAPED-SURFACE HEAT EXCHANGERS

Scraped-surface heat exchangers (SSHEs) involve the use of a rotor equipped with scraper blades that rotate inside a cylinder as shown in Figure 5.8. The food material is pumped inside while the heat transfer medium is circulated between the cylinder and a jacket. The blades, pushed against the surface due to centrifugal force, continuously scrape the inside of the cylinder wall. This scraping action assists in processing foods over a wide temperature range. In addition, highly viscous foods, as well as liquids containing discrete particles, can be processed. The rate of heat transfer between the heat transfer medium and the food product is controlled by two mechanisms: (1) the rate of heat conduction into the thin product layer at the heating surface and (2) the rate at which the thin product layer is removed and then mixed with the bulk liquid. The mixing step occurs as radial mixing, which enhances heat transfer. However, axial mixing with a certain amount of back mixing may also occur, which leads to a reduction in the driving force (the temperature difference between the heat transfer medium and the product) and thus to a decrease in the heat transfer coefficient.

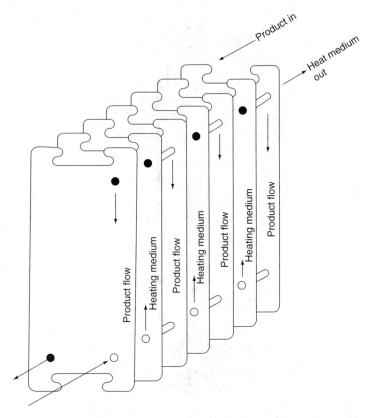

FIGURE 5.7 Schematic illustration of plate heat exchanger. (From Heldman, D.R. and Singh, R.P., 1981. *Food Process Engineering*. AVI Publishers, Westport, CT. With permission.)

In small-sized scraped-surface heat exchangers, there can be considerable back mixing, leading to poorer heat transfer than in larger units. Harrod (1987) has reviewed the flow mixing mechanisms and heat transfer in SSHEs.

Several investigators have used the penetration theory with surface renewal for mass transport (Higbie, 1935) to model heat transfer in SSHEs (Kool, 1958; Harlot, 1959; Latinen, 1959). The following dimensionless equation is suggested for determining the convective heat transfer coefficient:

$$N_{Nu} = 2_p^{-0.5} (N_{Re} N_{pr} n)^{0.5} \qquad (5.60)$$

Trommelin el al. (1971) suggest multiplying the value of N_{Nu} obtained from Equation 5.59 by a factor of less than 1 to accommodate for reduced heat transfer due to axial mixing and lack of complete equalization of temperature in the liquid film. Results from experimental studies on heat transfer in SSHEs used for heating, cooling, and freezing applications have been summarized by Harrod (1987). The experimentally obtained values of N_{Nu} are found to be 20 to 400% of the values predicted by Equation 5.60.

5.6 HEAT TRANSFER IN AGITATED VESSELS

Liquid foods are often heated or cooled in agitated vessels. These vessels are equipped with an agitator and contain either a coil or an enveloping jacket. Some common types of agitators, jackets, and coils are shown in Figure 5.9.

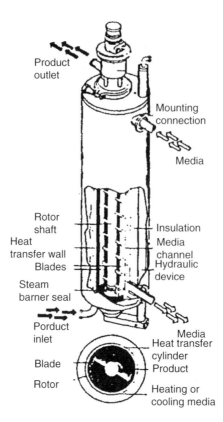

Product outlet

Mounting connection

Media

Rotor shaft
Heat transfer wall
Blades
Steam barner seal

Insulation
Media channel
Hydraulic device

Porduct inlet

Media
Heat transfer cylinder
Product

Blade

Rotor

Heating or cooling media

FIGURE 5.8   Schematic of a scraped-surface heat exchanger (Alpha Laval).

5.6.1 CONTINUOUS OPERATION

In a continuous operation, the following equation is used to determine the rate of heat transfer if the contents are at a constant temperature:

$$q = UA\Delta T \tag{5.61}$$

If the inlet and outlet temperature of the jacket medium vary, a log mean temperature difference is used and the equation becomes:

$$q = U\Delta T_{lm} \tag{5.62}$$

The overall heat transfer coefficient for a jacketed vessel is obtained from the following equation:

$$\frac{1}{U} = \frac{1}{h_i} + f_i + \frac{x}{k} + f_j + \frac{1}{h_j} \tag{5.63}$$

In the case of a heating coil, the overall heat transfer coefficient is obtained from the following equation:

$$\frac{1}{U_0} = \frac{1}{h_i} + f_i + \frac{x}{k}\frac{r_{co}}{r_{cm}} + \frac{1}{h_{ci}}\frac{r_{co}}{r_{ci}} + f_{ci} \tag{5.64}$$

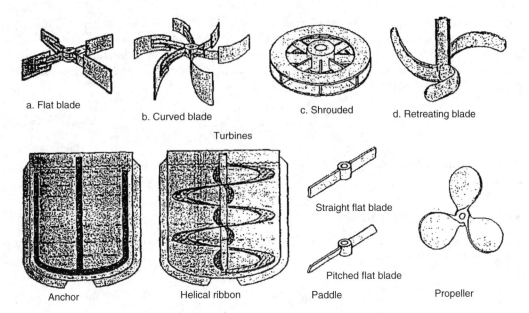

a. Flat blade

b. Curved blade

c. Shrouded

d. Retreating blade

Turbines

Straight flat blade

Pitched flat blade

Anchor

Helical ribbon

Paddle

Propeller

FIGURE 5.9 Different types of agitators used in heating and mixing applications. (From Bondy, R. and Lippa, S., 1983. *Chem. Eng.*, April 4: 62–71. With permission.)

5.6.2 BATCH OPERATIONS

Batch heating in an agitated vessel can be described by the following equation:

$$\ln\frac{T - T_1}{T - T_2} = \frac{UA}{\rho V c_{\mathrm{p}}}t \tag{5.65}$$

where T is the constant temperature of a coil or jacket medium, ρ the density of food material, V the volume of food material, T_1 the initial temperature, and, T_2 the final temperature after γ hours. If the jacket temperature is not constant during heating, then

$$\ln\frac{T - T_2}{T - T_2} = \frac{m_{\mathrm{j}}c_{\mathrm{j}}e^{UA/m_t c_t} - 1}{\rho V c_{\mathrm{p}}e^{UA/m_j c_j}}t \tag{5.66}$$

where m_{j} is the mass flow rate of fluid through jacket or coil, c_{f} the specific heat of fluid, and c_{p} the specific heat of vessel's contents.

5.6.3 CONVECTIVE HEAT TRANSFER COEFFICIENTS IN JACKETED VESSELS

Convective heat transfer coefficient may be calculated using dimensional correlations for different designs of agitating blades. Unless otherwise stated, most of the relationships given in this section are for Newtonian liquids. More details may be found in Bondy and Lippa (1983).

5.6.3.1 Flat-Blade Turbine

For six-bladed turbines, when the ratio of liquid depth to vessel diameter is 1 and ratio of impeller diameter to tank diameter is 1/3, the following relationships are available:

$$N_{\mathrm{Nu}} = C(N_{\mathrm{Re}})^{0.67}(N_{\mathrm{Pr}})^{0.33}\left(\frac{\mu}{\mu_{\mathrm{w}}}\right)^{0.14} \tag{5.67}$$

where $C = 0.54$ for $N_{Re} < 400$ and $C = 0.74$ for $N_{Re} > 400$. If viscosity is a strong function of temperature, a viscosity correction factor must be used in the relationship above. The following equation is useful to calculate the wall temperature using a trial-and-error solution. It is assumed that there is an equal heat flow through the jacket-side and vessel-side films and that the temperature drop across the metal of the vessel wall is negligible:

$$t_w = T - \frac{T - t}{1 + h_j A_0 / h_i A_i} \tag{5.68}$$

where A_0 is the jacketed area based on outside vessel radius and A_i is the area based on inside diameter.

5.6.3.2 Retreating-Blade Turbine

With a six-retreating-blade turbine in a jacketed and unbaffled vessel, the following relationship was developed by Ackley (1960):

$$N_{Nu} = 0.68(N_{Re})^{0.67}(N_{Pr})^{0.33}\left(\frac{\mu}{\mu_w}\right)^{0.14} \tag{5.69}$$

For baffled vessels, a similar relationship is available in Ackley (1960).

5.6.3.3 Helical Ribbon

The convective heat transfer coefficient is calculated using the following expressions:

$$N_{Nu} = 0.248(N_{Re})^{0.5}(N_{Pr})^{0.33}\left(\frac{\mu}{\mu_w}\right)^{0.14}\left(\frac{e}{D}\right)^{-0.22}\left(\frac{i}{D}\right)^{-0.28} \tag{5.70}$$

where $N_{Re} < 130, e$ is the clearance between vessel and impeller, and i is the agitator ribbon pitch, and

$$N_{Nu} = 0.238(N_{Re})^{0.67}(N_{Pr})^{0.33}\left(\frac{\mu}{\mu_w}\right)^{0.14}\left(\frac{i}{D}\right)^{-0.25} \tag{5.71}$$

where $N_{Re} > 130$ and i is the agitator ribbon pitch.

5.6.3.4 Propeller

For a 45° pitched four-bladed impeller, the convective heat transfer coefficient is calculated using the expression

$$N_{Nu} = 0.54(N_{Re})^{0.67}(N_{Pr})^{0.25}\left(\frac{\mu}{\mu_w}\right)^{0.14} \tag{5.72}$$

5.6.3.5 Paddle

For both baffled and unbaffled jacketed vessels,

$$N_{Nu} = 0.36(N_{Re})^{0.67}(N_{Pr})^{0.33}\left(\frac{\mu}{\mu_w}\right)^{0.14} \quad \text{for } N_{Re} > 4000 \tag{5.73}$$

and

$$N_{Nu} = 0.415(N_{Re})^{0.67}(N_{Pr})^{0.33} \left(\frac{\mu}{\mu_w}\right)^{0.24} \quad \text{for } 20 < N_{Re} < 4000 \tag{5.74}$$

5.6.3.6 Anchor

The following relationship is useful for anchor-type mixers:

$$N_{Nu} = 1.0(N_{Re})^{0.67}(N_{Pr})^{0.33} \left(\frac{\mu}{\mu_w}\right)^{0.18} \tag{5.75}$$

at $30 < N_{Re} < 300$ and anchor to wall clearance < 2.54 cm;

$$N_{Nu} = 0.38(N_{Re})^{0.67}(N_{Pr})^{0.33} \left(\frac{\mu}{\mu_w}\right)^{0.18} \tag{5.76}$$

at $300 < N_{Re} < 4000$ and anchor to wall clearance < 2.54 cm; and

$$N_{Nu} = 0.55(N_{Re})^{0.67}(N_{Pr})^{0.33} \left(\frac{\mu}{\mu_w}\right)^{0.14} \tag{5.77}$$

at $4000 < N_{Re} < 37,000$ and $2.54 <$ anchor to wall clearance > 14 cm.

5.6.3.7 Internal Coils (Turbine)

When internal helical coils are used in heating or cooling applications, and a six-bladed turbine is used for mixing, then

$$N_{Nu} = 0.17(N_{Re})^{0.67}(N_{Pr})^{0.37} \left(\frac{D}{D_T}\right)^{0.1} \left(\frac{d_\alpha}{D_T}\right)^{0.5} \left(\frac{\mu}{\mu_w}\right)^{\upsilon} \tag{5.78}$$

where η is a correction factor (Bondy and Lippa, 1983).

5.6.3.8 Internal Coils (Retreating Blades)

For an impeller with six retreating blades,

$$N_{Nu} = 1.4(N_{Re})^{0.62}(N_{Pr})^{0.33} \left(\frac{\mu}{\mu_w}\right)^{0.14} \tag{5.79}$$

5.6.3.9 Internal Coils (Propeller)

When a propeller is used for mixing and internal coils are used for heating or cooling, the Nusselt number is given by

$$N_{Nu} = 0.078(N_{Re})^{0.62}(N_{Pr})^{0.33} \left(\frac{\mu}{\mu_w}\right)^{0.14} \tag{5.80}$$

It is suggested by Bondy and Lippa (1983) that the expression (5.80) is based on limited data, and the calculated convective heat transfer coefficient should be divided by 1.3.

5.6.3.10 Paddle (Internal Coil)

When mixing is done with a paddle, and heating or cooling with internal coil,

$$N_{Nu} = 0.87(N_{Re})^{0.62}(N_{Pr})^{0.33}\left(\frac{\mu}{\mu_w}\right)^{0.14} \tag{5.81}$$

The fouling factors can contribute considerable heat resistance in food applications. A review by Lund and Sandu (1981) is recommended for further information on this topic.

NOMENCLATURE

A	Area
a	Constant in Equation 5.33
b	Constant in Equation 5.33
c_f	Specific heat of fluid
c_p	Specific heat capacity
d	Diameter of a pipe
D	Diameter
f_i	Fouling factor, inside vessel
f_j	Fouling factor, inside jacket
f_{ci}	Fouling factor on coil side referred to inside coil area
h	Convective heat transfer coefficient
h_{ci}	Convective heat transfer coefficient on coil side referred to inside coil area
h_r	Radiative heat transfer coefficient
J_1	Bessel function of first order
J_0	Bessel function of zero order
k	Thermal conductivity
L	Length
m_b	Consistency coefficient
m_f	Mass flow rate through jacket or coil
n	Flow behavior index
N_{Bi}	Biot number, dimensionless
N_{Fo}	Fourier number, dimensionless
N_{Gr}	Grashof number, dimensionless
N_{Gz}	Graetz number, dimensionless
N_{Nu}	Nusselt number, dimensionless
N_{pr}	Prandtl number, dimensionless
N_{Re}	Reynolds number, dimensionless
q	Rate of heat transfer
r	Radius
r_{ci}	Inner radius of coil
r_{co}	Outer radius of coil
r_{cm}	Log mean radius of coil
t	Time
T	Temperature
T_A	Temperature, absolute
T_m	Temperature, medium
T_o	Temperature, initial
T_s	Temperature, surface

T_w	Temperature, wall
T_∞	Temperature, surrounding
U	Overall heat transfer coefficient
V	Volume
x_1	Location in x direction
X_F	Mass fraction of fat
X_s	Mass fraction of solids
X_w	Mass fraction of water
X_i^w	Mass fraction of ith component
X_i^v	Volume fraction
α	Thermal diffusivity
δ	Boundary layer thickness
ε	Emissivity
Φ_{1-2}	View factor
μ	Viscosity
μ_b	Bulk viscosity
μ_w	Viscosity at wall
ρ	Density
σ	Stefan-Boltzmann constant

REFERENCES

Ackley, E.J. 1960. *Chem. Eng.*, Aug. 22, p. 133.

American Society of Heating, Refrigerating and Air Conditioning Engineers, In (1985). 1985 Fundamentals. ASHRAE, Atlanta, Georgia.

Bondy, F. and Lippa, S. 1983. Heat transfer in agitated vessels. *Chem. Eng.*, April 4: 62–71.

Chapman, A.J. 1974. *Heat Transfer*. Macmillan, New York.

Charm, S.E. 1971. *The Fundamentals of Food Engineering*, 2nd ed. AVI, Westport, Connecticut.

Charm, S.E. and Merrill, E.W. 1959. Heat transfer coefficients in straight tubes for pseudoplastic fluids in streamline flow. *Food Res.*, 24: 319.

Choi, Y. and Okos, M.R. 1983. The thermal properties of tomato juice concentrates. *Trans. ASAE*, 26: 305–311.

Choi, Y. and Okos, M.R. 1986. Effects of temperature and composition on the thermal properties of foods. In *Food Engineering and Process Applications*, Vol. 1, *Transport Phenomenon*, L. Maguer and P. Jelen (eds.). Elsevier, New York. pp. 93–101.

Crank, J. 1975. *The Mathematics of Differsion*, Clarendon Press, Oxford.

Dickerson, R.W. 1969. Thermal properties of foods. In *The Freezing Preservation of Foods*, Vol. 2, 4th ed., D.K. Tressler, W.B. Van Arsdel, and M.L. Copley (Eds.). AVI, Westport, Conn.

Foust, A.S., Wenzel, L.A., Clump, C.W., Mans, L., and Anderson, L.B., 1960. *Principles of Unit Operations*. John Wiley and Sons, New York.

Harriott, P. 1959. Heat transfer in scraped-surface exchangers. *Chem. Eng. Progr. Symp. Ser.*, 54: 137–139.

Harrod, M. 1987. Scraped surface heat exchanger: a literature survey of flow patterns, mixing effects, residence time distribution, heat transfer, and power requirements. *J. Food Proc. Eng.*, 9: 1–62.

Heldman, D.R. and Singh, R.P. 1981. *Food Process Engineering*. AVI Publishers, Westport, Connecticut, CT.

Higbie, R. 1935. The rate of absorption of a pure gas into a still liquid during short periods of exposure. *Trans. AlChE*, 31: 365–389.

Holman, J.P. 1997. *Heat Transfer*. McGraw-Hill, New York.

Jacob, M. 1949. *Heat Transfer*, Vol. 1. Wiley. New York.

Kool, J. 1958. Heat transfer in scraped vessels and pipes handling viscous materials. *Trans. Inst. Chem. Engrs.*, 36: 253–258.

Kopelman, I.J. 1966. Transient heat transfer and thermal properties in food systems. Ph.D. *Thesis*. Michigan State University, East Lansing, Michigan.

Kreith, F. and Black, W.Z. 1980. *Basic Heat Transfer*. Harper & Row, New York.

Latinen, G.A. 1959. Discussion of the paper Correlation of scraped film heat transfer in the Votator (A.H. Skelland), *Chem Eng. Sci.* 9: 263–266.

Lund, D.B. and Sandu, C. 1981. Slate-of-the-art of fouling: heat transfer surfaces. In *Fundamentals and Applications of Surface Phenomenon Associated with Fouling and Cleaning in Food Processing*, B. Hallstrom. D.B. Lund, and Ch. Tragardh (eds.). Proceedings, Div. of Food *Engineering, Lund University, Lund, Sweden.*

Martens, T. 1980. Mathematical model of heat processing in flat containers, Ph.D. thesis, Katholeike University, Leuven, Belgium.

Metzner. A.B., Vaugh, R.D., and Houghton, G.L. 1959. Turbulent flow of non-Newtonian systems. *AIChE J.*, 5: 189.

Nunge, R.J., Porta, E.W., and Gill, W.N. 1967. Axial conduction in the fluid stream of multistream heat exchanger. *Chem. Eng. Prog. Symp. Ser.* 77.66: 80–91.

Phipps, L.W. 1969. *The interrelationship of the viscosity, fat content and temperature of cream* between 40°C and 80°C. *J. Dairy Res.*, 36: 417–426.

Reidy, G.A. 1968. Thermal properties of foods and methods of their determination. M.S. thesis, Food Science Dept., Michigan State University.

Reidy, G.A. and Rippen, A.L. 1971. Methods for determining thermal conductivities of foods. *Trans. ASAE.*, 14: 248.

Riedel, L. 1949. Measurements of the thermal conductivity of sugar solutions, fruit juices and milk. *Chem. Ing. Tech.*, 21: 340 (in German).

Short, A.L. 1955. The temperature coefficient or expansion of raw milk. *J. Dairy Res.*, 22: 769.

Siebel, J.E. 1982. Specific heat of various products. *Ice Refrig.*, 2: 256–257.

Singh, R.P. 1982. Thermal diffusivity in food processing. *Food Technol.* 36: 87–91.

Skelland, A.H.P. 1967. *Non-Newtonian Flow and Heat Transfer*. John Wiley, New York.

Sweat, V.E. 1974. Experimental values of thermal conductivity of selected fruits and vegetables. J. *Food Sci.*, 39: 1080.

Trommelin, A.M., Beek, W.J., and Van De Westelaken, H.C. 1971. A mechanism for heat transfer in a Votator-type scraped surface heat exchanger. *Chem Eng Sci.*, 26: 1987–2001.

Wallapapan, K., Sweat, V.E., Diehl, K.C., and Engler, C.R. 1983. Thermal properties of porous foods. ASAE Paper 83-6575, American Society of Agricultural Engineering, St. Joseph, Michigan.

6 Food Freezing

Dennis R. Heldman

CONTENTS

6.1 INTRODUCTION

Freezing represents a preservation process for food where the product temperature is decreased to a temperature range resulting in the formation of ice crystals within the product structure. The purpose of the process is to reduce the temperature of the product as much as is economically feasible in an effort to reduce product quality deterioration reaction rates within the product. These steps result in an extension in the storage life of a perishable food. Most documented evidence of freezing as a preservation process has appeared within the last 150 years, and the process has become an integral part of food handling and distribution in most developed countries.

The engineering aspects of the freezing process are numerous. They deal with the computation of refrigeration requirements needed to accomplish the desired reductions in product temperature; these requirements involve removal of thermal energy in the form of both sensible and latent heats.

In addition, the design of processes for food freezing requires knowledge of the time needed to reduce product temperature to desired levels. Finally, the efficient design of frozen food storage requires knowledge of changes occurring within the food product during the time it is exposed to the environmental conditions within the storage systems. The design of these systems requires insight into the physical changes occurring within the product structure during freezing, the influence of these changes on product properties, the incorporation of these properties into computation of freezing times, and the use of these computed times in design of the freezing and product storage systems.

The objectives of this chapter are as follows:

1. To present the thermodynamics of food freezing in a manner that will illustrate physical changes occurring within the product during freezing and the relationship of these changes to frozen food properties
2. To present procedures as well as tabulated data required for the determination of frozen food properties needed in the calculation of freezing times and refrigeration requirements
3. To describe and discuss methods for computation of freezing times along with an analysis of the limitations of these procedures and to present guidelines for the selection of procedures to obtain the most acceptable results for different types of food products
4. To describe typical freezing systems and give an explanation of operating characteristics and typical applications of the systems
5. To present design calculations for refrigeration requirements to be used in selection of the capacity of food freezing and frozen food systems
6. To present and discuss food freezing processes and frozen food storage factors that influence frozen food quality

6.2 THERMODYNAMICS OF FOOD FREEZING

Thermodynamics can be used to describe the physical changes in water within a food product during the freezing process. As illustrated in Figure 6.1, the decrease in product temperature during

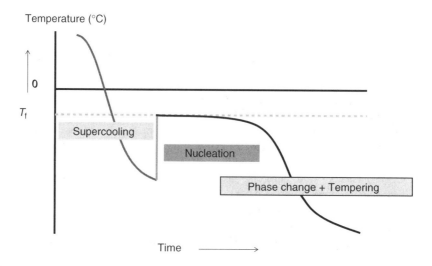

FIGURE 6.1 Changes in product temperature during freezing process.

freezing occurs gradually as the latent heat of fusion is removed from water within the product. The food freezing process has two unique characteristics compared to the freezing of pure water. First, the equilibrium temperature for initial formation of ice crystals is lower than the equilibrium temperature for ice crystal formation in pure water. Although supercooling may occur in the product before the initial ice crystal is formed, the temperature will be below that of a pure water system. The magnitude of the depression in equilibrium freezing temperature is a function of product composition. The second difference between freezing of the food product, as compared to pure water, occurs after the initial ice crystals are formed. In the food product, the removal of phase change energy occurs gradually over a range of decreasing product temperatures. The temperature–time relationship during phase changer is a function of the percent water frozen at any time during the freezing process.

The shape of the temperature–time curve during the freezing process will vary with product composition and with the location within the product structure. The gradual decrease in temperature with time will continue until reaching the eutectic temperature for a major product component. In practice, food products are not frozen to sufficiently low temperatures to reach these eutectic temperatures.

6.2.1 Freezing Temperature Depression

As previously indicated, the temperatures at which ice crystals are formed within a food product structure are depressed below that of a pure water system. In addition, it is evident that the magnitude of the temperature depression is a function of product composition. The relationship between product composition and temperature has been interpreted by Heldman (1974) and Schwartzberg (1976) in terms of the freezing temperature depression equation for an ideal solution, as follows:

$$\frac{\lambda}{R_g}\left[\frac{1}{T_{A_0}} - \frac{1}{T_A}\right] = \ln X_A \tag{6.1}$$

Equation 6.1 illustrates the relationship between the mole fraction (X_A) of water within the product and the equilibrium freezing temperature (T_A) and between the molar latent heat of fusion (λ) and the universal gas constant (R_g). The mole fraction of water within the product can be defined as follows:

$$X_A = \frac{m_A/m_A}{m_A/m_A + m_{si}/m_{si}} \tag{6.2}$$

In Equation 6.2 the mole fraction of water in the product is a function of product moisture content, expressed as a fraction (m_A), the molecular weight of water (M_A), the percentages of product components, expressed as a mass fraction (m_{si}), and the molecular weight of each product component (M_{si}). The applications of Equation 6.1 and Equation 6.2 may be direct for some food products, depending on the knowledge of the food product components and the influence of these components on the equilibrium freezing temperature. Since low molecular weight components of the product have a more dramatic influence on the magnitude of the mole fraction, knowledge of the mass fractions and molecular weights of these components are the most important.

Example The composition of orange juice is 88.3% water, 0.7% protein, 10.4% carbohydrate, 0.2% fat, and 0.4% ash (Table 6.1). Estimate the depression of the equilibrium freezing temperature of the product, based on composition.

TABLE 6.1
Composition of Foods

Product	Water g/100g	Protein g/100g	Carbohydrates			Lipid g/100g	Ash									
			Sugar g/100g	Starch g/100g	Fiber g/100g		g/100g	Ca mg/100g	Fe mg/100g	Mg mg/100g	P mg/100g	K mg/100g	Na mg/100g	Zn mg/100g	Cu mg/100g	Mn mg/100g
Beef (hamburger)	69.02	20	0	0	0	10	0.98	12	2.2	20	184	321	66	4.8	0.072	0.01
Beef (lean chuck)	74	21.45	0	0	0	3.56	0.99	18	1.86	23	204	339	77	5.68	0.09	0.01
Fish, Cod	81.22	17.81	0	0	0	0.67	1.16	16	0.38	32	203	413	54	0.45	0.03	0.02
Fish, Perch	79.13	19.39	0	0	0	0.92	1.24	80	0.9	30	200	269	62	1.11	0.15	0.7
Asparagus	93.22	2.2	1.88	3.88	2.1	0.12	0.58	24	2.14	14	52	202	2	0.54	0.19	0.16
Carrots, raw	88.29	0.93	4.54	9.58	2.8	0.24	0.97	33	0.3	12	35	320	69	0.24	0.05	0.14
Cucumbers	95.23	0.65	1.67	3.63	0.5	0.11	0.38	16	0.28	13	24	147	2	0.2	0.04	0.08
Onions, raw	88.54	0.92	4.28	10.11	1.4	0.08	0.35	22	0.19	10	27	144	3	0.16	0.04	0.13
Tomato	93	1.2	4	5.1	1.1	0.2	0.5	13	0.51	10	28	204	13	0.07	0.09	0.1
Peas, raw green	78.86	5.42	5.67	14.45	5.1	0.4	0.87	25	1.47	33	108	244	5	1.24	0.176	0.41

Spinach, raw	91.4	2.86	0.42	3.63 / 1.01	2.2	0.39	99	2.71	79	49	1.72 / 558	79	0.53	0.13	0.897
Blueberries, raw	84.21	0.74	9.96	14.49 / 2.13	2.4	0.33	6	0.28	6	12	0.24 / 77	1	0.16	0.057	0.336
Peaches, raw	88.87	0.91	8.39	9.54 / 0	1.5	0.25	6	0.25	9	20	0.43 / 190	0	0.17	0.07	0.06
Pears, raw	83.71	0.38	9.8	15.46 / 0	3.1	0.12	9	0.17	7	11	0.33 / 119	1	0.1	0.08	0.05
Plums, raw	87.23	0.7	9.92	11.42 / 0.1	1.4	0.28	6	0.17	7	16	0.37 / 157	0	0.1	0.057	0.052
Raspberries, raw	85.75	1.2	4.42	11.94 / 1.02	6.5	0.65	25	0.69	22	29	0.46 / 151	1	0.42	0.09	0.67
Strawberries, raw	90.95	0.67	4.66	7.68 / 1.02	2	0.3	16	0.42	13	24	0.4 / 153	1	0.14	0.048	0.386
Cherries, sweet	77.61	0.73	18.58	21.07 / 0	2.5	0.21	10	0.35	9	20	0.39 / 148	3	0.1	0.176	0
Egg, whites	87.57	10.9	0.71	0.73 / 0.02	0	0.17	25	0.69	22	29	0.63 / 151	1	0.42	0.09	0.67
Bread, white	36.7	8.2	4.31	49.5 / 42.89	2.3	3.6	108	3.03	24	94	1.9 / 119	27	0.62	0.126	0.383
Orange juice	88.3	0.7	8.4	10.4 / 0	0.2	0.2	11	0.2	11	17	0.4 / 200	1	0.05	0.04	0.01

Solution

1. The composition and molecular weight of product components are:

	Mass fraction	Molecular weight
Water	0.8830	18.02
Protein	0.0070	50,000.0
Carbohydrate		
Sugar(mono-sac.)	0.0570	180.2
Sugar(di-sac.)	0.0450	342.3
Fiber	0.0020	50,000.0
Lipid	0.0020	50,000.0
Ash	0.0040	37.75

The molecular weight of ash is a mass average of the nine minerals with highest composition in the orange juice, as given in Table 6.1 [USDA (2004)]. The molecular weights for proteins, lipids, and fiber are estimates, and these product components do not influence result.

2. Using Equation 6.2, the mole fraction is computed as follows:

$$X_A = \left(\frac{0.833}{18.02}\right)\left(\frac{0.833}{18.02} + \frac{0.007}{50,000} + \frac{0.057}{180.02} + \frac{0.045}{342.3} + \frac{0.002}{50,000} + \frac{0.002}{50,000} + \frac{0.004}{37.75}\right)^{-1}$$

and:

$$X_A = 0.98882$$

then:

$$\frac{(333.5)(18.02)}{8.3144}\left[\frac{1}{273} + \frac{1}{T_A}\right] = \ln(0.98882)$$

3. From Equation 6.1, the initial freezing temperature is:

$$T_A = 271.8456 \text{ K}$$

4. The predicted initial freezing temperature is very close to the experimental value given in Table 6.2.

6.2.2 Unfrozen Water Fraction

One of the unique characteristics of a frozen food is the relationship between unfrozen water fraction and temperature. This relationship is basic to the design of freezing systems and frozen food storage facilities and represents a key element in the establishment of frozen food storage stability. The unfrozen water fraction in a food product decreases gradually as temperature drops below the initial freezing temperature. The relationship can be described by the changes in unfrozen water fraction predicted by the freezing point depression equation. The estimation procedure requires

TABLE 6.2
Initial Freezing Temperature of Fruits, Vegetables, and Juices

Product	Water content (wt %)	Initial freezing temperature (°C)
Apple juice	87.2	−1.44
Apple juice concentrate	49.8	−11.33
Applesauce	82.8	−1.67
Asparagus	92.6	−0.67
Bilberries	85.1	−1.11
Bilberry juice	89.5	−1.11
Carrots	87.5	−1.11
Cherry juice	86.7	−1.44
Grape juice	84.7	−1.78
Onions	85.5	−1.44
Orange juice	89.0	−1.17
Peaches	85.1	−1.56
Pears	83.8	−1.61
Plums	80.3	−2.28
Raspberries	82.7	−1.22
Raspberry juice	88.5	−1.22
Spinach	90.2	−0.56
Strawberries	89.3	−0.89
Strawberry juice	91.7	−0.89
Sweet cherries	77.0	−2.61
Tall peas	75.8	−1.83
Tomato pulp	92.9	−0.72

Source: From Heldman D.R. and Singh, R.P. 1981. *Food Process Engineering*, 2nd ed., AVI Pub Co., Westport, CT.

the assumption that pure ice crystals are formed during freezing and that all solute is concentrated in the unfrozen water fraction.

Example   Estimate the percent unfrozen water in frozen strawberries at −10°C.

Solution

1. The composition of raspberries (Table 6.1) and molecular weights for each component are:

	Mass fraction	Molecular weight
Water	0.9095	18.02
Protein	0.0067	50,000.00
Carbohydrates		
Sugar (monosac.)	0.0336	180.20
Sugar (dissac.)	0.0432	342.30
Lipids	0.0030	50,000.00
Ash	0.0040	37.42

2. Using Equation 6.1, with T_A at $-10°C$ or 263 K:

$$(333.5)(18.0)/8.31441\{(1/273) - (1/263)\} = -0.10067 = \text{Ln}(X_A)$$
$$\text{and} \quad X_A = 0.904;$$

the mole fraction of unfrozen water in the partially frozen strawberries.

3. Next, the mole fraction of unfrozen water in the frozen product, and Equation 6.2 are used to compute the mass fraction of unfrozen water.

$$0.904 = \frac{m_u/18.02}{m_u/18.02 + 0.0067/50,000 + 0.0336/180.2 + 0.0432/342.3 + 0.003/50,000 + 0.004/37.42}$$

The mass fraction is 0.06642, and the unfrozen water is 7.3% at $-10°C$. This value is slightly lower than the experimental value given in Table 6.3.

The entire relationship for unfrozen water fracture vs. temperature for frozen strawberries over the temperature range -40 to $+5°C$ is illustrated in Figure 6.2. The predicted relationship is the solid curve and is compared to experimental data from Riedel (1951). This predicted relationship becomes an adequate input for computations needed for the design of food freezing systems and frozen food storage systems and for estimating the impact on freezing and storage on frozen food quality. The primary input parameters to the prediction are product composition and molecular weights of components. In some situations, initial freezing temperature data as presented in Table 6.2 may be useful. Typical data for unfrozen water fraction in foods are presented in Table 6.3 (Riedel data presented by Dickerson (1981). These relationships were analyzed by Chen (1985a, 1985b).

6.3 PROPERTIES OF FROZEN FOODS

The food product properties of interest when considering the freezing process include density, specific heat, thermal conductivity, enthalpy, and latent heat. These properties must be considered in the estimation of the refrigeration capacity for the freezing system and the computation of freezing times needed to assure adequate residence times. The approach to prediction of property magnitudes during the freezing process depends directly on the relationship between unfrozen water fraction and temperature, as described in Section 6.2.

The research literature contains numerous references to experimental data for properties of unfrozen and frozen foods. These data have been reviewed by Dickerson (1969), Woodams and Nowry (1968), Reidy (1968), Qashou et al. (1972), Mohsenin (1980), Heldman and Singh (1983), Lind (1991), Miles (1991), Becker and Fricke (1999), and Heldman (2001). It is generally agreed that the accuracy of experimental data as a function of product composition or measurement techniques is more evident for partially frozen products.

6.3.1 PRODUCT DENSITY

The influence of freezing on food product density is relatively small but a dramatic change does occur at and just below the initial freezing temperature. This change can be predicted by the following equation, as discussed by Heldman (2001):

$$\rho = \frac{1}{\Sigma(m_{si}/\rho_{si})} \tag{6.3}$$

TABLE 6.3
Enthalpy of Frozen Foods[a]

Product	Water content (wt %)	Mean specific heat[b] 4 to 32°C kJ/(kg·C)		Temperature (°C)																		
				40	-30	-20	-18	-16	-14	-12	-10	-9	-8	-7	-6	-5	-4	-3	-2	-1	0
Fruits and vegetables																					
Applesauce	82.8	3.73	Enthalpy (kJ/kg)	0	23	51	58	65	73	84	95	102	110	120	132	152	175	210	286	339	343
			% water unfrozen[c]	—	6	9	10	12	14	17	19	21	23	27	30	37	44	57	82	100	—
Asparagus, Peeled	92.6	3.98	Enthalpy(kJ/kg)	0	19	40	45	50	55	61	69	73	77	83	90	99	108	123	155	243	181
			% water unfrozen	—	—	—	—	—	5	—	—	7	8	10	12	15	17	20	29	58	100
Bilberries	85.1	3.77	Enthalpy (kJ/kg)	0	21	45	50	57	64	73	82	87	94	101	110	125	140	167	218	348	352
			% water unfrozen	—	—	—	7	8	9	11	14	15	17	18	21	25	30	38	57	100	—
Carrots	87.5	3.90	Enthalpy (kJ/kg)	0	21	46	51	57	64	72	81	87	94	102	111	124	139	166	218	357	361
			% water unfrozen	—	—	—	7	8	9	11	14	15	17	18	20	24	29	37	53	100	—
Cucumbers	95.4	4.02	Enthalpy (kJ/kg)	0	18	39	43	47	51	57	64	67	70	74	79	85	93	104	125	184	390
			% water unfrozen	—	—	—	—	—	—	—	—	5	—	—	—	—	—	14	20	37	100
Onions	85.5	3.81	Enthalpy (kJ/kg)	0	23	50	55	62	71	81	91	97	105	115	125	141	163	196	263	349	353
			% water unfrozen	—	5	8	10	12	14	16	18	19	20	23	26	31	38	49	71	100	—
Peaches without stones	85.1	3.77	Enthalpy (kJ/kg)	0	23	50	57	64	72	82	93	100	108	118	129	146	170	202	274	348	352
			% water unfrozen	—	5	8	9	11	13	16	18	20	22	25	28	33	40	51	75	100	—
Pears, Bartlett	83.8	3.73	Enthalpy (kJ/kg)	0	23	51	57	64	73	83	95	101	109	120	132	150	173	207	282	343	347
			% water unfrozen	—	6	9	10	12	14	17	19	21	23	26	29	35	43	54	80	100	—
Plums without stones	80.3	3.65	Enthalpy (kJ/kg)	0	25	57	65	74	84	97	111	119	129	142	159	182	214	262	326	329	333
			% water unfrozen	—	8	14	16	18	20	23	27	29	33	37	42	50	61	78	100	—	—
Raspberries	82.7	3.73	Enthalpy (kJ/kg)	0	20	47	53	59	65	75	85	90	97	105	115	129	148	174	231	340	344
			% water unfrozen	—	—	7	8	9	10	13	16	17	18	20	23	27	33	42	61	100	—
Spinach	90.2	3.90	Enthalpy (kJ/kg)	0	19	40	44	49	54	60	66	70	74	79	86	94	103	117	145	224	371
			% water unfrozen	—	—	—	—	—	—	6	7	—	—	9	11	13	16	19	28	53	100
Strawberries	89.3	3.94	Enthalpy (kJ/kg)	0	20	44	49	54	60	67	76	81	88	95	102	114	127	150	191	318	367
			% water unfrozen	—	—	5	—	6	7	9	11	12	14	16	18	20	24	30	43	86	100
Sweet Cherries Without Stones	77.0	3.60	Enthalpy (kJ/kg)	0	26	58	66	76	87	100	114	123	133	149	166	190	225	276	317	320	324
			% water unfrozen	—	9	15	17	19	21	26	29	32	36	40	47	55	67	86	100	—	—

(Continued)

TABLE 6.3
Continued

Product	Water content (wt %)	Mean specific heat[b] 4 to 32°C kJ/(kg·C)		Temperature (°C)																	
				−40	−30	−20	−18	−16	−14	−12	−10	−9	−8	−7	−6	−5	−4	−3	−2	−1	0
Tall peas	75.8	3.56	Enthalpy (kJ/kg)	0	23	51	56	64	73	84	95	102	111	121	133	152	176	212	289	319	323
			% water unfrozen	—	6	10	12	14	16	18	21	23	26	28	33	39	48	61	90	100	—
Tomato pulp	92.9	4.02	Enthalpy (kJ/kg)	0	20	42	47	52	57	63	71	75	81	87	93	103	114	131	166	266	382
			% water unfrozen	—	—	—	—	5	—	6	7	8	10	12	14	16	18	24	33	65	100
Eggs																					
Eggs white	86.5	3.81	Enthalpy (kJ/kg)	0	18	39	43	48	53	58	65	68	72	75	81	87	96	109	134	210	352
			% water unfrozen	—	—	10	—	—	—	—	13	—	—	—	18	20	23	28	40	82	100
Egg yolk	40.0	2.85	Enthalpy (kJ/kg)	0	19	40	45	50	56	62	68	72	76	80	85	92	99	109	128	182	191
			% water unfrozen	20	—	—	22	—	24	—	27	28	29	31	33	35	38	45	58	94	100
Whole egg with shell[d]	66.4	3.31	Enthalpy (kJ/kg)	0	17	36	40	45	50	55	61	64	67	71	75	81	88	98	117	175	281
Fish and meat																					
Cod	80.3	3.69	Enthalpy (kJ/kg)	0	19	42	47	53	66	74	79	84	89	96	105	118	137	177	298	323	—
			% water unfrozen	10	10	11	12	12	13	14	16	17	18	19	21	23	27	34	48	92	100
Haddock	83.6	3.73	Enthalpy (kJ/kg)	0	19	42	47	53	59	66	73	77	82	88	95	104	116	136	177	307	337
			% water unfrozen	8	8	9	10	11	11	12	13	14	15	16	18	20	24	31	44	90	100
Perch	79.1	3.60	Enthalpy (kJ/kg)	0	19	41	46	52	58	65	72	76	81	86	93	101	112	129	165	284	318
			% water unfrozen	10	10	11	12	13	14	15	16	17	18	20	22	24	31	40	55	95	100
Beef, lean Fresh[e]	74.5	3.52	Enthalpy (kJ/kg)	0	19	42	47	52	58	65	72	76	81	88	95	105	113	138	180	285	304
			% water unfrozen	10	10	11	12	13	14	15	16	17	18	20	22	24	31	40	55	95	100
Beef, lean Dried	26.1	2.47	Enthalpy (kJ/kg)	0	19	42	47	53	62	66	70	—	74	—	79	—	84	—	89	—	93
			% water unfrozen	96	96	97	98	98	100	—	—	—	—	—	—	—	—	—	—	—	—
Bread																					
White bread	37.3	2.60	Enthalpy (kJ/kg)	0	17	35	39	44	49	56	67	75	83	93	104	117	124	128	131	134	137
Whole wheat bread	42.4	2.68	Enthalpy (kJ/kg)	0	17	36	41	48	56	66	78	86	95	106	119	135	150	154	157	160	163

a Above −40°C.

b Temperature range limited to 20°C for meats and 20 to 40°C for egg yolk.

c Total weight of unfrozen water = (total weight of food) (% water content/100)(water unfrozen/100).

d Calculated for a weight composition of 58% white (86.5% water) and 32% yolk (50% water).

e Data for chicken, veal, and venison very nearly matched the data for beef of the same water content.

Source: Dickerson, R. W. Jr. 1981. In Handbook and Product Directory Fundamentals, Society of Heating, Refrigerating, and Air-Conditioning Engineers, Atlanta, GA.

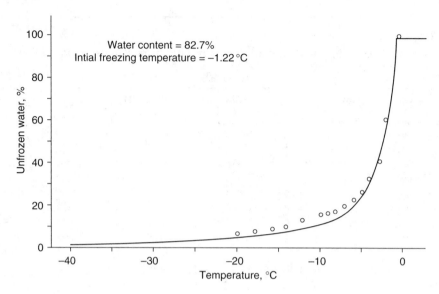

FIGURE 6.2 Relationship between unirozen water fraction and temperature of raspberries. (From Heldman, D.R. 1974. *Trans. ASAE* 17: 63–66. With permission.)

TABLE 6.4
Properties of Ice as a Function of Temperature

Temperature $t°Q$	Thermal conductivity (W/m-K)	Specific heat (kJ/kg-K)	Density (kg/m^3)
-101	3.50	1.382	925.8
-73	3.08	1.587	924.2
-45.5	2.72	1.783	922.6
-23	2.41	1.922	919.4
-18	2.37	1.955	919.4
-12	2.32	1.989	919.4
-7	2.27	2.022	917.8
	2.22	2.050	916.2

Source: Adapted from Dickerson, R. W. Jr. 1969. In *The Freezing Preservation of Foods*, 4th ed., Vol. 2, D.K. Tressler, W.B. Van Arsdel, and M.J. Copley (Eds), AVI Pub. Co., Westport, CT.

This expression allows for prediction of density changes during freezing from knowing the mass fraction of unfrozen water in product (m_u), mass fraction of individual product components (m_{si}), and mass fraction of frozen water in product (m_I). Density data for product components, such as liquid water (ρ_u) or ice (ρ_I) are available from standard tables. (See Table 6.4). Density data for food product components (protein, fat, carbohydrate, ash) were presented as a function of temperature by Choi and Okos (1986) and are presented in Table 6.5. Alternatively, the values can be predicted from the expressions in Table 6.6.

An example of the influence of freezing on density of strawberries is presented in Figure 6.3. Above the initial freezing temperature, the product density is relatively constant, At the initial freezing temperature, the product density decreases rapidly as the fraction of frozen water in the product increases. At temperatures below $-20°C$, the density of frozen strawberries becomes relatively

TABLE 6.5
Density of Pure Components Calculated from Literature Values of Liquid Foods

T (°C)	Density (kg/m³)					Data	Error	Standard no. of %
	Water	Protein	Fat	Carbohydrate	Ash			
20	997.6	1289.4	916.4	3424.6	1743.4	13	3.02270	0.28
30	995.2	1272.2	913.5	1413.3	1731.2	16	24.5498	2.29
40	991.2	1258.4	906.7	1399.2	1719.8	15	22.9114	2.14
50	986.8	1246.2	902.7	1386.4	1704.7	16	26.4704	2.48
60	983.3	1231.4	894.3	1369.5	1691.5	11	26.0143	2.45
70	978.2	1222.6	884.9	1358.2	1679.1	7	29.7849	2,82
80	971.5	1212.9	880.0	1346.4	1668.8	6	32.2467	3.04
90	965.0	1204.3	876.0	1337.2	1658.4	6	33.0122	3.15
100	958.0	1198.4	874.2	1331.7	1649.3	5-	32.0948	3.07

Source: From Choi, Y. and Okos, M.R. 1986. *Physical and Chemical Properties of Food*. Martin R. Okos (Ed.), ASAE, St. Joseph, MI. pp. 35–77.

constant. It should be noted that the overall change in product density between +5 and −40°C is less than 10%.

6.3.2 PRODUCT SPECIFIC HEAT

As illustrated by Heldman (2001), the specific heat capacity of a food product can be predicted, based on product composition and the specific heat capacity of individual product components. The following expression was proposed:

$$c_p = \sum (c_{psi} m_{si}) \tag{6.4}$$

where each factor on the right-side of the equation is the product of the mass fraction of a product component and the specific heat capacity of that component. The specific heat values for product components were estimated by Choi and Okos (1986) and are presented in Table 6.7 or from expressions in Table 6.6.

Equation 6.4 can be used to predict the specific heat capacity of product solids by removing the term for the water fraction. These specific heat magnitudes for the product solids can be used in the prediction of product enthalpy and apparent specific heat, as illustrated in sections to follow.

6.3.3 PRODUCT THERMAL CONDUCTIVITY

The thermal conductivity magnitudes of most food products are a function of water content and the physical structure of the product. Many models suggested for prediction of thermal conductivity are based on moisture content and do not consider structural orientation. The models proposed by Kopelman (1966) have been developed for three different types of structural orientation, as illustrated in Figure 6.4.

A model for the two-component homogeneous dispersion system (Figure 6.4a) is as follows:

$$k = k_L \left[\frac{1 - M_v^2}{1 - M_v^2(1 - M_v)} \right], \tag{6.5}$$

TABLE 6.6
Coefficients to Estimate Food Properties

Property	Component	Temperature function	Standard error	Standard % error
k (W/[m°C])	Protein	$k = 1.7881 \times 10^{-1} + 1.1958 \times 10^{-3}T - 2.7178 \times 10^{-6}T^2$	0.012	5.91
	Fat	$k = 1.8071 \times 10^{-1} - 2.7064 \times 10^{-3}T - 1.7749 \times 10^{-7}T^2$	0.0032	1.95
	Carbohydrate	$k = 2.0141 \times 10^{-1} + 1.3874 \times 10^{-3}T - 4.3312 \times 10^{-6}T^2$	0.0134	5.42
	Fiber	$k = 1.8331 \times 10^{-1} + 1.2497 \times 10^{-3}T - 3.1683 \times 10^{-6}T^2$	0.0127	5.55
	Ash	$k = 3.2962 \times 10^{-1} + 1.4011 \times 10^{-3}T - 2.9069 \times 10^{-6}T^2$	0.0083	2.15
	Water	$k = 5.7109 \times 10^{-1} + 1.7625 \times 10^{-3}T - 6.7063 \times 10^{-6}T^2$	0.0028	0.45
	Ice	$k = 2.2196 - 6.2459 \times 10^{-3}T + 1.0154 \times 10^{-4}T^2$	0.0079	0.79
α (m²/s)	Protein	$\alpha = 6.8714 \times 10^{-2} + 4.7578 \times 10^{-4}T - 1.4646 \times 10^{-6}T^2$	0.0038	4.50
	Fat	$\alpha = 9.8777 \times 10^{-2} - 1.2569 \times 10^{-4}T - 3.8286 \times 10^{-8}T^2$	0.0020	2.15
	Carbohydrate	$\alpha = 8.0842 \times 10^{-2} + 5.3052 \times 10^{-4}T - 2.3218 \times 10^{-6}T^2$	0.0058	5.84
	Fiber	$\alpha = 7.3976 \times 10^{-2} + 5.1902 \times 10^{-4}T - 2.2202 \times 10^{-6}T^2$	0.0026	3.14
	Ash	$\alpha = 1.2461 \times 10^{-1} + 3.7321 \times 10^{-4}T - 1.2244 \times 10^{-6}T^2$	0.0022	1.61
	Water	$\alpha = 1.3168 \times 10^{-1} + 6.2477 \times 10^{-4}T - 2.4022 \times 10^{-6}T^2$	0.0022×10^{-6}	1.44
	Ice	$\alpha = 1.1756 - 6.0833 \times 10^{-3}T + 9.5037 \times 10^{-5}T^2$	0.0044×10^{-6}	0.33
ρ (kg/m³)	Protein	$\rho = 1.3299 \times 10^3 - 5.1840 \times 10^{-1}T$	39.9501	3.07
	Fat	$\rho = 9.2559 \times 10^2 - 4.1757 \times 10^{-1}T$	4.2554	0.47
	Carbohydrate	$\rho = 1.5991 \times 10^3 - 3.1046 \times 10^{-1}T$	93.1249	5.98
	Fiber	$\rho = 1.3115 \times 10^3 - 3.6589 \times 10^{-1}T$	8.2687	0.64
	Ash	$\rho = 2.4238 \times 10^3 - 2.8063 \times 10^{-1}T$	2.2315	0.09
	Water	$\rho = 9.9718 \times 10^2 + 3.1439 \times 10^{-3}T - 3.7574 \times 10^{-3}T^2$	2.1044	0.22
	Ice	$\rho = 9.1689 \times 10^2 - 1.3071 \times 10^{-1}T$	0.5382	0.06
c_p (j/[kg°C])	Protein	$c_p = 2.0082 + 1.2089 \times 10^{-3}T - 1.3129 \times 10^{-6}T^2$	0.1147	5.57
	Fat	$c_p = 1.9842 + 1.4733 \times 10^{-3}T - 4.8006 \times 10^{-6}T^2$	0.0236	1.16
	Carbohydrate	$c_p = 1.5488 + 1.9625 \times 10^{-3}T - 5.9399 \times 10^{-6}T^2$	0.0986	5.96
	Fiber	$c_p = 1.8459 + 1.8306 \times 10^{-3}T - 4.6509 \times 10^{-6}T^2$	0.0293	1.66
	Ash	$c_p = 1.0926 + 1.8896 \times 10^{-3}T - 3.6817 \times 10^{-6}T^2$	0.0296	2.47
	Water[a]	$c_p = 4.0817 - 5.3062 \times 10^{-3}T + 9.9516 \times 10^{-4}T^2$	0.0988	2.15
	Water[b]	$c_p = 4.1762 - 9.0864 \times 10^{-5}T + 5.4731 \times 10^{-6}T^2$	0.0159	0.38
	Ice	$c_p = 2.0623 + 6.0769 \times 10^{-3}T$		

[a] For the temperature of −40 to 0°C.
[b] For the temperature of 0 to 150°C.

Source: From Choi, Y. and Okos, M.R. 1986. *Physical and Chemical Properties of Food.* Martin R. Okos (Ed.). ASAE, St. Joseph, MI, pp. 35–77.

where the volume fraction of the discontinuous phase within the product is M_v^3. In most food products, the discontinuous phase will be the product solids. Equation 6.5. is used when the thermal conductivity of the continuous phase or water is much larger than the thermal conductivity of the product solids. When the thermal conductivity of the continuous and discontinuous phases are similar, the following equation applies:

$$k = k_L \left[\frac{1 - Q}{1 - Q(1 - M_v)} \right] \tag{6.6}$$

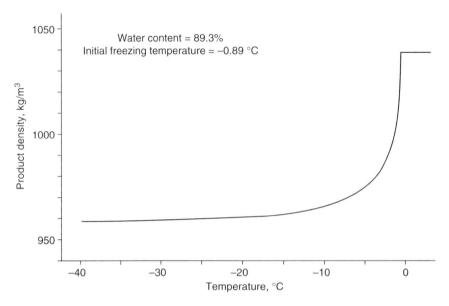

FIGURE 6.3 Influence of freezing on the predicted density of strawberries. (From Heldman, D.R. 1982. *Food Technol.* 36: 92–96. With permission.)

TABLE 6.7
Specific Heat of Pure Components Calculated from Literature Values of Liquid Foods

T (°C)	Specific heat (kJ/kg-°C)					No. of data	Standard error	Standard % error
	Water	Protein	Fat	Carbohydrate	Ash			
20	4.180	1.711	1.928	1.547	0.908	7	0.0614	1.59
30	4.172	1.765	1.953	1.586	0.937	7	0.0794	2.36
40	4.174	1.775	1.981	1.626	0.947	14	0.0681	1.81
50	4.176	1.842	2.004	1.639	0.976	13	0.0698	2.11
60	4.179	1.891	2.036	1.691	1,010	13	0.0702	2.07
70	4.185	1.914	2.062	1.734	1.025	14	0.0867	2.34
80	4.193	1.942	2.098	1.768	1.045	5	0.0943	2.85
90	4.199	1.967	2.124	1.787	1.057	5	0.0987	2.97
100	4.210	1.993	2.141	1.824	1.059	5	0.0972	2.98

Source: From Choi, Y. and Okos, M.R 1986. *Physical and chemical properties of Food.* Martin R. Okos (Ed.), ASAE, St. Joseph, MI. pp. 35–77.

where:

$$Q = M_v^2 \left[1 - \frac{k_s}{k_L} \right] \tag{6.7}$$

Equation 6.5 can be used for thermal conductivity prediction for many food systems, since the thermal conductivity of water is much larger than the thermal conductivity of food product solids. In addition, the structural orientation within these products does not influence the direction of transfer of thermal energy within the product. An exception would be low moisture food products, when

Food freezing

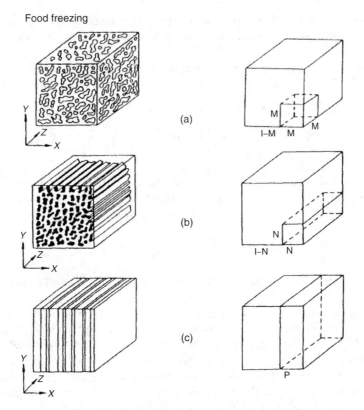

FIGURE 6.4 (a) Two-component homogeneous three-dimensional dispersion system, a, natural random state; b, rearrangements of the components, (b) Two-component homogeneous two-dimensional fibrous system, a, natural random state; b, rearrangements of the components, (c) Two-component homogeneous one-dimensional layered system, a, natural random state; b, rearrangement of the components. (From Kopelman, I.J. 1966. Transient heat transfer and thermal properties in food systems. Ph.D. Dissertation. Michigan State University. E. Lansing. With permission.)

water is not the continuous phase within the food system. For low moisture foods, Equation 6.6 and Equation 6.7 would be more appropriate prediction models.

If the food product structure includes fibrous or similar components that influence conduction of heat, the two-component anisotropic model in Figure 6.4b should be considered. For conduction of thermal energy parallel to the fibrous components, the following expression is proposed:

$$k_{11} = k_L \left[1 - N_v^2 \left(1 - \frac{k_s}{k_L} \right) \right] \tag{6.8}$$

where N_v^2 is the volume fraction of the discontinuous phase within the product. For prediction of thermal conductivity perpendicular to the fibrous components, the following equation is proposed:

$$k_\perp = k_L \left[\frac{1 - Q'}{1 - Q'(1 - N_v)} \right] \tag{6.9}$$

where

$$Q' = N_v \left(1 - \frac{k_s}{k_L} \right) \tag{6.10}$$

The expressions in the anisotropic model are used to differentiate between thermal conductivity magnitudes when heat transfer is parallel as compared to perpendicular to fibrous components in meats or similar foods.

The third model proposed in Figure 6.4c was developed for two-component, one-dimensional layered systems. For these situations, thermal conductivity parallel to the layer would be predicted from:

$$k_{11} = k_{L}\left[1 - P_{v}\left(1 - \frac{k_{s}}{k_{L}}\right)\right]$$

(6.11)

where P_v is the volume fraction of the discontinuous phase in the product. For prediction of thermal conductivity perpendicular to the product layer, the following equation is suggested:

$$k_{\perp} = k_{L}\left[\frac{k_{s}}{P_{v}k_{L} + k_{s}(1 - P_{v})}\right]$$

(6.12)

These expressions would be used in a variety of situations when the product contains two distinctly different components with a defined layer structure.

The complete derivations of Equation 6.5 to Equation 6.12 was presented and discussed by Kopelman (1966). The implications of the models when applied to food systems are discussed in Heldman and Singh (1981) and Heldman (2001). The application of these models to frozen food systems was first demonstrated by Heldman and Gorby (1975). Frozen foods are a three phase system (product solids, liquid phase water, and solid phase water), and applications of the models require recognition of changes in the food during the freezing process. An approach is to apply the expressions in sequence, with the first step involving the water/ice mixture. For this product component, the solid phase water (ice) is the discontinuous phase within the unfrozen water as a continuous phase. The second computational step applies the appropriate model with the water/ice mixture becoming the continuous phase and the product solids the discontinuous phase in the product system. An illustration of the relationship between thermal conductivity and temperature for frozen beef is presented in Figure 6.5, where it is possible to predict thermal conductivities parallel and

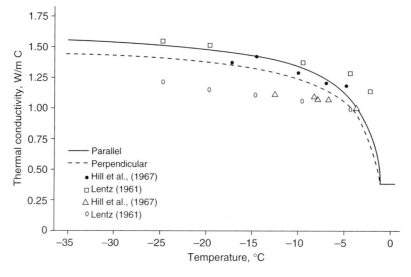

FIGURE 6.5 Thermal conductivity of frozen lean beef as a function of temperature. (From Heldman, D.R. and Gorby, D.P. 1975. *Trans. ASAE* 18: 740–744. With permission.)

perpendicular to the product fibers. The approach presented is acceptable during the early stages of the freezing process, when the volume fraction of liquid water exceeds the volume fraction of ice in the ice/water mixture.

The product solids can be considered homogeneous and disperse mixture of the basic components (protein, fat, carbohydrate, ash). This approach allows for the prediction of thermal conductivity for the product solids from the mass average of the component fractions and the thermal conductivity of each component. Thermal conductivities for each component of a food product, as estimated by Choi and Okos (1986), are presented in Table 6.8, and the expressions in Table 6.6.

6.3.4 PRODUCT ENTHALPY

The refrigeration requirements for food product freezing are directly dependent on the thermal energy content or enthalpy of the product. Magnitudes of enthalpy for frozen foods were measured by Riedel (1951, 1956, 1957a, 1957b), and a portion of the data is presented in Table 6.3. The enthalpy or thermal energy content is zero at $-40°C$, the reference temperature for refrigerants.

The enthalpy of a food product can be predicted, based on a reference temperature of $-40°C$, by using the following expression:

$$H = m_s c_{ps} \int_{-40}^{T_i} dT + m_u c_{pu} \int_{T_F}^{T_i} dT + \int_{-40}^{T_F} m_u(T)c_{pu}(T)dT + m_u(T)L + \int_{-40}^{T_F} m_I(T)c_{pI}(T)dT$$

(6.13)

The terms in Equation 6.13 include the sensible heat of product solids as a function of temperature (T_s). The second term of the equation accounts for sensible heat of unfrozen water when the temperature is above the initial freezing temperature (T_F) of the product. The sensible heat of unfrozen water in the frozen product is the third term of Equation 6.13, where the unfrozen water fraction (m_u) and specific heat of unfrozen water (c_{pu}) vary significantly with temperature. The fourth term of Equation 6.13 accounts for the contribution of phase change energy to the enthalpy and provides the influence of the unfrozen fraction (m_u), as it changes with temperature. The contribution of

---

TABLE 6.8
Thermal Conductivities of Pure Components Calculated from Literature Values of Liquid Foods

T (°C)	Thermal conductivity (W/m-°C)					No. of data	Standard error	Standard % error
	Water	Protein	Fat	Carbohydrate	Ash			
20	0.6012	0.1993	0.1765	0.2039	0.1356	20	0.0147	2.85
30	0.6191	0.2109	0.1759	0.2178	0.1402	26	0.0248	4.79
40	0.6332	0.2182	0.1737	0.2285	0.1430	23	0.0251	4.88
50	0.6464	0.2291	0.1724	0.2386	0.1480	23	0.0243	4.32
60	0.6542	0.2349	0.1708	0.2463	0.1543	23	0.0235	3.96
70	0.6643	0.2475	0.1686	0.2594	0.1577	23	0.0225	3.42
80	0.6712	0.2528	0.1669	0.2632	0.1619	15	0.0155	2.61
90	0.6768	0.2553	0.1656	0.2665	0.1642	11	0.0145	2.48
100	0.6827	0.2622	0.1645	0.2723	0.1645	11	0.0156	2.54

Source: From Choi, Y. and Okos, M.R. 1986. *Physical and Chemical Properties of Food.* Martin R. Okas (Ed.), ASAE, St Joseph, MI. pp. 35–77.

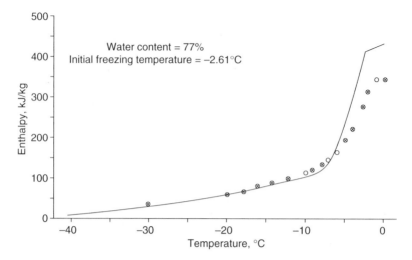

FIGURE 6.6 Enthalpy of sweet cherries as a function of temperature. (From Heldman, D.R. 1982. *Food Technol.* 36: 92–96. With permission.)

the frozen water fraction (m_I) to enthalpy is the final term of Equation 6.13. Analysis of similar approaches to prediction of frozen food enthalpy has been completed by Sastry (1984), Kerr et al. (1993), Pham (1996), and Fikiin and Fikiin (1999).

The mass fractions for unfrozen and frozen water can be predicted by procedures described in Section 6.2.1. The specific heat of product solids (c_{ps}) can be predicted as presented in Section 6.2.2. The specific heats of liquid and solid water are obtained from standard tables (see Table 6.4). A typical prediction curve for enthalpy of sweet cherries as a function of temperature is presented in Figure 6.6.

6.3.5 APPARENT SPECIFIC HEAT OF FOODS

By using the thermodynamic definition of specific heat, the derivative of enthalpy with respect to temperature produces an apparent specific heat function for frozen foods. The apparent specific heat,

$$c_{pA}(T) = \frac{dH}{dT} \tag{6.14}$$

is a significant and unique function of temperature for all frozen foods. By evaluating the derivative of enthalpy for sweet cherries over the temperature range -40 to $+5°C$, the function shown in Figure 6.7 is obtained. As illustrated, the apparent specific heat increases with increasing temperature until reaching the initial freezing temperature. The function increases rapidly as temperatures approach the initial freezing temperature, indicating the region where major portions of the phase change for the product occurs. At the initial freezing temperature, the apparent specific heat function reaches a near discontinuity and then decreases to magnitudes of enthalpy for the unfrozen product at temperatures above the initial freezing temperature.

6.3.6 APPARENT THERMAL DIFFUSIVITY

The thermal property most often incorporated into heat transfer equations is thermal diffusivity. For frozen foods, where properties are a function of temperature, the following definition would

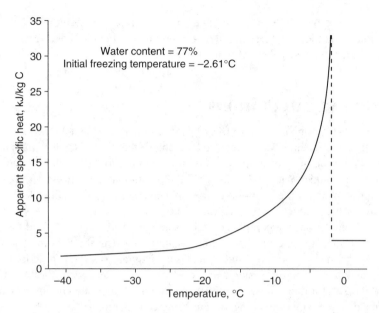

FIGURE 6.7 Predicted apparent specific heat of frozen sweet cherries as a function of temperature. (From Heldman, D.R. 1982. *Food Technol.* 36: 92–96. With permission.)

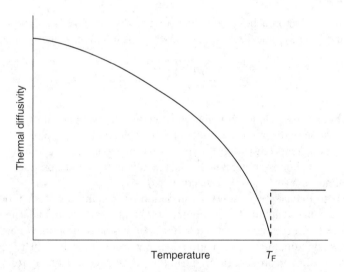

FIGURE 6.8 Relationship between thermal diffusivity and temperature during freezing of food product as predicted from initial freezing temperature assumption. (From Heldman, D.R. 1983. *Food Technol.* 37: 103–109. With permission.)

apply:

$$\alpha_A(T) = \frac{k(T)}{\rho(T)c_{pA}(T)} \tag{6.15}$$

and the thermal diffusivity magnitudes can be generated from expressions presented in previous sections. The apparent thermal diffusivity values as a function of temperature are illustrated in Figure 6.8. As is evident, the shape of the function at temperatures below the initial freezing temperature is similar to the shape of the thermal conductivity vs. temperature relationship. There is a near discontinuity at the initial freezing temperature, before the thermal diffusivity magnitudes become constant at temperatures above product freezing.

6.4 FREEZING-TIME CALCULATIONS

Freezing times are basic design criteria for freezing systems and represent the residence time for the food product within the freezing system required to achieve the desired level of freezing. The most widely accepted definition of freezing time is the time required to reduce the product temperature from some initial magnitude to an established final temperature at the slowest cooling location. An alternative definition changes the endpoint to the mass average enthalpy equivalent to the desired final temperature for the product.

Freezing-time calculations are completed as a first step in the design of a food freezing system. The freezing time establishes the residence time for the product in the system. The final product temperature is established as the magnitude needed to maintain optimum product quality during storage. For a continuous freezing system, the resident time is dependent on the rate of product moves through the system and on the length of the system. More specific characteristics of the design will depend on the type of freezing system being considered.

6.4.1 FREEZING-TIME EQUATIONS

Numerous equations and approaches to freezing-time prediction have been proposed and utilized. The best known and most used of the prediction methods is based on Planck's equation (1913):

$$t_F = \frac{\rho L}{t_F - T_\infty} \left[\frac{Pa}{h_c} + \frac{Ra^2}{k} \right] \tag{6.16}$$

where P and R are constants that depend on product geometry (see Table 6.9).

When the dimensions of the product are not infinite or spherical, Figure 6.9 can be used to evaluate the constants. The coefficients (β_1 and β_2) are the ratio of the maximum product (length) dimension to the minimum product dimension (thickness) and the ratio of the middle product dimension (width) to the minimum dimension (thichness), respectively.

The limitations to Planck's equation for estimation of freezing times for foods are numerous and have been discussed by Heldman and Singh (1981) and Ramaswami and Tung (1981). One of the concerns is selection of a latent heat magnitude (L) and an appropriate value for the thermal conductivity (k). In addition, the basic equation does not account for the time required for removal of sensible heat from unfrozen product above the initial freezing temperature or for removal of frozen product sensible heat.

There have been numerous attempts to modify Planck's equation or develop alternative expressions. The modifications include Nagaoka et al. (1955), Levy (1958), Charm and Slavin (1962),

---

TABLE 6.9
Constants for Plank's Equation

	P	R
Infinite slab	0.5	0.125
Infinite cylinder	0.24	0.0625
Sphere	0.167	0.04167

Tao (1967), Joshi and Tao (1974), Tien and Geiger (1967, 1968), Tien and Kuomo (1968, 1969), Mellor (1976), Gustschmidt (1964), and Mott (1964). In general, these modifications and alternatives provide improvements in the predictions, but limitations of various types still exist.

Cleland and Earle (1977, 1979a, 1979b, 1982) developed and presented a modification with sound empirical justification. These authors use Planck's equation in dimensionless form:

$$N_{\text{Fo}} = P\frac{1}{N_{\text{Bi}}N_{\text{Ste}}} + R\frac{1}{N_{\text{Ste}}} \tag{6.17}$$

$$N_{\text{Fo}} = \text{Fourier No.} = \frac{\alpha t}{d_c^2} \tag{6.18}$$

$$N_{\text{Bi}} = \text{Biot No.} = \frac{h_c d_c}{k} \tag{6.19}$$

$$N_{\text{Ste}} = \text{Stefan No.} = \frac{c_{\text{pF}}(T_\infty - T_{\text{Fo}})}{\Delta H} \tag{6.20}$$

The influence of sensible heat above freezing is incorporated by introducing Planck's number:

$$N_{\text{Pk}} = \frac{c_{\text{pu}}(T_i - T_F)}{\Delta H} \tag{6.21}$$

The values of the constants (P and R) are determined by using charts with relationships between Planck's number and Stefan's number. These charts are presented in Figure 6.10 and Figure 6.11.

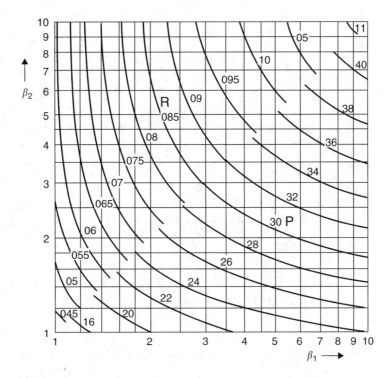

FIGURE 6.9 Chart providing P and R constants for Planck's equation when applied to a brick or block geometry. (From Ede, A.J. 1949. *Mod. Refrig.* 52: 53. With permission.)

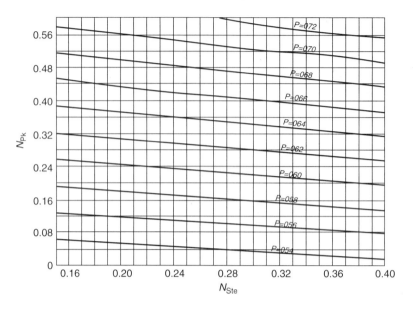

FIGURE 6.10 Chart showing the Planck number vs. the Stefan number for determination of different values of the empirical modification P. (From Cleland, A.C. and Earle, R.L. 1982. *Int. J. Refrig.* 5: 134–140. With permission.)

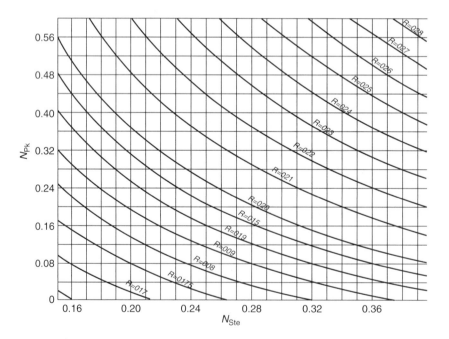

FIGURE 6.11 Chart showing the Planck's number vs. the Stefan number for determination of different values of the empirical modification P. (From Cleland, A.C. and Earle, R.L. 1982. *Int. J. Refrig.* 5: 134–140. With permission.)

Product shape is considered by an equivalent heat-transfer dimension (EHTD), as determined by:

$$EHTD = 1 + W_1 + W_2 \qquad (6.22)$$

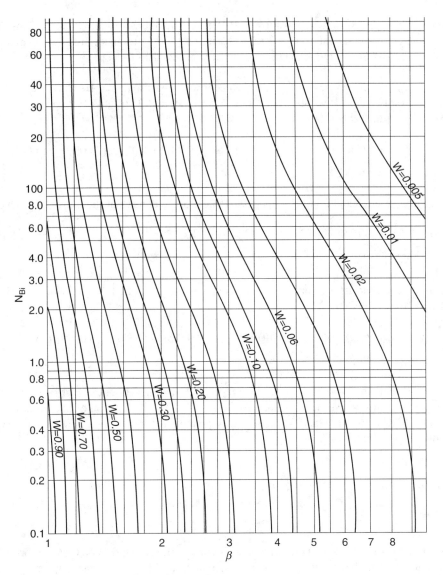

FIGURE 6.12 Chart showing the Biot number vs. the shape factor for determination of different values of W. (From Cleland, A.C. and Earle, R.L. 1982. *Int. J. Refrig.* 5: 134–140. With permission.)

The values of W_1 and W_2 are determined from Figure 6.12 using the Biot number and a shape factor (β). The factor (W_1) is determined using:

$$\beta_1 = d_1/2d_c \tag{6.23}$$

where d_1 is product width. The second factor (W_2) is obtained by using

$$\beta_2 = d_2/2d_c \tag{6.24}$$

where d_2 is product length.

Example   Beef steaks with dimensions of 0.1 m length, 0.06 m width, and 0.02 m thickness are frozen in an air-blast system at $-20°C$. The initial product temperature is $10°C$ and the final temperature is $-10°C$. Calculate the time required to freeze the product.

Solution   The product properties for unfrozen beef will be estimated at $10°C$. Based on Table 6.1, the composition is 74% water, 21.45% protein, 3.56% fat, and 0.99% ash. Using Equation 6.4 and the relations in from Table 6.6:

$$c_p = (0.74)(4.18) + (0.2145)(2.02) + (0.0356)(1.998) + (0.0099)(1.111)$$
$$= 3.538 \text{ kJ/kgC}.$$

Using Equation 6.3 and relationships from Table 6.6, the density of the unfrozen product is:

$$1/\rho = (0.74)(1/996.8) + (0.2145)(1/1324.7) + (0.0356)(1/921.4) + (0.0099)(1/2420.99)$$
$$\rho = 1055.9 \text{ kg/m}^3.$$

The thermal conductivity of unfrozen lean beef can be estimated based on composition and expressions from Table 6.6:

$$k = (0.74)(0.588) + (0.2145)(0.1995) + (0.0356)(0.1531) + (0.0099)(0.3433)$$
$$k = 0.483 \text{ W/mC}$$

In order to estimate the properties of the frozen product, the mass fractions of frozen and unfrozen water at $-10°C$ must be estimated. Using Equation 6.1, the mole fraction of unfrozen water in the product at $-10°C$ is found to be 0.904. Using this quantity and the following molecular weights for the product components:

Water	18.02
Proteins	50,000.00
Lipids	50,000.00
Ash	34.50

the mass fraction of unfrozen water in the lean beef at $-10C$ is found to be 0.04967, or 16.71% of the freezable water in the product when the mass fraction of 0.074 unfreezable water is included.

Using the composition of the frozen product (at $-10C$), the properties of the product can be estimated.

$$c_p = (0.12367)(4.18) + (0.61633)(2.0629) + (0.2145)(2.008)$$
$$+ (0.0356)(1.9839) + (0.0099)(1.0924)$$
$$= 2.3 \text{ kJ/kgC}$$
$$1/\rho = (0.12367)(1/986.8) + (0.61633)(1/918.2) + (0.2145)(1/1335.1) + (0.0356)(1/926.0)$$
$$+ (0.0099)(1/2426.6) = 9.997 \times 10^{-4}$$
$$\rho = 1000.3 \text{ kg/m}^3$$
$$k = (0.12367)(0.5528) + (0.61633)(2.167) + (0.2145)(0.1666)$$
$$+ (0.0356)(0.2083) + (0.0099)(0.3433)$$
$$k = 1.45 \text{ W/mC}$$

TABLE 6.10
Heat Transfer Coefficients

Condition	Heat transfer coefficient (W/m-K)
Naturally circulating	5
Air blast	22
Plate contact freezer	56
Slowly circulating brine	56
Rapidly circulating brine	85
Liquid nitrogen	
low side of horizontal plate	170
where gas blanket forms	
upper side of horizontal plate	425
Boiling water	568

The Biot number is determined using Equation 6.19:

$$N_{Bi} = (22)(0.01)/1.45 = 0.152$$

where the convective heat transfer coefficient for air blast freezing is obtained from Table 6.10. The Stefan number is calculated using Equation 6.20:

$$N_{Ste} = (2.3)[-0.8 - (-20)]/(0.61633)(333.5)$$
$$= 0.215$$

where enthalpy change is based on the fraction of water converted to frozen state. Planck's number is computed by using Equation 6.21:

$$N_{Pk} = (3.538)[10 - (-0.8)]/(0.61633)(333.5)$$
$$= 0.176$$

The shape factors for slices of lean beef are:

$$\beta_1 = 0.06/0.02 = 3$$
$$\beta_2 = 0.1/0.02 = 5$$

Using Figure 6.10 with $N_{Pk} = 0.176$ and $N_{Ste} = 0.215, P = 0.58$. Then Figure 6.11 is used with $N_{Pk} = 0.176$ and $N_{Ste} = 0.215$ to obtain $R = 0.187$. For $\beta_1 = 3$, and $N_{Bi} = 0.152$, Figure 12 is used to obtain $W_1 = 0.151$. Finally, Figure 6.12 is used with $\beta_2 = 5$ and $N_{Bi} = 0.152$ to obtain $W_2 = 0.061$.

Based on the above parameters, Equation 6.22:

$$EHTD = 1 + 0.151 + 0.061 = 1.212$$

Using Equation 6.17 to obtain the Fourier Number:

$$N_{Fo} = (0.58)[1/(0.152)(0.215)] + (0.187)[1/(0.152)]$$
$$= 18.978.$$

Then: $N_{Fo} = \alpha t_F/d_c^2$

$$t_F = (18.978)(0.01)^2(1000.3)(2.3)(1000)/(1.45)(1.212)$$
$$= 2484.5 \text{ sec} = 41.4 \text{ min.}$$

Pham (1986) presented an improvement of Planck's equation for the prediction of freezing times. The approach is based on the following equation:

$$t_F = \frac{d_c}{E_f h_c}\left[\frac{\Delta H_1}{\Delta T_1} + \frac{\Delta H_2}{\Delta T_2}\right]\left(1 + \frac{N_{Bi}}{2}\right) \tag{6.25}$$

where the following parameters are defined and must be evaluated.

E_f is a shape factor, with a value of 1 for an infinite slab, a value of 2 for an infinite cylinder, and a value of 3 for a sphere.

$$\Delta H_1 = \rho_u c_{pu}(T_i - T_{fm}) \tag{6.26}$$

and represents the change in volumetric enthalpy for a precooling period, where the end of the period is defined by a "mean freezing temperature" (T_{fm}). This temperature is defined as:

$$T_{fm} = 1.8 + 0.263T_C + 0.105T_\infty \tag{6.27}$$

and depends on the final temperature at the product center (T_C), and the freezing medium temperature (T_∞). This relationship is based on experimental data for a variety of foods.

The change in volumetric enthalpy is:

$$\Delta H_2 = \rho_F[L + c_{pF}(T_{fm} - T_C)] \tag{6.28}$$

and accounts for phase change at temperatures below the mean freezing temperature (T_{fm}).

The temperature gradients in Equation 6.25 are defined as follows:

$$\Delta T_1 = [(T_i + T_{fm})/2] - T_\infty \tag{6.29}$$
$$\Delta T_2 = T_{fm} - T_\infty \tag{6.30}$$

The solution of problem involving freezing time using the Pham approach requires evaluation of the parameters in Equation 6.26 to Equation 6.30, followed by the use of Equation 6.25.

Example Calculate the freezing time for the conditions presented in the previous example, using Pham's approach.

Solution The freezing time will be evaluated with the assumption that the geometry is an infinite slab and $E_f = 1$. The mean freezing temperature is:

$$T_{fm} = 1.8 + 0.263(-10) + 0.105(-20) = -2.93°C$$

$$\Delta H_1 = (1055.9)(3.538)[10 - (-2.93)](1000) = 48{,}303{,}560 \text{ J/m}^3$$

$$\Delta H_2 = (1000.3)\{(0.61633)(333.5) + (2.3)[-2.93 - (-10)]\}(1000)$$
$$= 221{,}873{,}590 \text{ J/m}^3$$

$$\Delta T_1 = [(10 + (-2.93))/2] - (-20) = 23.535 \text{ C}$$

$$\Delta T_2 = -2.93 - (-20) = 17.07 \text{ C}$$

$$N_{Bi} = 0.152 \text{(as computed in the previous example)}$$

Using Equation 6.25

$$t_F = [(0.01)/(22)][(48{,}303{,}560/23.535) + (221{,}873{,}590/17.07)][1 + (0.152/2)]$$

$$t_F = 7360.96 \text{ sec} = 2.05 \text{ h}$$

Pham's method (1986) provides additional relationships to account for the influence of product shape and geometry. The Pham approach uses the parameters (β_1) and β_2), as defined earlier, to determine inputs to a relationship for the shape factor (E_f);

$$E_f = G_1 + G_2 E_1 + G_3 E_2 \tag{6.31}$$

where the values for the coefficients (G_1, G_2, G_3) are obtained from Table 6.11. The parameter E_1 is determined from the following from:

$$E_1 = (X_1/\beta_1) + [1 - X_1](0.73/\beta_1^{2.5}) \tag{6.32}$$

and the value of X_1 is computed from:

$$X_1 = 2.32\beta_1^{-1.77}/[(2N_{Bi})^{1.34} + 2.32\beta_1^{-1.77}]. \tag{6.33}$$

The relationship for evaluation of E_2 is obtained from:

$$E_2 = (X_2/\beta_2) + [1 + X_2](0.73/\beta_2^{2.5}) \tag{6.34}$$

TABLE 6.11
Magnitudes of Shape Factors (G) for Different Shapes

Shape	G_1	G_2	G_3
Finite cylinder; height < diameter	1	2	0
Finite cylinder; height > diameter	2	0	1
Rectangular rod	1	1	0
Rectangular brick	1	1	1

with X_2 obtained from:

$$X_2 = 2.32\beta_2^{-1.77}/[(2N_{\text{Bi}})^{1.34} + 2.32\beta_2^{-1.77}] \tag{6.35}$$

Example   Estimate the freezing time for the previous example, while considering the impact of the shape factor.

Solution   The first step is to compute the shape factor (E_f). In order to determine E_f, the values of E_1 and X_1 must be computed.

$$X_1 = 2.32(3)^{-1.77}/\{[2(0.152)]^{1.34} + 2.32(3)^{-1.77}\} = 0.621$$

$$E_1 = 0.621/3 + [1 - 0.621](0.73/3^{2.5}) = 0.225$$

The values of E_2 and X_2 are obtained from:

$$X_2 = 2.32(5)^{-1.77}/\{[2(0.152)]^{1.34} + 2.32(5)^{-1.77}\} = 0.422$$

$$E_2 = 0.422/5 + [1 + 0.422](0.73/5^{2.5}) = 0.103$$

Then:

$$E_f = 1 + (1)(0.225) + (1)(0.103)$$

$$E_f = 1 + 0.225 + 0.103 = 1.328$$

And, the freezing time is computed using Equation 6.25:

$$t_F = 7360.96/1.328 = 5542.89 \text{ sec} = 92.38 \text{ min.}$$

6.4.2 NUMERICAL METHODS

The evolution of high-speed computing systems and appropriate numerical methods have provided opportunities to solve complex partial differential equations with temperature-dependent properties. Equations of the following form for one-dimensional heat transfer during freezing of the product can be solved to predict temperature distribution histories within the product:

$$\rho(T)c_{\text{pA}}(T)\frac{\partial T}{\partial t} = \frac{\partial}{\partial X}\left[k(T)\frac{\partial T}{\partial X}\right]. \tag{6.36}$$

The prediction of the product thermal properties by methods presented in Section 6.3 can be used to solve the equations by finite difference and/or finite element methods (Heldman, 1983; Cleland, 1990; Delgado and Sun, 2001). These solutions predict temperature distribution histories of the product, as illustrated in Figure 6.13. As indicated, the temperature at the center of the product decreases to a plateau at the initial freezing temperature of the product before decreasing to lower temperatures. At all locations closer to the surface, the temperature decreases more gradually during the early stages of the freezing process when most of the latent heat is being removed from the product. Numerical solutions for products with various geometries have been completed by Mannapperuma and Singh (1988, 1989). The simulation, for a range of geometries, has been incorporated into a software package for the prediction of conditions during food freezing and is available from the WFLO (2000).

Freezing times are established at the point when the temperature history curve passes through the temperature established for the storage of the frozen product. The numerical solutions can be

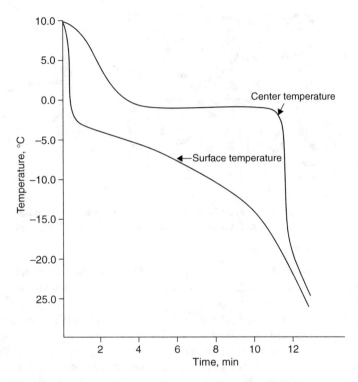

FIGURE 6.13 Predicted temperature history at the center and surface during freezing of a 2-cm strawberry $(T = 10°C; T_\infty = -35°C; h_c = 70 \text{ W/m}^2\text{K}; a = 2 \text{ cm})$. (From Heldman, D.R. 1983. *Food Technol.* 37: 103–109. With permission.)

used to compute enthalpy distributions and mass-average enthalpies. Often, mass-average enthalpies equivalent to the desired final temperature are used to establish the end of the freezing process.

6.5 FREEZING SYSTEMS

As indicated in Section 6.4, the reduction of product temperature from an initial value to the desired final temperature establishes the freezing time. The environment maintained during freezing of the product is established by a physical structure and a refrigeration system. The environment required to maintain the temperature and boundary conditions at the product surface are the primary factors that establish the effectiveness of the freezing system.

Freezing systems can be classified into two groups: direct-contact systems and indirect-contact systems. This classification is based on the type of contact between the product surface and the refrigeration medium.

As described in the following sections, the contact between product surface and refrigeration medium varies with the type of product package and the type of cooling medium.

6.5.1 DIRECT-CONTACT SYSTEMS

Any freezing system that brings a refrigeration medium into direct contact with the product surface would be classified as a direct contact system. As illustrated by Figure 6.14, these systems attempt to bring the cold medium into contact with the maximum product surface area. In general, these types of freezing systems would be expected to be highly efficient, since barriers to heat transfer are reduced to a minimum. It should be noted that the product surface may be covered by package film.

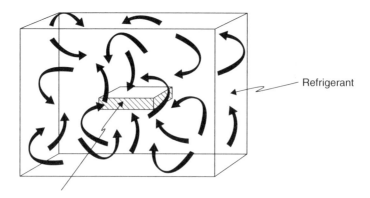

FIGURE 6.14 Schematic diagram of direct-contact freezing. (From Singh, R.P. and Heldman, D.R. 2001. *Introduction to Food Engineering*. Third Edition. Academic Press, New York. With permission.)

The refrigeration medium used for these systems would include low-temperature air moving over the product surface at high air speeds, as well as selected liquid refrigerants that may allow for phase change during the product freezing process.

An air-blast freezing system would be considered direct contact when low-temperature air is brought into direct contact with a product during freezing. The use of these types of systems is limited to situations where the residence time is low, in order to control moisture loss during freezing. When product dimensions are small, these systems are referred to as individual-quick-freezing (IQF) systems, with the individual small pieces of product exposed directly to low-temperature air.

The fluidized-bed freezing system illustrated in Figure 6.15 is a modified version of an IQF system. By maintaining the product pieces in a fluidized state, the movement of low-temperature air at the product surface created very high convective heat transfer coefficients. Since the product pieces must be relatively small in order to establish and maintain a fluidized bed, the freezing times will be short. The limits to the use of the process are based on efficiency: energy requirements necessary to maintain the fluidized condition. The primary product parameter infuencing energy required for fluidization is size or mass of the product particle.

A third type of direct-contact freezing system is the immersion freezer. In such systems the product is exposed to a liquid refrigerant that is undergoing phase change as the freezing process occurs. A schematic of the process is shown in Figure 6.16, where the movement of product through the refrigerant is illustrated. The common refrigerants used for immersion freezers — nitrogen, carbon dioxide, and Freon — must be approved for food product contact. A commercial immersion freezing system is shown in Figure 6.17. The product particles or pieces pass through a compartment filled with cold refrigerant vapor where the produce is exposed to a spray of liquid refrigerant. In general, very rapid freezing of product is achieved, resulting in superior product quality when rate of ice crystal formation influences quality. Overall process efficiency is influenced by the ability to recover expansive refrigerant as the freezing process is completed.

6.5.2 INDIRECT-CONTACT SYSTEMS

Most frozen foods are the result of using indirect-contact types of freezing systems. As illustrated schematically in Figure 6.18, the food is separated from the refrigerant by some type of barrier. These barriers would include product package surfaces as well as structural components of the freezing system.

A typical type of indirect-contact freezing system is the plate freezer. As indicated in Figure 6.19, the product may be maintained between plates during the freezing process. The plate separates the product from the refrigerant, although the product package may be a part of the barrier between

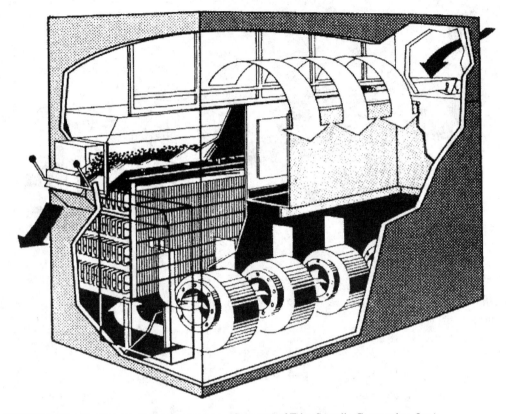

FIGURE 6.15   Fluidized-bed freezing system. (Courtesy of FrigoScandia Contracting, Inc.)

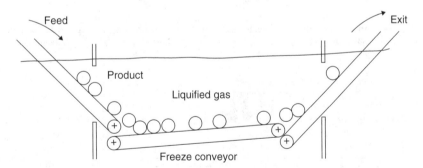

FIGURE 6.16   Schematic illustration of immersion freezing system. (From Singh, R.P. and Heldman, D.R. 2001. *Introduction to Food Engineering*. Third Edition. Academic Press, New York. With permission.)

product and package as well. The use of pressure tends to reduce the resistance to heat transfer and reduces freezing times. These types of systems may operate in a batch mode or continuously, as illustrated in Figure 6.20. Continuous plate freezers are designed with the plates holding the product moving in a manner that results in product movement from entrance to exit in an indexing fashion. Plate freezing systems are highly efficient but are limited to product shapes that fit the plate configuration.

For food products with unusual shapes, air-blast freezers of the type shown in Figure 6.21 are used. In these situations, the product package represents the barrier between product and refrigerant. Although a system of this type may function in a batch mode, continuous systems are most typical.

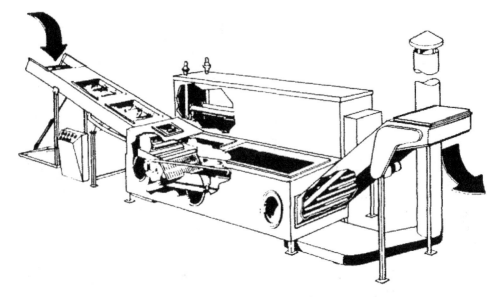

FIGURE 6.17 Individual quick freezing (IQF) using liquid refrigerant. (Courtesy of FrigoScandia Contracting, Inc.)

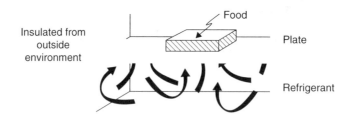

FIGURE 6.18 Schematic diagram of indirect-contact freezing system. (From Singh, R.P. and Heldman, D.R. 2001. *Introduction to Food Engineering*. Third Edition. Academic Press, New York. With permission.)

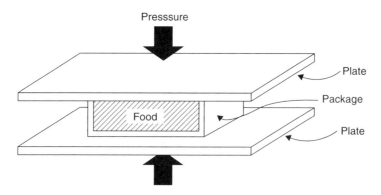

FIGURE 6.19 Schematic illustration of plate freezing system. (From Singh, R.P. and Heldman, D.R. 2001. *Introduction to Food Engineering*. 3rd ed. Academic Press, New York. With permission.)

Short freezing times are possible by maintaining high air velocities within the freezer compartment, low air temperatures, and good contact between package and product surface. In addition to the conveyor arrangement shown in Figure 6.21, product may be carried through the system on trays, spiral conveyors, and roller conveyors.

FIGURE 6.20   Plate contact freezing system. (Courtesy of Crepaco, Inc.)

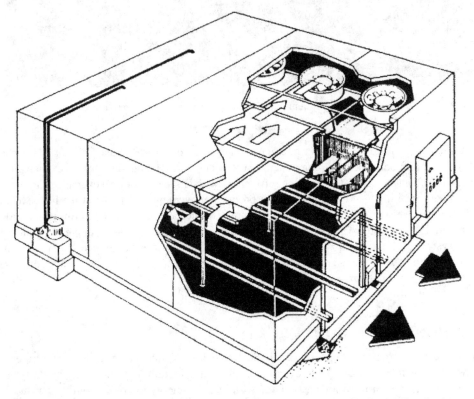

FIGURE 6.21   Continuous air-blast freezing system. (Courtesy of FrigoScandia Contracting, Inc.)

FIGURE 6.22   Continuous freezing system for liquid foods. (Courtesy of Cherry-Burrell Corp.)

The final type of indirect-contact freezing system to be discussed is used primarily for partial freezing of liquid foods. The basic component of the system is a scraped-surface heat exchanger with product within the tube separated from the refrigerant by the tube wall or heat exchange surface. The rotor component of the heat exchange system provides movement of product at the heat exchange surface to enhance heat transfer. The jacket surrounding the heat exchange surface is the evaporator for the refrigerant system used to maintain the desired temperature gradient. An example of a continuous freezing system for liquid food is shown in Figure 6.22. The use of this type of system results in the removal of 60 to 80% of the latent heat of fusion from a liquid food. The product leaves the system in the form of a frozen slurry.

6.6   DESIGN CALCULATIONS

The ultimate use of the properties discussed in Section 6.3 is to establish the capacity of refrigeration systems used for food freezing. The evaporator on the refrigeration system must be designed to accommodate the heat transfer rate associated with thermal energy removed from the product as it passes through the freezing system. By computation of the refrigeration requirement or difference

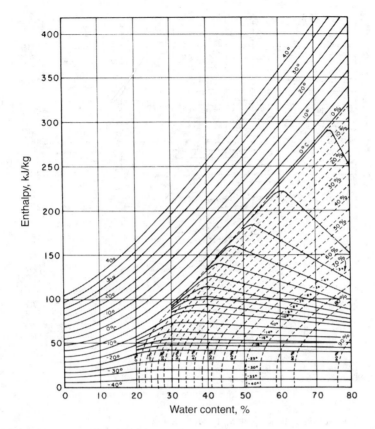

FIGURE 6.23   Enthalpy-composition chart for beef.

between initial and final product enthalpies, the first step in establishing a load on the refrigeration system is completed. The second step involves the rate at which the thermal energy is removed from product based on freezing times (from Section 6.2) and the residence time for product within the freezing system. Finally, the rate of thermal energy transfer to the refrigeration system evaporator is converted into refrigeration system compressor size, as well as other system sizes.

6.6.1 REFRIGERATION REQUIREMENTS

As illustrated in Section 6.3.4 and Figure 6.6, the enthalpy of a frozen food can be predicted based on the composition of the product. Based on the composition of the product, the specific heats of unfrozen and frozen food can be predicted, along with latent heat for changing the phase of water in the product frozen to a given temperature. For lean beef and fruits and vegetables, the enthalpies associated with various water contents and temperatures were presented in charts by Riedel (1956, 1957a, 1957b). The chart for lean beef is presented in Figure 6.23 and for fruit and vegetable juices, in Figure 6.24. To estimate the enthalpy requirement for food product freezing, the enthalpy magnitudes at the initial product temperature (before freezing) and the final product temperature are determined from Figure 6.23 or Figure 6.24. The difference between these two enthalpy values represents the thermal energy to be removed during freezing.

Example   Estimate the enthalpy change required during freezing of lean beef, when the initial temperature is $10°C$ and the final temperature is $-10°C$. The composition of lean beef is presented in the example in Section 6.4.1.

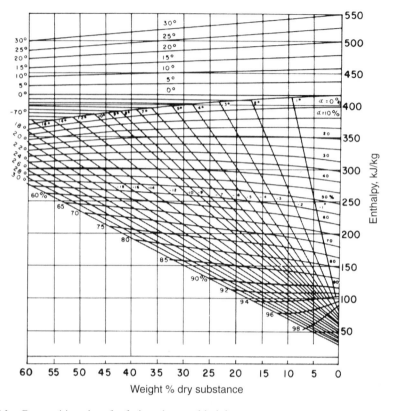

FIGURE 6.24 Composition chart for fruit and vegetable juices.

Solution   Given the composition from the previous example, the specific heats of the product can be predicted from the Choi and Okos (1986) relationships:

$$c_{pu} = 3.604 \text{ kJ/kg K (at } 5°C)$$
$$c_{ps} = 1.9635 \text{ kJ/kg K (at } -5°C)$$
$$c_{pl} = 2.032 \text{ kJ/kg K (at } -5°C)$$

(1) The enthalpy change above initial freezing temperature is:

$$\Delta H_{un} = (1)(3.604)[10 - (-0.8)] = 39.92 \text{ kJ/kg}$$

(2) Enthalpy change for the product solids fraction is:

$$\Delta H_s = (0.26)(1.9635)[(-0.8) - (-10)] = 4.70 \text{ kJ/kg}$$

(3) Enthalpy change for frozen fraction of water (ice):

$$\Delta H_l = (0.462)(2.032)[(-0.8) - (-10)] = 8.64 \text{ kJ/kg}$$

where the mass fraction of ice is 75% of the total change occuring when freezing to $-10°C$; 0.75 (0.616).

(4) Enthalpy change for unfrozen water fraction:

$$\Delta H_u = (0.278)(4.133)[(-0.8) - (-10)] = 10.57 \text{ kJ/kg}$$

where the mass fraction of unfrozen water (0.278) is 25% of the change in mass fraction of unfrozen water; 0.25 (0.74 − 0.12367) + 0.12367.

(5) Enthalpy change for change of phase:

$$\Delta H_{pc} = (0.61633)(333.5) = 205.55 \text{ kJ/kg}$$

(6) The estimated change in enthalpy for freezing the product is a totlal of the five components:

$$\Delta H = 33.16 + 4.7 + 8.64 + 10.57 + 205.55 = 266.28 \text{ kJ/kg}$$

This estimate can be compared to the results from Figure 6.23. The predicted enthalpy change is slightly higher than the value from the chart (Figure 6.23); a difference that is most likely related to the estimated values used for water and ice fractions below the initial freezing temperature. An expression for the estimation of the enthalpy change during freezing of fruits and vegetables was proposed by Riedel (1951):

$$\Delta H = \left(1 - \frac{X_{snj}}{100}\right)\Delta H_j + 1.21\left(\frac{X_{snj}}{100}\right)\Delta T, \tag{6.37}$$

where X_{snj} is the percentage of insoluble solids in the product. The enthalpy (ΔH_j) is obtained from Figure 6.24 using the appropriate product temperature and soluble solid content of the juice to determine the percentage of dry substance.

Example   Containers of cherries with a mass of 10 kg are being frozen in a continuous freezing system. The initial product temperature is 10°C and the final temperature is −10°C. Determine the enthalpy change during freezing for each container of cherries.

Solution   Based on Table 6.2, the water content of cherry juice is 86.7%, or soluble solids of 13.3%. In Table 6.1, the water content of sweet cherries is 77.61%, to give a total solids of 22.39%, 9.09% insoluble solids. Using Figure 6.24 and a soluble solids content of 13.3%, the initial enthalpy is 453 kJ/kg and the final enthalpy is 150 kJ/kg. Thus. $\Delta H = 453 - 150 = 303$ kJ/kg to be used in Equation 6.37. Using Equation 6.37 yields

$$\Delta H = \left(1 - \frac{9.09}{100}\right)\left(303\right) + 1.21\left(1 - \frac{9.09}{100}\right)\left(20\right)$$

$$\Delta H = 277.66 \text{ kJ/kg}$$

Since each container holds 10 kg of product, the total enthalpy change = (277.66 kJ/kg) (10 kg) = 2776.6 kJ

6.6.2 System Capacity

The second step in establishing the load on a refrigeration system for food product freezing incorporates the rate of thermal energy removal. In a continuous freezing system, this rate is a function of residence time within the system. In Section 6.4, the freezing time was defined as the time to reduce the product temperature at a defined location, from an initial value to the desired final temperature.

If the product freezing occurs within the freezing system, the residence time must be equal to or greater than the freezing time.

The establishment of residence time in a continuous system is a function of the linear distance that the product moves through the freezing system and the speed of product movement through the system. Freezing system design involves the establishment of refrigeration requirements, dimensions of the freezing system, and the speed of product movement in system.

Example   The containers of cherries described in the previous example are being frozen in a system with a length of 300 m, with each container occupying 0.5 m. Determine the refrigeration system capacity required if the product freezing time is 20 min.

Solution   Based on the results of the preceding example:

$$\text{Enthalpy change for freezing} = 2776.6 \text{ kJ/container}$$

Since the residence time for the product in the freezing system is 20 min, and the length of the system is 300 m, it is moving through the system at a speed of 15 m/min or 0.25 m/sec. Since there are 2 containers per meter in the system, the rate of product movement is 0.5 containers/sec.

$$\text{Then: Refrigeration capacity} = (2776.6 \text{ kJ/container})(0.5 \text{ containers/sec})$$
$$= 1388.3 \text{ kJ/sec} = 1388.3 \text{ kW}$$

This value represents the rate of thermal energy transfer through the refrigeration system evaporator surfaces, as well as the thermal energy to be absorbed by the refrigerant within the evaporator component of the system.

6.7 DESIGN OF FROZEN FOOD STORAGE

The storage facilities for frozen foods have significant influence on the quality of the frozen product. In order to maintain maximum quality, the facilities must be maintained at the optimum temperature, and any fluctuations in product temperature should be minimized.

Maintaining the temperature of the frozen food storage environment at the appropriate temperature depends on several factors including: (a) negligible heat transfer through the walls of the facility, (b) controlling the movement of ambient temperature air into the facility through openings, (c) minimizing the movement of elevated temperature product into the facility, and (d) continuous operation of the refrigeration system. Heat transfer through the walls of the facility is normally a minor factor due to the use of insulating materials within the wall sections. Since access to the frozen product occurs through openings, the movement of ambient temperature air into the storage environment through the opening cannot be completely eliminated. Various structural design features may be used to reduce the amount of air movement into the product storage space. The thermal capacity of frozen product within the storage space assists in keeping the temperature of the facility. The movement of the elevated temperature of the product into the storage space can cause temperature changes and must be controlled by careful management of the facility. The interruption of the refrigeration system operation has become a practice during peak electric power demand in an effort to control cost of operations but must be managed carefully.

Since the temperature of the storage environment is likely to vary from one location to another, frozen food products are likely to be exposed to less than optimum temperatures, and product temperatures may fluctuate over time. The impact of temperature and temperature fluctuations on frozen food quality has been evaluated. The most extensive experimental investigations of frozen food quality, as impacted by temperature, were described by van Arsdel et al. (1969). Similar results

TABLE 6.12
Shelf Life Parameters for Frozen Foods

Product	Activation energy coefficient (kJ/mole)	Reference shelf-life (days @ −18°C)
Asparagus	67.57	239
Beans, lima	67.45	379
Broccoli	67.45	379
Cod	58.45	163
Caulifower	67.45	379
Corn, cut	53.76	653
Haddock	53.26	282
Lobster	54.81	192
Peas	24.60	302
Shrimp	47.20	316
Spinach	56.40	474

Source: From Lai, D.J. and Heldman, D.R. 1982. *J. Food Process Engr.* 6: 179.

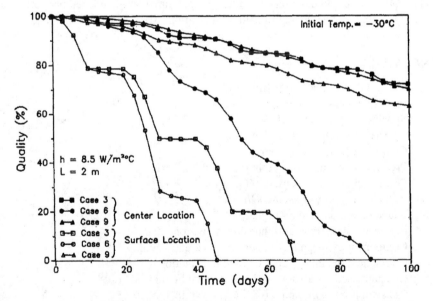

FIGURE 6.25 Effect of the magnitude of step changes in storage temperatures on the quality of strawberries. (Case 3: 10 days at both −30°C and −5°C; Case 6: 10 days at both −18°C and −5°C; Case 9; 10 days at both −18°C and −13°C. E_a = 182 KJ/mole; reference shelf life — 630 days at −18°C) [from Scott et al. (1989)].

have been presented by Jul (1984). The results of these investigations were reported in terms of frozen food shelf life as a function of storage temperature. Later, Lai and Heldman (1982) analyzed the experimental data, and present the results in terms of frozen food shelf life at a reference temperature, and a coefficient to describe the influence of temperature on shelf-life magnitude. Examples of reference shelf life values and activation energy constants are presented in Table 6.12. The reference shelf life values are Practical Shelf Life (PSL), or length of time that the frozen food can be held at a given temperature while retaining characteristic properties and suitable for consumption. The reference temperature is $-18°C$; the established standard for storage of frozen foods. Since the model used for analysis of experimental shelf life data is based on reaction kinetics, the temperature coefficient is referred to as an activation energy constant (E_A). The form of the model for influence of temperature on frozen food shelf life is as follows:

$$\ln t_Q = \ln B + \frac{E_A}{R_g T_A} \tag{6.38}$$

where B is a pre-exponential constant. The impact of temperature on shelf life at any other temperature (below the initial freezing temperature) can be predicted by using the model and the activation energy constants (E_A) and reference shelf life values from Table 6.12.

The impact of fluctuations in temperature of the storage environment on frozen food quality was described by Scott et al. (1989). The predictions of reduction in frozen food quality due to temperature fluctuations were obtained from a model similar to Equation 6.38, as follows:

$$t_Q = t_{QR} - \int_o^t \exp\left\{ -\frac{E_A}{R_g} \left[\frac{1}{T_A} - \frac{1}{T_{AR}} \right] \right\} dt \tag{6.39}$$

where the influence of variations in product temperature has been integrated over periods of storage time. The influence of the magnitudes of air temperature fluctuation has been evaluated and illustrated in Figure 6.25. Temperature fluctuations have significant impact on product quality near the surface of the package. In addition, fluctuations in storage temperature that include temperatures well above the reference temperature have more dramatic influence on frozen product shelf life.

NOMENCLATURE

a	Product dimension in Equation 6.16, m
B	Pre-exponential constant, in Equation 6.38
c_p	Specific heat capacity, kJ/kgK
c_{pA}	Apparent specific heat capacity, kJ/kgK
d_c	Characteristic dimension, m
d_1	Product dimension, as used in Equation 6.23, m
d_2	Product dimension, as used in Equation 6.24, m
E_A	Activation energy constant, kJ/mole
E_f	Shape factor, defined by Equation 6.31
EHTD	Equivalent heat transfer dimension, Equation 6.22
E_1	Component of shape factor, defined in Equation 6.32
E_2	Component of shape factor, defined in Equation 6.34
G_1	Constant used in Equation 6.31, and given in Table 6.11
G_2	Constant used in Equation 6.31, and given in Table 6.11
G_3	Constant used in Equation 6.31, and given in Table 6.11
h_c	Convective heat transfer coefficient, W/m^2 K
H	Enthalpy, kJ/kg
ΔH	Enthalpy change during freezing, kJ/kg
ΔH_1	Volumetric enthalpy, defined in Equation 6.26
ΔH_2	Volumetric enthalpy, defined in Equation 6.28
k	Thermal conductivity, W/mK
L	Latent heat of fusion, kJ/kg
M	Molecular weight
M_v^3	Volume fraction, defined in Equation 6.5
m	Mass, kg

N_v^2	Volume fraction, defined in Equation 6.8
N_{Bi}	Biot number
N_{Fo}	Fourier number
N_{Pk}	Planck number
N_{Ste}	Stefan number
P	Constant in Planck's equation, given in Table 6.5
P_v	Volume fraction, used in Equation 6.11
Q	Parameter defined in Equation 6.7
Q'	Parameter defined in Equation 6.10
R	Constant in Planck's equation, given in Table 6.5
R_g	Gas constant, J/mol K
T	Temperature, C
T_F	Equilibrium freezing temperature, C
T_{fm}	Mean freezing temperature, defined in Equation 6.27, C
T_A	Absolute temperature, K
T_{Ao}	Freezing temperature of water, K
ΔT	Temperature change, C
ΔT_1	Temperature gradient, defined in Equation 6.29, C
ΔT_2	Temperature gradient, defined in Equation 6.30, C
t	Time
t_F	Freezing time
t_Q	Frozen food shelf life
X	Mole fraction
X_1	Parameter defined in Equation 6.33
X_2	Parameter defined in Equation 6.35
W	Constant in Equation 6.22

GREEK SYMBOLS

α	Thermal diffusivity, m^2/sec
α_A	Apparent thermal diffusivity, m^2/sec
β	Constant used in Figure 6.9 and Figure 6.12.
β_1	Dimensionless ratio, defined in Equation 6.23
β_2	Dimensionless ratio, defined in Equation 6.24
χ	Insoluble solids, expressed as percentage
λ	Molar latent heat of fusion, kJ/mol
ρ	Density, kg/m^3

SUBSCRIPTS

A	Liquid component in solution
a	Ash component in product, mass fraction
B	Component in solution
C	Center or final
c	Carbohydrate component in product, mass fraction
F	Frozen condition
f	Fat component, mass fraction
I	Ice component, mass fraction
i	Initial condition
j	Juice component, mass fraction
L	Liquid component
p	Protein component, mass fraction

pc	Phase change
R	Reference condition
s	Product solids
si	Individual product components
snj	Solids-not-juice fraction
u	unfrozen component

REFERENCES

Becker, B.B. and Fricke, B.A. 1999. Food thermophysical property models. *Int. Comm. Heat & Mass Trans.* 26: 627–636.

Charm, S.E. and Slavin, J. 1962. A method for calculating freezing time of rectangular packages of food. *Annex Bull. Inst. Int. Froid* pp. 567–568.

Chen, C.S. 1985a. Thermodynamic analysis of the freezing and thawing of foods: Enthalpy and apparent specific heat. *J. Food Sci.* 50: 1158–1162.

Chen, C.S. 1985b. Thermodynamic analysis of the freezing and thawing of foods: Ice content and Mollier diagram. *J. Food Sci.* 50: 1163–1167.

Choi, Y. and Okos, M.R. 1986. Thermal properties of liquid foods: review. *Physical and Chemical Properties of Food.* Martin R. Okos (Ed.), ASAE, St. Joseph, MI. pp. 35–77.

Cleland, A.C. 1990. *Food Refrigeration Processes: Analysis, Desigh and Simulation.* Elsevier Science Pub. Co. New York.

Cleland, A.C. and Earle, R.L. 1977. A comparison of analytical and numerical methods of predicting the freezing times of foods. *J. Food Sci.* 42: 1390–1395.

Cleland, A.C. and Earte, R.L. 1979a. A comparison of methods for predicting the freezing times of cylindrical and spherical foodstuffs. *J. Food Sci.* 44: 958–963.

Cleland, A.C. and Earle, R.L. 1979b. Prediction of freezing times for foods in rectangular packages, *J. Food Sci.* 44: 964–970.

Cleland, A.C. and Earle, R.L. 1982. Freezing time prediction for foods: a simplified procedure. *Int. J. Refrig.* 5: 134–140.

Delgado, A.E. and Sun, D.-W. 2001. Heat and mass transfer models for freezing processes — a review. *J. Food Engr.* 47: 157–174.

Dickerson, R.W. Jr. 1969. Thermal properties of food. In *The Freezing Preservation of Foods,* 4th ed., Vol. 2, D.K. Tressler, W.B. Van Arsdel, and M.J. Copley (Eds), AVI Pub. Co., Westport, CT.

Dickerson, R.W. Jr. 1981. Enthalpy of frozen foods. In *Handbook and Product Directory Fundamentals,* Society of Heating, Refrigerating, and Air-Conditioning Engineers, Atlanta, GA.

Ede, A.J. 1949. The calculation of the freezing and thawing of foodstuffs. *Mod. Refrig.* 52: 53.

Fikiin, K.A and Fikiin, A.G. 1999. Predictive equations for thermophysical properties and enthalpy during cooling and freezing of food materials. *J. Food Engr.* 40: 1–6.

Gutschmidt, J. 1964. Cited in *Cooling Technology in the Food Industry,* 1975, A. Ciobanu, G. Lascu, V. Bercescu, and L. Niculescu, (Eds), Abacus Press, Turnbridge Wales, Kent.

Heldman, D.R. 1974. Predicting the relationship between unfrozen water fraction and temperature during food freezing using freezing point depression. *Trans. ASAE* 17: 63–66.

Heldman, D.R. 1982. Food properties during freezing. *Food Technol.* 36: 92–96.

Heldman, D.R. 1983. Factors influencing food freezing rates. *Food Technol.* 37: 103–109.

Heldman, D.R. 2001. Prediction models for thermophysical properties of foods. Chapter 1 In *Food Processing Operation Modeling: Design and Analysis.* J. Irudayaraj (Ed.), Marcel-Dekker, Inc. New York.

Heldman, D.R. and Gorby, D.P. 1975. Prediction of thermal conductivity in frozen food. *Trans. ASAE* 18: 740–744.

Heldman, D.R. and Singh, R.P. 1981. *Food Process Engineering,* 2nd ed., AVI Pub. Co., Westport, CT.

Heldman, D.R. and Singh, R.P. 1983. Thermal properties of frozen foods. *Physical and Chemical Properties of Foods.* Martin R. Okos (Ed.), ASAE, St. Joseph, MI, pp. 120–137.

Joshi, C. and Tao, L.C. 1974. A numerical method of simulating the axisymmetrical freezing of food systems. *J. Food Sci.* 39: 623.

Jul, M. 1984. *The Quality of Frozen Food.* Academic Press. London.

Kerr, W.L., Ju, J., and Reid, D.S. 1993. Enthalpy of frozen foods determined by differential compensated calorimetry. *J. Food Sci.* 58: 675–679.

Kopelman, I.J. 1966. Transient heat transfer and thermal properties in food systems. Ph.D. Dissertation. Michigan State University. E. Lansing.

Lai, D.J. and Heldman, D.R. 1982. Analysis of kinetics of quality change in frozen foods. *J. Food Process Engr.* 6: 179.

Levy, F.L. 1958. *J. Refrig. 1: 35.* Cited by Brennen el ah, 1976, *Food Engineering Operations*, 2nd ed., Chapter 14. Applied Science, London.

Lind, I. 1991. The measurement and prediction of thermal properties of food during freezing and thawing — a review with particular reference to meat and dough. *J. food Engr.* 13: 285–319.

Mannapperuma, J.D. and Singh, R.P. 1988. Prediction of freezing and thawing of foods using a numerical method based on enthalpy formulation. *J. Food Sci.* 53: 626–630.

Mannapperuma, J.D. and Singh, R.P. 1989. A computer-aided method for prediction of properties and freezing/thawing times of foods. *J. Food Engr.* 2: 275–304.

Mellor, J.D. 1976. Personal communications cited by A.C. Cleland and R.L. Earle, *J. Food Sci.* 44: 958.

Miles, C.A. 1991. The thermophysical properties of frozen foods. In *Food Freezing Today and Tomorrow.* W.B. Bald (Ed.), Springer Verlag London Limited. London, pp. 45–65.

Mohsenin, N.N. 1980. *Thermal Properties of Foods and Agricultural Materials.* Gordon and Breach, New York.

Mott, L.F. 1964. The prediction of product freezing time. *Aust. Refrig. Air Cond. Heat.* 18: 16.

Nagaoka, J., Takagi, S., and Hotani, S. 1955. Experiments on the freezing of fish in an air-blast freezer. *Proc. 9th Int. Congr. Refrig.*, Vol. 2, p. 4.

Pham, Q.T. 1986. Simplified equation for predicting the freezing time of foodstuffs. *J. Food Technol.* 21: 209–211.

Pham, Q.T. 1996. Prediction of calorimetric properties and freezing time of foods from composition data. *J. Food Engr.* 30: 95–107.

Qashou, M.S., Vachon, R.L, and Touloukian, V.A. 1972. Thermal conductivity of foods. *ASHRAE Trans.* 75: 165.

Ramaswamy, H.S. and Tung, M.A. 1981. Thermophysical properties of apples in relation to freezing. *J. Food Sci.* 46: 724–728.

Reidy, G.A. 1968. Thermal properties of foods and methods of their determination. M.S. Thesis. Food Science Department, Michigan State University.

Riedel, L. 1951. The refrigeration required to freeze fruits and vegetables. *Refrig. Eng.* 59: 670–673.

Riedel, L. 1956. Calorimetric investigations of the freezing of fresh meat. *Kaltetechnik* 8: 374–377 (in German).

Riedei, J. 1957a. Calorimetric investigations of the meat freezing process. *Kaltetechnik* 9: 38–40 (in German).

Riedel, J. 1957b. Calorimetric investigations of the freezing of egg whites and yolks. *Kaltetechnik* 9: 342–345 (in German).

Sastry, S.K. 1984. Freezing time prediction: an enthalpy-based approach. *J. Food Sci.* 49: 1121–1127.

Schwartzberg, H.G. 1976. Effective heat capacities for freezing and thawing of food. *J. Food Sci.* 41: 152–156.

Scott, E.P., Steffe, J.F., and Heldman, D.R. 1989. Frozen food quality improvements through storage temperature optimization. *Changing Food Technology II.* M. Kroger and A. Freed (Eds), Technomic Pub Co, Inc. pp. 189–208.

Singh, R.P. and Heldman, D.R. 2001. *Introduction to Food Engineering.* 3rd ed. Academic Press, New York.

Tao, L.C. 1967. Generalized numerical solutions of freezing a saturated liquid in cylinders and spheres. *AIChE J.* 13: 165.

Tien, R.H. and Geiger, G.E. 1967. A heat transfer analysis of the solidification of binary eutectic system, *J. Heat Transfer* 9: 230.

Tien, R.H. and Geiger, G.E. 1968. The unidimensional solidification of a binary eutectic system with a time-dependent surface temperature. *J. Heat Transfer* 9C: 27.

Tien, R.H. and Koumo, V. 1968. Unidimensional solidification of a subvariable surface temperature. *Trans. Metall. Soc. AIME* 242: 283.

Tien, R.H. and Koumo, V. 1969. Effect of density change on the solidification of alloys. *Am. Soc, Mech. Eng.* (Paper) 69-HT-45.

U.S. Department of Agriculture, Agricultural Research Service. 2004. USDA National Nutrient Database for Standard Reference, Release 17. Nutrient Data Laboratory Home Page, http://www.nal.usda.gov/fnic/foodcomp

van Arsdel, W.B., Copley, M.J., and Olson, R.D. 1969. *Quality and stability of frozen foods.* John Wiley and Sons. New York.

WFLO. 2000. Industrial Scale Food Freezing — Simulation and Process. http://www.wflo.org/hq/forms/freeze.pdf

Woodams, E.E. and Nowrey, J.E. 1968. Literature values of thermal conductivities of foods. *Food Technol.* 22: 150.

7 Mass Transfer in Foods

Bengt Hallström, Vassilis Gekas, Ingegerd Sjöholm, and Anne Marie Romulus

CONTENTS

7.1 INTRODUCTION

In the area of food engineering, mass transfer phenomena play an important role in many applications, including (1) water (liquid or vapor) in products and air in drying or similar operations, such as osmotic dehydration; (2) aroma components in products during drying and storage; (3) extraction/coagulation (sugar, tea, coffee); (4) salt in products (meat, cheese); and (5) gas and liquid permeation through packaging materials and membranes. The transfer of water is the dominant phenomenon, and most of this chapter is concerned with this and related problems. The terms "mass transfer" and "diffusion" are sometimes used without distinction. This is not correct. Mass transfer may take place in several modes, only one of which is diffusion. These different transfer modes are sometimes treated in the same way as diffusion mathematically (Fick's law), despite the fact that this is not correct from a phenomenological point of view. In such cases the corresponding coefficient is called the apparent or effective diffusion coefficient. Distillation is rare in the food industry and is not covered in this chapter.

7.2 PRINCIPLES AND THEORY OF DIFFUSION

Diffusion is the process by which matter is transported from one part of a system to another as a result of random molecular motion. Despite the fact that no molecule has a preferred direction of motion, a transfer of molecules from a region of higher concentration to a region of lower concentration is observed: as the number of molecules is higher in the former region, more molecules will pass toward the region of lower concentration than in the opposite direction.

Fick defines the diffusion coefficient by means of the concentration gradient

$$J_A = -D_A \frac{dc_A}{dx} \tag{7.1}$$

In the thermodynamic approach, the molar diffusion velocity of a component A is defined in terms of the chemical potential μ_c:

$$v_{A,x} - V_x = -\frac{D_A}{RT} \frac{d\mu_c}{dx} \tag{7.2}$$

where $v_{A,x}$ is the linear velocity of component A related to a fixed coordinate and V_x is the velocity of the bulk. Then the molar flux of A becomes

$$J_A = c_A(v_{A,x} - V_x) = -c_A \frac{D_A}{RT} \frac{d\mu_c}{dx} \tag{7.3}$$

Introducing

$$\mu_c = \mu_0 + RT \ln c_A \tag{7.4}$$

gives

$$J_A = -D_A \frac{dc_A}{dx} \left(\frac{kmol}{m^2 sec} = \frac{m^2}{sec} \cdot \frac{kmol}{m^3 \cdot m} \right) \tag{7.5}$$

This equation also gives the dimensions of D: m^2/sec.

Equation 7.4 is valid, of course, when the activity coefficient γ is equal to 1, that is, for dilute/ideal solutions.

For nonstationary conditions, Equation 7.5 leads to

$$\frac{\partial c_A}{\partial t} = D_A \frac{\partial^2 c_A}{\partial x^2} \tag{7.6}$$

For diffusion in all three directions,

$$\frac{\partial c_A}{\partial t} = D_A \left(\frac{\partial^2 c_A}{\partial x^2} + \frac{\partial^2 c_A}{\partial y^2} + \frac{\partial^2 c_A}{\partial z^2} \right) \tag{7.7}$$

If D_A is related to c_A,

$$\frac{\partial c_A}{\partial t} = \frac{\partial}{\partial x} \left(D_A \frac{\partial c_A}{\partial x} \right) \tag{7.8}$$

This defines ordinary diffusion. However, if A is diffusing into another component, B, B is simultaneously diffusing into A. This is called mutual or binary diffusion. It can be shown that

$$D_{AB} = D_{BA} \tag{7.9}$$

So far, a plane of no net molar transfer is assumed as the section defining D. If this section is moving, due to bulk transport or volume change related to the mixing process, the total flow in relation to fixed coordinates is given by

$$N_A = J_A + x_A(N_A + N_B) \tag{7.10}$$

where N_A is the flow of A in relation to fixed coordinates, J_A is diffusive flow, and $x_A(N_A + N_B)$ is the part of A in the bulk flow.

By means of labeled molecules it is possible to measure the diffusion of one component, A, in a uniform system, AB; this is termed self-diffusion, D_{AA}. Tracer-diffusion (also called intradiffusion) refers to diffusion of very small amounts of component A into another component B, or in a mixture AB: D_A^*.

If in Equation 7.2, the flux is used instead of velocities, then an interesting alternative approach arises, defining the so-called mass conductivity, L_A:

$$J_A = -L_A \frac{d\mu_A}{dx} \tag{7.11}$$

Combining 7.1 and 7.11 and using the chain property:

$$\frac{\partial c}{\partial x} = \frac{\partial \mu}{\partial x} \cdot \left(\frac{dc}{d\mu} \right)_T \tag{7.12}$$

This leads to a relationship between the D_A and the L_A:

$$D_A = \frac{L_A}{(dc/d\mu)_T} \tag{7.13}$$

(Isothermal notation is followed here for emphasis.)

If, now, we use the molar fraction X instead of the concentration C, $C = \rho X$, Equation 7.13 becomes analogous to the well-known heat transfer properties $\alpha = \lambda / C_p$ relationship:

$$D_A = \frac{L_A}{\rho (dX_A / d\mu)_T} \tag{7.14}$$

As we see in Section 7.3, the derivative in the denominator is closely related to the sorption isotherm if A is the water contained in a food.

7.3 WATER AND WATER VAPOR IN POROUS MATERIALS

7.3.1 VAPOR

7.3.1.1 Knudsen Diffusion (D_K)

The mean-free path of the molecules is long compared with the pore diameter (in general, negligible for pore diameters greater than 0.1 μm).

7.3.1.2 Ordinary or Binary Diffusion (D_{AB})

The mean-free path of the molecules is much less than the pore diameter. The transition region between Knudsen and ordinary diffusion is defined by

$$\frac{1}{D_P} = \frac{1}{D_K} + \frac{1}{D_{AB}} \tag{7.15}$$

7.3.1.3 Stephan Diffusion

Diffusion of vapor through a layer of stagnant air (e.g., water evaporating from a wet surface into bulk air [i.e., convective drying]):

$$N_A = \frac{D_P}{RT\,\Delta x} \ln \frac{P - p_{w\infty}}{p - p_{wo}} \quad \text{(deduced layer)} \tag{7.16}$$

A surface transfer coefficient may be defined:

$$k_g = \frac{D}{\Delta x} \tag{7.17}$$

7.3.1.4 Hydraulic Flow

$$N = -Ab\frac{dp}{dx} \tag{7.18}$$

7.3.1.4.1 Laminar Flow

$$b_{lam} = \frac{d^2}{32v} \tag{7.19}$$

where d is the pore diameter and v is the viscosity. Turbulent flow is seldom reached, due to the very small pore diameter.

7.3.1.5 Condensation Evaporation

Closed pores partially filled with water are subjected to a temperature gradient. Water is vaporized at the end of the pore with the higher temperature and condenses at the opposite end. Liquid water is transferred back along the wall of the pore.

7.3.2 WATER

7.3.2.1 Capillary Flow

Surface Diffusion: Surface diffusion is the transfer of water molecules adsorbed onto the wall in the pores.

7.3.2.2 Hydraulic Flow

Components: Components dissolved in water.

Effective or apparent diffusion coefficient:

$$D_{\text{eff}} = D\frac{\in}{\tau} \tag{7.20}$$

where $\in$ porosity and τ is the tortuosity factor ($\tau = l_e/l$, where l_e is the effective average path length, l is the shortest distance).

7.4 MASS TRANSFER WITHIN FOODSTUFFS

Most solid foodstuffs are capillary porous and colloidal materials. Several of the transport mechanisms mentioned in Section 7.3 are applicable. For some of these foodstuffs the type and rate of transfer (i.e., the effective or apparent diffusion coefficient) change during (thermal) processing. Examples are the baking of bread, cooking of meat, and drying of several products. These structural changes also often result in geometrical changes: swelling or shrinking. It is therefore often difficult to determine single values of the diffusion coefficient for this type of material.

The European project DOPPOF, a continuation of the previous COST 90 projects, calculated effective diffusivity data in various foods. A comprehensive list has been presented in a book by Saravacos and Maroulis (2001).

A knowledge base on the D_{eff} has been published by Doulia et al. (2000) and this publication contains a large portion of the relevant DOPPOF database.

7.4.1 DIFFUSION OF SALT IN CHEESE

Within the COST 90 bis project, Gros and Rüegg (1987) studied the diffusion of salt in agar gel and cheese at 25°C. The diffusivity was measured during unidirectional diffusion between a gel cylinder containing an initially uniform salt concentration and a gel cylinder containing a lower concentration of salt (the infinite couple). The diffusion coefficient was calculated using a fitting method. In a 3% w/w agar gel the diffusion coefficient was found to be between 1.36 and 1.46 × 10^{-9} m²/sec. This study was performed by a group of laboratories. The results for cheese (Sbrinz hard cheese with 35% water and 31% fat) are reported in Table 7.1.

7.4.2 DIFFUSION OF WATER IN MINCED MEAT

Motarjemi (1988) studied the diffusion of water in minced meat (beef: 76% water, 21% protein, 1 to 2% fat). She used two methods: drying and the infinite couple (or concentration profile curve)

TABLE 7.1
Diffusion Coefficients for Salt in Cheese at Different Temperatures

Temperature (°C)	Diffusion coefficient (10^{-10} m²/sec)
7	1.06
11	1.49
15	1.71
20	1.88

Source: From Gros, J.B. and Rüegg, M. 1987. *Physical Properties of Foods,* R, Jowitt et al. (eds.), Elsevier, London.

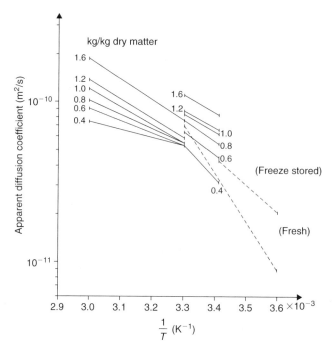

FIGURE 7.1   Apparent diffusivity of minced meat (beef). Comparison of the results obtained from the concentration profile curve with drying method. (——) = Drying; (- - - - -) = Concentration profile curve. (From Motarjemi, Y. 1988. A study of some physical properties of water in foodstuffs. Ph.D. thesis, Lund University, Lund, Sweeden. With permission.)

method. The results are shown in Figure 7.1. It is interesting to note that despite the use of different methods, the results agree very well. No previous publication has reported on such a comparison of methods.

7.4.3 Diffusion of Volatiles in Water

Voilley and Roques (1987) studied the diffusion of aromatic compounds in water. The same experimental technique as for salt in cheese was used (COST 90 bis project). Some of the results are presented in Figure 7.2.

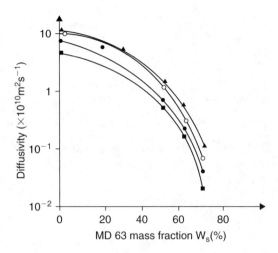

FIGURE 7.2 Diffusivity of volatiles as a function of sugar concentration. (▲) Acetone; (O) diacetyl; (●) *N* hexanol; (■) acetone. (From Voilley, A. and Roques, M. 1987. In *Physical Properties of Foods*, Vol. 2, R. Jowitt et al. (eds.), Elsevier, London. With permission.)

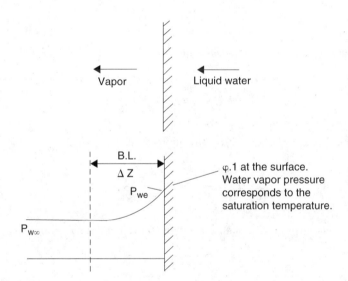

FIGURE 7.3 Vapor diffusion in the boundary layer.

7.5 VAPOR DIFFUSION IN STAGNANT AIR (STEPHAN DIFFUSION)

In Section 7.2 the following equation was given for a binary system:

$$N_A = J_A + x_A(N_A + N_B)$$

A and B are diffusing into each other simultaneously. Suppose that A is water vapor leaving the surface of a moist material (drying) and diffusing into an air boundary layer (B) (Figure 7.3).

The air is not moving perpendicular to the surface, and accordingly, $N_B = 0$.

$$N_A = \frac{1}{1 - x_A} J_A \tag{7.21}$$

$$J_A = -D_A \frac{dc_A}{dz} \tag{7.22}$$

c_A is expressed here as density. However, in this problem it is better to measure the concentration of water vapor either as mole fraction (x_A) or as partial pressure (p_w).

$$x_A = \frac{c_A}{c_A + c_B} \tag{7.23}$$

$$x_A = x_w = \frac{W/M_w}{W/M_w + L/M_L} \tag{7.24}$$

Dalton's law gives

$$\frac{W}{L} = \frac{M_w}{M_L} \frac{p_w}{P - p_w} \tag{7.25}$$

These equations are used to rewrite N_A, resulting in

$$N_A = -\frac{D}{RT} \frac{P}{P - p_w} \frac{dp_w}{dz} \tag{7.26}$$

Integrating over $p_{w\infty}$ (the air bulk) and p_{wo} (the moist surface) gives

$$N_A = \frac{D}{RT\Delta Z} \ln \frac{P - p_{w\infty}}{P - p_{wo}} \tag{7.27}$$

Often, a surface mass transfer coefficient, k_g, is defined (as mentioned before):

$$k_g = \frac{D}{\Delta Z} \tag{7.28}$$

$$N_A = k_g \frac{P}{RT} \ln \frac{P - p_{w\infty}}{P - p_{wo}} \tag{7.29}$$

Here N_A is expressed in kmol/m$^2 \cdot$ sec. Multiplication by the molecular weight of water gives the result in kg/m$^2 \cdot$ sec.

We note here, in advance, that Equation 7.29 is remarkably analogous to the film theory equation regarding the phenomenon of concentration polarization (Section 7.11).

7.6 MASS TRANSFER COEFFICIENT

Calculation of mass transfer coefficients is in most cases based on empirical results. By means of dimensional analysis, equations can be derived for different situations.

In the main, the following dimensionless numbers are used:

$$\text{Sherwood number:} \quad \text{Sh} = \frac{k_g d}{D} \tag{7.30}$$

$$\text{Reynold's number:} \quad \text{Re} = \frac{vd}{\nu} \tag{7.31}$$

$$\text{Schmidt number:} \quad \text{Sc} = \frac{\nu}{D} \tag{7.32}$$

$$\text{Grashof number:} \quad \text{Gr} = g\frac{\beta d^3 \Delta T}{\nu^2} \tag{7.33}$$

The general form of the Sherwood correlation is:

$$\text{Sh} = c_1 \cdot \text{Re}^{c_2} \cdot \text{Sc}^{c_3} \tag{7.34}$$

Usually, the exponent of the Sc number is 1/3, independent of the type of flow. The exponent of the Reynold number is usually 1/3 for laminar flow and 0.8 for turbulent flow in a circular tube or a flat channel.

Turbulent flow through a circular tube or a flat channel was examined by Gekas and Hallström in two subsequent papers in *The Journal of Membrane Science* (1987, 1988). Their remarks are not only applicable to the membrane area.

Laminar flow through a circular tube:

$$\text{Sh} = 1.86 \, \text{Re}^{0.8} \tag{7.35}$$

Forced conversion around a solid sphere:

$$\text{Sh} = 2.0 + 0.6 \, \text{Re}^{1/2} \, \text{Sc}^{1/3} \tag{7.36}$$

7.7 SORPTION ISOTHERMS

The presence and status of water in a material are described by the equilibrium between the water content inside the food and the water activity of the food. This is the situation for hygroscopic foodstuffs. For nonhygroscopic materials the vapor pressure above the surface is equal to the vapor pressure for pure water at the same temperature. For a hygroscopic material the vapor pressure above the surface is lower and the relationship P_w/P_{ws} is called water activity. The decreased vapor pressure above a hygroscopic material is mainly due to the structure and the porosity of the material. Important phenomena in food processing are related more to water activity than to the water content, as can be seen from Figure 7.4.

The equilibrium between the water content and the water activity of a material is described by the sorption isotherm (Figure 7.5). In most cases the sorption isotherm of a foodstuff is an S-shaped curve. The first part of the curve (a) is convex toward the water content axis, and this part is normally said to correspond to the monomolecular layer of adsorbed water. The middle part (b) is an almost straight line and corresponds to the multimolecular layer. In the third part (c) capillary condensation takes place.

Sorption isotherms for different foodstuffs are given in the literature, but it must be emphasized that the relationship sometimes varies with the origin and prior treatment of the product. Several attempts have been made in the literature to describe isotherms by means of equations or models.

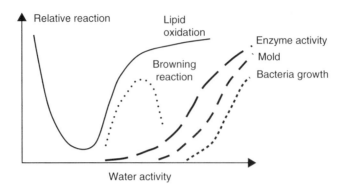

FIGURE 7.4 Importance of water activity.

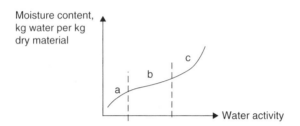

FIGURE 7.5 Principle of the sorption isotherm.

Some of these are presented below (Chirife and Iglesias, 1978; Bruin and Luyben, 1980; van den Berg and Bruin, 1981):

1. Linear relationship: $c_w = c\alpha_w + c_2$
2. Oswin: $c_w = c\alpha_w^n / (1 - \alpha_w)^n$
3. Kühn: $c_w = c(\ln \alpha_w)^n + c_2$
4. Fugassi: $c_w = c\alpha_w / c_2 \alpha_w (1 - \alpha_w) + c_3 \alpha_w$
5. BET: $c_w = c_{max} c\alpha_w / (1 + c\alpha_w - \alpha_w)(1 - \alpha_w)$
6. Halsey: $c_n \alpha_w = -c(c_2 / c_w)^n$
7. Chirife: $c_w + (c_w^2 + c_{w0.5})^{0.5} = \exp(b\alpha_w + d)$

7.8 DRYING AND OSMOTIC DEHYDRATION

7.8.1 Basic Theory of Drying

Drying is characterized by the simultaneous transfer of heat and mass (water). This makes the theoretical treatment complex, and engineering calculations of the process become complicated. During the first period of the drying, when the surface of the material is wet, all heat transferred from the air is used for the evaporation of water from the surface. Water from internal parts of the material is transferred rapidly to the surface. During this period — the constant-rate period — the surface maintains a constant temperature equal to the wet-bulb temperature. Accordingly, the water is evaporated at a constant rate, but the constant rate drying period represents a relatively short portion of the total drying time.

Gradually, the moisture content at the surface decreases and the internal resistance to water transport increases. The evaporation zone moves from the surface into the material and a crust or a skin is often formed. This means that the heat necessary for evaporation has to be transported

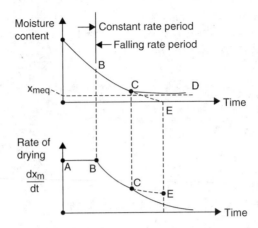

FIGURE 7.6 Principle drying curves.

from the air to the surface and then into the material to the evaporation zone. In the same way, the evaporated water has to make its way through the dry material to the surface and then into the air. A temperature gradient is required for heat transport to take place, and, accordingly, the surface temperature increases above the wet-bulb temperature. At the evaporation zone α_w is 1, or, expressed in terms of relative humidity, ϕ is equal to 100% at the surface to obtain water vapor transport to the surface, a potential is needed, and therefore α_w at the surface is decreased to below 1. This phase of drying is called the falling-rate period. Conventionally, the procedure is illustrated by means of drying curves (Figure 7.6). x_{mB} is called the critical moisture content.

When drying a hygroscopic material, there is a further breakpoint in the curve, C. This is when the part of the material with the highest moisture content reaches the maximum hygroscopic moisture content, corresponding to $\alpha_w = 1$. The drying rate is then decreased further and the moisture content of the material slowly reaches a value x_{meq} corresponding to the actual relative humidity of the air, ϕ. A hygroscopic material can never be dried to $x_m = 0$ as long as the relative humidity of the surrounding air is above 0. In a nonhygroscopic material, drying continues to point E, where the moisture content of the material reaches 0, irrespective of the value of the relative humidity of the air.

As already mentioned, during the falling-rate period the surface temperature increases above the wet-bulb temperature. Therefore, this period is especially dangerous when drying heat-sensitive materials (as most foodstuffs are), and special care must be taken. This is illustrated in Figure 7.7. It is interesting to note that in concurrent air flow drier a, the final material temperature is limited by the temperature of the air leaving the drier. In countercurrent flow, however, the material may reach the temperature of the incoming hot air (Figure 7.8). This may be avoided if the drier is provided with two drying chambers.

In this case, a high air temperature may be used during the constant-rate period while the material is maintained at the wet-bulb temperature. During the falling-rate period, the drying may take place in a separate chamber with more gentle treatment. In this way the energy consumption may be considerably decreased. This two-stage drying arrangement is often used in the spray drying of milk.

7.8.2 Constant-Rate Period

As discussed above, during this period there is an exchange of heat and mass only between the air and the surface. The internal parts of the material are not involved in the calculation of the drying rate. In most types of drying, heat is transferred by convection:

$$q = h(t_\infty - t_s)$$

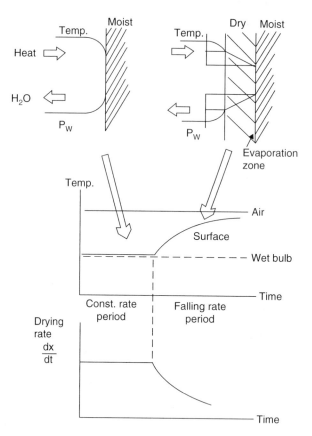

FIGURE 7.7   Temperature profiles during the constant- and falling-rate periods.

and in the same way the water is evaporated (transfer of mass):

$$m = k_g \frac{M_w P}{RT} \ln \frac{P - p_\infty}{p - p_s}$$

m and q are balanced via

$$m \cdot r = q$$

h and k_g may be calculated by means of the dimensionless numbers Nu and Sh.

However, with knowledge of the relationship between m and q it is necessary to know only one of the transfer coefficients. For the mixture of air and water there is also a relationship according to Lewis:

$$\frac{h}{k_g} = c_p \rho \left(\frac{\alpha}{D}\right)^{2/3}$$

7.8.3   Falling-Rate Period

The falling-rate period is difficult to treat theoretically. It is thus necessary to rely on experimental results. It is often possible to adapt the results to a simplified mathematical model. The model can then be used to estimate how the different parameters influence the drying procedure.

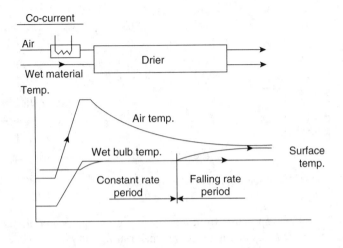

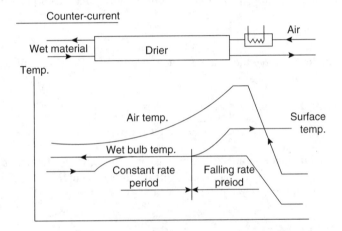

FIGURE 7.8   Temperature relationships during concurrent and countercurrent drying.

7.8.3.1   Porous Materials

According to the model most often used, evaporation takes place at an evaporation front inside the material, and the vapor formed is transferred through the pores of the dry material toward the surface. As evaporation progresses, the front moves away from the surface toward the interior of the material. This is called the shrinking core or receding-front model (Clement et al., 1991). Heat transfer to the evaporation front depends on both the heat transfer coefficient at the surface and the thermal conductivity of the dry material between the surface and the evaporation front:

$$q = \frac{1}{1/h + \Delta z/k}(T_\infty - T_z)$$

The transfer of vapor also has two resistances to overcome: diffusion in the dry material and the mass transfer coefficient at the surface:

$$m = \frac{M}{RT}\frac{1}{1/k_g + \Delta z/D}(P_{vz} - p_{w\infty})$$

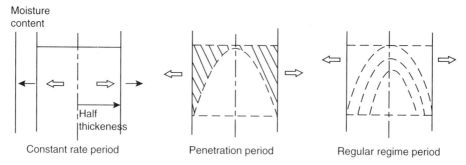

FIGURE 7.9 Moisture profiles inside the material during drying.

T_z corresponds to P_{vz} at the evaporation front. For a nonhygroscopic material this means that T_z is the saturation temperature at P_{vz}. For a hygroscopic material T_z and P_{vz} are related via the sorption isotherm. Inside the evaporation zone the transfer of water takes place as liquid diffusion.

7.8.3.2 Polymers

Fick's law is applicable. For a parallel plate geometry, constant diffusion coefficient, and sufficient drying time, the solution is

$$t = \frac{1}{D} \left(\frac{\Delta z}{\pi} \right)^2 \left(\ln \frac{8}{\pi^2} \right) \frac{x_0 - x_e}{x_t - x_e}$$

The calculation procedure is the same as that for nonstationary heat transfer and the same diagrams are useful. One drawback of this method is that the diffusion constant of liquid water in these materials varies considerably with the moisture content.

7.8.4 REGULAR REGIME MODEL

This method was developed to enable calculations concerning the drying process for materials with a diffusivity that is highly dependent on moisture content (Schoeber, 1976; Yamamoto et al., 2002). The drying process is divided into three periods: (1) the constant-rate period, (2) the penetration period, and (3) the regular regime period. The moisture profile inside the material to be dried determines the actual drying period. Figure 7.9 shows the profiles corresponding to the different periods, which will be explained more theoretically.

7.8.4.1 Constant-Rate Period

During the constant-rate period the water transport depends only on external transfer resistance. The mass transfer coefficient k_g is calculated by means of dimensionless numbers, as described previously.

7.8.4.2 Penetration Period

When a moisture profile starts to develop on the product, the penetration period begins. This changeover occurs at the critical moisture content. The penetration period proceeds until a stable moisture profile is developed. The moisture profile is described by the equations

$$F_{pp}E = \text{constant}$$

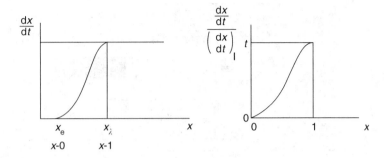

FIGURE 7.10 Normalized drying curves.

where $F_{pp} = m\rho_s \Delta z_s$ (ρ_s is the density of solids and Δz_s is the half-thickness of solids) and

$$E = \frac{x_0 - x}{x_0 - x_e}$$

A characteristic of the penetration period is that the drying rate is dependent on the rate in the first period.

7.8.4.3 Regular Regime Period

When a stable moisture profile moves toward the center of the product, the regular regime period begins. The equation for the regular regime period is

$$\text{FRR} = C \exp\left(\frac{G}{T}\right)(x - x_e)^B$$

where C, G, and B are constant and T is the temperature. The calculation procedure is described in Schoeber (1976).

7.8.5 NORMALIZED DRYING CURVES

For many problems it is very difficult and time consuming to analyze how the drying rate is influenced by different variables. In such cases a normalized drying curve may be used. The drying rate is divided by the drying rate of the constant-rate period. A dimensionless moisture content x is formed so that $x = 0$ equals the equilibrium moisture content and $x = 1$ the critical moisture content (Figure 7.10). Some authors maintain that such curves can be used generally for the same material. However, it is recommended that the curve should be experimentally determined when air conditions are changing. The normalized drying curves can then easily be used in a calculation program as Excel, Matlab, or FEMLAB Multiphysics Modeling to predict drying curves and drying times.

7.8.6 OSMOTIC DEHYDRATION

Fresh fruit and vegetables contain 75 to 95% water and one way to reduce the water content initially is to use osmotic dehydration. The difference in osmotic pressure of the immersion solution and the product is the driving force of the process. Often used solutes are sugar solutions, 30 to 50 Brix, or various salt solutions (Gekas et al., 1998). The osmotic dehydration step can remove up to 50% of the water in the original fruit or vegetable. The product will lose water and most often gain solutes from the immersion solution. To achieve a stable product with a long shelf life requires a final stage of convection airdrying, vacuum drying or microwave-assisted drying.

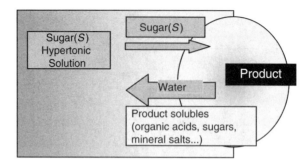

FIGURE 7.11   Mass transport in osmotic drying.

TABLE 7.2
Examples of Extraction Processes in the Food Industry

Raw material	Solvent	Product
Processes where solute is the valuable product		
Sugar beets	Water	Sugar juice
Coffee beans	Water	Coffee extract
Fish meal	Organic	Fish oil
Calf stomach	Water	Rennet
Processes where purified solid is the valuable product		
Potatoes	Water	Starch
Fermentation broth	Methyl isobutyl ketone	Penicillin

The mass transfer in a system when immersing pieces of fruit is a function of the temperature and osmotic pressures of the system, the reology of the immersion solution, the viability of the product, the geometry and stucture of the product, and agitation. The viability and eventual collapse of the cell structure during immersion will determine the status of the product before the final drying (Prothon et al., 2003).

The kinetics of mass transfer in osmotic drying is a two-way exchange of solutes and soluble components. The system is schematically described in Figure 7.11.

7.9   LIQUID–SOLID EXTRACTION

This process is also called leaching. A soluble component is extracted from a solid by means of a solvent. Some examples from the food industry are given in Table 7.2. The calculation of liquid–solid extraction is based on a material balance and the concept of an ideal or theoretical stage in which it is supposed that the phases are in equilibrium. The time required to reach equilibrium depends on the diffusion coefficient and the length of the diffusion path (i.e., the size of the solid's particles). The equilibrium time can be calculated by means of Fick's law:

$$\frac{\partial c}{\partial t} = D \frac{\partial c^2}{\partial x^2}$$

This equation may be solved in the same way as for Fourier's law (Chapter 5). However, for the comparison with heat transfer, $\Delta T / \Delta T_0$ is replaced by $\Delta c / \Delta c_0$ and $\alpha = k / \rho c_p$ is replaced by D.

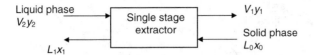

FIGURE 7.12   Material balance for single-stage extraction.

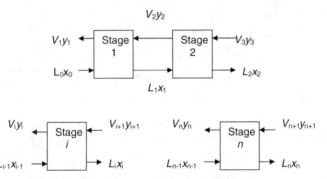

FIGURE 7.13   Material balance for multistage countercurrent extraction.

The material balance for a single-stage extraction system is illustrated in Figure 7.12.

The weight concentration of a component is denoted by y in the liquid phase, also called the overflow. The concentration in the solid phase, underflow, is termed x.

$$L_0 + V_2 = L_1 + V_1$$

$$L_0 x_0 + V_2 y_2 = L_1 x_1 + V_1 y_1$$

These equations have to be combined with the equilibrium relationships between the components. This may be done by means of triangular diagrams (for the procedure, the reader is referred to the relevant literature). This type of extraction is normally performed in a multistage countercurrent plant (Figure 7.13). The total material balance

$$L_0 + V_{n+1} = L_n + V_1$$

and the balance for the extractable component

$$L_0 x_0 + V_{n+1} y_{n+1} = L_n x_n + V_1 y_1$$

A stepwise procedure for calculating the ideal number of stages is as:

1. Calculate flow quantities and concentrations of all intermediate flows between stages.
2. Calculate the concentration for an ideal stage 1 using the material balance and the equilibrium characteristics.
3. Calculate the concentration for the following ideal stage by using the material balance equations given above.

The principle is demonstrated in Figure 7.14. In the figure, EF represents the locus of overflow compositions for the case in which the overflow stream contains inert solids. Lines GF, GL, and GM represent the loci of underflow compositions for the three different conditions indicated on

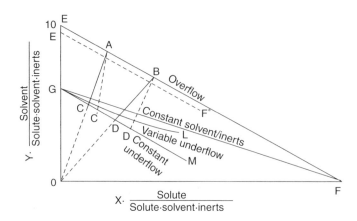

FIGURE 7.14 Calculation of the number stages. (From Perry, R.H. and Greene, D. 1984. *Perry's Chemical Engineering Hand Book*. McGraw-Hill, New York. With permission.)

the diagram. In the figure, the constant underflow line GM is parallel to EF, the hypotenuse of the triangle, whereas GF passes through the right-hand vertex, representing 100% solute.

The compositions of the overflow and underflow streams leaving the same stage are represented by the interaction of the composition lines for those streams with a tie line (AC, BD). Equilibrium tie lines (AC, BD) pass through the origin (representing 100% inerts). For nonequilibrium conditions with or without adsorption or for equilibrium conditions with selective adsorption, the tie lines are replaced (e.g., AC and BV). Point C′ is to the right of C if the solute concentration in the overflow solution is less than that in the underflow solution adhering to the solids. Unequal concentrations in the two solutions indicate insufficient contact time, preferential adsorption of one of the components on the inert solids, or both. Lines such as AC can be considered to be practical lines (i.e., they have been obtained experimentally under conditions simulating actual operation, particularly with respect to contact time, agitation, and the particle size of solids).

7.9.1 EXTRACTOR TYPES IN THE FOOD INDUSTRY*

The simplest type of extractor used is the single-stage extractor consisting of a tank containing the solid to be extracted (Table 7.3). The liquid phase flows over the solid and is drained from the bottom of the tank. This is a batch system, and after extraction the tank has to be emptied and refilled. Several such batch extractors may be connected in a series, with the solvent being pumped from one stage to the next, forming an extraction battery. The next development in terms of configuration is a continuous belt conveyor. Continuous systems of special design have been developed: Rotocal (Rosedowns, U.K.) is a rotary extractor and the DDS double-screw extractor (DDS, Denmark) was originally developed for sugar extraction from beets but is now also used in fruit juice extraction.

7.10 MASS TRANSFER THROUGH PACKAGING MATERIALS

A general equation for the calculation of the transfer of a constituent through a packaging film is

$$G = \frac{P}{s}At\Delta p$$

---

* The information in this section is from Brennan et al. (1976) and Lyndersen (1983).

TABLE 7.3
Continuous Extractors Used in the Food Industry

	Sugar conc.	Sugar beet	Fish meal	Soy	Coffee	Fruit juice	Oil seeds	Hops	Tea
Rotocal (Rosedowns, U.K.)	x		x	x	x		x	x	x
Carroucel (Extraktionstechnik, Germany)	x		x	x	x		x	x	x
DDS double-screw extractor (DDS, Denmark)		x				x			
Vertical screw tower		x				x			
Double-screw conveyor	x	x	x			x			x
Stationary cells, rotating feed			x			x			
Continuous belt	x	x			x	x			

TABLE 7.4
Examples of Transmission Data for Some Packaging Films

Packaging material	Density (kg/m³)	WVTR (g/m²·day)	GTR (cm³/m² · day · bar, 25 μm)		
			Oxygen	Nitrogen	Carbon dioxide
Cellophane	1,440	18–198	7.8–12.4	7.8–24.8	6.2–93
Cellulose acetate		2,480	1,813–2,235	465–620	13,300–15,500
Nylon	1,130	248–341	40–527	14–53	155–2,370
Polyethylene					
Low density	910–950	21.7	7,750	2,790	41,850
High density	940–970	4.6	2,667	651	8,990
Polyethylene-vinyl acetate		31–46	13,020	6,200	93,000
Polyvinyl chloride-acetate (plasticized)	1,230	77.5–124	310–2,325	155–930	1,085–12,400

Source: From Perry, R.H. and Greene, D. 1984. *Perry's Chemical Engineer's Handbook.* McGraw-Hall, New York. With permission.

Here P/s is the permeability constant of the film, normally given in g/m² · day · mmHg, s is the thickness of the film, A is the area of the package in m², t is the time in days, and Δp the difference in partial pressure. This results in an amount, G, of the mass transferred in grams.

For water losses through the package, the term "water vapour transmission rate" (WVTR) is often used. The WVTR is given in g H_2O/m² · day and is measured at 37°C and 90% relative humidity. For gases such as oxygen and carbon dioxide, the gas transmission rate (GTR) is used, measured as cm³/m² · day · bar and 25.4 μm thickness, at 25°C. Some data for various packaging films are given in Table 7.4.

The water permeability of different packaging materials influences the relative humidity in the head space around the product and thus the loss of water content and microbiological growth. In addition to water permeability in packing materials and coatings, the transport of aroma compounds is a high priority (Debeaufort et al., 1998).

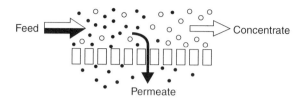

FIGURE 7.15 Principle of membrane separation.

TABLE 7.5
Pressure-Driven Membrane Processes

Process	Driving force	Cutoff (MW/particle size)
Reverse osmosis (RO)	Pressure, MPa	<MW 500
Ultrafiltration (UF)	Pressure, 0.1–10 MPa	MW 500–0.1 μm
Cross-flow microfiltration (CFMF)	Pressure, 0.01–0.1 MPa	0.1–10 μm

7.11 MASS TRANSFER IN MEMBRANE TECHNOLOGY

The principle of membrane processes is described in Figure 7.15. The feed passes tangentially over the membrane and some of the constituents (molecules) pass through the membrane. The feed is separated into two fractions: the permeate passing through the membrane and the concentrate or retentate rejected by the membrane.

Properties of the membrane such as porosity (tightness) determine which constituents pass through the membrane. The amount passing through depends further on the driving force — the chemical potential. Some processes are categorized in Table 7.5.

Two theories for transport through the membrane dominate in the literature at present: hydraulic flow and diffusion. In most cases, both are probably relevant: in RO, diffusion dominates, whereas in UF and CFMF, hydraulic flow is more applicable. For hydraulic flow of water, Poiseuelle's law can be written as

$$J_{\mathrm{w}} = \frac{k_{\mathrm{w}}}{\delta}\frac{\Delta P}{n}$$

where k_{w} is the permeability coefficient of water. For diffusion flow of water, the following equation is used:

$$J_{\mathrm{w}} = k_{\mathrm{Dw}}\frac{\Delta P - \Delta \Pi}{\delta} = \frac{\Delta P - \Delta \Pi}{R_{\mathrm{m}}}$$

where Π is the osmotic pressure and R_{m} the resistance of the membrane. However, this equation has to be modified depending on the permeability of the membrane to the solute, and therefore a reflection coefficient σ is introduced. Also, R_{m} is replaced by $1/L_{\mathrm{p}}$, giving the total flow

$$J = L_{\mathrm{p}}(\Delta P - \sigma \Delta \Pi)$$

If the membrane is permeable to the solvent but not to the solute, $\sigma = 1$; if the membrane is equally permeable to both the solute and the solvent, $\sigma = 0$.

The most serious problem in the membrane operations mentioned here is concentration polarization (CP) at the membrane surface. Due to the transport through the membrane of molecules/particles

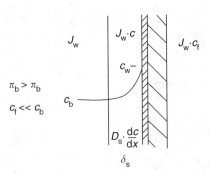

FIGURE 7.16   Concentration polarization.

below a certain size, molecules/particles above this size are concentrated on the feed side of the membrane surface. The phenomenon is illustrated in Figure 7.16. c_b is the concentration in the bulk, c_w the concentration at the membrane surface, c_f the concentration in the permeate, and δ_s the thickness of the polarized layer. In this layer there is a diffusive flow to the left due to the concentration gradient:

$$-D\frac{dc}{dx}$$

The flow toward the membrane brings an amount $J_w c$ of particles/molecules. The transport through the membrane is $J_w c_f$, which gives the following equation:

$$J_w c + D\frac{dc}{dx} = J_w c_f$$

Integration results in

$$\frac{J_w \delta_s}{D} = \ln\frac{c_w - c_f}{c_b - c_f}$$

or, if the retention is 100% (i.e., $c_f = 0$),

$$\frac{J_w \delta_s}{D} = \ln\frac{c_w}{c_b}$$

D/δ_s is equal to the mass transfer coefficient k_g and can be calculated by means of a dimensionless relationship. The relevance of such equations has been discussed by Gekas and Hallström (1987).

NOMENCLATURE

A	Area, m^2
a	Thermal diffusivity, m^2/sec
α_w	Water activity
c	Concentration, $kmol/m^3$, kg/m^3
c_p	Specific heat, Wsec/kg, K
D	Mass diffusivity, m^2/sec
d	Diameter, m
G	Mass flow (through package material), g
h	Heat transfer coefficient, W/m^2, K

J	Diffusive flow rate, kmol/m^2 · sec, kg/m^2 · sec
k	Thermal conductivity, W/m, K
k_g	Mass transfer coefficient, m/sec
k_w	Permeability coefficient, kg/m
L	Solid phase (extraction problems)
L	Mass of air, kg
L	Length, m
M	Molecular weight, kg/kmol
m	Mass flow rate, kg/m^2 · sec
N	Bulk flow rate, kmol/m^2 · sec, kg/m^2 · sec
P	Total pressure, N/m^2, Pa
p	Partial pressure, N/m^2, Pa
q	Heat flow rate, W/m^2
R	Gas constant, Wsec/m, K
R_m	Membrane resistance, l/sec
r	Heat of evaporation, Wsec/kg
T, t	Temperature, K, °C
t	Time, sec
V, v	Velocity, m/sec
V	Liquid phase (extraction problems)
W	Mass of water, kg
x, y, z	Coordinates, m
x, y	Concentration, kmol/kmol, kg/kg
β	Coefficient of thermal expansion, l/K
δ	Membrane thickness, m
ε	Porosity
η	Dynamic viscosity, Pa · sec
μ	Chemical potential, Wsec/m
ν	Kinematic viscosity, m^2/sec
Π	Osmotic pressure, Pa
ρ	Density, kg/m^3
σ	Reflection coefficient
τ	Tortuosity factor
ϕ	Air humidity

REFERENCES

Brennan, J.G. et al. 1976. *Food Engineering Operations.* Applied Science, London.

Bruin, S. and Luyben, C. 1980. Drying of food materials: a review on recent developments. In *Advances in Drying*, Vol. 1, A.S. Mujumdar (ed.). Hemisphere, New York.

Chirife, J. and Iglesias, H.A. 1978. Equations for fitting water sorption isotherms of foods, *Journal of Food Technology*, 13: 159.

Clement, K.H., Hallstroem, A., Dich, H.C., Le, C.M., Mortensen, J., and Thomsen, H.A. 1991. On the dynamic behavior of spray dryers. Dep. Chem. Chem. Eng., Eng. Acad. Denmark, Lyngby, Denmark, *Chemical Engineering Research and Design*, 69(A3): 245–252.

Debeaufort, F., Quezada-Gallo, J.-A., and Voilley, A. 1998. Mechanism of aroma transfer through edible and plastic packagings. ENSBANA, Lab. G.F.A.B., Dijon, Fr. Book of Abstracts, 215th ACS National Meeting, Dallas, March 29–April 2, 1998.

Doulia, D., Tzia, K., and Gekas, V. 2000. A knowledge base for the apparent mass diffusion Coefficient (DEFF) in foods, *International Journal of Food Properties*, 3(1): 14.

Gekas, V. and Hallström, B. 1987. Mass transfer in the membrane concentration polarization layer under turbulent cross flow, *Journal of Membrane Science,* 30: 153–170.

Gekas, V. and Mavroudis, N. 1998. Mass transfer properties of osmotic solutions. II. Diffusivities. Food Engineering Department, Center for Chemistry and Chemical Engineering, Lund University, Lund, Sweden, *International Journal of Food Properties*, 1(2): 181–195.

Gekas, V. and Oelund, K. 1988. Mass transfer in the membrane concentration polarization layer under turbulent cross flow. II. Application to the characterization of ultrafiltration membranes, *Journal of Membrane Science*, 37(2): 145–163.

Gros, J.B. and Rüegg, M. 1987. Determination of the apparent diffusion coefficient of sodium chloride in model foods and cheese. In *Physical Properties of Foods*, R. Jowitt et al. (eds.). Elsevier, London.

Lyndersen, A.L. 1983. *Mass Transfer in Engineering Practice*. Wiley, New York.

Motarjemi, Y. 1988. A study of some physical properties of water in foodstuffs. Ph.D. thesis, Lund University, Lund, Sweden.

Perry, R.H. and Greene, D. 1984. *Perry's Chemical Engineer's Handbook*. McGraw-Hill, New York.

Prothon, F., Ahrne, L., and Sjoeholm, I. 2003. Mechanisms and prevention of plant tissue collapse during dehydration: a critical review. Environment and Process Engineering, SIK — The Swedish Institute for Food and Biotechnology, Göteborg, Sweden. *Critical Reviews in Food Science and Nutrition*, 43(4): 447–479.

Saravacos, G.D. and Maroulis, Z.B. 2001. *Transport Properties of Foods*. Marcel Dekker, New York.

Schoeber, W.J.A.H. 1976. Regular regime in sorption processes. Ph.D. thesis, Technical University of Eindhoven, The Netherlands.

Van den Berg, C. and Bruin, S. 1981. Water activity and its estimation in food systems: theoretical aspects. In *Water Activity: Influences on Food Quality*, L.B. Rockland and G.F. Stewart (eds.). Academic Press, New York.

Voilley, A. and Roques, M. 1987. Diffusivity of volatiles in water in the presence of a third substance. In *Physical Properties of Foods*, Vol. 2, R. Jowitt et al. (eds.). Elsevier, London.

Yamamoto, S., Saeki, T., and Inoshita, T. 2002. Drying of gelled sugar solutions — water diffusion behavior. Department of Chemical Engineering, Yamaguchi University, Tokiwadai, Ube, Japan, *Chemical Engineering Journal* (Amsterdam, The Netherlands), 86(1–2): 179–184.

8 Evaporation and Freeze Concentration

Ken R. Morison and Richard W. Hartel

CONTENTS

8.1 INTRODUCTION

In many food processes water, or another solvent, needs to be removed from a dilute liquid to produce a concentrated or dried product. The processes that can be used to remove water include evaporation, freeze concentration, reverse osmosis (or other membrane separations), and drying (Figure 8.1). The choice depends on the extent of concentration required, the effect of the process on the product, available energy sources, and on the relative costs of the processes. Often a combination of processes is used.

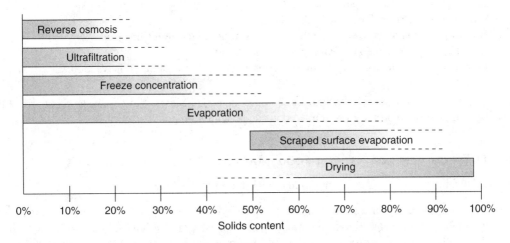

FIGURE 8.1 The range of solids content for alternative water removal processes.

Evaporation has several advantages over other water removal processes. Modern evaporation plants are very effective at utilizing small amounts of steam or electrical energy to generate large rates of evaporation. Techniques such as multiple-effect evaporation, thermal vapor recompression and mechanical vapor recompression greatly reduce the amount of energy required to give a certain degree of concentration. Another advantage of evaporation is the level of concentration attainable. Evaporation can concentrate most liquid feeds of any dilution up to 50% solids easily, while, in the extreme, sugar solutions for the production of hard candies (toffees) are evaporated to about 98% solids (Schwartzberg, 1989).

Freeze concentration may be used for thermally sensitive products or when loss of volatile components must be minimized, and concentrations of between 40 and 55% can be attained. Reverse osmosis is most suitable for dilute liquids and also has very little effect on product quality. Reverse osmosis should be considered for the preconcentration before evaporation of feed solutions containing less than 10% solids. Drying is energy intensive and normally requires a high concentration feed, so some form of preconcentration is required.

In this chapter we focus on evaporation and freeze concentration. Membrane separations are covered in Chapter 9, while drying is covered in Chapter 10.

8.2 OVERVIEW OF EVAPORATION

The primary objective of evaporation is to remove the solvent (usually water) from a solution so as to increase the concentration of the solute. A secondary objective is to do this with minimum total cost. The total cost will include the capital cost and operating costs such as the energy cost, product loss, and cleaning costs.

Usually the liquid being concentrated flows through a tube while heat is applied to the outside of the tube. The solvent boils and is separated from the concentrated liquid. Most foods are damaged by heat, so they are normally evaporated under vacuum conditions with a low boiling point. The latent heat of vaporization of water is high but by reusing energy in multiple stages or with vapor recycle, good energy efficiency can be obtained.

The main components of evaporators are:

A feed preheater to bring the feed close to the boiling point
A feed distribution system to distribute the feed equally between the tubes

An energy supply, usually steam or electricity
A method of heat transfer to the boiling liquid
Vapor/liquid separators to separate the vapor with minimal liquid carryover
A vacuum system to keep the boiling temperature low
A condenser to remove energy from the vapor and/or to help maintain the vacuum

Integration and optimization of each of these components is necessary to obtain the most efficient evaporation process to produce concentrated products with high quality.

8.2.1 TYPICAL APPLICATIONS IN THE FOOD INDUSTRY

Some typical applications of evaporation in the food industry include:

Fruit products. Concentrated fruit juices are made by evaporation at low temperatures to provide stability to the product, as well as to minimize storage and shipping volumes. In the 1995/96 season Florida and Brazil together produced about 10^{10} l of orange juice and about two thirds of this was evaporated from about 12% solids to produce frozen concentrated orange juice with about 65% solids (Brown and Lesser, 1995). Jams and jellies are also produced from fruit by evaporation.

Dairy products. Evaporation is used widely in the dairy industry to concentrate milk, whey, and lactose prior to drying. Some products are sold as concentrated liquids, for example, canned evaporated milk, but most are dried to produce products that have excellent stability, low volume, and low shipping costs.

Sugars. Refined sucrose from sugar beet and cane is made by extracting the sugar with hot water, evaporating off some of this water to give a concentrated syrup, and then using controlled evaporation to generate the supersaturation necessary for the crystallization process. Malt and glucose syrups are evaporated after enzymatic hydrolysis of barley and cornstarch.

Salt. In dry countries fresh water is produced by evaporation of seawater, which gives salt as a co-product. It is concentrated and crystallized in a manner similar to sugar.

Vegetables. Water is removed from vegetable juices to give textural advantages, as in purees and pastes.

Any food product in a concentrated or dried form, that is normally liquid, is likely to be made by using evaporation as one step.

8.3 TYPES OF EVAPORATORS

8.3.1 PHYSICAL TYPE

Many types of evaporators are currently in use for evaporation of foods. The choice of the evaporator that should be used for a particular process depends on the product, economics, and historical preferences of the particular industry. There are many references in which detailed descriptions of all types of evaporators may be found. These include Minton (1986), Perry et al. (1997), and Billet (1989) for general information, while Chen and Hernandez (1997) give more specific information about citrus juice evaporation, and Bouman et al. (1988) about evaporators in the dairy industry.

An evaporator is normally made up of more than one vessel containing many tubes. Each vessel is sometimes known as a body or a calandria, though some people use the term calandria for the bundle of tubes welded between two tube sheets inside a steam chest. If a number of vessels are operated at different pressures they are also known as effects, but sometimes many vessels are operated at the same pressure (as in a mechanical vapor recompression evaporator; see Section 8.3.2.4) and they form just a single effect. Sometimes the liquid is passed through a calandria more than once but each time through a different group of tubes giving rise to a multi-pass effect or a multi-pass calandria.

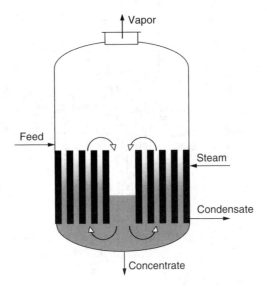

FIGURE 8.2   A short tube evaporator.

The tubes are welded to a tube sheet which is fixed into the vessel. The space created on the outside of the tubes is known as the steam chest or vapor chest.

8.3.1.1   Short-Tube Evaporator

One of the oldest types of evaporators used in the food industry is the short-tube vertical evaporator, sometimes called the Robert evaporator. Here there are a bundle of vertical tubes about 2.2 to 3.5 m long welded between a top and bottom tube sheet to form a calandria that is placed vertically in the larger shell, as seen in Figure 8.2. Steam is added to the calandria on the outside of the tubes and the fluid boils within the tubes, producing vapors that rise. The natural convection created by the vapor flow is sufficient, but internal agitators are used to promote flow, especially with more viscous fluids. The operation of these is discussed in detail by Hugot (1986) and a thorough literature review is given by Peacock (1999). There are numerous variations of feed, product, and vapor arrangements.

This type of evaporator can still be found in use, especially in the cane sugar refining industry for concentrating syrups, where typically five effects are connected in series to get energy efficiencies. Because the cane sugar industry has its own fuel supply from the cane, there is little incentive to change to more efficient evaporators. In contrast, the beet sugar industry uses coal, oil or gas as a fuel source and some companies have moved to falling film type evaporators. The short-tube evaporator can also be used for crystallization with the addition of a stirring propeller to maintain circulation and suspend crystals. Advantages of this type of evaporator include relatively high heat transfer rates at high temperature differences, ease of cleaning, and relatively low cost. However there is high liquid hold up giving a high residence time so heat-related product damage can occur.

8.3.1.2   Long-Tube Vertical Rising Film

Chronologically the rising film (or climbing film) evaporator followed the short-tube evaporator, in an effort to reduce the residence time. The feed is introduced at the bottom of each tube and is heated from the outside of the tube by steam. The evaporating vapors carry the liquid up the tube by shear force, creating a film on the inside of the tube. A typical long-tube vertical rising film is shown in Figure 8.3.

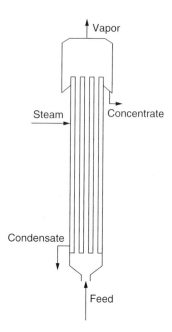

FIGURE 8.3   A long-tube vertical rising film evaporator.

   This type gives short residence times and good heat transfer. However a significant part of the temperature difference is required to provide the pressure difference to carry the liquid upwards, so these evaporators cannot operate with low temperature differences. Rising film evaporators have, in the main part, been superceded by falling film evaporators, except that they are used for handling high viscosity liquids and products with suspended solids where it is not possible to make a falling film. Products like honey, glucose syrup at 80° Brix, tomato paste, and saturated solutions of salts are example applications.

8.3.1.3 Long-Tube Vertical Falling Film

The falling film evaporator was first introduced by Wiegand Apparatebau GmbH of Karlsruhe, Germany, in 1953, and it is now the most common type of evaporator used in the food industry. Liquid is fed at the top of the evaporator at near boiling point (Figure 8.4) and after distribution passes down the inside of the tubes while being heated from the outside by steam. The vapor and concentrated liquid is separated and normally the liquid is pumped to the next stage, while the vapor is used to heat the next stage.

   The falling film evaporator offers the advantages of low residence time and is normally operated under vacuum to allow low temperatures and energy efficiency. They can operate with temperature differences as low as 2°C (Ward, 1994). Tubes are typically 5 to 15 m long. They can operate with liquids up to a viscosity of about 200 mPa·s, but the flow rate must be sufficiently great to ensure that the tube surface is kept wet. Singh and Heldman (1984) give the maximum viscosity as 3 Pa·s. Equation 8.33 given later relates the film thickness to the viscosity.

8.3.1.4 Forced Circulation

In this type of evaporator, liquid is heated by pumping through a heat exchanger (shell and tube or plate), but vaporization is not usually allowed to occur in the heat exchanger but rather in the flash chamber, where the vapor is then separated. An external heat exchange forced-circulation evaporator is shown in Figure 8.5.

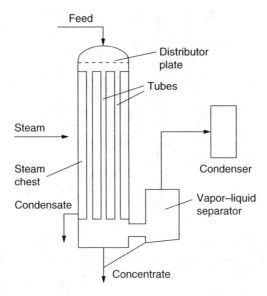

FIGURE 8.4 A long-tube vertical falling film evaporator.

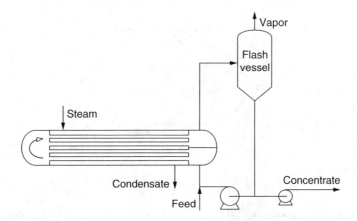

FIGURE 8.5 A forced circulation evaporator.

Due to the high recirculation rates through the heat exchanger the heat transfer coefficients are high. Viscous liquids and particulates can be handled easily in this way and the absence of evaporation in the tubes reduces fouling. A major disadvantage for the food industry is the high residence time due to the increased hold up volume. Forced circulation evaporators are used in the processing of tomato products as well as in the crystallization step of sugar refining.

8.3.1.5 Scraped Surface Thin Film

Scraped surface thin film evaporators (Figure 8.6) are very similar to scraped surface heat exchangers and are designed for the evaporation of highly viscous and sticky products that cannot be otherwise evaporated easily. A rotating blade wipes the surface to promote heat transfer, to prevent deposition or charring, and to allow less concentrated liquid to move to the surface. They are used for liquids that are very viscous (up to 20 Pa·sec), heat sensitive, or foul easily. Applications include tomato pastes, gelatine, and crystallizing whey. The residence time in this type of evaporator can be kept very short,

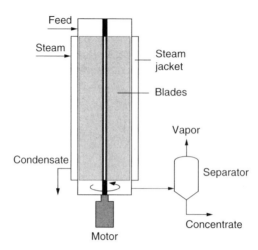

FIGURE 8.6   A scraped surface evaporator.

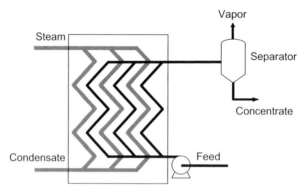

FIGURE 8.7   A plate evaporator.

making it useful for heat sensitive materials. Typically these evaporators have only a single tube and rotor, so they have a low surface area and a relatively expensive method of evaporation. Normally just a single scraped surface effect is used usually after concentration by another type of evaporator. Continuous falling film and rising film, as well as batch, scraped surface evaporators, have been designed.

8.3.1.6 Plate Evaporators

Plate evaporators are constructed and operate in a manner similar to plate heat exchangers (Figure 8.7). The liquid to be evaporated passes, either upwards or downwards, on one side of the plate while the heating medium (steam or hot water) passes on the other. Evaporation may take place either within the plate system or externally in a flash chamber, as in a forced-circulation evaporator (Figure 8.5).

Plate evaporators offer several advantages over other types. These include the ease and flexibility of operation, good heat transfer rates, low hold up time for heat sensitive products, and reduced fouling when high fluid velocities are maintained. This type of evaporator has a relatively low capacity (up to 40,000 L h^{-1} of water removal). The use of multiple effects, with the resulting energy saving, seems to be less common with plate evaporators possibly because of the lower throughput and because of the pressure drop through the plate channels. Plate evaporators find use in many processing applications, including dairy products, coffee concentrates, gelatin, soup broths, sugar syrups, and fruit juice concentrates. Viscous fluids can be efficiently concentrated in these evaporators, with concentrations of up to 50 to 60° Brix being reported for fruit purees and up to 97% for sugar and corn syrups (APV, 2001).

8.3.1.7 Thin-Film Spinning Cone

Another type of evaporator that has recently been redeveloped is the thin-film, spinning cone evaporator that uses centrifugal flow to produce a film about 0.1 mm thick on the underside of a hollow cone heated by steam (Flavourtech, 2003). Liquids with a viscosity up to 20 Pa·s can be processed with evaporation rates up to about 5000 $kg\,h^{-1}$. The technology allows very short residence times and hence very little heat effect on the product.

8.3.2 TYPES OF ENERGY USE

Each of these different evaporator types can be arranged in an evaporator train with different forms of vapor flow and reuse. There are at least four different types from this point of view.

Single effect
Multiple effect with direct steam expansion
Multiple effect with thermal vapor recompression
Mechanical vapor recompression

The aim of the different designs is to minimize energy usage. The efficiency of an evaporator is normally expressed as a specific energy consumption or as its inverse the energy economy ratio.

$$\text{Specific energy consumption} = \frac{\text{Energy input}}{\text{Energy required for evaporation achieved}} \qquad (8.1)$$

or

$$\text{Specific energy consumption} = \frac{Q_{\text{input}}}{\dot{m}_{\text{evap}} \Delta h_v} \qquad (8.2)$$

where Q_{input} is the rate of energy input, $\dot{m}_{\text{evap}}$ is the rate of evaporation, and Δh_v is the latent heat of vaporization. In the case where steam is the heat source, the rate of energy input is obtained from the steam flow and, after noting that the latent heat for the condensing steam, $\Delta h_{v,\text{cond}}$, is almost the same as for evaporation, an expression for energy consumption is obtained in terms of steam usage.

$$\text{Specific energy consumption} = \frac{\dot{m}_{\text{steam}} \Delta h_{v,\text{cond}}}{\dot{m}_{\text{evap}} \Delta h_{v,\text{evap}}} = \frac{\dot{m}_{\text{steam}}}{\dot{m}_{\text{evap}}} \qquad (8.3)$$

Often energy consumption is expressed as mass of steam required per mass of water evaporation, or more generally as energy input per mass of evaporation.

For the purposes of this initial discussion about relative efficiencies of different types of evaporators, heat losses from the evaporator and energy required for preheating are ignored. They will be considered in Section 8.4.2 on thermal design.

A falling film evaporator is used here in the diagrams as an example, but any type of evaporator can be configured with the different forms of energy reuse discussed below. The choice of the configuration of evaporator to be used is not straightforward and may require a complete analysis of the energy requirements of the site before a decision is made. A comparison of some is given by Zimmer (1980).

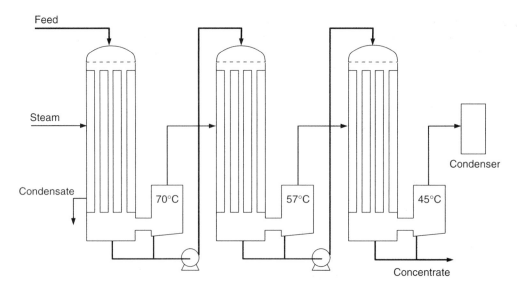

FIGURE 8.8   A simple multiple effect evaporator with possible evaporating temperatures shown.

8.3.2.1 Single Effect

A single effect falling film evaporator is shown in Figure 8.4. In a single effect evaporator the amount of steam required equals the amount of water evaporation required. Ignoring the heat losses, the specific energy consumption is equal to one kilogram of steam per kilogram of water evaporation. The vapor from the single effect is condensed in a condenser.

8.3.2.2 Direct Steam Expansion (DSE) Evaporators

In a simple multi-effect evaporator (Figure 8.8) there are two or more effects and each operates at a pressure and temperature lower than the previous. Steam is supplied to the first effect and the vapor from each effect is used to heat the next. The vapor from the final effect is condensed normally by direct or indirect cold water. These evaporators are also known as steam driven multi-effect evaporators (Schwartzberg, 1989). This type of evaporator requires a steam supply of at least 1 bar gauge. This could possibly come from a low quality boiler or be the exhaust from another process such as a steam turbine used to generate electricity. A vacuum pump may be required to establish and maintain a vacuum, but steam ejectors can be used to create a vacuum if there is a supply of 6 bar steam (see Section 8.4.3.7). A condenser will also be required to condense the vapor from the final effect, thus helping to maintain the vacuum.

Normally the product flow is co-current with the vapor flow as then heat damage to high concentration product is minimized by using the minimum possible temperature. However, counter-current flow is used with some products such as tomato juice so that the viscosity of the concentrated product is minimized by the high temperatures in the early effects (Schwartzberg, 1989).

Because the energy of the steam or vapor is reused in each of the n effects the specific energy consumption is approximately $1/n$, that is, 1 kg of steam gives about n kg of evaporation.

8.3.2.3 Thermal Vapor Recompression Evaporators

In a thermal vapor recompression (TVR) evaporator (Figure 8.9), vapor from one of the effects is effectively recompressed by adding new steam to it through a steam ejector. In this way some of the vapor is used again and better efficiencies are obtained. A TVR evaporator requires steam with a

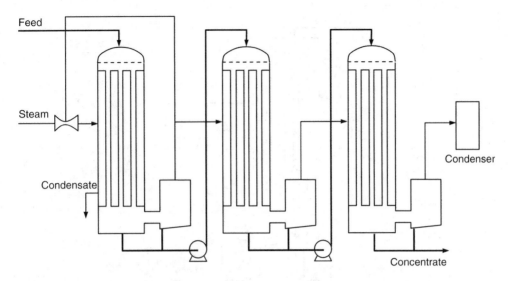

FIGURE 8.9 Thermal vapor recompression evaporator with recompression around the first effect.

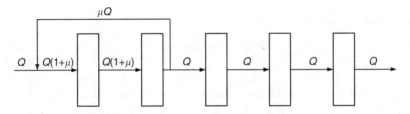

FIGURE 8.10 Energy flows in a five-stage thermal vapor recompression evaporator with recompression around the first two effects.

pressure of about 6 bar absolute, but higher pressures give a greater margin to enable good control. This type is always more efficient than a simple evaporator with the same number of effects.

The vapor recycle can be around the first effect only or from the second or third back to the first stage. The choice will be an optimization based on the number of effects and the operating pressures of the evaporator.

The extra efficiency of a TVR can be shown from a block diagram of heat flows (Figure 8.10).

Here we see that, if there is a recycle of energy at a rate μQ, over $n_{recycle}$ effects, and there are a total of n effects, then the energy flow Q is used $(n + \mu n_{recycle})$ times. Thus the specific energy consumption is $1/(n + \mu n_{recycle})$. The recycle ratio μ is generally known as the entrainment ratio.

$$\mu = \frac{\text{mass of vapour recycled}}{\text{mass of actuating steam}} \qquad (8.4)$$

The entrainment ratio is typically in the range 1 to 2 and depends on the steam ejector and the pressures of the various vapor flows. Methods for the calculation of the entrainment ratio are given below in Section 8.4.2.2.

The cost of a steam ejector is low so, if sufficient steam pressure is available, a TVR evaporator should always be used in preference to a direct steam expansion evaporator.

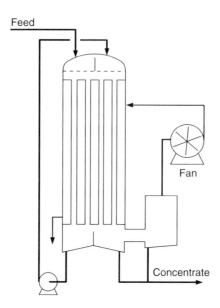

FIGURE 8.11   A single body, two pass evaporator with mechanical vapor recompression.

8.3.2.4 Mechanical Vapor Recompression

A mechanical vapor recompression (MVR) evaporator uses a compressor or fan, rather than steam, to recompress the vapor before returning it to the same stage. Once started, the energy input is the electricity used to drive the compressor or fan, so a reliable source of electricity is normally required. This could be supplied by a cogeneration system that produces steam and electricity. Alternatively the compressor could be driven by a gas turbine or engine. The MVR evaporator does not require much steam and does not require a condenser, but a vacuum pump is required to establish and maintain the vacuum by removing noncondensable gases.

The MVR evaporator generally consists of just one effect, which may be contained within one or more calandria. All of the product is at the same temperature (ignoring the small effect of boiling point elevation). Figure 8.11 shows a single effect, two pass MVR evaporator. The two passes could be contained in a single body. The thermodynamics and design of these are detailed further in Section 8.4.2.3. Typically MVR evaporators may give 20 J of evaporation per joule of electrical energy, but if the electricity has been produced in a thermal power plant with a 33% efficiency then the overall efficiency of an MVR evaporator is a little better than a good TVR evaporator. However a cogeneration system can make the overall efficiency and economics very attractive.

8.3.3 PHYSICAL, CHEMICAL, AND BIOLOGICAL PROPERTIES

The operation and design of all evaporators are constrained by the physical properties of the product, but for food evaporators, product degradation by heat and bacteria add extra constraints.

The physical properties that change the most during evaporation and which may limit the operating range of the evaporator are viscosity and boiling point elevation, while less variable but still important are thermal conductivity, heat capacity, and density. All of these affect the rate of heat transfer to a lesser or greater amount.

8.3.3.1 Heat Sensitivity

Many food components undergo thermally induced modifications during processing. These are generally negative to the food quality, but may at times be beneficial, as in the case of thermal deactivation of enzymes or microorganisms. Any intentional heat treatment during evaporation is usually carried out in the preheat sections where the residence time and temperature are easily controlled. In the main part of the evaporator, residence times are large so temperatures are kept sufficiently low to minimize thermal damage. Adverse effects that are often encountered include (1) protein denaturation (van Boekel and Walstra, 1995), which can cause loss of nutritional value and fouling, bloom loss in gelatins, and grade loss in pectins; (2) production of off-colors and off-flavors, as in browning of dairy products and sugar syrups, and cooked flavors in ultra-heat treated milk; and (3) chemical reactions, as in sucrose hydrolysis and vitamin degradation.

To minimize thermal reactions the reaction rate and/or the reaction time must be minimized. This is done by minimizing the temperature of the product or heating surfaces and by minimizing the residence time in the evaporator. The product temperature can be kept low by using evaporating pressures less than atmospheric pressure so that the boiling temperature is reduced. In general, most evaporation for the concentration of heat sensitive products, especially those containing proteins occurs at temperatures below 70°C (Knipschildt and Anderson, 1994). The heat surface temperatures are kept low by using low temperature driving forces, typically in the range 2 to 8°C. The residence time is minimized by design. Falling film, rising film, plate, and scraped surface film evaporators are all capable of low residence times, but when many effects are used to get thermal efficiency, residence times are often several minutes if not more.

8.3.3.2 Bacterial Growth

The temperatures required to minimize thermal degradation are sufficiently low to allow bacterial growth. Thermophilic bacteria and spores can accumulate during the normal operation of an evaporator (Murphy et al., 1999), especially in preheaters (Refstrup, 1998). Apart from the need for good design and manufacture of the evaporator, as discussed briefly in Section 8.5, the risk of bacterial growth adds the constraints that any product recycle must be avoided and hold up times minimized.

8.3.3.3 Viscosity

The product viscosity directly influences the heat transfer coefficient and at high concentrations most food liquids become so viscous that they do not flow well causing increased fouling and risk of blockage. There are numerous different equations for viscosity that may be grouped as:

Very dilute solutions
Solutions where the effective volume fraction of dissolved or suspended solids is less than about 70%
Concentrated Newtonian solutions with significant intermolecular interaction
Solutions with non-Newtonian behavior

Few foods can be characterized with certainty, notable exceptions being sugar solutions and milk products. Holdsworth (1993) carried out a review of models used for the prediction of flow properties of food products.

The viscosity of fruit juices can be quite variable and are dependent on the total solids, fruit species, the pulp content, and the pectin level (Crandall et al., 1982). The viscosity in an evaporator may vary from 0.5 mPa·sec in the first stage to 50 mPa·sec in the final stage (Schwartzberg, 1989).

Relative viscosity is commonly used in viscosity equations and is defined as:

$$\mu_{rel} = \frac{\mu}{\mu_o} \tag{8.5}$$

where μ and μ_o are the viscosity of the solution and solvent, respectively.

8.3.3.3.1 Dilute and low concentration solutions

For very dilute solutions, the viscosity is used to determine other properties of the molecules and Einstein's first order equation based on spherical molecules is often used:

$$\mu_{\text{rel}} = 1 + 2.5\phi \tag{8.6}$$

where ϕ is the effective volume fraction of the solute. The effective volume fraction of a hydrated solute in solution, ϕ, is given by

$$\phi = w\rho v_{\text{eff, solute}} \tag{8.7}$$

where $v_{\text{eff, solute}}$ is the effective specific volume of the solute ($\text{m}^3 \text{ kg}^{-1}$) including any attached solvent molecules. In general for a multi-solute system the effective volumes of the solutes are summed:

$$\phi = \rho \sum_i w v_{\text{eff},i} \tag{8.8}$$

However it is better to use Einstein's higher order equation as derived by Kunitz (1926) for solution for which ϕ is up to about 0.60.

$$\mu_{\text{rel}} = \frac{(1 + 0.5\phi)}{(1 - \phi)^2} \tag{8.9}$$

This equation has been largely ignored by the literature but works well.

8.3.3.3.2 Concentrated newtonian solutions

When the volume fraction of solute exceeds about 0.60, Einstein's equation becomes inaccurate as intermolecular interactions become significant. At higher concentrations exponential equations seem to work well.

$$\ln(\mu_{\text{rel}}) = \sum_{\text{solutes}} \left(a_i \frac{w_i}{w_{\text{solvent}}} + b_i \left(\frac{w_i}{w_{\text{solvent}}} \right)^2 \right) \tag{8.10}$$

where w is the mass fraction and a_i and b_i are constants for solute i.

8.3.3.3.2.1 Non-newtonian behavior

Non-Newtonian behavior arises from the interaction of long chain molecules such as proteins and polysaccharides in liquid food products. A review of concentrated whey solutions (Morison and MacKay, 2001) showed that the protein mass fraction in whey had to be greater than about 20% before non-Newtonian effects are significant.

Above this concentration the non-Newtonian behavior index, n, in whey solutions the index is less than 1.0 and the apparent viscosity depends on the shear rate being applied to the fluid. Equation 8.11 can be used to describe the behavior of many liquids (Holdsworth, 1993).

$$\mu_{\text{app}} = K\dot{\gamma}^{n-1} \tag{8.11}$$

Normally concentrations are low enough so non-Newtonian behavior is not significant, but an engineer must be aware of the possibility of pseudo plastic (shear thinning) and thixotropic (shear-time dependence) behavior. The viscosity of concentrated milk products is further confused by their age-related effects. In an evaporator age effects are minimal, and shear rates are difficult to calculate so Newtonian relationships are often used.

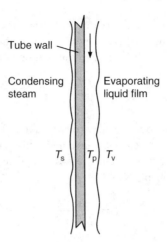

FIGURE 8.12 Steam, product and vapor temperatures in a falling film evaporator.

8.3.3.4 Boiling Point Elevation

Solutions boil at a temperature higher than the vapor above them. This becomes important in multiple effect evaporation where the rate of heat transfer depends on the product temperature (Equation 8.12 and Figure 8.12), but the usefulness of the vapor for heating a subsequent effect depends on the vapor temperature which is less than the product temperature (Equation 8.13).

$$Q = UA(T_s - T_p) \tag{8.12}$$

$$T_v = T_p - \Delta T_b \tag{8.13}$$

Boiling point elevation is a colligative property, so it is governed by some well-defined equations (Berry et al., 1980). It is given almost exactly by

$$\Delta T_b = \frac{-RT_{wb}^2 \ln a_w}{\Delta h_v} \tag{8.14}$$

where T_{wb} is the boiling temperature of water in kelvin, a_w is the water activity, Δh_v is the molar latent heat of vaporization of water and is a function of temperature. The molar latent heat of vaporization (Haar et al., 1984) can be described with adequate accuracy by:

$$\Delta h_v = 57,222 - 44.3T_{wb} \text{ J mol}^{-1} \tag{8.15}$$

Many authors follow the old derivation using molality which arose because the calculation of logarithms was difficult but which is less accurate at higher concentrations and hence of limited use for evaporation. Given the variability of most food products, except pure sugars, the water activity is best estimated assuming that the activity coefficient of milk is one. Thus the activity of water equals the mole fraction, x_w, of water and Equation 8.16 can be written as:

$$\Delta T_b = \frac{-RT_{wb}^2 \ln x_w}{\Delta h_v} \tag{8.16}$$

Significant contributions to the mole fraction of the solutes come from sugars and salts. In the case of milk, the data of Radewonuk et al. (1983) for freezing point elevation of skim milk concentrates was analyzed. It was found that good estimates of the mole fraction of water could be obtained using the lactose mass fraction (molecular mass 0.342 kg mol^{-1}) and by assuming that the ash in milk does not dissociate and has a molecular mass of about 0.067 kg mol^{-1}. Proteins and fat have very high effective molecular masses and hence very low mole fractions making negligible contribution to boiling point elevation. For other substances the molar concentration can be found from the freezing point depression, ΔT_f, which is easier to measure than boiling point elevation.

$$\Delta T_f = \frac{RT_{wf}^2 \ln x_w}{\Delta h_f} \tag{8.17}$$

where Δh_f is the molar heat of fusion of water, (6009.7 J mol^{-1} at 0°C). From this the effective average molecular mass of the solute can be calculated and then used to calculate boiling point elevation. The mole fraction of water, x_w, can be easily calculated from the mass fraction, w_i, and the molecular mass of each component, M:

$$x_w = \frac{w_w/M_w}{\sum_i w_i/M_i} \tag{8.18}$$

8.3.3.5 Density

Density is best calculated from the weighted mass fractions of specific volume (inverse density).

$$\frac{1}{\rho} = \sum_i \frac{w_i}{\rho_i} \tag{8.19}$$

For milk it is often sufficient to use only three components, water, fat, and nonfat solids (NFS).

$$\frac{1}{\rho_{milk}} = \frac{w_{water}}{\rho_{water}} + \frac{w_{fat}}{\rho_{fat}} + \frac{w_{NFS}}{\rho_{NFS}} \tag{8.20}$$

Písecký (1997) suggests an equation for the component densities in kg m^{-3} in terms of temperature (°C).

$$\rho_{fat} = 966.665 - 1.334T \tag{8.21}$$

$$\rho_{NFS} = 1635 - 2.6T + 0.01T^2 \tag{8.22}$$

The density of water can be calculated from temperature with good accuracy in the range 5 to 100°C using

$$\rho_{water} = 1000.35 + 0.004085T - 0.0057504T^2 + 1.50673 \times 10^{-5}T^3 \tag{8.23}$$

Choi and Okos (1986) have similar equations obtained for a range of foods that can be used with Equation 8.19. They suggest component densities as given in Table 8.1. The equation for water is not very good and is included only to give the complete set as recommended by the authors. Equation 8.23 above is much more accurate for pure water.

8.3.3.6 Thermal Conductivity

The thermal conductivity is used in the calculation of heat transfer coefficients. Choi and Okos (1986) suggest a volumetric average of component thermal conductivities that can be expressed as:

$$k = \frac{\sum_i \frac{w_i}{\rho_i} k_i}{\sum_i \frac{w_i}{\rho_i}} \tag{8.24}$$

For a wide range of food products they obtained component equations for thermal conductivity (Table 8.2) that were used with the component densities in Table 8.1. The standard error was typically 3–5%.

TABLE 8.1
Density Equations for Foods

Component	Component density $(kg\,m^{-3})$
Protein	$1330 - 0.52T$
Fat	$926 - 0.42T$
Carbohydrate	$1599 - 0.31T$
Ash	$2424 - 0.28T$
Water	$997.2 + 0.00314T - 0.00376T^2$

TABLE 8.2
Component Equations for Thermal Conductivity of Foods

Component	Component thermal conductivity $(W\,m^{-1}\,K^{-1})$
Protein	$0.1788 + 0.0012T - 2.7 \times 10^{-6}T^2$
Fat	$0.181 - 0.00276T - 1.77 \times 10^{-7}T^2$
Carbohydrate	$0.183 + 0.00139T - 4.33 \times 10^{-6}T^2$
Ash	$0.330 + 0.0014T - 2.91 \times 10^{-6}T^2$
Water	$0.571 + 0.00176T - 6.7 \times 10^{-6}T^2$

8.3.3.7 Specific Heat Capacity

Choi and Okos (1986) suggest a mass average of specific heat capacity:

$$C_p = \sum_i w_i C_{p,i} \tag{8.25}$$

For a wide range of food products they obtained component equations for specific heat capacity (Table 8.3). The standard error was typically 3–5%.

8.3.3.8 Heat Transfer Coefficient

In all designs an estimate of the heat transfer coefficient is required. Ward (1994) suggests 100–500 Btu h^{-1} ft^{-2}°F^{-1} (550–2800 W $m^{-2}K^{-1}$). In an analysis of an industrial falling film evaporator for whey permeate (lactose, minerals and water) (Morison, unpublished) the heat transfer coefficient was estimated to range from 1700 to 600 W $m^{-2}K^{-1}$ and for solid contents from 5 to 39%. Within the accuracy of the data the overall heat transfer coefficient was found to be a linear function of the mass fraction of solids.

$$U = 1940 - 3450w \; W\,m^{-2}\,K^{-1} \tag{8.26}$$

Jebson and Chen (1997) report a number of heat transfer coefficients for milk in falling film evaporators in the range 190 to 3100 W $m^{-2}\,K^{-1}$ but without sufficient other information to have confidence in the correlation given.

Chen and Hernandez (1997) gave overall heat transfer coefficients for a falling film orange juice ("TASTE") evaporator ranging from 1260 to 300 W $m^{-2}\,K^{-1}$ for concentration from 13 to 60° Brix. Estimates of viscosity are not given in this work so extrapolation to other situations is difficult.

TABLE 8.3
Component Equations for Specific Heat Capacity

Component	Component specific heat capacity (J kg^{-1} K^{-1})
Protein	$2008 + 1.21T - 0.00131T^2$
Fat	$1984 + 1.47T - 0.0048T^2$
Carbohydrate	$1549 + 1.96T - 0.00594T^2$
Ash	$1093 + 1.90T - 0.00368T^2$
Water	$4176 - 0.0909T + 0.00547T^2$

For Robert type short tube evaporators the sugar industry base heat transfer coefficients on historical overall averages. Peacock (1999) gives a review of numerous studies of heat transfer coefficents for sugar, while Schwartzberg (1989) reports average values of 3500, 2290, 1690, 1190, and 650 W m^{-2} K^{-1} for a five-effect Robert beet sugar evaporator.

Sangrame et al. (2000) evaluated the performance of a scraped surface evaporator for tomato pulp and obtained overall heat transfer coefficients in the range 480 to 940 W m^{-2} K^{-1} depending on flow rates and concentrations. Chawankul et al. (2001) give equations to evaluate the physical properties and heat transfer coefficients of tangerine orange juice in a scraped surface evaporator.

Extrapolation to other situations is most reliable if a more rigorous approach is taken. Normally film coefficients are calculated and combined the overall heat transfer coefficient. The overall heat transfer coefficient, in this case based on the outer diameter, is given in terms of three heat transfer resistances; the condensing vapor resistance outside the tube, the tube wall resistance, and the inside evaporation resistance:

$$\frac{1}{U} = \frac{1}{h_o} + \frac{t}{k} + \frac{D_o}{D_i h_i} \tag{8.27}$$

where h_o and h_i are the film heat transfer coefficients for the outside and inside of the tube, and t is the wall thickness. In most cases the diameter ratio D_o/D_i has a small effect compared with the uncertainty in the film coefficients.

Heat transfer coefficients are normally expressed as:

$$Nu = a Re^b Pr^c \tag{8.28}$$

where Re is the Reynolds number, Pr is the Prandtl number (both defined below), and the Nusselt number, Nu, is defined in standard conditions as:

$$Nu = \frac{hD}{k} \tag{8.29}$$

but for film heat transfer the diameter, D, is substituted by the average laminar film thickness (Chun and Seban, 1971):

$$Nu = \frac{h}{k} \left(\frac{\mu^2}{\rho^2 g} \right)^{1/3} \tag{8.30}$$

Often Equations 8.28 and Equation 8.30 are combined and the film coefficient is expressed as:

$$h = ak \left(\frac{\rho^2 g}{\mu^2} \right)^{1/3} Re^b Pr^c \tag{8.31}$$

The Reynolds number for film heat transfer is normally defined as Equation 8.32, but some authors omit the factor 4, causing potential confusion.

$$\text{Re} = \frac{4\Gamma}{\mu} \tag{8.32}$$

The tube wetting rate or the peripheral flow rate, Γ, is the flow rate per tube divided by the appropriate circumference of the tube (see Section 8.3.3.8 also).

$$\Gamma = \frac{\dot{m}}{\pi D} \tag{8.33}$$

The mean film thickness can be estimated from Equation 8.34 (Nusselt, 1916) but is not normally required.

$$\delta = \left(\frac{3\dot{m}\mu}{\rho^2 \pi Dg}\right)^{1/3} = \left(\frac{3\Gamma\mu}{\rho^2 g}\right)^{1/3} \tag{8.34}$$

The Prandtl number, Pr, is based as usual on the specific heat capacity, viscosity, and thermal conductivity:

$$\text{Pr} = \frac{C_p \mu}{k} \tag{8.35}$$

8.3.3.8.1 Outside film coefficient

The outside film coefficient changes as the quantity of condensed vapor flowing down the tube increases so average values are often given. A recent equation for the average film heat transfer coefficient (over the length) for condensing vapor on a vertical tube in a stagnant environment is from Chen et al. (1987):

$$h_o = k_l \left(\frac{\rho_l^2 g}{\mu_l^2}\right)^{1/3} [\text{Re}_L^{-0.44} + 5.82 \times 10^{-6} \text{Re}_L^{0.8} \text{Pr}_l^{1/3}]^{1/2} \tag{8.36}$$

which they claim was accurate to $\pm 10\%$ of experimental results. Subscript lower case l refers to the liquid properties, while upper case L refers to the total condensation at the base of the tube. The Reynolds number is based on the total wetting rate of condensate at the base on the tube which can be calculated from the evaporation flow rate from the calandria.

$$\text{Re} = \frac{4\Gamma_{o,L}}{\mu} \tag{8.37}$$

$$\Gamma_{o,L} = \frac{\dot{m}_{\text{evap}}}{\pi D_o n_{\text{tubes}}} \tag{8.38}$$

Using typical values for food evaporation the outside film coefficient is found to range from 7 to 8 kW m^{-2} K^{-1}. These estimates are a good starting point for more detailed calculations once the evaporation per effect and number of tubes is known.

8.3.3.8.2 Wall conduction

The conduction term is t/k, the thickness divided by the thermal conductivity. If stainless steel is used the thermal conductivity is about 17 W m^{-1} K^{-1} and the wall thickness is at most 1.5 mm. Thus this term is at most 10^{-4} K m^2 W^{-1} and does not significantly affect the overall coefficient.

8.3.3.8.3   Inside film coefficient
The inside falling film coefficient is much more difficult to determine with confidence and it has a strong influence on the design.

Schwartzberg (1989) confirmed, for water and 13% sucrose, with Re < 1600, the equations of Nusselt (1916)

$$h_L = 1.1 \text{Re}^{-1/3} k_L \left(\frac{\rho_L^2 g}{\mu_L^2} \right)^{1/3} \tag{8.39}$$

and for Re > 1600 Schwartzberg confirmed Chun and Seban (1971)

$$h_L = 0.0038 k_L \left(\frac{\rho_L^2 g}{\mu_L^2} \right)^{1/3} \text{Re}^{0.4} \text{Pr}^{0.65} \tag{8.40}$$

where the Reynolds number is defined as in Equation 8.31 with the factor of 4. However poorer agreement was obtained for runs with 30% sugar solutions (viscosity 1.2 mPa·sec).

Alhusseini et al. (1998) reported results for the evaporator of water from solutions with propylene glycol and gave a detailed equation that is best obtained from the original source. Bouman et al. (1993) give an overall heat transfer coefficient for whole milk in falling-film evaporators

$$h_i = 6.05 q^{0.47} \Gamma^{0.26} \mu^{-0.44} \tag{8.41}$$

where q is the heating flux in W m^{-2}, and all other quantities are in standard SI units, and for skim milk

$$h_i = 0.77 q^{0.69} \mu^{-0.41} \tag{8.42}$$

but at typical feed concentrations this gives predictions up to five times other estimates.

There is a clear need for more research giving interior evaporation film heat transfer coefficients for aqueous solutions a viscosity range from 1 to 500 mPa·sec. The apparent difference between the coefficients obtained for single tube pilot scale evaporators and industrial evaporators needs to be resolved.

8.3.3.9   Tube Wetting

The tube wetting rate (also known as the peripheral flow rate and the irrigation density) of a falling film, Γ, is the mass flow rate in a tube divided by the appropriate circumference of the tube (Equation 8.33). It is important in falling-film evaporators in which it must be sufficiently high to ensure that all of the tube remains wet. Heavy fouling and poor heat transfer will result if the tube is not always wet. Data from Paramalingam et al. (2000) showed clearly (Figure 8.13) the effect of the wetting rate (averaged over many tubes) on the heat transfer coefficient for water. The overall heat transfer coefficient for water was reduced proportionally approximately by the factor Γ / Γ_{min} when the wetting rate was less than $0.2 \text{ kg m}^{-1} \text{ sec}^{-1}$. Data given by Schwartzberg (1989) showed the same effect with a proportional decrease in heat transfer coefficients for water and 13% sucrose when the wetting rate dropped below about $0.08 \text{ kg m}^{-1} \text{ sec}^{-1}$.

Various formulae are given, both for the minimum liquid rate to maintain a wet surface and for the (higher) liquid rate required to wet a dry surface. Given the natural variations in flows seen in an evaporator it seems likely that one must always aim to rewet a dry surface.

Bouman et al. (1993) discuss the effect of tube length and diameter on wetting. They say that as the diameter is decreased or the length is increased there is more wetting. However, if the area is to be kept constant, reduction in diameter will have no theoretical effect on wetting rates as the number of tubes increases in proportion and the wetting rate (per tube) remains constant.

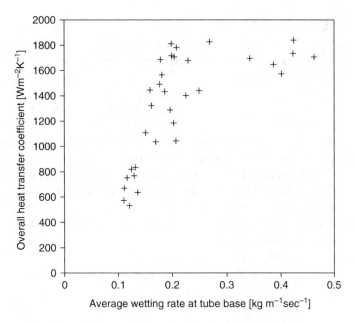

FIGURE 8.13 The effect of insufficient wetting on heat transfer. (After Paramalingam, S., Winchester, J., and Marsh, C. (2000), On the fouling of falling film evaporators due to film break-up, *Transactions of the Institution of Chemical Engineers*, **78C**, 79–84.)

Hartley and Murgatroyd (1964) gave Equation 8.43 for the minimum wetting for a fluid flowing over a surface. Minton (1986) gave the same equation but with θ set to 45°.

$$\Gamma_{\min} = 1.69 \left(\frac{\mu\rho}{g} \right)^{1/5} (\sigma(1 - \cos\theta))^{3/5} \tag{8.43}$$

The equation has been modified by a number of subsequent authors but was tested as shown for whole milk by Paramalingam et al. (2000). They found that the prediction was good at low concentrations but it overestimated the minimum wetting rate at higher concentrations. At 20°C for whole milk with concentrations from 5 to 40% by mass, they observed minimum wetting rates from 0.123 to 0.151 $kg\,m^{-1}\,sec^{-1}$. They noted the observation by Yang et al. (1991) that the minimum wetting rate is lower when proteins are present on the surface. Based on the work of Paramalingam et al., an initial design minimum wetting rate might be 0.2 $kg\,m^{-1}\,sec^{-1}$. This would allow for a variation in liquid distribution into the tubes of ±30%.

Schwartzberg (1989) recommended the minimum wetting rate ($kg\,m^{-1}sec^{-1}$) as:

$$\Gamma_{\min} = 0.085(\mu/\mu_1)^{0.2} \tag{8.44}$$

where μ_1 is the viscosity in the first stage, but to be safe he suggested a minimum of 0.25 $kg\,m^{-1}sec^{-1}$ in all stages.

Ward (1994) states "the evaporator designer normally selects the tube wetting rate based on personal experience" but, given that the wetting rate is often an active constraint in evaporator design, there is a need for more research on wetting of solutions under evaporation conditions (Schwartzberg, 1989).

8.4 DESIGN CALCULATIONS

Design calculations need, as a starting point, the required capacity, the composition of the feed material and the required total solid content for the product. In addition the physical properties listed in the previous section will be required.

8.4.1 MASS FLOWS

It is likely that the daily throughput of an evaporator will be given, but perhaps four hours must be allowed for cleaning every day, and so only about 20 h/day are available for evaporation. In other cases the factory may not run for 24 h/day or the evaporator may be required to handle multiple products in a day, so the available processing time is shorter. From the processing time available the continuous flow rate through the evaporator is easily calculated.

By mass balance of the total solid components the other flow rates and the amount of evaporation can be calculated. Conservation of mass gives:

$$\dot{m}_{feed} w_{feed} = \dot{m}_{conc} w_{conc} \tag{8.45}$$

or

$$\dot{m}_{conc} = \dot{m}_{feed} \frac{w_{feed}}{w_{conc}} \tag{8.46}$$

and

$$\dot{m}_{evap} = \dot{m}_{feed} - \dot{m}_{conc} = \dot{m}_{feed} \left(1 - \frac{w_{feed}}{w_{conc}} \right) \tag{8.47}$$

where $\dot{m}$ is mass flow rate of solution in kg sec^{-1} and w is mass fraction. The evaporation mass flow rate is the difference of the feed and concentrate flow rates.

In milk powder production it is possible to have three or four evaporators feeding a continuously operated dryer. If there are three evaporators, then at any point in time two will operate and the third will be cleaned. The flow rate through each evaporator corresponds to half of the dryer capacity.

At the design stage the feed specifications may be uncertain, or likely to change, and thus the evaporation capacity may change. An example of this is milk powder production, where depending on market demand, either whole milk powder or skim milk powder may be required. It is likely that the spray dryer will be the process bottleneck as this generally has the highest capital cost. Say for example, a dryer is able to dry 10,000 kg of water per hour from a feed at 50% total solids to a powder with 4% moisture. If the evaporator feed is whole milk with a total solid content of 13.5%, the evaporator will need to evaporate 51,900 kg of water per hour, but if the feed is skim milk with a total solid content of 9% the evaporator will need to evaporate 87,500 kg of water per hour. A decision could be made to ensure that the evaporator is sufficiently flexible to handle both circumstances. In particular the designer will need to ensure that there is sufficient wetting of the tube in both cases.

Mass flows within the evaporator cannot be calculated until the energy flows are known.

Example 8.1 Mass flowrates

An evaporator is required to concentrate 1,000,000 kg of whole milk per day from 13.0% total solids to 50% total solids. It is estimated that 4 h/day will be required for cleaning. Calculate the rate of evaporation.

The mass flow rate is simply calculated from the total volume and 20 h/day available for processing and is calculated to be 50,000 kg h^{-1}. Using Equation 8.46 and Equation 8.47

$$\dot{m}_{conc} = 50,000 \frac{0.13}{0.50} = 13,000 \ \text{kg h}^{-1}$$

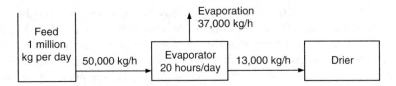

FIGURE 8.14 Flow rates around an evaporator.

and the evaporation required is

$$\dot{m}_{\text{evap}} = \dot{m}_{\text{feed}} - \dot{m}_{\text{conc}} = 50{,}000 - 13{,}000 = 37{,}000 \text{ kg h}^{-1}$$

These mass flows are shown in Figure 8.14.

Example 8.2 Mass flowrates

In another application about 70,000 kg h^{-1} of whey is concentrated by ultrafiltration to 20% total solids with a product flow rate of 2,500 kg h^{-1}. An evaporator is required to concentrate the whey to 50% total solids prior to drying.

By the same calculations we find that the evaporation is only 1,500 kg h^{-1}. In this case pretreatment with ultrafiltration reduces the evaporation requirements significantly.

8.4.2 ENERGY FLOWS

Energy flow calculations depend very much on the type of evaporator and calculations for each specific type are dealt with below. Here the examples are based on falling-film evaporators but some of the concepts apply to other physical types also.

Energy flow calculations are based on Equation 8.48 relating the energy flow rate, Q, to the evaporation rate and the heat of vaporization:

$$Q = \dot{m}_{\text{evap}} \Delta h_{\text{v}} \tag{8.48}$$

In addition the heat transfer is calculated by Equation 8.12 with Equation 8.13 and the mass flow Equation 8.45 through Equation 8.47.

8.4.2.1 Temperatures

Energy flows depend very much on the temperatures within an evaporator. The maximum and minimum temperatures that can be used are normally tightly constrained in a food evaporator. The maximum temperature for products containing proteins is typically 70°C while the minimum vapor temperature is constrained to about 43°C by the practicalities of maintaining a vacuum (Bouman et al., 1993; Knipschildt and Anderson, 1994). In the sugar industry temperatures range from 120 to 45°C (Schwartzberg, 1989). Chen and Hernandez (1997) give data for a citrus evaporator operating from about 90 to 40°C.

The overall temperature difference is the first effect product temperature, $T_{\text{p},1}$, less than the final effect vapor temperature, $T_{\text{v},n}$. When multiple effects are used this overall temperature difference can be broken down into the heat transfer temperature differences in each effect after the first and the boiling point elevations in all the effects.

$$\Delta T_{\text{overall}} = T_{\text{p},1} - T_{\text{v},n} = \sum_{i=2,n} \Delta T_i + \sum_{i=1,n} \Delta T_{\text{b},i} \tag{8.49}$$

Initially the values of the boiling point elevation may be unknown but, to enable calculations, an estimate of an average value of 1°C for milk or 3°C for juices and sugar should be adequate. It is also normal to assume that the temperature difference over each effect will be the same. So, if for example, a three-effect evaporator is used (Figure 8.8) with a maximum product temperature of 70°C, a minimum vapor temperature of 45°C, and an average boiling point elevation of 1°C, the temperature difference over each effect will be about $(70 - 45 - 3)/2 = 11°C$.

The temperature differences cannot be set directly but will be the result of the area of each effect, energy flow rate, and the heat transfer coefficient. A low temperature difference will result if the area is relatively large, but this may result in a low wetting rate that may be undesirable (Section 8.3.3.8).

8.4.2.2 Energy Losses

Heat losses are common to all types of evaporators and are considered next. When calculating heat flows it may be useful to ignore heat and temperature losses on the first iteration so that estimates can be obtained of various parameters.

Heat losses arise from different causes:

Convective heat losses from evaporator calandrias and separators
Convective heat losses from vapor ducts
Energy in vapor removed with noncondensable gases (Section 8.4.3.6)
Pressure drop in evaporator tubes and vapor ducts (Section 8.4.3.5)
Energy used to preheat the feed (Section 8.4.3.1)

Schwartzberg (1989, discussion) estimated that, in every stage, 2% of the energy in the vapor is lost from the vessel and duct walls. Kern (1964) gives an equation for natural convection from vertical walls from which an approximate heat transfer coefficient of 5 W m^{-2} K^{-1} can be calculated for losses.

8.4.2.3 Direct Steam Expansion (DSE) Design

Because the energy of the steam is reused in each effect (Section 8.3.2.2) the energy requirement for an n effect evaporator without vapor recompression is approximately

$$Q = \frac{\dot{m}_{evap} \Delta h_v}{n} \tag{8.50}$$

where Q is the energy flow rate [J sec^{-1} or W], $\dot{m}_{evap}$ is the total mass flow of evaporation required [kg sec^{-1}], and Δh_v is the heat of vaporization of water [kJ kg^{-1} K^{-1}]. Thus the specific energy consumption is 1 kg of steam for about n kg of evaporation. The annual cost of energy can be easily calculated from this, using the number of hours of operation per year and the cost of energy.

This calculation indicates that the number of stages should be maximized but as each stage is added another body, pump, and pipework are required. Further Bouman et al. (1988) warn that more stages with lower temperature differences, give rise to larger surface areas and more area for fouling. In some cases, they found in milk evaporators that the cost of fouling through lost product was about 40% of the total cost of evaporation. They showed that for low fouling the evaporator should be designed with a ratio of area to evaporation rate of 0.100–0.110 m^2 h^{-1} kg^{-1} (6.3–6.9 kW m^{-2}).

Example 8.3 Energy cost
Calculate the energy flows for Example 8.1.

TABLE 8.4
Energy Costs and Savings for Example 8.3

Number of effects	Energy cost (k$/year)	Energy cost saving from the extra effect (k$/year)
1	2305	
2	1152	1152
3	768	384
4	576	192
5	461	115
6	384	77
7	329	55
8	288	41

In Example 8.1 the evaporation flow rate is 37,000 kg h^{-1} (10.3 kg sec^{-1}). The heat of vaporization of water is about 2,360 kJ kg^{-1} at a typical temperature of 60°C. We do not need to consider the exact pressure of the evaporation at this stage, as it will have little influence on decision making.

For a single effect, the heat flow is thus:

$$Q = \dot{m}\Delta h_v = 10.3 \times 2,360 = 24,300 \text{ kJ sec}^{-1}$$

For 300 days operation at about 22 h/day (allowing some energy input for startup, shut down and cleaning), the total annual energy consumption is

$$\text{Annual energy consumption} = \frac{24,300 \times 300 \times 22 \times 3,600}{1,000,000} = 576,000 \text{ GJ year}^{-1}$$

If steam is used, then the energy will cost about 4 US$ GJ^{-1} (10 US$ tonne^{-1}), and the total energy cost is about US$2.4 million per year so there is clearly some incentive to improve efficiency.

If more effects are used, the energy is reused in each effect and the efficiency improves proportionally. As a first estimate we can estimate the energy use as shown in Table 8.4.

Table 8.4 shows that even with seven effects there is a significant gain to be made by adding an extra effect, and depending on the required investment criteria it may be worth spending US$250,000 to obtain this saving.

However as the number of effects increases the temperature driving force decreases so the area and capital cost increase. Also as the area increases the wetting rate in each tube decreases, potentially leading to increased fouling and decreased heat transfer.

Example 8.4 Energy cost
The same calculations can be repeated but for Example 8.2 giving the results in Table 8.5.

Here we see that the gains from increasing the number of effects are much more limited. Almost certainly it will be worth using two effects, thus saving US$50,000 per annum, but savings beyond that may not be justified by the extra capital cost.

8.4.2.4 TVR Design

The value of the entrainment ratio (Equation 8.4) depends upon the relative pressures of the fresh steam, P_s, the supply vapor, P_v, and the discharge vapor, P_d. Figure 8.15 from Power (1994) was

TABLE 8.5
Energy Costs and Savings for Example 8.4

Number of effects	Energy cost (k$/year)	Energy cost saving from the extra effect (k$/year)
1	99	
2	49	50
3	33	16
4	25	8
5	20	5

obtained from data from a number of ejector manufacturers, and it enables the determination of R_s, the inverse of the entrainment ratio (Equation 8.4). Power also gives a thermodynamic explanation of steam ejectors. Additional data are given by Billet (1989) for higher evaporation temperatures. This gives entrainment ratios 20–25% lower than those of Power. Minton (1986) provides the formula:

$$R_s = 0.40e^{4.6 \ln(P_d/P_v)/\ln(P_s/P_v)} \tag{8.51}$$

Clearly a final design must be based on the performance of a particular ejector.

In the context of TVR evaporation the motive pressure referred to in Figure 8.15 is the absolute steam pressure, and the suction pressure is the vapor pressure before the thermocompressor.

Example 8.5 TVR evaporator design
Calculate the specific heat consumption of a five effect TVR evaporator as shown in Figure 8.10 with temperatures from 70 to 40°C. Steam is available at 10 bar abs. but to allow for control of pressure disturbances and for fouling the design will be based on 6 bar abs.

We assume that the temperatures are evenly distributed. This is not necessarily the case but is a good first approximation. Thus the first effect vapor will be at 70°C and the second at 62.5°C. Ignoring boiling point elevation and pressure drops in tubes or vapor ducts, the temperature driving force is 7.5°C. We arbitrarily decide that the steam entering the first effect should provide the same driving force and thus its temperature should be 77.5°C; a higher temperature increases the rate of fouling. Thus we have the temperatures and pressures obtained from steam tables shown in Table 8.6.

The compression ratio is the ratio of discharge pressure to suction pressure and in this case is $0.427/0.223 = 1.91$. The expansion ratio is the ratio of steam pressure to suction pressure and in this case is $6.0/0.223 = 26.9$. From Figure 8.15 the value of R_s is about 1.03. Thus the entrainment ratio, μ, is $1/1.03 = 0.97$. By way of comparison, using Minton's equation gives $R_s = 0.99$, and from GEA (1998) the value of R_s is about 0.86 giving an entrainment ratio, of 1.16.

Hence the specific energy consumption is $1/(n + \mu n_{recycle}) = 1/(5 + 0.97 \times 2) = 0.144$ kg steam/kg evaporation. If the evaporator did not have TVR the specific energy consumption would have been about $1/5 = 0.2$. The effect of uncertainty in the entrainment ratio on these calculations is small.

Referring to Example 8.3 and Table 8.4, the required energy flow rate will be about $0.144 \times 24,300 = 3,500$ kJ sec^{-1} and the energy cost will be about $0.144 \times 2,305,000 = $US\$332,000 per year compared with US\$461,000 for a standard five-effect evaporator. The use of TVR would reduce the energy cost by about US\$133,000 per year. The additional capital cost of TVR is small. Even with some uncertainty in this initial estimate, it indicates that TVR evaporation offers significant savings.

If the vapor recycle was around the first effect only the entrainment ratio would have been about 1.95, and the specific energy consumption is again found to be 0.144. The reduced efficiency of vapor

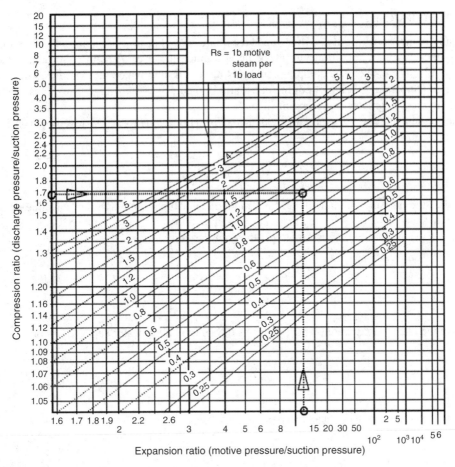

Compression ratio (discharge pressure/suction pressure)

Rs = 1b motive steam per 1b load

Expansion ratio (motive pressure/suction pressure)

FIGURE 8.15 Steam rates for thermocompressors and ejectors. (From Power, R.B. (1994), *Stream Jet Ejectors for the Process Industries*, McGraw-Hill Inc, New York.)

TABLE 8.6
Temperature and Pressures for Example 8.5

	Temperature (°C)	Pressure bar absolute
First effect chest	77.5	0.427
Second effect vapor = Second effect chest	70	0.312
Third effect vapor	62.5	0.223

recycle over only one effect is exactly offset by the increased efficiency of the thermocompressor operating at a lower compression ratio.

In a TVR evaporator more heat is transferred in the effects within the TVR loop, so the area in these must be greater than for the same effects in a standard evaporator. As a result the number of tubes must be greater and the wetting factors are lower in these effects. Conversely because there is relatively less heat transfer in the other effects, the wetting in these is greater. Generally this is an advantage.

8.4.2.5 MVR Design

Originally compressors were used for vapor recompression, but now suitable radial fans that can operate under vacuum conditions are available at a much lower cost (Knipschildt and Andersen, 1994). Typically fans achieve compression ratios up to 1.4, and hence they produce only a small increase in temperature, being a limiting constraint in the design. Two single stage fans can be put in series to provide greater compression.

The pressure increase for a fan can be calculated from Equation (8.42) (Davidson and von Bertele, 1996)

$$\Delta P = \psi \rho_{\text{inlet}} \frac{u_{\text{tip}}^2}{2} \tag{8.52}$$

where ψ is the pressure coefficient and for a modern centrifugal fan may it have a value up to 1.7. The fan tip speed, u_{tip}, is limited by material strength to about 265 m sec^{-1} at 65°C.

We can manipulate Equation 8.52 to get the compression ratio:

$$\frac{P_{\text{out}}}{P_{\text{inlet}}} = 1 + \frac{\psi \rho_{\text{inlet}} u_{\text{tip}}^2}{2 P_{\text{inlet}}} \tag{8.53}$$

and from the ideal gas equation, which applies sufficiently well to water vapor at low pressures,

$$\frac{\rho_{\text{inlet}}}{P_{\text{inlet}}} = \frac{M}{RT} \tag{8.54}$$

where M is the molar mass of water (0.018 kg/mol). Thus

$$\frac{P_{\text{out}}}{P_{\text{in}}} = 1 + \frac{\psi u_{\text{tip}}^2 M}{2RT} \tag{8.55}$$

Using the maximum values of 1.7 for ψ and 265 m sec^{-1} for u_{tip}, the maximum compression ratio is found to be 1.4 at 60°C and it is relatively independent of temperature. From steam tables the corresponding temperature rise of saturated water vapor is calculated to be 7.0°C. This temperature rise does not take into account any pressure drops in vapor ducts and the corresponding heat transfer temperature driving force is further reduced by boiling point elevation. In practice two fans may be used in series, achieving a maximum compression ratio of about 1.8 with a corresponding temperature rise of 13°C.

The power requirement, Po, of a fan can be approximated from

$$Po = \frac{Q \Delta P}{\eta} \tag{8.56}$$

where Q is the average volumetric flow rate calculated from the inlet and outlet densities. A more rigorous equation is given by Davidson and von Bertele (1996). The efficiency of a fan, η, is defined as:

$$\eta = \frac{\text{isentropic power requirement}}{\text{actual power requirement}} \tag{8.57}$$

and typically a value of 80% can be used.

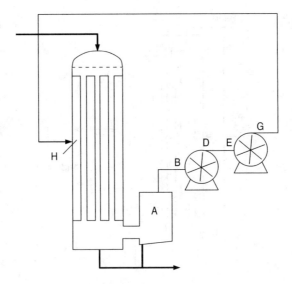

FIGURE 8.16 A single MVR stage with two fans for vapor recompression.

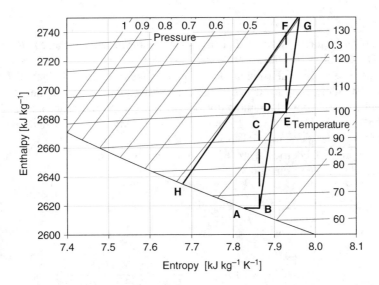

FIGURE 8.17 A Mollier diagram of MVR evaporator with two fans in series. Pressure is shown in bar and temperature in °C.

Alternatively, if the thermodynamic calculations are done, the power requirement is given by the enthalpy change of the vapor

$$Po = \dot{m}_{evap}\Delta h_{fan} \tag{8.58}$$

where Δh_{fan} is the change in the enthalpy of the vapor as it passes through the fan.

The MVR process, shown with two fans and a single stage in Figure 8.16, is best described by referring to a thermodynamic Mollier chart of water vapor enthalpy vs. entropy as shown in Figure 8.17. The points and process path marked in Figure 8.17 are described in Example 8.6. An understanding of entropy is not required to interpret the process.

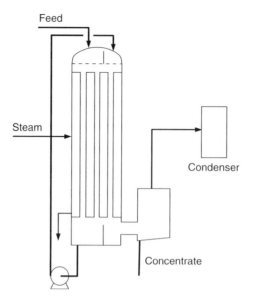

FIGURE 8.18   A two-pass evaporator effect.

Example 8.6   Calculations for Figure 8.17.
Calculate the power requirement for a single fan and two fans in series for an MVR evaporator with a total evaporation of 50,000 kg/h (13.89 kg sec^{-1}) at 65°C.

This example was calculated using thermodynamic equations given by Haar et al. (1984) and is presented here with more accuracy than can be obtained from the diagram alone.

From steam tables at 65°C the saturated vapor pressure is 25,009 Pa absolute and the enthalpy is 2618.2 kJ kg^{-1} (point A in Figure 8.16 and Figure 8.17). After the pressure drop of 2,000 Pa, at the same enthalpy the temperature reduces to 64.8°C.

We assume a pressure coefficient, ψ, of 1.7 and a fan tip speed of 265 m s^{-1}. The pressure out of the fan is calculated from Equation 8.53.

$$P_{out} = P_{in} \left(1 + \frac{\psi u_{tip}^2 M}{2RT} \right)$$

$$= 23,009 \left(1 + \frac{1.7 \times 265^2 \times 0.018}{2 \times 8.314 \times (273.15 + 64.8)} \right)$$

$$= 31,807 \text{ Pa}$$

By using Equation 8.53 (rather than assuming ideal gas behavior) the outlet pressure is found a little more accurately to be 31,887 Pa. Initially a constant entropy line is followed from point B to C and the new enthalpy of 2,671.3 kJ kg^{-1} is obtained. The enthalpy change of 53.1 kJ kg^{-1} is for the isentropic case, but if we have an isentropic efficiency of 80% the actual enthalpy change will be 53.1/0.8 = 66.4 kJ kg^{-1}. The actual outlet vapor has an enthalpy of 2,618.2+66.4 = 2,684.6 kJ kg^{-1} at the pressure already calculated as 31807 Pa (point D). The power requirement of the fan can be calculated from

$$Po = \dot{m}_{evap} \Delta h_{fan} = 13.89 \times 66.4 = 922 \text{ kW}$$

An additional 4% should be allowed for the inefficiency of the electrical motor (Walas, 1990), giving a requirement of 960 kW.

The calculation from D to G is the same as for A to D, and we find the pressure at G to be 40,400 Pa at a temperature of 136.4°C. For the second fan the power requirement is 970 kW.

From the fan exit (G), the vapor loses some pressure in the ducting. The vapor also loses its super-heat either by the addition of water, by heat loss from the ducting, or by contact with the evaporator tubes. Point H in the vapor chest is saturated but with a pressure of 38,400 Pa equal to the pressure at G minus the loss, assumed here to be 2,000 Pa. The saturated vapor temperature at this pressure is 74.9°C, and thus the available temperature driving force is 9.9°C less the boiling point elevation.

The vapor condenses on the outside of the tubes at point H and is discharged as liquid. An almost equal amount of vapor is evaporated from the product in the inside of the tubes. There is an energy path but not a material path from H to A.

The total power required is 1,930 kW to evaporate 13.89 kg sec^{-1}, that is, 140 kJ kg^{-1}. At the operating temperature of 65°C the latent heat of evaporation is about 2,350 kJ kg^{-1}. Thus 140 kJ is required to achieve evaporation of 2,350 kJ giving a specific energy consumption of 0.06 kJ of electricity per kJ of evaporation. The cost of a suitable pair of fans was about US$350,000 in 2001. In addition, induction motors are required and a variable speed drive is required for at least one of the fans. Variable speed drives cost about US$100 per kW. The total price for fans, motors and variable speed drives would be about US$650,000.

When the concentration is high the boiling point elevation can be significant and the effective temperature difference is reduced. This effect limits the application of MVR to solutions with a boiling point elevation of less than about 2°C, so one or two TVR effects are sometimes added after the MVR section. The combination allows a reduced area and better control of the final concentration (Knipschildt and Andersen, 1994).

In the MVR evaporator the vapor is condensed in the chest, so there is no need for a condenser or condenser water.

In contrast to a TVR evaporator, which must consist of several effects contained in several calandrias, an MVR evaporator has only one effect with possibly many passes within one calandria. Thus an MVR evaporator can be much more compact than a TVR evaporator.

MVR fans can be driven by an electric motor, a steam turbine, or a gas engine or turbine. Electric motors are the most common choice because of price, availability, and the maturity of the technology. Variable speed is possible with all the types of drives and variable speed drives are now available for motors up to 2 MW with a price in the year 2000 of about US$100 per kW.

The choice between TVR or MVR depends on many factors. Table 8.7 lists the circumstances that might favor each type. A complete plant-wide evaluation may be required to determine the best option.

8.4.2.6 Multipass Design

For any style of falling film evaporator, the tubes must be kept wet, but a simple initial design sometimes does not guarantee this. Consider, for example, the design calculations for a simple

TABLE 8.7
Conditions Appropriate for TVR and MVR Evaporators

TVR	MVR
Low cost steam	Low cost electricity
High cost electricity	On site cogeneration
Low throughput	High throughput
Low capital cost required	Small space requirement
No large motor expertise	

TABLE 8.8
Calculated Design with Single Pass Effects

Effect	Evaporation (kg sec^{-1})	Exit flow (kg sec^{-1})	Total solids (%)	Average HTC (W m^{-2} K^{-1})	Number of tubes	Wetting rate (kg m^{-1} sec^{-1})
1	2.05	11.8	15.3	1560	180	0.44
2	2.05	9.8	18.5	1480	200	0.33
3	2.05	7.7	23.4	1360	220	0.23
4	2.05	5.7	31.9	1150	275	0.14
5	2.05	3.6	50.0	700	525	0.05

TABLE 8.9
Calculated Design with Multiple Passes in Effects 4 and 5

Effect	Exit flow (kg sec^{-1})	Total solids (%)	Average HTC (W m^{-2} K^{-1})	Number of tubes	Wetting rate (kg m^{-1} sec^{-1})
1	10.7	11.7	1647	306	0.23
2	7.4	16.8	1519	324	0.15
3	5.8	21.6	1399	197	0.20
4a	5.0	25.2	1310	121	0.27
4b	4.1	30.2	1185	129	0.21
5a	3.4	36.7	1022	143	0.16
5b	2.9	43.7	848	130	0.15
5c	2.5	50.0	690	116	0.14

five-effect DSE evaporator shown in Table 8.8. As the liquid becomes more concentrated the heat transfer coefficient drops (a linear relationship was used here) and more area, that is, a greater number of tubes, is required to transfer the same amount of energy. Also with increased concentration the flow rate drops, and the combined effect is that the wetting rate (Section 8.3.3.9) reduces markedly. The wetting rate in the last two effects is lower than the recommended 0.2 kg m^{-1} sec^{-1}.

To overcome this difficulty, the design might be altered to have two passes through effect 4, (as shown in Figure 8.18) and three through effect 5. The passes are separated by partitions at the top and bottom of the tube bundle. The resulting design is shown in Table 8.9. The wetting rates obtained are better but still lower than the recommended level.

With TVR, MVR, and multiple passes, there are many possible ways to configure an evaporator. For example Knipschildt and Andersen (1994) show an APV Anhydro evaporator with 7 effects. It has a TVR loop around the first three effects and another around the fifth effect. There are 2 passes in the 3rd effect, 5th, 6th and 7th effects. The product flow goes through the effects in the order 1, 2, 3, 4, 5 (pass 1), 7, 6, and 5 (pass 2).

8.4.3 Design and Operation of Other Components

8.4.3.1 Preheat

The feed to the evaporator is generally preheated to:

Bring it up to the boiling temperature in the first effect
Destroy pathogenic bacteria

In the case of milk to denature the whey protein and alter functional properties
Deaerate the feed

The preheating section may be used as a legal pasteurization treatment as shown by Knipschildt and Andersen (1994, Figure 7). However in some countries this may also require a system to divert unpasteurized product in case the pasteurization conditions are not met. The diversion system would be designed but it may never be used, as control of an evaporator with a sudden change from product to water would be very difficult. It has been found easier in practice to pasteurize all the feed using a plate heat exchanger.

In milk powders the amount of heat treatment affects the functional properties of the powder. A low heat treatment is given for milk powder that is to be used in cheese making, for example, where undenatured protein is desired, but high heat treated milk can be more desirable for baking and chocolate applications.

A number of different methods are used for preheating:

Indirect heating by passing the feed tubes through each steam chest
Indirect heating with hot water in a heat exchanger
Direct steam injection
Direct contact preheat

A combination of these systems might also be used. For example one popular system has been indirect heating with hot water then direct steam injection.

Indirect heating in the condenser and steam chests has been more popular in the past than at present. The feed pipe spirals its way through the condenser, then the final effect chest and through each effect to the first or second. The preheater in the condenser helps condense some of the vapor and reduces the cooling water requirement. Preheating in the steam chests reduces the required capacity of the main preheater but it uses energy in the vapor that might otherwise be used for evaporation. It has the advantage of the energy flow to later effects being lower and hence wetting rates being higher. It has disadvantages of making the construction of the calandrias more complex and hence expensive, and it reduces the flexibility of the preheat system. The same system is also used but with an external preheater on each effect using the vapor from the steam chest of the effect (Figure 8.19). This system allows easier cleaning and alteration of the preheaters.

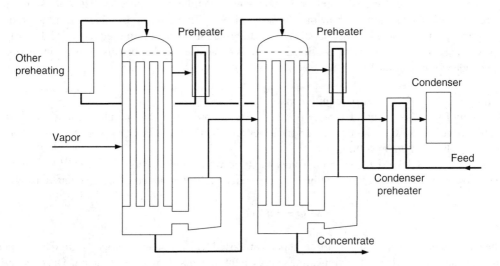

FIGURE 8.19   A two-effect evaporator with feed preheating.

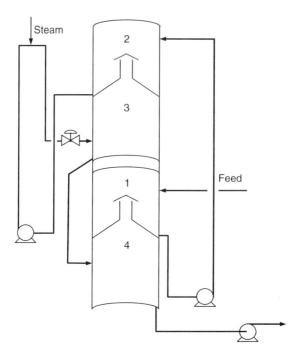

FIGURE 8.20 A direct contact preheat system. (After Refstrup, E. (1998), Heat treatment of milk prior to evaporation and drying, *International Dairy Congress*, Vol. 2, pp. 228–237.)

Indirect preheating with a heat exchanger is common. In MVR evaporation the condensate is often used to partially preheat the feed. In other evaporators feed can be heated by hot water in a plate heat exchanger or a tube-in-tube heat exchanger.

In the direct steam injection (DSI) process, clean filtered steam is injected directly into the product through a set of small holes. The liquid is heated without a heat exchange surface and fouling is less of a problem. DSI systems need to be designed to avoid burning the product in the region of the holes. This requires that the steam pressure be high enough to prevent product entering the steam holes and also that the heating rate is not too great as then local temperatures can become excessive.

The direct contact (or flash) preheat systems are becoming more popular. They are somewhat more complex. One system is shown by Refstrup (1998) (Figure 8.20). In this system the product flows from sections 1 to 4 in order but the temperatures are in the ascending sequence 1:2:4:3. Vapor flashes in section 3 and passes to section 2 where is condenses, and vapor also flashes in section 4 and condenses in section 1.

Example 8.7 Preheat energy requirements
Following Examples 8.1 and Example 8.3, calculate the preheat energy requirements if the feed is heated from 8 to 32°C in the condenser preheater and then to 68°C in chest preheaters and then to 100°C by direct steam injection.

From Section 8.3.3.6, the specific heat capacity of whole milk is about 3.9 kJ kg^{-1} K^{-1} and the mass flow rate is 13.9 kg sec^{-1}. The heat input of each preheater can be easily calculated using

$$Q = \dot{m}C_p(T_{\text{out}} - T_{\text{in}}) \tag{8.59}$$

The required energy inputs of the different parts are listed with the steam energy required for evaporation in Table 8.10. Clearly the preheating energy requirements are not insignificant. The condenser preheater uses energy that would otherwise be wasted, but the energy for the chest preheater and the

TABLE 8.10
Preheater Energy Flow Rates

Section	Energy flow rate (kJ sec^{-1})
Condenser preheater	1451
Chest preheaters	1814
DSI preheater	1680
Five effect TVR steam	3500

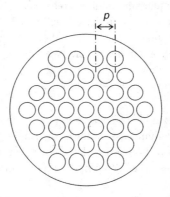

FIGURE 8.21   A typical tube sheet with triangular pitch.

direct steam injection must come from another source. The amount of energy transferred in the chest preheaters is one half of the energy of the steam. The calculations of Example 8.5 should be redone to include the preheating energy.

8.4.3.2 Tube Sheet Design

There is a tube sheet (or tube plate) at each end of the bundle of tubes to hold and seal the tubes. Both ends of the tubes are welded into a tube sheet and the tube sheet is welded or sealed into the calandria wall, thus forming a sealed steam chest around the tubes. The standard arrangement of tubes in the sheet is triangular as shown in Figure 8.21. The ratio of tube exterior cross-sectional area to plate area is designated k_e and is given by

$$k_e = \frac{A_{tube}}{A_{plate}} = \frac{\pi D_e^2/4}{\sqrt{3}p^2/2} = \frac{\pi}{2\sqrt{3}}\frac{D_e^2}{p^2} \tag{8.60}$$

where D_e is the exterior diameter of the tube, and p is distance between the tube centers (pitch). Thus

$$p = 0.952\frac{D_e}{\sqrt{k_e}} \tag{8.61}$$

For a Robert sugar evaporator (not a falling film) the value of k_e is typically 0.45–0.55. (Hugot, 1986). Falling film evaporators have an internal diameter to pitch ratio of 0.75 to 0.8 (Stork Friesland, 1992; Bouman et al., 1988). Thus using a 50.8 mm OD tube with a 1.5 mm wall thickness, the gap between the outer walls of the tubes will be 9 to 13 mm. The tube spacing is constrained by the need to have adequate metal in the tube sheet for strength and welding.

The question can arise as to whether a small tube pitch could cause a pressure drop as the steam travels into the center of the bundle. We can calculate the tube bundle dimensions and find that the gap between tubes would need to be less than about 1 mm before pressure drop causes even 0.1°C loss of temperature driving force so this does not seem to be a design constraint.

8.4.3.3 Liquid Distribution

The objectives of the feed distribution system are to:

Distribute the feed so that an equal amount flows into each tube
Distribute the liquid around the circumference of each tube
Allow flashed vapor to pass through to the tubes without affecting the distribution
Allow overflow when flow rates are high, especially during cleaning
Be cleanable especially on the under side
Not cause excess residence time

Bouman et al. (1988) reported inspections on evaporators after processing milk but before cleaning. They found that in many cases the liquid was not well distributed with the possibility of excessive deposit formation in tubes with a low flow. They show a distribution system similar to that which is described here. A variety of systems have been proposed with various levels of complexity, but fluid mechanics favors the simple design.

Minton (1986) offers a little advice and for an undisclosed application suggests a liquid height above the distributor of 150 to 300 mm. This would cause an excessive increase in residence time and is not appropriate for some food products.

It is normal to use a plate with holes in it where the holes are lined up to hit the tube sheet and not the tubes directly. As shown in Figure 8.22 each distributor hole will feed three tubes and each tube is surrounded by six holes, so there are twice as many holes as tubes. The exceptions to this are the outer holes and tubes. As shown the outer tubes will receive a greater share of the flow and there is a case for replacing each of the outer pairs of distributor holes with a single hole, that is, each of the outer tubes has only five surrounding holes. APV (1999) show a design with half as many distributor holes.

The hole sizing is straightforward. For good distribution without excessive holdup, the liquid level should be about 30–50 mm above each hole. A simple orifice equation, also known as Torricelli's equation, can be used to calculate the flow rate through each hole or alternatively calculate the size

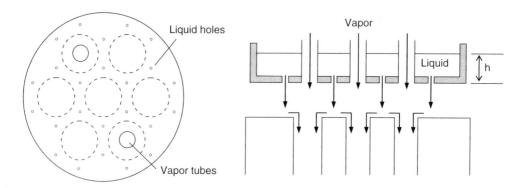

FIGURE 8.22   Plan view and cross-section of a distributor plate.

of each hole.

$$F = n_{holes} A_{hole} C_d \sqrt{2gh} \tag{8.62}$$

or

$$\frac{\pi D_{hole}^2}{4} = \frac{F}{n_{holes} C_d \sqrt{2gh}} \tag{8.63}$$

where the discharge coefficient, C_d, is typically 0.6, and F is the total flow rate (m^3 sec^{-1}).

For example, assume the total flow rate to a calandria is 0.0134 m^3 sec^{-1} (50,000 kg h^{-1}) and the design requires 266 tubes. From a tube layout it was found that there will be 540 distributor holes using the pattern in Figure 8.22. Using a desired liquid height of 30 mm the hole diameter is calculated from Equation 8.63 to be 8.3 mm.

Here the importance of the distributor plate being level and flat is apparent. If the plate is not installed level or if it becomes warped after years of use, the flow through each hole will vary. For example, a height variation of 6 mm over the plate will give a flow variation of 10%.

The feed to the evaporator must not flow directly onto the distributor plate as then more liquid will be forced through at the point of impact. Instead a deflection plate is installed under the inlet pipe to distribute the liquid onto the distribution plate.

The distributor plate must also allow a path for vapor that has flashed off at the top of each calandria above the distributor plate and needs to find its way down the tubes. Typically the liquid in one effect is 5 to 7°C hotter than the subsequent effect so some flashing will occur. Further, after preheating it is possible that the feed to the first effect may be much hotter than the first effect. To provide a path for the vapor, tubes referred to as "vapor tubes" are welded into the distributor plate as shown in Figure 8.22.

Flash calculations. Suppose 10.7 kg sec^{-1} of skim milk with 11.7% total solids is pumped from the first effect of an evaporator at 70°C to the second effect which has a temperature of 62.5°C. From the physical property equations above, the heat capacity of the milk is 3.91 kJ kg^{-1} K^{-1}, and so when the temperature drops from 70 to 62.5°C (ignoring boiling point elevation) the flow rate of enthalpy, Q, available is

$$Q = \dot{m} C_p \Delta T = 10.7 \times 3.91 \times (70 - 62.5) = 313.6 \text{ kJ sec}^{-1} \tag{8.64}$$

The energy required to evaporate water at 62.5°C is 2352 kJ kg^{-1}, so the rate of water evaporation, $\dot{m}_{evap}$ is

$$\dot{m}_{evap} = \frac{Q}{\Delta h_v} = \frac{313.6}{2352} = 0.133 \text{ kg sec}^{-1} \tag{8.65}$$

At this temperature the vapor density is 0.144 kg m^{-3} and thus the volumetric flow rate is

$$F = \frac{\dot{m}_{evap}}{\rho} = \frac{0.133}{0.144} = 0.93 \text{ m}^3 \text{ sec}^{-1} \tag{8.66}$$

This vapor must pass through or around the distribution plate with a pressure drop small enough so that an excessive amount of liquid is not forced through the plate. Normally vapor tubes are included for this purpose. Given that the liquid level is at least 30 mm, the maximum pressure should be about 5 mm water gauge or 50 Pa, and the vapor tubes need to be about 60 mm high so that in general liquid does not flow down them. (In periods of high liquid flow, for example, during cleaning, flow down the vapor tubes may be useful.) A low pressure drop can be calculated using incompressible fluid equations. There is a contraction, the tube and an expansion and the appropriate coefficients for these are:

$$K_{contraction} = 0.5, \quad K_{expansion} = 1.0, \quad f = 0.003 \text{ (estimate)}$$

The pressure drop is given by

$$\Delta P = \left(K_c + K_e + \frac{4fL}{D} \right) \rho_{vapour} \frac{u^2}{2} \tag{8.67}$$

or

$$u = \sqrt{\frac{2\Delta P_{max}}{\rho(K_c + K_e + 4fL/D)}} \tag{8.68}$$

which in this case, say, using a 25 mm OD (23 mm ID) vapor tube,

$$u = \sqrt{\frac{2 \times 50}{0.144(0.5 + 1.0 + 4 \times 0.003 \times 0.06/0.023)}}$$

$$\approx 21 \text{ m sec}^{-1}$$

This result is not very sensitive to the choice of tube diameter or the friction factor. The required hole area can be calculated from the volumetric flow rate and the velocity giving 0.044 m^2 which can be achieved with 106 holes with 23 mm diameter. Independently it was calculated that this effect would require 325 tubes of 50 mm diameter, with a calandria diameter of 1.32 m. Thus there should be one vapor tube for every three evaporator tubes. About 3.4% of the distributor plate area will be occupied by the vapor tubes, so some liquid may pass directly down the vapor tube rather than through the distributor holes, but given the small area of the tubes the amount would be acceptably low.

8.4.3.4 Separator Design

When the vapor leaves each effect it must be separated from any entrained liquid so that liquid product is not carried over with the vapor. If there is product carry-over it causes product loss, it potentially causes fouling of other parts of the evaporator, and it reduces the quality of the condensate. In falling-film evaporators liquid entrainment in the vapor is minimal as droplets produced during boiling are entrained by the film within the tubes. However in a short tube evaporator the separation occurs above the tubes and some form of de-entrainment is normally necessary. See Hugot (1986) for more details.

For falling-film evaporators, vertical cylindrical vessels with tangential entry have proved successful as vapor–liquid separators. As will be seen in the next section high vapor velocities must be avoided to ensure that pressure drops are low.

8.4.3.5 Vapor Pressure Drop

Any pressure drop in the vapor causes a drop in the vapor temperature and hence in the temperature difference available in the next effect. Vapor is taken from each separator to the next effect by a vapor duct and, in the case of an MVR evaporator, from each separator to the MVR fan and then from the fan back to each effect. The ducts are usually circular as this shape provides the greatest strength against collapse. Although no enthalpy is lost by friction, the pressure is reduced and thus the temperature driving force is reduced. A reduction of 0.1°C in vapor temperature will reduce heat transfer capacity by about 1.4%. At 70°C a temperature drop of 0.1°C will occur if the pressure drop is 135 Pa and at 50°C, the corresponding pressure drop is only 60 Pa.

The pressure drop is given by

$$\Delta P = \left(\Sigma K_{fittings} + \frac{4fL}{D} \right) \rho_{vapour} \frac{u^2}{2} \tag{8.69}$$

Example 8.8 Vapor duct pressure drop

An MVR evaporator processing 50,000 $kg\,h^{-1}$ of whole milk has a 1.0 m diameter vapor duct to and from the fan. The duct is 20 m long and includes four smooth bends. Calculate the pressure and temperature drop if the vapor temperature is 65°C.

At 65°C the vapor density is 0.16 $kg\,m^{-3}$ and from Example 8.1 the evaporation flow rate is 10.28 $kg\,sec^{-1}$ so the vapor volumetric flow rate is 63.7 $m^3\,sec^{-1}$. The velocity in the 1 m diameter duct will be 81 $m\,sec^{-1}$. The resistance coefficient for a smooth bend is about 0.25 and the friction factor is found to be 0.0028. The pressure drop is given by

$$\Delta P = \left(\Sigma K_{\text{fittings}} + \frac{4fL}{D} \right) \rho_{\text{vapour}} \frac{u^2}{2}$$

$$= \left(4 \times 0.25 + \frac{4 \times 0.0028 \times 20}{1.0} \right) 0.16 \frac{81^2}{2}$$

$$= 650 \text{ Pa}$$

The pressure thus decreases from 25,023 Pa (the saturation pressure at 65°C) to 24,373 Pa with a saturation temperature of 64.4°C. The drop in saturation temperature from the friction is 0.6°C, and given that a typical MVR evaporator has a temperature difference of at most 9°C, this would require an increase to area of 7% to compensate. If instead the duct diameter was increased to 1.5 m the temperature drop can be calculated to be only 0.1°C. Almost certainly, a greater capital investment in the ducting would lead to significant saving in the cost of the calandrias. This calculation tells us that the vapor duct system must be designed with care.

8.4.3.6 Noncondensable Gases

If noncondensable gases from air build up in the vapor chest, the partial pressure of water vapor is lowered and hence the temperature driving force is lowered. Further, when vapor condenses, the air that is carried with it will build up at the condensing surface, providing a physical barrier for heat transfer. Minton (1986) gives a simple correlation for the fouling heat transfer coefficient as:

$$h_f = 247/w_{\text{air}} \tag{8.70}$$

where h_f is the apparent fouling coefficient due to noncondensable gases ($W\,m^{-2}\,K^{-1}$) and w is the mass fraction of air in the steam chest. For example, if the concentration of noncondensable gases was 2% by mass, h_f has a value 12,350 $W\,m^{-2}\,K^{-1}$, and thus an overall heat transfer coefficient of say 1,000 $W\,m^{-2}\,K^{-1}$ is reduced to 925 $W\,m^{-2}\,K^{-1}$ giving a significant loss of capacity.

A vacuum pump is normally connected to each vapor chest and to the condenser to withdraw vapor and noncondensable gases. The required extraction rate depends on the amount of dissolved or entrained air in the feed. Air in the feed may be evaporated off in the first effect and hence accumulates in the steam chest of the second effect. Excess extraction gives excessive energy loss because useful vapor must be extracted with the noncondensable gases. Schwartzberg (1989) gave an estimate that 3% of the vapor from each vessel is vented from noncondensable gas removal.

8.4.3.7 Vacuum System

A vacuum system is required to evacuate the evaporator before startup, to maintain the vacuum and to remove noncondensable gases that otherwise accumulate. A vacuum can be produced by a steam ejector or a mechanical vacuum pump. It is likely that a mechanical vacuum pump will be more efficient. Minton (1986) gives the energy efficiency of a steam ejector at about 5%, whereas an electric liquid ring pump may be 40 to 50% efficient.

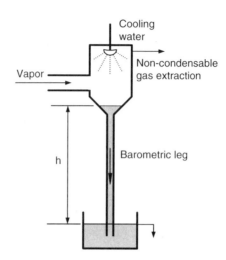

FIGURE 8.23   A simple direct contact condenser.

The normal choice of pump is a liquid ring vacuum pump as they can tolerate slugs of liquid while positive displacement vacuum pumps cannot. The liquid ring condenses any vapors and forms a vacuum seal.

The capacity of the vacuum pump depends on the required draw-down time when the evaporator is started up. From the draw-down time we can find the required actual volumetric flow rate required (Minton, 1986).

$$F = \frac{V_{\text{system}}}{t} \ln \left(\frac{P_{\text{initial}}}{P_{\text{final}}} \right) \tag{8.71}$$

For example, if the evaporator volume is 80 m^3 and the required pressure of 10,000 Pa abs. must be obtained in 30 min, then the actual flow rate is 368 m^3 h^{-1}.

The capacity required for the maintenance of the vacuum is likely to be much smaller than, this, so normally two vacuum pumps are used — both for draw-down and just one for maintenance of the vacuum. This would reduce the energy overall electrical power cost. If steam jets are used there are also two for the same reason.

8.4.3.8 Condensers

Condensers may be direct or indirect. In a direct condenser cooling water is sprayed into the condenser and it mixes directly with the vapor, immediately condensing it. Sometimes packing is used to help distribute the water and enhance contacting. The condenser requires water at a temperature sufficiently lower than the vapor temperature; Billet (1989) suggests that the outlet water temperature is 6–8°C below the required vapor pressure. A typical arrangement is shown in Figure 8.23.

Example 8.9   Water flow rate for a condenser
From the final effect there is a flow of 1.6 kg sec^{-1} of vapor at 40°C. Water is available at 25°C. How much water will be required?

To achieve the desired temperature of 40°C in the final effect, the water in the condenser must have a temperature of less than 40°C. We will allow 6°C, so the water must leave the condenser with a temperature, $T_{\text{w,out}}$, of 34°C. The heat balance over the condenser is

$$\dot{m}_v h_v + \dot{m}_w C_{p,w} T_{\text{w,in}} = (\dot{m}_v + \dot{m}_w) C_{p,w} T_{\text{w,out}} \tag{8.72}$$

from which

$$\dot{m}_{\mathrm{w}} = \frac{\dot{m}_{\mathrm{v}}(h_{\mathrm{v}} - C_{\mathrm{p,w}}T_{\mathrm{w,out}})}{C_{\mathrm{p,w}}(T_{\mathrm{w,out}} - T_{\mathrm{w,in}})} \tag{8.73}$$

At 40°C water vapor has an enthalpy, h_{v}, of 2575 kJ kg^{-1} and the specific heat capacity of water, $C_{\mathrm{p,w}}$ is 4.18 kJ kg^{-1}K^{-1}. Substitution of values gives a required water flow rate of 133 kg sec^{-1} or 481 m^3 h^{-1}. This is a large amount of water so some form of reuse is required. A cooling tower might be considered so that the water can be recirculated.

The pressure within the condenser, and hence within the final effect, is controlled by altering the flow rate of condenser water. If the flow rate is lowered, the temperature of the outlet water will increase and hence the condenser pressure will also increase.

The barometric leg shown in Figure 8.23 is just an open pipe that allows free passage of water out of the condenser without the need for a pump. If the leg is sufficiently long, the head of water, h, is enough to prevent the vacuum in the condenser from sucking the water back into the condenser. Equation 8.74 for static pressure applies.

$$P_{\mathrm{ambient}} - P_{\mathrm{condenser}} = \rho g h \tag{8.74}$$

At 40°C the density of water, ρ, is 992 kg m^{-3}, the evaporator pressure will be 7,380 Pa, and the maximum ambient pressure normally encountered is about 104,000 Pa so the required head calculated to be 9.9 m. A longer pipe can be used and the water will find a level in the pipe corresponding to the ambient and condenser pressures.

Noncondensable gases tend to accumulate in the condenser, so these must be extracted by a vacuum pump or ejector.

An indirect condenser might just be a shell and tube heat exchanger. Design calculations for such a heat exchanger are given by Kern (1964). A typical heat transfer coefficient for condensing vapor and water is given as 600 kW m^{-2} K^{-1}.

8.4.3.9 Pumps

Pumps are required to feed the evaporator, to transfer liquid between stages, and to remove the condensate. In each case the pumping head is the sum of the head due to the pressure difference between the vessels and the height change in the liquid. Pipe friction will generally be negligible if the liquid velocities are less than 2 m sec^{-1}.

It is desirable to maintain the level of liquid in the pipe from the separator down to the pump within a range to ensure that the pump does not cavitate, while at the same time avoiding hold up in the separator. The sizing of the interstage transfer pumps requires a little attention to achieve this. The maximum pumping head required, ΔH_{max}, as shown in Figure 8.24, is when the liquid is at the minimum level possible without cavitation

$$\Delta H_{\mathrm{max}} = \rho g \Delta h_{\mathrm{max}} - (P_i - P_{i+1})/\rho g \tag{8.75}$$

The minimum required pumping head, ΔH_{min}, corresponds to Δh_{min}. The pump should be designed to be able to pump within this range for any flow rate up to the likely maximum. At the same time the pump curve should be such that small changes in liquid level do not cause large changes in flow rate. Some friction in the system will help give a stable flow rate as then the operating point is less sensitive to fluctuations in the suction side liquid level. Thus the pump curve should be selected to be something like in Figure 8.25. Also if possible the efficiency at the nominal flow rate should be maximized.

The feed pump is often aided by the vacuum in the first effect, but a pressure drop (perhaps 0.4 bar) over the feed control valve should be included to ensure good control of the feed flow rate. The

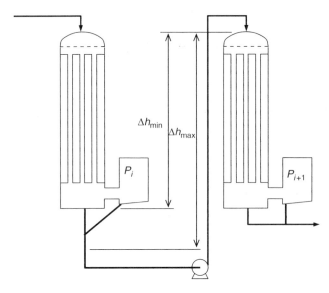

FIGURE 8.24   Operating range for a transfer pump.

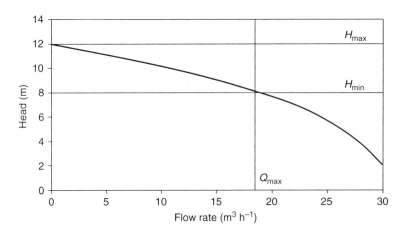

FIGURE 8.25   An ideal pump curve for an evaporator transfer pump.

final concentrate pump and its associated pipework must be designed to ensure that the evaporator vacuum is not lost.

The pumps will almost certainly have three phase induction motors in which case the power input will be

$$Po = \frac{Q\Delta P}{\eta} = \sqrt{3}VI \tag{8.76}$$

The voltage, V, is likely to be 400 V. Given a pump curve, the voltage, and a measurement of the motor current, it is usually quite easy to estimate the flow rate through the pump.

8.4.3.10   Condensate Return

In a simple evaporator, the condensate from the first effect should be uncontaminated and thus is ideal boiler feed water. Condensate recovery will reduce the need for boiler makeup water and its associated treatment, and it will recover some of the energy in the condensate. When TVR is used the condensate from effects other than the first may contain traces of volatile components or entrained liquid particles from the product and thus may be less suitable. Normally a conductivity meter and controller is installed to measure the concentration of dissolved solids and hence to decide whether condensate is acceptable as boiler feed water or not. The relatively low flows from each effect and the range of condensate temperatures often makes the recovery of heat from condensate uneconomic. The condensate from an MVR evaporator is likely to be contaminated by volatile or entrained product so it cannot be considered pure.

8.4.3.11 Materials of Construction

For food systems it is very likely that stainless steel 304 will be the material of choice as it resists corrosion and has a good surface finish that can be cleaned easily. If there are chlorides in the product, or in cleaning chemicals, 316 stainless steel is likely to be used.

8.4.4 FOULING PREVENTION, CLEANING, AND HYGIENE

Most evaporators become fouled over time. Fouling can limit operating times to as little as eight hours before cleaning becomes necessary. Often the loss of heat transfer is not a problem, but foulant provides a suitable environment for the growth of thermophilic bacteria which will limit the acceptable processing time. During preheating and in the main part of the evaporator, certain locations will have a temperature ideal for the growth of thermophiles, and if any spores adhere in this region growth is likely. Refstrup (1998) proposes direct contact preheating (Section 8.4.3.1) to reduce the area of surfaces at the ideal temperatures. The loss of processing time while the evaporator is cleaned requires that the evaporator be designed with a throughput perhaps 20% higher than that required for continuous operation.

Fouling is almost inevitable. Proteins will adhere to most transfer surfaces unless they have been strongly heat treated, but this is normally undesirable as it may damage the product. Minerals may precipitate as their concentration increases in the evaporator and some, e.g., calcium phosphate in dairy products, have reduced solubility as temperature increases. In some cases pretreatment with pH changes and temperature it is possible to precipitate minerals but such processes may be more expensive than the cost of fouling.

Measurement of fouling is an important step in the determination of conditions that enhance or reduce fouling. A crude measure is the pressure of steam in the first effect as this will be automatically or manually increased to achieve the required amount of evaporation. A more direct measure is the pressure difference across the first effect from the steam chest to the vapor in the separator. The temperature difference is directly related to the pressure difference and the overall heat transfer coefficient is inversely related to the temperature difference if the amount of evaporation is constant. In a well-controlled evaporator this pressure difference will increase slowly during each run.

Cleaning is often carried out by introducing cleaning chemicals such as sodium hydroxide and nitric acid into the feed tank. The rate of evaporation is reduced and the solutions are pumped through the evaporator. This method has a number of disadvantages. As the sodium hydroxide passes through the evaporator it becomes more concentrated, but contrary to intuition this decreases its cleaning effectiveness (Bird and Fryer, 1991). The need to pass from one effect to the next extends the cleaning time, especially in large evaporators. The pumps are designed to process only the normal flow rates, and the high flow rates desirable for cleaning might not achieved.

Normal food process design hygiene applies to evaporators. All fittings must be sanitary with no dead spaces. Care needs to be taken around the underside of any surfaces for example, the distributor plate, to ensure that these are cleaned effectively. The pipes to and from the interstage transfer pumps are normally left full as there is normally no means of drainage. This provides a potential site for bacterial growth between runs, which can be countered by using a sterilizing agent as the last rinse of the plant after cleaning.

8.4.5 CONTROL

General aspects of process control are not covered here as there are numerous textbooks that cover these for example, Ogunnaike and Ray (1994). Most of the evaporator controls are designed to keep the inputs of the evaporator constant. If all the inputs are constant, the product should be consistent, but small changes in control setpoints may be required as fouling occurs.

8.4.5.1 Evaporator Balance Tank Level

The level of the evaporator balance tank may be controlled by a float valve or by a modulating inlet valve on the incoming flow. If it is not constant it may cause variations in the feed flow rate. If a float valve is used no other control is required, but otherwise the control shown in Figure 8.26 applies.

8.4.5.2 Evaporator Feed Flow Rate

Variations in the flow rate through an evaporator will have a direct effect on the total solid content of the product. Disturbances can become amplified through the evaporator. A control valve is preferred over a variable speed pump for two reasons. First, the evaporator vacuum helps to suck the feed into the evaporator so some form of resistance may be required, and second a control valve may be required to maintain a suitable pressure for the preheating system.

8.4.5.3 Evaporator Preheat

There are many different schemes for evaporator preheat depending on the type of evaporator. In MVR evaporators the cold feed is usually heat exchanged with the condensate, as shown in Figure 8.26 to help maintain the heat balance. In DSE or TVR evaporators, the only temperature control points are likely to be around the direct or indirect steam preheater.

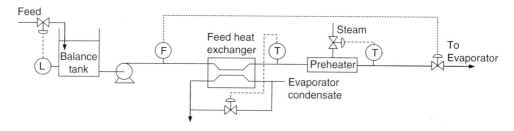

FIGURE 8.26   Possible evaporator feed and preheat control loops.

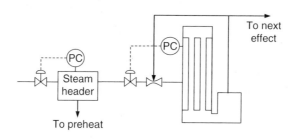

FIGURE 8.27   Main steam header control and one possible TVR pressure control using chest pressure.

8.4.5.4 Evaporator Heat Control

The control of the amount of energy entering the evaporator is the most important part of the control system. The amount of evaporation is directly related to the amount of energy transferred to the product so the rate of heat transfer should be controlled to remain constant. The rate of heat transfer, Q, is given by Equation 8.12 and over short periods (less than 1 h) the heat transfer coefficient can be considered to be constant. Thus to achieve a constant rate of heat transfer the temperature difference must be held constant. In vapor systems the pressure difference from the outside to the inside of tubes, ΔP, is directly related to the temperature difference. The pressure difference can be measured with more accuracy and resolution than temperature difference. The best control loop to control heat input is one in which the pressure difference is measured and controlled.

8.4.5.5 TVR and DSE Heat Input Control

The heat input into a TVR or DSE evaporator is controlled by controlling the steam supply to the evaporator. In all cases the final control element is the evaporator steam valve. A number of different selections can been made of the process variable that is measured. The traditional method has been to measure the steam pressure inside the steam chest (Figure 8.27). Another method is to control the steam pressure upstream of the TVR ejector. The control achieved using these two methods is affected by disturbances in the internal pressure (temperature) of the evaporator.

8.4.5.6 Main Steam Pressure Control

As part of the heat input control on TVR evaporators it is desirable to have a constant pressure of steam available as shown in Figure 8.27. This also helps to give consistent feed preheating control.

8.4.5.7 MVR Energy Input

The energy input into an MVR evaporator is altered by changing the fan speed and hence the electrical power is transferred to the vapor. Generally there is no measurement of energy input, but rather it is implicitly and incorrectly assumed that a specific fan speed will give a specific heat input. The MVR fan speed is changed in direct response to the concentrate total solids measurement as shown in Figure 8.28.

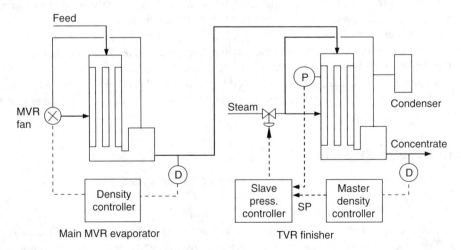

FIGURE 8.28 An effective control scheme for an MVR evaporator and a TVR finisher with separate controllers for each.

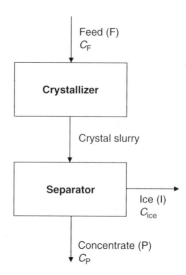

FIGURE 8.29 General process for freeze concentration.

8.4.5.8 Evaporator Total Solids Control

The total solids of the evaporator concentrate is usually measured by density. The relationship between total solids and density is discussed in Section 8.3.3.4. In some plants the density is converted into a total solids reading for the operator, but in many plants operators use the density value directly without difficulty.

In general the output of the total solids controller is the setpoint for the heat input controller using cascade control as shown in Figure 8.28.

8.4.5.9 Total Solids Control with Finishers

Some evaporators incorporate a finisher that has an independent steam supply and thus can be controlled independently with a relatively fast time response. In general the finisher produces only a small amount of the evaporation and the use of the finisher does not remove the need for good control in the main part of the evaporator. In some cases the position of the steam valve on the finisher can be used to alter the heat input control of the main part of the evaporator.

The ideal system is to separate completely the control of the main part of the evaporator from the control of the finisher by installing density control on each part as shown in Figure 8.28. Then the fast response of the finisher control can be used to remove some of the variation in total solids in the concentrate from the main evaporator. Feed forward control can be implemented on a finisher by using the feed total solids and the required product total solids to calculate the amount of steam required.

8.4.5.10 Evaporator Pressure

The evaporator pressure control depends on the type of vacuum system used.

8.4.5.10.1 TVR and DSE final effect pressure control
TVR and DSE evaporators may use a direct or indirect condenser to create the vacuum (Section 8.4.3.8). The temperature of water leaving a direct condenser is closely related to the evaporator final effect pressure and can be controlled by varying the flow. Often in practice it is desirable to have as low a pressure as possible, to maximize throughput and to minimize heat damage to the product, so

condensers may be operated at maximum water flow and thus no effective control is obtained. If this is the case, any changes in the condenser conditions are likely to affect the total solids control.

8.4.5.10.2 MVR pressure control

The pressure inside an MVR evaporator is closely related to the overall heat balance. Vacuum pumps are used to remove noncondensable gases, but some means is required to ensure that the evaporation temperature is controlled. The temperature and, hence the pressure of the MVR effect, can be controlled by adding small amounts of steam, by cooling with water or by changing the feed temperature.

8.4.5.11 Evaporator Separator Level

The level of liquid in each separator can be controlled through a control loop or by design. Normally the interstage pumps are designed so that the pumping flow rate will change depending on the level of liquid in the pipe down from the separator (see Section 8.4.3.9). Thus no control loop is required. Level control can also be achieved using variable speed pumps and differential pressure level sensors.

8.5 OVERVIEW OF FREEZE CONCENTRATION

Another method of concentration of aqueous foods is based on the separation of pure ice crystals from a freeze concentrated solution. In many cases, crystallization is a separation process, where ice crystals are essentially pure water, and freezing followed by removal of ice can be used to concentrate aqueous solvents.

The basic steps in the freeze concentration system are shown in Figure 8.29. The first step involves formation of ice crystals of the appropriate number, size, and shape from the feed product. These ice crystals are then separated from the freeze concentrated solution, which is the product concentrate. Controlling both ice crystal formation and separation is important to an efficient and economical process technology and has been the subject of considerable study over the years.

Compared to evaporation and membrane separations freeze concentration has some significant potential advantages. Freeze concentration has great potential for producing a product with high quality, because no elevated temperatures are used and no vapor–liquid interface exists so there is no loss of volatiles. Thus, the flavor and quality of freeze concentrated products is exceptionally high, especially relative to evaporation. These benefits make freeze concentration particularly suitable for concentration of such products as fruit juices, coffee and tea extracts, and aroma extracts. More recently, freeze concentration has been tested on heat-sensitive herbal extracts and nutraceuticals. Another benefit of freeze concentration is that it can be used to concentrate alcoholic beverages. Thus, freeze concentration has been used to concentrate beer and wine. A similar benefit makes freeze concentration particularly appropriate for concentration of vinegar.

Despite considerable development of technology and the substantial benefits, freeze concentration has been found to be a viable commercial operation in only limited applications. Probably the main limitation to commercial application of freeze concentration is the upper limit on concentration that can be attained due to the increased viscosity at freezing temperatures. Separation of ice from viscous, concentrated product becomes more and more difficult as concentration increases until, eventually, no further concentration effect can be achieved. According to Figure 8.1, the practical upper limit of freeze concentration is somewhere in the range of 40 to 50% solid content, although this is certainly dependent on the nature of the feed material. Furthermore, freeze concentration has not been a low cost alternative. Commercial systems utilize expensive scraped surface heat exchangers to provide cooling and require costly refrigeration systems to provide the needed cooling effect. New developments (Verschuur et al., 2002) strive to reduce these costs, but their efficiency remains to be seen. In some types of freeze concentration, solids losses in the ice stream may be an economic deterrent to the use of freeze concentration, although advances in separation technology can bring solids losses down to as low as 100 ppm.

8.6 FREEZE CONCENTRATION TECHNOLOGY

8.6.1 Types of Freeze Concentrators

Numerous types of freeze concentration units have been developed and tested over the years. These may be broadly classified according to the methods of freezing and subsequent separations (Heist, 1979; Deshpande et al., 1984; Chowdhury, 1988).

8.6.1.1 Freezer Types

In the crystallization section of freeze concentration, ice must crystallize in a form that is easily separated from the freeze concentrated solution and that is in a very pure state. Typically, freeze concentration involves freezing ice in the form of individual ice crystals of a size that are easily separated from the aqueous phase. Careful control of the crystal size distribution is needed to ensure efficient separation. Various types of freezing systems have been used for freeze concentration technologies.

Direct contact freezers provide for intimate mixing between the refrigerant, usually a boiling liquid like Freon, butane, or some other primary refrigerant, and the product being frozen. Expansion of the refrigerant from the high pressure liquid state to vapor phase provides a refrigeration effect and causes formation of ice crystals within the product. Alternatively, a high vacuum can be used to vaporize a portion of the water, which then provides the refrigeration effect to promote ice crystallization.

Formation of ice in layers on cold surfaces has also been evaluated for freeze concentration. In this case, a cold surface is immersed in the fluid to be concentrated (or alternatively, the fluid is pumped across the cold surface). Ice freezes at the cold temperatures of the cold surface and a layer of ice forms. The ice layer continues to grow as heat is removed through the ice layer and through the cold surface. Ice separation is very simple; the remaining concentrated fluid is allowed to drain off the cold surface. However, the partition coefficient for solutes between the ice and the liquid is fairly high in such a system and considerable solids are lost in the ice. Subsequent purification steps like washing, sweating, or pressing have been tested to facilitate removal of entrapped solutes (and liquid pockets), but the efficiency of solid-layer systems for freeze concentration of fluid foods has never been demonstrated.

Most freeze concentration systems have focused on indirect cooling to promote the formation of a slurry of high-purity ice crystals that can be efficiently separated from the remaining concentrate. Suspension systems can be cooled in various ways, from using a cooling jacket on a stirred vessel to recirculating liquid or suspending through a cooling device. In one system, for example, the slurry is pumped out of a crystallization vessel, through a heat exchanger to cool the slurry and then back into the crystallizer vessel to provide cooling for the remainder of the contents in the vessel. Agitation promotes separation of solutes from the ice so that the ice crystals are essentially pure water with no dissolved or entrapped solids. The efficiency of separation of the slurry from the remaining concentrate then determines the extent of solute lost.

8.6.1.2 Separation Devices

Once the ice has been formed in the crystallization unit, it must be separated from the concentrate as efficiently as possible. Separation devices that have been tested include presses, centrifuges, and wash columns Thijssen, 1974a, 1975). The capacity of separation increases with mean ice crystal size and the pressure differential applied across the crystal slurry, and decreases with increased viscosity of the liquid phase. Thus, for a given type of separator, larger ice crystals and lower concentrate viscosity are desired for most efficient separation.

A mechanical press can be used to expel the liquid from within and between the ice crystals. This method may be more appropriate for solid-layer crystallization than suspension crystallization, but the amount of solids remaining with the ice is generally too high for use in food freeze concentration.

A slurry of ice crystals can be separated from the concentrate in a centrifuge based on the difference in specific gravity between ice and concentrate. The efficiency of separation depends on the rotational speed of the centrifuge, the nature of the ice crystals (size, shape, distribution, and total mass), and the viscosity of the concentrated product. Rinsing with water may be necessary to remove solids, but then the wash stream containing dilute solids content must be processed back into the system. A vapor–liquid interface in centrifuge systems generally leads to loss of volatile flavors and aromas.

The best alternative found for separation of ice in freeze concentration is the wash column, which can yield a pure ice stream with as little as 100 ppm dissolved solids. Wash columns also operate continuously with ice and concentrate flowing countercurrent to enhance separation. The ice crystals are transported to the top of the column, either by pressure or natural buoyancy (or both), while the concentrate flows down the column. Pure ice is removed at the top of the column by a scraping device. A portion of this ice is melted and the water allowed to pass down the column, giving the washing effect. The fluid at the top of the column is thus diluted and has lower viscosity than the primary concentrate, which enhances separation. Concentrate is removed through a filter at the bottom of the column.

Thijssen (1974b) classified wash columns according to the force applied to promote countercurrent flow of ice and concentrate. These included buoyancy columns, screw conveyors (spiral column), and piston bed columns. In buoyancy bed columns, the driving force for separation is the density difference between ice and concentrate. Since ice is less dense than the concentrate, it floats upwards at a speed related to the difference in density and the viscosity of the concentrate. The capacity of buoyancy columns is limited in high-viscosity systems experienced in foods, which makes these columns generally unsuitable for commercial freeze concentration.

A screw conveyor may be used to assist the upward motion of the ice crystals in a column. The action is similar to that of a screw pump, where the helical motion of the screw lifts the ice crystals while allowing the concentrate to flow down through the column.

The most commonly used wash column for food freeze concentration is the piston bed column, which utilizes a reciprocating piston at the concentrate end (bottom) of the column to apply a force on the ice crystals. A packed bed of ice is formed, with ice moving upwards due to the applied force and concentrate draining down through the column. Application of wash water at the top of the column promotes this countercurrent operation. When working correctly, a piston bed wash column has a sharp wash front between the concentrate and purified ice segments of the column. Under correct operating conditions, recrystallization of ice crystals in dendritic form, which reduces bed permeability at this point, helps to stabilize the wash front (van der Malen and van Pelt, 1983).

8.6.2 PHYSICAL PROPERTIES

8.6.2.1 Freezing Point Depression

The dissolved solutes present in foods result in a freezing point (or more technically, melting point) lower than that for pure water. This freezing point depression is due primarily to low molecular weight solutes like salts and sugars. Equation 8.17 can be used to calculate the freezing point depression given the molar concentration of solutes present in the food.

$$\Delta T_f = \frac{RT_{wf}^2 \ln x_w}{\Delta h_f} \tag{8.17}$$

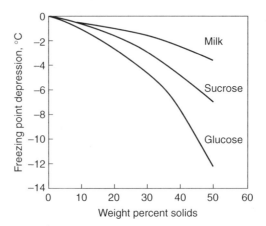

FIGURE 8.30   Freezing point depression curve for sucrose.

Equation 8.17 is often simplified to allow the calculation of freezing point depression based on the molality, m, of the solute, as given in Equation 8.77.

$$\Delta T_{\mathrm{f}} = \frac{R T_{\mathrm{wf}}^2 W_{\mathrm{A}} m}{1000 \Delta h_{\mathrm{f}}} \tag{8.77}$$

Here, W_{A} is the molecular weight of the solute. For pure binary mixtures, Equation 8.77 can be used directly to estimate freezing point depression. However, since foods often comprise several compounds, an average solute concentration and molecular weight must be determined to calculate freezing point depression. For example, both the lactose and salts in milk influence freezing point depression.

Equation 8.17 and Equation 8.77 give values of the initial freezing point when composition data for the initial food to be freeze concentrated are used. During the process of freeze concentration, the freezing point continually decreases as water is removed in the form of ice and the solutes become concentrated in the remaining solution. The freezing point at any increased concentration can also be found from Equation 8.17 with increasing solute concentration. An example of the decrease in freezing point with increased concentration is shown for sucrose in Figure 8.30.

8.6.2.2 Viscosity

Low temperature and high concentration lead to high viscosity in liquid foods, which can limit the maximum concentration attainable in freeze concentration. For example, the viscosity of sucrose at 0°C increases from 3.8 cP at 20% to 93.9 cP at 55% concentration (Pancoast and Junk, 1980). This increase in viscosity limits the maximum concentration attainable in freeze concentration since separation of ice from concentrate is inversely dependent on viscosity. For this reason, concentrations of 40 to 50% are typically the highest attainable in freeze concentration of foods.

To reduce viscosity during freeze concentration, pretreatment of the feed may be necessary. Pulp removal by centrifugation prior to freeze concentration of fruit juices results in greater efficiency of separation and lower solute losses due to decreased viscosity. Ultrafiltration to remove macromolecules (proteins, starches, etc.) prior to freeze concentration has also been suggested as a means of enhancing the efficiency of freeze concentration.

High viscosity also impacts ice crystallization through concentration of components that inhibit crystal formation and growth. The higher solute concentrations found in freeze concentration result in greater inhibition and slower crystallization, which also impacts the maximum concentration attainable.

8.6.3 Engineering Principles

8.6.3.1 Mass Balance

The general relationship between the amount of water removed as ice and the concentration of the solution phase is given by the mass balance equation. If an ice stream containing a low level of solutes is separated from the remaining concentrated phase, the mass balance can be written according to the system shown in Figure 8.29.

$$\frac{I}{F} = \frac{C_P - C_F}{C_P - C_{\text{ice}}} \tag{8.78}$$

Here, I is the amount (kg) of ice stream separated, F is the amount (kg) of feed, and C_P, C_F and C_{ice} represent the concentrations (weight percent) of solutes in the concentrated product, feed and ice streams, respectively. When the amount of solute retained in the ice stream is negligible, Equation 8.78 simplifies to:

$$\frac{I}{F} = 1 - \frac{C_F}{C_P} \tag{8.79}$$

According to Equation 8.79, to increase the concentration of the product by a factor of two over the feed, half of the original feed solution must be removed as ice.

8.6.3.2 Ice Crystallization

Control of ice formation and growth is critical to efficient freeze concentration. By controlling ice formation, the separation of pure ice from concentrated product can be done efficiently and easily. However, controlling ice formation to allow efficient separation has proven to be a relatively difficult challenge. Numerous methods have been tried over the years, but only a handful have found any commercial success.

A review of crystallization processes of importance in freeze concentration is given by Huige (1972) and more information on crystallization in foods can be found in Hartel (2001). The first step in controlling ice crystallization is nucleation or the formation of the crystalline phase from the liquid state. Nucleation involves water molecules coming together to form a crystal lattice when a necessary driving force is attained. In aqueous systems, a driving force for nucleation occurs when the temperature is below the melting point. The term subcooling, the temperature difference below the melting point, is often used to define the temperature driving force in aqueous systems. For pure water, the melting point occurs at 0°C, but in foods, the melting point is lowered by the presence of small molecular weight solutes (freezing point depression). According to the theory of homogeneous nucleation (Mullin, 1993), water molecules aggregate together to form molecular clusters that eventually reach a critical size where the cluster becomes a stable crystal nucleus. The critical size of a stable nucleus is given, according to the classical homogeneous nucleation theory, as the point where the maximum Gibb's free energy change occurs. In general, extremely high subcoolings are needed for homogeneous nucleation. In pure water with no agitation, homogeneous nucleation of ice occurs at about –40°C.

In practice, it is rare that homogeneous nucleation occurs since nucleation sites abound in food systems. Foreign particles like dust and the wall of the vessel often provide sufficient energy to allow nucleation to occur at reduced driving forces (lower subcooling). In heterogeneous nucleation, the foreign surface provides some of the energy necessary to form a stable cluster. In ice, nucleation at vessel walls occurs readily, and this is a problem for some freeze concentration systems where individual ice crystals are needed for efficient separation. In typical foods, heterogeneous nucleation occurs when the temperature is reduced to between −6 and −8°C. Thus, hetereogeneous nucleation occurs at temperatures well above those expected for homogeneous nucleation.

In agitated suspensions of ice crystals, additional nuclei can be generated by secondary nucleation. Secondary nucleation occurs under conditions of subcooling that would not by itself promote heterogeneous nucleation, so typically secondary nucleation is only important in systems that are close to the melting point. Contacts between growing crystals and other elements of the crystallizer (wall, impeller, other crystals, etc.) are thought to be responsible for generation of secondary nuclei. Thus, agitation speed plays an important role in the rate of secondary nucleation.

The rate of nucleation is dependent on both process conditions and ingredients in the food to be freeze concentrated. The most important of the process conditions is subcooling or the temperature at which nucleation occurs. Very high subcooling (10 to 20°C) promotes rapid nucleation whereas primary, heterogeneous nucleation is very slow at temperatures close to the melting point. In agitated suspensions, the rate of stirring and subcooling are both important process variables that affect nucleation rate. In foods, the solute molecules can significantly impact nucleation. In general, the presence of soluble components in foods inhibits ice nucleation, whereas the presence of insoluble components may promote nucleation by providing heterogeneous nucleation sites.

Once nuclei have formed, crystals grow according to the environmental conditions (process and formulation). Growth of ice crystals has been found to be governed by two steps: counterdiffusion of solute molecules and transfer of latent heat of fusion away from the growing ice crystal surface. Ice crystal growth rates increase linearly as subcooling increases, whereas increasing agitation promotes crystal growth only up to the point where heat and mass transfer are no longer limited. The presence of dissolved solutes inhibits ice crystal growth since more molecules must diffuse away from the crystal surface. The increased viscosity as solute concentration increases promotes this growth inhibition.

Under certain conditions, small ice crystals can agglomerate into larger crystals. Shirai et al. (1987) have used this approach to form large ice crystals in freeze concentration systems.

Ice crystals undergo a thermodynamic ripening process, particularly if the initial ice crystals are small and not spherical in shape, which is often the case for nucleated crystals in freeze concentration systems. The regions of high radius (sharp edges) are less thermodynamically stable (have a slightly higher melting point) that flat surfaces, so very small ($<10\ \mu$m) crystals can melt away at the same time larger ice crystals can grow. This phenomenon is often called ripening (or recrystallization) and is governed by the Gibbs–Thomson equation.

$$T_{f\infty} - T_{fd} = \frac{4\sigma T_{f\infty}}{\Delta h_f \rho_i d_p} \tag{8.80}$$

Here, $T_{f\infty}$ is the melting point for an infinitely flat surface, T_{fd} is the melting point for a crystal of size, d_p, Δh_f is the latent heat of fusion, σ is interfacial tension, and ρ_i is the density of ice. The temperature dependence of melting point on crystal size for smooth spherical crystals is shown in Figure 8.31. When particle size is greater than about 10 μm for ice in aqueous solutions, this type of ripening is negligible. However, for crystals than are irregular in shape, with numerous extensions and dendritic aspects, Equation 8.80 applies roughly for all the different aspects of the crystal. This type of ripening leads to the formation of large, spherical crystals from initial crystals that are irregularly shaped. Huige and Thijssen (1972) studied the use of a ripening vessel to generate large crystals for efficient separation in freeze concentration processes.

8.6.3.3 Ice Separation

Once ice crystals of the correct size, shape, and purity have been formed, an efficient separation process is needed to ensure minimal solute carryover and loss. Both presses and centrifuges present problems with carry-over of solutes into the ice stream, whereas wash columns have developed to the point where carryover has been reduced to less than 100 ppm in many cases.

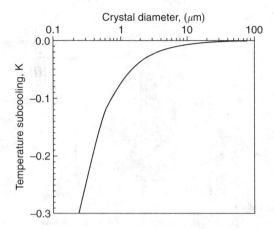

FIGURE 8.31 Effect of crystal size on equilibrium temperature for ice crystals in 30% sucrose solution.

Separation of ice in a wash column is dependent on applied pressure, ice crystal size, and the viscosity of the concentrate. The velocity, V, of fluid through a wash column has been given as (van Pelt and Swinkels, 1986):

$$V = \frac{E^3}{180(1 - E^2)} \frac{(k_s d_p)^2}{\mu} \frac{dP}{dZ} \tag{8.81}$$

where, E is the volume fraction of ice crystals in the column, k_s is the area shape factor of ice crystals, μ is liquid phase viscosity, and dP/dZ is the pressure gradient along the packed bed. Based on Equation 8.81, the separation capacity is governed by mean ice crystal size, ice crystal size distribution, the phase volume of ice in the column, the applied pressure, and the viscosity of the concentrate. For a given column (applied pressure), large ice crystals and low concentrate viscosity promote the best separations.

In a continuous wash column, a wash front is created between the concentrated product and the diluted wash stream applied to the top of the column. The stability of this wash front is enhanced by the recrystallization (ripening) process, due to the density differences at the front. Thijssen (1974a) gives the stability criterion for this wash front as

$$\frac{d_m^2}{\mu} > (10^{-15}) \text{m}^2/\text{Pa·sec} \tag{8.82}$$

where, d_m is mean ice crystal size (in m) and μ is concentrate viscosity (in Pa·sec). As viscosity goes up, particle size must increase accordingly to maintain efficient separation.

8.6.4 COMMERCIAL FREEZE CONCENTRATION

Numerous methods of freeze concentration have been evaluated over the years, but only one has reached the level of commercial importance. A system based on work by Thijssen (1974b) was developed commercially in the 1980s and is still functional today, as shown schematically in Figure 8.32. This system is based on separate nucleation and growth of ice crystals followed by separation of ice from concentrate in a pressurized wash column. Ice is formed in a scraped-surface freezer (SSF) to generate myriads of small crystals. The SSF is used as the cooling mechanism for an adiabatic ripening vessel. Liquid from the ripening vessel (after removal of ice crystals through a filter) is circulated through the SSF to generate ice crystal seeds. The cooled stream containing seed crystals is fed back into the ripening tank where the seeds are allowed to ripen adiabatically. Small, dendritic ice crystals from the SSF melt and ripen, leaving a few crystals to grow to a larger size in

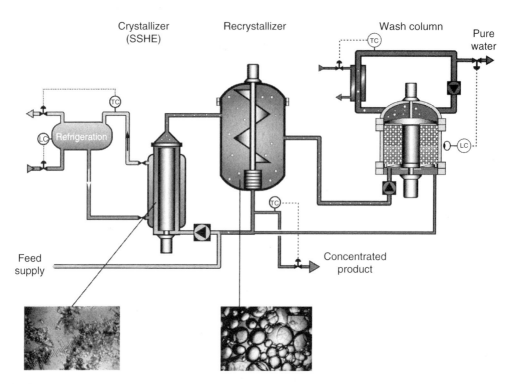

FIGURE 8.32 Commercial freeze concentration process showing the change in ice crystals from the scraped surface freezer to the recrystallization vessel. (Courtesy of Niro Process Technology.)

the recrystallizer. Larger, round crystals, about 200 to 300 μm in size, are formed in this process. The change in ice crystals from the SSF to the recrystallizer is also seen in Figure 8.32. The product from the ripening vessel is fed to the wash column, where purified water is separated at the top and concentrate recirculated to the ripening vessel. Product concentrate is removed from the bottom of the recrystallizer vessel.

Despite the technical success of this system and numerous commercial trials, freeze concentration currently finds limited application in the food industry primarily due to economic constraints. Although product quality is high with freeze concentration, commercial technology results in costs that are significantly higher than the competitive processes of evaporation and reverse osmosis. Economic comparisons in the 1980s suggested that freeze concentration had equivalent or slightly lower energy costs as a triple-effect MVR evaporator but higher energy costs than reverse osmosis. However, due to the high costs of scraped-surface freezer and the recrystallization vessel, capital costs of freeze concentration are considerably higher than for evaporation and reverse osmosis. The high capital costs have limited the application of freeze concentration systems in recent years.

Recent innovations have caused renewed interest in the potential for high quality products made by freeze concentration technology (Verschuur et al., 2002). In one modification, the SSF is replaced by a vacuum crystallizer. Here, small ice crystals are generated by the cooling effect caused by evaporation of water at low pressures, below the triple point of water (<600 Pa abs). The vacuum crystallizer replaces the SSF as the cooling recirculation loop from the recrystallization/ripening vessel. The product slurry from the recrystallizer is separated on a wash column as before. The combination of vacuum crystallizer and recrystallization vessel, however, still does not result in significant savings in capital costs. In a second modification, a slurry crystallizer is used to generate the crystal population for feeding to a wash column. In this case, the entire slurry is recirculated through an external device (SSF, or a shell and tube heat exchanger) to provide cooling. Although this process produces smaller crystals than the commercial freeze concentration system, good separation with the piston-type wash column can still be attained at significantly lower capital costs.

The use of ice-nucleating bacteria to promote ice formation has the potential of enhancing the efficiency of freeze concentration. Watanabe et al. (1996) added *Xanthomonas campestris* cells to soy sauce allowed freeze concentration at $-25°C$. In the absence of the ice-nucleating bacteria, soy sauce which was found to reach a temperature of $-30°C$ before nucleation occurred and freeze concentration was difficult. The addition of the bacteria improved the potential for freeze concentration although no commercial trials with ice-nucleating bacteria in freeze concentration have been reported.

Future developments will determine if these modifications allow freeze concentration to become competitive with evaporation and reverse osmosis as a concentration technique.

NOMENCLATURE

a_w	Water activity	
A	Area	m^2
C_d	Discharge coefficient	
C_p	Specific heat capacity	$J\,kg^{-1}\,K^{-1}$
D	Diameter	m
f	Friction fraction	
F	Volumetric flow rate	$m^3\,sec^{-1}$
g	Gravitational acceleration	$m\,sec^{-2}$
h	Enthalpy	$J\,kg^{-1}$
h	Height	m
h	Film heat transfer coefficient	$W\,m^{-2}\,K^{-1}$
Δh_v	Latent heat of vaporization	$J\,kg^{-1}$
Δh_f	Latent heat of freezing	$J\,kg^{-1}$
ΔH	Pumping head	m
I	Current	A
k	Thermal conductivity	$W\,m^{-1}\,K^{-1}$
k_e	Ratio is tube sheet design	
K	Non-Newtonian flow constant	
K	Resistance coefficient	
L	Length	m
$\dot{m}$	Mass flow rate	$kg\,sec^{-1}$
M	Molecular mass	$kg\,mol^{-1}$
n	Non-Newtonian flow behavior index	
n	Number of effects or tubes	
Nu	Nusselt number	
p	Tube pitch	m
P	Pressure	Pa
ΔP	Pressure difference	Pa
Po	Power	W
Pr	Prandtl number	
q	Energy flux	$W\,m^{-2}$
Q	Energy flow rate	W
R	Gas constant	$8.314\,J\,mol^{-1}\,K^{-1}$
R_s	Steam ratio	
Re	Reynolds number	
t	Time	sec
t	Wall thickness	m

T	Temperature	K or °C
T_{wb}	Boiling temperature of water	K
T_{wf}	Freezing temperature of water	K
ΔT	Temperature difference	K or °C
ΔT_b	Boiling point elevation	K or °C
ΔT_f	Freezing point depression	K or °C
u	Velocity	$m\,sec^{-1}$
U	Overall heat transfer coefficient	$W\,m^{-2}\,K^{-1}$
v_{eff}	Effective volume of a solute	$m^3\,kg^{-1}$
V	Volume	m^3
V	Voltage	V
w	Mass fraction	
x	Mole fraction	
$\dot{\gamma}$	Shear rate	sec^{-1}
δ	Film thickness	m
η	Efficiency	
Γ	Wetting rate or peripheral flow rate	$kg\,m^{-1}\,sec^{-1}$
μ	Entrainment ratio	
μ	Viscosity	Pa·sec
ρ	Density	$kg\,m^{-3}$
σ	Surface tension	$N\,m^{-1}$
ϕ	Effective volume fraction	
ψ	Fan pressure coefficient	
θ	Wetting angle	radians

ABBREVIATIONS

DSE	Direct steam expansion
MVR	Mechanical vapor recompression
TVR	Thermal vapor recompression

REFERENCES

Alhusseini, A.A., Tuzla, K., and Chen, J.C. (1998), Falling film evaporation of single component liquids, *International Journal of Heat and Mass Transfer*, **41**, 1623–1632.

APV (1999), *Evaporation Handbook*, from www.apv.invensys.com (accessed May 2003).

APV (2001), *The APV plate evaporator, product brochure*, APV Anhydro AS, Denmark.

Berry, R.S., Rice S.A., and Ross, J. (1980), *Physical Chemistry*, John Wiley & Sons, NY.

Billet, R. (1989), *Evaporation Technology: Principles, Applications and Economics*, VCH publishers, Weinheim, Germany.

Bird, M.R. and Fryer, P.J. (1991), An experimental study of the cleaning of surfaces fouled by whey proteins, *Trans IChemE Part C Food Bioprod Proc*, **69**, 13–21.

Bouman, S., Brinkman, D.W., de Jong, P., and Waalewijn, R. (1988), Multistage evaporation in the dairy industry: energy saving, product losses and cleaning, In: *Preconcentration and Drying of Food Materials*, S. Bruin, Ed., Elsevier Science, Amsterdam.

Bouman, S., Waalewijn, R., de Jong, P., and van der Linden, H.J.L.J. (1993), Design of falling-film evaporators in the dairy industry, *Journal of the Society of Dairy Technology*, **46**, 100–106.

Brown, M.G. and Lesser, P. (1995), *Florida Citrus Outlook*, 1995–96 Season, University of Florida report.

Chawankul, N., Chuaprasert, S., Douglas, P., and Luewisutthichat W. (2001), Simulation of an agitated thin film evaporator for concentrating orange juice using AspenPlus, *Journal of Food Engineering*, **47**, 247–253.

Chen, C.S. and Hernandez, E. (1997), Design and performance evaluation of evaporation, In: *Handbook of Food Engineering Practice*, Kenneth J. Valentas, Enrique Rotstein, R. Paul Singh, (Eds), CRC Press, Boca Raton, Florida.

Chen, S.L., Gerner, C.L., and Tien C.L., (1987), General film condensation correlations, *Experimental Heat Transfer*, **1**, 93–107.

Choi, Y. and Okos M.R. (1986), Effects of temperature and composition on the thermal properties of foods, In: *Food Engineering and Process Applications Vol 1 Transport Phenomena*, Elsevier Applied Science, London.

Chowdhury, J. (1988), CPI warmup to freeze concentration, *Chemical Engineering*, **95**, 24.

Chun, K.R. and Seban, R.A. (1971), Heat transfer to evaporating liquid films, Journal of Heat Transfer, *Journal of Heat Transfer, ASME Transfer*, **93**, 391–396.

Crandall, P.G., Chen, C.S., and Carter, R.D. (1982), Models for predicting viscosity of orange juice concentrate, *Food Technology*, **36**, 245–252.

Davidson, J. and von Bertele, O. (Eds), (1996), *Process Fan and Compressor Selection*, Mechanical Engineering Publications, London.

Deshpande, S.S., Cheryan, M., Sathe, S.K. and Salunkhe, D.K., (1984), Freeze concentration of fruit juices, *CRC Critical Reviews in Food Science and Nutrition*, **20**, 173.

Flavourtech (2003), Flavourtech website, www.flavourtech.com (accessed May 2003).

GEA (1998), *Jet Pumps and Gas Scrubbers*, GEA Jet Pumps GmbH, Ettlingen, Germany.

Haar, L, Gallagher, J.S., and Kell, G.S. (1984), *NBS/NRC Steam Tables*, Hemisphere Pub Corp, Washington.

Hartel, R.W. (2001), *Crystallization in Foods*, Aspen, Gaithersburg, MD.

Hartley, D.E. and Murgatroyd, W. (1964), Criteria for the break-up of thin liquid layers flowing isothermally over solid surfaces, *International Journal of Heat and Mass Transfer*, **7**, 1003–1015.

Heist, J.A. (1979), Freeze crystallization, *Chemical Engineering*, **86**, 72.

Holdsworth, S.D. (1993), Rheological models used for the prediction of the flow properties of food products: a literature review, *Trans IChemE*, **71C**, 139–179.

Hugot, E. (1986), *Handbook of Cane Sugar Engineering*, 3rd ed., Elsevier, Amsterdam.

Huige, N.J.J. (1972), Nucleation and growth of ice crystals from water and sugar solutions in continuous stirred tank crystallizers, Ph.D. Dissertation, Technical University of Eindhoven, The Netherlands.

Huige, N.J.J. and Thijssen, H.A.C. (1972), Production of large crystals by continuous ripening in a stirred tank, *Journal of Crystal Growth*, **13/14**, 483.

Jebson and Chen (1997), Performances of falling film evaporators on whole milk and a comparison with performance on skim milk, *Journal of Dairy Research*, **64**, 57–67.

Kern, D.Q. (1964), *Process Heat Transfer*, McGraw-Hill Kogakusha Ltd, Tokyo.

Knipschildt, M.E. and Andersen, G.G. (1994), Drying of milk and milk products, In: *Modern Dairy Technology*, Vol 1, 2nd ed., pp. 159–254, Robinson, R.K. (Ed.), Chapman and Hall, London.

Kunitz M. (1926), An empirical formula for the relation between viscosity of solution and volume of solute. *The Journal of General Physiology*, **9**, 715–725.

Minton, P.E., (1986), *Handbook of Evaporation Technology*, Noyes Publications, Park Ridge, New Jersey.

Morison, K.R. and Mackay, F.M. (2001), Viscosity of lactose and whey protein solutions, *International Journal of Food Properties*, **4**, 441–454.

Mullin, J.W. (1993), *Crystallization*, 3rd ed., Butterworth Heinemann, Oxford.

Murphy, P.M., Lynch, D., and Kelly, P.M. (1999), Growth of thermophilic spore forming bacilli in milk during the manufacture of low heat powders, *International Journal of Dairy Technology*, **52**, 45–50.

Nusselt, W. (1916), The surface condensation of water vapour, *Zeitschrift Des Vereines Deutscher Ingenieure*, **60**, 541–546.

Ogunnaike, B.A. and Ray, W.H. (1994), *Process dynamics, modeling, and control*. Oxford University Press, New York.

Pancoast, H.M. and Junk, W.R. (1980), *Handbook of Sugars, 2nd Ed.*, AVI Publishing, Westport, CT.

Paramalingam, S., Winchester, J., and Marsh, C. (2000), On the fouling of falling film evaporators due to film break-up, *Transactions of the Institution of Chemical Engineers*, **78C**, 79–84.

Peacock, S.D. (1999), *A Survey of the Literature Regarding Robert Evaporators*, Communications from the Sugar Milling Research Institute, No. 167, University of Natal, Durban, South Africa.

Perry, R.H., Green, D.W., and Maloney, J.O., (Eds), (1997), *Perry's Chemical Engineers' Handbook*, McGraw-Hill, New York.

Písecký, J. (1997), *Handbook of Milk Powder Manufacture*, Niro A/S, Copenhagen.

Power, R.B. (1994), *Steam Jet Ejectors for the Process Industries*, McGraw-Hill Inc, New York.

Radewonuk, E.R., Strolle E.O., and Craig, J.C. (1983), Freezing points of skim milk concentrates, *Journal of Dairy Science*, **66**, 2061–2069.

Refstrup, E. (1998), Heat treatment of milk prior to evaporation and drying, *Proceedings of the 25th International Dairy Congress*, Aarhus, Denmark, **2**, 228–237.

Sangrame, G., Bhagavathi, D., Thakare, H., Ali, S., and Das. H. (2000), Performance evaluation of a thin film scraped surface evaporator for concentration of tomato pulp, *Journal of Food Engineering*, **43**, 205–211.

Schwartzberg, H.G. (1989), Food property effects in evaporation, In: *Food Properties and Computer-Aided Engineering of Food Processing Systems*, R.P. Singh and A.G. Medina (Eds), pp. 443–470.

Shirai, Y. Sugimoto, T., Hashimoto, M., Nakanishi, K., and Matsuno, R. (1987), Mechanism of ice growth in a batch crystallizer with an external cooler for freeze concentration, *Agriculture and Biological Chemistry*, **51**, 2359.

Singh, R.P. and Heldman, D.R. (1984), *Introduction to Food Engineering*, Academic Press, Orlando.

Stork Friesland (1992), *Evaporation Technology*, Technical brochure, Stork Friesland B.V. (Ed.), Gorredijk, The Netherland.

Thijssen, H.A.C. (1974a), Fundamentals of concentration processes, In: *Advances in Preconcentration and Dehydration of Foods*, A. Spicer (Ed.), Applied Science, London, p. 14.

Thijssen, H.A.C. (1974b), Freeze concentration, In: *Advances in Preconcentration and Dehydration of Foods*, A. Spicer (Ed.), Applied Science, London, p. 115.

Thijssen, H.A.C. (1975), Current developments in the freeze concentration of liquid foods, In: *Freeze Drying and Advanced Food Technology*, S.A. Goldblith, L. Rey, and W.W. Rothmayr (Eds), Academic Press, New York, Chapter 30.

Van Boekel, M.A.J.S. and Walstra, P. (1995), In: *Heat-Induced Changes in Milk*, 2nd ed., Fox P.F. (Ed.), International Dairy Federation, Brussels.

Van der Malen, B.G.H. and Van Pelt, W.H.J.M. (1983), Multistage freeze concentration economics and potential, In: *Progress in Food Engineering*, C. Cantarelli and C. Peri (Eds), Foster-Verlag, Switzerland, p. 413.

Van Pelt, W.H.J.M. and Swinkels, W.J., (1986), Recent developments in freeze concentration, In: *Food Engineering and Process Applications*, Vol. 2, *Unit Operations*, M. Lamaguer and P. Jelen (Eds), p. 275.

Verschuur, R.J., Scholz, R., van Nistelrooij, M., and Schreurs, B., (2002), Innovations in freeze concentration technology, In: *Industrial Crystallization*, proceedings of International Symposium on Industrial Crystallization, Sorrento, Italy (in press).

Walas, S.M. (1990), *Chemical Process Equipment: Selection and Design*, Butterworth-Heinemann, Boston.

Ward, A. (1994), Consider mechanical recompression evaporation, *Chemical Engineering Progress*, **90**, 65–71.

Watanabe, M., Tesaki, S., and Aria, S., (1996), Production of low-salt soy sauce with enriched flavor by freeze concentration using bacterial ice nucleation, *Bioscience Biotechnol, and Biochemistry*, **60**, 1519–1521.

Yang J., McGuire, J., and Kolbe, E., (1991), Use of equilibrium contact angle as an index of surface cleanliness, *Journal of Food Protection*, **54**, 879–884.

Zimmer, A (1980), Developments in energy-efficient evaporation, *Chemical Engineering Progress*, Aug 1980, 50–56.

9 Membrane Concentration of Liquid Foods

Munir Cheryan

CONTENTS

9.1 INTRODUCTION

Membrane separation processes are based on the ability of semipermeable membranes of the appropriate physical and chemical nature to discriminate between molecules primarily on the basis of size, and to a lesser extent, on shape and chemical composition. A membrane's role is to act as a selective barrier, enriching certain components in a feed stream, and depleting it of others. In this regard, the phenomenon is very similar to osmosis, which has been observed and studied for more than 250 years, beginning with the efforts of the French scientist Abbe Nollet in 1748. However, there was little interest in the osmosis process outside the academic, medical, and photographic fields until the early 1950s, mostly because membranes capable of withstanding the high pressures necessary (>20 bar), and to give high flux and high rejections, were not available. Serious study of reverse osmosis (RO) as a practical tool for the production of potable water from brackish or saline water began in 1953 when the U.S. Department of the Interior, Office of Saline Water, began supporting research projects aimed at developing RO technology for desalination. The first breakthrough was made by Reid and Breton (1959) who, while screening membrane materials for desalination, discovered that cellulose acetate membranes gave high rejections and reasonable fluxes (dewatering rates). Shortly thereafter, Loeb and Sourirajan (1960, 1962) developed the casting procedure for asymmetric cellulose acetate membranes and demonstrated that flux could be greatly improved by making asymmetric rather than homogeneous membranes.

This landmark event is generally considered the birth of modern membrane separations technology. Originally termed the "surface skimming" of sea water or brackish water for the production of pure water by a nonthermal energy-efficient method, it led to a vast array of applications unmatched by any other processing technique in its variety and versatility. Desalination and water treatment by reverse osmosis is probably the earliest and best known application. The decade of the 1970s saw its increasing usage in the chemical process industries (paint, textiles, oil recovery, pulp, and paper), and in the 1980s, it was the food and biotechnology industries that benefited most from membrane technology, particularly ultrafiltration (UF). With the development and maturation of sister processes microfiltration (MF) and nanofiltration (NF), the applicability of membrane technology widened considerably, especially in water treatment (for potable, industrial, semiconductor, and biotechnology uses) and in waste treatment. Today, it is difficult to imagine a liquid phase process, especially in biological applications, that would not benefit from one or more of the many membrane technologies.

9.1.1 PRINCIPLES

The chemical nature and physical properties of the membrane control which components are retained and which permeate the membrane. Thus the distinction between the four pressure-driven

Membrane separations

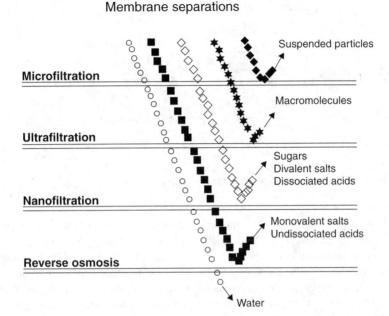

FIGURE 9.1 Characteristics of pressure-driven membrane processes.

processes — RO, NF, UF, and MF — is somewhat arbitrary although the mechanism of transport through each type of membrane may be different and, accordingly, so will the operating strategies for optimum performance. As shown in Figure 9.1, reverse osmosis (RO) can be used to concentrate or dewater components in a solution, while nanofiltration (NF) can be used to separate low molecular weight (<1000 MW) components from each other (e.g., sugars from salts). Large molecules (<1000 MW) can be separated from smaller molecules by ultrafiltration (UF) and microfiltration (MF) is used to clarify slurries or remove suspended matter. All these separation operations occur while simultaneously concentrating the retained components.

In all four cases, hydraulic pressure (through the pump) is used to provide the driving force for permeation. With UF and MF, the driving force is used to overcome the resistance of the membrane and the polarized layer of rejected macromolecules on the surface. In the case of RO and NF, it is primarily to overcome the chemical potential difference between the concentrate and the permeate, expressed in terms of the osmotic pressure (see Section 9.1.2). The pressure applied to the feed side increases the thermodynamic activities of the solutes and solvents by an amount proportional to their partial molar volumes. The difference in the thermodynamic activity on either side of the membrane is the driving force for the permeation of solutes and solvent. Relatively pure solvent is withdrawn as the permeate, leaving a concentrated solution (the retentate) on the high pressure side of the membrane.

Due to the high osmotic pressures of small soluble solutes retained by the RO membrane, pressures in RO are frequently of the order of 20–50 bars (300–750 psi). On the other hand, most NF membranes will allow substantial passage of small molecules less than ~300 MW, and effective osmotic pressures are lower on the retentate side. NF pressures are thus generally lower, for example, 10–30 bars (150–450 psi). Since UF and MF are designed to retain macromolecules and submicron sized particles, respectively, which exert little osmotic pressure, the pressures required are much less (1–7 bars, 15–100 psig), primarily to overcome hydraulic resistance of the polarized macromolecular layer on the membrane surface, a phenomenon known as concentration polarization. If the polarization phenomenon is serious enough, mass transfer could also be limiting, in which case high cross-flow velocities will also be required.

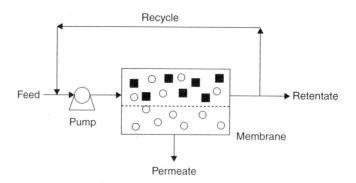

FIGURE 9.2   Flow schematic of a typical membrane process.

The most appealing feature of membrane technology is its simplicity. It involves only the bulk movement (i.e., pumping) of fluids using mechanical energy (Figure 9.2). In addition, they are among the few continuous molecular separation processes that do not involve a phase change or interphase mass transfer. The removal of solvent or water is accomplished without a change in its state from liquid to vapor (as in evaporation) or liquid to solid (as in freeze concentration). Membrane processes can thus also be operated at ambient temperatures if necessary. This avoids product degradation problems associated with thermal processing resulting in products with better functional and nutritional properties. However, membranes may be operated at higher temperatures to lower viscosity and reduce pumping costs, and to minimize microbial growth during processing.

Energy requirements are low compared to other dewatering processes. Typically, while open-pan evaporation may need over 600 kWh/1000 kg water removed and a 5 to 7 effect evaporator with mechanical vapor recompression (MVR) requires 37 to 53 kWh/1000 kg, reverse osmosis for desalination requires 5 to 20 kWh/1000 kg water removed. Energy consumption by RO for concentrating whole milk to 31% solids was only 6 to 7 kcal/kg of milk, compared to 70–90 kcal/kg using MVR evaporators and 330 kcal/kg by double-effect evaporators (Cheryan et al., 1987). In addition to energy savings, no complicated heat transfer or heat generating equipment is needed, and the membrane operation requires only electrical energy to drive the pump motor.

9.1.1.1   Electrodialysis and Pervaporation

Two other membrane processes are shown in Figure 9.3 and Figure 9.4. Although they are not normally considered as concentration processes, there is some enrichment and/or depletion of membrane-permeable components during electrodialysis (ED) and pervaporation (PV). As shown in Figure 9.3, there are two main types of ED. In conventional electrodialysis, flat-sheet membranes are stacked in alternating layers of cation-exchange and anion-exchange membranes. The feed is pumped through the "diluting" stream compartments, while the product is removed from the adjacent "concentrating" stream compartments. With the application of electrical current, cations migrate through the cation-exchange membranes towards the cathode, but the cations are stopped by anion-exchange membranes. Similarly, anions migrate towards the anode through anion-exchange membranes but are stopped by cation-exchange membranes. The result is a separation of ionic species from uncharged species or a concentration of the salts in the concentration compartment.

An extension of conventional ED uses bipolar membranes, which splits water into its component H^+ and OH^- ions. When used along with conventional cation- and anion-exchange membranes, it allows a salt stream to be converted into an acid and a base stream. This is particularly useful in downstream processing of organic acids such as citric, lactic, acetic, and gluconic acids, in that it produces the more desirable acid form of the compound while regenerating the alkali which is used in the fermentation vessel.

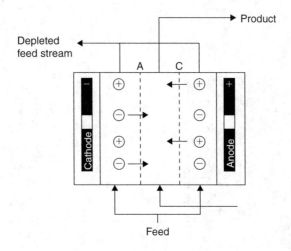

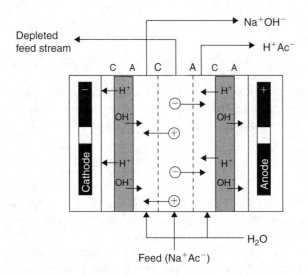

FIGURE 9.3 Electrodialysis. Top: Conventional electrodialysis. Bottom: Bipolar electrodialysis.

Like other membrane processes, ED is affected by concentration polarization and membrane fouling. In addition, ED also exhibits simultaneous water transport, which limits the maximum concentration that can be attained for permeable species. With bipolar membranes, fouling by divalent cations is quite severe, since they can precipitate with the hydroxyls encountered in the cationic membranes. ED is generally more expensive than pressure-driven membrane processes, primarily due to the electrical energy requirements. It is justified in some applications, such as water desalting, removing tartarate from wine, demineralization of protein solutions, and separation and concentration of organic acids (Bailly et al., 2001).

Pervaporation is also a pressure-driven process except that unlike all the others discussed so far, the permeate is a vapor and not a liquid. The solutes permeating the membrane are relatively more volatile than the solvent. As shown in Figure 9.4, a liquid stream containing the volatile component(s) that are to be separated is pumped past the appropriate membrane that selectively allows the desired components to permeate through as vapor. The driving force for transport is a chemical potential gradient that arises due to a decrease in activity of the permeating components. The activity decrease can be accomplished by pressure reduction, for example, a vacuum can be applied on the permeate

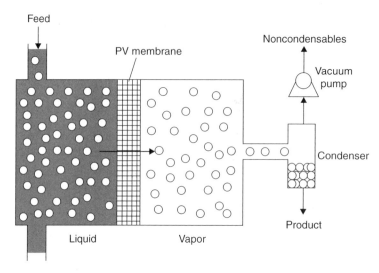

FIGURE 9.4 Pervaporation process.

side that causes the permeating solutes to pass through the membrane and emerge as vapor. The vapor is then condensed and the noncondensables are removed by the vacuum pump. Compared to other membrane technologies, PV has yet to emerge as a widely-used unit operation. The most common application of PV is purification of ethanol using ethanol-permeable membranes. In food processing, it could find a niche in recovery of flavor compounds (Rajagopalan et al., 1994; Rajagopalan and Cheryan, 1995).

9.1.1.2 Limitations

There are some limitations of using membranes as a liquid concentration process. It cannot take a liquid stream to very high solids concentrations, especially compared to an evaporator. In the case of RO and NF, it is the osmotic pressure of the concentrated solutes that limits the upper concentration that can be comfortably handled. For example, milk and whey exert an osmotic pressure of about 7 bar (100 psi) at room temperature. Since RO is commonly conducted at pressures of 40 bar (600 psi), it implies that a 4X concentration of milk and whey can be achieved. In ultrafiltration, it is rarely the osmotic pressure, but rather the low mass transfer rates and high viscosity of the concentrate that limits the process. Skim milk can be economically concentrated to 36–45% total solids by UF for cheese making applications.

Due to substantial improvements in membrane chemistry, systems engineering, fouling and cleaning protocols that occurred in the 1990s, good membranes in well-engineered systems that can meet the rigorous sanitation and safety demands of the food industry are available at reasonable cost. The primary focus in this chapter is reverse osmosis and nanofiltration, since these are the membrane technologies that could be used for concentrating or dewatering liquid foods. Microfiltration and ultrafiltration, on the other hand, are primarily clarification and fractionation techniques which have been covered elsewhere (Cheryan, 1998).

9.1.2 THERMODYNAMICS AND OSMOTIC PRESSURE

Osmotic pressure is a critically important property in reverse osmosis and thus warrants some detailed discussion. It is based upon Gibbs free energy which, on a molar basis, is called the thermodynamic or chemical potential (μ). It is an intensive quantity, that is, dependent on its nature and concentration, but independent of size of the system, and is best described as a driving force that describes

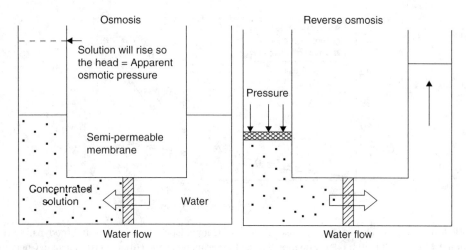

FIGURE 9.5 Osmosis and reverse osmosis. The two compartments are separated by a semipermeable membrane.

changes in free energy (G) when one mole of a component is added to or removed from the system. Therefore, an expression for chemical potential for a single component i at constant pressure (P) and temperature (T) is:

$$\mu_i = \left(\frac{\partial G}{\partial n_i}\right)_{P,T} \tag{9.1}$$

where n_i is the number of moles of component i. For multicomponent systems, each constituent has its own partial contribution to the overall chemical potential.

If a solvent (e.g., water, denoted as component 1) and a solution (e.g., of NaCl) are separated by a semipermeable membrane (defined as a barrier that allows passage of water and not of solutes), the mole fraction of water (X_1) is 1.0 and the chemical potential of the solvent in the pure solvent compartment is denoted as μ_1^o. On the solution side, however, $X_1 < 1$, and thus the chemical potential (μ_1) is less than 1. This difference in chemical potential of the water is the driving force for permeation of water from the high-potential side to the low-potential side. This phenomenon is known as *osmosis* and is shown in Figure 9.5. Thus the water would be transported from the water compartment to the solution compartment. With a U-shaped assembly shown in Figure 9.5, the increase in height of the liquid column in the left (solution) compartment would create a hydraulic pressure against the membrane, and the water would stop diffusing through the membrane when the pressure developed would just balance the chemical potential difference.

Assuming that (a) the solvent vapor behaves ideally and Raoult's law applies, and (b) the liquid is incompressible, the free energy equation can be written in terms of volume changes due to the presence of a solute at constant temperature and composition. A relationship between mole fraction of water and the chemical potential difference on either side of the membrane can be derived (Cheryan, 1998):

$$\mu_1^o - \mu_1 = -RT \ln X_1 \tag{9.2}$$

where μ_1^o is the chemical potential of the pure solvent and μ_1 is the chemical potential of the solution. To counter the diffusion of pure water down the chemical potential gradient, energy in the form of pressure can be applied to the solution side in order to raise its chemical potential. By definition, the pressure applied to balance the chemical potentials is the osmotic pressure (π). Thus, at equilibrium,

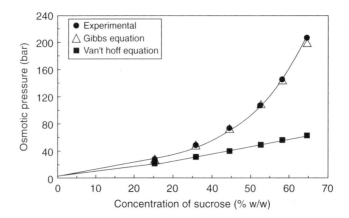

FIGURE 9.6 Osmotic pressure of sucrose solutions. Comparison of the Gibbs thermodynamic model (Equation 9.4), the van't Hoff model (Equation 9.5) and experimental data. (Adapted from data in Cheryan, M. 1998. *Ultrafiltration and Microfiltration Hndbook.* CRC Press, Boca Raton, FL.)

the governing equation will be:

$$\mu_1^o - \mu_1 = \pi \bar{V}_i = -RT \ln X_1 \tag{9.3}$$

$$\pi = -\frac{RT}{\bar{V}_1} \ln X_1 \tag{9.4}$$

where $\bar{V}_1$ is the partial molar volume of water which is the increase in volume per mole of water when an infinitesimal amount of water is added. van't Hoff also developed an osmotic pressure relationship by making several assumptions (Cheryan, 1998):

$$\pi = n_2 RT = ic_2 RT/M \tag{9.5}$$

where n_2 is the moles of solute per liter of solution, c_2 is concentration of solute in grams per liter of solution, M is molecular weight of solute, and i is the number of ions for ionized solutes.

The van't Hoff equation is a gross approximation and is valid only for dilute solutions under ideal conditions. It assumes that osmotic pressure increases linearly with molar concentration of the solute, while it actually increases exponentially in most food and biological systems, as shown in Figure 9.6 for sucrose solutions. To account for this relationship, the equation for osmotic pressure (π) as a function of solute concentration (C) is frequently written in terms of a power series:

$$\pi = A_1 C + A_2 C^2 + A_3 C^3 + \cdots \tag{9.6}$$

The virial coefficients for Equation 9.10 for a variety of solutes have been tabulated by Cheryan (1998). Some osmotic pressures of selected food products are shown in Table 9.1.

The physical significance of osmotic pressure in reverse osmosis is that it represents the minimum pressure that must be applied to a feed solution in order to obtain any permeation or flux. The basic model that relates applied and osmotic pressure to flow of solvent through a membrane is, like many transport processes, expressed as the flux (the rate of solvent transport per unit area per unit time) and a transport coefficient:

$$J = A(\Delta P_T - \Delta \pi) \tag{9.7}$$

TABLE 9.1

Osmotic Pressure of Foods at Room Temperature

Food	Concentration	Osmotic pressure (psi)
Apple juice	15% total solids	300
Coffee extract	28% total solids	500
Grape juice	16% total solids	300
Lactic acid	1% w/v	80
Lactose	5% w/v	55
Orange juice	11% total solids	230
Perilla anthocyanins	10.6% total solids	330
Skimmilk	9% total solids	100
Sodium chloride	1% w/v	125
Sweet potato waste	22% total solids	870
Whey	6% total solids	100

Note: 1 psi = 6.9 kPa.

Source: From Cheryan, M. 1998. *Ultrafiltration and Microfiltration Handbook.* CRC Press, Boca Raton, FL.

where A is the membrane permeability coefficient (the reciprocal of resistance to flow), ΔP_T is the transmembrane pressure, and $\Delta \pi$ is the difference in osmotic pressure between the feed solution and the permeate. For example, using the van't Hoff equation, the osmotic pressure of a 1% solution of NaCl with a molecular weight (MW) of 58.5 is about 125 psi. This means no flux will be obtained unless the pressure is above the theoretical 125 psi. This is shown in Figure 9.7, which shows typical reverse osmosis performance data with NaCl solutions. The higher the solute concentration, the higher the osmotic pressure and the higher the applied pressures needed to obtain the required flux (Figure 9.7a). Due to differences in their manner of transport, the solvent and solute move through the membrane at different rates (Figure 9.7a and Figure 9.7b) giving rise to the pattern of rejection observed in Figure 9.7d. The permeate quality will be better at higher pressures and lower concentrations of the solute (Figure 9.7c).

On the other hand, a 1% solution of lactose (MW = 342) will have an osmotic pressure of 10 psi and a 1% solution of bovine serum albumin (MW = 60,000) only 0.006 psi. Thus much higher pressures have to be applied with salt solutions than with protein solution. This is why osmotic pressures are of little or no consequence in UF or MF but important in RO and NF. This is shown in Figure 9.8, which shows typical flux behavior with model solutions of 1% NaCl, 1% lactose and a real liquid food (skim milk of 9.1% total solids with an osmotic pressure of 100 psi). As expected, no permeation was observed until the applied pressure was higher than the osmotic pressure. The slopes of the salt and sugar lines are almost the same as the water line. With skimmilk, however, there is a marked deviation from linearity at higher pressures. This is due to concentration polarization of rejected particles, and flux becomes controlled by the mass-transfer characteristics of the system. This explains why turbulence (in the form of higher velocities) has a beneficial effect with skimmilk, but not so much with salt or lactose where polarization is less significant under these conditions.

The osmotic pressure also tells us the approximate limits to a reverse osmosis process. For example, many RO membranes can tolerate only about 600 psi. This means that in theory, the maximum concentration that can be obtained by RO for the 1% salt solution is 600/125 = 4.8×, or 4.8%. Higher concentrations should be obtained with sugar protein solutions, as long as the feed solutions can be pumped and the solutes remain in solution. In practice, the maximum concentration

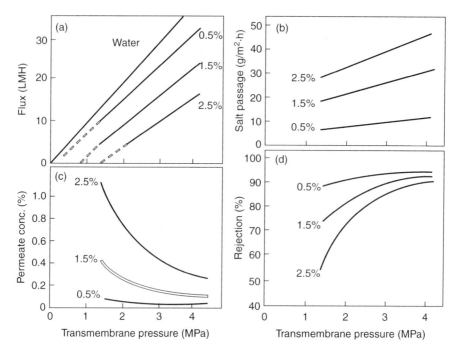

FIGURE 9.7 Reverse osmosis of 0 to 2.5% NaCl solutions with a cellulose acetate membrane (SEPA-CA97, Osmonics spiral). Data show effect of transmembrane pressure on (a) flux, (b) solute transport, (c) NaCl concentration in the permeate, and (d) rejection of NaCl. (Adapted from Cheryan (1989).)

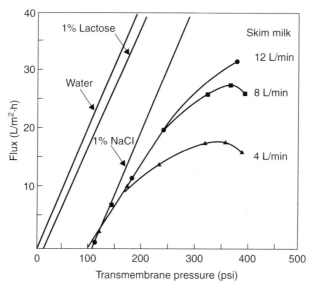

FIGURE 9.8 Reverse osmosis of model NaCl (1% w/v) and lactose (1% w/v) solutions, and of skim milk at 50°C with a thin-film composite RO membrane (PA-99, Osmonics spiral). (Adapted from Cheryan (1989).)

is limited by solubility of the solute, concentration polarization, fouling, and economics, that is, the flux at high concentrations may become too low to be practical.

9.2 THE MEMBRANES

The heart of any membrane process is obviously the membrane itself. A wide variety of membranes are commercially available (Table 9.2), although not all are suitable for food applications. The most important structural properties of a RO membrane are its chemical nature, its pore statistics (pore size, pore-size distribution and density, and void volume) and its degree of asymmetry. From a functional point of view, the most important are its permeability (measure of the rate at which a given molecule permeates) and its permselectivity (measure of the rate of permeation of one molecule relative to another). These characteristics are more commonly termed as "flux" and "rejection." Since these properties can change with time and environmental conditions, secondary properties such as resistance to compaction, temperature and chemical stability, resistance to microbial attack, tolerance to cleaning and disinfecting solutions and lack of toxicity of the contact materials, are also important. Additional requirements for food processing are good tolerance of cleaning and

---

TABLE 9.2
Membrane Materials

Material	MF	UF	NF	RO
Alumina	X			
Carbon-carbon composites	X			
Cellulose (regenerated)	X	X		
Cellulose acetate (CA)	X	X	X	X
Cellulose esters (mixed)	X			
Cellulose nitrate	X			
Cellulose triacetate (CTA)	X	X		X
CA/CTA blends				X
Ceramics and ceramic composites	X	X		
Composites, polymeric thin film			X	X
Polyacrylonitrile (PAN)	X	X		
Polyamide, aliphatic (e.g., nylon)	X			
Polyamide, aromatic (PA)			X	X
Polybenzimidazole (PBI)				X
Polycarbonate (track-etch)	X			
Polyetherimide (PEI)				X
Polyethersulfone (PES)	X	X		
Polyimide (PI)		X		X
Polyester (track-etch)	X			
Polypropylene (PP)	X			
Polysulfone (PS)	X	X		
Polytetrafluoroethylene (PTFE)	X			
Polyvinyl alcohol (PVA)	X	X		
Polyvinyl chloride (PVC)	X			
Polyvinylidene fluoride (PVDF)	X			
Stainless steel	X			

Source: From Cheryan, M. 1998. *Ultrafiltration and Microfiltration Handbook.* CRC Press, Boca Raton, FL.

TABLE 9.3
Methods of Manufacture of Synthetic Membranes

Process	Materials
Phase inversion by:	
solvent evaporation	Cellulose acetate, polyamide
temperature change	Polypropylene, polyamide
precipitant addition	Polysulfone, nitrocellulose
Stretching sheets of partially crystalline polymers	PTFE
Irradiation and etching	Polycarbonate, polyester
Molding and sintering of fine-grain powders	Ceramic (alumina, zirconia, titania)

Source: From Ripperger, S. and Schulz, G. 1986. *Bioprocess. Eng.* 1: 43–49.

disinfecting solutions, and lack of toxicity of the contact materials. Membrane preparation methods, transport properties, and performance characteristics have been discussed in the literature (Kesting, 1985; Lloyd, 1985; Cheryan, 1998).

9.2.1 First Generation Membranes

9.2.1.1 Cellulose Acetate

Cellulose acetate (CA) and its derivatives have been widely used since the development of the Loeb–Sourirajan asymmetric membrane. CA polymers are produced by acetylation of cellulose with acetic anhydride, acetic acid, and a catalyst such as sulfuric acid. Acetylation is carried out until modification of the three hydroxyl groups of each unit occurs (esterified by acetyl groups). This is called cellulose triacetate (CTA) with a degree of substitution (DS) of 3.0. DS values of 3.0 are rare and DS > 2.75 is considered to be CTA. CA polymers result from the deacetylation of CTA until a practical polymer of DS between 2.3 and 2.8 is obtained (37.5 to 43.4% acetyl). A DS of 2.4 is common among RO membranes (Kesting, 1985).

The chemical structure of the CA polymer has a marked influence on the behavior of the membrane. As discussed in Section 9.3, it is believed that the H-bonding between water and the hydroxyl sites results in high water permeability. The hydrophilic properties of the membrane are a function of the number of hydroxyl groups remaining after acetylation. Thus the DS must be carefully controlled. The hydrophobic acetyl groups act as crosslinks due to dipole–dipole interaction and restrict the swelling resulting in the increased permselectivity of the membrane, that is, water and not hydrated ions are allowed to pass the polymer micellar matrix. A balance between the hydroxyl and acetyl groups will determine the permeability and permselectivity of the membrane at this level (Kesting, 1985).

There are several ways of manufacturing membranes (Table 9.3). CA membranes are produced using the phase inversion process (Kesting, 1985; Cheryan, 1998). CA powder is first dissolved in a solvent such as acetone. A swelling agent (i.e., formamide or magnesium chloride) is added to form the casting dope or solution. The viscous, two phase system in which the polymer-solvent forms the continuous phase and the swelling agent the dispersed phase is formed into a thin film by some type of casting procedure and exposed to air (1–3 min). As the solvent evaporates, the surface concentration of polymer increases and the membrane skin layer forms. After the skin layer forms, solvent evaporation is much slower. Immersion of the film in cold water ($0°C$) results in the formation of a gel. Water replaces the remaining solvent and swelling agent and the gel solidifies to form the membrane. After gelation, the film is heat-treated or annealed at 60 to 90°C. During heating, the entire membrane shrinks and the skin layer becomes more dense. Thus, heating decreases the

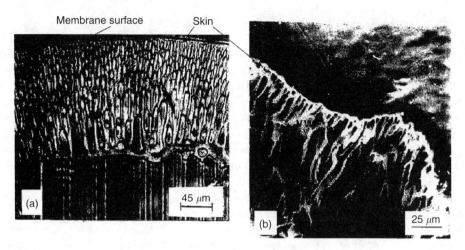

FIGURE 9.9 Views of an asymmetric membrane (Millipore PTGC ultrafiltration membrane). Left: (a) Light microscope view of a cross-section. Right: (b) Scanning electron micrograph of the surface and cross-section [v = voids; vw = void wall; mb = membrane bottom; ms = membrane surface]. (From Cheryan, M. 1998. *Ultrafiltration and Microfiltration Handbook.* CRC Press, Boca Raton, FL. With permission.)

membrane's permeability but increases its selectivity. This results in an asymmetric structure in which a very thin, dense layer (commonly termed the "skin") forms over a porous support layer (Figure 9.9). The total membrane thickness is 50–200 microns, while the skin — which is the critical part of the membrane — is only about 0.1–0.5 microns. Since the rate of permeation is inversely proportional to the thickness of the separating barrier — in this case, the skin — the importance of the asymmetry in obtaining good performance is obvious.

For RO membranes, the question of whether the skin layer is porous or nonporous brings up the question of the flow mechanism — is it flow through pores of the critical diameter or is it diffusive solubility flow so that pore flow (if any) is due to membrane defects? This is still a matter of controversy, even after almost five decades. In any case, the solubility and diffusivity of solute and solvent in the membrane determines its selectivity and permeability. For good separation, the solubility and diffusivity of the solute in the membrane must be less than those of the solvent.

9.2.1.2 Factors Affecting the Applicability of Cellulose Acetate

The asymmetric cellulose acetate membrane was quite a breakthrough for dewatering applications in that it had a good combination of high flux and high rejections of most solutes of interest in a wide variety of potential applications. Furthermore, CA membranes are fairly easy to manufacture, and a wide variety of pore sizes are available, from RO to MF. In addition, cellulose (the raw material) is an abundant and renewable resource. There are, however, several limitations to CA which restricts its use, especially in food and biotechnology applications where standards and requirements are quite rigorous. Factors to account for include temperature and pH of operation, use of acidic or basic cleaners, chlorine and other oxidizing agents used for sanitation, microbiological activity, and other mechanical influences such as pressure and high shear in the system.

CA is very temperature sensitive, which limits the maximum operating temperature to 30°C, with some blends of CA and CTA tolerating 35°C. This low temperature poses several problems: low flux, since viscosity, diffusivity, and solubility are aided by high temperatures; microbial growth may be a problem at these temperatures, which RO and NF will only make worse since all solutes are concentrated during the process, unless the process is operated at refrigeration temperatures) and cleaning is also more difficult since cleaning agents work best at higher temperatures.

CA is also pH sensitive; recommended pH limits are pH 2–8, preferably pH 3–6. CA will hydrolyze in water at either high or low pH. The rate of hydrolysis, which is lowest at pH 4.5–5.0, is temperature dependent with the recommended pH range decreasing with increasing temperature. This pH constraint limits the choices of cleaning solutions for CA membranes. Oxidation, especially by sanitizers such as chlorine, a sanitizer extensively used in the food industry, weakens the membrane and increases flux to a degree greater than that by hydrolysis. Most membrane manufacturers suggest that continuous exposure to 1 ppm chlorine is possible and a maximum single dose of 50 ppm. Mechanical defects such as scratches and cracks on the surface lead to overall poor performance.

Compaction of the membrane due to high pressures in RO tends to decrease the flux and increase rejection which becomes noticeable with time. This compaction phenomenon affects CA membranes to a larger degree than other membrane types and is accelerated at higher temperatures and pressures. This may be the reason for recommending pressures less than 30 bar (450 psi) for CA membranes to prevent loss of permeability due to compaction.

There is a strong interaction among the factors causing deterioration of CA membranes, that is, the effect of one factor (e.g., temperature) depends on the level of other factors (e.g., pH, pressure). Higher temperature accelerates hydrolysis. Higher pressure increases the rate of compaction. pH extremes are better tolerated at lower temperatures and lower pressures. The operating strategy with CA membranes must be optimized with respect to the specific application to use them most effectively.

The life of CA membranes used in the food industry is expected to be about 12 months. However, this can be shortened by adverse environmental conditions that lead to hydrolysis, solvation, and deterioration due to microorganisms. some microorganisms have been reported to enzymatically digest CA membranes due to their cellulosic nature. Colonies form on the membrane surface and pinhole defects are found underneath them, thus increasing flux and decreasing membrane life. Polar organic compounds with low molecular weights such as alcohols, ketones, and amides plasticize or swell CA membranes.

9.2.2 SECOND-GENERATION MEMBRANES: THIN-FILM COMPOSITES

Owing to several limitations of cellulose acetate as discussed above, there is a continuing search for better polymers. The general aim of such research is the development of hydrophilic polymers superior to CA. Membranes made of polyamides, polyamidehydrazides, polybenz-imidazole, and others looked very promising (Cheryan, 1998). The mechanical, chemical, and biological properties of polyamide (PA) membranes are generally superior to CA. These membranes are not as susceptible to hydrolysis or to microbial attack and they can tolerate alkaline conditions and temperatures up to 50°C, but are extremely sensitive to chlorine. In the latter respect they are much worse than CA, tolerating a maximum exposure of 0.1 ppm active chlorine. PA membranes exhibit higher permselectivities but lower water permeabilities as compared to CA membranes. In general, PA membranes are better than CA membranes for organic separation and have been used to separate aqueous solutions of alcohols, phenols, ethers, ketones, and aldehydes (Matsuura et al., 1974). Perhaps the best-known polyamide noncomposite membrane is the Dupont B-9 and B-10 Aramid membrane which were used extensively in the form of hollow fine fibers for desalination of brackish water and sea water in the 1970s and 1980s.

A major milestone was marked in the early 1970s with the development of thin-film composite membranes. Table 9.4 lists several methods of manufacturing thin-film composite membranes. These membranes consist of a submicron barrier layer of a noncellulosic polymer formed or coated on a microporous layer such as a polysulfone ultrafiltration membrane. This composite structure is then backed by a reinforced fabric made of various materials — nonwoven polyester, polyolefine, sailcloth. PA membranes, used for separations in their own right, are a good skin layer for thin-film composite membranes. Figure 9.10 shows some typical chemical structures of separating layers.

TABLE 9.4
Methods of Manufacture of Thin-Film Composite Membranes

Cast the ultrathin barrier membrane separately, then laminate to a porous support
Dip-coat a polymer solution onto a support film, followed by drying
Dip-coat a reactive monomer or prepolymer solution on to a support followed by heat or radiation curing
Deposit a barrier film from a gaseous phase monomer plasma onto the support
Interfacially polymerize a reactive set of monomers at the surface of the support

Source: From Cheryan, M. 1998. *Ultrafiltration and Microfiltration Handbook.* CRC Press, Boca Raton, FL.

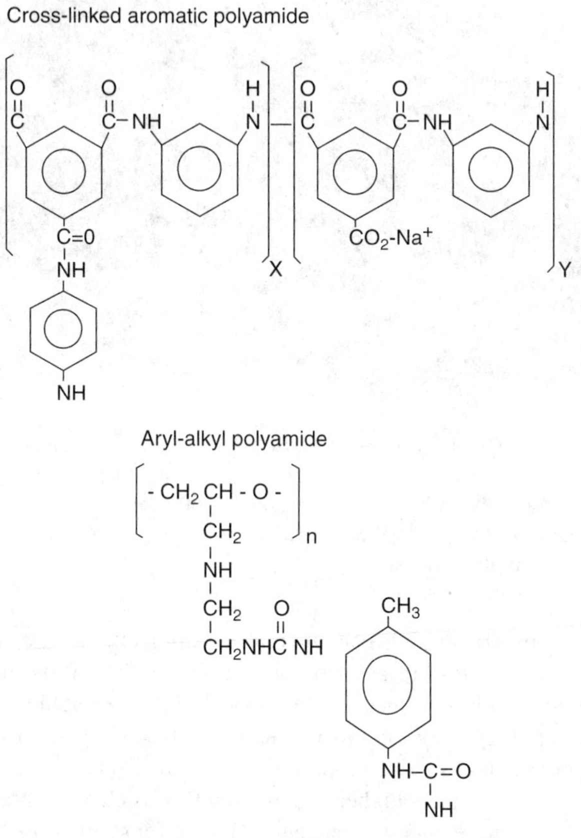

Cross-linked aromatic polyamide

Aryl-alkyl polyamide

FIGURE 9.10 Structure of the separating layers of commercial thin-film composite membranes. Top: Cross-linked aromatic polyamide. Bottom: Aryl-alkyl polyamide/polyurea. (Adapted from Allegrezza (1988). In *Reverse Osmosis Technology*, B.S. Parekh, Ed., Dekker, NY. pp. 53–120.)

The first commercially successful composite membrane is the Dow-FilmTec FT-30 membrane. It is composed of a polysulfone support cross-linked by interfacial polycondensation with the PA polymer. The skin layer may itself be composed of several layers which enhances its strength, flexibility, and abrasion resistance. The hydrophilic carboxyl groups are responsible for its relatively high water permeability, similar to the hydroxyl groups in CA membranes.

Thin-film composite membranes are available from almost every company that manufactures reverse osmosis membranes. Figure 9.11 shows the ultrastructure of a commercial thin-layer

Fluid system TFC membrane
TOP: PVA layer
Middle: Polytheramineurea TF layer
Bottom: Polysulfone support
Magnification:126,000

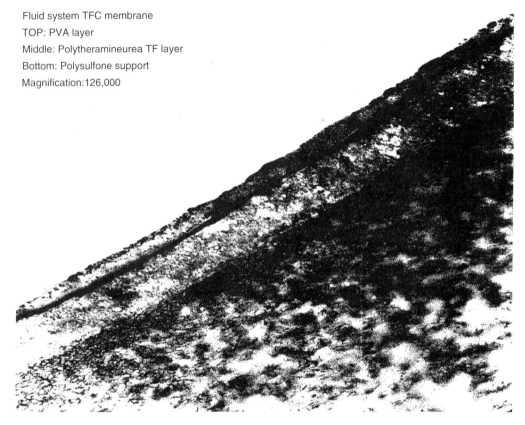

FIGURE 9.11 Electron micrograph of an asymmetric thin-film composite membrane. Three regions are visible: a top protective coating of polyvinyl alcohol, the polyether-amine-urea rejecting surface, and the porous polysulfone support. This transmission electron micrograph is of a fresh membrane, stained with RuO$_4$ and embedded in Medcast epoxy. Magnification 128,000. (Micrograph prepared by A.X. Swamikanni and provided courtesy of W.G. Light, UOP Fluid Systems.)

composite membrane from Koch/Fluid Systems that uses polyetherurea as the separating barrier. For the food industry, composites are available as spirals, as flat sheets in plate systems and in tubular form. It has been used in seawater desalination and has been field tested in food processing facilities for several years. Figure 9.12 compares the properties of a composite membrane with the cellulose acetate membrane. Its superior properties are obvious and are a real benefit in food processing. Salt rejection is better, flux is higher, and it outperforms CA membranes in COD and BOD reduction. Many composites have good tolerance to pH 3–11 for short periods. They can be cleaned with phosphoric or nitric acid (at low concentrations) and alkalis such as dilute solutions of sodium hydroxide or trisodium phosphate. These membranes resist compaction better than CA. Composites can retain many small organic molecules better than CA membranes, an important consideration in biotechnology and flavor applications.

The one drawback with composite membranes is poor chlorine tolerance. Most composite membranes cannot tolerate even 1 ppm chlorine for any length of time.

9.3 MECHANISMS OF MEMBRANE TRANSPORT

The exact mechanism of transport through reverse osmosis membranes is still a matter of controversy. Transport could be due to diffusion and/or convective flow through very fine pores in the skin of the

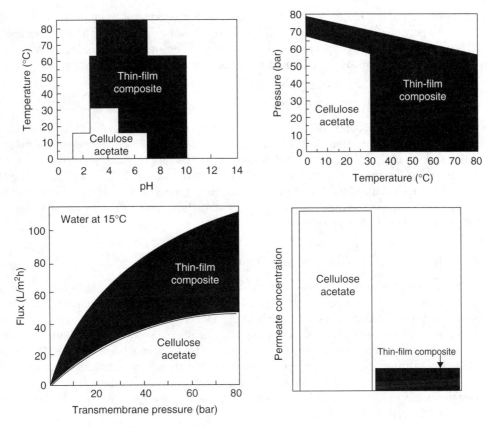

FIGURE 9.12 Properties of thin-film composite membrane. The data shown is for the ZF-99 membrane, manufactured by PCI, England. It is similar to the FT-30 membrane developed by Film-Tec. (Adapted from product literature, courtesy of PCI.)

asymmetric membrane. There is still some confusion among workers in this area about a true RO membrane being porous or nonporous. Several of the more significant mechanisms are discussed below.

9.3.1 SIEVE MECHANISM

One of the earliest and simplest mechanisms proposed was the sieve mechanism. According to this concept, a semipermeable membrane has pores intermediate in size between the solvent and solute molecules. Separation occurs because solute molecules are blocked out of the pores while smaller solvent molecules are able to enter the pores. The sieve mechanism probably accounts in part for the removal of large molecules in UF. However, for RO, in which molecules of approximately the same size are separated, a purely steric explanation for membrane rejection is less plausible.

9.3.2 HYDROGEN BONDING MECHANISM

The hydrogen bonding mechanism was developed specifically for cellulose acetate membranes (Reid and Breton, 1959; Reid, 1972). According to this mechanism, permeation occurs in the noncrystalline portions of the membrane and is much faster for molecules which can form hydrogen bonds with the membrane material. Flow is pictured as the migration of water molecules from one hydrogen bond site to the next. Orofino et al. (1969) proposed a similar mechanism in which water tends to cling to the membrane surface, by means of hydrogen bonding, as an absorbed film. This film blocks the membrane pores, preventing solute ions from entering. Solute molecules can move through the membrane only by displacing solvent molecules from absorption sites. Since displacement requires large amounts of energy, relatively few solute molecules are absorbed by the membrane.

9.3.3 SOLUTION–DIFFUSION MECHANISM

The solution–diffusion mechanism attributes solute rejection to large differences in the diffusivity of solvent and solute in the membrane and/or to differences in their solubility in the membrane material. According to this well-known mechanism (Lonsdale et al., 1965, 1971; Banks and Sharples, 1966), both solvent and solute dissolve in the homogeneous nonporous surface layer of the membrane and then diffuse through the membrane in an uncoupled manner. Thus, it is desirable to have membranes with completely nonporous surface layers in which the solubility and diffusivity of the solvent are much higher than those of the solute.

9.3.4 PREFERENTIAL SORPTION–CAPILLARY FLOW MECHANISM

According to the preferential sorption–capillary flow mechanism (Kimura and Sourirajan, 1967; Sourirajan, 1970, 1977), RO is governed by two distinct factors: (1) an equilibrium effect involving preferential sorption at the membrane surface and (2) a kinetic effect which is concerned with the movement of solute and solvent molecules through the membrane pores. The equilibrium effect is governed by repulsive or attractive potential force gradients at the membrane surface. The kinetic effect is governed by both potential force gradients and steric effects associated with the structure and size of the solute and solvent molecules relative to the membrane pores. Consequently, the chemical nature of the membrane surface, and the size, number and distribution of its pores will determine the success of an RO separation process.

For separation, at least one of the components in the feed solution must be preferentially sorbed at the membrane surface. This means that a concentration gradient, arising from the influence of surface forces, must exist at the membrane–solution interface. Preferential sorption at a membrane–solution interface is a function of solute–solvent–membrane material interactions, similar to those governing the effect of structure on the reactivity of molecules. Such interactions arise in general from the ionic, polar (hydrogen bonding), steric, and nonpolar character of each one of the components involved in the RO system. The overall result of these interactions determines the equilibrium condition under which solvent or solute, or neither, is preferentially sorbed at the membrane–solution interface. This equilibrium condition, together with the friction and shear forces affecting the relative mobility of the solute molecule in the membrane pores under the applied pressure gradient, determines the extent of solute separation and solvent flux obtainable in RO with a given membrane under specified operating conditions (Sourirajan and Matsuura, 1984).

For RO systems involving aqueous solutions and polymeric membranes, solutes can be characterized by their ionic, polar, steric, and non-polar parameters. Values of these parameters for many solutes and CA membranes are given by Sourirajan and Matsuura (1977, 1984). Unique correlations exist between the above parameters characterizing solutes and their rejection by RO membranes (Matsuura and Sourirajan, 1973).

Figure 9.13 is a schematic representation of the preferential sorption–capillary flow mechanism for RO separation of sodium chloride from an aqueous solution. The term "pore" or "capillary" refers to any void space (whatever its size or origin) connecting the high and low pressure sides of the membrane. The critical pore size for a given membrane-feed solution is not necessarily correlated with the size of the solute and solvent molecules: it is twice the thickness of the preferentially sorbed layer. Thus, this mechanism is not sieve filtration.

Polymeric membranes with low dielectric constants, such as CA, repel ions in the close vicinity of the surface which results in the preferential sorption of water. Theories of salt rejection based on the

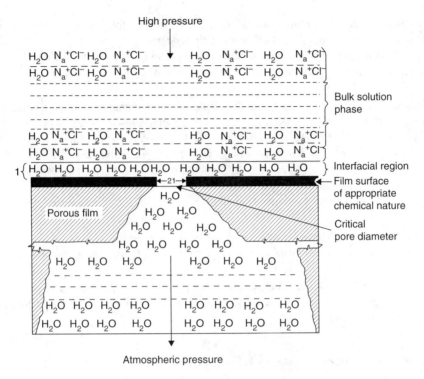

FIGURE 9.13 Preferential sorption-capillary flow model of membrane transport. (Suggested by Sourirajan, S. and Matsuura, T. 1984. *Reverse Osmosis and Ultrafiltration*, National Research Council of Canada, Ottawa, Canada.)

repulsion of ions by membranes with low dielectric constants have been proposed. Separation occurs when the preferentially sorbed layer is forced through the membrane capillaries under pressure.

This mechanism explains most of the variations in separation behavior that have been experimentally observed. In the radial direction, the convective transport of the two ions is unequal. Because the electrical current has to be zero, a potential gradient is generated which drives the ions opposite to the direction of convection. The overall result is a flow of water from the pore containing a smaller amount of salt than at the entrance of the pore. The membrane can reject salt only up to a maximum value for a given set of conditions. This maximum is primarily determined by the electrical properties of the pore walls and to a very large extent by the diffusivities of the coions and counterions. Since RO membranes are made from several different materials and physicochemical parameters are unique to each solution–membrane system, it is reasonable to assume that various rejection mechanisms are possible.

9.4 MEMBRANE TRANSPORT MODELS

A membrane transport model is necessary in order to relate the performance (expressed in terms of flux of solvent and solute) to the operating conditions and other measurable properties of the system (Dickson, 1988). To fully describe and understand the overall transport processes involved, the following relationships must be defined (Jonsson, 1980a; Cheryan and Nichols, 1992):

1. An equation relating permeate flux to the wall or boundary layer concentration (i.e., a membrane transport equation)
2. An equation relating the boundary layer concentration to a mass transfer coefficient

3. A model describing the mass transfer characteristics of the system in relation to feed solution properties and fluid mechanics of the system
4. Equations relating the fluid mechanics to the operating parameters and the flow regime (e.g., turbulent flow, laminar flow — entrance region or fully developed)

Several models attempting to describe the transport phenomena exist in the literature. In most cases, each model has been evaluated for one particular membrane–solute combination and thus the transport coefficients so obtained may not be applicable to other membrane–solute combinations. However, it is important to obtain an understanding of the forces involved and the mechanism by which solutes and solvent are transported through RO membranes, especially for development of suitable membranes and processing systems.

Two very different approaches have been taken in developing RO transport models:

1. Using irreversible thermodynamics to describe the process
2. Using physical-chemical-structural descriptions of the RO membrane–solution system

It should be kept in mind that to be useful, the transport model must be able to predict permeate flux and solute rejection from the transport coefficients characterizing the solution–membrane system and the operating parameters.

9.4.1 IRREVERSIBLE THERMODYNAMICS (IT) TRANSPORT MODELS

Irreversible thermodynamics provides a general description of membrane transport that can be used for both RO and UF, since it uses a black box approach leading to phenomenological equations which correlate rejection behavior with macroscopic properties that can be measured. However, since this approach is not model dependent, it gives no explanation of the flow and separation mechanisms.

In any irreversible process, the dissipation function or rate of entropy production (Φ) can be expressed as the sum of the products of the flows (fluxes) and the conjugated forces,

$$\Phi = \sum_{i=1}^{N} J_i X_i \tag{9.8}$$

The Onsager principle states that, for sufficiently slow processes or small deviations from equilibrium, there is a linear dependence between all fluxes and all forces. This relationship is expressed in the phenomenological equations:

$$J_i = L_{ii} X_i + L_{ij} X_j \tag{9.9}$$

The L's are the so-called phenomenological coefficients and the X's represent the thermodynamic driving forces. This linear law is a generalization of the well-known laws of Fourier, Fick, Poiseuille, Ohm, and D'arcy in which flux is proportional to a driving force or conjugate force (straight coefficients). In Onsager's phenomenological equations, additional contributions are assumed to be made by nonconjugated forces (denoted by the cross-coefficients) such that any flux (J_i) depends directly linearly on not only its conjugate force but also on the nonconjugate forces.

In an isothermal RO process, two solutions of the same solvent and solute are separated by a membrane. The driving forces are the chemical potential gradients across the membrane, and

Equation 9.8 can be rewritten as:

$$J_1 = -L_{11} \text{ grad } \mu_1 - L_{12} \text{ grad } \mu_1 \tag{9.10}$$

$$J_2 = -L_{21} \text{ grad } \mu_2 - L_{22} \text{ grad } \mu_2 \tag{9.11}$$

J_1 is the flux of solvent and J_2 is the flux of solute through the membrane. The external forces in RO are concentration and pressure gradients, both of which appear explicitly in the chemical potential (μ). Since $L_{12} = L_{21}$, Equation 9.11 can be written as:

$$J_2 = \frac{L_{12}}{L_{21}} J_1 - L_{22} \left[1 - \left(\frac{L_{12}}{L_{11}} \right)^2 \frac{L_{11}}{L_{22}} \right] \text{grad } \mu_2 \tag{9.12}$$

To use this approach, the L-coefficients or their equivalents (appropriate ratios) must be determined by experimentation or suitable theories. The L-coefficients should be independent of the fluxes and forces and may be functions of composition. L_{11} can be easily determined from measurements of J_1 when no solute is present but only when the presence of the solute does not change the membrane characteristics (Johnson et al., 1966).

Several transport models have been derived using the IT approach. These models differ somewhat in their choice of flows, their evaluation of the L-coefficients, their application of local equilibria, and other parameters. Four of the IT models are presented here.

9.4.1.1 Kedem and Katchalsky's Transport Model

Kedem and Katchalsky (1958) transformed the coupled phenomenological equations in more practical terms, using three transport coefficients that are easier to experimentally determine:

$$J_v = L_p(\Delta P_T - \sigma \Delta \pi) \tag{9.13}$$

$$J_2 = \omega \Delta \pi + (1 - \sigma) J_v (c_2)_{\ln} \tag{9.14}$$

where J_v is the volume flux through the membrane, J_2 is solute flux through the membrane, ΔP_T is pressure difference across the membrane, $\Delta \pi$ is osmotic pressure difference across the membrane, $(c_2)_{\ln}$ is log mean solute concentration difference across the membrane, L_p is hydrodynamic permeability coefficient, σ is reflection coefficient, and ω is the solute permeability coefficient.

Since the solute concentration gradient in the membrane is usually large and nonlinear, they used a log-mean solute concentration to account for the nonlinearity in their linear model.

9.4.1.2 Spiegler and Kedem's Transport Model

To account for a changing concentration profile at different volume fluxes, Spiegler and Kedem (1966) derived a set of differential flux equations in terms of a local solute permeability coefficient, P_2, a specific membrane permeability coefficient, P_h, and the reflection coefficient:

$$J_v = P_h \left(\frac{dp}{dy} - \sigma \frac{d\pi}{dy} \right) \tag{9.15}$$

$$J_2 = \bar{P}_2 \frac{dc_2}{dy} + (1 - \sigma) c_2 J_v \tag{9.16}$$

The equations can be integrated across the membrane thickness (λ). Volumetric flux, solute flux and rejection are:

$$J_v = \frac{P_h}{\lambda(\Delta P_T - \sigma \Delta \pi)} \tag{9.17}$$

$$j_2 = (1 - \sigma)j_v \left[\frac{c_2'' - c_2' \exp[J_v(1 - \sigma)\lambda/\bar{P}_2]}{1 - \exp[J_v(1 - \sigma)\lambda\bar{P}_2]} \right] \tag{9.18}$$

$$R = \sigma \left[\frac{1 - \exp[-J_v(1 - \sigma)\lambda/\bar{P}_2]}{1 - \sigma \exp[-J_v(1 - \sigma)\lambda/\bar{P}_2]} \right] \tag{9.19}$$

Thus, solute rejection is a function of volume flux and the parameters ($\bar{P}_2$ and σ), which are assumed to be constant since they are combinations of the L-coefficients and average solute concentrations which are considered to be constant for a given membrane. Being independent of solute concentration and applied pressure, they are thus independent of external forces. These parameters can be determined by independent osmotic experiments, in which no pressure is applied to the system, and used in Equation 9.18 to predict solute rejection. At extremely high pressures, $J_v \to \infty$ and, according to Equation 9.18, maximum solute rejection (R_{max}) is equal to the reflection coefficient (σ). R_{max} provides an indication of the degree of coupling of the solute and solvent fluxes. R approaches 100% as a limit if fluxes are not coupled, since solvent flux through the membrane is usually much larger than solute flux. In case of coupled fluxes, however, an increase in volume flux will increase solute flux proportionately, causing rejection to asymptotically approach a limiting value (R_{max}). The latter is frequently encountered in practice with small molecules (e.g., salt solutions), while the former is observed with large molecules.

9.4.1.3 Pusch's Linear IT Model

Pusch (1977) also based his equations on linear phenomenological equations:

$$J_v = L_p(\Delta P_T - \Delta \pi) \tag{9.20}$$

$$J_2 = \left(\frac{L_\pi}{L_p} - \sigma^2 \right) L_p \bar{c}_2 \Delta \pi + J_v(1 - \sigma)\bar{c}_2 \tag{9.21}$$

where L_p is the hydrodynamic permeability coefficient, L_π is the osmotic permeability coefficient, and σ is the reflection coefficient. A linear relationship between rejection and volume flux can be derived from these equations:

$$\frac{1}{R} = \frac{1}{R_{max}} + \frac{(L_\pi/L_p - \sigma^2)(1 - R_{max})L_p\pi'}{(1 - \sigma)R_{max}J_v} \tag{9.22}$$

When $R_{max} = \sigma$, Equation 9.22 reduces to:

$$\frac{1}{R} = \frac{1}{R_{max}} + \frac{(L_\pi/L_p - R_{max}^2)L_p\pi'}{R_{max}J_v} \tag{9.23}$$

Since Equation 9.23 was derived from the general principles of the thermodynamics of irreversible processes without using any particular membrane model, it should have broad applicability in RO where the transport coefficients are independent of pressure and concentration.

9.4.1.4 Bi-Layer and Extended IT Models

Spiegler and Kedem's equations were derived for a single layer membrane. Bi-layer IT models consider asymmetric cellulose acetate membranes as two homogeneous layers in series (Kedem and Katchalsky, 1963; Jagur-Grodzinski and Kedem, 1966; Elata, 1969; Groepl and Pusch, 1970; Jonsson, 1980b; Hwang and Pusch, 1981). The reflection coefficient for the whole membrane is related to the individual reflection coefficients and solute permeabilities of the two layers. This model now has six transport coefficients which can be experimentally determined by measuring the coefficients for one of the layers separately and then for the entire membrane. A homogeneous membrane analog of the porous layer, identical to the skin layer in thickness, can be prepared or removed from an asymmetric membrane by abrasion or hydrolysis. All bi-layer models predict that, at a given volume flux, solute rejection is completely determined by the reflection coefficient and solute permeability coefficient for the skin layer, provided the reflection coefficient for the porous layer is zero. Single layer models give a good description of rejection vs. permeate flux for asymmetric membranes under RO conditions.

The IT models presented so far are based on two rather restrictive assumptions: (1) The Onsager Reciprocal Relationships (ORR) are valid and (2) that the mechanism for transport is diffusion, that is, there is no convection flow term in these models. In RO, the validity of the ORR has frequently been questioned on the applicability of the linear laws and equilibrium conditions. Attempts to extend the model without assuming ORR to be valid and including convection terms have not been very successful (Lui, 1978; Soltanieh and Gill, 1981).

9.4.2 PHYSICAL-CHEMICAL-STRUCTURAL MODELS

All physical-chemical-structural models use some combination of three parallel transport mechanisms: (1) diffusion due to a concentration gradient, (2) diffusion due to a pressure gradient, or (3) hydrodynamic flow through pores. These models assume a mechanism for transport and then correlate rejection behavior with known or assumed structural properties of the membrane (porosity, pore size, tortuosity) and/or physical-chemical properties of the solution–membrane system (diffusivity, solubility, frictional interactions, etc.). This approach gives more information on the flow and separation mechanisms, depending on the suitability of the chosen model.

9.4.2.1 Solution–Diffusion Model

This is probably the most widely used model to describe RO. It assumes that the membrane is a nonporous diffusive barrier, in which all components dissolve in accordance with phase equilibrium considerations and diffuse by the same mechanisms that govern diffusion through liquids and solids (Lonsdale, 1972; Baker, 2000). It is assumed that the solution process is rapid enough for equilibrium distributions of solute and solvent between the solutions immediately adjacent to the membrane and the membrane surface to be maintained. Under these conditions, the rate of transport of water and solute is proportional to the chemical potential gradients across the membrane. Each component dissolves in the membrane on the high pressure side and diffuses through the membrane in response to concentration and pressure gradients, without any coupling between the individual fluxes.

The flux of each component (J_i) is given by the product of its mobility, concentration, and driving force:

$$J_i = -\frac{D_{im}c_{im}}{RT}\operatorname{grad}\mu_i \tag{9.24}$$

where D_{im} is the diffusion coefficient and c_{im} is the concentration of component i within the membrane. The driving force for solvent flux is $\bar{v}_1(\Delta P_T - \Delta\pi)$, and the solvent flux equation is:

$$J_i = -\frac{D_{im}c_{im}\bar{v}_i}{RT\lambda}(\Delta P_T - \Delta \pi) = A(\Delta P_T - \Delta \pi) \tag{9.25}$$

The influence of pressure on solute transport is neglected so that solute flux across the membrane is:

$$J_2 = -\frac{D_{2m}K_2}{\lambda}\Delta c_2 = B\Delta c_2 \tag{9.26}$$

where K_2 is the solute distribution coefficient.

According to this model, operating variables (transmembrane pressure, temperature, and feed concentration), membrane thickness, and the water and solute permeabilities ($D_{1m}c_{1m}\bar{v}_1$ and $D_{2m}K_2$) determine flux. At constant temperature, the solution–diffusion model uses two parameters, A, which is ($-D_{1m}c_{1m}\bar{v}_1/RT\lambda$) and B ($-D_{2m}K_2/\lambda$), to describe membrane performance. As a membrane constant, A is a measure of water flux through the membrane and the ratio A/B is a measure of the rejection capabilities of the membrane (Lonsdale, 1972). Solute rejection for the solution–diffusion model is:

$$R = \frac{A(\Delta P_T - \Delta \pi)}{A(\Delta P_T - \Delta \pi) + Bc_1'} \tag{9.27}$$

Both water and solute fluxes are inversely proportional to membrane thickness, but rejection of solute, given by the ratio of the fluxes, is independent of membrane thickness. According to this model, solvent flux increases linearly with net pressure, reaching a maximum rejection (R_{max}) of 1 while solute flux is independent of pressure. This is because no coupling of flows is assumed. This implies that high transmembrane pressures will produce more permeate of better quality. This has generally been observed in practice, although rejection never quite reaches 100% even at very high pressures.

The solution–diffusion model is limited in that it does not allow for convective flow. It has been extended to include a coupling effect and for cases where solvent flux is not linearly proportional to the applied pressure but increases exponentially (Lee, 1975; Paul, 1976; Bo and Stannett, 1976).

9.4.2.2 Pore Flow Model

Unlike the solution–diffusion model, which represents one extreme in that no coupling of solute and solvent is assumed, pore flow models represent the other extreme in which coupling is a characteristic of viscous flow. Volume flux for Merten's (1966) highly porous transport model is based on Poiseuille's equation and includes a tortuosity factor (t) which accounts for twisting of pores and increases in effective pore length. Solute flux for this model is the sum of solute flux due to convective or bulk movement within the pores and due to diffusion of solute through the membrane pores. Separation of solute and solvent is assumed to occur because the solute concentration in the pore fluid is not the same as that in the feed solution. Assuming chemical equilibria at the solution–membrane interfaces, solute concentration in the pore fluid at each interface is related to the solute concentration in the permeate and in the feed solution through equilibrium or partition coefficients.

For Merten's model, solute flux is:

$$N_2 = \frac{v_w}{\varepsilon}\left[\frac{K_2''c_2'' - K_2'c_2'\exp(v_w t\lambda/\varepsilon D_{21})}{1 - \exp(v_w t\lambda/\varepsilon D_{21})}\right] \tag{9.28}$$

Solute rejection is

$$R = 1 - \frac{K_2' \exp(v_w t\lambda/\varepsilon D_{21})}{K_2'' - \varepsilon + \varepsilon \exp(v_w t\lambda/\varepsilon D_{21})} \tag{9.29}$$

where ε is the porosity of membrane ($n\pi r^2$), n is number of pores per unit area, π is pi = 3.142, r is pore radius, K_2 is generalized solute permeability coefficient, and c_2 is the solute concentration.

According to these equations the highest rejections will be obtained at the highest volume flux (highest pressure) and the maximum rejection is:

$$R_{\max} = 1 - \frac{K_2'}{\varepsilon} \tag{9.30}$$

The existence of a distribution coefficient different from the water content of the membrane in these equations implies that the membrane affects the equilibrium properties of the pore solution without affecting transport within the pores. Any force (mechanical, chemical, etc.) that affects equilibrium properties should also affect transport. This picture of a highly porous membrane which excludes a solute but allows viscous flow may not be realized in practice, and a frictional pore model is more realistic for RO membranes (Merten, 1966).

9.4.2.3 Frictional Transport Models

In the IT models previously discussed, no attempt was made to define specific interactions between solute, solvent, and membrane relative to the phenomenological coefficients. The basic assumption in all frictional models is that, at mechanical equilibrium, the thermodynamic driving forces acting on each particle are exactly balanced by the frictional forces between the particle and all others in the surrounding medium. These frictional forces, F_{ij}, are proportional to the mean relative velocities of the components, that is,

$$F_{ij} = f_{ij}(v_i - v_j) \tag{9.31}$$

where f_{ij} is the friction coefficient between components i and j, and v_i and v_j are the mean linear velocities of components i and j. Belfort (1976) derived relationships between the frictional coefficients and the transport coefficients defined in Spiegler and Kedem's model. It appears that solvent–solute interactions within the membrane are much greater than those in the external solutions and that solute–membrane interactions are much greater than solvent–membrane interactions in hydrophilic membranes where the solvent is water and the solute an electrolyte. This concept of frictional forces inhibiting movement of solute and solvent within the membrane was used in the derivation of Merten's finely porous model and Jonsson and Boeson's viscous flow-frictional model (see below).

9.4.2.4 Finely Porous Model

This model was developed by Merten (1966) for RO membranes whose transport properties are intermediate between the solution–diffusion and the pore flow models. For dilute solutions, solute flux through the membrane is:

$$N_2 = -\frac{RT}{bf_{21}} \frac{dc_{2m}}{dy} + \frac{c_{2m}v_p}{b} \tag{9.32}$$

f_{21} is the coefficient for friction between the solute and the solvent, and b was defined as a drag or friction factor $(1 + f_{23}/f_{21})$. Solute–solvent and solute–membrane interactions are important but not solvent–membrane interactions. Volume flux through the membrane pores is given by a Poiseuille type equation. The total force acting on the pore fluid includes the frictional force between the solute and the membrane in addition to the applied pressure. Membrane performance is a function of several factors: the equilibrium solubility of the solute in the pore fluid, solute-membrane and solute-solvent interactions, and the velocity of the pore fluid. However, for a given membrane and feed solution, solute rejection depends only on the volumetric flow rate since most of the parameters in the model are assumed to be independent of solute concentration (Cheryan and Nichols, 1992). Maximum rejection for this model is described by a model similar to Equation 9.30.

9.4.2.5 Combined Viscous Flow-Frictional Model

This is based on the Finely Porous model, but the distribution or partition coefficient is taken to be the same on both sides of the skin layer (Jonsson and Boesen, 1975). In addition, this model was derived for the effective membrane thickness λ or the thickness of skin layer. The fractional pore volume was corrected for pore tortuosity so that the total pore volume is εt. The maximum solute rejection is similar to Equation 9.30:

$$R_{\max} = 1 - \frac{K_2}{b} \tag{9.33}$$

The parameters $(1 - K_2/b)$ and $(t\lambda/\varepsilon)$ which appear in the flux and rejection equations are determined from nonlinear parameter estimation procedures and should be independent of pressures and concentrations (Cheryan and Nichols, 1992). The parameter $(t\lambda/\varepsilon)$ should also be independent of the type of solute being processed and thus should be a constant (Jonsson, 1978). This is in accordance with the physical meaning of the parameter — the effective pore length in the membrane skin divided by the pore area.

9.4.3 COMBINATION MODELS

The physical-chemical-structural models discussed thus far differ in their description of flow through the membrane (i.e., either purely diffusive flow, purely convection flow through membrane pores, or diffusive and convective flow through membrane pores). Combination models do not make a distinction among the three types of flow but merely assume that both diffusive and pore flow or diffusive and convective flow are possible.

The solution–diffusion-imperfection model was developed by Sherwood et al. (1967) based on the assumption that water and solute cross the membrane by the parallel processes of diffusion and pore flow. Solute and solvent dissolve into the water swollen membrane on the high pressure side, diffuse across the membrane due to a chemical potential gradient and desorb from the membrane surface on the low pressure side. A fraction of the feed solution is assumed to pass through open channels or pores, with negligible change in solute concentration due to diffusion in the pore fluid. The rate of pore flow is proportional to the applied transmembrane pressure. In order to have high solute rejections, pore flow must be a small fraction of the total flux. The flux equations included transport coefficients for the diffusion of water, for pore flow, and for diffusion of solute which is proportional to the product of the diffusivity of the solute in the membrane and the distribution coefficient for the solute between the solution at the membrane surface and the membrane material (Cheryan and Nichols, 1992).

The diffusion-flow model (Yasada and Lamaze, 1971) assumes that water transport in water-swollen polymer membranes under an applied pressure can occur by both molecular diffusion and bulk flow, depending on the water content of the membrane or the degree of swelling. Solute rejection is due to transport depletion of solute in relation to water rather than the exclusion of the solute by the membrane. Rather than treat pores as capillaries in which the transport of water can be described by

Poiseuille's flow equation, they propose a dynamic membrane structure in which water movement is diffusive at some point in time, and later as viscous flow, when the structure changes to a more loose geometric arrangement. This model uses two types of water permeability coefficients — a hydraulic permeability coefficient (K_1) and a diffusive permeability coefficient (P_1^*). A parameter ϕ is defined as the ratio of the two permeability coefficients:

$$\phi = \frac{K_1^* RT}{P_1^* \bar{v}_1} \tag{9.34}$$

Thus membranes can be classified according to their mode of water transport using ϕ. If $\phi = 1$, it is a diffusion membrane and solute rejection can be described by an equation like the solution–diffusion model (Equation 9.27). If $\phi > 1$, it is a diffusion-flow membrane and if $\phi \gg 1$, it is considered as a flow membrane.

This model also assumes that whenever water moves through the membrane by diffusion, solute also moves in a diffusive mode. On the other hand, when water moves by viscous flow, there is no solute rejection.

9.4.4 PREFERENTIAL SORPTION–CAPILLARY FLOW MODEL

Two sets of transport equations quantitatively describe the preferential sorption-capillary flow (PSCF) mechanism for aqueous solutions developed originally by Sourirajan and Matsuura (1982). The first set is for systems in which water is preferentially sorbed at the membrane surface, similar to the basic capillary or pore flow model. Water transport is by viscous flow, and solute transport is by pore diffusion. The film theory is used to determine the boundary conditions. At a given temperature and pressure, the transport equations are:

$$J_w = A_w \Delta P_T \tag{9.35}$$

$$J_1 = A_w (\Delta P_T - \Delta \pi) \tag{9.36}$$

$$J_2 = \frac{D_{2m}}{K_2' \lambda} (c_2' - c_2'') \tag{9.37}$$

$$v_w = \frac{D_{2m}}{K_2' \lambda} \left(\frac{c_2'}{c_2''} - 1 \right) \tag{9.38}$$

$$v_w = k \ln \frac{c_2' - c_2''}{c_2^b - c_2''} \tag{9.39}$$

Permeate flux depends on the magnitude of the preferentially sorbed water layer, the effective thickness of the membrane, the size, number, and distribution of pores on the membrane surface, and the operating pressure, temperature, and flow conditions in the RO module.

The basic equations appear similar to the solution–diffusion model except for two differences: (1) the concentration at the membrane surface on the high pressure side (c_2') is related to the bulk concentration by the film theory, and (2) explicit expressions for the pure water permeability coefficient and the solute permeability coefficient in terms of the operating variables are required. This is one argument for claiming that the PSCF analysis is more realistic than the solution–diffusion model (Sourirajan and Matsuura, 1977). In the solution–diffusion model there is no pure water permeability constant (A_w); there is no mass transfer coefficient (k) for the high pressure side of the membrane; there is no means of calculating the boundary layer concentration on the high pressure side of the membrane, and the constant A in the solution–diffusion model cannot be a constant since it must depend on the nature of the solute, solute concentration, and feed flow conditions.

The second set of equations is for the surface force-pore flow model involving preferential sorption of either water or solute at the membrane solution interface. Pores on the membrane surface are treated as circular cylindrical pores with or without a pore size distribution. Solute–membrane interactions, relative to water, are expressed in terms of electrostatic or Lennard–Jones type surface potential functions. Solute–solvent transport through the membrane pore is governed by these surface forces together with pressure, friction and viscous shear forces. Based on the above considerations, transport equations can be derived in terms of several dimensionless parameters which are broadly applicable to any single-solute aqueous solution, whether RO or UF, at any given set of operating conditions provided the pore radius, osmotic pressure data, and surface force and friction force functions are known (Sourirajan and Matsuura, 1984; Cheryan and Nichols, 1992).

9.5 EQUIPMENT

There are two basic classes of membrane equipment which are distinguished by the manner in which concentration polarization is controlled:

1. By movement of the fluid, for example, by cross-flow or tangential-flow pumping of feed through a membrane module, or by stirring in a cylindrical vessel as practiced in dead-end cells. Within the subclass of cross-flow modules, there are four different types of equipment available for reverse osmosis: tubular, hollow fine fiber (external feed), spiral-wound, and plate systems.
2. By movement of the membrane, for example, rotary or vibratory modules.

In all likelihood, one or more of these designs will be able to handle almost any type of feed stream that can be pumped. Each design has its own special applications, advantages, and disadvantages.

9.5.1 FLUID MOVEMENT MODULES

9.5.1.1 Tubular Modules

The tubular configuration is one of the earliest designs. The membrane is cast inside a porous support tube, several of which are housed within one pressure vessel in a shell-and-tube arrangement (Figure 9.14). The usual flow is through the bore of the tubes which are usually connected in series with the appropriate end-cap. The permeate collects in the shell side from where it is removed. Most manufacturers use tubes of 12.5 mm (0.5″) internal diameter and lengths of 3.3–20 feet (1–6 m). The number of tubes per module varies from 7 to 130, with 18 to 19 being most common.

Tubular modules have several advantages: (1) high turbulence and Reynolds Numbers, which generally leads to higher flux with polarizing feed streams; (2) ability to handle suspended matter with particle sizes as large as 1 to 1.5 mm; and (3) easy to clean. The major disadvantages are: (1) low surface area-to-volume ratio (i.e., low packing density), as illustrated in Figure 9.15. This implies that, in comparison to other designs, a larger floor area is required; (2) high hold up in the modules; and (3) high capital cost and high pumping energy costs (in comparison to other designs).

9.5.1.2 Hollow Fibers

There are two types of hollow fiber devices in use. The hollow (fine) fibers have an internal diameter of 40–50 microns, an outer diameter of 85 microns, and a wall thickness of ~21 microns (Figure 9.16). They act like thick-walled pressure vessels and can remain unsupported under the high pressures encountered in RO systems. The feed is pumped from the shell side while permeate is withdrawn from the inside of the fibers. On the other hand, many hollow (large) fibers that are used primarily for process applications rather than for drinking water (e.g., from Romicon/Koch, Amicon/Millipore,

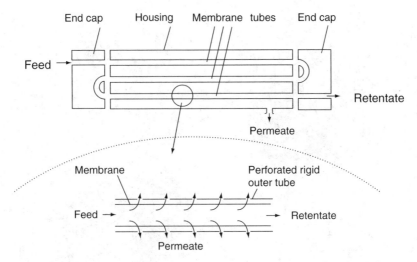

FIGURE 9.14 Schematic of tubular membrane. In the PCI design, the membrane is cast on a filter paper tube of 12.5 mm i.d., which is inserted into a 14 mm perforated stainless steel outer tube. These tubes are then arranged in a shell-and-tube configuration within the housing as shown on top.

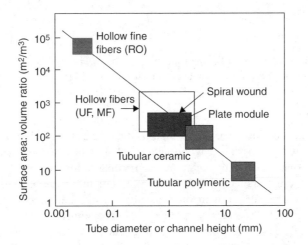

FIGURE 9.15 Relationship between channel size and surface area:volume ratio of membrane modules.

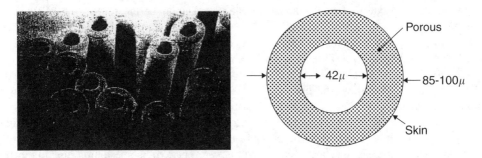

FIGURE 9.16 Ultrastructure of Dupont Aramid hollow fine fibers. (Adapted from Dupont product literature.)

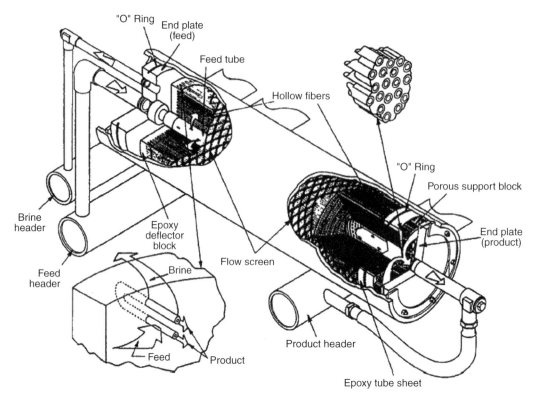

FIGURE 9.17 Schematic of Du Pont hollow (fine) fiber module. The feed is pumped from the shell side, permeate withdrawn from the tube side. (Adapted from Cheryan, M. 1998. *Ultrafiltration and Microfiltration Handbook.* CRC Press, Boca Raton, FL and Du Pont product literature.)

AGTechnology/Amersham Biosciences) have internal diameters of 0.2–3 mm and the feed is pumped through the tube side, with the permeate withdrawn from the shell side (Cheryan, 1998). The former are used for reverse osmosis, while the latter are used primarily for UF and MF.

Hollow fine fibers are manufactured by forcing the polymer solution through an orifice with a centered needle which delivers an inert fluid, usually dry nitrogen, into a coagulation bath. The resulting fibers are then treated chemically and thermally depending upon the final desired properties. After being wound to the necessary size, bundles of fibers are arrested in resin, while making sure that the fibers are not blocked, and the resin cured. Bundles may be sealed with epoxy into a straight configuration with open fiber ends on opposite sides of the module for interior feed flow, or U-shaped with both open fiber ends on one side of the module for exterior feed flow. After trimming the potted fibers, modules are assembled using appropriate pressure vessels.

An example of an exterior feed, U-shaped bundle is the Dupont permeator (Figure 9.17) in which feed is introduced from a central perforated tube and the permeate moves radially into the fibers, collects in the open fiber bore and exits at the fiber ends to join other permeate streams. It is obvious that this type of arrangement is quite prone to fouling/plugging by suspended matter in the feed and thus is appropriate only for clear solutions.

The advantages of hollow fine fibers are: (1) high packing density, resulting in the highest membrane area per module volume among the various designs (Figure 9.15) and (2) high resistance to compression, and thus very high pressures of up to 100 bar can be routinely used. This is a particular advantage in applications where osmotic pressure is the main limiting factor, as will be seen later with the FreshNote process for concentrating fruit juices. There are some limitations: (1) they are easily fouled by suspended matter which means that the feed stream must be prefiltered to remove

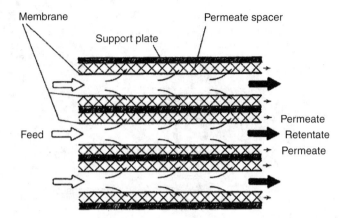

FIGURE 9.18 Schematic of flow paths in a typical plate module. Some designs incorporate a space/mesh in the feed channel. In others, the support plate may be hollow and perforated, so that the permeate passes into the support plate, and is removed through an opening in the periphery or center of the plate. The general arrangement of the membrane stack is like a plate and frame filter press. The plates may be rectangular, square, circular or oval in shape.

suspended matter, (2) they have a lower flux compared to other designs, which is probably related to the nonasymmetric nature of the membrane itself, and (3) individual membrane fibers cannot be replaced when damaged or leaking; thus replacement costs per module may be high.

9.5.1.3 Plate Systems

The plate-and-frame design utilizes flat membrane sheets on either side of a supporting plate (Figure 9.18). Many such membrane–plate combinations can be stacked together in either a vertical or horizontal manner to make up a module. Any flat-sheet membrane can be used in these systems, which makes for considerable flexibility in use. When membranes need replacing, the support plates are re-used, and so this type of arrangement can be economical in the long run.

Some advantages include: (1) fairly low hold up and moderately high packing density, intermediate between the hollow fine fibers and the tubular designs (Figure 9.15), (2) individual membranes can be replaced, (3) flexible with regard to membrane usage — flat sheet membranes from any manufacturer can be easily cut and used in the plate units, and (4) capacity can be increased by adding more membrane–plate units. The main disadvantages seem to be the relatively high initial capital cost and complexity in maintaining leak-free seals under the high pressures encountered in RO.

9.5.1.4 Spiral-Wound

Spiral wound modules also utilize flat-sheet membranes. As shown in Figure 9.19, two membrane sheets are sandwiched with a permeate spacer between them and three edges are sealed. The fourth is connected to a central perforated tube. A feed channel spacer is placed on top of one layer, and the membrane–screen composite is rolled into a spiral configuration around the central collection tube. The module is placed inside a tubular pressure vessel. Feed flows longitudinally in the feed channel, while permeate flows between the membrane sandwich and spirally around to the permeate collection tube. Several individual spiral modules may be placed in series in a housing with the permeate tubes interconnected (Figure 9.20).

Some advantages of the spiral wound system are: (1) relatively high packing density, depending primarily on the feed channel spacer thickness, (2) low cost per unit membrane area, (3) relatively easy replacement of modules from the pressure vessels, and (4) low energy consumption per unit energy. Some disadvantages are: (1) they may be difficult to clean especially with feeds containing

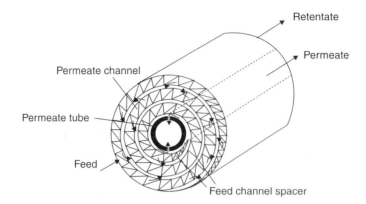

FIGURE 9.19 Spiral-wound membrane module.

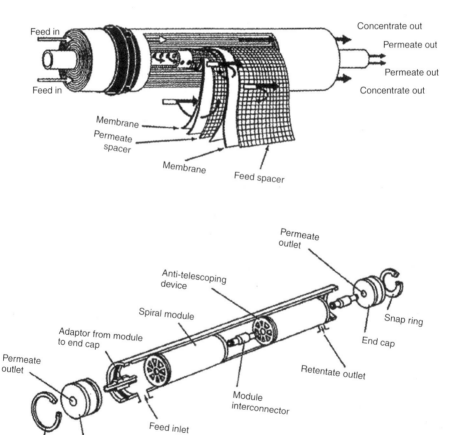

FIGURE 9.20 Spiral-wound module. Top: Cutaway of membrane showing flow path of feed, retentate and permeate. Bottom: Arrangement of spiral modules in housing. Upto six of these modules may be used in series, separated by anti-telescoping devices (ATD) with interconnectors in between.

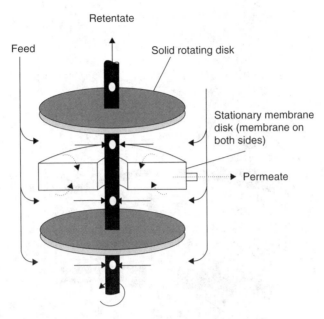

FIGURE 9.21 Schematic of a rotary module with alternating disks of stationary and rotating disks. The membrane is on the stationary disks.

a large amount of suspended matter and (2) the feed must be prefiltered well, perhaps to 1/20 to 1/50 the feed channel height (Cheryan, 1998).

9.5.2 MEMBRANE MOVEMENT MODULES

Rather than moving the fluid as in conventional modules, a special class of modules controls concentration polarization by moving the membrane itself. The two designs currently available are the rotary modules and the vibratory modules. Figure 9.21 shows the principle of the rotating flat-plate or disc module. The membrane is cast or fitted on to a hollow polymeric plate or disk with provision for removing the permeate. In one embodiment, the membrane disk does not rotate. Rather, solid disks are connected to a central rod and placed a few millimeters away from both sides of the membrane disks. The solid rotors rotate at speeds up to 6,000 RPM, resulting in shear rates up to 400,000 sec^{-1}, higher than cylindrical devices or conventional modules. High shear rates at the membrane surface help to keep it free of suspended matter and minimizes concentration polarization. This makes it particularly useful for fine separation and otherwise difficult fractionations, for example, separation of similarly sized proteins. However, if the membrane is truly fouled by solute–membrane interactions, rotating the membrane, even at high RPMs, will not be very helpful. There are also designs where the membrane disks themselves rotate at about 1,000 to 2,000 RPM (Cheryan, 1998).

Another variation on the moving membrane concept is the vibratory shear-enhanced module developed by New Logic International (Figure 9.22). One version is marketed by Pall as the "Pall-Sep VMF." The module is essentially a series of flat-sheet disk membranes separated from each other by a gasket of 1–2 mm thick. The units are constructed as a stack of these disk leaf elements, which may be 30–50 cm in diameter. Each leaf element has a membrane placed on a drainage spacer on both sides. Operating pressures can be up to 600 psi which is in the range of NF and RO.

Unlike rotating modules, the vibratory module vibrates in a torsional oscillation similar in action to a clothes washing machine. The stack rotates back and forth to a displacement of 1.5″ (3.75 cm) peak to peak at the rim of the stack. The oscillation frequency is 60 times a second: for an 18″ diameter stack, this results in shear rates up to 150,000 sec^{-1} at the rim. Since turbulence is generated by

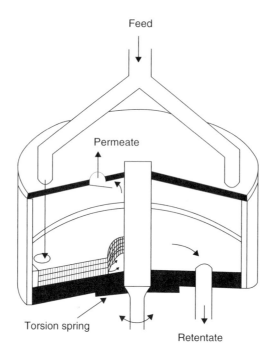

Feed

Permeate

Torsion spring

Retentate

FIGURE 9.22   Schematic of a vibratory module.

vibration and not cross flow, the unit can be operated in a single-pass mode, with no recirculation of the retentate.

9.5.3 SELECTING THE BEST MODULE

The final decision on the type of module to use in a particular application should be based on the following criteria: (1) physical properties of the feed stream and retentate, especially viscosity, (2) particle size of suspended matter in the feed, and (3) fouling potential of the feed stream. For example, most food materials containing macromolecules display non-Newtonian behavior. Their viscosity will increase dramatically above certain concentrations, making pumping difficult and reducing the mass transfer within the boundary layer. This will necessitate the use of modules that can withstand high pressure drops. The hollow fine fibers can withstand very high pressure and have extremely high packing densities (surface area: volume ratios), but the feedstream must be fairly clean with no suspended matter.

On the other hand, a general rule of thumb states that the feed should be prefiltered to remove all particles larger than one-tenth the channel height or diameter to prevent a physical blockage of the inlet of the module (Cheryan, 1998). This makes the large-bore tubular modules appropriate with many applications containing suspended particles, especially with waste streams where the additional cost of prefiltration required with spiral, hollow fiber and many plate designs may make it uneconomical. Thus large-bore tubular modules are more suitable for feed streams with large amounts of suspended matter such as citrus juices and animal waste streams, even though they have the lowest packing densities and highest energy consumption among all designs.

Modules utilizing flat sheets (spiral-wound and some plate unit) have a mesh-like spacer between sheets of membrane. This restricts their use to clear liquids or feed streams containing low levels of finely suspended matter. On the other hand, spiral-wound units are the lowest in capital costs and lowest in energy consumption, although this is somewhat offset by their generally lower flux with polarizing feed streams.

The rotary and vibratory modules are best utilized with feed streams containing suspended particles. Since osmotic pressure is usually the limiting factor with RO and NF, and osmotic pressure in turn is most affected by small molecules that are in solution, these modules may not offer any advantages. However, if concentration polarization is also significant, then there may be some advantage.

9.5.4 FOULING AND CLEANING

If there has been one single factor that has inhibited the large-scale use of membrane technology in the food industry, it is fouling of membranes. Fouling manifests itself as a decline in performance, usually a decline in flux under constant operating conditions and a possible change in the sieving properties of the membrane. This phenomenon is frequently confused in the literature with other flux-depressing phenomena, such as change in membrane properties with time due to deterioration, changes in feed solution properties and concentration polarization.

Minimizing membrane fouling must include consideration of the chemical nature of the membrane as well as physicochemical properties of the feed stream. This is clearly shown by cheese whey, which is one of the most notorious foulants of membranes. Although it contains less than half the solids and one-third the protein of milk, cheese whey fouls membranes much more severely than does milk. Furthermore, hydrophilic membranes (e.g., cellulosics) generally foul less than hydrophobic membranes (e.g., polysulfone). This is also true with microbial fouling: Ridgway (1988) observed that adhesion by *Mycobacterium* bacteria was much less with cellulose acetate membranes than a polyamide membrane.

Frequently, membrane fouling problems may really be cleaning problems. Cleaning and sanitation is a separate art and science by itself. Ideally, the membrane material, and all other food contact surfaces, should be compatible with normal food and dairy cleaners, such as mild caustic or acid solutions. Severe problems may require the use of enzyme detergents. It behooves the user to clearly understand the nature and extent of the interaction of the foulants with the membrane in order to establish the optimum cleaning regime.

9.6 ECONOMICS AND APPLICATIONS

In terms of energy consumption alone, reverse osmosis is the most economical among all the water removal processes (Table 9.5). RO is best used when the concentration of dissolved solutes is low, thus keeping the osmotic pressure low. Unlike most heat transfer processes, the performance of the RO system drops as the concentration of the solute increases. Thus RO may be most useful as a preconcentration step (Table 9.6 and Table 9.7).

RO was first used to demineralize sea water and brackish water, producing fresh water for municipal and industrial purposes. Of the 23 million cubic meters per day of desalination capacity in the world in 1997, as much as 44% is by RO; up from 24% in 1986. RO plants for brackish water treatment (<5000 ppm total dissolved solids) are being installed at more than five times the rate of installation of Multi-Stage Flash distillation units. Among the largest installations are the 25 million gallons per day (MGD) RO plant for producing drinking water from the sea in Tampa, Florida, the 72 MGD RO plant in Yuma, Arizona for treating brackish water from the Colorado river, and the 86 MGD NF plant in Mery-sur-Oise, France for treating river water.

9.6.1 DAIRY INDUSTRY

The use of reverse osmosis for concentration of milk has been studied since the early 1960s. The major attraction to the dairy industry was energy savings and the possibility of using temperatures lower than conventional evaporation. The osmotic pressure of milk is about 6 to 7 bar due largely to the lactose and dissolved salts. Figure 9.8 showed typical data obtained for the concentration

TABLE 9.5

Energy Economy of Moisture Removal Processes [TC = Thermal Compression; MSF = Multi-Stage Flash Distillation; MVR = Mechanical Vapor Recompression; RO = Reverse Osmosis]

	TC	MSF	MVR	RO
Economy (lb water/1000 BTU)	3–10	3–12	17–33	30–150[a]
Type of energy	Steam	Steam	Electricity	Electricity
Operating temperature (C)	50–80	85–115	80–95	30–50
Complexity[b]	3	2	4	1
Maintenance[b]	1	2	4	3

Notes:

[a] With pressure recovery.

[b] 1 = Least, 4 = Most.

Source: From Eisenberg, T.N. and Middlebrooks, E.J. 1986. *Reverse Osmosis of Drinking Water*, Butterworths, Stoneham, MA.

TABLE 9.6

Comparison of Energy Consumption and Surface Areas for Preconcentration of Whole Milk from 15 to 31% Total Solids at a Feed Rate of 1000 kg/h

Process	Area (m^2)	Energy (kcal/kg milk)
Thermal concentration		
Open-pan boiling	10.4	455
Evaporator		
Double-effect	25	209
MVR	32	136
Membrane process		
Batch, single pump	64	80
Batch, dual pump	64	7
Continuous, one-stage	206	16
Continuous, three-stage	93	7

Source: From Cheryan, M., Sarma, S.C., and Pal, D. 1987. *Asian J. Dairy Res.* 6: 143–153.

of milk in a spiral wound thin-layer composite membrane The asymptotic relationship between applied transmembrane pressure and flux is due to concentration polarization of proteins and fat, giving rise to increased hydrodynamic resistance to permeate flow and/or to higher osmotic pressure. The decrease in flux at higher pressures is probably due to a compaction of the polarized/fouling layer, giving rise to increased resistance, or to a much higher osmotic pressure (Cheryan et al., 1990).

Fouling (i.e., a decrease in flux with time with under constant operating conditions) is generally much less with milk than with whey. The nature of the deposit is mostly protein, but inorganic salts such as calcium phosphate play an important role. Operating conditions should be carefully

TABLE 9.7

Operating Cost ($) for Removal of 8500 kg/h of Water from Cheese Whey by Double-Effect Evaporation and by Reverse Osmosis

Item	Evaporator	RO
Steam (@$11/1000 kg)	346,000	48,000[a]
Cleaning chemicals	5,000	19,000
Labor	18,600	18,600
Electric power	10,400	14,000
Membrane replacement[b]	—	30,000
Annual operating cost	380,000	130,000

Annual savings ($250,000) exceeds cost of RO plant

Notes:

[a] To reheat whey to 32°C and for CIP.

[b] One year life.

Source: From Pepper, D. 1980. *Desalination* 35: 383–396.

controlled, especially with RO of whole milk. Although the high pressures per se pose no problem, the sudden release of pressure at the outlet of the module can damage fat globules. This can be overcome by proper design of the outlet high-pressure valve. RO of raw milk will give concentrates of poor bacteriological quality and increased free fat content. Suitable preheat treatment at a sub-pasteurization level (thermization) can usually avoid such problems, as will operating the RO plant at temperatures of 50–55°C. The maximum solids concentration in the retentate is limited largely by the osmotic pressure. Most commercial modules have a limit of about 3–4 MPa, which means a maximum of 3–4× concentration before the flux drops too low to be economical.

Perhaps the greatest potential for RO in the dairy industry is in bulk milk transport, especially in those countries which have large distances between producing and consumption areas. Considering that milk is more than 85% water, preconcentration of the milk prior to shipment to central dairies should result in considerable savings in transportation costs, as well as reducing chilling and storage costs. RO-milk products, when reconstituted with good quality water, are indistinguishable from unconcentrated milks in flavor and other quality attributes.

The removal of water from milk during the production of dried milk powder accounts for a significant portion of the product's final cost. Milk (either whole or skimmed) is usually concentrated to 45 to 50% total solids before spray drying. Because of the upper solids limitation mentioned above, RO cannot be used by itself as a complete substitute for conventional evaporation. RO, however, can be used as a preconcentration step ahead of the evaporators to reduce the time and energy demands of evaporation, or as a means to increase capacity of existing evaporators. Table 9.6 compares the energy consumption of RO and thermal evaporation for a 2.5X concentration of milk. RO can concentrate whey to 25% total solids prior to thermal evaporation and spray drying. Table 9.7 compares operating costs for dewatering cheese whey by RO and evaporation.

For equivalent capital costs, the operating cost of a RO plant is much less than a thermal evaporation plant (Cheryan et al., 1990). The potential energy savings are enormous. In 2000, there were 1.21 billion lb. of dried milk products and 0.58 billion lb. of condensed and evaporated milk products produced in USA (IDFA, 2001). This represents potentially about 8 billion lb. of water that can be removed by incorporating a 2.5X RO process in the evaporation plants. Even if only 25% of the condensed, evaporated, or dried milk products were processed with RO, it represents a savings of at least 100×10^9 kcal per year.

Desalting of whey permeate is perhaps the single most important dairy application of nanofiltration membranes. Salty cheese whey is a tremendous waste disposal problem in the dairy industries. It contains 4 to 6% NaCl and up to 6% whey solids. It has a biological oxygen demand (BOD) activity of 45,000 ppm. Hence, it can neither be discharged directly into sewers nor can it be mixed with normal cheese whey because of its salt content. In the Ultra-Osmosis^{TM} process for desalting whey, the NF system selectively removes the dissolved salts as permeate, which can be recycled or discharged, while the retentate can be added back to the normal whey. This has twin benefits. It not only solves the waste disposal problem but also increases profitability of the whey processor (Raman et al., 1994).

9.6.2 FRUIT JUICES

There are three primary areas where membranes can be applied in processing of fruit juices: (1) clarification, for example, in the production of sparkling clear beverages using microfiltration or ultrafiltration; (2) concentration, for example, using reverse osmosis to produce fruit juice concentrates of greater than 42° Brix; and (3) deacidification, for example, electrodialysis or nanofiltration to reduce the acidity in citrus juices. This section focuses on the second application (concentration): details on the other applications have been provided by Cheryan and Alvarez (1995), Cheryan (1998), and Girard and Fukumoto (2000).

Fruit juice concentration by reverse osmosis is potentially one of the largest applications of membrane technology. The major benefit of RO is that it avoids thermal damage of delicate aroma components, although these effects are more noticeable with some products.

9.6.2.1 Apple Juice

Early work by Merson and Morgan (1968) with apple juice showed that cellulose acetate membranes showed good sugar retention, but aroma/flavor compounds permeated the membrane, which lowered the quality of the concentrate (Demeczky et al., 1981). Aroma is determined by C2–C6 alcohols, C4–C8 esters, and C2–C6 aldehydes. Three compounds are especially important in providing apple-like characteristics: ethyl 2-methylbutyrate, hexanal, and 2-hexenal. High osmotic pressure limits concentration to 20 to 25 Brix with conventional RO. Pectins contributed to fouling and viscosity which also affect plant performance. Thus apple juice RO is usually preceded by depectinization by enzymes and clarification (usually by UF), which then makes it one of the easiest liquids to process by RO (Pepper, 1990). Cleaning and restoration of capacity at the end of a process cycle is no problem. Although this suggests that spiral modules are probably the best design for this application, tubular membranes are also used.

Sheu and Wiley (1983, 1984) reported permeate fluxes between 15 and 27 liters per square meter per hour (LMH) and 17 to 87% retention of apple flavor compounds for CA and PA membranes. Alvarez et al. (1997, 1998) reported on the flux and flavor retention characteristics of polyamide RO membranes. Polyamide RO membranes retained more of the flavors than cellulose acetate and also resulted in higher flux. Flux for the regular cellulose acetate membrane (with 97% salt rejection) was much higher than for a high retention (99% salt rejection) membrane, but the flavor retention was much poorer (Chua et al., 1988).

9.6.2.2 Orange Juice

With orange and other citrus juices, many of the flavor compounds reside in an oil phase present in the juice in the form of an emulsion. Since they are sparingly soluble in water, flavor retention should be much better when membrane-processed than with apple juice. Orange juice concentrates are today produced mostly by conventional multistage evaporation. Most of the citrus essences are stripped in the first effect, and in the later stages other compounds are destroyed or transformed into undesirable compounds (e.g., furfurals).

Cellulosic RO membranes can concentrate centrifuged juice up to 25° Brix at 5–7°C and 30–70 bar. Although a high retention of sugars, acids, phenolics, nitrogen compounds, and ash was obtained, the retention of volatile aroma compounds was lower (Peri, 1974). Reverse osmosis with a polyamide composite membrane can concentrate juice without a significant loss of aroma, sugar, or acids (Medina and Garcia, 1988). No losses of acids, vitamin C, limonene, or pectin were observed. Glucose and fructose amounts in permeate were 0.02–0.4% and there was no loss of sucrose. There were more aroma losses by vaporization than by permeation through the membrane (Braddock et al., 1988). However, conventional RO is limited by osmotic pressure and viscosity considerations to less than 30° Brix. Therefore RO can be used as a preconcentration step, with thermal evaporation completing the required concentration to 42° Brix. Adding RO ahead of the evaporators can increase evaporator capacity and reducing thermal treatment.

The first commercial RO plant used PCI's AFC-99 membranes with a feed rate of 4200–9200 liters per hour (L/h) and a water removal rate of 2000 L/h. The feed was pasteurized orange juice with a dissolved solids content of 10–12° Brix, pH 3.2–3.7 and a pulp content of 2–7.5%. It was necessary to do a quick flush with alkali every 4–6 h to remove the hesperidin that fouled the membranes (Pepper, 1990). The flavor of the concentrate was judged to be very good and comparable to single-strength fresh juice when rediluted.

In order to produce highly concentrated (42–60° Brix) fruit juices, SeparaSystems, a joint venture of Du Pont and Food Machinery Corporation in the 1980s, used a combination of high and low-retention RO membranes to develop the FreshNote process. The osmotic pressure of the juices such as orange juice containing 10–12% soluble solids (10–12° Brix) is about 250–300 psi. It increases to 1500 psi at 42° Brix and 3000 psi at 60° Brix. The resulting high viscosity and osmotic pressures would result in very high energy consumption and require modules capable of pressures beyond present-day state of the art.

The FreshNote process overcomes these limitations (Cross, 1988; Walker, 1990). As shown in Figure 9.23, the basic concept is a two-stage process. Ultrafiltration is first used with the juice to separate out the pulp (the bottom solids) from the serum which contains the sugars and flavor compounds. The bottom solids contains some soluble solids, all the insoluble solids, pectins, enzymes, orange oils, and the microorganisms that would affect the stability of the concentrate. The UF retentate, about 1/10 to 1/20th the feed volume, is subjected to a pasteurization treatment that destroys spoilage microorganisms and improves stability of the finished product when blended back with the concentrated UF permeate.

The serum (UF permeate), which amounts to about 90–95% of the feed volume, is concentrated by reverse osmosis using hollow fine fibers made of aromatic polyamide. Pressures are typically 1000–2000 psi. A multi-stage system is used with high-rejection membranes in the early stages and low rejection membranes (i.e., leaky, loose membranes) in later stages. When the serum concentration becomes high enough to make the effective driving force too low, the concentrate is pumped through low-rejection membranes. Some of the sugars leak through, but this results in a lower osmotic pressure difference across the membrane, which allows lower pressures to be used. Permeates with a nonzero sugar or flavor content is returned to stages containing high-rejection membranes. For particularly demanding applications, the water removed from the RO modules can be sent to a separate high-rejection RO module to remove any traces of sugar or flavor compounds. The concentrated serum can then be blended back with the pulp or bottom solids stream.

SeparaSytems reported that fruit juice concentrates of 45 to 55° Brix were obtained commercially and up to 70° Brix was obtained in pilot trials (Walker, 1990). The careful control of operating conditions is necessary. For example, the freshly extracted juice was blanketed with nitrogen and its temperature was controlled below 10°C throughout the remainder of the process. The flavor

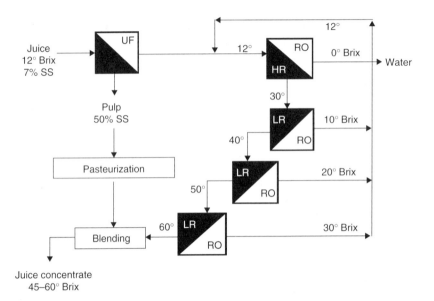

FIGURE 9.23 The *FreshNote* process for high-Brix concentrates of fruit juices. The Brix values for retentates and permeates are for illustrative purposes only. Exact values depend on the application and process design. LR = Low retention membrane; HR = High-retention membrane, SS = Suspended solids. (Adapted from Cheryan, M. and Alvarez, J.R. 1995. In *Membrane Separations Technology, Principles and Applications*, R.D. Noble and S.A. Stern (Eds.), Elsevier, Amsterdam, pp. 415–465.)

compounds in the serum were not subjected to any heat during processing, which also explained the high flavor scores for this product.

Watanabe et al. (1990) used a similar combination of high and low-rejection membranes (Nitto-Denko spirals NTR-7199 and NTR-7450 respectively) to concentrate 11° Brix apple juice to 45° Brix, using pressures of 7.5 MPa and 9.5 MPa, respectively. Similar concepts were used by Koseoglu et al. (1990) to produce concentrated and/or sterilized single-strength orange juice using hollow fiber UF and tubular membrane RO.

An interesting variation of this theme (to reduce the net osmotic pressure difference across the membrane) is direct osmosis. Milleville (1990) describes a system which uses 2-inch diameter tubes, each 10 feet long, with RO membranes of 100 MWCO. The fruit juice flows in the tubes at 31°C, while the osmotic agent (concentrated sugar solution) flows in the shell side at 45°C. The difference in concentration on either side of the membrane results in a chemical potential difference that will cause the water to flow from the dilute side to the concentrated side. It is claimed that concentrates of 42 to 60° Brix can be produced by this method. The reported advantage of this system is that no transmembrane pressure is applied, which reduces energy consumption and reduces fouling. The diluted osmotic agent (sugar syrup on the shell side) can be re-concentrated or used within the plant. The economics of this process will of course depend heavily on the latter factor.

9.6.2.3 Tomato Juice

Concentration of tomato juice presents a difficult problem, because it has a high pulp content (25% fiber) and a high viscosity (which behaves in a non-Newtonian manner). The rheological properties of the juice are affected by the method of juice manufacture. The break temperature, which is the temperature to which tomatoes are exposed after being chopped, can give different levels of enzyme inactivation which, in turn, affects final viscosity. RO-concentration of tomato juice is limited by both osmotic pressure and viscosity. If the fibers were separated, the remaining serum would behave as a water-like liquid.

Natural tomato juice is 4.5–5° Brix, commercial tomato sauces are 8–12° Brix and tomato pastes are 28–29° Brix. Because of the fiber content and particle size, tubular modules are probably the best; for example, little fouling has been reported with the PCI AFC-99 tubular membrane (Pepper et al., 1985; Merlo et al., 1986). In addition, flux was unaffected by the feed velocity in the range 0.3–4 m/sec. This has been attributed to a tubular pinch effect, that is, the solids are suspended in the middle of the flow channel and thus away from the wall where the membrane is located. Rejection of natural tomato soluble solids was 94–99%. Flux is 39.7 LMH at 72°C and 55.2 bar after 52 h of operation, and 41 LMH at 78°C and 41.4 bar after 717 h (Merlo et al., 1986).

The first RO plant for tomato juice has operated since 1984 with PCI AFC-99 tubular membranes, processing more than 250 tons per hour (during the season) of 4.5° Brix juice to 8° Brix. Retention of organic acids (citric and L-malic), sugars (glucose, fructose), mineral ions (K, Mg, Na, PO$_4$, Cl), and free amino acids was excellent (Gherardi et al., 1986). Some loss of low molecular weight volatiles (methanol, ethanol) was observed. The color was very good and showed none of the browning normally associated with evaporation. This color quality is retained when evaporated to 28 to 30 Brix paste because of the reduced time in the evaporator. This plant in Italy was operated for five seasons with the same set of membranes, despite aggressive cleaning (Pepper, 1980).

The original work was stimulated by potential energy savings. Thus further development work has been done to reach 15 to 20° Brix, although direct RO may not be the most economical way of reaching this concentration. Using the combination of HR (high rejection) and LR (low rejection) membranes as with the FreshNote process has not been successful so far because the reblended pulp and serum concentrate separate out on standing (Pepper, 1990).

9.6.3 OTHER APPLICATIONS

In the food industry, RO has been evaluated for concentrating egg whites (Conrad et al., 1993; Elaya and Gunasekaran, 2002), concentrating flavor compounds from citrus (Braddock et al., 1991; Kane et al., 1995), mango puree (Olle et al., 1997) and lobster (Jayarajah and Lee, 1999), and for processing coffee and tea extracts, sugar solutions, color pigments, and low-alcohol beer (Cheryan and Alvarez, 1995). Reverse osmosis of maple syrup started in the 1970s in response to increasing energy prices. After removal of about 60% of water from the maple sap, the concentrate is boiled in a conventional open pan evaporator to develop the characteristic color and flavor. For 90,000 gallons of maple sap, this results in a decrease of 33% in the processing cost compared to the all-thermal process (Sendak and Morselly, 1984). The number of installations increased from less than 5 prior to 1980 to more than 100 in 1985 (Gekas et al., 1985). RO has been successfully used in waste treatment and resource recovery applications such as effluents from potato starch processing, sucrose from candy-making waste water, protein from soy whey solutions, and anthocyanins from cranberry pulp wastes. Table 9.8 shows the benefits of using a hybrid process combining RO with evaporation for concentrating corn steep liquor.

To date, the largest application of NF membranes is the removal of hardness and dissolved organics from water. Water hardness is caused by sulfate and bicarbonate salts of calcium and magnesium. Since nanofiltration membranes contain negatively charged hydrophilic groups which are attached to a hydrophobic UF support membrane, they have higher water fluxes compared to RO membranes because of the favorable orientation of water dipoles. Due to the surface active groups, they also have improved fouling resistance against hydrophobic colloids, oils, and proteins. However, the presence of charged groups on the membrane surface can also be a disadvantage. Solutes with a charge opposite to that of the membrane would interact with the membrane, causing fouling. NF membranes are best in applications that utilize both the ability to reject uncharged molecules due to size exclusion and charged components due to electrostatic interaction.

Food and pharmaceutical industries generate many liquids containing NaCl and organics with molecular weights of 300 to 1000. The use of NF to desalt whey was discussed earlier. Baker's yeast is produced commercially by harvesting aerobically grown biomass from a molasses medium.

TABLE 9.8
Energy Requirements for Concentrating Corn Steep Liquor
[Basis: Capacity of 360,000 Gallons per Day, Concentrating
Steep Liquor from 6 to 50% Total Solids (TS), Electricity Cost
of 5¢/kWh, Steam Cost of $5/MMBTU]

Process	Electrical (kw)	Thermal (MMBTU/h)	Cost ($1000/year)
MVR Evaporator	1800	10	1200
RO + MVR Evaporator[a]	500	4	390

Note:
[a] 6–14% total solids by RO (57% of the water removed), 14–50% total solids by
MVR (31% of water removed).

Source: From Gienger, J.K. and Ray, P.J. 1988. *AIChE Symp. Ser.* 84: 168–177.

The process generates heavily polluting effluents; sugars, dark brown pigments, and several other BOD and COD causing species. The effluent is usually processed by evaporation and bacterial digestion to recover the sugars and the residual molasses and used as a cattle feed. Microfiltration was used to replace the centrifuge or filter press for cell harvesting while nanofiltration can be used to treat the wash water. A Nitto-Denko NF membrane gave high flux while rejecting 97% of the COD and 56% of the BOD-causing species, while achieving a 10-fold reduction in volume. NF was more economical than evaporation (Raman et al., 1994).

An RO plant with thin layer composite membranes was installed in a yeast factory in Europe in 1986 (Merry, 1991). However, the membranes were severely fouled and it then was retrofitted with CA membranes. It solved the fouling problem but resulted in bacterial growth due to the need for lower operating temperatures. With the development of nanofiltration, the CA membranes have now been replaced with NF membranes. The membrane is resistant to fouling and does not encourage bacterial growth. It also reduces evaporator corrosion by lowering the salt content.

Preconcentrating sugar solutions from cane or beets by NF increases the capacity of the evaporators and reduces sugar losses in the molasses by 10%. The retentate from the first NF unit goes into the evaporators, thus reducing the load on the evaporators to 65% of the conventional process. The NF permeate is used as a part of the diffusion water for extracting the sugar, without introducing non-sugars into the juice. A second NF unit is used to concentrate excess permeate to a Brix of 20°, which can be used for molasses dilution. RO and BF membranes have been used for the manufacture of pure oligosaccharides (Nakajima et al., 1990; Watanabe et al., 1991) and to recover them from steamed soybean waste water (Matsubara et al., 1996).

Ikeda et al. (1988) suggest the use of NF in the manufacture of amino acids. Amino acids are amphiphilic in nature (i.e., the molecule contains both negative and positive charges). The pH of the surrounding environment affects the charge on the amino acid. For example, aspartic acid, isoleucine, and ornithine have isoelectric points (IEP) of pH 2.8, 5.9, and 9.7 respectively. The rejection of these acids is a function of pH. At a pH lower than the IEP, the amino acids would pass through a negatively charged NF membrane but would be rejected at pH greater than the IEP. For example, at a pH of 5.0, rejection of aspartic acid is about 40%, while rejection of isoleucine and ornithine is less than 10%. Even though the differences in rejection are not very large, it has provided a method for separation of the nucleic acids.

NF is finding increasing application in the recovery of nutraceuticals from extracts of plants, for example, concentrating lutein and zeaxanthin from ethanol extracts of corn (Cheryan, 2001). These extracts tend to be very dilute (20–500 ppm) and the desired compounds are 300–1000 in molecular weight. This makes it ideal for NF or RO which work best with dilute solutions and for separating components in this size range.

NOMENCLATURE

A	Membrane permeability coefficient
A	Virial coefficient in expanded osmotic pressure relationship
b	Friction factor $= (1 + f_{23}/f_{21})$
B	Solute permeability coefficient
c	Concentration
$(c_2)_{\ln}$	Log mean concentration difference
D	Diffusion coefficient
f	Friction coefficient
i	Frictional force
G	Gibbs free energy
H	Hydraulic permeability for pore flow $(\varepsilon r^2/8\eta)$
i	Number of ions in osmotic pressure equation
J	Flux
k	Mass transfer coefficient
K	Distribution coefficient
K^*	Hydraulic permeability coefficient
K^{**}	Total water permeability coefficient
K_2	Generalized solute permeability coefficient
K'	Transport coefficient
L	Phenomenological coefficient
L_p	Hydraulic permeability coefficient
L	Osmotic permeability coefficient
M	Molecular weight
n	Number of moles of component i
N	Number of moles
p	Pressure
P	Vapor pressure
P^*	Diffusive permeability coefficient
P	Transmembrane pressure
P_2	Local solute permeability coefficient
P_h	Specific permeability coefficient
R	Universal gas constant
R	Solute rejection $[1 - (c_2''/c_2')]$
R_{max}	Maximum solute rejection
S	Entropy
t	Tortuosity factor
t	Time
T	Temperature
v	Transverse velocity of liquid
$\bar{v}$	Partial molar volume
X	Axial length
X	Mole fraction
X	Conjugated forces in Onsager relationship
y	Transverse length

GREEK LETTERS

ε	Fractional pore area
κ	Transport coefficient
λ	Effective membrane thickness
η	Viscosity
σ	Reflection coefficient $(\Delta P_T / \Delta \pi)_{Jv=0}$
μ	Solute permeability coefficient $(J_2 / \Delta \pi)_{Jv=0}$
μ	Chemical potential osmotic pressure dissipation function

SUPERSCRIPTS

$'$	Interface conditions on high pressure side of membrane
$''$	Interface conditions on low pressure side of membrane
b	Bulk fluid conditions

SUBSCRIPTS

1	Solvent
2	Solute
m	At the membrane surface
p	Pore
s	Solute
v	Volume
w	Water

REFERENCES

Allegrezza, A.E. 1988. Commercial reverse osmosis membranes and modules. In *Reverse Osmosis Technology*, B.S. Parekh (Ed.), Dekker, NY. pp. 53–120.

Alvarez, V., Alvarez, S., Riera, F.A., and Alvarez, R. 1997. Permeate flux prediction in apple juice concentration by reverse osmosis. *J. Membrane Sci.* 127: 25–34.

Alvarez, S., Riera, F.A., Alvarez, R., and Coca, J. 1998. Permeation of apple juice aroma compounds in reverse osmosis. *Separ. Purif. Technol.* 14: 209–218.

Banks, W. and Sharples, A. 1966. Studies on desalination by reverse osmosis. III. Mechanism of solute rejection. *J. Appl. Chem.* 16: 153–158.

Bailly, M., Roux-de Balmann, H., Aimar, P., Lutin, F., and Cheryan, M. 2001. Production processes of fermented organic acids targeted around membrane operations: Design of the concentration step by conventional electrodialysis. *J. Membrane Sci.* 191: 129–142.

Baker, R.W. 2000. *Membrane Technology and Applications*. McGraw-Hill, NY.

Belfort, G. 1976. A molecular frictional model for transport of uncharged solutes in neutral hyperfiltration and ultrafiltration membranes containing bound water. *Desalination* 18: 259–281.

Bo, F. and Stannett, V. 1976. On the salt rejection of non-ionic polymeric membranes. *Desalination* 18: 113–135.

Braddock, R.J., Nikdel S., and Nagy, S. 1988. Composition of some organic and inorganic compounds in reverse osmosis-concentrated citrus juices. *J. Food Sci.* 53: 508–512.

Braddock, R.J., Sadler, G.D., and Chen, C.S. 1991. Reverse osmosis concentration of aqueous-phase citrus juice essence. *J. Food Sci.* 56: 1027–1029.

Cheryan, M. 1989. Membrane separations: mechanisms and models. In *Food Properties and Computer-Aided Engineering of Food Processing Systems*, R.P.Singh and A.G.Medina (Eds.), Kluwer Academic Publishers, Dordrecht, The Netherlands. pp. 367–392.

Cheryan, M. 1998. *Ultrafiltration and Microfiltration Handbook*. CRC Press, Boca Raton, FL.

Cheryan, M. 2001. Method for extracting xanthophylls from corn. U.S. Patent 6, 169, 217.

Cheryan, M. and Alvarez, J.R. 1995. Food and beverage industry applications. In *Membrane Separations Technology. Principles and Applications*, R.D. Noble and S.A. Stern (Eds.), Elsevier, Amsterdam. pp. 415–465.

Cheryan, M. and Nichols, D.J. 1992. Modelling of membrane processes. In *Mathematical Modelling of Food Processes*, S. Thorne (Ed.), Elsevier, London, UK. pp. 49–98.

Cheryan, M., Sarma, S.C., and Pal, D. 1987. Energy considerations in the manufacture of khoa by reverse osmosis. *Asian J. Dairy Res.* 6: 143–153.

Cheryan, M., Veeranjaneyulu, B., and Schlicher, L.R. 1990. Reverse osmosis of milk with thin-film composite membranes. *J. Membrane Sci.* 48: 103–114.

Chua, H.T., Rao, M.A., Acree, T.E., and Cunningham, D.H. 1987. Reverse osmosis concentration of apple juice: flux and flavor retention by cellulose acetate and polyamide membranes. *J. Food Process Engr.* 9: 231–245.

Chua, H.T., Rao, M.A., Acree, T.E., and Cunningham, D.H. 1988. Reverse osmosis concentration of apple juice: flux and flavor retention by cellulose acetate and polyamide membranes. *J. Food Process Engr.* 9: 231–245.

Conrad, K.M., Mast, M.G., and Ball, H.R. 1993. Concentration of liquid egg white by vacuum evaporation and reverse osmosis. *J. Food Sci.* 58: 1017–1020.

Cross, S. 1988. Achieving 60 Brix with membrane technology. Presented at the 49th Annual Meeting, Institute of Food Technologists, New Orleans, June 19–22.

Demeczky, M., Khell-Wicklein, M., and Godek-Kerek, E. 1981. The preparation of fruit juice semi-concentrates by reverse osmosis. In *Developments in Food Preservation*. pp. 93–119.

Dickson, J.M. 1988. Fundamental aspects of reverse osmosis. In *Reverse Osmosis Technology*, B.S. Parekh (Ed.), Dekker, NY. pp. 1–52.

Elata, C. 1969. The determination of the intrinsic characteristics of reverse osmosis membranes. *Desalination* 6: 1–12.

Elaya, M.M.O. and Gunasekaran, S. 2002. Gelling properties of egg white produced using a conventional and a low-shear reverse osmosis process. *J. Food Sci.* 67: 725–729.

Eisenberg, T.N. and Middlebrooks, E.J. 1986. *Reverse Osmosis of Drinking Water*, Butterworths, Stoneham, MA.

Gekas, V., Hallstrom, B., and Tragardh, G. 1985. Food and dairy applications: the state of the art. *Desalination* 53: 95–127.

Gherardi, S., Bazzarini, R., Trifiro, A., Voi, A.L., and Palamas, D. 1986. Pre-concentration of tomato juice by reverse osmosis. *Int. Fruchtstaft-Union,Wiss.-Tech. Komm.,[Berlin].* 19: 241–252.

Gienger, J.K. and Ray, P.J. 1988. Membrane-based hybrid processes. *AIChE Symp. Ser.* 84: 168–177.

Girard, B. and Fukumoto, L.R. 2000. Membrane processing of fruit juices and beverages. *Crit. Rev. Biotechnol.* 20: 109–175.

Groepl, R. and Pusch, W. 1970. Asymmetric behavior of cellulose acetate membranes in hyperfiltration experiments as a result of concentration polarization. *Desalination* 8: 277–292.

Hwang, S. and Pusch, W. 1981. Asymptotic solute rejection in reverse osmosis. In *Synthetic Membranes: Volume I, Desalination*, A.F. Turbak (Ed.), ACS Symposium Series 153, American Chemical Society, Washington, DC. pp. 253–266.

IDFA. 2001. *Milk Facts 2001*. International Dairy and Food Association, Washington, DC.

Ikeda, K., Nakano, T., Ito, H., Kubota, T., and Yamamoto, S. 1988. New composite charged reverse osmosis membrane. *Desalination* 68: 109–119.

Jagur-Grodzinski, J. and Kedem, O. 1966. Transport coefficients and salt rejection in uncharged hyperfiltration membranes. *Desalination* 1: 327–341.

Jayarajah, C.N. and Lee, C.M. 1999. Ultrafiltration/reverse osmosis concentration of lobster extract. *J. Food Sci.* 64: 93–98.

Johnson, J.S., Dresner, L., and Kraus, K.A. 1966. Hyperfiltration (reverse osmosis). In *Principles of Desalination*, K.S. Spiegler (Ed.), Academic Press, NY. pp. 346–439.

Jonsson, G. 1978. Methods for determining the selectivity of reverse osmosis membranes. *Desalination* 24: 19–37.

Jonsson, G. 1980a. Overview of theories for water and solute transport in UF/RO membranes. *Desalination* 35: 21–38.

Jonsson, G. 1980b. The influence of the porous sublayer on the salt rejection and reflection coefficient of asymmetric cellulose acetate membranes. *Desalination* 34: 141–157.

Jonsson, G. and Boesen, C.E. 1975. Water and solute transport through cellulose acetate reverse osmosis membranes. *Desalination* 17: 145–165.

Kane, L., Braddock, R., and Sims, C.A. 1995. Lemon juice aroma concentration by reverse osmosis. *J. Food Sci.* 60: 190–194.

Kedem, O. and Katchalsky, A. 1958. Thermodynamic analysis of the permeability of biological membranes to non-electrolytes. *Biochimica et Biophysica Acta* 27: 229–246.

Kedem, O. and Katchalsky, A. 1963. Permeability of composite membranes. Part 3. Series array of elements. *Trans. Faraday Soc.* 59: 1941–1953.

Kesting, R.E. 1985. *Synthetic Polymeric Membranes: A Structural Perspective.* John Wiley & Sons, NY.

Kimura, S. and Sourirajan, S. 1967. Analysis of data in reverse osmosis with porous cellulose acetate membranes used. *AIChE J.* 13: 497–503.

Koseoglu S.S., Lawhon, J.T., and Lusas, E.W. 1990. Use of membranes in citrus juice processing. *Food Technol.* 44: 90–97

Lee, C.H. 1975. Theory of reverse osmosis and some other membrane permeation operations. *J. Appl. Polym. Sci.* 19: 83–95.

Lloyd, D.R. 1985. *Materials Science of Synthetic Membranes*, American Chemical Society, Washington, DC.

Loeb, S. and Sourirajan, S. 1960. Sea water demineralization by means of semi-permeable membranes. UCLA Engineering Report No. 60.

Loeb, S. and Sourirajan, S. 1962. Sea water demineralization by means of an osmotic membrane. *Advan. Chem. Ser.* 38: 117–132.

Lonsdale, H.K. 1972. Theory and practice of reverse osmosis and ultrafiltration. In *Industrial Processing with Membranes*, R.E. Lacey and S. Loeb (Eds.), Wiley-Interscience, NY. pp. 123–178.

Lonsdale, H.K., Merten, U., and Riley, R.L. 1965. Transport properties of cellulose acetate osmotic membranes. *J. Appl. Polym. Sci.* 9: 1341–1362.

Lonsdale, H.K., Cross, B.P., Graber, F.M., and Milstead, C.E. 1971. Permeability of cellulose acetate membranes to selected solutes. In *Permselective Membranes*, C.E. Rogers (Ed.), Dekker, NY.

Lui, K.W. 1978. Experimental investigation of intramembrane transport models in reverse osmosis. M.S. Thesis, State University of New York at Buffalo, NY.

Matsubara, Y., Iwasaki, K.I., Nakajima, M., Nabetani, H., and Nakao, S.I. 1996. Recovery of oligosaccharides from steamed soybean waste water in tofu processing by reverse osmosis and nanofiltration membranes. *Biosci. Biotechnol. Biochem. J.* 60: 421–428.

Matsuura, T. and Sourirajan, S. 1973. Reverse osmosis separation of organic acids in aqueous solutions using porous cellulose acetate membranes. *J. Appl. Polym. Sci.* 17: 3661–3682.

Matsuura, T., Blais, P., Dickson, J.M., and Sourirajan, S. 1974. Reverse osmosis separations for some alcohols and phenols in aqueous solutions using aromatic polyamide membranes. *J. Appl. Polym. Sci.* 18: 3671–3683.

Medina, B.G. and Garcia, A. 1988. Concentration of orange juice by reverse osmosis. *J. Food Process Engr.* 10: 217–230.

Merlo, C.A., Rose W.W., Petersen, L.D., White, E.M., and Nicholson, J.A. 1986. Hyperfiltration of tomato juice. Pilot-scale high temperature testing. *J. Food Sci.* 51: 403–407.

Merry, A.J. 1991. A case study of reverse osmosis applied to concentration of yeast effluent. In *Effective Industrial Membrane Processes — Benefits and Opportunities*, M.K. Turner (Ed.), Elsevier, London.

Merson, R.L. and Morgan, A.I. 1968. Juice concentration by reverse osmosis. *Food Technol.* 22: 97–100.

Merten, U. 1966. Transport properties of osmotic membranes. In *Desalination by Reverse Osmosis*, U. Merten (Ed.), MIT Press, Cambridge, MA.

Milleville, H.P. 1990. Direct osmosis concentrates juices at low temperature. *Food Process.* 51: 70–71.

Nakajima, M., Watanabe, A., Nabeya, H., Hiadaka, H., and Otusuka, R. 1990. Manufacture of highly pure fructo-oligosaccharides with fructosyltransferase and loose reverse osmosis. Japan Kokai, 90 050289.

Olle, D., Baron, A., and Lozana, Y.F. 1997. Microfiltration and reverse osmosis affect recovery of mango puree flavor compounds. *J. Food Sci.* 62: 1116–1119.

Orofino, T.A., Hopfenberg, H.B., and Stannett, V. 1969. Characterization of penetrant clustering in polymers. *J. Macromol. Sci. Phys.* B3: 777–778.

Paul, D.R 1976. The solution–diffusion model for highly swollen membranes. *Separation and Purification Methods* 5: 33–50.

Pepper, D. 1980. Membrane life, capacity and costs. *Desalination* 35: 383–396.

Pepper, D. 1990. RO for improved products in the food and chemical industries and water treatment. *Desalination* 77: 55–71.

Pepper, D., Orchard, A.C.J., and Merry, A.J. 1985. Concentration of tomato juice and other fruit juices by reverse osmosis. *Desalination* 53: 157–166.

Peri, C. 1974. Concentration of orange juice by reverse osmosis. *Sci.e.Tec. Degli Alimenti.* 4: 43–47.

Pusch, W. 1977. Determination of transport parameters of synthetic membranes by hyperfiltration experiments. Part I: Derivation of transport relationship from the linear relations of thermodynamics of irreversible processes. *Ber. Bunsenges. Physik. Chem.* 81: 269–276.

Rajagopalan, N., Cheryan, M., and Matsuura, T. 1994. Recovery of diacetyl by pervaporation. *Biotechnol. Tech.* 8: 869–872.

Rajagopalan, N. and Cheryan, M. 1995. Pervaporation of grape juice aroma. *J. Membrane Sci.* 104: 243–250.

Raman, L.P., Cheryan, M., and Rajagopalan, N. 1994. Consider nanofiltration for membrane separations. *Chem. Engr. Progr.* 90: 68–74.

Reid, C.E. 1972. Principles of reverse osmosis. In *Industrial Processing with Membranes*, R.E. Lacey and S. Loeb (Eds.), Wiley-Interscience, NY.

Reid, C.E. and Breton, E.J. 1959. Water and ion flow across cellulose acetate membranes. *J. Appl. Polymer Sci.* 1: 133–143.

Ridgway, H.F. 1988. Microbial adhesion and biofouling of reverse osmosis membranes. In *Reverse Osmosis Technology*, B.S. Parekh (Ed.), Dekker, NY. pp. 429–482.

Ripperger, S. and Schulz, G. 1986. Microporous membranes in biotechnical applications. *Bioprocess Engr.* 1: 43–49.

Sendak, P.E. and Morselly, M.F. 1984. Reverse osmosis in the production of maple syrup. *Forest Products J.* 34: 57–61.

Sherwood, T.K., Brian, P.L.T., and Fisher, R.E. 1967. Desalination by reverse osmosis. *Ind. Eng. Chem. Fund.* 6: 2–12.

Sheu, M.J. and Wiley, R.C. 1983. Preconcentration of apple juice by reverse osmosis. *J. Food Sci.* 48: 422–428.

Sheu, M.J. and Wiley, R.C. 1984. Influence of reverse osmosis on sugar retention in apple juice concentration. *J. Food Sci.* 48: 422–428.

Singh, N. and Cheryan, M. 1997. Membrane applications in corn wet milling. *Cereal Foods World.* 42: 520–525.

Soltanieh, M. and Gill, W.N. 1981. Review of reverse osmosis membranes and transport models. *Chem. Eng. Commun.* 12: 279–363.

Sourirajan, S. 1970. *Reverse Osmosis.* Academic Press, NY.

Sourirajan, S. 1977. Reverse osmosis — a general separation technique. In *Reverse Osmosis and Synthetic Membranes. Theory-Technology-Engineering*, S. Sourirajan (Ed.), National Research Council of Canada Publications, Ottawa, Canada.

Sourirajan, S. and Matsuura, T. 1977. Physicochemical criteria for reverse osmosis separations. In *Reverse Osmosis and Synthetic Membranes, Theory-Technology-Engineering*, S. Sourirajan (Ed.), National Research Council of Canada Publications, Ottawa, Canada.

Sourirajan, S. and Matsuura, T. 1982. Science of reverse osmosis — an essential tool for the chemical engineer. *Chem. Eng.* 385: 359–376.

Sourirajan, S. and Matsuura, T. 1984. *Reverse Osmosis and Ultrafiltration*, National Research Council of Canada, Ottawa, Canada.

Spiegler, K.S. and Kedem, O. 1966. Thermodynamics of hyperfiltration (reverse osmosis): criteria for efficient membranes. *Desalination* 1: 311–326.

Walker, J.B. 1990. Membrane process for the production of superior quality fruit juice concentrates. *Proceedings of the 1990 International Congress on Membranes and Membrane Processes*, Vol. 1, pp. 283–285.

Watanabe, S., Nabetani, H., Nakajima, M., Ohmori, T., Yamada, Y., and Isiguro, Y. 1990. Development of multi-stage RO combined system (MRC) for high concentration of apple juice. *Proceedings of the 1990 International Congress on Membranes and Membrane Processes*, Vol. 1, p. 282.

Watanabe, A., Nakajima, M., Nabeya, H., Otusuka, R., and Hirayama, M. 1991. Manufacture of highly pure oligosaccharides from a sugar mixture by loose reverse osmosis membranes. *Japan Kokai* 91: 77896.

Yasada, H. and Lamaze, C.E. 1971. Salt rejection by polymer membranes in reverse osmosis. I. Nonionic polymers. *J. Appl. Polym. Sci. Part A-2* 9: 1537–1551.

10 Food Dehydration

Martin R. Okos, Osvaldo Campanella, Ganesan Narsimhan,
Rakesh K. Singh, and A.C. Weitnauer

CONTENTS

10.1 INTRODUCTION

The purpose of drying food products is to allow longer periods of storage with minimized packaging requirements and reduced shipping weights. The quality of the product and its cost are greatly influenced by the drying operation. The quality of a food product is judged by the amount of physical and biochemical degradation occurring during the dehydration process. The drying time, temperature, and water activity influence the final product quality. Low temperatures generally have a positive influence on the quality but require longer processing times. Low water activity retards or eliminates the growth of microorganisms, but results in higher lipid oxidation rates. Maillard (nonenzymatic) browning reactions peak at intermediate water activities (0.6 to 0.7), indicating the need for a rapid transition from medium to high water activities (Franzen, 1988).

Many dried foods are rehydrated before consumption. The structure, density and particle size of the food plays an important role in reconstitution. Ease of rehydration is increased with decreasing particle size, and the addition of emulsifiers such as lecithin or surfactants. Processing factors which affect structure, density, and rehydration include puffing, vacuum, foaming, surface temperature, low temperature processing, agglomeration, and surface coating (King, 1974).

Storage stability of a food product increases as the water activity decreases, and the products that have been dried at lower temperatures exhibit good storage stability. Since lipid-containing foods are susceptible to lipid oxidation at low water activities, these foods must be stored in oxygen impermeable packages. Poor color retention has been a problem in the freeze-drying of coffee because the number of light-reflecting surfaces is decreased during rapid drying. This problem has been improved by slow freezing, partial melting, and refreezing to insure large ice crystal formation (King, 1971). Other food materials have different drying problems and specific solutions must be developed.

The following goals for drying foods have been summarized by King (1974):

1. *Product quality*
   Minimal chemical and biochemical degradation reactions
   Selective removal of water over other salts and volatile flavor and aroma substances
   Maintenance of product structure (for a structured food)

Control of density
Rapid and simple rehydration or redispersion
Storage stability: less refrigeration and packaging requirements
Desired color
Lack of contamination or adulteration
2. *Process economics*
Minimal product loss
Rapid rate of water removal (high capacity per unit amount of drying equipment)
Inexpensive energy source (if phase change is involved)
Inexpensive regeneration of mass separating agents
Minimal solids handling problems
Facility of continuous operation
Noncomplex apparatus (reliable and minimal labor requirement)
3. *Other*
Minimal environmental impact

10.2 DRYING FUNDAMENTALS

Drying is defined as a process of moisture removal due to simultaneous heat and mass transfer. Heat transfer from the surrounding environment evaporates the surface moisture. The moisture can be either transported to the surface of the product and then evaporated or evaporated internally at a liquid vapor interface and then transported as vapor to the surface.

The transfer of energy (heat) depends on the air temperature, air humidity, air flow rate, exposed area of food material, and pressure. The physical nature of the food, including temperature, composition, and in particular moisture content, governs the rate of moisture transfer. The dehydration equipment generally utilizes conduction, convection, or radiation to transfer energy from a heat source to the food material. The heat is transferred directly from a hot gas or indirectly through a metal surface.

The model equations for dryers cannot be discussed without a thorough understanding of the basic heat and mass transfer concepts. The typical drying cycle consists of three stages: heating the food to the drying temperature, evaporation of moisture from the product surface occurring at a rate proportional to the moisture content, and once the critical moisture point is reached, the falling of the drying rate. The critical moisture point depends greatly on the drying rate since high drying rates will raise the critical point and low drying rates will decrease them. Terminology and basic concepts associated with drying will be discussed to facilitate proper selection of dryers.

1. *Partial pressure of a liquid contained in a gas*

$$P_w V_w = NRT \tag{10.1}$$

where P_w is the partial pressure of moisture vapor, P_a, V_w is specific molar volume m^3, N is number of moles of the gas, R is gas constant, J/mol·K, and T is the temperature, K

2. *Humidity*

$$H = \frac{P_w}{P - P_w} \frac{m_w}{m_g} \tag{10.2}$$

where H is the humidity, m_w is mass of moisture vapor, kg, m_g is mass of the dry air, kg, P is total pressure, Pa, and P_w is the partial pressure of moisture vapor, Pa.

When the partial pressure of the vapor in the gas equals the vapor pressure of the liquid, the gas becomes saturated.

$$H_s = \frac{P_w^o}{P - P_w^o} \frac{m_w}{m_g}$$ (10.3)

where P_w^o is the saturated vapor pressure, Pa

3. *Relative humidity*: the measure of moisture saturation at a given temperature:

$$\Psi = \frac{P_w}{P_w^o}$$ (10.4)

10.2.1 UNSATURATED VAPOR-GAS MIXTURES

1. Dry bulb temperature, T_B: temperature determined by an ordinary thermometer.
2. Wet bulb temperature, T_W: temperature measured when a gas passes rapidly over a wet thermometer bulb. Used along with dry bulb temperature to measure relative humidity of a gas.
3. Dew point, T_{DP}: temperature at which a vapor condenses its first drop of liquid: At this temperature the dry bulb temperature equals the wet bulb temperature.
4. Moisture content, W: Amount of water contained in the product. It can be calculated either on a dry (W_d) or a wet basis (W_w) by dividing the weight of water contained in the sample by the weight of the dry or wet solid, respectively. Moisture contents in dry or wet basis can be related by the following equations.

$$W_w = \frac{W_d}{W_d + 1}$$ (10.5)

or

$$W_d = \frac{W_w}{1 - W_w}$$ (10.6)

where W_w is the weight of water per kilogram of wet material and W_d is the weight of water per kilogram of dry material.

5. Unbound moisture: moisture in excess of the equilibrium moisture content corresponding to saturated humidity.
6. Bound moisture: amount of moisture tightly bound to the food matrix with properties different from those of bulk water (also known as unfreezable water).
7. Free moisture content: amount of moisture mechanically entrapped in the void spaces of the system, having nearly all properties similar to those of bulk water.
8. Equilibrium moisture content: moisture content of a product in equilibrium with the surrounding temperature and humidity conditions.
9. Water activity: an index of the availability of water for chemical reactions and microbial growth (Banwart, 1981).
10. Enthalpy: the heat energy content of the air, a relative measure of internal energy and the flow work per unit mass.

10.2.2 EQUILIBRIUM MOISTURE CONTENT

Air is the gas most commonly used for drying foods and its humidity content is important when predicting drying conditions. When air is brought in contact with a wet food material, an equilibrium

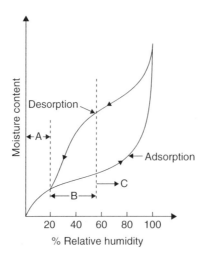

FIGURE 10.1 Sorption isotherm for typical food product. (From Fortes, M. and Okos, M.R. 1980. *Adv. Drying*, 119–150.)

between the water present in the air and contained in the food material is eventually reached. The moisture content of the food under such a condition is called the equilibrium moisture content. A plot of the equilibrium moisture content vs. the relative humidity of the air at different temperatures is often used to illustrate the effect of temperature. The equilibrium moisture content usually decreases with an increase in temperature and is dependent on whether the amount of humidity is changed from high to low (drying) or from low to high. Hence different curves are plotted for desorption and adsorption. The difference in the equilibrium moisture content between the adsorption and desorption curves is called *hysteresis*. Hysteresis is generally observed in most hygroscopic products. A product is hygroscopic if it is able to bind water when the vapor pressure is lowered. Water activity is a term commonly used in the food area. It can be calculated as the ratio of the vapor pressure of the water in the food and the vapor pressure of pure water at the temperature of the food. At equilibrium conditions the water activity of the food is equal to the relative humidity of the air in contact with the food. The degree of water interactions with the food is determined by moisture content and water activity.

A typical sorption isotherm for food products is shown in Figure 10.1. The curve is divided into three regions. In region A, the water is tightly bound to the food product. In region B, the water is less tightly held and usually present in small capillaries. In region C, the water is held loosely in large capillaries, or is free (Fortes and Okos, 1980). Foods high in protein, starch, and high-molecular-weight polymers have higher equilibrium moisture contents than those high in soluble solids, crystalline salts, and sugars. Sorption isotherms vary drastically from one food product to another. Shown in Figure 10.2 (Menon and Mujumdar, 1987) are equilibrium moisture contents vs. water activity ψ for a variety of foods.

10.3 WATER SORPTION ISOTHERMS OF FOODS AND FOOD COMPONENTS

Water sorption by foods is a process wherein water molecules progressively and reversibly combine with food solids via chemisorption, physical adsorption, and multilayer condensation (van den Berg and Bruin, 1981a). The sorption isotherm of a food material is a curve showing the equilibrium moisture content vs. the relative humidity or water activity of the vapor space surrounding the material. It usually presents type II isotherm which is shown in Figure 10.3. The concept of water activity that is used most commonly by researchers and processors in the food industry can be defined

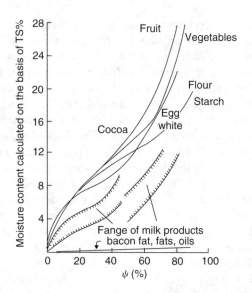

FIGURE 10.2 Typical equilibrium moisture content of food materials at 25°C. (From Menon, A.S. and Mujumdar, A.S. 1987. In *Handbook of Industrial Drying*, Marcel Dekker, New York, pp. 3–45.)

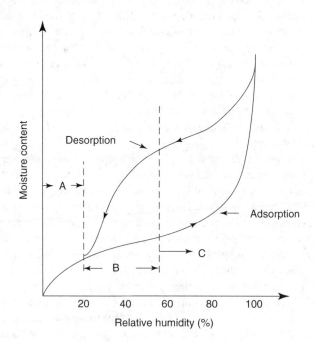

FIGURE 10.3 General type of sorption isotherm for food products.

by following equation:

$$a_w = \frac{P}{P_o} = \frac{\text{relative humidity } (\%)}{100} \tag{10.7}$$

where P = water vapor pressure in food material, P_o = vapor pressure of pure water at the temperature of the food, and a_W = water activity.

An isotherm can be divided into three regions as seen in Figure 10.3. The water in region A represents strongly bound water with an enthalpy of vaporization considerably higher than pure water. These first water molecules are adsorbed at hydrophilic, charged, and polar groups of food components (protein, polysaccharides), which include structural water (Hydrogen-bonded water), hydrophobic hydration water and monolayer water (Kinsella and Fox, 1986). The presence of monolayer water has been recently challenged (Karel and Lund, 2003) and although it has been recognized that it is unlikely that a water monolayer exists, the concept is still used for interpretation of fitting parameters associated with models that are used to describe sorption phenomena. Usually, the water in these regions is unfreezable and is not available for chemical reactions or as a plasticizer. In region B, water molecules bind less firmly than the first. The vaporization enthalpy is little greater than that for pure water. This water is available as a solvent for low-molecular-weight solutes and for some biochemical reactions. This class of constituent water can be looked upon as the continuous transition of the bound to the free type of water. The properties of water in region C are pretty close to those of pure water and no excess heat of binding can be detected. That water, also known as free water, may be held in voids, crevices, capillaries, and is loosely bound to the food material.

The process of increasing moisture content (water gain) is termed *adsorption*, and that of decreasing moisture content (water loss), *desorption*. The term describing the difference between adsorption and desorption is *hysteresis*. Generally, adsorption isotherms exhibit a lower moisture content than do desorption isotherms at a given water activity, possibly because of the structure of the food or because equilibrium had not been reached during experimental measurement. In foods, a variety of hysteresis loop shapes have been observed, depending on the type of food and the temperature (Wolf et al. 1972). In high sugar/high pectin foods such as air-dried apple, hysteresis occurs mainly in the monomolecular layer of water region (refer to Figure 10.4a). Although the total hysteresis is large, there is no hysteresis above $a_w = 0.65$. In high-protein foods such as pork, a moderate hysteresis begins at about $a_w = 0.85$ (i.e., in the capillary condensation region; Figure 10.4b) (Kapasalis, 1981) and extends over the rest of the isotherms to the zero water activity. In both adsorption and desorption, the isotherms retain the characteristic sigmoid shape for proteins. In starchy foods (Figure 10.4c), a large hysteresis loop occurs with a maximum at about $a_w = 0.70$, which is within the capillary condensation region. Increasing temperature decreases the total hysteresis (Wolfe et al., 1972). Desorption isotherms usually give a higher water content than do adsorption isotherms. In general, the type of changes encountered upon adsorption and desorption will depend on the initial state of the sorbent (amorphous vs. crystalline), the transitions taking place during adsorption, and the speed of desorption (Kapasalis, 1981). Hysteresis seems to be reproducible and persistent over many adsorption–desorption scans, especially at low temperatures and over relatively short periods of time (Benson and Richardson, 1955; Strasser, 1969). However, at higher temperatures, this may not be the case (Chung, 1966), due probably to denaturation of proteins. Elimination of hysteresis upon the second or subsequent cycles may take place for a variety of reasons, such as change in the crystalline structure when a new crystalline form persists upon subsequent cycles, swelling, and increased elasticity of capillary walls, resulting in a loss of water-holding capacity (Rao, 1939a, 1939b). Another factor involved in determining the shape of the curve is temperature. As discussed in Section 3.3, increased temperature decreases the moisture content. Also, as the moisture content decreases, the heat of vaporization increases.

In addition to the physical factors in sorption isotherms, chemical factors influence water activity. Hydrogen-bond formation is one such factor. Also, the presence of dissolved solutes affects water activity. Differences between electrolytic and nonelectrolytic solutions as well as amount of positively and negatively charged ions play a role in the sorption process.

10.3.1 BOUND WATER

Many experimental results show that some water molecules have different kinetic and thermodynamic properties from ordinary water (e.g., they exhibit a lower vapor pressure, lower mobility, greatly

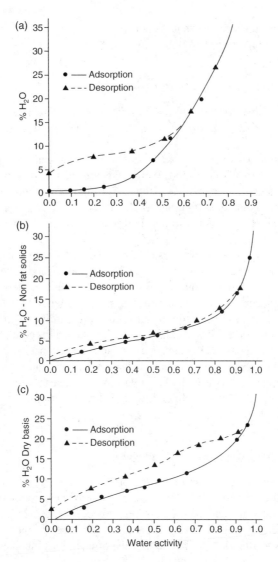

FIGURE 10.4 Examples of sorption hysteresis in foods.

reduced freezing point, etc.). Such water has frequently been referred to as bound water. This definition is an operational one that depends on the physical properties studied and the techniques used (Fennema, 1976). Bound water contents may be different for the same foods if different criteria are used for its identification (Leung, 1986). Nowadays it is widely recognized that bound water is a questionable term that has caused more confusion than understanding (Walstra, 2003).

Kuprianoff (1958) stated that bound water may best be determined by measuring the water that cannot be frozen at subfreezing temperatures. Under this definition, Toledo et al. (1968) measured the unfreezable water content in wheat flour with wide-line NMR. The results showed that the signals from all samples with different moisture content remained constant as the temperature decreased to less than 28°F. Below this temperature, the bound water fell to a common value that is equivalent to 25% (DB) moisture. Berlin et al. (1970) assessed the amount of unfreezable water in proteins (Bovine casein) through DSC and found such bound water was to be 50 to 60% of the dry weight of protein. From the water sorption isotherms of these materials, interestingly, it can be seen that unfreezable water of foods, in general, corresponds to equilibrium moisture content at water activities in a range 0.8 to 0.9. Thus unfreezable water is not strongly bound to food materials and is available for chemical reactions and microbial growth (Leung, 1986).

10.3.2 MEASUREMENT OF SORPTION ISOTHERMS

Many methods are available for determining water sorption isotherms (Gal, 1981). Basically, these methods can be classified under three categories: (1) gravimetric, (2) manometric, and (3) hygrometric.

Gravimetric methods involve the registration of weight changes upon changes in the relative humidity of the air in contact with the food. Weight changes can be determined both continuously and discontinuously in static and dynamic systems. Continuous methods employ the use of electrobalances or quartz spring balances (Guillard et al., 2004). In discontinuous systems, sulfuric acid or salt solution is placed in a vacuum with the food material to give a measure of the equilibrium relative humidity.

As the name implies, manometric methods involve the use of sensitive manometers. Manometric devices measure the vapor pressure of water in equilibrium with a food material at a given moisture content. To improve accuracy, the fluid selected for the manometer is often oil instead of mercury because the relative displacement of oil is an order of magnitude greater than that of mercury.

Hygrometric methods measure the equilibrium relative humidity of air in contact with a food material at a given moisture content. Dew-point hygrometers detect the condensation of cooling water vapor. Electric hygrometers measure the change in conductance or capacitance of hygrosensors. Most hygrosensors are coated with a hygroscopic salt, such as LiC1, which absorbs moisture from the food sample. Another hygrometric device is the hair hygrometer. Measurements are determined by the stretching of human hair as it is exposed to high water activities. Instruments based in this principle are uncommon in the field.

10.3.3 MATHEMATICAL DESCRIPTION OF ISOTHERMS

Experimental food isotherms can be fitted to one or more of a host of theoretical and empirical models. No single equation has been found to depict accurately the sorption isotherms of all types of foods in the entire range of water activity. Such being the case, isotherm data are studied individually and the model that describes the behavior most accurately is used to report the water activity and moisture content for a specific food. Following is a list of equations that have been studied, as well as a definition of the terms used in the models.

Bradley's equation (1936) is based on the theory of polarization. The dipoles induced in the first layer of absorbed molecules induce more dipoles, and so on. The constants included in the equation are temperature dependent and the equation is not very accurate above 0.7 to 0.85 a_w.

$$\ln \frac{1}{a_w} = B(2)B(1)^X \tag{10.8}$$

The BET equation (1938) is used most frequently to characterize isothermal of foods. The equation contains only one constant, C, which is related to the net heat of sorption. This equation best fits data in the low water activity region, 0.45 to 0.5 a_w. Above 0.5 a_w this theory no longer holds.

$$\frac{a_w}{(1 - a_w)X} = \frac{1}{X_M C} + \frac{a_w(C - 1)}{X_M C} \tag{10.9}$$

The GAB equation (1966) is a multilayer model that takes into account different properties of water in the multilayer region, where X_M is the monolayer moisture content. It is considered one of the best equations for fitting isotherms of many food materials. It has two constants and requires at least five experimental data points, but is good up to water activities of 0.94.

$$\frac{X}{X_M} = \frac{C_G K a_w}{(1 - K a_w)[1 + (C_G - 1)K a_w]} \tag{10.10}$$

Oswin's equation (1946) is a series expansion for S-shaped curves.

$$X = B(2) \left(\frac{a_w}{1 - a_w}\right)^{B(1)} \tag{10.11}$$

Smith's equation (1947) is useful in describing isotherms of various bipolymers and food products of 0.3 to 0.5 a_w and higher.

$$X = B(2) - B(1) \ln(1 - a_w) \tag{10.12}$$

Halsey's equation (1948) provides an expression for condensation multilayers at a relatively long distance from the surface. It can be used for a variety of foods and food components with water activities from 0.1 to 0.8.

$$a_w = \exp\left[-\frac{B(2)}{X^{B(1)}}\right] \tag{10.13}$$

Henderson's equation (1952) is good for describing globular proteins and is widely used as an empirical equation for fitting water sorption isotherms to foods.

$$1 - a_w = \exp\{-[B(2)X^{B(1)}]\} \tag{10.14}$$

Kuhn's equation (Labuza et al., 1972) exhibits inconsistencies at low a_w values.

$$X = \frac{B(1)}{\ln a_w} + B(2) \tag{10.15}$$

Iglesias and Chirife's equation (1978) is designed to describe high-sugar foods.

$$\ln[X + (X^2 + X_{0.5})^{1/2}] = B(1) + B(2) \tag{10.16}$$

Iglesias and Chirife's equation (1981):

$$X = B(1) \left(\frac{a_w}{1 - a_w}\right) + B(2) \tag{10.17}$$

Modified Halsey

$$a_w = \exp\left(\frac{B(2)}{X^{B(1)}}\right) \tag{10.18}$$

Modified Oswin

$$a_w = \frac{1}{\left(\frac{B(2)}{X}\right)^{B(1)} + 1} \tag{10.19}$$

Modified Henderson

$$1 - a_w = \exp(B(2)X^{B(1)}) \tag{10.20}$$

where a_w is the water activity, X is moisture content, dry basis, X_M is monolayer moisture content, $X_{0.5}$ is moisture content at 0.5 a_w, $B(1)$ is statistically determined constant, $B(2)$ is statistically determined constant, C_G is Guggenheim–Anderson constant, K is constant, and C is the constant.

Tests by Lomauro et al. (1985) showed that the GAB equation gave good fits to over 75% of food isotherms studied. The Oswin equation fits 57% of the food isotherms. The Iglesias and Chirife equations require knowledge of moisture content at $0.5a_w$. This value is difficult to determine; thus guesswork is involved when using the Iglesias and Chirife equations.

Peleg and Normand (1992) developed an equation based on a polynomial to calculate the moisture content of the material as a function of the water activity. The paper also discusses the use of that polynomial to estimate water activities of a mixture of ingredients. A website has been developed wherein calculations can be carried out in an excel® environment.

The website is: (http://www-unix.oit.umass.edu/~aew2000/wateractivity.html)

10.3.4 ISOTHERM DATA

Isotherm sorption data have been published for many different foods. Reports have been submitted in many journals for various types of foods. *Handbook of Food Isotherms: Water Sorption Parameters for Food and Food Components*, a text authored by Iglesias and Chirife (1982), is a source of information regarding many foods, ranging from beef to yogurt. All water sorption data contained therein are presented both graphically and statistically. Some of them are introduced in Table 10.1 for quick reference. A source of information regarding food isotherms is *Sorption Isotherms and Water Activity of Food Materials*. This book is a bibliography compiled by Wolf et al. (1985) and can be used to reference the concepts of the sorption process as well as sorption data of food products. An article by van den Berg and Bruin (1981) lists 77 mathematical models for food sorption isotherms. These constants were obtained from *Handbook of Food Isotherms* (Iglesias and Chirife, 1982). In addition, the GAB isotherm equation has been fitted to sorption isotherm data of a variety of food products. Table 10.2 lists the values of the constants.

Recently Bell and Labuza (2000) summarized most of the information concerning water isotherm sorption of selected food materials.

10.3.5 FACTORS AFFECTING WATER BINDING

10.3.5.1 Temperature

The amount of water adsorbed usually decreases as temperature increases. In the sorption isotherm, this effect shows a downward shift (Berlin et al., 1973). Noguchi (1981) reported that hydrophobic hydration of biopolymers (dextran) decreases rapidly with increasing temperature (e.g., the water content decreased from 4 mol per mole sorbate to 2 as temperature was increased from 20 to 40°C). There is a possibility that high temperature/high humidity conditions may cause degradation in the substrate.

For constant moisture contents, water activity increases with rising temperatures and the temperature effect on isotherms follows the Clausius–Clayperon equation:

$$\left[\frac{\partial \ln a_w}{\partial (1/T)}\right]_m = -\frac{E_b}{R} \tag{10.21}$$

where E_b is the binding energy of water in cal/mol or J/mol and R, gas constant, is 1.987 cal/mol·K or 8.314 J/mol·K. The binding energy is defined as the difference between the heat of adsorption of

TABLE 10.1
Constants for Isotherm Equations

Food product	Specifications	Average range	Eq.	B(1)	B(2)	Xm	Source
Almond, California	15°C des.	0.37–0.91	12	5.1008	1.2543	—	
	25°C des.	0.39–0.91	12	5.2167	0.9873	—	
	35°C des.	0.41–0.91	13	1.6457	9.4005	—	
Anise (seed)	5°C ads.	0.05–0.80	19	0.4970	8.9980	—	Soysal and Öztekin (2001)
	25°C ads.	0.05–0.80	19	0.5260	8.3740	—	
	45°C ads.	0.05–0.80	19	0.6350	7.4860	—	
Barley malt, Six-row	15°C sorp.	0.10–0.50	15	−47.1200	−17.5300	—	Barreiro et al. (2003)
	35°C sorp.	0.10–0.50	15	−45.8800	−17.1010	—	
	15°C sorp.	0.50–0.95	12	35.6270	−62.7630	—	
	35°C sorp.	0.50–0.95	12	17.7480	−92.7730	—	
Beef							
Raw	30°C des.	0.6–0.80	11	0.4292	10.8829	6.4	
	40°C ads.	0.10–0.80	8	0.8226	4.402	6.2	
	50°C ads.	0.10–0.80	14	2.0365	0.0113	5.1	
Raw minced	10°C des.	0.05–0.88	11	0.5480	93.6908	—	Lind and Rask (1991)
	40°C des.	0.05–0.88	11	0.5480	75.9443	—	
	70°C des.	0.05–0.88	11	0.5480	61.5559	—	Singh et al. (2001)
Black tea	60°C sorp.	0.10–0.90	11	0.5070	6.5400	—	Temple and van Boxtel (1999)
Cake dough	100°C sorp.	0.10–0.80	11	0.6480	7.7500	—	Bassal et al. (1993)
	120°C sorp.	0.10–0.80	11	0.5996	5.5500	—	
	140°C sorp.	0.10–0.80	11	0.5512	3.3500	—	
Cardamom	5°C des.	0.10–0.80	14	2.4577	0.0010	8.1	
	45°C des.	0.10–0.80	11	4.9129	9.1760	5.1	
Fruit	5°C ads.	0.05–0.80	19	0.3370	12.1710	—	Johnson and Brennan (2000)
	25°C ads.	0.05–0.80	19	0.3730	10.5120	—	
	45°C ads.	0.05–0.80	19	0.4740	8.7740	—	
	60°C ads.	0.05–0.80	19	0.4300	7.2510	—	
Celery	5°C ads.	0.10–0.80	11	0.6203	12.6920	6.3	
	25°C ads.	0.10–0.80	11	0.6581	12.0212	6.2	
	45°C ads.	0.10–0.80	13	0.88	4.4222	3.4	
	60°C ads.	—	—	—	—	3.2	
Chamomile (flower)	5°C ads.	0.05–0.80	13	1.3670	−18.8120	—	Soysal and Öztekin (2001)
	25°C ads.	0.05–0.80	13	1.3670	−18.8120	—	
	45°C ads.	0.05–0.80	13	1.0250	6.0210	—	
	60°C ads.	0.05–0.80	13	1.0830	4.8530	—	
Cheese	25°C ads.	0.10–0.80	13	1.1889	5.9967	3.3	
	45°C ads.	0.10–0.80	15	−2.5189	−0.1019	2.2	
Chicken, Cooked	5°C des.	0.10–0.80	14	2.2287	0.0019	8.4	
	45°C des.	0.10–0.80	13	1.7319	32.8144	5.0	
	60°C des.	0.10–0.80	15	−3.4730	1.7060	3.7	
Chickpea seed	5°C ads.	0.11–0.877	18	1.5823	1.1410	—	Menkov (2000)
	20°C ads.	0.11–0.877	18	1.5823	1.0220	—	

(Continued)

**TABLE 10.1
Continued**

Food product	Specifications	Average range	Eq.	$B(1)$	$B(2)$	X_m	Source
	40°C ads.	0.11–0.877	18	1.5823	0.8642	—	
	60°C ads.	0.11–0.877	18	1.5823	0.7062	—	
	5°C des.	0.11–0.877	18	1.7774	0.5809	—	
	20°C des.	0.11–0.877	18	1.7774	0.3906	—	
	40°C des.	0.11–0.877	18	1.7774	0.1376	—	
	60°C des.	0.11–0.877	18	1.7774	−0.1154	—	
Cinnamon	5°C des.		—	—	—	10.3	
	25°C des.	0.10–0.80	14	2.2918	0.0024	7.0	
	45°C des.	0.10–0.80	14	1.7941	0.0122	5.4	
Shell	5°C ads.	0.05–0.80	19	0.3180	13.5900	—	Soysal and Öztekin (2001)
	25°C ads.	0.05–0.80	19	0.3590	10.7840	—	
	45°C ads.	0.05–0.80	19	0.4480	8.1290	—	
	60°C ads.	0.05–0.80	19	0.4050	7.1170	—	
Clove (flower)	5°C ads.	0.05–0.80	19	0.3040	8.8800	—	
	25°C ads.	0.05–0.80	13	1.7370	−23.0360	—	
	45°C ads.	0.05–0.80	13	1.3560	−8.5290	—	
	60°C ads.	0.05–0.80	13	1.2620	−3.0020	—	
Coffee	20°C sorp.	0.10–0.60	13	1.1738	4.4409	2.5	
	30°C sorp.	0.10–0.60	13	0.7459	1.5689	1.6	
Coriander (seed)	5°C ads.	0.05–0.80	19	0.3830	8.9100	—	Soysal and Öztekin (2001)
	25°C ads.	0.05–0.80	19	0.3830	8.9100	—	
	45°C ads.	0.05–0.80	19	0.5450	6.5840	—	
Corn	22°C ads.	0.10–0.80	8	0.8325	6.3766	7.0	
	50°C ads.	0.10–0.80	11	0.3378	10.7490	6.0	
Daphne (leaf)	5°C ads.	0.05–0.80	19	0.3420	10.9630	—	Soysal and Öztekin (2001)
	25°C ads.	0.05–0.80	19	1.3890	12.8050	—	
	45°C ads.	0.05–0.80	19	0.5720	6.4340	—	
	60°C ads.	0.05–0.80	19	0.5050	4.7030	—	
Dough, White bread	10°C des.	0.05–0.88	11	0.3860	83.8475	—	Corzo and Fuentes (2004)
	40°C des.	0.05–0.88	11	0.3860	67.7619	—	
	70°C des.	0.05–0.88	11	0.3860	54.7622	—	
Eggs	10°C ads.	0.10–0.70	11	0.4805	8.0481	4.5	
	37°C ads.	0.10–0.70	11	0.5097	7.1133	3.9	
	60°C ads.	0.10–0.70	13	1.2875	6.4464	3.1	
	80°C ads.	0.10–0.70	11	0.5893	4.6593	2.5	
Fennel (seed)	5°C ads.	0.05–0.80	19	0.6650	5.5950	—	Soysal and Öztekin (2001)
	25°C ads.	0.05–0.80	19	0.6650	5.5950	—	
	45°C ads.	0.05–0.80	19	0.7010	5.0300	—	
Fufu	25°C sorp.	0.11–0.96	8	0.6806	12.7600	—	Sanni et al. (1997)
	32°C sorp.	0.11–0.96	8	0.6915	8.3990	—	
	45°C sorp.	0.11–0.96	8	0.6410	8.6110	—	
Ginger (rhizome)	5°C ads.	0.05–0.80	19	0.3410	12.1470	—	Soysal and Öztekin (2001)
	25°C ads.	0.05–0.80	19	0.3490	11.4220	—	

TABLE 10.1
Continued

Food product	Specifications	Average range	Eq.	$B(1)$	$B(2)$	X_m	Source
	45°C ads.	0.05–0.80	19	0.4580	8.3240	—	
Grapefruit	5°C ads.	—	—	—	—	6.5	
	25°C ads.	—	—	—	—	6.5	
	45°C ads.	0.10–0.80	14	0.6645	0.1519	6.5	
	60°C ads.	—	—	—	—	1.9	
Groundnut							
Roasted	27°C ads.	0.05–0.75	12	0.0814	0.0128	—	Sopade (2001)
	34°C ads.	0.05–0.76	12	0.0824	−0.0002	—	
	45°C ads.	0.05–0.77	12	0.5800	0.0400	—	
	27°C des.	0.05–0.78	12	0.0999	0.0184	—	
	34°C des.	0.05–0.79	12	0.0908	0.0122	—	
	45°C des.	0.05–0.80	12	0.0864	0.0044	—	
Unroasted	27°C ads.	0.05–0.75	12	0.0766	0.0142	—	
	34°C ads.	0.05–0.76	12	0.0789	−0.0016	—	
	45°C ads.	0.05–0.77	12	0.0802	−0.0123	—	
	27°C des.	0.05–0.78	12	0.0786	0.0287	—	
	34°C des.	0.05–0.79	12	0.0796	0.0152	—	
	45°C des.	0.05–0.80	12	0.0788	0.0049	—	
Horseradish	5°C des.	0.10–0.80	13	1.6670	55.3696	6.9	
	45°C des.	0.10–0.80	13	1.0553	6.9463	4.5	
Lima bean pre-cooked flour	18°C sorp.	0.10–0.75	14	1.2700	−12.9100	—	Corzo and Fuentes (2004)
	28°C sorp.	0.10–0.75	14	1.3000	−13.9200	—	
	38°C sorp.	0.10–0.75	14	1.0700	−10.0200	—	
	48°C sorp.	0.10–0.75	14	1.1400	−12.6400	—	
Marjoram (leaf)	5°C ads.	0.05–0.80	19	0.3530	11.2750	—	Soysal and Öztekin (2001)
	25°C ads.	0.05–0.80	13	1.4400	−16.3740	—	
	45°C ads.	0.05–0.80	13	1.0920	−5.2270	—	
	60°C ads.	0.05–0.80	13	1.0730	−3.3180	—	
Microcrystalline cellulose	80°C des.	0.05–0.40	11	1.0000	2.3290	—	Ferrasse and Lecomte (2004)
	100°C des.	0.05–0.41	11	1.0000	2.3602	—	
	120°C des.	0.05–0.42	11	1.0000	2.3652	—	
	100°C sorp.	0.10–0.80	11	0.5287	278.5550	—	Bassal et al. (1993)
	120°C sorp.	0.10–0.80	11	0.5704	333.0550	—	
	140°C sorp.	0.10–0.80	11	0.6122	387.5550	—	
Milk, whole	24°C ads.	—	—	—	—	3.1	
	34°C ads.	—	—	—	—	3.5	
Mixed meat (minced)	10°C des.	0.05–0.88	11	0.5650	26.3377	—	Lind and Rask (1991)
	40°C des.	0.05–0.88	11	0.5650	24.3614	—	
	70°C des.	0.05–0.88	11	0.5650	22.5334	—	
Morel (mushroom)	5°C des.	0.11–0.92	19	0.7170	0.1008	—	Johnson and Brennan (2000)
	15°C des.	0.11–0.92	19	0.7170	0.0988	—	
	25°C des.	0.11–0.92	19	0.7170	0.0969	—	
	35°C des.	0.11–0.92	19	0.7170	0.0949	—	

(Continued)

TABLE 10.1
Continued

Food product	Specifications	Average range	Eq.	$B(1)$	$B(2)$	X_m	Source
Muscat (fruit)	5°C ads.	0.05–0.80	19	0.2850	9.1170	—	
	25°C ads.	0.05–0.80	19	0.3430	7.8490	—	
	45°C ads.	0.05–0.80	19	0.4860	6.4060	—	
	60°C ads.	0.05–0.80	19	0.4720	4.9630	—	
Onion	17°C ads.	0.10–0.70	11	0.7070	16.1993	9.4	
	27°C ads.	0.10–0.70	11	0.7923	13.5043	9 5	
Shreds	30°C sorp.	0.15–0.85	20	2.4800	0.0113	—	Viswanathan et al. (2003)
	40°C sorp.	0.15–0.85	20	2.4800	0.0117	—	
	50°C sorp.	0.15–0.85	20	2.4800	0.0120	—	
Pasta	25°C sorp.	0.10–0.90	11	0.3557	0.1323	—	Xiong et al. (1991)
	35°C sorp.	0.10–0.90	11	0.4251	0.1148	—	
	45°C sorp.	0.10–0.90	11	0.4946	0.0973	—	
	50°C sorp.	0.10–0.90	11	0.5293	0.0886	—	
Peach	20°C = 30°C ads.	0.10–0.80	14	1.0096	0.0471	8.7	
Peanut							
Hull	5°C sorp.	0.10–0.95	19	0.4208	8.8600	—	Chen (2000)
	25°C sorp.	0.10–0.95	19	0.4208	8.3600	—	
	50°C sorp.	0.10–0.95	19	0.4208	7.7300	—	
	5°C sorp.	0.10–0.95	20	1.5936	0.0197	—	
	25°C sorp.	0.10–0.95	20	1.5936	0.0224	—	
	50°C sorp.	0.10–0.95	20	1.5936	0.0257	—	
Kernel	5°C sorp.	0.10–0.95	19	0.3633	6.1400	—	
	25°C sorp.	0.10–0.95	19	0.3633	5.7800	—	
	50°C sorp.	0.10–0.95	19	0.3633	5.3300	—	
	5°C sorp.	0.10–0.95	20	1.7939	0.0245	—	
	25°C sorp.	0.10–0.95	20	1.7939	0.0283	—	
	50°C sorp.	0.10–0.95	20	1.7939	0.0330	—	
Pod	5°C sorp.	0.10–0.95	19	0.3936	6.7300	—	
	25°C sorp.	0.10–0.95	19	0.3936	6.3800	—	
	50°C sorp.	0.10–0.95	19	0.3936	5.9400	—	
	5°C sorp.	0.10–0.95	20	1.6636	0.0270	—	
	25°C sorp.	0.10–0.95	20	1.6636	0.0307	—	
	50°C sorp.	0.10–0.95	20	1.6636	0.0353	—	
Oil	30°C	0.21–0.76	15	−0.0359	0.0468	—	
	80°C	0.21–0.85	15	−0.0416	0.0774	—	
Pepper							
Green	30°C ads.	0.10–0.9	13	1.1830	−0.0420	—	Kaymak-Ertekin and Sultanoğlu (2001)
	30°C des.	0.10–0.9	13	1.1640	−0.0620	—	
	45°C ads.	0.10–0.88	13	1.1660	−0.0350	—	
	45°C des.	0.10–0.88	13	1.1090	−0.0620	—	
	60°C ads.	0.10–0.84	13	1.1290	−0.0250	—	
	60°C des.	0.10–0.84	13	1.1240	−0.0300	—	
Red	30°C ads.	0.10–0.90	13	1.1630	−0.0600	—	
	30°C des.	0.10–0.90	13	1.1920	−0.0660	—	
	45°C ads.	0.10–0.88	13	1.0650	−0.0630	—	
	45°C des.	0.10–0.88	13	1.1490	−0.0660	—	
	60°C ads.	0.10–0.84	13	1.1360	−0.0400	—	
	60°C des.	0.10–0.84	13	1.0480	−0.0550	—	

(Continued)

TABLE 10.1
Continued

Food product	Specifications	Average range	Eq.	B(1)	B(2)	X_m	Source
Peppermint (leaf)	5°C ads.	0.05–0.80	19	0.3980	11.5230	—	Soysal and Öztekin (2001)
	25°C ads.	0.05–0.80	19	0.3980	11.5230	—	
	45°C ads.	0.05–0.80	19	0.5100	8.8480	—	
	60°C ads.	0.05–0.80	13	1.3170	−7.0410	—	
Pistachio nut paste	10°C sorp.	0.10–0.90	13	1.8160	−0.7910	—	Maskan and Göğüş (1997)
	20°C sorp.	0.10–0.90	13	1.2220	−0.6570	—	
	30°C sorp.	0.10–0.90	13	1.2480	−0.6570	—	
Plantain pretreated by moist infusion	40°C ads.	0.10–0.70	16	3.1700	1.3600	—	Johnson and Brennan (2000)
	50°C ads.	0.10–0.70	16	3.6200	1.2600	—	
	60°C ads.	0.10–0.70	16	3.1100	1.3800	—	
Freeze-dried	40°C ads.	0.10–0.70	14	1.6100	−35.9000	—	
	50°C ads.	0.10–0.70	14	1.7200	−53.4000	—	
	60°C ads.	0.10–0.70	14	1.3400	−32.3000	—	
Fresh	40°C des.	0.10–0.70	14	0.8450	−4.5300	—	
	50°C des.	0.10–0.70	14	0.7070	−4.5900	—	
	60°C des.	0.10–0.70	14	0.5010	−2.4500	—	
Potato	10°C sorp.	0.10–0.80	13	1.5068	27.4606	6.3	
	37°C sorp.	0.10–0.80	13	1.6622	33.2730	5.9	
	60°C sorp.	—	—	—	—	5.4	
	80°C sorp.	0.10–0.80	13	1.5638	17.1200	4.7	
Red chilli	25°C ads.	0.12–0.87	18	1.1526	10.4385	—	Kaleemullah and Kailappan (2004)
	35°C ads.	0.12–0.88	18	1.1526	8.5037	—	
	45°C ads.	0.12–0.89	18	1.1526	6.9275	—	
	25°C des.	0.12–0.90	18	1.0489	2.3742	—	
	35°C des.	0.12–0.91	18	1.0489	2.1957	—	
	45°C des.	0.12–0.92	18	1.0489	2.0172	—	
Rice							
Rough	0°C des.	0.20–0.90	8	0.8290	9.1228	—	
	20°C des.	0.20–0.90	14	2.4516	0.0013	—	
	30°C des.	0.20–0.90	11	2.3771	0.0018	—	
Medium Grain rough	19.7°C sorp.	0.371–0.897	20	2.1360	0.0026	—	Basunia and Abe (2001)
	29.7°C sorp.	0.371–0.898	20	2.1360	0.0030	—	
	37.8°C sorp.	0.371–0.899	20	2.1360	0.0033	—	
	51°C sorp.	0.371–0.900	20	2.1360	0.0038	—	
Ruziz date paste	5°C sorp.	0.10–0.90	16	1.1430	3.1390	—	Johnson and Brennan (2000)
	25°C sorp.	0.10–0.85	16	2.3510	2.4530	—	
	40°C sorp.	0.10–0.80	16	3.4490	1.8570	—	
Sausage, Smoked chicken	5°C sorp.	0.113–0.877	13	−1.8385	−5.1461	—	Singh et al. (2001)
	25°C sorp.	0.113–0.843	13	−1.3619	−2.8224	—	
	50°C sorp.	0.111–0.812	13	−0.6626	−0.5112	—	
	5°C sorp.	0.113–0.877	11	0.3855	3.0654	—	

TABLE 10.1
Continued

Food product	Specifications	Average range	Eq.	$B(1)$	$B(2)$	X_m	Source
	25°C sorp.	0.113–0.843	11	0.4968	2.4197	—	
	50°C sorp.	0.111–0.812	11	1.0251	1.4843	—	
Soup, Meat and vegetable	37°C sorp.	0.10–0.80	11	0.7883	9.7198	—	
Soybean	40°C sorp.	0.07–0.98	19	0.2688	14.5529	—	Aviara et al. (2004)
	50°C sorp.	0.07–0.98	19	0.2688	13.0429	—	
	60°C sorp.	0.07–0.98	19	0.2688	11.5329	—	
	70°C sorp.	0.07–0.98	19	0.2688	10.0229	—	
Starch	30°C	0.10–0.84	14	2.3241	0.0016	9.4	
	80°C	0.10–0.84	14	1.9557	0.0070	7.4	
Maize	25°C ads.	0.34–0.87	12	5.4823	8.8112	—	
Wheat	50°C ads.	0.32–0.87	8	0.8363	5.9465	—	
Sugar beet root	20°C des.	0.10–0.70	13	0.9645	7.5659	5.5	
	35°C des.	0.10–0.70	13	0.8553	5.0309	4.9	
	47°C des.	—	—	—	—	5.0	
	65°C des.	—	—	—	—	3.9	
Sweet potato	25°C des.	0.10–0.90	19	0.6373	13.5100	—	Chen (2002)
	25°C ads.	0.10–0.90	19	0.4481	12.9608	—	
	50°C des.	0.10–0.90	19	0.5285	10.2135	—	
	50°C ads.	0.10–0.90	19	0.5915	10.2395	—	
Tapioca	25°C des.	—	—	—	—	8.7	
	45°C des.	0.10–0.80	13	1.6565	0.0143	6.9	
	25°C sorp.	0.11–0.96	8	0.6881	12.0200	—	Sanni et al. (1997)
	32°C sorp.	0.11–0.96	8	0.6727	8.2390	—	
	45°C sorp.	0.11–0.96	8	0.6168	6.9800	—	
Thyme (leaf with small stem)	5°C ads.	0.05–0.80	13	0.6550	−30.6050	—	Soysal and Öztekin (2001)
	25°C ads.	0.05–0.80	13	1.4870	−17.5830	—	
	45°C ads.	0.05–0.80	13	1.2900	−9.0420	—	
	60°C ads.	0.05–0.80	13	1.4040	−9.0070	—	
Tomato	17°C des.	0.10–0.80	13	0.9704	10.0587	8.2	
	27°C des.	—	—	—	—	6.1	
Slice	30°C sorp.	0.15–0.85	20	2.9200	0.0052	—	Viswanathan et al. (2003)
	40°C sorp.	0.15–0.85	20	2.9200	0.0054	—	
	50°C sorp.	0.15–0.85	20	2.9200	0.0055	—	
Tortilla Chips	25°C des.	0.10–0.95	12	1.1062	0.0019	—	Kawas and Moreira (2001)
	48.8°C des.	0.10–0.95	12	1.0980	0.0019	—	
	68.8°C des.	0.10–0.95	12	1.0920	0.0019	—	
Trout							
Cooked	45°C des.	0.10–0.80	11	1.5526	20.3488	4.4	
Raw	5°C des.	0.10–0.80	11	0.3871	15.3339	8.8	
	45°C des.	0.10–0.80	11	0.3903	15.2898	8.8	
	60°C des.	0.10–0.80	17	5.3728	1.6769	3.5	
Turkey, Cooked	10°C des.	0.10–0.80	14	1.4720	0.0160	7.8	
	0°C des.	0.10–0.80	14	1.3296	0.0201	8.8	
Wheat	25°C des.	0.09–0.86	11	0.3205	13.0959	7.8	
	50°C des.	0.13–0.82	11	0.3906	10.4073	6.0	

(Continued)

TABLE 10.1
Continued

Food product	Specifications	Average range	Eq.	$B(1)$	$B(2)$	X_m	Source
Winter savory	25°C des.	0.10–0.80	13	2.3329	252.6355	7.0	
	45°C des.	—	—	—	—	3.5	
Yeast	16°C des.	0.10–0.75	17	6.3072	4.9435	5.6	
	27°C des.	0.10–0.75	17	6.0234	4.1318	4.8	
Yogurt	5°C des.	0.10–0.80	16	2.5442	1.9048	5.4	
	25°C des	0.10–0.80	13	1.0529	6.4806	4.2	

water and its latent heat of condensation. Integrating the Clausius–Clapeyron equation at a constant moisture, one obtains

$$\ln \frac{a_{w2}}{a_{w1}} = \frac{E_b}{R}\left(\frac{1}{T_1} - \frac{1}{T_2}\right) \tag{10.22}$$

where a_{w_1} and a_{w_2} are the water activities at temperatures T_1 and T_2, respectively at a fixed moisture content. From the experimental measurement of adsorption isotherms at different temperatures, the average binding energy over the temperature range can be determined from a plot of $\ln a_w$ vs. $1/T$ at different moisture contents. Over a relatively narrow temperature range, such plots should yield a straight line for each moisture content, the slope of the straight line being $-E_b/R$. The average moisture binding energy for various products is given in Table 10.3. These values were obtained from the adsorption isotherm data of these products at different temperatures reported in the *Handbook of Food Isotherms Water: Sorption Parameters for Food and Food Components* by Iglesias and Chirife (1982). For most of food materials, the binding energy is positive since moisture readily adsorbs onto the food. For hydrophobic materials such as peanut oil, for example (see Table 10.3), binding energy can be negative, implying thereby that water interacts more readily with other water molecules than with hydrophobic food.

Alternatively, the binding energy E_b can also be determined from Equation 10.21 by differentiating with respect to temperature the fitted adsorption isotherm equation at constant moisture content. As can be seen from Equation 10.21, the moisture binding energy is a function of both moisture content and temperature. The isotherm of extruded pasta at different temperatures was fitted to the following Oswin equation (Xiong et al., 1991):

$$m = (0.176 - 1.748 \times 10^{-3}T)\left(\frac{a_w}{1 - a_w}\right)^{(0.182 + 6.946 \times 10^3 T)} \tag{10.23}$$

where m is the moisture content (DB). A plot of E_b vs. moisture content is shown in Figure 10.5. The average binding energy values compare well with those predicted from the isotherms. It is worth noting that moisture binding energy is higher at lower moisture contents as well as at lower temperatures. Also, the binding energy tends to zero at high moisture contents since water eventually behaves as free water. Although in this example, the Oswin equation is used as an example to fit the isotherm data it is, however, preferable to use the GAB equation to fit the data since this equation has been found to fit better the isotherm data for food systems.

From a knowledge of moisture binding energy, the moisture sorption isotherm at a different temperature can be predicted.

TABLE 10.2
Data of GAB Parameters for Water Sorption for Water Sorption Isotherms of Various Foods and Related Products at Given Temperature and Applicable a_w Region

Food product	State of sorption[a]	a_w below	T (°C)	X_m (kg w/kg ds)	C_G	K	Source
Alligator meat	A,D	0.9	10	0.07436	1.6132	1.0549	Lopes Filho et al. (2002)
	A,D	0.9	15	0.07779	2.2691	0.9106	
	A,D	0.9	25	0.0311	3.4111	0.9278	
	A,D	0.9	35	0.1562	3.6589	1.0291	
Amioca	A,D	0.6	25	0.0954	24.191	0.6768	Vagenas and Karathanos (1991)
Amylose	A	0.90	20	0.0997	16.16	0.724	van den Berg (1984)
Apple	D	0.97	25	0.051	1.322	1.009	Prothon and Ahrné (2004)
	D	0.97	45	0.057	2.563	1.007	
	D	0.97	55	0.039	2.606	1.011	
	D	0.97	65	0.051	2.614	1.007	
Babusa	A	0.85	20	0.0786	−25.43	0.8094	Ahmed et al. (2004)
	A	0.85	30	0.0799	17.44	0.792	
	A	0.85	40	0.0594	−44.92	0.8333	
	A	0.85	50	0.0482	−19.2	0.8515	Ahmed et al. (2004)
Beef, lean	A, D	0.98	5	0.0522	4.27	0.9721	Trujillo et al. (2003)
	A, D	0.98	15	0.0497	3.23	0.9688	
	A, D	0.98	25	0.054	3.8212	0.975	
	A, D	0.98	40	0.0527	3.0732	0.9878	
Chicken (cooked)	D	0.8	19.5	0.0775	18.7	0.86	Timmermann et al. (2001)
Chicken meat	A,D	0.9	10	0.0758	1727.7	0.9579	Delgado and Sun (2002)
	A,D	0.9	20	0.06471	43.2	1.0339	
	A,D	0.9	30	0.07682	137.5	1.1099	
Chicken sausage (smoked)	D	0.877	5	0.11628	52.1157	−0.0225	Singh et al. (2001)
	D	0.843	25	0.05248	34.7762	−0.0302	
	D	0.812	50	0.02009	2.4871	−0.5869	
Chips (Doritos)	A	0.9	25	0.0326	11.18	1.01	Palou et al. (1997)
	A	0.9	35	0.0306	11.01	1.06	
	A	0.9	45	0.03	8.01	1.04	
Chips (Tostitos)	A	0.9	25	0.0371	7.23	0.95	
	A	0.9	35	0.0377	5.84	0.94	
	A	0.9	45	0.0345	6.33	0.97	
Cocoa beans nonfermented	A	0.90	30	0.1614	11.87	0.512	van den Berg (1984)
	A,D	0.95	25	0.0218	−76.8	0.99	Sandoval and Barreiro (2002)
Coffee extract	A	0.85	20	0.0285	3.01	1.009	van den Berg (1984)
	A,D	0.90	20	0.0623	2.85	1.023	
Collagen	A,D	0.8	25	0.115	17.3	0.8	Timmermann (2003)

(Continued)

TABLE 10.2
Continued

Food Product	State of sorption[a]	a_w below	T (°C)	X_m (kg w/kg ds)	C_G	K	Source
Cookie	A,D	0.85	20	0.034	5.866	0.982	Kim et al. (1998)
	A,D	0.85	30	0.032	4.204	0.997	
	A,D	0.85	40	0.03	2.979	1.01	
Animalitos	A	0.9	25	0.0441	4.2	1.02	Palou et al. (1997)
	A	0.9	35	0.0411	3.59	1.05	
	A	0.9	45	0.0397	3.55	1.05	
Corn	D	0.8	30	0.0978	18.8	0.633	Timmermann et al. (2001)
	D	0.91	25	0.115	14.3	0.58	Aguerre and Suarez (2004)
Corn bran	A	0.8	25	0.0721	9.8	0.76	Timmermann et al. (2001)
	A	0.9	5	0.062	22.046	0.845	Dural and Hines (1993)
	A	0.9	15	0.062	15.104	0.853	
	A	0.9	25	0.062	14.61	0.832	
	A	0.9	37	0.062	12.333	0.857	
Corn starch	D	0.75	25	0.101	24.3	0.69	Aguerre et al. (1996)
Cracker	A,D	0.85	20	0.05	9.012	0.957	Kim et al. (1998)
	A,D	0.85	30	0.044	5.051	0.974	
	A,D	0.85	40	0.04	3.378	0.987	
Eggalbumen (coag.)	A,D	0.8	25	0.0629	0.118	0.78	Timmermann (2003)
Ethylcellulose	A	0.9	9	1.1593	5.3597	0.9082	Velázquez de la Cruz et al. (2001)
	A	0.9	15	1.0867	4.6522	0.9035	
	A	0.9	20	0.9068	4.4003	0.9109	
	A	0.9	25	0.7885	3.7338	0.9067	
	A	0.9	35	0.7505	3.0674	0.9224	
Extruded pasta	D	0.90	55	0.0540	10.27	0.798	Wannanen (1990)
Garden mint leaves	D	0.9	30	0.073	36.953	0.902	Park et al. (2002)
	D		40	0.095	11.247	0.727	
Glucose-alginate gel	D	0.95	40	0.1200	2.78	1.000	van den Berg (1984)
Green beans	A,D	0.95	20	0.0734	2.58	0.94	Samaniego-Esguerra et al. (1991)
	A,D	0.92	30	0.0699	2.28	0.95	
	A,D	0.89	40	0.0668	2.03	0.96	
Ground coffee	A	0.80	20	0.0349	16.65	0.963	Weisser (1986)
Groundnut protein isolate	A	0.92	20	0.0620	14.55	0.751	van den Berg (1984)
Highly amylopectin powder	A	0.95	30	0.032	22.2	0.888	Al-Muhtaseb et al. (2003)
		0.95	45	0.032	15.3	0.882	

(Continued)

TABLE 10.2
Continued

Food Product	State of sorption[a]	a_w below	T (°C)	X_m (kg w/kg ds)	C_G	K	Source
		0.95	60	0.027	9.86	0.887	
	D	0.95	30	0.041	14.59	0.902	
		0.95	45	0.041	9.13	0.881	
		0.95	60	0.034	6.79	0.89	
Highly amylose powder	A	0.95	30	0.031	11.1	0.913	
		0.95	45	0.028	8.52	0.893	
		0.95	60	0.028	6.04	0.91	
	D	0.95	30	0.043	10.2	0.89	
		0.95	45	0.037	7.5	0.881	
		0.95	60	0.036	5.13	0.88	
Hylon 7	A,D	0.6	25	0.0904	15.906	0.7277	Vagenas and Karathanos (1991)
Jam	A,D	0.85	20	0.135	75.773	0.99	Kim et al. (1998)
	A,D	0.85	30	0.08	42.034	1.009	
	A,D	0.85	40	0.05	25.98	1.031	
Khudari date paste	A,D	0.9	5	0.148	7.22	0.69	Alhamdan and Hassan (1999)
	A,D	0.9	25	0.132	5.863	0.849	
	A,D	0.9	40	0.063	5.045	1.324	
Lentil seeds	A	0.8	5	0.07215	24.83	0.8056	Menkov (2000)
	D	0.8	5	0.09121	25.72	0.6341	
	A	0.8	20	0.07215	14.726	0.7782	
	D	0.8	20	0.09121	14.98	0.6125	
	A	0.8	40	0.07215	7.931	0.747	
	D	0.8	40	0.09121	7.8995	0.588	
	A	0.8	60	0.07215	4.601	0.7205	
	D	0.8	60	0.09121	4.498	0.5671	
Lupine	D	0.8	25	0.0497	2.7355	0.9682	Vasquez et al. (2003)
	D	0.8	35	0.0497	1.9345	0.9456	
	D	0.8	45	0.0497	1.3982	0.9249	
Macadamia nut	A	0.8	50	0.0317	12.44168	0.8774	Beristain et al. (1996)
	A	0.8	60	0.0299	10.2746	0.8766	
Manioc Starch	A	0.9	25	0.093	26	0.72	Aguerre and Suarez (2004)
Methylcellulose	A	0.9	9	3.9284	11.0592	0.9743	Velázquez de la Cruz et al. (2001)
	A	0.9	15	4.0573	8.6868	0.9564	
	A	0.9	20	3.5236	11.0973	0.9642	
	A	0.9	25	3.6727	5.0327	0.9259	
	A	0.9	35	4.8565	4.3591	0.8773	
Microcrystalline cellulose	A	0.95	20	0.0506	16.60	0.806	van den Berg (1984)
	D	0.9	35	0.004	15	0.9	Ferrasse and Lecomte (2004)
	D	0.9	45	0.003	53.4	0.92	

(Continued)

TABLE 10.2
Continued

Food Product	State of sorption[a]	a_w below	T (°C)	X_m (kg w/kg ds)	C_G	K	Source
	D	0.4	35	0.004	15	0.9	
	D	0.4	45	0.003	53.4	0.92	
Myosin	A,D	0.9	10	0.03642	12.962	0.924	Das and Das (2002)
	A,D	0.9	27	0.04906	5.333	0.897	
	A,D	0.9	45	0.073	2.412	0.851	
Oat fiber	A	0.9	5	0.0566	19.851	0.85	Dural and Hines (1993)
	A	0.9	15	0.0566	20.071	0.843	
	A	0.9	25	0.0566	18.463	0.844	
	A	0.9	37	0.0566	13.743	0.857	
Onion	A,D	0.95	20	0.0754	2.72	0.96	Samaniego-Esguerra et al. (1991)
	A,D	0.92	30	0.0739	2.29	0.97	
	A,D	0.89	40	0.0726	1.95	0.98	
Pea	A	0.85	25	0.0327	35.49	0.744	van den Berg (1984)
	D	0.6	20	0.0592	0.9499	0.8553	Rahman et al. (1997)
	D	0.6	40	0.0383	0.9831	8.8797	
	D	0.6	60	0.0235	10.1583	1.0129	
Pepper							
Green	A	0.9	30	0.082	3.306	0.858	Kaymak-Ertekin and Sultanoglu (2001)
	D	0.9	30	0.101	3.543	0.901	
	A	0.88	45	0.0487	6.563	0.95	
	D	0.88	45	0.082	3.961	0.945	
	A	0.84	60	0.0484	2.318	0.892	
	D	0.84	60	0.038	5.801	0.995	
Red	A	0.9	30	0.0996	3.587	0.884	
	D	0.9	30	0.113	3.229	0.901	
	A	0.88	45	0.0995	2.567	0.866	
	D	0.88	45	0.09	5.075	0.928	
	A	0.84	60	0.086	2.551	0.808	
	D	0.84	60	0.067	3.838	0.933	
Potassium caseinate	A	0.90	40	0.0785	5.30	0.780	van den Berg (1984)
Potato	A	0.95	30	0.035	17.6	0.907	Al-Muhtaseb et al. (2003)
		0.95	45	0.027	11.8	0.905	
		0.95	60	0.021	8.19	0.889	
	D	0.95	30	0.056	12.1	0.88	
		0.95	45	0.046	8.24	0.888	
		0.95	60	0.029	6.38	0.917	
	A	0.8	40	0.052	13.73	0.83	Beristain et al. (1996)
	A	0.8	50	0.048	13.41	0.82	
	A	0.8	60	0.036	21.18	0.86	
	A	0.8	70	0.029	17.75	0.9	

(Continued)

TABLE 10.2
Continued

Food Product	State of sorption[a]	a_W below	T (°C)	X_m (kg w/kg ds)	C_G	K	Source
Potato starch	A	0.90	20	0.1012	17.60	0.740	van den Berg (1984)
	D	0.90		0.1399	17.40	0.651	
	A	0.903	25	0.085	10.9	0.8	Aguerre and Suarez (2004)
Native	A	0.8	20	0.0979	20.4	0.75	Timmermann et al. (2001)
Potato	D		25	0.06344	67.29	0.946	Chen (2002)
	D		50	0.04627	153.01	0.968	
Prickly pear	A	0.9	30	0.1296	88.21	0.9299	Lahsasni et al. (2004)
	D	0.9	30	0.1917	35.28	0.8245	
	A	0.9	40	0.1296	58.25	0.9437	
	D	0.9	40	0.1917	24.52	0.8609	
	A	0.9	50	0.1296	39.46	0.9569	
	D	0.9	50	0.1917	17.44	0.8964	
Quinoa grain	A	0.851	20	0.0867	15.3	0.7	Tolaba et al. (2004)
	A	0.851	30	0.0851	11.91	0.68	
	A	0.851	40	0.059	9.72	0.8	
	D	0.851	20	0.0999	14.44	0.62	
	D	0.851	30	0.0866	11.77	0.67	
	D	0.851	40	0.0573	11.46	0.82	
Rice	A	0.8	25	0.11	19.2	0.58	Timmermann et al. (2001)
Fiber	A	0.9	5	0.037	7.96	0.918	Dural and Hines (1993)
	A	0.9	15	0.037	8.976	0.922	
	A	0.9	25	0.037	10.308	0.925	
	A	0.9	37	0.037	8.778	0.891	
Rough	D	0.9	25	0.079	44	0.75	Aguerre and Suarez (2004)
Skim milk	A	0.8	34	0.0427	38	0.876	Timmermann et al. (2001)
Sodium caseinate	A,D	0.92	25	0.0723	6.15	0.862	van den Berg (1984)
Sorghum	D	0.92	37.8	0.082	23.4	0.72	Aguerre and Suarez (2004)
Soy flour	A,D	0.9	30	0.05	25.3	0.96	Riganakos et al. (1994)
Soy protein isolate	A,D	0.9	20	0.075	60	0.906	Jovanovich et al. (2003)
Spaghetti	A	0.9	22	0.805	7.74	0.87	Lagoudaki et al. (1993)
Conventional	A	0.9	30	0.77	7.48	0.84	
	A	0.9	37	0.712	6.84	0.77	
	A	0.9	45	0.704	6.37	0.76	
Diet	A	0.9	22	0.66	7.18	0.88	
	A	0.9	30	0.589	6.54	0.84	

(Continued)

TABLE 10.2
Continued

Food Product	State of sorption[a]	a_w below	T (°C)	X_m (kg w/kg ds)	C_G	K	Source
	A	0.9	37	0.56	5.6	0.83	
	A	0.9	45	0.45	5.28	0.81	
Strawberry	A	0.9	30	0.051	3.5	1.16	Moraga et al. (2004)
	D	0.9	30	0.098	4.9	0.99	
Sunflower	A	0.90	23	0.0222	28.40	0.919	van den Berg (1984)
Tapioca starch	D	0.93	25	0.101	23.5	0.71	Aguerre et al. (1996)
Tomato	A	0.8	30	0.166	31.4	0.83	Timmermann et al. (2001)
Turkey (cooked)	A	0.8	22	0.0629	7.41	0.82	
	D	0.8	22	0.0692	7.5	0.79	
Wheat	A	0.00	25	0.0844	23.58	0.743	van den Berg (1984)
	D	0.90		0.0968	21.87	0.698	
	D	0.8	25	0.1024	18.6	0.62	Timmermann et al. (2001)
Bran	A	0.9	5	0.0566	10.198	0.895	Dural and Hines (1993)
	A	0.9	15	0.0566	9.425	0.887	
	A	0.9	25	0.0566	10.782	0.863	
	A	0.9	37	0.0566	11.602	0.849	
Flour	A,D	0.92	27	0.067	31.7	0.82	Aguerre et al. (1996)
	A,D	0.9	30	0.063	20.28	0.91	Riganakos et al. (1994)
Gluten	A	0.8	3	0.0638	16.4	0.78	Timmermann et al. (2001)
Semolina	A,D	0.92	27	0.077	30.5	0.76	Aguerre et al. (1996)
Starch	A	0.88	20	0.0982	27.30	0.681	van den Berg (1984)
	D	0.89		0.1288	21.10	0.584	
	A	0.91	20	0.0960	17.40	0.728	
	A	0.8	20	0.0989	26.7	0.68	Timmermann (2003)
Wool keratine	A	0.96	24.6	0.0915	8.83	0.696	van den Berg (1984)
	D	0.96		0.0991	16.16	0.679	
β-Lactoglobulin	A,D	0.8	25	0.0772	9.5	0.81	Timmermann (2003)

Notes: [a] A, adsorption; D, desorption.

TABLE 10.3
Average Moisture Binding Energy Values for Various Food Products

Product	Moisture content (g/100 g solid)	Average E_b (cal/mol)	a_w			
			25°C	35°C	45°C	
Skim	10	2,030	0.52	0.58	0.645	
Milk	15	2,065	0.73	0.76	0.845	
	20	1,167	0.83	0.895	0.94	
			30°C	50°C		
Almonds	10	2,673	0.38	0.50		
	15	645	0.73	0.78		
	20	110	0.88	0.89		
			30°C	40°C		
Beef	10	4,720	0.45	0.565	0.73	
(raw)	15	2,446	0.70	0.80	0.90	
	20	1,532	0.83	0.90	—	
			20°C	35°C	47°C	65°C
Sugarbeet	10	4,710	0.46	0.51	0.52	0.495
Root	15	1,006	0.57	0.62	0.60	0.565
	20	532	0.66	0.69	0.67	0.64
			5°C	45°C		
Cardamom	10	3,584	0.25	0.565		
	15	1,304	0.55	0.74		
			5°C	25°C	45°C	60°C
Celery	10	2,403	0.41	0.43	0.555	0.605
	15	1,109	0.59	0.60	0.675	0.70
	20	608	0.67	0.675	0.72	0.75
			25°C	45°C		
Cheese	5	3,359	0.42	0.60		
	10	1,573	0.66	0.78		
			5°C	45°C	60°C	
Chicken	5	11,451	—	0.15	0.34	
(cooked)	10	8,186	0.29	0.54	0.64	
	15	1,255	0.56	0.74	0.76	
			5°C	25°C	45°C	
Cinnamon	5	7,010	—	0.095	0.20	
	10	5,400	0.20	0.385	0.53	
	15	4,018	0.43	0.70	0.80	
			20°C	30°C		
Coffee	1	133,639	0.025	0.19		
	2	21,171	0.12	0.40		
	3	9,579	0.29	0.50		
			22°C	50°C		
Corn	10	1,825	0.355	0.465		
	15	480	0.68	0.73		
			10°C	37°C	60°C	80°C
Eggs	5	3,592	0.255	0.33	0.445	0.53
	10	410	0.61	0.65	—	—
			5°C	25°C	45°C	60°C
Grapefruit	10	3,885	0.46	0.46	0.50	0.66
	15	2,492	0.515	0.515	0.59	0.705
	20	2,002	0.60	0.60	0.65	0.75
			5°C	45°C		

(Continued)

TABLE 10.3
Continued

Product	Moisture content (g/100 g solid)	Average E_b (cal/mol)	a_w			
Horseradish	10	2,460	0.32	0.56		
	15	815	0.54	0.65		
	20	437	0.67	0.74		
			14°C	24°C	34°C	
Milk	5	−2,919	0.475	0.40	0.375	
(whole)	10	4,730	0.64	0.60	0.78	
	15	0	0.915	0.90	0.90	
			30°C	80°C		
Peanut	0.1	−3,388	0.51	0.23		
Oil	0.15	−625	0.695	0.60		
	0.2	−545	0.79	0.695		
			17°C	27°C		
Onion	10	2,808	0.34	0.40		
	20	857	0.59	0.62		
	30	843	0.70	0.735		
			20°C	30°C	90°C	50°C
Peach	10	6,846	0.40	0.40	0.50	0.65
	15	4,417	0.50	0.50	0.65	0.81
	20	4,965	0.60	0.60	0.78	—
			10°C	36°C	60°C	80°C
Potato	5	3,613	0.10	0.12	0.18	0.23
	10	3,613	0.40	0.47	0.60	0.62
	15	1,206	0.63	0.69	0.79	0.795
			0°C	20°C	30°C	
Rice	10	1,548	0.24	0.295	0.32	
(rough)	15	1,727	0.57	0.63	0.695	
	20	1,973	0.80	0.88	0.89	
			15°C	37°C		
Soup:	10	−311	0.53	0.51		
Meat and	15	−261	0.63	0.61		
Vegetable	20	−228	0.72	0.70		
			30°C	80°C		
Starch	5	2,258	0.10	0.17		
	10	2,404	0.27	0.475		
	15	949	0.60	0.75		
			25°C	40°C		
Starch	15	−722	0.66	0.62		
(maize)	20	277	0.88	0.90		
	25	−62	0.99	0.985		
			30°C	50°C		
Starch	15	1,608	0.39	0.46		
(potato)	20	140	0.69	0.7		
	23	0	0.81	0.81		
			30°C	50°C		
Starch	15	319	0.90	0.93		
(wheat)	20	60	0.805	0.81		
	25	780	0.6	065		
			25°C	45°C		

(Continued)

TABLE 10.3
Continued

Product	Moisture content (g/100 g solid)	Average E_b (cal/mol)	a_w		
Tapioca	5	3,591	0.14	0.205	
	10	3,672	0.325	0.48	
	15	1,452	0.60	0.70	
			17°C	27°C	
Tomato	10	1,006	0.65	0.69	
	20	705	0.585	0.61	
	30	5,101	0.325	0.44	
			45°C	60°C	
Trout	5	8,407	0.17	0.31	
(cooked)	10	1,838	0.57	0.65	
	15	571	0.72	0.75	
			5°C	45°C	60°C
Trout	10	12,482	0.25	0.25	0.61
(raw)	15	5,729	0.48	0.48	0.70
	20	2,371	0.65	0.65	0.77
			0°C	10°C	22°C
Turkey	10	3,864	0.36	0.4	0.53
(cooked)	20	1,271	0.68	0.74	0.81
	30	1,867	0.815	0.92	(>1)
			25°C	50°C	
Wheat	10	3,592	0.3	0.48	
	15	1,286	0.6	0.71	
	20	469	0.79	0.84	
			5°C	25°C	45°C
Winter	10	6,633	0.15	0.345	0.68
Savory	15	1,972	0.51	0.66	0.80
	20	869	0.72	0.80	—
			16°C	27°C	44°C
Yeast	10	600	0.485	0.51	0.53
	15	881	0.615	0.62	0.67
	20	622	0.70	0.71	0.75
			5°C	25°C	
Yogurt	10	1,858	0.455	0.57	
	15	1,372	0.58	0.685	
	20	569	0.70	0.75	

Source: Data from Iglesias, H.A. and Chirife, J. 1982. *Handbook of Food Isotherms.* Academic Press, New York.

Example 10.1 The average moisture binding energy of extruded pasta at moisture content of 10 g per 100 g of dry solid is 3000 cal/mol. If the water activity of extruded pasta at the moisture content stated is 0.35 at 35°C, predict the water activity of the sample at 50°C.

Solution From Equation 10.22.

$$\ln \frac{a_{w2}}{a_{w1}} = \frac{E_b}{R}\left(\frac{1}{T_1} - \frac{1}{T_2}\right)$$

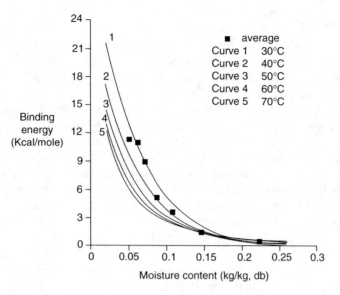

FIGURE 10.5 Plot of binding energy vs. moisture content for extruded pasta.

As $E_b = 3000$ cal/mol, $R = 1.987$ cal/mole·K, $T_1 = 298$ K, and $T_2 = 323$ K,

$$\ln \frac{a_{w_2}}{a_{w_1}} = \frac{3000}{1.987} \left(\frac{1}{298} - \frac{1}{323} \right)$$

$$= 0.392$$

$$\frac{a_{w_2}}{a_{w_1}} = 1.48; \quad a_{w_2} = 0.518$$

10.3.5.2 Pressure

The effect of pressure on the adsorption isotherm is relatively small and negligible at reasonable pressure levels. At constant moisture content, the variation of water activity with pressure is given by

$$\ln \frac{a_{w_2}}{a_{w_1}} = \frac{\bar{V}_L}{RT}(P_2 - P_1) \tag{10.24}$$

where $a_{w_1} = a_w$ at P_1, $a_{w_2} = a_w$ at P_2, $\bar{V}_L$ is molar volume of water $= 18$ cm³/mol, R is gas constant $= 82.05$ cm³ atm g/mol·K, T is temperature (K), and P_1, P_2 are total pressure (atm).

Example 10.2 If the water activity of extruded pasta at a moisture content of 10 g per 100 g of dry solid at 1 atm 25°C is 0.35, what will be the water activity of the sample at 10 atm pressure?

Solution

$$\ln \frac{a_{w_2}}{a_{w_1}} = \frac{18}{82.05 \times 298}(10 - 1)$$

$$= 0.00663$$

$$\frac{a_{w_2}}{a_{w_1}} = 1.0066; \quad a_{w_2} = 0.3523$$

10.3.5.3 Composition

The composition of the foodstuffs affects their water sorption properties. The basic components in a food system usually include proteins, lipids, polysaccharides, and other minor components. The binding energy of water depends on the nature of the water interaction with these food constituents. The stronger the interaction, the higher is the binding energy. Water-carbohydrate interactions are considered to consist of hydrophilic and hydrophobic interactions in addition to gel forming characteristics of these interactions. Numerous hydroxyl groups which make up carbohydrates hydrophilic interact with water molecules by hydrogen bonding, and this leads to solvation and/or solubilization of sugars and many other carbohydrate polymers. The structure of carbohydrates can greatly affect the rate of water binding and the amount of water bound (Whistler and Daniel, 1985). Starch gelatinization along with the minimum water content necessary for it to occur was studied by Donovan (1979) and Eliasson (1980). After that seminal research there was significant work in the area and it is now widely recognized that the phenomenon of gelatinization is a consequence of the order–disorder phase transition occurring when starch granules are heated in the presence of adequate water.

In regard to the interaction of water and proteins one factor that influences the extent and the manner that water interacts with proteins is the amino acid composition of the proteins. Bull and Breese (1968) studied the hydration properties of various globular proteins and found that protein hydration correlated strongly with the sum of the polar residues (hydroxyls, carboxyls, and basic groups) minus the amide groups. Amide groups were observed to inhibit water binding. Kuntz (1971) used a NMR technique to study the hydration of some synthetic polypeptides and showed that the bound water content decreased in the order: ionic groups > polar groups > nonpolar groups. Based on these measurements a formula that depended on the number of water molecules associated with the different amino acids forming the protein was developed: ($A = f_c + 0.4f_p + 0.2f_n$), where A represents the grams of water per gram of protein and f_c, f_p, and f_n are the fractions of ionic, polar, and nonpolar amino acids, respectively. This indicated that the approximate level of hydration of proteins may be estimated from their amino acid composition. However, this method can only be used at low to intermediate water activity range, because of the influences of molecular structure and the conformational features of proteins at high water activity.

Bushuk and Winkler (1957) measured the water sorption isotherms and the binding energies of wheat flour, starch, and gluten. They indicated that the sorptive capacity of these materials for water vapor were: starch > flour > gluten, and the effect of temperature on adsorption was greater at low water activity and became almost negligible as the saturation pressure was approached, but they did not investigate in detail how starch and gluten as individual components influence the water sorption in a food system containing these components.

Xiong et al. (1991) measured the desorption isotherms of extruded mixtures of starch and gluten of different compositions (see Figure 10.6) and found that the water binding ability of the samples decreased as the gluten content increased from a ratio of starch to gluten (s:g) of 1 to 1/3. This can be explained by the fact that gluten is a protein with a low content of ionic functional groups but rich in nonpolar amino acids, whereas starch is highly hydroxylated and thus very hydrophilic.

10.3.5.4 Structure

Literature on the structural influences of food component on water binding is rare. Water sorption (rate and amount) may be influenced by surface area and the porosity of food materials. The number and size of pores in the food matrix may influence the rate and extent of hydration (Kinsella and Fox, 1986). Food materials with a porous structure may, at high humidities, imbibe water in addition to the hydration of water forming multilayers (Watt, 1983). However, Gal (1983) indicated that the great majority of foods and food ingredients of biological origin have no definite surface area when exposed to water vapor because of the extremely strong hydrogen-bond-splitting ability of these surfaces; in other words, foodstuffs possess no rigid pores for water. Xiong et al. (1991) investigated the effect of structure through the experimental measurement of desorption isotherms of puffed and unpuffed pasta. Even though the two samples had very different pore structures, their desorption isotherms did not differ much. Similar results were found by King et al. (1968) for poultry and Shubin (1982) for wood. Xiong et al. (1990) also found that the amount of water adsorbed by unextruded materials was higher than the amount absorbed by extrudates measured at the same conditions of temperature and moisture content. This difference may be due to the influence of extrusion pressure and to the modification of the material surface as a result of the extrusion process and also of the surface hydration during extrusion. They also showed that pregelatinized samples exhibited a lower moisture binding ability than did ungelatinized ones.

10.3.6 EFFECTS OF WATER ACTIVITY AND MOISTURE CONTENT

Water activity and moisture content of the material are very important in food unit operations. During processing and storage, many chemical and physical factors are influenced by the water activity and moisture content level. Chemical changes that are enhanced by water activity include enzymatic reactions, nonenzymatic browning, and microbial activity (Duckworth, 1975). In many food products, enzymes are not inactivated during the heating process. Consequently, enzymatic reactions can take place at even low moisture contents. Water activity also affects the nonenzymatic browning reactions in foods. When water is present, carboxyl and amino compounds are involved as reactants, products, or catalysts in the browning process.

Bacterial growth is also affected at fairly high water activity levels. If water activity is maintained at a value below 0.90, most bacteria remain dormant. Most yeasts and molds, however, can grow and multiply at water activity levels as low as 0.80.

Physical changes such as texture and aroma can depend greatly on water activity. Textural changes are most often seen in freeze drying and subsequent storing of foods, particularly meats and fish. The water activity in dried foods can also affect the retention of aroma. Different foods that are stored together will be altered if their individual relative humidities (water activities) are different. Under the action of a driving force created by a difference in water activities changes in the moisture and water activities of these foods will follow their own isotherm curve until an equilibrium water activity is achieved.

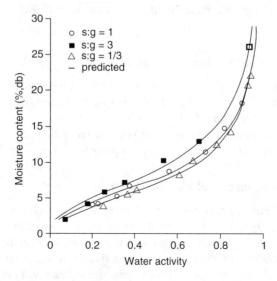

FIGURE 10.6 Desorption isotherms of extruded mixtures of starch and gluten.

With such factors dependent upon water activity, much work needs to be done to determine water sorption data accurately. Studies have been done to determine the effect of different constituents on water activity. Fat content, for example, apparently has no effect on water activity. Most other composition variables exhibit conflicting results when their importance to water activity is determined. Now that water sorption data exist, future work needs to concentrate on the prediction of water activity and its effects based on food composition, processing, and storage. One such study done by Crapiste and Rotstein (1982) showed the prediction of isotherms for potatoes, peas, beans, corn, and white rice with reasonably accurate results. The predictions were close to the experimental values but exhibited sensitivity to variation in composition. The method required knowledge of the composition and sorptional behavior of the basic components of the food. The work of Peleg and Normand (1992) represents another approach in which a simple numerical method implemented in a commercial spreadsheet was used for the determination of equilibrium activities of a mixture of dry powders.

10.4 GLASS TRANSITION TEMPERATURE

10.4.1 GLASS TRANSITION TEMPERATURE

The second order transition where a substance transforms from the glassy state to the rubbery state is known as the glass transition temperature (T_g) and can be used in conjunction with water activity to predict the self-life and stability of stored foods. Every amorphous polymer in its pure form has a particular glass transition point, below which it takes the form of a highly viscous, amorphous glassy matrix. Diffusion is slow in this state, and water is generally unavailable for biological processes. Above the glass transition temperature, a substance is in its rubbery state. While it is still amorphous, polymer chains in this state are more flexible and diffusion more rapid.

10.4.2 TESTING

10.4.2.1 Dilatometry

Dilatometry was one of the first methods used to find T_g by measuring a change in the volumetric thermal expansion coefficient. This can be found by comparing the specific volume of the system at different temperatures. For the measurement the sample is placed in contact with a nonplasticizing fluid such as mercury or silicon oil, where the dilatometer measures thermal expansion at a controlled temperature.

10.4.2.2 Dynamic Thermal Mechanical Analysis (DMTA)

DMTA measures changes on the mechanical properties of the material due to changes in temperature. In order to evaluate the mechanical properties of the sample it is required that the sample have a well-defined shape such as for example a uniform rectangular slab. Few foods are naturally found in these shapes, so they need to be ground and pressed into the required form. Unfortunately the sample forming process may very well affect the mechanical properties of the samples Biliaderis (2002).

10.4.2.3 Differential Scanning Calorimetry (DSC)

DSC is a commonly accepted method to measure glass transition of food materials Zimeri and Kokini (2003). In this test, the sample is sealed in an aluminum pan to prevent moisture loss and heated at a constant rate, usually 5 to 20°C/min. A control system is used to maintain either the temperature or the rate of heating/cooling of the sample. A heat flux calorimeter measures the temperature difference between the sample pan and the control pan, usually empty. This temperature difference is proportional to heat flux. Using a power-compensated calorimeter, the computer measures the

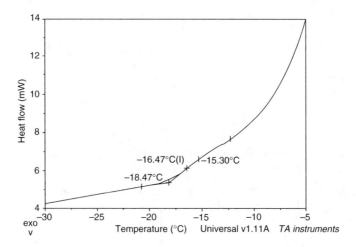

FIGURE 10.7 A DSC plot of temperature vs. heat flux for chicken meat. (From Delgado, A.E. and Sun, D. 2002. *J. Food Eng.,* 55: 1–8.)

amount of electrical energy needed to heat the sample at the same rate as the control pan. The calorimeter then reports the isotherm data as a function of temperature and heat flux as shown in Figure 10.7. The difference in heat flow necessary to achieve identical temperatures quantifies the thermal properties of the polymer. In a successful test a transition can be observed as a slight depression in the heat flow temperature curve. The range of glass transition temperatures is in general very close to the gelatinization temperature of starches. This event, which requires higher levels of energy than the glassy-rubbery transition can then mask the glass transition effect and make its detection difficult. It is then recommendable for foods that the temperature be ramped up and down prior to the determination of T_g. Since there is no change in latent heat, the glass transition is a second-order transition. Researchers take the mid-point of this depression as the glass transition temperature unless otherwise stated.

It should be noted that in addition to composition and moisture content, T_g is dependent on the rate of heating chosen for testing. T_g can decrease linearly by as much as $\pm 3°C$ with cooling rate. A standard of testing T_g using DSC has not been established, so the rate of cooling is chosen with respect to the equipment and type of sample used. Figure 10.8 shows how a fast cooling rate can affect the point at which the product makes the state transition.

10.4.3 PREDICTION OF T_g

All starch-containing materials have a glass transition temperature range that must be found individually due to variations in composition and qualities. Since the vast majority of foods are mixtures and not pure compounds, the effects of other ingredients must be taken into account. This presents a difficulty in predicting values of T_g. One of the simplest models was introduced by Fox and Flory (1950) and relates the glass transition temperature of a pure substance to its molecular weight:

$$T_g = T_{g\infty} - \frac{k}{M_W} \tag{10.25}$$

where $T_{g\infty}$ is a limiting value for T_g and k is a constant specific to the type of polymer whereas M_w is the average molecular weight of the sample. However simple, this equation is highly limited as to its practical application in foods. The Gordon–Taylor equation (1952) is the most commonly

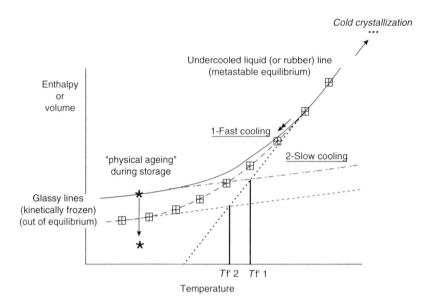

FIGURE 10.8 The effect of cooling rate on the enthalpy or volume on a sample polymer. (From Borde, B. et al. 2002. *Carbohydr. Polym.*, 48: 83–96.)

used method for calculating T_g of a mixture of two components. For foods one of the components is water.

$$T_g = \frac{w_1 T_{g_1} + k_{G-T} w_2 T_{g_2}}{w_1 + k_{G-T} w_2} \qquad (10.26)$$

where w_1 is the weight fraction of dry solids, w_2 is the weight fraction of water, T_{g_1} is the glass transition temperature of the sample at 0% moisture content, T_{g_2} is the glass transition temperature for glassy water ($-138°C$). k_{G-T} is a constant specific to the sample. With known moisture content and constant k_{G-T}, the glass transition curve can be predicted. Figure 10.9 shows an example of Greek honey data which has been fitted with the Gordon–Taylor equation. In this case, the constants are $k_{G-T} = 3.14$ and $T_{g_2} = 288.0$ K (Lazaridou, et al., 2004).

Additional models include the following models (Katkov and Levine, 2004):

Couchman–Karasz:

$$\ln(T_g) = \frac{\Delta C_{p_1} f_1 \ln(T_{g_1}) + \Delta C_{p_2} f_2 \ln(T_{g_2})}{\Delta C_{p_1} f_1 + \Delta C_{p_2} f_2} \qquad (10.27)$$

where $\Delta C_p = C_p^{\text{liquid}} - C_p^{\text{glass}}$ as the heat capacity of the sample changes from liquid to glass and f_i is the partial fraction of the ith component. The above equation is written for two components

modified Couchman–Karasz:

$$T_g = \frac{\Delta C_{p_1} f_1 T_{g_1} + \Delta C_{p_2} f_2 T_{g_2}}{\Delta C_{p_1} f_1 + \Delta C_{p_2} f_2} \qquad (10.28)$$

modified Couchman:

$$T_g = \frac{w_1 T_{g_1} + k_{C-K} w_2 T_{g_2}}{w_1 + k_{C-K} w_2} \qquad (10.29)$$

Fox:

$$1/T_g = w_1/T_{g_1} + w_2/T_{g_2} \qquad (10.30)$$

modified Fox:

$$1/T_g = \frac{\Delta C_{p_1} f_1 T_{g_1} + \Delta C_{p_2} f_2 T_{g_2}}{\Delta C_{p_1} f_1 + \Delta C_{p_2} f_2} \qquad (10.31)$$

Kwei

$$T_g = \frac{w_1 T_{g_1} + k_{G-T} w_2 T_{g_2}}{w_1 + k_{G-T} w_2} + q w_1 w_2 \qquad (10.32)$$

where q is an empirical constant the could evaluate the deviation from the water plasticization theory for homogenous water-polymer systems.

Table 10.4 provides food products and the model parameters that could be used to predict the glass transition vs. moisture content curves.

Simple empirical models can be used to predict T_g as well. Figure 10.10 shows a dataset for freeze-dried date paste which has been plotted and fitted with a linear model. Researchers in this case found the model to fit with an R^2 value of 0.95 (Ahmed, 2004). Peleg (1993a,b) used an equation based on the Fermi's equation to describe the behavior of amorphous food polymers near their glass transition.

10.4.4 RELATIONSHIP TO DRYING

When dried, a food product loses moisture to the ambient. This necessarily lowers its water activity, which is related to the glass transition temperature of the food. As water activity decreases, the glass transition temperature increases. The individual glass transition temperature of a food product affects its physical state during storage, shelf life, as well as cooking and consumption properties. T_g cannot be assumed to be constant over a range of moisture contents, leading to problems for producers and

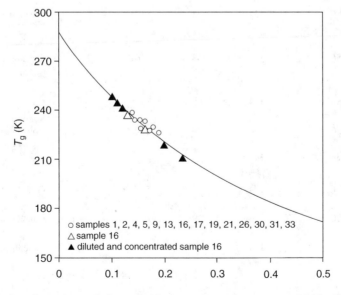

FIGURE 10.9 A plot of experimental glass transition temperatures vs. water weight fraction of Greek honey along with the predictions of the Gordon–Taylor equation. (From Lazaridou, A., et al., 2004. *J. Food Eng.*, 64: 9–21.)

TABLE 10.4
Glass Transition Models

Product	Measurement method	T_g (water) (°K)	T_g (food component) (°K)	k_{G-T}	q	R_2	Model	Source
Amylopectin	DSC	134	502	4.73	—	—	G–T	Zimeri and Kokini (2003)
Amylopectin β limit dextrin	DSC	134	542	0.155	—	1	Couchman	Bizot et al. (1997)
Apple	DSC	138	271.5	2.34	—	—	G–T	del Valle et al. (1998)
Apple	—	138	314	3.95	—	—	G–T	Bai et al. (2001)
Chitosan	DMTA	135	367.9	1.68	—	—	G–T	Lazaridou and Biliaderis (2002)
Cornstarch	DSC	138	551	0.176	—	—	G–T	Zhong and Sun (2005)
	DSC	138	341.5	4	—	—	G–T	Rahman (2004)
Date flesh (Barni variety)	DSC	138	330.6	3.2	—	—	G–T	Rahman (2004)
Dextran T500	DSC	134	462	0.246	—	1	Couchman	Bizot et al. (1997)
Honey, Greek	DSC	135	288	3.14	—	—	G–T	Lazaridou et al. (2004)
Inulin	DSC	134	393	2.98	—	—	G–T	Zimeri and Kokini, (2003)
Kiwifruit	DSC	138	313.5	4.88	—	—	G–T	Moraga et al. (2006)
	DSC	136	374	7.4	—	—	G–T	Ozmen and Langrish (2002)
Maize phytoglycogen, su-1	DSC	134	533	0.174	—	1	Couchman	Bizot et al. (1997)
Maize starch, waxy (cast 100°C)	DSC	134	558	0.161	—	1	Couchman	Bizot et al. (1997)
Pea amylase (cast 100°C)	DSC	134	605	0.145	—	1	Couchman	Bizot et al. (1997)
Phytoglycogen β-limit dextrin	DSC	134	539	0.15	—	1	Couchman	Bizot et al. (1997)
Pineapple	DSC	138	330.9	0.21	—	1	G–T	Telis and Sobral (2001)
Polydextrose	DSC	134	367	5.88	—	—	G–T	Ribeiro et al. (2003)
Potato starch (cast 90°C)	DSC	134	589	0.145	—	1	Couchman	Bizot et al. (1997)
Potato starch, lintnerized	DSC	134	410	0.238	—	1	Couchman	Bizot et al. (1997)
Pullulan	DMTA	138	396.6–408.7	3.03–4.16	—	0.96–0.99	G–T	Lazaridou et al. (2003)

(Continued)

**TABLE 10.4
Continued**

Product	Measurement method	T_g (water) (°K)	T_g (food component) (°K)	k_{G-T}	q	R_2	Model	Source
Pullulan	DSC	134	488	0.18	—	1	Couchman	Bizot et al. (1997)
Pullulan/starch	DMTA	135	353.5	1.72	—	—	G–T	Lazaridou and Biliaderis (2002)
Starch/chitosan	DMTA	135	363.7	1.97	—	—	G–T	Lazaridou and Biliaderis (2002)
Strawberry (entire tissue, adsorption)	DSC	138	300	4.14	—	—	G–T	Moraga et al. (2004a)
Strawberry (entire tissue, desorption)	DSC	138	301	4.14	—	—	G–T	Moraga et al. (2004a)
Strawberry (homogenized tissue, adsorption)	DSC	138	313	4.32	—	—	G–T	Moraga et al. (2004a)
Strawberry (homogenized tissue, desorption)	DSC	138	336	4.82	—	—	G–T	Moraga et al. (2004a)
Sucrose	DSC	139	338	5.7	—	—	G–T	Blond et al. (1997)
Tomato, air-dried	DSC	138	380	5.21	414	0.9	Kwei	Telis and Sobral (2002)
Tomato, freeze-dried	DSC	138	603.9	9.35	—	1	G–T	Telis and Sobral (2002)
Tomato, osmotically-treated	DSC	138	1958.7	83.3	583	0.9	Kwei	Telis and Sobral (2002)
Tortilla chip	DSC	138	660	5.2	—	—	G–T	Kawas and Moreira, (2002)
Trehalose	DSC	138	373	6.54	—	—	G–T	Katkov and Levine (2004)
Trehalose	DSC	138	373	3.56	—	—	C–K	Katkov and Levine (2004)
Tuna meat	DSC	138	368	2.89	—	—	G–T	Rahman et al. (2003)
Wheat durum semolina (cooling)	DSC	138	435	3.4	—	1	G–T	Cuq and Icard-Vernière (2001)
Wheat durum semolina (cooling)	DSC	138	546	9.5	346	1	Kwei	Cuq and Icard-Vernière (2001)

(Continued)

TABLE 10.4
Continued

Product	Measurement method	T_g (water) (°K)	T_g (food component) (°K)	k_{G-T}	q	R_2	Model	Source
Wheat durum semolina (heating)	DSC	138	449	4.2	—	0.9	G–T	Cuq and Icard-Vernière (2001)
Wheat durum semolina (heating)	DSC	138	518	9.6	277	1	Kwei	Cuq and Icard-Vernière (2001)
Wheat durum semolina (re-heating)	DSC	138	498	5.5	—	1	G–T	Cuq and Icard-Vernière (2001)
Wheat durum semolina (re-heating)	DSC	138	589	12.1	304	1	Kwei	Cuq and Icard-Vernière (2001)
Wheat gluten	DSC	138	446	5.1	—	—	G–T	Micard and Guilbert (2000)
Wheat gluten	DSC	138	446	6.54	184	—	Kwei	Micard and Guilbert (2000)
Wheat gluten protein (alkylated)	DSC	134	412	5	—	—	G–T	Noel et al. (1995)
Wheat gluten protein (high molecular weight)	DSC	134	411	4.3	—	—	G–T	Noel et al. (1995)
Wheat gluten protein (α-gliadin)	DSC	134	417	4.6	—	—	G–T	Noel et al. (1995)
Wheat gluten protein (γ-gliadin)	DSC	134	397	3.8	—	—	G–T	Noel et al. (1995)
Wheat gluten protein (ω-gliadin)	DSC	134	418	4.2	—	—	G–T	Noel et al. (1995)

Simple empirical models can be used to predict T_g as well. Figure 10.10 shows a dataset for freeze-dried date paste which has been plotted and fitted with a linear model. Researchers in this case found the model to fit with an R^2 value of 0.95.

Source: From Ahmed, J., Khan, A.R., and Hanan, A.S. 2004a. *J. Food Eng.*, 64: 187–192.

researchers. Testing for foods using the aforementioned methods must be done over a practical range of moisture contents if the results are to be of any use to producers. Knowledge of T_g can help extend the shelf life of dried and intermediate moisture foods since there is no moisture diffusion below this temperature.

Figure 10.11 shows a plot of glass transition temperature of cornstarch with respect to moisture content. Testing was performed using DSC (Zhong and Sun, 2005).

As the product dries, the glass transition temperature increases. Assuming that the product is being dried using a constant ambient temperature and T_g is sufficiently high, there will be a point at which the ambient temperature is equal to the glass transition temperature at a low moisture content.

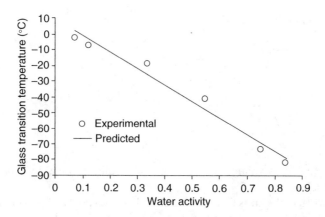

FIGURE 10.10 Data and graph showing the linear relationship between water activity and glass transition temperature for freeze-dried date paste samples. (From Ahmed, J., et al., 2004a. *J. Food Eng.*, 64: 187–192; Ahmed, J. et al., 2005. *J. Food Eng.*, 66: 253–258.)

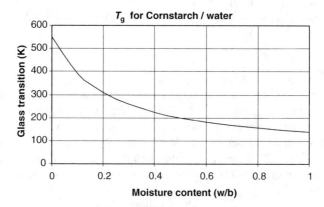

FIGURE 10.11 Plot of T_g vs. moisture content for cornstarch using parameters from Table 10.3. (From Zhong, Z. and Sun, X.S. 2005. *J. Food Eng.*, 69: 453–459.)

In addition, the temperature at which drying takes place affects the point where this change occurs, as does the composition of the food. Once the transition has been achieved, further drying and reduction on the sample moisture content implies that water mobility is severely limited, making it difficult to remove water from the food through diffusion. Thus, drying of foods is extremely important because normal food degradation processes are slowed in the absence of water; however, a state in foods without water is not possible for many food products due to the problems mentioned above.

The shelf life for products with low moisture contents can be extended by storing them below their glass transition temperatures. Due to changes in water mobility, there is significant benefit to drying a food above T_g and storing it at temperatures below T_g. T_g and a_w must be known and used together to precisely predict stability as neither one is sufficient on its own.

10.4.4.1 Porosity

During dehydration, water vapor is transported from the center of the material to the environment. Thus, the diffusivity of gases and liquids (mainly water) in the material is affected by the porosity. Fruits and vegetables are the most common porous solids. The porous structure complicates the mass transfer in the plant tissue (Barat et al., 2001). Porosity is defined as the ratio of the empty space volume to the total volume:

$$\varepsilon = \frac{V_a}{V_s + V_w + V_a} \tag{10.33}$$

and,

$$\varepsilon = 1 - \frac{\rho_b}{\rho_p} \tag{10.34}$$

where ε is the porosity, V_i is volume of i (a = air, s = solid, w = water), ρ_i is the density of i (b = bulk, p = particle).

Depending on whether the air is contained in closed or open pores, closed or open pores porosity can be calculated. In addition to the material contained inside a container the interparticle space should be accounted for and in that case bulk density is calculated. Porosity including both closed and open pores along the space between particles is defined as the total porosity of the system. The effective vapor diffusivity in the material increases with the porosity and average pore size (Huizenga and Smith, 1986; Karathanos and Saravacos, 1993), therefore, the study of porosity, pore size, and distribution are important parameters to be considered when designing a dehydration process.

10.4.4.2 Shrinkage

During drying, water evaporates out and causes a pressure gradient to the material. This pressure gradient produces contracting stresses and the material shrinks (Mayor and Sereno, 2004). Porous structure of the material is supported by a solid matrix. Changes in the solid network would change the porous structure. At a certain temperature, called collapse temperature, the solid matrix collapses and causes product shrinkage. Shrinkage is defined as the relative volume change to the original volume, and a shrinkage coefficient (SC) can be calculated as follows (Lozano et al., 1983):

$$SC = \frac{\text{Actual volume}}{\text{Initial volume}} = \frac{V}{V_o} \tag{10.35}$$

10.4.4.3 Factor Affecting Shrinkage

10.4.4.3.1 Glass Transition Temperature and Moisture Content
When the temperature of the material is above T_g, the solid network is in the rubbery state and becomes more mobile. Therefore, shrinkage of the material occurs. As mentioned before, high water content would lower the glass transition temperature. Therefore, for the same temperature, material with high moisture content would experience more shrinkage. The shrinkage in high moisture product decreases with moisture. Due to the shrinkage many times a phenomenon of casehardening occurs during the drying of foods. Due to casehardening the moisture movement is retarded and thus the rate of shrinkage decreases. This leads to a nonlinearity of the shrinkage profile in the final stage of convective drying and in vibrofluidized dryers (Achanta and Okos., 1996; Ramesh et al., 2003).

10.4.4.3.2 Drying Method
Drying method affects the extent of product shrinkage. Maskan (2001) studied the shrinkage of kiwifruits during hot air and microwave drying. They found that with combined hot air and microwave drying, the shrinkage is significantly reduced. Hot air drying method produces moderate shrinkage. This is because of the longer drying time which allows the product to have more time to shrink. Microwave drying produces a large pressure gradient and therefore the product shrinkage by this method is higher and rapid. The shrinkage during freeze-drying is usually very little (Karathanos

et al., 1996), because water is removed by sublimation at low temperatures and under vacuum and thus the solid matrix is at the glassy state below its glass transition temperature with a restricted deformation and reduced shrinkage.

10.4.4.4 Modeling Shrinkage

Work has been done in regard to the use of empirical and fundamental models to predict product shrinkage. Empirical models are classified as linear and nonlinear model. When the porosity development of the material is very little, the linear model can be used to describe the shrinkage. However, if porosity of the material increases significantly during drying, the shrinkage is generally not linear with the drying time . Therefore, a nonlinear model is more suitable to describe the shrinkage. Mayor and Sereno (2004) collected some of the linear and nonlinear empirical models for shrinkage of various food products. Since these are empirical models, experimental work should be used to ensure the applicability of the model to the material and process condition used.

Fundamental models for shrinkage of various food products are also collected in Mayor's review paper. Like the empirical model, there are linear and nonlinear fundamental models for shrinkage. Models in which the porosity term is explicitly expressed are also included.

10.5 PREDICTION OF DRYING RATES: RATE OF DRYING CURVES

To select a dryer, it is necessary to determine the drying rate at a specific air temperature and humidity. These data are scarce for food materials and must be obtained experimentally by plotting the free moisture content vs. drying time. This plot is converted to a drying rate curve by calculating the derivative of the curve over the time. Typical curves are shown in Figure 10.12. At time zero, the moisture content of a food is given by the point A if the food is at a cold temperature and by A' if it is at hot temperature. The drying curve is divided into two distinct portions. The first is the constant-rate period, in which unbound water is removed (line BC in Figure 10.12). Water evaporates as if there is no solid matrix present, and its rate of evaporation is not dependent on the solid matrix. This continues until water from the interior is no longer available at the surface of the food material. Point C distinguishes the constant-rate period from the falling-rate period and is called the critical moisture content. The surface of the food is no longer wet. The falling-rate period has two sections as seen in Figure 10.12. From point C to D, the wet areas on the surface become completely dry. When the surface is dry (point D), the evaporation will continue moving towards the center of the food. This is shown by the curve from D to E. The water that is removed from the center of the food moves to the surface as a vapor. Although the amount of water removed in the falling-rate period is relatively small, it can take considerably longer time than in the constant-rate period (Geankoplis, 2003). In general, increased air velocity and air temperature increase the drying rate, while increased humidity and food thickness decrease it.

The drying rate in the constant-rate period is determined by conditions external to the material being dried, including temperature, gas velocity, total pressure, and partial vapor pressure. Mass transfer during the constant-rate period involves diffusion of water vapor from the material surface through a boundary layer into the drying medium. During the falling-rate period, the drying rate decreases with time, and the rate of internal mass transfer to the material surface typically controls the process. A falling drying rate may be observed when internal mass transfer resistance is controlling and the surface vapor pressure of the solid matrix is decreasing as moisture content drops.

The importance of internal vs. external mass transfer resistance can be inferred from drying studies on samples of different size [i.e., varying slab thickness (l) or sphere and cylinder radii (r)]. The drying time required to reach a given moisture content will be proportional to l or r for external

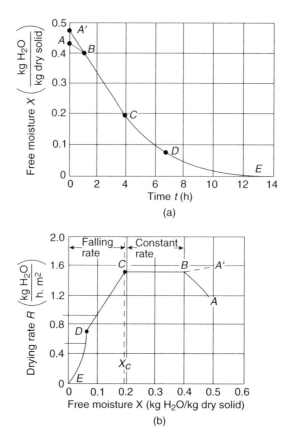

FIGURE 10.12 Typical drying rate curve: (a) free moisture vs. time; (b) drying rate vs. free moisture content. (From Geankopolis, C.J. 2003. *Transport Processes and Separation Process Principles.* 4th edn Prentice Hall, New Jersey.)

mass transfer control and proportional to l^2 or r^2 for control by internal diffusion. Andrieu and Stamatopoulos (1986a) analyzed drying data for pasta cylinders of different diameters and found internal mass transfer resistance to be controlling. Vaccarezza et al. (1974) analyzed the effect of nonisothermal drying conditions on the results obtained from samples of different thickness. When heat transfer effects were considered for sugarbeet root slabs, the thickness dependence of drying data was consistent with internal mass transfer control. Litchfield and Okos (1986) identified internal mass transfer control for drying of pasta slabs by varying the velocity of the drying medium and observing no change in drying rates.

The importance of internal vs. external mass transfer resistance can also be determined using the concept of an overall mass transfer coefficient.

$$\frac{1}{K} = \frac{1}{k_c} + \frac{L}{D_{eff}} \qquad (10.36)$$

where K is the overall mass transfer coefficient (m/sec), k_c the external mass transfer coefficient (m/sec), D_{eff} the internal moisture diffusivity (m^2/sec), and L the characteristic sample dimension (m). The overall mass transfer coefficient K can be determined from experimental drying data. An estimate of k_c can be obtained from general correlations found in mass transfer texts. For example,

Geankoplis (2003) gives the following correlation for laminar flow parallel to a flat plate:

$$\frac{k_c l}{D_{AB}} = 0.664 N_{Re,l}^{0.25} N_{Sc}^{1/3}$$

where l is the length of the plate in the direction of flow (m), $N_{Re,l}$ the Reynold's number $= lV\rho/\mu$, D_{AB} the molecular diffusivity of an air-water vapor mixture (m^2/sec), V the velocity of flowing gas (m/sec), ρ the gas density (kg/m^3), μ the gas viscosity (kg/m·sec), and N_{Sc} the Schmidt number $= \mu/(\rho D_{AB})$. If $1/K$ is approximately equal to $1/k_c$, external mass transfer control is suggested. If $1/K$ is much greater than $1/k_c$, internal mass transfer control can be inferred. If $k_c L/D_{eff} > 10$, external mass transfer resistance can be considered negligible (King, 1968). Therefore, the rate of drying can be predicted using

$$R_c = k_c M_B (H_W - h) \tag{10.37}$$

where R_c is the constant rate of drying (kg/sec·m^2), k_c is mass transfer coefficient (m/sec), M_B is molecular weight of air, 29 g/mol, H_w is humidity at wet-bulb temperature (kg/kg dry air), H is the humidity of external environment (kg/kg dry air).

Alternatively, the rate can also be predicted from heat transfer as

$$R_c = \frac{h(T - T_w)}{\lambda} \tag{10.38}$$

where h is the heat transfer coefficient, (W/m^2·K); T and T_W the dry- and wet-bulb temperature (K), respectively; and λ the latent heat of vaporization of water $= 2433$ kJ/kg at approximately 30°C. The heat transfer coefficient h can be predicted using the correlation (Geankopolis, 1983):

$$h = 0.0204 G^{0.8} \tag{10.39}$$

where $G = \rho v$ is the mass velocity of air (kg/sec·m^2).

In the falling-rate period, the internal resistance to mass transfer is controlling. The rate of drying can therefore be predicted by considering the rate of diffusion of moisture through the food product. Unlike external heat and mass transfer coefficients, which depend only on the external flow conditions, the diffusion of moisture through the food product depends on both the pore structure and the specific interactions of moisture with the food matrix and therefore is system specific. Detailed discussion on the experimental determination of the falling rate for different food products is given in the next section.

10.5.1 MOISTURE DIFFUSIVITIES IN FOODS

Moisture transfer in foods is a subject of considerable importance in industry today. The mechanisms of moisture transport are numerous and often complex. Transport phenomena are usually classified as resulting from pressure diffusion, thermal diffusion, forced diffusion, and ordinary diffusion (net transport of material without fluid movement) (Van Arsdel, 1963).

Often a diffusion transport mechanism is assumed, and the rate of moisture movement is described by an effective diffusivity value, D_{eff} no matter which mechanism is really involved in moisture movement. Even though this method is not quite sound theoretically, it is a very practical and convenient approach to describe moisture content change during processing. Parameters required in this approach are only sample dimensions and the effective diffusion coefficient while more complex analyses need permeability, liquid, and vapor conductivities, or various phenomenological coefficients which are difficult to determine experimentally.

In this section, therefore, food drying studies using a simple diffusion theory are reviewed and reported drying parameters are presented.

10.5.1.1 Solutions of Fick's Law

Fick's law is often used to describe a moisture diffusion process (Becker and Sallans, 1955; Fish, 1958; Saravacos and Charm, 1962; Del Valle and Nickerson, 1968; Vaccarezza et al., 1974; Steffe and Singh, 1980a,b; Suarez et al., 1980).

$$\frac{\partial m}{\partial t} = D_{\text{eff}} \frac{\partial^2 m}{\partial x^2} \tag{10.40}$$

where m is the local moisture content on a dry basis, t is time, and x is the spatial coordinate.

A simple application of Fick's law usually involves assuming that the transfer of mass is unidimensional, the food has an uniform initial moisture content, and that the internal moisture movement is its main resistance to moisture transfer.

Once the shape of the food product is determined or assumed, the corresponding solution of Fick's law is used to obtain the effective diffusion coefficient. The solution of Fick's law for a spherical food is given as follows:

$$\Gamma = \frac{m - m_{\text{s}}}{m_{\text{o}} - m_{\text{s}}} = \frac{6}{\pi^2} \sum_{n=1}^{\text{inf}} \frac{1}{n^2} \exp\left[n^2 \frac{D_{\text{eff}_t}}{r^2}\right] \tag{10.41}$$

where m is the moisture content, m_{o} is initial moisture content, m_{s} is surface moisture content, D_{eff} is effective diffusion coefficient, m^2/sec, r is radius, m, and t is the time, sec.

Examples of products modeled as spheres include wheat kernels (Becker and Sallans, 1955, 1956), soybeans (Kitic and Viollaz, 1984), and walnuts (Alves and Rumsey, 1985).

The solution of Fick's law for a slab is as follows:

$$\Gamma = \frac{m - m_{\text{s}}}{m_{\text{o}} - m_{\text{s}}} = \frac{8}{\pi^2} \sum_{n=0}^{\infty} \frac{1}{2n+1^2} \exp\left[(2n+1)^2 \frac{\pi^2}{4} \frac{D_{\text{eff}} t}{L^2}\right] \tag{10.42}$$

where L is the half thickness of slab, m.

Foods modeled using a slab geometry include tapioca root (Chirife, 1971), apple (Roman et al., 1979), turnip, oatmeal cookie, shredded wheat, and flour (Lomauro et al., 1985a,b).

The solution of Fick's law for a cylinder is as follows:

$$\Gamma = \frac{m - m_{\text{s}}}{m_{\text{o}} - m_{\text{s}}} = \frac{4}{a^2} \sum_{n=1}^{\infty} \left(\frac{1}{\beta_n^2}\right) \exp\left(-D_{\text{eff}} \beta_n^2 t\right) \tag{10.43}$$

where β_n^2 is the Bessel function roots of the first kind and zero order, and a is the radius of cylinder, m.

At sufficiently large times, only the leading term in the series expansion (Equation 10.41 to Equation 10.43) need be taken into account. At smaller times, however, more number of terms are to be taken. Potato samples (Lawrence and Scott, 1966), raisins (Lomauro et al., 1985a,b), and pasta (Andrieu and Stamatopoulos, 1986) have been modeled using the solution of Fick's law for a cylinder.

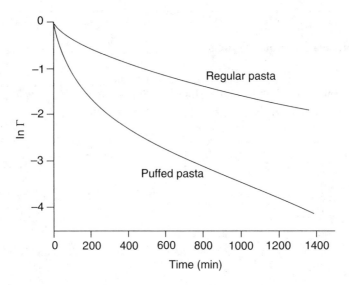

FIGURE 10.13 Typical drying curve for pasta.

Diffusion coefficients are typically determined by plotting experimental drying data in terms of $\ln \Gamma$ vs. time, where Γ is defined by Eqs 10.41 to 10.43. The slope of the linear segments will yield the effective diffusion coefficient. A typical drying curve for pasta is shown in Figure 10.13.

Rotstein et al. (1974) studied the effects of sample shape and size on the diffusivity constant. They concluded that the cross-sectional area should be incorporated into an analytical model to describe diffusivity. The shapes studied included: cardioid, circle, corrugated square, epitochoid, and hexagon. Their study showed that, in particular, the effect of the shape must be accounted for after 100 minutes drying.

Steffe and Singh (1980a,b) used a more complex model to incorporate shape effects. They assumed a rough rice kernel to consist of a spherical core surrounded by two concentric shells. A number of assumptions were made to implement this model; constant diffusivity was assumed, heat transfer and shrinkage were not considered and the components of the rice were assumed homogeneous. Table 10.5 shows effective diffusivity data reported for various food products.

These effective diffusivity data can be used to predict the drying curve of food products of different geometrics using Equation 10.41 to Equation 10.43. It is to be noted that these equations are strictly valid only for constant values of effective diffusivities and for the geometries considered by the equations. Consequently, they can predict the drying curves only for small enough changes in moisture since effective diffusivity changes with moisture. However, these equations can be employed with average effective diffusivity values as a good approximation to predict the drying curve.

10.5.1.2 Effective Diffusivity Expressions as a Function of Moisture and Temperature

The effective diffusivities are often expressed as a function of moisture content and temperature rather than constant values. The relationship between effective diffusivity and moisture content has been described by many different functional forms. However, this is often done in a purely empirical manner using curve fitting techniques.

TABLE 10.5
Effective Diffusivities for Food Products

Product	Temp (°C)	MC (DB) Initial	MC (DB) Final	MC (DB) Equilib.	MC (DB) RM (%)	D_{eff} (m^2/sec)	Reference
Alfalfa (ads.)		0.08	0.15			$7.5 - 10 - 1.26 \times 10^{-09}$	Fasina and Sokhansanj (1996)
Alfalfa (des.)		0.08	0.15			$2.0 - 09 - 4.12 \times 10^{-09}$	
Alfalfa stem	26			3.7		$2.610 - 10 - 2.6 \times 10^{-09}$	Bagnall et al. (1970)
				3.7		$2.6 - 10 - 2.6 \times 10^{-09}$	Saravacos and Charm (1962)
Axial	∼ 26	< 3.7				$\sim 2.6 \times 10^{-9}$	
Radial	∼ 26	< 3.7				$\sim 2.6 \times 10^{-10}$	
Almonds, roasted		0.01				1.6×10^{-12}	
Sliced	25	0.0102				1.608×10^{-11}	Hong et al. (1986)
Animal feed		0.01–0.15				$1.8 \times 10^{-11} - 2.8 \times 10^{-09}$	Luyben et al. (1980)
Apple	66				13	6.4×10^{-9}	Sarvacos and Charm (1962)
	71					1.6×10^{-09}	Alzamora (1979)
	65					1.16×10^{-09}	Labuza and Simon (1970)
	30 to 70	0.10–1.50				$1.0 \times 10^{-11} - 3.3 \times 10^{-09}$	Luyben et al. (1980)
	76					3.6×10^{-09}	Rostein et al. (1974)
	65					6.4×10^{-09}	
	30	0.13–0.15				4.9×10^{-11}	Roman et al. (1979)
	30					$2.2 \times 10^{-12} - 1.0 \times 10^{-11}$	Saravacos (1969)
Freeze-dried	30					$9.0 \times 10^{-11} - 7.0 \times 10^{-10}$	
	25	0.333				2.43×10^{-10}	Lomauro et al. (1985b)
Macintosh osmotically	66	0.06				1.1×10^{-09}	Labuza and Simon (1970)
Dried	38	0.01–4.00				$3.0 \times 10^{-11} - 3.5 \times 10^{-10}$	Bakalis et al. (1994)
	55	0.01–5.50				$2.5 \times 10^{-10} - 2.2 \times 10^{-09}$	Karathanos et al. (1994)
Puff dried	30					$2.0 \times 10^{-11} - 7.0 \times 10^{-10}$	Saravacos (1969)
Freeze-dried	25	0.3330				2.43×10^{-10}	Lomauro et al. (1985)
Granny Smith	30	> 0.15				2.6×10^{-10}	Roman et al. (1979)
	30	< 0.15				4.9×10^{-11}	Roman et al. (1980)
	70				10	3.617×10^{-11}	Rotstein et al. (1974)
MacIntosh	66	0.057			4–7	1.1×10^{-9}	Labuza and Simon (1970)

(Continued)

TABLE 10.5
Continued

Product	Temp (°C)	MC (DB) Initial	MC (DB) Final	MC (DB) Equilib.	MC (DB) RM (%)	D_{eff} (m^2/sec)	Reference
Avocado (~ 15% oil)	31 to 56					1.1×10^{-10}–3.3×10^{-10}	Alzamora et al. (1979)
Avocado (5–9% oil)	58					1.2×10^{-09}–1.8×10^{-09}	
Banana	50 to 70					2.8×10^{-10}–6.4×10^{-10}	Nogueira and Park, (1992)
	20 to 40	0.01–3.50				3.0×10^{-13}–2.1×10^{-10}	Kechau and Maalej (1994)
Banana chip	25	0.0096	0.0102			2.911×10^{-11}	Hong et al. (1986)
	25	0.03				2.9×10^{-12}	
Beef							
freeze dried	25	0.13				3.1×10^{-11}	Lomauro et al. (1985a)
Raw	30	0.01–0.10				1.0×10^{-11}	Saravacos and Stinchfield (1965)
freeze-dried	25	0.1268				3.07×10^{-11}	Lomauro et al. (1985)
raw ground							
freeze-dried raw	25	0.1268				3.07×10^{-11}	Lomauro et al. (1985a)
Beet	65					1.5×10^{-09}	Alzamora (1979)
Biscuit	20 to 100	0.10–0.60				8.6×10^{-10}–9.4×10^{-08}	Tong and Lund, 1990
Bran	10	0.02				1.9×10^{-11}	Sokhansanj (1982)
Bread	20 to 100	0.10–0.75				2.8×10^{-09}–9.6×10^{-07}	Tong and Lund, 1990
	22					3.9×10^{-09}	Zhou et al. (1994)
Carrot	30 to 70					2.3×10^{-10}–4.5×10^{-09}	Mulet et al. (1989)
	30 to 70	0.01–5.0				8.0×10^{-10}–8.0×10^{-09}	
	30 to 70					1.2×10^{-09}–5.9×10^{-09}	
	40 to 100	0.6				2.9×10^{-10}–9.9×10^{-10}	Bimbenet et al. (1985)
	64					6.4×10^{-09}	Saravacos and Charm (1962)
Vacuum dried)	80	0.03–5.0				1.1×10^{-12}–4.0×10^{-11}	Kompany et al. (1993)
Cube	40	0.6				6.75×10^{-11}	Bimbenet et al. (1985)
	60	0.6				12.1×10^{-11}	
	80	0.6				17.9×10^{-11}	
	100	0.6				24.1×10^{-11}	
	64				14	6.4×10^{-9}	Sarvascos and Charm (1962)
Catfish, 0.102% fat	30					3.61×10^{-11}	Jason (1965)
	45	0.6				4.6×10^{-11}	
	60	0.6				7.7×10^{-11}	

(Continued)

TABLE 10.5
Continued

Product	Temp (°C)	MC (DB) Initial	MC (DB) Final	MC (DB) Equilib.	MC (DB) RM (%)	D_{eff} (m^2/sec)	Reference
Coconut albumen	45 to 110	0.2–0.6				4.6×10^{-11}–6.6×10^{-10}	Bimbenet et al. (1985)
	70	0.6				8.2×10^{-11}	
	80	0.6				10.4×10^{-11}	
	90	0.6				10.8×10^{-11}	
	100	0.6				11.8×10^{-11}	
	110	0.6				12.8×10^{-11}	
	45	0.2				1.0×10^{-10}	
	80	0.2				3.3×10^{-10}	
	110	0.2				6.6×10^{-10}	
Cod, 0.05% fat	30					$3.40(\pm 0.36) \times 10^{-11}$	Jason (1965)
Coffee extra	30 to 70	0.08–1.5				1.0×10^{-11}–3.3×10^{-10}	Luyben et al. (1980)
Conger eel, 0.460% fat	30					2.28×10^{-11}	Jason (1965)
Cookie, Oatmeal	25	0.1776				3.97×10^{-11}	Lomauro et al. (1985)
Corn	40	0.05–0.2				1.0×10^{-12}–1.0×10^{-10}	Syarief et al. (1984)
	20 to 65	0.10–0.4				1.1×10^{-11}–5.3×10^{-10}	Ulku and Uckan (1986)
	70 to 90					2.7×10^{-11}–7.7×10^{-11}	Cabrera et al. (1984)
	10 to 27	0.04–0.4				3.5×10^{-12}–6.7×10^{-11}	Parti and Dugmanics (1990)
	40 to 70	0.10–0.2				3.6×10^{-11}–8.3×10^{-11}	Tolaba et al. (1991)
	25 to 40	0.04–0.4				5.3×10^{-12}–1.8×10^{-11}	Muthukumarappan and Gunasekaran (1994)
	50 to 70	0.05–0.35				2.8×10^{-11}–9.1×10^{-10}	Chu and Hustruhid (1968)
Corn (yellow-dent)	40	0.050–0.230				1.0×10^{-11}–1.0×10^{-09}	Litchfield and Okos (1992)
Corn, Gelatinized cooked	70					2.69×10^{-11}–4.53×10^{-11}	Cabrera et al. (1984)
	80					4.3×10^{-11}–4.02×10^{11}	
	85					4.81×10^{-11}–4.22×10^{-11}	
	90					7.56×10^{-11}–7.69×10^{-11}	
Cracker	40 to 90	0.03–0.14				1.4×10^{-04}–1.8×10^{-09}	Kim and Okos (1999)
Dab	30					2.94×10^{-11}	Jason (1965)
Dogfish							
4.0% fat	30					2.2×10^{-11}	
4.00% fat	30					1.3×10^{-11}	
6.40% fat	30					1.6×10^{-11}	
8.60% fat	30					8.3×10^{-11}	

(Continued)

TABLE 10.5
Continued

Product	Temp (°C)	MC (DB) Initial	MC (DB) Final	MC (DB) Equilib.	MC (DB) RM (%)	D_{eff} (m²/sec)	Reference
Egg							
Liquid	85 to 105					1.0×10^{-11}–1.5×10^{-11}	Kincall (1987)
Fresh	85	300	0.73			10×10^{-12}	
incubated	100	213	0.75			16×10^{-12}	
Liquid	100	300	1.05			15×10^{-12}	
	105	300	1.09			14×10^{-12}	
	105	213	0.75			17×10^{-12}	
	125	213	0.88			19×10^{-12}	
Fig	55 to 85					1.13×10^{-10}–6.48×10^{-10}	Babalis and Belessiotis (2004)
Fish (various)	30	0.05–0.3				1.3×10^{-11}–3.1×10^{-10}	Jason (1958)
Flour	25	0.0628				3.86×10^{-12}	Lomauro et al. (1985)
	25	0.1687				3.20×10^{-11}	
	25	0.06–0.17				3.9×10^{-12}–3.2×10^{-11}	Lomauro et al. (1985b)
Garlic	22 to 58	0.20–1.60				1.1×10^{-11}–2.0×10^{-10}	Pinaga et al. (1984)
Dry zone	45	0.140–0.250				3.37×10^{-11}	Pezzutti and Crapiste (1997)
	60	0.090–0.140				4.49×10^{-11}	Pezzutti and Crapiste (1997)
	75	0.06–0.130				5.85×10^{-11}	Pezzutti and Crapiste (1997)
Slice	61	2.4				1.1×10^{-11}	Saravacos and Charm (1962)
Wet zone	37	0.400–1.350				1.54×10^{-10}	Pezzutti and Crapiste (1997)
	49.7	0.240–1.310				2.35×10^{-10}	Pezzutti and Crapiste (1997)
	62.2	0.270–1.360				3.45×10^{-10}	Pezzutti and Crapiste (1997)
Slab	61	2.4		15		1.06×10^{-10}	Saravacos and Charm (1962)
Germ	100	0.02				7.38×10^{-11}	Sokhansanj (1982)
Glucose	30 to 70	0.08–1.50				4.5×10^{-12}–6.5×10^{-10}	Luyben et al. (1980)
Grain (maize)	30 to 90	0.53				3.99×10^{-08}–4.10×10^{-07}	Verma and Prasad (1999)
Grain sorghum	55	0.021				2.8×10^{-11}	Suarez et al. (1980)
	55	0.021				2.8×10^{-11}	
	40	0.021				1.75×10^{-11}	
	30	0.021				1.4×10^{-11}	
	20	0.21				9.0×10^{-12}	
Haddock 0.105% fat	30					3.25×10^{-11}	Jason (1965)

(Continued)

TABLE 10.5
Continued

Product	Temp (°C)	MC (DB) Initial	MC (DB) Final	MC (DB) Equilib.	MC (DB) RM (%)	D_{eff} (m²/sec)	Reference
Halibut 0.208% fat	30					2.49×10^{-11}	
Herring							
2.0% fat	30					1.9×10^{-11}	
12.5% fat	30					3.9×10^{-11}	
6.4% fat	30					9.5×10^{-11}	
Leek	38 to 64					1.1×10^{-09}–6.6×10^{-09}	Akbaba and Cakaloz, 1994
Lemon Sole 0.094% fat	30					2.63×10^{-11}	Jason (1965)
Lentil	30 to 50	0.10–0.20				2.8×10^{-11}–2.8×10^{-09}	Tang and Sokhansanj (1993)
Ling 0.047% fat	30					2.21×10^{-11}	Jason (1965)
Mackerel 0.694% fat	30					2.21×10^{-11}	
Meat ("sobrasada")	10 to 16	0.40–0.9				2.9×10^{-11}–5.4×10^{-11}	Mulet et al. (1992)
Milk foam, whole	35 to 55	0.2				8.5×10^{-10}–3.0×10^{-09}	Komanowsky et al. (1964)
	55					3×10^{-9}	
	50					2×10^{-9}	
	40					1.4×10^{-9}	Lomauro et al. (1985)
	35					8.5×10^{-10}	
Milk							
Dry (nonfat)	25	0.13				2.1×10^{-11}	Lomauro et al. (1985a)
Nonfat dry	25	0.1273				2.13×10^{-11}	Lomauro et al. (1985)
Skim	50 to 90	0.25–0.80				2.8×10^{-11}–3.1×10^{-10}	Ferrari et al. (1989)
Monkfish, 0.094% fat	30					3.06×10^{-11}	Jason (1965)
Muffin	20 to 100	0.10–0.95				8.5×10^{-10}–1.6×10^{-07}	Tong and Lund (1990)
Noodle, Japanese (fresh)	20 to 40	0.23–0.58				2.1×10^{-11}–3.7×10^{-11}	Inazu and Iwasaki (1999)
Oatmeal (cookie)	25	0.1549				3.97×10^{-11}	Lomauro et al. (1985)
Onion	60 to 80	0.05–18.7				2.3×10^{-10}–6.6×10^{-09}	Kiranoudis et al. (1992)
	62	6				1.6×10^{-10}	Saravacos and Charm (1962)
	62	6.00		14		4.17×10^{-11}	
Pasta	40	0.27				2.5×10^{-11}	Andrieu et al. (1986)
	40	0.23				1.8×10^{-11}	Waananen et al. (1996)

(Continued)

TABLE 10.5
Continued

Product	Temp (°C)	MC (DB) Initial	MC (DB) Final	MC (DB) Equilib.	MC (DB) RM (%)	D_{eff} (m²/sec)	Reference
	40	0.256				1.47×10^{-11}	Litchfield and Okos (1992)
		0.125				9.38×10^{-12}	
		0.108				5.83×10^{-12}	
		0.089				4.28×10^{-12}	
		0.06				1.55×10^{-12}	
	55	0.254				2.33×10^{-11}	
		0.12				1.85×10^{-11}	
		0.105				8.75×10^{-12}	
		0.095				2.60×10^{-12}	
		0.088				2.33×10^{-12}	
		0.079				2.07×10^{-12}	
		0.067				1.84×10^{-12}	
		0.058				1.66×10^{-12}	
		0.057				1.61×10^{-12}	
	70	0.239				3.50×10^{-11}	
		0.052				3.50×10^{-12}	
	85	0.238				4.84×10^{-11}	
		0.075				7.02×10^{-12}	
		0.066				6.26×10^{-12}	
		0.048				6.20×10^{-12}	
		0.037				3.96×10^{-12}	
		0.027				3.60×10^{-12}	
		0.021				2.85×10^{-12}	
		0.015				1.60×10^{-12}	
	44	0.14				1.32×10^{-11}	Xiong et al. (1991)
		0.15				1.46×10^{-11}	
		0.21				1.66×10^{-11}	
	55	0.13				2.61×10^{-11}	
		0.21				2.68×10^{-11}	
	71	0.21				3.55×10^{-11}	
Corn	40 to 80	0.10–0.50				2.8×10^{-11}–1.9×10^{-10}	Andrieu et al. (1988)
Dense	40 to 122	0.03–0.23				$2. \times 10^{-12}$–1.05×10^{-10}	Litchfield (1986)
	40 to 90	0.12–0.23				7.0×10^{-12}–4.39×10^{-11}	Andrieu and Stamatopoulos (1986)
	40 to 122	0.03–0.23				9.4×10^{-12}–1.06×10^{-10}	Waananen and Okos (1996)
	40 to 122	0.23				1.84×10^{-11}–6.21×10^{-11}	Xiong et al. (1991)
Dur. wheat	50 to 90	0.16–0.35				2.5×10^{-12}–5.6×10^{-11}	Piazza et al. (1990)
	40 to 90					1.0×10^{-11}–9.0×10^{-11}	Andrieu and Stamatopoulos (1986)

(Continued)

TABLE 10.5
Continued

Product	Temp (°C)	MC (DB) Initial	MC (DB) Final	MC (DB) Equilib.	MC (DB) RM (%)	D_{eff} (m^2/sec)	Reference
Porous	55 to 105	0.02–0.21				3.0×10^{-11}–3.41×10^{-10}	Waananen and Okos (1996)
	40	0.07–0.20				1.87×10^{-11}–3.0×10^{-11}	
	55 to 105	0.02–0.21				1.49×10^{-11}–1.80×10^{-10}	
Semolina	40 to 125	0.10–0.25				3.0×10^{-11}–1.5×10^{-10}	Okos et al. (1989)
	44 to 71	0.14–0.21				1.3×10^{-11}–3.6×10^{-11}	Xiong et al., 1991
	40 to 85	0.01–0.26				1.5×10^{-12}–4.8×10^{-11}	Litchfield and Okos (1992)
Drum wheat	40			> 0.27		0.25×10^{-10}	Andrieu and Stamtopoulos (1986)
	40			0.18–0.27		0.14×10^{-10}	
	40			0.136–0.18		0.08×10^{-10}	
	50			> 0.27		0.35×10^{-10}	
	50			0.18–0.27		0.19×10^{-10}	
	50			0.136–0.18		0.12×10^{-10}	
	60			> 0.27		0.41×10^{-10}	
	60			0.18–0.27		0.24×10^{-10}	
	60			0.136–0.18		0.15×10^{-10}	
	70			> 0.27		0.61×10^{-10}	
	70			0.18–0.27		0.36×10^{-10}	
	70			0.136–0.18		6.25×10^{-10}	
	80			> 0.27		0.72×10^{-10}	
	80			0.18–0.27		0.42×10^{-10}	
	80			0.136–0.18		0.28×10^{-10}	
	90			> 0.27		0.89×10^{-10}	
	90			0.18–0.27		0.50×10^{-10}	
	90			0.136–0.18		0.31×10^{-10}	
Peanut (roasted)	25	0.01–0.02				3.8×10^{-12}	Hong et al. (1986)
	25	0.0175	0.0143			3.800×10^{-11}	
Pear	65	6.5				9.6×10^{-10}	Saravacos and Charm (1962)
	66	6.5		18		9.63×10^{-10}	
Pepper (green)	60 to 80	0.04–16.2				3.8×10^{-10}–1.2×10^{-08}	Kiranoudis et al. (1992)
Pepperoni	12	0.16				4.7×10^{-11}–5.7×10^{-11}	Palumbo et al., (1977)
Pepperoni 13.3% fat		12				5.7×10^{-11}	Palumbo et al. (1970)
17.4% fat		12				5.6×10^{-11}	
25.1% fat		12				4.7×10^{-11}	
Pineapple	20 to 60	0.2–2.0				1.62×10^{-10}–1.2×10^{-09}	Rahman and Lamb (1991)
Pistachio nut	35 to 60	0.05–0.25				1.7×10^{-11}–6.3×10^{-10}	Karatas and Battalbey (1991)

(Continued)

TABLE 10.5
Continued

Product	Temp (°C)	MC (DB) Initial	MC (DB) Final	MC (DB) Equilib.	MC (DB) RM (%)	D_{eff} (m²/sec)	Reference
Pizza							
Cheese	191 to 218					7.0×10^{-11}	Dumas and Mittal (2002)
Crust	191 to 218					5.9×10^{-10}–6.2×10^{-10}	
Tomato paste	191 to 218					1.3×10^{-11}–1.5×10^{-11}	
Pork carcass (muscle)	1 to 33					1.2×10^{-11}–8.6×10^{-11}	Gou et al. (2004)
Potato	54.44					2.58×10^{-11}	Sarvacos and Charm, (1962)
	60					3.94×10^{-11}	
	65.55					4.37×10^{-11}	
	68.88					6.36×10^{-11}	
	65	0.15				2.0×10^{-10}	Aguilera et al., (1975)
	65					1.4×10^{-09}	Alzamora (1979)
	60 to 80					2.4×10^{-10}–2.6×10^{-10}	Gekas and Lamberg (1991)
	65					9.0×10^{-10}	Islam and Flink (1982)
	60 to 100	0.03–5.00				2.8×10^{-10}–5.3×10^{-09}	Kiranoudis et al. (1995)
	31					6.0×10^{-11}–1.6×10^{-10}	Lawrence and Scott (1966)
	30 to 70	0.05–1.50				2.0×10^{-11}–4.2×10^{-10}	Luyben et al. (1980)
	30 to 90					1.1×10^{-10}–4.5×10^{-10}	Mulet (1994)
	30 to 90					1.4×10^{-10}–8.2×10^{-10}	
	40					8.8×10^{-10}–1.2×10^{-09}	Ronald et al. (1995)
	65					4.4×10^{-10}	Saravacos and Charm (1962)
	60					1.8×10^{-10}	Rovedo et al. (1995)
	65	0.010–0.150				1.0×10^{-10}	Litchfield and Okos (1992)
Air-dried	30					3.0×10^{-12}–2.0×10^{-11}	Saravacos (1967)
Freeze-dried	30	3.39–4.96				1.3×10^{-10}–3.2×10^{-10}	
	30	0.01–0.10				8.3×10^{-12}	Saravacos and Stinchfield (1965)
Infrared-dried	1.5	0.10–1.00				6.0×10^{-11}–1.73×10^{-09}	Afzal and Abe (1998)
Puff-dried	30					2.0×10^{-11}–7.0×10^{-11}	Saravacos (1967)
Sugarbeet root	81	2.5–3.6				1.3×10^{-11}–7.0×10^{-11}	Vaccarezza et al. (1974)
Potato starch gel	25	0.2				2.4×10^{-11}	Fish (1958)

(Continued)

TABLE 10.5
Continued

Product	Temp (°C)	MC (DB) Initial	MC (DB) Final	MC (DB) Equilib.	MC (DB) RM (%)	Deff (m²/sec)	Reference
Potato strip (blanched)	180					6.61×10^{-09}–6.81×10^{-09}	Moyano and Berna, (2002)
	170					5.19×10^{-09}–5.29×10^{-09}	
	160					4.11×10^{-09}–4.17×10^{-09}	
Arran Banner	31					1.5×10^{-11}	Lawrence and Scott (1966)
Binjie variety	31					6.2×10^{-11}	
King Edward Variety	31					1.6×10^{-11}	
Pentland Crown Crown	31					1.4×10^{-11}	
Sweet	328	0.1–3.5				3.7×10^{-10}–4.35×10^{-10}	Biswal et al. (1997)
Raisin	25	0.27				4.167×10^{-11}	Lomauro et al. (1985)
	25	0.27				4.2×10^{-13}	Lomauro et al. (1985b)
	60	0.15–2.40				5.0×10^{-11}–2.5×10^{-10}	Saravacos and Raouzeos (1986)
	50	0.17–3.50				1.0×10^{-10}–4.0×10^{-10}	Raghavan et al. (1994)
Rice	60	0.18–0.36				1.3×10^{-11}–2.3×10^{-11}	Zuritz and Singh (1982)
	49 to 82	0.26–0.32				4.7×10^{-12}–1.6×10^{-11}	Steffe and Singh (1980)
	30 to 50	0.10–0.25				3.8×10^{-08}–2.5×10^{-07}	Kameoka et al. (1984)
Brown	20 to 60	0.06–0.34				1.7×10^{-12}–2.8×10^{-09}	Yamaguchi (1992)
	50 to 120	0.15				4.5×10^{-10}–3.9×10^{-09}	Bakshi and Singh (1980)
	35 to 36	0.32				9.27×10^{-11}	Yamaguchi et al. (1985)
	35.4 to 54	0.14–0.31				1.48×10^{-11}–4.1×10^{-11}	Steffe and Singh (1980)
Rough	50 to 120	0.15				1.3×10^{-10}–3.2×10^{-09}	Bakshi and Singh (1980)
	35.4 to 54	0.13–0.32				3.8×10^{-12}–2.0×10^{-11}	Steffe and Singh (1980)
	57 to 70.2	0.19–0.36				1.2×10^{-11}–3.2×10^{-11}	Zuritz and Singh (1982)
Brown	35–36	0.32			32–35	9.27×10^{-11}	Yamaguchi et al. (1985)
	35.4	0.283	0.162		64.4	1.57×10^{-11}	Steffe and Singh (1980)
	36.0	0.299	0.153		62.8	1.48×10^{-11}	
	40.8	0.279	0.163		70.8	1.803×10^{-11}	
	41.0	0.310	0.162		69.7	2.20×10^{-11}	
	45.7	0.247	0.153		69.1	1.96×10^{-11}	

(Continued)

TABLE 10.5
Continued

Product	Temp (°C)	MC (DB) Initial	MC (DB) Final	Equilib.	MC (DB) RM (%)	Deff (m²/sec)	Reference
	45.7	0.287	0.153		70.2	2.41×10^{-11}	
	49.0	0.280	0.136		65.9	3.02×10^{-11}	
	50.0	0.309	0.140		66.4	3.24×10^{-11}	
	54.4	0.306	0.138		70.1	4.07×10^{-11}	
	54.8	0.285	0.137		69.5	3.76×10^{-11}	
Rough	35.4	0.315	0.158		65.5	4.39×10^{-12}	
	35.4	0.320	0.16		65.8	3.79×10^{-12}	
	35.5	0.313	0.154		64.4	4.36×10^{-12}	
	40.2	0.303	0.154		68.8	5.56×10^{-12}	
	40.2	0.320	0.159		68.8	4.99×10^{-12}	
	40.1	0.312	0.15		66.1	5.56×10^{-12}	
	44.7	0.318	0.146		68.1	6.49×10^{-12}	
	45.0	0.323	0.144		67.3	7.43×10^{-12}	
	49.9	0.319	0.133		67.5	1.27×10^{-11}	
	49.9	0.313	0.137		68.0	1.18×10^{-11}	
	55.0	0.316	0.126		69.3	1.42×10^{-11}	
	54.8	0.317	0.131		69.9	1.58×10^{-11}	
	54.8	0.312	0.129		69.5	2.01×10^{-11}	
	52.0	0.3565				1.33×10^{-11}	Zuritz and Singh (1982)
	51.0	0.2660				1.33×10^{-11}	
	50.5	0.2340				1.00×10^{-11}	
	50.5	0.1990				1.33×10^{-11}	
	60.8	0.3607				2.33×10^{-11}	
	60.0	0.2412				2.00×10^{-11}	
	61.0	0.2332				1.67×10^{-11}	
	60.0	0.1862				1.67×10^{-11}	
	60.5	0.2323				1.33×10^{-11}	
	60.5	0.1915				1.67×10^{-11}	
	70.2	0.3565				2.67×10^{-11}	
	70.5	0.2602				3.17×10^{-11}	
	70.3	0.2347				1.17×10^{-11}	
	71.0	0.1972				2.00×10^{-11}	
	70.2	0.2321				1.67×10^{-11}	
White	35.3	0.298	0.160		62.5	6.88×10^{-11}	Steffe and Singh (1980)
	39.9	0.298	0.165		68.6	6.64×10^{-11}	
	47.5	0.309	0.155		68.1	9.07×10^{-11}	
	50.0	0.311	0.145		66.4	9.60×10^{-11}	
	54.8	0.331	0.144		69.3	1.09×10^{-10}	
	35.5	0.275	0.161		65.4	6.50×10^{-11}	
	39.8	0.289	0.166		69.6	6.78×10^{-11}	
	45.5	0.258	0.151		68.4	8.68×10^{-11}	
	49.8	0.269	0.140		66.5	9.79×10^{-11}	
	54.6	0.279	0 140		69.9	1.11×10^{-10}	
Saithe 0.111% fat	30					3.06×10^{-11}	Jason (1965)
Silica gel	25					3.0×10^{-06}–5.6×10^{-06}	Litchfield and Okos, (1992)

(Continued)

TABLE 10.5
Continued

Product	Temp (°C)	MC (DB) Initial	MC (DB) Final	MC (DB) Equilib.	MC (DB) RM (%)	Deff (m²/sec)	Reference
Skate 0.139% fat	30					3.28×10^{-11}	Jason (1965)
Soybean	50	0.638				1.0×10^{-10}	Kitic and Viollaz (1984)
	50	0.631				8.7×10^{-11}	
	63	0.872				2.2×10^{-10}	
	63.5	0.845				1.9×10^{-10}	
	78.4	0.523				2.6×10^{-6}	
	87.2	0.523				2.3×10^{-6}	
	30	0.07				7.5×10^{-13}–5.4×10^{-12}	Saravacos (1969)
	25	0.14				4.3×10^{-10}	Fukuoka et al., (1994)
	50 to 87.2	0.52–0.87				8.7×10^{-11}–2.6×10^{-10}	Kitic and Viollaz (1984)
Squid (mantle)	34.3	0.15–2.74				8.23×10^{-11}	Teixeira and Tobinga (1998)
Starch (granular)	25 to 140	0.10–0.50				4.0×10^{-10}–1.3×10^{-09}	Karathanos (1990);
Starch gel	25	0.10–0.30				1.0×10^{-12}–2.3×10^{-11}	Fish (1958)
	30 to 50	0.20–3.00				1.0×10^{-10}–1.2×10^{-09}	Saravacos and Raouzeos (1984)
	140	0.75–2.50				7.0×10^{-11}–1.5×10^{-09}	Karathanos (1990)
	60 to 100	0.10–1.00				1.4×10^{-11}–3.2×10^{-10}	Vagenas and Karathanos (1991)
	30 to 60	0.5–4.0				1.8×10^{-10}–2.1×10^{-09}	McMinn and Magee (1996)
	30	0.01–0.10				1.2×10^{-11}	Saravacos and Stinchfield (1965)
	30 to 50	0.2–1.7				7.3×10^{-09}–8.8×10^{-09}	Buvanasundaram et al. (1994)
	25	0.010–0.140				1.0×10^{-14}–3.6×10^{-12}	Litchfield and Okos (1992)
Sugarbeet	40 to 80	2.50–3.60				4.0×10^{-10}–1.3×10^{-09}	Vaccarezza and Chirife (1978)
Sugarbeet root	81	2.5–3.6				1.3×10^{-11}	Vaccarezza et al. (1974)
	60	2.5–3.6				7.0×10^{-11}	
	47	2.5–3.6				3.8×10^{-11}	
Swordfish	40 to 55	1.00–5.0				2.5×10^{-10}–8.9×10^{-10}	Del Valle and Nickerson (1968)
2–3% fat	40					3.0×10^{-11}	
	55					3.9×10^{-11}	
Saturated in Salt brine	40					2.6×10^{-11}	
	55					3.3×10^{-11}	

(Continued)

TABLE 10.5
Continued

Product	Temp (°C)	MC (DB)			MC (DB) RM (%)	Deff (m²/sec)	Reference
		Initial	Final	Equilib.			
Tapioca root	84					6.7×10^{-11}	Chirife (1971)
	74					4.8×10^{-11}	Chirife (1971)
	55					3.5×10^{-11}	Chirife (1971)
	55 to 100	0.16–1.95				3.3×10^{-10}–8.6×10^{-10}	Chirife (1971)
Tomato (concetr.)	60 to 100	5.7				1.7×10^{-10}–6.5×10^{-09}	Karatas and Esin (1994)
Turkey	22	0.04				8.0×10^{-14}	Margaritis and King (1971)
Turnip (freeze-dried)	25	0.28				7.6×10^{-12}	Lomauro et al. (1985)
	25	0.2812				7.61×10^{-11}	Lomauro et al. (1985)
Walnut	11 to 43	0.04–0.56				5.7×10^{-12}–9.3×10^{-11}	Alves-Filho and Rumsey (1985)
Ground English	32	0.564	0.155		47	1.908×10^{-11}	
	32	0.279	0.083		47	3.694×10^{-11}	
	32	0.166	0.081		31	6.167×10^{-11}	
	32	0.13	0.059		23	8.333×10^{-11}	
	32	0.114	0.068		48	6.139×10^{-11}	
	43	0.503	0.054		26	3.972×10^{-11}	
	43	0.295	0.052		28	9.333×10^{-11}	
	43	0.186	0.041		30	1.150×10^{-11}	
	43	0.161	0.042		20	1.267×10^{-11}	
	43	0.135	0.043		28	1.464×10^{-11}	
	16	0.042	0.111		91	6.139×10^{-11}	
	16	0.037	0.08		72	6.806×10^{-11}	
	11	0.059	0.113		89	5.694×10^{-11}	
	13	0.081	0.091		70	6.361×10^{-11}	
Wheat	21 to 80	0.12–0.30				6.9×10^{-12}–2.8×10^{-10}	Becker and Sallans (1955)
	20	0.13–0.20				3.3×10^{-10}–3.7×10^{-09}	Hayakawa and Rosser, (1977)
	4 to 50	0.05–0.30				5.1×10^{-10}–2.2×10^{-09}	Sun and Woods (1994)
	50	0.06–0.2				3.3×10^{-10}–5.1×10^{-08}	Jayas et al. (1991)
	100	0.02				5.25×10^{-11}	Sokhansanj (1982)
Bran	100	0.02				1.85×10^{-11}	
endosperm	100	0.02				2.13×10^{-24}	
Flour	25	0.06–0.17				3.3×10^{-10}–5.1×10^{-08}	Lomauro et al. (1985)
Germ	100	0.02				7.38×10^{-11}	Sokhansanj (1982)
hard winter	20	0.13				3.7×10^{-9}	Hayakawa and Rossen (1977)
	20		0.20			3.3×10^{-10}	
Shredded	25	0.1549				5.53×10^{-11}	Lomauro et al. (1985)

(Continued)

TABLE 10.5
Continued

Product	Temp (°C)	MC (DB) Initial	MC (DB) Final	MC (DB) Equilib.	MC (DB) RM (%)	D_{eff} (m^2/sec)	Reference
Western Canadian hard spring	20.8	0.840				6.9×10^{-11}	Becker and Sallans (1955)
	27	0.796				1.16×10^{-11}	
	42.5	0.62				4.40×10^{-11}	
	48.1	0.532				7.10×10^{-11}	
	67.8	0.282				2.210×10^{-11}	
	77.6	0.045				7.200×10^{-11}	
	24.7	0.823				9.7×10^{-11}	
	44.3	0.648				3.75×10^{-11}	
	45.5	0.588				5.36×10^{-11}	
	50.0	0.576				5.65×10^{-11}	
	52.8	0.558				6.35×10^{-11}	
	59.4	0.475				9.52×10^{-11}	
	61.0	0.363				1.580×10^{-11}	
	67.3	0.38				1.500×10^{-11}	
	79.5	0.23				2.770×10^{-11}	
Whiting 0.036% fat	30					2.72×10^{-11}	Jason (1965)

In those models, the dependence of the diffusivity on temperature is generally described by the Arrhenius equation as follows.

$$D_{eff} = D_o \exp(-E_a/RT) \tag{10.44}$$

where D_{eff} is the effective diffusivity, E_a is activation energy, and T is the absolute temperature.

The activation energy E_a can be determined from the plot of $\ln D_{eff}$ vs. $1/T$. The slope of the line is $-E_a/R$ and the intercept equals $\ln D_o$. For regular pasta the activation energy E_a is 5.2 Kcal/mole and $D_o = 6.39 \ 10^{-8} m^2/sec$ [Xiong et al. (1991)]. D_o depends only on the pore structure of the food material and therefore can be considered as structure parameter. E_a may depend on the type and the amount of solutes in water.

Values of E_a found in the literature are presented in Table 10.6. One trend that should be noted is that activation energy is inversely related to moisture content.

10.5.1.3 Prediction of Variation of D_{eff} with Moisture

As pointed out earlier, D_{eff} can be obtained from a drying curve by plotting log $(m - m_s)/(m_0 - m_s)$ vs. time. The slopes of the curves at different moistures yield D_{eff} as a function of moisture. Even though, one should obtain D_{eff} values from the drying curves involving small changes in moisture, Waananen (1989) demonstrated that D_{eff} values obtained from drying curves involving relatively large changes in moisture are reliable.

Typical values of D_{eff} at different moistures for extruded pasta are given in Table 10.7 and its variation with moisture and temperature is shown in Figure 10.14. D_{eff} is found to be smaller at lower moisture contents, increasing with moisture and eventually becoming constant at sufficiently high moistures.

Xiong et al. (1991) postulated that the decrease in the effective diffusivity at lower moisture contents is a result of a decrease in the availability of water molecules for diffusion, that is, D_{eff} can

TABLE 10.6
Activation Energy for Moisture Diffusivity

Product		Activation energy (kJ/mol)	Reference
Animal feed	(MC 0.01–0.15)	106.8–37.8	Luyben et al. (1980)
Apple	(MC 0.05–2)	110.5–51.1	Luyben et al. (1980)
Carrot		20.9	Bimbenet et al. (1985)
Coconut albumen	(MC > 0.4)	13.0	Bimbenet et al. (1985)
	(MC < 0.4)	33.9	Bimbenet et al. (1985)
Coffee extract	(MC 0.1–2)	108.0–23.0	Luyben et al. (1980)
Corn		46.0	Bimbenet et al. (1985)
Fish	(MC < 0.1)	36.8	Jason (1958)
Fish muscle (cod)		29.7	Jason (1958)
Fish, swordfish		15.1	Del Valle and Nickerson (1968)
Fish, salted swordfish		15.1	Del Valle and Nickerson (1968)
Glucose	(MC0.1–2)	24.7–23.9	Luyben et al. (1980)
Grain sorghum	(MC = 0.21)	31.4	Suarez et al. (1980)
Whole milk concentrate foam	(0.4 g/cm^3 density)	58.6	Komanowsky et al. (1964)
Potato		52.3	Sarvacos and Charm (1962)
	(MC 0.05–2)	108–31.4	Luyben et al. (1980)
Rice, starchy Endosperm	(MC = 0.13–0.34)	28.5	Steffe and Singh (1980)
Bran		44.8	Steffe and Singh (1980)
Semolina pasta		25.9	Litchfield and Okos (1986)
Corn-based pasta		48	Andrieu et al. (1988)
Skim milk	(MC 0.15–2)	13.0–10.9	Luyben et al. (1980)
Soybeans	(MC = 0.29)	36.4	Suarez et al. (1980)
	(MC = 0.42)	30.1	Kinc and Viollaz (1984)
	(MC = 0.62)	28.9	Kitic and Voillaz (1984)
Starch gel	(MC = 0.14)	26.4	Fish (1958)
Sugarbeet root		28.9	Vaccarezza et al. (1974)
Tapioca root		22.6	Chirife (1971)
Wheat		54.1–61.1	Becker and Sallans (1955)
Durum wheat pasta	Cylinder Slab	22.5 ± 0.524.5	Andrieu and Stamatopoulous (1986)

be related to the proportion of free water molecules in the porous food. This can be expressed as follows:

$$\frac{d[F]}{dt} = k_1[B] - k_2[F] = k_{1o}e^{-E_1/RT}[B] - k_{2o}e^{-E_2/RT}[F] \qquad (10.45)$$

When the free water [F] does not change more with time $d[F]/dt = 0$ and the above equation results in:

$$\frac{[F]}{[B]} = \frac{k_{1o}}{k_{2o}}e^{(-E_1+E_2)/RT} = Ke^{-E_b/RT} \qquad (10.46)$$

TABLE 10.7
Effective Diffusivities for Pasta

Temp. (°C)	Average moisture (DB)	$D_{\text{eff}}(\times 10^{12}\,\text{m}^2/\text{sec})$ Regular pasta	Puffed pasta
105	0.21	82.8	189.8
105	0.08	79.2	180.4
105	0.07	74.9	159.2
105	0.06	70.9	140.4
105	0.05	—	131.4
105	0.04	56.0	128.4
105	0.03	52.7	103.5
71	0.21	35.8	108.0
71	0.11	34.1	—
71	0.10	31.2	118.6
71	0.09	29.2	106.4
71	0.08	26.7	93.1
71	0.07	24.0	74.9
71	0.06	21.3	60.7
71	0.05	20.7	45.9
55	0.21	26.8	88.8
55	0.13	26.12	—
55	0.12	24.9	—
55	0.11	23.01	—
55	0.10	21.02	81.1
55	0.09	19.5	68.6
55	0.08	18.3	51.1
55	0.07	—	38.3
55	0.06	—	27.5
55	0.05	—	14.93
44	0.21	16.6	44.4
44	0.15	14.6	—
44	0.14	13.2	—
44	0.13	11.8	—
44	0.12	10.45	—
44	0.11	9.22	40.1
44	0.10	8.40	33.6
44	0.09	7.35	30.2
44	0.08	—	24.3
44	0.07	—	18.7

Upon a rearranging of the above equation

$$\frac{[F]}{[F]+[B]} = \frac{K\,e^{-E_b/RT}}{1+Ke^{-E_b/RT}} \tag{10.47}$$

$$\frac{D_{\text{eff}}}{D_o} = e^{-E_a/RT} \cdot \frac{Ke^{-E_b/RT}}{1+Ke^{-E_b/RT}} \tag{10.48}$$

Equation 10.45 is the rate of formation of free water, where $[F]$ is the concentration of free water and $[B]$ is that for bound water; Equation 10.46 shows that the process is at equilibrium state, where $-E_b$ is the binding energy. Equation 10.47 provides the proportion of free water to whole water

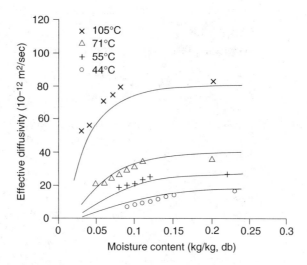

FIGURE 10.14 Effective diffusivity for the regular pasta.

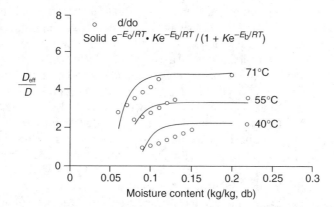

FIGURE 10.15 Comparison of experimental and predicted D/D^o vs. moisture content for regular pasta.

molecules. Equation 10.48 shows that the diffusivity is in proportion to the amount of free water, where $-E_a$ is the activation energy for diffusion at high moistures. A K value of 1032.6 was obtained from the regression of diffusivity data and binding energy data for regular pasta at 55°C according to Equation 10.48. When plotting $e^{-E_a/RT} \cdot K e^{-E_b/RT} / 1 + K e^{-E_b/RT}$ vs. moisture content and D_{eff}/D_o vs. moisture content at 40°C using K obtained at 55°C, the two curves agree reasonably well. The same is found to be true at 71°C (Figure 10.15). Equation 10.48 is insensitive to the K value, because K is much larger than the other terms in Equation 10.48. At high moisture level, the binding energy is near zero and this results in D_{eff}/D_o being constant. So, at high moisture levels, D_{eff} is independent of moisture content.

In order to use Equation 10.48 to predict the variation of D_{eff} with moisture for other products, the values of the constants D_o, E_a, E_b, and K have to be determined from experiments.

As pointed out earlier, D_o and E_a are obtained from the experimental measurement of D_{eff} at high moistures and at different temperatures. As discussed the slope of the plot of $\ln D_{eff}$ vs. $1/T$ should give E_b and the intercept D_o. Experimental measurements of moisture sorption isotherms at different temperatures can be also used in the Clausius–Clayperon (Equation 10.21) to determine E_b at different moisture contents. Finally the constant K can be determined by fitting the experimental measurements of D_{eff} at different moisture contents to Equation 10.48. In addition to diffusion,

TABLE 10.8
Proposed Mass Transfer Mechanisms

Vapor	Liquid
Mutual diffusion	Diffusion
Knudsen diffusion	Capillary flow
Effusion	Surface diffusion
Slip flow	Hydrodynamic flow
Hydrodynamic flow	
Stefan diffusion	
Poiseuille flow	
Evaporation/condensation	

several other mechanisms of internal mass transfer have been proposed in the drying literature including surface diffusion, hydrodynamic or bulk flow, and capillary flow. Table 10.8 lists internal mass transfer mechanisms that have been proposed for vapor and liquid phase water. Modeling is complicated because more than one mechanism may contribute to the total flow, and the contribution of different mechanisms may change as the drying process proceeds (Bruin and Luyben, 1980). The development of a generally applicable drying model requires the identification and inclusion of all contributing mechanisms. A diffusional internal mass transfer mechanism has been assumed in many modeling studies. A distinction can be made between liquid diffusion, vapor diffusion, and surface diffusion of adsorbed molecules. Van den Berg (1981) supports the concept of surface diffusion of water in starch molecules, suggesting that molecules hop from one adsorbed site to another.

Ceaglske and Hougen (1937a,b) postulated that capillary water movement was predominant in granular solids for all water above the saturation point. Comings and Sherwood (1934) used a capillary moisture transfer mechanism to qualitatively account for observed drying characteristics of a clay mix.

Internal mass transfer due to a gradient in total pressure has been postulated in materials ranging from food and wood products to chemical catalyst pellets. At temperatures approaching and exceeding the boiling point of water, rapid vapor generation may produce significant total pressure gradients in addition to partial vapor pressure gradients. Total pressure driven flow may occur in moderate temperature vacuum drying and high temperature convective and contact drying. Pounder and Ahrens (1987) suggest that bulk vapor flow during high intensity paper drying may reduce energy usage by removing liquid water through physical displacement rather than through evaporation.

Moyne and de Giovanni (1985) derived a drying model for a one-dimensional system which accounts for total pressure driven flow. Theoretical predictions were qualitatively compared with experimental results obtained for the drying of light concrete slabs in superheated steam. Predicted and experimental curves were shown to have similar shapes.

Cross et al. (1979) and Gibson et al. (1979) studied the theoretical development of pressure gradients within iron ore pellets during drying. Based on the results from the model, the porosity of the body was shown to be a significant factor in determining the maximum pressure gradient. The model predicted that there would be no pressure gradient in the material for porosities greater than 0.3. This conclusion was supported by manufacturer experience with dense silica shapes, but no experimental confirmation was reported.

Irreversible thermodynamics theory has been used to develop drying models that account for cross-effects between different driving forces. Phenomenological laws such as Fourier's lass, Fick's law, and Ohm's law are based on proportionalities between a flux and a driving force. When more than one driving force is present in a process, cross-effects can occur. For example, in coupled heat conduction and mass diffusion, a gradient in temperature can cause mass transfer (Soret effect) and a

---

TABLE 10.9
Factors That Influence Quality during Drying

Chemical	Physical	Nutritional
Browning reactions	Rehydration	Vitamin loss
	Solubility	Protein loss
Lipid oxidation	Texture	Microbial survival
	Aroma loss	
Color loss		

---

gradient in mass concentration can cause heat flow (Dufour effect). Application of the principles of irreversible thermodynamic enables consideration of coupled transport processes. Instead of separate differential equations for heat and mass transfer, a system of coupled equations is obtained.

Whitaker (1988a) performed a detailed analysis of the Soret and Dufour effects for heat and mass transfer in a porous medium. He concluded that the effects are negligible compared with coupling effects caused by classic equilibrium thermodynamic considerations. Fortes and Okos (1981c) developed a drying model for extruded corn meal using the irreversible thermodynamics framework, but observed that cross-effect terms were small compared to direct terms. Based upon this analysis, it does not appear practical to utilize the irreversible thermodynamics framework for development for porous solids.

10.6 QUALITY CHANGES IN FOOD DURING DRYING

The types of degradation common in food drying have been identified in Table 10.9. Numerous factors can influence to what degree the product is changed, and their effects on the degradation were reviewed as follows.

10.6.1 BROWNING REACTIONS

Browning reactions change color, decrease nutritional value and solubility, create off-flavors, and may induce textural changes. Browning reactions can be classified as enzymatic or nonenzymatic with the latter being more serious as far as the drying process is concerned. The two major types of nonenzymatic browning (NEB) are caramelization and Maillard browning.

In addition to the moisture level, temperature, pH, and the composition are all parameters which affect the rate of nonenzymatic browning. The Arrhenius relationship provides a good description of the temperature dependence of the browning reaction (Burton, 1954; Hendel et al., 1955; Labuza and Saltmarch, 1981; and Toribio and Lozano, 1984). It seems that browning follows zero order kinetics after an initial induction period (Hendel et al., 1955; Singh, 1983; and Petriella et al., 1985). The rate of browning has been shown to be most rapid in the intermediate moisture range and decreases at very low and very high moistures.

During drying, browning tends to occur primarily at the center. This may be due to migration of soluble constituents (sugars) towards the center of the food. Browning is also more severe near the end of the drying period when the moisture level is low (Hendel et al., 1955) and less evaporative cooling is taking place which cause the product temperature to rise.

There are several suggestions to reduce browning during drying. They all emphasize that the product should not experience unnecessary heat when it is in its critical moisture content range. Five different models describing nonenzymatic browning as a function of temperature and moisture were found in the literature. The form of each model is very similar where temperature effect is expressed

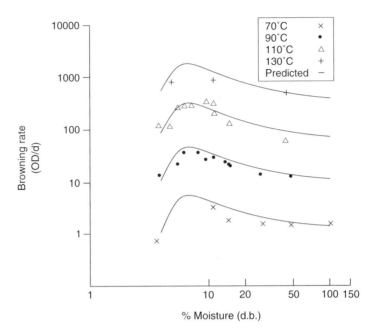

FIGURE 10.16   Predicted and experimental values of browning rate constant as a function of moisture contents at various temperatures (°C): x, 70; • 90; △, 110; +, 130; −, predicted. (From Franzen, K., et al., 1990. *J. Food Eng.*, 11: 225–239.)

by the Arrhenius relation and the coefficients of model take care of the moisture dependencies (Franzen et al., 1990).

10.6.1.1 Nonenzymatic Browning Kinetics of Milk

A model of the browning rate as a function of time and temperature is useful to food processors in determining the amount of browning a product will attain in a particular process. Such a model could also be used to evaluate new dryer designs with respect to quality. Franzen et al. (1990) determined the rates of NEB in skim milk at constant temperatures (ranging from 35 to 130°C) and at constant moistures (ranging from 3 to 50%) and developed a model to describe NEB as a function of time, temperature, and moisture (Figure 10.16).

Isotherms of humidified nonfat dry milk (NFDM) were measured for temperature range of 25 to 45°C and moisture content range of 1.5 to 35% (d.b.) to correlate water activities with experimental moisture contents and temperatures. The modified Henderson equation yielded the best fit with the following expression.

$$a_w = 1 - \exp[-2.482T^{-0.735}M^{5.392 \times 10^{-5}T^{1.771}}] \qquad (10.49)$$

The samples humidified to different moisture contents were heat-treated in an oil bath which was maintained at test temperatures. Browning of NFDM was obtained at predetermined time intervals from spectrophotometer readings in the form of optical density.

An initial induction period with a slower browning rate was found to precede a browning period in which the rate of browning linearly changed with time.

The browning rate of this latter period increased as temperature increased, as exposure time to heat treatment increased, and as moisture content increased until a certain browning-critical-moisture content was reached. The critical moisture content occurred between 4 and 11% (d.b.). Above the critical moisture content, the browning rate became primarily a function of temperature only. The maximum browning rate was found to occur at about 7% moisture content (d.b.)

A model describing the rate of nonenzymatic browning (NEB) in skim milk was developed as a function of temperature and moisture content for the range of 35 to 130°C and 3 to 5% (d.b.), respectively. It was found that the browning rate was zero order after an initial lag period, and that the temperature dependency of NEB satisfied the Arrhenius relation. The developed model was as follows.

$$\frac{dB}{dt} = k_o e^{-E_a/RT} \text{ with } k_o - \exp 38.53 + \frac{15.83}{m}, \frac{E_a}{R} = 13157.19 + \frac{90816.51}{m^3} \quad (10.50)$$

dB/dt is the rate of NEB, k_o is the Arrhenius constant, E_a is activation energy (kcal/mol), R is the gas constant, T is absolute temperature (K), and m is the moisture content (d.b.).

10.6.2 LIPID OXIDATION

Lipid oxidation is responsible for rancidity, development of off-flavors, and the loss of fat soluble vitamins and pigments in many foods, especially in dehydrated foods. Factors which affect oxidation rate include: moisture content, type of substrate (fatty acid), extent of reaction, oxygen content, temperature, presence of metals, presence of natural antioxidants, enzyme activity, UV light, protein content, free amino acid content, and other chemical reactions. Moisture plays an important part in the rate of oxidation. At water activities around $a_w \sim 0.3$, resistance to oxidation is the greatest.

The elimination of oxygen from foods can reduce oxidation, but the oxygen concentration must be very low to have an effect. The effect of oxygen on lipid oxidation is also closely related to the product porosity. Freeze-dried foods are more susceptible to oxygen because of their high porosity. Air-dried foods tend to have less surface area due to shrinkage, and thus they are not affected much by the presence of oxygen (Villota and Karel, 1980a).

Since the lipid oxidation reaction is auto-catalytic and the rate depends on the progress of reaction, it is difficult to model lipid oxidation. Minimizing oxygen level during processing and storage and addition of antioxidants as well as sequestrants have been recommended to prevent lipid oxidation.

10.6.3 COLOR LOSS

Carotenoids are fat soluble pigments present in green leaves and red and yellow vegetables. Their unsaturated chemical structure makes them susceptible to the same types of degradation that lipid undergo, namely oxidation (Stefanovich and Karel, 1982). Many studies indicate that the bulk of carotene destruction occurs during storage rather than as a result of the dehydration process. Thus, the composition of the food exerts a major effect on the oxidation reaction. The foods in this study, as well as other studies, followed first-order reaction kinetics.

Haralampu and Karel (1983) have developed an empirical model to predict carotene degradation. This model is most accurate at high water activity levels (Figure 10.17).

In a study on beetroots, pigment retention decreased as temperature increased and as moisture increased. Thus it was found that the beet pigments were most stable in the powders, then slices, and finally least stable in solution. Saguy et al. (1978) developed models describing the reaction kinetics of betanine (red) and vulgaxanthin I (yellow) pigment as a function of temperature and moisture.

10.6.4 REHYDRATION AND SHRINKAGE

The degree to which a dehydrated sample will rehydrate is influenced by structural and chemical changes caused by dehydration, processing conditions, sample preparation, and sample composition. Rehydration is maximized when cellular and structural disruption, such as shrinkage, are minimized.

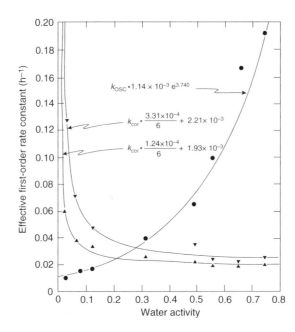

FIGURE 10.17 Observed and fitted values of the effective first-order degradation rate constants for ascorbic acid (k_{asc}), ß-carotene (k_{car}). The observed rates were determined in a 15-day storage test at 40°C. (From Haralampu, S.G. and Karel, M. 1983. *J. Food Sci.*, 48: 1872.)

Several researchers have found that freeze-drying causes fewer structural changes and fewer changes to the product's hydrophyllic properties than other drying processes (Hamm, 1960; Calloway, 1962; McIlrath et al., 1962). During drying most of the shrinkage occurs in the early drying stages where 40 to 50% shrinkage may occur. Thus, to minimize shrinkage low drying temperatures should be employed so that moisture gradients throughout the product are minimized (Van Arsdel, 1963).

10.6.5 SOLUBILITY

Many factors affect the solubility such as processing conditions, storage conditions, composition, pH, density, and particle size. It has been found that the increase in product temperatures is accompanied by an increase in protein denaturation which decreases solubility of the product.

A low bulk density (<0.4 g/ml) is required for good dispersibility of nonfat dry milk. It was found that particle agglomeration, which increases particle size, increased sinkability of this food product. However, Baldwin et al. (1980) found that larger particles were less soluble. This was attributed to the longer drying time required to dry large particles. Thus, more protein was denatured and solubility decreased. This shows that the heat treatment as well as the particle size must be considered when determining solubility.

10.6.6 TEXTURE

Factors that affect texture include moisture content, composition, variety, pH, product history (maturity), and sample dimensions. The chemical changes associated with textural changes in fruits and vegetables include crystallization of cellulose, degradation of pectins, and starch gelatinization. Lund (1983) found that starch gelatinization was too complex to model. Experiments must be carried out on each system in order to approximate it.

Texture is also dependent on the method of dehydration. If high temperatures are used during the drying process, case hardening results in the outer surface being hard or glassy. The hardening of fruit pieces can be limited by infusion with sugar prior to drying. The sugar helps maintain a soft, pliable texture at moisture contents as low as 5%.

10.6.7 AROMA AND FLAVOR

Volatile organic compounds responsible for aroma and flavor have boiling temperatures lower than water. As a result, they are often lost during dehydration. However, if a thin dry layer is formed over the product during the initial stage of drying, these components can be retained. This is because the thin layer of dried food material is selectively permeable to water only.

10.6.8 VITAMIN LOSS

Ascorbic acid is sensitive to high temperatures at high moisture contents (Mishkin et al., 1982). Several studies have shown that the maximum rate of ascorbic acid degradation occurs at specific (critical) moisture levels. The critical moisture level appears to vary with the product being dried and/or the dehydration process. Various models have been developed to describe the rate of ascorbic acid degradation in foods during dehydration and storage (Wanninger, 1972; Villota and Karel, 1980a, 1980b; and Haralampu and Karel, 1983).

To optimize ascorbic acid retention, the product should be dried at a low initial temperature when the moisture content is high since ascorbic acid is most heat-sensitive at high moisture contents. The temperature can then be increased as drying progresses and ascorbic acid is more stable due to a decrease in moisture (Mishkin et al., 1982).

10.6.9 PROTEIN LOSS

Several studies have shown that protein loss during drying is not a major nutritional problem.

10.6.10 MICROBIOLOGICAL QUALITY

Reducing the water activity of a product below 0.85 inhibits growth but does not result in a sterile product. The heat of the drying process does reduce their numbers, but the survival of food-spoilage organisms may give rise to problems in the reconstituted food (Gibbs, 1986). Recommendations for the control of microorganisms during processing are often very basic. The highest possible drying temperatures should be used to maximize thermal death even though low drying temperatures are best for maintaining organoleptic characteristics. If a process is optimized for other quality factors, it constrains the maximum allowable water content.

10.6.11 VISCOELASTIC PROPERTIES OF FOODS

Basic models which describe viscoelastic behavior of dried food materials were introduced with the equations for the corresponding stress-strain relationship. They include the Maxwell, Kelvin, Burger model, and the differential operator equation model. Like other properties, viscoelastic properties such as modulus and compliance functions of biological materials strongly depend on temperature and moisture content. A special type of temperature and moisture content effect on the viscoelastic properties called the thermorheological and hydrorheological simplicity was explained in great detail.

Characterizing the viscoelastic behavior of dried food materials is usually done by the measurement of relaxation modulus $E(t)$ and Poisson's ratio $v(t)$ from simple uniaxial tension or compression test. Other experimental techniques such as constant strain rate test, stress relaxation test, creep test,

dynamic test, and the Hertz and Boussinesq technique have been described in regards to their principles and data analysis in a book related to the mechanical properties of plant and animals (Mohsenin,
1986).

10.6.12 DRYING INDUCED STRESS CRACKS IN FOODS

During drying of foods, internal stress cracks or fissures are often produced. Serious quality deterioration can be caused due to cracks. For example, the presence of cracks or fissures increases breakage
of the dried products when they are subjected to mechanical stresses during handling; cracked grain
kernels lead to increased susceptibility to insect and microbial attack. The cause of stress crack
formation during drying is due to uneven volumetric changes resulting from uneven moisture and
temperature distribution (gradients) in the material. Temperature increase leads to product expansion, while moisture loss results in product contraction. Temperature-induced gradient stresses are
called thermal stresses and gradients caused by moisture gradients hydro stresses. Stresses depend on
transient moisture and temperature gradients in the material, but moisture gradient plays a dominant
role (Litchfield and Okos, 1988). Cracks form when the combined thermal and hydro stresses exceed
a critical value called failure stress which is normally determined by an axial tensile or compressive
test of the material. Failure stress depends on the physical and rheological properties of the material and is influenced significantly by moisture content (Liu et al., 1990). Crack formation is also
significantly affected by operational conditions. Drying tests of cylindrical food indicated that air
humidity influenced most strongly crack formation, followed by air temperature and initial moisture
content, and other factors in descending order of influence were surface heat transfer coefficient,
mass transfer coefficient, initial food diameter, and moisture diffusivity (Liu et al., 1997). Crack
initiation and propagation can be tracked using either NMR imaging technique (Song and Litchfield
1994) or X-ray microtomography (Leonard et al., 2004). Since experimental investigation on the
interaction between different factors is very costly and time consuming, mathematical models are
usually employed to quantify this interactive influence in order to get a proper process design which
minimizes stress cracks. A number of these models can be found in the literature (e.g., Litchfield
and Okos, 1988; Irudayaraj and Haghighi, 1993; Akiyama et al., 1997; Inazu et al., 2005).

10.7 DRYER DESIGN

10.7.1 CONVENTIONAL

10.7.1.1 Spray Dryer

In a spray dryer, foods are transformed from a pumpable liquid into a dry powder. The liquid is pumped
through a nozzle where it is atomized. The droplets are dried by hot air as they fall to the bottom of the
chamber. Spray drying is especially advantageous for heat sensitive products because the particles
are never subjected to a temperature higher than the wet bulb temperature of the drying air, and their
residence time is short, usually between three and thirty seconds (Dittman and Cook, 1977). The
spray drying operation is easily divided into three distinct processes: atomization, drying through the
contact between the droplets and the heated air, and collection of the product by separating it from
the drying air. A typical spray dryer configuration is shown in Figure 10.18. The dryer configuration
and the properties of the feed material determine the operating conditions necessary to provide a
high quality finished product. Each aspect of the dryer will be discussed in turn.

10.7.1.2 Atomization

The type of atomizer is important because it determines the energy required to form the spray,
the size and distribution of the droplets, available heat transfer area, drying rate, the droplet speed
and trajectory, and the final product size (Filkova and Mujumdar, 1987). The types of atomizers

most commonly used in spray drying are hydraulic (pressure) nozzles and rotary wheels. Two-fluid pneumatic nozzles have been used in special instances such as in the drying of thick slurries or pastes, but they are not efficient at high capacities (Williams-Gardner, 1971).

1. Wheel atomizer

   A wheel type atomizer is shown in Figure 10.19. Liquid is fed into the center of the spinning wheel under centrifugal force. The droplets are guided and shaped by vanes in the wheel. The droplets are projected horizontally away at 100 to 200 m/sec with angular velocities of 10,000 to 30,000 rpm (Filkova and Mujumdar, 1987). Disk diameters typically range from 2 to 18 in. Since wheel atomizers are not susceptible to clogging, they are often used for slurries or pastes (Williams-Gardner, 1971). Wheel atomizers produce a homogeneous spray and the mean particle diameter can be controlled by varying rotational speed. Wheel atomizers are widely used in the food industry because they can handle a wide range of liquid viscosities and physical properties (Filkova and Mujumdar, 1987).

2. Pressure nozzles

   Pressure nozzles are used to create droplets by forcing the liquid through a small orifice (0.4 to 4 mm). Nozzles have a maximum flow rate of 1 l/h. In situations which warrant higher flow rates, several nozzles are installed in the drying chamber. Typical pressures range from

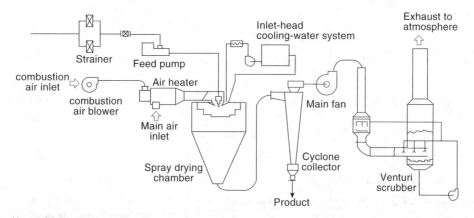

FIGURE 10.18 Spray-drying plant. (From Filkova, I. and Mujumdar, A.S. 1987. In *Handbook of Industrial Drying*, Marcel Dekker, New York, pp. 243–294.)

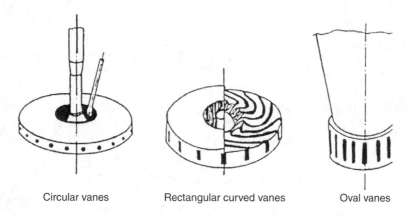

Circular vanes   Rectangular curved vanes   Oval vanes

FIGURE 10.19 Wheel atomizer for a spray dryer. (From Filkova, I. and Mujumdar, A.S. 1987. In *Handbook of Industrial Drying*, Marcel Dekker, New York, pp. 243–294.)

TABLE 10.10
Range of Droplet and Particle Sizes
Obtained in Spray Dryers (μm)

Rotating wheels	1–600
Pressure nozzles	10–800
Pneumatic nozzles	6–300
Milk	30–250
Coffee	80–400

Source: Filkova and Mujumdar (1987).

300 to 4000 psig. Dryers with pressure nozzles typically contain drying chambers that are narrow in diameter and tall in height. The small orifice size facilitates clogging, so pressure nozzles are seldom used when the feed is highly concentrated. The droplets produced have a narrow range of diameters and the dried product consists of hollow spheres (Filkova and Mujumdar, 1987). The operating cost of pressure atomizers is lower than that of wheel or pneumatic nozzles.

10.7.1.3 Droplet Size Determination

The first step in sizing a spray dryer is the determination of the droplet size. The droplet size will vary with nozzle type and feed material. An estimation of droplet size can be made with the following equations, but actual testing will be required before the exact dryer configuration is obtained. The range of particle size obtained by different systems is given in Table 10.10.

10.7.1.3.1 Rotating Wheel Atomizers
To estimate droplet size with the rotating atomizer, the wheel size and speed must be known. If unknown, it is best to estimate these values using a trial and error process, find the droplet size that can be dried at the lowest cost, and attempt to choose an atomizer which will produce that size of droplet. The equation which relates average particle size and nozzle operating parameter is (Dittman and Cook, 1977):

$$D_\mathrm{a} = 12.2 \times 10^4 r \left(\frac{\Gamma}{\rho_1 N r^2} \right)^6 \left(\frac{\mu}{\Gamma} \right)^2 \left(\frac{\alpha \rho_1 L}{\Gamma^2} \right)^1 \qquad (10.51)$$

D_a is the average particle size (μm), α is surface tension of the liquid (lb/min^2), ρ_1 is liquid density (lb/ft^3), r is the disk radius (ft), Γ is the spray mass velocity per foot of disk periphery (*lb/ft* min), N is the disk speed (rpm), and L is the disk periphery (ft), or (Filkova and Mujumdar, 1987)

$$D_\mathrm{a} = 1.62 \times 10^3 N^{-0.53} M^{0.21} (2r)^{-0.39} \qquad (10.52)$$

where D_a is the sauter mean diameter (m), N is rotational speed (rps), M is mass flow rate (kg/sec), and r wheel diameter (m).

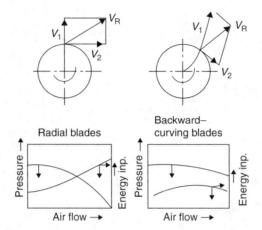

FIGURE 10.20 Typical radial and backward bending fan blades. (From Filkova, I. and Mujumdar, A.S. 1987. In *Handbook of Industrial Drying*, Marcel Dekker, New York, pp. 243–294.)

10.7.1.3.2 Pressure Nozzle

The calculation for estimating the droplet size from a pressure nozzle requires only the pressure drop across the nozzle as shown in the following equation (Dittman and Cook, 1977):

$$D_a = \frac{500}{\Delta P^{\frac{1}{3}}} \tag{10.53}$$

D_a is the mean particle diameter (μm), and ΔP is the pressure drop across the nozzle (psi) or (Filkova and Mujumdar, 1987):

$$D_a = \frac{9575}{\Delta P^{1/3}} \tag{10.54}$$

D_a is the mean particle diameter (μm), ΔP is the pressure drop across the nozzle (Pa).

10.7.1.4 Dryer Chamber Design

The chamber design depends on the type of atomizer selected, the air flow pattern, the production rate, and when drying a heat sensitive product, the temperature profile of the air in the chamber. Pilot scale tests determine the optimum size of the dryer chamber.

The shape of the drying chamber is a function of the trajectory angle of the droplets as they leave the atomizer. The chamber must be sized so that the largest droplet is dry before it contacts a wall.

10.7.1.5 Auxiliary Equipment

The auxiliary equipment will vary with the spray dryer design, but the most commonly employed auxiliary parts are air heaters and fans. The heater may be direct or indirect and fueled by steam, fuel oil, gas, electricity, or thermal fluids. The most common heater in the food industry is the steam type heater. Saturated steam at 150 to 200°C is used to heat the air up to 10°C below the steam temperature. Because they are able to produce high flow rates, centrifugal fans are used to control the air flow in most spray dryers. A two-fan system, with one fan positioned behind the powder recovery cyclones and the other at the drying chamber inlet, provides chamber pressure control. The pressure produced is a function of the blade design. Blades which have a backward bend as shown in Figure 10.20 are the most common.

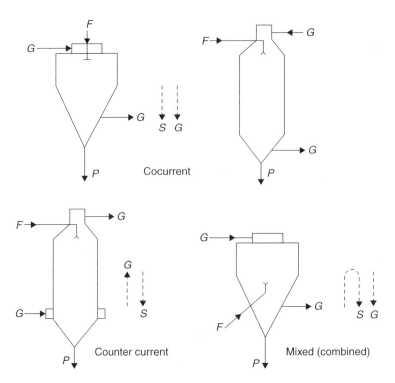

FIGURE 10.21 Airflow patterns in spray dryers. (From Filkova, I. and Mujumdar, A.S. 1987. In *Handbook of Industrial Drying*, Marcel Dekker, New York, pp. 243–294.)

10.7.1.6 Air Flow Patterns

There are three air flow patterns which are commonly used in spray drying: co-current, counter-current, and mixed flow (See Figure 10.21). The air pattern used most often with heat sensitive materials is co-current because the product temperature is lower than the inlet air temperature. When high density dried products of heat sensitive materials are required, counter-current flow is utilized. The drying air flows in the opposite direction of the falling particles. If the size of the dryer is limited, mixed flow patterns are used. The most economical spray drying systems have typically been mixed flow, but that flow pattern is not suitable for heat sensitive materials (Filkova and Mujumdar, 1987). To size a chamber, pilot tests are run to determine the percentage of solids allowable in the feed, the dryer inlet and outlet temperatures, air flow rate, and the configuration of the dryer. The scale up from the pilot test is not precise and requires experience and extensive knowledge of drying operations. Usually, calculations with the air flow rate and temperatures are used to predict the heat load. From the air flow rate and exposure time, the volume of the chamber is calculated (Dittman and Cook, 1977).

10.7.1.7 Calculation of Heat Input

$$Q = h_{\text{g}}(T_{\text{gas}} - T_{\text{s}})A_{\text{s}} \tag{10.55}$$

h_{g} is the gas film coefficient (J/m^2K), T_{gas} is temperature of the gas (K), T_{s} is temperature of the solid (K), A_{s} is the area of the solid surface (m^2).

TABLE 10.11
Typical Operating Parameters for Spray Dryers

Material	Inlet MC (%)	Outlet Mc (1%)	Atomizing device	Liquid air layout	Inlet temperature (°F)	Outlet temperature (°F)
Skim milk	48–55	4	Wheel	Cocurrent	<250	95–100
Whey	50	4	Wheel	Cocurrent	150–180	70–80
Milk	50–60	2.5	Wheel		170–200	90–100
			Pressure nozzle (100–140 bar)	Cocurrent		
Whole eggs	74–76	2.4	Wheel Pressure nozzle	Cocurrent	140–200	50–80
Coffee (instant)	75–85	3–3.5	Pressure nozzle	Cocurrent	270	110
Tea (instant)	60	2	Pressure nozzle	Cocurrent	190–250	90–100
Cream	52–60	4	Wheel	Cocurrent		50–y60
Processed cheese	6	3–4	Wheel	Cocurrent		
Whole eggs	74–76	2–4	Wheel			

The thermal efficiency η of a spray dryer can be determined using the following equation:

$$\eta = \frac{M_{CH} \cdot \lambda}{L_A(T_A - T_{WB})C_{PA} + M_F(T_F - T_{WB})C_{PF}} \tag{10.56}$$

M_{CH} is the chamber evaporation capacity (kg_{H_2O}/sec), λ is latent heat of evaporation of water (J/kg), L_A is air flow rate (kg/sec), M_F is feed flow rate (kg/sec), C_{PA} is heat capacity of the air (J/kg K), C_{PF} is heat capacity of the feed (J/kg K), T_F is temperature of the feed (°C), T_A is temperature of the air (°C), T_{WB} is the wet bulb temperature (°C).

Typically, the heat consumption required for the evaporation of 1 kg of water is 6000 kJ. This compares high with 430 kJ for a six-stage evaporator (Filkova and Mujumdar, 1987).

10.7.1.8 Product Collection

There are many ways to collect the dried product. If the product separates from the air at the bottom of the conical chamber, it is continuously removed through a rotary valve or screw conveyor. It is common for much of the product to remain entrained in the air stream; cyclones, followed by bag filters or wet scrubbers, are used to recover the product. The efficiency of a cyclone is 98 to 99%. It is sometimes desirable to follow the first cyclone with another to collect more of the product (Williams-Gardner, 1971).

10.7.1.9 The Use of Spray Dryers in the Food Industry

Although spray dryers are widely used in the food industry, each type of food product dried in a spray dryer has its own set of challenges. For example, food flavorings must be combined with edible gums and carbohydrates before drying to prevent the loss of volatile components. Table 10.11 shows the operating parameters for spray drying of food products.

10.7.1.10 Food Quality Factors

Volatile retention is a problem for most spray-dried food products. The loss of volatilized material is minimized by increasing the particle diameter (decreases the surface to volume ratio), decreasing the feed temperature (lowers the liquid phase diffusion coefficient), and decreasing the air temperature (minimize particle expansion).

Thermal degradation could be a problem for the droplets which remain in the hot portion of the dryer for too long. If the length of the falling rate period and the time the droplets remain at a higher temperature are long, thermal degradation is more likely. Experimental tests followed by optimization is the best approach for avoiding quality concerns.

10.7.1.11 Example Calculations

Pilot test results for a food flavoring: Initial moisture content 20%
Final moisture content 5%
Air inlet temperature 230°F
Air outlet temperature 100°F
Residence time 6 sec
Ambient air 70°F
Ambient air humidity 0.008(lb/lb dry air)
Output rate 500(lb/h)

Calculate chamber size required and capacity of steam heater:
Find feed rate:

$$\frac{500 \text{ lb}}{\text{h}} * 0.95 = 475 \text{ lb dry solids} \tag{10.57}$$

$$\frac{475}{0.20} = 2375 \text{ lb wet feed} \tag{10.58}$$

Evaporation rate:

$$2375 - 500 = 1875 \frac{\text{lb}}{\text{h}} = 31.25 \frac{\text{lb}}{\text{min}} \tag{10.59}$$

From enthalpy chart or psychrometric chart: Ambient air at 70°F and humidity of 0.008 has

$$\text{Enthalphy} = 18.4 \frac{\text{btu}}{\text{lb}} \tag{10.60}$$

$$\text{Specific Volume} = 13.5 \frac{\text{ft}^3}{\text{lb}} \tag{10.61}$$

After the air has been heated to 230°F, humidity remains 0.008.

$$\text{Enthalphy} = 59.5 \frac{\text{btu}}{\text{lb}} \tag{10.62}$$

$$\text{Specific Volume} = 17.6 \frac{\text{ft}^3}{\text{lb}} \tag{10.63}$$

Assume the amount of energy lost due to convection and other causes is 11% of the difference between the inlet and outlet air enthalpies.

$$(59.5 - 18.4) * 0.11 = 41.1 * 0.11 = 4.5 \frac{\text{btu}}{\text{lb}} \tag{10.64}$$

The resulting enthalpy will be:

$$59.5 - 4.5 = 55.0 \, \frac{\text{btu}}{\text{lb}} \tag{10.65}$$

From a psychrometric chart, the humidity is 0.035 lb/lb dry air and the specific volume is 14.9 $\frac{ft^3}{\text{lb}}$.

Each pound of dry air will acquire $0.035 - 0.008 = 0.027$ lb of moisture. The dry air requirement will be:

$$\frac{31.25}{0.027} = 1157 \, \frac{\text{lb}}{\text{min}} \tag{10.66}$$

The volume of the chamber should be sized according to residence time:

$$\frac{\left(\frac{1157 \, \text{lb}}{\text{min}} \right) \left(14.9 \frac{ft^3}{\text{lb}} \right) (6 \, \text{sec})}{\frac{60 \, \text{sec}}{\text{min}}} = 1724 \, ft^3 \tag{10.67}$$

The volume of air required to be heated is:

$$\left(13.5 \frac{ft^3}{\text{lb}} \right) \left(\frac{1157 \, \text{lb}}{\text{min}} \right) = 15,620 \, \frac{ft^3}{\text{min}} \tag{10.68}$$

This allows no safety factor. The heater should be designed to provide an air temperature of 270°F.

From psychrometric charts, the enthalpy at 270°F is $70 \frac{\text{btu}}{\text{lb}}$ (at a humidity of 0.008).

The heat transfer rate is:

$$\left(1157 \frac{\text{lb}}{\text{min}} \right) \left(70 \frac{\text{btu}}{\text{lb}} \right) \left(60 \frac{\text{min}}{\text{h}} \right) = 4,860,000 \, \frac{\text{btu}}{\text{h}} \tag{10.69}$$

Normal operational requirement:

$$\left(1157 \frac{\text{lb}}{\text{min}} \right) \left(59.5 - 18.4 \frac{\text{btu}}{\text{lb}} \right) \left(60 \frac{\text{min}}{\text{h}} \right) = 2,853,000 \, \frac{\text{btu}}{\text{h}} \tag{10.70}$$

10.7.2 FLUID BED DRYING

Fluid bed drying is commonly utilized in the food industry. Fluid bed drying permits continuous, large scale drying of foods without over-drying. The high heat transfer rates make it an economical process, and the lack of mechanical parts insure low maintenance costs. The rapid mixing in the bed provides nearly isothermal drying conditions. As with most other dryers prevalent in the food industry, the selection and design of fluid bed dryers depends on empirical knowledge.

A typical fluidized bed design consists of a cylindrical column supported on a grid. The grid must be fine enough to prevent the product from falling through it when the dryer is not in operation. Beneath the grid lies a gas distributor and air heater. The column is made tall enough to allow for expansion of the bed due to fluidizing and to prevent particles from being carried into the air exhaust system. The air exhaust system is connected to a dust recovery system and an exhaust fan. Ambient air is introduced into the drying chamber at its base. The air is heated by steam, electricity, or a combustion chamber. Control of the air temperature is based on the bed temperature or exit air temperature. The bed depth is usually not greater than the bed diameter.

10.7.2.1 Drying Theory

Granular particles are fluidized in a drying chamber by a hot gas, typically air. The gas is passed through a grid which supports the granules. The velocity of the gas determines the degree of fluidization. The degree of fluidization is shown graphically in Figure 10.22. When the pressure of the air equals the weight of the particles per area of bed, the layer of particles is incipiently fluidized. At this pressure, the layer undergoes moderate particle mixing. Velocities lower than this result in no particle mixing. Increasing the air velocity over that caused by incipient fluidizing results in rapid mixing of the particles. The additional fluidizing gas passes through the particle layer in bubbles. At higher gas velocities, the particles entrained in the fluidizing gas may be pneumatically conveyed

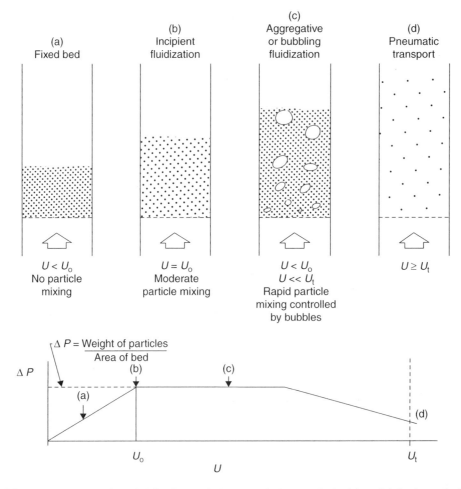

FIGURE 10.22   Regions of gas fluidization. U is the gas velocity, U_o the incipient fluidization velocity, U_t the thermal velocity, and ΔP the pressure drop. (From Hovmand, S. 1987. In: *Handbook of Industrial Drying*, Marcel Dekker, New York, pp. 165–221.)

out of the drying chamber. It is imperative that the product to be fluid bed dried is flowable. Hovmand (1987) listed the characteristics which generally describe the materials suitable for fluid bed drying:

1. The average particle size must be between $20\,\mu$ and $10\,mm$ to avoid channeling and slugging. Particles smaller than $20\,\mu m$ tend to lump together because of their large surface area.
2. The particle size distribution must be narrow to insure that the majority of the particles are fluidized and few are lost by entrainment in the air.
3. For proper fluidization, especially of larger particles, the particles should be spherically shaped.
4. If any lumps are present in the fluid material, they must break up readily once in the dryer to retain fluidity in the bed.
5. The particles must be strong enough to withstand the vigorous mixing in the bed.
6. The final product must not be sticky at the fluid bed exit temperature.

10.7.2.2 Equipment

The major parts of a fluid bed dryer are:

1. The reaction vessel which includes the fluidized bed and the disengaging, or free board, space
2. The gas distributor
3. The solids feeder
4. The product discharge mechanism
5. The instrumentation
6. The gas supply (typically air)

The bed height is determined by the space available and flow rate required, the product residence time, and the space required for internal heat exchanges. Bed heights typically range from one to fifty feet. The free board height is the distance between the top of the fluid bed and the air exit nozzle. It permits entrained particles to fall back to the bed. The gas distributor is designed to prevent the back flow of solids during normal operation. It consists of a perforated ring through which gas is supplied to the bed. The solid feed mechanism is usually a standard weighing and conveying device, such as a screw conveyor or dip pipe. The solids are discharged by an overflow weir or a flapper valve. Solids may be collected from the exit air in cyclones (Wells, 1973).

10.7.2.3 Design Parameters

One of the most fundamental parameters for the fluid bed dryer design is the incipient fluidization gas velocity. The following equations relate the pressure drop with this velocity:

$$\Delta P_o = (\rho_s - \rho_f)(1 - \varepsilon_o)H_o g \tag{10.71}$$

ΔP_o is the pressure drop across the bed, ρ_s is density of the solid, ρ_f is density of the gas, ε_o is bed voltage at incipient fluidization, H_o is bed height at incipient fluidization, and g is the acceleration due to gravity.

ε_o is estimated to be 0.4 for spherical particles, but no general relationship has been made between ε_o and a particle's shape factor. For most systems, values range from $0.4 < \varepsilon_o < 0.55$, with higher values for finer particles. For fine particles,

For fine particles, the Carman–Kozeny equation relates the pressure drop across the bed with the incipient fluidization point.

$$U_o = \frac{\varepsilon_o{}^3}{5(1 - \varepsilon_o)} \frac{\Delta P_o}{S \mu H_o} \tag{10.72}$$

U_o is minimizing fluidizing gas velocity, ε_o is bed voidage at incipient fluidization, ΔP_o is pressure drop across the bed, S is specific surface of the particle, μ is the viscosity of the fluid, and H_o is the bed height at incipient fluidization.

If the particles are spherical, $s = \frac{6}{d}$, and $\varepsilon_o = 0.4$

$$U_o = \frac{d^2(\rho_s - \rho_f)g}{1695\mu} \tag{10.73}$$

Another set of equations has been recommended by Grace and Richardson for particles which are not cohesive:

$$\mathrm{Re} = \left(c_1{}^2 + c_2 A_r\right)^{1/2} - c_1 \tag{10.74}$$

$$\mathrm{Re} = dU_o \frac{\rho_f}{\mu} \tag{10.75}$$

$$A_r = \frac{\rho_f(\rho_s - \rho_f)gd^3}{\mu^3} \tag{10.76}$$

Re is the Reynolds number at minimum fluidization, d is particle diameter, U_o is minimizing fluidizing gas velocity, ρ_s is density of the solid, ρ_f is density of the gas, g is acceleration due to gravity, A_r is Archimedes number, and μ and viscosity of the fluid.

A more exact determination of the incipient velocity can be made by pilot testing.

All of the mixing which occurs in fluid bed dryers is a result of the fluidizing air. The most vigorous mixing occurs just above the air distributor. Heat and mass transfer are efficient between the fluidizing air and the particles because of the large surface area of the particles. Because the air-particle heat and mass transfer are not the limiting elements of fluidized bed drying, a uniform temperature is achieved throughout the dryer. Rising bubbles are the cause of the mixing, and the vertical mixing rate is typically greater than the horizontal mixing rate. The mixing rate increases as the average particle size decreases.

10.7.2.4 Variations in Fluid Bed Design

1. Vibrated Fluid Bed

   If the product to be dried does not fluidize in a standard fluid bed dryer because the particle distribution is too wide or the particles break up due to their low strength, or if they are sticky, thermoplastic or pasty, a vibrated fluid bed dryer may be applicable. A long, rectangular narrow drying chamber is vibrated at 5 to 25 Hz. The air velocity within the dryer can be as low as 20% of the minimum fluidization velocity. The large particles are transported through the dryer by the vibration of the dryer. The vibration provides a gentler means of transportation than vigorous agitation in stationary fluid bed dryers. Vibrated fluid bed dryers typically dry products such as milk, whey, cocoa, and coffee. More detailed information about the operating characteristics for fluid bed dryers may be found in Gupta et al. (1980), Mujumdar (1983), Pakowski and Mujumdar (1982), Ringer and Mujumdar (1982), and Strumillo and Pakowski (1980).

2. Fluid Bed Granulation

A second variation of fluid bed drying is fluid bed granulation. A binding liquid is sprayed into the fluidized bed of granules causing the particles to agglomerate. The fluid bed process is typically batch, although some continuous units are in production (Hoebink and Rietema, 1980). It involves drying, cooling, reacting, mixing, agglomeration, and coating. The main parameters which determine the size distribution and bulk density of the product are how the new particles are formed, and how they grow in the bed (Hovmand, 1987). The process is advantageous because it is not dust producing. The average particle size produced ranges from 0.5 to 2 mm. The process has widespread use in the pharmaceutical industry (Story, 1981).

3. Spouted Bed Dryer

When the particles to be dried are larger than 5 mm and are not readily fluidized in a conventional fluid bed dryer, a spouted bed dryer is employed. The drying air enters the drying chamber at the center of the conical bottom. The particles move in a cyclic fashion through the dryer. As they travel upward in the center, they are carried by the incoming air stream and fall downward at the periphery of the chamber. The advantages of spouted bed dryers are the excellent solids mixing and heat transfer rates. Spouted bed dryers have successfully dried heat sensitive goods such as wheat and peas. Romankov and Raskovskaya (1968) discussed the drying of granular heat sensitive materials in spouted bed dryers.

4. Mechanical Agitated Fluid Bed Dryer

A combination fluid bed and flash dryer used to dry wet cakes was developed to conserve energy costs. The wet cake is fed directly into a dryer via a screw conveyor. Once in the drying chamber, a mechanical agitator breaks up the particles while air is introduced to fluidize the small particles. Dry particles are carried to the exhaust system by the fluidizing one (Hovmand, 1987). Pastes of pigments and dyes are industrially dried with this method (Ormos and Blickle, 1980).

5. Centrifugal Fluid Bed Dryers

Rapid predrying of sticky foods with a high moisture content has been done in centrifugal fluid bed dryers (Figure 10.22). Diced, sliced, and shredded vegetables which are difficult to fluidize, and too heat sensitive to dry in conveyor dryers, are also dried in centrifugal fluid bed dryers. The cylindrical dryer rotates horizontally while air flows into the chamber through the perforated wall. The solids alternate between fluid bed and fixed bed configuration as the dryer rotates (Hovmand, 1987).

6. Fluidized Spray Dryer

Another alternative for hygroscopic and thermoplastic foods is the fluidized spray dryer. A fluid bed chamber is installed directly in the spray drying chamber. The fluidizing air is led to the bottom of the drying chamber. The combination of partially dried and dried products allows agglomeration to take place. Small particles entrapped in the air are recycled from the exhaust system to the drying chamber. The combination of spray and fluid drying provides very efficient use of the drying chamber and produces agglomerated products with low bulk density and good instatizing characteristics (Hovmand, 1987).

10.7.2.5 Air Velocity

The air velocity in the bed must be large enough to promote mixing yet small enough to prevent excessive entrainment of small particles. An air velocity two to three times larger than the incipient fluidization velocity is commonly recognized as the proper fluidization velocity. The velocity at which the particles become entrained in the air is called the terminal velocity. When the flow is

TABLE 10.12
Typical Velocities of Particles

Average particle size	Velocity
100–300	0.2–0.4
300–800	0.4–0.8
800–2000	1.2–3.0

laminar, the terminal velocity is found by Stokes law:

$$U_t = \frac{(\rho_s - \rho_f)g\, d^2}{18\mu} \tag{10.77}$$

d is the particle diameter, U_t is terminal velocity for a particle, ρ_s is density of the solid, ρ_f is density of the gas, g is acceleration due to gravity, and μ is the viscosity of the fluid.

Typical velocities of particles with densities between 1000 and 2000 (kg/m^3), given by Hovmand (1987), are shown in Table 10.12.

10.7.2.6 Heat Transfer in Fluid Beds

Heat transfer, which is an important design parameter for all drying systems, depends on the heat capacity of the particles and the degree of particle circulation at the heat transfer surfaces.

1. Fluid Bed Drying
   Generalized equations which can be used to estimate the heat transfer rate have been suggested by Botterill (1975), Zabrodsky (1966), Gelperin and Einstein (1972), Schlünder (1980), and Martin (1980). Some generalization may be noted: heat transfer increases with vigorous bubbling, and the maximum temperature differential is maintained by replacement of the particles at the hot wall surfaces. The heat transfer coefficient goes through a maximum value as the gas fluidization velocity increases because of the increasing number of bubbles at the wall and because the volumetric heat capacity of the particles is greater than the gas heat capacity, which promotes heat transfer from particles to the interior of the bed.
2. Residence Time Distribution
   The residence time distribution of fluid bed dryers is determined by whether the dryer has back-mixed or plug flow (Figure 10.23). Broad residence time distributions are obtained in back-mixed fluid beds. In these dryers, the wet feed is introduced directly into the drying bed. The bed height is maintained by an overflow weir, and the drying chamber is usually taller than its width. The behavior in these beds is much like that of an agitated tank with an overflow weir; the vigorous mixing in the bed results in a nearly isothermal condition. The moisture content of the dried particles from back-mixed fluid bed dryers vary dramatically between particles. Approximately 40% of the product remains in the dryer for one half of the average residence time. Therefore, some granules may be underdried, while others are overdried (Hovmand, 1987). The stringent restrictions on the moisture content of dried foods prevent widespread use of back-mixed fluid bed drying. The advantage of back-mixed fluid bed drying is that feed material which is not easily fluidizable can be fed directly into the dryer. The incoming feed is rapidly dispersed among the other particles because of the vigorous mixing in the bed.

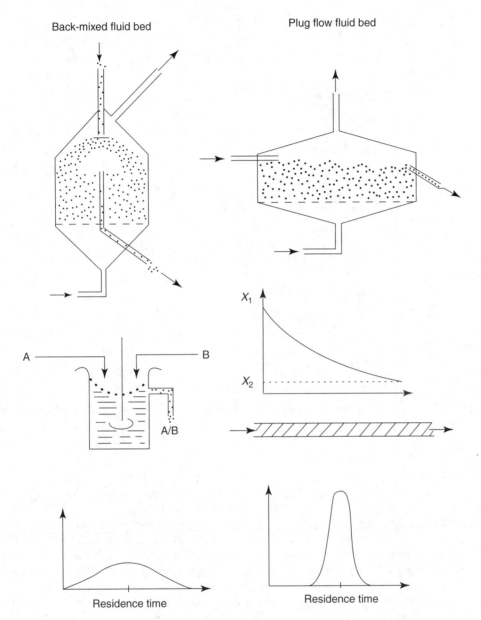

FIGURE 10.23   Residence-time distribution for back-mixed and pilot flow fluid-bed dryers. (From Hovmand, S. 1987. In: *Handbook of Industrial Drying*, Marcel Dekker, New York, pp. 165–221.)

A narrow residence distribution is obtained in plug flow fluid bed dryers which are wider than their height. A narrow residence time distribution can also be obtained by compartmentalizing the dryer (Figure 10.24). These dryers are used to control the residence time in the dryer. Feed is introduced directly into the center of the fluid bed, and the particles are forced to travel in a spiral pattern to the edge of the drying chamber, where they are discarded. At discharge, the product and drying gas are in equilibrium. Since the moisture content is easily controlled, plug flow fluid bed dryers are ideal for heat sensitive food materials (Hovmand, 1987).

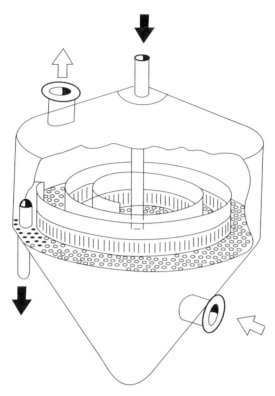

FIGURE 10.24 Compartmentalized dryer for narrow residence-time distribution. (From Hovmand, S. 1987. In: *Handbook of Industrial Drying*, Marcel Dekker, New York, pp. 165–221.)

3. Dryer Design

A small scale fluid bed drying test is used to construct a drying curve (residual volatiles vs. time). A typical drying curve is shown in Figure 10.25. The surface moisture evaporates rapidly until the critical moisture content is reached. After the critical moisture content is reached, drying rate decreases as it is limited by the rate of diffusion of moisture inside the particles. The drying air becomes saturated quickly due to the rapid heat and moisture transfer. The maximum drying air temperature is determined by the heat sensitivity of the product. The temperature limit can be found in pilot plant testing. Visual observation of the fluidization at different air velocities determines the best fluidization velocity with a minimum product loss. As stated by Williams-Gardner (1971), the steps in designing a continuous fluid bed dryer are:

1. Determine that the feed material can be fluidized
2. Determine the optimum fluidizing velocity, inlet air temperature and residence time in pilot tests
3. Construct the drying rate curve and equilibrium moisture content curve at the selected bed temperature
4. Determine the exit gas relative humidity
5. Solve heat and mass balance equations to determine the bed temperature and fluidization flow rate
6. Calculate the bed diameter

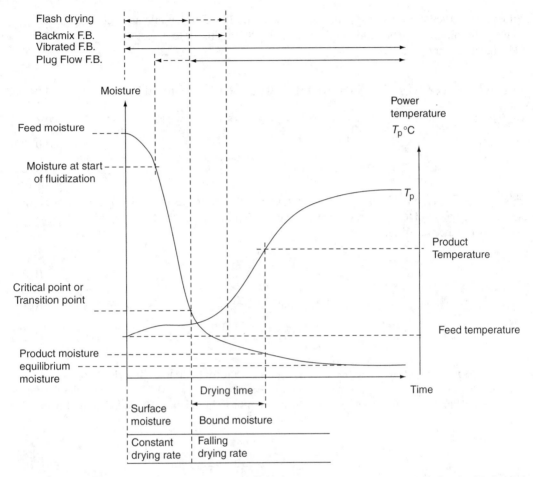

Flash drying
Backmix F.B.
Vibrated F.B.
Plug Flow F.B.

Moisture

Power
temperature
T_p °C

Feed moisture

Moisture at start
of fluidization

T_p

Product
Temperature

Critical point or
Transition point

Feed temperature

Product moisture
equilibrium
moisture

Drying time

Time

Surface
moisture

Bound moisture

Constant
drying rate

Falling
drying rate

FIGURE 10.25 Drying curve for granular product. (From Hovmand, S. 1987. In: *Handbook of Industrial Drying*, Marcel Dekker, New York, pp. 165–221.)

A glass tube of 3 to 4 in. diameter fed with air can be used to test the fluidization behavior of the product. Pilot-scale tests should be performed in a drying chamber at least 12 in. in diameter, and 5 or 6 feet high. A series of tests at different air temperatures and velocities should determine the combination which yields the desired final moisture content. The drying chamber should be shaped similarly to the industrial sized dryer. Because bed depth is usually a lot greater than bed diameter, tests at different bed depths will determine the minimum bed diameter. The evaporation rate per square foot of bed is used to scale up from pilot sized to full scale. The free board space should be at least 1.5 times the bed depth to insure small particles are disentrained (Williams-Gardner, 1971).

10.7.2.7 Example Calculation

A granulated food product is to be dried from 15 to 2% (dry basis) in a fluid bed dryer

Inlet air temperature: 400°F
Fluidizing velocity: 250 ft/min
Retention time: 7 min
Production rate: 8000 lb/h of wet feeds

Pilot plant studies show that bed temperature should be 170°F for an inlet temperature of 400°F. The equilibrium relative humidity of the exit air is 48% at this air temperature.
Heat balance using:

$$C_m G(T_{in} - T_{out}) = \lambda_{T_{out}} W_1 (X_{in} - X_{out}) - W[(t_{out} - t_{in})(C_p + C_1 X_{in})] \tag{10.78}$$

$$0.24 G(400 - 170) = 996(8000)\frac{(0.15)}{(1.15)}(0.15 - 0.02) + 8000\frac{(1.0)}{(1.15)}(170 - 60) \cdot (0.2 + (1.0)(1.5)) \tag{10.79}$$

$$G = 7300 \text{ lb/h} \tag{10.80}$$

T is the gas temperature, t is product temperature, C_m is average heat capacity of gas, C_p is average heat capacity of solid, C_1 is average heat capacity of liquid, $\lambda_{Y_{out}}$ is the heat of vapor at T_{out}.

Mass Balance on Water

$$Y_{out} = \frac{W}{G(X_{in} - X_{out})} + Y_{in} \tag{10.81}$$

$$Y_m = 0.01 \text{lb water/lb dry air} \tag{10.82}$$

$$Y_{out} = \frac{8000}{7300}(0.15 - 0.02)\frac{1.0}{1.15} + 0.01 \tag{10.83}$$

Y is the gas humidity lb water/lb dry gas, X is product moisture content lb water/lb dry solid, G is dry gas rate, lb dry gas/h, W is product rate lb dry solid/h, and Y_{out} is 0.1339 lb water/lb dry gas.
Relative humidity from psychrometric chart is 0.48 Minimum cross-sectional area of bed

$$D = \left[\frac{4G}{60_{ef} \pi V_f}\right]^{0.5} = \left[\frac{4(7300)}{60(0.0455)(\pi)(250)}\right]^{0.5} = 3.69 \text{ ft} \tag{10.84}$$

Bed height in settled state

$$V = \frac{M}{\rho_b} = \frac{W\tau}{\rho_b} = \frac{(8000)(7)}{(40)(60)} = 23.3 \text{ ft}^3 \tag{10.85}$$

V is the bed volume (ft^3), M is hold up in dryer, and ρ_b is the bulk density of settled bed (lb/ft^3).
Fluid bed heights

$$H = \frac{(v)(1.5)}{\pi D^2} = \frac{23.3(1.5)}{(\pi)(3.69)^2 4} = 3.27 \text{ ft} \tag{10.86}$$

W is the production rate (lb/h), τ is retention time (min), and H is the bed height (ft).

10.7.3 FREEZE DRYING

Food products which are too sensitive to withstand any heat are often freeze-dried. In freeze drying, the product is frozen, then the frozen solvent (typically water) is removed by sublimation under vacuum. The sublimed ice, now as a vapor, is pulled from the vacuum chamber by vacuum pumps or steam jet ejectors. The heat of sublimation is supplied by conduction or radiation. Frozen water sublimes at temperatures of 0°C or lower under pressures of 627 Pa or less (Millman et al., 1985).

It is well known that freeze-drying produces the highest quality food product. This is largely because the structure of the food is not severely damaged as in other drying processes. When water is removed from a material by sublimation, a porous, nonshrunken structure remains. Freeze-dried foods are easily rehydrated. Little or no loss of flavor and aroma occur during freeze drying. Product quality remains high because the low drying temperature is not conducive to most degradative processes such as nonenzymatic browning, protein deterioration and enzymatic reactions. The greatest disadvantage of freeze drying is the cost. The drying rate is slow and the use of a vacuum adds to the cost. The final product has low moisture content so some cost is saved by alleviating the refrigeration and storage costs (Liapis, 1987).

10.7.3.1 Process

Since water exists in a combined state in most biological materials, freeze-drying is performed at $-10°C$ to ensure that the water remains in a frozen state. An absolute pressure of 2 mm or less is common (Liapis, 1987). The heat of sublimation must be controlled to ensure that the ice sublimes without melting. Sublimation begins at the exterior surface of the frozen material and recedes toward the bottom. The porous structure left behind obstructs the sublimed vapors, and the drying rate slows as the layer thickens. Minimal heat is supplied because the sublimation is driven by the vacuum. 98 to 99% of the water is removed in this stage. The supply of heat is controlled to provide maximum sublimination rates without the product melting. The removal of the last 1 to 2% usually takes much more time, but the material can be allowed to reach room temperature. The drying rate is influenced by the thickness of the product and its composition. The thinner the product, the higher is the drying rate. Optimum rates are achieved at thicknesses of 1/2 to 3/4 of an inch. Foods with higher sugar contents have slower drying rates (Williams-Gardner, 1971).

The freeze drying operation has three steps: the freezing of the product, ice sublimation, and water vapor removal. The removal of water vapor from the chamber is the most expensive of those processes, and the feasibility of freeze drying often hinges on this step. A vapor trap is placed between the drying chamber and the vacuum pump or steam jet ejectors. The vapor trap has refrigerated surfaces and the vapors condense on the trap as they contact it. The efficiency of the vapor trap is dependent upon the pressure difference between the freeze drying chamber and the vapor trap area, the temperature of the trap, the thickness of ice built up on it, and the temperature difference between the trap surface and the evaporating refrigerant. The less efficient the vapor trap, the lower the temperature in the freeze drying chamber. The area of the vapor condenser is usually equal to the shelf area. Water is removed by the vacuum pump, and the vacuum pump also serves to maintain sub-atmospheric pressures in the drying chamber. The removal of noncondensable gasses reduces the resistance of the sublimed water vapors migrating to the condenser. The presence of noncondensables in the drying chamber greatly reduces the efficiency of the dryer. The vacuum pump should be able to reduce the pressure in the vacuum chamber to at least 5 μM (Powell, 1976).

10.7.3.2 Dryer Chamber Design

1. Pilot Scale
   Portable freeze dryers are used in the food industry in laboratories and in instances when very small amounts of product are required. These units are usually mobile and have self-contained refrigeration, heating, and vacuum pumping processes. Typical capacities range from 2 to 20 kg of frozen product (Liapis, 1987) or 6 to 36 square feet.
2. Scale-Up
   The factors that influence the sizing of a freeze dryer are: chamber size, vacuum pump and condenser capacity and plate area. A one to one scale up ratio is used. Scale up is accomplished by increasing the drying surface area to compensate for the increased food capacity. The vacuum pump and condenser are scaled up proportionally with the increase

in drying area. The thickness of the product is not typically increased above the optimum drying thickness found in pilot scale tests.

10.7.3.3 Design Equations

The heat of sublimation is 1220 btu/lb (2838 kJ/kg). It is conducted inward through the frozen food material. The vaporized ice is transferred through the layer of dry material. Heat and mass transfer occur simultaneously. The heat of sublimation is supplied from the vaporizing gasses to the sample surface. Design equations for predicting the energy requirements and drying time are given by Geankoplis (2003).

The heat flux to the surface occurs by convection and conduction to the sublimation surface:

$$q = h(T_e - T_s) = \frac{k}{L_2 - L_1}[T_s - T_f] \tag{10.87}$$

q is the heat flux (W or J/sec), h is external heat transfer coefficient (W/m^2), T_e is external temperature of gas (°C), T_s is surface temperature of dry solid (°C), T_f is temperature of sublimation front (ice layer), k is thermal conductivity of the dry solid (W/mK), and $L_2 - L_1$ is the thickness of dry layer (m).

The flux of water vapor from the sublimation front is given by:

$$N_a = \frac{D'}{RT(L_2 - L_1)}[p_{fw} - p_{sw}] = K_g(p_{sw} - p_{ew}) \tag{10.88}$$

where N_a is the flux of water vapor (kg mols/sm^2), D' is average effective diffusivity in the dry layer (m^2/sec), R is universal gas constant, T is average temperature in dry layer (°C), $L_2 - L_1$ is thickness of dry layer (m), p_{fw} is partial pressure of the water vapor in equilibrium with the sublimation ice front (atm), P_{sw} is partial pressure of the water vapor at surface in atm, K_g is external mass transfer coefficient (kg mols/sm^2 atm), and P_{ew} is the partial pressure of water vapor in external bulk gas phase (atm).

Arrangement of these equations gives:

$$q = \frac{T_e - T_f}{1/h + \frac{(L_2 - L_1)}{k}} \tag{10.89}$$

$$N\Delta = \frac{(\rho_{fw} - \rho_{ew})}{\frac{1}{K_g} + RT\frac{L_2 - L_1}{D'}} \tag{10.90}$$

where h and K_g are constants which are determined by the gas velocities and characteristics of the dryer. T_e and P_{ew} are set by external operating conditions. k and D' are determined by the nature of the dried material.

$$q = \Delta H_s N_a \tag{10.91}$$

ΔH_s is the latent heat of sublimation of ice (J/kg mol).

p_{fw} is uniquely determined by T_f the equilibrium vapor pressure of ice at that temperature), and we can combine Equations 10.86, 10.87, and 10.88 yielding:

$$\frac{1}{\frac{\Delta L}{k}}(T_s - T_f) = \Delta H_s \frac{1}{\frac{1}{K_g} + RT\frac{\Delta L}{D'}}(p_{fw} - p_{ew}) \tag{10.92}$$

As T_e and T_s are raised, the rate of drying is increased. T_s is limited by the heat sensitivity of the material, and T_f must be kept below its melting point.

To solve the equation, ΔL is set to $(1 - x)(L/2)$.

The rate of freeze-drying is related to N_a by

$$N_a = \frac{L}{2}\left(\frac{1}{M_a V_s}\right)\left(\frac{-dx}{dt}\right) \tag{10.93}$$

where L is the thickness of the solid material (m), V_s is volume of solid material occupied by a unit kg of water, initially $V_s - \frac{1}{X_o \rho_s}$, X_o is initial free moisture content (kg water/kg dry solid), and ρ_s is bulk density of the dry solid (kg/m^3).

By integration, the time for drying is given by:

$$t = \left(\frac{L^2 \Delta H_s}{4k V_s M_a (T_e - T_f)}\right)\left(x_1 - x_2 - \frac{x_1^2}{2} + \frac{x_2^2}{2}\right) \tag{10.94}$$

Integration from $t = 0$, $x_1 = 1.0$, $t = t$ at $x_2 = x_2$. h is assumed to be very large.

10.7.3.4 Industrial Freeze Dryers

1. Tray Freeze Dryer
   The most common type of freeze dryer in operation is the tray freeze dryer. The condensers are mounted in the same chamber as the tray-heater assembly or in a separate chamber joined by a wide tube. The size of these dryers range from 120 to 220 ft^2 and the product rests on trays within the drying chamber. If the required capacity of the dryer is great enough, several freeze dryers may be operated from a central tray heater, condenser, refrigeration, and vacuum pump system. The system would be programmed to stagger the dryer cycles and even out the load on each part of the system. Each dryer could be individually controlled by its own panel, yet the operation would be semicontinuous.
2. Tunnel Freeze Dryer
   This process consists of trays being loaded into the freeze dryer at one end and dried product being discharged from the other end. The process takes place in a large vacuum cabinet, and the trays are loaded and unloaded through vapor locks. The dryer is divided into five independent processing areas. It is cooled by an aqua-ammonia absorption refrigerator which can control the load more readily. This type of freeze dryer is advantageous because the flowrate can be increased as demand increases, although it is difficult to switch from one product to another.

10.7.3.5 Special Considerations for Freeze Drying in Foods

1. Meat and Fish
   Meat and fish are semisolids, consisting of muscular tissue which contains labile proteins, adipose tissue, glycogen, and approximately 75% water. The orientation of muscle is important for heat and vapor transfer. The bones, lean meat, and adipose tissue require quite different drying times. The bone cannot be completely dried, as the fat tissue will melt. Common practice dictates that the bones are packaged with a desicant inside the container. The desicant will remove moisture not removed in the drying process and protect the meat protein from dangerously high moisture levels. The adipose tissue contains 10 to 15% water compared with 70 to 75% for lean meat. The fat tissue may begin to melt before the lean meat is dry. Most of the fat should be trimmed off the meat before freeze drying.

Lean meat should be cut perpendicular to the grain of the muscle fiber. Skin should be cut 10 to 15 mm thick.

2. Fruit and Vegetables
    The cells which make up fruits and vegetables consist of protein, aqueous solutions and cell organs. Between cells is a pectic compound. The structure of plants is such that the orientation during freeze drying is not important. It is important that the structure is not damaged and that dehydration produces a firm, crisp product. Fruits to be freeze-dried should have a high solid content, and good color and flavor. To prevent nonenzymatic browning before freezing, apples and pears are dipped in a 0.1% sodium sulfite solution after they have been peeled, cored and diced. Strawberries and raspberries should be washed and plugged. If the skin of the fruit is impermeable to water vapor, the skin should be slit or perforated. Vegetables are washed, peeled, trimmed, cut, rewashed, and blanched before freeze drying. Vegetables vary in how long they can be stored before freezing. For example, peas should be prepared for freeze drying immediately after harvest, whereas potatoes may be stored for several months. Care must be taken to prevent product losses when the vegetables are hand or machine peeled. Continuous abrasive peeling or chemical peeling in hot caustic may reduce these losses. Root vegetables are diced, while leafy vegetables are shredded. Enzymes in the freshly cut material are decavitated by scalding in hot steam or water. Onions and leeks should not be blanched because enzyme activity is desirable. After blanching, the vegetables are cooled in air and then frozen (Mellor, 1978).

10.7.3.6 Quality Concerns for Freeze Drying Foods

The quality of freeze-dried foods is superior to conventionally dried products. The effect of major quality reducing reactions will be discussed.

1. Lipid Oxidation
    The oxidation of lipids is a major quality concern for freeze-dried products. The low moistures that are produced in freeze drying are conductive to lipid oxidation reactions. Free metals which catalyze the reaction are unbound upon water removal. To control the degradation of fatty acids, foods should be packaged in oxygen impermeable containers (King, 1971).

2. Nonenzymatic Browning
    Nonenzymatic browning causes food products to appear brown in color, have an off flavor, and lose nutritional value. It occurs at intermediate moisture contents. Freeze-dried foods are not in danger of enzymatic browning because the intermediate moisture content range is largely avoided. There is a rapid transition from high to low moisture content during freeze drying.

3. Protein Denaturation
    High temperature and salt content will cause protein denaturation. The denaturation Temperature for meats is estimated to be 40 to 60°C (King, 1971). Freeze-dried products are not subjected to this temperature range and are spared from protein degradation.

10.7.4 DRUM DRYERS

10.7.4.1 Introduction

A heated cylindrical drum which rotates around a horizontal axis can be implemented for the drying of slurries, pastes, or solutions. The material to be dried is spread onto the surface of the drum and heat from condensing steam inside the drum is transferred through the metal thickness of the drum to dry the adhering feedstock. The capacity of a drum dryer is a function of the drying rate of the thin layer of material and the amount of product which adheres to the drum surface. The drying rate depends on the type of feed device, steam pressure within the drum, and the drum speed. Preheating and preconcentration of the feed can reduce the drying load. The amount of pre-concentration is limited to the optimum concentration which can be effectively applied to the drum surface. Properties which affect drum adherence are viscosity, surface tension, and wetting power. The four variables which govern the operation of drum dryers are steam pressure, rotational speed, film thickness, and feed material characteristics. The steam pressure, or heating medium temperature, will regulate the drum's temperature. The rotational speed of the drum determines contact time (Moore, 1987). The wet material is applied to the drum from below by either splash feeders which splash the product onto the drums by rotary blades, dip feeders where the drum dips into a tank and the concentrated material adheres to the drum, or from below by a pendulum feed pipe.

10.7.4.2 Types of Drum Dryers

1. Double-Drum Dryer

   With the double-drum dryer, the thickness of the food product is determined by the distance between the drums (see Figure 10.26). The product dries as the drums rotate and is then scraped off by the knives. This type of drum dryer is advantageous because it can handle a wide range of products, has high production rates, and low labor requirements. Materials ranging from dilute solutions to heavy pastes can be effectively dried in double-drum dryers. Food products dried by this method include heat sensitive liquids and pastes which can be quickly rehydrated from the resulting flakes or powders. Applesauce, fruit purees, bananas, precooked breakfast cereals, and dry soup materials are manufactured in double-drum dryers (Moore, 1987).

2. Twin-Double-Drum

   In a twin double-drum dryer the drums rotate away from one another, and the wet feed is applied by splash feeders at the bottom of the drum (Figure 10.27). The dry material is removed by knives located 270° away from the rotary feed devices. This type of drum dryer is used for materials with solids that are dusty when dry, such as salt solutions or clay slips. The twin-drum dryer can be used as a pre-dryer when a top feed is installed. The material is removed from the dryer at a high moisture content and drying is completed with a rotary dryer. This drying method is economical only when increased capacity is obtained

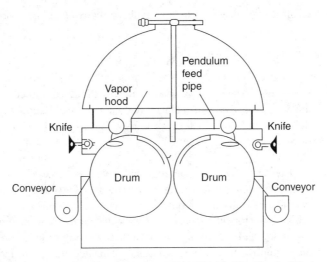

FIGURE 10.26 Double-drum dryer. (From Moore, J.G. 1987. In *Handbook of Industrial Drying*, Marcel Dekker, New York, pp. 227–242.)

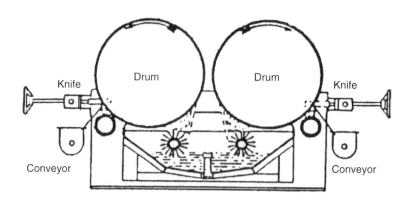

FIGURE 10.27   Twin-drum dryer. (From Moore, J.G. 1987. In *Handbook of Industrial Drying*, Marcel Dekker, New York, pp. 227–242.)

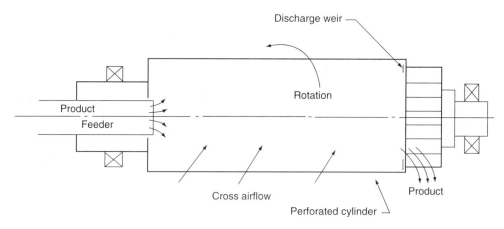

FIGURE 10.28   Centrifugal fluidized bed dryer. (From Hovmand, S. 1987. In: *Handbook of Industrial Drying*, Marcel Dekker, New York, pp. 165–221.)

by avoiding difficulties which would be encountered by using either dryer alone (Moore, 1987).

3. Single-Drum Dryer

   The application of wet material onto a single-drum dryer is done with applicator rolls. The rolls provide heavy product sheets resulting in a thick final product. The number of applicator rolls used determines the characteristics of the dry product sheet. Applicator rolls are advantageous when the wet material does not uniformly coat the drum surface. It is used for drying pastes and for food products high in starch such as potato flakes. Figure 10.28 shows a single-drum dryer.

4. Vacuum Drum Dryer

   A drum dryer can be enclosed and operated under vacuum (see Figure 10.29). In this way, materials which are heat sensitive can be dried without adversely affecting enzymes, vitamins, or proteins. The auxillary equipment necessary to run a vacuum drum dryer includes a wet dust collector, condenser, and vacuum pump. A screw conveyor is used to remove the dry product. Two product reservoirs are used in order to break the vacuum on one reservoir so the other can be discharged. The design, construction and excess equipment make the cost of vacuum drum dryers high (Harcourt, 1938). Typical drum dryer dimensions and operating parameters are shown in Table 10.13.

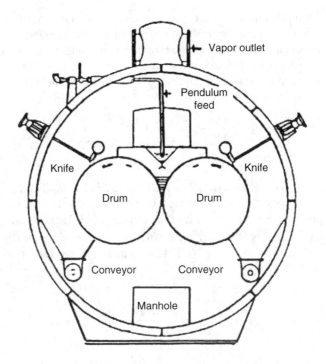

FIGURE 10.29   Vacuum drum dryer. (From Harcourt, G.N. 1938. *Chem. Metallurgical Eng.*, 45: 179–182.)

TABLE 10.13
Typical Drum Dryer Dimensions and Operating Parameters

Diameter	500–1500 mm (20–60 in.)
Length	1000–3000 mm (40–0120 in.)
(Wall) thickness	20–40 mm (3/4–1.5 in.)
Product thickness	0.1–0.5 mm
Revolution rate	5–30 rpm
Steam temperature	120–162°C (248–329°F)
Steam pressure	29–100 psi
Specific heat consumption	3000–3500 kJ/kg H_2O removed
Specific steam consumption	1.3–1.5 kg steam/kg H_2O removed
Specific Water evaporation	10–30 kg H2O/m^2·h

Source: Kessler (1981).

10.7.4.3   Design Equations

As discussed previously, the rate of drying for the thin film of product on a drum dryer is determined by the rate of heat transmission into the product from the drum. The removal of free water during the constant rate period is given by the following equation:

$$R = 2.45V^8 \Delta P \qquad\qquad (10.95)$$

R is the rate of drying (lb/h/ft^2), V is air velocity (ft/sec), ΔP is $p_s - p_a$, p_s is vapor pressure of the surface of evaporation (atm), and p_a is the vapor pressure of the mainstream of air (atm).

The rate of heat transfer is hindered by resistance from condensate formed on the inside of the drum, the drum wall, the food material, and by the outside surface. Heat flows to the surface by radiation, convection, and conduction. The rate of heat flow is given by:

$$\frac{Q}{A\theta} = \frac{\Delta T}{\frac{1}{h_w} + \frac{1}{h_m} + \frac{1}{h_p} + \frac{1}{h_c + h_r + h_s}} \qquad (10.96)$$

where Q is the Heat (btu), A is Area (ft^2), θ is Time (h), ΔT is $T_a - T_s$ (°F), T_u is Air temperature (°F), T_s is Temperature of the sheet (°F), h_w is Heat transfer coefficient of the condensate (Btu/h/ft^2°F), h_m is Heat transfer coefficient of the metal (Btu/h/ft^2°F), h_p is Heat transfer coefficient of the food material (Btu/h/ft^2°F), h_c is Heat transfer coefficient due to convection (Btu/h/ft^2°F), h_r is Heat transfer coefficient due to radiation (Btu/h/ft^2°F), and h_s is the Heat transfer coefficient due to vaporization (Btu/h/ft^2°F), and

$$h_s = \frac{2.45 V^8 \Delta p \lambda}{\Delta A t_f} \qquad (10.97)$$

where V is the Air velocity (ft/sec), Δp is Vapor pressure (atm), λ is Latent heat of vaporization (btu), and $\Delta A t_f$ is Temperature drop (°F).

10.7.4.4 Drying Characteristics and Heat Transfer

The layer of material on the heated drum is thin and presents no restriction to the vaporizing water. Three stages of heat transfer occur in this thin layer: the first stage consists of heating the thin layer up to its boiling point; in the second stage the water is vaporized and the material gradually changes from a liquid to solid state, and in a final stage the temperature of the product approaches that of the drum.

Drum drying involves heat transfer from condensing steam through the metal drum to the product layer. The rate of heat transfer depends on the resistance to the removal of water at lower moisture contents and product characteristics. When the time to remove the last few percent moisture becomes too great to be practical, the drum dryer can be used as a predryer for a more suitable drying technique. The capacity of the dryers often increase dramatically in these instances. Maximum evaporation rates for drum dryers can be as high as 18.5 lb/h/ft^2. This rate is attained with dilute solutions which evaporate easily. The heat transferred varies over the surface area of the drum. For example, the drum surface between the knife scraper and the feed applicator is not utilized. No heat is transferred at this point except for radiation loss. This is where the drum temperature is at its maximum. Once the wet feed material is applied, the surface temperature drops rapidly. Depending on the product's coefficient of heat transfer, the temperature will remain at or below the steam temperature. At the point of product removal, heat transfer rates are low due to the resistance of the dry product. The product will be approaching the drum temperature and the temperature difference will be quite small. Temperature of the drum surface of a double-drum dryer used for drying milk was studied by Roeser and Mueller (1930). The temperature of the milk was within 3°F of the metal temperature at the point of product removal. Calculations have shown that the overall heat transfer is 360 Btu/hr/ft^2°F in the area between the drums of a double-drum dryer, and 220 Btu/hr/ft^2°F close to the knives (Van Marle, 1938).

10.7.4.5 Example Problem

The drying of a sheet of food material is performed on a 4 ft steel drum. Steam is supplied at 200°F, the air temperature is 100°F at 40% humidity. The air velocity is 300 ft/min. Find the surface temperature of the sheet, the heat flow rate, and the evaporation rate:

Assume

$$t_s = 183°F$$

$$p_s = 0.547 \text{ atm}$$

$$p_a = 0.033 \text{ atm}$$

$$\Delta p = 0.514 \text{ atm}$$

$$\lambda = 988 \text{ btu/lb}$$

$$V = 5 \text{ ft/sec so } V^8 = 3.62$$

$$\Delta At_f = 183 - 100 = 83°F$$

$$h_w = 800 \text{ btu/hr/ft}^2°F$$

$$h_m = 1000 \text{ btu/hr/ft}^2°F$$

$$h_p = 800 \text{ btu/hr/ft}^2°F$$

$$h_s = 1.1 \text{ btu/hr/ft}^2°F$$

$$h_r = 1.5 \text{ btu/hr/ft}^2°F$$

From Equation 10.96:

$$h_s = \frac{(2.45)(3.62)(.514)(988)}{83} = 54.2 \tag{10.98}$$

$$\frac{Q}{A\Delta\theta} = \frac{\Delta T}{1/800 + 1/1000 + 1/800 + 1/(1.1 + 1.5 + 54.2)} = \frac{200 - 100}{0.0211} = 4740 \text{ btu/h/ft}^2 \tag{10.99}$$

and

$$\Delta t_f = \frac{0.0176}{0.0211} \cdot 100 = 83$$

and

$$t_s \text{ (calculated)} = 183°F$$

Since the calculated and assumed values of t_s are equal, the solution is correct.

$$R = (2.45)(3.62)(.514) = 4.55/\text{lb/h/ft}^2 \tag{10.100}$$

The heat transfer coefficients can be combined into an overall coefficient, u, and the basic equation is:

$$\frac{dw}{dt} = \frac{UA\Delta T_m}{L} \tag{10.101}$$

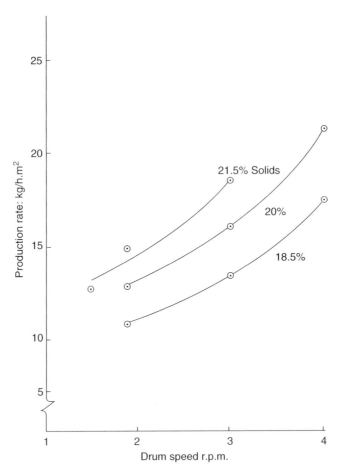

FIGURE 10.30 Influence of drum speed and solids content on potatoflake production rate. (From Heldman, D.A. and Singh, P.R. 1981. *Food Process Eng.*, 310–315.

where $\frac{dw}{dt}$ is the rate of water dehydration (kg/h), U is overall heat transfer coefficient W/m^2°C, A is surface area (m^2), ΔT_m and mean temperature difference between the roller surface, and L is the latent heat of vaporation (kJ/kg) and the product (°C).

U has been measured experimentally and ranges from 1000 to 2000 W/m^2°C. Solids content and drum speed also influence drying rate, as illustrated in Figure 10.30. These data for potato flakes show that the drying rate increases with increasing solids content and with drum speed.

10.7.4.6 Example

A drum dryer which dries a food product consisting of 12 to 96% solids is to be designed. The overall heat coefficient is 1700 W/m^2°C. The average temperature difference between the rollers and the product is 85°C. Find the surface area of the rollers to produce 20 kg of product per hour. Using equation:

$$\text{Feed rate} \frac{20\,\text{kg product}}{\text{h}} * 0.96 \frac{\text{kg solids}}{\text{kg feed}} = 19.2 \frac{\text{kg solids}}{\text{h}} \qquad (10.102)$$

Feed rate:

$$\frac{19.2\,\text{kg solids/hr}}{0.12\,\text{kg solids/kg}} = \frac{160\,\text{kg feed}}{h} \tag{10.103}$$

$$\text{Water Removal } 160 - 20\,\text{kg/hr} = 140\,\text{kg/hr} \tag{10.104}$$

If $L = 2420$ kJ/kg

$$\frac{140\,\text{kg water}}{h} = \frac{1700A(85)}{2420} \tag{10.105}$$

$A = 2.35\,\text{m}^2$

Since one third of the roller area is not used for heat transfer, drum size is adjusted to account for this discrepancy (Heldman and Singh, 1981).

10.7.4.6.1 Operating Parameters

Examples of operating parameters for atmosphere and vacuum drum dryers are given in Table 10.14, Table 10.15, and Table 10.16. Detailed reports of drum drying potatoes are given by Drazga and Eskew (1982), Cording et al. (1964), Moore and Samuel (1964), and Kozempel et al. (1985). A predictive model of the drying process is given by Kozempel and Sullivan. A description of instant applesauce flakes is given by Lazer and Morgan (1985).

10.7.5 CONVECTIVE DRYERS

10.7.5.1 Introduction

Mass production of dried foods is often accomplished through the use of convective dryers. In this type of dryer, a conveyor is used to transfer the foodstuffs through a tunnel of circulating gas, normally air. These systems require smaller equipment components but are larger in overall size due to the necessity of a drying tunnel replacing a drying chamber. The continuous and varied directions of the air flow through the tunnel, as well as into the bed, aid in the development of

---

TABLE 10.14
Drum Dryer Applications

Material (dried)	Type feed	Moisture in feed (%)	Steam pressure (lb/in.^{2})	Drum speed (rpm)	Vacuum (in. Hg)	Product (lb/h·ft^2)	Water in product (%)
Extract	Pan	59	35	8	27.9	7.74	4.76
Extract	Pan	59	35	6	27.7	2.76	1.92
Extract	Pan	59	36	4	Atm.	2.09	1.01
Extract	Pan	56.5	35	7 1/2	27.5	1.95	3.19
Extract	Pan	56.5	50	2 1/2	Atm.	1.16	0.75
Skim milk	Pan	65	10–20	4–5	—	2–3	2.5–3.2
Malted milk	Pan	60	30–35	4–5	—	2	2.6
Coffee	Pan	65	5–10	1–1.5	—	2–3	1.6–2.1
Malt extract	Spray	65	3–5	0.5–1	—	3–4	1.3–1.6
Taning extract	Pan	50–55	30–35	8–10	—	8–10	5.3–6.4
Vegetable glue	Pan	60–70	15–30	5–7	—	10–12	2–4

Source: From Harcourt, G.N. 1938. Effective drum drying by present-day methods, *Chem. Metallurgical Eng.*, 45: 179–182.

TABLE 10.15
Drum Dryer Applications

	Yeast cream	Stone slop	Starch solutions	Glaze	Zirconium silicate	Brewer's yeast	Clay slip
Feed solids (wt%)	16	40	36	64	70	25	75
Product moisture (% w/w basis)	5–7	0–2	5	0–2	0–2	5	9
Capacity (lb prod/h)	168	420	300–400	225–	1120	146	4000
Drye type (a), single; (b), twin; (c), double	(a)	(a)	(a)	(a)	(a)	(a)	(a)
Drum							
Diameter	4 ft 0 in.	2ft 6 in.	4i in.	18 in.	36 in.	28 in.	48 in.
Length	10 ft 0 in.	5 ft 0 in.	120 in.	36 in.	72 in.	60 in.	120 in.
Type of feed method	Top roller	Dip	Top roller	Side	Dip	Center nip	Side
Steam pressure (lb/in.^{2} gauge)	80	60	80	—	80	40	40
Atmospheric or vacuum	Atmos.	Atmos.	Atmos.	Atmos.	Atmos.	Atmos.	Atmos.
Steam consumption (lb/lb evaporated)	—	—	1–3	1–3	—	—	1–35
Average effective area (%)	—	—	86	—	—	—	65
Evaporation (ft^2/h)	6–5	4	5	9	8–4	6	8–4

Source: From Williams-Gardner, A. 1971. *Industrial Drying*. CRC Press, Cleveland, OH.

more uniform moisture contents (Sokhansanj and Jayas, 1987). Further advantages stem from the continuous automatically controlled conveyors. Products are constantly fed onto the conveyor or trays, and the machine automation controls the rest.

Disadvantages also exist in the use of these conveyor bands. Sticky materials tend to lodge in the chains and links of the conveyor band, blocking the passage of air. The use of wire mesh bands has decreased due to their tendency to distort and disintegrate in extreme heats (Williams-Gardner, 1971).

10.7.5.2 Preformers

Most products dried in these types of dryers need to be preformed. This is accomplished with the use of extruders. The basic concept behind an extruder is to force the product through a small diameter hole of constant area. As the material is compressed, the moisture rises to the surface and evaporates. This leaves a porous material that is much easier to dry (Williams-Gardner, 1971). The size and shape of the extrusions depend upon the nature of the material, rate of drying required, and the rate of extrusion possible. Different types of extruders are used for different products. A wiper blade extruder is suitable for materials which should not be subjected to high pressures. Roller extruders are generally used to granulate materials. Cam blade extruders are used for more dense materials that require extreme pressure to preform. These different styles of extruders are shown in Figure 10.31.

10.7.5.3 Drying Chamber Design

The chamber normally consists of insulated panels and doors held together with air tight joints. The thickness of these panels is generally less than 1.5 in. (Williams-Gardner, 1971). Several fans are

TABLE 10.16
Drum Drying Parameters for Several Examples

	Temp.		Moisture content		Drum speed	Drum dimensions and capacity	Drying time	Reference
	Wet	Dry	Initial	Final				
Fruit puree, double drum		85–90%	2–4%			12.7 cm diameter 23.4 cm long		Kitson and MaGregor (1982)
Plant juice, protein conc.		130°C		Below 20% wet basis	Through-put rates 3.3–6.4 kg/m²·h	Drum no. 2 mm Drum no. 8 cm/sec		Straub et al. (1979)
Yeast		25°C		4.45 g H_2O/g DM		15% solids	180 sec	Labuza (1971)
		25°C		7.23 g H_2O/g DM		4.5 lb/h		
Beef powder		123°C		3–0.8–4.0%		60 cm × 35 cm dia.	18 sec	Kopelman and Saguy (1977)
Cowpea puree			2 rpm	5.5%	2–4 rpm	Clearance between 0.016 in.	20 min	Drayemi 19
			4 rpm	6.5%	Heated by pressure	45 psi		
Potato slurry			81.5%	3.8%	2 rpm	2.11 lb/h	15 sec	Charm (1963)
			81.5%	4.7%	4 rpm	2.11 lb/h	15 sec	
Milk		155°C			24 rpm	24 in. × 36 in., 53–60 lb/h	33 min	Combs and Hubbard (1931)
		50–60°C			24 rpm	58 lb/h	45 min	
Mashed potatoes	Pressure 55–60 psi			6–6.5%	2–3 rpm	2 dia. 3 ft long, 4 rolls		Draza and Eskew (1962)
Applesauce		82.2–85°C		1.40%				Lazar and Hart (1968)
Tomato powder		70°C		5.3%	2.5 rpm	1.6 lb/h	10–17 s	Henig (1971)
		70°C		10.2%	4 rpm	6.5 in. dia. 6 in. long	10–17 sec	
Tomato flakes		33°F		3.2%	3 1/2 rpm	0.0106 gan. 3.5	4 h	Lazar and Miers (1971)

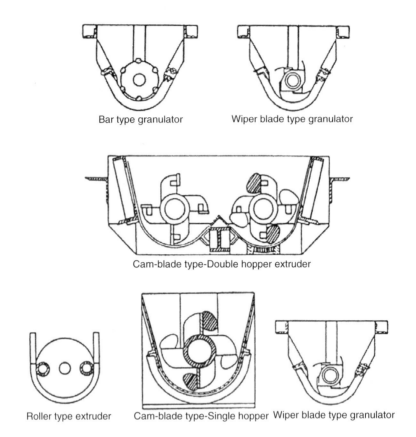

Bar type granulator   Wiper blade type granulator

Cam-blade type-Double hopper extruder

Roller type extruder   Cam-blade type-Single hopper   Wiper blade type granulator

FIGURE 10.31 Extruder types. (From Williams-Gardner, A. 1971. *Industrial Drying*. CRC Press, Cleveland, OH.)

present in the chamber to create the necessary air flow inside. Reheat coils are often used after each cross flow pass of air, in order to maintain a constant temperature within the chamber. The bands, aprons, trays, or flights are on a set of tracks or roller conveyor chains to facilitate the motion of the product through the drying chamber. Each of the product carriers is perforated to increase air flow through the product.

10.7.5.4 Pilot Tests and Scale Up

Pilot tests are the best resources for determining the size and specifications of a particular dryer. Current publications, vendors, and research are all viable means of gaining such information. After pilot tests determine the drying rates for specific loading, scale up factors can be implemented. Design calculations are not very accurate and should not be used (Williams-Gardner, 1971). Previous research experience can provide the necessary data to start pilot tests.

Once production rates, loading rates, and drying time have been estimated, the approximate size of the band to be used can be calculated:

$$A_{band} = \frac{\text{production rate}}{\text{loading of dry product/ft}^2} \cdot \text{drying time} \qquad (10.106)$$

Figure 10.32 shows the experimental drying time vs. the effective band area over a variety of production rates.

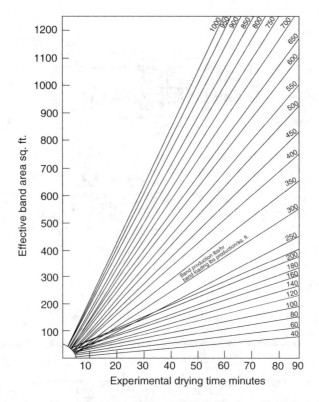

FIGURE 10.32   Estimation of band dryer size. (From Williams-Gardner, A. 1971. *Industrial Drying*. CRC Press, Cleveland, OH.)

10.7.5.5   Variations of Design

There are several major divisions of convective dryers. These include tunnel, through-circulation, conveyor band, and belt dryers. Each one has a unique characteristic that makes its use suitable only for specific foodstuffs. All of the dryers utilize varied air flows, which is an important aspect of the convective dryer.

10.7.5.6   Air Flow

A parallel air flow dryer has the air and the product travel in the same direction through the chamber. The hottest gas contacts the wettest solid first. This improves the control of moisture content in the solid and increases the rate of drying.

In a counter-current air flow dryer, the hottest gas first contacts the driest solid on the band. With this type, the air and the product travel in opposite directions. This results in a slower drying time but eventually produces a lower moisture content than parallel systems. The main disadvantage with this system is that the solid may carry out a lot of sensible heat, thus lowering the efficiency of the system (Treybal, 1980).

Combined air flow dryers usually consist of a series of tunnels, each one having a different air flow direction. These multi-stage dryers can almost achieve optimum conditions by varying the air flows. Often a three tunnel system is used, consisting of two wet tunnels and one dry tunnel. The product from both wet tunnels is loaded alternately into the dry tunnel after initial drying has occurred. The residence time in the dry tunnel is half that of the wet tunnel. Dry tunnels are operated at lower temperatures than the wet tunnels. In this manner, products are not in the presence of high temperatures for long periods of time, thus improving the drying process for heat sensitive products

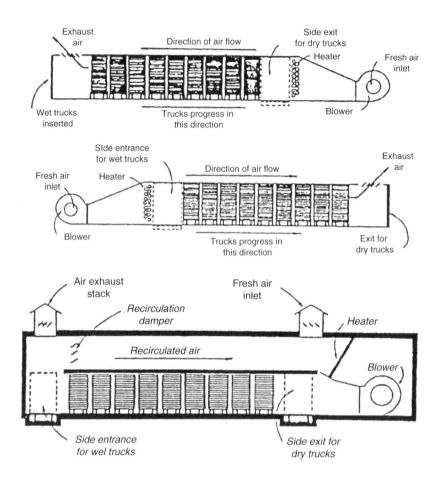

FIGURE 10.33 Airflow patterns in tunnel dryers. (From Sokhansanj, S. and Jayas, D.S. 1987. In: *Handbook of Industrial Drying*, Mujumdar, A.S., (Ed.), Marcel Dekker, New York, pp. 532–537.)

or foodstuffs. The temperature range for wet tunnels is 99 to 104°C, and for dry tunnels is 65 to 71°C. These temperatures are for drying root vegetables (Sokhansanj and Jayas, 1987) (Figure 10.33).

Each type of dryer also has its own characteristic additional air flow patterns created within the drying chamber.

In tunnel dryers, a cross flow is usually generated. This is produced by fans on the side of the tunnel. Food pieces of any shape and size can be handled on this dryer, and if solid trays are used, fluids can also be dried.

The through-circulation dryers also have an extra air flow. Air is forced onto the loaded pan and through the particles, which have a bed depth of 1.5 to 2 inches (Treybal, 1980).

In the conveyor band dryer, the product is loaded onto perforated plates. During initial drying, the air is blown up through a bed depth of 6 to 8 inches (Holdsworth, 1971). Later air is blown down through the semi-dry product.

The gas is usually recycled in these systems. The perforations in the trays usually run in the following sizes:

Perforation	Size
Round	Maximum diameter 1/16 in.
Elongated Slits	Maximum 3/32 × 1/2 in.

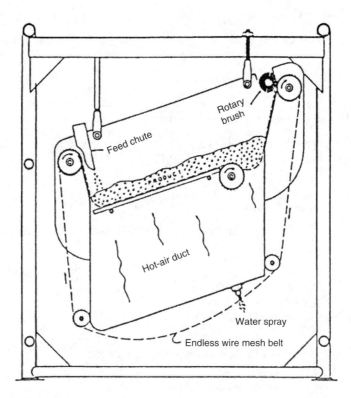

FIGURE 10.34 Bell dryer. (From Sokhansanj, S. and Jayas, D.S. 1987. In: *Handbook of Industrial Drying*, Mujumdar, A.S., (Ed.), Marcel Dekker, New York, pp. 532–537.)

Belt dryers are best suited for cut vegetables. In this dryer, the belt is a fine mesh band shaped into a trough. Hot air is blown up through the band to increase evaporation. Soft materials can be dried without damaging the existing form because the air flow provides a lift to the product. The material must consist of uniformly sized particles to insure proper lift. The belt is inclined at a 15 to 20 degree angle to help with gravity unloading of the dried product (Mujumdar, 1987) (Figure 10.34).

In these direct contact dryers, the drying time is lessened when perforated trays are used (Treybal, 1980). This illustrates the importance of using particulate products to allow efficient air flow through the beds.

10.7.5.7 Heat Input

There are two types of heat input available for these systems: direct fired and steam heated.

In a direct fired dryer, the combustion gases of either gas or oil burner fires are intimately mixed with the recirculating air. This heats up the chamber, providing temperatures high enough to produce proper drying. This mixing of gases maintains the heat requirements of each section of the chamber. The temperature of the mixed gas controls the fuel input to the burner. This produces a simple, effective, and closely controlled temperature system. Initially, this heating system is expensive, but is highly efficient thus lowering the overall cost.

Steam heated dryers rely upon heat transfer from tubes that are heated with steam. These tubes are located either above or below the band, on the side of the chamber, or in a combination of these arrangements. The area of the heater tubes will meet the thermal requirements for heating up and evaporating moisture from the product as it is relayed down the passage. The drying rate curve found from the pilot tests helps show the distribution of the heat load on each section.

10.7.5.8 Product Collection

Dried product is collected in a gravity unload situation. The dried product is generally unloaded off the band down a chute into a lower compartment. The band is cleaned on the return trip by scrubbers and sprays.

10.7.5.9 Use of Convective Dryers in Industry

Convective dryers are best suited for drying a wide range of vegetables. Each specific type is also good for other products, as shown in Table 10.17.

10.7.5.10 Example Calculation

A food product is to be dried from filter cake at 25% moisture to 0.5% moisture (by weight), in a convective dryer.

> Production rate: 1000 lb/h dry product
> Specific heat of material: 0.2 Btu/lb
> Bulk density of dry product: 24 lb/ft^3

Pilot plant studies show that the maximum product temperature is 220°F.
Filter cake is fed into an extruder at 40°F.

> Fed into heater:
> Ambient air   Temperature 60°F
>   Relative humidity 55%
> Exhaust from heater:
> Air   Temperature 180°F
>   Relative humidity 30%

Pilot tests show:
Roller extruder best suited for material drying characteristics of extrusions on a 1 ft 2 test tray at bed depth of 3 in. (8 lb material):

> Wet weight   8 lb
> Dry product (0.5%)   6 lb
> Air velocity   250 ft/min
> Air inlet temperature   250°F
> Material outlet temperature   205°F

Find the size of dryer required.
Over time, moisture content and evaporation rate were measured at the pilot conditions. A time of 40 min was required to dry the product to 0.5%.

> Evaporation rate is 327 lb/h
> Production of bone dry material is:
> (1000 lb/h) (0.005) = 5 lb H$_2$O/h
> 995 lb/h bone dry products.

> Final moisture product temperature 205°F
> Exhaust air temperature 180°F.

TABLE 10.17
Convective Drying

	Temp.		Moisture content		Drying time	Air velocity flow rate	Reference
	Wet	Dry	Initial	Final			
Beets		70°C	6.67 gH₂O/g solid	0.004	6 h	150 ft/min	Saguy et al. (1980)
Potatoes		73°C		0.05	3 1/2 h		Mishkin et al. (1983)
Grain		70°C	29%	5%		250 ft/min	Jayaraman et al. (1980)
HIST method		170–180°C	50–55%	20%	4–6 min	2000 ft/min at end	
Rice (bed Depth 7 cm)		38 ± 5°C	31.1%	25.4%	1/3 h	21.3 m/min	Stefe et al. (1979)
				24.3%	7/12 h	21.3 m/min	
Peanut kernels		130°C	25% WB	22.91%	3 h	21/3 m/min	Young et al. (1982)
		110°C	25%WB	22.97%	3 h	21.3 m/min	
Grain corn		3–18°C	20–30% WB	14–15% WB	144 h	11 m/min	Brown et al. (1979)
Carrots		65.6°C	30%	4%		200 ft/min	Holdsworth (1971)
Apples	100°F	65.6°C	20%	2%	5 h	350 ft/min	
Celery	95–100°F	54.4°C	35%	5% or lower	2.75 h	200 ft/min	
White beans		48.9°C	33%	20%	3 h	0.25 m/sec	Hutchinson an Otten (1983)
				17%	5 h		
Soybeans		49.4°C	22%	14%	3 h	0.58 m/sec	
				13%	5 hrs		
Sunflower		93°C	26% WB		1 h	0.3 m/sec	Syarief et al. (1984)
Seeds		49°C			1 h	0.3 m/sec	
Wheat		37.88°C			3 h	0.61 m/sec	Ramaswamy and Lo (1983)
Grain					3 h	0.41 m/sec	
					5 h	0.61 m/sec	
					9 h	0.61 m/sec	
Rice grain		59°C	29.8%	9.1%	12 h		Kunze (1979)
Chicken		50°C	65%	45%	4 h	240 m/min	Agarwal et al. (1972)
Yellow globe		65°C	6.8 kgH₂O/	0.6	1 h	8.1 m/min	Mazza and
Onions		50°C	kg DM	1.5	1 h	8.1 m/min	LeMaguer (1980)
Sweet		61°C	4.1 DB	2 DB	1 2/3 h	10 m/sec	Suarez et. al. (1984)
Corn		69.5°C	3.2 DB	1 DB	1 2/3 h	10 m/sec	
Field		61°C	1.35 DB	0.8	1 2/3 h	10 m/sec	
Corn				0.48		10 m/sec	
Shelled		175°C	25% WB	15% WB	1.25h	40 m/min	Brook and Baker-Arkema (1980)
Carrots	95°F	48.3°C		Reduced by 75%	2 1/2 h	750 ft³/min	Havighorst (1943)
Secondary dryer	98°F	70°C		4%	6 h	750 ft³/min	Havighorst (1943)
Peach halves	92°F	68.3°C		25–30%	12 h	950 ft/min	Phaff et al. (1945)
White	122°F	65°C		75%	2 h	1150 ft/min	Hendel et al. (1955)
Potatoes				8.7%	16 h	1150 ft/min	Hendel et al. (1955)
	140°F	75°C		60%	2 h	1150 ft/min	Hendel et al. (1955)
				8.3%	8 h	1150 ft/min	Hendel et al. (1955)

(Continued)

TABLE 10.17
Continued

	Temp.		Moisture content		Drying time	Air velocity flow rate	Reference
	Wet	Dry	Initial	Final			
Pineapple	Vacuum	65.6°C		2.5%	2 3/4 h	Pressure	Natter et al. (1958)
Juice	Shelf	25°C		3.7%	48 h	2 mmHg	
Sugarbeets		70°C				6 m/sec	Vaccarezza and Chirife (1975)
Tapioca root		84°C	2.0	0.4	45 min	1.16 m/sec	Chirife (1971)
			2.0	0	2 h	1.16 m/sec	
Soybeans		99–102°C		13.2% WB	48 h	5.18 ft^3/min·ft^2	Alam and Shove (1973)
Shelled corn		37.8°C	34.1%	16.3%	13 h	50–70 ft/min	Westerman et al. (1973)
			34.1%	21.7%	6 h	50–70 ft/min	
White shelled corn		71.1°C	28% DB	16% DB	1.5 h		White et al. (1973)
Shelled corn		37.8°C	34.1%	28.3%	3 h		Westerman et al. (1973)
Soybeans		37.8°C	29.7% DB	19%	3 h		Overhults et al. (1973)
			29.7	14%	12 h		
		71.1°C	29.7	16.5%	1 h		
			29.7	11%	3 h		
Alfalfa		76.7°C			0.15 h	100 ft^3/min	Rowe and Gunkel (1973)
					0.30 h	100 ft^3/min	
Egg powder	Enter at 104.4°C	Leave at 82.2°C		1.5–2%	2 sec	100 ft^3/min	Van Arsdel (1973)
Beat powder		93°C			2.5 h	250 ft/min	Kopelman and Saguy (1977)
		77°C			2.5 h		
		68°C			2.5 h		

Heat to solids:

$$\Delta H = mC_{\mathrm{p}}\Delta T$$
$$= 995 \times 0.2 \times (205 - 40) \quad = 32{,}800 \text{ Btu/h}$$

Heat to water:

$$= 332 \times 1 \times (180 - 40) \quad = 46{,}500 \text{ Btu/h}$$

Heat to evaporate water

$$= 327 \times 970 \quad \frac{317{,}000 \text{ Btu/h}}{396{,}300 \text{ Btu/h}}$$

Add 10% for convection and radiation losses

$$\text{Total} \quad \frac{39{,}630 \text{ Btu/h}}{435{,}930 \text{ Btu/h}}$$

With a production rate of 1000 lb/h and a product loading of $8 \times 0.75 = 6$ lb/ft^3 equivalent to 166.7 ft^2/h with drying time of 40 min, there would be 140 ft^2 of band within the dryer allowing a 25% scale up factor.

Assuming the dryer will have a band 6 ft wide, and each dryer section is 6 ft long, a four section dryer will be required to have an effective length of 24 ft.

10.7.6 NOVEL DRYING TECHNOLOGIES

Drying is an important energy intensive operation in many processes. It is recognized that a large amount of energy is consumed by drying in the United States. So even a small percentage saving in energy consumption will result in considerable overall improvement in energy efficiency. In addition, the final quality of the product is greatly influenced by the drying technique and strategy.

To investigate alternate novel drying technologies in order to arrive at more energy efficient processes with improved quality, a brief overview of some novel drying techniques is given below. This is followed by a detailed case study of application of supercritical fluid extraction and mechanical vapor recompression to drying.

10.7.6.1 Microwave Drying

High frequency radio waves of up to 30,000 MHz are utilized in microwave drying. A high frequency generator guides the waves and channels them into an oven designed to prevent the waves from leaving the chamber. Proper wavelength selection is necessary to insure thorough penetration into the food. Penetration is also affected by the depth of the material and type of material being exposed. It is important that each product be evaluated individually to insure proper wavelengths and dehydration. As the energy enters the foods, the molecules try to align in the electric field orientation. They oscillate around their axis, generating heat within the food, resulting in dehydration. The waves bounce from wall to wall, until eventually all of the energy is absorbed by the product. In this manner, the drying rate is increased greatly. There is a problem with uniformity of drying because of the penetration of the microwaves through the product. This type of heating is highly efficient, and power utilization efficiencies are generally greater than 70%. Important commercial aspects include the ability to maintain color and quality of the natural food. This has been found prevalently in potato chips. Cabbage and potato blocks were reduced in moisture content from 15 to 9% and 7 to 5% respectively, and the time required was about one fifth of that in a cross air blow dryer (Rushton, 1945).

Further work was done by Jeppson (1964), Nury and Salunkhe (1968), Huxsoll and Morgan (1966), Davis (1965), Wienecke (1989), Porter (1973), Blau et al. (1965), and O'Meara (1966). Microwave drying of fruits and vegetables is reviewed by Nijhuis et al. (1998) whereas Vega-Mercado et al. (2001) describes this operation. Specifics to the microwave drying operation are detailed in Chen et al., 2006a,b.

10.7.6.2 Ethyl Oleate

In the drying process, certain compounds can be used to enhance the drying of specific products. One such compound is ethyl oleate. This acts as a wetting agent, increasing the evaporation rate of water in the initial stages of drying. Many researchers believe that this results from a solvent reaction of the ethyl oleate on the skin wax and cell walls of certain products. Only high amylose starches, seedless grapes, maize grains, and certain porous fruits have been proven to be affected by this product. Also, it affects starch pastes but not gels.

Ethyl oleate acts as a surfactant by increasing the spreading of free water within the sample. This in turn increases the drying rate. On the skin of grapes, ethyl oleate causes a dissolving action on the waxy components of the skin. It is these components that offer high resistance to moisture transfer. By dissolving these, the moisture transfer is much easier. Figure 10.35 exemplifies this effect of ethyl oleate on the rate of air dried foods.

Further work has been done by Saravacos (1986), Suarez et al. (1984), Salas and Labuza (1968), Loncin and Roth (1985), Saravacos and Charm (1962), Ponting and McBean (1970), Raouzeos and Saravacos (1986), Riva and Masi (1986), and Chambers and Possingham (1963).

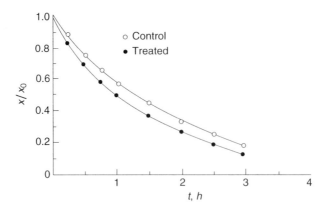

FIGURE 10.35   Effect of ethyl oleate on the rate of drying of unheated amylase pastes in air at 60°C and 2 m/sec. Control $X_O = 1.93$, treated (0.2% ethyl oleate) $X_O = 1.94$ kg water/kg dry solids.

10.7.6.3 Acoustic Drying

A recently developed technology in the drying field is acoustic drying. Products are dried at relatively low temperatures, 140 to 200°F, with intense low frequency sound waves. These strong sonic waves increase heat and mass transfer coefficients across the boundary layer of the product. This tends to promote liquid – solid separation. The drying rates of these dryers are three to ten times faster than conventional dryers. Its efficiency rate is around 1500 Btu/lbH$_2$O removed. Work has also been done in the ultrasonic region of sound waves.

The product enters the dryer at the top in a gravity fall situation. The product is atomized in order to enter the chamber. As the product hits the air and sound waves, it is dried in seconds. A cyclone baghouse helps collect the dried products. This dryer requires sound diminishing devices because of the loud noise produced during drying.

Foods that have been commonly classified as difficult to dry have been dried successfully in acoustic dryers. Liquids containing between 5 and 78% moisture have been dried to 0.5% moisture content. Products with high fat contents, up to 30%, have also been dried in these dryers. Other products that dry well are high fructose corn syrups, tomato pastes, lemon juice, and orange juice. Because this process is relatively quick and cool, the degradation of natural color, flavor, and nutrition is reduced. Reconstitution is easy and a good taste usually remains in these products.

Further information is available from the U.S. Development Corp., and Muralidhara et al. (1985), and Ensminger (1988).

10.7.6.4 Infrared Radiation Drying

This technique is often used in conjunction with freeze drying (in order to accelerate the sublimation process), batch drying, and continuous band drying. The infrared radiation is generated by heating the product to a high temperature, resulting in direct penetration of the surface by the radiation. The ideal products for this system are referred to as black bodies. This heat is usually generated by gas flames, electrical methods including reflector incandescent lamps (100–5000 W), quartz tubes, and resistance elements. The heat radiation is projected from plates arranged above the trays of product. Band systems are common because of the need for the product layers to be no thicker than 3 mm. Slurries and gels work best in this system, providing optimal penetration. This drying technique produces a high drying rate without burning.

Other research has been done by Biau (1986), Winecke (1989), Hagen and Drawert (1987), Sandu (1986), Orfeuil (1981), Sifauori and Perrier (1978), Hasatani et al. (1983 – 1984), and Kuts et al. (1982). Most of the major research in this field has been done in the U.S.S.R.

10.7.6.5 Electric and Magnetic Field Dewatering

The method of using electric and magnetic fields is mostly used as a separation technique with solid–liquid combinations. Once the product has been dewatered through this method, it is sent on to another conventional dryer to complete the dehydration process. The backbone of the process is the flow of a dc electric field applied to enhance dewatering. This influences surface characteristics of the solid/liquid such as zeta potential, dipole interactions, and hydrophobicity. This method works with two processes: electroosmosis and electrofiltration. Electroosmosis is the movement of water through the porous membranes of the product with the application of a dc electric potential. This is a surface diffusion process. Electrofiltration is the movement of charged particles towards electrodes in the presence of a dc electric field. The processes are carried out in a separation chamber before the product is sent on to another dryer. These processes are not used on a commercial scale yet because of the economics of the system, but they are becoming more prominent because of the growing environmental concern with supernatant discard.

Further research has been conducted by Bureau of Mines, USA; Commonwealth Scientific and Industrial Research Organization, Australia; Central Electricity Generating Board, U.K.; Battelle Memorial Institute, Switzerland; Fuji Electric Co., Ltd., Japan; Monsanto Enviro-Chem Systems, USA.

10.7.6.6 Superheated Steam

This method of drying is normally incorporated with a two belt dryer system. The product is put onto an upper belt in a uniform noncompact layer. As it passes through the dryer, steam is passed up through the belt and product. At the end of the first belt, a gravity fall system loads the product onto a second lower belt. The product is passed through the chamber once again. The steam is blown over the product in a parallel fashion. This is good for drying pulp products and alfalfa. Figure 10.36 shows a schematic diagram of a superheated steam dryer with vapor recompression.

No air is introduced into this type of system. For this reason it is important to minimize any leaks within the dryer. Superheated steam at atmospheric pressure is the only drying medium involved. This occurs because steam increases the efficiency of the system over air. This in turn reduces drying time and dryer surface area. The evaporation rate is controlled by the rate of heat exchange between

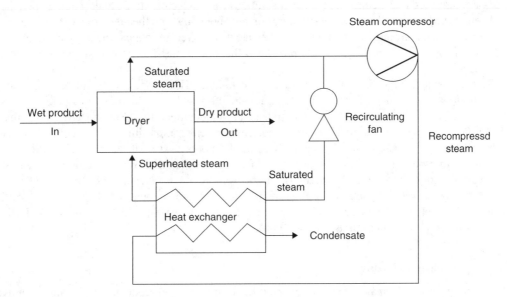

FIGURE 10.36   Superheated steam dryer.

the drying fluid and the product. At the end of the drying process, the steam is diffused from inside the product to the outside more easily than air would be. The steam leaves the dryer completely saturated, and upon exit, is compressed to condense out some of the water present in the steam. This steam is then expanded and recycled through the dryer.

Despite its initial high expense, this system has many advantages. There is no material loss, and no burning of the product. No pollution, a higher quality dry product, uniform drying, a sterile atmosphere, and a 50% energy savings in primary consumption are all benefits of the procedure. For further information see Garin et al. (1988).

10.7.6.7 Desiccant

Desiccants have been used primarily in the air conditioning industry for many years, but several people believe that this particular technique has some aspects applicable to the dehydration of food. The process is basically a dehumidification of the air followed by an adiabatic evaporative cooling period. The idea is that the water in the product would condense out through the pores of the desiccant and the latent heat of vaporization would be converted into sensible heat. The product is normally dried in a rotary type or batch dryer. The high speed of revolution of the drying drum plus a low air flow rate causes a great potential for mass transfer from air to desiccant. Cooling water is circulated continuously to maintain a constant desiccant temperature.

A solid gel or liquid desiccant can be used. The solid gel requires very high pressures to evaporate the water out. Higher temperatures are required for reactivating the used desiccant. There are significant operating costs involved here. The liquid desiccant has a lower vapor pressure than the water, which results in a dilution of the desiccant. This means the desiccant is absorbing the moisture from the product.

Further research has been done by Gandhidasan et al. (1988), Penney and Maclaine-Cross (1985), and Epstein and Grolmes (1983).

10.7.6.8 Osmotic Dehydration

There are several solutions used for this process. The main kind is a sugar syrup treatment. This causes removal of moisture by placing the food in contact with the sugar solution. The product slices are immersed in the concentrated sugar solutions for a range of 4 to 24 h, depending upon the food being dried. This will reduce the moisture content about 50% and then the product can be dried further by another conventional method. The other agents involved are sugars, sugar-starch mixtures, and sugar syrups. This candying leaves a product with a porous, crisp texture while retaining most of its original flavor.

Reverse osmosis is also used as a dewatering technique. This process is useful for fruit juice concentrations, and is much faster than conventional methods. The optimum temperature range is 20 to 40°C. The drying time is inversely proportional to the temperature. Studies have shown that raising the temperature range to 40 to 80°C shortens the dehydration time, but lowers the quality of the product. It also causes the cell membranes to plasmolyze. Also, at higher temperatures, the viscosity of the solution decreases as the water diffusion coefficient in osmotic solution increases. Figure 10.37 shows this effect of temperature on the osmotic processes in apples.

Further research has been done by Ponting et al. (1966), Adambounou and Castaigne (1983), Lenart and Lewicki (1988), Lee and Salunkhe (1966, 1967a, 1967b), Salunkhe and Do (1973), Bolin (1970), Morgan (1965), and Igarashi et al. (1988),

10.7.6.9 Explosion Puffing

This technique produces many of the desired qualities of freeze-dried product, but with significantly less expense and in a much shorter time with better reconstitution. The product is partially dehydrated in a preliminary stage, and then loaded into a closed rotating cylinder called a gun. The product

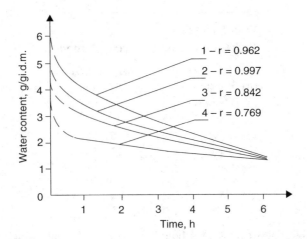

FIGURE 10.37 Effect of temperature on the course of osmotic dehydrationof apples in saccharose solution. Temperature: 1, 30°C; 2, 50°C; 3, 7°C; 4, 90°C. (From Lenart, A. and Lewicki, P. 1988. Osmotic Dehydration of Apples at High Temperature, Sixth International Drying Symposium.)

is heated inside until the internal pressure reaches a predetermined value. The food is instantly discharged to atmospheric pressure. As this occurs, some of the water is vaporized. The important fact is that the explosion produces a porous network within the particle. This porosity enables the final dehydration to be completed rapidly, approximately two times faster than conventional methods. The gun was redesigned some years ago by introducing superheated steam at 500°F and 55 psig into it. This prevents condensation and guarantees that the particles are exposed on all sides. This process is used mainly for fruits, vegetables, and grains.

Other research has been done on both drying and reconstitution by Strolle (1970), Isidro (1968), Eisenhardt et al. (1962, 1964, 1967, 1968), Wilson (1965), Sullivan and Cording (1969), Cording et al. (1964), Sullivan and Eskew (1964), Kozempel et al. (1989), Eskew and Gelber (1964), and Eskew et al. (1965).

10.7.6.10 Foam-Mat Drying

The conversion of liquid foods to foams has been found to be a less costly means of improving product quality. Foams dry more rapidly than liquids, allowing the use of lower temperatures and shorter residence times. The increased rate of drying is due to an increase in surface area and the relative ease of moisture transport through the porous dried foam structure as compared to the less porous structure of the dried liquid. Heat transfer is less efficient in a foam, but is adequate since the drying of food materials is predominately controlled by internal mass transfer. In addition to the reduction in thermal exposure to heat sensitive foods, dried foams retain a porous structure allowing rapid rehydration characteristics.

Several applications of foam drying have been developed including vacuum puff drying, foam spray drying, and foam-mat drying. Vacuum puff drying refers to processes which use a vacuum to induce product foaming. According to Holdsworth (1974), vacuum puff drying developed out of observations made during the freeze drying of orange juice concentrate. Some samples of the freeze-dried concentrate showed a significantly increased drying rate and a porous structure due to foaming of the liquid under a vacuum.

When gases are dissolved in the liquid feed under considerable pressure prior to spray drying, the process is called foam spray drying. The density of the foam spray dried products are typically reduced by half, and "whereas spray-dried particles are hollow spheres surrounded by thick walls of dried material, the foam process produces particles having many internal spaces and relatively thin walls" (Holdsworth, 1974). Crosby and Weyl (1977), discuss the general principles of foam spray drying.

TABLE 10.18
Cost Comparison of Drying Methods

Drying method	Approximate cost (cents/lb)
Drum	0.9–1.0
Air	1.0–1.5
Spray	0.7–1.5
Foam-mat	2.3–3.0
Vacuum-puff	3.0–4.0
Freeze	5.0–10.0

In foam-mat drying, the foaming of a liquid is due to surfactants which are either naturally occurring or added. A stable foam is spread out in a thin mat and dried with heated air. The dried product, examples of which include milk, mashed potatoes, and fruits, can be scraped off moving trays or belts and crumbled. The cost of foam-mat drying is higher than spray or drum drying, but less than freeze drying (Morgan, 1974). This is due to the large drying surface required to dry a thin film layer. Thicker films are unsuitable because the drying time exceeds the time of stability for most foams. Hertzendorf and Moshy (1970) present Table 10.18 comparing the cost of various drying operations. While absolute costs may have increased significantly since 1970, the relative costs are likely to be similar.

Recent developments in the foam-mat drying process and its application in the food industry can be traced to work conducted by Morgan and co-workers at the USDA Western Regional Research Laboratory around 1960 (Hertzendorf and Moshy, 1970). However, the basic concept of foam-mat drying is much older, as evidenced by patents issued to Campbell (1917), and Mink (1939, 1940) for processes involving milk, and egg whites, respectively.

10.7.6.11 Supercritical Fluid Extraction (SCF) and Its Application to Drying

Supercritical fluids possess unique properties that enable them to selectively extract components from a mixture. This ability has been investigated recently as an alternative to currently used extraction processes such as distillation or liquid extraction. Investigation of SCF extraction is motivated by the desire to find separation techniques with lower energy costs and improved health and safety standards.

SCF extraction exploits the properties that occur at or above the supercritical pressure and temperature. Figure 10.38 is a pressure–temperature phase diagram for carbon dioxide in which regions of solid, liquid, and gas are shown. The supercritical region is that part beyond the critical point. The properties exhibited by fluids in this region are intermediate between those of liquids and gases (Table 10.19). The property of greatest interest for SCF extraction is that of density. Figure 10.39 shows the relationship between pressure and density. With increasing pressure, a gas will become increasingly dense. Above the critical point, this increased density produces enhanced solvency, approaching that of a liquid. It is this solvency that makes SCF extraction a viable alternative. Mass transfer properties resembling that of gases are also a significant factor in SCF extraction.

Many patents involving SCF extraction have been granted since 1974 when the first was issued (U.S. Patent 3,843,824). Patents have involved extraction of hops, caffeine, spices, fatty acids, minerals, and aroma compounds. The most commonly used solvent in these SCF extractions is carbon dioxide. Carbon dioxide is popular because it is non-toxic, nonflammable, inexpensive, and has a relatively low critical temperature (304 K).

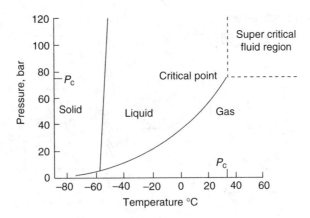

FIGURE 10.38 Carbon dioxide phase diagram.

TABLE 10.19
Typical Physical Propeties Associated with Different Fluid States

State of fluid	Density (g/cm^3)	Diffusivity (cm^2/sec)	Viscosity (g/cm-sec)
Gas			
$P = 1$ atm, $T = 15$–$30°C$	$(0.6$–$2) \times 10^{-3}$	0.1–0.4	$(1$–$3) \times 10^{-4}$
Liquid			
$P = 1$ atm, $T = 15$–$30°C$	0.6–1.6	$(0.2$–$2) \times 10^{-5}$	$(0.2$–$3) \times 10^{-2}$
Supercritical			
$P = P_c, T = T_c$	0.2–0.5	0.7×10^{-3}	$(1$–$3) \times 10^{-4}$
$P = 4P_c, T = T_c$	0.4–0.9	0.2×10^{-3}	$(3$–$9) \times 10^{-4}$

Source: From Hoyer, G.G. 1985. *Chemtech.* 15: 440–448.

An application of SCF extraction which has seemingly gone unexplored is to the drying of food products. Since moisture content influences texture, chemical reactions, and susceptibility to microbial spoilage, drying is a way to retain quality and prolong shelf life. A complication associated with drying of food products is that they may undergo changes which alter the physical or chemical structure, thus changing the integrity of the product. SCF extraction avoids this problem because it allows the food product to be dehydrated without undergoing a phase from liquid water to water vapor. Also, if a solvent such as supercritical carbon dioxide is used, it will not be necessary to heat the product above ambient temperatures.

10.7.6.11.1 Extraction

The process for SCF extraction consists of one basic step. The complexity of the system needed for any extraction will depend on the product and conditions of operation.

Figure 10.40 shows a basic system for the process. The product is first contacted with the SCF. The SCF is at conditions where maximum solubility exists between it and the extract. A continuous process is shown in this diagram, with the product entering the vessel through rotating locks. The temperature of the food has been adjusted so that it is the same as the temperature of the SCF. This precaution is taken so that the conditions of maximum solubility are not altered. The food then proceeds down the vessel, exiting through another rotating lock. The SCF containing the extract

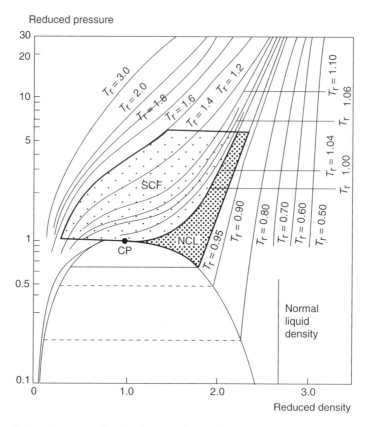

FIGURE 10.39 Reduced pressure–density diagram. Supercritical fluid (SCF) and near-critical liquid (NCL) regions as indicated. (From deFilippi, R.P. 1982. *Chem. Ind.*, 19: 390–394.)

leaves the vessel and is transported through an apparatus which will alter the conditions to decrease miscibility. Since miscibility is dictated by temperature and pressure, a heat exchanger which will decrease the temperature of the SCF/extract stream is illustrated. The mixture then enters a separator where the extractant is allowed to exit, while the solvent is recycled through the system.

Obviously, loss of SCF through the rotating lock must be considered. This problem may be solved by adding a SCF reservoir which would supply a continuous supply to the process. Figure 10.41 shows a flow scheme that not only supplies the fluid, but also generates and regulates the pressure. It should be noted that the pressure throughout the process is never lowered below the critical pressure. Should this occur, a phase change will take place, possibly changing the quality of the final product.

10.7.6.11.2 Application to Drying

The use of supercritical carbon dioxide in conjunction with food products has received attention because it is considered safe and convenient. It is already being used in the extraction of many food components, the most common being caffeine and hops. Its low critical temperature (304 K) is very important in that most foods experience any chemical and physical changes slowly at this condition. The SCF process itself avoids a phase change which is very important if the structural integrity of the product is to be retained. Current drying techniques may cause browning, marked shrinkage, and even loss of flavor and aroma. SCF extraction of water can avoid these often undesirable effects.

The solubility of water in carbon dioxide is an important consideration for drying by SCF extraction. Although water is considered to have poor solubility in carbon dioxide, it is not completely insoluble. Figure 10.42 shows the relationship between the composition of water in carbon dioxide

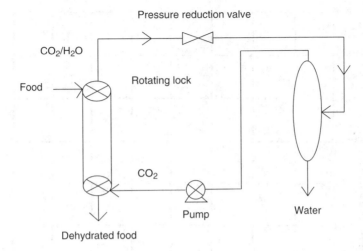

FIGURE 10.40   Schematic diagram of the SCF extraction process.

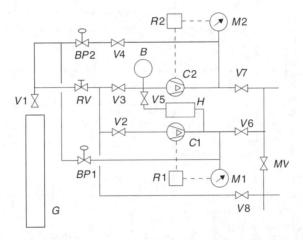

FIGURE 10.41   Pressure generation and regulation in a laboratory installation for pressures up to 3000 bar. G gas supply; RV, reducing valve; BP1, BP2, back-pressure regulators; MV, fine metering valve; V1–V8, out-off valves; B, buffer volume; C1, membrane compressor to 1000 bar; C2, membranecompressor to 3000 bar; R1, R2, switching relays; M1, M2, contactmanometers; H, heat exchanger. (From Stahl, E., et al., 1988. *Dense Gases for Extraction and Refining*. Springer-Verlag, Berlin Heidelberg, Germany.)

and pressure at various temperatures (Wiebe, 1941). The results are given in grams of water per liter of expanded carbon dioxide at the STP (273 K, 1 atm). This would indicate that the amount of water would be higher measured under supercritical conditions, where the density of carbon dioxide is increased. It is evident from this figure that a significant change in solubility occurs simply by changing the temperature.

10.7.6.11.3 Entrainer
As discussed, carbon dioxide is considered a weak solvent in relation to water when compared with its solvency to other compounds. The low solvency has the advantage of high selectivity but may cause limitations by dictating that the process occurs under extreme or uneconomical conditions. Improving the solvent capacity can be accomplished by using a third compound. This compound is referred to as an entrainer, cosolvent, auxiliary, or modifier.

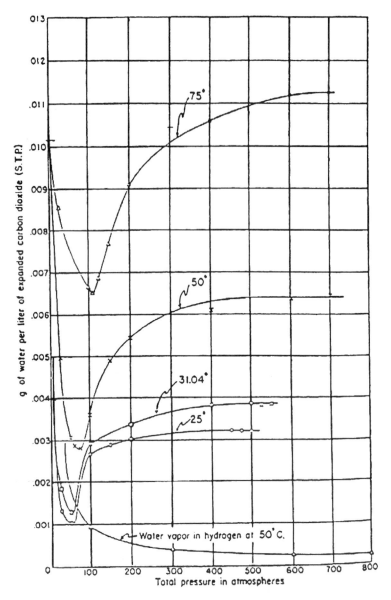

FIGURE 10.42 Composition of the phase rich in carbon dioxide asa function of pressure and temperature. (From Wiebe, R. 1941. *Chem. Rev.,* 29: 475.)

The purpose of an entrainer is to increase the solvent capacity in a system. The entrainer must be chosen so that the volatility lies between that of the SCF and the solute. Brunner and Peter (1982) present three criteria that an entrainer must fulfill.

1. The entrainer must enhance solubility, allowing the process to operate at lower pressures.
2. The entrainer must allow regeneration of the gases to be achieved by a temperature change only.
3. The entrainer must improve the separation factor.

A ternary phase diagram showing the relationship of a system with an entrainer is shown in Figure 10.43. It is evident that at the prescribed conditions, a miscibility gap exists between the

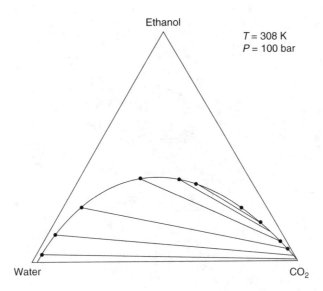

FIGURE 10.43   Ternary-phase diagram for ethanol–water–carbon dioxideat 308 K and 100 bar. (From Paulaitis, M.E., et al., 1983. *Rev. Chem. Eng.*, 1: 181–211.)

carbon dioxide and water, while no gap exists between the entrainer and water. The miscibility of the entrainer in the solvent can thus be used. When the entrainer dissolves in the solvent, it will carry the solute with it. This process can be controlled with miscibility changes. Figure 10.44 shows how changing temperature can alter miscibility. Brunner and Peter (1982) explain this diagram in detail. Figure 10.45 shows a flow scheme for this two step process along with the ternary phase diagrams corresponding to the steps. In the first column, which operates in the supercritical range (temperature is slightly above the critical temperature), a miscibility gap exists only between the SCF and the extract. The second column is operated at the same pressure, but a higher temperature. An additional miscibility gap now exists, causing the extractant to separate as its solubility is decreased.

Disadvantages of using an entrainer range from practical complications to added expenses. One complication is that a second separation step will be necessary for the entrainer. The loss of entrainer from this step must be replaced in order to maintain a constant concentration in the system.

Currently, few studies have been published concerning measurements for ternary systems. The addition of an entrainer to a binary system would seem to result in marked flexibility, but the exact nature of the changes it induces may not be known. Experimentation is necessary to examine the possibility of an advantageous entrainer for drying.

10.7.6.11.4   Feasibility

The major disadvantage in SCF extraction is the high capital investment necessary for pressure equipment. Since SCF extraction is a relatively new process, research for the design and implementation of efficient equipment has been limited. It has been predicted, though, that the attractiveness of SCF extraction will bring about increased interest in this area. With increased interest, developments will be made to decrease the cost of the process.

The cost for the process itself is dependent on the amount of extraction desired and the dilution of the extract. Generally, foods containing a large amount of water require a greater amount of energy to dry to completion than those with a lower water content. On the other hand, low dilutions of extract also require more energy. Figure 10.46 shows the relationship between the recovery efficiency and time. At lower solute concentrations, the time necessary to extract a small amount of solute is much greater than that necessary to extract the same amount at high concentrations. It is evident that efficiency will reach a maximum, at which point further extraction would prove uneconomical.

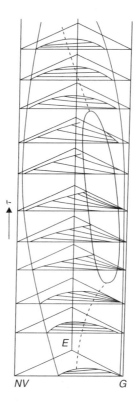

FIGURE 10.44   Temperature dependence of phase equilibrium of aquasi-ternary system consisting of substances of low volatility(NV), an entrainer (E), and a supercritical component (G). (From Stahl, E., et al., 1988. *Dense Gases for Extraction and Refining*. Springer-Verlag, Berlin Heidelberg, Germany.)

One method of separation involves stepwise extraction. A diagram for this is shown in Figure 10.47. The number of stages, as well as the time spent in each stage, is again, dictated by the extraction desired. The stagewise extraction allows for fractionation. If the processor requires a product at different degrees of dryness, this design would be beneficial. There is, though, no evidence of increased savings with this design. Longer time periods for complete drying, or a greater number of stages may not make this process feasible.

The relative throughput of the solvent can be increased. It has been found that for carbon dioxide a good extraction in which saturation occurs requires 5 to 20 kg gas/h/kg sample (Stahl et al., 1988).

Energy saving steps have also been suggested to make the process more feasible. For example, Figure 10.48 is a plot of the energy requirements necessary to raise the pressure from 60 to 600 bars. The process of isentropic compression which is accompanied by an increase in temperature is shown. It is evident that the two stage process, which includes intermediate cooling, saves over a quarter of the energy necessary for compression.

It has been reported that the energy requirements for SCF extraction are less than those required for distillation and are comparable or less than those for liquid extraction. This is, in part, due to fast extraction rates, low heats of absorption, and availability of inexpensive solvents.

The major operating cost of pressurizing and repressurizing the SCF can be minimized if the extraction can be controlled by temperature changes. Energy costs for pressurizing a system are high. If the solubility changes can be controlled by temperature changes, the high costs incurred with repressurizing the SCF can be avoided. It is extremely advantageous to use temperature changes with SCF carbon dioxide because of the moderate temperatures that can be utilized. Minimal energy can be used to make the solubility change that are necessary for drying if temperature changes can be employed.

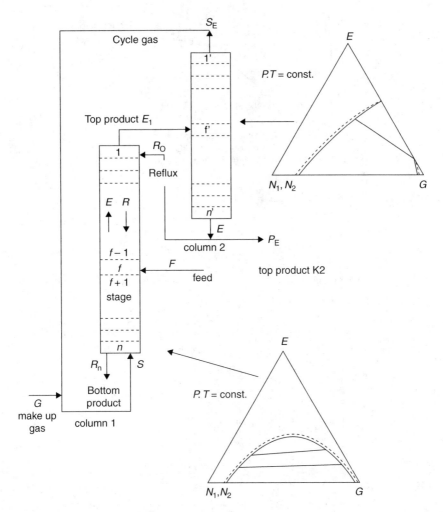

FIGURE 10.45 Flow diagram of gas extraction with an entrainer and related phase equilibria. E, entrainer; G, gas; N_1, N_2, nonvolatile substance 1 and 2. (From Brunner, G. and Peter, S. 1982. *Sep. Sci. Technol.*, 17: 199.)

The labor costs for SCF have been determined to be lower since only one step, extraction, is necessary as opposed to the two steps, extraction and evaporation, needed for current extraction processes. Maintenance costs have been found to be comparable.

Specific costs associated with SCF extraction are difficult to determine. Little research has been done on this topic, and the research that has been done is not widely available. The competition for employing this technique in the marketplace restricts the amount of published information. It is clear, however, that SCF extraction necessitates a high initial investment. The cost for operation is then dependent on the product. For drying, SCF extraction may prove to be feasible for high value products. In the future, improved design and efficiency may make it a viable alternative for all products.

10.7.6.11.5 Economic Analysis of Mechanical Vapor Recompression for a Drum Dryer

It was reported that for a regular drum drying process 1300 Btu was required to remove one pount of water. If the system was equipped with an one-effect mechanical vapor recompressor (MVR), only the thermal energy of 218 Btu/lb water removed would be needed, and 176 Btu/lb would be

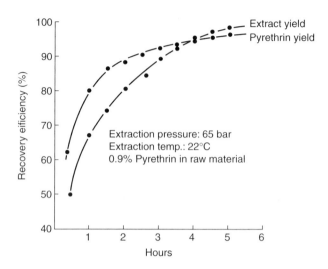

FIGURE 10.46 Efficiency of CO_2 extraction vs. time on stream. (Printed with permission; copyright 1987 Marc Sims.)

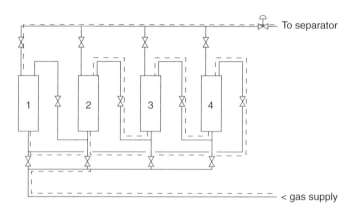

FIGURE 10.47 Scheme of a multiple-vessel countercurrent extraction. (From Stahl, E., et al., 1988. *Dense Gases for Extraction and Refining*. Springer-Verlag, Berlin Heidelberg, Germany.)

sufficient for a system with a two-effect MVR. If a drum dryer with MVR were operated at an optimal condition, more significant energy savings could be achieved.

Since this work is focused on the drum drying of food with an ordinary heat source, the recovery of low level heat will be the main area of reducing process cost. A heat pump and heat exchanger are widely used equipment for the recovery of thermal energy. The major limitation of the latter is that the temperature level of the heat receiver in all cases is lower than the level of the heat source. On the other hand, although a heat pump does not suffer this limitation, its high investment and operating costs will hinder its use in industry unless its use meets the economic criteria, mainly the payback period, the cash flow and the internal rate of return (Moser and Schnitzer, 1985).

According to Heap (1979), there are several operating cycles for a heat pump. Among the types widely used in industry are mechanical vapor recompression (MVR), Brayton heat pump, open-cycle, and close-cycle heat pump.

MVR is one of the most important process cycles. It works with the process vapor directly and does not limit the maximum obtainable temperatures because it is independent on a special heat transfer medium. Both the latent and sensible heat of the vapor is recovered in MVR and that provides a very efficient energy savings in the drying process. MVR can be applied to any contact dryer such as a drum dryer, tray dryer, rotary dryer, etc.

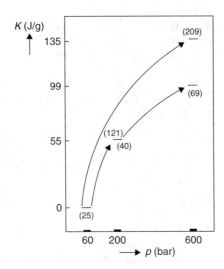

FIGURE 10.48 Scheme illustrating the energy requirement for isentropic compression of carbon dioxide with and without intermediate cooling. (Numbers in parentheses = temperature in °C). (From Stahl, E., et al., 1988. *Dense Gases for Extraction and Refining.* Springer-Verlag, Berlin Heidelberg, Germany.)

REFERENCES

Achanta, S. Okos, M.R. 1996. Predicting the quality of dehydrated foods and biopolymers–research needs and opportunities. *Drying Technol.*, 14: 1329–1368.

Acker, L.W. 1969. Water activity and enzyme activity. *Food Technol.* 23: 1257.

Adambounou, T.L. and Castaigne, F. 1983. *Lebensm. - Wiss. u. - Technol.* 16: 230.

Afzal, T.M. and Abe, T. 1998. Diffusion in potato during far infrared radiation drying. *J. Food Eng.*, 37: 353–365.

Agarwal, S.R., Heiligman, F., and Powers, E.M. 1972. Comparison of pre-cooked irradiated chicken and lamb with and without partial dehydration. *J. Food Sci.*, 37: 469–472.

Agrawal, K.K., Clary, B.L., and Nelson, G.L. 1969. Investigation into the theories of desorption isotherms for rough rice and peanuts. I. Paper No. 69–890, 1969. Winter Meeting American Society of Agricultural Engineers, Chicago.

Aguilera, J.M., Chirife, J., Flink, J.M., and Karel M. 1975. Computer simulation of non-enzymatic browning during potato dehydration. *Lebensm.-Wiss. u.-Technol.* 8: 128.

Ahmad, S.S., Morgan, M.T., and Okos, M.R. 2001. Effects of microwave on the drying, checking and mechanical strength of baked biscuits. *J. Food Eng.*, 50: 63–75.

Ahmed, J., Khan, A.R., and Hanan, A.S. 2004a. Moisture adsorption of an Arabian sweet (basbusa) at different temperatures. *J. Food Eng.*, 64: 187–192.

Ahmed, J. Ramaswamy, H.S., and Khan, A.R. 2005. Effect of water activity on glass transitions of date pastes. *J. Food Eng.*, 66: 253–258.

Akbaba, H. and Cakaloz, T. 1994. Air drying of leek. In: *Drying '94*, Vol. 2, Mujumdar, A.S., (Ed.), Hemisphere, McGraw-Hill, pp. 1137–1143.

Akiyama, T., Liu, H., and Hayakawa, K. 1997. Hygrostress-multicrack formation and propagation in cylindrical viscoelastic food undergoing heat and moisture transfer processes. *Int. J. Heat Mass Transfer*, 40: 1601–1609.

Al-Muhtaseb, A.H., McMinn, W.A.M., and Magee, T.R.A. 2003. Water sorption isotherms of starch powders: Part 1: mathematical description of experimental data. *J. Food Eng.*, 61: 297–307.

Alam, A. and Shove, G.C. 1973. Simulation of soybean drying. *Trans. ASAE*, 16: 134–136.

Alcaraz, E.C., Martin, M.A., and Marin, J.P. 1977. Metodo manometrico para medida de humedades de equilibrio. *Grasas y Aceites*, 28: 403.

Alhamdan, A.M. and Hassan, B.H. 1999. Water sorption isotherms of date pastes as influenced by date cultivar and storage temperature. *J. Food Eng.*, 39: 301–306.

Alves-Filho, O. and Rumsey, T.R. 1985. Thin layer drying and rewetting models to predict moisture diffusion in spherical agricultural products. *Drying '85*, 434–437.

Alzamora, S. 1979. Thesis. Facultad de Ciencias Exactas y Naturales. Universedad de Buenos Aires, Argentina.

Alzamora, S., Chirife, J., Viollaz, P., and Vaccarezza, L.M. 1979. Heat and mass transfer during air drying of avocado. In: *Developments in Drying*, Mujumdar, A.S., (Ed.), Science Press, NJ.

Amundson, C.H., Ishino, K., and Lindsay, R.C. 1980. Retention of free fatty acids during spray drying of cheddar cheese. *J. Food Process Eng.*, 4: 213–225.

Andrieu, J. and Stamatopoulos, A. 1986a. Durum wheat pasta drying kinetics. *Lebensmittel-Wissenschaft und-Technologie,* 19: 448–456.

Andrieu, J. and Stamatopoulos, A. 1986b. Moisture and heat transfer modeling during durum wheat pasta drying. In: *Drying '86*, Vol. 2, Mujumdar, A.S., (Ed.), Hemisphere Publ. Corp., Washington, DC, pp. 492–498.

Andrieu, J., Stamatopoulos, A., and Zafiropoulos, M. 1985. Equation for fitting desorption isotherms of durum wheat pasta. *J. Food Technol.*, 20: 651–657.

Andrieu, J., Stamatopoulos, A., and Zafiropoulos, M. 1986. Corn past water desorption isotherms. *Lebensm. - Wiss.' u-Technol.*, 19: 415–418.

Andrieu, J., Jallut, C., Stamatopoulos, A., and Zafiropoulos, M. 1988. Identification of water apparent diffusivities for drying of corn based extruded pasta. Proceedings of the Sixth International Drying Symposium. IDS '88 Versailles, France: 71–74, Sept. 5–8.

Ashworth, J.C. 1977. The Mathematical Simulation of the Batch-Drying of Softwood Timber, Ph.D Dissertation. University of Canterbury.

Auerbach, E., Wang, H., Maynard, N., Doty, D.M., and Kraybill, H.R., 1954. A histological and histochemical study of beef dehydration. V. Some factors influencing the rehydration level of frozen-dried muscle tissue. *Food Res.*, 19: 557–563.

Aviara, N.A., Ajibola. O.O., and Oni, S.A. 2004. Sorption equilibrium and thermodynamic characteristics of soya bean. *Biosystems Eng.*, 87: 79–190.

Azoubel, P.M. and Murr, F.E.X. 2004. Mass transfer kinetics of osmotic dehydration of cherry tomato. *J. Food Eng.,* 61: 291–295.

Azuara, E., Beristain, C.I., and Gutierrez, G.F. 1998. A method for continuous kinetic evaluation of osmotic dehydration. *Lebensm.-Wiss. U.-Technol.*, 31: 317–321.

Babalis, S.J. and Belessiotis, V.G. 2004. Influence of the drying conditions on the drying constants and moisture diffusivity during the thin-layer drying of figs. *J. Food Eng.*, 65: 449–458.

Bagnall, L.O., Millier, W.F., and Scott, N.R., 1970. Drying the alfalfa stem. *Trans. ASAE,* 13: 232–245.

Bai, Y., Rahman, M.S., Perera, C.O., Smith, B., and Melton, L.D. 2001. State diagram of apple slices: glass transition and freezing curves. *Food Res. Int.*, 34: 89–95.

Baik, O.D. and Marcotte, M. 2002. Modeling the moisture diffusivity in a baking cake. *J. Food Eng.,* 56: 27–36.

Bakalis, S., Karathanos, V., Maroulis, Z., Marinos-Kouris, D., and Saravacos, G. 1994. Moisture diffusivity in osmotically dehydrated fruits. In: *Drying '94*, Vol. 1, Mujumdar, A.S., (Ed.), Hemisphere, McGraw-Hill, pp. 857–862.

Baker, Christopher G.J. 1997. *Industrial Drying of Foods*. Blackie Academic & Professional, London, pp. 53–54.

Bakshi, A. and Singh, P. 1980. Kinetics of water diffusion and starch gelatinization during rice parboiling. *J. Food Sci.*, 45: 1387–1392.

Baloch, A.K., Buckle, K.A., and Edwards, R.A. 1977a. Effect of processing variables on the quality of dehydrated carrot II. Leaching losses and stability of carrot cuting dehydration and storage. *J. Food Technol.*, 12: 285–307.

Baloch, A.K., Buckle, K.A., and Edwards, R.A. 1977b. Stability of beta-carotene in model systems containing sulphite. *J. Food Technol.*, 12: 309–316.

Banwart, G.J. 1981. *Basic Food Microbiology*. AVI Publishing Company, Westport, CT, p. 78.

Barat, J.M., Fito, P., and Chiralt, A. 2001. Modeling of simultaneous mass transfer and structural changes in fruit tissues. *J. Food Eng.,* 49: 77–85.

Barbosa-Canovas, G.V. and Vega-Mercado, H. 1996. *Dehydration of Foods*. Chapman & Hall, New York.

Barrett, D.M. and Lund, D.B. 1989. Effect of oxygen on thermal degradation of 5-Methyl-5, 6, 7, 8-tetrahydrofolic acid. *J. Food Sci.*, 54: 146–149.

Barreiro, J.A., Fernández, S., and Sandoval, A.J. 2003. Water sorption characteristics of six row barley malt (Hordeum vulgare). *Lebensm.-Wiss. u.-Technol.*, 36: 37–42.

Bassal, A., Vasseur, J., and Loncin, M. 1993. Sorption isotherms of food materials above 100°C. *Lebensm.-Wiss. u.-Technol.*, 26: 505–511.

Basunia, M.A. and Abe, T. 2001. Moisture desorption isotherms of medium-grain rough rice. *J. Stored Prod. Res.*, 37: 205–219.

Becker, H.A. and Sallans, H.R. 1955. A study of internal moisture movement in the drying of the wheat kernel. *Cereal Chem.*, 32: 212–226.

Becker, H.A. and Sallans, H.R. 1956. A study of the desorption isotherms of wheat at 25 and 50°C. *Cereal Chem.*, 33: 79.

Benado, A.L. and Rizvi, S.S.H. 1985. Thermodynamic properties of water on rice as calculated from reversible and irreversible isotherms. *J. Food Sci.*, 50: 101–105.

Bell, L.N and Labuza, T.P. 2000. Moisture Sorption. Practical Aspects of Isotherm Measurement and Use.2nd Ed. AACC, St Paul, MN.

Bell, L.N. and Touma, D.E. 1996. Glass transition temperatures determined using a temperature-cycling differential scanning calorimeter. *J. Food Sci.* 61: 807–810.

Benson, S.W. and Richardson, R.L. 1955. A study of hysteresis in sorption of polar gases by native and denatured proteins. *J. Am. Chem. Soc.*, 70: 2585.

Beristain, C.I., Azuara, E., and Vernon-Carter, E.J. 2002. Effect of water activity on the stability to oxidation of spray-dried encapsulated orange peel oil using mesquite gum (*Prosopis Juliflora*) as wall material. *J. Food Sci.*, 67: 206–211.

Berlin, E., Anderson, B.A., and Pallansch, M.J. 1970. Effect of temperature on water vapor sorption by dried milk powders. *J. Dairy Sci.*, 53: 146.

Berlin, E., Anderson, B.A., and Pallansch, M.J., 1973. Water sorption by dried products stabilized with carboxymethyl cellulose. *J. Dairy Sci.*, 56: 685.

Berlin, E., Kliman, P.G., and Pallansch, M.J., 1970. Changes in state of water in proteinaceous systems. *J. Colloid Interface Sci.*, 34: 488.

Berlin, E., Kliman, P.G., and Pallansch, M.J. 1971. Calorimetry and thermogravimetry of bound water in dried milk and whey powders. *J. Dairy Sci.*, 54: 300–305.

Bhandari, B.R. and Howes, T. 1999. Implication of glass transition for the drying and stability of dried foods. *J. Food Eng.*, 40: 71–79.

Bhandari, B.R. and Howes, T. 2000. Implication of glass trasition for food stability. *Food Australia*. 52: 579–585.

Biau, D. 1986. *Les applications industrielles du Chauffage par rayonnement infrarouge*, Eyrolles, Paris.

Biliaderis, C.G. 1992. Structures and phase transitions of starch in food systems. *Food Technol*, 46: 98–100, 102, 104, 106, 108, 109, 145.

Bimbenet, J.J., Daudin, J.D., and Wolf, E. 1985. Air drying kinetics of biological materials. In: *Drying '85*, Mujumdar, A.S., (Ed.), Hemisphere, New York, pp. 178–185.

Birch, G.G. and Parker, K.J. 1974. *Vitamin C*. John Wiley & Sons, New York.

Biswal, R.N., Wilhelm, L.R., Rojas, A., and Mount, J.R., 1997. Moisture diffusivity in osmotically concentrated diced sweet potato during air drying. *T. ASAE.*, 40: 1383–1390.

Bizot, H., Le Bail, P., Leroux, B., Davy, J., Roger, P., and Buleon, A. 1997. Calorimetric evaluation of the glass transition in hydrated, linear and branched polyanhydroglucose compounds. *Carbohydr. Polym.*, 32: 33–50.

Blau, R., Powell M., and Gerling, J.E. 1965. Results of 2450 megacycle microwave treatments in potato chip finishing. Proceedings of the 28th Annual Conference and Exhibit of the Potato Chip Institute, Int., NY.

Blazek A. 1973. In: *Thermal Analysis*, Chalmers R.A. (Ed.), Van Nostrand Reinhold Co., London, pp. 152–159.

Blond, G., Simatos, D., Catté, M., Dussap, C.G., and Gros, J.B. 1997 Modeling of the water–sucrose state diagram below 0°C. *Carbohydr. Res.*, 298: 139–145.

Bolin, H.R. 1970. Fruit Juice Concentrates and Powders. I. Development of a New Concentrate Procedure. II. Physicochemical and Volatice Flavor Changes, Ph.D. Dissertation, Univ. Utah.

Bone, D.P. 1986. Practical applications of water activity and moisture relations in foods. In Water activity: Theory and applications to food, L.B. Rockland and L.R. Beuchat (Eds.), pp. 369–395.

Borde, B. Bizot, H., Vigier, G., and Buleon, A. 2002. Calorimetric analysis of the structural relaxation in partially hydrated amorphous polysaccharides. I. Glass transition and fragility. *Carbohydr. Polym.*, 48: 83–96.

Borenstein, B. 1968. Vitamins and amino acids. In: *Handbook of Food Additives*, Furia, T.E. (Ed.), CRC Press, Cleveland, pp. 107–137.

Botterill, J.S.M. 1975. *Fluid Bed Heat Transfer*. Academic Press, London.

Boudhrioua, N., Michon, C., Cuvelier, G., and Bonazzi, C. 2002. Influence of ripeness and air temperature on changes in banana texture during drying. *J. Food Eng.,* 55: 115–121.

Brailsford, A.D. and Major, K.G. 1964. The thermal conductivity of aggregates of several phases, includinf porous materials. *Brit. J. Appl. Phys.,* 15: 313–319.

Brake, N.C. and Fennema, O.R. 1999. Glass transition values of muscle tissue. *J. Food Sci.,* 64: 10–15.

Brakel, J.V. and Heertjes, P.M. 1974. Analysis of diffusion in macroporous media in terms of a porosity, a tortuosity and a constrictivity factor. *Int. J. Heat Mass Transfer.,* 17: 1093–1103.

Briggs, G.M. and Calloway, D.H. 1979. *Bogert's Nutrition and Physical Fitness*. W.B. Saunders Company, Philadelphia.

Brook, R.C. and Bakker-Arkema, F.W. 1980. Dynamic programming for process optimization 1. An algorithm for design of multi-stage grain dryers. *J. Food Process Eng.,* 2: 199–211.

Brown, R.B., Fulford, G.N., Daynard, T.B., Meiering, A.G., and Otten, L. 1979. Effect of drying method on grain corn quality. *Cereal Chem.,* 56: 529–532.

Bruin, S. and Luyben, K.Ch.A.M. 1980. Drying of food materials: A review of recent developments. In: *Advances in Drying*, Vol. 1, Mujumdar, A.S., (Ed.), Hemisphere Publishing Corp., New York, pp. 155–215.

Brunauer, S., Emmett, P.H., and Teller, E. 1938. The adsorption of gases in multimolecular layers. *J. Am. Chem. Soc.,* 60: 309.

Brunner, G. and Peter, S. 1982. On the solubility of glycerides and fatty acids in compressed gases in the presence of an entrainer. *Sep. Sci. Technol.,* 17: 199.

Bruygidyr, A.M., Rzepecka, M.A., and McConnel, M.B. 1977. Characterization and drying of tomato paste foam by hot air and microwave energy, *J. Inst. Can. Sci. Technol. Ailment.,* 10: 313–319.

Buera, M., Chirife, J., Resnik, S.L., and Wetzler, G. 1987b. Nonenzymatic browning in liquid model systems of high water activity: kinetics of color changes due to Maillard's reaction between different single sugars and glycine and comparison with caramelization browning. *J. Food Sci.,* 52: 1063–1067.

Buera, M., Chirife, J., Resnik, S.L., and Lozano, R.D., 1987a. Nonenzymatic browning in liquid model systems of high water activity: kinetics of color changes due to caramelization of various single sugars. *J. Food Sci.,* 52: 1059–1062, 1073.

Buera, M., Chirife, J., Resnik, S.L., and Lozano, R.D. 1987c. Nonenzymatic browning in liquid model systems of high water activity: kinetics of color changes due to reaction between glucose and glycine peptides. *J. Food Sci.,* 52: 1068–1070, 1073.

Buera, M.P., Jouppila, K., Roos, Y.H., and Chirife, J. 1998. Differential scanning calorimetry glass transition temperatures of white bread and mold growth in the putative glassy state. *Cereal Chem.,* 75: 64–69.

Bull, H.B. and Breese, K. 1968. Protein hydration. 1. Binding sites. *Arch. Biochem. Biophys.,* 128, 488–496.

Burin, L., Buera, M.P., Hough, G., and Chirife, J. 2002. Thermal resistance of β-galactosidase in dehydrated dairy model systems as affected by physical and chemical changes. *Food Chem.,* 76: 423–430.

Burton, H. 1954. Colour changes in heated and unheated milk. I. The browning of milk on heating. *J. Dairy Res.,* 21: 194–203

Burton, H. and Rowland, S.J. 1955. Colour changes in heated and unheated milk. III. The effect of variation in milk composition on the whitening and browning of separated milk on heating. *J. Dairy Res.* 22: 82–90.

Bushuk, W. and Winkler, C.A. 1957. Sorption of water vapor on wheat flour, starch, and gluten. *Cereal Chem.,* 34, 73–85.

Buvansaundaram, K., Mukai, N., Tsukada, T., and Hozawa, M. 1994. Experimental and simulation study on drying of food gel. In: *Drying '94*, Vol. 2, Mujumdar, A.S., (Ed.), Hemisphere, McGraw-Hill, pp. 1291–1298.

Cabrera, E., Pineda, J.C., de Bazula, C.D., Segurajaureregui, J.S., and Vernon, E.J. 1984. Kinetics of water diffusion and starch gelatinization during corn nixtamalization. In: *Engineering and Food*, Vol. 1, Mackenna, B.M., (Ed.), Elsevier Applied Science Publishers, Elsevier, London.

Cain, R.F. 1967. Water-soluble vitamins. Changes during processing and storage of fruit and vegetables. *Food Technol.,* 21: 998.

Calame, J.P. and Steiner, R. 1982. Carbon dioxide extraction in the flavour and perfumery industries. *Chem. Ind.,* 19: 399.

Calloway, D.H. 1962. Dehydrated foods. *Nutr. Rev.,* 20: 257–260.

Campbell, C.H. 1917. Drying milk. U.S. Patent 1250427.

Campbell, A.M., Penfield, M.P., and Griswold, R.M. 1979. *The Experimental Study of Food*. Houghton Mifflin Company, Boston, MA.

Caurie, M. 1981. Derivation of full range moisture sorption isotherms. In: *Water Activity: Influences on Food Quality*, Rockland, L.B. and Stewart, G.F., (Eds), Academic Press, New York, pp. 83–87.

Ceaglske, N.H. and Hougen, O.A. 1937. The drying of granular solids. *Trans. Am. Inst. Chem. Eng.*, 33: 283–312.

Cerrutti, P., Resnik, S.L., Seldes, A., and Fontan, C.F. 1985. Kinetics of deteriorative reactions in model food systems of high water activity: Glucose loss, 5-hydroxymethylfurfural accumulation and fluorescence development due to nonenzymatic browning. *J. Food Sci.*, 50: 627–630, 656.

Chace, E.M. 1942. The Present Status of Food Dehydration in the United States. IFT Proceedings. pp. 70–89.

Chambers, T.C. and Possingham, J.V. 1963. Studies of the fine structure of the wax layer of sultana grapes. *Aust. J. Biol. Sci.*, 16: 818–825.

Charms, S.E. 1963. *The Fundamentals of Food Engineering*. AVI Publ. Co., Westport, CT.

Chaudhary, D.R. and Bhandari, R.C. 1968. Heat-transfer through a 3-phase porous medium. Brit. *J. Appl. Phys., Ser. 2*, 1: 815–817.

Chen, C. 2000a. A rapid method to determine the sorption isotherms of peanuts. *J. Ag. Eng. Res.*, 75: 401–408.

Chen, C. 2002b. PH — pastharvest technology: sorption isotherms of sweet potato slices. *Biosyst. Eng.* 83: 85–95.

Cheng, W.M. Raghavan, G.S.V. Ngadi, M. Wang, N. 2006a. Microwave power control strategies on the drying process. I. Development and evaluation of new microwave drying system. *J. Food Eng.*, 76: 188–194.

Cheng, W.M. Raghavan, G.S.V. Ngadi, M. Wang, N. 2006b. Microwave power control strategies on the drying process. II. Phase-controlled and cycle-controlled microwave/air drying. *J. Food Eng.*, 76: 195–201.

Chilton, W.G. and Collison, R. 1974. Hydration and gelation of modified potato starches. *J. FD. Technol.*, 9, 87–93.

Chipley, J.R. and May, K.N. 1968. Survival of aerobic and anerobic bacteria in chicken meat during freeze-dehydration, rehydration, and storage. *J. Appl. Microbiol.*, 16: 445–449.

Chirife, J. 1971. Diffusional process in the drying of tapioca root. *J. Food Sci.*, 36: 327–329.

Chirife, J. and Iglesies, H.A. 1978. Equations for fitting water sorption isotherms of foods: Part 1 — a review. *Food Technol.*, 13: 159–174.

Choi, Y. 1985. Food thermal property prediction as affected by temperature and composition. Ph.D. Thesis, Purdue University.

Chu, S. and Hustrulid, A., 1968. Numeral solution of diffusion equation. *Trans. ASAE.* 11: 705–708.

Chung, D.S. and Pfost, H.B. 1967. Adsorption and desorption of water vapor by cereal grains and their products. *Trans ASAE*, 10: 549.

Chung, H.J., Woo, K.S., and Lim, S.T. 2004. Glass transition and enthalpy relaxation of cross-linked corn starches. *Carbohydr Polym.* 55: 9–15.

Clayton, J.T. and Huang, C.T. 1984. Porosity of extruded foods. In: *Engineering and Food*, Vol. 2, McKenna, B.M., (Ed.), Elsevier Applied Science Publishers, London and New York.

Clegg, K.M. 1964. Non-enzymatic browning of lemon juice. *J. Sci. Food Agic.*, 15: 878, 885.

Cohen, E. and Saguy, I. 1985. Statistical evaluation of arrhenius model and its applicability in prediction of food quality losses. *J. Food Proc. Preserv.*, 9: 273–290.

Cole, S.J. 1967. The maillard reaction in food products carbon dioxide production. *J. Food Sci.* 32: 245.

Collison, R. and Dickson, A. 1971. Heats of starch dehydration by differential thermal analysis. *Die Starch*, 23: 45.

Combs, W.B. and Hubbard, E.F. 1931. Some Factors Influencing the Capacity of the Atmospheric Drum Dryer. Paper No. 1035, Journal Series, Minnesota Agr. Exp. Sta., Univ. of MN.

Comings, E.W. and Sherwood, T.K. 1934. They drying of solids-VII, Moisture movement by capillarity in drying granular materials. *Ind. Eng. Chem.*, 25: 1096–1098.

Cooke R. and Kuntz I.D., 1974. The properties of water in biological systems. *Ann. Rev. Biophys. Bioeng.*, 3: 95.

Cording, J., Jr., Willard, M.J., Eskew, R.K., and Sullivan, J.F. 1964. Advances in the dehydration of mashed potatoes by the flake press, Eastern Regional Research Laboratory, Philadelphia.

Corzo, O. and Fuentes, A. 2004. Moisture sorption isotherms and modeling for pre-cooked flours of pigeon pea (Cajanus cajans L millsp) and lima bean (Canavalia ensiformis). *J. Food Eng.*, 65: 443–448.

Couchman, P.R. and Karasz, F.E. 1978. A classical thermodynamic discussion of the effect of composition on glass-transition temperatures. *Macromolecules.* 11: 117–119.

Coulter, S.T., Jennes, R., and Geddes, W.F., 1951. Physical and chemical aspects of the production, storage, and utility of dry milk products. *Adv. Food Res.,* 3: 45–118.

Coronel, P., Simunovic, J., and Sandeep, K.P. 2003. Temperature profiles within milk after heating in a continous-flow tubular microwave system operating at 915 MHz. *J. Food Sci.* 68: 1976–1980.

Corzo, O. and Gomez, E.R. 2004. Optimization of osmotic dehydration of cantaloupe using desired function methodology. *J. Food Eng.,* 64: 213–219.

Craig, J.C., Jr., Aceto, N.C., and Della Monica, E.S. 1961. Occurrence of 5-hydroxymethylfurfural in vacuum foam-dried whole milk and its relation to processing and storage. *J. Dairy Sci.,* 44: 1827–1835.

Crank, J. and Park, G.S. 1968. *Diffusion in Polymers.* Academic Press, London and New York.

Crapiste, G.H. and Rotstein, E. 1982. Prediction of sorptional equilibrium data for starch-containing foodstuffs. *J. Food Sci.,* 47: 1501–1507.

Crosby, E.J. and Weyl, R.W. 1977. Foam spray drying: general principles. *AIChE Symp. Ser.,* 73: 82–94.

Cross, M., Gibson, R.E., and Young, R.W. 1979. Pressure generation during the drying of a porous half-space. *Int. J. Heat Mass Transfer,* 22: 47–50.

Cuq, B. and Icard-Verniere, C. 2001. Characterisation of glass transition of durum wheat semolina using modulated differential scanning calorimetry. *J Cereal Sci.,* 33: 213–221.

D'Arcy, R.L. and Watt, I.C. 1981. Water vapor isotherms on macromolecular substrates. In: *Water Activity: Influences on Food Quality,* Rockland, L.B. and Stewart, G.F., (Eds), Academic Press, New York, pp. 111–142.

Daniels, T. 1973. *Thermal Analysis.* Hohn Wiley & Sons Inc. NY.

Das, M. and Das, S.K. 2002. Analysis of moisture sorption characteristics of fish protein myosin. *Int. J. Food Sci. Technol.,* 37: 223–227.

Davis, C.O. 1965. Microwave processing of potato chips, I, II, III, *Potato Chipper* 25: 2–4.

Day, LeRoy C. 1964. A device for measuring voids in porous materials. *Agri. Engineer.,* 45: 36–37.

deFilippi, R.P. 1982. Carbon dioxide as a solvent: application to fats, oils, and other materials. *Chem. Ind.,* 19: 390–394.

Delgado, A.E. and Sun, D. 2002. Desorption isotherms and glass transition temperature for chicken meat. *J. Food Eng.,* 55: 1–8.

Del Valle, F.R. and Nickerson, J.T.R. 1968. Salting and drying of fish. 3. Diffusion of water. *J. Food Sci.* 33: 499: 503.

Del Valle, J.M., Cuadros, R.M., and Aguilera, J.M. 1998. Glass transitions and shrinkage during drying and storage of osmosed apple pieces. *Food Res. Intern.,* 31: 191–204.

Dennison, D.B. and Kirk, J.R., 1982. Effect of trace mineral fortificaion on the storage stability of ascorbic acid in a dehydrated model food system. *J. Food Sci.,* 47: 1198–1200, 1217.

Diehl, K.C., Garwood, V.A., and Haugh, C.G. 1988. Volume measurement using the Air-Comparison Pycnometer. *Trans. ASAE,* 31: 284–287.

Di Matteo, P., Donsi, G., and Ferrari, G. 2003. The role of heat and mass transfer phenomena in atmospheric freeze-drying of foods in a fluidized bed. *J. Food Eng.,* 59: 267–275.

Dittman, F.W. and Cook, E.M. 1977. Establishing the parameters for a spray dryer. *Chem. Eng.,* 84: 108–112.

Donovan, J.W. 1979. Phase transitions of the starch-water system. *Biopolymers,* 18, 263–275.

Doymaz, I. 2004. Convective air drying characteristics of thin layer carrots. *J. Food Eng.,* 61: 359–364.

Drazga, F.H. and Eskew, R.K. 1962. Observations on drum drying mashed potatoes. *Food Technol.,* 103–105.

Drouzas, A.E. and Schubert, H. 1996. Microwave application in vacuum drying of fruits. *J. Food Eng.,* 28: 203–209.

Duckworth, R.B., 1975. *Water Relations in Foods,* Academic Press, London.

Dumas, C. and Mittal, G.S. 2002. Heat and mass transfer properties of pizza during baking. *Int. J. Food Prop.,* 5: 161–177.

Dural, N.H. and Hines, A.L. 1993. Adsorption of water on cereal-bread type dietary fibers. *J. Food Eng.,* 20: 17–43.

Dziezak, J.D. 1986. Innovative separation process finding its way into the food industry. *Food Technol.,* 40: 66.

Eckoff, S.R., Tuite, J.F., Foster, G.H., Kirles, A.W., and Okos, M.R. 1983. Microbial growth inhibition by SO_2 or SO_2 plus NH_3 treatments during the slow drying of corn. *Cereal Chem.,* 60: 185–188.

Eichner, K. and Karel, M. 1972. The influence of water content and water activity on the sugar-amino browning reaction in model systems under various conditions. *J. Agric. Food Chem.*, 20: 218–223.

Eisenhardt, N.H., Cording, J., Jr., Eskew, R.K., and Sullivan, J.F. 1962. Quick-cooking dehydrated vegetable pieces, I. Properties of potato and carrot products. *Food Technol.*, 16: 143.

Eisenhardt, N.H. et al., 1964. *Food Eng.*, 36: 53.

Eisenhardt, N.H. et al. 1967. USDA-ARS 73–54.

Eisenhardt, N.H. et al. 1968. USDA-ARS, 73–57.

Eliasson, A.-C. 1980. Effect of water content on the gelatinization of wheat starch. *Starch* 32: 270–272.

Elizalde, B.E. and Pilosof, A.M.R. 1999. Kinetics of physio-chemical changes in wheat gluten in the vicinity of the glass transition temperature. *J. Food Eng.*, 42: 97–102.

Engels, C., Hendrickx, M., and Tobback, P. 1987. Limited multilayer desorption of brown parboiled rice. *Int. J. Food Sci. Technol.*, 22: 219–223.

Ensminger, D. 1988. Acoustic and electroacoustic methods of dewatering and drying. *Drying Technol.*, 6: 473–499.

Epstein, M. and Grolmes, M.A. 1983. A Comparison Between Liquid and Solid Desiccant Cooling Systems Operating in the Ventilation Mode. Fauske & Associates, Inc., IL. Report NO. FAI/83–12.

Eskew, R.K. and Gelber, P. 1964. *Food. Process*, 25: 70.

Eskew, R.W., 1965. Proceedings of the first International Congress of Food Science and Technology, London.

Eucken, A. 1932. B3. Forschungshaft no. 353. (fr. Fernandez-Martin and Montes, 1977)

Farhat, I.A. 2000. Measuring and modelling the glass transition temperature. In: *Understanding and Measuring the Shelf-Life of Food*, Steele, R. (Ed.), Woodhead Publishing, Cambridge, 218–232.

Farrar, K.T.H. 1945. The thermal destruction of vitamin B1. I. The influence of buffer salts on the rate of destruction of aneurin at 100. *Biochem. J.*, 39: 128–132.

Farrar, K.T.H. 1955. The thermal destruction of vitamin B1 in foods. *Adv. Food Res.* 6: 257–311.

Fasina, O. and Sokhansanj, S. 1996. Estimation of moisture diffusivity coefficient and thermal properties of alfalfa pellets. *J. Ag. Eng. Res.* 63: 333–344

Fennema, O. 1976. Water and protein hydration. In: *Food Proteins*, Whitaker, J.R. and Tannebaum, S.R. (Eds), AVI Pub. Co., Westport, CT, pp. 50–90.

Fernandez, E., Schebor, C., and Chirife, J. 2003. Glass transition temperature of regular and lactose hydrolyzed milk powders. *Lebensm.-Wiss. U.-Technol.* 36: 547–551.

Ferrari, G., Meerdink, J., and Walstra, P. 1989. Drying kinetics for a single droplet of skim-milk. *J. Food Eng.*, 10: 215–230.

Ferrasse, J. and Lecomte, D. 2004. Simultaneous heat-flow differential calorimetry and thermogravimetry for fast determination of sorption isotherms and heat of sorption in environmental or food engineering. *Chem. Eng. Sci.* 59: 1374–1376.

Filkova, I. and Mujumdar, A.S. 1987. Industrial spray drying systems. In *Handbook of Industrial Drying*, Marcel Dekker, New York, pp. 243–294.

Fish, B.P. 1958. Diffusion and thermodynamics of water in potato starch gel. In *Fundamental Aspects of Dehydration of Foodstuffs*. Society of Chemical Industrialists. London: 143–157.

Fito, P., Chiralt, A., Barat, J.M., Spiess, W.E.L., and Bensnilian, D. 2001. *Osmotic Dehydration and Vacuum Impregnation*. Technomic, Lancaster.

Ford, J.E., Hurrell, R.F., and Finot, P.A., 1983. Storage of milk powders under adverse conditions. 2. Influence of the content of water-soluble vitamins. *Br. J. Nutr.*, 49: 355.

Fortes, M. and Okos, M.R. 1980. Drying theories: their bases and limitations as applied to foods, *Adv. Drying*, 119–150.

Fortes, M. and Okos, M.R. 1981c. Heat and mass transfer analysis of intra-kernal wheat drying and rewetting. *J. Argic. Eng. Res.*, 26: 109–125.

Fox, T.G. and Flory, P.J., 1950. Second-order transition temperatures and related properties of polystyrene. *J. Appl. Phys.* 21, pp. 581–591.

Franzen, K.A. 1988. Nonenzymatic Browning of Skim Milk During Dehydration. M.S. Thesis, Purdue University, West Lafayette, IN.

Franzen, K., Singh, R.K., and Okos, M.R. 1990. Kinetics of nonenzymatic browning in dried skim milk. *J. Food Eng.*, 11: 225–239.

Fukuoka, M., Watanbe, H., Mihori, T., and Shinada, S. 1994. Moisture diffusion in a dry soybean seed measured using pulsed-field-gradient NMR. *J. Food Eng.*, 23: 533–541.

Gailani, B.M. and Fung, D.Y.C. 1989. Critical review of water activities and microbiology of drying of meats. *CRC Crit. Rev. Food Sci. Nut.*, 25, 2 159–183.

Gal, S., 1981. Recent developments in techniques for obtaining complete sorption isotherms. In: *Water Activity: Influences on Food Quality*, Rockland, L.B. and Stewart, G.E. (Eds), Academic Press.

Gal, S., 1983. *Physical Properties of Foods*, Jowett, R., (Ed.), Applied Science, London, p. 13.

Gal, S., Arm, H., and Siguer, R., 1962. *Helv. Chim. Acta.*, 45: 751.

Gandhidasan, P., Goring, Q., and Myers, K. 1988. Design and Testing of a Rotary-Type Liquid Desiccant Dehumidifier, Sixth International Drying Symp.

Gane, R. 1943. Dried egg. VI. The water relations of dried egg. *J. Soc. Chem. Ind.*, 42: 185.

Garin, P., Boy-Marcotte, J.L., Roche, A., and Danneville, A. 1988. Superheated Steam Drying with Mechanical Stem Recompression, Sixth International Drying Symposium.

Geankopolis, C.J. 2003. *Transport Processes and Separation Process Principles.* 4th edn Prentice Hall, New Jersey.

Gekas, V. and Lamberg, I. 1991. Determination of diffusion coefficients in volume-changing systems-application in the case of potato drying. *J. Food Eng.*, 14: 317–326.

Gelperin, N.I. and Einstein, V.G. 1980. In: *Heat Transfer in Fluidized Beds, Fluidification,* Davidson, J.F. and Hamson, D. (Eds), Academic Press, London.

Genin, N. and Rene, F. 1996. Influence of freezing rate and the ripeness state of fresh courgette on the quality of freeze-dried products and freeze-drying time. *J. Food Eng.*, 29: 201–209.

Geurts, T.J., Walstra, P., and Mulder, H. 1974. Water binding to milk protein, with particular reference to cheese. *Neth. Milk Dairy J.*, 28: 46.

Gibbs, P.A. 1984. Microbiological quality of dried foods. In: *Concentration and Drying of Foods*. MacCarthy, D. (Ed.), Elsevier App. Sci. Publ., New York, pp. 89–111.

Gibson, R.D., Cross J., and Young, R.W. 1979. Pressure gradients generated during the drying of porous shapes. *Int. Ju. Heat Mass Transfer*, 22: 827–830.

Giraldo, G., Talens, P., Fito, P., and Chiralt, A. Influence of sucrose solution concentration on kinetics and yield during osmotic dehydration of mango. *J. Food Eng.*, 58: 33–43.

Glouannec, P., Lecharpentier, D., and Noel, H. 2002. Experimental survey on the combination of radiating infrared and microwave sources for the drying of porous material. *Appl. Thermal Eng.*, 22: 1689–1703.

Gorobtsova, N.Y.E. 1982. A method for describing and calculating sorption-desorption isotherms for a variety of materials. *Heat Trans. Soviet Res.*, 14: 92–96.

Gou, P., Comaposada, J., and Arnau, J. 2004. Moisture diffusivity in the lean tissue of dry-cured ham at different process times. *Meat Sci.*, 67: 203–209.

Greenkorn, R.A. and Kessler, D.P. 1972. Transfer Operations, McGraw Hill, New York.

Greenshields, R.N. and Macgillivray, A.W., 1972. Caramel-part 1. The browning reactions. *Process Biochem.*, 12: 11–16.

Greensmith, M. 1998. *Practical Deyhdryation.*, CRC Press, Cambridge, pp. 65–104.

Gregory, J.F. and Kirk, J.R., 1978. Assessment of storage effects on vitamin B 6 stability and bioavailability in dehydrated food systems. *J. Food Sci.*, 43: 1801–1815.

Guggenheim, E.A. 1966. *Applications of Statistical Mechanics.* Clarendon Press, Oxford.

Guillard, V., Broyart, B., Guilbert, S., Bonazzi, C., and Gontard, N. 2004. Moisture diffusivity and transfer modelling in dry biscuit. *J. Food Eng.*, 64: 81–87.

Gupta, R., Leung, P., and Mujumdar, A.S. 1980. *Drying '80*, Vol. 2, Mujumdar, A.S., (Ed.), Hemisphere, Washington, DC, pp. 201–207.

Hagen, W. and Drawert, F. 1987. Determination of water in hop cones and pellets by infrared drying. *Monatsschrift fuer Brauwissenschaft*, 40: 451–455.

Halsey, G. 1948. Physical adsorption in non-uniform surfaces. *J. Chem. Phys.*, 16: 931.

Hamdami, N., Monteau, J.-Y., and Le Bail, A. 2004. Transport properties of a high porosity model food at above and sub-freezing temperatures. Part 2: Evaluation of the effective moisture diffusivity from drying data. *J. Food Eng.*, 62: 385–392.

Hamm, R. 1960. Biochemistry of meat dehydration. *Adv. Food Res.*, 10: 355–463.

Hansen, E., Andersen, M.L., and Skibsted, L.H. 2003. Mobility of solutes in frozen pork studied by electron spin resonance spectroscopy: evidence for two phase transition temperatures. *Meat Sci.*, 63: 63–67.

Haralampu, S.G. and Karel, M. 1983. Kinetic models for moisture dependence of ascorbic acid and *b-carotene degradation in dehydrated sweet potato. *J. Food Sci.*, 48: 1872.

Harcourt, G.N. 1938. Effective drum drying by present-day methods, *Chem. Metallurgical Eng.*, 45: 179–182.

Harper, J.M. 1981. *Extrusion of Foods*, CRC Press, Inc., Florida.

Harris, R.S. and von Loesecke, H. (Ed.) 1960. *Nutritional Evaluation of Food Processing*. John Wiley & Sons Inc., New York, pp. 1–4

Hasatani, M., 1983–1984. Drying of optically semitransparent materials by combined radiative — convective heating. *Drying Technol.*, 1: 193–214.

Havighorst, C.R. 1943. New dehydration plant handles record quantities. *Food Ind.*, 82–85.

Hayakawa, K.I. and Rossen, J.L., 1977. Simultaneous heat and moisture transfer in capillary-porous material in a moderately large time range. *Lebensmittel-Wissenschaft und-Technologie*, 10: 273–278.

Hayakawa, K.I., Matas, J., and Hwant, P. 1978. Moisture sorption isotherms of coffee products. *J. Food Sci.*, 43: 1026–1027.

Heap, R.D. 1979. *Heat Pumps*, John Wiely & Sons Inc, New York.

Heldman, D.A. and Hohner, G.A., 1974. An analysis of atmospheric freeze drying. *J. Food Sci.*, 39: 147–155.

Heldman, D.A. and Singh, P.R. 1981. Food dehydration. *Food Process Eng.*, 310–315.

Hendel, C.E., Silveira, V.G., and Harrington, W.O. 1955. Rates of nonenzymatic browning of white potato during dehydration. *Food Technol.*, 9: 433.

Henderson, S.M. 1952. A basic concept of equilibrium moisture. *Agri. Eng.*, 33: 29–32.

Hernandez, J.A., Pavon, G., and Garcia, M.A. 2000. Analytical solution of mass transfer equation considering shrinkage for modeling food-drying kinetics. *J. Food Eng.*, 45: 1–10.

Hertzendorf, M.S. and Moshy, R.J. 1970. Foam drying in the food industry. *CRC Cri. Rev. Food Technol.*, 1: 25–70.

Hills, B.P., Godward, J., and Wright, K.M. 1997. Fast radial NMR microimaging studies of pasta drying. *J. Food Eng.*, 33: 321–335.

Hodge, J.E. 1953. Dehydrated foods: chemistry of browning reactions in model systems. *J. Agric. Food Chem.*, 1: 928–943.

Hoebink, J.H.B.J. and Rietema, K. 1980. *Chem. Eng. Sci.*, 35: 2135–2140.

Holdsworth, S.D. 1971a. *Encyclopedia of Food Technol.*, 282–290.

Holdsworth, S.D. 1971b. *Food Technol.*, 6: 331–370.

Holdsworth, S.D. 1974. Dehydration. *Ency. of Food Technol.*, AVI Publishing Co., AVI Publishing Co., Westport, CT, p. 288.

Holdsworth, S.D. 1985. Optimisation of thermal processing — a review. *J. Food Eng.*, 4: 89–116.

Hong, Y.C., Bakshi, A.S., and Labuza, T.P., 1986. Finite element modeling of moisture transfer during storage of mixed multicomponent dried foods. *J. Food Sci.*, 51: 554–558.

Houger, O.A. 1940. Typical dryer calculations. *Chem. Metallurgical Eng.*, 47: 15–18.

Hovmand, S. 1987. Fluidized bed drying. In: *Handbook of Industrial Drying*, Marcel Dekker, New York, pp. 165–221.

Hoyer, G.G. 1985. Extraction with supercritical fluids: why, how, and so what. *Chemtech.*, 15: 440–448.

Hsineh, F., Acott, K., and Labuza, T.P. 1976a. Death kinetics of pathogens in a pasta product. *J. Food Technol.*, 41: 516–519.

Hsineh, F., Acott, K., and Labuza, T.P. 1976b. Prediction of microbial death during drying of a macaroni product. *J. Milk Food Technol.*, 39: 619–623.

Hui, Y.H., Chazala, S., Grahm, D.M., Murrell, K.P., and Nip, W.-K. 2004. *Handbook of Vegetable Preservation and Processing*. Marcel Dekker, New York, pp. 350–370.

Huizenga, D.G. and Smith, D.M. 1986. Knudsen diffusion in random assemblages of uniform spheres. *AICHE J.*, 32, 1–6.

Hurrell, R.F., Finot, P.A., and Ford, J.E. 1983. Storage of milk powders under adverse conditions. 1. Losses of lysine and of other essential amino acids as determined by chemical microbiological methods. *Br. J. Nutr.*, 49: 343.

Husain, A., Chen, C.S., Clayton, J.T., and Whitney, L.F. 1972. Mathematical simulation of mass and heat transfer in high moisture foods. *Trans. ASAE*, 25: 732.

Hussain, M.A., Shafiur Rahman, M., and Ng, C.W. 2002. Prediction of pores formation (porosity) in foods during drying: generic models by the use of hybrid neural network. *J. Food Eng.*, 51: 239–248.

Hutchinson, D. and Otten, L. 1983. Thin-layer air drying of soybeans and white beans. *J. Food Technol.*, 18: 507–522.

Huxsoll, C.C. and Morgan, A.I. 1966. Use of microwaves in the food industry, 26th Institute of Food Technologists Annual Mtg., OR.

Hynd, J. 1980. Drying of whey. *J. Soc. Dairy Technol.*, 33: 2.

Igarashi, S., Matsubara, M., and Tanaka, S. 1988. Dehydration method and dehydration system. U.S. Patent 4-793-072.

Iglesias, H.A. and Chirife, J. 1995. An alternative to the Guggenheim, Anderson and De Boer model for the mathematical description of moisture sorption isotherms of foods. *Food Res. Intl.* 28: 317–321.

Iglesias, H.A., Chirife, J., and Fontan, C.F. 1986. Temperature dependence of water sorption isotherms of some foods. *J. Food Sci.*, 51: 551–553.

Iglesias, H.A., Chirife, J., and Lombardi, J.L. 1975a. Water sorption isotherms in sugar beet root. *J. Food Technol.*, 10: 299.

Iglesias, H.A. and Chirife, J. 1976. Prediction of the effect of temperature on water sorption isotherms of food material. *J. Food Technol.*, 11: 109–116.

Iglesias, H.A. and Chirife, J. 1978. An empirical equation for fitting water sorption isotherms of fruits and related products. *Can. Inst. Food Sci. Technol. J.*, 11: 12.

Iglesias, H.A. and Chirife, J. 1981. An equation for fitting uncommon water sorption isotherms in foods. *Lebensum. Wiss-u. Technol.*, 14: 105.

Iglesias, H.A. and Chirife, J. 1982. *Handbook of Food Isotherms.* Academic Press, New York.

Inazu, T., Iwasaki, K., and Furuta, T. (2005). Stress and crack prediction during drying of Japanese noodle (udon). *Int. J. Food Sci. Technol.*, 40: 621–630.

Irani, C.A. and Funk, E.W. 1977. *Recent Developments in Separation Science*, Vol. III, part A. CRS Press, W. Palm Beach, FL.

Irudayaraj, J. and Haghighi, K. (1993). Stress analysis of viscoelastic materials during drying: I-Theory and finite element formation. *Drying Technology*, 11: 901–927.

Isidro, D.S. 1968. Research Report No. 20, M.S.U. Agric. Exp. Station, East Lansing, MI.

Islam, M.N. and Flink, J.M. 1982. Dehydration of potato II. Osmotic concentration and its effects on air drying behavior. *J. Food Tech.* 17: 373–385.

Jackson, S.F., Chichester, C.O., and Joslyn, M.A. 1960. The browning of ascorbic acid. *Food Res.* 25: 484–490.

Jadhav, S., Steele, L., and Hadziyev, D. 1975. Vitamin C losses during production of dehydrated mashed potatoes. Lebensmittel-Wissenschaft + Technologie, 8: 225–230.

Jagannath, J.H., Nanjappa, C., Das Gupta, D.K., and Arya, S.S. 2001. Crystallization kinetics of precooked potato starch under different drying conditions (methods). *Food Chem.*, 75: 281–286.

Jason, A.C. 1958. A study of evaporation and diffusion processes in the drying of fish muscle, *Fundamental Aspects of the Dehydration of Foodstuffs*. Metchim & Son Ltd. Soc. Chem. Ind. London, pp. 103–135.

Jason, A.C. 1965. Effects of fat content on diffusion of water in fish muscle. *J. Sci. Food Agrc.*, 16: 281.

Jayaraman, K.S., Gopinathan, V.K., and Ramanathan, L.A. 1980. Development of quick-cooking dehydrated pulses by high temperature short time phenumatic drying. *J. Food Technol.*, 15: 217–226.

Jayas, D.S., Cenkowski, S., Pabis, S., and Muir, W.E. 1991. Review of thin-layer drying and wetting equations. *Dry. Technol.*, 9: 551–588.

Jensen, A. 1969. Tocopherol content of seaweed and seaweed meal. III. Influence of processing and storage on the content of tocopherols, carotenoids and ascorbic acid in seaweed meal. *J. Sci. Food Agric.*, 20: 622–626.

Jensen, K.N., Jorgensen, B.M., and Nielsen, J. 2003. Low-temperature transitions in cod and tuna determined by differential scanning calorimetry. *Lebensm.-Wiss. U.-Technol.* 36: 369–374.

Jeppson, M.R. 1964. Techniques of continuous microwave food processing. *Cornell Hotel and Restaurant Admin.Q.*, 5: 60.

Johnson, P-N.T. and Brennan, J.G. 2000. Moisture sorption isotherm characteristics of plantain (Musa, AAB). *J. Food Eng.*, 44: 79–84.

Jopelman, I.J. 1966. Transient heat transfer and thermal properties in food systems. Ph.D. Thesis. Michigan State Univ.

Joslyn, M.A. 1957. Role of amino acids in the browning of orange juice. *Food Res.*, 22: 1–13.

Jovanovich, G., Puppo, M.C., Giner, S.A., and Añón, M.C. 2003. Water uptake by dehydrated soy protein isolates: Comparison of equilibrium vapour sorption and water imbibing methods. *J.Food Eng.*, 56: 331–338.

Kaleemullah, S. and Kailappan, R. 2004. Moisture sorption isotherms of red chillies. *Biosyst. Eng.*, 88: 95–104.

Kawas, M.L. and Moreira, R.G. 2001. Characterization of product quality attributes of tortilla chips during the frying process. *J. Food Eng.*, 47: 97–107.

Kameoka, T., Hosokawa, A. and Morishima, H. 1984. Simulation of Heat and Mass Transfer During Through-Drying Process of Rough Rice. *Proceedings of The Fourth International Drying Symposium*, Kyoto International Conference Hall, Kyoto, Japan, 625–632.

Kamman, J.F., Labuza, T.P., and Warthesen, J.J. 1981. Kinetics of thiamin and riboflavin loss in pasta as a function of constant and variable storage conditions. *J. Food Sci.*, 46: 1457–1461.

Kanner, J., Mendel, H., and Budowski, P. 1979. Carotene oxidation factors in red pepper fruits (Capsicum annuum L.): oleoresin-cellulose solid model. *J. Food Sci.*, 43: 709–712.

Kapsalis, J.G., 1981. Moisture sorption hysterisis. In: *Water Activity: Influences on food Quality*. Rockland, L.B. and Stewart, G.E. (Eds), Academic Press. New York.

Karatas, S. and Battalbey, F. 1991. Determination of moisture diffusivity of pistachio nut meat during drying. *Lebensm.-Wiss. u.-Technol.* 24: 484–487.

Karatas, S. and Esin, A. 1994. Determination of moisture diffusivity and behavior of tomato concentrate droplets during drying in air. *Drying Technol.*, 12: 799–822.

Karathanos, V.T., Kostaropoulos, A.E., and Saravacos. G.D. 1995. Diffusion and equilibrium of water in dough/raisin mixtures. *J. Food Eng.*, 25: 113–121.

Karathanos, V., Reppa, A., and Kostaropoulos, A. 1994. Air-drying kinetics of osmotically dehydrated fruits. In: *Drying '94*, Vol. 1, Mujumdar, A.S., (Ed.), Hemisphere, McGraw Hill, pp. 871–878.

Karathanos, V.T. and Saravacos, G.D. 1993. Porosity and pore size distribution of starch materials. *J. Food Eng.*, 18, 259–279.

Karathanos, V.T., Kostaropoulos, A.E., and Saravacos, G.D. 1995. Diffusion and equilibrium of water in dough/raisin mixtures. *J. Food Eng.*, 25: 113–121.

Karathanos, V.T., Kanellopoulos, C.N.K., and Belessiotis, V.G. 1996. Development of porous structure during air drying of agricultural plant products. *J. Food Eng.*, 29: 167–183.

Karel, M. 1984. Control of lipid oxidation in foods. In: *Concentration and Drying of Foods*, MacCarthy, D. (Ed.), Elsevier App. Sci. Publ., New York, pp. 37–68.

Karel, M. and Nickerson, J.T.R. 1964. Effects of relative humidity, air, and vacuum on browning of dehydrated orange juice. *Food Technol.*, 18: 1214–1218.

Karel, M. and Labuza, T.P. 1968. Nonenzymatic browning in model systems containing sucrose. *J. Agric. Food Chem.*, 16: 717–719.

Karel, M. and Lund, D.B. 2003. *Physical Principles of Food Preservation*. Second edition. Marcel Dekker, Inc, New York.

Katkov, I.I. and Levine, F. 2004. Prediction of the glass transition temperature of water solutions: comparison of different models. *Cryobiology*. 49: 62–82.

Kawas, M.L. and Moreira, R.G. 2001. Characterization of product quality attributes of tortilla chips during the frying process. *J. Food Eng.*, 47: 97–107.

Kaymak-Ertekin, F. and Sultanoglu, M. 2000. Modelling of mass transfer during osmotic dehydration of apples. *J. Food Eng.*, 46: 243–250.

Kaymak-Ertekin, F. and Sultanoðlu, M. 2001. Moisture sorption isotherm characteristics of peppers. *J. Food Eng.*, 47: 225–231.

Kaymak-Ertekin, F. and Gedik, A. 2004. Sorption isotherms and isosteric heat of sorption for grapes, apricots, apples, and potatoes. *Lebensm.-Wiss. u.-Technol.* 37: 429–438.

Keattch, C.J. 1969. Fric. *An Introduction to Thermogravimetry*. Heyden & Son Ltd., Great Britain

Kechau, N. and Maalej, M. 1994. Evaluation of diffusion coefficient in the case of banana drying. In: *Drying '94*, Vol. 1. Mujumdar, A.S., (Ed.), Hemispere, McGrawHill. 841–848.

Keey, R.B. 1980. *Advances in Drying*, Vol. 1. Hemisphere Publishing Co., Washington, p. 14.

Keey, R.B. 1972. *Drying Principles and Practice*. Pergamon Press, New York.

Keey, R.B. 1992. *Drying of Loose and Particulate Materials*. Hemisphere, New York.

Kessler, H.G. 1981. *Food Engineering Dairy Technology*, Verlag, Germany.

Kessler, H. and Fink, R. 1986. Changes in heated and stored milk with an interpretation by reaction kinetics. *J. Food Sci.*, 51: 1105–1111, 1155.

Kim, M.H. and Okos, M.R. 1999. Some physical, mechanical, and transport properties of crackers related to the checking phenomenon. *J. Food Eng.*, 40: 189–198.

Khalloufi, S., El-maslouhi, Y., and Ratti, C. 2000a. Mathematical model for prediction of glass transition temperature of fruit powders. *J. Food Sci.*, 65: 842–848.

Khalloufi, S., Giasson, J., and Ratti, C. 2000b. Water activity of freeze dried mushrooms and berries. *Can. Agric. Eng.*, 42: 7.1–7.13.

Khalloufi, S. and Ratti, C. 2003. Quality deterioration of freeze-dried foods as explained by their glass transition temperature and internal structure. *J. Food Sci.*, 68: 892–902.

Kim, M.H. and Okos, M.R. 1999. Some physical, mechanical, and transport properties of crackers related to the checking phenomenon. *J. Food Eng.*, 40: 189–198.

Kim, S.S., Kim, S.Y., Kim, D.W., Shin, S.G., and Chang, K.S. 1998. Moisture Sorption Characteristics of Composite Foods Filled with Strawberry Jam. *Lebensm.-Wiss. u.-Technol.* 31: 399–401.

Kincall, N.S. 1987. Transport properties of liquid egg related to spray drying behavior. *J. Food Eng.*, 6: 467–474.

King, C.J. 1968a. Rates of moisture sorption and desorption in porous, dried foodstuffs. *Food Technol.*, 22: 165–171.

King, C.J. 1968b. Rates of moisture sorption and desorption in porous, dried foodstuffs. *Food Technol.*, 22: 509.

King, C.J. 1971. *Freeze-Drying of Foods*. CRC Press, Cleveland, OH.

King, C.J. 1974. Understanding and conceiving chemical processes. AIChE Monograph Series, *Am. Inst. Chem. Eng.*, New York, p. 70.

King, C.J. and Clark, J.P. 1968. Convective heat transfer for freeze drying of foods. *Food Technol.*, 22: 33–37.

King, M.B. and Bott, T.R. 1982. Problems associated with the development of gas extraction and similar processes. *Sep. Sci. Technol.*, 17: 119–150.

King, C.J., Lam, W.K., and Sandall, O.C. 1968. Physical properties important for freeze-drying poultry meat. *Food Technol.*, 22: 1302–1308.

Kinsella, J.E. and Fox, P.F. 1986. Water sorption by proteins: Milk and Whey Proteins. *CRC Crit. Rev. Food Sci. Nut.*, 24: 91.

Kiranoudis, C.T. 1998. Design and operational performance of conveyer-belt drying structures. *Chem. Eng. J.*, 69: 27–38.

Kiranoudis, C.T. and Markatos, N.C. 2000. Pareto design of conveyer-belt dryers. *J. Food Eng.*, 46: 145–155.

Kiranoudis, C.T., Maroulis, Z.B., and Marinos-Kouris, D. 1992. Model selection in air drying of foods. *Drying Technol.*, 10: 1097–1106.

Kiranoudis, C.T., Maroulis, Z.B., and Marinos-Kouris, D. 1995a. Design and production planning for multiproduct dehydration plants. *Computers Chem. Eng.* 19: 581–606.

Kiranoudis, C.T., Maroulis, Z.B., and Marinos-Kouris, D. 1995b. Heat and mass transfer model building in drying with multiresponse data. *Int. J. Heat Mass Trans.*, 38: 463–480.

Kiranoudis, C.T., Maroulis, Z.B., and Marinos-Kouris, D. 1996. Drying of solids: selection of some continuous operation dryer types. *Computers Chem. Eng.*, 20: S177-S182.

Kirk, J., Dennison, D., Kokoczka, P., and Heldman, D. 1977. Degradation of ascorbic acid in a dehydrated food system. *F. J. Food Sci.*, 42: 1274–1279

Kitic, D. and Viollaz, P.E. 1984. Comparison of drying kinetics of soybeans in thin layer fluidized beds. *J. Food Technol.*, 19: 399–408.

Kitson, J.A. and MacGregor, D.R. 1982. Technical note: drying fruit purees on an improved pilot plant drum-dyrer. *J. Food Technol.*, 17: 285–288.

Kliman, P.G. and Pallansch, M.J. 1968. Chemical changes in spray-dried skimmilk held near dryer outlet temperatures. *J. Dairy Sci.*, 51: 498–502.

Komanowsky, M., Sinnamon, H.I., and Aceto, N.C. 1964. Mass transfer in the cross-circulation drying of foam. *I & EC Process Design and Development,* 3: 193–197.

Kompany, E., Benchimol, J., Allaf, K., Ainseba, B., and Bouvier, J.M. 1993. Dehydration kinetics and modeling. *Drying Technol.*, 11: 451–470.

Kopelman, I.J. and Saguy, I. 1977. Drum dried beet powder. *J. Food Technol.*, 12: 615–621.

Kouassi, K. and Roos, Y.H. 2000. Glass transition and water effects on sucrose inversion by invertase in a lactose-sucrose system. *J. Agric. Food Chem.* 48: 2461–2466.

Kozempel, M.F., Sullivan, J.F., Craig, J.C., and Heiland, W.K. 1985. Drum Drying Potato Flakes — A Predictive Model. Eastern Regional Research Laboratory, Philadelphia.

Kozempel, M.F., Sullivan, J.F., Craig, J.C., and Konstance, R.P. 1989. Explosion puffing of fruits and vegetables. *J. Food Sci.*, 54: 772–773.

Krokida, M.K., Foundoukidis, E., and Maroulis, Z. 2004. Drying constant: literature data compilation for foodstuffs. *J. Food Eng.,* 61: 321–330.

Krokida, M.K., Karathanos, V.T., Maroulis, Z.B., and Marinos-Kouris, D. 2003. Drying kinetics of some vegetables. *J. Food Eng.,* 59: 391–403.

Kuntz, I.D., Jr. 1971. Hydration of macromolecules. 3. Hydration of polypeptides. *J. Am. Chem. Soc.,* 93, 514–516.

Kuntz, I.D., Jr. and Kauzmann, W. 1974. Hydration of proteins and polypeptides. *Adv. Protein Chem.,* 28: 239–245.

Kunze, O.R. 1979. Fissuring of the rice grain after heated air drying. *Trans. ASAE,* 25: 1197–1201.

Kuprianoff, J. 1958. Bound water in food. In: *Fundamental Aspects of the Dehydration of Foodstuffs.* Soc. Chemical Industry, London, pp. 2–23.

Kuts, P. 1982. Mathematical Modelling of Heat and Mass Transfer Processes in Thermoradiative — Convective Drying of Polymers Coatings. Third Intl Drying Symposium, pp. 439–448.

Laaksonen, T.J. and Roos, Y.H. 2003. Water sorption and dielectric relaxations of wheat dough (containing sucrose, NaCl, and their mixtures). *J. Cereal Sci.,* 37: 319–326.

Labuza, T.P. 1971. Kinetics of lipid oxidation in foods. *CRC Crit. Rev. Food Technol.,* 2: 355–405.

Labuza, T.P. 1972. Nutrient losses during drying and storage of dehydrated foods. *CRC Crit. Rev. Food Technol.,* 3: 217.

Labuza, T.P. 1975. Sorption Phenomena in Foods: Theoretical and Practical Aspects. Theory, Determination and Control of Physical Properties of Food Materials. pp. 197–219.

Labuza, T.P. 1984. Moisture Sorption : Practical Aspects of isotherm. Measurement and Use. Am. Assoc. Cereal Chem., Minnesota, 16.

Labuza, T.P. 1968a. Sorption phenomenon in foods. *Food Technol.,* 22: 15–24.

Labuza,T.P. 1968b. Sorption phenomena in foods. *Food Technol.,* 22: 263.

Labuza, T.P. and Simon, I.B. 1970. Air drying of apple slices. *Food Technol.* 24: 712–715.

Labuza, T.P. and Santos, D.B. 1971. Concentration and drying of yeast for human food: effect of evaporation and drying on cell viability and SCP quality. *Trans. ASAE,* 14: 701–705.

Labuza, T.P. and Chou, H.E. 1974. Decrease of Linoleate Oxidation Rate Due to Water at Intermediate Water Activity. *J. Food Sci.,* 39: 112–113.

Labuza, and Saltmarch. 1981. Kinetics of browning and protein quality loss in whey powders during steady state and nonsteady state storage conditions. *J. Food Sci.,* 47: 92–96.

Labuza, T.P. and Ragnarsson, J.O. 1983. Kinetic history effect on lipid oxidation of methyl linoleate in a model system. *J. Food Sci.,* 50: 145–174.

Labuza, T.P., Tannenbaum, S.R., and Karel, M. 1970a. Water content and stability of low-moisture & intermediate-moisture foods. *Food Technol.,* 24: 543.

Labuza, T.P., Tannenbaum, S.R., and Karel, M. 1970b. Water content and stability of low-moisture and intermediate-moisture foods. *Food Technol.,* 24: 35–42.

Labuza, T.P., Mizrahi, S., and Karel, M. 1972a. Mathematical models for optimization of flexible film packaging of foods for storage. *Trans. ASAE,* 15: 150.

Labuza, T.P., Cassil, S., and Sinskey, J. 1972b. Stability of intermediate moisture foods. 2. Microbiology. *J. Food Sci.,* 37: 160–162.

Labuza, T.P., McNally, L., Gallagher, D., Hawks, J., and Hurtad, F. 1972c. Stability of intermediate moisture foods. 1. Lipid oxidation. *J. Food Sci.,* 37: 154.

Labuza, T.P., Acott, K., Tatini, S.R., and Lee, R.Y. 1976. Water Activity Determination: A Collaborative Study of Different Methods. *J. Food Sci.,* 41: 910–917.

Lagoudaki, M., Demertzis, P.G., and Kontominas, M.G. 1993. Moisture Adsorption Behaviour of Pasta Products. *Lebensm.-Wiss. u.-Technol.* 26: 512–516.

Lahsasni, S., Kouhila, M., and Mahrouz, M. 2004. Adsorption–desorption isotherms and heat of sorption of prickly pear fruit (Opuntia ficus indica). *Energy Convers. Manage.* 45: 249–261.

Laignelet, B. 1983. Lipids in Pasta. In: *Lipids in Cereal Technology,* Barnes J.P. (Ed.), Academic Press, New York

Laing, B.M., Schlueter, D.L., and Labuza, T.P. 1978. Degradation kinetics of ascorbic acid at high temperature and water activity. *J. Food Sci.,* 43: 1440–1443.

Langrish, T.A.G. and Fletcher, D.F. 2001. Spray drying of food ingredients and applications of CFD in spray drying. *Chem. Eng. Proc.* 40: 345–354.

Lawrence, J.C. and Scott, R.P. 1966. Determination of the diffusivity of water in biological tissue. Nature 210: 301–303.

Lazar, M.E. and Hart, M.R. 1968. Densified instant applesauce. *Food Technol.*, 22: 39–40.

Lazar, M.E. and Miers, J.C. 1971. Improved drum-dried tomato flakes are produced by a modified drum dryer. *Food Technol.*, 25: 72–74.

Lazar, M.E. and Farkas, D.F. 1980. *Drying '80*, Vol. 1, (Ed. Mujumdar, A.S.). Hemisphere, Washington, DC, pp. 242–246.

Lazar, M.E. and Morgan, A.I., Jr. 1985. Instant applesauce. *Food Technol.*, pp. 179–181.

Lazaridou, A. and Biliaderis, C.G. 2002. Thermophysical properties of chitosan, chitosan–pullulan films near the glass transition. *Carbohydr. Polym.* 48: 19–190.

Lazaridou, A., Biliaderis, C.G., and Kontogiorgos, V. 2003. Molecular weight effects on solution rheology of pullulan and mechanical properties on its films. *Carbohydr. Polym.* 52: 151–166.

Lazaridou, A., Biliaderis, C.G., Bacandristos, N., and Sabatini, A.G. 2004. Composition, thermal and rheological behavior of selected Greek honeys. *J. Food Eng.*, 64: 9–21.

Lee, C.Y. and Salunkhe, D.K. 1966. Effects of gamma radiation on freeze dried apples. Nature, 210 (5039), 971.

Lee, C.Y. and Salunkhe, D.K. 1967a. Sucrose penetration in osmo-freeze dehydrated apple slices. Curr. Sci. 37(10): 297.

Lee, C.Y. and Salunkhe, D.K. 1967b. Effects of dehydration process on color and rehydration of fruits. J. Sci. Food Agric., 18: 566.

Lee, F.A. 1983. *Basic Food Chemistry*, 2nd edn. The AVI Publishing Company, Inc. Westport, CT, pp. 288–302.

Lee, S.H. and Labuza, T.P. 1975. Destruction of ascorbic acid as a function of water activity. *J. Food Sci.*, 40: 370

Lee, W.H., Staples, C.L., and Olson J.C., Jr. 1975. Staphylococcus aureus growth and survival in macaroni dough and the persistence of enterotoxins in the dried products. *J. Food Sci.*, 40: 119–120.

LeMeste, M., Huang, V.T. Panama, J., Anderson, G., and Lentz, R. 1992. Glass transition of bread. *Cereal Foods World.* 37: 264–267.

Le Meste, M., Champion, D., Roudaut, G., Blond, G., and Simatos, D. 2002. Glass transition and food technology: a critical appraisal. *J. Food Sci.* 67: 2444–2458.

Lenart, A. and Lewicki, P. 1988. Osmotic Dehydration of Apples at High Temperature, Sixth International Drying Symposium.

Leonard, A., Blacher, S., Marchot, P., Pirard, J.P., and Crine, M. 2004. Measurement of shrinkage and cracks associated to convective drying of soft materials by X-ray microtomography. *Drying Technol.*, 22: 1695–1708.

Lerici, C.R., Mastrocola, D., and Nicoli, M.C. 1988. Use of direct osmosis as fruit and vegetables dehydration. *Acta Alimentaria Polonica.*, 14: 35–40.

Leung, H.K. 1983. Water activity and other colligative properties of foods. ASAE Paper No. 83–6508, Chicago.

Leung, H.K., 1986, Water activity and other colligative properties of foods, In: *Physical and Chemical Properties of Foods*, M.R. Okos (Ed.), *Am. Society of Agri. Eng.*, Michigan, p. 138.

Lewicki, P.P. 2000. Raoult's law based food water sorption isotherm. *J. Food Eng.*, 43: 31–40.

Lewin, L.M. and Mateles, R.I. 1962. Freeze drying without vaccum: a preliminary investigation. *Food Technol.*, 16: 94–96.

Li, Y., Kloeppel, K.M., and Hsieh, F. 1998. Texture of glassy corn cakes as a function of moisture content. *J. Food Sci.*, 63: 869–872.

Liapis, A.I. 1987. Freeze drying. In: *Handbook of Industrial Drying*, Marcel Dekker, New York, pp. 295–326.

Lievonen, S.M., Laaksonen, T.J., and Roos, Y.H. 2002. Nonenzymatic browning in food models in the vicinity of the glass transition: effects of fructose, glucose, and xylose as reducing sugar. *J. Agric. Food Chem.*, 50: 7034–7041.

Lind, I. and Rask, C. 1991. Sorption isotherms of mixed minced meat, dough, and bread crust. *J. Food Eng.*, 14: 303–315

Litchfield, J.B. 1986. Analysis of mass transfer for the drying of extruded durum semolina. Ph.D. Dissertation. Purdue University, West Lafayette, IN.

Litchfield, J.B. and Okos, M.R. 1986. Moisture Diffusivity in Pasta During Drying. ASAE paper no. 86–651g. Presented at the Winter Meetin of ASAE. Chicago, IL December 16–19.

Litchfield, J.B. and Okos, M.R. 1988. Prediction of corn kernel stress and breakage induced by drying, tempering, and cooling. *Trans. ASAE*, 31: 585–594.

Litchfield, J.B. and Okos, M.R. 1992. Moisture diffusivity in pasta during drying. *J. Food Eng.*, 17: 117–142.

Liu, M., Haghighi, K., Stroshine, R.L., and Ting, E.C. 1990. Mechanical properties of the soybean cotyledon and failure strength of soybean kernels. *Trans. ASAE*, 33: 559–566.

Liu, H., Zhou, L. and Hayakawa, K. 1997. Sensitivity analysis for hygrostress crack formation in cylindrical food during drying. *J. Food Sci.*, 62: 447–450.

Lomauro, C.J., Bakshi, A.S., and Labuza, T.P. 1985a. Evaluation of food moisture sorption isotherm equations. II: Milk, coffee, tea, nuts, oilseeds, spices, and starchy foods.*Lebensm.-Wiss. u.-Technol.* 18: 118–124.

Lomauro, C.J., Bakshi, A.S., and Labuza, T.P. 1985b. Moisture transfer properties of dry and semimoist foods. *J. Food Sci.*, 50: 397–400.

Loncin, M., Bimbenet, J.J., and Lenges, J. 1968. Influence of the activity of water on the spoilage of foodstuffs. *J. Food Technol.*, 3: 131–142.

Loncin, M. and Roth, T. 1985. *Drying '85*, Hemisphere, New York, pp. 97–101.

Longan, B.J., Hruzek, G.A., and Burns, E.E. 1974. Effect of processing variables on volatile retention of freeze-dried carrots. *J. Food Sci.*, 39: 1191–1194.

Lopes Filho, J.F., Romanelli, F.P., Barboza, S.H.R., Gabas, A.L., and Telis-Romero, J. 2002. Sorption isotherms of alligator's meat (Caiman crocodilus yacare). *J. Food Eng.*, 52: 201–206.

Lozano, J.E., Rotstein, E., and Urbicain, M.J. (1983). Shrinkage, porosity and bulk density of food stuffs at changing moisture contents. *J. Food Sci.*, 48, 1497–1502, 1553.

Luikov, A.V. 1966a. *Heat and Mass Transfer in Capillary-porous Bodies*. Pergamon Press, Oxford.

Lund, D.B. 1983. Physical changes in foods. ASAE Paper No. 83–6514, ASAE, St. Joseph, MI.

Luyben, K.C.A.M., Olieman, J.J., and Bruin, S. 1980. Concentration dependent diffusion coefficients derived from experimental drying curves. In *Drying '80*. Mujumdar, A.S. (Ed.) Proc. 2nd Int. Drying Symp., Montreal Vol. 2, pp. 233–243. Hemisphere Publ. Corp., Washington, DC.

MacCarthy, D., (Ed.) 1985. Concentration and drying of foods. Elsevier, London. "Water Activity" Cornelis Van den Berg, pp. 11–36.

Mackenzie, R.C. 1969, Nomenclature in Thermal Analysis. *Talanta*, 16: 1227–1230.

Margaritis, A. and King, C.J. 1971. Measurement of rates of moisture transport in porous media. *Ind. Eng. Chem. Fund.*, 10: 510.

Maltini, E., Torreggiani, D., Venir, E., and Bertolo, G. 2003. Water activity and the preservation of plant foods. *Food Chem.*, 82: 79–86.

Maroulis, Z.B. and Saravacos, G.D. 2003. *Food Process Design*. Marcel Dekker, New York.

Maroulsi, Z.B., Kiranoudis, C.T., and Marinos-Kouris, D. 1995. Heat and Mass Transfer in Modeling in Air Drying of Foods. *J. Food Eng.*, 26: 113–130.

Marouze, C., Giroux, F., Collignan, A., and Rivier, M. 2001. Equipment design for osmotic treatments. *J. Food Eng.*, 49: 207–221.

Martin, H. 1980. Wärme- und Stoffübertragung in der Wirbelschicht. *Chem. Ing. Technik.*, 52: 199–209.

Mascou, P. and Lub, S. 1981. Practical use of mercury porosimetry in the study of porous solids. *Powder Technol.*, 29: 45.

Maskan, M. 2001. Drying, shrinkage and rehydration characteristics of kiwifruits during hot air and microwave drying. *J. Food Eng.*, 48: 177–182.

Maskan, M. and Göðüþ, F. 1997. The fitting of various models to water sorption isotherms of pistachio nut paste. *J. Food Eng.*, 33: 227–237.

Mathur, K.B. and Epstein, M. 1974. *Spouted Beds*, Academic Press, New York.

Mauro, M.A. and Menegalli, F.C. 2003. Evaluation of water and sucrose diffusion coefficients in potato tissue during osmotic concentration. *J. Food Eng.*, 57: 367–374.

Mauro, M.A., Tavares, D.Q., and Menagalli, F.C. 2002. Behavior of plant tissue in osmotic solutions. *J. Food Eng.*, 56: 1–15.

Mayor, L. and Sereno, A.M. 2004. Modelling shrinkage during convective drying of food materials: a review. *J. Food Eng.*, 61: 373–386.

Mazza, G. and LeMaguer, M. 1980. Dehydration of onion: some theoretical and practical considerations. *J. Food Technol.*, 15: 181–194.

McCord, J.D. and Kilara, A. 1983. Control of enzymatic browning in processed mushrooms (IAgaricus bisporusR). *J. Food Sci.*, 48: 1479–1483

McCormick, P.Y. 1973. Solids drying fundamentals. In: *Chemical Engineers' Handbook*, fifth edition, Perry, R.H. and Chilton, C.H., (Eds), McGraw-Hill, New York.

McIlrath, W.J., Dekazos, E.D., and Johnson, K.R. 1962. Rehydration characteristics of freeze-dried plant tissue. *Conference on Freeze-Drying of Foods*, Washington, DC.

McMinn, W.A.M. and Magee, T.R.A. 1996. Moisture transport in starch gels during convective drying. *T. I. Chem. Eng.-Lond.*, 74: 3–12.

McWeeny, D.J., Biltcliffe, D.O., Powell, R.C.T., and Spark, A.A. 1969. The Maillard reaction and its inhibition by sulfite. *J. Food Sci.*, 34: 641–643.

Meade, R.E. 1971. Develops novel dual dryer. *Food Eng.*, pp. 88–89.

Mellor, J.D. 1978. *Fundamentals of Freeze Drying*, Academic Press, New York, pp. 257–262.

Menkov, N.D. 2000a. Moisture sorption isotherms of chickpea seeds at several temperatures. *J. Food Eng.*, 45: 189–194.

Menkov, N.D. 2000b. Moisture sorption isotherms of lentil seeds at several temperatures. *J. Food Eng.*, 44: 205–211.

Menon, A.S. and Mujumdar, A.S. 1987. Drying of solids: principles, classification, and selection of dryers. In *Handbook of Industrial Drying*, Marcel Dekker, New York, pp. 3–45.

Micard, V. and Guilbert, S. 2000. Thermal behavior of native and hydrophobized wheat gluten, gliadin and glutenin-rich fractions by modulated DSC. *Int. J. Biol. Macromol.* 27: 229–236.

Miller, D.L, Goepfert, J.M., and Amunds, C.H. 1972. Survival of Salmonellae and Escherichia coli during the spray drying of various food products. *J. Food Sci.*, 37: 828–830.

Millman, M.J., Liapis, A.I., and Marchello, J.M. 1985. An analysis of the liophilization process using a sorption-sublimation model and various operational policies. *AIChE Journal*, 31: 1594–1604.

Minemoto, Y., Adachi, S., and Matsuno, R. 1997. Comparison of oxidation of methyl linoleate encapsulated with gum arabic by hot-air-drying and freeze-drying. *J. Agric. Food Chem.*, 45: 4530–4534.

Mink, L.D. 1939. Egg material treatment. U.S. Patent 2183516.

Mink, L.D. 1940. Treatment of egg whites. U.S. Patent 2200963.

Mishkin, M., Karel, M., and Saguy, I. 1982. Applications of optimization in food dehydration. *Food Technol.*, 36: 101–109.

Mishkin, M., Saguy, I., and Karel, M. 1983. Dynamic optimization of dehydration processes: minimizing browning in dehydration of potatoes. *J. Food Sci.*, 48: 1617–1621.

Mishkin, M., Saguy, I., and Karel, M. 1984a. A dynamic test for kinetic models of chemical changes during processing: ascorbic acid degradation in dehydration of potatoes. *J. Food Sci.*, 49: 1267–1274.

Mishkin, M., Saguy, I., and Karel, M. 1984b. Optimization of nutrient retention during processing: Ascorbic acid in potato dehydration *J. Food Sci.*, 49: 1262–1266.

Mitsuiki, M., Mizuno, A., and Motoki, M. 1999. Determination of molecular weight of agars and effect of the molecular weight on the glass transition. *J. Agric. Food Chem.*, 47: 473–478.

Mizrahi, S., Labuza, T.P., and Karel, M. 1970a. Computer-aided predictions of extent of browning in dehydrated cabbage. *J. Food Sci.*, 35: 799.

Mizrahi, S., Labuza, T.P., and Karel, M. 1970b. Feasibility of accelerated tests for browning in dehydrated cabbage. *J. Food Sci.*, 35: 804.

Mohsenin, N.M. 1986. *Physical Properties of Plant and Animal Materials*. Gordon Breach, New York.

Moore, J.G. 1987. Drum dryers. In *Handbook of Industrial Drying*, Marcel Dekker, New York, pp. 227–242.

Moraga, G., Martinez-Navarrete, N., and Chiralt, A. 2004a. Water sorption isotherms and glass transition in strawberries: influence of pretreatment. *J. Food Eng.*, 62: 315–321.

Moraga, G., Martinez-Navarrete, N., Chiralt, A. 2006. Water sorption isotherms and phase transitions in kiwifruit. *J. Food Eng.*, 72: 147–156.

Moreira, R. and Sereno, A.M. 2003. Evaluation of mass transfer coefficients and volumetric shrinkage during osmotic dehydration of apple using sucrose solutions in static and non-static conditions. *J. Food Eng.*, 57: 25–31.

Morgan, A.J. 1965. Reverse osmosis. *Food Technol.*, 19: 52.

Morgan, A.J. 1974. Foam-mat drying. Ency. *Food Technol.*, AVI Publishing Co., Westport, CT, p. 432.

Mortensen, S. and Hovmand, S. 1983. Fluidized-bed spray granulation. *Chem. Eng. Prog.*, 37–42.

Moser, F. and Schnitzer, H. 1985. *Heat Pumps in Industry*. The Netherlands.

Mossel, D.A.A. and Shennan, J.L. 1976. Micro-organisms in dried foods: their significance, limitation and enumeration. *J. Food Technol.*, 11: 205–220.

Moyano, P.C. and Berna, A.Z. 2002. Modeling water loss during frying of potato strips: effect of solute impregnation. *Dry. Technol..* 20: 1303–1318.

Moyne, C. and de Giovanni, A. 1985. Importance of gas phase momentum equation in drying above the boiling point of water. In: *Drying '85*, Mujumdar, A.S. and Toei, R. (Eds), Hemisphere Publishing Corp., Washington, DC, pp. 109–115.

Mowlah, G., Tamano, K., Kamoi, I., and Obara, T. 1982. Effects of drying method on water sorption and color properties of dehydrated whole banana powders. *J. Agric. Sci. Tokyo* 27: 145–155.

Mujumdar, A.S. 1983. *Lat. Am. J. Heat Mass. Trans.*, 7: 99–110.

Mujumdar, A.S. 1987. *Handbook of Industrial Drying*, Marcel Dekker, New York.

Mujumdar, A.S. (Ed.) 2000. *Drying Technology in Agriculture and Food Sciences*. Science Publishers, Enfield.

Mulet, A. 1994. Drying modeling and water diffusivity in carrots and potatoes. *J. Food Eng.*, 22: 329–348.

Mulet, A. Berna, A., and Rossello, C. 1989. Drying of carrots. I: Drying models. *Drying Technol.*, 7: 537–557.

Mulet, A. Berna, A., Rossello, C., Canellas, J., and Lopez, N. 1992. Influence of fat content on the drying of meat products, In *Drying '92*, Vol. 1, Mujumdar, A.S., (Ed.), Elsevier Science Publishers B.V. pp. 844–853.

Mulet, A., García-Pascual, P., Sanjuán, N., and García-Reverter, J. 2002. Equilibrium isotherms and isosteric heats of morel (Morchella esculenta) *J. Food Eng.*, 53: 75–81.

Mulley, E.A., Stumbo, C.R., and Hunting, W.M. 1975a. Kinetics of thiamine degradation by heat. *J. Food Sci.*, 40: 985.

Mulley, E.A., Stumbo, C.R., and Hunting, W.M. 1975b. Kinetics of thiamine degradation by heat. Effect of pH and form of the vitamin on its rate of destruction. *J. Food Sci.*, 40: 989.

Muralidhara, H.S. 1988. Combined Fields Dewatering Techniques, Sixth International Drying Symposium.

Muralidhara, H.S., Ensminger, D., and Putnam, A. 1985. Acoustic dewatering and drying (low and high frequency): state of the art review. *Drying Technol.*, 3: 529–566.

Muthukumarappan, K. and Gunasekaran, S. 1994. Moisture diffusivity of corn kernel components during adsorption Part I: Germ. *Trans. ASAE*, 37: 1263–1268.

Nadeau, J.P., Puiggali, J.R., Aregba, W., and Quintard, M. 1988. *Infrared Tunnel Dryer: A Kinetics Study*, Sixth International Drying Symposium.

Namiki, M. 1988. Chemistry of maillard reactions: recent studies on the browning reaction mechanism and the development of antioxidents and mutagens. *Adv. Food Res.*, 32: 115–185.

Natter, G.K., Taylor, D.H., and Brekke, J.E. 1958. Pineapple Juice Powder. Western Utilization Research and Development Division, ARS-USDA, presented at IFT meeting, Pittsburgh.

Navankasattusas, S. and Lund, D.B. 1982. Thermal destruction of vitamin B 6 vitamers in buffer solutions and cauliflower puree. *J. Food Sci.*, 47: 1512–1518.

Nawar, W.W. 1985. Lipids. In: *Food Chemistry*, Fennema, O.R., (Ed.), Marcel Dekker, Inc., New York, pp. 139–244.

Nelson, V. 1948. The color of evaporated milk with respect to time and temperature of processing. *J. Dairy Sci.*, 31: 415–419.

Ngoddy, P.O. and Bakker-Arkema, F.W. 1970. A generalized theory of sorption phenomena in biological materials (Part I: The isotherm equation). *Trans. ASAE*, 13: 612–617.

Ngoddy, P.O. and Bakker-Arkema, F.W. 1975. A theory of sorption hysteresis in biological materials. *J. Agric. Eng. Res.*, 20: 109–121.

Nickerson, J.T. and Sinskey, A.J. 1972. *Microbiology of Foods and Food Processing*. Elsevier Publishing Co., Inc. New York.

Nieto, A.B., Salvatori, D.M., Castro, M.A., and Alzamora, S.M. 2004. Structural changes in apple tissue during glucose and sucrose osmotic dehydration: shrinkage, porosity, density and microscopic features. *J. Food Eng.*, 61: 269–278.

Nijhuis, H.H., Torringa, H.M., Muresan, S., Yuksel, D., Leguijt, C., and Kloek, W. 1998. Approaches to improving the quality of dried fruit and vegetables. *Trends Food Sci. Tech.*, 9: 13–20

Nikolaidis, A. and Labuza, T.P. 1996. Glass Transition State Diagram of a Baked Cracker and Its Relationship to Gluten. *J. Food Sci.*, 61: 803–806.

Noel, T.R., Parker, R., Ring, S.G., and Tatham, A.S. 1995. The glass transition behaviour of wheat gluten proteins. *Int. J. Biol. Macromol.*, 17: 81–85.

Noguchi, H., 1981. *Water Activity: Influence on Food Quality*, Rockland, L. and Stewart, G. (Eds.), Academic Press, New York, p. 281.

Nogueira, R. and Park, K. 1992. Drying parameters to obtain "banana-passa." In: *Drying '92*, Vol. 1. Mujumdar, A.S. (Ed.), Elsevier Science Publishers BV. 873–883.

Nursten, H.E. 1984. Maillard browning in dried foods. In: *Concentration and Drying of Foods*, MacCarthy, D. (Ed.), Elsevier Applied Science Publishers, New York.

Nury, F.S. and Salunkhe, D.K. 1968. Effect of microwave dehydration on components of apples. U.S. Dept. Agric. ARS. Spec. Bull., pp. 74–45.

Okos, M. et al. 1989. Design and control of energy efficient food drying processes with specific reference to quality. DOE/ID/12608–4, DE910099999. National Technical Information (NTIS). Springfield, VA.

O'Meara, J.R. 1966. Progress report on microwave drying. Proceedings of the 29th Annual Conference and Exhibit of the Potato Chip Institute, NY.

Odom, J.W. and Low, P.F. 1983. A kinetic method for determining desorption isotherms of water on clay. *Soil Sci. Soc. Am. J.*, 47: 1039–1041.

Olmos, A., Trelea, I.C., Poligne, I., Collignan, A., Broyart, B., and Trystram, G. 2004. Optimal operating conditions calculation for a pork meat dehydration-impregnation-soaking process. *Lebensm.-Wiss. U.-Technol.*

Onayemi, O. and Potter, N.N. 1976. Cowpea powders dried with methionine: preparation, storage stability, organoleptic properties, nutritional quality. *J. Food Sci.*, 41: 48–53.

Orent-Keiles, E., Hewston, E.M., and Butler, L. 1945. Effects of different methods of dehydration on vitamins and mineral value of meats. *Food Res.*, 11: 486.

Orfeuil, M. 1981. *Electrotermie Industrille: Fours at Equipements Thermiques Electriques Industrilles*. Dunod, Paris.

Ormos, Z. and Blickle, T. 1980. *Drying '80*, Vol. 1, Mujumdar, A.S., (Ed.), Hemisphere, Washington, DC, pp. 200–204.

Oswin, C.R. 1946. *J. Chem. Ind.*, London, 65: 419.

Overhults, D.G., White, G.M., Hamilton, H.E., and Ross, I.J. 1973. Drying soybeans with heated air. *Trans. ASAE*, 112–113.

Ozmen, L. and Langrish, T.A.G. 2002. Comparison of glass transition temperature for skim milk powder. *Drying Technol.*, 20: 1177–1192.

Pakowski, Z. and Mujumdar, A.S. 1982. Proceedings of the Third International Symposium on Drying, Mujumdar, A.S., (Ed.), Birmingham, England, pp. 149–155.

Palou, E., López-Malo, A., and Argaiz, A. 1997. Effect of temperature on the moisture sorption isotherms of some cookies and corn snacks. *J. Food Eng.*, 31: 85–93.

Palumbo, S.A., Komanowsky, M., Metzger, V., and Smith, J.L. 1977. Kinetics of pepperoni drying. *J. Food Sci.*, 42: 1029–1033.

Park, K.J., Bin, A., Brod, F.P.R., and Park, T.H.K.B. Osmotic dehydration kinetics of pear D'anjou (*Pyrus communis* L.). *J. Food Eng.*, 52: 293–298.

Park, K.J., Vohnikova, Z., and Brod. F.P.R. 2002. Evaluation of drying parameters and desorption isotherms of garden mint leaves (Mentha crispa L.). *J. Food Eng.*, 51: 193–199.

Parti, M. and Dugmanics, I. 1990. Diffusion coefficient for corn drying. *Trans. ASAE*, 33: 1652–1656.

Patton, S. 1955. Browning and associated changes in milk and its products: a review. *J. Dairy Sci.*, 38: 457–478.

Paulaitis, M.E., Krukonis, V.J., Kurnik, R.T., and Reid, R.C. 1983. Supercritical fluid extraction. *Rev. Chem. Eng.*, 1: 181–211.

Peleg, M. 1993a. Glass transition and the physical stability of food powders. In: The Glassy State in Foods. Eds. J. M. V. Blanshard and P. J. Lilliford, pp 435–451.

Peleg, M. 1993b. Mapping the stiffness-temperature-moisture relationship of solid biomaterials at and around their glass transition. *Rheolica Acta*, 32: 575–580.

Peleg, M. and Normand, M.D. 1992. Estimation of the water activity of multicomponent dry mixtures. *Trends Food Science and Technololgy*, 3: 157–160.

Pence, J.W. et al., 1950. Characterization of wheat gluten. II. Amino Acid Composition. *Cereal Chem.*, 27: 335.

Penney, T.R. and Maclaine-Cross, I. 1985. Promising Advances in Desiccant Cooling. U.S. Department of Energy. No. DE-AC02–83CH10093.

Pere, C. and Rodier, E. 2002. Microwave vacuum drying of porous media: experimental study and qualitative considerations of internal transfers. *Chem. Eng. Process.*, 41: 427–436.

Pereira, P.M. and Oliveira, J.C. 2000. Measurement of glass transition in native wheat flour by dynamic mechanical thermal analysis (DMTA). *Int. J. Food Sci. Technol.*, 35: 183–192.

Peri, C. and DeCesari, L. 1974. Thermodynamics of water sorption on Sacc. cerevisiae and cell viability during spray-drying. *Lebensm. Wiss. Technol.*, 7: 56.

Peters, M.S. and Timmerhaus, K.D. 1980. *Plant Design and Economics for Chemical Engineers*, McGraw-Hill, New York.

Peterson, E.E. and Lorentzen, J. 1973. Influence of freeze-drying parameters on the retention of flavor compounds of coffee. *J. Food Sci.*, 39: 119–122.

Petriella, C., Resnik, S.L., Lozano, R.D., and J. Chirife. 1985. Kinetics of deteriorative reactions in model food systems of high water activity: color changes due to nonenzymatic browning. *J. Food Sci.*, 50: 622–626.

Pezzutti, A. and Crapiste G.H. 1997. Sorptional equilibrium and drying characteristics of garlic. *J. Food Eng.*, 31: 113–123.

Phaff, H.J., Mrak, E.M., Perry, R.L., and Fisher, C.D. 1945. New methods produce superior dehydrated cut fruits. *Food Ind.*, 84: 634–637.

Piazza, L., Riva, M. and Masi, P. 1990. Modeling pasta during drying processes. In *Engineering and Food*, Vol. 1. Spiess, W.E.L. and Schubert, H. (Eds), Elsevier Applied Science, 592–602.

Pinaga, F., Carbonel, J.V., Pena, J.L., Miquel, J.J. 1984. Experimental simulation of solar drying of garlic using an adsorbent energy storage bed. *J. Food Eng.*, 3: 187–208.

Pisecky, J. 1983. *Dairy Industries Int.* April 21.

Pitombo, R.N.M. and Lima, G.A.M.R. 2003. Nuclear magnetic resonance and water activity in measuring the water mobility in Pintado (*Pseudoplatystoma corruscans*) fish. *J. Food Eng.*, 58: 59–66.

Pixton, S.W. and Henderson, S. 1979. Moisture relations of dried peas, shelled almonds and lupins. *J. Stored Prod. Res.*, 15: 59.

Ponting, J.D., 1966. Osmotic dehydration of fruits. *Food Technol.*, 20: 125.

Ponting, J.D. and McBean, D.M. 1970. *Food Technol.*, 24: 1403–6.

Pope, M.I. and Judd, M.D., 1977. *Differential Thermal Analysis*. Heyden & Son Ltd., N.J.

Porter, V.L. 1973. Microwave finish drying of potato chips, *J. Food Sci.*, 38: 583.

Pounder, J.R. and Ahrens, F.W. 1987. A mathematical model of high intensity paper drying. *Drying Technol.*, 5: 213–243.

Powell, H.R. 1976. Trends in freeze-drying equipment and materials. *Stokes Division*, Pennwalt Corporation.

Prakash, S. Jha, S.K., and Datta, N. 2004. Performance evaluation of blanched carrots dried by three different driers. *J. Food Eng.*, 62: 305–313.

Priop, B.A. 1979. Measurement of water activity in foods: a review. *J. Food Prot.*, 42: 668–674.

Prothon, F. and Ahrné, L.M. 2004. Application of the Guggenheim, Anderson and De Boer model to correlate water activity and moisture content during osmotic dehydration of apples. *J. Food Eng.*, 61: 467–470.

Quast, D.G. and Teixeira Neto, R.O. 1976. Moisture Problems of Foods in Tropical Climates. *Food Technol.*, May, pp. 98–105.

Raghavan, G., Tulasidas, T., Sablani, S., and Ramaswamy, H. 1994. Concentration dependent moisture diffusivity in Drying of shrinkable commodities. In *Drying '94*, Vol. 1. Mujumdar, A.S. (Ed.), Hemisphere, McGrawHill. 277–290.

Rahman, M.S. 2004. State diagram of date flesh using differential scanning calorimetry (DSC). *Int. J. Food Prop.*, 7: 407–428.

Rahman, M.S. and Lamb, J. 1991. Air drying behavior of fresh and osmotically dehydrated pineapple. *J. Food Process Eng.*, 14: 163–171.

Rahman, M.S., Perera, C.O., and Thebaud, C. 1997. Desorption isotherm and heat pump drying kinetics of peas. *Food Res. Int.*, 30: 485–491.

Rahman, M.S., Kasapis, S., Guizani, N., and Al-Amri, O.S. 2003. State diagram of tuna meat: freezing curve and glass transition. *J. Food Eng.*, 57: 321–326.

Ramaswamy, H.S. and Lo, K.V. 1983. Simplified mass transfer relationships for diffusion-controlled air dehydration of regular solids. *Can. Agr. Eng.*, 25: 143–148.

Ramesh, M.N. 2003. Moisture transfer properties of cooked rice during drying. *Lebensm-Wiss. U.-Technol.* 36: 245–255.

Rao, K.S. 1939a. Hysteresis in the sorption of water on rice. *Curr. Sai.*, 8: 256.

Rao, K.S. 1939b. Hysterisis loop in sorption. *Curr. Sai.*, 8: 468.

Rao, S.S. 1984. *Optimization: Theory and Applications*, 2nd ed., Halsted Press, New York.

Raouzeos, G.S. and Saravacos, G.D. 1986. *Drying '86*, Vol. 2, Hemisphere, Washington, DC, pp. 487–91.

Ratti, C., Crapiste, G.H., and Rotstein, E. 1989. A new water sorption equilibrium expression for solid foods based on thermodynamic considerations. *J. Food Sci.,* 54: 3.

Ravindra, M.R. and Cattopadhyay, P.K. 2000. Optimisation of osmotic preconcentration and fluidised bed drying to produce dehydrated quick-cooking potato cubes. *J. Food Eng.,* 44: 5–11.

Reimer, J. and M. Karel. 1978. Shelf-life studies of vitamin C during food storage: prediction of L-ascorbic acid retention in dehydrated tomato juice. unknown.

Resnik, S. and J. Chirife. 1979. Effect of moisture content and temperature on some aspects of nonenzymatic browning in dehydrated apple. *J. Food Sci.,* 44: 601–605.

Reynolds, T.M. 1963. Chemistry of nonenzymic browning. I. The reaction between aldoses and amines. *Adv. Food Res.,* 12:

Reynolds, T.M. 1965. Chemistry of nonenzymic browning. II. *Adv. Food Res.,* 14: 168.

Ribeiro, C., Zimeri, J.E., Yildiz, E., and Kokini, J.L. 2003. Estimation of effective diffusivities and glass transition temperature of polydextrose as a function of moisture content. *Carbohydr. Polym.,* 51: 273–280.

Richards, E.L. 1963. A quantitative study of changes in dried skim-milk and lactose-casein in the "dry" state during storage. *J. Dairy Res.,* 30: 223–234.

Riganakos, K.A., Demertzis, P.G., and Kontominas, M.G. 1994. Water sorption by wheat and soy flour: comparison of three methods. *J. Cereal Sci.* 20: 101–106.

Ringer, D.U. and Mujumdar, A.S. 1982. Proceedings of the Third International Symposium on Drying, Birmingham, England, pp. 107–114.

Riva, M. and Masi, P. 1986. *Drying '86*, Vol. 1, Hemisphere, Washington, DC, pp. 454–60.

Rizvi, S.S.H. and Benado, A.L. 1984. Thermodynamic Properties of Dehydrated Foods. *Food Technol.,* 83–92.

Rizvi, S.S.H., Benado, A.L., Zollweg, J.A., and Daniels, J.A. 1986. Supercritical fluid extraction: Fundamental principles and modeling methods. *Food Technol.,* 40: 65.

Rizvi, S.S.H., Santos, J., and Nigogosyan, N. 1984. An Accelerated Method for Adjustment of Equilibrium Moisture Content of Foods. *J. Food Eng.,* 3: 3–11.

Rizzolo, A., Nani, R.C., Viscardi, D., Bertolo, G., and Torreggiani, D. 2003. Modification of glass transition temperature through carbohydrates addition and anthocyanin and soluble phenol stability of frozen blueberry juices. *J. Food Eng.,* 56: 229–231.

Roberts, R.L., Carlson, R.A., and Farkas, D.F. 1979. Application of a continuous centrifugal fluidized bed drier to the preparation of quick-cooking rice products. *J. Food Sci.,* 44: 248–250.

Roberts, I.S., Tong, C.H., and Lund, D.B. 2002. Drying kinetics and time-temperature distribution of pregelatinized bread. *Inst. Food Technol.,* 67: 1080–1087.

Rockland, L.B. and Nishi, S.K. 1980. influence of water activity on food product quality and stability. *Food Technol.,* April, pp. 42–50.

Roeser, W.F. and Mueller, E.F. 1930. Bureau of Standards, Research Paper 231.

Roman, G.N., E. Rotstein and Urbicain, M.J. 1979. Kinetics of water vapor desorption from apples. *J. Food Sci.,* 44: 193–197.

Romankov, P.G. and Rashkovskaya, N.B. 1968. Drying in a fluidized bed, Khimiya, Leningrad.

Ronald, T., Magee, A., and Wilkinson, P.D. 1992. Influence of process variables on the drying of potato slices. *Int. J. Heat Mass Transfer.,* 27: 541–549.

Rootare, H.M. and Prenzlow, C.F. 1967. Surface area from mercury porosimeter measurements. *J. Phys. Chem.,* 71: 2733.

Rotstein, E., Laura, P.A., and Cemborain, M.E. 1974. Analytical prediction of drying performance in unconventional shapes. *J. Food Sci.,* 39: 627.

Rotstein, E. and Cornish, A.R.H. 1977. Prediction of the sorptional equilibrium relationship for the drying of foodstuffs. *Drying: Principles and Technology,* pp. 493–502.

Rotstein, E., Laura, P.A., and Cemborain, M.E. 1974. Analytical prediction of drying performance in nonconvectional shapes. *J. Food Sci.,* 39: 627–631.

Rovedo, C., Suarez, C., and Viollaz, P. 1995. Drying of foods: evaluation of a drying model. *J. Food Eng.,* 26: 1–12.

Row, R.J. and Gunkel, W.W. 1972. Simulation of temperature and moisture content of alfalfa during thin-layer drying. *Trans. ASAE,* pp. 805–810.

Rushton, E. 1945. Compressed dehydrated vegetable blocks, the application of high frequency heating. *Chem. Ind.,* 35: 274.

Sablani, S.S., Rahman, M.S., and Al-Sadeiri, D.S. 2002. Equilibrium distriubution data for osmotic drying of apple cubes in sugar-water solution. *J. Food Eng.,* 52: 193–199.

Saguy, I., Kopelman, I.J., and Mizrahi, S. 1978a. Extent of nonenzymatic browning in grapefruit juice during thermal and concentration processes: Kinetics and prediction. *J. Food Proc. Preserv.,* 2: 175–184.

Saguy, I., Mizrahi,S., Villota, R., and Karel, M. 1978b. Accelerated method for determining the kinetic model of ascorbic acid loss during dehydration. *J. Food Sci.,* 43: 1861–1864.

Saguy, I., Kopelman, I.J., and S. Mizrahi. 1979. Simulation of ascorbic acid stability during heat processing and concentration of grapefruit juice. *J. Food Proc. Eng.,* 2: 213–225.

Saguy, I., Kopelman, I.J., and S. Mizrahi. 1978. Computer-aided prediction of beet pigment *Journal of Food Science,* 43: 124–127.

Sahasrabudhe, M.R., Larmond, E., and Nunes, A.C. 1976. Sulphur dioxide in instant mashed potatoes. *Can. Inst. Food Sci. Technol. J.,* 9: 207–211.

Saidi, B. and Warthesen, J.J. 1983. Influence of pH and light on the kinetics of vitamin B 6 degradation. *J. Agric. Food Chem.,* 31: 876–880.

Salas, F. and Labuza, T.P. 1968. *Food Technol.,* 22: 1576–80.

Saltmarch, M., Vagnini-Ferrari, M., and Labuza, T.P. 1981. Theoretical basis and application of kinetics to browning in spray-dried whey food systems. *Prog. Food Nutr. Sci.,* 5: 331–344.

Salunkhe, D.K. and Do, J.Y. 1973. Developments in technology and nutritive value of dehydrated fruits, vegetables, and their products, *CRC Crit. Rev. Food Technol.,* 153–192.

Samaniego-Esguerra, C.M., Boag, I.F., and Robertson, G.L. 1991. Comparison of regression methods for fitting the GAB model to the moisture isotherms of some dried fruit and vegetables. *J. Food Eng.,* 13: 115–133.

Sandoval, A.J. and Barreiro, J.A. 2002. Water sorption isotherms of non-fermented cocoa beans (Theobroma cacao). *J. Food Eng.,* 51: 119–123.

Sandu, C. 1986. Infrared radiative drying in food engineering: a process analysis. *Biotechnol. Progress,* 2: 109–119.

Sanga, E.C.M., Mujumdar, A.S., and Raghavan, G.S.V. 2002. Simulation of convection-microwave drying for a shrinking material. *Chem. Eng. Proc.* 41: 487–499.

Sanjuan, N., Bon, J., Clemente, G., and Mulet, A. 2004. Changes the quality of dehydrated broccoli florets during storage. *J. Food Eng.,* 62: 15–21.

Sanni, L.O., Atere, C., and Kuye, A. 1997. Moisture sorption isotherms of fufu and tapioca at different temperatures. *J. Food Eng.,* 34: 203–212.

Sano, Y. and S. Yamamoto. 1986. Calculation method of concentration dependent mutual diffusion coefficient based on the assumption of similar concentration distribution. *Drying* 86: 85.

Saravacos, G.D. 1965. Freeze-drying rates and water sorption of model food gels. *Food Technol.,* 193–197.

Saravacos, G.D. 1967. Effect of the drying method on the water sorption of dehydrated apple and potato. *J. Food Sci.,* 32: 81–84.

Saravacos, G.D. 1969. Sorption and diffusion of water in dry soybeans. *Food Technol.,* 23: 145–147.

Saravacos, G.D. 1986. Mass transfer properties of food. In *Engineering Properties of Foods,* Marcel Dekker, New York, pp. 89–132.

Saravacos, G.D. and Charm, S.E. 1962. A study of the mechanism of vegetable and fruit dehydration. *Food Technol.,* 16: 78–81.

Saravacos, G.D. and Stinchfield, R.M. 1965. Effect of temperature and pressure on the sorption of water vapor by freeze-dried food materials. *J. Food Sci.,* 30: 779–786.

Sarncova, V. and J. Davidek. 1972. Reaction of pyridoxal and pyridoxal-5-phosphate with proteins. *J. Food Sci.,* 37: 310–312.

Saravacos, G.D. and Raouzeos, G.S. 1984. Diffusivity of moisture in air drying of starch gels. In *Engineering and Food,* Vol. 1. McKenna, B.M. (Ed.), Elsevier, London, pp. 499–507.

Saravacos, G.D., Raouzeos, G.S. 1986. Diffusivity of moisture in air drying of raisins. In *Drying '86,* Vol. 2. Mujumdar, A.S. (Ed.), Hemisphere. McGrawHill, pp. 487–491.

Saravacos, G.D., Marousis, S.N., and Raouzeos, G.S. 1988. Effect of ethyl oleate on the rate of air-drying foods, *J. Food Eng.,* 7: 263–270.

Schebor, C., Buera, M.P., and Chirife, J. 1996. Glassy state in relation to the thermal inactivation of the enzyme invertase in amorphous dried matrices of trehalos, malodextrin, and PVP. *J. Food Eng.,* 30: 269–282.

Schebor, C., Buera, M.P., Karel, M., and Chirife, J. 1999. Color formation due to non-enzymatic browning in amorphous, glassy, anhydrous, model systems. *Food Chem.* 65: 427–432.

Schlunder, E.V. 1980. *Verfahrenstechnik*, 14: 459–467.

Schwimmer, S. 1981. *Source Book of Food Enzymology*, AVI Publishing Co., Westport, Conn.

Shaw, T.M. 1944. The surface area of crystalline egg albumen. *J. Chem. Phys.*, 12: 391.

Shimada, Y., Roos, Y., and Karel, M. 1991. Oxidation of methy linoleate encapsulated in amorphous lactose-based food model. *J. Agric. Food Chem.*, 39: 637–641.

Shotton, E. and Harb, N. 1965. The effect of humidity and temperature on the equilibrium moisture content of powders. *J. Pharm. Pharmacol.*, 17: 504.

Shubin, G.S. 1982. Thermal moisture conductivity of colloidal capillary-porous bodies. *Heat Transfer.*, 14: 65.

Sifaouri, M.S. and Perrier, A. 1978. Caracterisation de l'evaporation profonde. *Int. J. Heat Mass Transfer*, 21.

Simal, S., Deya, E., Frau, M., and Rossello, C. 1997. Simple modeling of air drying curves of fresh and osmotically pre-dehydrated apple cubes. *J. Food Eng.*, 33: 139–150.

Singh, R.K. 1983. Kinetics and computer simulation of storage stability in intermediate moisture foods. Ph.D. thesis, University of Wisconsin-Madison, WI.

Singh, R.R.B., Rao, K.H., Anjaneyulu A.S.R., and Patil, G.R. 2001. Moisture sorption properties of smoked chicken sausages from spent hen meat. *Food Res. Int.* 34: 143–148.

Slade, L. and Levine, H. 1995. Water and the glass transition — dependence of the glass transition on composition and chemical structure: special implications for flour functionality. *J. Food Eng.*, 24: 431–509.

Smith, S.E. 1947. The sorption of water vapor by high polymers. *J. Am. Chem. Soc.*, 69: 646.

Snedekar, R.A. 1955. Ph.D. thesis, Princeton University.

Soekarto, S.T. and Steinberg, M.P., 1981. *Water Activity: Influences on Food Quality*, Rockland, L. and Stewart,G. (Eds.), Academic Press, New York, p. 265.

Sokhansanj, S. 1982. Drying induced stresses in food grains — a finite element approach. pp. 214–219. In *Drying '82*. Mujumdar, A.S. (Ed.), Hemisphere. McGrawHill. 256–262.

Sokhansanj, S. and Jayas, D.S. 1987. In: *Handbook of Industrial Drying*, Mujumdar, A.S., (Ed.), Marcel Dekker, New York, pp. 532–537.

Song, H.P. and Litchfield, J.B. (1994). Measurement of stress cracking in maize kernels by magnetic-resonance-imaging. *J. Agric. Eng. Res.*, 57: 109–118.

Sopade, P.A. 2001. Criteria for an appropriate sorption model based on statistical analysis. *Int J. Food Prop.*, 4: 405–418.

Sopade, P.A., Halley, P., Bhandari, B., D'Arcy, B., Doebler, C., and Caffin, N. 2002. Application of the Williams-Landel-Ferry model to the viscosity-temperature relationship of Australian honeys. *J. Food Eng.*, 56: 67–75.

Soysal, Y. and Öztekin, S. 2001. PH — Postharvest technology: sorption isosteric heat for some medicinal and aromatic plants. *J. Agric. Eng. Res.* 78: 160–162.

Spanyar, P. and Kevei, E., 1963. Uber die stabilisierung von vitamin C in lebensmitteln. *Z. Lebensm.* Unters. Forsch. 120: 1–17.

Spiazzi, E. and Mascheroni, R. 1997. Mass transfer model for osmotic dehydration of fruits and vegetables — I. Development of the simulation model. *J. Food Eng.*, 34: 387–410.

Stahl, E., Quirin, K.W., and Gerard, D., 1988. *Dense Gases for Extraction and Refining*. Springer-Verlag, Berlin Heidelberg, Germany.

Stateler, E.S. (Ed.) 1944. Study shows trends in dehydration. *Food Ind.*, 11: 902.

Stefanovich, A.F. and Karel, M., 1982. Kinetics of beta-carotene degradation at temperatures typical of air drying of foods. *J. Food Proc. Preserv.*, 6: 227–242.

Steffe, J.F. and Singh, R.P. 1980a. Liquid diffusivity of rough rice components. *Trans. ASAE*, 23: 767–774.

Steffe, J.F. and Singh, R.P. 1980b. Parameters required in the analysis of rough rice drying. In *Drying '80*, Vol. 2, Mujumdar, A.S., (Ed.), Hemisphere. McGrawHill, pp. 256–262.

Steffe, J.F., Singh, R.P., and Bakshi, A.S. 1979. Influence of tempering time and cooling on rice milling yields and moisture removal. *Trans. ASAE*, 1214–1224.

Story, M.J. 1981. *Int. J. Pharm. Technol. Prod. Mfg.* 2: 19–23.

Straatsma, J., Van Houwelingen, G., Steenbergen, A.E., and De Jong, P. 1999a. Spray drying of food products: 1. Simulation model. *J. Food Eng.*, 42: 67–72.

Straatsma, J., Van Houwelingen, G., Steenbergen, A.E., and De Jong, P. 1999b. Spray drying of food products: 2. Prediction of insolubility index. *J. Food Eng.*, 42: 73–77.

Strasser, J. 1969. Detection of quality changes in freeze-dried beef by measurement of the sorption isobar hysterisis. *J. Food Sci.*, 34: 18.

Straub, R.J., Tung, J.Y., Koegel, R.G., and Bruhn, H.D. 1979. Drum drying of plant juice protein concentrates. *Trans. ASAE*, 484–486.

Strolle, E.O. 1970. *J. Food Sci.,* 35: 338.

Strumillo, C. and Pakowski, Z. 1980. *Drying '80*, Vol. 1, Mujumdar, A.S., (Ed.), Hemisphere, Washington, DC, pp. 211–225.

Strumillo, C. and Kudra, T. 1986. *Drying: Principles, Applications and Design*. Gordon and Breach, New York.

Strumillo, C., Markowski, A. and Kaminski, W. 1983. Modern developments in drying of pasta-like materials. *Advances in Drying*. Hemisphere Publishing Corp., Washington.

Stuchly, S.S. and Stuchly, M.A. 1983. Microwave drying potential and limitations. *Adv. Drying*, 2: 53–71.

Suarez, C. Loncin, M., and Chirife, J. 1984. A preliminary study on the effect of ethyl oleate dipping treatments on drying rate of grain corn. *J. Food Sci.,* 49: 236–238.

Suarez, C. and Viollaz, P.E. 1982. Effect of pressurized gas freezing pre-treatment of carrot dehydration in air flow. *J. Food Technol.,* 17: 607–613.

Suarez, C., Viollaz, P., and Chirife, J., 1980. Diffusional analysis of air drying of grain sorghum. *J. Food Technol.,* 15: 523–531.

Sullivan, J.F. and Cording, J. 1969. *Food Eng.,* 41: 90.

Sun, D.W. and Woods, J.L. 1994. Low temperature moisture transfer characteristics of wheat in thin layers. *Trans. ASAE*. 37: 1919–1926.

Swientek, Robert J. 1986. Sonic technology applied to food drying. *Food Process.,* 62–63.

Syarief, A.M., Gustafson, R.J., and Vance, M. 1984. Moisture diffusion coefficients for yellow dent corn. ASAE Paper NO. 84–3551, Presented at the Winter Meeting of the ASAE, December 11–14. New Orleans, Louisiana, ASAE, St. Joseph, MI.

Tannenbaum, S.R., Young, V.R., and Archer, M.C. 1985. Vitamins and Minerals, In: *Food Chemistry,* Fennema, O.R. (Ed.), Marcel Dekker, Inc., New York, pp. 477–544.

Tang, J. and Sokhansanj, S. 1993. Moisture diffusivity in laird lentil seed components. *Trans. ASAE*. 36: 1791–1798.

Taylor, A.A. 1961. Determination of moisture equilibria in dehydrated foods. *Food Technol.,* 15: 536–540.

Teixeira, M.B.F. and Tobinga, S. 1998. A diffusion model for describing water transport in round squid mantle during drying with a moisture-dependent effective diffusivity. *J. Food Eng.,* 36: 169–181.

Telis, V.R.N. and Sobral, P.J.A. 2001. Glass transitions and state diagrams for freeze-dried pineapple. *Lebensm.-Wiss. U.-Technol.* 34: 199–205.

Telis, V.R.N. and Sobral, P.J.A. 2002. Glass transitions for freeze-dried and air-dried tomato. *Food Res. Intl.,* 35: 435–443.

Temple, S.J. and van Boxtel, A.J.B. 1999. Equilibrium moisture content of tea. *J. Agric. Eng. Res.,* 74: 83–89.

Teoh, H.M., Schmidt, S.J., Day, G.A., and Faller, J.F. 2001. Investigation of cornmeal components using dynamic vapor sorption and differential scanning calorimetry. *J. Food Sci.,* 66: 434–440.

Teunou, E., Fitzpatric, J.J., and Synnott, E.C. 1999. Characterisation of food powder flowability. *J. Food Eng.,* 39: 31–37.

Thed, S.R. and Lillard, D.A. 1989. Antioxidative effect of Maillard reaction Products on Lipid Oxidation. Paper presented at IFT annual meeting, Chicago-June 25–29.

Thijssen, H.A.C. 1979. Optimization of process conditions during drying with regard to quality factors. *Lebensm.-Wiss. u. -Technol.,* 12: 308–317.

Thijssen, H.A.C. and Kerkhof, P.J.A.M. 1977. Effect of temperature and water concentration during processing on food quality. *J. Food Proc. Eng.,* 1: 129–147.

Thompson, S.S., Harmon, L.G., Stine, C.M. 1978. Survival of selected organisms during spray drying of skim milk and storage of nonfat dry milk. *J. Food Prot.,* 41. 16–19.

Timmermann, E.O. 2003. Multilayer sorption parameters: BET or GAB values? *Colloids Surf.,* A. 220: 245.

Timmermann, E.O., Chirife, J., and Iglesias, H.A. 2001. Water sorption isotherms of foods and foodstuffs: BET or GAB parameters? *J. Food Eng.,* 48: 19–31.

Toledo, R., Steinberg, M.P., and Nelson, A.I. 1968. Quatitative determination of bound water by NMR. *J. Food. Sci.,* 33,315.

Tolaba, M.P., Suarez, C., and Viollaz, P.E. 1991. Diffusion coefficient estimation for shelled corn. *Lebensm.-Wiss. u.-Technol.* 24: 303–306.

Tolaba, M.P., Peltzer, M., Enriquez, N., and Pollio, M.L. 2004. Grain sorption equilibria of quinoa grains. *J. Food Eng.,* 61: 368.

Tong, C.H. and Lund, D.B. 1990. Effective moisture diffusivity in porous material as a function of termperature and moisture content. *Biotechnol. Progr.*, 6: 67–75.

Toribio, J.L. and Lozano, J.E. 1984. Nonenzymatic browning in apple juice concentrate during storage. *J. Food Sci.*, 49: 889–892.

Torringa, Erik, Esveld, Erik, Scheewe, Ischa, van den Berg, Robert, and Bartels, Paul. 2001. Osmotic dehydration as a pre-treatment before combined microwave-hot-air drying of mushrooms. *J. Food Eng.*, 49: 185–191.

Treybal, R.E. 1980. *Mass Transfer Operations*, 3rd ed., McGraw Hill Co.

Trujillo, F.J., Yeow, P.C., and Pham, Q.T. 2003. Moisture sorption isotherm of fresh lean beef and external beef fat. *J. Food Eng.*, 60: 357–366.

Uddin, M.B., Ainsworth, P., and Ibanoglu, S. 2004. Evaluation of mass exchange during osmotic dehydration of carrots using response surface methodology. *J. Food Eng.*

Ulku, S. and Uckan, G. 1986. Corn drying in fluidized. In *Drying '86*, Vol. 1. Mujumdar, A.S., (Ed.), Hemisphere. Spirnger-Verlag, NY, London, pp. 531–536.

Ulrich, G.D. 1984. *A Guide to Chemical Engineering Process Design and Economics*. Wiley, New York.

Vámos–Vigyázó, L., 1981. Polyphenol oxidase and peroxidase in fruits and vegetables. *CRC Crit. Rev. Food Sci. Nutr.*, 15: 49–127

Vaccarezza, L.M. and Chirfe, J. 1975. On the mechanism of moisture transport during air drying of sugar beet root. *J. Food Sci.*, 40, 1286–1289.

Vaccarezza, L.M., Lombardi, J.L., and Chirife, J. 1974. Kinetics of moisture movement during air drying of sugar beet root. *Food Technol.*, 9: 317.

Vaccarezza, L.M. and Chirife, J. 1978. On the application of Fick's law for the kinetic analysis of fair drying of foods. *J. Food Sci.*, 43: 236–238.

Vagenas, G.K. and Karathanos, V.T. 1991. Prediction of moisture diffusivity in granular materials, with special applications to foods. *Biotechnol. Prog.*, 7: 419–426.

Vagenas, G.K. and Karathanos, V.T. 1993. Prediction of the effective moisture diffusivity in gelatinized food systems. *J. Food Eng.*, 18: 159–179.

Valentas, K.J., Rotstein, E., and Singh, R.P. (Eds.) 1997. *Handbook of Food Engineering*. CRC Press, Boca Raton.

Van Arsdel, W.B. 1963a. *Food Dehydration*, Vol. 1 The AVI Publ. Co., Inc., Westport, CT.

Van Arsdel, W.B. 1963b. *Some Theoretical Characteristics of Other Types of Drying*. AVI Publishing Co., Inc. Westport, CT.

Van Arsdel, W.B. 1973. *Food Dehydration*. AVI Publishing Co., Inc. Westport, CN.

Van Arsdel, W.B., Copley, M.J., and Morgan, A.I. 1973. *Food Dehydration Vol. 2 — Practices and Applications*, 2nd ed. The AVI Publishing Company, Inc., Westport, Connecticut

Van Brakel, J., Modry, S., and Svata, M. 1981. Mesury porosimetry: State of the art. *Powder Technol.*, 29: 1.

Van den Berg, C. 1981. Vapor sorption equilibria and other water-starch interactions; a physico-chemical approach. Doctoral thesis, Agr. Univ. Wageningen, Netherlands.

Van den Berg, C. 1986. Water activity. *Concentration and Drying of Foods*, ed. Diarmuid MacCarthy. New York. pp. 11–36.

Van den Berg, C. and Bruin, S, 1981a. *Water Activity : Influence on Food Quality*, Rockland, L. and Stewart, G., (Eds), Academic Press, New York.

Van den Berg, C. and Bruin S. 1981b. Water activity and its estimation in food systems. theoretical aspects. *Water Activity: Influences on Food Quality*. Louis B.R. and George F.S. (Eds), Academic Press, New York, pp. 34–61.

Van Marle, D.J. 1938. Drum drying. *Ind. Eng. Chem.*, 30: 1006–1008.

Vazquez, A. and Calvelo, A. 1980. Gas particle heat transfer coefficient in fluidized pea deds. *J. Food Process Eng.*, 4: 53–70.

Vázquez, G., Chenlo, F., and Moreira, R. 2003. Sorption isotherms of lupine at different temperatures. *J. Food Eng.*, 60: 449–452.

Vega-Mercado, H., Gongora-Nieto, M.M., and Barbosa-Canovas, G.V. 2001. Advances in dehydration of foods. *J. Food Eng.*, 49: 271–289.

Velázquez de la Cruz, G., Torres, J.A., and Martín-Polo, M.O. 2001. Temperature effect on the moisture sorption isotherms for methylcellulose and ethylcellulose films. *J. Food Eng.*, 48: 91–94.

Verhey, J.G.P. 1973. Vacuole formation in spray powder particles. 3. Atomization and droplet drying. *Neth. Milk Dairy J.*, 27: 3–18.

Verma, R.C. and Prasad, S. 1999. Kinetics of adsorption of water by maize grains. *J. Food Eng.*, 39: 395–400.

Villota, R. and Hawkes, J.G. 1983. Effect of processing on kinetics of nutrients and organoleptic changes in foods. Paper presented at the 1983 winter meeting ASAE, ASAE, St. Joseph, MI.

Villota, R. and M. Karel. 1980a. Prediction of ascorbic acid retention during drying. I. Moisture and temperature distribution in a model system. *J. Food Proc. Preserv.*, 4: 111–134.

Villota, R. and M. Karel. 1980b. Prediction of ascorbic acid retention during drying. II. simulation of retention in a model system. *J. Food Proc. Preserv.*, 4: 141–159.

Viswanathan, R., Jayas, D.S., and Hulasare, R.B. 2003. Sorption isotherms of tomato slices and onion shreds. *Biosystems Eng.*, 86: 465–472.

Vojnovich, C. and Pfeifer, V.F. 1970. Stability of ascorbic acid in blends with wheat flour, CSM, and infant cereals. *Cereal Sci. Today* 15: 317.

Volman, D.H., Simons, J.W., Seed, J.R., and Sterling, C. 1960. *J. Ploym. Sci.*, 46: 355.

Von Loesecke. 1943. *Drying and Dehydration of Foods*. Reinhold Publishing Corp., NY.

Vullioud, M., Márquez, C.A., and Michelis, A.D. 2004. Desorption isotherms for sweet and sour cherry. *J. Food Eng.*, 63: 15–19.

Waananen. 1986. Analysis of mass transfer mechanisms during drying of extruded pasta. Ph.D. Thesis, Purdue University, W. Lafayette, IN.

Waananen, K.M. and Okos, M.R. 1996. Effect of porosity on moisture diffusion during drying of pasta. *J. Food Eng.*, 28: 121–137.

Waletzko, P. and Labuza, T.P. 1976. Accelerated shelf-life testing of an intermediate moisture food in air and in an oxygen-free atmosphere. *J. Food Sci.*, 41: 1338–1344.

Walsh, D.E. 1975. The influence of spaghetti extrusion, drying and storage on survival of Staphloococcus aureus. *J. Food Sci.*, 40: 714–716.

Wang, Z.H. and Chen, G. 2000. Heat and mass transfer in batch fluidized-bed drying of porous particles. *Chem. Eng. Sci.* 55: 1857–1869.

Wanninger, L.A. 1972. Mathematical model predicts stability of ascorbic acid in food products. *Food Technol.*, 26: 42

Warmbier, H.C., Schnickels, R.A., and Labuza, T.P. 1976. Effect of glycerol on nonenzymatic browning in a solid intermediate moisture model food system. *J. Food Sci.*, 41: 528–531.

Washburn, E.W. 1921. *Phys. Rev.*, 17: 273.

Watt, I. 1983. *Physical Properties of Foods*, Jowett, R. ed., Applied Science, London, 27.

Wedzicha, B.L. 1984. *Chemistry of Sulphur Dioxide in Foods*. Elsevier Science Publishing Co., Inc. New York.

Weisser, H. 1986. Influence of Temperature on Sorption Isotherms. In *Food Engineering and Process Applications — Vol. I. Transport Phenomena*, Le Maguer, M. and Jelen, P. (Eds).

Wells, D.F. 1973. Fluidized Bed Systems. *Chemical Engineers Handbook*, Perry, R. and Chilton, C. (Eds), McGraw-Hill, New York, 20: 64–74.

Westerman, P.W., White, G.M., and Ross, I.J. 1973. Relative humidity effect on the high-temperature drying of shelled corn. *Trans. ASAE*, pp. 1136–1139.

Whistler, R.L. and Paschall, E.F. 1967. *Starch : Chemistry and Technology*, Academic Press, New York, Chap. VII.

Whistler, R.L. and Daniel, J.R. 1985. Carbohydrates. In: *Food Chemistry*, 2nd ed., Fennema,O.R., (Ed.), Marcel Dekker, Inc., New York and Basel.

Whitaker, S. 1988a. The role of irreversible thermodynamics and the Onsager ralations in the analysis of drying phenomena. Preceedings of the Sixth International Drying Symposium. IDS '88, Versailles, Sept. 5–8.

White, G.M., Ross, I.J., and Westerman, P.P. 1973. Drying rate and quality of white shelled corn as influenced by dew point temperature. *Trans. ASAE*, pp. 118–120.

Wiebe, R. 1941. The binary system: Carbon dioxide-water under pressure. *Chem. Rev.*, 29: 475.

Wienecke, F. 1989. Process and device for drying products of plant or animal origin by means of microwave and IR radiation. German (FRG) Patent DE-37-21-412-A1.

Wilkinson, S.A., Earle, M.D., and Cleland, A.C. 1982. Effects of food composition, pH, and copper on the degradation of vitamin A in beef liver puree during heat processing *J. Food Sci.*, 47: 844.

Williams, D.F. 1981. Extraction with supercritical gases. *Chem. Eng. Sci.*, 36: 1769.

Williams-Gardner, A. 1971. *Industrial Drying*. CRC Press, Cleveland, OH.

Wilson, C.W. 1965. *Food Technol.*, 19: 1280.

Winslow, D.N. 1984. Advances in experimental techniques for mercury intrusion porosimetry. In: *Surface and Colloid Science*, Matijevic, E. and Good, R.J. (Eds), Plenum Press, New York and London.

Wisakowsky, E.E. and Burns, E.E. 1977. Factors affecting the quality of freeze-dried compressed spinach. *J. Food Sci.*, 42: 782–783, 794.

Wolf, J.C. and Thompson, D.R. 1977. Initial losses of available lysine in model systems. *J. Food Sci.*, 42: 1540–1544.

Wolf, M., Walker, J.E., Jr., and Kapsalis, J.G. 1972. Water vapor sorption hysteresis in dehydrated foods. *J. Agric. Food Chem.*, 20: 1073.

Wolf, W., Spiess, W.E.L., and Jung, G. 1973. Die Wasserdampfsorptionsisothermen einiger in der Literatur bislangwening berucksichtigter lebensmittel. Lebensm. *Wiss. Technol.*, 6: 94 (190).

Wolf, W., Spiess, W.E.L., and Jung, G. 1985. *Sorption Isotherms and Water Activity of Food Materials*. Science and Technology Publishers, Ltd., England.

Wolfrom, M.L., Kashimura, N., and Horton, D., 1974. Factors affecting the Maillard browning reaction between sugars and amino acids. Studies on the nonenzymic browning of dehydrated orange juice. *J. Agric. Food Chem.*, 22: 796.

Woodroof, J.G. and Luh, B.S. 1975. *Commercial Fruit Processing*. The AVI Publishing Co. Inc., Westport, Connecticut.

Woollard, D.C. and Edmiston, A.D. 1983. Stability of vitamins in fortified milk powders during a two-year storage period. *NZ J. Dairy Sci. Technol.*, 18: 21–26.

Wursch, P., J. Rosset, B. Kollreutter, and A. Klein. 1984. Crystallization of B-lactose under elevated storage temperature in spray-dried milk powder. *Milchwissenschaft* 39: 579–582.

Xiong, X., Narsimhan, G., and Okos, M.R. 1991. Effect of composition and pore structure on binding energy and effective diffusivity of moisture in porous food. *J. Food Eng.*, 15: 187–208.

Xu, F., Wang, Z., Xu, S., Sun, D.W. 2001. Cryostability of frozen concentrated orange juices produced by enzymatic process. *J. Food Eng.*, 50: 217–222.

Yamaguchi, S. 1992. Temperature and moisture-dependent diffusivity of moisture in rice kernel. In *Drying '92*, Vol. 2. Mujumdar, A.S., (Ed.), Elsevier Science Publishers B.V. 1389–1398.

Yamaguchi, S., Wakabayashi, K., and Yamazawa, S. 1985. Properties of brown rice kernels for calculation of drying stress. *Drying '85*. pp. 438–444.

Young, J.H., Whitaker, T.B., Blankenship, P.D., Brusewitz, G.H., Troeger, J.M., Steele, J.L., and Person, N.K. 1982. Effect of oven drying time on peanut moisture determination. *Trans. ASAE*, 491–496.

Zabik, M.E. and Figa, J.E. 1968. Comparison of frozen, foam-spray dried, freeze-dried, and spray-dried eggs. *Food Technol.*, 22: 119–125.

Zabrodsky, S.S. 1966. *Hydrodynamics and Heat Transfer in Fluidized Beds*. MIT Press, MA.

Zanoni, B., Pierucci, S., and Peir, C. 1994. Study of the bread breaking process — II. Mathematical modeling. *J. Food Eng.*, 23: 321–336.

Zarkarian, J.A. and King, C.J. 1978. Acceleration of limited freeze-drying in conventional dryers. *J. Food Sci.*, 43: 998–1001.

Zhong, Z. and Sun, X.S. 2005. Thermal characterization and phase behavior of cornstarch studied by differential scanning calorimetry. *J. Food Eng.*, 69: 453–459.

Zhou, L., Puri, V.M., and Anantheswaran, R.C. 1994. Effect of temperature gradient on moisture migration during microwave heating. *Drying Technol.*, 12: 777–798.

Zimeri, J.E. and Kokini, J.L. 2003. Phase transitions of inulin–waxy maize starch systems in limited moisture environments. *Carbohydr. Polym.*, 51: 183–190.

Zuritz, C.A. and Singh, R.P. 1982. Simulation of rough rice drying in a spouted-bed. *Drying '82*, pp. 239–247. Mujumdar, A.S., (Ed.), Hemisphere. McGrawHill. 862–867.

11 Thermal Processing of Canned Foods

Arthur Teixeira

CONTENTS

11.1 INTRODUCTION

Thermal processing of canned foods has been one of the most widely used methods of food preservation during the twentieth century and has contributed significantly to the nutritional well-being of much of the world's population. Thermal processing consists of heating food containers in pressurized retorts at specified temperatures for prescribed lengths of time. These process times are calculated on the basis of achieving sufficient bacterial inactivation in each container to comply with public health standards and to ensure that the probability of spoilage will be less than some minimum. Associated with each thermal process is always some degradation of heat-sensitive vitamins and other quality factors that is undesirable. Because of these quality and safety factors, great care is taken in the calculation of these process times and in the control of time and temperature during processing to avoid either under- or over-processing. The heat transfer considerations that govern the temperature profiles achieved within the container of food are critical factors in the determination of time and temperature requirements for sterilization.

An understanding of two distinct bodies of knowledge is required to appreciate the basic principles involved in thermal process calculation. The first of these is an understanding of the thermal inactivation kinetics (heat resistance) of food spoilage causing organisms. The second body of knowledge is an understanding of heat transfer considerations that govern the temperature profiles achieved within the food container during the process, commonly referred to in the canning industry as heat penetration.

Figure 11.1 conceptually illustrates the interdependence between the thermal inactivation kinetics of bacterial spores and the heat transfer considerations in the food product. Thermal inactivation of bacteria generally follows first-order kinetics and can be described by logarithmic reduction in the concentration of bacterial spores with time for any given lethal temperature, as shown in the upper family of curves in Figure 11.1. These are known as survivor curves. The decimal reduction time,

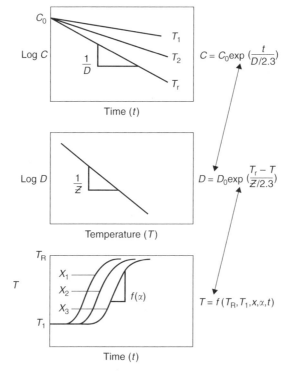

FIGURE 11.1 Time-temperature dependence of thermal inactivation kinetics of bacterial spores in thermal processing.

D, is expressed as the time in minutes to achieve one log cycle of reduction in concentration, C. As suggested by the family of curves shown, D is temperature dependent and varies logarithmically with temperature, as shown in the second graph. This is known as a thermal death time curve and is essentially a straight line over the range of temperatures employed in food sterilization. The slope of the curve that describes this relationship is expressed as the temperature difference, Z, required for the curve to transverse one log cycle. The temperature in the food product, in turn, is a function of the retort temperature (T_R), initial product temperature (T_I), location within the container (x), thermal diffusivity of the product (α), and time (t) in the case of a conduction-heating food.

Thus, the concentration of viable bacterial spores during thermal processing decreases with time in accordance with the inactivation kinetics, which are a function of temperature. The temperature, in turn, is a function of the heat transfer considerations, involving time, location, thermal properties of the product, and initial and boundary conditions of the process. This interrelationship is illustrated by the functional expressions given in Figure 11.1. The reason that temperature profiles can be computed independently of the reaction rates is the lack of a heat of reaction; this distinguishes sterilization from other reactions or phase-change situations.

The topics to be covered in this handbook chapter are organized as follows. Section 11.1 is the Introduction. Section 11.2 Commercial Equipment Systems describes commercially available canned food sterilization equipment systems, along with an explanation of their operating principles and typical applications. Section 11.3 Thermal Inactivation Kinetics of Bacterial Spores and how they are used to specify a process lethality value as a target objective for process calculations. Section 11.4 Heat Transfer in Canned Foods describes methods for obtaining heat penetration data on canned foods and methods of analyzing these data for subsequent process calculations. Section 11.5 Process Calculations describes current and accepted methods for calculating thermal processes, including the process lethality delivered by a specific process, as well as the process time required at a given temperature to deliver a specified lethality value. Section 11.6 Numerical Simulation of Heat Transfer describes the development of mathematical heat transfer models for thermal process simulation. Section 11.7 Process Optimization shows use of numerical models to find optimum process conditions that maximize quality retention without compromise of sterility assurance. Section 11.8 On-Line Computer Control shows use of numerical models for off-line evaluation of process deviations and on-line correction of process deviations. Section 11.9 Aseptic Processing describes aseptic processing and packaging systems, and methods for process calculation involving ultra-high temperature (UHT) product sterilization through heat exchangers and hold tubes. Section 11.10 FDA Low-Acid Canned Food Regulations summarizes the compliance requirements of the low acid canned food regulations.

11.2 COMMERCIAL STERILIZATION SYSTEMS

This section describes briefly some of the commercial equipment systems that are used in the food canning industry to accomplish thermal processes efficiently on a production scale. Just as with most industrial processing operations, both batch and continuous systems are available. As the name implies, batch systems are made up of individual batch retorts that operate intermittently. Scheduling of the retorts is skillfully staggered so that workers move from retort to retort, manually unloading and reloading each retort as its scheduled process cycle comes to an end. In continuous systems, cans are automatically fed into and out of retort systems that operate continuously over one or more working shifts.

11.2.1 BATCH RETORTS

The vertical still cook batch retort shown schematically in Figure 11.2 is, perhaps, the grandfather of all batch retorts. Hardly any food science pilot plant or laboratory is complete without

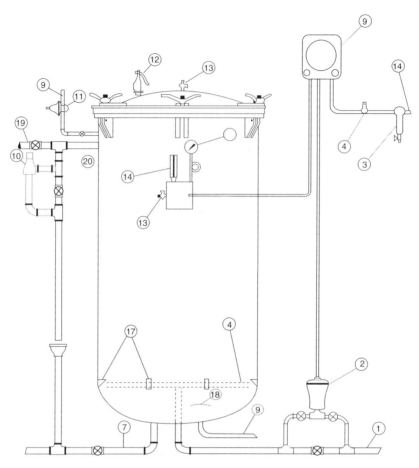

FIGURE 11.2 Vertical still-cook retort. (From Lopez, A., 1987, *Basic Information on Canning*, 11th ed. Courtesy CTI Publications, Inc.)

one. A typical production unit will measure 42 in. in diameter by 8 or 9 ft in height. Cans are loaded in crates that are handled by chain hoist for lifting and lowering into the retort. Most retorts are designed to hold either three or four crates, with a total capacity of more than 1000 No. 2 cans per batch, or 400 No. 10 cans. Although the basic design of these retorts has changed little since the turn of the century, they are still quite popular and can be found operating in many food canneries. Part of the reason for this continued popularity is the simplicity of their design and operation and their versatility to accommodate virtually all can sizes and shapes.

Although the unloading and reloading operations are labor intensive, a well-managed cook room can operate with surprising efficiency. The cook room is the room or area within a food canning plant in which the retorts are located. Some cook rooms are known to have more than 100 vertical still cook retorts operating at full production. Although each retort is a batch cook operation, the cook room as a whole operates as a continuous production system in that filled and sealed unsterilized cans enter the cook room continuously from the filling line operations, and fully processed sterilized cans leave the cook room continuously. Within the cook room itself, teams of factory workers move from retort to retort to carry out loading and unloading operations, while retort operators are responsible for a given number or bank of retorts. These operators carefully

FIGURE 11.3 Agitating horizontal retort. FMC orbitort pressure sterilizer. (Courtesy FMC Corporation, Food Processing Machinery Division.)

monitor the operation of each retort to make sure that the scheduled process is delivered for each batch.

An alternative to the vertical still cook retort is the horizontal still cook retort. In general, all operations are the same as with vertical retorts, except that crates are usually moved into and out of horizontal retorts on trolley tracks instead of chain hoists. For convection heating products that benefit from mechanical agitation during processing, agitating batch retorts are available. The FMC Orbitort® shown in Figure 11.3 is a batch horizontal agitating retort, that accomplishes axial rotation of the cans. Other types of agitating batch retorts are available that accomplish end-over-end agitation by rotating the entire crate during processing operations.

11.2.2 CONTINUOUS RETORT SYSTEMS

Continuous retort operations require some means by which filled, sealed containers are automatically and continuously moved from atmospheric conditions into a pressurized steam environment, held or conveyed through that environment for the specified process time, and then returned to atmospheric conditions for further handling operations. The best-known commercially available systems that accomplish these requirements are the crateless retort, the continuous rotary cooker, and the hydrostatic sterilizer. Two other systems, which operate on different principles but accomplish the same purpose for special products, are the Flash "18" system and Steriflamme system.

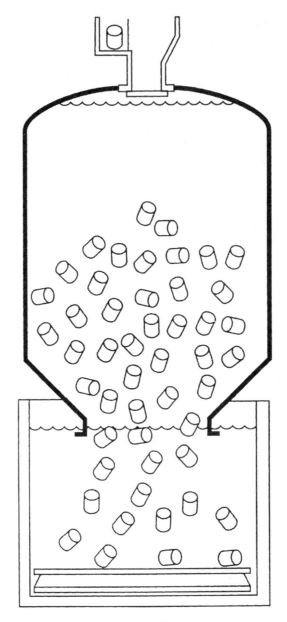

FIGURE 11.4   Illustration of crateless retort. (Courtesy FMC Corporation.)

11.2.3 CRATELESS RETORTS

A crateless retort system is, in a sense, an automatic cook room in that the system is made up of a series of individual retorts, each operating in a batch mode, with loading, unloading, and process scheduling operations all carried out automatically without the use of crates. An individual crateless retort is illustrated in Figure 11.4. When ready to load, the top hatch opens automatically, and cans fed from an incoming conveyor literally fall into the retort, which is filled with hot water to cushion the fall. Once fully charged, the hatch is closed and steam entering from the top displaces the cushion water out the bottom. When the cushion water has been fully displaced, all vales are closed and processing begins. At the end of the process time, the retort is refilled with warm water and the bottom hatch, which lies beneath the

FIGURE 11.5 Continuous crateless retort system. (Courtesy FMC Corporation, Food Processing Machinery Division.)

water level in the discharge cooling canal, is opened to let the cans fall gently onto the moving discharge conveyor in the cooling canal. After all cans are discharged, the bottom hatch is reclosed and the retort is ready to begin a new cycle. A commercial system of crateless retorts would consist of several such retorts in a row sharing a common infeed and discharge conveyor system to achieve continuous operation of any design capacity. Such a system is shown in Figure 11.5.

11.2.4 CONTINUOUS ROTARY COOKERS

The continuous rotary pressure sterilizer or cooker is a horizontal rotating retort through which the cans are conveyed while they rotate about their own axis through a spiral path and rotating reel mechanism as illustrated in the cutaway view of Figure 11.6. Residence time through the sterilizer is controlled by the rotating speed of the reel, which can be adjusted to achieve the required process time. This, in turn, sets the line speed for the entire system. Cans are transformed from an incoming can conveyor through a synchronized feeding device to a rotary transfer valve, which indexes the cans into the sterilizer while preventing the escape of steam and loss of pressure. Once cans have entered the sterilizer, they travel in the annular space between the reel and the shell. They are

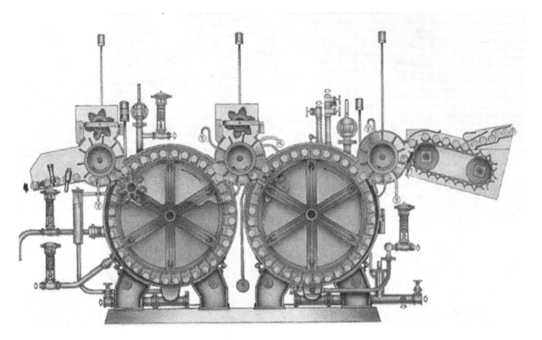

FIGURE 11.6   Cutaway view of continuous horizontal rotary cooker/sterilizer. (Courtesy FMC Corporation.)

held between spines on the reel and a helical or spiral track welded to the shell. In this way the cans are carried by the reel around the inner circumference of the shell, imparting a rotation about their own axes, while the spiral track in the shell directs the cans forward along the length of the sterilizer by one can length for each revolution of the reel. At the end of the sterilizer, cans are ejected from the reel into another rotary valve and into the next shell for either additional cooking or cooling.

Most common systems require at least three shells in series as shown in Figure 11.7 to accomplish controlled cooling through both a pressure cool shell and an atmospheric cool shell following the cooker or sterilizer. For cold-fill products that require controlled preheating, as many as five shells may be required in order to deliver an atmospheric preheat, pressure preheat, pressure cook, pressure cool, and atmospheric cool. By nature of its design and principle of operation, a continuous rotary sterilizer system is manufactured to accommodate a specific can size and cannot easily be adapted to other sizes. For this reason it is not uncommon to see several systems in operation in one food canning plant, each system dedicated to a different can size.

11.2.5 HYDROSTATIC STERILIZERS

These systems are so named because steam pressure is controlled hydrostatically by the height of a leg of water. Because of the height of water leg required, these sterilizers are usually installed outdoors adjacent to a canning plant. They are self-contained structures with the external appearance of a rectangular tower, as shown in Figure 11.8. They are basically made up of four chambers: a hydrostatic bring-up leg, a sterilizing steam dome, a hydrostatic bring-down leg, and a cooling section.

The principle of operation for a hydrostatic sterilizer can be explained with reference to the schematic flow diagram in Figure 11.9. Containers are conveyed through the sterilizer on carriers connected to a continuous chain link mechanism that provides positive line speed control and thus residence-time control to achieve specified process time in the steam dome. Carriers are loaded

FIGURE 11.7   Continuous rotary sterilizer system. (Courtesy FMC Corporation.)

automatically from incoming can conveyors and travel to the top of the sterilizer, where they enter the bring-up water leg. They travel downward through this leg as they encounter progressively hotter water. As they enter the bottom of the steam dome, the water temperature will be in equilibrium with steam temperature at the water seal interface. In the steam dome, the cans are exposed to the specified process or retort temperature controlled by the hydrostatic pressure for the prescribed process time controlled by the carrier line speed. When cans exit the steam dome, they again pass through the water seal interface at the bottom and travel upward through the bring-down leg as they encounter progressively cooler water until they exit at the top. Cans are then sprayed with cooling water as the carriers travel down the outside of the sterilizer on their return to the discharge conveyor station. Pressure and temperature profiles that are experienced by the water and steam in the various chambers, as well as by the cans themselves in a typical hydrostatic sterilizer system, are shown in Figure 11.10.

11.2.6 FLASH 18

The Flash "18" process is unique in that the product is brought to sterilizing temperature prior to filling through steam injection heating and then pumped while at sterilizing temperature to a hot fill operation carried out under pressure to accomplish sterility at the product-can wall interface. Conventional filling equipment and steam-flow can sealers are housed in a pressurized room or tank maintained at 18 lb of air pressure, as shown in Figure 11.11a. Hot product enters the tank at a sterilizing temperature of 265°F. It then flash cools to 255°F (the boiling point at 18 lb of air pressure). The filled and sealed cans are then processed through a continuous horizontal retort to accomplish a controlled hold time at 255°F to sterilize the inside can surfaces and deliver the required process time before final cooling and release to the outside through pressure seal can valves. This system is used primarily for large institutional size cans that would otherwise require such long retort processes that the resulting product quality would be unacceptable.

FIGURE 11.8   Exterior view of continuous hydrostatic sieiilizer. (Courtesy FMC Corporation.)

11.2.7 STERIFLAMME SYSTEM

The Steriflamme process is shown in Figure 11.11b. After closing under a high vacuum, cans are first preheated in steam and then further heated by rotating rapidly over direct contact with flames from a gas burner. After a necessary holding time to ensure sterilization, the cans are cooled by means of a water spray, as shown in the schematic diagram of the process in Figure 11.12. A high vacuum is important to prevent distortion of can seams, since the cans themselves become their own retort pressure vessels when heated by the gas flames. The process is often used for canned vegetables such as corn, peas, carrots, and mushrooms when minimum brine content is required.

11.3   THERMAL INACTIVATION KINETICS OF BACTERIAL SPORES

The precise mechanism by which a bacterial spore is rendered inactive when exposed to a lethal heat treatment is not yet fully understood. However, when a homogeneous population of viable spores is exposed to a lethal temperature, the number of spores that remain viable is observed to decrease exponentially (or logarithmically) with time; and the rate of this exponential decrease varies with

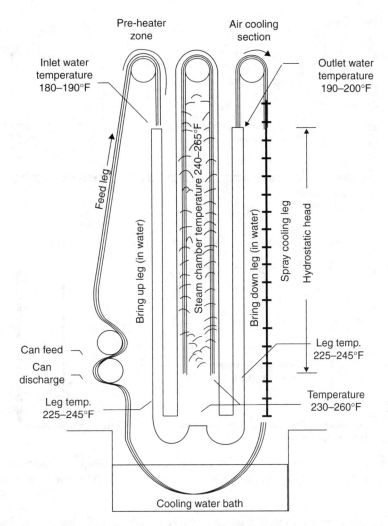

Pre-heater zone

Air cooling section

Inlet water temperature 180–190°F

Outlet water temperature 190–200°F

Feed leg

Steam chamber temperature 240–265°F

Bring up leg (in water)

Bring down leg (in water)

Spray cooling leg

Hydrostatic head

Can feed

Can discharge

Leg temp. 225–245°F

Leg temp. 225–245°F

Temperature 230–260°F

Cooling water bath

FIGURE 11.9 Flow diagram of a hydrostatic sterilizer for canned foods. (From Lopez, A., 1987, *Basic Information on Canning*, 11th ed. Courtesy CTI Publications, Inc.)

temperature within the lethal range. This is precisely the behavior of a first-order reaction in which the reactant (viable spores) is depleted exponentially as the reaction proceeds over time. Mathematically, this reaction can be described by the general rate equation for a first-order reaction in terms of a rate constant at a reference temperature and an activation energy for use of the Arrhenius equation to describe the temperature dependency of the rate constant. This method of describing reaction kinetics is presented in detail in Chapter 2 of this handbook and is applied to the thermal inactivation of bacterial spores in this chapter for continuity and completeness. However, this will be followed immediately by a presentation of the more traditional use of D and Z values based on common logarithms that is found in the food science literature in describing the thermal death time behavior of bacterial spores.

11.3.1 RATE CONSTANT AND ARRHENIUS RELATIONSHIP

The general rate equation,

$$-\frac{dC}{dt} = kC \tag{11.1}$$

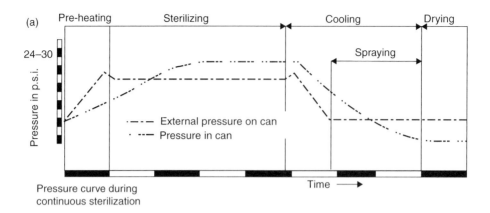

Pressure curve during
continuous sterilization

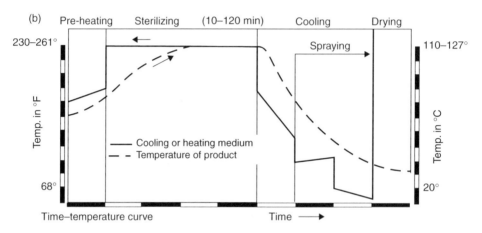

Time–temperature curve

FIGURE 11.10 Pressure and temperature histories during thermal processing in a typical hydrostatic process. (From Lopez, A., 1987, *Basic Information on Canning*, 11th ed. Courtesy CTI Publications, Inc.)

which applies to any first-order reaction process, can be used to describe the thermal inactivation of bacterial spores when C represents the concentration of viable spores, k is the rate constant, and t is time. After rearranging terms and integrating over time, the solution of this differential equation becomes

$$C = C_0 \exp(-kt) \tag{11.2}$$

which can also be expressed in natural logarithms as

$$\ln \frac{C_0}{C} = kt \tag{11.3}$$

Thus if a semilog plot is constructed in which the natural logarithm of the concentration of viable spores is plotted against time of exposure to a lethal temperature, a straight line will be produced as shown in Figure 11.13, which intercepts the ordinate axis at the natural log of initial concentration with a slope of $-k$ (the rate constant). Since the slope of a straight line is always given as the rise over the run, the units for the rate constant are cycles per unit of time or reciprocal of time (t^{-1}). At different lethal temperatures, similar straight lines would be produced but with different slopes, as shown by the family of curves in Figure 11.14, in which T_1, T_2, and T_3 represent increasingly higher lethal temperatures with corresponding rate constants, $-k_1$, $-k_2$, and $-k_3$.

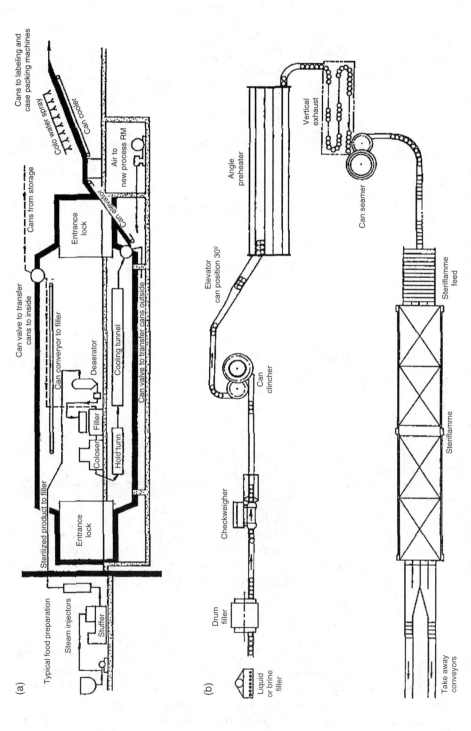

FIGURE 11.11 (a) Flash "18" process; (b) Steriflammie process. (From Lopez, A.., 1987, *Basic Information on Canning*, 11th ed. Courtesy CTI Publications, Inc.)

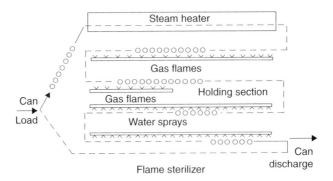

FIGURE 11.12   Flame sterilizer system. (From Lopez, A., 1987, *Basic Information on Canning*, 11th ed. Courtesy CTI Publications, Inc.)

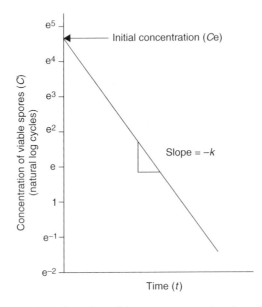

FIGURE 11.13   Semilog plot of number of surviving spores per unit volume (concentration) vs. time for viable bacterial spores subjected to a lethal temperature.

The temperature dependency of the rate constant is also an exponential function that can be described by a straight line on a semilog plot when the natural log of the rate constant is plotted against the reciprocal of absolute temperature as shown in Figure 11.15. The equation describing this straight line is known as the Arrhenius equation (discussed in Chapter 2):

$$\ln \frac{k}{k_0} = -\frac{E_\alpha}{R}\left[\frac{T_0 - T}{T_0 T}\right] \tag{11.4}$$

where k is the rate constant at any temperature T, and k_0 is the reference rate constant at a reference temperature T_0. The slope of the line produces the term E_α/R, in which E_α is the activation energy, and R is the universal gas constant. Thus once the activation energy is obtained in this way, Equation 11.4 can be used to predict the rate constant at any other temperature. Once the rate constant is known for a specified temperature, Equation 11.3 can be used to determine the time required for exposure at that temperature to reduce the initial concentration of viable bacterial spores by any number of log cycles. Alternatively, when the time of exposure to a given lethal temperature is specified, the same equation can be used to determine the number of log cycles by which the initial concentration

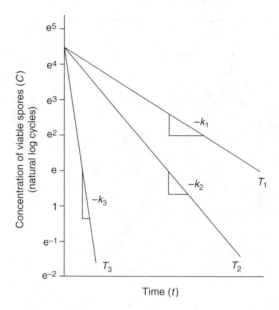

FIGURE 11.14 Family of spore survivor curves on semilog plot showing viable spore concentration vs. time at different lethal temperatures.

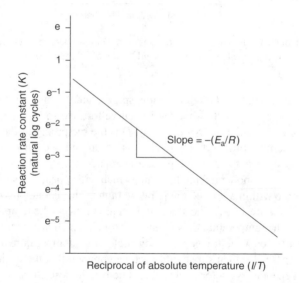

FIGURE 11.15 Arrhenious plot showing logarithmic temperature dependency of rate constant (k) for thermal inactivation of bacterial spores, where E_a is the activation energy and R is the universal gas constant.

of viable spores was reduced. This log cycle reduction can be translated into a sterilizing value or degree of lethality that will be discussed later.

11.3.2 THERMAL DEATH TIME RELATIONSHIP

The first-order rate process describing the thermal inactivation kinetics of bacterial spores is more commonly referred to in much of the food science literature as the logarithmic order of death and is described mathematically with the use of common (base 10) logarithms. The semilog plot used to show the logarithmic reduction in the number of surviving spores over time when exposed to a lethal temperature is known as a survivor curve, as shown if Figure 11.16. In this curve, the log cycles

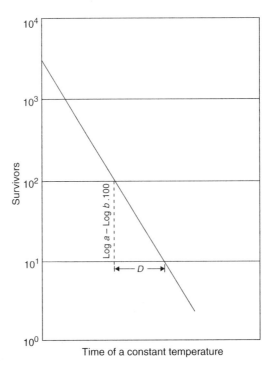

FIGURE 11.16 Survivor curve showing logarithmic order of death for bacterial spores subjected to a constant lethal temperature, and showing D value as time required for a tenfold reduction in spore population.

represent a tenfold decrease in the number of viable spores, and the rate constant called the decimal reduction time (D) is expressed as the time for the curve to traverse one log cycle or the time for one log cycle reduction in population. Numerically, D is the reciprocal of the rate constant k when multiplied by 2.3 (the number of natural log cycles per common log cycle).

Just as with the rate constant k, D is temperature dependent and will take on different values at different temperatures in an exponential relationship, which will appear as a straight line on a semilog plot of D vs. temperature within a limited temperature range. This is known as a thermal death time (TDT) curve, shown in Figure 11.17. The slope of this curve reflects the temperature dependency of D and is used to derive the temperature dependency factor Z, which is expressed as the temperature difference required for the curve to traverse one log cycle or the temperature difference required for a 10-fold change in the D value. Thus it should become apparent that the decimal reduction time D and the temperature dependency factor Z play the same role in describing the thermal inactivation kinetics of bacterial spores as do the rate constant k and the activation energy E_a from the Arrhenius relationships discussed in the preceding section.

Once the TDT curve has been established for a given microorganism, it can be used to calculate the time–temperature requirements for any idealized thermal process. For example, assume a process is required that will achieve a six-log-cycle reduction in the population of bacterial spores whose kinetics are described by the TDT curve in Figure 11.6, and that a temperature of 235°F has been chosen for the process. The TDT curve shows that the D value at 235°F is 10 min. This means that, at that temperature, 10 min will be required for each log cycle reduction in population. If a six-log-cycle reduction is required, a total of 60 min is needed for the process. If a temperature of 270°F had been chosen for the process, the D value at that temperature is approximately 0.1 min, and only 0.6 min (or 36 sec) would be required at that temperature to accomplish the same six-log-cycle reduction.

Since the TDT curve is a straight line on a semilog plot, all that is needed to specify such a curve is its slope and a single reference point on the curve. The slope of the curve is specified by the Z value,

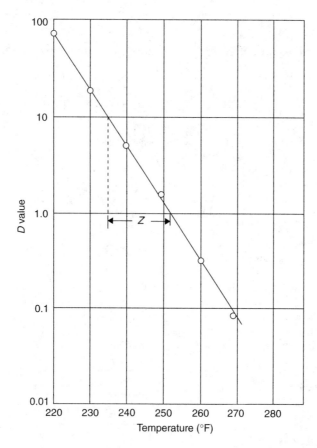

FIGURE 11.17 Thermal death time (TDT) curve showing temperature dependency of D value given by temperature change (Z) required for tenfold change in D value.

and the reference point is the D value at a reference temperature. For sterilization of low-acid foods (pH above 4.5), in which thermophilic spores of relatively high heat resistance are of concern, this reference temperature is usually taken to be 250°F. For high-acid foods or pasteurization processes in which microorganisms of much lower heat resistance are of concern, lower reference temperatures are used, such as 212 or 150°F. In specifying a reference D value for a microorganism, the reference temperature is shown as a subscript, such as D_{250}. For example, the TDT curve in Figure 11.17 can be specified by a Z value of 18°F degrees and a D_{250} value of 1.16 min.

Ranges of D values for different classification of bacteria are given in Table 11.1, and D_{250} values for specific organisms in selected food products are given in Table 11.2.

11.3.3 PROCESS LETHALITY

The example process calculations carried out in the preceding section using the TDT curve in Figure 11.17 showed clearly how two widely different processes (60 min at 235°F and 0.6 min at 270°F) were equivalent with respect to their ability to achieve the same log-cycle reduction in spore population (sterilizing value). In fact, the straight line drawn between these two points plotted on the TDT graph would lie parallel to the TDT curve and would represent all possible combinations of time and temperature that would accomplish a six-log-cycle reduction for that microorganism. Therefore, for a given Z value, the specification of any one point on this line is sufficient to specify the sterilizing value of any process combination of time and temperature on that line. The reference point that has been adopted for this purpose is the time in minutes at the reference temperature

TABLE 11.1
D Values for Different Classifications of Foodborne Bacteria

Bacterial groups	D-value
Low-acid and semiacid foods (pH above 4.5)	D_{250}
Thermophiles	
Flat-sour group (B. stearothermophilus)	4.0–5.0
Gaseous-spoilage group (C. thermosaccharolyticum)	3.0–4.0
Sulfide stinkers (C. nigrigicans)	2.0–3.0
Mesophiles	
Putrefactive anaerobes	
C. botulinum (types A and B)	0.10–1.20
C. sporogenes group (including P.A. 3679)	0.10–1.5
Acid foods (pH 4.0–4.5)	
Thermophiles	
B. coagulans (facultatively mesophilic)	0.01–0.07
Mesophiles	D_{212}
B. polymyxa and B. macerans 0.10-0.50	
Butyric anaerobes (C. pasteurianum)	0.10–0.50
High-acid foods D_{150}	
Mesophilic nonspore-bearing bacteria	
Lactobacillus spp., Leuconostoc spp., yeast and molds	0.50–1.00

Source: From Stumbo, C.R. 1965. *Termobacteriology in Food Processing.* Academic Press, New York.

TABLE 11.2
Comparison of D_{250} Values for Specific Organisms in Selected Food Substrates

Organism	Substrate	TDT method	D_{250}
P.A. 3679	Cream-style corn	Can	2.47
P.A. 3679	Whole-kernel corn (1)	Can	1.52
P.A. 3679	Whole-kernel corn (2)	Can	1.82
P.A. 3679	Phosphate buffer	Tube	1.31
F.S. 5010	Cream-style corn	Can	1.14
F.S. 5010	Whole-kernel corn	Can	1.35
F.S. 1518	Phosphate buffer	Tube	3.01
F.S. 617	Whole milk	Can	0.84
F.S. 617	Evaporated milk	Tube	1.05

Source: From Stumbo, C.R. 1965. *Termobacteriology in Food Processing.* Academic Press, New York.

of 250°F, or the point in time where the equivalent process curve crosses the vertical axis drawn at 250°F, and is known as the F value for the process. F is often referred to as the *lethality* of a process, and since it is expressed in minutes at 250°F, the *unit of lethality* is 1 min at 250°F. Thus if a process is assigned an F value of 6, it means that the integrated lethality achieved by whatever time–temperature history is employed by the process must be equivalent to the lethality achieved by 6 min of exposure to 250°F, assuming an idealized process of instantaneous heating to 250°F and instantaneous cooling from 250°F.

To illustrate, the example process calculation using the TDT curve in Figure 11.17 will be repeated by specifying the F value for the required process. Recall from that example that the process was required to accomplish a six-log-cycle reduction in spore population. All that is required to specify the F value is to determine how many minutes at 250°F will be required to achieve that level of log-cycle reduction. The D_{250} value is used for this purpose, since it represents the number of minutes at 250°F to accomplish one log-cycle reduction. Thus the F value is equal to D_{250} multiplied by the number of log cycles required in population reduction, or

$$F = D_{250}(\log a - \log b) \tag{11.5}$$

where a is the initial number of viable spores and b is the final number of viable spores (or survivors).

In this example, $D_{250} = 1.16$ min as taken from the TDT curve in Figure 11.17, and $(\log a - \log b) = 6$. Thus $F = 1.16(6) = 7$ min, and the sterilizing value for this process has been specified as $F = 7$. This is normally the way in which a thermal process is specified for subsequent calculation of a process time at some other temperature. In this way information regarding specific microorganisms or numbers of log-cycles reduction can be replaced by the F value (lethality) as a process specification.

Note also that this F value serves as the reference point to specify the equivalent process curve discussed earlier. By plotting a point at 7 min on the vertical line passing through 250°F in Figure 11.17, and drawing a curve parallel to the TDT curve through this point, the line will pass through the two equivalent process points that were calculated earlier (60 min at 235°F, and 0.6 min at 270°F). Alternatively, the equation of this straight line can be used to calculate the process time (t) at some other constant temperature (T) when F is specified.

$$F = 10^{[(T-250)/Z]t} \tag{11.6}$$

Equation 11.7 becomes important in the general case when the product temperature varies with the time during a process, and the F value delivered by the process must be integrated mathematically,

$$F = \int_0^t 10^{[(T-250)/Z]t} \tag{11.7}$$

At this point Equation 11.5 and Equation 11.7 have been presented as two clearly different mathematical expressions for the process lethality, F. It is most important that the distinction between these two expressions be clearly understood. Equation 11.5 is used to determine the F value that should be *specified* for a process and is determined from the log cycle reduction in spore population by considering factors related to safety and wholesomeness of the processed food, as discussed in the following section. Equation 11.7 is used to determine the F value *delivered* by a process as a result of the time–temperature history experienced by the product during the process. Another observation is that Equation 11.5 makes use of the D_{250} value in converting log cycles of reduction into minutes at 250°F, while Equation 11.7 makes use of the Z value in converting temperature–time history into minutes at 250°F. Because a Z value of 18°F (10°C) is so commonly observed or assumed for most thermal processing calculations, F values calculated with a Z of 18°F and reference to D_{250} are designated F_o.

11.3.4 SPECIFICATION OF PROCESS LETHALITY

Establishing the sterilizing value to be specified for a low-acid canned food is undoubtedly one of the most critical responsibilities taken on by a food scientist or engineer acting on behalf of a food company in the role of a competent thermal processing authority. In this section we outline briefly the steps normally taken for this purpose, as presented by Graves (1987) and reported in Stumbo (1965).

There are two types of bacterial populations of concern in canned food sterilization. First is the population of organisms of public health significance. In low-acid foods with pH above 4.5, the chief organism of concern is *Clostridium botulinum*. A safe level of survival probability that has been accepted for this organism is 10^{-12}, or one survivor in 10^{12} cans processed. This is known as the 12 D concept for botulinum cook. Since the highest D_{250} value known for this organism in foods is 0.21 min, the minimum lethality value for a botulinum cook assuming an initial spore load one organism per can is

$$F = 0.21 \times 12 = 2.52$$

Essentially all low-acid foods are processed far beyond the minimum botulinum cook in order to deal with spoilage-causing bacteria of much greater heat resistance. For these organisms, acceptable levels of spoilage probability are usually dictated by economic considerations. Most food companies accept a spoilage probability of 10^{-5} from mesophilic spore-formers (organisms that can grow and spoil food at room temperature). The organism most frequently used to characterize this classification of food spoilage is a strain of *C. sporogenes*, known as PA 3679, with a maximum D_{250} value of 1.00. Thus a minimum lethality value for a mesophilic spoilage cook assuming an initial spore of load of one spore per can is

$$F = 1.00 \times 5 = 5.00$$

Where thermophilic spoilage is a problem, more severe processes may be necessary because of the high heat resistance of thermophilic spores. Fortunately, most thermophiles do not grow readily at room temperature and require incubation at unusually high storage temperatures (110 to 130°F) to cause food spoilage. Generally, foods with no more than 1% spoilage (spoilage probability of 10^{-2}) upon incubation after processing, will show less than the accepted 10^{-5} spoilage probability in normal commerce. Therefore, when thermophilic spoilage is a concern, the target value for the final number of survivors is usually taken as 10^{-2}, and the initial spore load needs to be determined through microbiological analysis since contamination from these organisms varies greatly. For a situation with an initial thermophilic spore load of 100 spores per can, and an average D_{250} value of 4.00, the process lethality required would be

$$F = 4.00(\log 100 - \log 0.01)$$
$$= 4.00(4) = 16$$

The procedural steps above are only first-cut guidelines for average conditions, and often need to be adjusted up or down in view of the types of contaminating bacteria that may be present, the initial level of contamination or bioburden of the most resistant types, the spoilage risk accepted, and the nature of the food product from the standpoint of its ability to support the growth of the different types of contaminating bacteria that are found. Table 11.3 contains a listing of process lethalities (F_o) specified for the commercial processing of selected canned foods (Lopez, 1987).

11.4   HEAT TRANSFER IN CANNED FOODS

In the previous sections on thermal inactivation kinetics of bacterial spores, frequent reference was made to an idealized process in which the food product was assumed to be heated instantaneously to a lethal temperature, then cooled instantaneously after the required process time temperature. These idealized processes are important to gain an understanding of how the kinetic data can be used

TABLE 11.3
Lethality Values (F_0) for Commercial Sterilization of Selected Canned Foods

Product	Can sizes	Lethality value, F_0
Asparagus	All	2–4
Green beans, brine packed	No. 2	3.5
	No. 10	3.5
Chicken, boned	All	6–8
Corn, whole kernel,	All	9
brine packed	No. 10	15
Cream style corn	No. 2	5–6
	No. 10	2.3
Dog food	No. 2	12
	No. 10	6
Mackerel in brine	301×411	2.9–3.6
Meat loaf	No. 2	6
Peas, brine packed	No. 2	7
	No. 10	11
Sausage, Vienna, in brine	Various	5
Chili con carne	Various	6

Source: From Lopez, A., *Basic Information on Canning*, 11th ed., 1987. Courtesy of American Can Company, Inc.

directly to determine the process time at any given lethal temperature. There are in fact, commercial sterilization processes for which this method of process-time determination is applicable. These are high temperature/short time (HTST) sterilization processes for liquid foods that make use of steam injection heaters and flash cooling chambers for instantaneous heating and cooling. The process time is accomplished through the residence time in the holding tube between the heater and cooler as the product flows continuously through the system. This method of product sterilization is most often used with aseptic filling systems, discussed later in this chapter.

In traditional thermal processing of most canned foods, the situation is quite different from the idealized processes described above. Cans are filled with relatively cool unsterile product, sealed after headspace evacuation, and placed in steam retorts, which apply heat to the outside can wall. The product temperature can then only respond in accordance with the physical laws of heat transfer and will gradually rise in an effort to approach the temperature at the wall followed by a gradual fall in response to cooling at the wall. In this situation, the lethality delivered by the process will be the result of the transient time–temperature history experienced by the product at the slowest-heating location in the can; this is usually the geometric center. Therefore, the ability to determine this time–temperature history accurately is of paramount importance in the calculation of thermal processes. In this section we review the various modes of heat transfer found in canned foods and describe methods of temperature measurement and recording and how these data are treated to obtain important heat penetration parameters for subsequent use in various methods of thermal process calculation.

11.4.1 MODES OF HEAT TRANSFER

11.4.1.1 Conduction Heating

Solid-packed foods in which there is essentially no product movement within the container, even when agitated, heat largely by conduction heat transfer. Because of the lack of product movement and the low thermal diffusivity of most foods, these products heat very slowly and exhibit a nonuniform temperature distribution during heating and cooling caused by the temperature gradient that is set up between the can wall and geometric center. For conduction-heating products, the geometric center is the slowest heating point in the container. Therefore, process calculations are based on the temperature history experienced by the product at the can center. Solid-packed foods such as canned fish and meats, baby foods, pet foods, pumpkin, and squash fall into this category. These foods are usually processed in still cook or continuous hydrostatic retorts that provide no mechanical agitation.

11.4.1.2 Convection Heating

Thin-bodied liquid products packed in cans such as, milk, soups, sauces, and gravies will heat by either natural or forced convection heat transfer, depending on the use of mechanical agitation during processing. In a still cook retort that provides no agitation, product movement will still occur within the container because of natural convective currents induced by density differences between the warmer liquid near the hot can wall and the cooler liquid near the can center. An extensive study and analysis of natural convective heat transfer in canned foods is given in Datta and Teixeira (1987, 1988). The rate of heat transfer in nearly all convection-heating products can be increased substantially by inducing forced convection through mechanical agitation. For this reason, most convection-heating foods are processed in agitating retorts designed to provide either axial or end-over-end can rotation. Normally, end-over-end rotation is preferred and can be provided in batch retorts, while continuous agitating retorts can provide only limited axial rotation.

Unlike conduction-heating products, because of product movement in forced convection-heating products, the temperature distribution throughout the product is reasonably uniform under mechanical agitation. In natural convection the slowest heating point is somewhat below the geometric center and should be located experimentally in each new case. The two basic mechanisms of conduction and convection heat transfer in canned foods are illustrated schematically in Figure 11.18.

11.4.1.3 Broken Heating

There are also broken-heating canned food products that exhibit a break between these two modes of heat transfer and will heat part of the time by convection and part of the time by conduction. The more common of these foods are those that initially heat by convection; then, because of starch gelatinization or other thickening agent activity, they set up or thicken and proceed to heat by conduction. Less common are products that begin heating first by conduction then for the remainder of the period heat by convection. Generally, these are products with solid pieces in liquid brine that

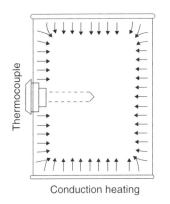
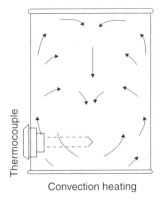

FIGURE 11.18   Conduction and convection heat transfer in solid and liquid canned foods, respectively. (From Lopez, A., 1987, *Basic Information on Canning*, 11th ed. Courtesy CTI Publications, Inc.)

settle and pack into the lower two-thirds or so of the container when placed in the retort. After some time of heating, when convective currents become sufficiently strong, the solid pieces are lifted and disperse to begin moving with the liquid phase.

11.4.2 HEAT PENETRATION MEASUREMENT

The primary objective of heat penetration measurements is to obtain an accurate recording of the product temperature at the can cold spot over time while the container is being treated under a controlled set of retort processing conditions. This is normally accomplished through the use of copper–constantan thermocouples inserted through the can wall so as to have the junction located at the can geometric center. Thermocouple lead wires pass through a packing gland in the wall of the retort for connection to an appropriate data acquisition system in the case of a still cook retort. For agitating retorts, the thermocouple lead wires are connected to a rotating shaft for electrical signal pick up from the rotating armature outside the retort. Specially designed thermocouple fittings are commercially available for these purposes (Ecklund, 1949), and further details on instrumentation, equipment, and methodology for conducting heat penetration tests can be found in Lopez (1987), NFPA (1980), and Stumbo (1965).

Also, the discussion of heat penetration in this chapter is limited to the use of saturated steam under pressure as the heat exchange medium in the retort. Hot water with overriding air pressure is usually used when processing foods in glass jars (water cook), and steam-air mixtures with overriding air pressure are also gaining in popularity for the thermal processing of foods in flexible packages. Details on how heat penetration studies and retort operations differ from saturated steam for these process methods can also be found in Lopez (1987) and NFPA (1980).

The precise temperature–time profile experienced by the product at the can center will depend on the physical and thermal properties of the product, size and shape of the container, and retort operating conditions. Therefore, it is imperative that test cans of product used in heat penetrations tests be truly representative of the commercial product with respect to ingredient formulation, fill weight, headspace, can size, and so on. In addition, the laboratory or pilot plant retort being used must accurately simulate the operating conditions that will be experienced by the product during commercial processing on the production-scale retort systems intended for the product. If this is not possible, heat penetration tests should be carried out using the actual production retort during scheduled breaks in production operations.

11.4.3 HEAT PENETRATION CURVES

During a heat penetration test, both the retort temperature history and product temperature history at the can center are measured and recorded over time. A typical test process will include venting of the retort with live steam to remove all atmospheric air, then closing the vents to bring the retort up to operating pressure and temperature. This is the point at which process time begins, and the retort temperature is held constant over this period of time. At the end of the prescribed process time, the steam is shut off and cooling water is introduced under overriding air pressure to prevent a sudden pressure drop in the retort. This begins the cooling phase of the process, which ends when the retort pressure returns to atmosphere and the product temperature in the can has reached a safe low level for removal from the retort.

A typical temperature–time plot of these data is shown in Figure 11.19 and illustrates the degree to which the product center temperature in the can lags behind the retort temperature during both heating and cooling. The product center temperature history can be taken directly from this plot to perform a process calculation by numerical integration of Equation 11.7 and will be discussed in further detail later.

As mentioned earlier, the response of the product temperature at the can center to the steam temperature applied at the can wall is governed by the physical laws of heat transfer and can be

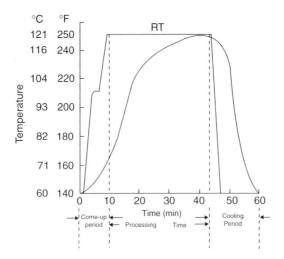

FIGURE 11.19 Typical heat penetration curve for canned foods showing retort temperature and temperature of food at slowest heating point within can. RT, retort temperature; can temp., temperature of can contents at slowest heating point. (From Lopez, A., 1987, *Basic Information on Canning*, 11th ed. Courtesy CTI Publications, Inc.)

expressed mathematically. This mathematical expression serves as the basis of the Ball formula method (Ball and Olson, 1957) of process calculation, explained later. It also serves as a basis for obtaining effective values for thermal properties of canned foods in order to use numerical computer models that are capable of simulating the heat transfer in thermal processing of canned foods.

A heat balance between the heat absorbed by the product and the heat transferred across the can wall from the steam retort could be expressed as follows for an element of food volume facing the can wall of surface area A and thickness L:

$$\rho L A C_p \frac{dT}{dt} = \frac{k}{L} A (T_r - T) \tag{11.8}$$

where T is product temperature, T_r is retort temperature, and ρ, C_p, and k are density, specific heat, and thermal conductivity of the product, respectively. Because of high surface heat transfer coefficient of condensing steam at the can wall and high thermal conductivity of the metal can, overall surface resistance to heat transfer can be assumed negligible, in contrast to the product's resistance to heat transfer. After rearranging terms, expression (11.8) can be written in the form of an ordinary differential equation:

$$\frac{dT}{dt} = \frac{k}{\rho C_p} L^2 (T_r - T) \tag{11.9}$$

By letting the thermal diffusivity (α) represent the combination of thermal and physical properties ($k/\rho C_p$), and letting T_o represent the initial product temperature, the solution to Equation 11.9 becomes

$$\frac{T_r - T}{T_r - T_o} = \exp\left(\frac{\alpha}{L^2} t\right) \tag{11.10}$$

Thus the product center temperature can be seen to be an exponential function of time; a semilog plot of the temperature difference ($T_r - T$) against time would produce a straight line sloping downward, having a slope related to the product's thermal diffusivity and can dimensions. Since

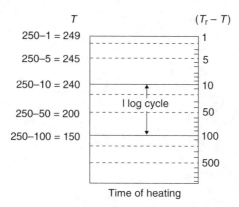

FIGURE 11.20 Technique for plotting heat penetration data on semilog graphing paper with the ordinate labeled for can temperature at a retort temperature of 250°F. (From Toledo, R.T., 1980, *Fundamentals of Food Process Engineering*. AVI, Westport, CT. With permission.)

the can center temperature itself rises upward during heating and not downward, a straight-line semilog plot of the center temperature increasing over time can be made with a minimum of data manipulation by rotating the semilog graph paper 180° (upside down) and labeling the top cycle 1° below the retort temperature, as shown in Figure 11.20. With the graph paper oriented in this manner, the difference between retort temperature and internal can temperature $(T_R - T)$ is shown increasing downward logarithmically on the right-hand scale, while the actual product temperature can be plotted against the left-hand scale, for the case in which the retort temperature is 250°F. A heat penetration curve plotted on such a graph, in which the ordinate axis on inverted semilog graph paper is labeled 1° below retort temperature at the top, with each log cycle representing a tenfold decrease in temperature proceeding downward, is shown in Figure 11.21. Since the center temperature can never theoretically reach the retort temperature, but only approaches that value asymptotically, this method of plotting allows the curve to extend as a straight line for any length of heating time desired.

Some of the traditional methods for calculating thermal processes described in the following chapter subsection are based on the mathematical equation of the straight-line portion of the heat penetration curve when plotted in this fashion. These methods make use of the heating curve slope factor, f_h, which is taken as the time in minutes for the straight-line portion of the heat penetration curve to traverse one log cycle; and the heating curve lag factor, j_{ch}, which is taken as the ratio of the difference between the retort temperature (T_r) and pseudo-initial temperature (T_o), the temperature at which an extension of the straight-line portion of the heating curve intersects the ordinate axis $(T_r - T_o)$ over the difference between retort temperature and actual initial product temperature $(T_r - T_i)$. The parameter g represents the difference between retort temperature and the maximum product temperature reached at the can center.

Since the heat penetration rate factor, f_h, represents the slope of the heat penetration curve, it is related to the product's thermal diffusivity and container dimensions. For a finite cylinder, with all parameters expressed in English units, the following relationship can be used to obtain the thermal diffusivity, α, from a heat penetration curve (Stumbo, 1965):

$$\alpha = \frac{0.398}{1/R^2 + (0.427/H^2)f_h} \tag{11.11}$$

where R is the can radius in inches, H is one-half the can height in inches, f_h the heating curve slope factor in minutes, and α the product thermal diffusivity in square inches per minute. This relationship is also useful to determine the heating curve slope factor, f_h, for the same product in a different-sized

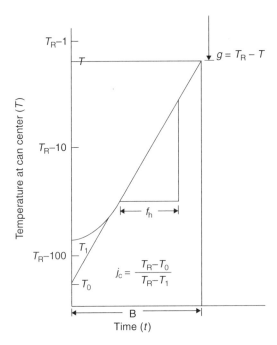

FIGURE 11.21 Typical heat penetration curve describing the temperature–time profile at the can center plotted on inverted semilog scale for use with Ball's formula method of calculating thermal process time.

container, since the thermal diffusivity is a combination of physical properties that characterize the product and its ingredient formulation and remains unaffected by different container sizes.

11.5 PROCESS CALCULATIONS

Once a heat penetration curve has been obtained from laboratory heat penetration data or predicted by a computer model, there are essentially two widely accepted methods for using these data to perform thermal process calculations. The first of these is the general method of process calculation described by Bigelow et al. (1920) and the second is the Ball formula method of process calculation (Ball, 1923). Both methods are described.

11.5.1 GENERAL METHOD

As the name implies, the general method is the most versatile method of process calculation because it is universally applicable to essentially any type of thermal processing situation. It makes direct use of the product temperature history at the can center obtained from a heat penetration test or predicted by a computer model to evaluate the integral shown in Equation 11.7 for calculating the process lethality delivered by a given temperature–time history. A straightforward numerical integration of Equation 11.7 can be expressed as follows with reference to Figure 11.22:

$$F_o = \sum_{i=1}^{n} \Delta F_i = \sum_{i=1}^{n} 10^{[(T_i - 250)/Z]\Delta t} \tag{11.12}$$

Figure 11.22 is a direct plot of the can center temperature experienced during a heat penetration test. Since no appreciable lethality can occur until the product temperature has reached the lethal temperature range (above 220°F), Equation 11.12 need only be evaluated over the time period during

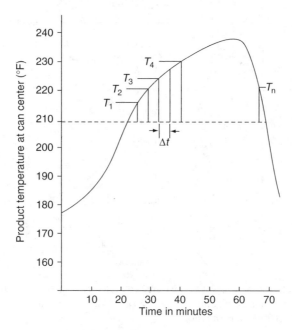

FIGURE 11.22 Temperature history at center of canned food during thermal process for calculation of process lethality by general method.

which the product temperature remains above 220°F. By dividing this time period into small time intervals (Δt) of short duration as shown in Figure 11.22, temperature T_i at each time interval can be read from the curve and used to calculate the incremental lethality (ΔF_i) accomplished during that time interval. Then the sum of all these incremental sterilizing values equals the total lethality, F_o, delivered by the test process. To determine the process time required to deliver a specified lethality, the cooling portion of the curve in Figure 11.22 is shifted to the right or left and the integration is repeated until the delivered sterilizing value so calculated agrees with the value specified for the process.

When introduced, this method was sometimes referred to as the graphical trial-and-error method because the integration was performed on specially designed graph paper to ease the tedious calculations that were required. The method was also time consuming and soon gave way in popularity to the more convenient Ball formula method described in the following section. With the current widespread availability of low-cost programmable calculators and desktop computers, these limitations are no longer of any consequence, and the general method is currently the method of choice because of its accuracy and versatility.

In fact, the general method is particularly useful in taking maximum advantage of computer-based data logging systems used in connection with heat penetration tests. Such systems are capable of reading temperature signals received directly from thermocouples monitoring both retort and product center temperature and processing these signals through the computer. Through programming instructions, both retort temperature and product center temperature are plotted against time without any data transformation. This allows the operator to see what has actually happened throughout the duration of the process test. As the data are being read by the computer, additional programming instructions call for calculation of the incremental process lethality (ΔF_i) at each time interval between temperature readings and summing these over time as the process is under way. As a result, the accumulated lethality (F) is known at any time during the process and can be plotted on the graph along with the temperature histories to show the final value reached at the end of the process.

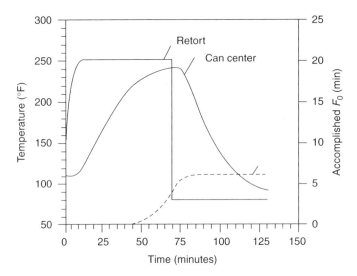

FIGURE 11.23 Computer-generated plot of measured retort temperature and calculated center temperature and accomplished F_o for a given thermal process. (From Datta, A.K., Teixeira, A.A., and Manson, J.E., 1986, *J. Food Sci.* 51: 480–483, 507. With permission.)

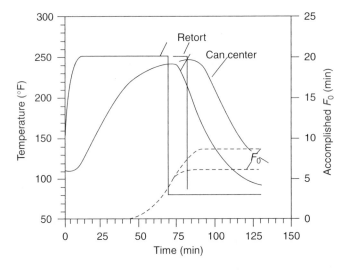

FIGURE 11.24 Computer-generated plot of retort temperature, can center temperature, and accomplished lethality (F_o) over time for two different process times.

   An example of the computer printout from such a heat penetration test is shown in Figure 11.23. Another test can be repeated quickly for a longer or shorter process time with instant results on the F_o achieved. By examining the results from both tests, the desired process time for the target F value can be closely estimated and then quickly tested for confirmation. The results of two such heat penetration tests are shown superimposed on each other in Figure 11.24. These results show that test 1, with a process time of 68 min, produced an F value of 6; test 2, with a process time of 80 min, produced an F value of 8, suggesting that a target F value of 7 will be achieved by an intermediate process time. This can be confirmed by running a test at the suggested process time and examining the resulting F value. Rather than carrying out repeated heat penetration tests in the laboratory for such purposes, numerical computer models can be used at the computer terminal to generate the

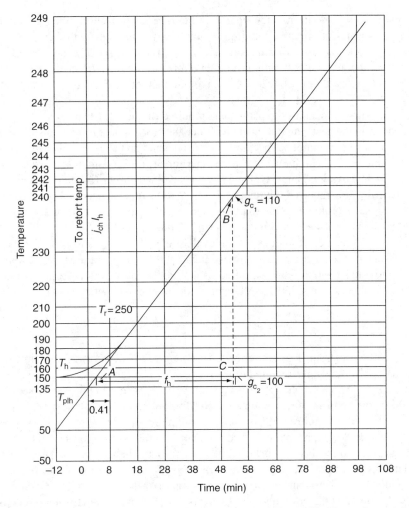

FIGURE 11.25 Semilog plot of heating curve. (From Stumbo, C.R., 1965, *Thermobacteriology in Food Processing*. Academic Press, New York. With permission.)

heat penetration data for different process times once the product thermal diffusivity is known, as described previously.

11.5.2 BALL FORMULA METHOD

Ball's formula method for calculating the process time at a given retort temperature is based on a mathematical equation for the straight-line portion of the temperature–time profile at the can center when plotted on inverted semilog graph paper. A heat penetration curve of this type is shown in Figure 11.25 for a retort temperature of 250°F. The ordinate axis is labeled 1° below retort temperature at the top, with each log cycle representing a tenfold decrease in temperature proceeding downward, as explained earlier in the section on heat penetration curves. This method of data transformation is a straightforward mathematical technique and allows Ball's formula to take on a simple expression that obeys standard heat conduction theory.

The equation that Ball derived for this straight-line heating curve can be expressed as follows:

$$B = f_h(\log j_c I - \log g) \tag{11.13}$$

where B is the process time in minutes when no time is required to bring the retort to processing temperature, f_h is time, in minutes, required for the straight-line portion of the heating curve to traverse one log cycle, j_c is heating lag factor $(T_r - T_o)/(T_r - T_i)$, I is temperature difference between retort temperature and the initial product temperature $(T_r - T_i)$, and g is the temperature difference between retort temperature and the maximum temperature reached by the food at the can center.

With the exception of g, all of the parameters required on the right-hand side of Equation 11.13 can be obtained from the heat penetration curve. Since the process time is not known, the endpoint of the heat penetration curve cannot be specified but must depend on the degree of bacterial inactivation required of the process. The value for g can be determined from a complex series of relationships based on thermal inactivation kinetics (D and Z values defined earlier for the organism of concern), the degree of sterilization required (number of log cycle reductions in bacterial spore concentration), the f_h value, and the retort temperature to be specified for the process. These relationships also take into account the additional sterilization that takes place during cooling.

In practice, g is obtained from a set of tables that relate g to the ratio f_h/U for the specific Z value of the organism of concern. U is the time required at retort temperature to accomplish the same degree of bacterial inactivation that is required of the process at the reference temperature used to determine the D value (usually, 250°F). Mathematically, U can be expressed as

$$U = D_R \cdot \log(C_o/C) \cdot 10^{[(T_r - T_R)/Z]} \tag{11.14}$$

where D_R is D value measured at reference temperature (T_R), C_o/C is ratio of initial over final concentration of bacterial spores required of the process, T_r is reference temperature at which D_R is determined for the organism of concern (usually, 250°F), T_R is retort temperature specified for the process, and Z is the temperature dependency factor for the organism of concern.

Figure 11.26 through Figure 11.28 contain a set of graphs showing f_h/U vs. $\log g$ for z values of 14, 18, and 22, respectively. The families of curves on these graphs also show how these relationships vary slightly as a function of the cooling lag factor j_{cc}. This factor helps to account for the contribution of cooling lethality to the process calculation and is derived from a semilog plot of the cooling curve shown in Figure 11.29. Note that for a cooling penetration curve, the semilog graph paper is kept in normal orientation and the bottom log cycle is labeled 1° above cooling water temperature, with each log cycle representing a tenfold increase in temperature moving upward. The cooling lag factor is the ratio of the difference between the temperature at which the extended straight-line portion of the cooling curve intersects the ordinate axis (T_{pic}) and the cooling water temperature (T_w) over the difference between the product temperature when cooling begins (T_{ic}) and the cooling water temperature (T_w), or mathematically,

$$j_{cc} = \frac{T_{pic} - T_w}{T_{ic} - T_w} \tag{11.15}$$

Thus for the cooling curve shown in Figure 11.29, the cooling lag factor would be

$$j_{cc} = \frac{340 - 70}{240 - 70} = \frac{270}{170} = 1.6 \tag{11.16}$$

Example 11.1 Suppose that for processing a particular product, the following heat penetration data are given:

$$T_R = 250°F \qquad j_{ch} = 2 \qquad f_h = 25 \text{ min}$$
$$T_I = 170°F \qquad j_{cc} = 1.4$$

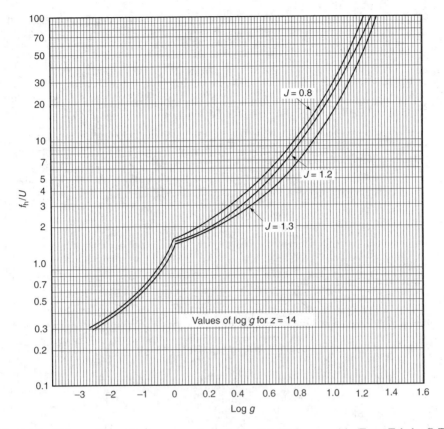

FIGURE 11.26 Values of $\log g$ for corresponding values of f_h/U for $z = 14$. (From Toledo, R.T., 1980, *Fundamentals of Food Process Engineering*. AVI, Westport, CT. With permission.)

What process time (B) will be required to achieve a specified lethality value of $F_0 = 5.6$ assuming that $Z = 14°F$?

Solution Since the retort temperature specified for this process is the same as the universal reference temperature for unit lethality, $250°F$, the factor u has the same value as F_0; otherwise, u would be calculated using Equation 11.14 for a different retort temperature (T_R). Referring to the Ball formula, we have

$$B = f_h(\log j_{ch}I - \log g)$$
$$I = (T_R - T_I) = (250 - 170) = 80$$
$$j_{ch}I = 2.00 \times 80 = 160$$
$$\log j_{ch}I = 2.778$$

Find $\log g$ from the f_h/u vs. $\log g$ graph for $Z = 14$ on Figure 11.26.

$$\frac{f_h}{u} = \frac{25}{5.6} = 4.46$$

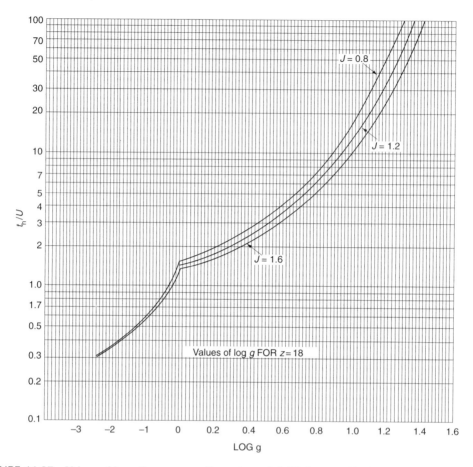

FIGURE 11.27   Values of $\log g$ for corresponding values of f_h/U for $z = 18$. (From Toledo, R.T., 1980, *Fundamentals of Food Process Engineering*. AVI, Westport, CT. With permission.)

By locating this value midway between the curves for $j_{cc} = 1.2$ and $j_{cc} = 1.6$, read the value for log $g = 0.625$; then

$$B = 25(2.778 - 0.625) = 25(2.153) = 54 \text{ min}$$

11.6 NUMERICAL COMPUTER SIMULATION OF HEAT TRANSFER

Another purpose for obtaining the thermal diffusivity of the product from a heat penetration curve is to make use of numerical computer models capable of simulating the heat transfer in canned foods. One of the primary advantages of these models is that once the thermal diffusivity has been determined, the model can be used to predict the product temperature history at any specified location within the can for any set of processing conditions and container size specified. Thus with the use of such models, it is unnecessary to carry out repeated heat penetration tests in the laboratory or pilot plant to determine the heat penetration curve for a different retort temperature or can size. A second advantage of even greater importance is that the retort temperature need not be held constant but can vary in any prescribed manner throughout the process, and the model will predict the correct product temperature history at the can center. Use of these models has become invaluable for simulating the

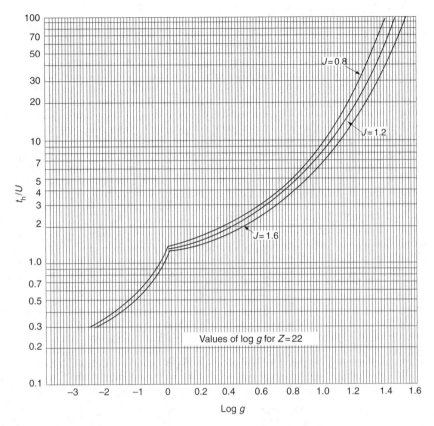

FIGURE 11.28 Values of $\log g$ for corresponding values of f_h/U for $z = 22$. (From Toledo, R.T., 1980, *Fundamentals of Food Process Engineering*. AVI, Westport, CT. With permission.)

process conditions experienced in continuous sterilizer systems, in which containers pass from one chamber to another, experiencing a changing boundary temperature as they pass through the system. Another important application of these models is in the rapid evaluation of an unscheduled process deviation, such as when an unexpected drop in retort temperature occurs during the course of the process. The model can quickly predict the product center temperature profile in response to such a deviation and calculate the delivered lethality value, F_o, for comparison with the lethality value specified for the product.

The use of a numerical computer model for stimulating the thermal processing of canned foods was first described by Teixeira et al., (1969) and continually improved upon over the years (Teixeira et al., 1975; Teixeira and Manson, 1982; Datta et al., 1986; Teixeira et al., 1999). The model makes use of a numerical solution by finite differences of the two-dimensional partial differential equation that describes conduction heat transfer in a finite cylinder. During conduction heating, heat is applied only at the can surface; temperatures will rise first only in regions near the can walls, while temperature near the can center will begin to respond only after a considerable time lag. Mathematically, the temperature is a distributed parameter in that at any point in time during heating, the temperature takes on a different value with location in the can; and in any one location, the temperature changes with time as heat gradually penetrates the product from the can walls toward the center.

The mathematical expression that describes this temperature distribution pattern over time is shown in Figure 11.30 and lies at the heart of the numerical computer model. This expression is the classic partial differential equation for two-dimensional unsteady heat conduction in a finite cylinder and can be written in the form of finite differences for numerical solution by digital computer, as

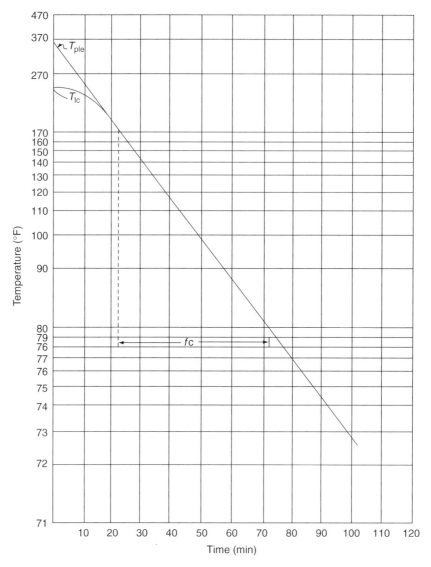

FIGURE 11.29 Semilog plot of cooling curve. (From Stumbo, C.R., 1965, *Thermobacteriology in Food Processing*. Academic Press, New York. With permission.)

shown in Figure 11.31. The finite differences are discrete increments of time and space defined as small intervals of process time and small increments of container height and radius (Δt, Δh, and Δr, respectively).

As a framework for computer iterations, the cylindrical container is imagined to be subdivided into volume elements that appear as layers of concentric rings having rectangular cross sections, as illustrated in Figure 11.32 for the upper half of the container. Temperature nodes are assigned at the corners of each volume element on a vertical plane, as shown in Figure 11.33, where I and J are used to denote the sequence of radial and vertical volume elements, respectively. By assigning appropriate boundary and initial conditions to all the temperature nodes (interior nodes set at initial product temperature and surface nodes set at retort temperature), the new temperature reached at each node can be calculated after a short time interval (Δt) that would be consistent with the thermal diffusivity of the product obtained from heat penetration data (f_h). This new temperature distribution

$$\frac{\partial T}{\partial t} = \alpha \left[\frac{\partial^2 T}{\partial r^2} + \frac{1}{r}\frac{\partial T}{\partial r} + \frac{\partial T}{\partial h^2} \right]$$

Where: T = temperature
t = time
α = thermal diffusivity*
r = radial position in cylinder
h = vertical position in cylinder

$$^*\alpha = \frac{0.398}{[1/R^2 + 0.427/H^2]f_h}$$ Where: R = can radius
H = one-half can height
f_h = slope of heat penetration curve.

FIGURE 11.30 Classical heat conduction equation for a finite cylinder. (From Teixeira, A.A. and Manson, J.E., 1982, *Food Technol.* 36: 85. With permission.)

$$T_{(ij)}^{(t+\Delta t)} = T_{(ij)}^{(t)} + \frac{\alpha\,\Delta t}{\Delta r^2}[T_{(i-1,j)} - 2T_{(ij)} + T_{(i+1,j)}]^{(t)}$$

$$+ \frac{\alpha\,\Delta t}{2r\,\Delta r}[T_{(i-1,j)} - T_{(i+1,j)}]^{(t)}$$

$$+ \frac{\alpha\,\Delta t}{\Delta h^2}[T_{(i,j-1)} - 2T_{(i,j)} + T_{(i,j+1)}]^{(t)}$$

Where: Δt, Δr, Δh = discrete increments of time, radius, and height, and i and j denote sequence of radial and vertical increments away from can wall and mid-plane.

FIGURE 11.31 Heat conduction equation for a finite cylinder expressed in the form of finite differences for numerical solution by computer iterations. (From Teixeira, A.A. and Manson, J.E., 1982, *Food Technol.* 36: 85. With permission.)

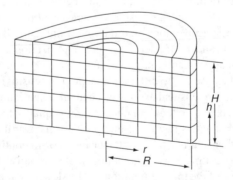

FIGURE 11.32 Subdivision of a cylindrical container for application of finite differences. (From Teixeira, A.A., Dixon, J.R., Zahradnik, J.W., and Zinsmeister, G.E., 1969, *Food Technol.* 23: 137. With permission.)

is then taken to replace the initial one, and the procedure repeated to calculate the temperature distribution after another time interval. In this way, the temperature at any point in the container at any instant in time is obtained. At the end of process time, when steam is shut off and cooling water is admitted to the retort, the cooling process is simulated by simply changing the boundary conditions from retort temperature T_R to cooling temperature T_C at the surface nodes and continuing with the computer iterations described above.

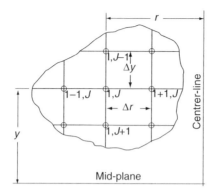

FIGURE 11.33 Labeling of grid nodes in matrix of volume elements on a vertical plane for application of finite differences.

The temperature at the can center can be calculated after each time interval to produce a predicted heat penetration curve upon which the process lethality, F, can be calculated. When the numerical computer model is used to calculate the process time required at a given retort temperature to achieve a specified lethality, F, the computer follows a programmed search routine of assumed process times that quickly converges on the precise time at which cooling should begin in order to achieve the specified F value. Thus the model can be used to determine the process time required for any given set of constant or variable retort temperature conditions.

11.7 PROCESS OPTIMIZATION

11.7.1 Objective Functions

The principal objective of thermal process optimization is to maximize product quality, minimize undesirable changes, minimize cost, and maximize profits. At all times, a minimal process must be maintained to exclude the danger from microorganisms of public health and spoilage concern. Five elements common to all optimization problems are performance or objective function (nutrients, texture, and sensory characteristics), decision variables (retort temperature and process time), constraints (practical limits for temperatures and required minimal lethality), mathematical model (analytical, finite differences, and finite element), and optimization technique (search, response surface, and linear or nonlinear programming).

Optimization theory makes use of the different temperature sensitivity of microbial and quality factor destruction rates. Microorganisms have lower decimal reduction time (less resistant to heat) and a lower Z-value (more sensitive to temperature) than most quality factors. Hence, a higher temperature will result in preferential destruction of microorganisms over the quality factor. Especially applied to liquid product either in a batch in-container mode or in continuous aseptic systems, the higher temperature with shorter time offers a great potential for quality optimization. However, for conduction heating foods, one of the major limitations is the slower heating. All higher temperatures do not necessarily favor the best quality retention because they also expose the product nearer the surface to more severe temperature than the product at the center, which might result in diminished overall quality.

11.7.2 Thermal Degradation of Quality Factors

Optimum combinations of retort temperature and process time that maximize quality or nutrient retention (or minimize process time) can be found if the kinetic parameters describing the thermal degradation kinetics of the quality factors are known. Using the numerical computer simulation models described earlier, process times needed at different retort temperatures to achieve the same

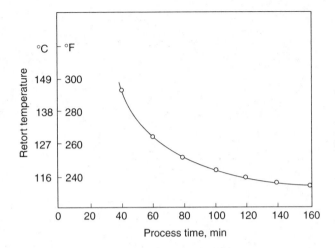

FIGURE 11.34 Iso-lethality curve showing combinations of retort temperature and process time that deliver the same level of lethality for pea puree in No. 2 cans. (From Teixeira, A.A., Dixon, J.R., Zahradnik, J.W., and Zinsmeister, G.E., 1969, *Food Technol.* 23: 137. With permission.)

TABLE 11.4

Kinetic Parameters for Thermal Degradation of Quality Factors in Selected Thermally Processes Foods

Quality factor in food systems	$D_{121°C}$ (min)	$K_{121°C}$ (min^{-1})	$Z(°C)$	E_a (kcal/mol)
Thiamine in Beans	329.77	6.9837×10^{-3}	27.95	25.416
Lysine in Beans	178.28	9.051×10^{-2}	25.44	27.32
Texture in Beans	101.68	2.260×10^{-2}	20.62	35.44

Source: From Thermobacteriology Laboratory, Food Science Department, College of Food Engineering, UNICAMP, Campinas, Sao Paulo, Brazil.

process lethality can be quickly calculated over a range of retort temperatures that falls within the operating performance limitations of the retort. A plot of these equivalent retort temperature–process time combinations produces an isolethality curve such as the one shown in Figure 11.34 for the case of pea puree in No. 2 cans.

The total level of nutrient/quality retention can be quickly calculated for each set of equivalent process conditions by replacing the kinetic parameters for microbial inactivation with those for quality degradation in the model. Table 11.4 gives examples of such kinetic parameters for the thermal degradation of selected quality factors in specific food systems. A plot of nutrient retention vs. equivalent process conditions reveals the range of process conditions that result in maximum nutrient retention, as shown in Figure 11.35 for the case of pea puree in No. 2 cans. Note that the same exercise is also useful when seeking to minimize process time, because these results reveal the price that is paid in lower quality retention caused by the higher surface temperatures needed to allow for shorter process time.

11.7.3 Volume Average Determination of Quality Retention

Quality retention in thermally processed conduction-heated foods is a nonuniformly distributed parameter. Relatively long exposure to the higher temperatures near the product surface causes

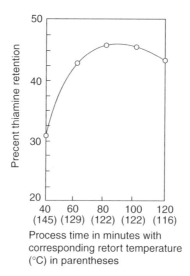

FIGURE 11.35   Optimization curve showing percent thiamine retention for pea puree in No. 2 cans after various retort temperature and process time combinations that deliver the same level of lethality. (From Teixeira, A.A., Dixon, J.R., Zahradnik, J.W., and Zinsmeister, G.E., 1969, *Food Technol.* 23: 137. With permission.)

much more quality degradation in product near the surface than in the product near the cold spot or center. This is because temperature distribution throughout the food container is nonuniformly distributed as heating and cooling proceed during the process. For this reason, quality retention must be calculated by volume integration of the different levels of retention at different locations. This is done by taking advantage of the finite element feature of the numerical simulation model. As the computer iterations make each sweep across the finite element nodes in carrying out the heat transfer calculations, the small change in nutrient concentration that occurs in that time interval can be calculated from the momentary value of rate constant which prevails at the local temperature at that time. When the process simulation is ended, a different final nutrient/quality concentration will exist within each volume element. Recall that the volume elements are in the shape of concentric rings with known dimensions from which the volume of each different-size ring can be calculated. Total nutrient retention within each ring is calculated by multiplying the final nutrient concentration within the ring by the volume of that ring. Total nutrient retention in the product is the summation of final retention in all the rings.

11.8   ON-LINE COMPUTER CONTROL

Traditional control of thermal process operations has consisted of maintaining specified operating conditions that have been predetermined from product and process development research, such as the process calculations for the time and temperature of a batch cook. Sometimes unexpected changes can occur during the course of the process operation or at some point upstream in a processing sequence such that the pre-specified processing conditions are no longer valid or appropriate, and off-specification product is produced that must be either reprocessed or destroyed at appreciable economic loss. These types of situations can be of critical importance in food processing operations, because the physical process variables that can be measured and controlled are often only indicators of complex biochemical reactions that are required to take place under the specified process conditions.

Because of the important emphasis placed on the public safety of canned foods, processors operate in strict compliance with the Food and Drug Administration's Low-Acid Canned Food Regulations. Among other things, these regulations require strict documentation and record-keeping of all critical

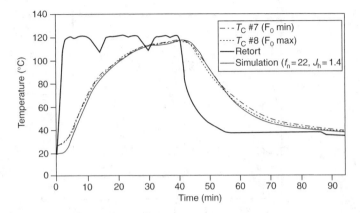

FIGURE 11.36 Comparison of internal cold spot temperatures predicted by model simulation with profiles measured by thermocouples in response to multiple retort temperature deviations during a heat penetration test with 5% bentonite suspension in 6-ounce tuna cans. (From Teixeira, A.A., Balaban, M.O., Germer, S.P.M., Sadahira, M.S., Teixeira-Neto, R.O., and Vitali. 1999. *J. Food Sci.* 64: 488–493. With permission.)

control points in the processing of each retort load or batch of canned product. Particular emphasis is placed on product batches that experience an unscheduled process deviation, such as when a drop in retort temperature occurs during the course of the process which may result from loss of steam pressure. In such a case, the product will not have received the established scheduled process and must be either destroyed, fully reprocessed, or set aside for evaluation by a competent processing authority. If the product is judged to be safe, then batch records must contain documentation showing how that judgment was reached. If judged unsafe, then the product must be fully reprocessed or destroyed. Such practices are costly.

In recent years food engineers knowledgeable in the use of engineering mathematics and scientific principles of heat transfer have developed computer models capable of simulating thermal processing of conduction-heated canned foods as described earlier. These models make use of numerical solutions to mathematical heat transfer equations capable of predicting accurately the internal product cold spot temperature in response to any dynamic temperature experienced by the retort during the process. Accuracy of such models is demonstrated in Figure 11.36 from Teixeira et al. (1999), which compares internal cold spot temperatures predicted by model simulation with profiles measured by thermocouple in response to multiple retort temperature deviations during a heat penetration test. The accomplished lethality (F_o) for any thermal process is easily calculated by numerical integration of the predicted cold spot temperature over time as explained previously. Thus, if the cold spot temperature can be accurately predicted over time, so can accumulated process lethality.

Computer-based intelligent online control systems make use of these models as part of the decision-making software in a computer-based online control system. Instead of specifying the retort temperature as a constant boundary condition, the actual retort temperature is read directly from sensors located in the retort and is continually updated with each iteration of the numerical solution. Using only the measured retort temperature as input to the control system, the model operates as a subroutine calculating the internal product cold spot temperature at small time intervals for computer iteration in carrying out the numerical solution to the heat conduction equation by finite differences. At the same time, the model also calculates the accomplishing process lethality associated with cold spot temperature in real time as the process is under way. At each time step, the subroutine simulates the additional lethality that will be contributed by the cooling phase if cooling were to begin at that time. In this way, the control system decision of when to end heating and begin cooling is withheld until the model has determined that final target process lethality will be reached at the end of cooling.

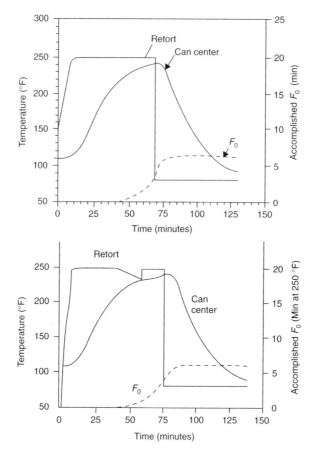

FIGURE 11.37   Output documentation of computer-based on-line control system showing scheduled heating time of 68 min for normal process (above), and heating time extended automatically to 76 min in compensation for unscheduled temporary loss of retort temperature, or process deviation (below).

By programming the control logic to continue heating until the accumulated lethality has reached some designated target value, the process will always end with the desired level of lethality (F_o) regardless of an unscheduled process temperature deviation. At the end of the process, complete documentation of measured retort temperature history, calculated center temperature history, and accomplished lethality (F_o) can be generated in compliance with regulatory record-keeping requirements. Such documents are shown in Figure 11.37 for a normal process (above) and for the same intended process with an unexpected deviation (below).

11.9   ASEPTIC PROCESSING

Aseptic canning systems have rapidly developed in recent years, primarily to allow for the marketing of shelf-stable foods in novel or more economical packaging systems that cannot withstand normal retort processing conditions. The primary goal in earlier development work on aseptic canning systems was to minimize quality losses that occur in slow-heating foods processed in conventional retorting systems. In either case, aseptic canning circumvents the need for retort operations by sterilizing the product outside the container through heat exchanger systems before it is filled aseptically into separately sterilized containers or packaging systems.

Expectations of quality improvement from aseptic canning stem from the realization that the temperature dependency of the first-order reaction rate constant describing the thermal degradation of

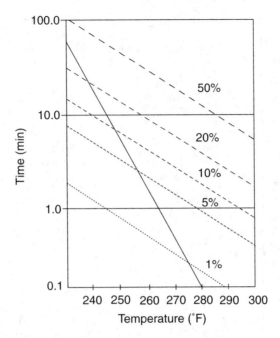

FIGURE 11.38 Time–temperature sterilization curve for bacteria (F_0 = 6 min) compared with time-temperature destruction curves for 1, 5, 10, 20, and 50% loss of thiamine. (From Joslyn, M.A. and Heid, J.L., 1963, *Food Processing Operations*, Vol. II. AVI, Westport, CT. With permission.)

most heat-sensitive quality factors is much less severe than that for the thermal inactivation of bacteria. The thermal degradation kinetics for thiamine (vitamin B_1) are often used to illustrate this point because thiamine is generally accepted as one of the most heat-sensitive nutrients of importance in thermally processed foods. The significance of this difference in temperature dependency is revealed in Figure 11.38 from Joslyn and Heid (1963). In this figure a family of thermal destruction time curves representing different levels of thiamine degradation are shown superimposed on one thermal sterilization (TDT) curve representing lethality value of F_0 = 6 min. Since the figure is a semilog plot of decimal reduction time versus temperature, all the points on any one line represent combinations of time and temperature that are equivalent in achieving the level of destruction represented by that line. Assuming that the TDT curve for bacterial spores is a process constraint, only time–temperature combinations that fall on this curve can be allowed. This curve intersects with curves representing lower levels of thiamine destruction as conditions of higher temperature with correspondingly shorter times are approached. Then, by observation, the optimum process is the highest temperature possible with the associated minimum process time. These differences between thiamine degradation and bacterial inactivation kinetics formed the basis for optimization studies on retort operating conditions reported by Teixeira et al. (1969, 1975).

The benefits of high temperature/short time processing have been known for a number of years and have led to the development of aseptic processing and filling systems wherever possible. These methods generally apply only to fluid products that can be pumped through heat exchangers capable of applying ultrahigh-temperature/short time (UHT) heating conditions to the product before it is filled and sealed aseptically. The general types of heat exchangers commonly used with aseptic canning systems fall into the two basic categories of direct or indirect heating. In direct heating, the product is brought into direct contact with live steam through either steam injection or steam infusion heating, the product contacts the heated metal surfaces of a heat exchanger, which separate the product from direct contact with the heat exchange medium. Either plate, tubular, or swept-surface heat exchangers are most often used for this purpose. The residence time experienced by

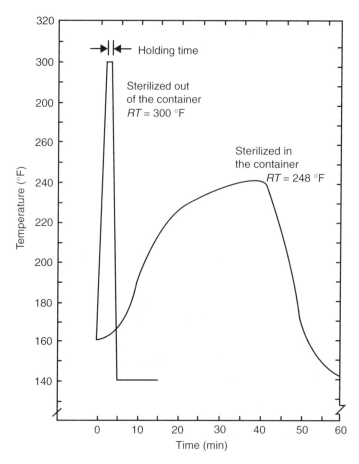

FIGURE 11.39 Comparison of product heating curves for thermal processes of equivalent lethality in an aseptic system at 300°F and in a conventional still retort at 248°F. (From Lopez, A., 1987, *Basic Information on Canning*, 11th ed., The Canning Trade, Baltimore, MD. Courtesy CTI Publications, Inc.)

the product as it flows through an insulated holding tube or holding section between heating and cooling accomplishes the necessary process time for delivering the specified sterilizing value and is controlled by flow rate. A comparison of product heating curves for thermal processes of equivalent sterilizing value in an aseptic system at 300°F and in a conventional still-cook retort at 248°F are shown in Figure 11.39.

11.9.1 COMMERCIAL ASEPTIC PROCESSING SYSTEMS

Among the first commercially successful aseptic canning systems is the Dole system, illustrated schematically in Figure 11.40. The system was designed to fill conventional steel cans aseptically and made use of superheated steam chambers to sterilize empty can bodies and covers as they were slowly conveyed to the filling chamber. The filling chamber was also maintained sterile by superheated steam under positive pressure and received cool sterile product from the heat exchangers in the product sterilizing subsystem. The entire system was sterilized prior to operation by passing superheated steam through the can tunnel, cover and closing chamber, and filling chamber for a prescribed startup program of specified times and temperatures. The product sterilizing line was

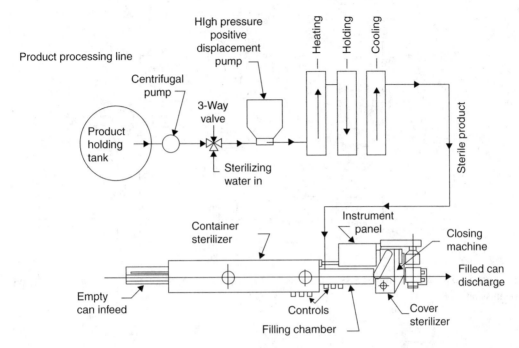

FIGURE 11.40 Aseptic canning line. (From Lopez, A., 1987, *Basic Information on Canning*, 11th ed., The Canning Trade, Baltimore, MD. Courtesy CTI Publications, Inc.)

presterilized by passing pressurized hot water through the cooling heat exchanger (with coolant turned off), product filling line, and filler heads. This startup procedure had to be repeated every time a compromise in sterility occurred at any system component. Obviously, careful monitoring and control by skillful and highly trained operators is a must for such an intricately orchestrated system.

Although the Dole system continues to be the mainstay for aseptic filling into metal cans, recent regulatory approval for the use of chemical sterilants such as hydrogen peroxide to sterilize the surfaces of various paper, plastic, and laminated packaging materials has opened the door to a wide array of commercially available aseptic filling systems to produce shelf-stable liquid foods in a variety of gable-topped, brick-packed, and other novel package configurations. Filling machines designed for these packaging systems are usually based on the use of form-fill-seal operations. In these machines, the packaging material is fed from either precut blanks or directly from roll stock, passed through a chemical sterilant bath or spray treatment, formed into the final package shape while being filled with cool sterile product from the product sterilizing system, and then sealed and discharged al within a controlled aseptic environment.

Another important commercial application of aseptic processing technology is in the storage and handling of large bulk quantities of sterilized food ingredients, such as tomato paste, fruit purees, and other liquid food concentrates that need to be purchased by food processors, or institutional end users for use as ingredients in further processed prepared foods. The containers for such applications can range in size from the classic 55-gallon steel drum to railroad tank cars or stationary silo storage tanks. Specially designed aseptic transfer valves and related handling systems make it possible to transfer sterile product from one such container to another without compromising sterility. A schematic flow diagram of a typical aseptic processing system for 55-gallon steel drums is shown in Figure 11.41, with a cutaway view of a typical drum-filling station in Figure 11.42. A system such as this is capable of filling 15 drums per hour, with each drum containing nearly 500 lb of product (Wagner, 1982; Lopez, 1987).

Furman canning's aseptic operation

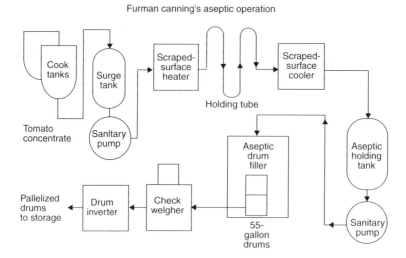

FIGURE 11.41   Aseptic filling of 55-gallon drums. (From Wagner, J.N. 1982. *Food Eng.* 54: 120–121. With permission.)

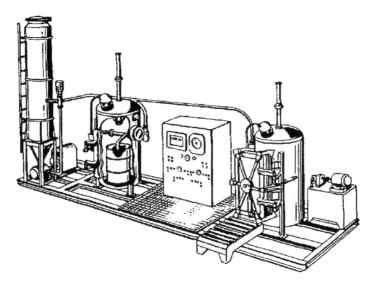

FIGURE 11.42   Aseptic filling system for 55-gallon drums. "No-Bac 55" Filler Drum filling and sterilizing retorts with sterile product holding surge tank. (Courtesy of Cherry-Burrell Corp.; from Lopez, A., 1987, *Basic Information on Canning*, 11th ed. Courtesy CTI Publications, Inc.)

11.9.2 UHT PROCESS CALCULATIONS

The heat exchangers used to provide the ultrahigh-temperature (UHT)/short time sterilization treatments for aseptic filling processes accomplish such rapid heating and cooling that for most practical design purposes, the process sterilizing value is assumed to be delivered during the product residence time in the holding tube or holding section between the heater and cooler. Since little temperature is lost through a well-insulated holding tube, an idealized process can be assumed; and the process time can be determined directly from the TDT curve, as in the example calculation given in Section 11.2.3. To account for any heat losses that may occur in the holding tube, it is always a good practice to use the temperature measured near the end of the holding tube, for this purpose. Once the process time has been determined in this way, the design objective of the food engineer is to determine the length of holding tube required to deliver this process time as residence time in the holding tube for any given product flow rate. An understanding of fluid mechanics and the flow behavior properties of Newtonian and non-Newtonian fluids is required for this purpose.

For a given flow rate and hold tube diameter, bulk average fluid velocity can easily be calculated to specify a hold tube length that will deliver an average residence time. However, depending on the governing flow regime (turbulent or laminar flow), not all liquid particles will travel through the tube at the average velocity. Some may travel much faster than the average velocity and escape through the holding tube with insufficient residence time, thus resulting in underprocessed product. Holding tube design based strictly on bulk average velocity is applicable only in those cases when turbulent flow is known to be fully established throughout the sterilizing length of the hold tube, and that documented conditions are met that are known to sustain turbulent flow adequately. Turbulent flow is characterized by a combination of physical product properties and flow conditions that result in a Reynolds number greater than 10,000. The Reynolds number is the ration of the product of the fluid velocity, density, and pipe diameter over the fluid viscosity. Thus high Reynolds numbers are achieved when thin-bodied fluids of relatively low viscosity are pumped through relatively large diameter pipes at relatively high flow rates. Although these conditions can sometimes be found with products such as fluid milk and single-strength fruit juices, they are usually the exception and not the rule with most liquid food products.

Holdsworth (1969) reported that most food products subject to continuous sterilization are high apparent viscosity, non-Newtonian fluids in laminar flow. Fluids in laminar flow exhibit a nonuniform velocity profile that is a function of the flow behavior index s. When $s = 1$, the fluid is Newtonian and exhibits a parabolic velocity profile, as shown in Figure 11.43, in which the maximum velocity at the center streamline is twice the bulk average velocity (Charm, 1971). For non-Newtonian fluids that are pseudoplastic (shear thinning), s is less than 1 and the velocity profile is less pronounced. For such fluids, which are typical of many foods, the maximum velocity at the center streamline is closer to the bulk average velocity. When s is greater than 1, the fluid is dilatant (shear thickening) and will exhibit an elongated velocity profile approaching a conical shape, in which the maximum velocity at the center streamline can reach as much as three times the bulk average velocity in the limiting case where s goes to infinity (Palmer and Jones, 1976). This type of flow behavior is rare in most food products (Heldman and Singh, 1975). Many researchers have studied the heat transfer during heating and cooling of non-Newtonian fluids with different flow behavior characteristics. Vocadlo and Charles (1973) characterized the laminar flow of viscoplastic fluids, while Palmer and Jones (1976), and Simpson and Williams (1974) studied the continuous sterilization of power-law fluids and Guariguata et al. (1979) did a similar analysis of Bingham plastic fluids. Rao and Anantheswaran (1982) reviewed and summarized much of this work.

In designing the holding tube for continuous sterilization processes, it is sufficient for process evaluation to consider only the lethal effect achieved while the product is resident in the holding tube. This simplifies the heat-transfer model by allowing the assumption of isothermal conditions

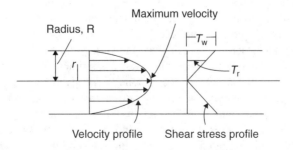

FIGURE 11.43 Velocity profile and shear stress distribution for a Newtonian liquid in laminar flow through a tube. (From Teixeira A.A. and Manson J.E., 1982, *Food Technol.* 36: 85.)

in an insulated holding tube. Because of laminar flow, the resulting parabolic velocity distribution imposes a residence-time distribution (i.e., all parts of the product do not remain in the holding tube for the same length of time). Normally, the process requirements are determined by calculating the fastest particle velocity (hence minimum holding time) from viscometric data. More often, however, processors are encouraged to make a worst-case assumption that their product behaves as a Newtonian fluid and choose a maximum velocity that is twice the bulk average velocity as a safety factor in complying with good manufacturing practice (Palmer and Jones, 1976). The lethal effects on those portions of the product having a residence time greater than the fastest particle are not considered and add to a further safety factor.

In this way, process evaluation is based on single-point lethality (fastest-moving particle), analogous to evaluating a retort sterilization process on the basis of lethality F_0 received at the can center. Charm (1966) first suggested the concept of integrating lethality across the holding tube to account for the different residence times of various streamlines from the centerline to the tube wall; he later revised the mathematical expression for this integral (Charm, 1971) on the basis of comments by Beverloo (1967). This concept is analogous to Stumbo's (1965) integrated lethality F_s across a can of conduction-heated food from can center to can wall.

11.9.3 Two-Phase Flow with Solid Particulates

Of growing interest in the food processing industry is the desire to apply aseptic processing to fluid foods that contain chunks of meat and vegetables, such as soups and stews. Thermal process calculations for hold tube design in these applications become quite challenging. During the residence time through the hold tube, heat must transfer from the carrier fluid (broth, sauce, or gravy) across the boundary layer interface between the fluid and particle surface and then penetrate the particle by conduction heat transfer to reach the cold spot at the particle center for sufficient time to achieve the necessary sterilizing value at the particle center.

Since no known methods are available to monitor the particle center temperature experimentally as it travels with the carrier fluid in an actual process, food engineers must resort to mathematical heat transfer models, like the models described in Section 11.6. In this case, the model is used to represent the solid particle instead of a cylindrical can, with the carrier fluid temperature serving in the role of retort temperature. However, unlike the case with condensing steam on the can wall, the boundary layer effect at the fluid-particle surface interface can be expected to cause significant impedance to heat transfer across this surface interface, characterized by a relatively low surface convective heat transfer coefficient. Mathematical heat transfer models can be modified to account for this surface heat transfer coefficient, as suggested by Manson and Cullen (1974) and demonstrated by Sastry (1986). Difficulty often arises, however, in choosing a realistic value for this coefficient. This coefficient is not a physical property but a parameter that represents mathematically the physical effect of a combined set of conditions. Among other things, these conditions include fluid viscosity and, most important, the relative velocity between the fluid and particle. A combination of high fluid viscosity (as in a sauce or gravy) with a low relative velocity produces a worse-case situation in which the surface convective heat transfer coefficient takes on a relatively low value reflecting a low rate of heat transfer across the particle surface.

Because of the complex factors that influence the surface heat transfer coefficient, it is common engineering practice to determine these coefficients experimentally under laboratory conditions that closely simulate the actual process. This is done by conducting a heat penetration test on a "particle" of known regular geometry and known thermal properties when immersed in the carrier fluid under conditions of known and controlled relative velocity. The problem that becomes evident to food engineers in this situation is the difficulty in determining a realistic relative velocity for such experiments. Although it is often tempting to choose to relate this relative velocity in some way to the fluid carrier flow rate, this approach is not at all advised. Regardless of fluid flow rate and any mixing action induced by swept surface heat exchangers, it is important to remember that the solid particles are always being carried by the fluid, much like a leaf or twig floating down-stream. Thus, with the exception of momentary inertia effects, it is more likely that true relative velocities between carrier fluid and suspended particles are quite low. Therefore, until new methods are developed that may demonstrate otherwise, it would be most prudent to assume little or no relative velocity in determining realistic values for the surface convective heat transfer coefficient. Under these considerations, a safe value for this coefficient can be obtained from heat penetration tests in which the instrumented particle is immersed in a still bath of heated carrier fluid.

11.10 LOW-ACID CANNED FOOD REGULATIONS

Food engineers involved with thermal processing operations should be familiar with all federal regulations applicable to sterilization of low-acid canned foods. The specific provisions for regulating the low-acid canned food industry are contained in Title 2, Part 113 of the U.S. Code of Federal Regulations entitled "Thermally Processed Low-Acid Foods Packaged in Hermetically Sealed Containers." These regulations are also published in detail in *The Almanac of the Canning, Freezing, Preserving Industries* (Judge, 1986). The purpose of this concluding section is to acquaint the food engineer with the scope of compliance activities required to initiate and sustain commercial food canning operations under these regulations. In the broadest sense, the regulation directs the attention of low-acid canned food processors to four operational levels:

1. Adequacy of equipment and procedures to perform safe processing operations
2. Adequacy of record keeping to prove safe operations
3. Justification of the adequacy of time-and-temperature processes used
4. Qualifications of supervisory staff responsible for thermal processing and container closure operations

The requirements of the regulation can be further broken down into 11 specific compliance activities described below.

1. *Plant registration.* This compliance activity requires that every plant producing low-acid canned foods and selling these foods in the United States be registered with the Food and Drug Administration (FDA). This is accomplished by the submission of necessary forms (FD 2541), which require such information as:

Name of company
Place of business
Location of plant
Processing method: type of equipment used
List of food products processed

Although most processors normally provide this type of information regularly for their trade associations and for various business accounting purposes, technical and administrative personnel need to exercise care in such matters as choosing appropriate definitions for the type of equipment and processing method they use, and in defining each product for the list of products required. If a plant has to close for reasons other than seasonal operations or labor disputes, the regulation requires notification to the FDA within 90 days of closing.

2. *Process filing.* This compliance activity requires all processors to file Form 2541a for each product with the FDA within 60 days of plant registration and prior to packing any new product or adopting any change in process for an existing product. The type of information required on each form may include:

Name of product and container size
Processing method used and type of retort
Minimum initial product temperature (IT)
Time and temperature of processing
Sterilizing value of the process or equivalent scientific evidence of process adequacy
Critical factors affecting heat penetration
Authoritative source used and date of establishment of the process

One form containing all of this information is required for each product in each size container for all product-container size combinations processed in any given plant.

3. *Personnel training.* This compliance activity requires that supervisors of operators of retort processing systems and container closure inspectors must have attended a school approved by the FDA and have satisfactorily completed the prescribed course of instruction. These "Better Process Control and Container Closure" schools are sponsored jointly on a regular basis by the FDA and the National Food Processors Association (NFPA). They are held in conjunction with the food science departments at a number of colleges and universities across the country to bring them within reasonable proximity to most canned food processors. The curriculum is presented in a short-course format over 4-1/2 days, including examinations of the material presented and the awarding of certificates of completion.

4. *Equipment and procedures.* This compliance activity requires all processors to make certain that equipment related to the thermal processing operations is maintained in compliance with established specifications. For still cook retort operations, these requirements relate to such items as:

Mercury and glass thermometers
Temperature recorders or recorder-controllers
Steam pressure controllers and gages
Steam pressure controllers and gages
Steam inlet size, headers, and location in retort
Steam spreaders and bleeders
Crates (baskets), crate supports, and separators
Vents, size and location, venting times and temperature
Water-level indicators
Level indication for retort headspace in pressure cooking
Air supply to pneumatic controllers

5. *Product preparation.* This compliance activity requires each processor to have documented policies and procedures for this product preparation, production, and sanitation, delineating proper procedures to be followed in such areas as:

Raw material testing and certification, including proper storage and inventory control
Blanching and cooling operations
Filling operations, including frequent monitoring of critical factors such as initial product temperature, fill weight, headspace, product density, viscosity, and pH
Exhausting of headspace air prior to closing by heat, vacuum, steam injection, hot brine, and so on
All areas of plant, equipment, and materials-handling sanitation

This compliance activity forces all processors to review thoroughly existing quality control and sanitation policies and procedures or develop appropriate policies and procedures where none had previously existed.

6. *Establishing scheduled processes.* This compliance activity requires all processors to document and file the following information in support of establishing the scheduled process for any new product or product-process change that is to be filed with the FDA:

Source of qualified expert knowledge used in establishing the scheduled process

Heat penetration tests, microbial death time data, and thermal process calculations used to establish the scheduled process

Specification of all critical control factors affecting the scheduled process

Verification of the scheduled process through inoculated packs or incubation of product samples from initial production runs

7. *Thermal process operations.* This compliance activity specifies minimum requirements that processors have to meet with respect to operations that take place in the retort room or cook room, where the filled and sealed cans or jars are sterilized under pressure in steam or water-air override retorts. Some of these requirements include:

Posting of scheduled processes

Use of heat-sensitive indicators

Review of data on all critical control factors to make certain that they fall within specifications for the scheduled process prior to sterilization

Calibration of thermometers, recorders, controllers, and timing devices

Control of retort operations to assure compliance with specified venting procedures and time-temperature conditions for the established thermal process

Use of a fail-safe traffic control pattern to make certain that no unprocessed product can be mistaken for processed product, or vice versa

8. *Process deviations.* This compliance activity specifies what processors have to do in the event of a process deviation, such as drop in temperature caused by a sudden loss of steam pressure or a reduced cook time caused by a faulty timer, which would suggest that the product received a process less than the scheduled process. In the event of such a process deviation, the regulation specifies that the processor must either (a) reprocess the product according to the established scheduled process and retain all records of such event; or (b) put the product on hold and have the deviate process evaluated for its public health significance by a recognized processing authority with qualified expert knowledge. Such a processing authority may "clear" the deviation if it is judged to pose no significant risk to public health. Again, records containing documentation in support of such an evaluation have to be retained on file.

Since most canned food products cannot tolerate a second exposure to the heat sterilization process without serious degradation in physical quality, processors generally prefer to put products on hold while process deviations are evaluated. Large processors clear deviations quickly with appropriate documentation. Other processors may rely on outside services provided by trade associations, can manufacturers, or consultants.

9. *Container closure and coding.* This compliance activity specified the inspection and testing required to assure that all containers are properly closed and coded prior to sterilization. Some of the activities specified include:

Visual inspection of can top seams (or glass jar closures), at a minimum frequency of once every 30 min, with documentation for records retention

Complete seam teardown with measurement of critical dimensions taken under optic magnification (or coldwater vacuum tests for glass jars) at a minimum frequency of once every 4 h, along with documentation for records retention

Periodic testing of cooling water to check concentration of residual chlorine

Proper code on each container for:

Identity of product
Where packed (plant)
When packed (date)
Who packed (shift or line)

Use of proper postprocessing can handling systems to minimize damage to can seams or closures prior to labeling and case packing.

10. *Records and storage.* This compliance activity requires all processors to prepare, review, and retain all records from each product packed for at least one full year or packing season at the processing plant itself, followed by retention of these records for at least 2 years at some other location, so that all records will be readily available for inspection over a minimum period of three full years from the date the product was packed. The records themselves include all documents and recordings of data, test results, inspections, critical control factors, and so on, required by all of the individual compliance activities described previously. This means that on a continuing basis, essentially all processors must have procedures in place at each plant to:

Review all records for completeness
Collate and arrange records in an organized file for each product batch code
Store the records in sequence with a systematic file system for future retrieval

11. *Recall planning.* This final compliance activity requires that all processors have on hand a plan for recalling any product through primary distribution, plus a plan for each distributor to use in recalling the product from further distribution channels downstream. Some processors have adopted the practice of conducting drills to test the effectiveness of their recall plans. Such drills are not specifically required by the regulation but are strongly advised as part of this compliance activity.

NOMENCLATURE

A	Area through which heat transfer occurs (m^2)
a	Initial number of viable bacterial spores at the beginning of a thermal process
B	Process time for the Ball formula method of thermal process calculations (min)
b	Final number of viable bacterial spores (survivors) at the end of a thermal process
C	Concentration of primary component in a first-order reaction, quantity per unit mass or volume (e.g., spores/mL)
C_O	Initial concentration of primary component at beginning of reaction (spores/mL)
C_p	Specific heat or heat capacity (kJ/kg·°C)
D	Decimal reduction time, time for one log cycle reduction in population during exposure to constant lethal temperature (min)
D_r	Decimal reduction time at a specified reference temperature (T_r) (min)
D_{250}	Decimal reduction time at temperature of 212°F (121°C) (min)
D_{212}	Decimal reduction time at temperature of 212°F (100°C) (min)
D_{150}	Decimal reduction time at temperature of 150°F (65°C) (min)
E_a	Activation energy (kJ/kg)
e	Base of natural (Naperian) logarithms
F	Process lethality (min) at any specified temperature and z-value, applied to destruction of microorganisms

F_o Lethality value applied to destruction of microorganisms with z-value of 10°C (18°F), min at 250°F

f_h Heat penetration factor, time for straight-line portion of semilog heat penetration curve to traverse one log cycle (min)

f_c Cooling penetration factor, time for straight-line portion of semilog cooling curve to traverse one log cycle (min)

g Difference between retort temperature and final temperature reached by any point within a food container at the end of heating

g_c Difference between retort temperature and final temperature reached at the geometric can center at the end of heating

H Half-height of cylindrical food can (m)

h Any distance along vertical dimension (height) from the midplane in a cylindrical can (m)

I Difference between retort temperature and initial product temperature at beginning of heating or cooling (°C)

I_h Same as above but specific for heating phase of process (°C)

j_c Lag factor at geometric center of food container (dimensionless)

j_{ch} Heating lag factor at geometric center of food container (dimensionless)

j_{cc} Cooling lag factor at geometric center of food container (dimensionless)

k Thermal conductivity (W/m·k), with reference to heat transfer; also, rate constant in kinetic reactions (1/time)

k_0 Reference rate constant at reference temperature (T_r) (1/time)

L Length or thickness (m)

R Universal gas constant (kJ/kg·K); also can radius (m)

r Any distance along can radius from centerline (m)

T Temperature (°C)

T_c Temperature of cooling medium (°C)

T_I Initial product temperature (°C)

T_o Pseudo-initial temperature, the temperature at which an extension of the straight line portion of the heat penetration curve intersects the ordinate axis (°C)

T_R Retort temperature (°C)

T_r Reference temperature at which D_r is measured; also, retort temperature (°C)

T_{ih} Initial product temperature at beginning of heating (°C)

T_{pih} Pseudo-initial product temperature at beginning of heating (see also T_o) (°C)

T_{ic} Initial product temperature at beginning of cooling (°C)

T_{pic} Pseudo-initial product temperature at beginning of cooling, that is, temperature at which extension of straight-line portion of semilog cooling curve intersects ordinate axis (°C)

$T_{(ij)}$ Temperature at any grid node (i,j) in finite difference solution to heat transfer equation (°C)

$T_{(ij)}^{(t)}$ Temperature at any grid node (i,j) at time (t) (°C)

$T_{(ij)}^{(t+\Delta t)}$ Temperature at any grid node (i,j) at time $(t + \Delta t)$, or one time interval (Δt) later °C

T_w Cooling water temperature (°C)

t Time (min)

U Time required at retort temperature to accomplish the same lethality that is required of the process at the reference temperature used to determine the D value, (min)

x Spatial location within a food container (m)

Z Temperature-dependency factor in thermal inactivation kinetics, temperature difference required for ten-fold change in decimal reduction time (D value) (°C)

α Thermal diffusivity (m²/sec)

ΔF_i Incremental lethality accomplished over time interval (Δt) (min)
Δh Vertical height of incremental volume element ring in finite difference solution to heat transfer equation (m)
Δr Radial width of incremental volume element ring in finite difference solution to heat transfer equation (m)
Δt Time interval between computational iterations in finite difference solution to heat transfer equation (min)
ρ Product density (kg/m^3)

REFERENCES

Ball, C.O. 1923. Thermal Process Time for Canned Foods. *Bull. Natl. Res. Council*, 7, Part 1, No. 37, 76 pp.

Ball, C.O. and Olson, F.C.W. 1957. *Sterilization in Food Technology*. McGraw-Hill, New York.

Beverloo, W.A. 1967. Survival of microorganisms in continuous HTST processes: an error and additional observations. *Food Technol.* 21: 964.

Bigelow, W.D., Bohart, G.S., Richardson, A.C., and Ball, C.O. 1920. *Heat Penetration in Processing of Canned Foods*. Natl. Canners Assoc. Bull. 16L.

Charm, S.E. 1966. On the margin of safety in canned foods. *Food Technol.* 20: 97.

Charm, S.E. 1971. *Fundamentals of Food Engineering*, 2nd ed., AVI, Westport, Conn.

Datta, A.K. and Teixeira, A.A. 1987. Numerical modeling of natural convection heating in canned foods. *Trans. ASAE.* 30: 1542–1551.

Datta, A.K. and Teixeira, A.A. 1988. Numerically predicted transient temperature and velocity profiles during natural convection heating of canned liquid foods. *J. Food Sci.* 53: 191–196.

Datta, A.K., Teixeira, A.A., and Manson, J.E. 1986. Computer-based retort control logic for on-line correction of process deviations. *J. Food Sci.* 51: 480–483, 507.

Graves, R. 1987. Determination of the sterilizing value (FO) requirement for low acid canned foods. Presentation to *Inst. Thermal Process Spec. Annu. Conf.* November 4–6, Washington, DC.

Guariguata, C., Barreiro, J.A., and Guariguata, G. 1979. Analysis of continuous sterilization processes for Bingham plastic fluids in laminar flow. *J. Food Sci.* 44: 905.

Heldman, D.R. and Singh, R.P. 1975. *Food Process Engineering*. AVI, Westport, CT.

Holdsworth, S.D. 1969. Processing on non-Newtonian foods. *Process Biochem.* 4: 15.

Joslyn, M.A. and Heid, J.L. 1963. *Food Processing Operations*, Vol. II. AVI, Westport, CT.

Judge, E.E. 1986. *The Almanac of the Canning, Freezing, Preserving Industries*. Edward E. Judge & Sons, Westminster, MD.

Lopez, A. 1987. A complete course in canning, Book 1, *Basic Information on Canning*, 11th ed., The Canning Trade, Baltimore, MD.

Manson, J.E. and Cullen, J.F. 1974. Thermal process simulation for aseptic processing of foods containing discrete particulate matter. *J. Food Sci.* 39: 1084.

NFPA. 1980. *Laboratory Manual for Food Canners and Processors*, Vol. 1. AVI, Westport, CT.

Palmer, J.A. and Jones, V.A. 1976. Prediction of holding times for continuous thermal processing of power-law fluids. *J. Food Sci.* 41: 1233.

Rao, M.A. and Anantheswaran, R.C. 1982. Rheology of fluids in food processing. *Food Technol.* 36: 116.

Rice, J. 1987. International trends in food packaging. *Food Process.* 48: 86–91.

Sastry, S.K. 1986. Mathematical evaluation of process schedules for aseptic processing of low-acid foods containing discrete particles. *J. Food Sci.* 51: 1323–1328.

Simpson, G.S. and Williams, M.C. 1974. Analysis of high temperature/short time sterilization during laminar flow. *J. Food Sci.* 39: 1047.

Stumbo, C.R. 1965. *Thermobacteriology in Food Processing*. Academic Press, New York.

Teixeira, A.A. and Manson, J.E. 1982. Computer control of batch retort operations with on-line correction of process deviations. *Food Technol.* 36: 85.

Teixeira, A.A., Dixon, J.R., Zahradnik, J.W., and Zinsmeister, G.E. 1969. Computer optimization of nutrient retention in thermal processing of conduction-heated foods. *Food Technol.* 23: 137.

Teixeira, A.A., Zinsmeister, G.E., and Zahradnik, J.W. 1975. Computer simulation of variable retort control and container geometry as a possible means of improving thiamine retention in thermally processed foods. *J. Food Sci.* 40: 656.

Teixeira, A.A., Balaban, M.O., Germer, S.P.M., Sadahira, M.S., Teixeira-Neto, R.O., and Vitali. 1999. Heat Transfer Model Performance in Simulation of Process Deviations. *J. Food Sci.* 64: 488–493.

Toledo, R.T. 1980. *Fundamentals of Food Process Engineering*. AVI, Westport, CT.

Vocadlo, J.J. and Charles, M.E. 1973. Characterization and laminar flow of fluid-like viscoplastic substances. *Can. J. Chem. Eng.* 51: 116.

Wagner, J.N. 1982. Aseptic drum processing. *Food Eng.* 54: 120–121.

12 Extrusion Processes

Leon Levine and Robert C. Miller

CONTENTS

12.1 INTRODUCTION

Extrusion processing of foods has been practiced extensively for many years. The earliest examples were the forming of pasta products (macaroni, spaghetti, etc.) and production of pellets for conversion into ready-to-eat cereals in subsequent processing. Both of these processes are simple operations that result in little or no significant structural changes to the molecular components of the material being processed. And both employed the simplest type of extruder: a single screw machine operating at low temperature, nearly isothermal conditions.

Today the application of extrusion processes and the types of extruders are considerably more varied and complex. Both single-screw and twin-screw extruders are used for commercial production of a wide variety of food products, ranging from snack half-products, textured vegetable protein, animal feed (including pet foods), expanded ready-to-eat cereals, and flat breads. In general, these processes are carried out under conditions where the extruder introduces significant quantities of energy into the extrudate. This results in many physical and chemical changes in the product, including expansion (puffing), starch gelatinization, and protein denaturation, Modern extruders, particularly those of the corotating, twin-screw type, have such flexible physical configurations that the range of characteristics of the products produced in them seems limited only by the skill and imagination of the product scientists and engineers working with them.

Our objective is not to provide the reader with information how to make particular products with extruders. The capabilities are so wide, and the potential raw material supplies so large as to make the discussion of the chemistry and techniques used for control of product attributes impossible to discuss meaningfully in a handbook. Those who are interested in these subjects should explore the literature (Van Zuileichem et al., 1975; Ibave and Harper, 1982; Molay and Harper, 1982; Mueser et al., 1982, 1984a, 1987; Meuser and Van Lengerich, 1984a, 1984b; Miller, 1985; Fletcher et al., 1985; Megard et al., 1985; Bhattacharya and Hanna, 1986; Hagan and Villota, 1986; Yacu, 1987b, 1987c; Chinnaswamy and Hanna, 1988; Falcone and Phillips, 1988; Linko, 1989; Naguchi 1989; Stanley, 1989; Eerikainen and Linko, 1989; Meuser and Weidmann, 1989; Colonna et al., 1989). Our intent is to provide the food engineer with a quantitative understanding of the performance of various extrusion devices. The reader will find that even this limited discussion of the extrusion process is a formidable subject.

12.2 RHEOLOGY OF EXTRUDATES

Before embarking on a discussion of the performances of the various extruder types, it is essential that the reader have some understanding of the rheological properties of food extrudates. The analyses of extruder behavior rest on the foundation of extrudate rheology. For those not familiar with this subject, it is suggested that they review Chapter 1 of this handbook.

The literature (Cervone and Harper, 1978; Remsen and Clark, 1978; Jao et al., 1978; Morgan et al., 1978; Harper, 1979, 1981; Levine, 1982; Baird and Reed, 1989) indicates that food extrudates are, as would be expected, highly non-Newtonian, exhibiting shear-thinning behavior, where the apparent viscosity of a material decreases with increasing shear. This is commonly described with a power law (Ostwald-de Waele) model:

$$\tau = m\dot{\gamma}^n \tag{12.1}$$

For a shear-thinning material, the flow index, n, is less than unity.

In addition to viscosity being a strong function of the shear environment, the literature also indicates that formulation (primarily water content) and temperature have important roles in the viscous behavior of the extrudates. To account for these effects, Equation 12.1 is usually modified:

$$\tau = m_0 \, e^{A/T} \, e^{BM_{DB}} \dot{\gamma}^n \tag{12.2}$$

Table 12.1 summarizes some published values for the constants appearing in Equation 12.1 and Equation 12.2. These values should be used only for preliminary estimation of process performance because the actual rheological properties may change greatly with only minor changes in composition. These changes may be the result of agricultural variety and growing conditions. An additional difficulty with using these values, as the literature (Remsen and Clark, 1978; Morgan et al., 1978; Baird and Reed, 1989) points out, is that the rheological properties are functions of the path used to create the extrudates, possibly resulting in different physical and chemical states even when their final compositions and temperatures are identical. This is discussed in some detail in the literature.

Some authors (Morgan et al., 1918; Remsen and Clark, 1978) have attempted to include time–temperature history into the viscous model by assuming that the flow consistency is controlled

TABLE 12.1
Reported Power Law Models (Equation 12.2) for Food Extrudates

Material	m_0	n	Temperature range (°C)	Moisture range (%)	A (K)	B $(1/\%M_{DB})$	Reference
Cooked cereal dough (80% corn grits, 20% oat flour)	78.5	0.51	67–100	25–30	2500	−7.9[a]	Harper et al., 1971
Pregelatinized corn flour	36.0	0.36	90–150	22–35	4390	−14	Cervone and Harper, 1978
Soy grits	0.79	0.34	35–60	32	3670	—	Remson and Clark, 1978
Hard wheat dough	1,885	0.41	35–52	27.5–32.5	1800	−6.8	Levine, 1982
Corn grits	28,000	~0.5	177	13	—	—	Van Zuilichem et al., 1974
	17,000	~0.5	193	13	—	—	
	7,600	~0.5	207	13	—	—	
Full-fat soybeans	3,440	0.3	120	15–30	—	—	Fricke et al., 1977
Moist food products	223	0.78	95	35	—	—	Tsao et al., 1978
Pregelatinized corn flour	17,200	0.34	88	32	—	—	Hermann and Harper, 1974
Sausage emulsion	430	0.21	15	63	—	—	Toledo et al., 1977
Semolina flour	20,000	0.5	45	30	—	—	Nazarov et al., 1971
Defatted soy	110,600	0.05	100	25	—	—	Jao et al., 1978
	15,900	0.40	130	25	—	—	
	671	0.75	160	25	—	—	
	78,400	0.13	100	28	—	—	
	23,100	0.34	130	28	—	—	
	299	0.65	160	28	—	—	
	28,800	0.19	100	35	—	—	
	28,600	0.18	130	35	—	—	
	17,800	0.16	160	35	—	—	
Wheat flour	4,450	0.35	33	43	—	—	Launay and Bure, 1973
Defatted soy flour	1,210	0.49	54	25	—	—	Luxenburg et al., 1985
	868	0.45	54	50	—	—	
	700	0.43	54	75	—	—	
	1,580	0.37	54	85	—	—	
	2,360	0.31	54	100	—	—	
	2,270	0.31	54	110	—	—	

[a] Wet basis moisture used.

by a first-order equation. They suggest correlation of the data in the following form:

$$m = m_0 \dot{\gamma}^n e^{A/T} e \int_0^t K_\infty e^{E/RT} \, dt \qquad (12.3)$$

It is not clear how universally valid this model is. Some of the reported cooking curves for starch and proteinaceous materials (Baird and Reed, 1989) illustrate very complicated changes in rheology during cooking, suggesting that the first-order model cannot be valid universally.

From this discussion one can conclude that the prediction of rheology is very uncertain, suggesting that determination of rheological models via capillary rheometry or some other suitable method is necessary for virtually every new formulation. One may (Levine, 1982) take a different approach by using viscosity measurements of the in-process material. This is done by using the extruder die as an in-process rheometer. The technique provides "quick-and-dirty" determinations of rheology but is not rigorous in that complications such as die inlet and outlet effects are ignored. A more rigorous technique has recently been described: it uses a slot rheometer attached to the extruder in such a way as to draw out in-process material. Unfortunately, it is a very expensive procedure, and incorporation into commercial or even pilot-plant extruders may not always be feasible.

12.3 NEWTONIAN MODELS OF SINGLE-SCREW EXTRUDER PERFORMANCE

A quantitative model for the performance of the melt zone (the portion of the screw where the extrudate flows like a very viscous fluid) of a single-screw extruder has existed since the middle of the century (Carley and McKelvey, 1953; Carley and Strub 1953a, 1953b; Carley et al., 1953; McKelvey, 1953; Mallouk and McKelvey, 1953; Jepson, 1953). Like many food unit operations, the earliest studies of this type are found in the plastics and rubber literature. It is important to understand the development of this model since the analysis of all other extruder designs and problems originate at least in part from it. Detailed analyses of the flow may be found in many fine texts (Schenkel, 1966; Tadmor and Klein, 1970; Middleman, 1977; Harper, 1979, 1981; Tadmor and Gogos, 1979; Stevens, 1985; Rauwendaal, 1986). The reader interested in details of the various derivations should consult these sources. We present a summary of the analyses.

All of the derivations begin with an assumption that the extrudate is Newtonian (a gross simplification) and that there is no slip at the boundaries (the screw surface and barrel wall). They continue with another assumption that the channel formed by the screw root and flights, and the barrel wall is shallow and that the gap between the tips of the screw flights and the barrel wall are negligible (no flow occurs over the flights). These assumptions are often very good approximations. The assumption of a shallow channel allows the curved shape of the screw to be replaced by a flat, more easily analyzed geometry, illustrated in Figure 12.1.

This is still a complex situation, requiring solution of the Navier–Stokes equation in two dimensions. A further simplification, assuming the pitch of the screw, like the diameter, is much larger than the channel depth reduces the analysis to a one-dimensional problem. After integration of the resulting differential equation, extruder output becomes:

$$Q = \frac{V_z WH}{2} + \frac{WH^3}{12\mu} \frac{\partial P}{\partial Z} \qquad (12.4)$$

Equation 12.4 is composed of two distinct groups of terms that represent mechanisms of material transport. The first group is a function of the geometry and speed of the screw. It represents the drag of material through the screw channel and is given the name drag flow. Physically, it is the maximum

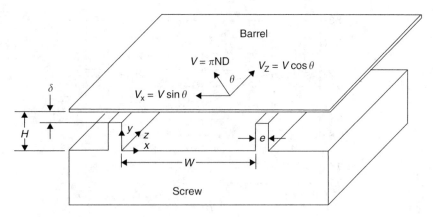

FIGURE 12.1 Geometry of a screw channel.

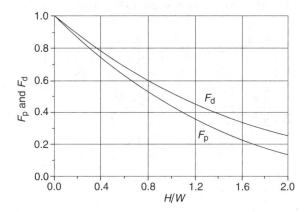

FIGURE 12.2 Drag and pressure flow corrections for channel depth. (Adapted from Tadmor, Z. and Klein, I. 1970. *Engineering Principles of Plasticating Extrusion.* Van Nostrand Reinhold, New York.)

theoretical output of the screw: that which would be expected if a die were not present at the discharge of the screw and the screw were fed in such a way as to keep up with this flow (no easy task!). The second group is a function of the discharge pressure, screw geometry, and viscosity of the extrudates and represents an *imaginary* flow that opposes the drag flow. The reader may recognize that this term is identical to the pressure-driven flow through a slot. In terms of these two flows, Equation 12.4 may be written as

$$Q = Q_d + Q_p \tag{12.5}$$

To account for the simplifying assumptions, Equation 12.4 and Equation 12.5 are modified in several ways. First, corrections must be applied to incorporate the fact that the depth of channel/width of channel ratio is not negligible. This is accomplished by the application of correction factors to the drag and pressure flow terms:

$$Q = \frac{V_z WH}{2} F_d + \frac{WH^3}{12\mu} \frac{\partial P}{\partial z} F_p \tag{12.6}$$

These corrections are derived by the solution of the flow problem in two dimensions rather than one. Figure 12.2 illustrates the corrections as a function of channel depth/width ratio. Note that for most screw designs, the factors are close to unity.

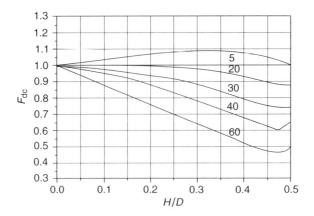

FIGURE 12.3 Drag flow correction for curvature. (Adapted from Tadmor, Z. and Klein, I. 1970. *Engineering Principles of Plasticating Extrusion.* Van Nostrand Reinhold, New York.)

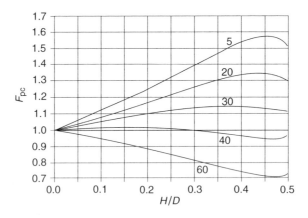

FIGURE 12.4 Pressure flow correction for curvature. (Adapted from Tadmor, Z. and Klein, I. 1970. *Engineering Principles of Plasticating Extrusion.* Van Nostrand Reinhold, New York.)

The second corrections for Equation 12.4 and Equation 12.5 account for the fact that the channel depth/screw diameter ratio is not negligible. As a consequence, the transformation of the problem from curved to flat geometry may not be precisely correct. Again, correction factors are applied to the pressure and drag flow terms in Equation 12.4 to Equation 12.6. Equation 12.6 now becomes

$$Q = \frac{V_z WH}{2} F_d F_{dc} + \frac{WH^3}{12\mu} \frac{\partial P}{\partial z} F_p F_{pc} \qquad (12.7)$$

Figure 12.3 and Figure 12.4 illustrate the effect of the channel depth/screw diameter ratio on the correction factors. Once again the factors are nearly unity.

One last, often neglected correction must be applied: the model assumes that the channel is lengthwise rectangular rather than trapezoidal. As a result, end corrections are required. Equation 12.6 now becomes

$$Q = \frac{V_z WH}{2} F_c F_{dc} F_{de} + \frac{WH^3}{12\mu} \frac{\partial P}{\partial z} F_p F_{pc} F_{pe} \qquad (12.8)$$

Figure 12.5 and Figure 12.6 illustrate the end corrections for pressure and drag flow.

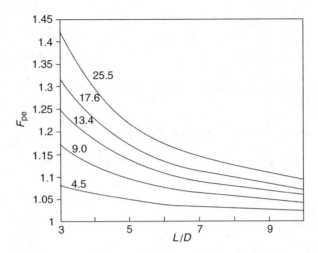

FIGURE 12.5 Pressure flow end correction (helix angle as a parameter). (Adapted from Tadmor, Z. and Klein, I. 1970. *Engineering Principles of Plasticating Extrusion.* Van Nostrand Reinhold, New York.)

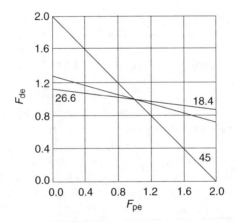

FIGURE 12.6 Drag flow end correction factor (helix angle as a parameter). (Adapted from Tadmor, Z. and Klein, I. 1970. *Engineering Principles of Plasticating Extrusion.* Van Nostrand Reinhold, New York.)

12.4 NON-NEWTONIAN MODELS OF SINGLE-SCREW EXTRUDER PERFORMANCE

All of the above equations assume that the extrudate exhibits Newtonian behavior. Since most food extrudates are highly pseudoplastic in nature, this is not valid. There are several approaches to dealing with the non-Newtonian (power law) behavior of real extrudates. Some sources (Harper, 1981; Martelli, 1983; Stevens, 1985) suggest that these equations may still be used after applying an approximation for non-Newtonian behavior. The technique assumes that Equation 12.4 is still valid if an apparent viscosity of the non-Newtonian material is used in place of the Newtonian viscosity. It is recommended that this be accomplished by approximating the shear rate within the extruder by

$$\dot{\gamma}_{app} \approx \frac{\pi ND}{H}$$

(12.9)

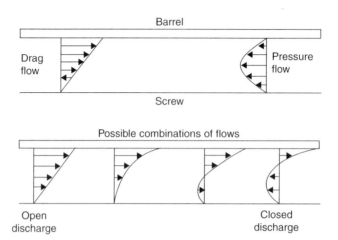

FIGURE 12.7   Possible velocity profiles within the screw channel.

Recognize that this is only an approximation (discussed further later in the chapter). An apparent Newtonian viscosity is then given by

$$\mu_{app} \approx m\dot{\gamma}_{app}^{n-1} \tag{12.10}$$

This viscosity is then used in Equation 12.4 to Equation 12.8 to determine the output of the extruder at any pressure.

The sources indicate that this method tends to predict higher extruder output than when using a rigorous calculation. There is a reasonable explanation for this difference. Consider Figure 12.7, which illustrates the in-channel fluid velocity profiles for drag flow, pressure flow, and the combined flows. It is clear that when pressure flow is significant, shear rates at the barrel wall and barrel surfaces will always be higher than for drag flow alone. The approximation used for shear in Equation 12.9 is an estimate of the shear rate due to drag flow alone. Since actual shear rate is always higher than this value, the apparent viscosity will always be lower than that indicated by Equation 12.10.

Inspection of Equation 12.4 reveals that at a fixed discharge pressure the lower extrudate viscosities result in greater pressure flows and hence reduced extruder outputs. We can infer that this particular approximation is most accurate when pressure flow is small compared to drag flow. This may be true for certain situations, such as for filled screws having a long length/diameter ratio or for screws operating at low discharge pressures.

The problem with the above approximation is a result of the nonlinear relationship between shear rate and viscosity, described in Equation 12.1. Introduction of this model into the fundamental differential equations renders them nonlinear. As a result, the solution is not that found in Equation 12.4. For rigor, the governing differential equations must be solved numerically (Middleman, 1977). The results are normally presented in dimensionless form as shown in Figure 12.8 (without shape correction factors). The dimensionless parameter on the horizontal axis of the figure is:

$$G_z = \frac{P_2 - P_1}{L} \frac{H^{n+1}\sin\theta}{m(\pi DN)^n} \frac{1}{\cos\theta} \tag{12.11}$$

Since the figure does not include shape correction factors, it cannot be used directly. Drag flow is estimated by reading the zero-pressure intercept (where $G_z = 0$) for the appropriate flow index from the graph. Pressure flow is found by reading the dimensionless net output from the graph and calculating pressure flow as the difference between the calculated drag flow and the dimensionless net output. Drag and pressure flows thus calculated may be used in Equation 12.5 along with flow

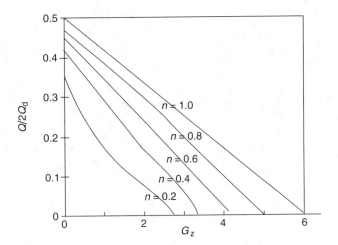

FIGURE 12.8 Dimensionless plot of extruder output for power law materials (flow index as a parameter). (Adapted from Griffith, R.M. 1962. *Ind. Eng. Chem. Fundam.* I: 180–181.)

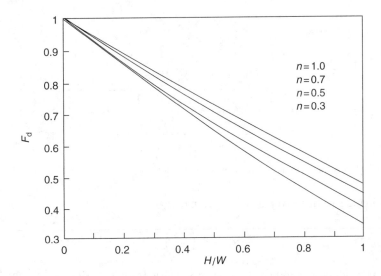

FIGURE 12.9 Drag flow correction factors for power law materials (flow index as a parameter). (Adapted from Tadmor, Z. and Klein, I. 1970. *Engineering Principles of Plasticating Extrusion.* Van Nostrand Reinhold, New York.)

correction factors. The flow correction factors for non-Newtonian fluids (Tadmor and Klein, 1970) are somewhat different from those presented in Figure 12.2. Figure 12.9 and Figure 12.10 provide the non-Newtonian corrections for the drag flow and pressure flow terms.

Rauwendaal (1986) gives a useful, easy to use linear approximation that accounts for non-Newtonian behavior and two dimensional flow. For screw pitch angles between 15 and 25 degrees, and flow indices between 0.2 and 1.0, the following approximation, with an error of less than 10% over the stated flow index and pitch range, for the extruder output is provided.

$$Q_{net} = \frac{4+n}{10} WHV_z - \frac{1}{1+2n} \frac{WH^3}{4\mu_a} \frac{dP}{dz} \qquad (12.12)$$

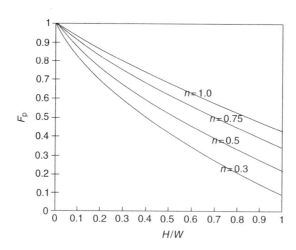

FIGURE 12.10   Pressure flow correction factors for power law materials (flow index as a parameter). (Adapted from Tadmor, Z. and Klein, I. 1970. *Engineering Principles of Plasticating Extrusion.* Van Nostrand Reinhold, New York.)

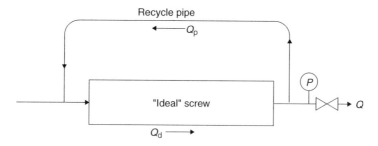

FIGURE 12.11   Highly simplified visualization of an extruder.

The viscosity in the equation given above is given by,

$$\mu_a = m \left(\frac{V_z}{H} \right)^{n-1} \tag{12.13}$$

A fourth, very simplified approach has been suggested (Levine, 1982; Levine and Rockwood, 1985). These papers suggest viewing the extruder as a perfect positive-displacement pump surrounded by a bypass, illustrated in Figure 12.11. Flow through this ideal extruder is always equal to the drag flow (at any discharge pressure). Since the net output of a real extruder is, as indicated in Equation 12.5, the difference between drag and pressure flows, flow through the recycle line must be equal to the pressure flow. Geometry in this system is simplified even further than that indicated in Figure 12.1: all geometric corrections and the pitch angle of the screw are neglected. Drag flow in the extruder is thus approximated by:

$$Q_d = \frac{\pi N D W H}{2} \tag{12.14}$$

The recycle line is a helically wound slot with a cross section indicated in Figure 12.12, and a length of:

$$L_{eq} = \frac{\pi D L}{W + e} \sqrt{1 + \left(\frac{W + e}{\pi D} \right)^2} \tag{12.15}$$

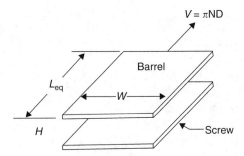

FIGURE 12.12 Geometry used for highly simplified extrusion model.

with an effective diameter of:

$$D_{eq} = 2R = 2\left[\frac{3n+1}{2\pi(2n+1)}\right]^{1/3+n}(WH^{2+n})^{1/3+n} \qquad (12.16)$$

Flow through the recycle line is now calculated with the friction factor-generalized Reynolds number relationship used for pipe flow:

$$f = \frac{16}{Re_f} \qquad (12.17)$$

where:

$$f = \frac{D_{eq}(P/4L_{eq})}{\rho v_r/2} \qquad (12.17a)$$

$$Re_f = \frac{D_{eq}^n v_r - n\rho}{(m/8)(6n + 2/n)^n} \qquad (12.17b)$$

For a specified output pressure the average recycle pipe velocity, v_r, may be calculated through the use of Equation 12.17. Pressure flow is calculated from the average velocity in the recycle pipe with:

$$Q_p = V_r HW \qquad (12.18)$$

There are numerous simplifying assumptions in this model. It has been pointed out that this solution is equivalent to solving the governing nonlinear differential equations through the use of superposition of the flow components, a method valid only for linear equations. It can be shown (Tadmor and Klein, 1970; Middleman, 1977) that superposition would be correct for the limiting cases of zero pressure flow or zero extruder output. The former is sometimes observed in practice. In fact, for this case this technique is identical to the simplified technique described at the beginning of this section.

Despite all of these simplifications, after slight modification to Equation 12.17, the technique appears to reasonably predict the behavior of real extrusion screws, operating at realistic operating conditions. As illustrated in Figure 12.13, Equation 12.17 should be modified to:

$$f \approx \frac{7}{Re_f} \qquad (12.19)$$

Figure 12.13 indicates that this model predicts extruder behavior over a wide range of operating conditions and extruder sizes.

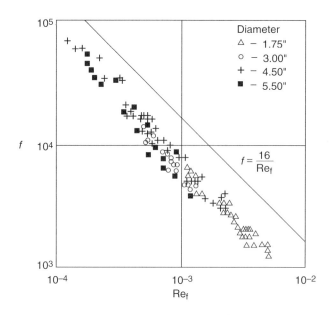

FIGURE 12.13 Dimensionless correlation for extruder output. (Adapted from Levine, L. 1982. *J. Food Process Eng.* 6: 1–13.)

12.5 SINGLE-SCREW EXTRUDER LEAKAGE FLOWS

The models that have been discussed thus far neglect extruder leakage flows. Similar to pressure flow, these flows tend to decrease extruder output. They are discussed in detail in the literature (Schenkel, 1966; Tadmor and Klein, 1970; Harper, 1979, 1981; Stevens, 1985; Rauwendaal, 1986). We shall attempt to summarize the findings of the literature.

Two kinds of leaks occur in single-screw extruders: flow over the flight tips in the gaps between them and the barrel surface; and flow in slots usually machined into the barrel wall to prevent slipping of the extrudate along the barrel surface or to encourage heat generation.

Leakage over the flights occurs because a machining tolerance must be built into the construction of the machine and, wear of the screw and barrel causes this gap to increase. Net extruder flow is not only reduced by a pressure flow over the flight but by a drag flow component as well — the shear imparted by the relative motion of the screw tip and barrel surface creates a drag flow analogous to that in the screw channel but in the opposite direction. Analysis is complicated because the temperature at the wall may be different from the bulk temperature of the extrudate. To correct Equation 12.8 for leakage flow, we can add correction factors for leakage. Equation 12.8 then becomes:

$$Q = p\frac{V_z WH}{2}F_d F_{dc} F_{de} F_{dl} + p\frac{WH^3}{12\mu}\frac{\partial P}{\partial z}F_p F_{pc} F_{pe} F_{pl} \tag{12.20}$$

The drag flow correction factor is given by

$$F_{dl} = 1 - \frac{\delta}{H} \tag{12.21}$$

The pressure flow correction factor is given by

$$P_{pl} = 1 + f \tag{12.22}$$

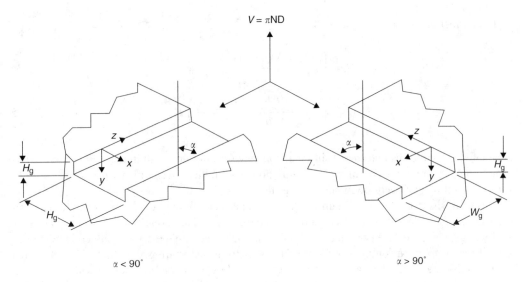

$$V = \pi ND$$

$\alpha < 90°$ $\alpha > 90°$

FIGURE 12.14   Geometry of barrel grooves.

where

$$f = G_3 \frac{\mu}{\mu_\delta} + G_4 \left[\left(\frac{G_5 \mu N}{P_1 - P_2} + G_6 \right) \bigg/ \left(1 + \frac{G_7 \mu_\delta}{\mu} \right) \right] \qquad (12.22a)$$

$$G_3 = \left(\frac{\delta}{H} \right)^3 \frac{e}{W} \qquad (12.22b)$$

$$G_4 = 1 + \frac{e}{W} \qquad (12.22c)$$

$$G_5 = -\frac{6 L \pi D (H - \delta)}{H^3 \tan \theta} \qquad (12.22d)$$

$$G_6 = \frac{1 + (e/W)}{\tan^2 \theta} \qquad (12.22e)$$

$$G_7 = \left(\frac{H}{\delta} \right)^3 \frac{e}{W} \qquad (12.22f)$$

Examination of the pressure flow correction term will reveal that the leakage flow increases with the cube of the clearance. This is very significant. A new screw may be very tight and exhibit very little leakage flow. But after only a small amount of screw or barrel wear, leakage flow can become significant. When that occurs, net output (or pumping efficiency) declines, increasing the specific mechanical energy imparted to the product, and raising its temperature.

Analysis of flow in barrel grooves is analogous to leakage over the flights. Flow in the grooves is the sum of a pressure flow and a drag flow induced by the motion of the screw land past the slot. Figure 12.14 describes the geometry of the situation. Equation 12.20 now becomes

$$Q = p \frac{V_z W H}{2} F_d F_{dc} F_{de} F_{dl} + p \frac{W H^3}{12 \mu} \left(\frac{\partial P}{\partial z} \right) F_p F_{pc} F_{pe} F_{pl} + Q_{gd} + Q_{gp} \qquad (12.23)$$

The correction factors are described by

$$Q_{gd} = j \left(\frac{\pi D N W_g H_g \cos \alpha}{2} \right) F_d \qquad (12.24)$$

$$Q_{gp} = j \left(\frac{W_g H_g^3 \sin \alpha}{12 \mu_g} \frac{P_1 - P_2}{L} \right) F_p \qquad (12.25)$$

Equation 12.24 indicates that there will be a tendency for drag flow within the grooves to increase extruder output. In reality, any increase in drag flow is more than overcome by pressure flow in the grooves, resulting in a decrease in net extruder flow.

This analysis assumes normal laminar flow with perfect adhesion of the extrudate to the screw and barrel surfaces. When shear stress at the boundaries exceed adhesive forces, slipping can occur, usually at the barrel surface (which has a smaller surface area than the screw, with correspondingly larger shear stresses). Net output can drop significantly from the resulting reduction in drag flow. This is why single-screw extruder barrels are usually grooved to provide more traction with the product. Twin-screw barrels are not normally grooved, because their asymmetrical geometry makes slipping less likely. But they can also exhibit slip under some conditions (Kalyon et al., 1999).

12.6 THE EXTRUDER DIE AND ITS INTERACTION WITH EXTRUDER BEHAVIOR

Before moving on to other extruder design considerations, we must look at the behavior of the extrusion die(s) and its interaction with extruder performance. No simple extruder the authors are aware of operates at free discharge — resistance to flow is an integral part of the extrusion operation (accomplished with relatively small die orifices, or internal resistances along complex screw profiles). Equations for flow through extruder dies may be found in almost any basic text on fluid mechanics. The following equations (Schenkel, 1966) may be found for the relationships between flow, viscosity, geometry, and pressure drop for three of the most common shapes: tube, slot, and narrow annulus. Discussion and estimation of the pressure flow relationships for more complex die designs may also be found (e.g., Wilkinson, 1960; Schenkel, 1966; Middleman, 1977; Bird et al., 1977; Michaeli, 1992).

The relationship between geometry, viscosity, pressure drop, and laminar flow through a tube (or circular cross-section die orifice) is given by:

$$Q = \frac{n \pi R^3}{3n + 1} \left(\frac{R \Delta P}{2Lm} \right)^{1/n} \qquad (12.26)$$

For flow through a narrow slot it is:

$$Q = \frac{n W h^2}{2 (2n + 1)} \left(\frac{h \Delta P}{2Lm} \right)^{1/n} \qquad (12.27)$$

And the relationship between geometry, viscosity, pressure drop, and flow through a thin annular orifice is:

$$Q = \frac{n \pi \bar{R} h^2}{2n + 1} \left(\frac{h \Delta P}{2Lm} \right)^{1/n} \qquad (12.28)$$

Examination of these or any other flow equations in the literature reveals that the general form for the relationship for flow through any die is in the form:

$$Q = k_p \Delta P^{1/n} \tag{12.29}$$

The equations above neglect such complications of entry and exit effects. And normally, there is more than one die hole at the screw discharge. Equation 12.29 still applies to multiple die applications, provided the flow rate used in the equation is the flow rate per hole.

Also, the die is not normally the only resistance encountered by the product at the screw discharge. A series of additional resistances are present in the form of pipe connections, kneading plates, screens, and transition pieces. Each of these may be described by the fundamental die equation, Equation 12.29. The flow resistance seen at the screw discharge is the sum of all these resistances plus the die resistance: resistances in series are additive. For multiple resistances Equation 12.28 becomes:

$$Q = \left[\frac{1}{\sum(1/K_{DL})^n} \right]^{1/n} \Delta P^{1/n} \tag{12.30}$$

This analysis of die resistance assumes normal laminar flow, in which the fluid velocity is zero at the die surfaces. In many cases, this is valid, but there are exceptions in food processing where the adhesive forces between the product and the constraining solid surfaces are insufficient to keep it from slipping. This is most noticeable in low-moisture or high-fat formulations, or for very smooth die surfaces, and is actually encouraged for making products such as pasta, where surface texture is an important quality parameter (Donnelly, 1982; Dintheer, 1993). In these cases, Teflon is commonly used to line the die surfaces to provide a very smooth, nonadhesive surface, especially in large high-velocity, high-pressure applications (Maldari and Maldari, 1993). In other cases, slip can be detrimental to good forming and textural development.

Incorporating a degree of slip at the die surfaces changes the flow equation, increasing the flow rate for a particular pressure, sometimes significantly (Lawal et al., 2000). In correcting for this change, an additional flow component is calculated by applying Navier's slip law (note: in some references, the reciprocal of the Navier's slip coefficient, β, is used):

$$V_s = \beta \tau_s \tag{12.31}$$

In terms of volumetric flow rate and pressure, this equation becomes:

$$Q_s = k_s \beta \Delta P \tag{12.32}$$

where the die constant for slipping flow is an expression of cross-sectional geometry (Miller, 1998). For circular dies, it is:

$$k_s = \frac{\pi R^3}{2L} \tag{12.32a}$$

For annular sections, it is:

$$k_s = \frac{2\pi \bar{R} h^2}{L} \tag{12.32b}$$

And for rectangular shapes, it is:

$$k_s = \frac{W^2 h^2}{2L\,(W + h)} \tag{12.32c}$$

The flow rates calculated with Equation 12.32 must be added to those for simple laminar flow in Equation 12.26 to Equation 12.28. Since there is a different relationship between cross-sectional dimensions and flow for slipping than for laminar flow, when slipping occurs, it is expected to be proportionally greater in smaller dies (Miller, 1998).

Another exception to the normal laminar flow model is encountered when solid flow occurs, as in the unusual case of flowing compacted powders. Here the pressure drop along the die path is not linear but exponential (Miller, 1998), so the die length dimension becomes much more critical — increasing it creates a disproportionally large increase in total pressure drop.

Equation 12.30 acts in unison with the equations governing the extruder output to define the extruder's operating state. Equation 12.8 can be rewritten as follows:

$$Q = k_f N + k_p \Delta P \tag{12.33}$$

Equation 12.29 and Equation 12.33 have been plotted in Figure 12.15 which describes the interaction between the screw and die designs for a Newtonian extrudate (similar, nonlinear curves could be drawn for non-Newtonian materials) Since the output of the screw equals the flow through the die, the operating output and pressure of the extruder are found at the intersection of the two curves, where the flows and pressures of each are identical. This type of analysis is used to understand various aspects of screw and die design. Figure 12.16 illustrates the operating points for two screws having different channel (thread) depths and two dies having different flow resistances. The operating points are indicated. For a die with low flow resistance, the screw with a deep channel can operate at higher outputs than the screw with a shallow channel. For a die with a high resistance, the converse is true. The implications of different channel depths are explored in more detail for twin-screw extruders in a later section. Figure 12.17 illustrates the effect of increasing screw speed when using a particular die. Figure 12.18 illustrates the effect of increasing the length of the screw. Note that the output of the screw increases rapidly with screw length, and that the sensitivity of screw output to pressure variations is reduced as screw length increases. This is the primary justification for designing screws with large length/diameter ratios. Of course, this improved stability and increased screw output is paid for in increased screw power consumption.

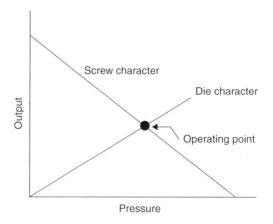

FIGURE 12.15   Determination of the screw's operating point.

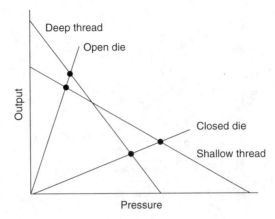

FIGURE 12.16   Effect of die design and channel depth on the operating point.

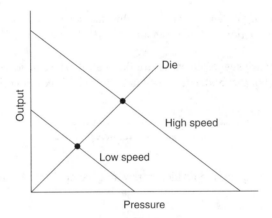

FIGURE 12.17   Effect of screw speed on the operating characteristic.

12.7   SCREW POWER CONSUMPTION

Aside from the requirement of specifying a drive motor, there are several reasons for being able to predict the power consumption of an extruder. Generally, a desired extrudate temperature is specified. Usually, most of the heat generated in an extruder is the direct result of viscous dissipation of motor power. In addition, some of the literature (Meuser et al., 1982, 1984a, 1984b, 1986, 1987; Meuser and Van Lengerich, 1984a, 1984b; Kuhle, 1986; Dreiblatt, 1987; Della Valle, 1989; Meuser and Weidmann, 1989) suggests that the specific mechanical energy input (watt hours/gram) by the screw to the extrudate is a critical determinant of extrudate properties.

For a Newtonian material, power consumption in a simple screw is readily estimated (Schenkel, 1966; Tadmor and Klein, 1970; Harper, 1981; Stevens, 1985) by:

$$p_t = \frac{(\pi ND)^2 L}{\sin \theta} \left[\mu \frac{W}{H} (\cos^2 \theta + 4 \sin^2 \theta) + \mu_\delta \frac{e}{\delta} \right] + \frac{\pi NDWH}{2} \Delta P \cos \theta \qquad (12.34)$$

For a non-Newtonian extrudates, problems in integrating the nonlinear differential equation again arise. One suggestion (Harper, 1981; Martelli, 1983; Stevens, 1985) follows the earlier approach

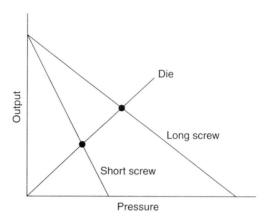

FIGURE 12.18   Effect of screw length on the operating characteristic.

used for prediction of output. The shear rate within the screw and the apparent Newtonian viscosity is estimated by Equation 12.9 and Equation 12.10. That estimated viscosity is then used in Equation 12.34.

The simplified approach suggested earlier for flow can also be considered (Levine, 1982; Levine and Rockwood, 1985). For the model described in Figure 12.11, power consumption of the screw is given by

$$p_t = p_s + Q_D \Delta P \tag{12.35}$$

The first term, viscous dissipation in an ideal screw, may be predicted from

$$N_p = \frac{1}{Re_s} \tag{12.36}$$

where

$$Re_s = \frac{\rho (DN)^{2-n} H^n}{m \pi^{2+n}} \tag{12.36a}$$

$$N_p = \frac{p_s}{\rho N^3 D^4 L} \tag{12.36b}$$

Note that once again all geometric corrections have been neglected. Nonetheless, Figure 12.19 illustrates that this model does a remarkably good job of predicting power consumption over a wide range of extruder operating conditions and geometries.

The qualitative effect of speed, screw design, and rheology on power dissipation is easy to extract from the models presented above. To a first approximation, viscous dissipation increases with rpm^{1+n} while the output increases approximately with rpm^n. So increasing output by increasing the screw speed will always result in increased specific mechanical energy. For geometrically similar screws operating at the same discharge pressure, power increases with the cube of screw diameter, as does the output of the screw. So increasing the capacity of the extruder by increasing its diameter, while proportionately increasing all other dimensions, does not affect the specific energy input if the discharge pressure and screw speed are held constant.

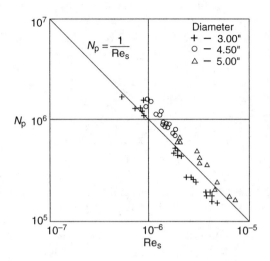

FIGURE 12.19 Dimensionless correlation for extruder power consumption. (Adapted from Levine, L. 1982. *J. Food Process Eng.* 6: 1–13.)

12.8 NONISOTHERMAL SCREW OPERATION

The introduction of nonisothermal operating conditions adds many complications to the models that have been described. The most analyzed situation is that of adiabatic extrusion of a Newtonian extrudate (Tadmor and Klein, 1970; Middleman, 1977). Even in this special case, assumptions must be made about the dependency of extrudate viscosity on temperature. Since data are often not available, and discussion of this situation is very convoluted, we will briefly discuss the consequences of adiabatic operation and refer the reader to the literature for a more detailed analysis.

Modeling of adiabatic operation can be summarized via a modification of Equation 12.8 (Tadmor and Klein, 1970; Middleman, 1977). Net flow in the extruder is still the difference between the drag and pressure flows in the channel, but the pressure flow is modified by a viscosity correction.

$$(Q_p)_{\text{adiabatic}} = f(Q_p)_{\text{isothermal}} \tag{12.37}$$

The correction factor is a function of the log mean viscosity,

$$f = \frac{1 - \mu_o/\mu_z}{\ln(\mu_z/\mu_o)} \tag{12.38}$$

Since viscosity of the extrudate decreases with increasing temperature, the output of an extruder operated adiabatically will always be less than that of the same extruder operating isothermally at the same discharge pressure. This difference is very small at low extruder output pressures and grows as the extruder's output approaches zero, as illustrated in Figure 12.20. It has been stated (Schenkel, 1966) that all real extruders operate between isothermal and adiabatic extremes, so we are able to find actual conditions within those limits. No satisfactory solutions exist for the non-Newtonian extrudates. It has been suggested (Middleman, 1977) that the same approach be applied but using a correction factor based on log mean flow consistency rather than log mean viscosity.

Another important aspect of nonisothermal performance is a temperature gradient in the product between the barrel and screw surfaces, introduced by heat transfer. Cored screws and barrels with heating/cooling jackets or electrical heaters are common.

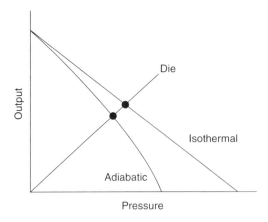

FIGURE 12.20   Comparison of isothermal and adiabatic operating curves.

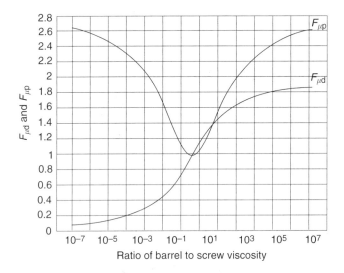

FIGURE 12.21   Drag and pressure flow correlations as a function of the barrel/screw viscosity ratio. (Adapted from Tadmor, Z. and Klein, I. 1970. *Engineering Principles of Plasticating Extrusion*. Van Nostrand Reinhold, New York.)

Analysis of this situation is quite complex. The literature (Tadmor and Klein, 1970; Harper, 1981) considers one highly simplified case that is enlightening: Assume that the extrudate is Newtonian, that its viscosity changes exponentially with temperature, and that a temperature profile is imposed on the region between the screw and barrel. The results are presented in Figure 12.21 as corrections to the drag flow and pressure

$$Q = Q_\mathrm{d}F_{\mu\mathbf{d}} + Q_\mathrm{p}F_{\mu\mathbf{p}} \tag{12.39}$$

Examination of Figure 12.21 reveals that as the viscosity at the barrel surface is increased relative to that at the screw surface, drag and pressure flows both increase. As long as the ratio is less than 10, the drag flow increases at a faster rate than the pressure flow, resulting in an increase in net output. Beyond a viscosity ratio of 10 the opposite is true, suggesting that careful attention to the barrel and screw temperatures can be used to maximize extruder output.

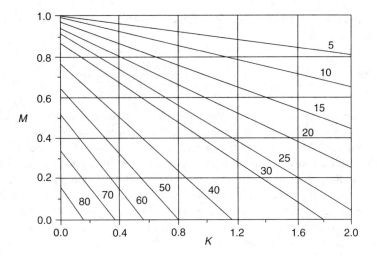

FIGURE 12.22 Figure for the estimation of solids conveying angle. (Adapted from Schenkel, O. 1966. *Plastics Extrusion Technology and Theory*. American Elsevier, New York.)

12.9 THE FEED ZONE

Feeding of the ingredients to the extruder can raise significant issues. The feed material is usually in a much different form than that leaving the die. In most cases, it is in the form of powder, pellets, or wetted balls of material with physical properties very different from those we find in the screw. Most obviously their bulk densities are less than that of the extrudate. This is why the extrusion screw normally has a deeper channel in the feed zone than at the discharge end. More importantly, the feed materials do not behave like fluids. They consist of discreet solid particles, having more (the preferred situation) or less free flowing properties. It is absolutely essential that the feed zone be properly designed. If not, the metering end of the screw will be starved, destabilizing the flow (Schenkel, 1966; Rauwendaal, 1986). Ideally, the feed zone of the extruder will be sized to meet any flow demand at the discharge end.

Detailed discussions of the flow of solids in screws may be found in the literature (Schenkel, 1966; Tadmor and Klein, 1970; Tadmor and Gogos, 1979; Harper, 1981; Rauwendaal, 1986). Unlike fluids, solids flow as a plug. Here, the most important physical property is not viscosity but the coefficients of friction against the various materials it contacts. In a key equation that may be extracted from the literature (assuming that operation of the feed zone is pressureless — usually true for the first few turns of the screw in the feed zone), the conveying angle of the solids is given by:

$$\cos\theta_\sigma = K\sin\theta_\sigma + \frac{f_s}{f_b}\sin\theta(K+\cot\theta)\left(1+\frac{2H}{W}\right) \tag{12.40}$$

where

$$K = \frac{\sin\theta + f_s\cos\theta}{\cos\theta - f_s\sin\theta} \tag{12.41}$$

Equation 12.36 and Equation 12.37 can be rewritten as

$$\cos\theta_\sigma = K\sin\theta_\sigma + M \tag{12.42}$$

Figure 12.22 provides the relationship between the conveying angle and the factors K and M.

Once the conveying angle is known, the volumetric conveying capacity of the screw is readily calculated from

$$Q_\sigma = \pi^2 NDH(D-H)\frac{\tan\theta_\sigma \tan\theta}{\tan\theta_\sigma + \tan\theta}\frac{W}{W+e} \qquad (12.43)$$

Note that, as expected, the deeper the screw, the greater its conveying capacity. The important physical factors are the coefficients of friction, f_b and f_s, between the feed material and the barrel wall and screw surfaces. The ideal situation would be for the screw to be frictionless and the barrel to have a high coefficient of friction. This is why screws are often polished and the barrel in the feed zone grooved. Since heating the feed material may result in changing its coefficient of friction, separate heating or cooling jackets may be used to manipulate surface temperature and to improve feed zone performance. Coefficients of friction tend to increase with temperature, so heating the barrel could have a positive effect on the performance of the feed zone, but would have a negative effect on the performance of the metering zone.

Since we would like to ensure that the feed zone provides sufficient material to the extruder, there is a temptation to cut very deep channels in the feed zone screw to ensure that this is always true. This may result in the feed zone overfeeding the extruder. To ensure stability, the literature (Schenkel, 1966; Levine and Rockwood, 1985; Levine et al., 1987a) suggests that this be avoided.

12.10   BEHAVIOR OF MORE COMPLEX SINGLE-SCREW DESIGNS

Most food extruders are not composed of a simple metering section (constant depth channel). In general, there is a deeper channel feed zone (for the reasons expressed above) followed by what is normally called a compression zone. This region is normally a constant pitch screw with linearly decreasing channel depth, as illustrated in Figure 12.23. The ratio of the channel depth at the feed end of the section to that at the discharge end of the screw is normally called the compression ratio. Typical values range from 2:1 to 5:1. Analysis of these screws for Newtonian flow have been reported (Schenkel, 1966; Tadmor and Klein, 1970; Harper, 1981). The output of this type of screw is given by

$$Q = \frac{WH_2 V_z}{2}\frac{2}{1+H_2/H_1} - \frac{WH_2^2}{12\mu}\frac{P_2 - P_1}{L}\frac{2}{H_2/H_1(1+H_2/H_1)} \qquad (12.44)$$

Note that Equation 12.44 has the same general form as that of all the other extruder equations. The first term represents drag flow and the second term is a pressure flow. Both terms are multiplied by correction factors for the compression ratio. The literature (Harper, 1981) suggests that the previously reported channel shape correction factors be used to correct Equation 12.40 for finite-sized channels by using the average channel depth for calculating the width/depth ratio.

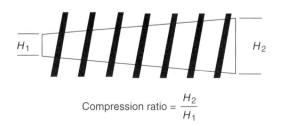

Compression ratio $= \dfrac{H_2}{H_1}$

FIGURE 12.23   Compression screw.

Note that the output of a compression screw will always be larger than for a straight screw having a channel depth equal to that of the discharge end of the compression screw and operating at the same discharge pressure. This is a consequence of the feed end of the compression screw having a higher capacity than the discharge end. As a result, a point of maximum pressure may exist within the compression zone. On the rising-pressure side of the maximum pressure, pressure flow is directed toward the feed end of the extruder, as in a normal screw. On the decreasing-pressure side of the peak, the pressure flow is directed toward the discharge end of the screw, resulting in an output higher than the drag flow for this section of the screw. Figure 12.24 illustrates the pressure and velocity profiles within the screw.

The literature on the behavior of compression screw for non-Newtonian extrudates is limited. One paper (Levine and Rockwood, 1985) compares the performance of three different compression screws: channel depth compression, screw diameter compression, and screw pitch compression. The three types of screws are illustrated in Figure 12.25. These screws were evaluated using the highly simplified approach suggested earlier. The results were presented as corrections to the dimensionless equations, Equation 12.17 and Equation 12.36, derived for a constant channel depth screw. These corrections were functions of the compression ratio and the dimensionless delivery of the screw (output expressed as a fraction of drag flow at the screw discharge). The tables presented are quite long. The reader is directed to the original paper for a complete list of these correction factors.

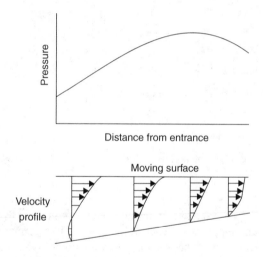

FIGURE 12.24   Pressure and velocity profiles in a compression screw.

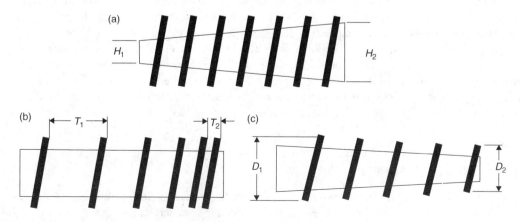

FIGURE 12.25   Three types of compression screws.

TABLE 12.2
Maximum Output of Depth, Diameter, and Pitch Compression Screws[a]

Flow index	Depth compression		Diameter compression		Pitch compression	
	Comp. ratio	Output	Comp. ratio	Output	Comp. ratio	Output
0.2	1.50	1.24	1.50	1.27	1.50	1.21
	2.00	1.47	2.00	1.56	2.00	1.38
	3.00	1.90	2.50	1.87	2.50	1.52
	4.00	2.36	3.00	2.17	3.00	1.65
0.4	1.50	1.23	1.50	1.26	1.50	1.21
	2.00	1.46	2.00	1.55	2.00	1.37
	3.00	1.87	2.50	1.85	2.50	1.51
	4.00	2.27	3.00	2.16	3.00	1.63
0.6	1.50	1.22	1.50	1.26	1.50	1.21
	2.00	1.45	2.00	1.55	2.00	1.37
	3.00	1.85	2.50	1.84	2.50	1.5
	4.00	2.22	3.00	2.14	3.00	1.61
0.8	1.50	1.22	10.50	1.26	1.50	1.21
	2.00	1.43	2.00	1.53	2.00	1.36
	3.00	1.82	2.50	1.83	2.50	1.47
	4.00	2.14	3.00	2.13	3.00	1.54
1.0	1.50	1.22	1.50	1.26	1.50	1.20
	2.00	1.43	2.00	1.53	2.00	1.36
	3.00	1.78	2.50	1.83	2.50	1.46
	4.00	2.12	3.00	2.13	3.00	1.53

[a] Output is relative to a metering screw.

Source: From Levine, L. and Rockwood, J. 1985. *Biotechnol. Prog.* 1: 189–199.

Table 12.2 and Table 12.3 describe two of the key conclusions drawn from computer simulation. Table 12.2 lists the maximum output (free discharge) of the various screws designs as a function of the flow index of the material. All of the designs provide a higher maximum output than that of a simple screw (maximum dimensionless output = 1.0). The increase in maximum output for channel depth and diameter compression screws is quite similar. The increase in output obtained with pitch compression screws is smaller. Table 12.3 lists the maximum pressure (closed discharge) that can be obtained with the different compression designs as compared to a simple screw. The maximum pressure is highest for a diameter compression screw and least for a pitch compression screw.

These two tables allow one to rough out an operating characteristic for a given screw since they provide the two extremes of the screw's performance. We will reserve, for a later section, discussion of how different screw designs, such as the combination of a compression and a metering screw work in tandem to produce the operating characteristic of a complete screw.

12.11 MULTIPLE-SCREW EXTRUDERS

Thus far we have limited our discussion to single-screw extruders. In recent years twin-screw extruders, of the intermeshing type, have become important in commercial food extrusion. Although they are more complicated, their performance is related to that of single-screw extruders.

TABLE 12.3
Maximum Pressure of Depth, Diameter, and Pitch Compression Screws[a]

Flow index	Depth compression		Diameter compression		Pitch compression	
	Comp. ratio	Pressure	Comp. ratio	Pressure	Comp. ratio	Pressure
0.2	1.50	1.00	1.50	1.28	1.50	0.77
	2.00	1.00	2.00	1.58	2.0	0.65
	3.00	1.00	2.50	1.89	2.50	0.56
	4.00	1.00	3.00	2.21	3.00	0.51
0.4	1.50	1.00	1.50	1.34	1.50	0.77
	2.00	1.00	2.00	1.72	2.00	0.64
	3.00	1.00	2.50	2.13	2.50	0.55
	4.00	1.00	3.00	2.56	3.00	0.50
0.6	1.50	1.00	1.50	1.41	1.50	0.76
	2.00	1.00	2.00	1.88	2.00	0.63
	3.00	1.00	2.50	2.40	2.50	0.55
	4.00	1.00	3.00	2.97	3.00	0.49
0.8	1.50	1.00	1.50	1.48	1.50	0.76
	2.00	1.00	2.00	2.05	2.00	0.63
	3.00	1.00	2.50	2.72	2.50	0.54
	4.00	1.00	3.00	3.47	3.00	0.49
1.0	1.50	1.00	1.50	1.55	1.50	0.75
	2.00	1.00	2.00	2.24	2.00	0.62
	3.00	1.00	2.50	3.08	2.50	0.53
	4.00	1.00	3.00	4.06	3.00	0.47

[a] Pressure is relative to a metering screw.

Source: From Levine, L. and Rockwood, J. 1985. *Biotechnol. Prog.* 1: 189–199.

12.11.1 NONINTERMESHING TWIN SCREW EXTRUDERS

After the single-screw extruder, the next most complex design is the counterrotating nonintermeshing twin-screw type. The nonstaggered screw alignment is illustrated in Figure 12.26. The primary advantage here is improved performance in the feed zone, a consequence of the nip formed at the flight tips, and a large feed throat. Because the screws do not intermesh, as a first approximation this extruder may be seen as two parallel single-screw devices — a simplification that is not quite true. The literature (Tadmor and Gogos, 1979; Rauwendaal, 1986) provides several approaches toward describing actual performance. We provide only the simplest analysis.

Figure 12.27 shows an end view of the twin-screw extruder. Several flows control performance: first, a drag flow that is closely related to that occurring in a single-screw extruder; second, a pressure flow similar to that in a single-screw extruder; third, a leakage flow between the screw flights and the barrel walls; and fourth, a leakage flow in the apex formed by the extruder shell and the two screws.

A key parameter needed to describe this situation is the fraction of the barrel surface in proximity to the screw tips, where the drag mechanism operates. This is given by:

$$f = \frac{\alpha_a}{\pi} \tag{12.45}$$

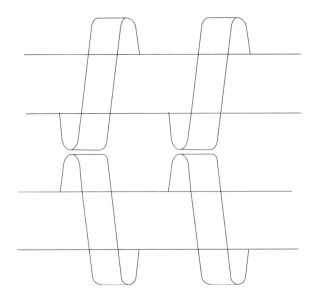

FIGURE 12.26   Nonintermeshing copunterrotating, twin screw extruder (nonstaggered configuration).

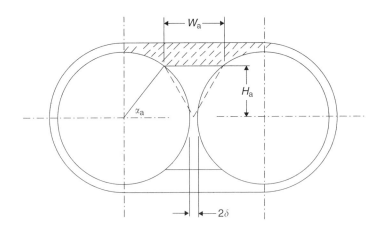

FIGURE 12.27   Apex region of nonintermeshing twin screw extruder.

As before, the output of the extruder can be written as the sum of drag and pressure-induced flows. For one screw this is:

$$Q = \frac{1}{2}fWHV_z + \frac{WH^3}{12\mu}\frac{dP}{dz} + Q_{11} + Q_{12} \tag{12.46}$$

where

$$Q_{11} = \frac{\pi D\delta^3 \cos\theta}{12\mu_\delta w}\left(\pi Dg_z\cos\theta + \frac{6\mu v_{bx}W}{H^2}\right) + \frac{w\delta^3}{12\mu_\delta}\frac{dP}{dz} \tag{12.47}$$

$$Q_{12} = 0.017(H_a - 0.61)\frac{W_a^3}{\sin\theta}\frac{W+e}{\mu e}\frac{dP}{dz} \tag{12.48}$$

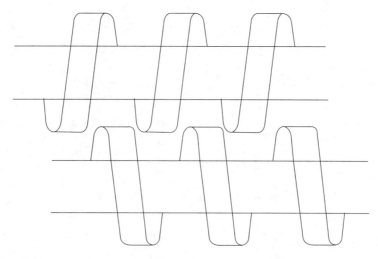

FIGURE 12.28   Nonintermeshing counterrotating, twin-screw extruder (staggered configuration).

For the staggered screw alignment in Figure 12.28, the output of *one* screw is given by

$$Q = \frac{1}{2}fWHV_z + \frac{WH^3}{12\mu}\frac{dP}{dz} + Q_{11} + Q_{ld} + Q_{lp} \qquad (12.49)$$

where

$$Q_{ld} = \frac{1}{2}(1-f)\pi D(H + 0.5W_a)v\left[\tan\theta + \left(1 + 0.5\frac{W_a}{H}\right)^2 \sin\theta\right] \qquad (12.50)$$

$$Q_{lp} = \frac{0.5\pi D(1-f)(H + 0.5W_a)^3}{12\mu\sin\theta}\frac{dP}{dz} \qquad (12.51)$$

Equation 12.49 predicts that the staggered configuration would produce less output, at the same output pressure, than the unstaggered screw configuration, Equation 12.46. This is to be expected since the leakage flow path is more open in this configuration. Neither Equation 12.46 nor Equation 12.49 includes channel shape correction factors for the drag and pressure flows. The corrections presented for a single screw should be applicable to these screw configurations.

12.11.2   INTERMESHING COUNTERROTATING TWIN-SCREW EXTRUDERS

The intermeshing counterrotating twin-screw configuration illustrated in Figure 12.29 is the closest approximation of a positive-displacement pump that can be obtained with a conventional extruder design. During each rotation of the screws a C-shaped space, as illustrated in Figure 12.30, is displaced in the forward direction by the screw. Without leaks, the output of this screw is (Janssen, 1978; Rauwendaal, 1986):

$$Q = 2NV \qquad (12.52)$$

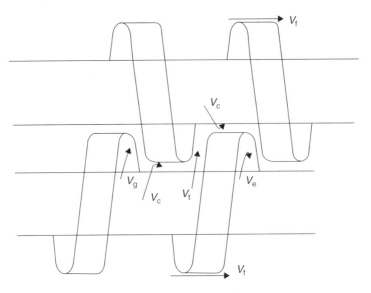

FIGURE 12.29 Intermeshing, counterrotating twin-screw extruder.

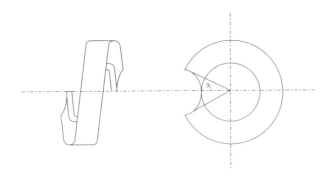

FIGURE 12.30 C-shaped section in an intermeshing, counterrotating twin-screw extruder.

The volume of the C-shaped chamber is given by:

$$V = \frac{\pi DHW_m}{\cos\theta} - \frac{w_m D^2 (2\alpha_i - \sin 2\alpha_i)}{4\cos\theta} \tag{12.53}$$

where

$$w_m = w + (D - H)\tan\varphi \tag{12.54}$$

$$W_m = \pi(D - H)\sin\theta - w_m \tag{12.55}$$

In this type of extruder there is no pressure flow such as that found in single-screw extruders. There are, however, several leakage flows that reduce the output of the extruder below that predicted by Equation 12.52. These leakages are illustrated in Figure 12.29 and may be described as follows:

1. A leakage (V_f) between the flight tips and the barrel, analogous to the leakage flows in single-screw devices.

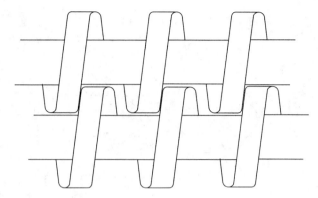

FIGURE 12.31 Intermeshing corotating twin-screw extruder.

2. A calender leakage (V_c) between the root of one screw and the flight tip of the other. This leakage is the result of pumping by the nip formed by these components and is analogous to pumping in a two-roll mill.
3. Interscrew leakage (V_t) through the gap between the flights flanks in the radial direction. This leakage occurs in the tetrahedral gap.
4. A tangential leakage (V_s) through the side gap.

Net output is the displacement volume flow less the sum of these four leakage flows:

$$Q = 2NV - 2(V_c + V_s) - V_f - V_t \tag{12.56}$$

Calculating all of these leakage flows is extremely complex and will not be reproduced here. The reader is referred to the literature (Janssen, 1978; Rauwendaal, 1986) for more detail.

12.11.3 INTERMESHING COROTATING TWIN-SCREW EXTRUDERS

Corotating, intermeshing extruders, illustrated in Figure 12.31, are difficult to analyze. The screws exhibit the properties of single-screw extruders in that, unlike intermeshing counterrotating twin-screw extruders, they provide a continuous flow path that allows pressure flow to occur. But analysis is complicated by the fact that the channel depth changes with axial and cross-channel positions. Discussion of this geometric problem may be found in the literature (Rauwendaal, 1986). And there is a small degree of positive impetus to flow in the overlapping region between the two screws. Some analyses (Yacu, 1985, 1987a; Tayeb et al., 1988, 1989) apparently ignore this conveying action.

Figure 12.32 is a cutaway view of the screws. Region A is the completely intermeshing area. The often-neglected forward flow generated by this region is given by

$$Q_a = A_a V \tan \theta \tag{12.57}$$

Pressure and drag flows occur in the regions marked B in Figure 12.32. The drag flow is given by

$$Q_d = \frac{1}{2} F_d H_{max} W V_{bz} (2p - 1) \tag{12.58}$$

The pressure flow is given by

$$Q_p = \frac{F_p H_{max}^3 W}{12\mu} \frac{dP}{dz} (2p - 1) \tag{12.59}$$

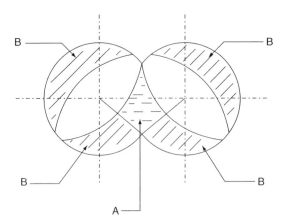

FIGURE 12.32   End view of intermeshing corotating twin-screw extruder.

The maximum channel depth is the difference between the screw centerline separation distance and the screw diameter:

$$H_{max} = D - L_c \tag{12.60}$$

The total output of these screws is the difference between the sum of displacement flow, Equation 12.53, and drag flow, Equation 12.54, and the pressure flow, Equation 12.55. Leakage flows are neglected in this analysis but could easily be estimated by using the principles applied to single-screw extruders.

The shape factors are difficult to calculate because of the variable channel geometry. They are given, for the drag flow correction by:

$$F_d = \int_{-w/2}^{w/2} \frac{H(x)\,dx}{WH_{max}} \tag{12.61}$$

and for the pressure flow correction:

$$F_p = \int_{-w/2}^{w/2} \frac{H(x)dx}{W} H_{max}^3 \tag{12.62}$$

The channel depth as a function of position is given by:

$$H(x) = \frac{D}{2}\left(1 - \cos\frac{2x}{D\sin\theta}\right) - \left(L_C^2 - \frac{1}{4}D^2\sin^2\frac{2x}{D\sin\theta}\right)^{1/2} \tag{12.63}$$

12.11.4   EMPIRICAL ANALYSIS OF THE TWIN SCREW FLOW EQUATIONS AND THE EFFECT OF KNEADING BLOCKS

White (1991) summarizes the work of Karian (1985) and Todd (1989) who performed empirical studies on APV twin-screw extruders. These works included a study of the performance of kneading blocks. Blocks, in a nonneutral position (90 degrees) approximate screws, with a pitch that is determined by the width of the block and the staggering angle. As a consequence they contribute a drag flow. White summarizes Karian and Todd's work in the form of an equation which encompasses

---

TABLE 12.4

Drag and Pressure Flow Parameters for a 50 mm Twin-Screw Extruder

Paddles/kneading blocks	Staggering angle, degrees	Paddle length/diameter	α (cm^3)	β' (cm^3)
	30	0.250	51.1	0.508
	45	0.125	18.7	0.198
		0.250	31.1	0.348
		0.500	36.4	0.603
	60	0.125	5.7	0.228
		0.250	17.9	0.336
		0.500	22.9	0.487
	90	0.250	0	0.429

Screws	Helix angle, degrees	Paddle length/diameter	α (cm^3)	β' (cm^3)
	18	N/A	42.0	0.112
	6.1	N/A	14.0	0.011

Source: From Frame, N.D. (Ed.) 1994. In *The Technology of Extrusion Cooking.* Blackie, New York.

---

drag and pressure flow.

$$Q = \alpha N - \frac{\beta'}{\mu L} \Delta P \qquad (12.64)$$

This equation is directly analogous to Equation 12.7. The drag and pressure flow terms are closely related to Equation 12.58 and Equation 12.59. The values determined for α and β' are shown in Table 12.4. Note that the kneading blocks with a small stagger angle can be effective in generating drag flow.

The values listed in Table 12.4 are only valid for 50 mm APV screws. A question remains about how they can be corrected for other diameter screws or screws having other manufacturers' geometries, such as different relative channel depths and pitch. Levine (1999a, 1999b, 1999c, 2000, 2001a, 2001b, 2001c) has discussed this issue in a series of short notes.

Examination of Equation 12.54 and Equation 12.55 reveal what the correction factors are. In geometrically similar screws, both the drag and pressure flow terms are simply proportional to the cube of the screw diameter. Therefore, for a particular manufacturer's geometry, the effect of screw diameter is readily estimated. But the effect of relative channel depth is more complicated. Drag flow is directly proportional to the ratio of depth to diameter. And pressure flow is proportional to the cube of that ratio. Levine (2002) has developed Table 12.5 to illustrate corrections that must be applied to the pressure and drag flow constants in White's equation to project from the APV extruder data to other geometries.

The pressure and drag flow terms may also be corrected for pitch of the screw by inclusion of the geometric relationships between channel width and down-channel velocity and the screw pitch. Drag flow is proportional to the product of the cosine and sine of the pitch angle, and pressure flow is proportional to the square of the sine of the pitch angle. The correction factors for various screw pitches (Levine, 2000) are summarized in Table 12.6. They provide a reasonable prediction of performance of a smaller pitch screw in Table 12.6 from the data given for a larger pitch screw.

TABLE 12.5

Approximate Corrections to Flow Parameters for Different Channel Depths, by Manufacturer

Manufacturer	Model	Drag flow factor	Pressure flow factor
APV		1.00	1.00
Coperion (W&P)	ZSK	0.71	0.36
	Continua	0.86	0.54
Wenger	TX	0.51	0.13
	Magnum	0.71	0.36
Buhler		0.82	0.54
Clextral		0.76	0.44
Readco		1.18	1.62

Source: From Levine, L. 2002. The role of rheology. Presented at the *AACC Short Course on Extrusion*, Stuttgart, Germany, February, 2002.

TABLE 12.6

Correction Factors for Various Screw Pitches

Screw pitch, degrees	Drag flow factor	Pressure flow factor
32	1.57	3.14
26	1.35	2.01
18	1.00	1.00
9	0.54	0.29
6	0.36	0.12

Source: From Levine, L. 2000. *Cereal Foods World.* 45: 223

12.11.5 NON-NEWTONIAN ANALYSIS OF TWIN-SCREW EXTRUDERS

Little has been reported on the corrections necessary to make the above models applicable to the more realistic non-Newtonian flows. All non-Newtonian analyses (Yacu, 1985; Tayeb et al., 1988, 1989) of flow in twin-screw extruders begin with an assumption that the apparent viscosity may be used in the equations developed for Newtonian fluids. In the main flow channel the apparent viscosity is estimated through the use of a gross shear rate: the apparent shear rate based on the speed of the screw and the screw channel depth. It is identical to the approximation suggested by Equation 12.9 and Equation 12.10.

Two of the papers (Tayeb et al., 1988, 1989) take these approximations one step further by differentiating between the apparent shear rate in the main flow channel from that in the intermeshing regions of the screw. The shear rate in this region is estimated as the relative velocity of the surfaces divided by the clearance. One paper (Yacu, 1985) estimates the shear rate as the product of the apparent shear rate due to the relative motion of screw to the barrel, Equation 12.9, times the apparent shear rate induced in a Newtonian fluid as it flows in an annulus formed by two concentric pipes (the screw shaft and the barrel). Although this model has produced good agreement with experimental data, the method of calculating the shear rate is not readily justifiable.

Levine (2001a) has suggested that the concept proposed by Rauwendaal, Equation 12.12 and Equation 12.13, for simplifying the analysis of single screw extruders could be applied to twin screw geometry, since the same kind of two-dimensional flow exists in most of the channel in a twin screw extruder. This implies that Equation 12.64 can also be corrected for non-Newtonian behavior, leading to a modified form:

$$Q = \frac{4+n}{5}\alpha N - \frac{3}{1+2n}\beta' \frac{\Delta P}{\mu L} \tag{12.65}$$

12.12 PARTIALLY FILLED SCREWS

All of the equations presented thus far deal with a completely filled screw, and this is the main condition of interest: Only in filled screw sections is heat transferred efficiently, pressure developed, and significant energy dissipated.

But in practice, many screws are only partially filled. This is particularly true in twin-screw extruders that are not flood fed. That is to say that the feed to the extruder is not equal to the conveying capacity of the melt conveying section of the screw. This may also occur in a single-screw extruder with a poorly designed feed section or where the bulk density of the feed material is very low. As a consequence, the feed section cannot keep up with the discharge end of the screw and the screw starves.

There are several circumstances where screws are starved by design:

1. When a low-pressure zone within the extruder is desired in order to vent the extruder to remove volatiles, or a noncompacted region is desired to ease the addition of steam or liquid ingredients.
2. When the feed of dry ingredients is controlled by a feeder external to the extruder itself: As a result, the fill of the screw is not controlled by a balance between the extruder's feed zone and discharge end. This is normally the case in the design of twin-screw extruders.

The degree of fill of a starved extrusion screw is a direct result of the following facts:

1. The ability of a screw(s) to develop pressure is directly proportional to the length of the screw. This is illustrated in Figure 12.18 and is a direct conclusion from Equation 12.4.
2. At steady state the pressure drop through the discharge resistance of the screw (the die) must be identical to the pressure rise in the screw. The output through this resistance must be identical to the feed rate (throughput) of the screw.

Using these two facts allows one to combine the equation for the die and the equations describing screw output to solve for the filled length of the screw. For Newtonian materials the literature (Janssen, 1989) provides the following equations: For a single screw or corotating twin screws,

$$L_t = \frac{\Delta P}{6\mu} \frac{H^3 W F_p}{W V_z H F_d - 2Q} \tag{12.66}$$

For an intermeshing counterrotating twin-screw extruder the number of filled chambers is,

$$\nu = \frac{BkQ}{(A - 2pV)N + Q} \tag{12.67}$$

The literature (Levine et al., 1987) provides the following equation for an estimate of the non-Newtonian flow situation:

$$\frac{L}{L_{\text{full}}} = \left[\frac{k_p/k_d}{(1/N_D) - 1} \right]^n \tag{12.68}$$

The delivery number is a dimensionless group that describes the loading of the screw. It is defined by

$$N_D = \frac{Q}{k_f N} \tag{12.69}$$

Equation 12.68 is presented graphically in Figure 12.33 for a value of the flow index of 0.5. We can draw the following important conclusions about the filled length of the screw.

1. The filled length of the screw is a complicated function of the feed rate to, and the rotational rate of, the screw(s). Lower feed rates or higher speeds will decrease the filled length of the screw.
2. The filled length of the screw decreases as the resistance of the die decreases (less open die area) or the flow resistance of the screw increases (shallower channels, tighter intermeshing, etc.).

Consider the physical consequences of these statements.

The system does not behave intuitively. Unlike a pump or a flood-fed extruder, the residence time does not vary in a simple manner with the speed of the screw or feed rate. For example, in a flood-fed extruder, doubling the rpm will roughly double the output and halve the residence time. In a starved extrusion screw, doubling the feed rate results in more of the screw being filled than before, and the residence time will be more than half its previous value. Halving or doubling the screw speed at the same feed rate will result in a more filled or less filled screw, but not double or half the previous fill.

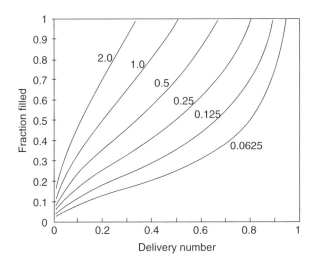

FIGURE 12.33 Filled length as a function of delivery number for a flow index of 0.5 (screw/die resistance ratio as a parameter).

12.13 ANALYZING COMPLEX SCREWS

The equations thus far presented consider only a simple screw with a simple die at its discharge. The concept of the operating point, which is illustrated graphically in Figure 12.33, may be used to examine the effect of complicated screw designs. This concept is discussed at various points in the literature (Harper, 1981; Stevens, 1985; Levine and Rockwood, 1985; Schenkel, 1986; Tadmor and Klein, 1970; Levine, 1988).

The concept is based on the fact (Janssen, 1989; Levine, 1989) that all the equations which describe the operation of screws, including multiple, intermeshing, nonintermeshing, counterrotating, and reverse pitch designs, may for Newtonian materials be generalized to:

$$Q = k_f N - k_p \Delta P \tag{12.70}$$

And the equation for the performance of any die must reduce, for a Newtonian material, to:

$$Q = k_D \Delta P \tag{12.71}$$

Equation 12.70 applies to any section of any screw. In addition, continuity requires that flow through the die must equal flow through any screw section and, of course, also the extruder feed rate. Conservation of energy requires that the sum of pressure rises and falls in the various screw elements must equal the pressure drop through the die. With these principles in mind, the operating point concept can be extended to a screw of any degree of complexity.

For example, consider the screw illustrated in Figure 12.34, made up of feed, compression, and metering zones. For the purposes of illustration, assume that the feed zone is always capable of providing the required flow to the transition and metering zones. Figure 12.35 shows assumed operating curves of the transition and metering zones. We have assumed the transition zone is shorter than the metering zone. The dashed lines in the figure represent lines of equal flow rate through each zone. Note that at low operating pressures, the output of the transition zone exceeds the drag flow output of the metering zone. Although not obvious, this is a physically realizable situation. The output of any screw can exceed its drag flow output if the zone experiences a pressure drop rather than the normal pressure rise. As a consequence, the induced pressure and leakage flows tend to enhance rather than retard screw output. The performance of the screw in this region may be estimated by extending the operating curve into the negative-pressure region, as illustrated in Figure 12.36.

Return to Figure 12.35, if one reads across the lines of equal flow rate through each zone and sums the pressure rise (or fall at low pressures) in each section, a new operating curve may be drawn, as illustrated in Figure 12.37. On this figure the operating curve for two different die designs may be superimposed, as was done previously for the simple screw design. Two die curves are illustrated (1) a die of low resistance results in an operating condition of the screw which exceeds the drag flow

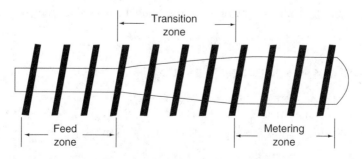

FIGURE 12.34 Screw made up of a feed, compression and metering zones.

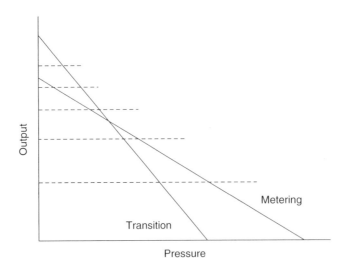

FIGURE 12.35 Construction of a composite operating curve.

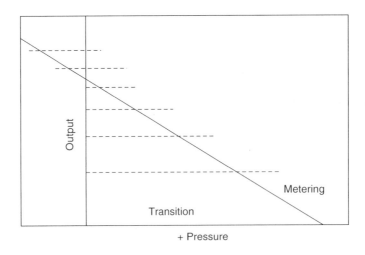

FIGURE 12.36 Extrapolation of the metering zone curve to pressure fall.

output of the metering zone and (2) a normal operating condition, where the output is less than the drag flow output of the metering zone. In all cases, the performance of two screw sections operating in tandem exceeds the output of either zone acting alone. In a similar fashion the analysis may be extended to screws of any arbitrary configuration, including the starved operating condition (Levine, 1988).

12.14 HEAT TRANSFER IN EXTRUDERS

Control of product temperature is an important issue in the extruder design. Unfortunately, the literature is very sparse. One article (Levine and Rockwood, 1986) provides the following correlation for the prediction of barrel-side heat transfer coefficients in single-screw extruders:

$$Nu = 2.2Br^{0.79} \tag{12.72}$$

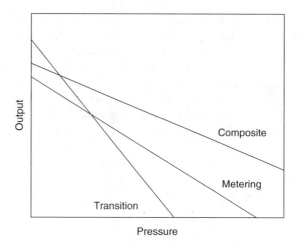

FIGURE 12.37 Operating curve for the composite of two screw solutions.

The Nusselt number is defined by

$$Nu = \frac{hH}{k} \tag{12.73}$$

The Brinkman number is defined by

$$Br = \frac{m(\pi ND)^{n+1}}{k(T_0 - T_B)H^{n-1}} \tag{12.74}$$

This correlation predicts heat transfer coefficients in the range 30 to 85 Btu/h-ft²-°F. Mathematical modeling (Mohamed et al., 1988) of the heat transfer problem suggests that the correlation suggested by Equation 12.72 should include at least one other dimensionless group. The results of that mathematical model are not presented in the form of an equation, so the results are not readily usable. Nonetheless, the model predicts heat transfer coefficients in the same range as reported above.

There is very limited additional information available. For single-screw extruders, a heat transfer coefficient of approximately 40 Btu/h-ft²-°F is quoted without reporting a source (Harper, 1981). Another model of heat transfer in single-screw extruders has been published (Tadmor and Klein, 1970). For the same range of data used in the correlation given above, predicted heat transfer coefficients were 5 to 10 times greater than actually measured.

Data for heat transfer in twin-screw extruders is even more limited. One source (Yacu, 1985) uses a value of 88 Btu/h-ft²-°F for the heat transfer coefficient in the melt conveying section and 5 Btu/h-ft²-°F in the feed section (solids conveying) of a corotating twin screw extruder. Unfortunately, the experimental source of this value is not provided. One would expect that, in this type of extruder, the improved mixing of the material and continuous wiping of films would result in higher heat transfer coefficients in the melt conveying section than have been reported for single-screw extruders. One paper (Todd, 1988) suggests correlating the heat transfer coefficient for corotating extruders through a Nusselt-type correlation. This paper reports on two sets of data. One set shows heat transfer coefficients in the melt conveying section of approximately 40 to 100 Btu/h-ft²-°F. The other data have a range of approximately 20 to 60 Btu/h-ft²-°F. They suggest the following correlation for the data:

$$\frac{hD}{k} = 0.94 \left(\frac{D^2 N \rho}{\mu}\right)^{0.28} \left(\frac{C_p \mu}{k}\right)^{0.33} \left(\frac{\mu}{\mu_w}\right)^{0.14} \tag{12.75}$$

Another paper (Larsen and Jones, 1988) suggests a model for heat transfer in twin-screw extruders with data presented in the form of tables rather than as an equation. Heat transfer coefficients predicted by the model were 83 to 163 Btu/h-ft^2-°F, while actual measurements on two different extruders were 55 to 157 Btu/h-ft^2-°F. One should not be misled by the similarity in these ranges: a reasonable correlation between the two does not appear to exist.

The most recent work on heat transfer in twin screw corotating extruders (Mohamed and Ofoli, 1989) uses an approach similar to that described earlier for single screws. As a result of both theoretical and experimental analysis, the suggested correlation for heat transfer coefficient is:

$$\bar{\text{Nu}} = 0.0042 G_z^{1.406} \text{Br}^{0.851} \tag{12.76}$$

The Brinkman number is defined by:

$$Br = \frac{K_0 L^2 \dot{\gamma}_a^{(n+1)}}{k(T_0 - T_w)} \exp\left[\frac{-\Delta E}{RT_0}\right] \tag{12.77}$$

And the Graetz number is:

$$G_z = \frac{\rho Q C_p}{KL} \tag{12.78}$$

This correlation predicts heat transfer coefficients in the range of 34–144 Btu/h-ft^2-°F which compares very favorably to the actual values of 34–135 Btu/h-ft^2-°F.

In addition to heat conducted through extruder surfaces, other thermal inputs are employed in food processing to augment (or minimize) heat generation by viscous dissipation of mechanical energy (shear). These include preheating of liquid inlet streams (using the high heat capacity of water to deliver a concentrated amount of energy) and preconditioning of the dry stream by mixing it with steam (and sometimes water) in a separate operation preceding the extruder. Steam may also be injected directly into the extruder through a barrel port. It then mixes with and condenses upon the cooler product, quickly raising its temperature.

These methods are especially important in large machines where the relative area for heat transfer diminishes: an important scale-up problem. Indeed, in very large machines, it is not possible to deliver a significant amount of energy by conduction, making alternate methods necessary. When excess energy must be removed by conductive cooling, the problem is worse because fewer alternatives exist. So larger machines often run at a higher temperature than do smaller ones, requiring changes in the basic process or equipment design to control product quality.

Both of the steam contacting methods may be analyzed with simple mass-energy balances. The overall heat balance in a food extruder is given by Harper (1981) as:

$$\frac{E_t}{\Delta t} = Q\rho \left[\int_{T_1}^{T_2} C_p dT + \int_{P_1}^{P_2} \frac{dP}{\rho} + \Delta H^\circ + \Delta H_{S_1}\right] \tag{12.79}$$

where $E_t/\Delta t$ is the total energy flux input (from conduction, dissipation of mechanical energy and condensation of steam, where used). Energy is consumed mainly by changes in sensible heat, the first term in the bracket. The second term is the change in fluid energy, and is usually ignored because it is relatively small, as is the fourth term, latent heats of fusion (unless there is a significant quantity of material such as solid fat in the formula that will change state in process). The third term is heat of reaction, which can be a significant but still a small component. It is included in rigorous calculations. The main reaction found in food processing is the endothermic gelatinization of starch, which is about 14 kJ/kg of starch and occurs at a temperature range of about 62 to 96°C, depending on the particular starch.

When the heat source is condensing steam (by injection or preconditioning), the energy input is from the change in enthalpy of the steam, leading to a simplification of the basic energy balance:

$$m_s(h_g - h_f) = \sum m_i C_{pi} \Delta T_i + m_r \Delta H^\circ \tag{12.80}$$

where m_s is the mass (or mass flow rate) of steam, m_r is the mass of reactive components (i.e., starch — usually ignored), and m_i is the mass of an input stream (dry mix, water, etc.). The enthalpies of steam at its inlet pressure and water from condensed steam at final in-process temperature are h_g and h_f. ΔT_i is the temperature rise in an input stream to the final temperature, and C_{pi} is the heat capacity of that stream. Heat capacities can be accurately estimated from product composition for any temperature range by weight averaging methods available in the literature (Harper, 1981; Choi and Okos, 1986).

By manipulating flow rates, any temperature up to the saturation temperature of the steam at the pressure in the process can be achieved. Most preconditioners operate at ambient atmospheric pressure, so the maximum possible final temperature is that of the boiling point of water, although in practice lower temperatures (about 90°C maximum) are typical due to problems in transferring preconditioned material into the extruder if it is too wet from the condensed steam. Some units operate at elevated pressure, requiring special equipment for ease of transfer. Most have a residence time of 2 to 4 min, which is sufficient time for fine particles to approach thermal equilibrium. Moisture takes longer to penetrate and is generally still more concentrated near the surface leading to the handling problem.

Since thermal diffusivity is so much greater than moisture diffusivity, the thermal effect can be ignored when analyzing moisture penetration — assume that the product is at a constant (final) temperature as moisture more slowly penetrates the particles with nonsteady-state water transfer as the controlling mechanism (Bouvier, 1995). Residence times up to about 10 min are sometimes employed for better moisture uniformity.

We can understand the mechanism of moisture penetration by analogy to well-established models of heat penetration into solid shapes (Foust et al., 1960), where data are first transformed into dimensionless form. The analogous dimensionless forms for moisture penetration are:

$$\Theta = \frac{w_e - w}{w_e - w_o} \tag{12.81}$$

for moisture content, where w is fraction of moisture (dry basis) with subscript e for equilibrium (final) moisture content and o for initial moisture content; and:

$$\tau = \frac{Dt}{R^2} \tag{12.82}$$

for time where D is the diffusivity of water in the product, and R is its radius. Dimensionless time, τ, is also called the Fourier Number. We expect to see a somewhat exponential decrease in Θ (approaching zero) with τ, with parallel lines at different radii within the particle having a slope determined by the Biot Number (a ratio of internal resistance to external or surface film resistance — also called the Sherwood number in mass transfer):

$$Bi = \frac{k_m R}{D} \tag{12.83}$$

When there is little surface resistance, we expect to see a large mass transfer coefficient, k_m, at the particle surface speeding the moisture rate up to a maximum controlled by internal resistance only. Large Biot Numbers (also a function of radius), therefore, make the process faster up to a limit. For that reason, high speed mixing is used in preconditioners for best steam-product contacting (Bouvier, 1995). Unfortunately, we do not usually know the diffusivity or mass transfer coefficient, and the particles are not truly spherical, so this relationship is not precisely known. But it can be used to interpret experimental data.

For high Biot numbers, constant physical parameters, and spherical particles, we can estimate how the average moisture changes with time for Fourier numbers greater than about 0.1, as the moisture approaches its end point to be (Miller, 1999):

$$\bar{w} = w_e - \frac{6}{\pi^2}(w_o - w_e)e^{-\pi^2 Dt/R^2} \tag{12.84}$$

which is identical to one equation used for analysis of drying, a very similar but reverse process (Heldman and Singh, 1981). Equation 12.84 is in a convenient form for fitting of experimental data:

$$y = c + a\,e^{bt} \tag{12.84a}$$

At least three experimental values are required to determine the three constants and to estimate an equivalent diffusivity and particle radius for the system.

An additional complication in preconditioner analysis is their wide residence-time distributions so that particles have varying exposure times. High speed mixing, for high Biot numbers also creates more axial dispersion (wider residence time distribution), a function of Froude number (Levine et al., 2002) which increases with mixer angular velocity, ω:

$$Fr = \omega R/g \tag{12.85}$$

Below a Froude number of unity, little dispersion occurs. One manufacturer has introduced a two-stage unit with a high Froude number dispersion section followed by a low speed holding section with little dispersion.

Steam injection uses the same heat balance relationship, Equation 12.80, but follows very different dynamics. It takes place at much higher pressures, usually in the range of 100 psi (about 700 kPa) in the extruder screw channel, which must be designed to produce at least that pressure leading into the contacting area and then to create enough empty volume (with starved screws of longer pitch) around the product to allow good flow of steam. Steam flows through a barrel port near the upstream end of the extruder where the extrudate is cool. A large temperature differential makes the steam condense very rapidly, leaving insufficient time (only a few seconds) for equilibration of moisture or temperature in the product.

So this process, unlike preconditioning, is controlled by the rate of heat penetration into the product. In an exercise like that discussed above for moisture penetration, one study (Miller, 1990; Miller, 1998b) found the same model based on heat penetration to fit steam injection data:

$$\frac{T_f - T}{T_f - T_o} = \frac{6}{\pi^2} \sum_{n=1}^{\infty} \frac{1}{n^2} e^{-n^2 \pi^2 \tau} \approx \frac{6}{\pi^2} e^{-\pi^2 \alpha t/R^2} \tag{12.86}$$

In that study, R^2/α was found empirically equal to 6.88 sec. With an estimated thermal diffusivity, based on the literature (Wallapapan et al., 1986), of about 0.145 mm²/sec, a particle radius of 0.98 mm was calculated, showing that the mechanism of heat penetration was based on flow of steam through a particulate bed in the extruder channel. In no case did the temperature reach equilibrium (equal to the steam temperature) but approached it generally to about 95%.

Venting is the inverse of steam injection and a means of quickly reducing product temperature by evaporative cooling. As in the case of injection, it is necessary to select extruder screws to assure that the flow is starved in the vent section to prevent extrudates from rising through the vent port. This is even more critical than in steam injection, because many extrudates tend to foam as volatiles are removed. Equation 12.80 still applies to this situation (without the reaction term) and with simplifications becomes:

$$m_i C_p (T_i - T_o) = m_{evap} \lambda \qquad (12.87)$$

where m_i is the total mass flow into the vent and C_p is the average heat capacity of that stream in the temperature range of T_i to T_o, the temperatures into and out of the vent section (assume at equilibrium with ambient pressure — sometimes vents are run in a partial vacuum). The steam heat of vaporization, λ, is that at the vent pressure. The total extrudate flow moving on past the vent is reduced by the mass of steam evaporated.

Many products flash off moisture as they leave the extrusion die: this is the mechanism of making puffed products. This is thermodynamically equal to a vent, except that more moisture usually leaves the product as it is exposed more completely to ambient air that is normally relatively dry. A typical final product temperature after flashing off its superheated moisture is typically about 90°C.

12.15   EXTRUDER RESIDENCE-TIME DISTRIBUTIONS

Since the extrudate undergoes chemical and physical changes during its transit through the extruder, there is a great deal of interest in the residence-time distributions in different extruder types. Residence-time distributions are discussed in many places in the literature (see Levine and Miller, 2003 for a more complete discussion on this subject). The cumulative age distribution function may be approximated by

$$F(\theta) = 1 - \exp \left\{ - \left[\left(\frac{1}{1-P} \right) (\theta - P) \right] \right\} \qquad (12.88)$$

Parameter P represents the fraction of the total residence seen as pure dead time. Typical values of P are approximately 0.75 (Bruin et al., 1978) for a single-screw extruder, 0.50 (Altomare and Ghossi, 1986) for a twin-screw intermeshing corotating extruder, and 0.91 (Lin and Armstrong, 1989) for an intermeshing counterrotating twin-screw extruder. These distribution functions, along with those for a perfectly stirred vessel, a laminar flow pipe, and plug flow reaction, are illustrated in Figure 12.38.

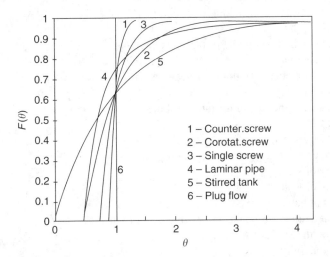

FIGURE 12.38   Residence-time distributions for various extruders.

The degree of mixing that occurs is least for the counterrotating extruder and most for a corotating extruder. The former approaches a plug flow reaction, while the latter approaches the residence-time distribution in a laminar flow pipe reactor.

One would expect that the residence-time distribution would change significantly with changed screw configuration in a twin-screw extruder. This does not appear to be the case for either of the twin-screw configurations (Altomare and Ghossi, 1986; Lin and Armstrong, 1989).

NOMENCLATURE

A	Empirical constant
A_a	Intermeshing area of corotating screws
B	Empirical constant
Br	Brinkman number
C_p	Heat capacity
D	Screw diameter, moisture diffusivity
D_{eq}	Equivalent diameter of screw channel
e	Flight thickness
E	Activation energy
E_t	Total energy input
F_d	Drag flow correction for channel depth
F_p	Pressure flow correction for channel depth
F_{dc}	Drag flow correction for channel curvature
F_{pc}	Pressure flow correction for channel curvature
F_{de}	Drag flow correction for channel end effect
F_{pe}	Drag flow correction for channel end effect
F_{dl}	Drag flow correction for leakage
F_{pl}	Pressure flow correction for leakage
$F_{\mu d}$	Drag flow correction for barrel/screw viscosity ratio
$F_{\mu p}$	Pressure flow correction for barrel/screw viscosity ratio
f	Channel friction factor for reverse flow; fraction of screw perimeter covered by barrel; leakage factor; correction factor for adiabatic operation
f_s	Coefficient of friction between extrudate and screw
f_b	Coefficient of friction between extrudate and barrel
G_z	Graetz number or dimensionless discharge pressure
H	Channel depth
H_1	Channel depth at beginning of tapered channel
H_2	Channel depth at discharge of tapered channel
H_a	Height of apex between nonintermeshing screws
H_g	Depth of barrel grooves
H_{max}	Maximum channel depth for intermeshing corotating screws
H_{S_1}	Heat of fusion
ΔH^o	Heat of reaction
h	Die gap; convective heat transfer coefficient
h_f	Enthalpy of steam
h_g	Enthalpy of water
j	Number of barrel grooves
K	Constant defined by Equation 12.36; used in Figure 12.22
k	Extrudate thermal conductivity
k_D	Die conductivity
k_f	Screw drag flow constant

k_p	Screw pressure flow conductivity
k_m	Surface mass transfer coefficient for water
L	Screw length, die land length
L_c	Shaft center-to-center distance
L_{eq}	Length of helical channel
L_f	Filled length
M	Constant defined by Equation 12.36, used in Figure 12.22
M_{DB}	Extrudate moisture, % dry basis
m	Power law flow consistency
m_{evap}	Mass of moisture evaporated
m_i	Mass of inlet stream
m_0	Power law flow consistency constant
m_r	Mass of reactant component
m_s	Mass of steam
N	Rotational rate of screw
N_p	Screw power number
Nu	Nusselt number
n	Power law flow index
P	Pressure; fraction of mean residence time seen as a pure delay
P_1	Pressure at inlet to channel
P_2	Pressure at discharge of channel
p	Number of thread starts
p_t	Screw power consumption
p_s	Power consumption of idealized screw
Q	Volumetric flow rate
Q_d	Drag flow
Q_p	Pressure flow
Q_{gd}	Drag flow in grooves
Q_{gp}	Pressure flow in grooves
R	Universal gas constant; screw radius; die radius, particle radius
$\bar{R}$	Average die radius in annular cross-section
Re_f	Channel reverse flow Reynolds number
Re_p	Screw rotational Reynolds number
T	Temperature, absolute temperature
$\bar{T}$	Average temperature
T_B	Barrel temperature
T_o	Initial temperature
ΔT_i	Temperature rise in an inlet stream
T_0	Extrudate temperature
T_f	Equilibrium temperature = temperature of heating/cooling medium
t	Time
V	Volume of C-shaped chamber
V_s	Slipping velocity at die surface
V_z	Surface velocity in down-channel direction
v_r	Average channel reverse velocity
W	Channel width
W_a	Width of apex between nonintermeshing screws
W_g	Groove width
x	Cross-channel direction
y	Direction normal to channel
z	Down-channel direction

Greek Letters

α	Angle of grooves relative to channel direction; drag flow geometry factor; thermal diffusivity
α_a	Angle of uncovered screw
α_i	Angle of twin-screw intermesh
β	Navier's slip coefficient
β'	Kneading block pressure flow geometry factor
$\dot{\gamma}$	Shear rate
δ	Screw clearance
θ	Screw helix angle; dimensionless time solids conveying angle
Θ	Dimensionless unaccomplished moisture change
λ	Heat of vaporization of water
μ	Viscosity
μ_0	Viscosity at screw entrance
μ_b	Viscosity at barrel wall
μ_z	Viscosity at screw discharge, viscosity in screw clearance
v	Number of filled chambers
ρ	Density
τ	Shear stress, Fourier number
τ_s	Shear stress at die surface
ψ	Flight flank angle

REFERENCES

Altomare, R.F. and Ghossi, P. 1986. An analysis of residence time distribution patterns in a twin screw extruder. *Biotechnol. Prog.*, 2: 157–163.

Baird, D.G. and Reed, C.M. 1989. Transport properties of food doughs. In *Cooking Extrusion*, C. Mercier, P. Linko, and J.M. Hamer (Eds), American Association of Cereal Chemists St. Paul, MN, pp. 205–234.

Bhattacharya, M. and Hanna, M.A. 1986. Mathematical modeling of food extruder. *Lebensm, Wiss. Technol.*, 19: 34–38.

Bird, R.B., Armstrong, R.C., and Hassager, O. 1977. *Dynamics of Polymeric Fluids*, Vol. I. John & Wiley, New York.

Bouvier, J.M. 1995. Preconditioning in the extrusioncookin process. *Int. Milling Flour & Feed J.*, 34–38.

Bruin, S., Van Zuilchem, D.J., and Stolp, W. 1978. Fundamental and engineering aspects of extrusion of biopolymers in a single screw extruder. *J. Food Process Eng.*, 2: 1–37.

Carley, J.F. and McKelvey, I.M. 1953. Extruder scale-up theory and experiments. *Ind. Eng. Chem.*, 45: 989–991.

Carley, J.F. and Strub, R.A. 1953a. Basic concepts of extrusion. *Ind. Eng. Chem.*, 45: 970–973.

Carley, J.F. and Strub, R.A. 1953b. Application of theory to design of screw extruders. *Ind. Eng. Chem.*, 45: 978–988.

Carley, J.F., Mallouk. R.S., and McKelvey, I.M. 1953. Simplified flow theory for screw extruders. *Ind. Eng. Chem.*, 45: 974–977.

Cervone, N.W. and Harper, J.M. 1978. Viscosity of an intermediate moisture dough. *J. Food Process Eng.*, 2: 83–95.

Chinnaswamy, R. and Hanna, M.A. 1988. Optimum extrusion-cooking conditions for maximum expansion of corn starch. *J. Food Sci.*, 53: 834–840.

Choi, Y. and Okos, M.R. 1986. Effects of temperature and composition on the thermal properties of foods. In *Food Engineering and Process Applications*, Vol. 1, *Transport Phenomina*, M. LeMaguer and P. Jelen (Eds), Elsevier, New York.

Colonna, P., Tayeb, J., and Mercier, C. 1989. Extrusion cooking of starch and starchy products. In *Cooking Extrusion*, C. Mercier, P. Linko, and J.M. Harper (Eds), American Association of Cereal Chemists, St. Paul, MN, pp. 247–320.

Della Valle, G., Kozlowski, C.P., and Tayeb, J. 1989. Starch transformation estimated by the energy balance on a twin screw extruder. *Lebesm. Wiss. Technol.* (in press).

Dintheer, W. 1993. Pasta extrusion. *Extrusion Communique*. 16–18.

Donnelly, B.J. 1982. Teflon and non-Teflon lined dies: effect on spaghetti quality. *J. Food Sci.*, 47: 1055–1058, 1069.

Dreiblatt, A. 1987. Accuracy in extruder scale-up. Paper presented at the *Natl. Meet. AIChE*, Minneapolis.

Eerikainen, T. and Linko, P. 1989. Extrusion cooking modeling, control and optimization. In *Cooking Extrusion*, C. Mercier, P. Linko, and J.M. Harper (Eds), American Association of Cereal Chemists, St. Paul, MN. pp. 91–156.

Eise, K., Herrmann, H., Werner, H., and Burkhardt, U. 1981. An analysis of twin-screw extruder mechanisms. *Adv. Plast. Technol.*, 1: 1–22.

Falcone, R.O. and Dixon P.R. 1988. Effects of feed composition, feed moisture, and barrel temperature on the physical and rheological properties of snack-like products prepared from cowpea and sorghum flours by extrusion. *J. Food Sci.*, 53: 1464–1469.

Fletcher, S.I., Richmond, P., and Smith, A.C. 1985. An experimental study of twin-screw extrusion-cooking of maize grits. *J. Food Eng.*, 4: 291–312.

Foust, A.S., Wenzel, L.A., Clump, C.W., Mans, L., and Anderson, L.B., 1960. *Principles of Unit Operations*. John Wiley & Sons, New York.

Frame, N.D. (Ed.). 1994. Operational characteristics of the co-rotating twin-screw extruder. In *The Technology of Extrusion Cooking*, Blackie, New York.

Fricke, A.L., Clark, J.P., and Mason, T.F. 1977. Cooking and drying of fortified cereal foods: extruder design. *AIChE Symp. Ser.*, 73: 134–141.

Griffith, R.M. 1962. Fully developed flow in screw extruders. *Ind. Eng. Chem. Fundam.* I: 180–181.

Hagan, R.C. and Villota, F.T. 1986. Texturization of coprecipitated soybean and peanut proteins by twin-screw extrusion. *J. Food Sci.*, 5I: 367–370.

Harmann, D.V. and Harper, J.M. 1974. Modeling a forming foods extruder. *J. Food Sci.*, 39: 1039–1044.

Harper, J.M. 1979. Food extrusion. *CRC Crit. Rev. Food Sci. Nutr.* 11155–11215.

Harper, J.M. 1981. *Extrusion of Foods*. CRC Press, Boca Raton, FL.

Harper, J.M., Rhodes, T.P., and Wanninger, L.A. 1971. Viscosity model for cooked cereal dough. *AIChE Symp. Ser.*, 67: 40–43.

Heldman, D.R. and Singh R.P. 1981. *Food Process Engineering*. AVI Publishing Co., Westport, CT.

Holay, S.H. and Harper, J.M. 1982. Influence of the extrusion shear environment on plant protein texturization. *J. Food Sci.*, 47: 1869–1874.

Ibave, J.L. and Harper, J.M. 1983. Textured soy protein dependence on extrusion parameters. Paper presented and the *Natl. Meet. AIChE*, Denver.

Janssen, L.P.B.M. 1978. *Twin Screw Extrusion*. Elsevier, New York.

Janssen, L.P.B.M. 1989. Engineering aspects of food extrusion. In *Cooking Extrusion*, C. Mercier, P. Linko, and J.M. Harper (Eds), American Association of Cereal Chemists, St. Paul, MN, pp. 39–56.

Jao, Y.C., Chen, A.H., Leandowski, D., and Irwin, W.E. 1978. Engineering analysis of soy dough. 1. *Food Process Eng.*, 2: 97–112.

Jepson, C.H. 1953. Future extrusion studies. *Ind. Eng. Chem.*, 45: 992–993.

Kalyon, D.M., Lawal, A., Yazici, R., Yaras, P., and Railcar, S. 1999. Mathematical modeling and experimental studies of twin-screw extrusion of filled polymers. *Polym. Eng. Sci.*, 39: 1139–1151.

Karian, H.G. 1985. *J. Vinyl Tech.*, 7: 154.

Kuhle, R. 1986. Continuous dough manufacturing system. Paper presented at the *Nail. Meet. AIChE*, Miami.

Larsen, H. and Jones, A. 1988. Heat transfer in twin screw extrusion, ANTEC 88 Conference Proceedings, Society of Plastics Engineers, *46th Annu, Tech. Conf Exhibit*, pp. 67–709.

Launay, B. and Bure, J. 1973. Application of a viscometric method to the study of wheat flour doughs. *J. Texture Stud.*, 4: 82–101.

Lawal, A., Railkar, S., and Kalyon, D.M., 2000. Mathematical modeling of three-dimensional extrusion and die flows of viscoplastic fluids with wall slip. *J. Reinf. Plast. Compos.*, 19: 1483–1492.

Levine, L. 1982. Estimating output and power of food extruders. *J. Food Process Eng.*, 6: 1–13.

Levine, L. 1988. Understanding extruder performance. *Cereal Foods World*, 33: 963–970.

Levine, L. 1989. Scale-up, experimentation and data evaluation. In *Cooking Extrusion*, C.

Mercier, P. Linko, and J.M. Harper (Eds), American Association of Cereal Chemists, St. Paul, MN, pp. 57–90.

Levine, L. and Miller, R.C. 2003. Extrusion system residence time distribution. In *Encyclopedia of Agricultural, Food and Biological Engineering*. D.R. Heldman, (Ed.), Marcel Dekker, New York.

Levine, L. and Rockwood, J. 1985. Simplified models for estimating isothermal operating characteristics of food extruders. *Biotechnol. Prog.*, 1: 189–199,

Levine, L. and Rockwood, J. 1986. A correlation of heat transfer coefficients in food extruders. *Biotechnol. Prog.*, 2: 105–108.

Levine, L., Symes, S., and Weimer, J. 1987a. A simulation of the effect of formula variations on the transient output of single screw extruders. *Biotechnol. Prog.*, 3: 212–220.

Levine. L., Symes, S., and Weimer, J. 1987b. A simulation of the effect of formula and feed rate variations on the transient behavior of starved extrusion screws. *Biotechnol. Prog.*, 3: 221–230.

Levine, L. 1999a. The effect of differing geometries on extruder screw performance. *Cereal Foods World*, 44: 162.

Levine, L. 1999b. The effect of differing geometries on extruder screw performance II. *Cereal Foods World*, 44: 426–427.

Levine, L. 1999c. The effect of differing geometries on extruder screw performance III. *Cereal Foods World*, 44: 681–682.

Levine, L. 2000. The effect of differing geometries on extruder screw performance IV. *Cereal Foods World*, 45: 223.

Levine, L. 2001a. The effect of differing geometries on extruder screw performance V. *Cereal Foods World*, 46: 169.

Levine, L. 2001b. The effect of differing geometries on extruder screw performance VI. *Cereal Foods World*, 46: 248–249.

Levine, L. 2001c. The effect of differing geometries on extruder screw performance VII. *Cereal Foods World*, 44: 442–443.

Levine L. 2002. The role of rheology. Presented at the *AACC Short Course on Extrusion*, Stuttgart, Germany, February, 2002.

Levine, L., Bouvier, J.-M., Brent, J.L. and Miller, R.C. 2002. An analysis of preconditioner residence time and residence time distribution. *Cereal Foods World*. 47: 142–148.

Lin, J.K. and Armstrong, D.J. 1989. Process variables affecting residence time distributions of cereal in intermeshing counter-rotating twin-screw extruder. *Trans. ASAE* (in press).

Linko, P. 1989. Extrusion cooking in bioconversions. In *Cooking Extrusion*, C. Mercier, P. Linko, and J.M. Harper (Eds), American Association of Cereal Chemists, St. Paul, MN, pp. 235–246.

Luxenburg, L.A., Baird, D.O., and Joseph, E.O. 1985. Background studies in the modeling of extrusion cooking processes for soy flour doughs. *Biotechnol. Prog.*, I: 33–38.

Maldari, D. and Maldari, C. 1993, Design and perfomance of pasta dies. *Cereal Foods World*, 38: 807–810.

Mallouk, R.S. and McKelvey, J.M. 1953. Power requirements of melt extruders. *Ind, Eng. Chem.*, 45: 987–988.

Martelli. F.G. 1983. *Twin Screw Extruders: A Basic Understanding.* Van Nostrand Reinhold, New York.

McKelvey, J.M. 1953. Experimental studies of melt extrusion. *Ind. Eng. Chem.*, 45: 982–986.

Megard, D., Kitabatake, N., and Cheftel, J.C. 1985. Continuous restructuring of mechanically deboned chicken meat by HTST extrusion-cooking. *J. Food Sci.*, 50: 1364–1369.

Meuser, P. and Van Lengerich, B. 1984a. System analytical model for the extrusion of starches. In *Thermal Processing and Quality of Foods*, P. Zeuthen, J.C. Cheftel, M. Jul, H. Leniger, P. Linko, F. Varela, and G. Vos (Eds), Elsevier, New York, pp. 175–179.

Meuser, F. and Van Lengerich, B. 1984b. Possibilities of quality optimization of industrially extruded flat breads. In *Thermal Processing and Quality of Foods* P. Zeuthen, J.C. Cheftel, M. Jul, H. Leniger, P. Linko, F. Varela, and G. Vos (Eds), Elsevier, New York, pp. 180–184.

Meuser, F. and Weidmann. W. 1989. Extrusion plant design. In *Cooking Extrusion*, C. Mercier, R. Linko, and J.M. Harper (Eds), American Association of Cereal Chemists, St. Paul, MN, pp. 91–156.

Meuser, F., Von, Van Lengerich, B., and Kohler, F. 1982. Einflub der Extrusion Extntsionsparameter auf funktionell Eigenschaften von Weizenstarke. *Starch*, 34: 366–372.

Meuser, F., VanLengerich, B., and Groneick, E. 1984a. The use of high temperature short time extrusion cooking of malt in beer production. In *Thermal Processing and Quality of Foods*, P. Zeuthen, J.C. Cheftel, M. Jul, H. Leniger, P. Linko, F. Varela, and G. Vos (Eds), Elsevier, New York, pp. 121–136.

Meuser, F.V., Van Lengeiich, B., and Rheimers, H. 1984b. Kochextrusion von Starken, *Starch*, 36: 194–199.

Meuser, F., Van Lengerich, H., and Kohler, F. 1986. Extrusion cooking of protein and dietary fiber enriched cereal products: nutritional aspects. Undated manuscript received in personal correspondence with B. Van Lengerich, Werner Pfleiderer Corporation, Ramsey, NJ.

Meuser, F., Van Lengerich, H., Pfaller, A.E., and Harmuth-Hoene, A.E. 1987. The influence of HTST-extrusion cooking of the protein nutritional value of cereal-based products, In *Extrusion Technology for the Food industry*, Part II, *Aspects of Technology*, P. Colonna (Ed), Elsevier, New York, pp. 35–53.

Middleman, S. 1977. *Fundamentals of Polymer Processing.* McGraw-Hill, New York.

Michaeli, W. 1992. *Extrusion Dies for Plastics and Rubber*, 2nd edn. Oxford University Press, New York.

Miller, R.C. 1985. Low moisture extrusion: effects of cooking moisture on product characteristics. *J. Food Sci.*, 50: 249–253.

Miller, R.C. 1990. Twin-screw extrusion: dynamics of steam injection. Paper #162, IFT Annual Meeting, Anaheim, CA.

Miller, R.C. 1999. Breakfast Cereal Cooking: Batch and Continuous, Presented at the *AACC Short Course on Breakfast Cereal Technology*, September, 1999.

Miller, R.C. 1998. Principles of die and cutter design, Presented at the *AACC Short Course on Extrusion*, Minneapolis, May, 1998.

Miller, R.C. 1998b. Extruder Heat Transfer, Presented at the *AACC Short Course on Extrusion*, Minneapolis, May 1998.

Mohamed, I.O. and Ofoli, R.Y. 1989. Average heat transfer coefficients in twin screw extruders, *Biotechnol. Prog.*, 5: 158–163.

Mohamed, I.O., Morgan, R.G., and Ofoli, R.Y. 1986. Average convective heat transfer coefficients in single screw extruders of non-Newtonian food materials. *Biotechnol. Prog.*, 4: 68–78.

Morgan, R.G., Suter, O.A., and Sweat, V.F. 1978. *J. Food Process Eng.*, 2: 65–81.

Naguchi, A. 1989. In *Cooking Extrusion*, C. Mercier, P. Linko, and J.M. Harper (Eds), American Association of Cereal Chemists, St. Paul, MN, pp. 343–370.

Nazarov, N.1., Azarov, B.M., and Chaplin, M.A. 1971. Capillary viscometry of macaroni dough. *Izv. Vyssh. Uchebn. Zaved. Pishch. Tekhnol.* 1971: l49.

Rauwendaal, C. 1986. *Polymer Extrusion*. Carl Hanser, New York.

Ravindran. C. and Ottino, J.M. 1985. Fluid mechanics of mixing in a single screw extruder. *Ind. Eng. Chem. Fundam.*, 24: 170–180.

Remsen, C.H. and Clark, P.J. 1978. Viscosity model for a cooking dough. *Food Process Eng.*, 2: 39–64.

Schenkel. O. 1966. *Plastics Extrusion Technology and Theory.* American Elsevier, New York.

Stanley, D.W. 1989. Protein reactions during extrusion cooking. In *Cooking Extrusion*, C. Mercier, P. Linko, and J.M. Harper (Eds), American Association of Cereal Chemists. St. Paul, MN, pp. 321–342.

Stevens, M.J. 1985. *Extruder Principles and Operation.* Elsevier, New York.

Tadmor, Z. and Gogos, C.G. 1979. *Principles of Polymer Processing.* Wiley-Interscience, New York.

Tadmor, Z. and Klein, I. 1970. *Engineering Principles of Plasticating Extrusion.* Van Nostrand Reinhold, New York.

Tayeb, J., Vergnes, H., and Della Valle, G. 1988. Theoretical computation of the isothermal flow of the reverse screw element of a twin screw extrusion cooker. *J. Food Sci.*, 53: 616–625.

Tayeb, J., Vergnes, B., and Della Valle, G. 1989. A basic model for a twin screw extruder. *J. Food Sci.*, 54: l047–1056.

Todd. D. 1988. Heat transfer in twin screw extrusion. *ANTEC '88 Conf Proc.*, Society of Plastics Engineers, 46th Annual Technical Conference and Exhibit, pp. 54–57.

Todd, D.B. 1989. *SPE ANTEC Tech. Papers*, 35: 168.

Toledo, R., Cabot, J., and Brown, D. 1977. Relationship between composition, stability and rheological properties of rat comminuted meat batters. *J. Food Sci.*, 42: 726.

Tsao, T.F., Harper, J.M., and Repholz, K.M. 1978. The effects of screw geometry on the extruder operational characteristics. *AIChE Symp. Ser.*, 74: 142–147.

Van Zuilichem, D.J., Buisman, G., and Stolp, W. 1974. Shear behavior of extruded maize. Paper presented at the *4th Int. Cong. Food Sci. Technol.*, International Union of Food Science and Technology, Madrid.

Van Zuilichem, D.J., Lammers, G., and Stolp, W. 1975. Influence of process variables on the quality of extruded maize. *Proc. 6th European Symp. Eng. Food Quality*, Cambridge, pp. 380–406.

Wallapapan, K., Sweat, V.E., Diehl, K.C., and Engler, C.R. 1986. Thermal properties of porous foods. In *Physical and Chemical Properties of Food*, M.R. Okos (Ed.), ASAE, St Joseph, MN.

White, J.L. 1990. *Twin Screw Extrusion: Technology and Principles.* Hanser Pub., New York.

Wilkinson, W.L. 1960. *Non-Newtonian Fluids.* Pergamon Press, Elmsford. New York.

Yacu, W.A. 1985. Modeling of a twin screw extruder. *J. Food Eng.*, 8: 1–21.

Yacu, W.A. 1987a. Energy balance in twin screw co-rotating extruders. Paper presented at the *AACC Short Course on Extrusion*, San Antonio.

Yacu, W.A. 1987b. Extrusion cooking analysis. I. Processing aspects of twin screw corotating extruders. Paper presented at the *AACC Short Course on Extrusion*, San Antonio.

Yacu, W.A. 1987c. Extrusion cooking analysis. II. Extrudate physical and functional properties. Paper presented at the *AACC Short Course on Extrusion*, San Antonio.

13 Food Packaging

John M. Krochta

CONTENTS

13.1 INTRODUCTION

13.1.1 ROLE OF FOOD PACKAGING

The goal of Food Engineers is to provide a broad variety of safe, convenient, high-quality, nutritious foods to consumers at the lowest prices possible. To accomplish this, it is necessary to understand how food deteriorates due to a combination of biological, chemical, and physical factors. Appropriate application of the food preservation principles presented in this Handbook prevents food deterioration and results in quality foods with long shelf life. The final step is packaging the food so that the preserved nature of the food is maintained and the package promotes product sale and provides customer convenience.

The goal of this chapter is to present the functions, terminology, development process, materials, properties, manufacture, uses, trends, safety, environmental issues, and regulation of food packaging so that food engineers can effectively work with food product developers and package engineers to select appropriate packaging. Of the several functions of packaging, the main focus of this chapter will be on the protective function of food packaging.

13.1.2 HISTORY OF FOOD PACKAGING

One can use the history of food packaging to study the development of civilization (Sacharow and Brody, 1987; Soroka, 2002a). At any time, food packaging reflects evolution in science, technology, art, psychology, sociology, politics, and law. The earliest packaging consisted of leaves, shells, gourds, animal skins and bladders, and even human skulls to contain and transport foods short distances from harvest. Woven baskets, leather bags, clay pots, glass containers, wood barrels, cloth sacks, and paper wraps reflected later development of human skills, tools, and discoveries about materials. These later packages still functioned mainly to contain and transport food but also provided increased protection.

In the early food industries, there was little packaging of individual units. Food products were made available to local stores and markets in bulk containers. Customers would buy butter from a large block, and it would be wrapped in paper. Liquid and solid foods would be kept in barrels, and customers would bring their own containers (Driscoll and Paterson, 1999). Only in relatively recent history have the functions of protection and convenience been highly developed in individually-packaged foods.

In the mid-18th century, the Dutch began preserving roast beef and then salmon by placing hot food into tin-plated iron cans, covering with hot fat, and quickly soldering on the lid (Robertson, 2006a). Nicolas Appert separately developed the concept of heat-processed food sealed in a glass bottle in response to a challenge by Napoleon in 1800 to develop more appetizing, stable, and convenient foods for his military. Peter Durand bought Appert's patent for use in England and added use of metal cans to the patent. Heat-preserved foods in tin cans were first made in 1812 (Sacharow and Brody, 1987). Other packaging innovations in the 19th century and early 20th century included mechanized production of paper bags and tubes, paperboard cartons (set-up and folding), corrugated paperboard, barrels, wood veneer produce baskets and berry boxes, tinfoil, wax paper, and glass bottles (Sacharow and Brody, 1987). Growing availability of refrigeration at the beginning of the 20th century stimulated the development by Clarence Birdseye of quick-frozen foods packaged in waxed paper and paperboard to prevent moisture loss. The first half of the 20th century saw the development of cellophane, aluminum cans, aluminum foil, and aluminum-foil-laminated paperboard. During this period, the form-fill-seal concept for in-line formation of flexible packaging was developed. World War II speeded the development of polyethylene (PE), polyvinyl chloride (PVC) and polyvinylidene chloride (PVDC).

In the second half of the 20th century, development of coextrusion using ethylene vinyl alcohol (EVOH) copolymer, PVDC and/or metalized films allowed production of multilayer barrier flexible and semi-rigid packaging for dried foods, aseptically-processed foods, wine, and other oxygen-sensitive foods. Development of polyethylene terephthalate (PET) revolutionized the bottling of beverages and packaging of many other products. Use of heat resistant polypropylene (PP) and PET allowed development of high-quality convenient microwaveable (PP) and dual-ovenable (PET) frozen, aseptically-processed and retorted foods in plastic containers. Availability of polymer films with varying permeabilities allowed development of modified atmosphere packaging of convenient prewashed and cut fruits and vegetables with extended shelf-life.

Packaging continues to evolve in response to developments in other advanced food processing techniques, such as irradiation and high-pressure processing. Developments in material properties and package design have improved the integrity of packaging, reduced the amount of packaging material required, and improved package appearance and convenience. In addition, developments in package law and regulation have increased the safety awareness of packaging manufacturers and increased the amount and quality of information provided to consumers on the package label.

Modern food preservation and packaging protect foods so well that less than 2% of all food is wasted in developed countries. In developing countries, lack of adequate preservation and packaging technology results in 30 to 50% waste of food (Driscoll and Paterson, 1999). Table 13.1 summarizes important developments in food packaging over the past several decades (Sacharow and Brody, 1987; Downes, 1989; Brody, 2004).

13.1.3 FUNCTIONS OF FOOD PACKAGING

Generally, packaging is discussed in terms of providing four basic functions (Yam et al., 1992; Marsh, 2001; Robertson, 2006b):

1. *Containment.* The containment function involves the ability of the packaging to maintain its integrity during the handling involved in filling, sealing, processing (in some cases, such as retorted, irradiated, and high-pressure-processed foods), transportation, marketing, and dispensing of the food.

2. *Protection.* The need for protection depends on the food product but generally includes prevention of biological contamination (from microorganisms, insects, rodents), oxidation (of lipids, flavors, colors, vitamins, etc.), moisture change (which affects microbial growth, oxidation rates, and food texture), aroma loss or gain, and physical damage (abrasion, fracture, and/or crushing). Protection can also include providing tamper evident features on the package (Rosette, 1997). In providing protection, packaging maintains food safety and quality achieved by refrigeration, freezing, drying, heat processing, and other preservation of foods.

3. *Communication.* The information that a package provides involves meeting both legal requirements and marketing objectives. Food labels are required to provide information on the food processor, ingredients (including possible allergens in simple language), net content, nutrient contents, and country of origin. Package graphics are intended to communicate product quality and, thus, sell the product. Bar codes allow rapid check-out and tracking of inventory. Other package codes allow determination of food production location and date. Various open dating systems inform the consumer about the shelf life of the food product. Plastic containers incorporate a recycling code for identification of the plastic material.

4. *Convenience.* Providing convenience (sometimes referred to as utility of use or functionality) to consumers has become a more important function of packaging. Range of sizes, easy handling, easy opening and dispensing, reclosability, and food preparation in the package are examples of packaging providing convenience to the consumer.

In addition to these four basic functions, three other functions have been defined that involve additional requirements of packaging:

1. *Production efficiency.* Another function often included is production efficiency (or machinability or economy), because of the requirement that packages perform well in rapid filling, closing,

TABLE 13.1
Important Developments in Food Packaging

Decade	Developments
1960s	• Spiral-wound composite juice cans with tear-off aluminum ends
	• Boil-in-bags
	• Saran-coated cellophane for snacks
	• Cheese in polypropylene film
	• Plastic tubs for cottage cheese
	• HDPE gallon milk jugs, margarine tubs, and mayonnaise jars
	• Tamper-resistant/evident closures for milk jugs
	• Polyethylene-coated milk cartons
	• Plastic-foam egg cartons
	• Clear PVC bottles for beverages
	• Plastic cans with full-panel end for ham
	• Aluminum beer cans with easy-open ends
	• Plastic loop carriers for beer cans
	• Steel coffee cans with plastic reseal lids
	• Shrink-wrapped corrugated fiberboard trays for canned goods
	• Screw-off closures for beer bottles
	• Bulk palletizing for glass bottles
1970s	• Large bottles for soft drinks
	• Metallized pouches for coffee
	• Bag-in-box for wine
	• PET soft-drink bottles
1980s	• Aseptic carton introduced in United States
	• Co-extruded ketchup bottle
	• Tamper-evident closures
	• Microwavable polymers
	• Microwave susceptors for browning and crisping
	• Thermal processing and heat sealing of rigid plastics
	• Modified atmosphere packaging
1990s and 2000s	• Increased use of PET for soft drinks, water, peanut butter, oils, etc.
	• Stand-up and resealable pouches
	• Retortable plastic pouches, trays, tubs and paperboard cartons
	• Microwavable PP and PP-coated-paperboard trays
	• Dual-ovenable CPET and CPET-coated-paperboard trays
	• Microwavable-defrosting HDPE frozen juice cans
	• Resealable aluminum bottles
	• Irradiation and high-pressure processing of plastic packaging
	• Active packaging (moisture absorbers, oxygen absorbers, etc.)
	• Edible films and coatings
	• Intelligent (communicative and responsive) packaging
	• Sous-vide and cook/chill packaging for home meal replacements
	• Case-ready, modified-atmosphere-packaged meats
	• High-barrier coatings (e.g., amorphous C, silica) for PET and PP
	• High-barrier polymers (e.g., liquid crystal polymers, PEN)
	• Controlled permeability materials (e.g., mineral-filled PP, microperforated films)
	• Shaped, embossed, and/or plastic-lined metal cans
	• Ring-pull easy-open metal-can lids
	• Self-heating and self-cooling cans
	• Hot-fill and retortable plastic bottles and jars

(Continued)

TABLE 13.1
Continued

1990s and 2000s	Developments
	• Hot-fill and aseptically-filled plastic tubes
	• Aseptic processing and packaging for low-acid particulate foods
	• Shelf life modeling and prediction

Source: Adapted from Sacharow, S. and Brody, A.L. (1987), *Packaging: An Intorduction.* M.N. Duluth (Ed.), Harcourt Brace Jovanovich Publications, Inc. 35–77; Downes, T.W. (1989). *Food Technology* 43: 228–229, 232–236, 238–240; and Brody, A.L. (2004). Personal communication.

handling, transportation, and storage operations (Paine and Paine, 1992b; Coles, 2003; Steven and Hotchkiss, 2003b).

2. *Minimal environmental impact.* Minimal environmental impact is an additional function that reflects the need to consider, at the time of package design, the used package disposal (Coles, 2003; Steven and Hotchkiss, 2003b). Reducing the amount of packaging (source reduction) to reduce packaging cost and waste has been a long-time aspect of package development. To further reduce the use of natural resources, package development has increasingly considered additional approaches to minimizing the amount of packaging that ends up in landfills, such as package recyclability and energy recovery.

3. *Package safety.* Package safety is a function that involves the need to consider any possibility that the package might cause contamination of the food product (Coles, 2003; Steven and Hotchkiss, 2003b). Increasing use of plastic packaging materials has brought increased attention to the possibility of migration of plastic monomers and additives into packaged foods. Thus, food packaging is highly regulated to ensure that packaging components do not migrate into food to produce a safety problem for consumers.

13.1.4 PACKAGING TERMINOLOGY

Several categories of packaging terminology exist which aid in considering packaging alternatives.

1. *Package form.* The package form describes the degree of package rigidity. Rigid packaging (e.g., glass jar, metal can) does not change shape upon filling and cannot be deformed without damage. Semi-rigid packaging (e.g., plastic water bottle) can be deformed to some degree without damage, returning to the original shape. Semi-rigid packaging can experience a small change in shape upon filling. Flexible packaging (e.g., plastic pouch for breakfast cereal) does not take form until it is filled. When the food has been dispensed, the package loses its filled shape.

2. *Package level.* The package level describes the food proximity and use of the package. The primary package is in direct contact with the food product (e.g., glass beverage bottle, metal food can, paperboard juice carton, plastic milk jug) and provides the main protection against the environment. Primary packages are also referred to as retail packages or consumer units, because they provide important functions in retail sale and home use (Fellows, 2000; Chinnan and Cha, 2003). The secondary package is the next layer of packaging and generally serves to add protection against physical abuse. The secondary packaging can serve as part of the retail package, by working with the primary package (e.g., a semi-rigid paperboard box that contains a flexible pouch of breakfast cereal) and/or by unitizing two or more primary packages (e.g., a paperboard carton that unitizes beverage cans or bottles). Secondary packaging is also sometimes defined as the distribution or shipping container (e.g., a corrugated box) for number of primary packages (Bourque, 2003). Secondary

packaging is sometimes designed for use in display of primary packages. Tertiary packaging and quaternary packaging are generally used in the distribution of the packaged food product and not seen by the consumer (e.g., stretch-wrapped pallet of boxes and large metal intermodal shipping containers, respectively) (Brighton, 1997; Robertson, 2006b). Tertiary and quaternary levels of packaging are also referred to as logistics packaging, distribution packaging or shipping containers, because they are utilized to contain and protect the product during storage, transport and distribution but have no marketing or consumer use (Twede, 1997; Fellows, 2000; Twede and Harte, 2003). Quaternary packaging is also referred to as a unit load, since a number of distribution packages are unitized into a single unit for handling, storage, and shipping (Soroka, 2002h).

3. *Preformed vs. in-line formed packaging.* Packages are manufactured at very different times, related to the filling of food into the package. Preformed packages (e.g., glass jars, metal cans) are produced at a separate manufacturing facility and then transported to the food processing facility. In-line-formed packages are formed immediately before filling (e.g., milk cartons formed from flat carton blanks) or are formed around the food product (e.g., form-fill-seal flexible pouches formed from rolls of plastic film).

4. *Integral vs. nonintegral packaging.* Whether a package is integral vs. nonintegral depends on whether or not the package is essential to the definition of the packaged product. Whipped cream, vacuum-packed peanuts, carbonated soft drinks, and heat-sterilized soup all rely on packaging for definition. After these products are removed from their packaging, their shelf lives are severely compromised. Thus, their packaging is integral. However, a flexible pouch used to package breakfast cereal in air is not integral to the product. The cereal could be removed from the package and poured into a plastic dispenser and not suffer any loss in shelf life.

13.1.5 FOOD PACKAGE DEVELOPMENT

Successful development of packaging for a new food product requires consideration of several factors (Twede and Downes, 1997; Brody and Lord, 2000):

- Marketplace analysis (i.e., targeted consumers, merchandizing, package-convenience)
- Food product assessment (i.e., needs for protection against physical, microbiological, and/or chemical deterioration)
- Packaging material comparison (for satisfying food protection needs)
- Package design that effectively considers all the functions of packaging
- Package manufacture and testing
- Food shelf life prediction and determination
- Food packaging law
- Market testing

Packaging system selection must also take into account (Paine and Paine, 1992b; Robertson, 2006b):

- Compatibility of the package with the method of preservation (e.g., heat processing or freezing)
- Hazards of food storage and distribution, including environmental conditions (temperature and relative humidity) to which the food will be exposed
- Potential for food–package interactions, including migration of packaging components into the food and absorption of food components into the packaging
- Packaging equipment considerations (i.e., packaging cleaning, filling, closing, labeling)

The package development process can be seen from the perspective of three phases: (1) planning, including identifying package concepts, feasibility assessment, and consumer concept- and usage-testing, (2) proving functionality by package testing, and (3) package launch, including production start-up and performance monitoring (DeMaria, 2000). Packaging designer's checklist, Gantt charts, and dividing the package design process into technical, manufacturing and engineering, marketing, and purchasing and traffic responsibilities help organize the multidisciplinary responsibilities involved (Soroka, 2002b). Integration of all of the design activities in a concurrent or simultaneous manner can allow achievement of goals in a shorter time period (Raper and Borchelt, 1997). Package design can be aided by utilization of design software such as Computer Assisted Packaging Evaluation (CAPE Systems, Inc.) and Total Optimization Packaging Software (TOPS Engineering Corp.). The concept of Quality Function Deployment can also be applied to package design to efficiently incorporate the views of customers and prioritize their needs (Raper and Borchelt, 1997).

13.2 PROTECTIVE FUNCTION OF FOOD PACKAGING

The expenses of purchasing the highest-quality raw materials and then using the most advanced food preservation methods are wasted if the appropriate protective packaging is not utilized. Destruction and inhibition of microorganisms are the main concerns of food preservation approaches, related to preventing food spoilage and food-borne illness, along with providing the highest food quality (appearance, aroma, taste, texture, etc.) possible. Packaging works with the preservation approach by preventing microbial contamination, inhibiting microbial growth, and minimizing quality loss. However, after packaging it is possible that foods can deteriorate from one or a combination of biological, chemical, and physical reasons (Brown and Williams, 2003; Tucker, 2003; Singh and Anderson, 2004; Robertson, 2006d). The packaging must be designed to protect foods against environmental interactions that affect these types of deterioration.

All packaging aims at preventing contamination of foods by providing a barrier to soils, microorganisms, insects, and/or rodents. Depending on the food product, packaging is also designed to control other environmental interactions, including oxygen uptake, moisture loss or gain, aroma loss or gain, food component absorption by the packaging material (scalping), package-component migration into the food, and light transmission. Figure 13.1 shows the interactions possible among food, package, and the atmosphere (Linssen and Roozen, 1994). Packaging is also designed to minimize the effect of physical stresses (compression, shock, and vibration) on the food. Knowledge of the food product's vulnerability to all of these environmental factors is critical to package design. Several references discuss the packaging requirements of particular foods (Rizvi, 1981; Paine and Paine, 1992a; Brody, 1997; Driscoll and Paterson, 1999; Petersen et al., 1999; Robertson, 2006c).

13.2.1 PACKAGED FOOD INTERACTION WITH THE SURROUNDING ATMOSPHERE

13.2.1.1 Oxygen, Nitrogen, and Carbon Dioxide

Exposure to oxygen can cause deterioration of many foods due to oxidation of lipids and other oxygen-sensitive components such as aromas, colors, and vitamins (Fennema, 1996a; Gordon, 2004; Robertson, 2006d). These foods benefit from packaging that can maintain a vacuum or nitrogen atmosphere and provides a barrier to oxygen. Oxygen diffusion resistance in the food can also affect the oxidation rates (Karel and Lund, 2003a). Table 13.2 shows amounts of oxygen with which foods can react before unacceptable changes in quality occur (Salame, 1974; Robertson, 1993b, 2006e).

Foods such as fresh meat, poultry, bakery and pasta products, and chilled prepared foods benefit from packaging that can maintain either a vacuum or a targeted low concentration of oxygen and high concentration of carbon dioxide to prevent oxidation and control microbial growth. High concentration of oxygen combined with high concentration of carbon dioxide maintains color of fresh red meat while delaying microbial spoilage (Zagory, 1997). Thus, these products benefit from packaging with low permeation to oxygen, carbon dioxide, and nitrogen.

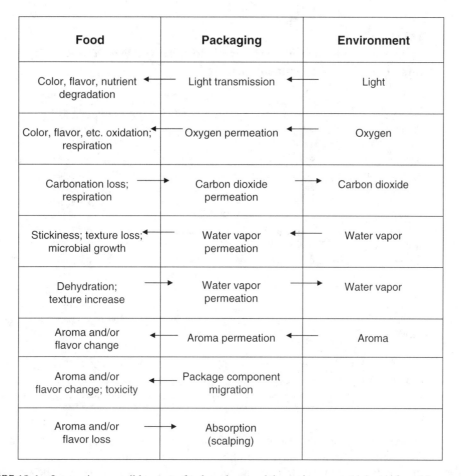

Food	Packaging	Environment
Color, flavor, nutrient degradation	Light transmission	Light
Color, flavor, etc. oxidation; respiration	Oxygen permeation	Oxygen
Carbonation loss; respiration	Carbon dioxide permeation	Carbon dioxide
Stickiness; texture loss; microbial growth	Water vapor permeation	Water vapor
Dehydration; texture increase	Water vapor permeation	Water vapor
Aroma and/or flavor change	Aroma permeation	Aroma
Aroma and/or flavor change; toxicity	Package component migration	
Aroma and/or flavor loss	Absorption (scalping)	

FIGURE 13.1 Interactions possible among food, package and the environment. (Adapted from Linssen, J.P.H. and Roozen, J.P. (1994). Food flavour and packaging interactions. In *Food Packaging and Preservation.* M. Mathlouthi, Ed, New York, Blackie Academic and Professional, 48–61.)

Fresh fruits and vegetables are respiring and thus need packaging that allows permeation of oxygen in and carbon dioxide out at appropriate rates. Proper design of fruit and vegetable packaging takes into account the different respiration rates of different fruits and vegetables and controls package-head-space oxygen and carbon dioxide concentrations to targeted levels that reduce product respiration rate and increase shelf life (Zagory, 1997).

13.2.1.2 Water Vapor

The most common environmental interaction that packaging is designed to control is food moisture loss or gain through desorption or absorption of water vapor, respectively. Table 13.2 shows the approximate changes in water content that various foods can experience before unacceptable changes in quality occur (Salame, 1974; Robertson, 1993b, 2006e).

In terms of food stability and food properties, it is appropriate to use water activity, a_w, as a measure of the degree of water association with the food's nonaqueous constituents. For a food at equilibrium with its environment (Fennema, 1996b):

$$a_w \sim p/p_o = \%ERH/100 \tag{13.1}$$

TABLE 13.2
Approximate Amounts of Oxygen and Moisture with Which Foods Can Interact before Unacceptable Change

Food or beverage	Maximum O_2 gain (ppm)	Maximum water gain or loss
Canned milk, vegetables, flesh foods, baby foods, soups, and sauces	1–5	3% loss
Beers and wines	1–5	3% loss
Instant coffee	1–5	2% gain
Canned fruits	5–15	3% loss
Dried foods	5–15	1% gain
Dry nuts and snacks	5–15	5% gain
Fruit juices, drinks and carbonated soft drinks	10–40	3% loss
Oils, shortenings, and salad dressings	50–200	10% gain
Jams, jellies, syrups, pickles, olives, and vinegars	50–200	3% loss
Liquors	50–200	3% loss
Condiments	50–200	1% gain
Peanut butter	50–200	10% gain

Source: Adapted from Salame, M. (1974). *Permeability of Films and Coatings.* H.B. Hopfenberg (Ed.), New York, Plenum Publ. Corp., 275; Robertson, G.L. (2006e). *Food Packaging — Principles and Practice.* Boca Raton, CRC Taylor & Francis Group, 225–254.

where p is the water vapor partial pressure of the food, p_o is the vapor pressure of pure water, and %ERH is the percent equilibrium relative humidity of the environment.

The water activity affects food stability in a number of different ways (Harte and Gray, 1987; Fennema, 1996b; Esse and Saari, 2004; Robertson, 2006d). Figure 13.2 shows a typical relationship between food water activity and moisture content (moisture isotherm) and the relative rates for a number of chemical reactions, enzyme activities, and microorganism growths that lead to food deterioration (Fennema, 1976). Table 13.3 gives additional details on the effect of water activity on the growth of microorganisms, oxidation, and Maillard browning in foods, along with examples of foods that fall within the listed water activity ranges (Mossel, 1975).

Food water content in the Zone 1 a_w range of Figure 13.2 is strongly associated with polar sites of the food constituents. Thus, it does not allow sufficient molecular mobility to produce food deterioration caused by enzyme activity, nonenzymic browning, hydrolytic reactions or microorganism growth. Interestingly, lack of water molecules at very low a_w allows greater vulnerability of lipids to oxidation. The oxidation rate then drops as a_w increases in Zone I, possibly due to water molecules associating with initial lipid oxidation products (hydroperoxides) or hydrating catalytic metal ions to inhibit the oxidation process. However, oxidation rate then increases as a_w increases further (Zone II) to increase oxygen solubility and expose more catalytic sites by the swelling of the food (Karel and Yong, 1981). Water in Zone II is more mobile than Zone I water, because its associations are by hydrogen bonding to available food constituent sites, strongly-associated Zone I water, and solute molecules. The resulting increased molecular mobility and swelling of the food increases the rates of most reactions. Zone III water is bulk-phase water that further increases molecular mobility and, thus, the rates of many reactions and the growth of microorganisms (Fennema, 1996b). The rates of some reactions decrease, possibly because additional water inhibits water-producing reactions and/or dilutes reactants (Eichner, 1975; Labuza and Saltmarch, 1981).

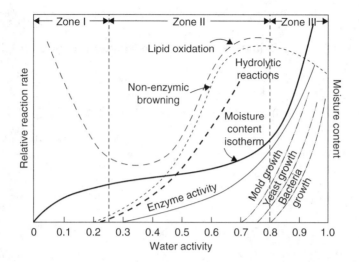

FIGURE 13.2 Relationship between food water activity and moisture content (moisture isotherm) and the relative rates for a number of chemical reactions, enzyme activities and microorganism growths that lead to food deterioration. (From Fennema, O.R., Ed., 1976. In *Principles of Food Science — Part I — Food Chemistry*. New York, Marcel Dekker, Inc. Copyright 1976 from *Principles of Food Science — Part I — Food Chemistry* by O.R. Fennema. Reproduced by permission of Routledge/Taylor & Francis Group, LLC.)

The zones shown in Figure 13.2 also relate to food texture (Fennema, 1996b). Water in Zone I behaves as if it is part of the solid food. Thus, it does not have a plasticizing (softening) effect on the food. As water content of dry food increases through Zone II and into Zone III, the food is increasingly plasticized and its crisp texture is eventually lost (Figure 13.2) (Katz and Labuza, 1981; Robertson, 2006f). In contrast, fresh fruits and vegetables ($a_w \sim 0.9$–1.0) have high cell turgor and, thus, are quite crisp. Loss of moisture reduces cell turgor and reduces crispness. Loss of moisture can also affect the character of preserved high-moisture foods and beverages.

One can see from Figure 13.2 that destruction of enzymes and microorganisms is necessary for shelf-stable high-moisture ($a_w = \sim 0.85$–1.0) foods. For such foods, hermetically-sealed packaging that prevents contamination with microorganisms is required. Fresh, high-moisture foods require refrigeration to slow enzymic and nonenzymic reactions and microbial growth, as well as packaging that prevents moisture loss. Freezing temperatures further reduce the rates of reactions and microbial growth to produce relatively long shelf life. Frozen, high-moisture foods have longer shelf life due to lower temperature and also require packaging that prevents moisture loss (freezer burn). Dry foods ($a_w < \sim 0.55$) require packaging that prevents moisture gain that causes loss of texture and allows growth of microorganisms. Intermediate-moisture foods ($a_w = \sim 0.55$–0.85) require packaging that prevents moisture gain that enhances microbial growth and moisture loss that produces loss of soft texture. Change in a_w can affect other physical changes besides crispness, including crystallization, stickiness, caking, and collapse (Karel and Lund, 2003b).

13.2.1.3 Aromas

Undesirable interactions of food with the environment include the possibility of loss or gain of aromas. Loss of food aromas to the environment reduces the fresh character of food. Gain of aromas from the environment can include engine fuel and exhaust vapors, as well as the aromas of other products such as cosmetics and cleaning agents. Thus, packaging that retains food aromas and excludes foreign aromas is important for maintaining food quality.

TABLE 13.3
Effect of Water Activity on Oxidation, Maillard Browning, and Growth of Microorganisms

a_W	Phenomena	Food examples
1.00		*Water-rich foods* ($0.90 < a_W < 1.0$)
~0.99		Fresh produce, meat, chicken, fish
0.95		Foods with 40% sucrose or 7% NaCl, for example, cooked sausages, bread crumbs
0.90	Lower limit for bacterial growth (general)	Foods with 55% sucrose or 12% NaCl, for example, dry ham, medium age cheese
		Intermediate moisture foods ($0.55 < a_W < 0.90$)
0.85	Lower limit for growth of most yeasts	Foods with 65% sucrose or 15% NaCl, for example, salami, "old" cheese
0.80	Lower limit for activity of most enzymes Lower limit for growth of most molds	Flour, rice (15–17% water), fruit cake, sweetened condensed milk
0.75	Lower limit for halophilic bacteria Maximum heat resistance of vegetative bacterial cells	Foods with 26% NaCl (saturated), for example, marzipan (15–17% water), jams
0.70	Lower limit for growth of most xerophilic ("dry loving") molds	
0.65	Maximum velocity of Maillard reactions	Rolled oats (10% water)
0.60	Lower limit for growth of osmophilic or xerophilic yeasts and molds	Dried fruits (15–20% water), toffees, caramels (8% water)
0.55	Lower limit for microbial growth	
		Dried foods ($0 < a_W < 0.55$)
0.50		Noodles (12% water), spices (10% water)
0.40	Minimum oxidation velocity	Whole egg powder (5% water)
0.30		Crackers, bread crusts (3–5% water)
0.25	Maximum heat resistance of bacterial spores	
0.20		Whole milk powder (2–3% water), dried vegetables (5% water), corn flakes (5% water)
0.00	Maximum oxidation velocity	

Source: Adapted from Mossel, D.A.A. (1975). *Water Relations of Foods* R.B. Duckworth (Ed.), New York, Academic Press Inc. (London) LTD., 347–361.

13.2.2 PACKAGED FOOD INTERACTION WITH LIGHT

Depending on food composition, light can catalyze a number of reactions that lead to chemical deterioration of the food. Light in the high ultraviolet (2900–4000 Å) and low visible (4000–4500 Å) wavelengths catalyzes lipid, color, flavor, and vitamin degradation (Robertson, 2006d). The degree of effect for a given reaction depends on the particular wavelength (Bossett et al., 1994).

Solid foods are least sensitive to light, because the penetration of light into the food decreases exponentially (Karel and Lund, 2003a). However, the situation is different for liquid foods. Diffusion in the liquid exchanges light-sensitive food components between the surface and interior, so that light-degraded compounds are replaced with nondegraded compounds at the surface that are subsequently degraded. The light-degraded compounds can also interact with compounds in the interior to cause further degradation.

The sensitivity of a particular food to light will determine the selection of packaging material. Foods that are vulnerable to light benefit from packaging that prevents light transmission over all or a portion of the wavelengths of concern. Thus, a trade-off between food visibility and light blocking is often necessary.

13.2.3 PACKAGED FOOD INTERACTION WITH PHYSICAL STRESSES

Food physical deterioration can result from bruising, deformation, breakage, or abrasion due to subjection of food to compression, shock, or vibration. Bruising of fresh fruits, vegetables, meat, poultry, and seafood can lead to chemical and biological deterioration. Deformed, fragmented, or abraded food is viewed as inferior by consumers. Additionally, any resulting increase in surface area increases the food's vulnerability to interactions with the atmosphere and/or packaging material.

Rigid and semi-rigid packages protect food from compression damage to the extent they maintain their integrity under compression. Flexible packaging provides little or no protection against compression damage. Thus, primary flexible packages of food are often placed in semi-rigid or rigid secondary packages. All packages, including flexible packages, limit shock and vibration damage to the extent they restrict movement of the food. Any cushioning that packaging materials provide reduces the effect of shock and vibration on the food. Fragile foods are often protected with cushioning materials added to the package (Peache, 1997; Soroka, 2002c; Karel and Lund, 2003a).

Beyond protecting food from physical deterioration, the packaging must maintain its integrity to provide its other functions. Failure of the packaging material will result in food contamination from soils and microorganisms, as well as increased interactions with the atmosphere.

13.2.4 PACKAGED FOOD INTERACTION WITH PACKAGING MATERIAL

Migration involves a component of a packaging material transferring to the food product and possibly to the environment external to the package (Selke, 1997c). In scalping (sorption), a component of a food product is sorbed by the packaging material without transfer to the surrounding atmosphere (Giacin, 1995; Giacin and Hernandez, 1997). To varying degrees, all materials used for food packaging have been found to interact with food in one or both ways (Katan, 1996). Possible migrating substances include plastic monomers and plasticizers (Figge, 1996), paper coating and adhesive components (Soderhjelm and Sipilainen-Malm, 1996), metals and metal coatings (Murphy, 1996) and glass component ions (Tingle, 1996). The greatest concern is with migration of low molecular weight substances from polymeric plastic materials in contact with food. The existence of these substances in packaging does not necessarily produce migration. Low levels can be totally or partially immobilized due to strong interactions with the packaging material (Miltz, 1992). The greatest concern with scalping is also with polymeric plastic materials, with resulting loss in food quality. The migration and scalping phenomena are very important to food safety and quality. Thus, much has been written about them in recent years (Gray et al., 1987; Hotchkiss, 1988; Risch and Hotchkiss, 1991; Linssen and Roozen, 1994; Giacin, 1995; Katan, 1996; Hernandez and Giacin, 1998; Risch, 2000). Because of concerns related to migration, food packaging is subject to rigorous laws and regulations to ensure food safety. (See Section 13.8.)

13.3 PACKAGING MATERIALS

Over $100 billion dollars are spent annually on food packaging in the United States., $\sim$40% on paper and paperboard packaging, $\sim$18% on plastic film and $\sim$12% on semi-rigid/rigid plastic packaging, $\sim$14% for aluminum, $\sim$6% for steel packaging, and $\sim$10% for glass containers (Brody, 2003b).

13.3.1 GLASS

Glass is one of the oldest manufactured materials and one of the first manufacturing businesses in the New World. Nonetheless, glass still serves as an important packaging material for food.

13.3.1.1   Advantages and Disadvantages

Glass is the most inert of the packaging materials and provides a total barrier to gases, water vapor and aromas, has good strength under compression, possesses good heat resistance to allow thermal processing of foods, allows viewing of the product, and is microwaveable. Other advantages of glass containers are customer perception that they add value to the food product and their recyclability. Because of their inertness, glass containers have the potential of being returned by the customer and refilled by the food manufacturer. This practice used to be common for milk, soft drink, and beer, and is practiced to a limited extent. However, the weight of glass and the return distance to increasingly more centralized food manufacturers work against returning and refilling. The disadvantages of glass include its weight and vulnerability to fracture from thermal shock (rapid temperature change) and physical shock. In recent years, advances in the science and technology of glass have resulted in lighter, stronger glass containers. For those food products vulnerable to light-catalyzed reactions, glass's transparency to light is another disadvantage. Use of light-absorbing colorants in the glass, as well as glass container labels and direct printing on the glass, will affect the transmission of light. The advantages and disadvantages of glass containers are summarized in Table 13.4.

13.3.1.2   Glass Composition and Properties

Glass is made by mixing several naturally-occurring inorganic compounds at a temperature above their melting points. The molten mixture is then cooled to produce a noncrystalline, amorphous solid. The main ingredient is silica (sand) (SiO_2) that serves as the network-forming backbone of the glass. However, silica has a very high melting temperature, and molten silica has high viscosity that makes it difficult to form into shapes. Adding soda (Na_2O) modifies the silica network by disrupting some of the Si-O bonds, with resulting lower melting temperature and viscosity but reduced resistance to dissolving in water. Thus, lime (CaO) is added as a network stabilizer, with the result that durability is increased but tendency to crystallize is also increased. Finally, alumina (Al_2O_3) is added as an intermediate to resist crystallization (Bayer, 2003; Robertson, 2006g). Minor amounts of colorants are added to produce colored glass, including chromium oxide for green, cobalt oxide for blue, nickel oxide for violet, selenium for red, and iron plus sulfur and carbon for amber. Amber provides the best protection for light-sensitive foods and beverages, transmitting very little light with wavelength shorter than 450 nm. Table 13.5 gives typical composition and properties of glass (Bayer, 2003).

TABLE 13.4
Advantages and Disadvantages of Glass Containers

Advantages	Disadvantages
Inert	Heavy
Total barrier to	Vulnerable to fracture
Gas	From thermal shock
Water vapor	From physical shock
Aroma	No protection from light
Good compression resistance	(unless colored)
Good heat resistance	
Allow viewing of product	
Microwavable	
Customer perception of high quality	
Reclosable	
Recyclable	
Refillable	

TABLE 13.5
Typical Composition and Desirable Properties of Glass

Oxide	Weight percent	Desirable properties
SiO_2	70–72	Moderate cost
Na_2O	13–15	Easily shaped
CaO	12	Chemically durable
Al_2O_3	2	Inherently strong
Minors	1	Low thermal expansion
		Nonpermeable
		Tasteless and odorless
		Transparent (flint)
		Light protection (amber)

Source: From Bayer, R. (2003). *Glass Packaging Essentials: A Multimedia Resource CD-ROM.* Alexandria, VA, Glass Packaging Institute.

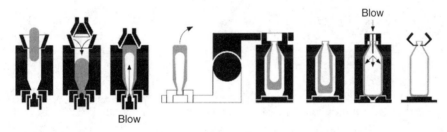

FIGURE 13.3 Blow and blow glass molding process. (Copyright 2003 from *Food Packaging Technology*, R. Coles, D. McDowell, and M.M. Kirwan, Eds. Courtesy of Simon Morgan, Rockware Glass Ltd., West Yorkshire, UK.)

Another important ingredient in glass making is crushed recycled glass, called cullet. Ability to use cullet allows glass recycling, with resulting diversion from landfills and reduced use of raw materials. Cullet also reduces melting temperature, saves energy, reduces corrosion of the heating furnace, and reduces atmospheric emissions (Bayer, 2003). Cullet can be added in any amount, with some states requiring a minimum amount to encourage use and enhance recycling.

13.3.1.3 Glass Package Manufacture

Glass can be molded into vials, tumblers, jars, bottles, jugs (large bottles with handles), and carboys with a wide variety of custom shapes and colors. Gobs of ~2100°F molten glass with desired shape and weight are fed into blank molds to be pressed or blown into thick, hollow, partially-formed containers called blanks, preforms, or parisons. The semi-molten parisons are then transferred to blow molds to be blown into the final container shape. Making the container in two steps allows greater control of glass thickness over different parts of the container. Figure 13.3 through Figure 13.5 show the glass molding processes used for different-shape containers. After molding, glass containers are passed through an annealing oven, where they are reheated to remove stresses in the glass, and then cooled slowly to prevent fracturing. Chemical treatments of the inner and outer surfaces of the glass containers can be used to provide greater chemical resistance and reduced coefficient of

FIGURE 13.4 Press and blow glass molding process for wide-mouth containers. (Copyright 2003 from Food Packaging Technology, R. Coles, D. McDowell, and M.M. Kirwan, Eds. Reproduced by permission of Blackwell Publishing Ltd, Oxford, UK, courtesy of Simon Morgan, Rockware Glass Ltd., West Yorkshire, UK.)

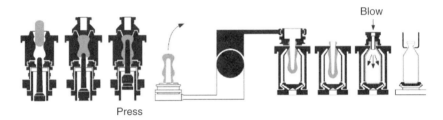

FIGURE 13.5 Press and blow glass molding process for narrow-neck containers. (Copyright 2003 from Food Packaging Technology, R. Coles, D. McDowell, and M.M. Kirwan, Eds. Reproduced by permission of Blackwell Publishing Ltd, Oxford, UK, courtesy of Simon Morgan, Rockware Glass Ltd., West Yorkshire, UK.)

friction, respectively. Several quality control tests are used. These include assessment of container dimensions, glass temper number, thermal shock resistance, and internal pressure resistance.

Details on the design and manufacture of glass containers can be found in several references (Yamato, 1990; Paine, 1991; Paine and Paine, 1992a; Cavanagh, 1997; Hambley, 1997; Hanlon et al., 1998a; Twede and Goddard, 1998; Soroka, 2002g; Barron and Burcham, 2003a; Bayer, 2003; Girling, 2003; Robertson, 2006g). Figure 13.6 shows a typical glass container. The terminology used in designing and describing these containers is shown. In addition to the parts shown, many glass containers also include a neck (e.g., all bottles), which is a relatively straight section between the shoulder and the bottom of the transfer bead or neck ring parting line.

13.3.1.4 Glass Package Closures

Several different closure options exist for glass containers (Nairn and Norpell, 1997; Soroka, 2002e). The traditional crown (crimp) closure used for glass beer and soft drink bottles does not allow for reclosure. However, the threaded and snap-on nature of many glass-container closures allows for resealing after opening. Figure 13.6 shows the continuous thread (for screw caps) and press-on/twist-off finish options for glass containers.

13.3.1.5 Glass Packaging Uses

Figure 13.7 shows glass packaging production by market sector (Bayer, 2003). Bottling of beer accounts for over half of glass container manufacture, with most of this glass colored amber or green. The heat resistance of glass is utilized in the packaging of heat-processed foods, with use of approximately one-fourth of glass containers. Other beverages account for most of the rest of glass packaging use.

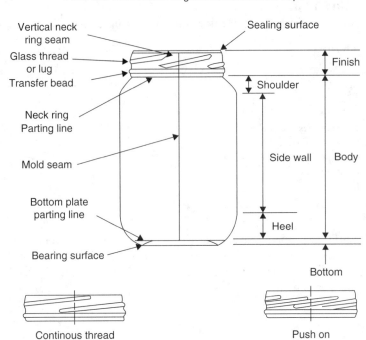

Parts and construction of glass containers and caps

FIGURE 13.6   Basic parts of a glass container. (Courtesy of Dr. Bradley Taylor, Food Processors Institute, Washington, DC).

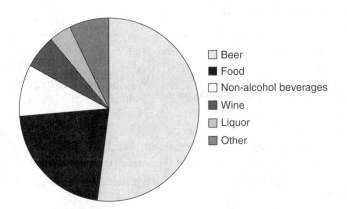

FIGURE 13.7   Glass packaging production by food market sector. (From Bayer, R., 2003. *Glass Packaging Essentials: A Multimedia Resource CD-ROM*. Alexandria, VA, Glass Packaging Institute.)

13.3.1.6   Glass Packaging Trends

A number of important trends have helped glass containers maintain their share of the food packaging market (Rowan, 2001; Bayer, 2003). The most significant trend in glass packaging in recent years has been lightweighting, the ability to manufacture glass containers that are increasingly lighter while not sacrificing strength or other properties. For example, beer bottles have been reduced in weight by over 40% since 1966. Development of the Narrow Neck Press and Blow forming process has allowed better distribution of glass throughout the bottle, so that thickness varies with need for strength. Also,

TABLE 13.6
Advantages and Disadvantages of Metal Containers

Advantages	Disadvantages
Total barrier to	Not inert, must be coated
Gas	Moderately heavy
Water vapor	Multi-step manufacture
Aroma	Do not allow viewing of product
Good compression resistance	
Good heat resistance	
Good thermal and physical shock resistance	
Light protection	
Recyclable	

improvements in surface treatments have reduced scratching that reduces container strength. Glass packaging has also benefited from innovative container shapes, labeling and decorating techniques, such as shrink-wrap labeling, shrink-sleeve (whole-body) labeling, ceramic ink (silk-screening) labeling, acid etching, and embossing.

13.3.2 METALS

Tin-plated iron cans have been used since the mid-18th century to preserve heat-processed foods. Since that time, great advances in the production of steel and aluminum, coating of these metals, and forming them into containers have resulted in their being important packaging materials.

13.3.2.1 Advantages and Disadvantages

Like glass, steel and aluminum are total barriers to gases, water vapor and aromas. Both also have good heat resistance and can withstand physical and thermal shock. Because of steel's greater strength, it is used more often in the thermal processing of foods. Neither steel nor aluminum is as inert as glass; thus both must be coated to avoid interactions with the foods they contain. Tin or chromium is used to coat steel, usually followed by a coating with a polymeric lacquer (enamel). Aluminum is coated directly with a lacquer. Other advantages of metal containers are exclusion of light from food products that are light-sensitive and their recyclability. The disadvantages of metal containers include their multi-step manufacture, weight (particularly steel), and (for some foods) lack of transparency. In recent years, advances in the science and technology of these metals have resulted in lighter, stronger metal containers. The advantages and disadvantages of metal containers are summarized in Table 13.6.

13.3.2.2 Tin- and Chromium-Coated Steel Composition and Properties

Steel is an alloy of iron and carbon, consisting of $\sim99.5\%$ iron with a small amount of carbon (~0.02–0.3%). It is produced by reduction of iron ore (iron oxide, Fe_2O_3) with coke (carbon) to remove the oxygen as CO_2. Small amounts of contaminants such as manganese, silicon, sulfur and excess carbon are reacted with oxygen to form oxides that are removed as slag and CO_2. Small adjustments to the composition give the steel more ductility (D) or higher strength and stiffness (N), or make the steel appropriate for moderately-corrosive (MR) or strongly-corrosive (L) products. Slabs of steel are formed from molten steel and the slabs are hot rolled into sheets, which are then cold-rolled (CR) to approximately the thickness desired for containers. At this point, the sheets are

TABLE 13.7
Primary Steel Alloys and Their Applications

Type	Properties	Application
L	High purity; low in residual elements.	Used where high internal corrosion resistance is required.
MR	Similar to L, but Cu and P content are raised. Most widely used tinplate steel.	Vegetable and meat packs where internal corrosion resistance is not critical.
N	Nitrogenized steel with up to 0.02% N to increase strength.	Used where high strength and rigidity required; for example, can ends and aerosol domes.
D	Stabilized steel and therefore nonaging. Less C than other tinplate steels.	Used for severe drawing operations; for example, D&I cans.

Source: Adapted from Robertson, G.L. (2006a). *Food Packaging — Principles and Practice.* Boca Raton, CRC Taylor & Francis Group, 121–156.

TABLE 13.8
Relationship between Steel Temper and Application

Temper classification	Rockwell hardness	Applications
T50	46–52	Nozzles, spouts and closures; deep drawn parts
T52	50–56	Shallow-drawn and specialized can parts
T57	54–63	Can ends, bodies; large diameter closures and crowns
T65	62–68	Stiff can ends and bodies for noncorrosive products
T70	67–73	Very stiff applications
DR8	70–76	Round can bodies and can ends
DR9	73–79	Round can bodies and can ends
DR9M	74–80	Beer and carbonated beverage can ends

Source: Adapted from Robertson, G.L. (1993a). *Food Packaging — Principles and Practice.* New York, Marcel Dekker, Inc., 173–203.

approximately 0.01 in. thick. The cold-rolling also increases the steel sheet strength and stiffness. The sheets are then heat-annealed to increase ductility and then, possibly, cold-rolled again (2CR or DR) to increase strength and stiffness (i.e., increase temper). Tables 13.7 and Table 13.8 relate steel composition and treatment to properties and uses of the resulting steel (Robertson, 1993a, 2006a).

To protect the steel sheets from corrosion, they are electrolytically coated with either tin or chromium. The coating thickness can be made different on the two sides of the sheet, with the thicker side commonly on the side facing the food product. When the steel sheets are coated with tin, the total tin thickness on the two sides of the sheet is approximately 1% of the sheet thickness. The resulting structure is often called tin-plate. Depending on the food to be stored, tin-plate steel is coated with an organic lacquer (enamel) to prevent interaction between the tin and food that could produce undesirable color or flavor change. Tin-free steel (TFS), also called Electrolytic Chromium-Coated Steel (ECCS), has chromium coatings that are thinner than tin coatings which always have to be coated with a lacquer. Table 13.9 gives information on the types of lacquers available and

TABLE 13.9
Metal Can Lacquers and Their Uses

Resin	Flexibility	Sulfide stain resistance	Typical uses
Oleo-resinous	Good	Poor	Acid fruits
Sulfur-resistant oleo-resinous (added zinc oxide)	Good	Good	Vegetables, soups (on can or as topcoat over epoxy-phenolic)
Phenolic	Moderate	Very good	Meat, fish, soups, vegetables
Epoxy-phenolic	Good	Poor	Meat, fish, soups, vegetables, beer, beverages (top coat)
Epoxy-phenolic with zinc oxide	Good	Good	Vegetables, soups (especially can ends)
Aluminized epoxy-phenolic	Good	Very good	Meat products
Vinyl solution	Excellent	N/A	Spray on can bodies, roller coat on ends, topcoat for beer and beverages
Vinyl organosol or plastisol	Good	N/A	Beer & beverage topcoat on ends, bottle closures, drawn cans
Acrylic	Very good (some ranges)	Very good (pigmented)	Vegetables, soups, prepared foods containing sulfide stainers
Polybutadiene	Moderate to poor	Very Good (if zinc)	Beer and beverage first coat, vegetables and, soups if with ZnO

Source: Adapted from Robertson, G.L. (1993a). *Food Packaging — Principles and Practice.* New York, Marcel Dekker, Inc., 173–203; Soroka, W. (2002f). *Fundamentals of Packaging Technology.* Naperville, IL, Institute of Packaging Professionals, 155–178.

their uses (Robertson, 1993a; Soroka, 2002f). Thermoplastic materials such as nylon have also been developed to coat directly on steel sheets, as a replacement for tin and chromium (Karel and Lund, 2003a).

13.3.2.3 Aluminum Composition and Properties

Aluminum (Al) is made by electrolytic reduction of alumina (Al_2O_3) that is separated from bauxite. Small amounts of other elements are added to produce aluminum alloys which have different desirable formability and corrosion-resistance properties. Table 13.10 gives the main aluminum alloy types and uses (Robertson, 1993a; Soroka, 2002f; Robertson, 2006a). Aluminum is lighter and weaker than steel but is more easily formed into cans. Slabs made from molten aluminum are hot-rolled into sheets for can-making. The same lacquers lists in Table 13.9 are used to coat the aluminum to prevent interaction with foods.

13.3.2.4 Metal Package Manufacture

Tin- and chromium-coated steel are used to make three-piece cans, whereas both coated steel and aluminum can be made into two-piece cans (Kraus and Tarulis, 1997; Reingardt and Nieder, 1997). Aluminum is also formed into trays, pans, and foils, as well as formed as coatings on plastics (metalized plastics).

Figure 13.8 shows the process used to manufacture *three-piece cans* with a welded side seam (Page et al., 2003). Tin- or chromium-coated steel sheets are cut into body blanks of appropriate size

---

TABLE 13.10
Main Aluminum Alloys and Their Applications

Alloy type	Application
1050	Foils and flexible tubes
3004	Beverage closures and D&I can bodies
5182	Easy open beverages

Source: Adapted from Soroka, W. (2002f), *Fundamentals of Packaging Technology.* Naperville, IL, Institute of Packaging Professionals, 155–178; Robertson, G.L. (2006a). *Food Packaging — Principles and Practice.* Boca and Raton, CRC Taylor & Francis Group, 121–156.

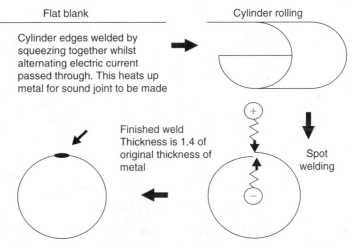

Flat blank

Cylinder rolling

Cylinder edges welded by squeezing together whilst alternating electric current passed through. This heats up metal for sound joint to be made

Finished weld
Thickness is 1.4 of original thickness of metal

Spot welding

Necking, flanging, beading as for DWI can

FIGURE 13.8 Steps in the manufacture of the 3-piece can. (From Page, B., Edwards, M., and May, N., 2003. *Food Packaging Technology.* R. Coles, D. McDowell, and M. Kirwan, Eds., Boca Raton, FL, CRC Press, 120–151. Copyright 2003 from *Food Packaging Technology*, R. Coles, D. McDowell and M.M. Kirwan, Eds. Reproduced by permission of Blackwell Publishing Ltd, Oxford, UK.)

to form the body of the desired can. The body blank is rolled into a cylinder with slight overlap. A side seam is formed along the overlap by welding. For ECCS, a strip of chromium at the overlap must be removed to allow welding. After pressure testing of the seamed body, a flange is formed on both ends of the can body, in order to allow seaming with the can ends. If intended for retorting the can body is rolled to form beads that reinforce the can against pressure difference which could produce collapse. Circular ends are stamped from sheets of coated steel and the edges are curled and coated with a sealing compound to allow seaming with the can body (Heck, 1997). One end is double seamed onto the can body, in a two-step operation shown in Figure 13.9 (Barron and Burcham, 2003b). The other end is double-seamed onto the can, after it is filled at a food processing plant. The double seams on both ends are examined on a regular basis to make sure all components conform to the exact dimensions that give a hermetic seal. Sealed cans are also checked for proper vacuum by tapping or by using a mechanical or optical technique to assess whether the can lid has the proper shape associated with desired vacuum.

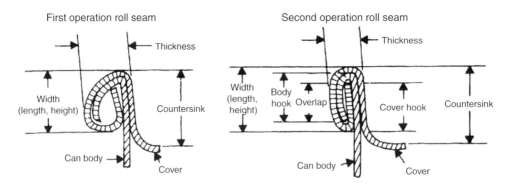

FIGURE 13.9 Metal double seam operations. (From Barron, 2003; Copyright 2003 from *Encyclopedia of Agricultural, Food, and Biological Engineering* by D.R. Heldman, Ed. Reproduced by permission of Routledge/Taylor & Francis Group, LLC.)

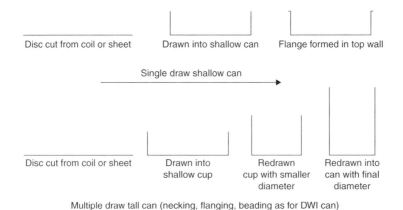

FIGURE 13.10 Steps in the manufacture of the draw and redraw (DRD) 2-piece can. (From Page, B., Edwards, M., and May, N., 2003. *Food Packaging Technology.* R. Coles, D. McDowell, and M. Kirwan, Eds., Boca Raton, FL, CRC Press, 120–151. Copyright 2003 from *Food Packaging Technology*, R. Coles, D. McDowell, and M.M. Kirwan, Eds. Reproduced by permission of Blackwell Publishing Ltd, Oxford, UK.)

Two-piece cans can be made from both coated steel and aluminum. Cans that have height less than diameter can be drawn (stamped) from a circular blank of tin-plate steel, ECCS or aluminum in one step through a die. If the can height is greater than the diameter, a second and possibly a third drawing step is necessary to achieve the desired can diameter and force more of the metal from the bottom of the originally-drawn form to the can side. The thickness of the metal on the can bottom and side is the same as the metal blank. This latter process, shown in Figure 13.10, is called draw and redraw (DRD) (Page et al., 2003). Lacquers can be coated onto the metal before the drawing operation(s), since they withstand the shaping process. Two-piece DRD cans are usually produced as sanitary food cans, since a thick side-wall is needed to withstand the pressure changes in heat processing. Tests similar to those used for three-piece cans are used to ensure leak-proof bodies and proper sealing and vacuum.

Two-piece cans can also be made with a process where the first step is similar to the first step of a DRD process. The resulting cup is redrawn to achieve the desired can diameter and force more metal to the can side, but then the side wall is thinned by forcing the redrawn can through rings that gradually iron out the can side-wall to the desired thickness. This process, called draw and iron (D&I) or draw and wall iron (DWI), is shown in Figure 13.11 (Page et al., 2003). The side-walls of

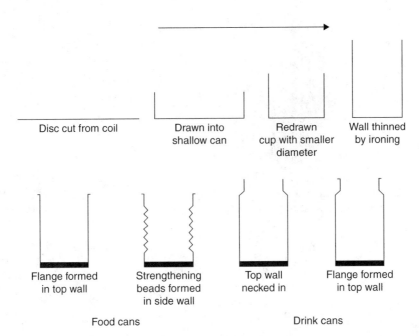

FIGURE 13.11   Steps in the manufacture of the draw and wall iron (DWI) 2-piece can. (From Page, B., Edwards, M., and May, N., 2003. *Food Packaging Technology*. R. Coles, D. McDowell, and M. Kirwan, Eds., Boca Raton, FL, CRC Press, 120–151. Copyright 2003 from *Food Packaging Technology*, R. Coles, D. McDowell, and M.M. Kirwan, Eds. Reproduced by permission of Blackwell Publishing Ltd, Oxford, UK.)

D&I cans are weaker than those of DRD cans and, thus, are not used for heat processing of food. However, D&I cans are well suited for containing carbonated beverages, where the internal pressure enhances the side-wall strength. D&I cans are also used for noncarbonated juices, for which nitrogen is injected to pressurize the can. Because laquers cannot endure the wall ironing process, they are sprayed onto D&I containers after being formed.

Details on the design and manufacture of metal containers can be found in several references (Matsubayashi, 1990; Paine, 1991; Paine and Paine, 1992a; Kraus and Tarulis, 1997; Reingardt and Nieder, 1997; Silbereis, 1997; Hanlon et al., 1998a; Twede and Goddard, 1998; Soroka, 2002d; Barron and Burcham, 2003b; Page et al., 2003; Robertson, 2006a).

13.3.2.5 Metal Package Closures

Can closure is generally accomplished by double seaming of a can end onto the can body. Easy-open options such as a perforated pull-ring lid and membrane lids have made cans more convenient (Page et al., 2003). Paperboard lids and/or plastic wraps are used on aluminum trays and pans. Aluminum-coated plastic films can be heat-sealed into pouches or heat-sealed as lidding on a variety of packages.

13.3.2.6 Metal Packaging Uses

Figure 13.12 shows metal packaging production by market sector (CMI, 2005). In the United States., over 100 billion aluminum cans are produced each year, almost entirely for soft drinks (68%) and beer (32%). The heat resistance of metal is utilized in the packaging of heat-processed foods. Approximately 31 billion steel cans are used each year in the United States., with 77% devoted to human food and 23% used for pet food. Of this amount, over 50% are now two-piece cans. The most

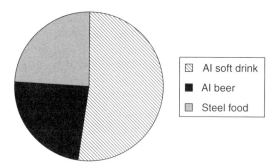

FIGURE 13.12   Metal packaging production by food market sector. (From CMI (2005). *Historical CMI Can Shipments*. Washington DC, Can Manufacturers Institute. http://www.cancentral.com/content.cfm.)

common uses for steel cans are vegetables and vegetable juices, fruits and fruit juices, soups, meat, poultry and seafood, dairy products, baby food, and coffee.

13.3.2.7   Metal Packaging Trends

Continuing developments have improved the effectiveness of metal containers (Narasimhan, 2001; Page et al., 2003). As with glass containers, lightweighting has been an important goal. Use of DR steel has allowed thinner-wall steel cans. Gradually, three-piece steel food cans are being replaced with DWI two-piece cans, which benefit from faster line speeds, lower metal cost, and greater container integrity (2003a). Additional metal saving is being achieved with new lightweight food cans now being produced (2003b). The thin walls require internal pressurization with nitrogen to maintain can shape. Cans with different sizes and shapes, including cans with bulged, concave, and fluted bodies has visual and handling appeal to consumers (Davis, 1998). The flexible square steel can also has reduced weight, while providing opportunities for improved food quality, new decoration approaches, and 20% savings in shelf space (Sonneveld, 2000). Easy-open ring-pull and peelable membrane lids are providing more convenience. ECCS is becoming more widely used because of economies in production and improved adhesion to lacquers (Hanlon et al., 1998a). Practically all metal beverage containers are now easy-open, stay-on-tab two-piece aluminum cans, and necking-in of the cans has reduced the diameter and cost of the can lid. Aluminum beverage cans were reduced in weight by 25% between 1972 and 1990 (Marsh, 1994), and they continue to be lightweighted. Aluminum bottles are becoming available for beverages, with the advantage that they are recloseable. Both steel and aluminum cans have benefited from new printing technology such as digital imaging which allows high-speed printing of photographic quality images directly on the cans.

13.3.3   PLASTICS

Plastics are high molecular weight polymers that can be molded into desired shapes such as films, trays, bottles, and jars using heat and pressure. Two broad categories of plastics exist, those based on thermplastic polymers and those based on thermoset polymers.

Thermoplastic polymers are linear or branched, but with no crosslinks between polymer chains. These polymers soften and become molten when heated, returning to their original condition upon cooling. Thus, they can be molded or extruded repeatedly. This property allows thermplastics to be recycled for many uses after their use in food packaging. Thermoplastic polymers constitute the most important category of plastics used for food packaging. These plastics exhibit a wide range of mechanical, optical, barrier, and thermal properties, depending on the specific polymer, polymer processing, and polymer additives. Additives can include plasticizers to improve plastic flexiblity, stabilizers to improve polymer resistance to degradation by heat and light, and antistatic agents to prevent plastics from clinging to packaging equipment.

Thermoset polymers crosslink into a set network when heated, often with addition of a crosslink-ing agent. Thus, after taking on their original cast shape, they cannot be reheated for molding into new shapes. Thermoset polymers play an important role in food packaging, often used for making package closures.

13.3.3.1 Advantages and Disadvantages

The most commonly used thermoplastic polymers are inexpensive, and their conversion into food packaging is also relatively inexpensive. These plastics can be molded or extruded into a wide range of flexible, semi-rigid and rigid containers that are lightweight, noncorrodible, shock-resistant, and heat-sealable. Most are transparent and some are microwaveable. Certain plastics have high enough heat resistance that they can be hot-filled, retorted and/or used in a conventional oven. Finally, the most commonly used plastic semi-rigid and rigid containers are recyclable. Similar to glass and metal, plastic properties have improved over the years so that less material is necessary for making containers with acceptable integrity.

However, unlike glass and metal, plastics do not provide a total barrier to gases, water vapor, and aromas. The permeabilities of a given plastic material to water vapor, oxygen, carbon dioxide, and aromas depend on the particular polymer composition and structure. This must be considered when selecting a plastic for a specific application and desired shelf life. Plastics are often combined in layers, to take advantage of the unique barrier properties of each polymer. Similar to glass, plastic container transparency to light can be detrimental to foods vulnerable to light-catalyzed reactions. Pigmenting, labeling or direct printing of plastic containers can reduce this problem for sensitive food products. Plastic materials do not have the compressive strength of glass or metal, and only a few plastics have high enough heat resistance for heat processing or preparation of foods. Plastic additives and any residual monomers have potential for migrating into foods. Thus, much attention and testing are devoted to minimizing this possibility. On the other hand, food components such as aromas and flavors can sorb into plastic packaging, with resulting loss of food quality. Finally, most plastic materials used in food packaging are not recyclable. Fortunately, these are used in lower quantities than recyclable plastic containers. The advantages and disadvantages of plastic containers are summarized in Table 13.11.

13.3.3.2 Plastic Materials and Properties

The properties of a plastic polymer are influenced by its chemical composition, structure, additives, processing, and conditions of use (Kondo, 1990; Miltz, 1992; Jasse et al., 1994; Hernandez, 1997b; Kirwan and Strawbridge, 2003; Robertson, 2006h). The structure of thermoplastic polymers can include both organized crystalline regions, where polymer chains are parallel and closely packed, and disorganized amorphous regions, where the greater free volume results in lower polymer density than the crystalline regions. The relative amounts of crystalline and amorphous regions depend on the polymer structure and polymer processing. Thermoplastic polymers with amorphous regions have a characteristic glass transition temperature, T_g, at which the amorphous polymer regions transition from a stiff glassy state to a more flexible rubbery state. If the polymer has crystalline regions, they have a melting temperature T_m at $\sim$1.5–2 T_g (°K). The T_g and T_m of a polymer have an important influence on the properties and uses of the polymer for food packaging. Table 13.12 lists the T_g and T_m of several common food packaging polymers (Armeniades and Baer, 1977; Robertson, 2006h). Note that some polymers are totally amorphous in nature.

Table 13.13 lists the most common plastics, along with their properties and common uses (Tice, 2002b, a; Leadbitter, 2003; Tice, 2003; APC, 2005).

TABLE 13.11
Advantages and Disadvantages of Plastic Containers

Advantages	Disadvantages
Inexpensive materials	Permeable to
Inexpensive conversion to packaging	Gas
Versatile	Water vapor
Flexible	Aroma
Rigid	Potential migration of
Semi-rigid	Monomers
Moldable	Additives
Light-weight	Food components can sorb into plastic
Noncorrodible	Low compressive strength
Shock-resistant	Lack heat resistance (some)
Heat-sealable	Not recyclable (some)
Transparent	
Can be pigmented	
Microwavable (some)	
Good heat resistance (some)	
Recyclable (some)	

TABLE 13.12
T_g and T_m of Common Plastics Used in Food Packaging

Polymer	$T_g(°C)$	$T_m(°C)$	$T_m/T_g(K)$
Polyethylene			
High-density	-125	137	2.67
Low-density	-25	98	1.50
Polypropylene	-18	176	1.76
Polyethylene terephthalate	69	267	1.57
Polystyrene (isotactic)	100	240	1.38
Polyvinylchloride	87	212	1.34
Polyvinylidene chloride)	-35	198	1.97
Poly(hexamethylene adipamide) (nylon 6,6)	50	265	1.66
Poly(hexamethylene sebacamide) (nylon 6,10)	40	277	1.59

Source: Adapted from Armeniades, C.D. and Baer, E. (1977). *Introduction to Polymer Science and Technology.* H.S. Kaufman and J.J. Falcetta (Eds), New York, John Wiley & Sons, Inc., Chapter 6; Robertson, G.L. (2006h). *Food Packaging — Principles and Practice.* Boca Raton, CRC Taylor & Francis Group, 9–42.

13.3.3.2.1 Polyethylene (PE)

Polymerization of ethylene gas produces the simplest, least expensive, and most widely used plastic, PE:

$$nCH_2 = CH_2 \rightarrow -[CH_2-CH_2]-_n$$

PE can be manufactured as a highly branched polymer, low density polyethylene (LDPE), a lightly branched polymer, linear low density polyethylene (LLDPE) or a linear polymer, high density

TABLE 13.13

Properties and Uses of Common Plastics

Plastic material	Properties	Selected uses
High Density Polyethylene (HDPE)	• Excellent moisture barrier	• Bottles and jugs for milk, water and juice
	• Poor O_2 and aroma barrier	• Cups and tubs for cottage cheese, yogurt, butter and margarine spread
	• Strong	• Bags for carrying groceries
Low Density Polyethylene and Linear Low Density Polyethylene (LDPE and LLDPE)	• Excellent moisture barrier	• Bags for fresh-produce and baked-goods
	• Poor O_2 and aroma barrier	• Pouches for frozen foods
	• Tough	• Moisture-barrier and/or heat-sealing layer/coating on multilayer cartons
Polypropylene (PP)	• Excellent moisture barrier	• Bottles for ketchup, syrup and oils
	• Poor O_2 and aroma barrier	• Cups and tubs for cottage cheese, yogurt, butter and margarine spread
	• Good heat resistance	• Overwraps for produce, baked goods and confectionery products
		• Pouches for snack foods
		• Trays for microwaveable foods
		• Coating for microwaveable paperboard cartons and trays
Polyvinyl Chloride (PVC)	• Good oil barrier	• Bottles for vegetable oils
	• Good moisture, O_2 and aroma barrier	• Overwraps for produce
	• Good stretch and cling	• Overwraps for meat
Polyvinylidene Chloride (PVDC)	• Excellent moisture, O_2 and aroma barrier	• Barrier layer or coating in multilayer containers
Polystyrene (PS)	• Poor moisture, O_2 and aroma barrier	• Clear trays and cartons for baked goods, fresh produce and meat
	• Glossy and clear	• Foamed trays and cartons for fresh produce, meats, poultry, fish, and eggs
	• Strong and stiff	• Clear and foamed cups and plates
	• Expanded foam is good cushioner and insulator	• Clear or pigmented cutlery
Polyethylene Terephthalate (PET)	• Good moisture, O_2 and aroma barrier	• Bottles for carbonated and noncarbonated beverages
	• Glossy and clear	• Bottles for oils, dressings, ketchup, sauces, and syrups
	• Strong and durable	• Jars for peanut butter, mustard, etc.
	• Excellent heat resistance	• Trays and lidding for dual-ovenable applications
		• Paperboard coatings for dual-ovenable applications
		• Retort pouches
		• Boil/microwave-in-bag pouches
Ethylene-vinyl alcohol copolymer (EVOH)	• Poor moisture barrier	• Barrier layer (sandwiched between moisture-barrier layers) in retort pouches, tubs and cans and aseptic packages
	• Excellent O_2 and aroma barrier when protected from moisture	

(Continued)

TABLE 13.13
Continued

Plastic material	Properties	Selected uses
Polyamide (PA) (Nylon)	• Poor moisture barrier	• Barrier layer (sandwiched between moisture-barrier layers) in retort pouches, tubs and cans and aseptic packages
	• Excellent O_2 and aroma barrier when protected from moisture	
	• Tough	
	• Good heat resistance	

Source: From Tice, P. 2002a. *Packaging Materials: 2. Polystyrene for Food Packaging Applications.* Brussels, ILSI Europe: 20; Tice, P. 2002b. *Packaging Materials: 3. Polyproplylene as a Packaging Material for Foods and Beverages.* Brussels, ILSI Europe: 24; Tice, P. 2003. *Packaging Materials: 4. Polyethylene for Food Packaging Applications.* Brussels, ILSI Europe: 24; Leadbitter, J. 2003. *Packaging Materials: 5. Polyvinyl Chloride (PVC) for Food Packaging Applications.* Brussels, ILSI Europe: 20; and APC (2005). Resin identification codes — plastic recycling codes. www.americanplasticscouncil.org.

polyethylene (HDPE). These polymers have low T_g values ($\sim-125°C$) and moderate T_m values ($\sim100–140°C$). The low T_g makes them quite flexible and resilient for use in packaging frozen foods as well as foods stored at ambient conditions. The moderate T_m makes them easily heat-sealed.

Because of its nonpolar nature, PE is an excellent moisture barrier. However, it is a poor barrier to O_2, CO_2, and aromas, which are also nonpolar and thus readily adsorb and then diffuse through PE. Because of its linear structure, HDPE is more crystalline and thus stiffer, stronger, less transparent, and a somewhat better barrier than LDPE.

LDPE is used extensively for bags (e.g., fresh produce and bread), pouches (e.g., frozen foods), coatings on paperboard cartons (e.g., refrigerated milk, frozen food), layers in LDPE/paperboard/LDPE/aluminum foil/LDPE laminate cartons (e.g., shelf-stable milk and juices) and coatings for other plastics that require a moisture barrier or heat-sealing layer. HDPE is used most often for bags (e.g., grocery bags), bottles and jugs (e.g., water, milk, and juice) and cups and tubs (e.g., yogurt, cottage cheese, margarine). The chemistry, properties, manufacture, applications, regulatory, safety, and environmental aspects of polyethylene have been summarized (Tice, 2003).

13.3.3.2.2 Polypropylene (PP)

The production, cost and properties of polyproylene (PP) are similar to LDPE, except that it is more glossy and stiff and has higher T_g ($\sim-20°C$) and T_m ($\sim175°C$).

$$n\ CH_2\!=\!CH \rightarrow -[CH_2\!-\!CH\ ]-_n$$
$$|\qquad\qquad\qquad |$$
$$CH_3\qquad\qquad\ CH_3$$

The barrier, mechanical, and optical properties of PP film are improved by orientation, which is accomplished by stretching the film while still semi-molten to produce better alignment of the polymer chains. If the stretching is done in one direction, the film is referred to as oriented polypropylene (OPP). If it is stretched in two directions, the film is biaxially oriented PP (BOPP).

Because of its relatively high T_g, PP does not have the resilience of LDPE for frozen foods. However, because of its high T_g and T_m, it is quite useful for hot-filled, retorted, and microwaveable food products.

Like PE, PP is an excellent moisture barrier and a poor O_2, CO_2, and aroma barrier. OPP is used widely for overwrap films (e.g., fresh produce, baked goods, confectionery products), pouches (e.g., chips, cookies, other snack items) and coatings for paperboard cartons and trays (e.g., microwaveable meals). PP is also formed into bottles, cups, and tubs for the same uses as for HDPE. The chemistry, properties, manufacture, applications, regulatory, safety, and environmental aspects of PP have been summarized (Tice, 2002b).

13.3.3.2.3 Polyvinylchloride (PVC or V)

PVC is another inexpensive polymer, but it has much more limited use in food packaging. Most PVC is used for nonfood applications, such as piping, house siding, and rain gutters and downspouts.

$$n CH_2 = \underset{\overset{|}{Cl}}{CH} \rightarrow -[CH_2 - \underset{\overset{|}{Cl}}{CH}]-_n$$

Strong polar, nonhydrophilic interaction between the C and Cl of adjacent PVC chains produces a very stiff, brittle material. Addition of plasticizer improves the flexibility and resilience of PVC but renders the polymer a moderately good barrier to moisture, O_2, CO_2, and aromas. Plasticized PVC film has good stretch and cling and is often used as an overwrap for fresh produce (which must exchange O_2 and CO_2 with the environment) and fresh meat (for which O_2 is necessary for red color). PVC is also used to make trays, bottles (e.g., vegetable oils), and jars (e.g., coffee creamer). The chemistry, properties, manufacture, applications, regulatory, safety and environmental aspects of PVC have been summarized (Leadbitter, 2003).

13.3.3.2.4 Polyvinylidene Chloride (PVDC)

PVDC has similar structure to PVC but with an additional Cl atom on each monomer:

$$n CH_2 = \underset{\overset{|}{Cl}}{\overset{\overset{Cl}{|}}{C}} \rightarrow -[CH_2 - \underset{\overset{|}{Cl}}{\overset{\overset{Cl}{|}}{C}}]-_n$$

PVDC also has strong polar, nonhydrophilic interactions between the C and Cl of adjacent polymer chains. Copolymerization with PVC and addition of plastizicer produces good mechanical properties and excellent H_2O, O_2, CO_2, and aroma barrier properties in the resulting PVDC/PVC copolymer. PVDC/PVC also has excellent stretch and cling properties. The consumer version of PVDC/PVC is known as Saran® wrap. The expense of PVDC/PVC copolymer generally limits its use to a coating, lamination, or co-extruded layer, where the PVDC/PVC provides the barrier properties, and another polymer provides the strength and stiffness.

13.3.3.2.5 Polystyrene (PS)

The bulky side group of PS prevents close interaction among polymer chains.

$$n CH_2 = \underset{\overset{|}{C_6H_5}}{CH} \rightarrow -[CH_2 - \underset{\overset{|}{C_6H_5}}{CH}]-_n$$

The resulting PS is totally amorphous, with a high T_g ($\sim 100°C$) and poor barrier properties. However, PS is a versatile polymer which can be made into a glossy, clear stiff material that can be formed into clear trays and cartons for baked goods, fresh produce, and meat. It can also be pigmented and used to form cups and tubs for dairy products. PS is easily foamed to make expanded polystyrene (EPS) useful for cushioning trays and cartons for fresh produce, meats, poultry, fish, and eggs. Clear and foamed PS are also used to make plastic cups and plates. Clear or pigmented PS is use to make disposable cutlery. The chemistry, properties, manufacture, applications, regulatory, safety, and environmental aspects of PS have been summarized (Tice, 2002a).

13.3.3.2.6 Polyethylene Terephthalate (PET or PETE)

PET is a more complicated polymer made by reacting the dicarboxylic terephthalic acid with the di-alcohol ethylene glycol to make a polyester:

$$n(HOOC-C_6H_4-COOH + HOCH_2-CH_2OH) \rightarrow -[OOC-C_6H_4-COO-CH_2CH_2]-_n + nH_2O$$

Oriented PET is a low-cost polymer which is strong, resilient and a good barrier to H_2O, O_2, and CO_2. It also has excellent clarity and gloss that make it resemble glass. PET is most commonly used for carbonated and noncarbonated beverage bottles. However, it is being increasingly used to make bottles for food products like vegetable oils and salad dressings and jars for products like peanut butter and mustard. The barrier properties of PET bottles and jars can be improved by coating with silicon or aluminum oxide, or by adding an excellent oxygen-barrier film such as ethylene vinyl alcohol copolymer (EVOH). PET has quite high T_g ($\sim 80°C$) and T_m ($\sim 270°C$), which allows PET bottles to be hot-filled or pasteurized, as well as used for PET trays and PET-coated paperboard trays that can be used in both microwave and convection ovens (i.e., dual-ovenable). For use as dual-ovenable trays, crystallization of the PET structure is increased in the forming process. The resulting crystallized PET (CPET) is heat stable at temperatures up to $\sim 225°C$. Biaxially orienting PET film to improve its barrier, mechanical and heat-resistance properties allows it to be used for "boil-in-bag" or "microwave-in-bag" pouches, retort pouches and dual-ovenable lidding. Because of its high heat resistance, PET has poor heat sealability. It must be coated with PE or PVDC where heat-sealing (e.g., a pouch) is desired. The chemistry, properties, manufacture, applications, regulatory, safety, and environmental aspects of PET have been summarized (Matthews, 2000).

13.3.3.2.7 Ethylene-Vinyl Alcohol Copolymer (EVOH)

EVOH is an expensive polymer that is made by reacting ethylene with vinyl acetate and then hydrolyzing the resulting polymer to EVOH.

$$n(CH_2 = \underset{\underset{OOCCH_3}{|}}{CH} + CH_2 = CH_2 \rightarrow -[CH_2- \underset{\underset{OOCCH_3}{|}}{CH} -CH_2-CH_2]-_n$$

$$\rightarrow -[CH_2- \underset{\underset{OH}{|}}{CH} -CH_2-CH_2]-_n + nHOOCCH_3$$

The relative amounts of ethylene and vinyl acetate affect the final properties of the polymer.

Because of its polar character, EVOH is an excellent O_2, CO_2, and aroma barrier. However, because it is hydrophilic, EVOH is sensitive to moisture and not a good moisture barrier. To take advantage of its O_2, CO_2, and aroma barrier properties, EVOH is sandwiched between layers of a nonpolar moisture barrier such as PP for manufacture of retort pouches, tubs (cans) and aseptic packages, or layers of PET for manufacture of beverage bottles.

13.3.3.2.8 Polyamides

Polyamides are made by reacting a dicarboxylic acid (e.g., adipic acid) with a diamine (e.g., hexamethylene diamine):

$$n(HOOC-(CH_2)_4-COOH + H_2N-(CH_2)_6-NH_2)$$

$$\rightarrow -[OC-(CH_2)_4-CO-NH-(CH_2)_6-NH]-_n + nH_2O$$

Polyamides are commonly referred to as Nylons, a term combining the names of New York and London. They are strong, heat resistant and have barrier properties similar to EVOH. They are also moisture sensitive and, thus, are often sandwiched between layers of a nonpolar, moisture-barrier polymer such as polypropylene.

13.3.3.3 Plastic Package Manufacture

Plastic polymers are quite versatile, as they can be formed into flexible, semi-rigid and rigid packaging.

13.3.3.3.1 Flexible Plastic Film Packaging

Flexible plastic films can be a single-layer structure, or they can be coated, laminated, or co-extruded structures (Dunn, 1997; Selke, 1997d; Hernandez et al., 2000; Soroka, 2002i; Robertson, 2006i).

The most commonly used materials for flexible packaging films are LDPE, LLDPE, HDPE, PP, and PVC (2002). Single-layer films are generally made by *extrusion*, in which plastic pellets are heat-softened sufficiently to melt and flow, and then the molten plastic is forced through either a slit (slot) die or a circular (tubular) die (Gibbons, 1997). The semi-molten film exiting from a slit die is cooled with a quenching water bath or chilled casting rolls (Figure 13.13) (Soroka, 2002i). The film can then be reheated and stretched in the machine direction and/or transverse to the machine direction to orient the polymer chains in the film to improve strength, barrier, and shrink properties (Kirwan and Strawbridge, 2003). One-direction orientation is called uni-axial orientation, while two-direction is called bi-axial orientation. From a circular die, the film can be blown up like a bubble to give transverse orientation while the film is being pulled to also give orientation in the machine direction (Figure 13.14) (Soroka, 2002i). The resulting films can be used as food wraps or heat-sealed into bags and pouches.

A polymer film can be solution-coated or extrusion-coated with another polymer to produce a bilayer film with improved strength, barrier, heat-sealability, appearance, and/or printability properties. Solution coating involves coating with a solution or dispersion of another polymer and then evaporating the solvent. In extrusion coating, a semi-molten film emerging from an extruder is deposited directly on the previously formed film. Plastic films, most often PP or PET, can also be coated with a thin layer of aluminum or glass. The aluminum is vaporized in a vacuum and then condenses onto the film surface (vacuum metallization). Coatings of SiOx can be formed onto plastic films by sputtering, evaporation, or plasma-enhanced chemical vapor deposition (Hill, 1997).

Two or more previously formed single-layer films can be laminated to give a multi-layer film with improved properties. The layers can be bonded by applying an adhesive between the films and

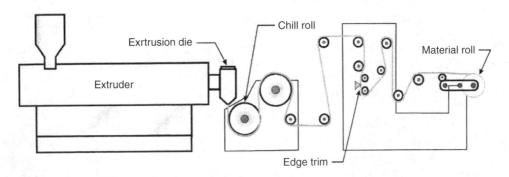

FIGURE 13.13 Extrusion of cast plastic film using a slit die. (From Soroka, W., 2002i. *Fundamentals of Packaging Technology*. Naperville, IL, Institute of Packaging Professionals, 223–259. Copyright 2002. Reproduced by permission of Institute of Packaging Professionals, Naperville, IL.)

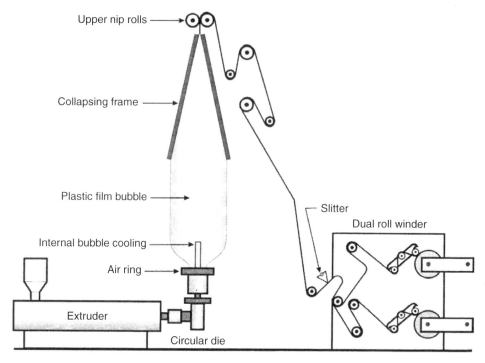

FIGURE 13.14   Extrusion of blown plastic film using a circular die. (From Soroka, W., 2002i. *Fundamentals of Packaging Technology.* Naperville, IL, Institute of Packaging Professionals, 223–259. Copyright 2002. Reproduced by permission of Institute of Packaging Professionals, Naperville, IL.)

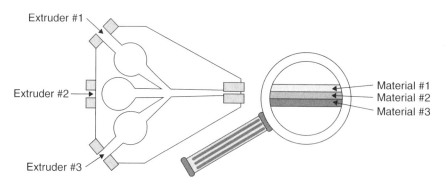

FIGURE 13.15   Co-extrusion of multilayer plastic film. (From Soroka, W., 2002i. *Fundamentals of Packaging Technology.* Naperville, IL, Institute of Packaging Professionals, 223–259. Copyright 2002. Reproduced by permission of Institute of Packaging Professionals, Naperville, IL.)

then passing the laminate structure between pressure rollers (adhesive laminating). The layers can also be bonded by extrusion coating one of the films and then immediately pressing the second film against the still-molten layer (extrusion laminating). Polymer films can also be laminated with paper and/or aluminum foil to combine the properties of each material into a package structure.

Co-extrusion is another way to form a multi-layer plastic film. It involves simultaneous extrusion of two or more different polymers from separate extruders to a common die (Figure 13.15) (Soroka, 2002i). After entering the die through different entry ports, the semi-molten polymers are brought

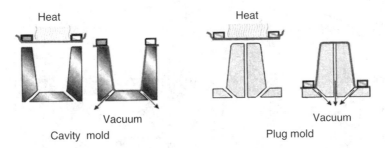

FIGURE 13.16 Thermoform vacuum molding of a plastic container over cavity and plug molds. (From Soroka, W., 2002i. *Fundamentals of Packaging Technology*. Naperville, IL, Institute of Packaging Professionals, 223–259. Copyright 2002. Reproduced by permission of Institute of Packaging Professionals, Naperville, IL.)

together in the die to form a multi-layer film. The multi-layer film then exits the slit or circular die. To achieve strong adhesion, the polymers should have similar chemical structures and flow properties. Co-extruded films tend to be less expensive than laminated films of the same composition, because there is no production of separate films that must be wound, unwound and then adhered to each other with an adhesive layer. In addition, thinner multi-layer films are possible with co-extrusion, and it is less likely that the film layer will separate.

All the above films can be used as food wraps. They can also be formed into bags or pouches, either preformed in a bag or pouch manufacturing facility, or (more often) in-lined formed in a form-fill-seal operation (Bardsley, 1997; Moyer, 1997). In the vertical form-fill-seal operation for pouches, the film is taken from a roll, folded over and sealed lengthwise to form a side-seal, then sealed horizontally, perpendicular to the movement of the film. After the side and bottom seals are formed, food or beverage can be filled into the pouch, and the top seal can be formed. The filled pouch is cut from the continuous film through the middle of the seal, leaving a bottom seal for the next pouch.

13.3.3.3.2 Semi-Rigid and Rigid Plastic Packaging

Depending on the type of semi-rigid or rigid plastic container desired, several different manufacturing methods are available for molding plastic into trays, tubs, cups, lids, jars, bottles, and jugs (Brody and Marsh, 1997; Selke, 1997d; Hernandez et al., 2000; Soroka, 2002i; Robertson, 2006i). Such containers generally have wall thickness greater than 75–150 μm, depending on the plastic material. Design of plastic containers includes many steps, including selection of appropriate manufacturing (Mandel, 1997).

Thermoforming involves heat-softening a previously-extruded plastic sheet and then forcing the sheet into or over a mold by vacuum (Figure 13.16) (Soroka, 2002i). Air pressure and/or mechanical means can also be used to form the softened sheet (McKinney et al., 1986). Food product applications include trays, tubs, and cups from PS or PET, trays, and cartons from EPS, and retortable and dual-oven-able container/dishware from PET (Huss, 1997). For the high temperature applications, a nucleating agent can be added to the PET sheet, resulting in crystallization of the PET structure in the thermoform mold. The resulting crystallized PET (CPET) is heat stable at temperatures up to ~225°C.

Injection molding involves heat-softening plastic pellets in an extruder and then injection of the molten plastic under pressure into a cool mold (Carter, 1997). The two halves of the mold then open to eject the solid container. PE, PP, and PS are the most commonly used materials to manufacture plastic tubs, cups, and lids by injection molding (2002). Retortable and microwaveable PP trays are also made by the injection molding process.

Blow molding includes several processes to produce a plastic container, each of which has a blowing step (Irwin, 1997). Injection blow molding of plastic containers is quite similar to

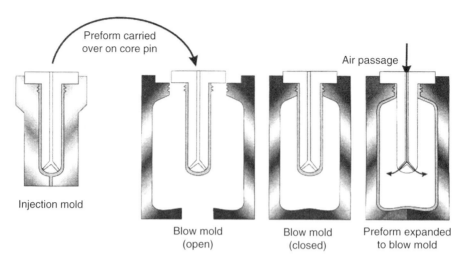

FIGURE 13.17 Injection blow molding of a plastic bottle. (From Soroka, W., 2002i. *Fundamentals of Packaging Technology*. Naperville, IL, Institute of Packaging Professionals, 223–259. Copyright 2002. Reproduced by permission of Institute of Packaging Professionals, Naperville, IL.)

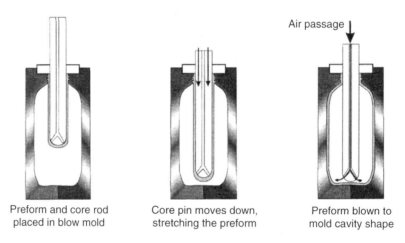

FIGURE 13.18 Stretch blow molding of a plastic bottle. (From Soroka, W., 2002i. *Fundamentals of Packaging Technology*. Naperville, IL, Institute of Packaging Professionals, 223–259. Copyright 2002. Reproduced by permission of Institute of Packaging Professionals, Naperville, IL.)

the two-step process for making glass bottles and jars (Figure 13.17) (Soroka, 2002i). A preform (parison) is first made in an injection mold around a blowing stick. While still semi-molten, the preform is transferred to a second mold, where the preform is blown to the final container shape, with resulting transverse polymer orientation. PE, PP, PVC, and PET are commonly used in this manner to make bottles and jars for food products. Co-injection blow molding involves use of two or more injection units to produce a multi-layer preform for blowing into a multi-layer bottle or jar with improved properties. An example is a multi-layer retortable container including an EVOH oxygen- and aroma-barrier layer between PP moisture-barrier structural layers.

Injection stretch blow molding is similar to injection blow molding. A preform is also made in an injection mold. However, in the blowing step, a rod is used to stretch the preform longitudinally at the same time as it is being blown transversally (Figure 13.18) (Soroka, 2002i). The resulting biaxial

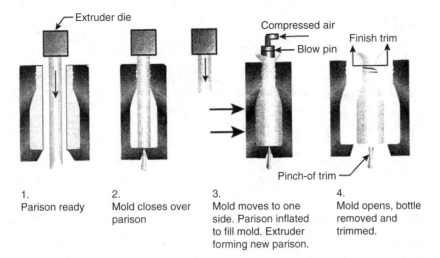

FIGURE 13.19   Extrusion blow molding of a plastic bottle. (From Soroka, W., 2002i. *Fundamentals of Packaging Technology*. Naperville, IL, Institute of Packaging Professionals, 223–259. Copyright 2002. Reproduced by permission of Institute of Packaging Professionals, Naperville, IL.)

orientation improves strength, barrier, and optical properties. The polymer that is most commonly injection stretch blow molded is PET for production of bottles intended for both carbonated and noncarbonated beverages. Other polymers that are sometimes molded in this manner include PVC and PP.

Extrusion blow molding involves extruding a heat-softened hollow tube (parison), quickly closing the two halves of a mold around the tube, and blowing the still-soft parison against the sides of the mold (Figure 13.19) (Soroka, 2002i). The newly formed bottle or jug is held in the mold until cool. The mold then opens, and the bottle or jug is removed and trimmed above the finish and at the bottom where the mold pinches the parison tube. The polymers that are most commonly extrusion blow molded are PE, PP, and PVC. This process also lends itself to forming bottles and jugs from multi-layer parisons made by co-extrusion. Common co-extruded blow molded containers include EVOH or Nylon as interior layer(s) sandwiched between layers of PE, PP, or PET. Adhesive (tie) layers bond the EVOH or Nylon to the outer layers. Co-extrusion blow molding also allows use of recycled plastic sandwiched between layers of virgin plastic, to protect food from any contaminant in the recycled plastic.

More details on plastic polymers plastic container design and manufacture can be found in a number of sources (Jenkins and Harrington, 1991; Paine, 1991; Paine and Paine, 1992a; Brody and Marsh, 1997; Selke, 1997d; Hanlon et al., 1998a; Twede and Goddard, 1998; Giles and Bain, 2000; Hernandez et al., 2000; Ehrenstein, 2001; Giles and Bain, 2001; Soroka, 2002i; Kirwan and Strawbridge, 2003; Robertson, 2006i).

13.3.3.4 Plastic Package Closures

The formability and thermoplastic nature of plastic allow the greatest selection of closures among the packaging materials (Guglielmini, 2001; Soroka, 2002e). Threaded and snap-on finishes allow reclosing after opening. Heat-sealed plastic pouches often include a convenient resealable feature.

13.3.3.5 Plastic Packaging Uses

Packages based on plastic or including a plastic layer are used for every food category. Extruded LDPE, LLDPE, HDPE, and PP films are converted into a broad range of flexible packaging for

TABLE 13.14

Processes and Amounts of Major Plastics Used in Packaging

Process	North America Markets (2000), Million lb						
	LDPE	LLDPE	HDPE	PP	PS	PVC	PET
Extruded film (<0.3 mm)	2066	2610	386	1296		1683[a]	326[b]
Food	973	480	288				
Nonfood	753	938	98				
Stretch/Shrink	340	1192					
Extruded carryout bags	90	224					
Extruded grocery sacks			825				
Extruded sheet (>0.3 mm)	96	34	740	330	1360		
Extruded expandable sheet					216		
Extrusion coating	943	35		38		436	
Injection molded containers	302	580	2240	1550	580		
Pails			902				
Tubs and containers			268				
Crates and totes			338				
Other			732				
Blow molded containers	57	12	4490	150	156		3062
Liquid food bottles			1380				
Household, etc. bottles			1230				
Motor oil bottles			170				
Industrial drums			327				
Other			1383				

[a] Includes extruded and calendered film and sheet.

[b] Includes extruded film and sheet.

Source: From (2001). Resin 2001 — Tables. *Modern Plastics* February: 42–43.

beverages and dry, frozen, and heat-processed foods (Table 13.13). These plastics are also used as the heat-seal layer in combination with other materials in flexible and semi-rigid packaging. In addition, LDPE and HDPE bags are often used to transport purchased foods from the store to home. PET (carbonated beverages and water) and HDPE (milk, water, and juices) are the most commonly used plastics for blow molding of beverage containers. HDPE, PP, and PS are injected molded to form cups and tubs for dairy products and other foods. PS and PET extruded sheets are thermoformed into trays for fresh produce, meat, and poultry. PP and PET trays are used for microwavable and dual-oven-able frozen foods, respectively.

Table 13.14 gives information on the processes used and amounts of major plastic polymers converted for the various types of plastic packaging (2001).

13.3.3.6 Plastic Packaging Trends

Because of weight, volume, simplicity of production, durability, and cost advantages, plastic containers have replaced glass and metal containers for many beverages and food products (Bain and Giles, 2000; Streeter, 2000). In addition, plastic properties have improved over the years so that less material is necessary for making containers with acceptable barrier and integrity (Marsh, 1994). Polyethylene continues to be used in food packaging in larger amounts than any other plastic, because of its low cost, versatility, and ease of conversion to a wide variety of packaging (Tice, 2003). Because of the unique combination of its properties, including recyclability, PET has found increasing applications for bottles, jars, dual-oven-able trays, and film wraps and pouches (Matthews, 2000). The properties of PP, including microwaveability, have also led to its increased use in food packaging (Tice, 2002b). Thinner stronger films with lower permeability, more reliable sealing, resealability, and stand-up design have made flexible plastic pouches an attractive option (Weinberg, 1998; Louis, 1999; Brody, 2000b; Shellhammer, 2003). Thin high-barrier coatings on plastics have opened up new applications, including beer, while achieving source reduction and maintaining recyclability (Ferrante, 1997; Sonneveld, 2000; Reynolds, 2002).

13.3.4 PAPER

More paper is used in food packaging than any other material. It can be found in all levels of packaging (primary, secondary, tertiary, and quaternary).

13.3.4.1 Advantages and Disadvantages

Paper is a quite versatile material, utilized in flexible, semi-rigid, and rigid packaging. It is made into a wide variety of single- and multi-wall bags. It can also be made into a thicker stronger structure (>0.012 in./0.03 cm) called paperboard (Pb) Which is made into cartons and boxes that provide mechanical protection for many foods. The paperboard can be converted to an even stronger material called corrugated paperboard that is converted into boxes used for logistics (tertiary and quaternary packaging). Most types of paper provide a partial or complete barrier to light. It can also be manufactured into transparent and clear materials. The starting material of paper, wood, is a renewable resource, and paper is recyclable and biodegradable.

The main disadvantages of paper are that it provides negligible barrier against water vapor, and oxygen, and that it is not heat-sealable. These disadvantages can be overcome by coating or laminating the paper or paperboard with wax, polyethylene, or other polymers (sometimes metalized) to improve the barrier properties and allow heat-sealability. Coated or laminated paper and paperboard cannot be recycled in most municipalities. But, the technology for separating the layers exists and is gradually being adopted for recycling. Table 13.15 summarizes the advantages and disadvantages of paper in packaging.

13.3.4.2 Paper Composition and Properties

Paper and paperboard (paper with thickness ≥0.012 in./0.3 mm) are made from paper pulp that is produced from wood by either the acid bisulfite process or the alkaline sulfate (Kraft) process.

TABLE 13.15
Advantages and Disadvantages of Paper Packaging

Advantages	Disadvantages
Versatile	Negligible resistance to
Rigid	Water vapor
Semi-rigid	Aromas
Flexible	Gas
Mechanical protection	
Logistics functions	Not heat sealable
Barrier to light	Not recyclable when coated or laminated
Renewable resource	
Recyclable	
Biodegradable	

Both processes remove the lignin and much of the hemicellulose in wood to give a pulp that is approximately 80% cellulose and 20% hemicellulose. The Kraft process is most commonly used, because it has less effect on the strength of the cellulose fibers. Lower quality pulp can be made in a mechanical process that involves grinding of wood chips into mechanical pulp that makes weaker papers and paperboards. A semi-chemical process is sometimes used that combines short acid or alkaline digestion of wood chips followed by grinding to give a pulp intermediate in quality. The pulp can then either be bleached to give white paper or left unbleached for production of brown paper or paperboard. The pulp can also be captured on a molded screen to produce cushioning pulpboard cartons for eggs, or trays for fresh fruits, or vegetables.

The next operation in the production of paper is beating of the pulp in a 5 to 7% pulp-in-water slurry. The beating flattens out the fibers, reduces space between fibers, and produces small fibrils on the fibers. These effects combine to increase cohesion among the fibers due to increase in hydrogen bonding. The result is paper that is stronger.

A number of compounds are then added to the pulp-water slurry that affect the properties of the paper. These include sizing compounds such as starch and casein that increase paper strength, stiffness, and smoothness. The sizing compounds also close gaps between fibers to improve paper resistance to water and oils and to reduce blurring of printing inks. Mineral fillers such as titanium oxide are often added to improve paper brightness, opacity, smoothness, and ink receptivity. Finally, pigments can be added to produce colored papers.

Paper sheets are produced by capturing the pulp and additives on a fine wire mesh. Most of the water from the mixture of pulp and additives (stock suspension) flows through the wire mesh, so that the water content drops from ∼99.5% to 80–90%. The wire mesh can be in the form of a continuous moving belt (Fourdrinier machine) or a rotating cylinder machine under vacuum that is partially submerged in the stock suspension. The sheet is then transferred to a felt blanket and carried through press rolls that reduce the moisture content to 60–70%. Next, the sheet goes through a drying oven that reduces moisture content to ∼10%. The dry sheet is usually then calendered (ironed) between rollers that smooth the paper. Finally, the paper sheet can be surface-treated with the same kinds of compounds added to the stock suspension. In this case, the paper must go through an additional drying step. Additional details on paper manufacture can be found in several sources (Paine, 1991; Paine and Paine, 1992a; Hanlon et al., 1998a; Twede and Goddard, 1998; Soroka, 2002j; Kirwan, 2003; Robertson, 2006j).

Different kinds of paper can be made in this process (Sikora, 1997). Both unbleached and bleached Kraft papers are produced for bags. Vegetable parchment is made by treating pulp in a bath of concentrated sulfuric acid to swell and partially dissolve the paper fibers. The pulp is then washed, resulting in precipitation of the dissolved fiber with fewer gaps and more consolidation. The resulting paper is less porous and has improved wet strength and resistance to grease and oils. Greaseproof paper is produced by extending the time of pulp beating to increase fibrillation and hydration of the cellulose fibers. The resulting paper is translucent and higher in density. This type of paper is suitable for packaging of foods such as pastries, fried foods, and butter, because of its greater resistance to oil and fat. Glassine is a paper that is more transparent, glossy, dense, and resistant to oils and fats than greaseproof paper, made by adding additional steps of dampening and rolling through a series of steam-heated rollers (super calendering). Tissue paper is thin, lightweight paper that can be used as cushioning wraps for fruits and vegetables. It can also be coated with wax to produce wax paper. It can also be laminated with LDPE and aluminum foil (Al) to achieve a better barrier. Thicker laminating paper can also be used. Pouch paper, which is also useful in coating, laminating, or printing, is made stronger with super calendering. Many other papers are made by modifications in the processing or additives of paper (Soroka and Zepf, 1998).

Different kinds of paperboard (also called boxboard, cartonboard, or cardboard) are also made (Attwood, 1997). Bleached Kraft board (sulfate board) intended for food use is called white board or food board, useful for many food product applications. High content of sizing compounds improves moisture resistance. Paperboard coated with LDPE is called liquid-packaging board, useful for packaging liquids such as milk and juices. Paperboard coated with PP is useful for packaging of foods that are microwaveable in the package. Paperboard coated with PET can withstand the higher temperatures of a convection oven and, thus, is dual-ovenable. Paperboard can also be combined with layers of LDPE and aluminum foil to produce packaging useful for aseptic packaging. A common structure is LDPE/Pb/LDPE/Al/LDPE.

Duplex board is unbleached paperboard with a thin layer of glued bleached liner paper. Chipboard is a kind of paperboard produced from recycled paper fibers. It is acceptable for contact with dry, nonfatty foods, but otherwise functions in secondary packaging. A thin layer of paper can be glued to the chipboard to make lined chipboard. The liner is a better printing surface than the chipboard. Many other types of paperboard are available (Soroka and Zepf, 1998).

Cellophane is a form of paper which was the first flexible transparent film. To make cellophane, sulfite pulp is refined to increase the cellulose content to ~93%. The pulp is converted to a dissolved cellulosic compound through treatment with alkali and then carbon disulfide to form cellulose xanthate, which dissolves in the alkali to form a viscous colloidal dispersion called viscose:

$$-[C_6-H_7-O_2-(OH)_3]-_n + nNaOH \rightarrow -[C_6-H_7-O_2-(OH)_2-ONa]-_n + nH_2O$$

$$-[C_6-H_7-O_2-(OH)_2-ONa]-_n + nCS_2 \rightarrow -[C_6-H_7-O_2-(OH)_2-OCS_2Na]-_n$$

The viscose is then forced through a slit die into an acid/salt bath, where the cellulose is regenerated as thin sheets of cellophane from the cellulose xanthate:

$$-[C_6-H_7-O_2-(OH)_2-OCS_2Na]-_n + nHCl \rightarrow -[C_6-H_7-O_2-(OH)_3]-_n + nCS_2 + nNaCl$$

The cellophane formed is passed through a bath of glycerol and then dried. Absorption of glycerol from the bath plasticizes the cellophane, which is otherwise too brittle for use. The nonporous, polar nature of cellophane makes it an excellent oxygen and aroma barrier at low to intermediate RH. However, because of its hydrophilic nature, cellophane is a poor moisture barrier, and its oxygen and aroma barrier properties are diminished at high RH. An additional disadvantage of cellophane is that it is not heat-sealable. However, coatings were developed to improve the moisture-barrier property and provide heat-sealability. Nitrocellulose lacquer coating allows heat-sealing while maintaining the biodegradability of the cellophane. Cellophane is also often coated with PVDC, which improves barrier properties and provides heat-sealability. Cellophane has also been laminated with PP and aluminum foil to improve functionality. Cellophane has been largely replaced by synthetic polymers. The main replacement has been OPP film, which has mechanical and optical properties similar to Cellophane, lower cost, and heat-sealability, without sensitivity to moisture. However, ~1 billion lbs of cellophane are still produced annually in the world.

13.3.4.3 Paper Packaging Manufacture

Paper is a quite versatile packaging material, as it can be formed into flexible, semi-rigid, and rigid packaging (Miltz, 1992; Twede and Goddard, 1998; Soroka, 2002k; Kirwan, 2003; Robertson, 2006j).

13.3.4.3.1 Flexible Packaging
Papers such as greaseproof paper can be used simply as interleavers between slices of meat or cheese or as wraps around sticks of butter or margarine. Paper can be made into single-wall bags (e.g., grocery bags) and multi-wall bags/sacks (e.g., flour and sugar sacks) by cutting a form (blank), folding, and gluing. Because of their lack of any barrier properties, these bags function mainly to contain and to protect the product(s) from contamination and physical damage. Paper can be combined with layers of LDPE and aluminum to make sealed bags and pouches that provide a better barrier to moisture and oxygen for the food product.

13.3.4.3.2 Semi-Rigid Pulpboard and Paperboard Packaging

Pulpboard containers are made by capturing paper pulp on a mold and then drying the molded form. Paperboard packaging involves a number of different production methods and designs, including folding cartons, set-up boxes, tubs, and trays (Soroka, 2002k). Paperboard boxes/cartons are made by first cutting and scoring paperboard to make a form (blank) (Lynch and Anderson, 1997). The desired box can be preformed by folding and joining the paperboard blank at a box manufacturing facility, thus producing a preformed set-up box. A folding carton can be in-line formed (folded and joined) from a paperboard blank just before filling of the carton at the food facility (Obolewicz, 1997). Paperboard coated with LDPE provides a much-improved moisture barrier and can be heat-sealed into a carton. Addition of an aluminum layer provides excellent protection from oxygen. Cartons useful for shelf-stable, aseptic products can be made from either multi-layer LDPE/Pb/LDPE/Al/LDPE collapsed blanks with preformed side-seams, or from a role of multi-layer LDPE/Pb/LDPE/Al/LDPE using a form-fill-seal procedure similar to that used for in-line formation of pouches.

Composite cans consist of a body with paper components and ends made with metal (Eubanks, 1997). The body can be a simple combination of layers of printing paper/Pb/glassine useful for dry foods such as beverage powders. The body can also be a more complex combination such as LDPE/Pb/LDPE/Al/LDPE useful for frozen liquid products.

13.3.4.3.3 Semi-Rigid/Rigid Corrugated Board Containers

Single-face corrugated board is made by gluing a layer of corrugated (fluted) paperboard to a flat layer (liner) of paperboard. Single-wall corrugated, which is the most common form of corrugated, is made by gluing one layer of corrugated paperboard between flat layers (liners) of paperboard. Most corrugated containers are made by cutting and scoring single-wall corrugated to make a form (blank) that can later be folded and joined to make a box. The more layers of alternating corrugated and liner paperboard, the more rigid and strong the resulting box. Corrugated boxes are used mainly as tertiary or quaterrnary, logistics/distribution packaging (Foster, 1997).

More details on paper packaging design and manufacture can be found in a number of sources (Sumimoto, 1990; Paine, 1991; Miltz, 1992; Paine and Paine, 1992a; Hanlon et al., 1998a; Twede and Goddard, 1998; Soroka, 2002k; Kirwan, 2003; Robertson, 2006j).

13.3.4.4 Paper Package Closures

Sealing of paper bags and pouches requires addition of an adhesive or coating with a heat-sealing layer. Cartons and boxes made of paperboard coated with PE or PP can be heat-sealed. Beverage cartons usually have a small hole with a LDPE/AL/LDPE membrane that can be punctured with a straw. Large juice cartons often have a plastic, recloseable spout. Boxes used as tertiary (logistics/distribution) packaging can be sealed with an adhesive, tape and/or staples.

13.3.4.5 Paper Packaging Uses

Packages based on paper or including a paper layer are used for every food category. In addition, paper bags are also often used to transport purchased foods from the store to home. Heat-sealable LDPE-coated paper is used to make pouches for a wide variety of dry foods. An aluminum layer is added for additional protection of the food from oxygen.

Shock-absorbing, molded pulpboard cartons and trays are used for fragile foods such as eggs, fruits and vegetables. Pulpboard trays are also used to hold meats and fish.

Paperboard boxes and cartons are often used as secondary packaging for mechanical protection of foods, such as dry breakfast cereal or pasta. LDPE-coated paperboard cartons are used as primary packaging for milk, juices and other beverages (Robertson, 2002). LDPE-coated paperboard boxes are used for frozen foods. The LDPE coating must be replaced by PP to allow microwaving of the frozen food in the box. PET-coated paperboard boxes are necessary for dual-ovenable frozen foods. Multi-layer LDPE/Pb/LDPE/Al/LDPE cartons are used for aseptically processed juices and other pumpable food products. Composite cans with paperboard bodies and metal ends have wide use for packaging of dry and frozen foods.

The strength and shock-absorbing character of corrugated boxes makes them ideal logistics/distribution containers for most foods.

13.3.4.6 Paper Packaging Trends

A number of improvements have resulted in more effective paper-based packaging (Vakevainen, 2000). Development of stronger and lighter paperboard provides more effective packages at lower cost. Innovative paperboard multi-pack secondary packaging is being used more often for unitizing two or more primary packages (Becton and Braselton, 2004). The multi-packs make shopping, handling, and storage more convenient. Some also provide convenient dispensing of the primary packaged product. Improvements in polymer coatings have allowed thinner paperboard coatings compared to laminated layers, resulting in packages that can be more easily recycled. PP-coated paperboard has found application in microwavable foods. PET-coated paperboard is more heat resistant, allowing dual-ovenable food products. Many innovations for the aseptic carton have been introduced, including reclosable tops, larger sizes, new shapes, and a microwaveable version that replaces the aluminum layer with a barrier plastic (Nielaender, 1996; Seidel, 2001). Retortable cartons have been developed that replace the LDPE used in the aseptic carton with PP (Robertson, 2002).

13.3.5 PACKAGING MATERIAL COMBINATIONS

Many uses of packaging material combinations have been mentioned earlier in this chapter. These include the complementary materials used in primary, secondary, and logistics packaging, in multi-layer laminates of paper, plastic and aluminum, and in multi-layer laminates of different plastic materials. The composite can, with body made of combinations of paperboard, plastic film, and aluminum foil and ends made of metal, is another example. In many instances, these combinations of paper, plastic, and aluminum foil have provided alternatives to traditional glass bottles and metal cans. Following is a list of examples:

- LDPE-coated Pb cartons for refrigerated milk and juices
- PET/Al/PP pouches for retorted foods
- PET/EVOH/PP pouches and trays for retorted foods
- LDPE/Pb/LDPE/Al/LDPE cartons for aseptically processed beverages and foods
- PP/EVOH/PP spigoted bag-in-corrugated-box combination for wines
- PET/EVOH/PET bottles for beer

New developments in food packaging often involve new combinations of packaging materials that provide better food protection at lower cost and with greater convenience.

13.4 QUANTIFICATION OF PACKAGING MATERIAL PROPERTIES

Ability to quantify the properties of packaging materials and packages manufactured from those materials is critical to development and design of packaging that will serve the intended functions (Gaynes, 1997; Barron and Burcham, 2003c).

13.4.1 MECHANICAL PROPERTIES OF GLASS, METAL, PLASTIC, AND PAPER PACKAGING

13.4.1.1 Properties of Packaging Materials

The ability of a package to maintain integrity is determined by the packaging material mechanical properties and the package closure's effectiveness. Thus, knowledge of mechanical properties is important for all packaging materials, since they reflect the ability of the package to maintain its protective functions under physical stress. A number of tests have been established that can assess the packaging material strength under tension, compression, bursting, tearing, or impact forces (Figure 13.20) (Karel and Lund, 2003a). The strength determined by each test is defined as the amount of force/area necessary to cause failure. The most commonly measured mechanical properties of packaging materials are the tensile properties, which include the material strength at break under tension, Young's (elastic) modulus (proportional to stiffness or rigidity), and elongation at break (Figure 13.21) (Robertson, 1993e). These properties are determined by determining the relationship between stress (force/area) and strain (elongation) when the material is stretched at a set rate (distance/time). Tough materials display a large area under the stress-strain curve, whereas brittle materials show a small area. Figure 13.22 shows typical results for a variety of materials having different tensile properties (Miltz, 1992). Table 13.16 gives a list of standard tests used to determine mechanical, including tensile, properties of packaging materials (ASTM, 2002, 2003).

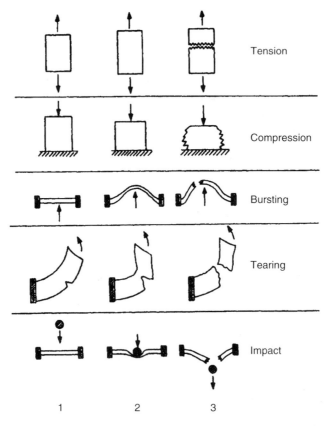

FIGURE 13.20 Tests that assess packaging material strength under tension, compression, bursting, tearing or impact forces. (From Karel, M. and Lund, D.B., 2003a; Copyright 2003 from *Physical Principles of Food Preservation* by Marcus Karel and Daryl B. Lund. Reproduced by permission of Routledge/Taylor & Francis Group, LLC.)

Other specialized tests are used to determine properties relevant to specific packaging materials functions (Gaynes, 1997). A test unique to can maker's quality steel is the Rockwell 30-T hardness test. The Rockwell hardness is determined by the degree of penetration of a hardened steel ball under given force into a sheet of steel. The hardness reflects the relative degree of cold rolling (stiffening) and annealing (softening) of the steel. The temper classifications given to manufactured steels are related to the Rockwell hardness values, as shown in Table 13.8.

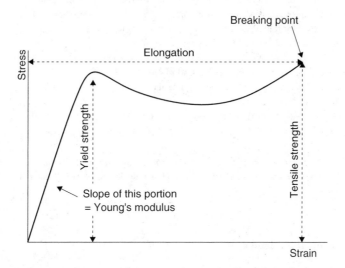

FIGURE 13.21   Tensile properties determined for a plastic material from the stress (force/area) vs. strain (elongation) relationship determined when the material is stretched at a set rate (distance/time). (From Robertson, G.L., 1993e. *Food Packaging — Principles and Practice.* New York, Marcel Dekker, Inc., 63–72. Copyright 1993 from *Food Packaging — Principles and Practice* by G.L. Robertson. Reproduced by permission of Routledge/Taylor & Francis Group, LLC.)

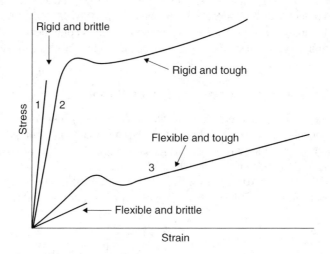

FIGURE 13.22   Stress (force/area) vs. strain (elongation) relationship when the material is stretched at a set rate (distance/time) for a variety of plastic materials having different tensile properties. (From Miltz, J., 1992. In *Handbook of Food Engineering.* D.R. Heldman and D.B. Lund, Eds., New York, Marcel Dekker, Inc., 667–718. Copyright 1992 from *Handbook of Food Engineering* by D.R. Heldman and D.B. Lund. Reproduced by permission of Routledge/Taylor & Francis Group, LLC.)

TABLE 13.16
Standard Tests for Mechanical Properties of Packaging Materials

ASTM D685 — Practice for conditioning paper and paper products for testing
ASTM D774/D774M — Test method for bursting strength of paper
ASTM D828 — Test method for tensile properties of paper and paperboard using constant-rate-of-elongation apparatus
ASTM D882 — Test method for tensile properties of thin plastic sheeting
ASTM D1596 — Test method for dynamic shock cushioning characteristics of packaging material
ASTM D1922 — Test method for propagation tear resistance of plastic film and thin sheeting by pendulum method
ASTM D 2176 — Test method for folding endurance of paper by the M.I.T. tester
ASTM F0392 — Test method for flex durability of flexible barrier materials
TAPPI T411 — Thickness (Caliper) of paper, paperboard and combined board
TAPPI T414 — Internal tearing resistance of paper (elmendorf-type method)
TAPPI T423 — Folding endurance of paper (Schopper type tester)
TAPPI T803 — Puncture test of container board
TAPPI T810 — Bursting strength of corrugated and solid fiberboard
TAPPI T811 — Edgewise compressive strength of corrugated fiberboard (Short column test)

Source: From ASTM (2002). *Consumer and Healthcare Packaging Standards.* Philadelphia, American Society for Testing and Materials, 366; (2003). *Selected ASTM Standards on Packaging.* Philadelphia, American Society for Testing and Materials, 493.

13.4.1.2 Properties of Packages

Mechanical properties of manufactured packages are certainly related to the mechanical properties of the packaging materials utilized. However, additional tests are used to assess package performance and detect package imperfections. Depending on the package material, these can include tests of package and/or seal dimensions; resistance to temperature, compression, impact or internal pressure; presence of desired vacuum, pressure or atmosphere; absence of leaks; and strength and integrity of seals. A number of leak test methods are available, depending on the package type (Arndt, 1997; Johnson and Demorest, 1997). Table 13.17 and Table 13.18 list standardized tests for mechanical properties of primary and distribution packages, respectively (ASTM, 2002, 2003). Additional tests are used to assess the integrity of packages. A test unique to glass containers is ASTM C 149 which that involves determination of resistance to thermal shock of glass bottles and jars.

13.4.2 LIGHT TRANSMISSION OF GLASS AND PLASTIC PACKAGING

Light transmission is an important property for glass and plastic packaging. As discussed earlier, various food components are sensitive to light, with the effect dependent on particular wavelength in the high UV and low visible ranges. Besides wavelength, the extent of light's effect on food component degradation depends on the intensity and time of exposure (Bossett et al., 1994).

The intensity of light absorbed by a packaged food depends on the light transmission and reflection properties of the packaging material and the reflection properties of the food (Fellows, 2000):

$$I_a = I_i T_p \left[\frac{1 - R_f}{1 - R_f R_p} \right] \tag{13.2}$$

where I_a is the intensity of light absorbed by the food, I_i is the intensity of light incident on the package, T_p is the fraction of light transmitted by the packaging material, R_f is the fraction of light reflected by the food, and R_p is the fraction of light reflected by the packaging material.

TABLE 13.17
Standard Tests for Mechanical Properties of Packages

ASTM D2561 — Test method for environmental stress-crack resistance of blow-molded polyethylene containers
ASTM D3078 — Test method for determination of leaks in flexible packaging by bubble emission
ASTM D4332 — Practice for conditioning containers, packages, or packaging components for testing
ASTM D4577 — Test method for compression resistance of a container under constant load
ASTM D4991 — Test method for leakage testing of empty rigid containers by vacuum method
ASTM D5094 — Test methods for gross leakage of liquids from containers with threaded or lug-style closures
ASTM D5276 — Test method for drop test of loaded containers by free fall
ASTM D5277 — Test method for performing programmed horizontal impacts using an inclined impact tester
ASTM D5487 — Test method for simulated drop of loaded containers by shock machines
ASTM D6537 — Practice for instrumented package shock testing for determination of package performance
ASTM F0088 — Test method for seal strength of flexible barrier materials
ASTM F1921 — Test methods for hot seal strength hot tack of thermoplastic polymers and blends comprising the sealing surfaces of flexible webs
ASTM F2054 — Test method for burst testing of flexible package seals using internal air pressurization within restraining plates
ASTM F 2095 — Test methods for pressure decay leak test for nonporous flexible packages with and without restraining plates

Source: From ASTM (2002). *Consumer and Healthcare Packaging Standards.* Philadelphia, American Society for Testing and Materials, 366; (2003). *Selected ASTM Standards on Packaging.* Philadelphia, American Society for Testing and Materials, 493.

The fraction of light transmitted by the packaging material can be assumed to follow the Beer–Lambert Law:

$$T_p = \frac{I}{I_i} = e^{-kx} \tag{13.3}$$

where I is the intensity of light transmitted by the packaging material, k is absorbance of the packaging material, and x is the thickness of the packaging material. The value of k depends on the packaging material and the light wavelength.

Figure 13.23 shows the effect of wavelength on the light transmission of common flint (clear) glass and the effect of various coloring agents (Robertson, 1993c, 2006g). Packages that transmit 10% or less of incident length at all wavelengths between 2900 and 4000 Å are defined as light-resistant. Table 13.19 compares the effect of wavelength on the light transmission properties of several plastic materials with translucent paper and clear glass (Robertson, 1993b; Karel and Lund, 2003a). The light transmission properties of plastic materials and glass can be modified by adding coloring agents or coatings that absorb light, depending on wavelength.

13.4.3 PERMEABILITY, MIGRATION AND SCALPING PROPERTIES OF PLASTIC PACKAGING

13.4.3.1 Permeability

Increased use of plastic packaging materials has depended on ability to quantify permeability, modify or combine polymers to control permeability, and use permeability coefficients in the design of packaging (Krochta, 2003).

TABLE 13.18

Standard Tests for Mechanical Properties of Distribution Packages

ASTM D642 — Test method for determining compressive resistance of shipping containers, components, and unit loads

ASTM D880 — Test method for impact testing for shipping containers and systems

ASTM D999 — Methods for vibration testing of shipping containers

ASTM D4003 — Test methods for programmable horizontal impact test for shipping containers and systems

ASTM D4169 — Practice for performance testing of shipping containers and systems

ASTM D4279 — Test methods for water vapor transmission of shipping containers-constant and cycle methods

ASTM D4728 — Test method for random vibration testing of shipping containers

ASTM D5276 — Test method for drop test of loaded containers by free fall

ASTM D5277 — Test method for performing programmed horizontal impacts using an inclined impact tester

ASTM D5331 — Test method for evaluation of mechanical handling of unitized loads secured with stretch wrap films

ASTM D5415 — Test method for evaluating load containment performance of stretch wrap films by vibration testing

ASTM D5487 — Test method for simulated drop of loaded containers by shock machines

ASTM D6537 — Practice for instrumented package shock testing for determination of package performance

Source: From ASTM (2002). *Consumer and Healthcare Packaging Standards.* Philadelphia, American Society for Testing and Materials, 366; (2003). *Selected ASTM Standards on Packaging.* Philadelphia, American Society for Testing and Materials, 493.

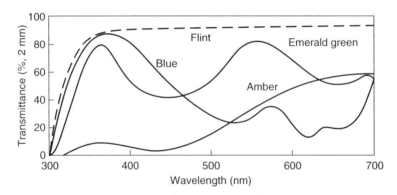

FIGURE 13.23 Wavelength effect on transmission properties of glasses. (From Robertson, G.L., 1993c. *Food Packaging — Principles and Practice.* New York, Marcel Dekker, Inc., 232–251; (2006g). *Food Packaging — Principles and Practice.* Boca Raton, CRC Taylor & Francis Group, 157–174. Copyright 1993. Reproduced by permission of Routledge/Taylor & Francis Group, LLC.)

13.4.3.1.1 Mechanism

Permeability is a process unique to plastic polymers, since glass and metal are impervious, and transport of gas and vapor through paper pores occurs by mechanisms different from permeability (Geankoplis, 1983; Vieth, 1991; Hernandez, 1997a; Johnson and Demorest, 1997).

Permeability is a three-step process. It includes adsorption (dissolution) of vapor or gas onto the side of a plastic material exposed to a high partial pressure of the vapor or gas, diffusion through the material, and then desorption from the side of the material exposed to a low partial pressure of the vapor or gas. The permeability of a plastic material to vapors and gases determines its usefulness for a particular food packaging application.

The diffusion step of permeability can be represented by Fick's first law of diffusion, in which permeant mass flux through a section of isotropic material is related to concentration gradient measured

TABLE 13.19

Wavelength Effect on Light Transmission Properties of Packaging Materials

Material	Thickness (mm)	% Transmission Wavelength (Å)		
		3000	4000	5000
HDPE	0.089	5	22	36
Vinylchloride/vinylidene chloride copolymer	0.028	4	85	86
Polyester	0.036	0	81	86
LDPE	0.038	75	85	85
Translucent wax paper	0.089	20	50	59
Clear glass	0.29	0	80	80

Source: Adapted from Robertson, G.L. (1993b). *Food Packaging—Priciples and Practice.* New York, Marcel Dekker, Inc., 252–302; Karel, M. and Lund, D.B. (2003a). *Physical Principles of Food Preservation.* New York, Marcel Dekker Inc., 514–592.

normal to the section by a diffusion coefficient (Crank, 1975):

$$F = -D\frac{\partial C}{\partial x} \tag{13.4}$$

where F is permeant flux or transmission rate (quantity of permeant per unit time per unit area), C is permeant concentration, $\partial C/\partial x$ is permeant concentration gradient in the x direction over thickness ∂x, and D is the diffusion coefficient. At steady state with D constant through a sheet of thickness L:

$$F = D(C_1 - C_2)/L \tag{13.5}$$

where C_1 and C_2 are the concentrations at surfaces 1 and 2 of the sheet.

If a linear relationship between concentration in the polymer material and the surrounding atmosphere is assumed, Henry's Law applies (Robertson, 2006k):

$$C = Sp \tag{13.6}$$

where p is partial pressure of the permeant in equilibrium with the polymer containing permeant with concentration C. The factor relating p and C is the solubility coefficient, S. If the solubility coefficient is constant, Equation 13.6 can be combined with Equation 13.5 to obtain:

$$F = DS(p_1 - p_2)/L \tag{13.7}$$

The product DS is defined as the permeability coefficient (or permeability), P:

$$F = P(p_1 - p_2)/L = P\Delta p/L \tag{13.8}$$

To determine P for a given polymer material, one can set the partial pressures of the permeant on each side of the sheet, p_1 and p_2, and then allow the sheet surfaces to come into equilibrium with the set partial pressures. Then, one can determine F, the transmission rate, by measuring the quantity of permeant (q) that permeates a film of a given area (A) over a specified time (t) at steady state:

$$F = q/A \cdot t \tag{13.9}$$

Equation 13.8 is generally correct when D and S are both constants. If D and S are not constant, P is an average or effective permeability coefficient that applies only over the selected partial pressure range p_1 to p_2 in relating F to $(p_1 - p_2)/L$. Depending on the permeant-polymer pair, P may be constant or not. D, S, and P are relatively constant when little interaction occurs between the permeant and polymer. An example of this would be water vapor, which is polar, permeating through polyethylene, which is nonpolar. However, when the permeant interacts with the polymer sufficiently to plasticize the polymer, the result is that permeant concentration has an effect on D, S, and P. Thus, they would not be constants (Rogers, 1985). An example of this would be water vapor permeating through ethylene-vinyl alcohol (EVOH) copolymer, which is polar.

It is important to distinguish among several terms in the literature which refer to the permeation process described in Equation 13.8 (Giacin and Hernandez, 1997; Hernandez, 1997a; Hernandez and Giacin, 1998). Sometimes, the terms transmission rate, thickness-normalized transmission rate, permeability coefficient (or permeability), and permeance are used interchangeably, in which case careful attention must be made to the actual units used.

13.4.3.1.2   Factors Affecting Permeability

The factors which affect permeability can be grouped into compositional factors and environmental factors (Pascat, 1986; Hernandez, 1997b; Hernandez and Giacin, 1998). Compositional factors include permeant size and shape, polymer morphology, polymer additives, and permeant-polymer interaction (Ashley, 1985; Jasse et al., 1994; Delassus, 1997). Temperature effect on the permeability coefficient reflects the temperature effect on the solubility and diffusion coefficients. An Arrhenius-type equation describes the relationship (Robertson, 2006k):

$$P = DS = D_o S_o \exp\{-(E_d + \Delta H_s)/RT\} = P_o \exp(-E_p/RT) \tag{13.10}$$

ΔH_s is the heat of solution of the permeant in the polymer. The ΔH_s is small and positive for permanent gases like oxygen, with the result that S increases slowly with temperature. However, the ΔH_s is negative and larger for condensable vapors like water, with the result that S decreases with temperature. E_d is the activation energy (necessary for hole to appear for a diffusion jump) for diffusion of the permeant in the polymer. Thus, E_d is always positive and D always increases with temperature. E_p is an apparent activation energy for permeability of the permeant in the polymer. Thus, depending on whether ΔH_s is positive or negative, and depending on the relative size of ΔH_s and E_d, the permeability coefficient theoretically may increase or decrease with temperature. However, for all known permeant-polymer pairs, the permeability coefficient increases with temperature (Delassus, 1997). Depending on the polymer material, the relative humidity can also affect permeability. Humidity has no effect on nonhydrophilic polymers. However, the permeability of hydrophilic polymers increases with the amount of moisture absorbed, since the water acts as a plasticizer for these polymers. Some polymers display a small decrease in permeability with absorption of moisture, including polyethylene terephthalate and amorphous nylons (Delassus, 1997).

Properties of different polymers can be combined by the formation of multi-layer films. If the permeability coefficients of the individual layers are independent of the permeant partial pressure and of the water-vapor partial pressure (if a gradient of the latter exists across the film), the permeability coefficient for the resulting multi-layer film, P_T, can be calculated using the following equation (Paine and Paine, 1992d; Hernandez, 1997a; Robertson, 2006k):

$$L_T/P_T = (L_1/P_1) + (L_2/P_2) + \cdots (L_n/P_n) \tag{13.11}$$

L_T is the resulting total thickness of the multi-layer film, and the permeability coefficients and thickness of the individual layers are designated by $1, 2, \ldots, n$. If the permeability coefficients of the individual layers are dependent on permeant concentration, the individual permeability coefficients

TABLE 13.20

Standard Tests for Water Vapor and Gas Transmission Properties of Packaging Materials and Manufactured Packages

ASTM D895 — Test method for water vapor permeability of packages
ASTM D1251 — Test method for water vapor permeability of packages by cycle method
ASTM D1434 — Test method for determining gas permeability characteristics of plastic film and sheeting
ASTM D3079 — Test method for water vapor transmission of flexible heat-sealed packages for dry products
ASTM D3199 — Test method for water vapor transmission through screw-cap closure liners
ASTM D3985 — Test method for oxygen gas transmission rate through plastic film and sheeting using a coulometric sensor
ASTM E96 — Test methods for water vapor transmission of materials
ASTM E0171 — Specification for standard atmospheres for conditioning and testing flexible barrier materials
ASTM F0372 — Test method for water vapor transmission rate of flexible barrier materials using an infrared detection technique
ASTM F1115 — Test method for determining the carbon dioxide loss of beverage containers
ASTM F1249 — Test method for water vapor transmission rate through plastic film and sheeting using a modulated infrared sensor
ASTM F1307 — Test method for oxygen transmission rate through dry packages using a coulometric sensor
ASTM F1769 — Test method for the measurement of diffusivity, solubility, and permeability of organic vapor barriers using a flame ionization detector
ASTM F1770 E01 — Test method for evaluation of solubility, diffusivity, and permeability of flexible barrier materials to water vapor
ASTM F1927 E01 — Test method for determination of oxygen gas transmission rate, permeability and permeance at controlled relative humidity through barrier materials using a coulometric detector

Source: From ASTM (2002). *Consumer and Healthcare Packaging Standards.* Philadelphia, American Society for Testing and Materials, 366; (2003). *Selected ASTM Standards on Packaging.* Philadelphia, American Society for Testing and Materials, 493.

will depend on the thickness and positioning of the layers. Other, more complicated composite structures are also possible (Karel and Lund, 2003a).

13.4.3.1.3 Measurement of Transmission Rates and Permeability Coefficients

Table 13.20 lists accepted standard methods for measuring the transmission and permeability properties of both films and formed packages (ASTM, 2002, 2003). The permeability coefficients of most common interest are the water vapor permeability (WVP), oxygen permeability (O_2P), carbon dioxide permeability (CO_2P), and organic compounds (e.g., aroma) (Johnson and Demorest, 1997). Alternative methods have been developed for measurement of O_2P and CO_2P (Gilbert and Pegaz, 1969; Demorest et al., 2000), as well as for measurement of aroma permeability (Hernandez et al., 1986; Hatzidimitriu et al., 1987; DeLassus, 1988; Hernandez et al., 1989; Seeley, 1997; Miller and Krochta, 1998; Risch et al., 2000). Modified methods have been developed for determination of WVP for hydrophilic films (McHugh et al., 1993; Gennadios et al., 1994).

For measuring film permeability, the film is sealed in a permeability test cell at a controlled constant temperature and pressure, and each side of the film is exposed to controlled water vapor, oxygen, carbon dioxide, or organic compound partial pressure to create a defined Δp. In the case of O_2P, CO_2P and organic compound permeability, the relative humidity must also be defined and controlled if the polymer is plasticized by absorption of water vapor. After the film has reached equilibrium with the conditions of the test and steady-state has been achieved, the quantity of permeant transferring through the film is measured by some method and then converted into a permeability coefficient according to Equation 13.8.

TABLE 13.21
Permeability Coefficients of Common Plastic Film Materials Used in Packaging

	$P \times 10^{11}$ [mL(STP) cm cm^{-2} sec^{-1} (cm Hg^{-1})]		
Polymer	O_2 (30°C)	CO_2 (30°C)	H_2O (90% RH, 25°C)
Low density polyethylene	55	352	800
High density polyethylene	11	35	130
Polypropylene	23	92	680
Polystyrene	11	88	12,000
Nylon 6	0.38	1.6	7,000
Poly(ethylene terephthalate)	0.22	1.5	1,300
Poly(vinylidene chloride)	0.053	0.29	14

Source: Adapted from Stannett, V., Szware, M., Bhargava, R.L., Meyers, J.A., Myers, A. W., and Roger, C.E. (1962). *Permeability of Plastic Films and Coated Papers to Gases and vapors.* New York, Technical Association of the Pulp and Paper Industry, 105; Robertson, G.L. (1993d). *Food Packaging — Principles and Practice.* New York, Marcel Dekker, Inc., 73–110.

13.4.3.1.4 Permeability Coefficient Values

Polymer permeability coefficients can be found in a number of references (Paine and Paine, 1992d; Delassus, 1997; Giacin and Hernandez, 1997; Hernandez, 1997a; Johnson and Demorest, 1997; Robertson, 2006k). Table 13.21 lists O_2P, CO_2P and WVP values for a number of common polymers (Stannett et al., 1962; Robertson, 1993d). Many combinations of units are used in the polymer literature. Tables are available for conversion from one set of units to another (Delassus, 1997; Hernandez, 1997a; Robertson, 2006k). Nonetheless, one must be careful in converting from one set of units to another. Furthermore, literature values of permeability should be used only for comparisons of different polymers and rough design estimates, and only for the conditions of permeability measurement. Considerable variation in the permeability of a given material can result from differences in polymer molecular structure and weight, additives, and polymer-product- (e.g., film-) formation conditions (Selke, 1997d). Permeability for a selected commercial polymer and polymer product should be obtained from the supplier and/or measured by the user.

The accuracies of various methods for estimating partition, solubility, and permeability coefficients of organic molecules in several polymers have been determined (Baner, 2000).

13.4.3.2 Migration and Scalping

Migration and scalping (sorption) are similar to permeability in that compounds are either gained by (migration) or lost from (scalping) the food. However, rather than the atmosphere being the source or sink of the compounds, migration and scalping involve food interaction with the packaging material itself (Linssen and Roozen, 1994). Many factors affect permeability, migration, and sorption in food-package systems (Hernandez and Giacin, 1998; Linssen et al., 2003). Standard tests for measuring migration and scalping in packaging materials are listed in Table 13.22 (ASTM, 2002, 2003).

13.4.3.2.1 Migration

An example of an area of concern about migration is in microwave heating of foods in packaging containers. Especially when aluminum susceptors are added to plastics to aid in crisping and browning of the food, high temperatures are created that could result in polymer breakdown and release of additives into the food. In-package pasteurization and sterilization using microwave heating are also

TABLE 13.22
Standard Tests for Migration and Scalping in Packaging Materials

ASTM F0034 — Practice for construction of test cell for liquid extraction of flexible barrier materials

ASTM F 0151 — Test method for residual solvents in flexible barrier materials

ASTM F 0874 — Test method for temperature measurement and profiling for microwave susceptors

ASTM F 1308 — Test method for quantitating volatile extractables in microwave susceptors used for food products

ASTM F 1317 — Test method for calibration of microwave ovens

ASTM F 1349 — Test method for nonvolatile ultraviolet UV absorbing extractables from microwave susceptors

ASTM F 1479 — Terminology relating to microwave food packaging

ASTM F 1500 — Test method for quantitating nonUV-absorbing nonvolatile extractables from microwave susceptors utilizing solvents as food simulants

ASTM F 1519 — Test method for qualitative analysis of volatile extractables in microwave susceptors used to heat food products

ASTM F 1884 — Test method for determining residual solvents in packaging materials

ASTM F 2013 — Test method for determination of residual acetaldehyde in polyethylene terephthalate bottle polymer using an automated static head-space sampling device and a capillary GC with a flame ionization detector

Source: From ASTM (2002). *Consumer and Healthcare Packaging Standards.* Philadelphia, American Society for Testing and Materials, 366; (2003). *Selected ASTM Standards on Packaging.* Philadelphia, American Society for Testing and Materials, 493.

used (Ohlsson, 2000). Possible migration of oligomers and additives from the packaging material must be considered in selecting appropriate packaging materials for in-package microwave processing and preparation (Ozen and Floros, 2001). Unpolymerized monomer can also migrate into foods that are not microwaved (Linssen and Roozen, 1994).

Mathematical modeling is of critical importance to research on migration, design of food-package systems, and development of package-safety regulations (Chatwin, 1996). Migration of a package component into a food can be modeled mathematically using a number of different approaches (Chang et al., 1988). For many food-package systems, components of the food or food simulant are assumed not to absorb in the packaging material. In addition, diffusion of the migrant in the packaging material is assumed to control the rate of migration into the food. The initial concentration of the migrating substance in a sheet of packaging material is assumed to be uniform. Additionally, the sheet surface concentration of the migrant can be assumed to drop to zero upon exposure of the sheet to a food simulant for which the migrant has affinity and when the simulant has large volume relative to the amount of migrant. Such conditions allow use of the following equation based on Fick's second law (Crank, 1975):

$$\frac{M_t}{M_\infty} = 1 - \frac{8}{\pi^2} \sum_{n=0}^{\infty} \frac{1}{(2n+1)^2} \exp\{-D(2n+1)^2\pi^2 t/l^2\} \tag{13.12}$$

where M_t is the amount of migrating substance that has diffused out of the sheet as a function of time t, M_∞ is the amount that would diffuse out at very large time (when the sheet has reached equilibrium with the simulant), l is the thickness of the sheet, and D is the diffusion coefficient for the migrant in the packaging material. The value of D for the migrant in the packaging material can determined by fitting the migration data to the solution.

The value of D can be estimated from the following approximation to Equation 13.12 for the initial stages of migration (Crank, 1975):

$$M_t/M_\infty = 4(Dt/\pi l^2)^{0.5} \tag{13.13}$$

where $M_t/M_\infty < \sim 0.6$.

The following equation can also be used to estimate the diffusion coefficient (Crank, 1975):

$$D = 0.049 \, l^2 / t_{1/2} \tag{13.14}$$

where $t_{1/2}$ is the time when half of the migrant has transferred to the food simulant (i.e., $M_t / M_\infty = 0.5$).

In the cases where the volume of food simulant is not large compared with the amount of migrating substance, the assumption that the sheet surface concentration of migrant drops to zero when exposed to the simulant is not appropriate. Then, more complicated mathematical analysis is necessary (Crank, 1975; Miltz, 1992).

The assumptions of simulant sorption into the packaging material having no effect on migrant diffusion and of migrant diffusion in the packaging material controlling the rate of migration do not apply to many food-package systems. Non-Fickian diffusion in the packaging material or migrant diffusion being controlled by the simulant phase may apply. In these cases, different mathematical approaches must be taken (Chang et al., 1988).

An equation for predicting diffusion coefficients of hydrocarbons in polyolefin plastics has been developed (Brandsch et al., 2000).

13.4.3.2.2 Scalping

In scalping (sorption), a component of a food product is sorbed by the packaging material without transfer to the surrounding atmosphere (Giacin, 1995; Giacin and Hernandez, 1997). The extent of scalping depends on properties of both the food and the packaging material (Halek and Luttmann, 1991). An important example of scalping is sorption of flavor/aroma compounds into the inner low-density polyethylene (LDPE) layer of multi-layer juice boxes which include an adjacent aluminum layer that prevents further transfer of the compound out of the LDPE layer. HDPE can also scalp volatile compounds from drinks (Linssen and Roozen, 1994). Another example is sorption of CO_2 from carbonated beverages by polyethylene terephthalate (PET) bottles with little transfer to the external environment (Koros, 1990). The latter type of scalping can be explained by the concept of dual-mode absorption, where one population of molecules sorbs so strongly onto sites in a polymer matrix that they are fixed or move very slowly (Stern and Trohalaki, 1990). Methods to quantify sorption of aroma compounds by food-contact polymers have been developed (Roland and Hotchkiss, 1991). Besides producing loss in food quality, scalping can compromise the package integrity and/or barrier properties (Harte and Gray, 1987; Hirose et al., 1988).

13.5 TRENDS IN FOOD PACKAGING

Trends in food packaging related to advances in glass, metal, plastic, paper, and combination materials were discussed earlier in this chapter. Additional trends reflect increased understanding of factors that improve food quality, including new processing techniques. Thus, these advances support the protection function of food packaging. Other trends reflect social and cultural changes that place more expectations on packaging, involving the communication, convenience and minimal environmental impact functions of packaging.

13.5.1 FLEXIBLE PACKAGING

Developments of new plastic materials and improved methods of converting them to packaging have supported rapid growth of the flexible packaging industry. Flexible packaging can be based on a single layer of plastic material or a combination of materials that can include several different plastics, paper, and aluminum. Flexible packaging is an attractive alternative to more traditional rigid and semi-rigid containers because of several advantages, including packaging material source reduction, storage space reduction, convenience for consumers, and visual and handling appeal (Shellhammer, 2003). The significant package weight reduction when using flexible packaging has to be balanced against the nonrecyclability of multilayer flexible packaging. Flexible packaging has several unique design criteria, including sealing, slip, stiffness, environmental impact, and food safety (Dixon, 2000). The stand-up pouch has replaced many applications of the lay-down pillow pouch, as well as opened up new categories of products such as liquid foods (Greely, 1997). Stand-up pouches with their greater visibility and new dispensing/resealing approaches are attractive to consumers.

Flexible packaging has had an important role in the development of several concepts that have led to improvement of food quality, including retortable pouches, aseptic processing and packaging, resealable pouches of individual quick frozen (IQF) foods, and modified atmosphere packaging of fresh and dry foods (Dixon, 2000; Shellhammer, 2003). Several of these are discussed in greater detail below.

13.5.2 RETORTABLE POUCHES, TRAYS, TUBS (CANS), AND CARTONS

Flexible pouches, semi-rigid/rigid plastic trays and cans, and paperboard-based cartons have been developed as alternatives to heat processing (retorting) in rigid metal cans or glass containers (Duxbury, 1997; Robertson, 2002; Blakistone, 2003). The pouches, trays, and tubs are always multi-layer laminate structures that contain different polymers which provide heat resistance, strength, and toughness (PET), pierce and pinhole resistance (nylon), oxygen barrier (EVOH, nylon or PVDC) and (for the pouches and trays) heat sealability (PP). An aluminum foil layer often serves as the moisture and oxygen barrier in pouches. The retortable paperboard cartons have external and internal PP layers that are impermeable to liquid and allow heat sealing, along with an internal aluminum layer that provides a gas and light barrier (Robertson, 2002).

Retortable pouches can be either preformed or in-line formed using form/fill/seal equipment. Common pouch structures are PET/nylon/foil/PP and PET/nylon/EVOH or PVDC/PP. In the U.S., the military has been a prime user of retortable pouches in the form of meals-ready-to-eat (MREs). However, retortable pouches are used by hotels, restaurants, and other institutions. Retail consumer products such as tuna, salmon, chicken patties, chipped beef, chili, and ground beef in retortable pouches have become available.

Retortable trays have a semi-rigid or rigid body and a sealable flexible lid. The trays are generally made from coextruded laminate such as PET/EVOH/PP by thermoforming. Retortable tubs are made from similar multi-layer laminates. An easy-open scored metal lid with pull ring is double seamed onto the tub body.

The advantage of retortable pouches and trays is that they have thinner profile than conventional metal or glass containers. The results are shortened process times, reduced energy consumption, and improved food quality due to more rapid and even heat transfer. In addition, retort pouches, trays, and tubs are convenient because of easy transport (due to shape and light weight) and easy opening. Plastic (with no foil layer) pouches, trays, and tubs are microwaveable. Trays and tubs can be used as serving dishes and bowls, respectively. Food products in retortable tubs are convenient items that can be made available in a dispensing machine with nearby microwave oven. Retortable rectangular cartons are lighter, take up less pallet and shelf space, have more surface area for information, and are easier to open than conventional cylindrical metal cans (Estrada, 2004).

The main disadvantage of retortable pouches, trays, tubs, and cartons is more difficult recycling. Pouch integrity and sealing have also been concerns that are addressed through vigorous package inspection and regulation. National Food Processors Association (NFPA) recommends several tests, including squeeze test, burst test, and seal tensile strength (Blakistone, 2003). Seals can also be tested using a dye penetration test or headspace gas composition test. Retorting of pouches and trays must include overpressure and critical control of pressure changes to prevent seal failure. Also, special racks or trays are incorporated in the retort to restrain pouches to a defined thickness for consistent heat transfer.

13.5.3 ASEPTIC PROCESSING AND PACKAGING

Aseptic processing and packaging is an alternative to conventional retorting of foods in metal, glass or plastic containers (Mabee, 1997; Robertson, 2002, 2006l). Common aseptically processed products include juices, milk and milk products, liquid eggs, tomato products, puddings, soups, and sauces. The advantage of aseptic processing is that pumpable foods can be heat sterilized at high temperature more quickly and evenly, compared to retorting, by passing through the narrow chambers of a heat exchanger. Heat is transferred by conduction and convection from a heat transfer medium (usually steam) through the walls of the heat exchanger to the food. After reaching and holding at the appropriate temperature, the food is then cooled quickly in another heat exchanger. The result is a high temperature short time (HTST) process that produces higher food quality. The concept is dependent on the fact that microorganisms are more vulnerable to temperature increase than the flavors, aromas, colors, and textures that comprise food quality. Thus, a HTST process that gives the same required microbial death as a lower temperature longer time process produces a higher quality food product. In addition, the process is faster and consumes less energy. A disadvantage of aseptic processing is the complexity of the equipment and control system. Only pumpable foods can be aseptically processed, and low-acid particulate foods present a greater challenge in terms of ensuring adequate heating of the particles due to uneven heat exchange.

New thermal processing methods have been developed, including electrical resistance (ohmic) heating, high-frequency heating, and microwave heating. These direct-heating methods can replace use of conventional conductive and convective heat transfer in the sterilization of foods in an aseptic processing system. The ohmic-, high-frequency-, or microwave-sterilized foods can then be filled aseptically into the same aseptic packaging used for foods that have been heat processed using conventional heat exchangers. The advantage is that the foods, including those with particles, are heated more uniformly and with less energy.

After cooling, the sterile food product is pumped to an aseptic packaging system where the food is filled and hermetically sealed into previously sterilized containers. Aseptically processed foods can be packaged in the same types of containers used for retorted foods. However, another advantage of aseptically processed foods is that they can be packaged in containers that do not have to survive the conditions of a retort. These include LDPE/Pb/LDPE/Al/LDPE laminate cartons and multilayer plastic flexible packaging that has cost and convenience advantages. The disadvantage of these packages is that they are not as easily recycled as metal and glass containers. Aseptic filling systems have also been developed for HDPE and PET bottles (Ammann, 2001). Aseptic filling of PET containers may have a cost advantage over hot filling of heat-set PET containers (Thompson, 1999). Another advantage of aseptically processed foods is that they can be filled into drums, railroad tank cars, tank trucks and silos that have been previously sterilized with steam. The food can be later reprocessed and packaged to meet market demands.

The sterilization agents available for aseptic packaging include heat, chemical treatment with hydrogen peroxide and high energy irradiation (UV light or ionizing (gamma) irradiation) (Ansari and Datta, 2003). A combination of hydrogen peroxide and mild heat is most commonly used with plastic and paperboard-based laminate packaging.

13.5.4 NONTHERMAL FOOD PROCESSING TECHNIQUES

Nonthermal technologies are available that can preserve food with little or no loss of food quality (Goddard, 1995; Bolado-Rodriguez et al., 2000; Gould, 2000). Ionizing radiation has been approved for several foods, including uncooked poultry, meat, and ground meat (Potter and Hotchkiss, 1995;

Thayer, 2003). Levels of allowed radiation are not sufficient to sterilize foods but can destroy vegetative food-spoilage and pathogenic microorganisms. To prevent recontamination, the irradiation process is performed on prepackaged foods. Irradiation has been found to affect the properties of glass, plastic, and paper packaging materials (El Makhzoumi, 1994; Ozen and Floros, 2001). The effect on most plastic polymers such as PE, PP, and PS is crosslinking. However, chain scission has been observed with some polymers. The result can be modification of the mechanical, barrier, and optical properties of the plastic material, as well as the strength of heat seals. Thus, it is necessary to be aware of these possible changes in packaging due to the irradiation process and select packaging materials that are compatible with the process. As a result, a limited number of plastic packaging materials have been approved for use in food irradiation.

High pressure processing has also been found effective for inactivation of vegetative microorganisms with little effect on food quality (Knorr, 2000; Balasubramaniam, 2003). High pressure processing is generally performed on foods that have already been filled into packages. Effects on properties of the packaging materials studied, including multilayer laminates, have generally been found to be small (Ozen and Floros, 2001). No changes in the structure of the materials studied have been observed, and high-pressure-treated packages have been found acceptable (Lanmbet et al., 2000). However, selecting compatible polymers for multilayer laminates is necessary to avoid delamination due to differential compressibility of polymers or blistering due to flashing of polymer-solubilized condensed vapors upon decompression (Sadler et al., 2005).

Several other nonthermal processing techniques are being developed (Barbosa-Canovas and Gould, 2000; Lozano et al., 2000). In each case, the packaging involved must be evaluated to ensure food safety and quality.

13.5.5 Modified Atmosphere Packaging (MAP)

The quality and shelf life of many foods have been improved due to packaging that maintains an atmosphere in the package headspace that is different from air (Ooraikul and Stiles, 1991; Smith et al., 1992; Riquelme et al., 1994; Perdue, 1997; Zagory, 1997; Blakistone, 1999; Ahvenainen, 2003a; Brody, 2003a; Mullan and McDowell, 2003; Robertson, 2006m). The modified atmosphere compliments refrigeration to retard chemical and/or microbiological deterioration of the food. In the case of fresh meat and poultry, fresh pasta and baked products, and fresh-prepared foods, plastic films that are good gas barriers maintain the atmosphere provided at the time of packaging. For fresh fruits and vegetables, the appropriate plastic film in combination with the respiring product creates and then maintains the desired levels of oxygen and carbon dioxide. In some cases, active packaging concepts (discussed below) such as oxygen absorbers, carbon dioxide absorbers, and/or moisture regulators work with the packaging film to maintain the desired atmosphere.

The shelf life of red meat can be extended by packaging in a vacuum, so that oxidation reactions and growth of aerobic bacteria such as malodorous *Pseudomonas* bacteria are inhibited. Vacuum packaging of poultry is difficult because of the irregular shapes and sharp edges. Packaging atmospheres of nitrogen and 20 to 30% carbon dioxide are used to retard oxidation and microbial growth in both red meat and poultry. Packaging of red meat in atmospheres of 40 to 80% oxygen and 20 to 30% carbon dioxide provides desirable red color and microbial inhibition, but with increased rates of oxidative rancidity.

Fresh pasta and baked products are also vulnerable to oxidative rancidity and microbial degradation, especially mold growth. Vacuum packaging can inhibit both, but package headspace atmospheres of 50 to 100% carbon dioxide and 0 to 50% nitrogen are more common.

Chilled, prepared foods such as pasta, pizza, precooked meats and complete dishes are increasingly packaged in a modified atmosphere with no oxygen and >25% carbon dioxide, with the remainder nitrogen. Quality and shelf life are enhanced due to reduction of oxidation and inhibition of aerobic microorganisms. A separate category of chilled, prepared foods is *Sous Vide*. Preparation starts with vacuum packaging of the food in a flexible barrier-film package, followed by cooking in a hot water bath, moist steam, or pressure cooker. The cooked food is then rapidly cooled and then refrigerated. The combination of cooking under vacuum and then rapid cooling achieves higher quality and longer shelf life.

Packaging many fresh-whole and fresh-cut fruits and vegetables in a modified atmosphere with selected low oxygen content and elevated carbon dioxide content reduces respiration, with resulting increase in shelf life. An example is the 5 to 9% oxygen and 1 to 5% carbon dioxide atmosphere that is beneficial to oranges (Singh and Mannapperuma, 2000). MAP design for fruits and vegetables requires selection of a plastic film with proper oxygen permeability (P_o), carbon dioxide permeability (P_c) and ratio between permeabilities (P_c/P_o) that will give the desired atmosphere. The design must take into account the desired weight of product (W) and achieve the targeted atmosphere with a film of reasonable thickness and sealed package of reasonable surface area. (See MAP model later in chapter.)

13.5.6 Active Packaging

Active packaging has been defined as performing some desired role other than providing an inert (passive) barrier to external conditions (Rooney, 1995; Rooney, 1997). Active packaging can then be considered as correcting some deficiency of passive packaging or enhancing the performance of the packaging. With consumer interest in ever higher quality and safety in foods, active packaging is a field of continuing interest and development (Labuza and Breene, 1989; Labuza, 1996; Hernandez and Giacin, 1998; Vermeiren et al., 1999; Rooney, 2000; Brody et al., 2001; Ahvenainen, 2003b; Day, 2003; de Kruijf and van Beest, 2003; Han, 2005; Robertson, 2006n).

13.5.6.1 Protective Active Packaging

Most active packaging concepts enhance the protective function of food packaging, thus improving quality, shelf life and safety (Ahvenainen, 2003a). MAP of fresh-whole and fresh-cut fruits and vegetables is often regarded as being a form of active packaging. In a MAP application, rather than being the best gas barrier, the appropriate film regulates oxygen and carbon dioxide transfer to achieve desirable levels in the packaging headspace. Other protective active packaging concepts include moisture regulating agents that complement the packaging moisture-barrier property (Powers and Calvo, 2003). In addition, oxygen-scavenger-incorporated sachets, labels, closure liners, films, and containers are available that complement the package oxygen-barrier property (Harima, 1990; Idol, 1997; Vermeiren et al., 2003). Other systems have been developed to regulate package carbon dioxide levels, absorb ethylene to slow ripening and absorb off-aromas (Linssen et al., 2003; Vermeiren et al., 2003).

Packaging systems that incorporate antimicrobial agents to reduce food microbial contents also enhance the protective function of packaging (Labuza and Breene, 1989; Rooney, 1995; Han, 2000; Appendini and Hotchkiss, 2002; Cutter, 2002; Quintavalla and Vicini, 2002; Vermeiren et al., 2002; Han, 2003; Steven and Hotchkiss, 2003a; Suppakul et al., 2003; Han, 2005; LaCoste et al., 2005). A number of concepts have been demonstrated for inhibiting microbial growth by modification of the package headspace composition. CO_2-generating sachets can produce high enough levels of carbon dioxide to inhibit microbial growth on food surfaces. Ethanol-releasing sachets and films have also been studied for inhibition of microorganisms on food surfaces. Combination CO_2-generating/O_2-scavenging sachets and ethanol-releasing/O_2-scavenging sachets are available. Allyl isothiocyanate and other volatile plant components have been emitted successfully to inhibit microorganisms. A number of approaches have been explored for producing antimicrobial films that inhibit microbial growth through direct contact with the food surface. Some concepts involve incorporation in the film matrix of an antimicrobial agent that migrates to the surface of the food. In this case, the agent must be stable to the conditions necessary to produce the film. To avoid this problem, previously formed films can be solution coated with a polymer that acts as a carrier for an antimicrobial agent that migrates to the food surface. Other approaches include using inherently antimicrobial polymers, immobilization of an antimicrobial agent on the polymer or on the film surface, and film surface modification by electron irradiation (Steven and Hotchkiss, 2003a). Several factors must be considered in development of an antimicrobial film, including effect of processing conditions for producing the film, interaction between the antimicrobial agent and the film polymer matrix, mass transfer of the agent from the film to the food surface, effect of the agent on the physical properties of the film, and properties of the targeted foods (Suppakul et al., 2003). The concept of nonmigrating bioactive polymers is also being developed (Steven and Hotchkiss, 2003a).

Uses of packaging for controlled release of antioxidants and flavors have also been proposed, including development of smart polymer blending to target desired release rates (LaCoste et al., 2005).

13.5.6.2 Convenience Active Packaging

A number of active packaging concepts enhance the convenience of packaged foods. Packaging that is stable to the conditions of a microwave or conventional oven (dual-ovenable) can serve as a convenient container for food preparation, service, and consumption. Incorporation of susceptors in microwaveable packaging allows crisping and browning of the food. The aerosol can for whipped cream and self-heating and self-cooling cans for beverages are other examples of convenience active packaging.

13.5.6.3 Edible Films and Coatings

Edible films and coatings formed from polysaccharides, proteins, lipids, resins, and/or waxes fall within the active packaging definition, since they can enhance the protective function, provide convenience, and minimize package environmental impact (Guilbert, 1986; Kester and Fennema, 1986; Gontard and Guilbert, 1994; Krochta et al., 1994; Guilbert and Gontard, 1995; Martin-Polo, 1995; Anker, 1996; Guilbert et al., 1997; Krochta, 1997a, b; Krochta and De Mulder-Johnston, 1997; Cuq et al., 1998; Guilbert, 2000; Haugaard et al., 2000, 2001; Park et al., 2001; Gennadios, 2002; Krochta, 2002; Han, 2005; Robertson, 2006o). Edible films placed or formed between components of a packaged food control transfer of moisture, oils, etc. over which the package has no control. Edible coatings or edible film pouches (as a primary package) work to complement the protective function of the nonedible (secondary) package. Such coatings and films can act as barriers to the external environment and maintain food integrity, thus reducing the amount of packaging required. Edible film pouches carrying premeasured amounts of ingredients can provide the convenience of placing pouch with ingredients into the food formulation. Edible coatings can also carry antimicrobials that can inhibit microbial growth at both the food-coating interface and the coating outer surface (Han, 2000; Franssen and Krochta, 2003; Han, 2003; Suppakul et al., 2003; Cagri et al., 2004).

A number of food applications of edible films and coatings have been explored (Krochta and De Mulder-Johnston, 1997; Haugaard et al., 2000, 2001). Several polysaccharide-, sucrose-ester-, lipid- and resin-based edible coating formulations are available commercially to control moisture loss and respiration in fresh fruits and vegetables. Starch, hydroxypropyl methylcellulose (HPMC), zein, gelatin, and shellac coatings are available for confectionery and other food products. Edible collagen casings and wraps for meat and HPMC pouches for dry foods are available commercially. A large number of foods would benefit from development of suitable edible films or coatings (Krochta and De Mulder-Johnston, 1997; Haugaard et al., 2000, 2001).

13.5.6.4 Intelligent Packaging

Intelligent packaging, which is also referred to as smart packaging, can be divided into two types (Goddard et al., 1997; Rodrigues and Han, 2003). Simple intelligent packaging contains components that sense the environment and communicate information important to proper handling of the food product. Interactive or responsive intelligent packaging has additional capability in that it can respond to environmental change and thereby prevent deterioration of the food product (Brody, 2000a; Karel, 2000).

Several intelligent packaging concepts involve sensors that provide information related to food quality (Goddard et al., 1997; de Kruijf et al., 2002; Ahvenainen, 2003a; Rodrigues and Han, 2003; Yam et al., 2005; Robertson, 2006n). One category includes temperature sensors that indicate whether the package has been exposed to temperatures above or below a critical limit. Time–temperature indicators are available that provide time-integrated information about the entire temperature history of the product (Taoukis and Labuza, 2003). Such indicators allow more accurate assessment of the remaining product shelf life (Johnson, 1997). Other sensors are available that indicate the composition of the package headspace. These include oxygen sensors that provide information about oxygen permeation or leakage into the package. Such indicators can be used with oxygen scavengers to signal oxygen level in the package. Indicators are also available to monitor carbon dioxide levels, useful to show whether desired levels are being maintained. Fruit ripeness indicators which detect headspace ethylene and aromas have also been studied. Another category of sensors includes indicators of food freshness and contamination (Smolander, 2003). These include indicator concepts that warn about food chemical deterioration or microbial growth. Shock abuse indicators are also available.

An example of responsive intelligent packaging for preventing food deterioration is plastic film whose permeability dependency on temperature is controlled through polymer structure response to temperature. Such film is useful for MAP of fruits and vegetables, since respiration rate is generally more affected by temperature than polymer permeability.

Another category of intelligent packaging includes components that range from bar codes to radio frequency transmitters (i.e., radio frequency identification (RFID) devices) that allow accurate tracking of product for improved supply chain management and rapid traceability (Barthel, 1997).

Intelligent packaging has been proposed for a future smart kitchen (Yam, 2000). The cooking appliance system would read a bar code, that includes information on optimum cooking conditions, and appropriately adjust the oven. The system could also read a time–temperature indicator to alert the consumer to spoiled food. An allergen alert component could also be added. Other applications and a research roadmap for intelligent packaging have been proposed (Yam et al., 2005).

13.5.7 CONSUMER-FRIENDLY PACKAGING

All of the packaging trends discussed above can be said to include consumer friendliness, to the degree they satisfy consumer desire for convenience, maximum food quality, and minimum packaging waste. Packaging innovation aimed at increasing convenience is becoming more important than decreasing package costs (Ferrante, 1997). Consumers want packaging that provides a high level of food security, has an easy-to-read label, is easy to open and resealable, provides an easy-to-prepare meal, and uses a minimum amount of material that is recyclable. In a global market, package design must also respond to cultural and demographic differences. Packaging will have to be adapted for electronic control of global distribution, the electronic purchasing of future smart shopping, and the electronic control of the future smart kitchen (Louis, 1999; Brody, 2000a; Sonneveld, 2000; Yam, 2000). Improvements in packaging materials, design, and intelligence will be necessary to achieve these goals (Goddard, 1995; Reynolds, 2002).

13.6 FOOD SHELF LIFE AND PACKAGE SELECTION

Food shelf life is generally understood to be the elapsed time between time of packaging of the food and time that the food becomes unacceptable to consumers. The shelf life of a food is influenced by three factors: (1) characteristics of the food (e.g., sensitivity to temperature, light, moisture and oxygen), (2) environment (e.g., temperature, light, RH and po_2) to which the food will be exposed, and (3) characteristics of the package. Accurate assessment of shelf life requires that an acceptance criterion be established by sensory evaluation and/or instrumental analysis. Several approaches are used to predict the packaged food shelf life (Marsh, 1997; Karel and Lund, 2003a; Steele, 2004; Robertson, 2006e). These include:

- Assuming the same shelf life as that of a similar food product in the same packaging
- Determining the shelf life by either exposure to real-world conditions of warehousing, shipping, and retailing, or to long-term storage at warehouse conditions
- Determining the shelf life by storage at controlled typical ambient conditions
- Calculating the ambient shelf life based on results of accelerated shelf life testing (ASLT)
- Calculating the shelf life by use of a mathematical model that takes into account the properties of the food, package, and environment

The first approach can result in under- or over-packaging of the product. The conditions of real-world or warehouse testing in the second approach are difficult to control; thus, the results are difficult to use for prediction.

13.6.1 Shelf Life from Storage Test at Controlled Typical Ambient Conditions

With the third approach, determination of shelf life by storage at typical ambient conditions will take a long time, if the test is run to actual time of unacceptable product. However, if a criterion of food quality (e.g., a color, flavor, or vitamin that can be identified by instrumental analysis) has been established, the kinetics of quality change can be determined by measurement over a relatively short time. This involves determination of the order (usually zero or first order), and the reaction rate constant (k_T) for the temperature studied. The reaction rate constant can then be used to calculate the time to reach the end of shelf life. Thus, for the conversion of attribute "A" to degradation product "D" (i.e., $A \rightarrow D$), the shelf life (t_S) before A drops to an original level (A_O) to an unacceptable level (A_S) or D increases from an original level (D_O) to an undesirable level (D_S) can be determined from the following equations, depending on the order of the reaction:

$$\text{Zero order kinetics:} \quad t_S = (A_O - A_S)/k_T \qquad (13.15)$$

$$t_S = (D_S - D_O)/k_T \qquad (13.16)$$

$$\text{First order kinetics :} \quad t_S = \ln(A_O/A_S)/k_T \qquad (13.17)$$

Assuming zero order kinetics in the determination of k_T and then in the calculation of t_S gives a conservative predicted shelf life, because with zero order kinetics the rate of quality change does not drop with time.

Using data from storage at one temperature to calculate shelf life in the manner just described has limited use. It is possible to calculate shelf life at only the temperature of the testing. There is no way from this data to predict shelf time at another temperature or if the food product experiences temperature changes. Thus, use of ASLT or a mathematical model has distinct advantages.

13.6.2 Shelf Life from Accelerated Shelf Life Testing (ASLT)

ASLT involves determination of kinetics of packaged food quality at controlled conditions that accelerate deterioration of the relevant quality factor (Taoukis and Labuza, 1996; Marsh, 1997; Karel and Lund, 2003a; Mizrahi, 2004; Robertson, 2006e). Using this approach requires determination of deterioration rate at three or more accelerating conditions to predict deterioration rate at a normal condition. Usually, kinetics of deterioration are measured at elevated temperatures and the Arrhenius equation or a linear model is used to predict the rate of deterioration at temperature(s) of normal handling and storage. For foods whose deterioration is determined by transport of moisture or oxygen, elevated levels of RH or p_{O_2} can be used to accelerate shelf life determination. A relationship between deterioration rate and RH or p_{O_2} must be established to predict deterioration at normal conditions.

Use of the Arrhenius relationship is the most rigorous approach to relating deterioration rates at high temperatures to the lower temperatures of normal storage:

$$k_T = k_0 \exp(-E/RT) \tag{13.18}$$

where k_T is the rate constant at temperature T ($°K$), k_0 is the Arrhenius constant for deterioration of the selected quality factor, E is the activation energy for the degradation reaction, and R is the ideal gas constant. If the rate constants determined from testing at three elevated temperatures fall in a straight line on a plot of $\log k$ vs. ($1/T°K$), the Arrhenius relationship applies. One can then interpolate or extrapolate on this plot to predict the rate constants of the degradation reaction at any other temperature. Also, the values of k_0 and E can be determined from the intercept and slope of the plot, respectively, and used in the Arrhenius equation to predict k_T for any temperature. Shelf life can then be determined by use of Equation 13.15 or Equation 13.17, depending on the order of the reaction.

A more empirical approach is to use a simplified linear equation to relate deterioration rates at high temperatures to the lower temperatures of normal storage:

$$k_T = k_1 \exp[b(T - T_1)] \tag{13.19}$$

where k_1 is the rate constant at T_T ($°C$) and b is a constant for the reaction. If the rate constants determined from testing at three elevated temperatures fall in a straight line on a plot of $\log k$ vs. $T°C$, the linear equation can be used. One can then interpolate or extrapolate on this plot to predict the rate constants of the degradation reaction at any other temperature. However, the error in extrapolated values is likely to increase with the amount of extrapolation. Also, the values of k_1 and b can be determined from the intercept and slope of the plot, respectively, and used in the linear Equation 13.19 to predict k_1 for any temperature. Again, possibility of error in k_T increases as the difference in temperature from the temperatures of the ASLT increases. Shelf life can then be determined by use of Equation 13.15 or Equation 13.17, depending on the order of the reaction.

A useful term in quantifying shelf life is the temperature quotient, Q_{10}, which indicates how much more rapidly a reaction proceeds when the temperature is raised by 10°C:

$$Q_{10} = k_{T+10}/k_T \tag{13.20}$$

It can be shown that:

$$k_T t_{S(T+10)} = k_{T+10} t_{S(T)} \tag{13.21}$$

where $t_{S(T)}$ and $t_{S(T+10)}$ are shelf lives at $T°C$ and $T + 10°C$, respectively. Thus:

$$Q_{10} = t_{S(T)}/t_{S(T+10)} \tag{13.22}$$

It can be shown that when using the linear model:

$$Q_{10} = \exp(10b) \tag{13.23}$$

Q_{10} values depend on the nature of the food, preservation process, and packaging. Values of 1.1 to 4, 1.5 to 10, and 3 to 40 have been reported for canned, dehydrated and frozen foods, respectively (Robertson, 2006e). Assuming a value of Q_{10} (e.g., $Q_{10} = 2$) can lead to large errors in predicted shelf life.

13.6.3 Shelf Life from Food-Package System Models

In cases when the packaging determines the food shelf life, food-package system models can be employed to either predict shelf life from properties of the package or select package properties based on desired shelf life.

13.6.3.1 Moisture-Sensitive Foods in Plastic Packaging

The end of shelf life for a food is often defined by the food reaching a critical moisture content. As an example, this would be the case for a dry snack food for which an acceptable crispness does not exist above a certain critical moisture content (m_S). Assuming that moisture-content change of the package head space is insignificant compared to the food and resistances of headspace and food to moisture diffusion are negligible, the unsteady-state material balance for moisture content of the food is

$$\frac{dm}{dt} W = P_w \frac{A}{L}(p_{w1} - p_{w2}) \tag{13.24}$$

where m is the initial moisture content of the food on a dry weight basis (db), W is the dry weight of the food, P_w is the water vapor permeability of the plastic packaging material, A is the package surface area, L is the package film thickness, p_{w1} is the water vapor partial pressure outside the package, p_{w2} is the water vapor partial pressure inside the package, and t is the time since packaging of the food. Assuming constant temperature and p_{w1} and that the food moisture content is in equilibrium with p_{w2}, it can be shown that (Taoukis et al., 1988; Karel and Lund, 2003a; Robertson, 2006e):

$$\ln \frac{[(m_e - m_i)}{(m_e - m_s)]} = \frac{P_w A p_{wo}}{bWL} t_s \tag{13.25}$$

where m_i is the initial moisture content on a dry weight basis (db), m_e is the moisture content (db) the food would reach if allowed to come into equilibrium with the atmosphere external to the package, m_s is the critical moisture content (db) that defines end of shelf life, b is the slope of the best-fit linear isotherm for the food over the range of moisture contents of interest, p_{wo} is the vapor pressure of pure water, and t_s is the time (i.e., shelf life) required to reach m_s. Equation 13.24 can be used to either predict shelf life (t_s) given certain package parameters (P_w, L, and A), or it can be used to determine the package parameters necessary to achieve a selected shelf life.

13.6.3.2 Oxygen-Sensitive Foods in Plastic Packaging

Depending on a food's critical quality component, it can react with a certain quantity of oxygen before the food becomes unacceptable due to oxidative rancidity. Table 13.2 shows amounts of oxygen with which various foods can react before end of shelf life is reached. Assuming that resistances of headspace and food to oxygen diffusion are negligible, the unsteady-state material balance for oxygen content of the package headspace is

$$\frac{dp_{o2}}{dt} \frac{V}{p_T} = \frac{P_o A (p_{o1} - p_{o2})}{L} - r_o W \tag{13.26}$$

where p_T is the total pressure, p_{o2} is the oxygen partial pressure inside the package, p_{o1} is the oxygen partial pressure outside the package, t is the time since packaging of the food, P_o is the oxygen permeability of the plastic packaging material, V is the package headspace volume, A is the package surface area, L is the package film thickness, W is the weight of the food, and r_o is the oxygen reaction rate per unit weight of the food. Assuming constant temperature and p_{o1}, constant moisture content, equilibrium of the food with p_{o2}, and an equation relating r_o to p_{o2}, Equation 13.25 can be solved to determine p_{o2} as a function of t (Khanna and Peppas, 1982). The simplest solution is obtained when r_o can be related to p_{o2} with a linear equation. The r_o can then be integrated over time, utilizing the relationship between r_o and p_{o2} and the relationship between p_{o2} and t, to determine how long it will take to reach the quantity of oxygen (Q_{os}) that produces an unacceptable food (i.e., shelf life).

A simpler approach that does not require an equation relating r_o to p_{o2} and that produces a conservative shelf life is to assume that oxygen reacts immediately upon permeating the packaging material, p_{o2} is zero, and steady state is reached quickly. Equation 13.25 then simplifies to:

$$\frac{P_o A (p_{o1})}{L} = r_o W \tag{13.27}$$

Recognizing that:

$$r_o = Q_{os}/t_s \tag{13.28}$$

Then:

$$t_s = \frac{Q_{os} W L}{P_o A p_{o1}} \tag{13.29}$$

This latter approach gives a conservative value of t_s, because it assumes the most rapid permeation possible (i.e., $p_{o2} = 0$).

13.6.3.3 Shelf-Life Extension of Fresh Fruits and Vegetables Using MAP

Fresh fruits and vegetables continue to respire after harvesting, consuming oxygen and producing carbon dioxide. A low oxygen ($\sim$2–4%) and somewhat elevated carbon dioxide ($>\sim$1%) environment reduces respiration rate in many fruits and vegetables, thus slowing ripening and senescence. Design of packaging for such produce can produce the desired environment in the package headspace. Several approaches of varying complexity have been developed for describing this modified atmosphere packaging (MAP) system.

One of the earliest approaches assumed steady state between oxygen consumption rate of the fruit or vegetable and oxygen permeation rate through the package (Jurin and Karel, 1963). The approach requires knowledge of the oxygen consumption rate as a function of the package headspace oxygen content:

$$R_o = f\{p_{o2}\} \tag{13.30}$$

where R_o is the oxygen consumption rate per unit weight of produce.
At steady state:

$$R_o W = f\{p_{o2}\} W = \frac{P_o A (p_{o1} - p_{o2})}{L} \tag{13.31}$$

Equation 13.30 allows determination of the p_{o2} achieved by a given package, either mathematically or graphically (Karel and Lund, 2003a). If the p_{o2} falls in the range of values that produce desirable reduced respiration rate without the produce going into anaerobic respiration, the selected packaging parameters (P_o, A, L) are satisfactory. Otherwise, different packaging parameters must be selected. Assuming a respiratory quotient (RQ) of 1, $R_c = R_o$, where R_c is the carbon dioxide production rate per unit weight of produce. Knowing R_c allows determination of the steady-state concentration of carbon dioxide (p_{c2}) at the determined steady-state value of oxygen concentration (p_{o2}) from the following equation:

$$R_c W = \frac{P_c A (p_{c2})}{L} \tag{13.32}$$

where P_c is the permeability of the selected packaging film to carbon dioxide.

Another approach that assumes steady-state conditions breaks the analysis into two steps (Singh and Mannapperuma, 2000). Again assuming $RQ = 1$, one can show that:

$$P_o(p_{o1} - p_{o2}) = P_c(p_{c2}) \tag{13.33}$$

Thus:

$$p_{c2} = \frac{P_o}{P_c}(p_{o1} - p_{o2}) = \frac{1}{\beta}(p_{o1} - p_{o2}) \tag{13.34}$$

By substituting the desired values of p_{o2} and p_{c2} for achieving reduced respiration into equation 13.33, the ratio of P_c to P_o(ß) can be determined for the packaging material that will achieve these values. This can also be done graphically on a plot of p_{c2} vs. p_{o2}. By drawing a line from the point representing $p_{o2} = p_{o1}$ and $p_{c2} = 0$ through a box determined by the desirable ranges of p_{o2} and p_{c2}, one can determine ß from the slope or intercept of the line according to Equation 13.33. After determination of ß, the other package parameters can be determined from Equation 13.30 or Equation 13.31. If reasonable combinations of W, A, and L cannot be achieved with the packaging material selected, another packaging material with similar ß must be tried.

Many fruits and vegetables need packaging with a ß value not available with existing polymer films. An approach for overcoming this limitation is the use of microperforated films. The model described above can be extended to include consideration of microperforated films (Singh and Mannapperuma, 2000). This approach widens application of MAP modeling to a wider range of products.

13.7 FOOD PACKAGING AND THE ENVIRONMENT

Accessing raw materials, producing packaging materials, converting packaging materials to containers, transporting packaging, and dealing with packaging waste all have impact on the environment (Selke, 1994; Selke, 1997a; Selke, 1997b). Life cycle analysis is an approach that has been used to take into account all the resources consumed in the creation, life, and disposal of a package (Brown, 1993; Franklin et al., 1997; Robertson, 2006p). The goal is to identify areas of concern and select the packaging having least impact. However, comparing the environmental impacts of competing packaging is complex. Tradeoffs between different environmental impacts are generally necessary when making comparisons. Often, there is no clearly superior packaging choice (Allen and Bakshani, 1992).

One focus has been on minimizing packaging waste that ends up in landfills. Packaging appears to occupy about one-third of municipal solid waste volume (Rowatt, 1993; Fearncombe, 1995;

Borchardt, 1997). Landfilling and incineration without energy recovery are not seen as support-able methods of waste management (McCormack, 2000). The approaches to minimizing this waste constitute the 4 Rs of packaging, source *r*eduction, package *r*euse, package *r*ecycling, and energy *r*ecovery (Selke, 1997a).

13.7.1 SOURCE REDUCTION

There is a constant economic incentive to reduce the amount and cost of packaging. As pointed out earlier in this chapter for each packaging material, advances in packaging material properties and in package design have reduced the weight of packaging material used for each container. Depending on material, packaging weights were reduced 12 to 50% between ~1975 and 1990 alone (Brown, 1993; Rowatt, 1993). Also, replacing glass and metal containers with lighter containers, including plastic containers, flexible pouches, and plastic/paperboard/plastic/foil/plastic cartons, has contributed greatly to source reduction (McCormack, 2000). Overall, this has resulted in a reduction in per-capita consumption of packaging, while other solid waste continues to increase (Marsh, 1994).

13.7.2 PACKAGE REUSE

When each town had its own dairy and beverage bottler, return of glass containers for reuse was practical. However, with larger regional food manufacturers, returnable bottles were generally no longer economic, especially because reusable bottles must have thicker walls to endure the increased handling. However, use of returnable bottles still exists in certain areas. Also, to the degree the consumer can reuse a package, it has greater value and reduces consumption of other materials.

13.7.3 PACKAGE RECYCLING

Three types of recycling are possible for packaging: mechanical, chemical, and biological (Brown, 1993; Dent, 2000; Robertson, 2006p).

13.7.3.1 Mechanical Recycling

The most common type is mechanical recycling, involving reprocessing of recycled materials through physical steps that can include cleaning, shredding/grinding, separating, and reforming. These steps result in metal and glass containers that are acceptable for use with foods. However, they generally do not ensure removal of all possible contaminants from paper and plastic materials to allow use of the recycled-content package with foods involving long-term contact. The FDA reviews food-contact applications of these recycled materials on a case-by-case basis that includes consideration of source control to ensure cleanliness, recycling process ability to remove possible contaminants, and the proposed food-contact application(s) (Thorsheim and Armstrong, 1993). FDA has approved several food-contact applications of mechanically-recycled plastics, including HDPE grocery bags, PS egg cartons, HDPE and PP crates for transporting fresh fruits and vegetables, and PET pint and quart baskets for fresh fruits and vegetables. All these applications involve a limited time and area of food contact at ambient and refrigerated temperatures, along with expectation that the food is normally cleaned before use or that the food is protected by a barrier (e.g., egg shell) (Thorsheim and Armstrong, 1993). FDA has also approved use of recycled plastic when it is co-extruded with a virgin layer of the plastic that is the food-contact surface.

13.7.3.2 Chemical Recycling

Chemical recycling involves depolymerization of plastic polymers to monomers or oligomers and then repolymerization to the polymer. This process allows removal of all possible contaminants, with the repolymerized polymer identical to virgin polymer. Several processes have been developed for chemical recycling of PET (Thorsheim and Armstrong, 1993; Borchardt, 1997). An ideal plastics recycling process would take mixtures of plastic and convert them at high temperature and pressure to an economical petrochemical process stream (Borchardt, 1997).

13.7.3.3 Biological Recycling

Increasing interest in sustainable systems has motivated investigation of renewable and biodegradable polymers for food packaging (Gontard and Guilbert, 1994; Guilbert and Gontard, 1995; Martin-Polo, 1995; Anker, 1996; Krochta and De Mulder-Johnston, 1996; Bastioli, 1997; Guilbert et al., 1997; Krochta and De Mulder-Johnston, 1997; Stratton, 1998; Petersen et al., 1999; Guilbert, 2000; Haugaard et al., 2000; Weber, 2000; Bastioli, 2001; Haugaard et al., 2001; Tharanathan, 2003). Polymers based on renewable resources can be grouped into three categories (Petersen et al., 1999; Guilbert, 2000):

- Polysaccharides and proteins extracted from plant, marine or animal sources (e.g., starch, chitosan and whey protein)
- Polymers synthesized from renewable, bio-derived monomers (e.g., polylactate)
- Polymers produced by microorganisms (e.g., polyhydroxyalkanoates)

Many of the same polysaccharides and proteins being explored for edible films and coatings are also candidates for biodegradable packaging.

For biodegradable polymers to replace synthetic nonbiodegradable polymers in food packaging, their mechanical, optical, and barrier properties must be comparable at competitive cost. At the present time, biodegradable polymers are generally more expensive than synthetic polymers (Petersen et al., 1999). An additional challenge is achieving controlled lifetime. Biodegradable packaging must be stable and function properly at the conditions of use, so as not to compromise the quality and safety of the food, and then biodegrade efficiently upon exposure to the appropriate microorganism(s) and environment (Kaplan et al., 1993). Biological recycling must also compete with mechanical and chemical recycling concepts that allow reusing materials rather than degrading them. Finally, biodegradable polymers must be made easily distinguishable from nonbiodegradable polymers so as not to interfere with the mechanical and chemical recycling processes. Reasons for considering use of a biodegradable polymer for packaging include (Guilbert, 2000):

- Life of the packaged product is short
- Mechanical or chemical recycling is not feasible
- Biological recycling is favored by consumers
- Biodegradability is legally mandated

Biodegradable packages have been developed for a number of food uses, including (Haugaard et al., 2000; Barry, 2001; Bastioli, 2001; Haugaard et al., 2001; Tullo, 2005):

- Pulp containers for fruits
- Woodpulp-starch trays for fresh beef and chicken
- Nitrocellulose-coated cellophane films for cheese, fruits, and confections
- Starch-based foamed shells for hamburgers and sandwiches
- Starch-based grocery bags
- Polylactic acid (PLA) bottles for water and tubs for fresh-cut produce and salads

Other possible food applications being explored for biodegradable polymers include fast-food containers, cups, plates, and cutlery. Biodegradability of all fast-food restaurant waste appears to be a feasible goal. Biodegradable polymers are also available for nonfood uses, such as refuse (composting) bags, loosefill packaging, agricultural mulches, and potting containers (Guilbert, 2000).

Widespread use of biodegradable polymers will require reductions in production costs, establishment of dedicated composting facilities, and increase in fossil resource costs. However, it seems inevitable that sustainable approaches to the production of packaging materials will be necessary.

13.7.4 ENERGY RECOVERY

Paper and plastic packaging materials, which consume more landfill volume than glass and metal packaging, have energy content that can be captured by incineration to produce electricity or steam. Energy recovery, sometimes called thermal recycling, is an attractive alternative for mixed plastic and mixed plastic/paper wastes that cannot be easily recycled (Brown, 1993). Waste incineration with energy recovery is more common in Europe and Asia but is increasing in the United States (Marsh, 1994).

13.8 FOOD PACKAGING LAWS AND REGULATIONS

Unfortunately, no single set of laws and regulations apply to food packaging. Laws and regulations exist at the federal, state, and local levels of government. There exist many packaging terms with specific legal definitions (Greenberg, 1996). Generally, packaging laws and regulations fall into three categories: safety of packaging materials, labeling of packages, and environmental impact of packaging (Greenberg, 1996; Simmons, 1997; White and Tice, 1997; Curtis, 2005).

13.8.1 SAFETY OF PACKAGING MATERIALS

In the United States., the Food and Drug Administration (FDA) has primary responsibility for ensuring that food packaging does not contaminate or adulterate food in violation of the Federal Food, Drug and Cosmetic Act (Simmons, 1997). If a component of a food packaging material migrates into a food, regulatory approval of the migrant as an indirect food additive must be obtained from FDA through a food-additive petition. This process requires an estimate of the amount of the substance that will enter the diet and demonstration that the amount is safe. However, the substance may be exempted from the FDA food-additive regulations if it (Simmons, 1997):

- Has received prior sanction for its intended use by the FDA or US Dept. of Agriculture (USDA) before the Food Additive Amendment of 1958
- Is generally recognized as safe (GRAS) by qualified experts and is thus listed as GRAS by FDA or can be self-determined as GRAS by a manufacturer of the packaging material
- Transfers to food at levels no higher than 50 ppb, or is a component of packaging used for dry, nonfat foods, or is a component of packaging intended for repeated use with bulk quantities of food

The latter exemption has not been formally adopted but has been allowed by FDA. Also, the latter exemption has not been applied to substances that are of special toxicological concern (e.g., heavy metals), or that are known to be carcinogens, or that result in toxic reactions in the diets of humans or animals at 40 ppm or less. If there is concern over the safety of the substance or if the substance will be exposed to food products for which there is high use (e.g., milk) it must be shown that less that 10 ppb or as low as 1 ppb migrates into the food.

Obtaining data to obtain an exemption based on transfer of an insignificant amount of the substance is a challenge. Extraction studies that simulate the intended use with a food must be performed on the packaging material. Because of the complexity of foods, solvents that simulate the intended food are generally used. More detailed information on migration can be found in (Robertson, 2006q).

Recycled plastic packaging materials are a potential food safety concern, because they may have been exposed to hazardous compounds that they absorbed. In the United States., FDA approval is not specifically required for using recycled plastics in food packaging. However, the FDA requires that all food-contact surfaces be suitably pure for their intended use. Furthermore, all packaging, virgin or recycled, must adhere to food additive regulations. Therefore, the packaging industry practice for using recycled plastic is to seek a no objection letter from FDA, based on proof that any potential contaminants would produce less than the "threshold" dietary level of 0.5 ppb (Simmons, 1997). In other countries, regulations on use of recycled plastic materials for food packaging vary. However, international bodies like the European Union (EU) have worked to develop cross-country legislation and guidance on food packaging materials (White and Tice, 1997).

13.8.2 LABELING OF FOOD PACKAGING

Food packaging laws in the United States derive mainly from the Food, Drug and Cosmetic Act (FDCA), the Fair Packaging and Labeling Act (FPLA), the Nutrition Labeling Education Act (NLEA), and the Food Allergen Labeling and Consumer Protection Act (FALCPA) (Simmons, 1997; Hanlon et al., 1998b; Blanchfield, 2000; Cramer, 2004). Food packaging laws in other countries also generally derive from a number of statutes, with international bodies such as the European Union (EU) working to bring uniformity (Paine and Paine, 1992c; White and Tice, 1997; Blanchfield, 2000). In the United States., the FDA regulates most food packaging, with the exception of the USDA which has responsibility for foods with greater than 2% cooked meat or poultry or greater than 3% raw meat or poultry (Storlie and Brody, 2000). USDA regulations generally conform to FDA policy.

In the U.S., package labels are required to appropriately communicate several items:

- Food product identity
- Net quantity of contents
- Manufacturer identity
- Ingredients in descending order of amount
- Nutritional facts
- Country of origin

The nutrient profile of a food is provided on a Nutrition Facts panel of the package label. The nature of the nutrient profile continues to evolve with understanding of the relationship between diet and health. Several format and display options are possible (Storlie and Hare, 1997). The most recent labeling requirement is easy-to-understand information on allergen ingredients, including declaration of allergens present in flavoring, coloring, or incidental additives (Cramer, 2004).

If the manufacturer wishes to make an absolute or comparative nutrient content claim on the food label, only certain defined terms that meet specific criteria may be used (Storlie and Brody, 2000). Defined absolute terms are "free", "low", "very low", "high", "source of", "healthy", "lean", and "extra lean". Defined comparative terms are "light", "lite", "reduced", "less", and "more". The FDA has also approved 11 health claims, with each involving well-defined criteria linking food components to lower health risk. In addition, "qualified health claims" not fully supported by scientific evidence can be made on a label, as long as it is accompanied by a disclaimer (Joy, 2005). Depending on the product, the label may be required to include the word "imitation", a percent juice statement, special handling instructions, or a statement alerting the consumer of certain processing techniques, ingredients, or possible exposure to foods not listed as ingredients (Storlie and Brody, 2000). Food products that require special label statements include unpasteurized juice products and irradiated foods.

13.8.3 ENVIRONMENTAL IMPACT OF PACKAGING

Regulations in the United States. governing packaging waste disposal have originated from state and local governments (Simmons, 1997). State laws have involved a range of approaches dealing with plastic waste, including required recycling rates, mandated recycle content, advance disposal fees, and landfill bans (Raymond, 1997b). Laws involving packaging waste are an evolving area, as manufacturers, consumers and lawmakers search for reasonable approaches. In the mean time, advances in material properties and container design continue to reduce the per-capita amount of packaging used by consumers (Marsh, 1994). Around 10 states require a beverage container deposit that is returned to the consumer when the container is returned to an authorized recycle location. Many cities have recycle programs, involving either regular curbside pickups or sorting of municipal solid waste (MSW), which include collection of paper, metal, glass, and plastic containers. Generally, the only plastic materials included are PET and HDPE. Such programs, along with separate collection of yard waste for composting, have allowed cities to meet state requirements for reducing waste disposal in landfills. Although there is no national solid-waste-reduction program in the United States., the Environmental Protection Agency (EPA) set a 25% MSW recycling goal that was met on schedule in 1995 and set a 35% MSW recycling goal for 2005 (Hanlon et al., 1998c).

Other countries have taken a wide range of approaches to environmental protection. Canada has largely left packaging waste legislation to provincial and local governments. However, national environmental goals have been established (Hanlon et al., 1998c). Europe has taken a more aggressive approach to reducing packaging waste, including required recovery and recycle rates for packaging (Raymond, 1997a; White and Tice, 1997; Hanlon et al., 1998c). Asian countries are developing policies similar to those instituted in Europe.

13.9 PROFESSIONAL AND BUSINESS ASSOCIATIONS

There are a number of professional and business institutes and associations that support individuals and companies that work in the packaging area. Many of these associations have publications and meetings that cover advances and trends in packaging. Several sponsor annual awards to outstanding packaging and packaging professionals. A list of mainly U.S. institutes and associations follows, along with their web addresses. (The author makes no claim that the following list is complete.)

- American Foil Container Manufacturers Association (AFCMA): www.afcma.rog
- American Forest and Paper Association (AF&PA): www.afandpa.org
- American Plastics Council (APC): www.americanplasticscouncil.org
- American Society for Testing and Materials (ASTM): www.astm.org
- Association of Independent Corrugated Converters (AICC): www.aiccbox.org
- Can Manufacturers Institute (CMI): www.cancentral.com
- Closure Manufacturers Association (CMA): www.cmadc.org
- Composite Can and Tube Institute (CCTI): www.cctiwdc.org
- Corrugated Packaging Council (CPC): www.corrugated.org
- Fibre Box Association (FBA): www.fibrebox.org
- Flexible Packaging Association (FPA): www.flexpack.org
- Food Service and Packaging Institute (FPI): www.fpi.org
- Food Processors Institute (FPI): www.fpi-food.org
- Food Processing Machinery Association (FPMA): www.foodprocessingmachinery.com
- Food Products Association (FPA): www.fpa-food.org
- Glass Packaging Institute (GPI): www.gpi.org
- Institute of Food Technologists (IFT): www.ift.org
- Institute of Packaging Professionals (IoPP): www.iopp.org
- International Corrugated Case Association (ICCA): www.iccanet.org

- International Molded Pulp Environmental Packaging Ass'n (IMPEPA): www.impepa.org
- National Paperbox Association (NPA): www.paperbox.org
- National Wood Pallet and Container Association (NWPCA): www.nwpca.org
- Packaging Machinery Manufacturers Institute (PMMI): www.pmmi.org
- Paperboard Packaging Council (PPC): www.ppcnet.org
- Society of Plastics Engineers (SPE): www.4spe.org
- Society of the Plastics Industries (SPI): www.socplas.org
- Technical Association of the Pulp and Paper Industry (TAPPI): www.tappi.org
- Tube Council: www.tube.org
- Women in Packaging (WP): www.womeninpackaging.org
- World Packaging Organization (WPO): www.packaging-technology.com/wpo

ACKNOWLEDGMENTS

The author wishes to acknowledge the literature searching and summarizing work of Ann Dragich, the figure production work of Jody Renner-Nantz, and the editing of Veronica Hernandez and May Janjarasskul.

REFERENCES

(2001). Resin 2001 — Tables. *Modern Plastics* February: 42–43.

(2002). Resin 2002 — Tables. *Modern Plastics* February: 27–28.

(2003a). DWI — Making the most of metal. *The Canmaker* 16(February): 31–32, 34.

(2003b). Green light for food cans. *The Canmaker* 16(February): 23–24.

Ahvenainen, R. (2003a). Active and intelligent packaging. In *Novel Food Packaging Techniques*. R. Ahvenainen (Ed.) Boca Raton, FL, CRC Press, 5–21.

Ahvenainen, R. (Ed.) (2003b). *Novel Food Packaging Techniques*. Boca Raton, FL, CRC Press. 590.

Allen, D.T. and Bakshani, N. (1992). Environmental impact of paper and plastic grocery sacks. *Chemecial Engineering Education* 26: 82–86.

Ammann, R. (2001). Aseptic filling of HDPE- and PET-bottles. *Fruit Process.* 11: 449–451.

Anker, M. (1996). *Edible and Biodegradable Films and Coatings for Food Packaging — A Literature Review*. Gotegorg, Sweden, SIK. 112.

Ansari, M.I.A. and Datta, A.K. (2003). An overview of sterilization methods for packaging materials used in aseptic packaging systems. *Food and Bioproducts Processing* 81: 57–65.

APC (2005). Resin identification codes — plastic recycling codes. www.americanplasticscouncil.org.

Appendini, P. and Hotchkiss, J.H. (2002). Review of antimicrobial food packaging. *Innovative Food Science & Emerging Technologies* 3: 113–126.

Armeniades, C.D. and Baer, E. (1977). Transitions and relaxations in polymers. In *Introduction to Polymer Science and Technology*, H.S. Kaufman and J.J. Falcetta (Eds), New York, John Wiley & Sons, Inc., Ch. 6.

Arndt, G.W., Jr. (1997). Leak testing. In *The Wiley Encyclopedia of Packaging Technology*, A.L. Brody and K.S. Marsh (Eds), New York, John Wiley & Sons, Inc., 558–561.

Ashley, R.J. (1985). Permeability and plastics packaging. *Polymer Permeability*. J Comyn. London, Elsevier Applied Science, 269–307.

ASTM (2002). *Consumer and Healthcare Packaging Standards*. Philadelphia, American Society for Testing and Materials. 366.

ASTM (2003). *Selected ASTM Standards on Packaging*. Philadelphia, American Society for Testing and Materials. 493.

Attwood, B.W. (1997). Paperboard. In *The Wiley Encyclopedia of Packaging Technology*. A.L. Brody and K.S. Marsh (Eds), New York, John Wiley & Sons, Inc., 717–723.

Bain, D.R. and Giles, G.A. (2000). Technical and commercial considerations. In *Materials and Development of Plastics Packaging for the Consumer Market*. G.A. Giles and D.R. Bain (Eds), Boca Raton, FL, CRC Press LLC, 1–14.

Balasubramaniam, V.W. (2003). High pressure food preservation. In *Encyclopedia of Agricultural, Food, and Biological Engineering*. D.R. Heldman (Ed.), New York, Marcel Dekker, Inc., 490–496.

Baner, A.L. (2000). The estimation of partition coefficients, solubility coefficients, and permeability coefficients for organic molecules in polymers. In *Food Packaging*. S.J. Risch (Ed.), Washington DC, American Chemical Society, 37–56.

Barbosa-Canovas, G.V. and Gould, G.W. (Eds) (2000). *Innovations in Food Processing*. Food Preservation Technology Series. Lancaster, PA, Technomic Publishing Co., Inc. 260.

Bardsley, R.F. (1997). Form/fill/seal, horizontal. In *The Wiley Encyclopedia of Packaging Technology*. A.L. Brody and K.S. Marsh (Eds), New York, John Wiley & Sons, Inc., 465–468.

Barron, F.H. and Burcham, J.D. (2003a). Glass containers. In *Encyclopedia of Agricultural, Food, and Biological Engineering*. D.R. Heldman (Ed.), New York, Marcel Dekker, Inc., 436–439.

Barron, F.H. and Burcham, J.D. (2003b). Metal containers. In *Encyclopedia of Agricultural, Food, and Biological Engineering*. D.R. Heldman (Ed.), New York, Marcel Dekker, Inc., 636–642.

Barron, F.H. and Burcham, J.D. (2003c). Package properties. In *Encyclopedia of Agricultural, Food, and Biological Engineering*. D.R. Heldman (Ed.), New York, Marcel Dekker, Inc., 727–733.

Barry, C. (2001). Bio-based plastics still say 'green.' *Food & Drug Packaging* October: 29–33.

Barthel, H. (1997). Code, bar. In *The Wiley Encyclopedia of Packaging Technology*. A.L. Brody and K.S. Marsh (Eds), New York, John Wiley & Sons, 225–228.

Bastioli, C. (1997). Biodegradable materials. In *The Wiley Encyclopedia of Packaging Technology*. A.L. Brody and K.S. Marsh (Eds), New York, John Wiley & Sons, Inc., 77–83.

Bastioli, C. (2001). Global status of the production of biobased packaging materials. *Starch/Starke* 53: 351–355.

Bayer, R. (2003). *Glass Packaging Essentials: A Multimedia Resource CD-ROM*. Alexandria, VA, Glass Packaging Institute.

Becton, L. and Braselton, V. (2004). Multi-pack offers more. *Dairy Foods* 105: 76–77.

Blakistone, B. (2003). Retortable pouches. In *Encyclopedia of Agricultural, Food, and Biological Engineering*. D.R. Heldman (Ed.), New York, Marcel Dekker, Inc., 846–851.

Blakistone, B.A. (Ed.) (1999). *Principles and Applications of Modified Atmosphere Packaging of Foods*. Gaithersburg, MD, Aspen Publishers Inc.

Blanchfield, J.R. (Ed.) (2000). *Food Labeling*. Boca Raton, FL, CRC Press. 320.

Bolado-Rodriguez, S., Gongora-Nieto, M.M., Pothakamury, U., Barbosa-Canovas, G.V., and Swanson, B.G., (2000). A review of nonthermal technologies. In *Trends in Food Engineering*. J.E. Lozano, C. Anon, E. Parada-Arias, and G.V. Barbosa-Canovas (Eds), Lancaster, PA, Technomic Publishing Co. Inc., 227–265.

Borchardt, J.K. (1997). Recycling. In *The Wiley Encyclopedia of Packaging Technology*. A.L. Brody and K.S. Marsh (Eds), New York, John Wiley & Sons, 799–805.

Bossett, J.O., Gallmann, P.U., and Sieber, R. (1994). Influence of light transmittance of packaging materials on the shelf-life of milk and dairy product — a review. In *Food Packaging and Preservation*. M. Mathlouthi (Ed.), New York, Chapman & Hall Inc., 222–268.

Bourque, R.A. (2003). Secondary packaging. In *Encyclopedia of Agricultural, Food, and Biological Engineering*. D.R. Heldman (Ed.), New York, Marcel Dekker, Inc., 873–879.

Brandsch, J, Mercea, P., and Piringer, O. (2000). Modeling of additive diffusion coefficients in polyolefins. In *Food Packaging*. S.J. Risch (Ed.), Washington DC, American Chemical Society, 27–36.

Brighton, T.B. (1997). Film, stretch. In *The Wiley Encyclopedia of Packaging Technology, Second Edition*. A.L. Brody and K.S. Marsh. New York, John Wiley & Sons, Inc., 434–445.

Brody, A. and Marsh, K. (Eds) (1997). *The Wiley Encyclopedia of Packaging Technology*. New York, John Wiley & Sons Inc.

Brody, A.L. (1997). Packaging of food. In *The Wiley Encyclopedia of Packaging Technology*. A.L. Brody and K.S. Marsh (Eds), New York, John Wiley & Sons Inc., 699–704.

Brody, A.L. (2000a). Smart packaging becomes intellipac.[tm] *Food Technology* 54: 104–107.

Brody, A.L. (2000b). What's ahead in food packaging. *Food Technology* 54: 193.

Brody, A.L. (2003a). Modified atmosphere packaging. In *Encyclopedia of Agricultural, Food, and Biological Engineering*. D.R. Heldman (Ed.), New York, Marcel Dekker, Inc., 666–670.

Brody, A.L. (2003b). Personal communication. Packaging/Brody Inc., Duluth, GA.

Brody, A.L. (2004). Personal communication. Packaging/Brody Inc., Duluth, GA.

Brody, A.L. and Lord, J.B. (2000). *Developing New Food Products for a Changing Marketplace*. Lancaster, PA, Technomic Publishing Co. 496.

Brody, A.L., Strupinsky, E.R., and Kline, L.R. (2001). *Active Packaging For Food Applications*. Lancaster, PA, Technomic Publishing Co. 218.

Brown, D. (1993). Plastic packaging of food products: the environmental dimension. *Trends in Food Science and Technology* 4: 294–300.

Brown, H. and Williams, J. (2003). Packaged product quality and shelf life. In *Food Packaging Technology*. R. Coles, D. McDowell, and M. Kirwan (Eds), Boca Raton, FL, CRC Press, 65–94.

Cagri, A., Ustunol, Z., and Ryser, E.T. (2004). Antimicrobial edible films and coatings. *Journal of Food Protection* 67: 833–848.

Carter, R. (1997). Injection molding. In *The Wiley Encyclopedia of Packaging Technology*. A.L. Brody and K.S. Marsh (Eds), New York, John Wiley & Sons, Inc., 503–511.

Cavanagh, J. (1997). Glass container manufacturing. In *The Wiley Encyclopedia of Packaging Technology*. A.L. Brody and K.S. Marsh (Eds), New York, John Wiley & Sons, Inc., 475–484.

Chang, S.-S., Guttman, C.M., Sanchez, I.C., and Smith, L.E. (1988). Theoretical and computational aspects of migration of package components to food. In *Food and Packaging Interactions*. J.H. Hotchkiss (Ed.), Washington DC, American Chemical Society, 106–117.

Chatwin, P.C. (1996). Mathematical modelling. In *Migration from Food Contact Materials*. L.L. Katan (Ed.), London, UK, Blackie Academic & Professional, 26–49.

Chinnan, M.S. and Cha, D.S. (2003). Primary packaging. In *Encyclopedia of Agricultural, Food, and Biological Engineering*. D.R. Heldman (Ed.), New York, Marcel Dekker, Inc., 781–784.

CMI (2005). *Historical CMI Can Shipments*. Washington DC, Can Manufacturers Institute. http://www.cancentral.com/content.cfm.

Coles, R. (2003). Introduction. In *Food Packaging Technology*. R. Coles, D. McDowell, and M. Kirwan (Eds), Boca Raton, FL, CRC Press, 1–31.

Cramer, M.M. (2004). The time has come for clear food allergen labeling. *Food Safety Magazine* 10: 18–22.

Crank, J. (1975). *The Mathematics of Diffusion*. New York, Oxford University Press. 414.

Cuq, B., Gontard, N., and Guilbert, S. (1998). Proteins as agricultural polymers for packaging production. *Cereal Chemistry* 75: 1–9.

Curtis, P.A. (2005). In *Guide to Food Laws and Regulations*. Ames, I.A. (Ed.), Blackwell Publishing Professional, 265.

Cutter, C.N. (2002). Microbial control by packaging: a review. *Critical Reviews in Food Science and Nutrition* 42: 151–161.

Davis, R. (1998). The can can. *Food Processing* 67: 9–10.

Day, B.P.F. (2003). Active packaging. In *Food Packaging Technology*. R. Coles, D. McDowell, and M. Kirwan (Eds), Boca Raton, FL, CRC Press, 282–302.

de Kruijf, N., van Beest, M., Rijk, R., and Sipilainen-Malm, T. (2002). Active and intelligent packaging: applications and regulatory aspects. *Food Additives and Contaminants* 19: 144–162.

de Kruijf, N. and van Beest, M.D. (2003). Active packaging. In *Encyclopedia of Agricultural, Food, and Biological Engineering*. D.R. Heldman (Ed.), New York, Marcel Dekker, Inc., 5–9.

Delassus, P. (1997). Barrier polymers. In *The Wiley Encyclopedia of Packaging Technology*. A.L. Brody and K.S. Marsh (Eds), New York, John Wiley & Sons, Inc., 71–77.

DeLassus, P.T. (1988). Barrier expectations for polymer combinations. *Tappi Journal* 71: 216–219.

DeLassus, P.T., Strandburg, G., and Howell, B.A. (1988). Flavor and aroma permeation in barrier film: the effects of high temperature and high humidity. *Tappi Journal* 71: 177–181.

DeMaria, K. (2000). *The Packaging Development Process — A Guide for Engineers and Project Managers*. Boca Raton, FL, CRC Press, 101.

Demorest, R.L., Mayer, W.N., and Mayer, D.W. (2000). New test methods for highly permeable materials. In *Food Packaging*. S.J. Risch (Ed.), Washington, DC, American Chemical Society, 115–124.

Dent, I.S. (2000). Recycling and reuse of plastics packaging for the consumer market. In *Materials and Development of Plastics Packaging for the Consumer Market*. G.A. Giles and D.R. Bain (Eds), Boca Raton, FL, Sheffield Academic Press, 177–202.

Dixon, J. (2000). Development of flexible plastics packaging. In *Materials and Development of Plastics Packaging for the Consumer Market*. G.A. Giles and D.R. Bain (Eds), Boca Raton, FL, CRC Press LLC, 79–104.

Downes, T.W. (1989). Food packaging in the IFT era: five decades of unprecedented growth and change. *Food Technology* 43: 228–229, 232–236, 238–240.

Driscoll, R.H. and Paterson, J.L. (1999). Packaging and food preservation. In *Handbook of Food Preservation*. S.M. Rahman (Ed.), New York, Marcel Dekker Inc., 687–733.

Dunn, T.J. (1997). Multilayer flexible packaging. In *The Wiley Encyclopedia of Packaging Technology*. A.L. Brody and K.S. Marsh (Eds), New York, John Wiley & Sons, Inc., 659–665.

Duxbury, D.D. (1997). Retortable flexible and semirigid packages. In *The Wiley Encyclopedia of Packaging Technology*. A.L. Brody and K.S. Marsh (Eds), New York, John Wiley & Sons, Inc., 808–811.

Ehrenstein, G.W. (2001). *Polymeric Materials: Structure-Properties-Applications*. Cincinnati, OH, Hanser/Gardner Publications. 277.

Eichner, K. (1975). The influence of water content on non-enzymic browning reactions in dehydrated foods and model systems and the inhibition of fat oxidation by browning intermediates. In *Water Relations of Foods*. R.B. Duckworth (Ed.), London, Academic Press, 417–434.

El Makhzoumi, Z. (1994). Effect of irradiation of polymeric packaging material on the formation of volatile compounds. In *Food Packaging and Preservation*. M. Mathlouthi (Ed.), New York, Chapman & Hall Inc., 88–99.

Esse, R. and Saari, A. (2004). Shelf-life and moisture management. In *Understanding and Measuring the Shelf-Life of Food*. R. Steele (Ed.), Boca Raton, FL, CRC Press LLC, 24–41.

Estrada, R. (2004). When it's hip to be square. *The Filling Business* October: 14–15.

Eubanks, M.G. (1997). Cans, composite. In *The Wiley Encyclopedia of Packaging Technology*. A.L. Brody and K.S. Marsh (Eds), New York, John Wiley & Sons, Inc., 134–137.

Fearncombe, J. (1995). Design strategies help packagers overcome environmental hurdles. *Packaging Technology and Engineering* 4: 22–27.

Fellows, P.J. (2000). *Food Processing Technology — Principles and Practice*. Boca Raton, FL, CRC Press. 462–510.

Fennema, O.R. (1976). Water and ice. In *Principles of Food Science — Part I — Food Chemistry*. O.R. Fennema (Ed.), New York, Marcel Dekker, Inc.

Fennema, O.R. (Ed.) (1996a). *Food Chemistry*. New York, Marcel Dekker, Inc. 1067.

Fennema, O.R. (1996b). Water and ice. In *Food Chemistry*. O.R. Fennema (Ed.), New York, Marcel Dekker, Inc., 17–94.

Ferrante, M.A. (1997). Packaging for the next millennium. *Food Engineering International* 22: 28–32, 34.

Figge, K. (1996). Plastics. In *Migration from Food Contact Materials*. L.L. Katan (Ed.), London, UK, Blackie Academic & Professional, 77–108.

Foster, G.A. (1997). Boxes, corrugated. In *The Wiley Encyclopedia of Packaging Technology*. A.L. Brody and K.S. Marsh (Eds), New York, John Wiley & Sons, Inc., 100–108.

Franklin, W.F., Boguski, T.K., and Fry, P. (1997). Life-cycle assessment. In *The Wiley Encyclopedia of Packaging Technology*. A.L. Brody and K.S. Marsh (Eds), New York, John Wiley & Sons, 563–569.

Franssen, L.R. and Krochta, J.M. (2003). Edible coatings containing natural microbials for processed foods. In *Natural Antimicrobials for the Minimal Processing of Foods*. S. Roller (Ed.), Baca Raton, FL, CRC Press LLC, 306.

Gaynes, C. (1997). Testing, packaging materials. In *The Wiley Encyclopedia of Packaging Technology*. A.L. Brody and K.S. Marsh (Eds), New York, John Wiley & Sons, Inc., 890–895.

Geankoplis, C.J. (1983). *Transport Processes and Unit Operations*. Englewood Cliffs, NJ, Prentice-Hall, Inc., 451–455.

Gennadios, A, (Ed.) (2002). *Protein-Based Films and Coatings*. Washington DC, CRC Press. 650.

Gennadios, A., Weller, C.L., and Gooding, C.H. (1994). Measurement errors in water vapor permeability of highly permeable, hydrophilic edible films. *Journal of Food Engineering* 21: 395–409.

Giacin, J.R. (1995). Factors affecting permeation, sorption and migration processes in package-product systems. In *Foods and Packaging Materials–Chemical Interactions*. P. Ackermann, M. Jagerstad, and T. Ohlsson (Eds), Cambridge, UK, The Royal Society of Chemistry, 12–22.

Giacin, J.R. and Hernandez, R.J. (1997). Permeability of aromas and solvents in polymeric packaging materials. In *The Wiley Encyclopedia of Packaging Technology*. A.L. Brody and K.S. Marsh (Eds), New York, John Wiley & Sons, Inc., 724–733.

Gibbons, J.A. (1997). Extrusion. In *The Wiley Encyclopedia of Packaging Technology*. A.L. Brody and K.S. Marsh (Eds), New York, John Wiley & Sons, 370–378.

Gilbert, S. and Pegaz, D. (1969). Find new way to measure gas permeability. *Packaging Engineering* 14: 66–69.

Giles, G.A. and Bain, D.R. (Eds) (2000). *Materials and Development of Plastics Packaging for the Consumer Market*. Boca Raton, FL, CRC Press LLC.

Giles, G.A. and Bain, D.R. (Eds) (2001). *Technology of Plastics Packaging for the Consumer Market*. Boca Raton, FL, CRC Press LLC.

Girling, P.J. (2003). Packaging of food in glass containers. In *Food Packaging Technology*. R. Coles, D. McDowell, and M. Kirwan (Eds), Boca Raton, FL, CRC Press LLC, 152–173.

Goddard, R.R. (1995). Packaging — a view of the future. *Packaging Technology and Science* 8: 119–126.

Goddard, N.D.R., Kemp, R.M.J., and Lane, R. (1997). An overview of smart technology. *Packaging Technology and Science* 10: 129–143.

Gontard, N. and Guilbert, S. (1994). Bio-packaging: technology and properties of edible and/or biodegradable material of agricultural origin. In *Food Packaging and Preservation*. M. Mathlouthi (Ed.), London, Blackie Academic & Professional, 159–179.

Gordon, M.H. (2004). Factors affecting lipid oxidation. In *Understanding and Measuring the Shelf-Life of Food*. R. Steele (Ed.), Boca Raton, FL, CRC Press LLC, 128–141.

Gould, G.W. (2000). Emerging technologies in food preservation and processing in the last 40 years. In *Innovations in Food Processing*. G.V. Barbosa-Canovas and G.W. Gould (Eds), Lancaster, PA, Technomic Publishing Co., Inc., 1–11.

Gray, I.J., Harte, B.R., and Miltz, J. (Eds) (1987). *Food Product — Package Compatibility*. Lancaster, PA, Technomic Publishing Co. 286.

Greely, M.J. (1997). Standup flexible pouches. In *The Wiley Encyclopedia of Packaging Technology*. A.L. Brody and K.S. Marsh (Eds), New York, John Wiley & Sons, 852–856.

Greenberg, E.F. (1996). *Guide to Packaging Law — A Primer for Packaging Professionals*. Herndon, VA, Institute of Packaging Professionals. 115.

Guglielmini, B. (2001). Plastic closures used in the consumer packaging market. *Technology of Plastics Packaging for the Consumer Market*. Boca Raton, FL, Sheffield Academic Press, 245–255.

Guilbert, S. (1986). Technology and application of edible protective films. In *Food Packaging and Preservation: Theory and Practice*. M. Mathlouthi (Ed.), New York, Elsevier Applied Science Publishers, 371–394.

Guilbert, S. (2000). Edible films and coatings and biodegradable packaging. *Bulletin of the International Dairy Federation* 346: 10–16.

Guilbert, S., Cuq, B., and Gontard, N. (1997). Recent innovations in edible and/or biodegradable packaging materials. *Food Additives and Contaminants* 14: 741–751.

Guilbert, S. and Gontard, N. (1995). Edible and biodegradable food packaging. In *Foods and Packaging Materials — Chemical Interactions*. P. Ackermann, M. Jagerstad, and T. Ohlsson (Eds), Cambridge, UK, The Royal Society of Chemistry, 159–168.

Halek, G.W. and Luttmann, J.P. (1991). Sorption behavior of citrus-flavor compounds in polyethylenes and polypropylenes: effects of permeant functional groups and polymer structure. In *Food and Packaging Interactions II*. S.J. Risch and J.H. Hotchkiss (Eds), Washington, DC, American Chemical Society, 212–226.

Hambley, D.L. (1997). Glass container design. In *The Wiley Encyclopedia of Packaging Technology*. A.L. Brody and K.S. Marsh (Eds), New York, John Wiley & Sons, Inc., 471–475.

Han, J.H. (2000). Antimicrobial food packaging. *Food Technology* 54: 56–65.

Han, J.H. (2003). Antimicrobial food packaging. In *Novel Food Packaging Techniques*. R. Ahvenainen (Ed.), Boca Raton, FL, CRC Press, 50–70.

Han, J.H. (Ed.) (2005). *Innovations in Food Packaging*. London, Elsevier Academic Press. 517.

Hanlon, J.F., Kelsey, R.J., and Forcinio, H.E. (1998a). *Handbook of Package Engineering*. Lancaster, PA, Technomic Publishing Co., 698.

Hanlon, J.F., Kelsey, R.J., and Forcinio, H.E. (1998b). *Handbook of Package Engineering*. Lancaster, PA, Technomic Publishing Co., 607–654.

Hanlon, J.F., Kelsey, R.J., and Forcinio, H.E. (1998c). *Handbook of Package Engineering*. Lancaster, PA, Technomic Publishing Co., 655–676.

Harima, Y. (1990). Frozen food and oven-proof trays. In *Food Packaging*. T. Kadoya (Ed.), New York, Academic Press, 229–252.

Harte, B.R. and Gray, J.I. (1987). The influence of packaging on product quality. In *Food Product-Package Compatibility*. J.I. Gray, B.R. Harte, and J. Miltz (Eds), Lancaster, PA, Technomic Publishing Co., Inc., 17–29.

Hatzidimitriu, E, Gilbert, S.G., and Loukakis, G. (1987). Odor barrier properties of multi-layer packaging films at different relative humidities. *Jounal of Food Science* 52: 472–474.

Haugaard, V.K., Udsen, A.-M., Mortensen, G., Hoegh, L., Petersen, K., and Monahan, F. (2000). Food biopackaging. In *Biobased Packaging Materials for the Food Industry — Status and Perspectives*. C.J. Weber (Ed.), Frederiksberg, Denmark, KVL Dept. of Dairy and Food Science, 45–84.

Haugaard, V.K., Udsen, A.-M., Mortensen, G., Hoegh, L., Petersen, K., and Monahan, F. (2001). Potential food applications of biobased materials. An EU-concerted action project. *Starch/Starke* 53: 189–200.

Heck, O.L. (1997). Can seamers. In *The Wiley Encyclopedia of Packaging Technology*. A.L. Brody and K.S. Marsh (Eds), New York, John Wiley & Sons, Inc., 128–132.

Hernandez, R.J. (1997a). Food packaging materials, barrier properties and selection. In *Handbook of Food Engineering Practice*. K.J. Valentas, E. Rotstein, and R.P. Singh (Eds), New York, CRC Press, 291–360.

Hernandez, R.J. (1997b). Polymer properties. In *The Wiley Encyclopedia of Packaging Technology*. A.L. Brody and K.S. Marsh (Eds), New York, John Wiley & Sons, Inc., 758–765.

Hernandez, R.J. and Giacin, J.R. (1998). Factors affecting permeation, sorption, and migration processes in package-product systems. In *Food Storage Stability*. I.A. Taub and R.P. Singh (Eds), New York, CRC Press, 269–330.

Hernandez, R.J., Giacin, J.R., and Baner, A.L. (1986). The evaluation of the aroma barrier properties of polymer films. *Journal of Plastic Film and Sheeting* 2: 187–211.

Hernandez, R.J., Giacin, J.R., and Baner, A.L. (1989). The evaluation of the aroma barrier properties of polymer films. *Plastic Film Technology*. Lancaster, PA, Technomic Publishing Co., Inc., 107–131.

Hernandez, R.J., Selke, S.E.M., and Culter, J.D. (2000). *Plastics Packaging*. Cincinnati, OH, Hanser/Gardner Publications Inc., 425.

Hill, R.J. (1997). Film, transparent glass on plastic food-packaging materials. In *The Wiley Encyclopedia of Packaging Technology*. A.L. Brody and K.S. Marsh. (Eds), New York, John Wiley & Sons, Inc., 445–448.

Hirose, K., Harte, B.R., Giacin, J.R., Miltz, J., and Stine, C. (1988). Sorption of *d*-limonene by sealant films and effect on mechanical properties. In *Food and Packaging Interactions*. J.H. Hotchkiss (Ed.), Washington, DC, American Chemical Society, 28–41.

Hotchkiss, J.H. (Ed.) (1988). *Food and Packaging Interactions*. ACS Symp. Series 365. Washington, DC, American Chemical Society. 305.

Huss, G.F. (1997). Microwavable packaging and dual-ovenable materials. In *The Wiley Encyclopedia of Packaging Technology*. A.L. Brody and K.S. Marsh (Eds), New York, John Wiley & Sons, Inc., 642–646.

Idol, R.C. (1997). Oxygen scavengers. In *The Wiley Encyclopedia of Packaging Technology*. A.L. Brody and K.S. Marsh (Eds), New York, John Wiley & Sons, 687–692.

Irwin, C. (1997). Blow molding. In *The Wiley Encyclopedia of Packaging Technology*. A.L. Brody and K.S. Marsh (Eds), New York, John Wiley & Sons, Inc., 83–93.

Jasse, B., Seuvre, A.M., and Mathlouthi, M. (1994). Permeability and structure in polymeric packaging materials. In *Food Packaging and Preserveration*. M. Mathlouthi (Ed.), New York, Blackie Academic & Professional, 1–22.

Jenkins, W.A. and Harrington, J.P. (1991). *Packaging Foods with Plastics*. Lancaster, PA, Technomic Publishing Co., 326.

Johnson, B. and Demorest, R. (1997). Testing, permeation and leakage. In *The Wiley Encyclopedia of Packaging Technology*. A.L. Brody and K.S. Marsh (Eds), New York, John Wiley & Sons, Inc., 895–900.

Johnson, D.L. (1997). Indicating devices. In *The Wiley Encyclopedia of Packaging Technology*. A.L. Brody and K.S. Marsh (Eds), New York, John Wiley & Sons, 498–503.

Joy, D. (2005). The FDA agenda for 2005. *Food Processing* 66: 17.

Jurin, V. and Karel, M. (1963). Studies on control of respiration of mcintosh apples by pakcaging methods. *Food Technology* 17: 104–108.

Kaplan, D.L., Mayer, J.M., Ball, D., McCassie, J., Allen, A.L., and Stenhouse, P. (1993). Fundamentals of biodegradable polymers. In *Biodegradable Polymers and Packaging*. C. Ching, D. Kaplan, and E. Thomas (Eds), Lancaster, PA, Technomic Publishing Co., Inc., 1–42.

Karel, M. (2000). Tasks of food technology in the 21st century. *Food Technology* 54: 56–64.

Karel, M. and Lund, D.B. (2003a). *Physical Principles of Food Preservation*. New York, Marcel Dekker Inc., 514–592.

Karel, M. and Lund, D.B. (2003b). *Physical Principles of Food Preservation*. New York, Marcel Dekker Inc., 117–179.

Karel, M. and Yong, S. (1981). Autooxidation-initiated reactions in food. In *Water Activity: Influences on Food Quality*. L.B. Rockland and G.F. Stewart (Eds), New York, Academic Press, 511–529.

Katan, L.L. (Ed.) (1996). *Migration from Food Contact Materials*. New York, Blackie Academic & Professional. 303.

Katz, E.E. and Labuza, T.P. (1981). Effect of water activity on sensory crispness and mechanical deformation of snack food products. *Journal of Food Science* 46: 403–409.

Kester, J.J. and Fennema, O.R. (1986). Edible films and coatings: a review. *Food Tech.* 40: 47–59.

Khanna, R. and Peppas, N.A. (1982). Mathematical analysis of transport properties of polymer films for food packaging: III. Moisture and oxygen diffusion. *AIChE Symposium Series 218* 78: 185–191.

Kirwan, M.J. (2003). Paper and paperboard packaging. In *Food Packaging Technology*. R. Coles, D. McDowell, and M. Kirwan (Eds), Boca Raton, FL, CRC Press, 241–282.

Kirwan, M.J. and Strawbridge, J.W. (2003). Plastics in food packaging. In *Food Packaging Technology*. R. Coles, D. McDowell, and M. Kirwan (Eds), Boca Raton, FL, CRC Press, 174–240.

Knorr, D. (2000). Process aspects of high-pressure treatment of food systems. In *Innovations in Food Processing*. G.V. Barbosa-Canovas and G.W. Gould (Eds), Lancaster, PA, Technomic Publishing Co., Inc., 13–30.

Kondo, K. (1990). Plastic containers. In *Food Packaging*. T. Kadoya (Ed.), New York, Academic Press, 117–145.

Koros, W.J. (1990). Barrier polymers and structures: overview. In *Barrier Polymers and Structures*. W.J. Koros (Ed.), Washington DC, American Chemical Society, 1–21.

Kraus, F.J. and Tarulis, G.J. (1997). Cans, steel. In *The Wiley Encyclopedia of Packaging Technology*. A.L. Brody and K.S. Marsh (Eds), New York, John Wiley & Sons, Inc., 144–155.

Krochta, J.M. (1997a). Edible protein films and coatings. In *Food Proteins and their Applications in Foods*. S. Damodaran and A. Paaraf (Eds), New York, Marcel Dekker, Inc., 529–549.

Krochta, J.M. (1997b). Films, edible. In *The Wiley Encyclopedia of Packaging Technology, 2nd Edition*. A.L. Brody and K.S. Marsh (Eds), New York, John Wiley & Sons, Inc.

Krochta, J.M. (2002). Proteins as raw materials for films and coatings: definitions, current status and opportunities. In *Protein-Based Films and Coatings*. A. Gennadios (Ed.), Boca Raton, FL, CRC Press LLC, 1–42.

Krochta, J.M. (2003). Package permeability. In *Encyclopedia of Agricultural, Food, and Biological Engineering*. D.R. Heldman (Ed.), New York, Marcel Dekker, Inc., 720–726.

Krochta, J.M., Baldwin, E.A., and Nisperos-Carriedo, M. (Eds) (1994). *Edible Coatings and Films to Improve Food Quality*. Lancaster, PA, Technomic Publishing Company, Inc., 379.

Krochta, J.M. and De Mulder-Johnston, C.L.C. (1996). Biodegradable polymers from agricultural products. In *Agricultural Materials as Renewable Resources Nonfood and Industrial Applications*. G. Fuller, T.A. McKeon, and D. Bills (Eds), Washington, DC, American Chemical Society, 120–140.

Krochta, J.M. and De Mulder-Johnston, C.L.C. (1997). Edible and biodegradable polymer films: challenges and opportunities. *Food Technology* 51: 61–73.

Labuza, T.P. (1996). An introduction to active packaging for foods. *Food Technology* 50: 68–71.

Labuza, T.P. and Breene, W.M. (1989). Applications of active packaging for improvement of shelf-life and nutritional quality of fresh and extended shelf-life foods. *Journal of Food Processing and Preservation* 13: 1–69.

Labuza, T.P. and Saltmarch, M. (1981). The nonenzymatic browing reaction as affected by water in foods. In *Water Activity: Influences on Food Quality*. L.B. Rockland and G.F. Stewart (Eds), New York, Academic Press, 605–650.

LaCoste, A., Schaich, K.M., Zumbrunnen, D., and Yam, K.L. (2005). Advancing controlled release packaging through smart blending. *Packaging Technology and Science* 18: 77–87.

Lanmbet, Y., Demazeau, G., Bouvier, J.M., Laborde-Croubit, S., and Cabannes, M. (2000). Packaging for high-pressure treatments in the food industry. *Packaging Technology and Science* 13: 63–71.

Leadbitter, J. 2003. *Packaging Materials: 5. Polyvinyl Chloride (PVC) for Food Packaging Applications*. Brussels, ILSI Europe: 20.

Linssen, J.P.H. and Roozen, J.P. (1994). Food flavour and packaging interactions. In *Food Packaging and Preservation*. M. Mathlouthi (Ed.), New York, Blackie Academic and Professional, 48–61.

Linssen, J.P.H, van Willige, R.W.G., and Dekker, M. (2003). Packaging-flavour interactions. In *Novel Food Packaging Techniques*. R. Ahvenainen (Ed.), Boca Raton, FL, CRC Press, 144–171.

Louis, P.J (1999). Review paper — food packaging in the next millennium. *Packaging Technology and Science* 12: 1–7.

Lozano, J.E., Anon, C., Parada-Arias, E., and Barbosa-Canovas, G.V (Eds) (2000). *Trends in Food Engineering*. Food Preservation Technology. Lancaster, PA, Technomic Publishing Co., Inc., 347.

Lynch, L. and Anderson, J. (1997). Cartons, rigid, paperboard. In *The Wiley Encyclopedia of Packaging Technology*. A.L. Brody and K.S. Marsh (Eds), New York, John Wiley & Sons, Inc., 108–110.

Mabee, M.S. (1997). Aseptic packaging. In *The Wiley Encyclopedia of Packaging Technology*. A.L. Brody and K.S. Marsh (Eds), New York, John Wiley & Sons, Inc., 41–45.

Mandel, A.S. (1997). Bottle design, plastic. In *The Wiley Encyclopedia of Packaging Technology*. A.L. Brody and K.S. Marsh (Eds), New York, John Wiley & Sons, Inc., 93–100.

Marsh, K (1994). *Package Design, Materials, and Related Environmental Issues*. Food Safety: A Comprehensive View, Boston, MA, Research and Development Associates for Military Food and Packaging Systems, Inc.

Marsh, K.S. (1997). Shelf life. In *The Wiley Encyclopedia of Packaging Technology*. A.L. Brody and K.S. Marsh (Eds), New York, John Wiley & Sons, 831–835.

Marsh, K.S. (2001). Looking at packaging in a new way to reduce food losses. *Food Technology* 55: 48–52.

Martin-Polo, M.O. (1995). Biopolymers in the fabrication of edible and biodegradable materials for food preservation. In *Food Preservation by Moisture Control*. G.V. Barbosa-Canovas and J. Welti-Chanes (Eds), Lancaster PA, Technomic Publishing Co., 849–868.

Matsubayashi, H. (1990). Metal containers. In *Food Packaging*. T. Kadoya (Ed.), New York, Academic Press, 85–104.

Matthews, V. (2000). *Packaging Materials: 4. Polyethylene Terephthalate (PET) for Food Packaging Applications*. Brussels, ILSI Europe: 16.

McCormack, T. (2000). Plastics packaging and the environment. In *Materials and Development of Plastics Packaging for the Consumer Market*. G.A. Giles and D.R. Bain (Eds), Boca Raton, FL, CRC Press LLC, 152–176.

McHugh, T.H., Avena-Bustillos, R., and Krochta, J.M. (1993). Hydrophilic edible films: modified procedure for water vapor permeability and explanation of thickness effects. *Journal of Food Science* 58: 899–903.

McKinney, L., Kent, W., and Roe, R. (1986). Thermoforming. In *The Wiley Encyclopedia of Packaging Technology*. M. Bakker and D. Eckroth (Eds), New York, John Wiley & Sons Inc., 668.

Miller, K.S. and Krochta, J.M. (1998). Measuring aroma transport in polymer films. *Transactions of the ASAE* 41: 427–433.

Miltz, J. (1992). Food packaging. In *Handbook of Food Engineering*. D.R. Heldman and D.B. Lund (Eds), New York, Marcel Dekker, Inc., 667–718.

Mizrahi, S. (2004). Accelerated shelf-life tests. In *Understanding and Measuring the Shelf-Life of Food*. R Steele (Ed.), Boca Raton, FL, CRC Press LLC, 317–339.

Mossel, D.A.A (1975). Water and micro-organisms in foods — a synthesis. In *Water Relations of Foods*. R.B. Duckworth (Ed.), New York, Academic Press Inc. (London) Ltd., 347–361.

Moyer, G.R. (1997). Form/film/seal, vertical. In *The Wiley Encyclopedia of Packaging Technology*. A.L. Brody and K.S. Marsh (Eds), New York, John Wiley & Sons, Inc., 468–470.

Mullan, M. and McDowell, D. (2003). Modified atmosphere packaging. In *Food Packaging Technology*. R. Coles, D. McDowell, and M. Kirwan (Eds), Boca Raton, FL, CRC Press, 303–339.

Murphy, T.P. (1996). Metals. In *Migration from Food Contact Materials*. L.L. Katan (Ed.), London UK, Blackie Academic & Professional, 111–143.

Nairn, J.F. and Norpell, T.M. (1997). Closures, bottle and jar. In *The Wiley Encyclopedia of Packaging Technology*. A.L. Brody and K.S. Marsh (Eds), New York, John Wiley & Sons, 206–220.

Narasimhan, K.R. (2001). Global trends and policies in food packaging. *Indian Food Industry* 20: 57–59.

Nielaender, G. (1996). Innovations of the aseptic carton package. *Fruit Processing* 6: 240–241.

Obolewicz, P. (1997). Cartons, folding. In *The Wiley Encyclopedia of Packaging Technology*. A.L. Brody and K.S. Marsh (Eds), New York, John Wiley & Sons, Inc., 181–187.

Ohlsson, T. (2000). Minimal processing of foods with thermal methods. In *Innovations in Food Processing*. G.V. Barbosa-Canovas and G.W. Gould (Eds), Lancaster PA, Technomic Publishing Company Inc., 141–148.

Ooraikul, B. and Stiles, M.E. (Eds) (1991). *Modified Atmosphere Packaging of Food*. Belmont, CA, Thompson International Publishing.

Ozen, B.F. and Floros, J.D. (2001). Effects of emerging food processing techniques on the packaging materials. *Trends in Food Science & Technology* 12: 60–67.

Page, B., Edwards, M., and May, N. (2003). Metal cans. *Food Packaging Technology*. R. Coles, D. McDowell, and M. Kirwan (Eds), Boca Raton, FL, CRC Press, 120–151.

Paine, F.A. (Ed.) (1991). *The Package User's Handbook*. New York, Van Nostrand Reinhold. 596.

Paine, F.A. and Paine, H.Y. (1992a). *A Handbook of Food Packaging*. New York, Van Nostrand Reinhold. 497.

Paine, F.A. and Paine, H.Y. (1992b). *A Handbook of Food Packaging*. New York, Van Nostrand Reinhold. 1–32.

Paine, F.A. and Paine, H.Y. (1992c). *A Handbook of Food Packaging*. New York, Van Nostrand Reinhold. 477–490.

Paine, F.A. and Paine, H.Y. (1992d). *A Handbook of Food Packaging*. New York, Blackie Academic & Professional. 390–425.

Park, H.J., Testin, R.F., Chinnan, M.S., and Park, J.W. (Eds) (2001). *Active Biopolymer Films and Coatings for Food and Biotechnological Uses*. Seoul, Korea, Laboratory of Packaging Engineering, Korea University. 250.

Pascat, B. (1986). Study of some factors affecting permeability. In *Food Packaging and Preservation*. M Mathlouthi (Ed.), London, Elsevier Applied Science Publishers, 7–24.

Peache, R. (1997). Cushioning, design. In *The Wiley Encyclopedia of Packaging Technology*. A Brody and K Marsh (Eds), New York, John Wiley & Sons Inc., 287–293.

Perdue, R. (1997). Vacuum packaging. In *The Wiley Encyclopedia of Packaging Technology*. A.L. Brody and K.S. Marsh (Eds), New York, John Wiley & Sons, 949–955.

Petersen, K., Nielsen, P.V., Bertelsen, G., Lawther, M., Olsen, M.B., Nilsson, N.H., and Mortensen, G. (1999). Potential of biobased materials for food packaging. *Trends in Food Science and Technology* 10: 52–68.

Potter, N.N. and Hotchkiss, J.H. (1995). *Food Science*. New York, Chapman & Hall. 608.

Powers, T. and Calvo, W.J. (2003). Moisture regulation. In *Novel Food Packaging Techniques*. R. Ahvenainen (Ed.), Boca Raton, FL, CRC Press, 172–185.

Quintavalla, S. and Vicini, L. (2002). Antimicrobial food packaging in meat industry. *Meat Science* 62: 373–380.

Raper, S.A. and Borchelt, R. (1997). Integrated packaging design and development. In *The Wiley Encyclopedia of Packaging Technology, Second Edition*. A.L. Brody and K.S. Marsh (Eds), New York, John Wiley & Sons, Inc., 514–519.

Raymond, M. (1997a). Environmental regulations, international. In *The Wiley Encyclopedia of Packaging Technology*. A.L. Brody and K.S. Marsh (Eds), New York, John Wiley & Sons, 348–351.

Raymond, M. (1997b). Environmental regulations, North America. *The Wiley Encyclopedia of Packaging Technology*. A.L. Brody and K.S. Marsh (Eds), New York, John Wiley & Sons, 351–355.

Reingardt, T. and Nieder, N.F. (1997). Cans, aluminum. In *The Wiley Encyclopedia of Packaging Technology*. A.L. Brody and K.S. Marsh (Eds), New York, John Wiley & Sons, Inc.

Reynolds, P. (2002). Technology to take us forward. *Packaging World* 70–76.

Riquelme, F., Pretel, M.T., Martinez, G., Serrano, M., Amoros, A., and Romojaro, F. (1994). Packaging of fruits and vegetables: recent results. In *Food Packaging and Preservation*. M. Mathlouthi (Ed.), New York, Chapman & Hall Inc., 88–99.

Risch, S.J. (Ed.) (2000). *Food Packaging — Testing Methods and Applications*. ACS Symp. Series 753. Washington, DC, American Chemical Society. 166.

Risch, S.J. and Hotchkiss, J.H. (Eds) (1991). *Food and Packaging Interactions II*. ACS Symp. Series 473. Washington, DC, American Chemical Society. 262.

Risch, S.J., Mayer, W.N., and Mayer, D.W. (2000). Prediction vs. equilibrium testing for permeation of organic volatiles through packaging materials. In *Food Packaging*. S.J. Risch (Ed.), Washington, DC, American Chemical Society, 141–150.

Rizvi, S.S.H. (1981). Requirements for foods packaged in polymeric films. *Critical Reviews in Food Science and Nutrition* 14: 111–134.

Robertson, G.L. (1993a). *Food Packaging — Principles and Practice*. New York, Marcel Dekker, Inc., 173–203.

Robertson, G.L. (1993b). *Food Packaging — Principles and Practice*. New York, Marcel Dekker, Inc., 252–302.

Robertson, G.L. (1993c). *Food Packaging — Principles and Practice*. New York, Marcel Dekker, Inc., 232–251.

Robertson, G.L. (1993d). *Food Packaging — Principles and Practice*. New York, Marcel Dekker, Inc., 73–110.

Robertson, G.L. (1993e). *Food Packaging — Principles and Practice*. New York, Marcel Dekker, Inc., 63–72.

Robertson, G.L. (2002). The paper beverage carton: past and future. *Food Technology* 56: 46–52.

Robertson, G.L. (2006a). *Food Packaging — Principles and Practice*. Boca Raton, CRC Taylor & Francis Group. 121–156.

Robertson, G.L. (2006b). *Food Packaging — Principles and Practice*. Boca Raton, CRC Taylor & Francis Group. 1–8.

Robertson, G.L. (2006c). *Food Packaging — Principles and Practice*. Boca Raton, CRC Taylor & Francis Group. 550.

Robertson, G.L. (2006d). *Food Packaging — Principles and Practice*. Boca Raton, CRC Taylor & Francis Group. 193–224.

Robertson, G.L. (2006e). *Food Packaging — Principles and Practice*. Boca Raton, CRC Taylor & Francis Group. 225–254.

Robertson, G.L. (2006f). *Food Packaging — Principles and Practice*. Boca Raton, CRC Taylor & Francis Group. 417–446.

Robertson, G.L. (2006g). *Food Packaging — Principles and Practice*. Boca Raton, CRC Taylor & Francis Group. 157–174.

Robertson, G.L. (2006h). *Food Packaging — Principles and Practice*. Boca Raton, CRC Taylor & Francis Group. 9–42.

Robertson, G.L. (2006i). *Food Packaging — Principles and Practice*. Boca Raton, CRC Taylor & Francis Group. 79–102.

Robertson, G.L. (2006j). *Food Packaging — Principles and Practice*. Boca Raton, CRC Taylor & Francis Group. 103–120.

Robertson, G.L. (2006k). *Food Packaging — Principles and Practice*. Boca Raton, CRC Taylor & Francis Group. 55–78.

Robertson, G.L. (2006l). *Food Packaging — Principles and Practice*. Boca Raton, CRC Taylor & Francis Group. 255–270.

Robertson, G.L. (2006m). *Food Packaging — Principles and Practice*. Boca Raton, CRC Taylor & Francis Group. 313–330.

Robertson, G.L. (2006n). *Food Packaging — Principles and Practice*. Boca Raton, CRC Taylor & Francis Group. 285–312.

Robertson, G.L. (2006o). *Food Packaging — Principles and Practice*. Boca Raton, CRC Taylor & Francis Group. 43–54.

Robertson, G.L. (2006p). *Food Packaging — Principles and Practice*. Boca Raton, CRC Taylor & Francis Group. 504–527.

Robertson, G.L. (2006q). *Food Packaging — Principles and Practice*. Boca Raton, CRC Taylor & Francis Group. 473–502.

Rodrigues, E.T. and Han, J.H. (2003). Intelligent packaging. In *Encyclopedia of Agricultural, Food, and Biological Engineering*. D.R. Heldman (Ed.), New York, Marcel Dekker, Inc., 528–535.

Rogers, C.E. (1985). Permeation of gases and vapours in polymers. *Polymer Permeability*. J Comyn. New York, Elsevier Aplied Science, 11–74.

Roland, A.M. and Hotchkiss, J.H. (1991). Determination of flavor — polymer interactions by vacuum — microgravimetric method. In *Food and Packaging Interactions II*. S.J. Risch and J.H. Hotchkiss (Eds), Washington, DC, American Chemical Society, 149–160.

Rooney, M.L. (Ed.) (1995). *Active Food Packaging*. New York, Blackie Academic & Professional.

Rooney, M.L. (1997). Active packaging. In *The Wiley Encyclopedia of Packaging Technology*. A.L. Brody and K.S. Marsh (Eds), New York, John Wiley & Sons, 2–8.

Rooney, M.L. (2000). Plastics in active packaging. In *Materials and Development of Plastics Packaging for the Consumer Market*. G.A. Giles and D.R. Bain (Eds), Boca Raton, FL, CRC Press LLC.

Rosette, J.L. (1997). Tamper-evident packaging. In *The Wiley Encyclopedia of Packaging Technology*. A.L. Brody and K.S. Marsh (Eds), New York, John Wiley & Sons, 879–882.

Rowan, C. (2001). Innovation in glass packaging. *Food Engineering and Ingredients* 26: 30–31, 34.

Rowatt, R.J. (1993). The plastics waste problem. *Chemical Technology* 23: 56–60.

Sacharow, S. and Brody, A.L. (1987). *Packaging: An Introduction*. Duluth MN, Harcourt Brace Jovanovich Publications, Inc., 35–77.

Sadler, G.D., Koutchma, T.N., and Setikaite, I. (2005). *Packaging Requirements for High Pressure Sterllization Processes — Abstract 39-4*. Inst. of Food Technologists Annual Meeting, New Orleans, LA.

Salame, M. (1974). The use of low permeation thermoplastics in food and beverage packaging. In *Permeability of Films and Coatings*. H.B. Hopfenberg (Ed.), New York, Plenum Publ. Corp., 275.

Seeley, D. (1997). Aroma barrier testing. In *The Wiley Encyclopedia of Packaging Technology*. A.L. Brody and K.S. Marsh (Eds), New York, John Wiley & Sons, Inc., 39–41.

Seidel, S. (2001). A change artist among packaging. *Fruit Processing* 11: 446–448.

Selke, S.E. (1997a). Environment. In *The Wiley Encyclopedia of Packaging Technology*. A.L. Brody and K.S. Marsh (Eds), New York, John Wiley & Sons, Inc., 343–348.

Selke, S.E.M. (1994). *Packaging and the Environment — Alternatives, Trends and Solutions*. Lancaster, PA, Technomic Publishing Co., Inc., 179.

Selke, S.E.M. (1997b). *Understanding Plastics Packaging Technology*. Cincinnati, OH, Hanser/Gardner Publications Inc., 173–198.

Selke, S.E.M. (1997c). *Understanding Plastics Packaging Technology*. Cincinnati, OH, Hanser/Gardner Publications Inc., 151–172.

Selke, S.E.M. (1997d). *Understanding Plastics Packaging Technology*. Cincinnati, OH, Hanser/Gardner Publications Inc., 206.

Shellhammer, T.H. (2003). Flexible packaging. In *Encyclopedia of Agricultural, Food, and Biological Engineering*. D.R. Heldman (Ed.), New York, Marcel Dekker, Inc., 333–336.

Sikora, M (1997). Paper. In *The Wiley Encyclopedia of Packaging Technology*. A.L. Brody and K.S. Marsh (Eds), New York, John Wiley & Sons, Inc., 714–717.

Silbereis, J (1997). Metal cans, fabrication. In *The Wiley Encyclopedia of Packaging Technology, Second Edition*. A.L. Brody and K.S. Marsh (Eds), New York, John Wiley & Sons, Inc., 615–629.

Simmons, R.A. (1997). Laws and regulations, United States. In *The Wiley Encyclopedia of Packaging Technology*. A.L. Brody and K.S. Marsh (Eds), New York, John Wiley & Sons, 552–558.

Singh, R.P. and Mannapperuma, J.D. (2000). Minimal processing of fruits and vegetables. *Trends in Food Engineering*. J.E. Lozano, C. Anon, E. Parada-Arias, and G.V. Barbosa-Canovas (Eds), Lancaster, PA, Technomic Publishing Co., Inc., 191–203.

Singh, R.P. and Anderson, B.A. (2004). The major types of food spoilage: an overview. In *Understanding and Measuring the Shelf-Life of Food*. R Steele (Ed.), Boca Raton, FL, CRC Press LLC, 3–23.

Smith, J.P., Simpson, B., and Ramaswamy, H. (1992). Packaging, Part IV: Modified atmosphere packaging — principles and applications. *Encyclopedia of Food Science and Technology*. Y.H. Hui. New York, John Wiley & Sons Inc., 1982–1992.

Smolander, M. (2003). The use of freshness indicators in packaging. In *Novel Food Packaging Techniques*. R Ahvenainen (Ed.), Boca Raton, FL, CRC Press, 127–143.

Soderhjelm, L. and Sipilainen-Malm, T. (1996). Paper and board. In *Migration from Food Contact Materials*. L.L. Katan (Ed.), London, UK, Blackie Academic & Professional, 159–179.

Sonneveld, K. (2000). What drives (food) packaging innovation? *Packaging Technology and Science* 13: 29–35.

Soroka, W. (2002a). *Fundamentals of Packaging Technology*. Naperville, IL, Institute of Packaging Professionals. 1–21.

Soroka, W. (2002b). *Fundamentals of Packaging Technology*. Naperville, IL, Institute of Packaging Professionals. 527–556.

Soroka, W. (2002c). *Fundamentals of Packaging Technology*. Naperville, IL, Institute of Packaging Professionals. 437–463.

Soroka, W. (2002d). *Fundamentals of Packaging Technology*. Naperville, IL, Institute of Packaging Professionals. 600.

Soroka, W. (2002e). *Fundamentals of Packaging Technology*. Naperville, IL, Institute of Packaging Professionals. 293–316.

Soroka, W. (2002f). *Fundamentals of Packaging Technology*. Naperville, IL, Institute of Packaging Professionals. 155–178.

Soroka, W. (2002g). *Fundamentals of Packaging Technology*. Naperville, IL, Institute of Packaging Professionals. 179–197.

Soroka, W. (2002h). *Fundamentals of Packaging Technology*. Naperville, IL, Institute of Packaging Professionals. 407–436.

Soroka, W. (2002i). *Fundamentals of Packaging Technology*. Naperville, IL, Institute of Packaging Professionals. 223–259.

Soroka, W. (2002j). *Fundamentals of Packaging Technology*. Naperville, IL, Institute of Packaging Professionals. 107–127.

Soroka, W. (2002k). *Fundamentals of Packaging Technology*. Naperville, IL, Institute of Packaging Professionals. 129–154.

Soroka, W.G. and Zepf, P.J. (1998). *The IoPP Glossary of Packaging Terminology*. Herndon, VA, Institute of Packaging Professionals. 380.

Stannett, V., Szwarc, M., Bhargava, R.L., Meyers, J.A., Myers, A.W., and Roger. C.E. (1962). *Permeability of Plastic Films and Coated Papers to Gases and vapors*. New York, Technical Association of the Pulp and Paper Industry. 105.

Steele, R. (2004). *Understanding and Measuring the Shelf-Life of Food*. Boca Raton, FL, CRC Press LLC. 407.

Stern, S.A. and Trohalaki, S. (1990). Fundamentals of gas diffusion in rubbery and glassy polymers. In *Barrier Polymers and Structures*. W.J. Koros (Ed.), Washington, DC, American Chemical Society, 22–59.

Steven, M.D. and Hotchkiss, J.H. (2003a). Non-migratory bioactive polymers (NMBP) in food packaging. In *Novel Food Packaging Techniques*. R. Ahvenainen (Ed.), Boca Raton, FL, CRC Press, 71–102.

Steven, M.D. and Hotchkiss, J.H. (2003b). Package functions. In *Encyclopedia of Agricultural, Food, and Biological Engineering*. D.R. Heldman (Ed.), New York, Marcel Dekker, Inc., 716–719.

Storlie, J. and Brody, A.L. (2000). Mandatory food package labeling in the United States. In *New Food Products for a Changing Marketplace*. A.L. Brody and J.B. Lord (Eds), Lancaster, PA, Technomic Publishing, 409–438.

Storlie, J. and Hare, K. (1997). Nutrition labeling. In *The Wiley Encyclopedia of Packaging Technology*. A.L. Brody and K.S. Marsh (Eds), New York, John Wiley & Sons, 674–681.

Stratton, K. (1998). Packaging materials get the green light. *Food Engineering International* 23: 43–48.

Streeter, A. (2000). The changing image of plastics packaging in the marketplace. In *Materials and Development of Plastics Packaging for the consumer Market*. G.A. Giles and D.R. Bain (Eds), Boca Raton, FL, Sheffield Academic Press, 130–151.

Sumimoto, M. (1990). Paper and paperboard containers. *Food Packaging*. T. Kadoya (Ed.), New York, Academic Press, 53–84.

Suppakul, P., Miltz, J., Sonneveld, K., and Bigger, S.W. (2003). Active packaging technologies with an emphasis on antimicrobial packaging and its applications. *Journal of Food Science* 68: 408–420.

Taoukis, P. and Labuza, T.P. (1996). Summary: integrative concepts. In *Food Chemistry*. O.R. Fennema (Ed.), New York, Marcel Dekker, Inc., 1013–1042.

Taoukis, P.S., Meskine, A.El., and Labuza, T.P. (1988). Moisture transfer and shelf life of packaged foods. In *Food and Packaging Interactions*. J.H. Hotchkiss (Ed.), Washington, DC, American Chemical Society, 243–261.

Taoukis, P.S. and Labuza, T.P. (2003). Time-temperature indicators (TTIs). In *Novel Food Packaging Techniques*. R. Ahvenainen (Ed.), Boca Raton, FL, CRC Press, 103–126.

Tharanathan, R.N. (2003). Biodegradable films and composite coatings: past, present and future. *Trends in Food Science and Technology* 14: 71–78.

Thayer, D.W. (2003). Ionizing irradiation, treatment of food. In *Encyclopedia of Agricultural, Food, and Biological Engineering*. D.R. Heldman (Ed.), New York, Marcel Dekker, Inc., 536–539.

Thompson, L. (1999). The aseptic revolution containers for immediate filling. *Fruit Processing* 9: 386–388, 390.

Thorsheim, H.R. and Armstrong, D.J. (1993). Recycled plastics for food packaging. *Chemical Technology* 23: 55–58.

Tice, P. (2002a). *Packaging Materials: 2. Polystyrene for Food Packaging Applications*. Brussels, ILSI Europe: 20.

Tice, P. (2002b). *Packaging Materials: 3. Polyproplylene as a Packaging Material for Foods and Beverages*. Brussels, ILSI Europe: 24.

Tice, P. (2003). *Packaging Materials: 4. Polyethylene for Food Packaging Applications*. Brussels, ILSI Europe: 24.

Tingle, V. (1996). Glass. In *Migration from Food Contact Materials*. L.L. Katan (Ed.), London, UK, Blackie Academic & Professional, 145–158.

Tucker, G.S. (2003). Food biodeterioration and methods of preservation. In *Food Packaging Technology*. R Coles, D. McDowell, and M. Kirwan (Eds), Boca Raton, FL, CRC Press, 32–64.

Tullo, A. (2005). Polylactic acid redux. *Chemical and Engineering News* 83: 26.

Twede, D. (1997). Logistical/distribution packaging. In *The Wiley Encyclopedia of Packaging Technology*. A.L. Brody and K.S. Marsh (Eds), New York, John Wiley & Sons, 572–579.

Twede, D. and Downes, T.W. (1997). Economics of packaging. *The Wiley Encyclopedia of Packaging Technology*. A.L. Brody and K.S. Marsh (Eds), New York, John Wiley & Sons, Inc., 325–331.

Twede, D. and Goddard, R. (1998). *Packaging Materials*. Surrey, England, Pira International. 230.

Twede, D. and Harte, B. (2003). Logistical packaging for food marketing systems. *Food Packaging Technology*. R. Coles, D. McDowell, and M. Kirwan (Eds), Boca Raton, FL, CRC Press, 95–119.

Vakevainen, J. (2000). Advanced packaging solutions. *The European Food and Drink Review*: 87–88.

Vermeiren, L., Devlieghere, F., and Debevere, J. (2002). Effectiveness of some recent antimicrobial packaging concepts. *Food Additives and Contaminants* 19: 163–171.

Vermeiren, L., Devlieghere, F., van Beest, M., de Kruijf, N., and Debevere, J. (1999). Developments in the active packaging of foods. *Trends in Food Science and Technology* 10: 77–86.

Vermeiren, L., Heirlings, L., Devlieghere, F., and Debevere, J. (2003). Oxygen, ethylene and other scavengers. In *Novel Food Packaging Techniques*. R. Ahvenainen (Ed.), Boca Raton, FL, CRC Press, 22–49.

Vieth, W.R. (1991). *Diffusion In and Through Polymers*. New York, Hanser Publishers. 322.

Weber, C.J. (Ed.) (2000). *Biobased Packaging Materials for the Food Industry Status and Perspectives*. Frederiksberg Denmark, KVL Department of dairy and Food Science. 136.

Weinberg, H. (1998). New trends in food packaging and labeling. *International Food Marketing and Technology* 12: 58–60.

White, R.M. and Tice, P.A. (1997). Laws and regulations, Europe. In *The Wiley Encyclopedia of Packaging Technology*. A.L. Brody and K.S. Marsh (Eds), New York, John Wiley & Sons, 541–552.

Yam, K.L. (2000). Intelligent packaging for the future smart kitchen. *Packaging Technology and Science* 13: 83–85.

Yam, K.L., Paik, J.S., and Lai, C.C. (1992). Packaging, Part I: General considerations. In *Encyclopedia of Food Science and Technology*. Y.H. Hui (Ed.), New York, John Wiley & Sons, Inc., 1971–1975.

Yam, K.L., Takhistov, P.T., and Miltz, J. (2005). Intelligent packaging: concepts and applications. *Journal of Food Science* 70: R1–R10.

Yamato, Y. (1990). Glass containers. In *Food Packaging*. T. Kadoya (Ed.), New York, Academic Press, 105–116.

Zagory, D. (1997). Modified atmosphere packaging. In *The Wiley Encyclopedia of Packaging Technology*. A.L. Brody and K.S. Marsh (Eds), New York, John Wiley & Sons, 651–656.

14 Cleaning and Sanitation

Erwin A. Plett and Albrecht Graßhoff

CONTENTS

14.1 FUNDAMENTALS

Hygiene considerations are of utmost importance in the food manufacturing process. In this chapter, plant hygiene is discussed to provide the reader with a basic understanding of the mass and heat transfer phenomena involved, to outline design considerations, and to provide practical user advice.

Cleaning and sanitation procedures have to be considered as an integral part of food production. On the one hand, the ability of equipment to produce high-quality foods depends significantly on hygienic conditions, and on the other hand, the intensity of the necessary cleaning and sanitation procedures depends on the previous soiling or fouling phenomena. Figure 14.1 shows this interdependence from the reactions point of view; cleaning and fouling are diametrically opposed processes with common starting/ending points (Plett and Loncin, 1985). No matter how well the production plant may be designed, if cleaning and sanitation are not possible or not done properly, it is impossible to achieve high quality products.

14.1.1 GOALS OF CLEANING AND SANITIZING

To set up goals for cleaning and sanitizing, different types of contamination sources have to be addressed: (1) dirty raw material; (2) foreign bodies, including foreign odors and flavors; and (3) dirty production equipment. In the first case, cleaning of raw materials will depend on its physical state (fluid or solid). For fluid raw materials screening (e.g., at inlet pipes), filtering (with and without filter aids), adsorption (e.g., on activated carbon), and centrifugation are used. For solids, settling, screening, and sorting are used.

Using all kinds of mechanical separation processes for cleaning should not divert attention from other commonsense points (Harper, 1972):

1. Use of more bulk handling systems to avoid the problems association with bags (strings, paper) and cans (metal swarf bottles, wooden boxes, and so on). This avoids such foreign-body sources as wood, paper, strings, nails, and wire.
2. When cleaning raw materials that are naturally dirty, such as vegetables, carryover of this dirt must be avoided.

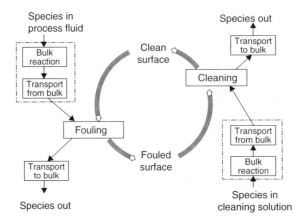

FIGURE 14.1 Fouling-cleaning cycle. (From Plett, E.A., and Loncin, M. 1985. In *Jahrbuch 1984 des Forschungskreises der Ernährungsindustrie*, Hannover, p. 139. With permission.)

3. Sieves for powders should not be constructed with stainless steel wire. Although this material is durable, it tends to be nonmagnetic, and therefore, if sieves break up, there is wholesale contamination, which will not be magnetically detectable.
4. Sieves must be of appropriate size (e.g., white flour can be sieved through an 850-μm sieve, which will remove adult and larval insects).
5. The opening of cans has caused major problems of swarf contamination. Punching openers should be used, not those based on a cutting action. After can-opening operations, magnets become a necessity in subsequent handling equipment.

Another source of contamination is the equipment surface. In this case the aim of cleaning and disinfection procedures is to achieve a hygienic unobjectionable final state of the surface. The surface must be free of soil, pathogenic microorganisms, food spoiling microorganisms, and detergents and/or disinfectant agents.

With the present state of knowledge about interactions between cleaning and fouling, another condition must be added to the final state of the cleaned surface. An antifouling treatment (e.g., passivation) must be applied to the clean surface in order to inhibit the first absorption step (induction phase) during the subsequent production phase.

14.1.2 TYPES OF SOIL

Soil is simply substance in the wrong place. It can be either simple product rests or fouling products in the form of a fouled layer and/or microorganisms. The various kinds of fouling mechanisms will produce different types of fouled layers containing one or more of the components described in Table 14.1. There are many possible classifications of soil, based on different criteria. One important classification of food processing is based on a microbiological or nonmicrobiological type of contamination. From the cleaning difficulties' point of view it is useful to classify typical soils according to their nature and structure, because this provides important information about the necessary cleaning agent to choose (Harper, 1972; Tamplin, 1980).

14.1.3 CLEANLINESS CRITERIA AND MEASUREMENT

In the search for objective procedures to control whether or not the goals of the cleaning/sanitation operation have been achieved, cleanliness criteria must be developed, as well as suitable measurement procedures. A clean surface is a surface free from residual film or soil (FID-IDF, 1979) (and

TABLE 14.1
Typical Components in Food Process Fouled Layers

Component	Solubility		Removal	Heat alterations
Sugar	Water:	Soluble	Easy	Caramelization
Fat	Water:	Insoluble	Difficult	Polymerization
	Alkali:	Poor	(good with	
	Acid:	Poor	surfactants)	
Protein	Water:	Poor	Difficult	Denaturation
	Alkali:	Good	Good	
	Acid:	Medium	Difficult	
Mineral salts				
Monovalent	Water:	Soluble	Easy	
	Acid:	Soluble		
Polyvalent	Water:	Insoluble	Difficult	Precipitation
	Acid:	Soluble		

TABLE 14.2
Methods of Determination of Remaining Soil after Cleaning

Method	Procedure
Visual	Presence of soil
	Wettability
	Adherence of gas bubbles
Optical	Adsorption (transparent surfaces)
	Reflectance
Microbiological	*In situ*
	Sterile cotton-wool swab
	Rinsing with nutrient solution
	Ultrasonic
Radiological	Incorporation of labeled tracers to product, microorganisms, or fouled layer
Gravimetric	Weight of remaining soil
Chemical	Dissolution of remaining soil and chemical analysis

in addition, free from cleaning agents) that will not contaminate food products in contact with it. If this criterion is strictly applied, we have to recognize that an absolutely clean surface will never be achieved. As will be described in Section 14.3, the removal of soil, microorganisms, and other potential contaminants is of logarithmic nature (i.e., it is impossible to reach zero remaining soil in finite time). The criteria are in this way unfortunately coupled with the sensitivity of the corresponding technical measurement procedure. Standards for so-called acceptable cleanliness have to be defined for each method (and this means frequently: the remaining soil amount must be below the method's sensitivity). If the primary concern is about microbiological recontamination, an upper limit in the number of remaining microorganisms per unit of surface will be set. This particular case is the only example, where international standards have been widely accepted, defining limits of acceptability for farm bulk tanks as 50 microorganisms/cm^2 of surface (American Public Health Association, 1960, 1972) and for industrial dairy equipment as 2 microorganisms/cm^2 (American Public Health Association, 1960; Dommet, 1968; Mckinnon and Cousins, 1969). If the cleanliness is measured through limitations in the performance of an equipment, criteria of physical nature will be used as, for example, maximum tolerable heat transfer resistance or pressure drop, or minimum acceptable flux through membranes. If these limits are reached, the cleaning has to be done much more drastically as during the routine procedures.

The effectiveness of a cleaning procedure can be judged by the amount of soil remaining after cleaning. In Table 14.2 there are some methods listed which include direct determination of total amount left (e.g., through weighing), determination of a physical property of the surface related to the soil still present (e.g., wettability, reflectance, color) and determination of some characteristic soil components (e.g., radiological count, chemical analysis). Reviews of these methods can be found in Jennings (1965), Kulkarni et al. (1975), and Corrieu (1981). The evaluation of the amount of soil remaining can be done only if internal surfaces are accessible, which means that these methods will be used mostly for laboratory cleaning tests, for periodical inspection of containers, or in disassembled equipment. Some standardized tests are described in Section 14.2.

14.2 THE CLEANING PROCESS

14.2.1 CLEANING PROCEDURE

The cleaning and sanitation procedure consists of a series of cleaning cycles with alternate composition of the medium used. When only water is used, the operation is called rinsing, and the cleaning

TABLE 14.3
Basic Steps in the Cleaning Procedure

Step	Function
Prerinse	Rinsing with water to remove grass, loose soil
Cleaning	Removal of residual soil by a suitable detergent
Interrinse	Removal and rinsing away of detergent and soil from plant items
Sanitizing	Destruction of most microorganisms by the application of chemicals with or without heat
Postrinse	Removal of sterilizing chemicals with water (of a suitable bacteriological quality) from the system

effect is provided by the mechanical action and the solventing power of water. If detergent solutions or sanitizers are used, the process is called cleaning or sanitizing, respectively. In Table 14.3, the basic steps in the cleaning procedure are shown. Some of these steps could be repeated several times, and others could be omitted during the procedure, depending on the cleaning strategy adopted.

14.2.2 Cleaning Techniques

Depending primarily on the type of the equipment to be used, several cleaning techniques have been developed. We must distinguish between the open and closed rinsing techniques (Plett and Loncin, 1985).

14.2.2.1 Open Cleaning Techniques

External surfaces, large storage tanks, transport tanks, and machine parts with complicated design will be cleaned and sanitized with relatively high labor involvement through one of the following:

COP: *Cleaning out of place*, which means dismantling of parts, soaking, and brushing.

CHP: *Central high pressure cleaning*, with pressures ranging between 15 and 120 bar, is based on shear forces applied by a spraying fluid jet and needs relatively little cleaning liquid.

CLP: *Central low pressure cleaning*, with pressures normally below 5 bar, needs high flow rates of cleaning agent; the detergent used becomes an important element.

CFS: *Central foaming system*, which is used mostly for sanitation purposes; a stable foam is applied to all surfaces and assures long contact times between detergent and/or sanitizer and the surface to be treated.

14.2.2.2 Closed Cleaning Techniques

In closed equipment consisting of pipe systems, tanks, processing apparatus, and so on, the single steps of the cleaning procedure can be carried out by pumping the different cleaning/sanitizing solutions through the equipment.

In cleaning-in-place (CIP) systems, the solutions are used only once in the single use CIP system; they are recovered and stored in the reuse CIP system. The multiuse CIP system is a combination of both systems, in which parts of equipments can be cleaned locally by smaller cleaning units located closer to the equipment. These smaller units are fed with detergents from a centrally placed reuse storage unit.

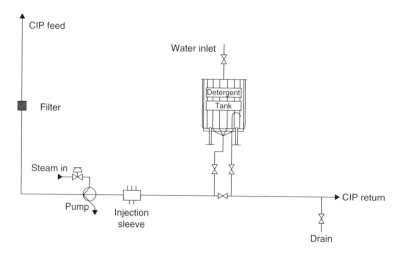

FIGURE 14.2   Flowchart of a single-use CIP system.

14.2.2.2.1 Single-use CIP system

Little space and a decentralized arrangement within the plant are specific features of a single-use system (Figure 14.2). Single-use systems manufactured in series and preassembly of all components ensure low cost, and quick installation. The principal components of a single-use system are:

1. A circulation station consisting of a frame on which the following parts are assembled: a tank with level electrodes, pneumatically controlled valves to inject steam, to admit water, and to regulate the circuit, including discharge, overflow, and throughflow.
2. A high capacity pump; a switchboard with temperature controller, solenoid valves to control the pneumatic valves, and a pressure and temperature recorder to check the progress of cleaning.
3. A set of dosing pumps for lye and acid.
4. A diaphragm dosing pump for additives.
5. A program unit. Box containing the entire program control for a fully automatic cleaning sequence.

The standard cleaning program for one tank cleaning covers, for example, a 20-min program sequence, which runs as follows:

1. Three prerinses of 20 sec with intervals of 40 sec each to remove the solid contaminations by means of a return pump to discharge the water to drain.
2. Dosage of cleaning media and injection of steam to provide the preadjusted temperature direct in the circuit. This circuit is maintained for about 20 to 12 min, after which the spent lye is discharged to drain.
3. A water circuit and injection of organic acid to lower the pH value to 5 to 4.5. Circulation of this cold solution for about 3 min and discharge to drain.

Examples of cleaning media:

1. An alkaline cleaning medium; dosage 0.65%
2. A product with a number of emulsifying, surface-active and antifoam substances; dosage 0.08% (depending on the make)
3. A liquid chloric cleaning medium; dosage 0.15% (depending on the make)

4. An acid product, consisting of four different organic acids; dosage 0.12% (depending on the make)

A simple way of lowering the water consumption in a single-use CIP system is to add a second tank to recover the rinsing water and use it for the prerinsing of the next cleaning cycle.

14.2.2.2.2 Reuse CIP system

The essential components of a CIP plant for a reuse system for lye and acid tanks; a freshwater tank; a return water tank, if required; a heating system; and pressure and return pumps. The piping system of a cleaning plant is fixed in the plant and equipped with remote controlled valves and measuring devices. A program control unit ensures automatic progress of the predetermined cleaning operations. The cleaning liquid is conveyed from the cleaning plant through the production plant, plant section, or machines to be cleaned. This requires some specific engineering when the production plant is planned.

Figure 14.3 shows an example layout of a CIP plant for a reuse system. In this case the cleaning liquid is heated in a plate heat exchanger (PHE) (Plett and Loncin, 1985). Two lye tanks are provided for the different lye concentrations, the low one for cleaning the tanks and pipes, the higher one for cleaning the pasteurizers. The concentrations of lye and acid are automatically adjusted by means of dosing pumps, which feed corresponding portions of detergent concentrate from containers into the tanks. Neutralizing tanks with automatically adjusted acids concentrations are provided to neutralize the effluents.

The plant of Figure 14.3 comprises two cleaning systems, as can be seen from the two pressure pumps (FID-IDF, 1979) and the two heating devices provided. The cleaning systems can be started independent of each other. In practice, a cleaning system is required to work on the individual cleaning circuits about 15 to 20 times a day. A cleaning system is distinguished by the liquid capacity and the cleaning program. The cleaning circuits in automated plants are switched on and off by means of remote-controlled valves from a central switchboard.

The CIP reuse unit, corresponding to Figure 14.3, incorporates some remarkable features (1) a valve combination for recirculation of hot water, (2) a return water tank, and (3) a leakage separation between the ranges of cleaning liquid and fresh water. Due to the recirculation of the hot water phase and to the return water tank, the water consumption of a reuse system can be optimized. The choice of a plate heat exchanger enables greater flexibility for different temperature requirements, allows full utilization of tank capacity, and enables versatile heating of water or other chemicals. A plate heat exchanger bypass valve can be fitted so that sterilizing solutions do not have to pass through the plate pack. The CIP reuse unit is designed to fill, empty, recirculate, heat, and dose contents automatically.

The automatic mode of the system is shown in the following examples.

1. *Prerinse operations*
   a. Either from water recovery (usual) or mains supply.
   b. Variable periods available at various temperatures.
   c. Prerinse can be either directed to drain or diverted via recirculatory loop for a timed period and then sent to drain.
   d. If required, chemical injection can be initiated to aid prerinsing efficiency.
2. *Detergent recirculation*
   a. Recirculation can be established either via the detergent vessel or the recirculatory loop.
   b. Recirculation times are variable and chemical injection can either boost the strength or utilize any combination of chemicals, as required.
   c. Recirculation can be either via the plate heat exchanger or PHE bypass loop. If a bypass valve is installed, variable-temperature programming allows for total

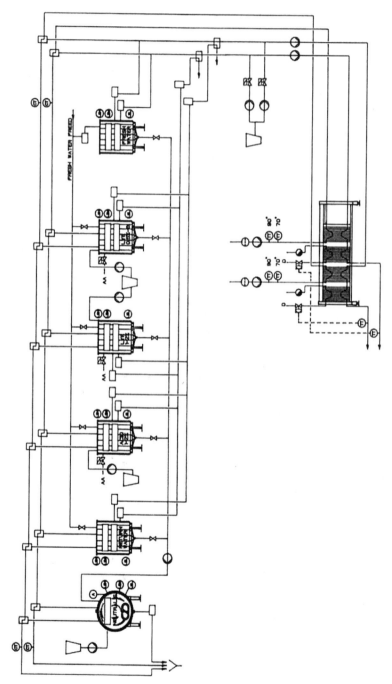

FIGURE 14.3 Flowchart of a reuse CIP system. (From FID-IDF. 1979. *Document 117: Design and Use of CIP-Systems in the Dairy Industry.* FED-IDF General Secretariat Brussels, Belgium. With permission.)

detergent tank heating or only heating the recirculatory loop, with the main detergent vessel maintained at ambient or different temperature.

  d. Chemicals can either be recovered or sent to drain.

3. *Intermediate rinse.* This is variable but similar in principle to the prerinse except that chemicals would not normally be required at this stage.

4. *Acid recirculation* (if necessary)

  a. This is variable but similar in principle to the detergent recirculation step with or without an acid tank. The recirculatory loop is established on mains water either through the PHE or via the PHE bypass loop (if included) in this case. Acid is injected to preset strength based on timing for a specific circuit volume.

  b. Recirculation time and temperature are both variable.

5. *Sterilant recirculation.* This is similar to the acid injection sequence except that heating would not be required under normal circumstances.

6. *Hot water sterilization*

  a. Recirculation loop established on freshwater tank via the PHE.

  b. Variable time and temperature available.

  c. Contents can be sent either to drain or water recovery.

7. *Final water rinse.* Mains water pumped through the CIP route and directed to water recovery; both times and temperatures are variable.

The cleaning programs for pasteurizers as well as tanks and pipes include a final flushing phase at 90°C. This has proved an efficient step to obtain the sanitary conditions necessarily required for food equipment. Another method consists of adding some disinfectant to the final flushing water.

The reuse CIP system requires higher investments compared with single-use CIP systems but lowers the running costs in complex installations and has the advantage of better personal safety by placing one unit together with detergent supply located away from the process area.

14.2.2.2.3 Multiuse CIP system

A centralized reuse system has the disadvantage of long pipework for supply and return to be installed. A large volume of liquids, heat losses, and dilution of residual product and detergents due to large mixing phases are the result. These disadvantages can be overcome by the use of decentralized multiuse CIP systems in which local small standard units are placed close to groups of equipment to be cleaned, but they are fed with detergents from a centrally placed storage unit. After having been used in the local multiuse CIP unit, the detergent is returned to storage.

The decentralized multiuse CIP system is a volume-controlled CIP unit incorporating a batch tank to which cleaning media is added. The volume of each cleaning medium corresponds to the required volume of each cleaning circuit. The batch of cleaning media is pumped through a plate heat exchanger, where it is heated and then circulated through this short cleaning circuit at a high flow rate up to 50,000 L/h.

The mode of operation enables rinse product as well as cleaning media to be effectively recovered. As the quantity of cleaning media is exactly the amount required, the consumption of water, steam, detergents, and pumping energy are reduced to a minimum. Although quite small quantities of liquids are used in this system (only a few hundred liters for even the largest vessel), it is quite common practice to retain the final rinse water for prerinse of the next clean.

When cleaning pipelines, the batch tank is used as a header. If the capacity of the batch tank is insufficient, liquid is added to it while the circuit is being filled. Once full, the circuit is cleaned in a similar manner to that of tank cleaning.

The multiuse CIP unit is designed for cleaning in place (CIP) of tanks and pipelines. It works with automatically controlled programs comprising different combinations of cleaning sequences, in which water, lye, acid, and highly diluted acid solution (acidified rinse) are circulated through the cleaning object for varying lengths of time and at varying temperatures. All the cleaning liquids

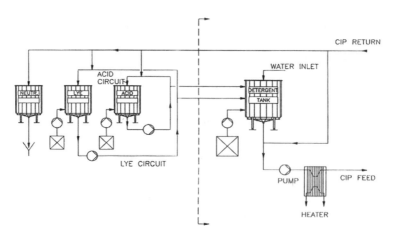

FIGURE 14.4   Flowchart of a multiuse CIP system.

are not included in each cleaning program. The decentralized multiuse CIP system is suitable for extensions of existing reuse units. Figure 14.4 shows a somewhat simplified flowchart. The multiuse unit is controlled and operated from a control panel.

The principal components of the decentralized multiuse CIP system are:

1. A batch tank in which the volume of cleaning liquid required is metered (the volume contained in the cleaning circuit plus the volume of the cleaning object).
2. A plate heat exchanger in which incoming water may be preheated and the cleaning liquid is brought to the temperature desired.
3. An acids injector by means of which a weak, acid solution is produced for the acidified rinse.
4. A circulation pump.
5. Valves of various kinds.
6. Transmitters, controllers, and so on, comprising the control system.
7. A control panel with its program unit.
8. An air blow valve for emptying (blowing clear) the cleaning circuit.
9. A lye plant with an insulated tank for prepared lye solution, a pump, and preferably tank level control and lye concentration control devices.
10. An acid plant with tank and pump for concentrated acid.
11. Piping and unions and/or valves for connecting the cleaning objects.

14.2.2.2.4 Two phase liquid/air CIP system

The conventional product pipelines cleaning method is to flush them completely, and pump cleaning and rinsing liquids through the circuit until all kind of soils that could contaminate subsequent product batches are removed. A prerequisite is that the volume of cleaning and rinsing liquids is such that the whole system is filled and the design of the circulation pumps is such that sufficient flow velocities are attained. However, there are special cases where the diameter of the pipelines has to be larger than that required for product flow. Cleaning them with the conventional CIP method would not be economical.

One of these special cases is the milkline of dairy farms . Unlike conventional product conveying pipelines in food producing plants, the milkline has to fulfil two tasks during the production phase: the first task is to transport the milk only by gravitation force, without additional pumping, from the milking unit to a central collecting tank without flushing the pipeline; the second is to feed the constant vacuum of 45 kPa[U], which is required for the milking process, through this very pipeline

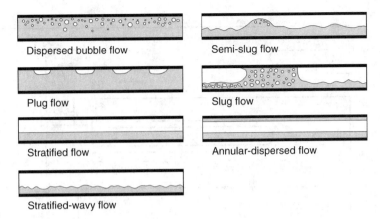

FIGURE 14.5 Diagrammatic view of the different flow images of a gas/liquid 2-phase flow in a horizontal pipeline (classified according to an increasing ratio of gas–liquid flow): dispersed bubble flow; plug flow; stratified flow; stratified wavy flow; semislug flow; slug flow; annular dispersed flow.

to the milking device. Milklines pipelines milklines may have considerable lengths (up to 100 m) and diameters of up to 100 mm. Cleaning such systems with the conventional circulating method requires extremely high volumes of cleaning and rinsing liquids as well as correspondingly designed circulating pumps, which are normally not mounted in a dairy farm.

The problem may be solved by applying the two-phase gas liquid flow system in the section of the (over-sized) milkline. It is known that a horizontally constructed pipeline forms slugs under defined conditions (Figure 14.5), for example, Brauer (1971), Hetsroni (1982), Reinemann and Mein (1994). Filling the total pipeline, these slugs rush at high velocities through the pipeline system. Due to the high velocity, the slugs have considerable amounts of kinetic energy which can be used as the component mechanics in the form of wall frictions. The underpressure produced by the vacuum system of the milking machine (Reinemann und Mein, 1995) can be used as driving force.

Graßhoff and Reinemann (1993), and Reinemann and Graßhoff (1994) described in detail the physical behavior of a slug in a horizontal pipeline, the developed forces, the moving velocity within the pipeline as well as the induced wall shear stress. Slug lengths between 1 and 7 m, moving velocities between 5 and 17 m/sec, and local wall shear stresses between 50 and 250 N/m^2 were detected in experimental measurements. Figure 14.6 displays the sketch of the experimental plant, consisting of two straight pipelines of nominal width of 73 mm each with a length of 36 m, a turn round bend of 180° in the centre, mounted with a slope of 1%, and a 27-L-receiver jar. During the experiments it has been possible to reduce the volumes of the cleaning liquids of the system (total volume 450 L) to less than 125 L. Compared to the conventional circulating cleaning method with total flush of the cleaning circuit, this means considerable savings of cleaning liquids for disposal and of energy costs. Reinemann et al. (1999) showed that the application of the two-phase gas/liquid flow technique at the cleaning of the milkline can lead to perfect cleaning results.

14.2.3 DETERGENTS

Choosing the right type of cleaner is very important. The choice depends on the chemical and physical composition of the soil (low or high viscosity, solid, nonfat, water/alkali or acid soluble, etc.) to be removed from the equipments by the CIP method. Not all the different types of chemicals available for cleaning are equally suited. Whereas acids attack mineral deposits, an alkaline medium is required for degrading organic or protein-containing substances. In food production, the soil in plants and apparatus after production are in most cases not exclusively mineral or organic. Therefore, different types of chemicals are used in as sequences of the common CIP methods. The chemicals used should trigger a chemical conversion of the deposits to be removed from the surfaces of the production

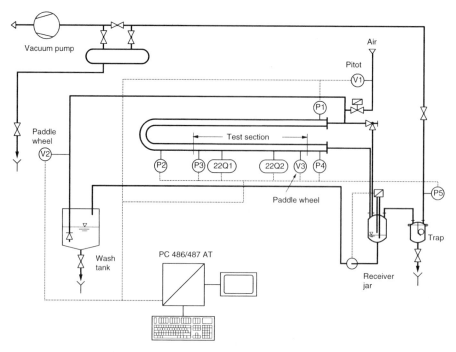

FIGURE 14.6 Experimental arrangement (plant) for cleaning a milkline with a two-phase gas/liquid flow.

equipments in such a way that the interactions with the deposits in the bulk phase of the cleaning solutions become stronger than the interactions of the deposits with the solid surface to be cleaned.

One possibility would be to induce a complete dissolution of the soil in the chemical solution as it happens, for example, with mineral salts under contact with impact with acid solutions. Another possibility would be to cover individual molecules or molecule aggregates with an electric double layer and to detach them by spacing by a few nanometers away from the solid surface or their solids bond. This process can reduce the adhesiveness between foreign substance and surface to such an extent that relatively small quantities of flow-mechanical energy, as it occurs, for example, in a cleaning solution flowing at low velocity (cm/sec), are sufficient for macroscopically transporting this chemically prepared material. The water of the chemical solutions largely contributes to the destabilization of the soil aggregates. This is due to the fact that water dipoles are incorporated into the complex protective matrix, thus inducing its loss in solidity, weakening its solid bond with the surface. Apart from this function, the water is an indispensable component of all cleaning solutions:

- As solvent for the chemicals required for soil degradation
- As transport medium for the detached and suspended soil
- As transmitter of thermal and chemical energy

An adequate selection and composition of chemicals should ensure that the soil on the cleanable surfaces are completely removed, whilst the surfaces of the equipments and apparatus are not attacked. This is the reason for using chemicals in highly diluted aqueous solution in a concentration range between 0.25 and 3% depending on their active substance.

14.2.3.1 Single Chemicals and Formulated Products

Cleaning chemicals are sold in various physical forms, for example, powder or liquid concentrate, pure substance or complex mixtures of several substances. The mixed products can be subdivided into fully formulated products and so-called component products made up of a basic substances and ingredient concentrates as a second component to be added to. According to Wildbrett (1996) alkaline component combinations have been successful in the beverage industry, for example, for the mechanical cleaning of bottles because

- Alkali and ingredient concentrates can be dosed independently of each other according to specific situations, for example, water quality, soil level, or cleaning plant
- It is more advantageous to buy the alkaline component separately than in formulated products
- The concentrates of the active substances can be stored more easily as single products than in a complex mixture

14.2.3.1.1 Alkalies
Alkaline cleaners may contain the following ingredients (Graßhoff, 1988).

- Strong alkalies like sodium hydroxide (NaOH) and/or potassium hydroxide (KOH) for degradation of organic soils (e.g., denatured protein).
- Weak alkalies like sodium carbonate (Na_2CO_3) for swelling slightly dried soil and keeping-in-solution solubilized soil.
- Complexing agents like potassium tripolyphosphate ($K_5P_3O_{10}$), sodium tripolyphosphate ($Na_5P_3O_{10}$), nitrilotriacetate NTA ($N-(CH_2-COONa)_3$), ethylen diamine tetraacetate EDTA [$N_2(CH_2)_2-(CH_2-COONa)_4$], gluconate ($CH_2OH-(CHOH)_4-COONa$), phosphonate ($H_2PO_3-C-(CH_2-COONa)_3$), polyacrylate [for further common complexing agents and their complexing properties see Graßhoff and Potthoff-Karl (1996)] for bonding water hardness (Ca- and Mg salts) and avoiding the formation of insoluble precipitations; additionally, substances as NTA or EDTA have a synergetic effect with the alkaline basic substances regarding cleaning of minerals containing organic soils. In this process Ca molecules are complexed (scaled off) out of the matrix, thus forming new surfaces within the matrix to be attacked by the alkalies one-phase cleaning process.
- Tensides (anionic: soaps, alkyl sulfates and alkyl benzene sulfates; cationic: for example, tetra alkyl ammonium- or N-alkyl pyridinium chloride; nonionic: alkyl or alkyl phenol polyethylene glycolether, fatty acid alkyloamides, and saccharose fatty acid ester) for lowering interface tension, emulsifying fat containing soils, and adjusting foaming properties (desired in foam cleaners, not desired under circulation cleaning).
- Stabilizers (e.g, phosphonates) for hot stored solutions.
- Oxidizers, for example, hydrogen peroxide (H_2O_2), sodium hypochlorite (NaOCl) for intensifying cleaning action.
- Corrosion inhibitors, for example, sodium silicate ($Na_2O:SiO_2$), sodium gluconate ($CH_2OH-(CHOH)_4-COONa$).
- Solubilizers, for example, calium compounds for stabilization of concentrates.

14.2.3.1.2 Acids
Product specific residues are formed in some sectors of the food industry, like beer stone — deposits created by proteins, hop resins, and particularly by Ca-oxalate — or milk stone mainly consisting of tricalcium phosphate. Additionally water stones form during warm water production in sites with hard water. Such deposits require acid cleaners. The efficiency of acids against mineral deposits is based on the fact that they transform the originally insoluble salts into water-soluble salts, for example:

$$Ca_3(PO_4)_2 + 4HNO_3 \rightarrow Ca(H_2PO_4)_2 + 2Ca(NO_3)_2$$

Acid cleaners may contain:

- Strong acids like nitric acid (HNO$_3$) or phosphoric acid (H$_3$PO$_4$) (without chloride, iron, and arsenic because of the risk of corrosion) for eluting mineral compounds from a complex protective matrix
- Inhibitors like urine or hydrazine compounds for binding nitrous gases
- Disperging agents for avoiding precipitations during the neutralizing process at the end of the cleaning phase
- Nonionogenic surfactants/tensides

Please note: For the acid cleaning of UHT plants only such cleaners are authorized that do not produce harmful disintegration products occurring under higher temperatures (e.g., nitrous gases for cleaners containing nitric acid).

14.2.3.1.3 Enzymatic Cleaners

In order to make the cleaning processes in the food industry as material-protective as possible, and to minimize the cleaning chemicals to be eliminated with the waste water, new cleaning processes have been developed. When using enzymes, soil degradation occurs with considerably reduced chemical amounts and almost in the neutral pH range. The detergent industry for textiles could reduce the use of chemicals and apply lower temperatures (energy saving) in the washing solutions using enzymes. In the dish washing formulae, the use of enzymes has been successful too.

In cleaning trials on milk heaters with the enzymes *Savinase*® and *Properase 1600 L*® at a concentration of 0.025% in aqueous solution, adjusted to pH 9.5 via low quantities of NaOH, Graßhoff (1999) detected a cleaning effect equalling that of 0.025% NaOH (see Figure 14.7). *Savinase*® and *Properase 1600 L*® are proteases obtained by submerse fermentation of Bacillus strains, which are used in washing agent formulae for eliminating proteinic soiling, for example, grass stains, blood, mold, faeces, and other stains like egg and sauce. Their effect consists in their ability of hydrolizing water-insoluble proteins in the stains into peptides which become completely soluble afterwards in

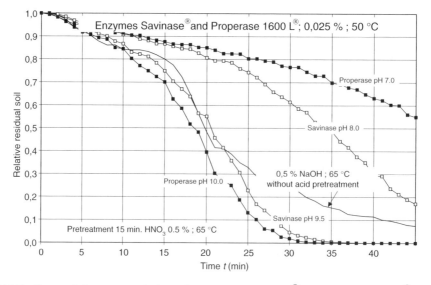

FIGURE 14.7 Removal via aqueous solutions of the enzymes *Savinase*® and *Properase 1600 L*® of encrusted fouling deposits produced during heating of milk on test elements from stainless steel. Pretreatment of the deposits for 15 min in nitric acid 0.5%, 65°C.

the washing solution. The results obtained in numerous laboratory trials with enzymatic cleaners could be confirmed for the cleaning of milk heating plants in the dairy sector.

Potthoff et al. (1997) describe an enzymatic process successfully developed and applied in the cold milk sector of a dairy (*Paradigm 2000*®).

14.2.4 SANITIZERS

There are three ways to reduce the number of living microorganisms: thermal deactivation, chemical sanitizing, and radiation. For sanitizing food equipment normally the first two methods are used. Thermal deactivation can be applied separately during a rinsing operation, for example, or can be combined with the action of a sanitizer. The basic components of detergents (alkalies and acids) also have bactericidal properties, which means that they reduce the number of living microorganisms. But for specific sanitizing processes they will be combined with other sanitizers, such as:

1. *Halogens*: active chlorine compounds (sodium hypochlorite, chlorinated trisodium phosphate), organic chlorine-release agents (dichlorodimethyl hydantoin, sodium dichloroisocyanurate), iodophors.
2. *Surface-active agents*: cation surfactants [quaternary ammonium compounds (quats), such as alkyldimethyl benzyl ammonium chloride or cetyl trimethyl ammonium bromide], amphoteric surfactants.
3. *Other components*: peroxides (hydrogen peroxide, peracetic acid), formaldehyde, ethylene oxide, monobrome acetic acid esters, ozone, labile germizide (P-propiolatone).

A great deal of knowledge is available regarding formulations, so it is advisable to consult detergent and sanitizer suppliers for tailored products in each application.

14.2.5 ENVIRONMENTAL ASPECTS

After having done their job as cleaners, the used, soiled chemical solutions have to be eliminated as waste water. This may occur in two possible ways:

- Direct discharge of the waste water into the sewage — after in-house treatment
- Indirect discharge of the waste water treatment in the municipal waste water treatment plant

14.2.5.1 Legal Regulations

For both systems the modalities on discharge are regulated in international framework agreements, for example, in the directives 91/271/EEC and 98/15/EC, whereby defining the admissible waste load is presently the task (state; December 2000) of the countries involved in the agreement. The framework agreement aims at determining internationally homogeneous and binding maximum values up to which waste water can be contaminated.

On the European level, pursuant to directive 91/271/EEC, direct discharge of (bio-degradable!) industrial waste water from industries of the sectors:

- Milk-processing
- Manufacture of fruit and vegetable products
- Manufacture and bottling of soft drinks
- Potato-processing
- Meat industry
- Breweries
- Production of alcohol and alcoholic beverages

- Manufacture of animal feed from plant
- Manufacture of gelatine and glue from hides, skins, and bones
- Malt-houses
- Fish-processing industry

requires the authorization of the local administration. The authorization can be granted only if the legal prerequisites are fulfilled (Table 14.4). In Germany, the conditions are fixed in § 7a of the Wasser-haushaltsgesetz (water management law), (German Federal Official Journal 1999, Bundesgesetzblatt, 1999).

The requirements listed in the table for directly discharged waste water can only be fulfilled by pretreatment within the own water treatment plant. If such a plant does not exist the criteria valid for indirect discharge have to be observed. In this case, the disposal of waste waters occurs via discharge into the local collecting system and treatment in the municipal waste water treatment plant.

In Germany, for example, the local authorities have fixed branch-specific reference values for indirect discharge, which follow the regulations of the ATV*-Arbeitsblatt (process flow sheet) A115 (version December 1994) "Discharge of industrial waste water into a public waste water treatment plant" (*ATV = Abwassertechnische Vereinigung — German Association for Water, Wastewater and Waste). The requirements for, for example, dairy effluents are determined in the ATV-fact sheet (explanatory note) M 708 of December 1994 (Table 14.5).

TABLE 14.4
Maximum Requirements in Germany for Directly Discharged Waste Water into the Drainage from Milk Processing Plants with a Waste Load of ≥ 3 kg BOD_5/d

Measuring parameter	Unit	Value
Chemical oxygen demand COD	mg/l	≤ 110
Biological oxygen demand BOD_5	mg/l	≤ 25
Ammonium-N (NH_4-N)	mg/l	≤ 10
Total phosphorus (P)	mg/l	≤ 2

TABLE 14.5
Requirements in Germany for Dairy Effluents at Discharge into the Municipal Waste Water Treatment Plants

Measuring parameter/ingredient	Quantity/unit
pH	6.5 to 10
Temperature	$\leq 35°C$
Nonvolatile lipophile substances (saponifiable fats)	≤ 250 mg/L
Directly precipitable hydrocarbons	≤ 50 mg/L
Absorbable organic halogen compounds (AOX as Cl)	≤ 1 mg/L
Nitrogen from ammonium and ammonia (N)	
at ≤ 5000 equiv. inhabitants	≤ 100 mg/L
at >5000 equiv. inhabitants	≤ 200 mg/L
Total phosphorus (P)	≤ 15 mg/L

The criteria listed in the above mentioned table have to be satisfied at the latest on the point of discharge in to the municipal collecting system. This means that, on company level, the solutions used during the individual cleaning phases (alkaline/acid) have first to be converted by neutralization into the pH-range between 6.5 and 10. This means that from the point of view of ecological assessment the compounds created during neutralization, and not the initial products, are relevant. These compounds mainly consist of sodium carbonate and sodium hydrogen carbonate like the sodium salts of the nitric and phosphorous acid. Calcium nitrate formed by the acidic impact on the deposits in the pasteurizer plates and sodium phosphate formed from phosphate-containing alkaline cleaners or from disintegration of the often used polyphosphates also enter waste waters directly. Indications of the load of phosphorus and nitrogen in waste waters from dairies were shown by a field study of Wildbrett and Böhner (1990a,b).

14.2.5.2 Biological Degradability

The behavior of the aforementioned substances in the water treatment plant is quite different (see Schindler-Stokar, 1992). Sodium carbonate goes through a two-step water treatment plant almost unchanged and enters the drain. Sodium phosphate is eliminated to approx. one-third in the mechanical–biological water treatments plants. Subsequently, 80–90% of the remaining 65% can be eliminated from the waste water in a separate chemical phosphate precipitation step. Sodium nitrate, the salt of the nitric acid, is degraded to a certain level (40–70%) and serves as a nitrogen source for the microorganisms in the water treatment plants.

An efficient complex-forming agent represents an essential component of a chemical cleaner for the milk pasteurizer. The polyphosphates are most frequently used for cost cutting reasons. Substitutes for the polyphosphates are searched due to the risk of water eutrophication (for water treatment plants having no phosphate precipitation step for treating phosphate-containing waste waters). The environmental behavior of the actually used phosphate substitutes is still quite controversially discussed:

- Ethylendiamintetraacetic acid (EDTA) is hardly biodegradable and thus enters the drain unchanged. Due to its excellent complex-forming properties it remobilizes heavy metals both in the sludge and in the sediment of the drain and conveys them into the waters. The chemical industry undertakes intensive efforts for detecting EDTA-substitutes (Graßhoff and Potthoff-Karl, 1995). In a study performed at the university GH Paderborn/Germany it has been possible to isolate an EDTA-degrading microorganisms with which EDTA-containing waste waters from the cleaning of the pasteurizer could be biodegraded to approx. 90% (Henneken et al., 1993). For the time being, the method is further developed and the technical applicability investigated (Otto et al., 2000).
- Nitrilotriacetic acid (NTA) has good biodegradable properties (Gousetis, 1991). In Germany, the Federal Ministry for the Environment, Nature Conservation and Nuclear Safety recommends NTA as a biodegradable substitute for EDTA. Fears that increased use of NTA could have negative effects on the environment have not been confirmed up to now. Nevertheless, research activities focus on detecting substitutes for NTA.
- Methylglycindiacetic acid (MGDA) is an easily biodegradable complex-forming agent newly developed on the basis of iminodiacetic acid (Potthoff-Karl, 1994). Graßhoff and Potthoff-Karl (1996) detected an effect similar to NTA in a study on of the efficiency of complex-forming agents being relevant for the formulation of alkaline cleaners for milk pasteurizers. For the time being, MGDA is tested on applicability by several manufacturers of industrial cleaners.
- Citric acid has an excellent biodegradability. It shows a good complex-forming ability for many metal ions and is thus getting increasingly important as an ecologically and physiologically harmless component of cleaning, deliming and rust-removing agents

as well as of water treatment products. As its complex-forming ability decreases under high temperatures, the citric acid is of no importance for the formulation of cleaners for pasteurizers.

- Gluconic acid has also a good biodegradability. It is a mild organic acid naturally occurring in foods (e.g., wine, honey) and belongs to the chemical group of the polyhydroxycarbon acids. The gluconic acid is said to have a strong complex-forming ability in the alkali medium as to calcium, iron, copper, and other heavy metals. In alkaline solution up to 15% NaOH it is resistant against boiling temperatures and is thus often used as a component for enhancing the effects of phosphate and polyphosphates in liquid formulated alkaline cleaners.

- Phosphonic acids and their derivates have a relatively low biodegradability. At temperatures up to 100°C they show a good complex-forming ability, whilst the formed complexes are considerably more stable than those of polyphosphates. Therefore, they are used, among others, for stabilizing of peroxi compounds and for formulating liquid alkaline cleaning concentrates, and have proved to be adequate complex-forming agents for CIP solutions which are to be hot-stacked stored for a longer period. Phosphonic acids can be used in the drainage by different algae as a phosphorus source. A remobilization of heavy metals cannot be excluded.

To intensify the cleaning effect, confectionated cleaners often contain additives with highly oxidizing effects (see above). In this context Gutknecht (1993) mentions the special risk for the waste water load by cleaning and disinfecting agents of the food industry at using AOX-producing products as chlorinated bleaching solutions and chlorine-containing cleaners. Therefore, he recommends the use of active oxygen products.

14.2.5.3 Regeneration of Cleaning Solutions

A significant decrease in the consumption of chemicals may be expected from methods which regenerate soiled cleaning solutions. In a bottle washing plant, during a 9-month permanent trial Ackermann et al. (1987) could reduce the waste water volume by 80% only by sedimentation of the solution during the week-end, disposal of the sedimented sludge (20%), and reuse of the carrier medium (80%). Thus, for the above mentioned process, compared to the practice of disposing the whole solution after a 1-week use, they observed a reduction of the waste water volume by 80% and of the consumption of chemicals by 20%. Under the influence of progressively rising costs for fresh water and for waste water disposal (imposition of load-proportional fees) flocculation and membrane methods become increasingly important.

Titz (1993) investigated the suitability of different flocculation agents for accelerating sludge sedimentation and reducing sludge volume in bottle-washing and evaporation solutions from dairy factories. She attained very good results with a number of commercial flocculation agents. With the development of temperature-resistant (up to 80°C) and chemical-resistant membranes (NaOH, KOH, H_3PO_4, HNO_3 in the concentration range 2–10%, pH-range 0–14) with high permeation performance the recycling of cleaning solutions with membrane methods is an interesting alternative. Apart from the insoluble soiled components and suspended substances this method allows to remove more than 90% of the dispersed and soluble organic soiling (proteins, carbohydrates, fat, etc.) from the used cleaning solution. Thus, the described method (after supplementing the chemicals used by reaction) can be applied practically without any limits. The consequence is a reduction of the waste water volume and thus a reduction of the fresh water and energy needs. The suitability and profitability of crossflow-filtration methods for treating cleaning solutions was thoroughly investigated by Henck (1993) and positively assessed.

Hofer (1993) made trials with CIP-solutions from dairies for detecting the regeneration efficiency. These solutions were liberated and cleaned of dirt and suspended solids by ultrafiltration. Hofer did not find a measurable reduction of the cleaning efficiency on the test deposits on plate pasteurizers in comparison with the fresh original solution. Only during the first phase of scaling-off, an analogy to the decrease of the pore width of the filtration membrane was stated. Presumably, membranes with a low separation limit induce a partial separation of higher molecular surfactant and complex-forming substances and thus a minimal reduction of the scaling-off activity. Trägardh and Johansson (1998) performed membrane filtration trials with different alkaline cleaning solutions (with synthetic and genuine dirt loads) and found that no NaOH was retained in any of the cases, whilst part of the additives brought to completion (e.g. complex-forming agents) were depleted. In permeate a 50% reduction of COD was detected. Novalic et al. (1998) found a reduction of the COD between 87 and 98% by nanofiltrating highly soiled cleaning solutions (acid and alkaline) coming from the evaporator or pasteurizer cleaning at dairies. From the alkaline solution, 90 to 100% of the NaOH could be recycled under flux rates 0.9 to 1.4×10^{-5} m^3/m^2 sec. The values of flux and COD retention ability obtained by nanofiltration were unsatisfactory if an acid solution (2% HNO_3) was used *prior* to the alkaline solution during the cleaning process, whereas in an acid solution used *after* the cleaning with the alkaline solution a reduction of the COD reaching up to 93% under good flux values was observed. Today, membrane filtration plants with chemical and temperature resistant membranes of different types are available on the market.

14.3 CLEANING KINETICS

The net mass flux (dm/dt) in fouling and in cleaning of a surface can be considered as the difference between a total deposition mass flux and a total removal mass flux (Kern–Seaton equation). The way cleaning is technically achieved consists simply of enlarging the removal term drastically while reducing the deposition or redeposition term. This is normally done on the one hand, by replacing the product (which is prone to foul) by water or a detergent solution and, on the other hand, by physically, physicochemically, or chemically changing the fouled layer properties using detergents, higher temperature, and/or mechanical forces to assist the mechanisms responsible for detaching the soil.

14.3.1 GENERAL REACTION SCHEME

The cleaning reaction is a heterogeneous reaction between the detergent solution and the fouled layer (Figure 14.8). As a heterogeneous reaction it can be divided into mass transport and reaction processes. There are reactions occurring in the bulk and in the fouled layer. For better understanding of the cleaning kinetics, the sequence of mechanisms acting on the stream of cleaning solution molecules entering the apparatus to be cleaned will be discussed here up to the point where they leave the apparatus again through convective transport.

14.3.1.1 Bulk Reaction

The detergents in the bulk can react with soil dispersed or dissolved in the bulk or with the water hardness. If this reaction is extensive, a large amount of cleaning agent will be consumed in the bulk and will not be available for cleaning. This means especially that a good-quality potable water supply must be used and that the detergent formulation must include complexing and sequestering agents selected for the specific water hardness conditions existing on site.

14.3.1.2 Transport to Surface

The mass transport of detergents through the boundary layer will be due to turbulence or concentration gradients (eddy and molecular diffusion, respectively) and adsorption on the liquid–solid interface.

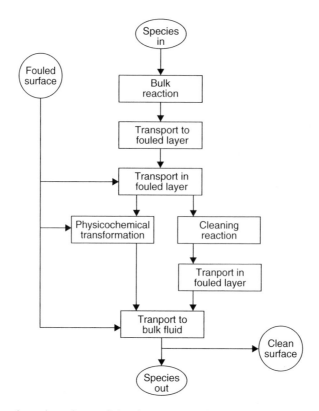

FIGURE 14.8 General reaction scheme of cleaning seen as an heterogeneous chemical reaction.

14.3.1.3 Transport into Fouled Layer

Depending on the structure of the fouled layer and the properties of the cleaning agents, the detergents will be transported into the fouled layer, building a whole reaction zone rather than reacting only at the interface. The detergents can be transported into the fouled layer by capillary or molecular diffusion. Surface-active agents have the ability to penetrate through pores and crevices due to the lower surface tension and adsorption characteristics and to adsorb on the surface weakening the soil–surface bonds.

14.3.1.4 Cleaning Reaction

The processes occurring when the cleaning agents come in contact with the fouled layer can be subdivided into the physicochemical transformations and chemical reactions. The physical and physicochemical transformations occurring are, among others melting, mechanical and thermal stress, wetting, soaking, swelling, shrinking, solvatation (as dissolution of soluble substances), emulsification, deflocculation, and desorption. The chemical reactions involved are hydrolization, peptization, saponification (to a lesser extent), solubilization (as forming soluble salts through chemical reaction), dispersion, chelation, sequestering, and suspending. These reactions will help to overcome the cohesion forces of soil–soil bonds and adhesion forces of soil–surface bonds (or detergent–surface bonds).

14.3.1.5 Transport Back to Interface

Smaller molecules formed in the reaction zone will diffuse through the fouled layer due to concentration gradients.

14.3.1.6 Transport to Bulk

The cleaning reaction products will be transferred through the boundary layer due to concentration gradients (back diffusion) or turbulence (eddy diffusion). When the mechanical and thermal stress and other physical, physicochemical and/or chemical processes have weakened the soil–surface or soil–soil bonds, larger sections of the fouled layer can be detached through spalling. On the other hand, the flow of the cleaning solution under turbulent conditions will cause erosion of the fouled layer. This mechanical cleaning mechanism will be enhanced if the soil–soil bonds are weakened through physical changes or due to the reactions mentioned in Section 14.3.1.4.

Listing this sequence of six steps does not imply that they always have to occur. In specific cases, some steps will be bypassed (see Figure 14.5). When cleaning, equipment with a fat soil, for example, it might be necessary to melt the fat (thereafter, the oil could simply be eroded by the hot water), or add some detergent to emulsify the oil or dissolve it with a solvent.

14.3.2 Parameters Influencing the Cleaning Rate

The rate at which the soil is eliminated during cleaning has been classically described as being logarithmic in nature (Jennings, 1965), which would suggest a first-order reaction. Recent research work on cleaning of industrial systems showed better accuracy by using a kinetic model with a coupled reaction (Plett and Loncin, 1981a, 1981b; Gallot-Lavallee 1982; Gallot-Lavallee et al., 1982; Corrieu et al., 1980):

$$\text{soil} + \text{detergents} \underset{k_{-1}}{\overset{k_{+1}}{\rightleftharpoons}} \text{complex} \overset{k_1}{\Rightarrow} \text{suspension} \tag{14.1}$$

In this kinetic model, the soaking and solvation steps are considered reversible first-order reactions, while the removal term is postulated as an irreversible first-order reaction. If the second reaction (suspending, erosion) is the total rate-limiting step, we can conclude that the cleaning rate will be logarithmic (i.e., the loosening of soil is governed by a given probability and it will be impossible to reach an absolute zero soil amount in a limited time). This is similar to the thermal deactivation of microorganisms, the sterilization process, where there is no absolute zero remaining contamination, but there is a certain probability of surviving microorganisms. The conclusion of this behavior is that for practical purposes cleanliness has to be defined as a maximum tolerable amount of remaining soil. Thus the cleaning time will be the time necessary to reach this preset limit of remaining soil and will depend on the cleaning rate.

When discussing the influence of different parameters on the cleaning rates, as reported in the literature, it must be kept in mind that individual authors use particular mathematical models to express their results. This means that the comparison can be done only qualitatively and will be made in this section according to the mathematical relations published by several researchers. The parameters influencing the cleaning rates are classified in three categories:

1. System parameters
   a. Nature of fouled layer
   b. Conditions of fouled layer
   c. Initial soil amount
   d. Water hardness
   e. Soil in the cleaning solution
2. Operational parameters
   a. Detergent nature
   b. Detergent concentration

      c. Temperature
      d. Mechanical action
   3. Design and construction parameters
      a. Nature of surface
      b. Finishing and condition of surface
      c. Equipment design

14.3.2.1 System Parameters

As stated previously, the nature of the fouled layer and of the surface will determine the strength of surface–soil bonds. In the same way, the nature and conditions of the fouled layer will determine soil–soil cohesion forces. This is the point where the interrelation between fouling and cleaning becomes more evident. The intensity of the cleaning operation will depend on the nature and conditions of the fouled layer. Efforts to prevent, diminish, or alter the fouled layer can therefore be more effective than any improvement in cleaning procedures. On the other hand, incomplete cleaning can shorten production time because of premature fouling (no fouling induction period). According to the nature and composition of soil, specific types of detergents must be used (see Table 14.1) to achieve the desired cleaning reactions. In this area much more applied experimental work is still needed because normal production fouling conditions are seldom used in laboratory experiments, so that the extrapolation of cleaning results is almost impossible. The discrepancies in the results published are certainly due to different soiling procedures and conditions rather than because of the various measuring techniques used.

If a first-order kinetics can be applied, no influence of the initial soil amount S_0 is expected. Nevertheless, Gallot-Lavallee (1982) found a linear decrease of the first order cleaning rate k_1 with increasing initial milk soil amount. Schlüssler (1970) reports no influence of the soil amount on k_0 for concentrated milk and buttermilk but linear increasing of k_0 for cacao and milk soils. He explains this behavior with different drying characteristics of these products when preparing the soiled samples. During industrial operations, the increase in the fouled layer thickness goes along with a change in the soil properties because of two reasons: first, them are time-dependent aging reactions going on among the adhered species, and second, if the amount of heat transferred has to be maintained constant (e.g., process controlled by the product exit temperature), the surface temperature and hence the temperature in the layer will be raised to compensate the increasing heat transfer resistance. These reactions can change the adherence of the soil, as it can be changed by drying of the soil. From the thermodynamic point of view, the adhesion forces that have to be overcome by the cleaning procedure are chemical and physical bonds. The physical bonds can be subdivided into material bridge, and gravitational, electrostatic, van der Waals, and capillary forces. Gravitational and capillary forces will be given by the radii of these particles, whereas electrostatic and van der Waals forces depend largely on the media involved and thus can be influenced during the cleaning procedure. Regarding electrostatic forces, the opposite sign will result in attraction and the same sign will result in repulsion.

Electrostatic forces have a wider range than van der Waals forces, thus in the case of the same sign charges, the particles will be repulsed in a distance range where the van der Waals' attraction forces are negligible. The effect of the medium between the surface and particles on the van der Waals forces can be seen in Table 14.6. If soil is dried, the water molecules between the surface and particles evaporated, resulting in high attraction forces, which will make soil removal difficult. This means, on the other hand, that laboratory experiments in which soiling is achieved by drying of liquids are not good models for fouling layers generated under industrial conditions.

Another system parameter to be considered is water hardness. Calcium and magnesium salts will react with the detergents in a bulk reaction, thus lowering the concentration of detergents available for cleaning the surface. On the other hand, water hardness will cause precipitation fouling, which also has to be removed; this is especially important regarding the rinsing procedure. For rinsing, only soft

TABLE 14.6
Van der Waals Forces between Surface and Particles
Separated by Water and by Air

		Particle	
		Hydrophilic (ions)	Hydrophobic (protein, fat)
Medium: water			
Surface	Hydrophilic (metal)	(Electrostatic)	Small
	Hydrophobic (plastic)	Small	High
Medium: air			
Surface	Hydrophilic (metal)	(Electrostatic)	High
	Hydrophobic (plastic)	High	High

Source: Data from Nassauer, J. 1985a. *Habilitationsschrift Techn.* Universität München, Nassauer, J. 1985b. *Grundlagen-Anwendung-Probleme.* Freising-Weihenstephan, Eigenverlag.

water of acceptable bacteriological standards may be used. Water hardness will increase detergent costs, because complexing and sequestering agents have to be added to the detergent formulation.

A high amount of soil in the cleaning solution can cause further decrease in the concentration of available detergent, because detached soil will participate in the bulk reaction. In reuse CIP it is advisable to separate soil flakes mechanically to avoid wasting detergent in dissolving these soil particles.

14.3.2.2 Operational Parameters

The reactants in the cleaning reaction are soil and detergents. The detergent could therefore be considered a system parameter, too, but since it can easily be changed by the plant operator, it makes more sense to classify it as an operational parameter. The detergent nature will he chosen according to the fouled layer composition: roughly, alkalies for organic soil and acids for mineral solids. This is the point where the know-how of the detergent manufacturer/supplier is very valuable. Formulated detergents will consist of a blend of those components cited in Section 14.2.3 to enhance wetting, dispersing, suspending, sequestering, emulsifying, dissolving, peptizing, or hydrolizing characteristics, since no single chemical compound can optimize all these properties. Some surfactants are able to penetrate between surface and soil, loosening the fouled layer adhesion so that larger soil flakes will detach without having to be dissolved. Detergents provide the necessary chemical energy that must be supplied to the system to reverse the natural state of a fouled surface. One way to increase this form of energy supply, besides choosing the right deterrent, is to raise the detergent concentration. The effect of the detergent concentration (mostly alkalies) on the cleaning rate has been a subject of study for a variety of researchers.

In Table 14.7 some of the results are reported, these can be summarized to indicate that initially an almost linear increase of cleaning rates with raise in concentration, passes over to a region where a maximum rate is obtained. In some cases even a still unexplained decline in cleaning rates has been observed when the concentration is too high. In this case, more will not yield to better. Reviewing the mechanisms involved in cleaning, as shown in Figure 14.1, a decrease in the overall Figure 14.1, a decrease in the overall cleaning rate at high detergent concentrations can only be due to transport inhibition. A moderate increase in the solution viscosity and/or a change in molecular weight of the cleaning reaction products have no significant effect on the molecular diffusion

TABLE 14.7
Influence of Detergent (C_{OH^-}) Concentration on Cleaning Rates

Model*	References
$k_1 = a\,C_{OH^-}$	Jennings, 1959b, 1963b, 1965
k_0 nonlinear with maximum	Schl üssler, 1970
$\ln k_{01} = a + b\,C_{OH^-}$	Gallot-Lavallee, 1982
$\ln k_{12} = a + b\,C_{OH^-} - c\,C_{OH^-}^2$	Gallot-Lavallee et al., 1984
$\ln k_1 = a + b\,C_{OH^-}$	

* a, b, c are positive constants.

TABLE 14.8
Effect of Temperature on Cleaning Rates

Model[a]	Q_{10}	Temperature (°C)	Reference
$\ln k_1 = a - b/T$	1.6	46–82	Jennings, 1959b
$\ln k_0 = a - b/T \quad T < T\,\text{max}$	2–4	5–100	Schlüssler, 1970
$\ln k_1 = a - b/T$	1.1	40–80	Hoffmann and Reuter, 1984b
$\ln k_{01}, \ln k_{12}, \ln k_1 = a + bT$	1.7	55–95	Gallot-Lavallee, 1982

[a] a and b, are positive constants.

coefficient. Most transport is due to the fast and extensive swelling of the outer layers, of soil at high detergent concentrations, thus diminishing the effective cross area for molecular diffusion in the soil matrix.

The second form of energy that can he supplied to the cleaning process is heat. Raising the detergent solution temperature will generally enhance the cleaning rates up to a point where other reactions cause heat alterations that will cause difficult cleaning (see Table 14.6).

The temperature effect on cleaning rates is usually expressed in the form of an Arrhenius-type equation, seen in Table 14.8. The overall heat sensitivity of the cleaning reaction rates with the broad Q_{10} value range reported (between 1 and 4) is situated between the activation energy E_a of diffusion processes (10 to 25 kJ/mol) and that of chemical reactions with E_a up to 150 kJ/mol. This leads to the conclusion that depending on the specific experimental conditions in some cases the overall reaction-rate-controlling step is diffusive transport or physicochemical transformations (low E_a) and, under different conditions, the chemical reaction between detergent and soil becomes limiting (high E_a).

The presence of an optimum cleaning temperature beyond which the cleaning rates decline due to heat alterations of some soils implies that the simple Arrhenius model is not valid over the whole range of applicable cleaning temperatures. A thermal degradation process will be overlapping the cleaning process. This is an area where much more systematic research is needed to understand the mechanism involved in the cleaning process.

The third form of energy input into the system to be cleaned is represented by the mechanical energy. The mechanical action is provided by either shear forces acting on the fouled layer or by the turbulent burst action of circulating detergent solutions in the case of CIP. In Cleaning Out of Place, the bristle brush has been the main device to scratch and shear the soil, despite some sporadic

TABLE 14.9
Models Expressing the Mechanical Effect on Cleaning Rates

Relation[a]	τ_W wall shear stress range (Pa)	Reference
$k_1 = a + b\mathrm{Re}, \quad \mathrm{Re} > 25{,}000$	>0.4 Pa	Jennings et al., 1957
$\ln k_{01} = a + bV$	$(0.3 < V < 1.9$ m/sec)	Gallot-Lavallee, 1982
$\ln k_{12} = a + bV + cV^2$		Gallot-Lavallee, 1982
$\ln k_1 = a + bV$		Gallot-Lavallee, 1982
$k_1 = a + b\tau_W$	1–14	Timperley, 1981
$\ln k_0 = a + b\tau_W$	0.19–7.5	Graßhoff, 1983b
$\ln S = a - b\ln \tau_W, \tau_W > 0.8$ Pa	0.41–1.3	Jackson and Low, 1982
$\ln S = a - b\tau_W$	1–50	Hoffmann, 1984b

[a] a, b, and c are positive constants.

use of ultrasound for small items (Schlüssler, 1976). Loncin and Merson (1979) state that in spray cleaning the kinetic energy of droplets reaching the wall is important in perpendicular impact, while droplet momentum is decisive in tangential impact on the wall. Although all researchers found an improvement in cleaning rates with increasing mechanical action, how to characterize this effect better has been the subject of controversy. The first role of thumb established was to advise a cleaning fluid velocity higher as approximately 1.5 m/sec (Loncin and Merson, 1979; FID-IDF, 1979). Jennings et al. (1957) postulated a minimum Reynolds number of 25,000, above which the mechanical effect is remarkable. Hankinson and Carver (1968), and Schlüssler (1970) also use the Reynolds number. Timperley (1981) shows that the number of remaining microorganisms after cleaning in place of stainless steel pipes is much better correlated with the mean velocity and hence the wall shear stress, than with the Reynolds number. The same opinion is expressed by Graßhoff (1983b), who relates the remaining fat soil amount to the wan shear stress in acrylic pipes. Jackson and Low (1982) found a shear stress threshold of 0.8 Pa, from which point on there is a remarkable mechanical effect in CIP of a plate heat exchanger [the Re = 25,000 threshold found in Jennings et al. (1957) for pipes corresponds to a shear stress of 0.4 Pa].

Some of the mathematical models developed to correlate the effect of mechanical action on the cleaning rates reported in the literature were summarized in Table 14.9. The tendency goes clearly toward the use of wall shear stress, which can be related in a better way to the mean velocity than to Reynolds numbers (Timperley, 1981). High shear stresses can be induced through turbulent flow conditions (i.e., that turbulence promoting design will improve the cleaning effect). Based on considerations regarding the time fluctuating characteristics of turbulent flow, it might be possible that average wall shear stress does not represent the best parameter to express abrasive effects of turbulent bunts on the fouled layer but that the amplitude and frequency of shear stress peaks play a much more significant role. Graßhoff (1983a) reports an increase in cleaning rates by a factor of 9 when the flow direction over the fouled layer is inverted every 5 or 15 sec. The pressure loss measurements and corresponding calculations revealed pressure oscillations after direction switching, with a maximum amplitude between 19 and 4 times larger than the average pressure loss during one-directional flow. These superimposed shear stress peaks can accelerate the soil removal after the fouled layer cohesion has been weakened by the cleaning reaction.

The cleaning time has not been considered here as an operational parameter, although it is in principle a free variable for a plant operator or the programmer of an automatized cleaning procedure. In addition, the necessary cleaning time is regarded here as the result of cleaning dynamics (determined by all parameters influencing cleaning kinetics), in order to achieve the cleaning result desired, which is measured through a proper cleanliness criterion.

14.3.2.3 Design and Construction Parameters

These influencing parameters have to be considered very carefully when planning and building equipment because they determine not only capital costs but also operation costs, due to their effect on fouling and cleaning dynamics. Their importance is of such magnitude that a great deal of work has been done to define and find what a hygienic design for all processing equipments, product storage, and transportation means, this will be explained separately in Section 14.4.

14.4 HYGIENIC DESIGN

14.4.1 PRINCIPLES OF HYGIENIC DESIGN

The prerequisite for obtaining perfect cleaning results with a cleaning process is that the object to be cleaned — ranging from the cleaning of the simple pipeline to that of the complex filling line — the whole processing and treatment plant have to be designed in such a way that they are cleanable. Cleanability means that the cleaning process should clean the object in such a way that all sources of microbial or chemical contamination (from lat.: contaminare = intensively contacting, soiling) are eliminated and that a product in direct or indirect contact with the object does not undergo (unintended) alteration immediately or later on. The expression *hygienic design* describes the compliance with the requirement to cleanability.

The requirement for the design of equipments fulfilling the criteria of cleanability in the nutritional technical sector had first been expressed in the 1920s when, in the USA, the Milk Industry Foundation (MIF) and the International Association of Food Industry Suppliers (IAFIS) together with the International Association of Milk, Food, and Environmental Sanitarians (IAMFES) initiated a concept but without obligation to be observed in the design of equipments — the so-called 3-A standard, which has been permanently amended up to the present day. By changing the manual cleaning methods by the automated CIP technique in the sixties the legislators of the industrial nations felt compelled, worldwide, to make regulations for design and operation of technical equipments in the food sector to protect the consumer and his basic right on body health in the consumption of food. In Germany, for example, it has been determined in the German Act on Food and Commodities (LMBG) of 1974, new version 1993 (Bundesgesetzblatt, 1993) in Article 5 (1) and Article 31 (1) that

> "Artifacts which are to be used in manufacturing, processing, handling and offered for sale or supply of foods to the consumer, and thus come into contact with the foods or affect them, have to be used in such a way that no constituent migrate into the food or its surface, except when considered to be safe with regard to consumers' health and acceptable with regard to affecting food odor and taste and only in such amounts which are technically unavoidable".

On the national level similar regulations were issued by most of the industrial nations (World Health Organization 1981, 1982). However, they were not all that helpful to constructors of equipment as they had a so-called exclusion character, that is, they determined what has not to be. None of these regulations — neither the Council Directive 93/43/EEC — contain provisions *how* such cleaning modules have to be constructed. Therefore, initiated by mechanical engineers, mainly empirical experiences have been gathered over the years and summarized in standards, for example, in the (more than 70) 3-A standards, in the DIN standard 11 480 (1978), in the ISO standard 14 159 (1999) in the DIN standard 1672-2 (1997) or in the bulletin No. 218 (1987) of the International Dairy Federation (IDF) especially dedicated to the dairy industry.

Due to the lack of practicable standards for (preliminary) monitoring of the hygienic design of components and of complete equipment for the production process an international expert panel with representatives of the mechanical industry, of users, of representatives of the legislative power, and of scientific research institutes was created in 1985. It is called European Hygienic Engineers Design Group (EHEDG) with 23 subgroups (January 2001):

Design Principles Group	Group for Monitoring of Plant Conditions,
Test Methods Group	Inspection and Maintenance
Materials Group	Air Handling Group
Group for Continuous Heat Treatment of	Conveyors Group
Food Products	Cooling and Chilling Equipment Group
Packing Machines Group	Dry Materials Handling Group
Group for Pumps, Homogenizers, and	Electrical Installation Group
Dampening Devices	Food Transport by Road Group
Valves Group	Hygienic Systems Integration Group
Pipe Couplings Group	Lubricants Group
Training Group	Mechanical Seals Group
Sensors Group	Process Water Group
Welding Group	Risk Assessment Group
Building Design Group	Slaughterhouses Group

The goal of EHEDG with its subgroups is to issue standards on the basis of secured knowledge and in cooperation with other international institutions like CEN (Comité Européen de Normalization), ISO (International Standards Organization), ANSI (American National Standards Institute), and the affiliated organisms FDA (Food and Drug Administration), NSF (National Sanitation Foundation) and USDA (United States Department of Agriculture), and, if required, to initiate specific scientific research studies.

14.4.1.1 Nature of Surface — Materials

The nature and conditions of the surface, among other factors, will determine the magnitude of soil-surface adhesion forces. Schlüssler (1970) compared the mean cleaning rates (k_o) of milk deposits on different materials. Relative ratings for glass, polished aluminum, and stainless steel were 100%, while tin had 92%, brass 52%, and copper 39%, respectively. The corrosion resistance of the construction materials has to be taken into account, otherwise the surface properties and conditions will deteriorate with time, producing more active sites for fouling. Nassauer (1985) could correlate the amount of quaternary ammonium compounds left after rinsing with the Galvani potential difference measured between the water and the duct wall made of different metals and plastics.

14.4.1.1.1 Metallic Surfaces — Stainless Steel

Product contact surfaces, that is, all the surfaces exposed to direct contact with the product as well as indirectly impacted surfaces from which sprayed product components, condensate, liquid or solid particles may run off, drop off, or may fall in the product in some other way should basically be made of materials meeting the following requirements:

- Corrosion proof to the product and to common cleaning and disinfecting chemicals
- Nontoxic, physiological, and organoleptic indifference
- Suitable for surface treatment, shaping, and welding ability
- Temperature proof
- Availability

Since the first stainless steel variety producible on a large technical scale had been patented in Germany in the year 1912 under the name Nirosta V2A more than 500 different stainless steel varieties have been developed worldwide. Approx. 60 varieties are standard steel types used internationally,

TABLE 14.10
Most Important Varieties of Stainless Steel with Their National Standard Designation

Material no. DIN 17 007	Euro standard 88-71	Germany DIN 17 440	France AFNOR 35-572-75	USA AISI 176-75	U.K. BS 970	Sweden SIS — 14
1.4301	X6 Cr Ni 18 10	X5 Cr Ni 18 9	Z6 CN 18.09	304	304 S 15	2333
1.4303	X8 Cr Ni 18 12	X5 Cr Ni 19 11	Z8 CN 18.12	305	305 S 19	—
1.4305	X10 Cr Ni S 18 10	X10 Cr Ni S 18 9	Z10 CNF 18.09	303	303 S 21	2346
1.4306	X3 Cr Ni 18 10	X2 Cr Ni 18 9	Z2 CN 18.10	304 L	304 S 12	2352
1.4310	X12 Cr Ni 17 7	X12 Cr Ni 17 7	Z12 CN 17.07	301	301 S 21	2331
1.4401	X6 Cr Ni Mo 17 12 2	X5 Cr Ni Mo 18 10	Z6 CND 17.11	316	316 S 16	2347
1.4404	X3 Cr Ni Mo 17 12 2	X2 Cr Ni Mo 18 10	Z2 CND 17.12	316 L	316 S 12	2348
1.4435	X3 Cr Ni Mo 17 13 3	X2 Cr Ni Mo 18 12	Z2 CND 17.13	316 L	316 S 12	2353
1.4436	X6 Cr Ni Mo 17 13 3	X5 Cr Ni Mo 18 12	Z6 CND 17.12	316	316 S 16	2343
1.4541	X6 Cr Ni Ti 18 10	X10 Cr Ni Ti 18 9	Z7 CNT 18.10	321	321 S 12	2337
1.4550	X6 Cr Ni Nb 18 10	X10 Cr Ni Nb 18 9	Z6 Cn Nb 18.1	347	347 S 17	2338
1.4571	X6 Cr Ni Mo Ti 17 12 2	X10 Cr Ni Mo Ti 18 10	Z6 CNDT 17.12	320	320 S 17	2344

TABLE 14.11
Chemical Composition of the Stainless Steel Types Mentioned in Table 14.10 (Indications in %)

Material no.	C max.	Si max.	Mn max.	P max.	S max.	Cr	Mo	Ni	Other elements
1.4301	0.07	1.0	2.0	0.045	0.030	17.0/19.0	—	8.5/10.5	—
1.4303	0.07	1.0	2.0	0.045	0.030	17.0/19.0	—	11.0/13.0	—
1.4305	0.12	1.0	2.0	0.060	0.15/0.35	17.0/19.0	—	8.0/10.0	—
1.4306	0.03	1.0	2.0	0.045	0.030	18.0/20.0	—	10.0/12.5	—
1.4310	0.12	1.0	2.0	0.045	0.030	16.0/18.0	—	6.5/9.5	—
1.4401	0.07	1.0	2.0	0.045	0.030	16.5/18.5	2.0/2.5	10.5/13.5	—
1.4404	0.03	1.0	2.0	0.045	0.030	16.5/18.5	2.0/2.5	11.0/14.0	—
1.4435	0.03	1.0	2.0	0.045	0.025	17.0/18.5	2.5/3.0	12.5/15.0	—
1.4436	0.07	1.0	2.0	0.045	0.025	16.5/18.5	2.5/3.0	11.0/14.0	—
1.4541	0.08	1.0	2.0	0.045	0.030	17.0/19.0	—	9.0/12.0	Ti > 5 x %C < 0.8
1.4550	0.08	1.0	2.0	0.045	0.030	17.0/19.0	—	9.0/12.0	Nb > 10 x %C < 1.0
1.4571	0.10	1.0	2.0	0.045	0.030	16.6/18.5	2.0/2.5	10.5/13.5	Ti > 5 x %C < 0.8

of which only the austenitic chrome-nickel or chrome-nickel-molybdenum steels are used for the construction of equipment and machining in the food industry. The most important varieties are shown with their national standard designation in Table 14.10 and their chemical composition in Table 14.11.

14.4.1.1.2 Rubber and Elastomers
The choice of elastic sealing materials is particularly important under the hygiene aspect in the food technology sector:

- Sealing components are wearing parts which become unusable much earlier by chemical, thermal, and mechanical stress than sealed metal components.

- Rubbed-off parts or parts of the sealing which have been eroded will stain the food in contact. However, they should by no means modify it.
- A sealing material which does not stand up to the chemical and physical requirements embrittles or looses its elasticity.
- Embrittled or creviced seals facilitate adherence of dirt and bacteria, there exists the risk of permanent product and process contamination due to insufficient cleaning and disinfecting conditions.
- Partly destroyed sealings favor crevice corrosion in stainless steel because nonrinsable liquid residues may subsist between damaged sealings and adjacent metal.

Natural rubber, as well as many other synthetic elastomers, may be used as material for elastic sealings. However, nearly exclusively synthetic products are used in the food processing technology as the natural rubber resources would by no means be sufficient for satisfying the rubber needs in the total technical sector, and as the synthetic elastomers can specifically be adapted to the various requirements (chemical, temperature proof, etc.). The following materials are used for elastic sealing materials:

Material	Common abbreviation	Used temperature range
Natural rubber	NR	−60 to +80°C
Acryl nitril butadiene rubber	NBR	−35 to +120°C
Silicone rubber	VMQ	−70 to +200°C
Ethylene propylene dien Methylene	EPDM	−60 to +135°C
Chloroprene	CR	−40 to +230°C

with the resistance characteristics showed in the following table:

Contact medium	NR	NBR	VMQ	EPDM	CR	FKM
Hot water (120°C)	−	++	++	++	++	++
Hot water (145°C)	−	−	++	++	−	++
NaOH (5%; 90°C)	+	++	++	++	+	++
NaOH (5%; 140°C)	−	−	−	+	(+)	++
H_3PO_4 (2%; 90°C)	−	++	++	++	++	++
H_3PO_4 (2%; 140°C)	−	−	−	+	−	++
HNO_3 (1%; 70°C)	−−	−	+	+	+	+

(++ = unlimited resistance; + = limited resistance; (+) = only short contact; − = nonresistent; −− = absolutely nonresistent)

The fact that the manufacturers of the materials often use different formulations, the characteristics in the aforementioned table may differ. In borderline cases (extremely high temperatures at heat sterilizing of the equipment, high chemical concentration, fats, oils, and solvents) binding statements can only be made after the performance of tests under working conditions. According to Wildbrett (1983), clyclic alternate tests (product/chemical solution) are compulsory in additional to the usual resistance tests (by immersion) made today. In Addition, the mechanical stress has to be taken into account. Guidelines for making general stress tests can be found in national standards for food technology, for example:

- DIN 53 521   Testing of rubber and elastomers, determination of resistance to liquids, vapors and gases;
- ASTM D 471   Change to properties of elastomeric vulcanizates resulting from immersion in liquids;
- BS 903 Part. A16   The resistance of vulcanized rubbers to liquids;
- ISO 1817   Vulcanized rubbers — the resistance to liquids — test method.

14.4.1.1.3 Plastics

Plastics is the collective name for a large group of macromolecular, organo-chemical materials, which are produced either by conversion of high molecular natural products, or fully synthetically. Starting materials are natural or low molecular substances from coal, natural gas, or mineral oil in connection with lime, common salt, water, and air. They are linked by specific methods to macromolecules or polymers. According to the applied linking methods, plastics can be subdivided into three groups:

- Polymerizates
- Polycondensates
- Polyadducts

Some of the reasons why they are widely used in the construction of apparatus for the food industry are the variety of the materials with special material characteristics like extraordinarily high resistance against numerous chemicals, the diversity of their mechanical properties covering the wide range from very solid to highly elastic material, their electric and dielectric properties as compared with metallic materials, and also their lower heat conductivity and low energy consumption during machining. A limited use in the food technology sector mainly results from their different temperature resistance due to material specificities. In order to safely exclude foreign bacteria the food contact surface has to show an absolute resistance against the necessary cleaning, sanitizing, and sterilizing processes, that is, the material has to be resistant up to 90°C, and (for several hours) hot water and/or steam resistant up to 140°C against cleaning and sterilizing chemicals. Thus, the number of possible plastics from a list of more than 1000 plastics is reduced to only a few products:

- Glass-reinforced polyamide (GF — PA)
- Glass-fiber-reinforced polycarbonate (GF — PC)
- Polyvinyldienefluoride (PVDF)
- Polytetra-fluorethylene (PTFE)

From these fours materials only polytetra-fluorethylene (PTFE) is totally chemical-resistant.

Polyamide is resistant against weak alkaline solutions, chlorinated hydrocarbons, organic solvents, fats and oils but not against acid. The tendency for crack formation due to tension is low.

Glass-fiber-reinforced polycarbonate is permanently resistant against low acids, and against fat and oils, but is not against alkaline solutions, chlorinated carbohydrates and solvents. In hot water, PC is jeopardized by crack formation.

Polyvinyldienefluoride (PVDF) is permanently resistant against weak alkaline solutions and acids in the temperature range up to 50°C. At higher temperatures, resistance may decrease, and strong alkaline solutions and acids attack PVDF.

Polytetra-fluorethylene (PTFE) is resistant against almost all chemicals. Only elementary fluor, chlorotrifluoride and molten alkali metals attack PTFE. The utility temperature range extends from $-100°C$ to $+250^{circ}C$. The material is water-reppelent and practically not wettable. It is a high-quality basic material for the production of sealings, membranes, bellows, and special casings and is well suited for covering corrosion-endangered metal surfaces. As far as the mechanic hardness properties allow it, PTFE (glass-reinforced or not) its use in the food technology sector is unlimited. For further properties, cf. Merkel and Thomas (2000).

14.4.1.2 Finishing and Condition of Surface

Before starting food production all apparatus surfaces involved into processing and treatment of foods (e.g., container walls, pipes, valves, shaped parts, measuring probes, pumps, shafts, agitators, static and dynamic sealings, etc.) have to be cleaned of impurities and foreign substances and brought into such condition that the subsequent processes can occur in a scheduled way. Events like product adhesion, cleaning, and disinfecting are influenced by a number of material- and processing-specific criteria, which is described in detail below:

- *Surface tension* Deposits of foreign materials, soil retaining ability, and cleaning behavior of the surface of solids are the complex product of their material-specific properties, of their physical processing state, and of the interactive events at the contact surface, as well as of the foreign materials being either solid, fluid or even gaseous. The adhesive forces are a function, among others, of surface energy, and with weaker adhesive forces the solid surface can be cleaned more easily from adhering soil. Thus, a solid surface can only be wetted by a fluid if the surface energy of the solid surface is larger than that of the fluid. Whereas mineral and metallic surfaces free of foreign materials are generally wettable, many plastics are not, especially PTFE. This is due to the low surface energy (18 mN/m) of PTFE (water at 20°C has 72.8 mN/m, by adding surface-active substances or surfactants, reducible to min. 50 to 40 mN/m). Therefore PTFE has the lowest soil adhering ability of all the technically relevant materials.

- *Surface potential* According to Nassauer (1985a) the surface potential considerably influences the adhesion ability of electrically charged product components. Thus, adhesion of positively charged particles on a solid wall decreases with increasing positive potential difference. For negatively charged particles, like those generally occurring in microbiological substrates and suspended cell cultures, the adhesive tendency increases accordingly. Temperature and pH value of the fluid phase are the decisive parameters for the potential difference between wall and charged solution ingredient. For the austenitic chrome-nickel-steel being the most important construction material in the field of food technology, the surface potential can be modified to a certain degree by electrochemical treatment. Nassauer could demonstrate that on an electropolished chrome-nickel-steel surface the fouling by microorganisms was significantly lower than on a comparable surface from the same material, the surface having been shifted to the positive range after cleaning with highly oxidizing disinfectants. This observation is important in connection with those surfaces in food technological apparatuses that reveal a less pronounced flow-mechanical component during cleaning processes, for example, at the walls of large tanks or containers.

- *Surface roughness* Concerning the production of technical apparatus for food processing the essential criterion for a surface as regards product adhesion ability and cleanability is roughness. The general rule is that the smoother surface is easier to clean. For apparatuses and plants in dairies, there exists a number of indicative values for the minimum requirements to the surface finish. Based on practical experience the minimum requirements for visible surfaces and food contact surfaces of plant components made from stainless steel have been determined for the dairy sector in the DIN standard 11480 as regards threshold values for the average roughness depth R_a and the maximum roughness depth R_t, and subdivided into five requirement levels. According to requirement levels 0, 1, and 2 for supporting and carrying constructions as well as for visible surfaces without product contact surface R_t-values between 6 μm and 2 μm, R_a between 0.8 and 3.3 μm are sufficient. For equipment used in transport, storage, production, processing and heat treatment of fluid and cheese factory milk, curd, cream, sour milk, mixed milk drinks, and ice-cream the surfaces should meet the requirement level 3 (medium roughness values between 0.1 and 0.4 μm — obtainable by fine grinding, lapping, electropolishing or glass bead blasting. The highest requirement levels to surface quality (requirement level 4 with R_a values <0.1 μm — polishing, lapping) are made to product contact surfaces of equipment used for aseptic processes and pure culture breeding. The surface roughness and the parameters for an objective classification of the numerical values are also specified in the ISO standard 468 (1982). Within an investigation on the cleanability in function of the surface roughness in the most favorable case as regards the flow-technical aspect, that is, the not branched tube, the (empirically) requested threshold values for dairy equipment were confirmed by Hoffmann (1983) as being completely satisfactory. Special steel surfaces in contact with foods are being analyzed within a comprehensive study of von Schmauderer, Werlein and Otto (1998) under the aspect of different surface treatment and cleaning processes, and assessed under their efficiency and cost aspect.

14.4.1.2.1 Mechanical Processes of Surface Treatment

• *Grinding* Grinding belongs to the surface-cutting treatment methods. The material is treated with abrasives of high hardness (e.g., silicone carbide (SiC), abrasive bodies or disks, or applied to belts. During grinding the cutting edges of the abrasive grain impact the material through pressure and temperature up to a flowing state, thus resulting in chip separation. In this process, very high temperatures and deformations occur for a short period of time leading to texture modifications, and to formation of strong punctual tensile and compressive strains. Due to immediate subsequent cooling, these strains are frozen into the surface. Possible consequences are microcrevices at the grain interfaces as well as the occurrence of impurities caused by tool abrasion, grinding and polishing agents, fat, forging scales, and oxides up to a considerable depth of the surface near material layers (up to 30 μm) with high impact on the corrosive behavior of the material.

When grinding stainless steel, please consider the following:

- Every type of stainless steel can be grounded but not all can be polished.
- Only iron-free grinding agents should be used.
- The treated work pieces should be kept as cold as possible.
- When changing the grain size of the abrasive agent the work pieces should be thoroughly cleaned to avoid scratches caused by rougher grains from the preceding work stage.
- During grinding it is important that grinding occurs in parallel direction.
- Grounded surfaces should be protected by bonded sheet or other appropriate agents in case of local successive treatment.
- Before starting grinding, graduation of the abrasive agent, velocity of the disk or the belts, and the lubricating and cooling stuff should be tested to ensure they produce the desired finish.

Prescribing an objective characterization of the finish attainable by grinding is not possible as, due to foreign substance incrustations on the grinding disk, and to the individual operating technique of the grinder or the grinding system differences that may occur as regards the grinding quality. Thus, the investigation of the surface profile on a sample (work piece 1.4401) ground with silicone carbide (grain size 150 mesh/in.) with an electric brush analyzer led to highly differing R_a- and R_t-values, depending on whether analyzing occurred parallel or transversal to the grinding direction (parallel: $R_t = 2$ μm, $R_a = 0.3$ μm; transversal: $R_t = 6$ μm, $R_a = 0.5$ μm).

• *Lapping* Like grinding, lapping is a cutting method using loose grains of geometrically undefined shape. The lapping grain is suspended in a fluid (e.g., oil). At lapping the surfaces are preserved with high measuring precision (0.1 to 0.2 μm) and very low roughness values $R_a < 0.2$ μm. In the case of corresponding grain sizes with subsequent polishing, even lower values may be obtained. Depending on the type of lapping equipment, smooth surfaces of highest quality or cylindric surfaces can be preserved. Due to the cutting effect of the lapping grain a texture modification of the treated surface is unavoidable up to a depth of a few micrometers.

- *Honing* This process consists of the finest honing where a honing tool performs rotations with simultaneous oscillation. Thus, extremely smooth surfaces ($R_t = 0.5$–1 μm) can be produced with high measuring precision (up to 1–3 μm). Honing belongs to the cutting surface treatment methods, and produces, as grinding, a change or even a damage up to a depth of 5μm.

- *Polishing* This process consists of a further treatment of precision-ground surfaces until they reach high polish or luster (mirror finish), whereby the limits between grinding and polishing are difficult to determine. With polishing wheels in connection with abrasive pastes, it is possible to obtain surfaces with a similar appearance than that of ground surfaces. With abrasive papers of finest grains, and by using appropriate oils, surfaces, which are similar to polished surfaces, can be obtained.

  In contrast to grinding with a cutting effect on the surface, theoretically polishing does not remove material. Nevertheless, bruises, chamfers, and furrows are levelled and closed to a large extent. The metallic surface becomes slightly plastic, even liquefied by the produced heat due to the pressure of the polishing equipment. During the polishing process, this crystalline film comparable to a viscous fluid flows over and fills up scratches, furrows, and small bruises. The action of the surface forces aims at a relative surface evenness similar to that of a fluid. The surface roughness obtainable by polishing lies under the roughness limit obtained with mechanical brush analyzers. The R_a-values may be under 0.05 μm. Polishing occurs in three steps: brushing, prepolishing, and polishing. The materials used for polishing wheels (for stainless steels) are sisal, fibre, cotton, and nettle. For stainless steels polishing the following agents are used: melted alumina, electrocorund, and fused pure chromoxide (Cr_2O_3), the latter consisting of very fine particles of high hardness.

  Steels stabilized with titan or niobum cannot be high-polished due to the formation of hard titan or niobum carbide or nitrides torn from the surface. Therefore, in order to comply with the requirement for mirror finish and consistency, steel qualities with reduced carbon content should be selected as austenitic steels.

- *Roller burnishing* Roller burnishing creates plastic deformation by displacing the material in the peaks which cold flows under the pressure of a smooth, hardened roll into the valleys, thereby generating a series of plateaus in the surface contact plane. The result is an accurately sized part with a mirror-like finish and a tough, work-hardened, wear-, and corrosion-resistant surface. Roller burnishing is particularly well suited for the treatment of rotational-symmetric surfaces (e.g., stirring shafts, axially driven valve spindles), also weldings in sheet metal can be finished by rolling after rough grinding. In case of roller burnishing it is important to exclude extremely high moments of flexion, so that the deformation forces may enter the material surface symmetrically. For avoiding undesired surface damages by persistent material and foreign particles the rolling surface should be permanently rinsed during the rolling process (oil bath).

- *Abrasive blasting* Abrasive blasting is applied for obtaining optically uniform larger surfaces, for example, for apparatus construction, mechanical end machining of geometrically complex surfaces and also as a mechanical prestep of chemical deforging/etching. Abrasive agents for special stainless steels are quartz sand, corund, glass beads, or special steel granulate (ordinary steel granulate should not be used due to the risk of potential foreign rust formation), which are applied via compressed air/water to the surfaces to be polished. Due to the mechanical impact power and/or effect of the steel grain when it strikes the surface at a high velocity, local compressive strains are produced on the surface, thus possibly compensating existing tensile stress caused by preceding cold forming. In the case of austenitic steel the risk of stress corrosion cracking can be reduced by abrasive blasting. Blasted surfaces have a matt appearance whereby the roughness depth can be influenced by the choice of the blasting grain. Average-peak-to-valley rates of <0.3 μm are attainable. Chemical etching and passivation should be applied for removing remaining residues of the blasting agent or residues stroke into the surface.

14.4.1.2.2 Chemical surface treatment

• *Pickling* For obtaining chemical resistance in stainless steels a metallic clean surface with a low roughness depth is required. The scaling layers created during warm moulding or welding (mainly chromoxides) and/or heat blooming paints have to be removed as they impede the formation of the corrosion inhibiting passive layer. The most popular method is pickling with chloride-free acids [nitric acid (HNO_3), sulfuric acid (H_2SO_4), and hydrofluoric acid (HF)] by immersion in a pickling bath, or application of a pickling agent (e.g., weldment joints). The strength of the layers to be removed by chemical etching is limited and lies normally under 5 μm. The pickling process has to be carefully monitored so that no etching defects (e.g., etching pores due to baths being too cold, used and hydrochloric-acid-containing, etching crevices in hardened or tempered steels) occur. After pickling, the acid residues have to be removed by careful rinsing with water, according to the requirements, by neutralization with diluted sodium hydroxide solution (NaOH) or lime milk $Ca(OH)_2$, and subsequently by rinsing with water.

• *Passivation* After the pickling treatment it is usual to immerge stainless steels for passivation in diluted nitric acid (10–20 parts per volume conc. HNO_3, 90–80 parts per volume H_2O), and to rinse subsequently. Although these varieties of steels with a metallic clear surface passivate already by staying in the air, this procedure has found large acceptance due to its successful corrosion resistant performance. The oxygen offers support to the rapid formation of the corrosion-inhibiting passive layer at the surface. The treatment in the passivating bath is a must in case undesired foreign particles, for example, abrasive swarf or rubbed-off particles from tools are on the surface of the prefabricated parts and may induce corrosion by extraneous rust. After passivation rinsing with water is necessary.

• *Electropolishing* Electropolishing belongs to the electrically removing finishing techniques and is essentially the opposite of galvanizing removed (Pießlinger-Schweiger, 1983). The material is removed without any kind of mechanical stress, only by electrochemical dissolution of the material surface connected to an anode within an electrolyte [based on concentrated mixtures of phosphoric acid (H_3PO_4) and sulphuric acid (H_2SO_4)] under the influence of external DC currents. The voltage does not exceed 20 V, and the produced current strengths are between 500 and 3,000 A/m^2. The electropolishing process lasts 15 min, and during this time, 15 to 25 μm are removed.

The material removal occurs under levelling conditions at the whole material surface, whereby the higher current density increases the removal at edges and corners. The smoothing effect of the method is due to the properties of the polishing acid. During electropolishing, the polishing acid forms a viscous dry layer — the so-called polishing film — under the influence of current above the metal surface representing a diffusion resistance for the escaping metal ions being much higher than the difference in metal removal at the individual, active spots at the surface. The removal rate is higher for peaks only covered with a thin polishing film than for those being better covered. Thus, a smoothing effect of the surface is obtained without a grain boundary attack, which is characteristic for pickling. In the micro area, the electropolished surface is smooth and generally without cracks and pores. In the macro area, it shows a certain remaining ripple the size of which depends on roughness, electropolishing intensity, and on the fineness of the texture. Electropolished surfaces show a crystalline basic texture with unfalsified qualities/properties. They are free of tensile or pressure stress, texture modifications, chemical reaction products, and have enough individual alloying constituents. Electropolished surfaces have optimum corrosion resistance of the corresponding material as, due to their metallic purity and homogeneity, they have the most favorable prerequisites for a perfect formation of the passive layer. During electropolishing, oxygen is released at the material surface. This oxygen is enriched at the above positioned electrolyte layer. Right after the switching-off of the polishing stream, the high oxygen quantity induces the formation of the passive layer. Thus a subsequent passivation, as required after pickling, is no more required.

Not all stainless steels, for example, materials stabilized with titan like 1.4541 (Euro standard 88-71: X6 Cr Ni Ti 18 10) or 1.4571 (Euro standard 88-71: X6 Cr Ni Mo Ti 17 2 2) are equally suited for electropolishing. As titan carbide is not attacked during electropolishing, carbide particles remain at the electropolished surface, and are thus the cause for an increased surface roughness being higher than that of the untreated surface.

14.4.1.3 Equipment Design

In general, machines and equipment for the manufacture, processing, and treatment of foods are more or less complex combinations of individual aggregates, and differentiation is based on the kind of processing they undergo:

- An open processing with open access to all machine food contacting surfaces, with the possibility of manual cleaning and visual inspection of the cleaning results is better.
- A closed processing without access to the machine food contacting surfaces, which are exclusively to be cleaned with CIP methods, and do not allow a direct control of the cleaning success are necessary.

Depending on the individual case, the machines and equipment used for these food production processes, particular constructional principles have to be taken into account, if they are attributed the label "hygienic design."

14.4.1.3.1 Open processing
A good example for an equipment used in open processing is the container with stirring device and outlet represented in Figure 14.9 [according to Hauser (1995) and DIN EN 1672-2]. On the left side of

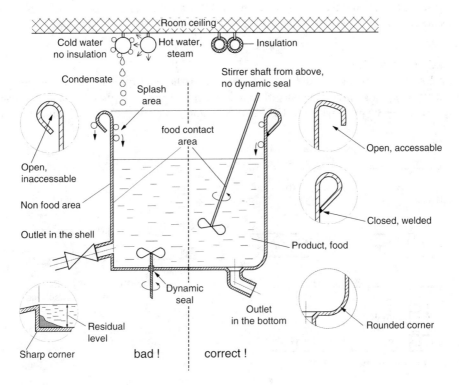

FIGURE 14.9 Hygienic design of an open stirrer container.

the figure some critical zones under the cleaning-technological aspect are displayed, and on the right side possible solutions for a hygienic design (upper container either completely open (cleanable), or hermetically welded instead of a simple bead, where soil may stick; outlet in the container bottom (complete drainage mut be possible) instead of outlet in the side shell (no complete drainage); rounded edges (good cleanability) at the container bottom instead of sharp corners, where soil may stay; stirrer shaft from above instead of from below or from the side in the container wall (dynamic seals are problematic)

It should be taken into account that also external factors may lead to a product contamination in the open processing. In the represented figure, pipes installed above the open container at the ceiling are displayed. On heat-uninsulated cold water pipes, or on pipes conducting cooled products, the condensate formed may drop into the open container. A heat insulation of the pipes and/or a gutter mounted under the pipes may solve the problem. Hauser (1999) mentions the hygienic risks in open food processing due to constructional defects in the building design and suggests remedial measures.

In general, equipment and aggregates cannot be made of a single piece of material. There are junctions and joints/connections, which represent hygienic risks if they are not correctly executed. As far as possible, permanent constructions (e.g., welded joints) should be preferred to detachable connections (e.g., screw-type connections). Nevertheless, some basic rules should always be observed (Figure 14.10) to have a hygienically perfect result.

The connections between the elements of the aggregate often have to be detachable for inspection and maintenance reasons (e.g., installation of sensors, etc.). In the above mentioned case the material transitions have to be absolutely free of crevices to avoid the collection of product residues and soil in these crevices. It has to be excluded that, when metal is in direct contact with metal, due to surface roughness some micrometers wide, undefined devices are formed, in which microorganisms may settle being unreachable to cleaning and disinfection. Inserting elastic intermediate layers is a remedial, whereby constructive measures should guarantee that the elastic element is not damaged by uncontrolled high surface pressure (beyond the elasticity limit of the elastomere). Thus, the metallic limit stop should be behind the sealing element. Models for the hygienic design of elastic sealing elements are presented in the DIN standard EN 1672-2 and in the ISO/DIS standard 14 159 (1999).

14.4.1.3.2 Closed Processing

Closed plants for the treatment and processing of foods have to fulfil higher cleaning-technical requirements compared to the manually cleanable plants for open processing. First, because their cleaning occurs exclusively with the CIP method without any possibility of individual, manual

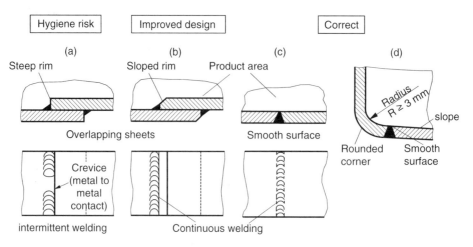

FIGURE 14.10   Material connections. (According to Hauser, G., 1995. *Trends Food Sci. Technol.* 6: 305.)

afterwards cleaning. Second, because the possibilities of checking the cleaning success are practically not given, at least not without a partly dismantling the equipment.

Examples for closed processing plants range from the simple, nonbranched pipe joining two equipment parts up to extensively branched pipe networks, from the simple open/close valve to the automatic valve nodes with hundreds of multiple-way-reverse valves in different executions, from the simple, closed container to the complicated sterile tank with aseptic intermediate storing of sterile products. Closed processing plants have pumps of different patterns, centrifuges, filtrating, heating and cooling devices, filling and dosing stations, etc. all equipped with the required sensors for process control of level, volume flow, temperature, pressure, pH, conductivity, density, etc.

This nearly endless variety of apparatuses, machines, and measuring devices have to bear no risk of process contamination induced by insufficient cleanability. Proposals, mainly based on empiric values, for the hygienic design of the components of closed plants for the manufacture of liquid foods are given in a paper of the EHEDG (European Hygienic Engineers Design Group, 1993). Scientifically founded examinations, from which concrete constructional indications, and necessary process parameters for the guaranteed cleanablility can be deduced, only exist for a small number of part components of the closed processing: as the comparably simplest cleaning task the flow-mechanical cleaning of straight, unbranched pipes were systematically investigated by Hoffmann (1983). Graßhoff (1980, 1983b) described the fluid dynamics and the stationary cleaning behavior of dead spaces in branchy pipe systems (Figure 14.11) and analyzed the behavior of the dynamic seal in spindle valves in a further study (1985).

Special attention must be paid to the possible formation of dead spaces in apparatuses and pipelines (Figure 14.12 and Figure 14.13). Graßhoff (1980, 1983b) has made extensive investigations on the cleanability of cylindrical dead ends and showed that the mechanical cleaning effect almost ceased in a relative depth L/d bigger than 4. He proposed a simple solution, introducing a stream divider to force part of the flow into the dead end (Graßhoff, 1983b). In industrial heat exchangers it is common to use parallel tubes (in shell-and-tube heat exchangers) or channels (in plate heat exchangers) to increase the throughput. In this case, it is of fundamental importance to achieve a homogeneous flux distribution through all conduits. The channels with lower velocities will be prone to foul more heavily because of reduced erosion of the fouled layer among other factors during the food processing procedure, and afterward, during the cleaning procedure, the mechanical action will be lower than in conduits with higher velocities, resulting in the necessity of much longer cleaning

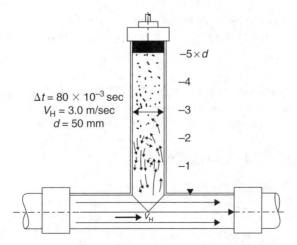

FIGURE 14.11 Fluid motion in a branched pipe. Marking the migration of particles which were photographed by a high speed cine camera within 80 msec. Frequency 500 frames/sec. (From Graßhoff, A. 1980. *Kieler Milchwirtschaftliche Forschungsberichte* 32: 273. With permission.)

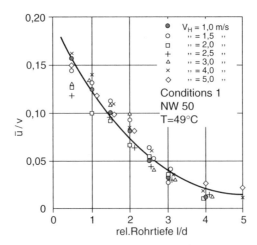

FIGURE 14.12 Relative motion in the deadend branch of a pipe connection, nominal width 50 mm, flow tangential to the orifice of the dead space. (From Graßhoff, A. 1980. *Kieler Milchwirtschaftliche Forschungsberichte* 32: 273. With permission.)

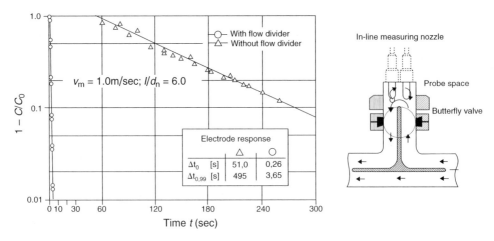

FIGURE 14.13 Solving the dead space problem exemplary at a in-line-measuring-nozzle by a flow divider. (From Graßhoff, A. 1992. Hygienic design — the basis for computer controlled automation. *Trans IChemE*, Vol. 70, Part C: 69. With permission.)

times. Those channels with higher-than-average flux will be cleaned unnecessarily and wastefully for long. The flow distribution in parallel channels of plate heat exchangers with both U-type and Z-type arrangements has been studied extensively by Bassiouny and Martin (1984).

14.5   CLEANABILITY TEST METHODS

The manufacturer has to give a practical demonstration on site, after mounting, to prove whether the delivered complex apparatus is cleanable. In general, he uses conventional methods, that is, after starting the equipment, a production phase is performed. Subsequently, a (partial) cleaning is done and the equipment is opened or dismantled and investigated for soil by:

- Visual inspection (eye control, UV-radiations, endoscopies for difficultly accessible zones)

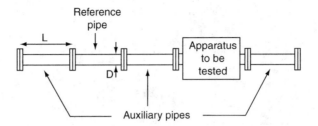

FIGURE 14.14 Test unit apparatus component with reference pipe.

- Wipe tests with clean wood pulp papers, physical or chemical tracing of product residues on the used wipe-papers
- Proof samples with sterile wood pulp papers or gauze pads, microbiological analysis in total bacterial count and bacterial spectrum
- Wipe tests with sterile gauze pads, detection of ATP (bioluminescent method)

The examinations performed with the aforementioned methods largely concentrate on (empiric) cleaning critical zones of an equipment and serve as a direct or indirect detection of the cleaning status of the assembled equipment after their starting. A previous detection of the cleanability of the components mounted in the equipment is thus not possible.

14.5.1 EHEDG — Cleanability Test (I)

The EHEDG (European Hygienic Engineers Design Group) — Cleanability Test (Holah et al., 1992) is based on a method first described by Galesloot et al. (1967), and modified by the Unilever Research Laboratorium (Lelieveld, 1985). The cleanability test is partly conceived to detect hygienically critical zones in food equipment, where product residues or microorganisms are protected from the cleaning process. However, individual components of the equipment should be assessed in a comparative test as regards their in-place-cleanability. The method is based on comparing (in the laboratory) the cleanability of a test item with that of a straight piece of pipe. For this, a degree of cleanliness is determined, which is based on the relative removal of a test soil still containing bacterial spores after cleaning. The test conditions are such that on using a relatively mild detergent a minimum of the initial soil remains detectable in the reference object (straight piece of tube). Thus, an individual assessment of the cleanability is possible.

The method is intended as a screening test for hygienic equipment design of individual equipment components and is not indicative of the performance of industrial cleaning processes. Nevertheless, an analogy of the cleaning results obtained in the comparative test to those obtained in practice is allowed.

The apparatus components to be assessed are joined with an appropriate adapter and a reference pipe with known surface roughness ($R_a = 0.5$ μm, according to ISO 468-1982) to form one unit (Figure 14.14). Afterwards, they are soiled in a circuit (Figure 14.15) by containing (*Bac. stearothermophilus*) sour milk suspension under a pressure of 5 bar. Movable parts (e.g., valve spindles, shafts, etc.) are to be activated continuously or at least 10× during pressurizing — also as a simulation of the operating conditions. After soiling, the test unit is dried with dry, filtered air for a few days and subsequently cleaned in a CIP subsequently cleaned in a CIP process (Figure 14.8):

- 1 min rinsing with water (41 ± 1°C)
- 10 min with a mild detergent solution (1%, 63 ± 2°C)
- 1 min rinsing with cold water

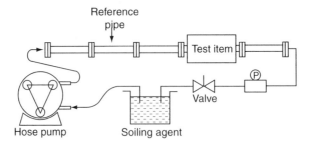

FIGURE 14.15   Soiling of the test unit (P = pressure measuring).

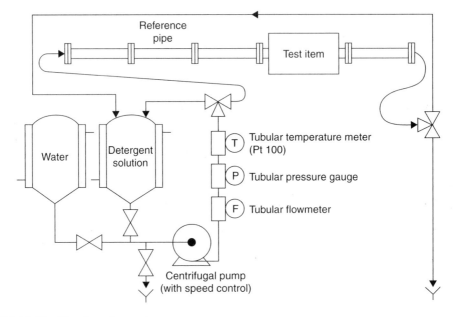

FIGURE 14.16   Cleaning of test unit.

The local flow velocity of the cleaning and rinsing solutions should be approx. 1.5 m/sec (Figure 14.16).

A liquefied nutritive agar (Shapton and Hindes, 1963) is applied after cleaning on the inside surfaces of the test item and of the reference pipe (e.g., by rotation of the elements until cooling down and solidifying of the agar) and incubated for 24 h at 58°C (incubator). Afterwards, the agar-coated surfaces are examined on yellow staining and/or colonies in the purple-red SHA agar.

The yellow zones indicate spores and microorganisms, which had not been removed by the cleaning process. In the reference pipe, the surface of the yellow zones as indicator of the normal course of the cleaning process should range between 5 and 30%. The cleaning ability of the test item is determined in percentages of the yellow surfaces in relation to that of the reference pipe. Repeated experiments should clarify whether the distribution of the yellow zones is random or whether they appear on the same spot of the test item. In case they do, they indicate zones of poor hygienic design.

14.5.2 EHEDG — CLEANABILITY TEST (II)

The EHEDG-Cleanability-Test (Venema-Keur et al., 1997), modified in 1997, is less time-consuming and labor cost-intensive, but not as sensitive as the a.m. method. With an apparative mounting more or

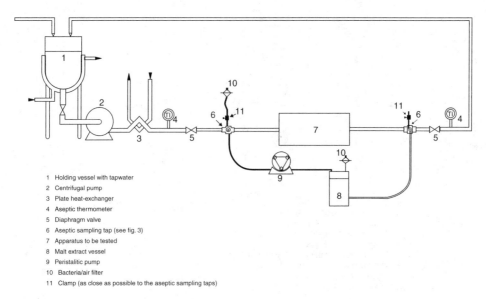

1 Holding vessel with tapwater
2 Centrifugal pump
3 Plate heat-exchanger
4 Aseptic thermometer
5 Diaphragm valve
6 Aseptic sampling tap (see fig. 3)
7 Apparatus to be tested
8 Malt extract vessel
9 Peristalitic pump
10 Bacteria/air filter
11 Clamp (as close as possible to the aseptic sampling taps)

FIGURE 14.17   Test rig for the pasteurizability test.

less identical to test (I) the soiling of test surfaces is not induced by bacteria-containing suspensions but by pressurizing with a ß-carotene colored fat emulsion. The degree of cleanliness of the test item surfaces is assessed by means of wiping samples performed with white wood pulp pads, which show the amount of soil remaining, in comparison with wiping samples from the inside surface of the reference pipe.

14.5.3   EHEDG — Pasteurizability Test

For a hygienic production of foods, the product-contact surfaces of a plant have not only to be physically, but also microbiologically, clean in order to avoid a contamination with undesired microorganisms. This can be induced by chemical disinfection of the plant or by treatment with hot water (90°C). In case of thermal disinfection, it has be ensured that all the product-contact surfaces reach a temperature, under which a sufficient reduction of bacteria is guaranteed. For checking the pasteurizability of a production equipment or individual components, the EHEDG — subgroup "Test methods" has proposed a special pasteurizability test (Venema-Keur et al., 1993).

The component/equipment to be tested is dismantled to such an extent that the product-contact surfaces are accessible. A spore containing suspension (*Neosartorya fisheri* var. *glabra*) is applied with a sterile brush and subsequently dried for 2 h at ambient temperature. The component/equipment is mounted and inserted into a test rig (see Figure 14.17). A treatment with hot water follows (90°C ± 0.5°C, measured at the flow runback — 30 min).

After cooling (by sprinkling the outer surfaces with cold water) the component/equipment is filled with a sterile nutrient broth. For at least 2 weeks a hose pump circulates the nutrient broth at 23 to 25oC via a buffer vessel. Ascospores, which survived the pasteurization process germinate and cause an initially light turbidity of the nutrient broth. Afterwards, a white mold-mycelium forms on the surface of the liquid and on the walls of the buffer vessel. The experiment has to be performed 3 times. If no mould is formed, the component/apparatus can be classified as pasteurizable at 90°C.

14.5.4   EHEDG — Sterilizability Test

The highest requirements for the hygienic design of production plants are procedures for sterile production, filling, or packaging of foods. It has to be guaranteed that both the product-contact

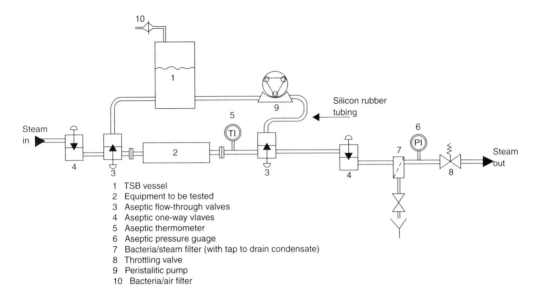

FIGURE 14.18   Test rig for the EHEDG In-line sterilizability test.

1 TSB vessel
2 Equipment to be tested
3 Aseptic flow-through valves
4 Aseptic one-way vlaves
5 Aseptic thermometer
6 Aseptic pressure guage
7 Bacteria/steam filter (with tap to drain condensate)
8 Throttling valve
9 Peristalitic pump
10 Bacteria/air filter

surfaces and the gas/air volumes in the apparatus are sterile before starting production and remain sterile during the production period for hours. For example, valves have to be hermetically sealed with aseptic bellow constructions against the insterile ambient atmosphere, and shaft seals have to be steam sterilizable. Through further construction measures, it has to be ensured that germ-infested ambient air does not penetrate into the inside of the equipment and so on. For checking the aseptic functionability of a production plant or of individual components the EHEDG subgroup Test methods proposed a special in-line sterilizability test (Timperley et al., 1993).

As in the pasteurizability test the apparatus/component to be tested is dismantled, degreased with alcohol, and manually cleaned, and, if required, mineral encrusting is removed by an aqueous acetic acid solution. Afterwards, all the individual components are sterilized in steam autoclaves (30 min at 120°C), or sterilized after reinstallation by direct treatment with saturated steam. In the cleaned and sterilized apparatus/component, all the inside surfaces, including seals and sealing grooves are humidified with a spore-containing (*Bacillus subtilis*) suspension. After a 2-hour drying at ambient temperature, the apparatus/component is mounted, inserted into the test rig (Figure 14.18), and sterilized (30 min, 120°C). For a period of at least 5 days a hose pump conveys a sterile nutrient broth (Trypticase Soy Broth – TSB – 15 g/L) through the apparatus/component at an ambient temperature of 25°C. Movable parts like valve spindles, shafts, etc., have to be actuated in regular intervals during this period. If the conveyed nutrient broth becomes turbid during the trial of several days (indicator of growth of microorganisms), it has to be investigated within a systematic error analysis. The cause of the turbidity can be growth of microorganisms, poor hygienic design of the tested apparatus/component, experiment execution errors, weak points of the experimental set-up, etc. If the TSB nutrient broth remains clear after 5 days and during at least three subsequent trials the tested apparatus/component can be classified as being in-line sterilizable and thus fit for the use in aseptic production processes.

REFERENCES

Ackermann, R., Winkler, K., Schumann, G., and Reinsch, V. 1987. Rezirkulationsverfahren für Flaschenwaschmaschinen-Reinigungslauge in Getränkebetrieben — eine einschätzung aus hygienischer Sicht. *Lebensmittelindustrie* 34: 28.

American Public Health Association. 1960. *Standard Methods for the Examination of Dairy Products*. APHA, New York.

American Public Health Association. 1972. *Standard Methods for the Examination of Dairy Products*, 13th ed., APHA, New York.

Amtsblatt der Europäischen Gemeinschaften, 19.07.1993, Nr. L 175:1. Richtlinie 93/43/EWG des Rates vom 14. Juni 1993 über Lebensmittelhygiene.

Arbeitsblatt ATV-A 115 1994. Einleiten von nicht häuslichem Abwasser in eine öffentliche Abwasseranlage. Oktober 1994. ISBN 3-933693-04-7.

Amtsblatt der Europäischen Gemeinschaften, 7.03.1998, Nr. L 67:29. Richtlinie 98/15/EG des Rates vom 27. Februar 1998 über die Behandlung von kommunalem Abwasser.

Amtsblatt der Europäischen Gemeinschaften, 30.05.1991, Nr. L 135:40. Richtlinie 91/271/EWG des Rates vom 21. Mai 1991 über die Behandlung von kommunalem Abwasser.

Bassiouny, M.K. and Martin, H. 1984. Flow distribution and pressure drop in plate heat exchanger. *Chem. Eng. Sci.* 39: 693.

Bourne, M.C. and Jennings, W.G. 1961. Some physicochemical relationships in cleaning hard surfaces. *Food Technol.* 15: 495.

Bourne, M.C. and Jennings, W.G. 1963a. Existence of two species in detergency investigations. *Nature* 197: 1003.

Bourne, M.C. and Jennings, W.G. 1963b. Kinetic studies of detergency. I. Analysis of cleaning curves. *J. Am. Oil Chem. Soc.* 40: 517.

Bourne, M.C. and Jennings, W.G. 1963c. Kinetic studies of detergency, II. Effect of age, temperature and cleaning time on rates of soil removal. *J. Am. Oil Chem. Soc* 40: 523.

Bourne, M.C. and Jennings, W.G. 1963d. Definition of detergent.*J. Am. Oil Chem. Soc.* 40: 212.

Brauer, H. 1971. In *Grundlagen der Einphasen- und Mehrphasenströmungen*. Verlag Sauerländer Aarau und Frankfurt A.M.

Bundesgesetzblatt 16.07.1993, Teil 1, Nr. 36: 1169. Neufassung des Lebensmittel- und Bedarfsgegenständege-setzes vom 8.07.1993.

Bundesgesetzblatt 18.02.1999, Teil 1, Nr. 6: 86. Verordnung über die Anforderungen an das Einleiten von Abwasser in Gewässer vom 9.02.1999.

Cheow, C.S. and Jackson, A.T. 1982. Circulation cleaning of a plate heat exchanger fouled by tomato juice. I. Cleaning with water. II. Cleaning with caustic soda solution. *J. Food Technol.* 17: 417–431.

Corrieu, G., Lalande, M., and Ferret, F. 1980. New monitoring equipment for the control and automation of milk pasteurization plants. In *Food Process Engineering*, Vol. 1, P. Linko, Y. Malkki, J. Olkku, and J. Larinkari (Eds), *Applied Science*, London, p. 1965.

Corrieu, G. 1981. *State-of-the-art of* cleaning surfaces. In *Fundamentals and Applications of Surface Phenomena Associated with Fouling and Cleaning in Food Processing*, B. Hallström, D.B. Land, and Ch. Träghardh (Eds), Lund University Reprocentralen, Lund, Sweden, p. 90.

Deutsches Institut für Normung (DIN) e.V., 1978. DIN 11 480, *Milchwirtschaftliche Maschinen in Molkereibe-trieben — Oberflächen*.

Deutsches Institut für Normung e.V., DIN EN 1672-2 1997. *Nahrungsmittelmaschinen — Allgemeine Gestaltungsgrundsätze — Teil 2: Hygienische Anforderungen*. Beuth-Verlag, D. 10787 Berlin/Germany.

Dommet, T.W. 1968. Bacteriological assessment of cheese factory hygiene. *Aust. J. Dairy Technol.* 176.

EHEDG 1993. Hygienic design of closed equipment for the processing of liquid food. *Trends Food Sci. Technol.* 4: 375.

FID-IDF. 1979. *Document 117: Design and Use of CIP-Systems in the Dairy Industry*. FED-IDF General Secretariat Brussels, Belgium.

Galesloot, Th.E., Radema, L.M., Kooy, E.G., and Hup, G. 1967. A sensitive method for the evaluation of cleaning processes, with a special version adapted to the study of the cleaning of tanks. *Neth. Milk Dairy J.* 21: 214.

Gallot-Lavallee, Th. 1982. *Contribution à l' étude de la cinétique du nettoyage des pasteurisateurs de lait*. Diss. ENSIA, Massy, France.

Gallot-Lavallee, Th., Lalande, M., and Corrieu, G. 1982. An optical method to study the kinetics of cleaning milk deposits by sodium hydroxide. *J. Food Process Eng.* 5: 131.

Gallot-Lavallee, Th., Lalande, M., and Corrieu, G. 1984. Cleaning kinetics modeling of holding tubes fouled during milk pasteurization. *J. Food Process Eng.* 7: 123.

Gousetis, C. 1991. Nitriloacetic acid. *Ullmann's Encyclopedia of Industrial Chemistry,* Vol. A 17: 377.

Graßhoff, A. 1980. Untersuchungen zum Strömungsverhalten von Flüssigkeiten in zylindrischen Toträumen von Rohrleitungssystemen. *Kieler Milchwirtschaftliche Forschungsberichte* 32: 273.

Graßhoff, A. 1983a. Modellversuche zur Ablösung festverkrusteter Milchbeläge von Erhitzerplatten im Zirkulationsreinigungsverfahren. *Kiel. Milchwirtsch. Forschungsber.* 35: 493.

Graßhoff, A. 1983b. Die örtliche Flüssigkeitsbewegung und deren Einfluss auf den Renigungsprozess in zylindrischen Toträumen. *Kiel. Milchwirtsch. Forschungsber.* 35: 471.

Graßhoff, A. 1986. Produktverschleppung und Rekontamination an bewegten Ventilspindeln. Kieler Milchwirtschaftliche Forschungsberichte 38: 3.

Graßhoff, A. 1988. Zum Einfluß der chemischen Komponeten alkalischer Reiniger auf die Ablösung festverkrusteter Beläge aus Milchbestandteilen von Erhitzerplatten. Kieler Milchwirtschaftliche Forschungsberichte 40: 139

Graßhoff, A. 1989. Environmental aspects of the use of alkaline cleaning soultions. *Proc. 3rd Int. Conf. on Fouling and Cleaning.* June 5–7, 1989 in Prien, Germany, (Kessler, H.G. and Lund, D., eds, printed by Walch, Augsburg, Germany).

Graßhoff, A. 1992. Hygienic design — the basis for computer controlled automation. *Trans IChemE,* Vol. 70, Part C: 69.

Graßhoff, A. 1999. Laborversuche zur Reinigung von Milcherhitzern mit enzymatischen Reinigungsmitteln. *Kieler Milchwirtschaftliche Forschungsberichte* 51: 295.

Graßhoff, A. and Reinemann, D.J. 1993. Zur Reinigung der Milchsammelleitung mit Hilfe einer 2-Phasen-Strömung. *Kieler Milchwirtschaftliche Forschungsberichte* 45: 205.

Graßhoff, A. and Potthoff-Karl, B. 1996. Komplexbildner in alkalischen Reinigern. *Tenside Surfactants Detergents* 33: 278.

Gutknecht, J. 1993. Gefahr durch Reinigungs- und Desinfektionsmittel. *die ernährungsindustrie* 5/95: 58.

Hankinson, D.J. and Carver, C.E. 1968. Fluid dynamic relationship involved in circulation cleaning. *J. Dairy Sci.* 51: 1761.

Harper, W.J. 1972. Sanitation in dairy food plants, In *Food Sanitation.* R.K. Guthrie (ed.), AVI, Westport, Corn., p. 112.

Hauser, G., 1995. Hygienic design of equipment for open processing. *Trends Food Sci. Technol.* 6: 305.

Hauser, G., 1999. Hygienic building design. *New Food* 2: 17.

Henck, M. 1993. Recycling von Reinigungslaugen mit Hilfe der Crossflow-Filtration in der Milchwirtschaft. Dissertation Eidgen. Techn. Hochschule Zürich Nr. 10190.

Henneken, L., Klüner, T., Nörtemann, B., and Hempel, D.C. 1993. Reaktionstechnische Untersuchungen zum mikrobiellen Abbau von Ethylendiamintetraacetat (EDTA) in Submers- und Fließbettreaktoren. *Chemie-Ing. Technik* 65: 1068.

Hetsroni, G. 1982. In *Handbook of Multiphase Systems.* Hemisphere Publishing Cooperation, Washington, New York, London.

Hofer, O. 1993. Aufbereitung von CIP-Laugen mit Hilfe der Ultrafiltration. Diplomarbeit am Inst. F. Lebensmittelwissenschaft der Eidgen. Techn. Hochschule Zürich/Schweiz.

Hoffmann, W. 1983. Zirkulationsreinigen (CIP) von geraden Rohren in Abhängigkeit von Oberflächenrauheit und anderen Einflußfaktoren. Dissertation Universität Kiel.

Hoffmann, W. and Reuter, H. 1984a. Zirkulationsreinigen (CIP) von geraden Rohren in Abhängigkeit von der Obefächenrauheit. *Milchwissenschaft* 39: 416.

Hoffmann, W. and Reuter, H. 1984b. Zirkulationsreinigung (CIP) von geraden Rohren in Abhängigkeit on physikalischen Einflussfaktoren. *Milchwissenschaft* 39: 594.

Holah, J.T. et al. 1992. A method for assessing the in-place-cleanability of food processing equipment. *Trends Food Sci. Technol.* 3: 325.

International Dairy Federation (IDF), Bulletin Nr. 218 1987. Hygienic Design of Dairy Processing Equipment. IDF, Brussels/Belgium ISSN 0250 – 5118.

ISO/DIS 14159, Febr. 1999. *International Organization for Standardization*, DIN, Beuth-Verlag, D 10787 Berlin/Germany.

Jackson, A.T. and Low. M.W. 1982. Circulation cleaning of a plate heal exchanger fouled by tomato juice.III. The effect of fluid flow rate on cleaning efficiency. *J. Food Technol.* 17:745.

Jennings, W.G. 1959a. Circulation cleaning. II. Effects of entrained air. *J. Dairy Sci.* 42: 476.

Jennings, W.G. 1959b. Circulation cleaning. III. The kinetics of a simple detergent system. *J. Dairy Sci.* 42: 1763.

Jennings, W.G. 1959c. Effective in-place cleaning. *Food Eng.* 31: 98.

Jennings, W.G. 1960. Effect of pressure on circulation cleaning. *Food Technol.* 14: 591.

Jennings, W.G. 1961. A critical evaluation of in vitro radioactive phosphorus additions for estimating soil deposits. *J. Dairy Sci.* 44: 258.

Jennings, W.G. 1963a. An interpretive review of detergency for the food technologist. *Food Technol.* 17: 53.

Jennings, W.G. 1963b. A kinetic approach to detergent synergism. *J. Am. Oil Chem. Soc.* 40: 17.

Jennings, W.G. 1965. Theory and practice of hard surface cleaning. *Adv. Food Res.* 14: 325.

Jennings. W.G., McKillop, A.A., and Luick, J.R. 1957. Circulation cleaning. *J. Dairy Sci.* 40: 1471.

Johnson, R. (ed.) 1981. World Health Organization. *Public Health in Europe-14-Foodsafety Services* (Copenhagen).

Kernforschungszentrum Karlsruhe GmbH (Hrsg.) 1990. Aquatische Umweltverträglichkeit von Nitrilotriessigsäure". *Abschlußkolloquium der vom Bundesminister für Forschung und Technologie und vom Bundesminister für Umwelt, Naturschutz und Reaktorsicherheit geförderten Sondervorhaben am 6. Juni 1990 in Bad Godesberg/Germany.*

Kulkarni, S.M., Arnold, R.G., and Maxcy R.B. 1975. Reuse limits and regeneration of solutions for cleaning dairy equipment. *J. Dairy Sci.* 58:1095.

Lelieveld, H.L.M. 1985. Hygienic design and test methods. *Journal of the Soc. of Dairy Technology* 38: 14.

Loncin, M. and Merson, R.L. 1979. *Food Engineering.* Academic Press, New York.

Loncin, M. and Plett, E.A. 1985. Cleaning and disinfection in the food industry. *Kagaku Kogaku, (Chem. Eng.)* 49: 186.

McKinnon, C.H. and Cousins, C.M. 1969. Automatic spray of bulk tanks: the results of a field trial. *J. Soc. Dairy Tecnol.* 22: 227.

Merkblatt ATV-M 708 1994. Abwasser bei der Milchverarbeitung. Dezember 1994. ISBN 3-933693-64-0

Merkel, M., Thomas, K.-H. 2000. In *Taschenbuch der Werkstoffe.* Fachbuchverlag Leipzig im Carl Hanser Verlag, München - Wien. ISBN 3-446-21410-0

Nakanishi, K. and Kessler, H.G. 1985. Stability and rinsing behavior of deposited layers. *J. Food Sci.* 50.

Nassauer, J. 1985a. Adsorption und Haftung an Oberflächen und Membranen. *Habilitationsschrift Techn.* Universität München.

Nassauer, J. 1985b. Adsorption und Haftung an Oberflächen and Membranen.*Grundlagen-Anwendung-Probleme.* Freising-Weihenstephan, Eigenverlag.

Novalic, S., Dabrowski, A., and Kulbe, D. 1998. Nanofiltration of caustic and acidic cleaning solutions with high COD. *J. Food Eng.* 38: 125.

Otto, P., Nörtemann, B., and Hempel, C. 2000. Vermeidung von Präzipitatanreicherungen in Bioreaktoren. *Umwelt Technik Aktuell* 1/2000: 52.

Pießlinger-Schweiger, S. 1983. Elektropolieren hochwertiger funktioneller Edelstahloberflächen. *Chemie-Technik* 12: 91.

Plett, E.A. 1984. Rinsing kinetics of fluid food equipment. In *Engineering and Food, Vol. 2 Processing Applications,* B. McKenna (ed.), Applied Science, New York, p. 659.

Plett, E.A. 1985a. Zur Kinetik des Entfernens von Reinigungs- und Desinfektionsmitteln aus lebensmittelverarbeitenden Anlagen. *Fortschr. Ber. VDI Z. 3.*

Plett, E.A. 1985b. Relevant mass transfer mechanisms during rinsing. In *Fouling and Cleaning in Food Processing,* D.B. Lund, E.A. Plett, and C. Sandu (eds), University of Wisconsin-Madison Extension Duplicating, Madison, Wis.

Plett, E.A. 1986. Optimization of rinsing during CIP-procedure. In *Engineering and Food 1985.* M. Le Maguer and P. Jelen (eds), Applied Science, New York.

Plett, E.A., and Loncin, M. 1981a. Das Nachspülen bei der CIP-Reinigung lebensmittel-verarbeitenden Anlagen. *Int. Z. Lebensm. Technol. Verfahrenstech.* 32: 117.

Plett, E.A. and Loncin, M. 1981b. The rinsing problem in the food industry. In *Fundamentals and Applications of Surface Phenomena Associated with Fouling and Cleaning in Food Processing,* B. Hallström, B.D. Lund, and Ch. Träghardh (eds), Lund University Reprocentralen, Lund, Sweden, p. 365.

Plett, E.A. and Loncin, M. 1984. Entfernen von Reinigungs und Desinfektionsmitteln aus Rohren und Plattenwämetauschern. *Chem. Ing. Tech.* 56: 306.

Plett, E.A. and Loncin, M. 1985. Zur Problematik des Reinigens. In *Jahrbuch 1984 des Forschungskreises der Ernährungsindustrie,* Hannover, p. 139.

Potthoff, A., Serve, W., and Macharis, P. 1997. The cleaning revolution. *Dairy Industries International* 62: 25.

Potthoff-Karl, B. 1994. Neue biologisch abbaubare Komplexbildner. *Seifen-Öle-Fette-Wachse-Journal* 120: 104.

Reinemann, D.J. and Graßhoff, A. 1994. Two-Phase Cleaning Flow Dynamics in Air Injected Milklines. *Trans ASAE* 37: 1531.

Reinemann, D.J. and Mein, G.A. 1994. Transition from Stratified to Slug Flow in Milklines. *Trans. ASAE* 37: 655.

Reinemann, D.J. and Mein, G.A. 1995. Sizing Vacuum Pumps for Cleaning Milking Systems. *Proc. National Mastitis Council Annual Meeting,* Febr. 19–22, Fort Worth, Texas.

Reinemann, D.J., Wong, A.C.L., Muljadi, A., and Graßhoff, A. 1999. Efficacy Assessment of CIP Processes in Milking Machines. *Proc. Int. Conference of Fouling and Cleaning in Food Processing*, 6–8 April 1998, Cambridge, UK, ISBN 92-828-5609-7.

Schindler-Stokar, M. 1992. Abwasserbelastung durch Reinigung und Desinfektion. *International Seminar 'Oekologie in der Milchwirtschaft',* 3./4. Sept. 1992 in Zollikofen/Schweiz.

Schlüssler, H.J. 1970. Zur Reinigung fester Oberflächen, in der Lebensmittelindustrie. *Milchwissenschaft* 25: 133.

Schlüssler H.J. 1976. Zur Kinetik on Reinigungsvorgängen an festen Oberflächen. *Brauwissenschaft* 29: 263.

Schmauderer, E., Werlein, H.-D., and Otto, F., 1998, 1999. Edelstahloberflächen im Kontakt mit Lebensmitteln — Aspekte der Auswahl der Oberflächenvergütung und der Reinigungs- und Sanatisierungsmethoden nach Effizienz und Kostenaufwand. *Deutsche Lebensmittelrundschau* 94: 329; 94: 397 und 95: 7

Shapton, D.A. and Hindes, W.R. 1963. The standardization of a spore count technique. *Chemistry and Industry*, February 9: 230.

Smith, G.A. 1957. Efficient CIP cleaning: cold milk equipment. *Milk Prod. J.* 48: 14.

Tamplin, T.C. 1980. CIP-technology, detergents, and sanitizers. In *Hygienic Design and Operation of Food Plant*, R. Jowitt (ed.), AVI, Westport, Conn., p. 183.

Thor, W. and Loncin, M. 1978a. Reinigen, Desinfizieren und Nachspülen in der Lebensmittel-industrie. *Chem. Ing. Tech.* 50:188.

Thor, W. and Loncin, M. 1978b. Untersuchungen zum Nachspülen bei der CIPReinigung in der Milchindustrie. *Milchwissenschaft* 33: 665.

Timperley. D. 1981. The effect of Reynolds-number and mean velocity of flow on the cleaning in-place of pipelines. In *Fundamentals and Applications of Surface Phenomena Associated with Fouling and Cleaning in Food Processing*, B. Hallström, B.D. Lund, and Ch. Trägardh, (eds), Lund University Reprocentralen, Lund, Sweden, p. 402.

Timperley, D.A. 1984. Surface finish and spray cleaning of stainless steel. In *Profitability of Food Processing*, Institution of Chemical Engineers (eds), Symposium Series 84. Pergamon Press, Oxford, p. 31.

Timperley, D.A. and Lawson, G.B. 1980. Test rigs for evaluation of hygiene in plant design, In *Hygienic Design and Operation of Food Plant,* R. Jowitt (ed.), AVI, Westport, Conn., p. 79.

Timperley, A.W., Axis, J., Graßhoff, A., Hodge, C.R., Holah, J.T., Kirby, R., Maingonnat, J.F., Träghardh, C., Venema-Keur, B.M., and Cerf, O. 1993. A method for the assessment of in-lin steam sterilizability of food processing equipment. *Trends Food Sci. Technol.* 4: 80.

Tissier, J.-P. 1984. *Etude du rincage d'un reservoir de stockage: Aptitude au rincage de produits laitier, du saccharose et de solutions de nettoyage.* Diss. Univ. Clermont-Ferrand 11, France.

Tissier, J.-P., Corrieu, G., and Lalande, M. 1981. Global and kinetic aspects of pre-rinse and cleaning of a milk storage tank. In *Fundamentals and Applications of Surface Phenomena Associated with Fouling and Cleaning in Food Processing*, B. Hallström, D.B. Lund, and Ch. Trägardh, (eds), Lund University Reprocentralen, Lund, Sweden, p. 413.

Tissier, J.-P., Corrieu, G., and Lalande, M. 1983a. Etude du nettoyage de tanks de stockage, I. Description d'une methode de suivi des différentes etapes du nettoyage: essai de quantification des phénomènes. *L.W.T.* 17: 294.

Tissier, J.-P., Corrieu, G., and Lalande, M. 1983b. Etude de nettoyage de tanks de stockage. II. Evaluation de quelques peparamétres suceptibles d'influencer la cinétique de pré-rincage — modélisation. *L.W.T.* 17: 300.

Titz, M. 1993. Aufarbeitung von Reinigungslaugen aus Molkereibetrieben. Diplomarbeit an der Fakultät für Thermischen Maschinenbau der Technischen Universität Magdeburg/Germany.

Trägardh, G. and Johansson, D. 1998. Purification of alkaline cleaning solutions from the dairy industry using membrane separation technology. *Desalination* 119: 21.

Venema-Keur, B.M., Axis, J., Graßhoff, A., Hodge, C.R., Holah, J.T., Kirby, R., Maingonnat, J.F., Täghardh, C., and Cerf, O. 1993. A method for the assessment of in-line pasteurization of food processing equipment. *Trends Food Sci. Technol.* 4: 52.

Venema-Keur, B.M., Horan, S.P., Axis, J., Graßhoff, A., Kastelein, J., Ramsey, C., Haugan, K., Cerf, O., Bénézech, C., Trägardh, C., Mattila-Sandholm, T., Kirby, R., Ronner, U., and Königsfeld, P. 1997. A method for the assessment of in-place cleanability of moderately-sized food processing equipment. *Trends Food Sci. Technol.* 8: 54.

Wildbrett, G. 1983. Gummidichtungen und ihr Verhalten gegenüber von Reinigungs- und Desinfektionsmitteln. *Deutsche Molkerei Zeitung* 104: 212.

Wildbrett, G. 1996. In *Reinigung und Desinfektion in der Lebensmittelindustrie*. B. Behr's Verlag GmbH & Co, D-22085 Hamburg/Germany, ISBN 3-86022-232-5.

Wildbrett, G. and Böhner, B. 1990a. Abschätzung der Phosphor- und Stickstoffbelastungen in Abwässern aus Milcherzeugerbetrieben. *dmz - Lebensmittelindustrie und Milchwirtschaft* 111: 1080.

Wildbrett, G. and Böhner, B. 1990b. Phosphor- und Stickstoffbelastung in Abwässern aus milchverarbeitenden Betrieben. *dmz - Lebensmittelindustrie und Milchwirtschaft* 111: 1246.

World Health Organization, 1982, *European Cooperation of Environmental Health Aspects of the Control of Chemicals* (Copenhagen).

Appendix

TABLE A.1(a)
Physical Properties of Water at the Saturation Pressure

Temperature t (°C)	T (K)	Density ρ (kg/m^3)	Coefficient of volumetric thermal expansion β ($\times 10^{-4}$ K^{-1})	Specific heat c_p (kJ/[kg K])	Thermal conductivity k (W/[m K])	Thermal diffusivity α ($\times 10^{-6}$ m^2/s)	Absolute viscosity μ ($\times 10^{-6}$ Pa s)	Kinematic viscosity ν ($\times 10^{-6}$ m^2/s)	Prandtl number N_{Pr}
0	273.15	999.9	−0.7	4.226	0.558	0.131	1793.636	1.789	13.7
5	278.15	1000.0	—	4.206	0.568	0.135	1534.741	1.535	11.4
10	283.15	999.7	0.95	4.195	0.577	0.137	1296.439	1.300	9.5
15	288,15	999.1	—	4.187	0.587	0.141	1135.610	1.146	8.1
20	293.15	998.2	2.1	4.182	0.597	0.143	993.414	1.006	7.0
25	298.15	997.1	—	4.178	0.606	0.146	880.637	0.884	6.1
30	303.15	995.7	3.0	4.176	0.615	0.149	792.377	0.805	5.4
35	308.15	994.1	—	4.175	0.624	0.150	719.808	0.725	4.8
40	313.15	992.2	3.9	4.175	0.633	0.151	658.026	0.658	4.3
45	318.15	990.2	—	4.176	0.640	0.155	605.070	0.611	3.9
50	323.15	988.1	4.6	4.178	0.647	0.157	555.056	0.556	3.55
55	328.15	985.7	—	4.179	0.652	0.158	509.946	0.517	3.27
60	333.15	983.2	5.3	4.181	0.658	0.159	471.650	0.478	3.00
65	338.15	980.6	—	4.184	0.663	0.161	435.415	0.444	2.76
70	343.15	977.8	5.8	4.187	0.668	0.163	404.034	0.415	2.55
75	348.15	974.9	—	4.190	0.671	0.164	376.575	0.366	2.23
80	353.15	971.8	6.3	4.194	0.673	0.165	352.059	0.364	2.25
85	358.15	968.7	—	4.198	0.676	0.166	328.523	0.339	2.04
90	363.15	965.3	7.0	4.202	0.678	0.167	308.909	0.326	1.95
95	368.15	961.9	—	4.206	0.680	0.168	292.238	0.310	1.84
100	373.15	958.4	7.5	4.211	0.682	0.169	277.528	0.294	1.75
110	383.15	951.0	8.0	4.224	0.684	0.170	254.973	0.268	1.57
120	393.15	943.5	8.5	4.232	0.685	0.171	235.360	0.244	1.43
130	403.15	934.8	9.1	4.250	0.686	0.172	211.824	0.226	1.32
140	413.15	926.3	9.7	4.257	0.684	0.172	201.036	0.212	1.23
150	423.15	916.9	10.3	4.270	0.684	0.173	185.346	0.201	1.17
160	433.15	907.6	10.8	4.285	0.680	0.173	171.616	0.191	1.10
170	443.15	897.3	11.5	4.396	0.679	0.172	162.290	0.181	1.05
180	453.15	886.6	12.1	4.396	0.673	0.172	152.003	0.173	1.01
190	463.15	876.0	12.8	4.480	0.670	0.171	145.138	0.166	0.97
200	473.15	862.8	13.5	4.501	0.665	0.170	139.254	0.160	0.95

(Continued)

TABLE A.1(a)
Continued

Temperature		Density	Coefficient of volumetric thermal expansion	Specific heat	Thermal conductivity	Thermal diffusivity	Absolute viscosity	Kinematic viscosity	Prandtl number
t	T	ρ	β	c_p	k	α	μ	ν	
(°C)	(K)	(kg/m^3)	($\times 10^{-4}$ K^{-1})	(kJ/[kg K])	(W/[m K])	($\times 10^{-6}$ m^2/s)	($\times 10^{-6}$ Pa s)	($\times 10^{-6}$ m^2/s)	N_{Pr}
210	483.15	852.8	14.3	4.560	0.655	0.168	131.409	0.154	0.92
220	493.15	837.0	15.2	4.605	0.652	0.167	124.544	0.149	0.90
230	503.15	827.3	16.2	4.690	0.637	0.164	119.641	0.145	0.88
240	513.15	809.0	17.2	4.731	0.634	0.162	113.757	0.141	0.86
250	523.15	799.2	18.6	4.857	0.618	0.160	109.834	0.137	0.86

Source: Adapted from Raznjevic, K. 1978. *Handbook of Thermodynamic Tables and Charts.* Hemisphere Pub. Corp. Washington, DC.

TABLE A.1(b)
Physical Properties of Dry Air at Atmospheric Pressure

Temperature		Density	Volumetric coefficient of expansion	Specific heat	Thermal conductivity	Thermal diffusivity	Viscosity	Kinematic viscosity
t	T	ρ	β	c_p	k	α	μ	ν
(°C)	(K)	(kg/m^3)	($\times 10^{-4}$ K^{-1})	(kJ/[kg K])	(W/[m K])	($\times 10^{-6}$ m^2/s)	($\times 10^{-6}$ N s/m^2)	($\times 10^{-6}$ m^2/s)
−20	253.15	1.365	3.97	1.005	0.0226	16.8	16.279	12.0
0	273.15	1.252	3.65	1.011	0.0237	19.2	17.456	13.9
10	283.15	1.206	3.53	1.010	0.0244	20.7	17.848	14.66
20	293.15	1.164	3.41	1.012	0.0251	22.0	18.240	15.7
30	303.15	1.127	3.30	1.013	0.0258	23.4	18.682	16.58
40	313.15	1.092	3.20	1.014	0.0265	24.8	19.123	17.6
50	323.15	1.057	3.10	1.016	0.0272	26.2	19.515	18.58
60	333.15	1.025	3.00	1.017	0.0279	27.6	19.907	19.4
70	343.15	0.996	2.91	1.018	0.0286	29.2	20.398	20.65
80	353.15	0.968	2.83	1.019	0.0293	30.6	20.790	21.5
90	363.15	0.942	2.76	1.021	0.0300	32.2	21.231	22.82
100	373.15	0.916	2.69	1.022	0.0307	33.6	21.673	23.6
120	393.15	0.870	2.55	1.025	0.0320	37.0	22.555	25.9
140	413.15	0.827	2.43	1.027	0.0333	40.0	23.340	28.2
150	423.15	0.810	2.37	1.028	0.0336	41.2	23.732	29.4
160	433.15	0.789	2.31	1.030	0.0344	43.3	24.124	30.6
180	453.15	0.755	2.20	1.032	0.0357	47.0	24.909	33.0
200	473.15	0.723	2.11	1.035	0.0370	49.7	25.693	35.5
250	523.15	0.653	1.89	1.043	0.0400	60.0	27.557	42.2

Source: Adapted from Raznjevic, K. 1978. *Handbook of Thermodynamic Tables and Charts.* Hemisphere Pub. Corp. Washington, DC.

TABLE A.2
Power Law Parameters of Various Foods

Products	m (Pa sec^n)	n	R^2	m' (Pa sec$^{n'}$)	n'	R^2	λ (sec)
Apple butter	222.90	0.145	0.99	156.03	0.566	0.99	8.21×10^{-2}
Canned frosting	355.84	0.117	0.99	816.11	0.244	0.99	2.90×10^0
Honey	15.39	0.989	—	—	—	—	—
Ketchup	29.10	0.136	0.99	39.47	0.258	0.99	4.70×10^{-2}
Marshmallow cream	563.10	0.379	0.99	185.45	0.127	0.99	1.27×10^3
Mayonnaise	100.13	0.131	0.99	256.40	-0.048	0.99	2.51×10^{-1}
Mustard	35.05	0.196	0.99	65.69	0.136	0.99	2.90×10^0
Peanut butter	501.13	0.065	0.99	3785.00	0.175	0.99	1.86×10^5
Stick butter	199.28	0.085	0.99	3403.00	0.398	0.99	1.06×10^3
Stick margarine	297.58	0.074	0.99	3010.13	0.299	0.99	1.34×10^3
Squeeze margarine	8.68	0.124	0.99	15.70	0.168	0.99	9.93×10^{-2}
Tube margarine	106.68	0.077	0.99	177.20	0.353	0.99	5.16×10^1
Whipped butter	312.30	0.057	0.99	110.76	0.476	0.99	1.61×10^{-2}
Whipped cream cheese	422.30	0.058	0.99	363.70	0.418	0.99	8.60×10^{-2}
Whipped dessert topping	35.98	0.120	0.99	138.00	0.309	0.99	3.09×10^1

Source: Dickie, A.M. 1982. Predicting the spreadability and thickness of foods from time dependent viscoelastic rheology. M.S. Thesis, Rutgers University, New Brunswick, NJ.

TABLE A.3
Properties of Fruit and Vegetables

Product	Total solids (%)	Temp. (°C)	n	m (Pa.sec^n)	π_y (Pa)	Shear rate ranges (sec^{-1})
APPLE						
Pulp	—	25.0	0.084	65.03	—	—
Sauce	—	23.8	0.645	0.50	—	—
Sauce	—	23.8	0.408	0.66	—	—
Sauce	—	—	0.470	5.63	58.6	—
Sauce	—	20.0	0.302	16.68	—	3.3–530
Sauce + 12.5% water	—	25.0	0.438	2.39	—	0.1–1.1
Sauce	11.6	27.0	0.28	12.7	—	160–340
	11.0	30.0	0.30	11.6	—	5–50
	11.0	82.2	0.30	9.0	—	5–50
Sauce	10.5	26.0	0.45	7.32	—	0.78–1260
	9.6	26.0	0.45	5.63	—	0.78–1260
	8.5	26.0	0.44	4.18	—	0.78–1260
APRICOT						
Puree	17.7	26.6	0.29	5.4	—	—
	23.0	26.6	0.35	11.2	—	—
	41.4	26.6	0.35	54.0	—	—
	44.3	26.6	0.37	56.0	—	0.5–80
	51.4	26.6	0.36	108.0	—	0.5–80
	55.2	26.6	0.34	152.0	—	0.5–80
	59.3	26.6	0.32	300.0	—	0.5–80

(Continued)

TABLE A.3
Continued

Product	Total solids (%)	Temp. (°C)	n	m (Pa.sec^n)	π_y (Pa)	Shear rate ranges (sec^{-1})
Reliable, conc., green	27.0	4.4	0.25	170.0	—	3.3–137
	27.0	25.0	0.22	141.0	—	3.3–137
Reliable, conc., ripe	24.1	4.4	0.25	67.0	—	3.3–137
	24.1	25.0	0.22	54.0	—	3.3–137
Reliable, conc., ripened	25.6	4.4	0.24	85.5	—	3.3–137
	25.6	25.0	0.26	71.0	—	3.3–137
Reliable, conc., overripe	26.0	4.4	0.27	90.0	—	3.3–137
	26.0	25.0	0.30	67.0	—	3.3–137
BANANA						
Puree A	—	23.8	0.458	6.5	—	—
Puree B	—	23.8	0.333	10.7	—	—
Puree (17.7 brix)	—	22.0	0.283	107.3	—	28–200
BLUEBERRY						
Pie Filling	—	20.0	0.426	6.08	—	3.3–530
CARROT						
Puree	—	25.0	0.228	24.16	—	—
GREEN BEAN						
Puree	—	25.0	0.246	16.91	—	—
GUAVA						
Puree (10.3 brix)	—	23.4	0.494	38.98	—	15–400
MANGO						
Puree (9.3 brix)	—	24.2	0.334	20.58	—	15–1000
ORANGE JUICE CONCENTRATE						
Hamlin, early (42.5 brix)	—	25.0	0.585	4.121	—	0–500
	—	15.0	0.602	5.973	—	0–500
	—	0.0	0.676	9.157	—	0–500
	—	−10.0	0.705	14.255	—	0–500
Hamlin, late (41.1 brix)	—	25.0	0.725	1.930	—	0–500
	—	15.0	0.560	8.118	—	0–500
	—	0.0	0.620	1.754	—	0–500
	—	−10.0	0.708	13.875	—	0–500
Pineapple, early (40.3 brix)	—	25.0	0.643	2.613	—	0–500
	—	15.0	0.587	5.887	—	0–500
	—	0.0	0.681	8.938	—	0–500
	—	−10.0	0.713	12.184	—	0–500
Pineapple, late (41.8 brix)	—	25.0	0.532	8.564	—	0–500
	—	15.0	0.538	13.432	—	0–500
	—	0.0	0.636	18.584	—	0–500
	—	−10.0	0.629	36.414	—	0–500
Valencia, early (43.0 brix)	—	25.0	0.538	5.059	—	0–500
	—	15.0	0.609	6.714	—	0–500
	—	0.0	0.622	14.036	—	0–500
	—	−10.0	0.619	27.16	—	0–500

(Continued)

TABLE A.3
Continued

Product	Total solids (%)	Temp. (°C)	n	m (Pa·sec^n)	π_y (Pa)	Shear rate ranges (sec^{-1})
Valencia, late (41.9 brix)	—	25.0	0.538	8.417	—	0–500
	—	15.0	0.568	11.802	—	0–500
	—	0.0	0.644	18.751	—	0–500
	—	−10.0	0.628	41.412	—	0–500
Naval (65.1 brix)	—	−18.5	0.71	39.2	—	—
	—	−14.1	0.76	14.6	—	—
	—	−9.3	0.74	10.8	—	—
	—	−5.0	0.72	7.9	—	—
	—	−0.7	0.71	5.9	—	—
	—	10.1	0.73	2.7	—	—
	—	19.9	0.72	1.6	—	—
	—	29.5	0.74	0.9	—	—
PAPAYA						
Puree (7.3 brix)	—	26.0	0.528	9.09	—	20–450
PEACH						
Pie Filling	—	20.0	0.46	20.22	—	0.1–140
Puree	10.9	26.6	0.44	0.94	—	—
	17.0	26.6	0.55	1.38	—	—
	21.9	26.6	0.55	2.11	—	—
	26.0	26.6	0.40	13.4	—	80–1000
	29.6	26.6	0.40	18.0	—	80–1000
	37.5	26.6	0.38	44.0	—	—
	40.1	26.6	0.35	58.5	—	2–300
	49.8	26.6	0.34	85.5	—	2–300
	58.4	26.6	0.34	440.0	—	—
Puree	11.7	30.0	0.28	7.2	—	5–50
	11.7	82.2	0.27	5.8	—	5–50
	10.0	27.0	0.34	4.5	—	160–3200
PEAR						
Puree	15.2	26.6	0.35	4.3	—	—
	24.3	26.6	0.39	5.8	—	—
	33.4	26.6	0.38	38.5	—	80–1000
	37.6	26.6	0.38	49.7	—	—
	39.5	26.6	0.38	64.8	—	2–300
	47.6	26.6	0.33	120.0	—	0.5–10
	49.3	26.6	0.34	170.0	—	—
	51.3	26.6	0.34	205.0	—	—
	45.8	32.2	0.479	35.5	—	—
	45.8	48.8	0.477	26.0	—	—
	45.8	65.5	0.484	20.0	—	—
	45.8	82.2	0.481	16.0	—	—
	14.0	30.0	0.35	5.6	—	5–50
	14.0	82.2	0.35	4.6	—	5–50
PLUM						
Puree	14.0	30.0	0.34	2.2	—	5–50
	14.0	8.2	0.34	2.0	—	5–50
	—	25.0	0.222	5.7	—	—

(Continued)

TABLE A.3
Continued

Product	Total solids (%)	Temp. (°C)	n	m (Pa.sec^n)	π_y (Pa)	Shear rate ranges (sec^{-1})
SQUASH						
Puree A	—	25.0	0.149	20.65	—	—
Puree B	—	25.0	0.281	11.42	—	—
TOMATO						
Juice Concentrate	5.8	32.2	0.590	0.223	—	500–800
	5.0	48.8	0.540	0.27	—	500–800
	5.8	65.5	0.470	0.37	—	500–800
	12.8	32.2	0.430	2.00	—	500–800
	12.8	48.8	0.430	1.88	—	500–800
	12.8	65.5	0.340	2.28	—	500–800
	12.8	82.2	0.350	2.12	—	500–800
	16.0	32.2	0.450	3.16	—	500–800
	16.0	48.8	0.450	2.77	—	500–800
	16.0	65.5	0.400	3.18	—	500–800
	16.0	82.2	0.380	3.27	—	500–800
	25.0	32.2	0.410	12.9	—	500–800
	25.0	48.8	0.420	10.5	—	500–800
	25.5	65.5	0.430	8.0	—	500–800
	25.0	82.2	0.430	6.1	—	500–800
	30.0	32.2	0.400	18.7	—	500–800
	30.0	48.8	0.420	15.1	—	500–800
	30.0	65.5	0.430	11.7	—	500–800
	30.0	82.2	0.450	7.9	—	500–800
Ketchup	—	25.0	0.27	18.7	32	10–560
	—	45.0	0.29	16.0	24	10–560
	—	65.0	0.29	11.3	14	10–560
	—	95.0	0.253	7.45	10.5	10–560
Puree	—	25.0	0.236	7.78	—	—
	—	47.7	0.550	1.08	2.04	—

Source: Steffe, J.F., Mohamed, I.O., and Ford, E.W., 1986, Rheological properties of fluid foods: data compilation. In: *Physical and Chemical Properties of Food*, M.E. Okos, Ed., ASAE Publications.

TABLE A.4
Properties of Apple and Grape Juice Concentrates

	°Brix	η_0 (Pa sec)	E_a kcal/gmole	Temp range (°C)
Apple juice concentrate	45.1	3.394×10^{-7}	6.0	−5 to 40
(from McIntosh apples)	50.4	1.182×10^{-7}	6.9	−10 to 40
	55.2	2.703×10^{-9}	9.4	−15 to 40
	60.1	3.935×10^{-10}	10.9	−15 to 40
	64.9	7.917×10^{-12}	13.6	−15 to 40
	68.3	1.156×10^{-12}	15.3	−15 to 40
Grape juice concentrate	43.1	8.147×10^{-8}	7.0	−5 to 40
(from concord grapes)	49.2	1.074×10^{-8}	8.5	−10 to 40
	54.0	9.169×10^{-8}	10.3	−15 to 40
	59.2	1.243×10^{-10}	11.8	−15 to 40
	64.5	1.340×10^{-10}	12.3	−15 to 40
	68.3	6.086×10^{-12}	14.5	−15 to 40

Source: Rao, M.A., Cooley, H.J., and Vitali, A.A., 1984. Flow properties of concentrated juices at low temperatures, *Food Technology*, 38: 113–119.

TABLE A.5
Properties of Meat, Fish and Dairy Products

Product	Total solids (%)	Temp. (°C)	n	m (Pa.sec^n)	π_y (Pa)	Shear rate ranges (sec^{-1})
CREAM						
10% Fat	—	40	1.0	0.00148	—	—
	—	60	1.0	0.00107	—	—
	—	80	1.0	0.00083	—	—
20% Fat	—	40	1.0	0.00238	—	—
	—	60	1.0	0.00171	—	—
	—	80	1.0	0.00129	—	—
30% Fat	—	40	1.0	0.00395	—	—
	—	60	1.0	0.00289	—	—
	—	80	1.0	0.00220	—	—
40% Fat	—	40	1.0	0.00690	—	—
	—	60	1.0	0.00510	—	—
	—	80	1.0	0.00395	—	—
FISH						
Minced paste	—	3–6	0.91	8.55	1600	0.7–238
MEAT						
Raw comminated batters						

% Fat	% Prot.	% MC						
15.0	13.0	66.8	—	15	0.156	639.3	1.53	300–500
18.7	12.9	65.9	—	15	0.104	858.0	0.28	300–500
22.5	12.1	63.2	—	15	0.209	429.5	0.00	300–500
30.0	10.4	57.5	—	15	0.341	160.2	27.80	300–500

(Continued)

TABLE A.5
Continued

Product			Total solids (%)	Temp. (°C)	n	m (Pa.sec^n)	π_y (Pa)	Shear rate ranges (sec^{-1})
33.8	9.5	54.5	—	15	0.390	103.3	17.90	300–500
45.0	6.9	45.9	—	15	0.723	14.0	2.30	300–500
45.0	6.9	45.9	—	15	0.685	17.9	27.60	300–500
67.3	28.9	1.8	—	15	0.205	306.8	0.00	300–500
MILK								
Homogenized			—	20	1.0	0.00200	—	—
			—	30	1.0	0.00150	—	—
			—	40	1.0	0.00110	—	—
			—	50	1.0	0.00095	—	—
			—	60	1.0	0.00078	—	—
			—	70	1.0	0.00070	—	—
			—	80	1.0	0.00060	—	—
Raw			—	0	1.0	0.00344	—	—
			—	5	1.0	0.00305	—	—
			—	10	1.0	0.00264	—	—
			—	15	1.0	0.00231	—	—
			—	20	1.0	0.00199	—	—
			—	25	1.0	0.00170	—	—
			—	30	1.0	0.00149	—	—
			—	35	1.0	0.00134	—	—
			—	40	1.0	0.00123	—	—
WHOLE SOYBEAN								
7% Soy Cotyledon Solids			—	10	0.85	0.0640	—	0–1300
7% Soy Cotyledon Solids			—	20	0.84	0.0400	—	0–1300
7% Soy Cotyledon Solids			—	30	0.80	0.0400	—	0–1300
7% Soy Cotyledon Solids			—	40	0.81	0.0330	—	0–1300
7% Soy Cotyledon Solids			—	50	0.82	0.0270	—	0–1300
7% Soy Cotyledon Solids			—	60	0.83	0.0240	—	0–1300
4.9% Soy Cotyledon Solids			—	25	0.90	0.0187	—	0–1300
6.2% Soy Cotyledon Solids			—	25	0.85	0.0415	—	0–1300
7.2% Soy Cotyledon Solids			—	25	0.84	0.0665	—	0–1300
8.1% Soy Cotyledon Solids			—	25	0.78	0.1171	—	0–1300
9.0% Soy Cotyledon Solids			—	25	0.76	0.2133	—	0–1300
10.2% Soy Cotyledon Solids			—	25	0.71	0.4880	—	0–1300

Source: Steffe, J.F., Mohamed, I.O., and Ford, E.W., 1986, Rheological properties of fluid foods: data compilation. In: *Physical and Chemical Properties of Food*, M.E. Okos, Ed., ASAE Publications.

TABLE A.6
Properties of Oils and Miscellaneous Products

Product	Total solids (%)	Temp. (°C)	n	m (Pa.s^n)	π_y (Pa)	Shear rate ranges (sec^{-1})
CHOCOLATE						
Melted	—	46.1	0.574	0.57	1.16	—
HONEY						
Buckwheat	18.6	24.8	1.0	3.86	—	—
Golden rod	19.4	24.3	1.0	2.93		—
Sage	18.6	25.9	1.0	8.88		—
Sweet clover	17.0	24.7	1.0	7.2		—
White clover	18.2	25.0	1.0	4.8		—
MAYONNAISE	—	25.0	0.55	6.4	—	30–1300
	—	25.0	0.54	6.6	—	30–1300
	—	25.0	0.60	4.2	—	40–1100
	—	25.0	0.59	4.7	—	40–1100
MUSTARD	—	25.0	0.39	18.5	—	30–1300
	—	25.0	0.39	19.1	—	30–1300
	—	25.0	0.34	27	—	40–1100
	—	25.0	0.28	33	—	−40–1100
OILS						
Castor	—	10.0	1.0	2.42	—	—
	—	30.0	1.0	0.451	—	—
	—	40.0	1.0	0.231	—	—
	—	100.0	1.0	0.0169	—	—
Corn	—	38.0	1.0	0.0317	—	—
	—	25.0	1.0	0.0565	—	—
Cottonseed	—	20.0	1.0	0.0704	—	—
	—	38.0	1.0	0.0386	—	—
Linseed	—	50.0	1.0	0.0176	—	—
	—	90.0	1.0	0.0071	—	—
Olive	—	10.0	1.0	0.1380	—	—
	—	40.0	1.0	0.0363	—	—
	—	70.0	1.0	0.0124	—	—
Peanut	—	25.0	1.0	0.0656	—	—
	—	38.0	1.0	0.0251	—	—
	—	21.1	1.0	0.0647	—	0.32–64
	—	37.8	1.0	0.0387	—	0.32–64
	—	54.4	1.0	0.0268	—	0.32–64
Rapeseed	—	0.0	1.0	2.530	—	—
	—	20.0	1.0	0.163	—	—
	—	30.0	1.0	0.096	—	—
Safflower	—	38.0	1.0	0.0286	—	—
	—	25.0	1.0	0.0922	—	—
Sesame	—	38.0	1.0	0.0324	—	—
Soybean	—	30.0	1.0	0.0406	—	—
	—	50.0	1.0	0.0206	—	—
	—	90.0	1.0	0.0078	—	—
Sunflower	—	38.0	1.0	0.0311	—	—

Source: Steffe, J.F., Mohamed, I.O., and Ford, E.W., 1986, Rheological properties In: *Physical and Chemical Properties of Food*, M.E. Okos, Ed., ASAE Publications.

TABLE A.7
Vapor Pressure of Ice, p_i, and Water, p_u, Vapor Volume, V_v, and Heat of Sublimation, ΔH_s at Temperatures Below 0°C

T (°C)	p_i (mbar)	p_u (mbar)	V_v (m³/Kg)	ΔH_s (J/g)
−98	0.00002			
−90	0.00009			
−80	0.00053			
−76	0.00103		908600	
−70	0.00259			
−65	0.00500		191600	
−60	0.01077		98110	
−55	0.02106		50260	
−50	0.03939		25760	
−45	0.07265		15030	
−40	0.12876			
−36	0.20088		5472	
−30	0.38110		2800	
−25	0.63451			
−20	1.03441	1.25323		
−15	1.65425	1.91419		
−10	2.59935	2.86462		
−5	4.01633	4.21628		
0	6.10381	6.10381		2834

TABLE A.8
Vapor Pressure, p_A^0, Temperature, T, and Corresponding Latent Heat of Vaporization, $\Delta_{vap}H$, and Volume, V_{vap}, of Vapor

p_A^0 (mbar)	T (°C)	V_{vap} (m³/kg)	$\Delta_{vap}H$ (kJ/kg)
10	6.98	129.21	2485
20	17.51	67.01	2460
30	24.09	45.69	2445
40	28.97	34.82	2433
50	32.89	28.20	2424
60	36.18	23.74	2416
70	39.02	20.53	2409
80	41.54	18.10	2403
90	43.79	16.20	2398
100	45.84	14.67	2393
150	54.00	10.02	2373
200	60.09	7.650	2359
250	64.99	6.206	2346
300	69.12	5.231	2336
400	75.88	3.995	2319
500	81.34	3.241	2305
600	85.95	2.732	2294
700	89.96	2.365	2283
800	93.51	2.087	2274
900	96.71	1.869	2266
1000	99.63	1.694	2258
1500	111.4	1.159	2226
2000	120.2	0.8853	2202
2500	127.4	0.7184	2181
3000	133.5	0.6056	2163
4000	143.6	0.4623	2133
5000	151.8	0.3747	2108
6000	158.8	0.3155	2085
7000	164.9	0.2727	2065
8000	170.4	0.2403	2047
9000	175.4	0.2148	2030
10000	179.9	0.1943	2014
15000	198.3	0.1316	1945
20000	212.4	0.09952	1889
25000	223.9	0.07990	1839
30000	233.8	0.06663	1794
40000	250.3	0.04975	1713
50000	263.9	0.03943	1640

TABLE A.9
Coefficients to Estimate Food Properties

Property	Component	Temperature function	Standard error	Standard % error
k (W/[m°C])	Protein	$k = 1.7881 \times 10^{-1} + 1.1958 \times 10^{-3}T - 2.7178 \times 10^{-6}T^2$	0.012	5.91
	Fat	$k = 1.8071 \times 10^{-1} - 2.7064 \times 10^{-3}T - 1.7749 \times 10^{-7}T^2$	0.0032	1.95
	Carbohydrate	$k = 2.0141 \times 10^{-1} + 1.3874 \times 10^{-3}T - 4.3312 \times 10^{-6}T^2$	0.0134	5.42
	Fiber	$k = 1.8331 \times 10^{-1} + 1.2497 \times 10^{-3}T - 3.1683 \times 10^{-6}T^2$	0.0127	5.55
	Ash	$k = 3.2962 \times 10^{-1} + 1.4011 \times 10^{-3}T - 2.9069 \times 10^{-6}T^2$	0.0083	2.15
	Water	$k = 5.7109 \times 10^{-1} + 1.7625 \times 10^{-3}T - 6.7063 \times 10^{-6}T^2$	0.0028	0.45
	Ice	$k = 2.2196 - 6.2459 \times 10^{-3}T + 1.0154 \times 10^{-4}T^2$	0.0079	0.79
α (m²/s)	Protein	$\alpha = 6.8714 \times 10^{-2} + 4.7578 \times 10^{-4}T - 1.4646 \times 10^{-6}T^2$	0.0038	4.50
	Fat	$\alpha = 9.8777 \times 10^{-2} - 1.2569 \times 10^{-4}T - 3.8286 \times 10^{-8}T^2$	0.0020	2.15
	Carbohydrate	$\alpha = 8.0842 \times 10^{-2} + 5.3052 \times 10^{-4}T - 2.3218 \times 10^{-6}T^2$	0.0058	5.84
	Fiber	$\alpha = 7.3976 \times 10^{-2} + 5.1902 \times 10^{-4}T - 2.2202 \times 10^{-6}T^2$	0.0026	3.14
	Ash	$\alpha = 1.2461 \times 10^{-1} + 3.7321 \times 10^{-4}T - 1.2244 \times 10^{-6}T^2$	0.0022	1.61
	Water	$\alpha = 1.3168 \times 10^{-1} + 6.2477 \times 10^{-4}T - 2.4022 \times 10^{-6}T^2$	0.0022×10^{-6}	1.44
	Ice	$\alpha = 1.1756 - 6.0833 \times 10^{-3}T + 9.5037 \times 10^{-5}T^2$	0.0044×10^{-6}	0.33
ρ (kg/m³)	Protein	$\rho = 1.3299 \times 10^3 - 5.1840 \times 10^{-1}T$	39.9501	3.07
	Fat	$\rho = 9.2559 \times 10^2 - 4.1757 \times 10^{-1}T$	4.2554	0.47
	Carbohydrate	$\rho = 1.5991 \times 10^3 - 3.1046 \times 10^{-1}T$	93.1249	5.98
	Fiber	$\rho = 1.3115 \times 10^3 - 3.6589 \times 10^{-1}T$	8.2687	0.64
	Ash	$\rho = 2.4238 \times 10^3 - 2.8063 \times 10^{-1}T$	2.2315	0.09
	Water	$\rho = 9.9718 \times 10^2 + 3.1439 \times 10^{-3}T - 3.7574 \times 10^{-3}T^2$	2.1044	0.22
	Ice	$\rho = 9.1689 \times 10^2 - 1.3071 \times 10^{-1}T$	0.5382	0.06
c_p (j/[kg°C])	Protein	$c_p = 2.0082 + 1.2089 \times 10^{-3}T - 1.3129 \times 10^{-6}T^2$	0.1147	5.57
	Fat	$c_p = 1.9842 + 1.4733 \times 10^{-3}T - 4.8006 \times 10^{-6}T^2$	0.0236	1.16
	Carbohydrate	$c_p = 1.5488 + 1.9625 \times 10^{-3}T - 5.9399 \times 10^{-6}T^2$	0.0986	5.96
	Fiber	$c_p = 1.8459 + 1.8306 \times 10^{-3}T - 4.6509 \times 10^{-6}T^2$	0.0293	1.66
	Ash	$c_p = 1.0926 + 1.8896 \times 10^{-3}T - 3.6817 \times 10^{-6}T^2$	0.0296	2.47
	Water[a]	$c_p = 4.0817 - 5.3062 \times 10^{-3}T + 9.9516 \times 10^{-4}T^2$	0.0988	2.15
	Water[b]	$c_p = 4.1762 - 9.0864 \times 10^{-5}T + 5.4731 \times 10^{-6}T^2$	0.0159	0.38
	Ice	$c_p = 2.0623 + 6.0769 \times 10^{-3}T$		

[a] For the temperature of -40 to $0°C$.
[b] For the temperature of 0 to $150°C$.

Source: From Choi, Y. and Okos, M.R. 1986. *Physical and Chemical Properties of Food.* Martin R. Okos (Ed.). ASAE, St. Joseph, MI, pp. 35–77.

TABLE A.10
Storage Requirements and Properties of Perishable Products

Commodity	Storage Temp. (°C)	Relative humidity (%)	Approximate storage life[a]	Water content (%)	Highest freezing (°C)	Specific heat above freezing[b] (J/kg ·°C)	Specific heat below freezing[b] (J/kg ·°C)	Latent heat[c] (J/kg)
			Vegetables[d]					
Artichokes								
Globe	0	95	2 weeks	84	−1.2	3.651	1.892	280.18
Jerusalem	0	90 to 95	5 months	80	−2.5[b]	3.517	1.842	266.84
Asparagus	0 to 2	95	2 to 3 weeks	93	−0.6	3.952	2.005	310.20
Beans								
Snap or green	4 to 7	90 to 95	7 to 10 days	89	−0.7	3.818	1.955	296.86
Lima	0 to 4	90 to 95	3 to 5 days	67	−0.6	3.081	1.679	223.48
Dried	10	70	6 to 8 months	11		1.206	0.975	
Beets								
Roots	0	95 to 100	4 to 6 months	88	−0.9	3.785	1.943	293.52
Bunch	0	95	10 to 14 days		−0.4			
Broccoli	0	95	10 to 14 days	90	−0.6	3.852	1.968	300.20
Brussels sprouts	0	95	3 to 5 weeks	85	−0.8	3.684	1.905	283.52
Cabbage late	0	95 to 100	5 to 6 months	92	−0.9	3.919	1.993	306.87
Carrots								
Topped, immature	0	98 to 100	4 to 6 weeks	88	−1.4	3.785	1.943	293.52
Topped, mature	0	98 to 100	5 to 9 months	88	−1.4	3.785	1.943	293.52
Cauliflower	0	95	2 to 4 weeks	92	−0.8	3.919	1.993	306.87
Celeriac	0	95 to 100	3 to 4 months	88	−0.9	3.785	1.943	293.52
Celery	0	95	1 to 2 months	94	−0.5	3.986	2.018	313.54
Collards	0	95	10 to 14 days	87	−0.8	3.751	1.930	290.19
Corn, sweet	0	95	4 to 8 days	74	−0.6	3.316	1.767	246.83
Cucumbers	10 to 13	90 to 95	10 to 14 days	96	−0.5	4.053	2.043	320.21
Eggplant	7 to 10	90 to 95	7 to 10 days	93	−0.8	3.952	2.005	310.20
Endive (escarole)	0	95	2 to 3 weeks	93	−0.1	3.952	2.005	310.20
Frozen vegetables	−23 to −18		6 to 12 months					
Garlic, dry	0	65 to 70	6 to 7 months	61	−0.8	2.880	1.604	203.46
Greens, leafy	0	95	10 to 14 days	93	−0.3	3.952	2.005	310.20
Horseradish	−1 to 0	95 to 100	10 to 12 months	75	−1.8	3.349	1.779	250.16
Kale	0	95	3 to 4 weeks	87	−0.5	3.751	1.930	290.19
Kohlrabi	0	95	2 to 4 weeks	90	−1.0	3.852	1.968	300.20
Leeks, green	0	95	1 to 3 months	85	−0.7	3.684	1.905	283.52
Lettuce, head	0 to 1	95 to 100	2 to 3 weeks	95	−0.2	4.019	2.031	316.87
Mushrooms	0	90	3 to 4 days	91	−0.9	3.885	1.980	303.53
Okra	7 to 10	90 to 95	7 to 10 days	90	−1.8	3.852	1.968	300.20
Onions								
Greens	0	95	3 to 4 weeks	89	−0.9	3.818	1.955	296.86
Dry and onion sets	0	65 to 75	1 to 8 months	88	−0.8	3.785	1.943	293.52

(Continued)

TABLE A.10
Continued

Commodity	Storage Temp. (°C)	Relative humidity (%)	Approximate storage life[a]	Water content (%)	Highest freezing (°C)	Specific heat above freezing[b] (J/kg ·°C)	Specific heat below freezing[b] (J/kg ·°C)	Latent heat[c] (J/kg)
Parsley	0	95	1 to 2 months	85	−1.1	3.684	1.905	283.52
Parsnips	0	98 to 100	4 to 6 months	79	−0.9	3.483	1.830	263.50
Peas								
Green	0	95	1 to 3 weeks	74	−0.6	3.316	1.767	246.83
Dried	10	70	6 to 8 months	12		1.239	0.988	
Peppers								
Dried	0 to 10	60 to 70	6 months	12		1.239	0.988	
Sweet	7 to 10	90 to 95	2 to 3 weeks	92	−0.7	3.919	1.993	306.87
Potatoes								
Early	10 to 13	90		81	−0.6	3.550	1.855	270.18
Main crop	3 to 10	90 to 95	5 to 8 months	78	−0.7	3.450	1.817	260.17
Sweet	13 to 16	85 to 90	4 to 7 months	69	−1.3	3.148	1.704	230.15
Pumpkins	10 to 13	70 to 75	2 to 3 months	91	−0.8	3.885	1.980	303.53
Radishes								
Spring	0	95	3 to 4 weeks	95	−0.7	4.019	2.031	316.87
Winter	0	95 to 100	2 to 4 months	95	−0.7	4.019	2.031	316.87
Rhubarb	0	95	2 to 4 weeks	95	−0.9	4.019	2.031	316.87
Rutabagas	0	98 to 100	4 to 6 months	89	−1.1	3.818	1.955	296.86
Salsify	0	98 to 100	2 to 4 months	79	−1.1	3.483	1.830	263.50
Seed, vegetable	0 to 10	50 to 65	10 to 12 months	7 to 15		1.206	0.976	036.69
Spinach	0	95	10 to 14 days	93	−0.3	3.952	2.005	310.20
Squash								
Acorn	7 to 10	70 to 75	5 to 8 weeks		−0.8			
Summer	0 to 10	85 to 95	5 to 14 days	94	−0.5	3.986	2.018	313.54
Winter	10 to 13	70 to 75	4 to 6 months	85	−0.8	3.684	1.905	283.52
Tomatoes								
Mature green	13 to 21	85 to 90	1 to 3 weeks	93	−0.6	3.952	2.005	310.20
Firm, ripe	7 to 10	85 to 90	4 to 7 days	94	−0.5	3.986	2.018	313.54
Turnips								
Roots	0	95	4 to 5 months	92	−1.1	3.919	1.993	306.87
Greens	0	95	10 to 14 days	90	−0.2	3.852	1.968	300.20
Watercress	0	95	3 to 4 days	93	−0.3	3.952	2.005	310.20
Yams	16	85 to 90	3 to 6 months	74		3.316	1.767	246.83
Fruits and Melons[d]								
Apples	−1 to 4	90	3 to 8 months	84	−1.1	3.651	1.892	280.18
Dried	0 to 5	55 to 60	5 to 8 months	24		1.641	1.139	
Apricots	0	90	1 to 2 weeks	85	−1.1	3.684	1.905	283.52
Avocados	4 to 13	85 to 90	2 to 4 weeks	65	−0.3	3.014	1.654	216.81
Bananas	d	85 to 95	d	75	−0.8	3.349	1.779	250.16
Blackberries	−0.5 to 0	95	3 days	85	−0.8	3.684	1.905	283.52
Blueberries	−1 to 0	90 to 95	2 weeks	82	−1.6	3.584	1.867	273.51
Cantaloupes	2 to 4	90 to 95	5 to 15 days	92	−1.2	3.919	1.993	306.87

(Continued)

TABLE A.10
Continued

Commodity	Storage Temp. (°C)	Relative humidity (%)	Approximate storage life[a]	Water content (%)	Highest freezing (°C)	Specific heat above freezing[b] (J/kg ·°C)	Specific heat below freezing[b] (J/kg ·°C)	Latent heat[c] (J/kg)
Cherries								
Sour	−1 to 0	90 to 95	3 to 7 days	84	−1.7	3.651	1.892	280.18
Sweet	−1	90 to 95	2 to 3 weeks	80	−1.8	3.517	1.842	266.84
Casaba melons	7 to 10	85 to 95	4 to 6 weeks	93	−1.1	3.952	2.005	310.20
Cranberries	2 to 4	90 to 95	2 to 4 months	87	−0.9	3.751	1.930	290.19
Currants	−0.5 to 0	90 to 95	10 to 14 days	85	−1.0	3.684	1.905	283.52
Dates, cured	−18 or 0	75 or less	6 to 12 months	20	15.7	1.507	1.089	66.71
Dewberries	−1 to 0	90 to 95	3 days	85	−1.3	3.684	1.905	283.52
Figs								
Dried	0 to 4	50 to 60	9 to 12 months	23		1.608	1.126	76.72
Fresh	−1 to 0	85 to 90	7 to 10 days	78	−2.4	3.450	1.817	260.17
Frozen fruits	−23 to −18	90 to 95	6 to 12 months					
Gooseberries	−1 to 0	90 to 95	2 to 4 weeks	89	−1.1	3.818	1.955	296.86
Grapefruit	10 to 16	85 to 90	4 to 6 weeks	89	−1.1	3.818	1.955	296.86
Grapes								
American	−1 to 0	85 to 90	2 to 8 weeks	82	−1.6	3.584	1.867	273.51
Vinifera	−1	90 to 95	3 to 6 months	82	−2.1	3.584	1.867	273.51
Guavas	7 to 10	90	2 to 3 weeks	83		3.617	1.880	276.85
Honeydew melons	7 to 10	90 to 95	3 to 4 weeks	93	−0.9	3.952	2.005	310.20
Lemons	0 or 10 to 14[e]	85 to 90	1 to. 6 months	89	−1.4	3.818	1.955	296.86
Limes	9 to 10	85 to 90	6 to 18 weeks	86	−1.6	3.718	1.918	286.85
Mangoes	13	85 to 90	2 to 3 weeks	81	−0.9	3.550	1.855	270.18
Nectarines	−0.5 to 0	90	2 to 4 weeks	82	−0.9	3.584	1.867	273.51
Olives, fresh	7 to 10	85 to 90	4 to 6 weeks	75	−1.4	3.349	1.779	250.16
Oranges	0 to 9	85 to 90	3 to 12 weeks	87	−0.8	3.751	1.930	290.19
Papayas	7	85 to 90	1 to 3 weeks	91	−0.8	3.885	1.980	303.53
Peaches	−0.5 to 0	90	2 to 4 weeks	89	−0.9	3.818	1.955	296.86
Dried	0 to 5	55 to 60	5 to 8 months	25		1.675	1.151	
Pears	−1.6 to −0.5	90 to 95	2 to 7 months	83	−1.6	3.617	1.880	276.85
Persian melons	7 to 10	90 to 95	2 weeks	93	−0.8	3.952	2.005	310.20
Persimmons	−1	90	3 to 4 months	78	−2.2	3.450	1.817	260.17
Pineapples, ripe	7	85 to 90	2 to 4 weeks	85	−1.0	3.684	1.905	283.52
Plums	−1 to 0	90 to 95	2 to 4 weeks	86	−0.8	3.718	1.918	286.85
Pomegranates	0	90	2 to 4 weeks	82	−3.0	3.584	1.867	273.51
Prunes								
Fresh	−1 to 0	90 to 95	2 to 4 weeks	86	−0.8	3.718	1.918	286.85
Dried	0 to 5	55 to 60	5 to 8 months	28		1.775	1.189	
Quinces	−1 to 0	90	2 to 3 months	85	−0.2	3.684	1.905	283.52
Raisins				18		1.440	1.063	
Raspberries								
Black	−0.5 to 0	90 to 95	2 to 3 days	81	−1.1	3.550	1.855	270.18
Red	−0.5 to 0	90 to 95	2 to 3 days	84	−0.6	3.651	1.892	280.18

(Continued)

TABLE A.10
Continued

Commodity	Storage Temp. (°C)	Relative humidity (%)	Approximate storage life[a]	Water content (%)	Highest freezing (°C)	Specific heat above freezing[b] (J/kg·°C)	Specific heat below freezing[b] (J/kg·°C)	Latent heat[c] (J/kg)
Strawberries	−0.5 to 0	90 to 95	5 to 7 days	90	−0.8	3.852	1.968	300.20
Tangerines	0 to 3	85 to 90	2 to 4 weeks	87	−1.1	3.751	1.930	290.19
Watermelons	4 to 10	80 to 90	2 to 3 weeks	93	−0.4	3.952	2.005	310.20
Seafood (Fish)[d]								
Haddock cod, perch	−1 to 1	95 to 100	12 days	81	−2.2	3.550	1.855	270.17
Hake, whiting	0 to 1	95 to 100	10 days	81	−2.2	3.550	1.855	270.17
Halibut	−1 to 1	95 to 100	18 days	75	−2.2	3.349	1.779	250.16
Herring								
Kippered	0 to 2	80 to 90	10 days	61	−2.2	2.880	1.604	203.46
Smoked	0 to 2	80 to 90	10 days	64	−2.2	2.981	1.641	213.47
Mackerel	0 to 1	95 to 100	6 to 8 days	65	−2.2	3.014	1.654	216.81
Menhaden	1 to 5	95 to 100	4 to 5 days	62	−2.2	2.914	1.615	206.80
Salmon	−1 to 1	95 to 100	18 days	64	−2.2	2.981	1.641	213.47
Tuna	0 to 2	95 to 100	14 days	70	−2.2	3.182	1.717	233.49
Frozen fish	−29 to −18	90 to 95	6 to 12 months					
Seafood (Shellfish)[d]								
Scallop meat	0 to 1	95 to 100	12 days	80	−2.2	3.517	1.842	266.84
Shrimp	−1 to 1	95 to 100	12 to 14 days	76	−2.2	3.383	1.792	253.50
Lobster, American	5 to 10	In sea water	Indefinitely in sea water	79	−2.2	3.483	1.830	263.50
Oysters, clams (meat and liquor)	0 to 2	100	5 to 8 days	87	−2.2	3.751	1.930	290.19
Oyster in shell	5 to 10	95 to 100	5 days	80	−2.8	3.517	1.842	266.84
Frozen shellfish	−29 to −18	90 to 95	3 to 8 months					
Meat (Beef)[d]								
Beef, fresh, average Carcass	0 to 1	88 to 92	1 to 6 weeks	62 to 77	−2.2 to −1.7[f]	2.914 to 3.426	1.616 to 1.804	206.80 to 256.83
Choice, 60% lean	0 to 4	85 10 90	1 to 3 weeks	49	−1.7	2.478	1.453	163.44
Prime, 54% lean	0 to 1	85	1 to 3 weeks	45	−2.2	2.345	1.403	150.10
Sirloin cut (choice)	0 to 1	85	1 to 3 weeks	56		2.713	1.541	186.79
Round cut (choice)	0 to 1	85	1 to 3 weeks	67		3.081	1.679	223.48
Dried, chipped	10 to 15	15	6 to 8 weeks	48		2.445	1.440	160.10
Liver	0 to I	90	1 to 5 days	70	−1.7	3.182	1.717	233.48

(Continued)

TABLE A.10
Continued

Commodity	Storage Temp. (°C)	Relative humidity (%)	Approximate storage life[a]	Water content (%)	Highest freezing (°C)	Specific heat above freezing[b] (J/kg·°C)	Specific heat below freezing[b] (J/kg·°C)	Latent heat[c] (J/kg)
Veal, 81% lean	0 to 1	90	1 to 7 days	66		3.048	1.666	220.14
Beef, frozen	−23 to −18	90 to 95	9 to 12 months					
Meat (Pork)[d]								
Pork, fresh, average	0 to 1	85 to 90	3 to 7 days	32 to 44	−2.2 to −2.7[f]	1.909 to 2.311	1.239 to 1.390	106.74 to 146.76
Carcass, 47% lean	0 to 1	85 to 90	3 to 5 days	37		2.077	1.302	123.41
Bellies, 33% lean	0 to 1	85	3 to 5 days	30		1.842	1.214	100.06
Backfat, 100% fat	0 to 1	85	3 to 7 days	8		1.105	0.938	
Shoulder, 67% lean	0 to 1	85	3 to 5 days	49	−2.2[f]	2.478	1.453	163.44
Pork, frozen	−23 to −18	90 to 95	4 to 6 months					
Ham								
74% lean	0 to 1	80 to 95	3 to 5 days	56	−1.7[f]	2.713	1.541	186.79
Light cure	3 to 5	80 to 85	1 to 2 weeks	57		2.746	1.553	190.12
Country cure	10 to 15	65 to 70	3 to 5 months	42		2.244	1.365	140.09
Frozen	−23 to-18	90 to 95	6 to 8 months					
Bacon								
Medium fat class	3 to 5	80 to 85	2 to 3 weeks	19		1.474	1.076	63.37
Cured, farm style	16 to 18	85	4 to 6 months	13–20		1.273 to 1.507	1.001 to 1.088	43.46 to 66.71
Cured, packer style	1 to 4	85	2 to 6 weeks					
Frozen	−23 to −18	90 to 95	4 to 6 months					
Sausage								
Links or bulk	0 to 1	85	1 to 7 days	38		2.110	1.315	126.75
Country, smoked	0	85	1 to 3 weeks	50	−3.9	2.512	1.465	166.78
Frankfurters								
average	0	85	1 to 3 weeks	56	−1.7	2.713	1.541	186.79
Polish style	0	85	1 to 3 weeks	54		2.646	1.516	180.12
Meat (Lamb)[d]								
Fresh, average	0 to 1	85 to 90	5 to 12 days	60 to 70	−2.2 to −1.7[f]	2.847 to 3.182	1.591 to 1.717	200.01 to 233.48
Choice, 67% lean	0	85	5 to 12 days	61	−1.9	2.880	1.604	203.47

(Continued)

TABLE A.10
Continued

Commodity	Storage Temp. (°C)	Relative humidity (%)	Approximate storage life[a]	Water content (%)	Highest freezing (°C)	Specific heat above freezing[b] (J/kg·°C)	Specific heat below freezing[b] (J/kg·°C)	Latent heat[c] (J/kg)
Leg, choice, 83% lean	0	85	5 to 12 days	65		3.014	1.654	216.81
Frozen	−23 to −18	90 to 95	8 to 10 months					
Meat (Poultry)[d]								
Poultry, fresh, average	0	85 to 90	1 week	74	−2.8	3.316	1.767	246.83
Chicken, all classes	0	85	1 week	74	−2.8	3.316	1.767	246.83
Turkey, all classes	0	85	1 week	64	−2.8	2.981	1.641	213.47
Duck	0	85	1 week	69	−2.8	3.148	1.704	230.15
Poultry, frozen	−23 to −18	90 to 95	8 to 12 months					
Meat (Miscellaneous)[d]								
Rabbits, fresh	0 to 1	90 to 95	1 to 5 days	68		3.115	1.691	226.81
Dairy Products[d]								
Butter	4	75 to 85	1 month	16	−20 to −0.6	1.373	1.038	53.37
Butter, frozen	−23	70 to 85	12 months					
Cheese								
Cheddar, long storage	−1 to 1	65 to 70	18 months	37	−13.3	2.077	1.302	123.41
Cheddar, short storage	4.4	65 to 70	6 months	37	−13.3	2.077	1.302	123.41
Cheddar, processed	4.4	65 to 70	12 months	39	−7.2	2.143	1.327	130.08
Cheddar, grated	4.4	60 to 70	12 months	31		1.876	1.227	103.40
Ice cream, 10% fat	−29 to −26		3 to 23 months	63	−5.6	2.948	1.629	210.14
Milk								
Whole, pasteurized grade A	0 to 1.1		2 to 4 months	87	−0.56	3.751	1.930	290.19
Dried, whole	21	Low	6 to 9 months	2		0.904	0.862	66.71
Dried, nonfat	7 to 21	Low	16 months	3		0.938	0.895	10.01
Evaporated,	4		24 months	74	−1.4	3.316	1.767	246.83
Evaporated, unsweetened	21		12 months	74	−1.4	3.316	1.767	246.83
Condensed, sweetened	4		15 months	27	−15	1.742	1.176	90.06
Whey, dried	21	Low	12 months	5		1.005	0.900	16.68
Poultry Products[d]								
Eggs								
Shell	−2 to 0[g]	80 to 85	5 to 6 months	66	−2.2[f]	3.048	1.666	220.14

(Continued)

TABLE A.10
Continued

Commodity	Storage Temp. (°C)	Relative humidity (%)	Approximate storage life[a]	Water content (%)	Highest freezing (°C)	Specific heat above freezing[b] (J/kg ·°C)	Specific heat below freezing[b] (J/kg ·°C)	Latent heat[c] (J/kg)
Shell, farm cooler	10 to 13	70 to 75	2 to 3 weeks	66	−2.2[f]	3.048	1.666	220.14
Frozen, whole	−18 or below		1 year plus	74		3.316	1.767	246.83
Frozen, yolk	−18 or below		1 year plus	55		2.680	1.528	183.45
Frozen, white	−18 or below		1 year plus	88		3.785	1.943	293.52
Whole egg solids	2 to 4	Low	6 to 12 months	2 to 4		0.938	0.875	10.01
Yolk solids	2 to 4	Low	6 to 12 months	3 to 5		0.972	0.888	13.34
Flake albumen solids	Room	Low	1 year plus	12 to 16		1.306	1.013	46.70
Dry spray albumen solids	Room	Low	1 year plus	5 to 8		1.055	0.919	21.68
Candy[d]								
Milk chocolate	−18 to 1.1	40	6 to 12 months	1		0.871	0.850	03.34
Peanut brittle	−18 to 1.1	40	1.5 to 6 months	2		0.904	0.862	06.67
Fudge	−18 to 1.1	65	5 to 12 months	10		1.172	0.963	33.35
Marshmallows	−18 to 1.1	65	3 to 9 months	17		1.407	1.051	56.70
Miscellaneous[d]								
Alfalfa meal	−18 or below	70 to 75	1 year plus					
Beer								
Keg[d]	2 to 4		3 to 8 weeks	90	−2.2[f]	3.852	1.968	300.20
Bottles and cans	2 to 4	65 or below	3 to 6 months	90				
Bread[d]	−18		3 to 13 weeks	32 to 37		1.993	1.271	106.74 to 123.41
Canned goods	0 to 16	70 or lower	1 year					
Cocoa	0 to 4	50 to 70	1 year plus					
Coconuts	0 to 2	80 to 85	1 to 2 months	47	−0.9	2.412	1.428	156.77
Coffee, green	2 to 3	80 to 85	2 to 4 months	10 to 15		1.172 to 1.340	0.962 to 1.026	033.36 to 050.03
Fur and fabrics	1 to 4	45 to 55	Several years					
Honey	below 10		1 year plus	17		1.407	1.051	056.70
Hops	−2 to 0	50 to 60	Several months					
Lard								
(without	7	90 to 95	4 to 8 months	0				
antioxidant)	−18	90 to 95	12 to 14 months	0				
Maple syrup				33		1.943	1.252	110.07
Nuts	0 to 10	65 to 75	8 to 12 months	3 to 6		0.938 to 1.038	0.875 to 0.913	010.01 to 020.01
Oil, vegetable, salad	21		1 year plus	0				

(Continued)

TABLE A.10
Continued

Commodity	Storage Temp. (°C)	Relative humidity (%)	Approximate storage life[a]	Water content (%)	Highest freezing (°C)	Specific heat above freezing[b] (J/kg ·°C)	Specific heat below freezing[b] (J/kg ·°C)	Latent heat[c] (J/kg)
Oleomargarine	2	60 to 70	1 year plus	16		1.372	1.038	053.37
Orange juice	−1 to 2		3 to 6 weeks	89		3.818	1.955	296.86
Popcorn, unpopped	0 to 4	85	4 to 6 weeks	10		1.172	0.963	033.36
Yeast baker's, compressed	−0.6 to 0			71		3.215	1.729	236.82
Tobacco								
Hogshead	10 to 18	50 to 55	1 year					
Bales	2 to 4	70 to 85	1 to 2 years					
Cigarettes	2 to 8	50 to 55	6 months					
Cigars	2 to 10	60 to 65	2 months					

[a] Storage life is not based on maintaining nutritional value.

[b] Calculated by Siebel's *formula and converted* to Sf units. For values below freezing, specific heat in kJ/kg·°C (Btu/lb·°F) = $0.0355a + 0.8374(0.008a + 0.20)$.
For values below freezing, specific heat in kJ/kg·°C (Btu/lb·°F) = $0.0126a + 0.8374(0.003a + 0.20)$. Seibel's formula is not very accurate in the frozen region, because foods are not simple mixtures of solids and liquids.

[c] Values for latent heat in kJ/kg were calculated by multiplying the percentages of water content by the latent heat of fusion of water.

[d] More specific information is available in the commodity chapters of the *ASHRAE Applications Handbook*.

[e] Lemons stored in production areas for conditioning are held at 12.8 to 14.4°C, but sometimes at 0°C.

[f] Average freezing point.

Source: From ASHRAE. 1982. *ASHRAE Handbook: Applications*. American Society of Heating, Refrigerating, and Air-Conditioning Engineers, Atlanta. GA.

TABLE A.11
Space, Weight, and Density Data for Commodities Stored in Refrigerated Warehouses

Commodity	Type of package	Outside dimensions of package, (in.)	Avg gross wt of pkg (lb)	Avg net wt mdse (lb)	Avg gross wt density (lb/cf)	Avg net wt density (lb/cf)
Apples	Wood box					
	Northwestern	$19\frac{1}{2} \times 11 \times 12\frac{3}{16}$	50	42	33.1	27.8
	Fiber tray carton	$20\frac{1}{2} \times 12\frac{1}{2} \times 13\frac{1}{4}$	$46\frac{3}{4}$	43	23.8	21.9
	Fiber master carton	$22\frac{1}{2} \times 12\frac{1}{2} \times 13$	$44\frac{3}{4}$	41	21.2	19.4
	Fiber bulk carton	$19 \times 12\frac{1}{2} \times 13$	$44\frac{3}{4}$	41	25.0	22.9
	Pallet box	$47 \times 47 \times 30$	1030	900	6.9	23.5
Beef						
Boneless	Fiber carton	$28 \times 18 \times 6$	146	140	83.4	80.0
Fores	Loose					22.2
Hinds	Loose					22.2
Celery	Wirebound crates	$20\frac{1}{4} \times 16 \times 9\frac{3}{4}$	60	55	32.8	30.0
	Fiber carton	$16 \times 11 \times 10$	36	32	35.4	31.4
Cheese	Hoops	$16 \times 16 \times 13$	84	78	43.6	40.5
	Wood, export	$17 \times 17 \times 14$	87	76	37.1	32.5
Cheese, Swiss	Wheels	$32\frac{1}{2} \times 32\frac{1}{2} \times 7$		171		40.0
Chili peppers	Bags	$45 \times 21 \times 26$	234	229	16.5	16.1
Citrus fruits						
Oranges	Box	$12\frac{1}{8} \times 13\frac{1}{4} \times 26\frac{1}{4}$	77	69	31.5	28.3
	Bruce box	$13 \times 11 \times 26\frac{1}{4}$	88	83	40.5	38.2
	Pallet, 40 cartons	$40 \times 48 \times 58\frac{1}{2}$	1690	1480	26.0	22.8
California oranges	Fiber carton	$16\frac{3}{8} \times 10\frac{1}{16} \times 10\frac{1}{2}$	40	37	38.0	35.2
Florida oranges	Fiber carton	$19\frac{1}{4} \times 12\frac{1}{4} \times 8$	45	37	41.3	33.9
Lemons	Fiber carton	$16\frac{3}{8} \times 10\frac{1}{16} \times 10\frac{1}{2}$	40	37	40.0	37.0
Grapefruit	Fiber carton	$19\frac{1}{4} \times 12\frac{1}{4} \times 8$	40	38	36.7	34.9
Coconut, shredded	Bags	$38 \times 18\frac{1}{2} \times 8$	101	100	31.0	30.7
Cranberries	Fiber carton	$15\frac{3}{4} \times 11\frac{1}{4} \times 10\frac{1}{2}$	26	24	24.1	22.2
Cream	Tins	$12 \times 12 \times 14$	$52\frac{3}{4}$	50	45.2	42.9
Dried fruit	Wood box	$15\frac{1}{2} \times 10 \times 6\frac{1}{2}$	$26\frac{1}{2}$	25	45.4	42.9
Dates	Fiber carton	$14 \times 14 \times 11$	32	30	25.7	24.0
Raisins, prunes, figs, peaches	Fiber carton	$15 \times 11 \times 7$	32	30	47.9	44.9
Eggs, shell	Wood cases	$26 \times 12 \times 13$	55	45	23.4	19.1
Eggs, frozen	Cans	$10 \times 10 \times 12\frac{1}{2}$	32	30	44.2	41.5
Frozen fishery products						
Blocks	$4/13\frac{1}{2}$-lb carton	$20\frac{3}{4} \times 12\frac{1}{8} \times 6\frac{3}{4}$	56	54	57.0	55.0
	$4/16\frac{1}{2}$-lb carton	$19\frac{3}{4} \times 10\frac{3}{4} \times 11\frac{1}{4}$	68	66	49.2	47.8
Filets	12/16-oz carton	$12\frac{3}{4} \times 8\frac{5}{8} \times 3\frac{13}{16}$	13.5	12	55.8	49.6
	10/5-lb carton	$14\frac{1}{2} \times 10 \times 14$	52.25	50	44.6	42.7

(Continued)

TABLE A.11
Continued

Commodity	Type of package	Outside dimensions of package, (in.)	Avg gross wt of pkg (lb)	Avg net wt mdse (lb)	Avg gross wt density (lb/cf)	Avg net wt density (lb/cf)
	5/10-lb carton	$14\frac{1}{2} \times 10 \times 14$	52.2	50	44.5	42.7
Fish sticks	12/8-oz carton	$11 \times 8\frac{3}{8} \times 3\frac{7}{8}$	6.9	6	33.6	29.3
	24/8-oz carton	$16\frac{7}{16} \times 8\frac{5}{16} \times 4\frac{5}{8}$	13.8	12	37.8	32.9
Panned fish portions	None, glazed	Wooden boxes				35.0
	2-, 3-, 5-, and 6-lb carton	Custom packing				29–33
Round ground fish	None, glazed	Stacked loose				33–35
Round Halibut	None, glazed	Wooden box, loose				30–35
		Stacked loose				38.0
Round salmon	None, glazed	Stacked loose				33–35
Shrimp	$2\frac{1}{2}$- and 5-lb cartons	Custom packing				35.0
Steaks	1-, 5-. or 10-lb packages	Custom packing				50–60
Frozen fruits, juices, vegetables						
Asparagus	24/12-oz carton	$13\frac{1}{2} \times 11\frac{3}{4} \times 8\frac{1}{4}$	21	18	27.7	23.8
Beans, green	36/10-oz carton	$12\frac{1}{2} \times 11 \times 8$	$25\frac{1}{2}$	$22\frac{1}{2}$	40.1	35.3
Blueberries	24/12-oz carton	$12 \times 11\frac{1}{2} \times 8$	20	18	31.3	28.2
Broccoli	24/10-oz carton	$12\frac{1}{2} \times 11\frac{1}{2} \times 8\frac{1}{2}$	$18\frac{1}{2}$	15	26.2	21.2
Citrus concentrates	Fiber carton 48/6 oz	$13 \times 8\frac{3}{4} \times 7\frac{1}{2}$	27	26	54.7	52.7
Peaches	24/1-1b carton	$13\frac{1}{2} \times 11\frac{1}{4} \times 7\frac{1}{2}$	27	24	41.0	36.4
Peas	6/5-1b carton	$17 \times 11 \times 9\frac{1}{2}$	32	30	31.1	28.2
	48/12-oz carton	$21\frac{1}{2} \times 8\frac{1}{2} \times 12\frac{1}{2}$	38	36	28.7	27.2
Potatoes, french fries	12/16-oz carton					28.6
	24/9-oz carton					24.0
Spinach	24/14-oz carton	$12\frac{1}{2} \times 11 \times 8\frac{1}{2}$	24	21	35.5	31.0
Strawberries	30-lb can	$12\frac{1}{2} \times 10 \times 10$	32	30	44.2	41.5
	24/1–1b carton	$13 \times 11 \times 8$	28	24	42.3	36.2
	450-lb barrel	$35 \times 25 \times 25$		450		35.5
Grapes, California	Wood lug box	$6\frac{1}{2} \times 15 \times 18$	31	28	32.4	29.2
Lamb, boneless	Fiber box	$20 \times 15 \times 5$	57	53	65.7	61.0
Lard (2/28 1b)	Wood export box	$18 \times 13\frac{1}{4} \times 7\frac{3}{4}$	64	56	59.8	52.5
Lettuce, head	Fiber carton	$20\frac{1}{2} \times 13\frac{1}{2} \times 9\frac{1}{2}$	$37\frac{1}{2}$	35	24.7	
	Fiber carton	$21\frac{1}{2} \times 14\frac{1}{4} \times 10\frac{1}{2}$	45–55	42–52	26.9	25.2
	Pallet, 30 cartons	$42 \times 50 \times 66$	1350	1170	16.8	14.6
Milk, condensed	Barrels	$35 \times 25\frac{1}{2} \times 25\frac{1}{2}$	670	600	50.9	45.6
Nuts						
Almonds, in shell	Sacks	$24 \times 15 \times 33$	$91\frac{1}{2}$	90	13.3	13.1

(Continued)

TABLE A.11
Continued

Commodity	Type of package	Outside dimensions of package, (in.)	Avg gross wt of pkg (lb)	Avg net wt mdse (lb)	Avg gross wt density (lb/cf)	Avg net wt density (lb/cf)
Almonds, shelled	Cases	$6\frac{3}{4} \times 23\frac{1}{2} \times 11$	32	28	31.7	27.7
English walnuts, in shell	Cases	$25 \times 11 \times 31$	103	100	20.9	20.3
English walnuts, shelled	Fiber carton	$14 \times 14 \times 10$	27	25	23.8	22.0
Peanuts, shelled	Burlap bag	$35 \times 10 \times 15$	127	125	39.2	38.6
Pecans, in shell	Burlap bag	$35 \times 22 \times 12$	$126\frac{1}{2}$	125	23.7	23.4
Pecans, shelled	Fiber carton	$13 \times 13 \times 11$	32	30	29.8	27.9
Peaches	3/4 bushel	$16\frac{7}{8}$ top dia	41	48	43.9	40.7
	1/2 bushel	$14\frac{1}{2}$ top dia	28	25	45.0	40.2
	Wirebound crate	$19 \times 11\frac{3}{4} \times 11\frac{1}{8}$	42	38	29.2	26.4
	Wood lug box	$18\frac{1}{8} \times 11\frac{1}{2} \times 5\frac{3}{4}$	26	23	38.0	33.1
Pears	Wood box	$8\frac{1}{2} \times 11\frac{1}{2} \times 18$	52	48	51.0	47.1
Pears, place pack	Fiber carton	$18\frac{1}{2} \times 12 \times 10$	52	46	40.5	35.6
Pork						
Bundle bellies	Bundles	$23\frac{1}{2} \times 10\frac{1}{2} \times 7$	57	57	57.0	57.0
Loins (regular)	Wood box	$28 \times 10 \times 10$	60	54	37.0	33.3
Loins (boneless)	Fiber box	$20 \times 15 \times 5$	57	52	65.7	59.9
Potatoes	Sack	$33 \times 17\frac{1}{2} \times 11$	101	100	27.5	27.2
Poultry, fresh (eviscerated)						
Fryers, whole, 24–30 to pkg.	Wirebound crate	$24 \times 10 \times 7$	65	60	27.5	25.4
Fryer parts Poultry, frozen (eviscerated)	Wirebound crate	$17\frac{3}{4} \times 10 \times 12\frac{1}{2}$	54	50	42.1	38.9
Ducks, 6 to pkg.	Fibei carton	$22 \times 16 \times 4$	$32\frac{1}{2}$	31	39.9	38.0
Fowl, 6 to pkg.	Fiber canon	$20\frac{3}{4} \times 18 \times 5\frac{1}{2}$	$33\frac{1}{2}$	31	28.2	26.1
Fryers, cut up, 12 to pkg.	Fiber. carton	$17\frac{1}{4} \times 15\frac{3}{4} \times 4\frac{1}{4}$	$30\frac{1}{2}$	28	45.4	41.7
Roasters, 8 to pkg.	Fiber carton	$20\frac{3}{4} \times 18 \times 5\frac{1}{2}$	$32\frac{1}{2}$	30	27.3	25.2
Turkeys,						
3–6 lb, 6 to pkg.	Fiber carton	$21 \times 17 \times 6\frac{1}{2}$	30	27	22.5	20.1
6–10 lb, 6 to pkg.	Fiber canon	$26 \times 21\frac{1}{2} \times 7$	$52\frac{1}{2}$	48	23.3	21.2
10–13 lb, 4 to pkg.	Fiber carton	$26\frac{1}{2} \times 16 \times 7\frac{1}{2}$	50	46	27.2	25.0
13–16 lb, 4 to pkg.	Fiber carton	$29 \times 18\frac{1}{2} \times 9$	$67\frac{1}{2}$	62	24.2	22.2
16–20 lb, 2 to pkg.	Fiber carton	$17 \times 16 \times 9$	39	36	27.7	25.4
20–24 lb, 2 to pkg.	Fiber carton	$19 \times 16\frac{1}{2} \times 9\frac{1}{2}$	$47\frac{1}{2}$	44	27.6	25.5
Tomatoes						
Florida	Fiber carton	$19 \times 10\frac{7}{8} \times 10\frac{3}{4}$	43	40	33.3	31.10
	Wirebound crate	$18\frac{3}{4} \times 11\frac{15}{16} \times 11\frac{15}{16}$	64	60	41.3	38.7
California	Wood lug box	$17\frac{1}{2} \times 14 \times 7\frac{3}{4}$	34	30	30.9	27.3
Texas	Wood lug box	$17\frac{1}{2} \times 14 \times 6\frac{5}{8}$	34	30	36.2	31.9
Veal (boneless)	Fiber carton	$20 \times 15 \times 5$	57	53	65.7	61.0

Source: From ASHRAE. 1982. *ASHRAE Handbook: Applications.* American Society of Heating, Refrigerating, and Air-conditioning engineers, Atlanta, GA.

TABLE A.12
Composition of Foods

Product	Water g/100g	Protein g/100g	Carbohydrates g/100g	Sugar g/100g	Starch g/100g	Fiber g/100g	Lipid g/100g	Ca mg/100g	Fe mg/100g	Mg mg/100g	P mg/100g	Ash g/100g	K mg/100g	Na mg/100g	Zn mg/100g	Cu mg/100g	Mn mg/100g
Beef (hamburger)	69.02	20	0	0	0	0	10	12	2.2	20	184	0.98	321	66	4.8	0.072	0.01
Beef (lean chuck)	74	21.45	0	0	0	0	3.56	18	1.86	23	204	0.99	339	77	5.68	0.09	0.01
Fish, Cod	81.22	17.81	0	0	0	0	0.67	16	0.38	32	203	1.16	413	54	0.45	0.03	0.02
Fish, Perch	79.13	19.39	0	0	0	0	0.92	80	0.9	30	200	1.24	269	62	1.11	0.15	0.7
Asparagus	93.22	2.2	3.88	1.88	0	2.1	0.12	24	2.14	14	52	0.58	202	2	0.54	0.19	0.16
Carrots, raw	88.29	0.93	9.58	4.54	1.43	2.8	0.24	33	0.3	12	35	0.97	320	69	0.24	0.05	0.14
Cucumbers	95.23	0.65	3.63	1.67	0.83	0.5	0.11	16	0.28	13	24	0.38	147	2	0.2	0.04	0.08
Onions, raw	88.54	0.92	10.11	4.28	0	1.4	0.08	22	0.19	10	27	0.35	144	3	0.16	0.04	0.13
Tomato	93	1.2	5.1	4	0	1.1	0.2	13	0.51	10	28	0.5	204	13	0.07	0.09	0.1
Peas, raw green	78.86	5.42	14.45	5.67	3.68	5.1	0.4	25	1.47	33	108	0.87	244	5	1.24	0.176	0.41

Food																	
Spinach, raw	91.4	2.86	0.42	3.63	1.01	2.2	0.39	99	2.71	79	49	1.72	558	79	0.53	0.13	0.897
Blueberries, raw	84.21	0.74	9.96	14.49	2.13	2.4	0.33	6	0.28	6	12	0.24	77	1	0.16	0.057	0.336
Peaches, raw	88.87	0.91	8.39	9.54	0	1.5	0.25	6	0.25	9	20	0.43	190	0	0.17	0.07	0.06
Pears, raw	83.71	0.38	9.8	15.46	0	3.1	0.12	9	0.17	7	11	0.33	119	1	0.1	0.08	0.05
Plums, raw	87.23	0.7	9.92	11.42	0.1	1.4	0.28	6	0.17	7	16	0.37	157	0	0.1	0.057	0.052
Raspberries, raw	85.75	1.2	4.42	11.94	1.02	6.5	0.65	25	0.69	22	29	0.46	151	1	0.42	0.09	0.67
Strawberries, raw	90.95	0.67	4.66	7.68	1.02	2	0.3	16	0.42	13	24	0.4	153	1	0.14	0.048	0.386
Cherries, sweet	77.61	0.73	18.58	21.07	0	2.5	0.21	10	0.35	9	20	0.39	148	3	0.1	0.176	0
Egg, whites	87.57	10.9	0.71	0.73	0.02	0	0.17	25	0.69	22	29	0.63	151	1	0.42	0.09	0.67
Bread, white	36.7	8.2	4.31	49.5	42.89	2.3	3.6	108	3.03	24	94	1.9	119	27	0.62	0.126	0.383
Orange juice	88.3	0.7	8.4	10.4	0	0.2	0.2	11	0.2	11	17	0.4	200	1	0.05	0.04	0.01

TABLE A.13
Enthalpy of Frozen Foods[a]

Product	Water content (wt %)	Mean specific heat[b] 4 to 32°C kJ/(kg-C)		40	-30	-20	-18	-16	-14	-12	-10	-9	-8	-7	-6	-5	-4	-3	-2	-1	0
Fruits and vegetables																					
Applesauce	82.8	3.73	Enthalpy (kJ/kg)	0	23	51	58	65	73	84	95	102	110	120	132	152	175	210	286	339	343
			% water unfrozen[c]	—	6	9	10	12	14	17	19	21	23	27	30	37	44	57	82	100	—
Asparagus, Peeled	92.6	3.98	Enthalpy(kJ/kg)	0	19	40	45	50	55	61	69	73	77	83	90	99	108	123	155	243	181
			% water unfrozen	—	—	—	—	—	5	6	—	7	8	10	12	15	17	20	29	58	100
Bilberries	85.1	3.77	Enthalpy (kJ/kg)	0	21	45	50	57	64	73	82	87	94	101	110	125	140	167	218	348	352
			% water unfrozen	—	—	—	7	8	9	11	14	15	17	18	21	25	30	38	57	100	—
Carrots	87.5	3.90	Enthalpy (kJ/kg)	0	21	46	51	57	64	72	81	87	94	102	111	124	139	166	218	357	361
			% water unfrozen	—	—	—	7	8	9	11	14	15	17	18	20	24	29	37	53	100	—
Cucumbers	95.4	4.02	Enthalpy (kJ/kg)	0	18	39	43	47	51	57	64	67	70	74	79	85	93	104	125	184	390
			% water unfrozen	—	—	—	—	—	—	—	—	5	—	—	—	—	—	14	20	37	100
Onions	85.5	3.81	Enthalpy (kJ/kg)	0	23	50	55	62	71	81	91	97	105	115	125	141	163	196	263	349	353
			% water unfrozen	—	5	8	10	12	14	16	18	19	20	23	26	31	38	49	71	100	—
Peaches without stones	85.1	3.77	Enthalpy (kJ/kg)	0	23	50	57	64	72	82	93	100	108	118	129	146	170	202	274	348	352
			% water unfrozen	—	5	8	9	11	13	16	18	20	22	25	28	33	40	51	75	100	—
Pears, Bartlett	83.8	3.73	Enthalpy (kJ/kg)	0	23	51	57	64	73	83	95	101	109	120	132	150	173	207	282	343	347
			% water unfrozen	—	6	9	10	12	14	17	19	21	23	26	29	35	43	54	80	100	—
Plums without stones	80.3	3.65	Enthalpy (kJ/kg)	0	25	57	65	74	84	97	111	119	129	142	159	182	214	262	326	329	333
			% water unfrozen	—	8	14	16	18	20	23	27	29	33	37	42	50	61	78	100	—	—
Raspberries	82.7	3.73	Enthalpy (kJ/kg)	0	20	47	53	59	65	75	85	90	97	105	115	129	148	174	231	340	344
			% water unfrozen	—	—	7	8	9	10	13	16	17	18	20	23	27	33	42	61	100	—
Spinach	90.2	3.90	Enthalpy (kJ/kg)	0	19	40	44	49	54	60	66	70	74	79	86	94	103	117	145	224	371
			% water unfrozen	—	—	—	—	—	—	6	7	7	8	9	11	13	16	19	28	53	100
Strawberries	89.3	3.94	Enthalpy (kJ/kg)	0	20	44	49	54	60	67	76	81	88	95	102	114	127	150	191	318	367
			% water unfrozen	—	—	5	—	6	7	9	11	12	14	16	18	20	24	30	43	86	100
Sweet Cherries Without Stones	77.0	3.60	Enthalpy (kJ/kg)	0	26	58	66	76	87	100	114	123	133	149	166	190	225	276	317	320	324
			% water unfrozen	—	9	15	17	19	21	26	29	32	36	40	47	55	67	86	100	—	—

Temperature (°C)

Food			Property																			
Tall peas	75.8	3.56	Enthalpy (kJ/kg)	0	23	51	56	64	73	84	95	102	111	121	133	152	176	212	289	319	323
			% water unfrozen		6	10	12	14	16	18	21	23	26	28	33	39	48	61	90	100	—
Tomato pulp	92.9	4.02	Enthalpy (kJ/kg)	0	20	42	47	52	57	63	71	75	81	87	93	103	114	131	166	266	382
			% water unfrozen	—	—	—	—	5	—	6	7	8	10	12	14	16	18	24	33	65	100
Eggs																						
Eggs white	86.5	3.81	Enthalpy (kJ/kg)	0	18	39	43	48	53	58	65	68	72	75	81	87	96	109	134	210	352
			% water unfrozen	—	—	10	—	—	—	—	13	—	—	—	18	20	23	28	40	82	100
Egg yolk	40.0	2.85	Enthalpy (kJ/kg)	0	19	40	45	50	56	62	68	72	76	80	85	92	99	109	128	182	191
			% water unfrozen	20	—	—	22	—	24	27	28	29	31	31	33	35	38	45	58	94	100
Whole egg with shell[d]	66.4	3.31	Enthalpy (kJ/kg)	0	17	36	40	45	50	55	61	64	67	71	75	81	88	98	117	175	281
Fish and meat																						
Cod	80.3	3.69	Enthalpy (kJ/kg)	0	19	42	47	53	66	74	79	84	89	96	105	118	137	177	298	323	
			% water unfrozen	10	10	11	12	13	13	14	16	17	18	19	21	23	27	34	48	92	100
Haddock	83.6	3.73	Enthalpy (kJ/kg)	0	19	42	47	53	59	66	73	77	82	88	95	104	116	136	177	307	337
			% water unfrozen	8	8	9	10	11	12	13	13	14	15	16	18	20	24	31	44	90	100
Perch	79.1	3.60	Enthalpy (kJ/kg)	0	19	41	46	52	58	65	72	76	81	86	93	101	112	129	165	284	318
			% water unfrozen	10	10	11	12	13	14	15	16	17	18	20	22	24	31	40	55	95	100
Beef, lean Fresh[e]	74.5	3.52	Enthalpy (kJ/kg)	0	19	42	47	52	58	65	72	76	81	88	95	105	113	138	180	285	304
			% water unfrozen	10	10	11	12	13	15	16	17	18	20	22	24	31	40	55	95	100	
Beef, lean Dried	26.1	2.47	Enthalpy (kJ/kg)	0	19	42	47	53	62	66	70	74		79	84						93
			% water unfrozen	96	96	97	98	98	100	—											
Bread																						
White bread	37.3	2.60	Enthalpy (kJ/kg)	0	17	35	39	44	49	56	67	75	83	93	104	117	124	128	131	134	137
Whole wheat bread	42.4	2.68	Enthalpy (kJ/kg)	0	17	36	41	48	56	66	78	86	95	106	119	135	150	154	157	160	163

[a] Above −40°C.

[b] Temperature range limited to 20°C for meats and 20 to 40°C for egg yolk.

[c] Total weight of unfrozen water = (total weight of food) (% water content/100)(water unfrozen/100).

[d] Calculated for a weight composition of 58% white (86.5% water) and 32% yolk (50% water).

[e] Data for chicken, veal, and venison very nearly matched the data for beef of the same water content.

Source: Dickerson, R. W. Jr. 1981. In *Handbook and Product Directory Fundamentals,* Society of Heating, Refrigerating, and Air-Conditioning Engineers, Atlanta, GA.

TABLE A.14
Properties of Ice as a Function of Temperature

Temperature $t°Q$	Thermal conductivity (W/m-K)	Specific heat (kJ/kg-K)	Density (kg/m³)
−101	3.50	1.382	925.8
−73	3.08	1.587	924.2
−45.5	2.72	1.783	922.6
−23	2.41	1.922	919.4
−18	2.37	1.955	919.4
−12	2.32	1.989	919.4
−7	2.27	2.022	917.8
	2.22	2.050	916.2

Source: Adapted from Dickerson, R. W. Jr. 1969. In *The Freezing Preservation of Foods*, 4th ed., Vol. 2, D.K. Tressler, W.B. Van Arsdel, and M.J. Copley (Eds), AVI Pub. Co., Westport, CT.

TABLE A.15
Heat Transfer Coefficients

Condition	Heat transfer coefficient (W/m-K)
Naturally circulating	5
Air blast	22
Plate contact freezer	56
Slowly circulating brine	56
Rapidly circulating brine	85
Liquid nitrogen	
low side of horizontal plate where gas blanket forms	170
upper side of horizontal plate	425
Boiling water	568

TABLE A.16
Examples of Transmission Data for Some Packaging Films

Packaging material	Density (kg/m³)	WVTR (g/m²·day)	GTR (cm³/m² · day · bar, 25 μm) Oxygen	Nitrogen	Carbon dioxide
Cellophane	1,440	18–198	7.8–12.4	7.8–24.8	6.2–93
Cellulose acetate		2,480	1,813–2,235	465–620	13,300–15,500
Nylon	1,130	248–341	40–527	14–53	155–2,370
Polyethylene					
Low density	910–950	21.7	7,750	2,790	41,850
High density	940–970	4.6	2,667	651	8,990
Polyethylene-vinyl acetate		31–46	13,020	6,200	93,000
Polyvinyl chloride-acetate (plasticized)	1,230	77.5–124	310–2,325	155–930	1,085–12,400

Source: From Perry, R.H. and Greene, D. 1984. *Perry's Chemical Engineer's Handbook.* McGraw-Hall, New York. With permission.

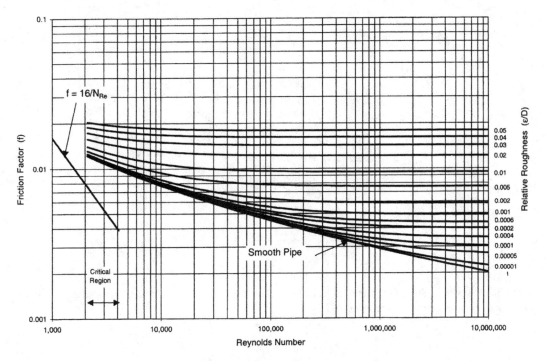

FIGURE A.1 The Moody diagram for the Fanning friction factor. Equivalent roughness for new pipes (ϵ in meters): cast iron, 259×10^{-6}; drawn tubing, 1.5235×10^{-6}, galvanized iron, 152×10^{-6}; steel or wrought iron, 45.7×10^{-6}. (Based on L.F. Moody, 1944. *Trans ASME*, 66, 671.)

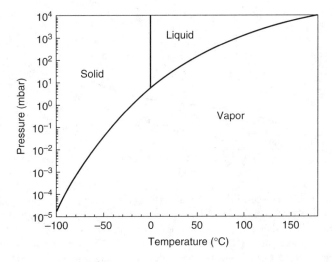

FIGURE A.2 Phase diagram of water. The triple point of water is at $0.0099°C$ and 6.104 mbar. In food materials the vapor pressure curve of ice is shifted to lower pressure values because of other compounds present.

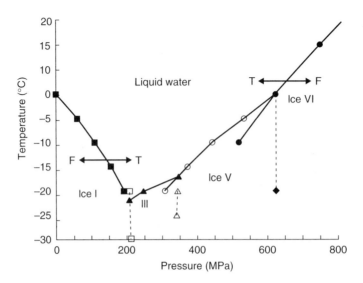

FIGURE A.3 Phase diagram of water showing the pressure-dependence of ice melting temperature for various pressure-dependent polymorphic forms of ice.

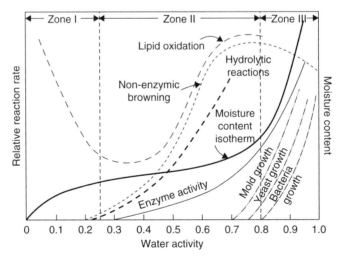

FIGURE A.4 Relationship between food water activity and moisture content (moisture isotherm) and the relative rates for a number of chemical reactions, enzyme activities and microorganism growths that lead to food deterioration. (From Fennema, O.R. (Ed.) (1976). In *Principles of Food Science — Part I — Food Chemistry*. New York, Marcel Dekker, Inc. Copyright 1976 from *Principles of Food Science — Part I — Food Chemistry* by O.R. Fennema. Reproduced by permission of Routledge/Taylor & Francis Group, LLC.)

REFERENCES

ASHRAE. 1982. ASHRAE Handbook Applications. American Society of Heating, Refrigeration and Air Conditioning Engineers. Atlanta, GA.

Choi, Y. and Okos, M.R. 1986. Effects of temperature and composition on the thermal properties of foods. In M. LeMaguer and P. Jelen (Ed) Food Engineering and Process Applications. Vol 1. Transport Phenomena. Elsevier Applied Science Publishers. London. pp 93–101.

Dickerson, R. W., Jr. 1969. Thermal properties of food. In The Freezing Preservation of Foodst 4th ed., Vol. 2, D. K. Tressler, W, B. van Arsdel, and M. J. Copley (eds.). AVI Pub. Co., Westport, CT.

Dickie, A.M. and Kokini, J.L. 1982. Use of Bird-Leider equation in food rheology. J. Food Process Engr. 5:157-174.

Fennema, O.R. (1976). Water and Ice. Principles of Food Science - Part I. Food Chemistry. OR Fennema. New York, Marcel Dekker, Inc.

Moody, L.F. 1944. Friction factors for pipe flow. Trans. ASME. 66:671.

Perry, R.H. and Greene, D. 1984. Perry's Chemical Engineer's Handbook. McGraw-Hill Publishing Co, Inc. New York.

Rao, M.A., Cooley, H.J. and Vitali. 1984. Flow properties of concentrated juices at low temperatures. Food Technol. 38:113-119.

Raznjevic, K. 1978. Handbook of Thermodynamic Tables and Charts. Hemisphere Pub. Corp. Washington, DC.

Steffe, J.F., Mohamed, I.O. and Ford, E.F. 1986. Rheological properties of fluid foods: data compilation. In: M.E. Okos, Ed. ASAE Publications. St. Joseph, MI

Index

A

Absorptivity, 410–411
Accelerated shelf life testing (ASLT), 905
Acids injector, 938
Activated-complex/transition state theory, 141
Activation energy, 47, 141–144, 252, 265, 465–466,
    658–661, 665, 755, 758–760, 794, 840, 894,
    906, 952
Activity coefficient, 472, 509
Adiabatic operation, 817, 840
Adsorption, 488, 606–609, 612, 619, 629–630, 637,
    892, 930, 932, 947–948
Aerosol can, 903
Agents
    complexing, 941
    sequestering, 947, 951
    surface–active, 2, 934, 943, 948, 959
Agitator, 357, 364–367, 392, 419–422, 499, 679, 959
Air blow, 669, 705, 938
Air velocity, 641, 676, 678–679, 692–693, 702–704
Airflow, 672, 690, 700
Alkaline, 129–130, 180–190, 250, 566, 883–884, 934,
    939, 941, 945–947, 958
α crystals, 331
Aluminum, 371, 632, 851, 859, 864, 876–877, 886, 896,
    898
    aluminum cans, 849, 866, 868–870
    aluminum foil, 849, 874, 878, 884–885, 887, 899
Amino acids, 179–181, 196–197, 243, 248, 251, 266,
    593–594, 630, 665
Amorphous sugar, 288, 300, 303, 338, 340–341
Amylopectin, 82, 321, 322, 324, 327, 338, 342, 621, 636
Amylose, 129, 288, 321–327, 342, 620, 705
Anaerobic respiration, 909
Angle
    dynamic, 371, 375, 393
    of friction, effective, 375, 393
    of internal friction, 371–374, 393
    of repose, 371, 374, 380, 393
Anisotropic model, 441, 442
Annealing, 302, 307, 317–319, 322–325, 331–332, 335,
    861, 889
Anthocyanins, 143, 203, 240–245, 254, 561, 593
Antifouling, 931
Appert, Nicolas, 849
Aroma(s), 250, 472, 489, 541, 543, 590–591, 602–603,
    631, 663, 667, 685, 710, 712, 850, 854–855,
    860, 864, 872–875, 885, 895, 898
Arrhenius, 47, 141–143, 180, 202, 251–265, 303,
    663–665, 658, 744, 755, 758, 760, 894, 906, 952

Arrhenius equation, 141, 142, 143, 180, 202, 254, 658,
    755, 758, 906
ASAE standards, 368, 370
Ascorbic acid, 127, 129, 133–134, 140, 143, 146, 179,
    185, 196–197, 241, 243, 245, 253–254,
    666–667; *see also* Vitamin C
    oxidation of, 145, 147, 250
Aseptic processing, 2, 202, 263, 747, 784–787, 790,
    852, 899–900
α-tocopherols, 196–197
Atomization, 668
Atomizer, 668–671

B

Bacteria, 198, 253, 262–263, 356, 480, 506–507, 526,
    537–538, 549, 594, 631, 746–747, 754–756,
    764, 774, 785, 794, 857, 858, 901, 957–958,
    967, 969–970
Baffled vessels, 422
Ball formula method, 768, 770–775, 794
Bar codes, 850, 904
Barrel groves, 811, 840
Batch, 2, 101–105, 356, 376–377, 457, 502, 588, 679,
    706, 708, 746, 780, 782–783, 794, 937, 938
    heating, 421
    retort, 747–750, 766
    system, 488, 747, 749
Beer-Lambert Law, 891
β-carotene, degradation of, 135, 192–194, 246–247,
    665–666
β crystals, 331–332
β-lactoglobulin, denaturation of, 327–328
BET equation, 480, 610
Binary diffusion, 470, 473–474
Binding energy, 612, 619, 626, 628, 630, 661
Bingham
    fluid, 789
    model, 42, 360
Biodegradable, 891, 883, 911–912, 945
Biopolymeric materials, 2
Biotin, 130, 190–191
Biot number, 412, 414, 424, 449, 451, 467, 837–838
Bird-Carreau model, 26, 60, 62, 71–75, 104,
    107–111
Bird-Leider equation, 39–40
Birdseye, Clarence, 849
Birefringence, 321–322
Black bodies, 706